TRIGONOMETRIC SERIES

VOLUMES I AND II

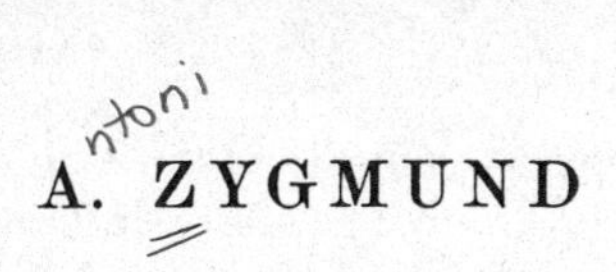

A. ZYGMUND

TRIGONOMETRIC SERIES

VOLUME I

The right of the
University of Cambridge
to print and sell
all manner of books
was granted by
Henry VIII in 1534.
The University has printed
and published continuously
since 1584.

CAMBRIDGE UNIVERSITY PRESS

Cambridge

New York Port Chester

Melbourne Sydney

Published by the Press Syndicate of the University of Cambridge
The Pitt Building, Trumpington Street, Cambridge CB2 1RP
40 West 20th Street, New York, NY 1011, USA
10 Stamford Road, Oakleigh, Melbourne 3166, Australia

First published in Warsaw 1935
Second edition published by Cambridge University Press 1959
Reprinted with corrections and some additions 1968
Reprinted 1977, 1979
First paperback edition 1988
Reprinted 1990

Printed in Great Britain at the University Press, Cambridge

ISBN 0 521 07477 0 hard covers
ISBN 0 521 35885 X paperback

DEDICATED TO THE MEMORIES OF

A. RAJCHMAN AND J. MARCINKIEWICZ

MY TEACHER AND MY PUPIL

CONTENTS

CHAPTER I

TRIGONOMETRIC SERIES AND FOURIER SERIES. AUXILIARY RESULTS

CHAPTER II

FOURIER COEFFICIENTS. ELEMENTARY THEOREMS ON THE CONVERGENCE OF $S[f]$ AND $\tilde{S}[f]$

CHAPTER III

SUMMABILITY OF FOURIER SERIES

CHAPTER IV

CLASSES OF FUNCTIONS AND FOURIER SERIES

CHAPTER V

SPECIAL TRIGONOMETRIC SERIES

CHAPTER VI

THE ABSOLUTE CONVERGENCE OF TRIGONOMETRIC SERIES

CHAPTER VII

COMPLEX METHODS IN FOURIER SERIES

CHAPTER VIII

DIVERGENCE OF FOURIER SERIES

CHAPTER IX

RIEMANN'S THEORY OF TRIGONOMETRIC SERIES

PREFACE

The first edition of this book was written almost twenty-five years ago. Since then the theory of trigonometric series has undergone considerable change. It has always been one of the central parts of Analysis, but now we see its notions and methods appearing, in abstract form, in distant fields like the theory of groups, algebra, theory of numbers. These abstract extensions are, however, not considered here and the subject of the second edition of this book is, as before, the classical theory of Fourier series, which may be described as the meeting ground of the Real and Complex Variables.

This theory has been a source of new ideas for analysts during the last two centuries, and is likely to be so in years to come. Many basic notions and results of the theory of functions have been obtained by mathematicians while working on trigonometric series. Conceivably these discoveries might have been made in different contexts, but in fact they came to life in connexion with the theory of trigonometric series. It was not accidental that the notion of function generally accepted now was first formulated in the celebrated memoir of Dirichlet (1837) dealing with the convergence of Fourier series; or that the definition of Riemann's integral in its general form appeared in Riemann's *Habilitationsschrift* devoted to trigonometric series; or that the theory of sets, one of the most important developments of nineteenth-century mathematics, was created by Cantor in his attempts to solve the problem of the sets of uniqueness for trigonometric series. In more recent times, the integral of Lebesgue was developed in close connexion with the theory of Fourier series, and the theory of generalized functions (distributions) with that of Fourier integrals.

A few words about the main problems of the present-day theory of trigonometric series. It has been decisively influenced by the methods of Lebesgue integration. These helped to solve the problem of the representation of functions by their Fourier series. This problem, stated in terms of summability of Fourier series, is now essentially a closed chapter (in spite of a large number of papers still written on the subject). The same holds for the problem of convergence of Fourier series at individual points. As regards the convergence or divergence almost everywhere, however, much still remains to be done. For example the problem of the existence of a continuous function with an everywhere divergent Fourier series is still open. One may argue that, owing to old-fashioned habits of thinking, one attaches too much importance to the notion of convergence as a method of summing Fourier series, and that, for example, the method of the first arithmetic mean is much more relevant; but there seems to be little doubt that the methods needed for the solution of the problem will be of considerable interest and value for the theory of functions.

Two other major problems of the theory also await their solution. These are the structure of the sets of uniqueness and the structure of the functions with absolutely convergent Fourier series; these problems are closely connected. General methods of solving them are still lacking and in a search for solutions we shall probably have to go beyond the domain of the theory of functions, in the direction of the theory of numbers and Diophantine approximation.

Among the outstanding problems of the theory one may also mention that of the behaviour of trigonometric series on sets of positive measure, and that of further developments of complex methods.

Another domain is that of multiple Fourier series. Here we have barely begun. Routine extensions from the case of a single variable are easy, but significant results are comparatively few. The field is vast and promising and at present we probably do not realize the shape of its problems, though the results here may be even more important for applications than in the case of a single variable.

Thinking of the extent and refinement reached by the theory of trigonometric series in its long development one sometimes wonders why only relatively few of these advanced achievements find an application. Part of the explanation is that in many problems of Analysis we can go far enough with more economical tools. For example, where in the past, to obtain a rigorous solution of a problem we had to prove the uniform convergence, or at least convergence almost everywhere, today we use convergence in norm, which effectively bypasses earlier difficulties. Other examples of a similar nature can be given. More subtle results of the theory, however, if we look at them in proper perspective, can give far-reaching applications. To give examples: extensions of methods dealing with conjugate functions of a single variable to the case of several variables can be an important tool in the theory of partial differential equations of elliptic type; results about the boundary behaviour of harmonic functions of two variables can be used to study boundary behaviour of analytic functions of several complex variables, etc.

We conclude with a remark about the character of the book. The first four chapters of it may serve as an introduction to the theory (part of the material contained herein can be omitted in this case; for example, the real-variable proof of the existence of the conjugate function, rearrangement of functions, linear operations). The material contained in subsequent chapters can be read, using cross-references, in practically any order. The 'miscellaneous theorems and examples' at the end of chapters are mostly accompanied by hints and are intended as possible exercises for the interested reader. Numbers in square brackets stand for items of the Bibliography at the end of the book. Notes at the end of each volume contain bibliographic references and additional information about the results of the text.

Practically all the manuscript of the book was read by Professor J. E. Littlewood and Mr P. Swinnerton-Dyer, and I have greatly profited from their criticism and suggestions. They, as well as Professor R. P. Boas, Mr T. G. Madsen, and Professors Guido and Mary Weiss, also helped me in the long and tedious process of proof-reading. Without this assistance many misprints and actual errors could have remained unnoticed, and I am grateful for this aid. I also appreciate the understanding and patience shown me by the Cambridge University Press. Finally I owe a great debt to my friend Professor R. Salem with whom I have collaborated over many years. The subject-matter of the book was often a topic of our discussions and in a considerable number of places any merits of the presentation are due to him.

A. Z.

CHICAGO,
AUGUST, 1958

NOTE ON THE 1968 IMPRESSION

This reprint has given me the opportunity of correcting a number of errors which slipped through in the preparation of the second edition and of including the more comprehensive index which several reviewers felt would be useful. I am most grateful to Drs L. Gordon, S. Lasher and L. Ziomek for preparing the new index.

A. Z.

CHICAGO,
JULY, 1968

NOTE ON THE 1977 IMPRESSION

The present reprinting is identical with that of 1968, except for the correction of misprints previously overlooked. In particular, we have not attempted to deal with the remarkable transformation of perspective in the field of almost everywhere convergence of Fourier Series which was brought about by Carleson through the proof of his celebrated theorem on almost everywhere convergence of the Fourier Series of L^2 functions, a result subsequently extended by Hunt to L^p, $p > 1$, functions. We refer the reader to the articles: L. Carleson, On convergence and growth of partial sums of Fourier Series, *Acta Mathematica* **116** (1966), 135–57, and R. A. Hunt, On the convergence of Fourier Series, *Proc. Conference Edwardsville, Ill.*, 1967, pp. 235–55.

A. Z.

CHICAGO
APRIL, 1977

LIST OF SYMBOLS

(Only symbols which are used systematically throughout the book are listed here. The numbers in parentheses refer to the pages where the symbols are defined; italics give pages of the second volume.)

(General symbols) $\in$, $\notin$, $\subset$, $\supset$ (9); o, O (14); $r'\left(=\frac{r}{r-1}\right)$ (17); sign x (3); $[x]$ (80); $\langle x \rangle$ (317); x^+, x^- (138); $\log^+ x$ (6); $\sim$ (5, 14); $\simeq$ (14); Σ' (41); $\sum'_{\nu \leqslant \omega}$ (55).

(Sets) $|E|$ (9); $E(f>y)$ (29).

(Functions) $\phi_x(t)$, $\phi_x(t;f)$; $\psi_x(t)$, $\psi_x(t;f)$ (50); $\chi_x(t)$ (*59*); $\Phi(h)=\Phi_{x_0}(h,f)$, $\Psi(h)=\Psi_{x_0}(h,f)$ (65); $\omega(\delta)$, $\omega(\delta;f)$ (42); $\omega_p(\delta)$, $\omega_p(\delta;f)$ (45).

(Classes of functions) L, L^r, $L^r(a,b)$, $L^\alpha(\log^+ L)^\beta$, L_ϕ, $L_\phi(a,b)$, $\phi(L)$ (16); $L^{r,\mu}$ (*94*); L^*_Φ (170); H^p, N (271); H^p, N, N_α (*316*); l^p (163); B, C, A, V, S (136).

(Norms) $\mathfrak{M}_r[f]$, $\mathfrak{A}_r[f]$, $\mathfrak{S}_r[a]$ (16); $\|x\|$, $\|x\|_p$ (163); $\|x\|_\Phi$ (170); $\|f\|_{r,\mu}$ (*94*).

(Kernels) $P(r,x)$, $Q(r,x)$ (1); $D_n(x)$, $\tilde{D}_n(x)$ (2); $D_n^*(x)$, $\tilde{D}_n^*(x)$ (50); $K_n(t)$ (88); $K_n^\alpha(t)$ (94); $D_{\mathbf{n}}(\mathbf{t})$, $K_{\mathbf{n}}(\mathbf{t})$, $P(\mathbf{r},\mathbf{t})$ (*302-3*).

(Trigonometric series and polynomials, Fourier series and integrals) $\tilde{S}$ (1); $A_n(x)$, $B_n(x)$ (3); a_ν, b_ν, c_ν (7); $c_{\mathbf{n}}$ (*301*); $S[f]$, $\tilde{S}[f]$ (7); $S[dF]$ (11); $S^{(k)}[f]$ (40); $S_n(x)$ $(=S_n(x;f))$, $\tilde{S}_n(x)$, $S_n^*(x)$, $\tilde{S}_n^*(x)$ (49, 50); $\sigma_n(x)$ $(=\sigma_n(x;f))$ (89); $\sigma_n^\alpha(x)$ $(=\sigma_n^\alpha(x;f))$ (94); $\tilde{\sigma}_n^\alpha(x;f)$ (95); $f(r,x)$, $\tilde{f}(r,x)$ (96); $\tilde{f}(x)$, $\tilde{f}(x;\epsilon)$ (51); $E_n(f)$ (115); $\tilde{f}(x)$ (247); $\sigma_{\mathbf{n}}(\mathbf{x})$, $f(\mathbf{r},\mathbf{x})$ (*302*); $I_n(x)$, $I_n[f]$, $I_n(x,f)$ (*5*); $I_{n,\nu}(x;f)$ (*6*); $a_\nu^{(n)}$, $b_\nu^{(n)}$, $c_\nu^{(n)}$ (*6*); $E_n(x)$, $E_n(x,f)$ (*13*); $J_n(x)$, $J_n(x,f)$ (*21*); $s(\theta)$ $(=s_\delta(\theta,F))$ (*207*); $g(\theta)$, $g(\theta,F)$ (*210*).

(Numbers) A_n^α, S_n^α, $\sigma_n^\alpha(x)$ (76).

(Methods of summability) (C, α) (76); A (80); T* (203); R (319); L (321); R_r (*69*); C (*66*).

(Sets of type) N (235); H (317); $H^{(m)}$ (346); U, M (344); $U(\varepsilon)$ (351).

CHAPTER I

TRIGONOMETRIC SERIES AND FOURIER SERIES. AUXILIARY RESULTS

1. Trigonometric series

These are series of the form

$$\tfrac{1}{2}a_0+\sum_{\nu=1}^{\infty}(a_\nu\cos\nu x+b_\nu\sin\nu x). \tag{1·1}$$

Here x is a real variable and the *coefficients* $a_0, a_1, b_1, \ldots$ are independent of x. We may usually suppose, if we wish, that the coefficients are real; when they are complex the real and imaginary parts of (1·1) can be taken separately. The factor $\frac{1}{2}$ in the constant term of (1·1) will be found to be a convenient convention.

Since the terms of (1·1) are all of period 2π, it is sufficient to study trigonometric series in an interval of length 2π, for example in $(0, 2\pi)$ or $(-\pi, \pi)$.

Consider the power series

$$\tfrac{1}{2}a_0+\sum_{\nu=1}^{\infty}(a_\nu-ib_\nu)z^\nu \tag{1·2}$$

on the unit circle $z=e^{ix}$. The series (1·1) is the real part of (1·2). The series

$$\sum_{\nu=1}^{\infty}(a_\nu\sin\nu x-b_\nu\cos\nu x) \tag{1·3}$$

(with zero constant term), which is the imaginary part of (1·2), is called the series *conjugate* to (1·1). If S is the series (1·1), its conjugate will be denoted by $\tilde{S}$. The conjugate of $\tilde{S}$ is, except for the constant term, $-S$.

A finite trigonometric sum

$$T(x)=\tfrac{1}{2}a_0+\sum_{\nu=1}^{n}(a_\nu\cos\nu x+b_\nu\sin\nu x)$$

is called a *trigonometric polynomial* of order n. If $|a_n|+|b_n|\neq 0$, $T(x)$ is said to be strictly of order n. Every $T(x)$ is the real part of an ordinary (power) polynomial $P(z)$ of degree n, where $z=e^{ix}$.

We shall often use the term 'polynomial' instead of 'trigonometric polynomial'.

The fact that trigonometric series are real parts of power series often suggests a method of summing them. For example, the series

$$P(r,x)=\tfrac{1}{2}+\sum_{\nu=1}^{\infty}r^\nu\cos\nu x, \quad Q(r,x)=\sum_{\nu=1}^{\infty}r^\nu\sin\nu x \quad (0\leqslant r<1)$$

are respectively the real and imaginary parts of

$$\tfrac{1}{2}+z+z^2+\ldots=\frac{1}{2}\frac{1+z}{1-z},$$

where $z=re^{ix}$. This gives

$$P(r,x)=\frac{1}{2}\frac{1-r^2}{1-2r\cos x+r^2}, \quad Q(r,x)=\frac{r\sin x}{1-2r\cos x+r^2}.$$

Similarly, from the formula

$$\log\frac{1}{1-z}=z+\tfrac{1}{2}z^2+\dots \quad (0\leqslant r<1),$$

we get
$$\sum_{\nu=1}^{\infty}\frac{\cos\nu x}{\nu}r^\nu=\tfrac{1}{2}\log\frac{1}{1-2r\cos x+r^2},\quad \sum_{\nu=1}^{\infty}\frac{\sin\nu x}{\nu}r^\nu=\arctan\frac{r\sin x}{1-r\cos x},\tag{1·4}$$

with $\arctan 0=0$.

Let us now consider the series

$$\tfrac{1}{2}+\sum_{\nu=1}^{\infty}\cos\nu x,\quad \sum_{\nu=1}^{\infty}\sin\nu x,$$

which are obtained by writing 1 for r in $P(r,x)$ and $Q(r,x)$, and let us denote by $D_n(x)$ and $\tilde{D}_n(x)$ the nth partial sums of these series. Arguing as before, we get

$$D_n(x)=\tfrac{1}{2}+\sum_{\nu=1}^{n}\cos\nu x=\frac{\sin(n+\frac{1}{2})x}{2\sin\frac{1}{2}x},\quad \tilde{D}_n(x)=\sum_{\nu=1}^{n}\sin\nu x=\frac{\cos\frac{1}{2}x-\cos(n+\frac{1}{2})x}{2\sin\frac{1}{2}x}.$$

A slightly simpler method of proving the formula for $D_n(x)$ is to multiply $D_n(x)$ by $2\sin\frac{1}{2}x$ and replace the products $2\sin\frac{1}{2}x\cos\nu x$ by differences of sines. Then all the terms except the last cancel. Similarly for $\tilde{D}_n(x)$.

These formulae show that $D_n(x)$ and $\tilde{D}_n(x)$ are uniformly bounded, indeed are absolutely less than $\operatorname{cosec}\frac{1}{2}\epsilon$, in each interval $0<\epsilon\leqslant x\leqslant 2\pi-\epsilon$.

Many trigonometric expressions have a term $2\sin\frac{1}{2}x$ or $2\tan\frac{1}{2}x$ in the denominator, and in this connexion we often use the inequalities

$$\sin u\leqslant u,\quad \sin u\geqslant\frac{2}{\pi}u,\quad \tan u\geqslant u\quad (0\leqslant u\leqslant\tfrac{1}{2}\pi).$$

Expressing the cosines and sines in terms of exponential functions, we write the nth partial sum $s_n(x)$ of (1·1) in the form

$$\tfrac{1}{2}a_0+\frac{1}{2}\sum_{\nu=1}^{n}\{(a_\nu-ib_\nu)e^{i\nu x}+(a_\nu+ib_\nu)e^{-i\nu x}\}.$$

If we define a_ν and b_ν for negative ν by the conditions

$$a_{-\nu}=a_\nu,\quad b_{-\nu}=-b_\nu\quad (\nu=0,1,2,\dots)$$

(thus in particular $b_0=0$), s_n is the nth *symmetric* partial sum, that is to say, the sum of the $2n+1$ central terms, of the Laurent series

$$\sum_{\nu=-\infty}^{+\infty}c_\nu e^{i\nu x}\quad (c_\nu=\tfrac{1}{2}(a_\nu-ib_\nu)),\tag{1·5}$$

where, if the a_ν and b_ν are real,

$$c_{-\nu}=\bar{c}_\nu\quad (\nu=0,1,2,\dots).\tag{1·6}$$

Conversely, any series (1·5) satisfying (1·6) may be written in the form (1·1) with a_ν and b_ν real. The series (1·5) satisfying (1·6) is a cosine series if and only if the c_ν are real; it is a sine series if and only if the c_ν are purely imaginary.

Whenever we speak of convergence or summability (see Chapter III) of a series (1·5), we are always concerned with the limit, ordinary or generalized, of the *symmetric* partial sums.

It is easily seen that the series conjugate to (1·1) is

$$-i\sum_{\nu=-\infty}^{+\infty}(\operatorname{sign}\nu)\,c_\nu e^{i\nu x}, \tag{1·7}$$

where the symbol 'sign z' is defined as follows:

$$\operatorname{sign}0=0,\quad \operatorname{sign}z=z/|z|\quad (z\neq 0).$$

Each of the forms (1·1) and (1·5) of trigonometric series has its advantages. Where we are dealing with (1·1) we suppose, unless the contrary is stated, that the a's and b's are real. Where we are dealing with (1·5), on the other hand, it is convenient to leave the c's unrestricted. The result is then that if (1·1) *has* complex coefficients and is of the form S_1+iS_2, where S_1 and S_2 have real coefficients, then the series conjugate to (1·1) is $\tilde{S}_1+i\tilde{S}_2$.

The following notation will also be used:

$$A_0(x)=\tfrac{1}{2}a_0,\quad A_n(x)=a_n\cos nx+b_n\sin nx,\quad B_n(x)=a_n\sin nx-b_n\cos nx\quad (n>0),$$

so that (1·1) and (1·3) are respectively

$$\sum_{n=0}^{\infty}A_n(x),\quad \sum_{n=1}^{\infty}B_n(x).$$

We shall sometimes write (1·1) in the form

$$\sum_{n=0}^{\infty}\rho_n\cos(nx+\alpha_n),\quad \text{where}\quad \rho_n=(a_n^2+b_n^2)^{\frac{1}{2}}\geqslant 0.$$

If $c_\nu=0$ for $\nu<0$, (1·5) will be said to be of *power series type*. For such series, $\tilde{S}$ is, except for the constant term, $-iS$. Obviously, S is the power series $c_0+c_1z+c_2z^2+\dots$ on the unit circle $|z|=1$.

In view of the periodicity of a trigonometric series it is often convenient to identify points x congruent mod 2π and to accept all the implications of this convention. Thus, generally, we shall say that two points are distinct if they are not congruent mod 2π; a point x will be said to be outside a set E if it is outside every set congruent to E mod 2π; and so on. This convention amounts to considering points x as situated on the circumference of the unit circle. If on occasion the convention is not followed the position will be clear from the context.

2. Summation by parts

This is the name given to the formula

$$\sum_{\nu=1}^{n}u_\nu v_\nu=\sum_{\nu=1}^{n-1}U_\nu(v_\nu-v_{\nu+1})+U_nv_n, \tag{2·1}$$

where $U_k=u_1+u_2+\dots+u_k$ for $k=1,2,\dots,n$; it is also called *Abel's transformation*.

(2·1) can be easily verified; it corresponds to integration by parts in the theory of integration. The following corollary is very useful.

(2·2) THEOREM. *If $v_1, v_2, \dots, v_n$ are non-negative and non-increasing, then*

$$|u_1v_1+u_2v_2+\dots+u_nv_n|\leqslant v_1\max_k|U_k|. \tag{2·3}$$

For the absolute value of the right-hand side of (2·1) does not exceed

$$\{(v_1-v_2)+(v_2-v_3)+\ldots+(v_{n-1}-v_n)+v_n\}\max|U_k|=v_1\max|U_k|.$$

The case when $\{v_\nu\}$ is non-negative and non-decreasing can be reduced to the preceding one by reversing the sequence. The left-hand side of (2·3) then does not exceed

$$v_n\max|U_n-U_{k-1}|\leqslant 2v_n\max|U_k|.$$

A sequence $v_0, v_1, \ldots, v_n, \ldots$ is said to be of *bounded variation* if the series

$$|v_1-v_0|+|v_2-v_1|+\ldots+|v_n-v_{n-1}|+\ldots$$

converges. This implies the convergence of $(v_1-v_0)+\ldots+(v_n-v_{n-1})+\ldots=\lim(v_n-v_0)$, and so every sequence of bounded variation is convergent.

The following result is an immediate consequence of (2·1).

(2·4) THEOREM. *If the series $u_0(x)+u_1(x)+\ldots$ converges uniformly and if the sequence $\{v_\nu\}$ is of bounded variation, then the series $u_0(x)v_0+u_1(x)v_1+\ldots$ converges uniformly.*

If the partial sums of $u_0(x)+u_1(x)+\ldots$ are uniformly bounded, and if the sequence $\{v_\nu\}$ is of bounded variation and tends to 0, then the series $u_0(x)v_0+u_1(x)v_1+\ldots$ converges uniformly.

The series (1·1) converges, and indeed uniformly, if $\Sigma(|a_\nu|+|b_\nu|)$ converges. Apart from this trivial case the convergence of a trigonometric series is a delicate problem. Some special but none the less important results follow from the theorem just stated. Applying it to the series

$$\tfrac{1}{2}a_0+\sum_{\nu=1}^{\infty}a_\nu\cos\nu x,\qquad \sum_{\nu=1}^{\infty}a_\nu\sin\nu x, \tag{2·5}$$

and taking into account the properties of $D_n(x)$ and $\tilde{D}_n(x)$ we have:

(2·6) THEOREM. *If $\{a_\nu\}$ tends to 0 and is of bounded variation (in particular, if $\{a_\nu\}$ tends monotonically to 0) both series (2·5), and so also the series $\Sigma a_\nu e^{i\nu x}$, converge uniformly in each interval $\epsilon\leqslant x\leqslant 2\pi-\epsilon$ $(\epsilon>0)$.*

As regards the neighbourhood of $x=0$, the behaviour of the cosine and sine series (2·5) may be totally different. The latter always converges at $x=0$ (and so everywhere), while the convergence of the former is equivalent to that of $a_1+a_2+\ldots$. If $\{a_\nu\}$ is of bounded variation but does not tend to 0, the uniform convergence in Theorem (2·6) is replaced by uniform boundedness.

Transforming the variable x we may present (2·6) in different forms. For example, replacing x by $x+\pi$ we have:

(2·7) THEOREM. *If $\{a_\nu\}$ is of bounded variation and tends to 0, the series*

$$\tfrac{1}{2}a_0+\sum_{\nu=1}^{\infty}(-1)^\nu a_\nu\cos\nu x,\qquad \sum_{\nu=1}^{\infty}(-1)^\nu a_\nu\sin\nu x$$

converge uniformly for $|x|\leqslant\pi-\epsilon$ $(\epsilon>0)$.

By (2·6), the series $\Sigma\nu^{-1}\cos\nu x$ and $\Sigma\nu^{-1}\sin\nu x$ converge for $x\neq 0$ (the latter indeed everywhere). Using the classical theorem of Abel which asserts that if $\Sigma\alpha_\nu$ converges to s then $\Sigma \alpha_\nu r^\nu \to s$ as $r \to 1-0$ (see Chapter III, § 1, below), we deduce from (1·4) the formulae

$$\left.\begin{aligned} \sum_{\nu=1}^{\infty} \frac{\cos \nu x}{\nu} &= \log \frac{1}{|2 \sin \frac{1}{2}x|} \\ \sum_{\nu=1}^{\infty} \frac{\sin \nu x}{\nu} &= \tfrac{1}{2}(\pi - x) \end{aligned}\right\} \quad (0 < x < 2\pi). \tag{2·8}$$

3. Orthogonal series

A system of real- or complex-valued functions $\phi_0(x)$, $\phi_1(x)$, $\phi_2(x)$, ..., defined in an interval (a, b), is said to be *orthogonal* over (a, b) if

$$\int_a^b \phi_m(x) \overline{\phi}_n(x)\, dx = \left\{\begin{matrix} 0 & \text{for} & m \neq n \\ \lambda_m > 0 & \text{for} & m = n \end{matrix}\right\} \quad (m, n = 0, 1, 2, \ldots). \tag{3·1}$$

In particular,

(i) the functions $|\phi_m(x)|^2$ are all integrable over (a, b);

(ii) no $\phi_m(x)$ can vanish identically (for that would imply $\lambda_m = 0$).

If in addition $\lambda_0 = \lambda_1 = \lambda_2 = \ldots = 1$, the system is said to be *normal*. A system orthogonal and normal is called *orthonormal*. If $\{\phi_n(x)\}$ is orthogonal, $\{\phi_n(x)/\lambda_n^{\frac{1}{2}}\}$ is orthonormal.

The importance of orthogonal systems is based on the following fact. Suppose that

$$c_0 \phi_0(x) + c_1 \phi_1(x) + \ldots + c_n \phi_n(x) + \ldots,$$

where $c_0, c_1, \ldots$ are constants, converges in (a, b) to a function $f(x)$. If we multiply both sides of the equation $f = c_0\phi_0 + c_1\phi_1 + \ldots$ by $\overline{\phi}_n$ and integrate term by term over (a, b), we have, after (3·1),

$$c_n = \frac{1}{\lambda_n} \int_a^b f(x) \overline{\phi}_n(x)\, dx \quad (n = 0, 1, 2, \ldots). \tag{3·2}$$

The argument is purely formal, although in some cases easily justifiable, e.g. if the series defining $f(x)$ converges almost everywhere, its partial sums are absolutely dominated by an integrable function, and each ϕ_n is bounded. It suggests the following problem. Suppose that a function $f(x)$ is defined in (a, b). We compute the numbers c_n by means of (3·2), and we write

$$f(x) \sim c_0\phi_0(x) + c_1\phi_1(x) + \ldots. \tag{3·3}$$

We call the numbers c_n the *Fourier coefficients of f*, and the series in (3·3) the *Fourier series of f*, with respect to the system $\{\phi_n\}$. The sign '$\sim$' in (3·3) only means that the c_n are connected with f by the formulae (3·2), and conveys no implication that the series is convergent, still less that it converges to $f(x)$. The problem is: *in what sense, and under what conditions, does the series* (3·3) *'represent' f(x)?*

This book is devoted to the study of a special but important orthogonal system, namely the trigonometric system (see § 4), and the theory of general orthogonal systems will be studied only in so far as it bears on this system. It may, however, be observed here that if an orthogonal system $\{\phi_n\}$ is to be at all useful for developing functions it must be *complete*, that is to say, whenever a new function ψ is added to it the new system is no longer orthogonal. For otherwise there would exist a function (namely, the function ψ), not vanishing identically, whose Fourier series with respect to the system $\{\phi_n\}$ would consist entirely of zeros.

If the functions ϕ_n are real-valued, we may drop the bars in (3·1) and (3·2).

The following system $\{\phi_\nu\}$, orthonormal over $(0, 1)$, is instructive. Let $\phi_0(x)$ be the function of period 1, equal to $+1$ for $0<x<\frac{1}{2}$ and to -1 for $\frac{1}{2}<x<1$; and let $\phi_0(0)=\phi_0(\frac{1}{2})=0$. Let

$$\phi_n(x)=\phi_0(2^n x) \quad (n=0, 1, 2, \ldots).$$

The function $\phi_n(x)$ takes alternately the values ± 1 inside the intervals

$$(0, 2^{-n-1}), \quad (2^{-n-1}, 2.2^{-n-1}), \quad (2.2^{-n-1}, 3.2^{-n-1}), \quad \ldots.$$

That $\{\phi_n\}$ is orthogonal follows from the fact that if $m>n$ the integral of $\phi_m\phi_n$ over any of these intervals is 0. The system is obviously normal. It is not complete, since the function $\psi(x)=1$ may be added to it (see also Ex. 6 on p. 34). The functions ϕ_n are called *Rademacher's functions*. Clearly,

$$\phi_n(x)=\operatorname{sign}\sin 2^{n+1}\pi x.\dagger$$

For certain problems the following extension of the notion of orthogonality is useful. Let $\omega(x)$ be a function non-decreasing over (a, b), and let $\phi_0, \phi_1, \phi_2, \ldots$ be a system of functions in (a, b) such that

$$\int_a^b \phi_m(x)\,\overline{\phi}_n(x)\,d\omega(x)=\begin{cases} 0 & \text{for} \quad m\neq n \\ \lambda_m>0 & \text{for} \quad m=n \end{cases} \quad (m, n=0, 1, \ldots), \tag{3·4}$$

where the integral is taken in the Stieltjes sense (Stieltjes-Riemann or Stieltjes-Lebesgue). The system $\{\phi_n\}$ is then called *orthogonal over (a, b) with respect to $d\omega(x)$*. If $\lambda_0=\lambda_1=\ldots=1$, the system is *orthonormal*. The Fourier coefficients of any function f with respect to $\{\phi_n\}$ are

$$c_n=\frac{1}{\lambda_n}\int_a^b f(x)\,\overline{\phi}_n(x)\,d\omega(x), \tag{3·5}$$

and the series $c_0\phi_0+c_1\phi_1+\ldots$ is the Fourier series of f. If $\omega(x)=x$, this is the same as the old definition. If $\omega(x)$ is absolutely continuous, $d\omega(x)$ may be replaced by $\omega'(x)\,dx$ and the functions $\phi_n(x)\sqrt{\{\omega'(x)\}}$ are orthogonal in the old sense. The case when $\omega(x)$ is a step function is important for trigonometric interpolation (see Chapter X).

4. The trigonometric system

The system of functions

$$e^{inx} \quad (n=0, \pm 1, \pm 2, \ldots) \tag{4·1}$$

is orthogonal over any interval of length 2π since, for any real α,

$$\int_\alpha^{\alpha+2\pi} e^{imx}\,e^{-inx}\,dx=\begin{cases} 0 & (m\neq n), \\ 2\pi & (m=n). \end{cases}$$

With respect to (4·1) the Fourier series of any function $f(x)$ defined, say, in the interval $(-\pi, \pi)$ is

$$\sum_{-\infty}^{+\infty} c_\nu e^{i\nu x}, \tag{4·2}$$

† The values $\phi_n(x)$ are closely related to the dyadic development of x. If $0<x<1$, x is not a dyadic rational and has dyadic development $\cdot d_1 d_2 \ldots d_n \ldots$, where the d_n are 0 or 1, then

$$d_n=d_n(x)=\tfrac{1}{2}\{1-\phi_{n-1}(x)\}.$$

where
$$c_\nu = \frac{1}{2\pi}\int_{-\pi}^{\pi} f(t)\, e^{-i\nu t}\, dt. \tag{4·3}$$

Let us set
$$a_\nu = \frac{1}{\pi}\int_{-\pi}^{\pi} f(t)\cos \nu t\, dt, \quad b_\nu = \frac{1}{\pi}\int_{-\pi}^{\pi} f(t)\sin \nu t\, dt \quad (\nu = 0, 1, \ldots) \tag{4·4}$$

(thus $b_0 = 0$), so that
$$c_\nu = \tfrac{1}{2}(a_\nu - ib_\nu), \quad c_{-\nu} = \tfrac{1}{2}(a_\nu + ib_\nu) \quad (\nu \geqslant 0). \tag{4·5}$$

Bracketing together in (4·2) the terms with $\pm\nu$, we write the series in the form
$$c_0 + (c_1 e^{ix} + c_{-1} e^{-ix}) + \ldots + (c_n e^{inx} + c_{-n} e^{-inx}) + \ldots,$$

or, taking into consideration (4·5),
$$\tfrac{1}{2}a_0 + (a_1 \cos x + b_1 \sin x) + \ldots + (a_n \cos nx + b_n \sin nx) + \ldots. \tag{4·6}$$

Since the orthonormality of a pair of functions ϕ_1, ϕ_2 implies the orthogonality of the pair $\phi_1 \pm \phi_2$, it is easily seen that the system
$$\frac{1}{2}, \quad \frac{e^{ix}+e^{-ix}}{2}, \quad \frac{e^{ix}-e^{-ix}}{2i}, \quad \ldots, \quad \frac{e^{inx}+e^{-inx}}{2}, \quad \frac{e^{inx}-e^{-inx}}{2i}, \quad \ldots,$$

or, what is the same thing, the system
$$\tfrac{1}{2}, \quad \cos x, \quad \sin x, \quad \cos 2x, \quad \sin 2x, \quad \ldots, \tag{4·7}$$

is orthogonal over any interval of length 2π.

The numbers λ (see §3) for this system are $\frac{1}{2}\pi, \pi, \pi, \ldots$, so that, in view of (4·4), (4·6) is the Fourier series of a function $f(x)$, $-\pi \leqslant x \leqslant \pi$, with respect to the system (4·7).

If the function $f(x)$ is even, that is, if $f(-x) = f(x)$, then
$$a_\nu = \frac{2}{\pi}\int_0^{\pi} f(t)\cos \nu t\, dt, \quad b_\nu = 0; \tag{4·8}$$

and if $f(x)$ is odd, that is, if $f(-x) = -f(x)$, then
$$a_\nu = 0, \quad b_\nu = \frac{2}{\pi}\int_0^{\pi} f(t)\sin \nu t\, dt. \tag{4·9}$$

The set of functions (4·7) is called the *trigonometric system*, and (4·1) the *complex trigonometric system*. The numbers a_ν, b_ν will be called the *Fourier coefficients* (the adjective *trigonometric* being understood), and the numbers c_ν the *complex Fourier coefficients*, of f. Finally, (4·6) is the *Fourier series* and (4·2) the *complex Fourier series*, of f. When no confusion can arise we shall simply speak of the *coefficients* of f and the *series* (or *development*) of f.

The Fourier series of f in either of the forms (4·2) and (4·6) will be denoted by
$$\mathsf{S}[f],$$
and the series conjugate to $\mathsf{S}[f]$ by $\tilde{\mathsf{S}}[f]$.

The series (4·2) and (4·6) are merely variants of each other, and in particular the partial sums of the latter are symmetric partial sums of the former. For real-valued functions we shall in this book use the forms (4·2) and (4·6) interchangeably. For complex-valued functions, in principle, only the form (4·2) will be used. However, for many problems of Fourier series (e.g. the problem of the representation of f by $\mathsf{S}[f]$) the limitation to real-valued functions is no restriction of generality.

Since the terms of (4·2) and (4·6) have period 2π, it is convenient to assume (as we shall always do in what follows) that the functions whose Fourier series we consider are defined not only in an interval of length 2π but for all values of x, by the condition of periodicity

$$f(x+2\pi)=f(x).$$

(This may necessitate a change in the value of f at one of the end-points of the interval, if initially the function had distinct values there.) In particular, when we speak of the Fourier series of a continuous function we shall always mean that the function is periodic and continuous in $(-\infty, +\infty)$. Similarly if we assert that a periodic f is integrable, of bounded variation, etc., we mean that f has these properties over a period. By *periodic* functions we shall always mean functions of period 2π.

If $\psi(x)$ is periodic, the integral of $\psi(x)$ over any interval of length 2π is always the same. In particular, since $f(x)$ is now defined everywhere and is periodic, the interval of integration $(-\pi, +\pi)$ in (4·3) and (4·4) may be replaced by any interval of length 2π, for instance by $(0, 2\pi)$.

(4·10) THEOREM. *If* (4·6) *or* (4·2) *converges almost everywhere to* $f(x)$, *and its partial sums are absolutely dominated by an integrable function, the series is* $S[f]$; *in particular, the conclusion holds if the series converges uniformly.*

That a_ν, b_ν are given by (4·4) follows by the same argument which led to (3·2) and which is now justified.

A function $f(x)$ defined in an interval of length 2π (and continued periodically) has a uniquely defined $S[f]$. With a function $f(x)$ defined in an interval (a, b) of length less than 2π we can associate various Fourier series, for we may define $f(x)$ arbitrarily in the remaining part of an interval of length 2π containing (a, b). The case $(a, b)=(0, \pi)$ is of particular interest. If we define $f(x)$ in $(-\pi, 0)$ by the condition $f(-x)=f(x)$, so that the extended f is even, we get a cosine Fourier series. If the extended f is odd, we have a sine Fourier series. These two series are respectively called the *cosine* and *sine* Fourier series of the function $f(x)$ defined in $(0, \pi)$.

By a linear change of variable we may transform the trigonometric system into a system orthogonal over any given finite interval (a, b). For example, the functions

$$\exp\{2\pi inx/(b-a)\} \quad (n=0, \pm 1, \pm 2, \ldots),$$

form an orthogonal system in (a, b), and with any $f(x)$ defined in that interval we may associate the Fourier series

$$\sum_{-\infty}^{+\infty} c_n \exp\frac{2\pi inx}{\omega}, \quad \text{where} \quad c_n=\frac{1}{\omega}\int_a^b f(t)\exp\left(-\frac{2\pi int}{\omega}\right)dt, \quad \omega=b-a.$$

By a change of variable the study of such series reduces to the study of ordinary Fourier series. The case of functions $f(x)$ of period 1 is particularly important. Here

$$f(x)\sim\sum_{n=-\infty}^{+\infty} c_n e^{2\pi inx}, \quad \text{where} \quad c_n=\int_0^1 f(t)\, e^{-2\pi int}\,dt.$$

The notion of Fourier coefficients a_ν, b_ν, c_ν has a parallel notion, that of *Fourier transforms*

$$\alpha(\nu)=\frac{1}{\pi}\int_{-\infty}^{+\infty} f(x)\cos\nu x\,dx, \quad \beta(\nu)=\frac{1}{\pi}\int_{-\infty}^{+\infty} f(x)\sin\nu x\,dx, \quad \gamma(\nu)=\frac{1}{2\pi}\int_{-\infty}^{+\infty} f(x)\, e^{-i\nu x}\,dx. \tag{4·11}$$

The function f here is defined over an infinite interval, and in general is not periodic; ν is a continuous variable ranging from $-\infty$ to $+\infty$. Unless f is a function absolutely integrable over $(-\infty, \infty)$, in which case the integrals (4·11) converge absolutely and uniformly for all ν, one must specify the sense in which these integrals are taken. Fourier integrals occur sporadically in the theory of Fourier series, but a more detailed discussion of them is postponed to a later chapter (see Chapter XVI).

The problems of the theory of Fourier series are closely connected with the notion of integration. In the formulae (4·4) we tacitly assumed that the products $f\cos \nu t$, $f\sin \nu t$ were integrable. Thus we may consider Fourier-Riemann, Fourier-Lebesgue, Fourier-Denjoy, etc., series, according to the way in which the integrals are defined. *In this book*, except when otherwise stated, *the integrals are always Lebesgue integrals*. It is assumed that the reader knows the elements of the Lebesgue theory. Proofs of results of a special character will be given in the text, or the reader will be referred to standard text-books.

Every integrable function $f(x)$, $0 \leqslant x \leqslant 2\pi$, has its Fourier series. It is even sufficient for f to be defined almost everywhere in $(0, 2\pi)$, that is to say, everywhere except in a set of measure zero. Functions $f_1(x)$ and $f_2(x)$ which are equal almost everywhere have the same Fourier series. Following the usage of Lebesgue, we call them equivalent (in symbols, $f_1(x) \equiv f_2(x)$), and we do not distinguish between equivalent functions.

Throughout this book the following notations will consistently be used:

$$x \in A, \quad x \notin A, \qquad A \subset B, \quad B \supset A.$$

The first means that x belongs to the set A; the second that x does not belong to A; the third and fourth that A is a subset of B.

The Lebesgue measure of a set (in particular, of an interval) E will be denoted by $|E|$. The sets and functions considered will always be measurable, even if this is not stated explicitly.

By a *denumerable* set we always mean a set which is either finite or denumerably infinite.

We list a few Fourier series which are useful in applications. Verifications are left to the reader.

(i) Let
$$\phi(x) = \tfrac{1}{2}(\pi - x) \quad \text{for} \quad 0 < x < 2\pi, \quad \phi(0) = \phi(2\pi) = 0.$$

Continued periodically $\phi(x)$ is odd and

$$\phi(x) \sim \sum_{\nu=1}^{\infty} \frac{\sin \nu x}{\nu} = \frac{1}{2} \sum_{\nu=-\infty}^{+\infty}{}' \frac{e^{i\nu x}}{i\nu} \tag{4·12}$$

(see also (2·8) above).

The function $\phi(x)$ can be used to remove discontinuities of other functions. For it is continuous except at $x = 0$, where it has a jump π. Thus, if $f(x)$ is periodic and at $x = x_0$ has a jump $d = f(x_0 + 0) - f(x_0 - 0)$, the difference

$$\Delta(x) = f(x) - (d/\pi)\,\phi(x - x_0)$$

is continuous at x_0, or may be made so by changing the value of $f(x_0)$.

(ii) Let $s(x) = +1$ for $0 < x < \pi$ and $s(x) = -1$ for $-\pi < x < 0$. Then

$$s(x) \sim \frac{4}{\pi} \sum_{\nu=1}^{\infty} \frac{\sin (2\nu - 1)x}{2\nu - 1}. \tag{4·13}$$

(iii) Given $0<h<\pi$, let $\chi(x)=1$ in $(-h,h)$, $\chi(x)=0$ at the remaining points of $(-\pi,\pi)$. The function χ is even and

$$\chi(x)\sim\frac{2h}{\pi}\left[\frac{1}{2}+\sum_{\nu=1}^{\infty}\frac{\sin\nu h}{\nu h}\cos\nu x\right]=\frac{h}{\pi}\sum_{\nu=-\infty}^{+\infty}\frac{\sin\nu h}{\nu h}e^{i\nu x}, \tag{4·14}$$

where the value of $(\sin\nu h)/\nu h$ when $\nu=0$ is taken to be 1.

(iv) Let $0<k\leqslant h$, $k+h\leqslant\pi$, and let $\mu(x)=\mu_{h,k}(x)$ be periodic, continuous, even, equal to 1 in $(0,h-k)$, 0 in $(h+k,\pi)$, linear in $(h-k,h+k)$. Then

$$\mu(x)\sim\frac{2h}{\pi}\left[\frac{1}{2}+\sum_{\nu=1}^{\infty}\left(\frac{\sin\nu h}{\nu h}\right)\left(\frac{\sin\nu k}{\nu k}\right)\cos\nu x\right]=\frac{h}{\pi}\sum_{-\infty}^{+\infty}\left(\frac{\sin\nu h}{\nu h}\right)\left(\frac{\sin\nu k}{\nu k}\right)e^{i\nu x}. \tag{4·15}$$

The νth coefficient of μ does not exceed a fixed multiple of ν^{-2}, so that $S[\mu]$ converges absolutely and uniformly. Using Theorem (6·3) below we see that the sign '$\sim$' in (4·15) can be replaced by the sign of equality.

For $k=0$ the series (4·15) go into the series (4·14).

(v) The special case $h=k$ of (4·15) deserves attention. Then

$$\lambda_h(x)=\mu_{h,h}(x)\sim\frac{2h}{\pi}\left[\frac{1}{2}+\sum_{\nu=1}^{\infty}\left(\frac{\sin\nu h}{\nu h}\right)^2\cos\nu x\right]=\frac{h}{\pi}\sum_{-\infty}^{+\infty}\left(\frac{\sin\nu h}{\nu h}\right)^2 e^{i\nu x}. \tag{4·16}$$

The function $\lambda_h(x)$ is even, decreases linearly from 1 to 0 over the interval $0\leqslant x\leqslant 2h$, and is zero in $(2h,\pi)$. It is useful to note that the coefficients of $S[\lambda]$ are non-negative. Using the remark made above that $S[\lambda]$ converges to λ and setting $x=0$, we have the formula

$$\sum_{\nu=-\infty}^{+\infty}\left(\frac{\sin\nu h}{\nu h}\right)^2=\frac{\pi}{h}, \tag{4·17}$$

which will be applied later.

The functions μ can be expressed in terms of the λ's:

$$\mu_{h,k}=\frac{1}{2}\left(\frac{h}{k}+1\right)\lambda_{\frac{1}{2}(h+k)}-\frac{1}{2}\left(\frac{h}{k}-1\right)\lambda_{\frac{1}{2}(h-k)}. \tag{4·18}$$

Since both sides here are even functions of x and represent polygonal lines, it is enough to check the formula for the values of x corresponding to the vertices, that is, for $x=0$, $h\pm k,\pi$.

λ is often called the *triangular*, or *roof*, function and μ the *trapezoidal* function.

(vi) Considering the Fourier series of the function $e^{-i\alpha x}$, $0<x<2\pi$, where α is any real or complex number, but not a real integer, we obtain the development

$$\frac{\pi}{\sin\pi\alpha}e^{i(\pi-x)\alpha}\sim\sum_{n=-\infty}^{\infty}\frac{e^{inx}}{n+\alpha} \tag{4·19}$$

This degenerates to (4·12) when $\alpha\to 0$.

5. Fourier-Stieltjes series

Let $F(x)$ be a function of bounded variation defined in the closed interval $0\leqslant x\leqslant 2\pi$. Let us consider the series $\sum_{-\infty}^{+\infty}c_\nu e^{i\nu x}$ with coefficients given by the formula

$$c_\nu=\frac{1}{2\pi}\int_0^{2\pi}e^{-i\nu t}dF(t)\quad(\nu=0,\pm 1,\pm 2,\ldots), \tag{5·1}$$

the integrals being Riemann-Stieltjes integrals. The numbers c_ν will be called the *Fourier-Stieltjes coefficients of F*, or the *Fourier coefficients of dF*. We write

$$dF(x) \sim \sum_{\nu=-\infty}^{+\infty} c_\nu e^{i\nu x}$$

and call the series here the *Fourier-Stieltjes series of F* or the *Fourier series of dF*; we denote it by $\mathsf{S}[dF]$. If $F(x)$ is absolutely continuous, then $\mathsf{S}[dF] = \mathsf{S}[F']$. We may also write $\mathsf{S}[dF]$ in the form (4·6) with

$$a_\nu = \frac{1}{\pi}\int_0^{2\pi} \cos \nu t \, dF(t), \quad b_\nu = \frac{1}{\pi}\int_0^{2\pi} \sin \nu t \, dF(t).$$

It is convenient to define $F(x)$ for all x by the condition

$$F(x+2\pi) - F(x) = F(2\pi) - F(0), \tag{5·2}$$

and this can be done without changing the values of F for $0 \leqslant x \leqslant 2\pi$. In the formulae for Fourier-Stieltjes coefficients we may then integrate over any interval of length 2π.

If we change $F(x)$ in a denumerable set, and if the new function is still of bounded variation, the numbers (5·1) remain unchanged. Thus we can assume once for all that the F we consider have no removable discontinuities.

The function $F(x)$ defined for all x by (5·2) is periodic if and only if $F(2\pi) - F(0)$ vanishes, i.e. if $c_0 = 0$. *The difference*

$$\Delta(x) = F(x) - c_0 x$$

is always periodic. For

$$\Delta(x+2\pi) - \Delta(x) = F(x+2\pi) - F(x) - 2\pi c_0 = 0.$$

A function $F(x)$ of bounded variation satisfying (5·2) may be called a *mass distribution* (of positive and negative masses, in general) on the circumference of the unit circle. If (α, β) is an arc on this circumference and $0 < \beta - \alpha \leqslant 2\pi$, then $F(\beta) - F(\alpha)$ is, by definition, the mass situated on the semi-open arc $\alpha < x \leqslant \beta$. The series

$$\sum_{-\infty}^{+\infty} e^{inx} = 2(\tfrac{1}{2} + \cos x + \cos 2x + \ldots)$$

is the Fourier-Stieltjes series of a mass 2π concentrated at the point $x = 0$ of the circumference.

6. Completeness of the trigonometric system

This theorem is a simple corollary of results we shall obtain later, but the following elementary proof, due to Lebesgue, is of interest in itself.

Let $f(x)$ be an integrable function whose coefficients $a_0, a_1, b_1, \ldots$ all vanish, so that

$$\int_{-\pi}^{\pi} f(x) T(x) \, dx = 0 \tag{6·1}$$

for any trigonometric polynomial $T(x)$. We have to show that $f(x) \equiv 0$. Let us assume first that $f(x)$ is continuous and not identically zero. There is then a point x_0 and two positive numbers ϵ, δ such that $|f(x)| > \epsilon$, say $f(x) > \epsilon$, in the interval $I = (x_0 - \delta, x_0 + \delta)$.

It will be enough to show that there is a sequence $T_n(x)$ of trigonometric polynomials such that

(i) $T_n(x) \geqslant 0$ for $x \in I$;

(ii) $T_n(x)$ tends uniformly to $+\infty$ in every interval I' interior to I;

(iii) the T_n are uniformly bounded outside I.

For then the integral in (6·1), with $T = T_n$, may be split into two, extended respectively over I and over the rest of $(-\pi, \pi)$. By (i), the first integral exceeds

$$\epsilon \,|\, I' \,|\, \min_{x \in I'} T_n(x),$$

and so, by (ii), tends to $+\infty$ with n. The second integral is bounded, in view of (iii). Thus (6·1) is impossible for $T = T_n$ with n large.

If we set

$$T_n(x) = \{t(x)\}^n, \quad t(x) = 1 + \cos(x - x_0) - \cos\delta,$$

then $t(x) \geqslant 1$ in I, $t(x) > 1$ in I', $|\,t(x)\,| \leqslant 1$ outside I. Conditions (i), (ii) and (iii) being satisfied, the theorem is proved for f continuous.

Suppose now that f is merely integrable, and let $F(x) = \int_{-\pi}^{x} f\,dt$. The condition $a_0 = 0$ implies that $F(x + 2\pi) - F(x) = 0$, so that $F(x)$ is periodic. Let $A_0, A_1, B_1, \ldots$ be the coefficients of F and let us integrate by parts the integrals

$$\int_{-\pi}^{\pi} F(x) \cos \nu x\, dx, \quad \int_{-\pi}^{\pi} F(x) \sin \nu x\, dx$$

for $\nu = 1, 2, \ldots$. Owing to the periodicity of F, the integrated terms vanish, and the hypothesis $a_1 = b_1 = a_2 = \ldots = 0$ implies that $A_1 = B_1 = A_2 = \ldots = 0$. Let $A_0', A_1', B_1', \ldots$ be the coefficients of $F(x) - A_0$. Obviously $A_0' = A_1' = B_1' = \ldots = 0$. Thus $F(x) - A_0$, being continuous, vanishes identically and $f \equiv 0$. This completes the proof. As corollaries we have:

(6·2) THEOREM. *If $f_1(x)$ and $f_2(x)$ have the same Fourier series, then $f_1 \equiv f_2$.*

(6·3) THEOREM. *If $f(x)$ is continuous, and $S[f]$ converges uniformly, its sum is $f(x)$.*

To prove (6·2) we observe that the coefficients of $f_1 - f_2$ all vanish, so that $f_1 - f_2 \equiv 0$. To prove (6·3), let $g(x)$ denote the sum of $S[f]$. Then the coefficients of $S[f]$ are the Fourier coefficients of g (see (4·10)). Hence $S[f] = S[g]$, so that $f \equiv g$ and, f and g being continuous, $f = g$. (For a more complete result see Chapter III, p. 89.)

7. Bessel's inequality and Parseval's formula

Let $\phi_0, \phi_1, \ldots$ be an orthonormal system of functions over (a, b) and let $f(x)$ be a function such that $|\,f(x)\,|^2$ is integrable over (a, b). We fix an integer $n \geqslant 0$, set

$$\Phi = \gamma_0 \phi_0 + \gamma_1 \phi_1 + \ldots + \gamma_n \phi_n$$

and seek the values of the constants $\gamma_0, \gamma_1, \ldots, \gamma_n$ which make the integral

$$J = \int_a^b |\,f - \Phi\,|^2\, dx \tag{7·1}$$

a minimum.

If we observe that

$$\int_a^b |\Phi|^2 dx = \int_a^b (\Sigma\gamma_\mu\phi_\mu)(\Sigma\bar{\gamma}_\nu\bar{\phi}_\nu)\,dx = \Sigma|\gamma_\nu|^2,$$

$$\int_a^b f\bar{\Phi}\,dx = \int_a^b f.(\Sigma\bar{\gamma}_\nu\bar{\phi}_\nu)\,dx = \Sigma c_\nu\bar{\gamma}_\nu,$$

where $c_0, c_1, \ldots$ are the Fourier coefficients of f with respect to $\{\phi_\nu\}$, we have

$$J = \int_a^b (f-\Phi)(\bar{f}-\bar{\Phi})\,dx = \int_a^b |f|^2 dx + \int_a^b |\Phi|^2 dx - 2\mathscr{R}\int_a^b f\bar{\Phi}dx$$

$$= \int_a^b |f|^2 dx + \Sigma|\gamma_\nu|^2 - 2\mathscr{R}\Sigma c_\nu\bar{\gamma}_\nu.$$

Adding and subtracting $\Sigma|c_\nu|^2$ we get

$$J = \int_a^b |f-\Phi|^2 dx = \int_a^b |f|^2 dx - \sum_{\nu=0}^{n}|c_\nu|^2 + \sum_{\nu=0}^{n}|c_\nu-\gamma_\nu|^2. \tag{7·2}$$

It follows that J attains its minimum if $\gamma_\nu = c_\nu$ for $\nu = 0, 1, \ldots, n$. Thus

(7·3) THEOREM. *If $|f(x)|^2$ is integrable over (a, b) and if $\Phi = \gamma_0\phi_0 + \gamma_1\phi_1 + \ldots + \gamma_n\phi_n$, where $\phi_0, \phi_1, \ldots$, form an orthonormal system over (a, b), the integral (7·1) is a minimum when Φ is the n-th partial sum of the Fourier series of f with respect to $\{\phi_\nu\}$.*

On account of (7·2) this minimum, necessarily non-negative, is

$$\int_a^b |f|^2 dx - \sum_{\nu=0}^{n}|c_\nu|^2. \tag{7·4}$$

Hence

$$\sum_{\nu=0}^{n}|c_\nu|^2 \leqslant \int_a^b |f|^2 dx.$$

This inequality is called *Bessel's inequality*. If $\{\phi_\nu\}$ is infinite we may make n tend to infinity, when Bessel's inequality becomes

$$\sum_{\nu=0}^{\infty}|c_\nu|^2 \leqslant \int_a^b |f|^2 dx. \tag{7·5}$$

Since the system $\{e^{i\nu x}/(2\pi)^{\frac{1}{2}}\}$ is orthonormal over $(0, 2\pi)$, we have

$$\sum_{\nu=-\infty}^{+\infty}|c_\nu|^2 \leqslant \frac{1}{2\pi}\int_0^{2\pi}|f|^2 dx,$$

where the c_ν are defined by (4·3). If f is real-valued this gives

$$\tfrac{1}{2}a_0^2 + \sum_{\nu=1}^{\infty}(a_\nu^2 + b_\nu^2) \leqslant \frac{1}{\pi}\int_0^{2\pi} f^2 dx.$$

It follows that *the Fourier coefficients a_ν, b_ν, c_ν tend to 0 with $1/\nu$, provided that $|f|^2$ is integrable.*

In some cases the sign '$\leqslant$' in (7·5) can be replaced by '$=$'. (From the preceding argument it follows that this is certainly the case if the Fourier series of f with respect to $\{\phi_\nu\}$ converges uniformly to f and (a, b) is finite.) The equation we then get is called *Parseval's formula*. It will be shown in Chapter II, § 1, that Parseval's formula holds for the trigonometric system.

Remark. If the functions ϕ_ν form on (a, b) an orthonormal system with respect to a non-decreasing function $\omega(x)$, Theorem (7·3) remains valid provided we replace the integral (7·1) by

$$\int_a^b |f-\Phi|^2 d\omega(x).$$

This remark will be useful in trigonometric interpolation (see Chapter X).

8.† Remarks on series and integrals

Let $f(x)$ and $g(x)$ be two functions defined for $x > x_0$, and let $g(x) \neq 0$ there. The symbols

$$f(x) = o(g(x)), \quad f(x) = O(g(x))$$

mean respectively that $f(x)/g(x) \to 0$ as $x \to +\infty$, and that $f(x)/g(x)$ is bounded for x large enough. The same notation is used when x tends to a finite limit or to $-\infty$, or even when x tends to its limit through a discrete sequence of values. In particular, an expression is $o(1)$ or $O(1)$ if it tends to 0 or is bounded, respectively.

Two functions $f(x)$ and $g(x)$ defined in the neighbourhood of a point x_0 (finite or infinite) are called *asymptotically equal* if $f(x)/g(x) \to 1$ as $x \to x_0$. We write then

$$f(x) \simeq g(x) \quad (x \to x_0).$$

If the ratios $f(x)/g(x)$ and $g(x)/f(x)$ are both bounded in the neighbourhood of x_0, we say that $f(x)$ and $g(x)$ are *of the same order* as $x \to x_0$, and write

$$f(x) \sim g(x) \quad (x \to x_0).$$

Let $u_0, u_1, u_2, \ldots$ be a sequence of numbers and let

$$U_n = u_0 + u_1 + \ldots + u_n \quad (n = 0, 1, \ldots).$$

A similar notation will be used with other letters. Let a be finite, and let $f(x)$ be a function defined in a finite or infinite interval $a \leqslant x < b$ and integrable over every interval (a, b'), $b' < b$. We shall write

$$F(x) = \int_a^x f(t)\, dt \quad (a \leqslant x < b).$$

(8·1) THEOREM. *Suppose that $f(x)$ and $g(x)$ are defined for $a \leqslant x < b$ and integrable over each (a, b') $(b' < b)$, that $g(x) \geqslant 0$, and that $G(x) \to +\infty$ as $x \to b$. Then, if $f(x) = o(g(x))$ as $x \to b$, we have $F(x) = o(G(x))$.*

Suppose that $|f(x)/g(x)| < \frac{1}{2}\epsilon$ for $x_0 < x < b$. For such x,

$$|F(x)| \leqslant \int_a^{x_0} |f|\, dt + \int_{x_0}^x |f|\, dt \leqslant \int_a^{x_0} |f|\, dt + \tfrac{1}{2}\epsilon G(x).$$

Since $G(x) \to \infty$, the last sum is less than $\epsilon G(x)$ for x close enough to b; and since ϵ is arbitrary, the result follows.

In this theorem the roles played by a and b can obviously be reversed. If $a = 0$ and $b = +\infty$, it has the following analogue for sums.

† The remainder of this chapter is not concerned with trigonometric series. It contains a concise presentation of various points from the theory of the real variable which will be frequently used later. Many of the results are familiar and we assemble them primarily for easy reference. We do not attempt to be complete. Some of the theorems, moreover, will be used later in a form more general than that in which they are proved here, but only when the general proof is essentially the same. (To give a typical example, the inequalities of Hölder and Minkowski will be applied to Stieltjes integrals although we prove them here only for ordinary Lebesgue integrals.) The material is not for detailed study, but only for consultation as required.

(8·2) THEOREM. *Let $\{u_\nu\}$ and $\{v_\nu\}$ be two sequences of numbers, the latter positive. If $u_n = o(v_n)$ and $V_n \to +\infty$, then $U_n = o(V_n)$.*

The proof is the same as that of (8·1).

(8·3) THEOREM. *Suppose that the series Σv_ν converges, that the v's are positive, and that $u_\nu = o(v_\nu)$. Then*

$$u_n + u_{n+1} + \ldots = o(v_n + v_{n+1} + \ldots).$$

This is obvious.

(8·4) THEOREM. *Let $f(x)$ be a positive, finite, and monotone function defined for $x \geqslant 0$, and let*

$$F(x) = \int_0^x f\,dt, \quad F_n = f(0) + f(1) + \ldots + f(n).$$

Then (i) *if $f(x)$ decreases, $F(n) - F_n$ tends to a finite limit;*

(ii) *if $f(x)$ increases,* $\quad F(n) \leqslant F_n \leqslant F(n) + f(n).$

To prove (i) we note that $f(k) \leqslant F(k) - F(k-1) \leqslant f(k-1)$ implies

$$0 \leqslant F(k) - F(k-1) - f(k) \leqslant f(k-1) - f(k) \quad (k = 1, 2, \ldots). \tag{8·5}$$

Since $\Sigma\{f(k-1) - f(k)\}$ converges, so does the series $\sum_1^\infty \{F(k) - F(k-1) - f(k)\}$; and it is enough to observe that its nth partial sum is $F(n) - F_n + f(0)$.

Case (ii) is proved by adding the obvious inequalities

$$f(k-1) \leqslant F(k) - F(k-1) \leqslant f(k) \quad (k = 1, 2, \ldots, n).$$

(8·6) THEOREM. *Let $f(x)$ be positive, finite and monotone for $x \geqslant 0$. If either* (i) *$f(x)$ decreases and $F(x) \to \infty$, or* (ii) *$f(x)$ increases and $f(x) = o(F(x))$, then*

$$F_n \simeq F(n).$$

This follows from (8·4).

(8·7) THEOREM. *Let $f(x)$, $x \geqslant 0$, be positive, monotone decreasing and integrable over $(0, +\infty)$, and let*

$$F^*(x) = \int_x^\infty f\,dt, \quad F_n^* = f(n) + f(n+1) + \ldots.$$

Then

$$F_{n+1}^* \leqslant F^*(n) \leqslant F_n^*.$$

If in addition $f(x) = o(F^*(x))$, then $F_n^* \simeq F^*(n)$.

It is enough to add the inequalities $f(k+1) \leqslant F^*(k) - F^*(k+1) \leqslant f(k)$ for $k = n$, $n+1, \ldots$.

Examples. From (8·6) and (8·7) it follows that

$$\sum_{k=1}^n k^\alpha \simeq \frac{n^{\alpha+1}}{\alpha+1}, \quad \sum_{k=n}^\infty k^{-\beta} \simeq \frac{n^{1-\beta}}{\beta-1} \tag{8·8}$$

for $\alpha > -1$, $\beta > 1$.

Taking $f(x) = 1/(1+x)$, $n = m-1$, we obtain from (8·4) that *the difference*

$$1 + \frac{1}{2} + \frac{1}{3} + \ldots + \frac{1}{m} - \log m$$

tends to a finite limit C (Euler's constant) as $m \to \infty$.

A more precise formula is sometimes needed, namely,

$$1+\frac{1}{2}+\dots+\frac{1}{m}=\log m+C+O\left(\frac{1}{m}\right). \tag{8·9}$$

To prove this, we observe that, for $f(x)=1/(1+x)$, the right-hand side of (8·5) is $1/k(k+1)$. Hence the mth partial sum of the series with terms $F(k)-F(k-1)-f(k)$ differs from the sum of the whole series by less than

$$\frac{1}{(m+1)(m+2)}+\frac{1}{(m+2)(m+3)}+\dots=\frac{1}{m+1},$$

and, arguing as in the proof of (8·4) (i), we get (8·9).

9. Inequalities

Let $\phi(u)$ be a non-negative function defined for $u \geqslant 0$. We say that a function $f(x)$ defined in an interval (a,b) *belongs to the class* $L_\phi(a,b)$, in symbols $f\in L_\phi(a,b)$, if $\phi(|f(x)|)$ is integrable over (a,b). If there is no danger of confusion, the class will be denoted simply by L_ϕ. In particular, if f is periodic, $f\in L_\phi$ will mean $f\in L_\phi(0,2\pi)$. If $\phi(u)=u^r$, $r>0$, L_ϕ will be written L^r. More generally, we shall occasionally write $\phi(L)$ for L_ϕ; thus, for example, $L^\alpha(\log^+ L)^\beta$ will denote the class of functions f such that $|f|^\alpha(\log^+|f|)^\beta$ is integrable.†

We shall also systematically use the notation

$$\mathfrak{M}_r[f;a,b]=\left\{\int_a^b |f(x)|^r\,dx\right\}^{1/r}, \quad \mathfrak{A}_r[f;a,b]=\left\{\frac{1}{b-a}\int_a^b |f(x)|^r\,dx\right\}^{1/r}.$$

If (a,b) is fixed we may simply write $\mathfrak{M}_r[f]$ and $\mathfrak{A}_r[f]$. Unlike $\mathfrak{M}_r$, $\mathfrak{A}_r$ is defined only if (a,b) is finite.

Similarly, given a finite or infinite sequence $a=\{a_n\}$ and a finite sequence

$$b=\{b_1,b_2,\dots,b_N\},$$

we write

$$\mathfrak{S}_r[a]=\{\Sigma\,|a_n|^r\}^{1/r},$$

Instead of L^1, $\mathfrak{M}_1$, $\mathfrak{A}_1$, $\mathfrak{S}_1$ we write L, $\mathfrak{M}$, $\mathfrak{A}$, $\mathfrak{S}$.

Let $\phi(u)$, $u\geqslant 0$, and $\psi(v)$, $v\geqslant 0$, be two functions, continuous, vanishing at the origin, strictly increasing, and inverse to each other. Then for $a,b\geqslant 0$ we have the following inequality, due to W. H. Young:

$$ab\leqslant \Phi(a)+\Psi(b), \quad \text{where} \quad \Phi(x)=\int_0^x \phi\,du, \quad \Psi(y)=\int_0^y \psi\,dv. \tag{9·1}$$

This is obvious geometrically, if we interpret the terms as areas. It is easy to see that *we have equality in* (9·1) *if and only if* $b=\phi(a)$. The functions Φ and Ψ will be called *complementary functions* (in the sense of Young).

On setting $\phi(u)=u^\alpha$, $\psi(v)=v^{1/\alpha}$ $(\alpha>0)$, $r=1+\alpha$, $r'=1+1/\alpha$, we get the inequality

$$ab\leqslant\frac{a^r}{r}+\frac{b^{r'}}{r'} \quad (a,b\geqslant 0), \tag{9·2}$$

where the 'complementary' exponents r, r' both exceed 1 and are connected by the relation

$$1/r+1/r'=1.$$

† By $\log^+|f|$ we mean $\log|f|$ wherever $|f|\geqslant 1$, and 0 otherwise.

This notation will be used systematically, so that, e.g., p' will denote the number such that

$$1/p+1/p'=1.$$

If $r=r'=2$, (9·2) reduces to the familiar inequality $2ab\leqslant a^2+b^2$. Clearly, either $r\leqslant 2\leqslant r'$ or $r'\leqslant 2\leqslant r$. If $r\to 1$, then $r'\to\infty$, and conversely. The connexion between r and r' may also be written

$$r'=\frac{r}{r-1}.$$

Integrating the inequality

$$|fg|\leqslant\Phi(|f|)+\Psi(|g|)$$

over $a\leqslant x\leqslant b$, we see that *fg is integrable over* (a,b) *if* $f\in \mathrm{L}_\Phi(a,b)$, $g\in \mathrm{L}_\Psi(a,b)$.

In particular, *fg is integrable if* $f\in \mathrm{L}^r$, $g\in \mathrm{L}^{r'}$.

Let us now consider (real or complex) sequences $A=\{A_n\}$, $B=\{B_n\}$, $AB=\{A_nB_n\}$, and let us assume that $\mathfrak{S}_r[A]=\mathfrak{S}_{r'}[B]=1$, $r>1$. If we sum the inequalities

$$|A_nB_n|\leqslant\frac{|A_n|^r}{r}+\frac{|B_n|^{r'}}{r'}$$

for $n=1,2,\ldots$, we get $\mathfrak{S}[AB]\leqslant 1$.

Now let $a=\{a_n\}$ and $b=\{b_n\}$ be any two sequences such that $\mathfrak{S}_r[a]$ and $\mathfrak{S}_{r'}[b]$ are positive and finite, and let us set $A_n=a_n/\mathfrak{S}_r[a]$, $B_n=b_n/\mathfrak{S}_{r'}[b]$ for all n. Then $\mathfrak{S}_r[A]=\mathfrak{S}_{r'}[B]=1$, so that $\mathfrak{S}[AB]\leqslant 1$. In other words,

$$\Sigma\,|a_nb_n|\leqslant\mathfrak{S}_r[a]\,\mathfrak{S}_{r'}[b], \tag{9·3}$$

and *a fortiori*

$$|\Sigma a_nb_n|\leqslant\mathfrak{S}_r[a]\,\mathfrak{S}_{r'}[b]. \tag{9·4}$$

These inequalities are called *Hölder's inequalities*. They are trivially true if $\mathfrak{S}_r[a]=0$ or $\mathfrak{S}_{r'}[b]=0$.

Hölder's inequality for integrals is

$$\left|\int_a^b fg\,dx\right|\leqslant\mathfrak{M}_r[f]\,\mathfrak{M}_{r'}[g], \tag{9·5}$$

and its proof is similar to that of (9·4), summation being replaced by integration. If $r=r'=2$, (9·4) and (9·5) reduce to the familiar Schwarz inequalities.

The remark concerning the sign of equality in (9·1) shows that we have equality in (9·2) if and only if $a^r=b^{r'}$. Hence if we assume that $\mathfrak{S}_r[a]$ and $\mathfrak{S}_{r'}[b]$ are distinct from 0 the proof of (9·3) shows that the sign of equality holds there if and only if $|A_n|^r=|B_n|^{r'}$ for all n; or, again, if and only if $|a_n|^r/|b_n|^{r'}$ is independent of n, with the understanding that a ratio 0/0 is to be disregarded. If $\mathfrak{S}_r[a]=0$, or $\mathfrak{S}_{r'}[b]=0$, we have automatic equality in (9·3), and at the same time $|a_n|^r/|b_n|^{r'}$ is 'independent of n', so that the rule is in this case also valid. Taking into account that the left-hand sides of (9·3) and (9·4) are equal if and only if $\arg(a_nb_n)$ is constant for all n for which $a_nb_n\neq 0$, we come to the following conclusion:

(9·6) THEOREM. *A necessary and sufficient condition for equality in* (9·4) *is that both sequences* $|a_n|^r/|b_n|^{r'}$ *and* $\arg(a_nb_n)$ *be independent of* n *(disregarding forms* 0/0 *and* $\arg 0$).

An argument similar in principle shows that

(9·7) Theorem. *The sign of equality holds in* (9·5) *if and only if* (i) *the ratio* $|f(x)|^r/|g(x)|^{r'}$ *is constant for almost all x for which it is not* 0/0, (ii) $\arg\{f(x)g(x)\}$ *is constant for almost all x for which* $fg \neq 0$.

The inequality (9·5) (and similarly (9·4)) may be extended as follows:

(9·8) Theorem. *If* $r_1, r_2, \ldots, r_k$ *are positive numbers such that* $1/r_1 + 1/r_2 + \ldots + 1/r_k = 1$, *and if* $f_i \in L^{r_i}(a, b)$ *for* $i = 1, 2, \ldots, k$, *then*

$$\left|\int_a^b f_1 f_2 \ldots f_k dx\right| \leqslant \mathfrak{M}_{r_1}[f_1]\,\mathfrak{M}_{r_2}[f_2] \ldots \mathfrak{M}_{r_k}[f_k].$$

The proof (by induction) is left to the reader.

A number M is called the *essential upper bound* (sometimes the *least essential upper bound*) of the function $f(x)$ in the interval (a, b) if (i) the set of points for which $f(x) > M$ is of measure 0, (ii) for every $M' < M$ the set of points for which $f(x) > M'$ is of positive measure. Similarly we define the *essential lower bound*. If both bounds are finite, $f(x)$ is said to be *essentially bounded*. (An equivalent definition is that $f(x)$ is essentially bounded if it is bounded outside a set of measure 0, or, again, that $f \equiv g$ where g is bounded.)

(9·9) Theorem. *If M is the essential upper bound of* $|f(x)|$ *in a finite interval* (a, b), *then*

$$\mathfrak{M}_r[f; a, b] \to M \quad as \quad r \to +\infty,$$

We may suppose that $M > 0$. Let $0 < M' < M$, and let E be the set of points where $|f(x)| > M'$. Then $|E| > 0$,

$$\mathfrak{M}_r[f] \geqslant M' \, |E|^{1/r},$$

so that $\liminf_{r\to\infty} \mathfrak{M}_r[f] \geqslant M'$. Hence $\liminf \mathfrak{M}_r[f] \geqslant M$. In particular, the theorem is proved if $M = +\infty$. This part of the proof holds even if $b - a = +\infty$.

Suppose then that $M < +\infty$. Since $\mathfrak{M}_r[f] \leqslant M(b-a)^{1/r}$, we have $\limsup \mathfrak{M}_r[f] \leqslant M$, and this, with the inequality $\liminf \mathfrak{M}_r[f] \geqslant M$ above, proves the theorem.

If $b - a = +\infty$, (9·9) is still true provided we assume that $\mathfrak{M}_r[f]$ is finite for some $r = r_0 > 0$. (Otherwise the result is false; take, for instance, $a = 2, b = +\infty, f(x) = 1/\log x$.) We have to show that $\limsup \mathfrak{M}_r[f] \leqslant M < +\infty$. Dividing by M, we may assume that $M = 1$. In order to show that $\limsup \mathfrak{M}_r[f] \leqslant 1$, we write $(a, b) = I + R$, where I is a finite subinterval of (a, b) so large that $\int_R |f|^{r_0} dx < 1$. Since $|f| \leqslant 1$ almost everywhere,

$$\int_a^b |f|^r dx = \int_I |f|^r dx + \int_R |f|^r dx \leqslant |I| + \int_R |f|^{r_0} dx \leqslant |I| + 1$$

for $r \geqslant r_0$. Hence $\limsup \mathfrak{M}_r[f] \leqslant 1$.

Since any sequence $a_0, a_1, \ldots$ may be treated as a function $f(x)$, where $f(x) = a_n$ for $n \leqslant x < n+1$, we see that $\mathfrak{S}_r[a]$ *tends to* $\max|a_n|$ *as* $r \to \infty$, *provided that* $\mathfrak{S}_r[a]$ *is finite for some* $r > 0$.

In virtue of (9·9), it is natural to define $\mathfrak{M}_\infty[f; a, b]$ as the essential upper bound of $|f(x)|$ in (a, b). By L^∞ we may denote the class of essentially bounded functions. The inequality (9·5) then remains meaningful and true for $r = \infty$, $r' = 1$.

Let $a = \{a_n\}$, $b = \{b_n\}$ be two sequences of numbers, and let $a + b = \{a_n + b_n\}$. The inequality

$$\mathfrak{S}_r[a+b] \leqslant \mathfrak{S}_r[a] + \mathfrak{S}_r[b] \quad (r \geqslant 1) \tag{9·10}$$

is called *Minkowski's inequality*. To prove it for $r>1$ (it is obvious for $r=1$), we write

$$\Sigma\,|a_n+b_n|^r \leqslant \Sigma\,|a_n+b_n|^{r-1}\,|a_n| + \Sigma\,|a_n+b_n|^{r-1}\,|\,b_n|,$$

and apply Hölder's inequality, with exponents r' and r, to the sums on the right. We get

$$\mathfrak{S}_r^r[a+b] \leqslant \mathfrak{S}_r^{r-1}[a+b]\,\mathfrak{S}_r[a] + \mathfrak{S}_r^{r-1}[a+b]\,\mathfrak{S}_r[b],$$

from which (9·10) follows, provided $\mathfrak{S}_r[a+b]$ is finite. Hence (9·10) holds when $\{a_n\}$ and $\{b_n\}$ are finite, and so also in the general case by passing to the limit.

A similar argument proves *Minkowski's inequality for integrals*

$$\mathfrak{M}_r[f+g] \leqslant \mathfrak{M}_r[f] + \mathfrak{M}_r[g] \quad (r \geqslant 1), \tag{9·11}$$

which implies that if f and g belong to L^r so does $f+g$.

Let $h(x,y)$ be a function defined for $a \leqslant x \leqslant b$, $c \leqslant y \leqslant d$. An argument similar to that which leads to (9·10) and (9·11) also gives the inequality

$$\left\{\int_a^b \left|\int_c^d h(x,y)\,dy\right|^r dx\right\}^{1/r} \leqslant \int_c^d \left\{\int_a^b |\,h(x,y)\,|^r\,dx\right\}^{1/r} dy \quad (r \geqslant 1) \tag{9·12}$$

which may be considered as a generalized form of Minkowski's inequality since it contains (9·10) and (9·11) as special cases. For if $(c,d)=(0,2)$, $h(x,y)=f(x)$ for $0 \leqslant y<1$, $h(x,y)=g(x)$ for $1 \leqslant y \leqslant 2$, (9·12) reduces to (9·11). If $(c,d)=(0,2)$, $(a,b)=(0,+\infty)$, and if for $n \leqslant x < n+1$ we set $h(x,y)=a_n$ or $h(x,y)=b_n$, according as $0 \leqslant y<1$ or $1 \leqslant y \leqslant 2$ $(n=0,1,\ldots)$, (9·12) gives (9·10).

The inequality (9·12) can also be written slightly differently. Let

$$H(x)=\int_c^d h(x,y)\,dy.$$

Then

$$\mathfrak{M}_r[H(x)] \leqslant \int_c^d \mathfrak{M}_r^x[h(x,y)]\,dy,$$

where $\mathfrak{M}_r^x$ means that integration is with respect to x.

If $0<r<1$, (9·10) and (9·11) cease to be true, but we have then the substitutes

$$\mathfrak{S}_r^r[a+b] \leqslant \mathfrak{S}_r^r[a] + \mathfrak{S}_r^r[b], \quad \mathfrak{M}_r^r[f+g] \leqslant \mathfrak{M}_r^r[f] + \mathfrak{M}_r^r[g] \quad (0<r \leqslant 1). \tag{9·13}$$

These are corollaries of the inequality $(x+y)^r \leqslant x^r + y^r$, or, what is the same thing,

$$(1+t)^r \leqslant 1+t^r \quad (t \geqslant 0,\ 0<r<1).$$

To prove the latter we observe that $(1+t)^r-1-t^r$ vanishes for $t=0$ and has a negative derivative for $t>0$.

In this connexion we may note in passing the inequality (a consequence of the last one)

$$|\,\Sigma a_n\,|^r \leqslant \Sigma\,|\,a_n\,|^r \quad (0<r \leqslant 1).$$

(9·14) THEOREM. *Given any function $F(x)$, $a \leqslant x \leqslant b$, and a number $1 \leqslant r < +\infty$, we have*

$$\mathfrak{M}_r[F;\,a,b] = \sup_G \left|\int_a^b FG\,dx\right|, \tag{9·15}$$

where the sup is taken over all G with $\mathfrak{M}_{r'}[G;\,a,b] \leqslant 1$. The result holds if $\mathfrak{M}_r[F]=+\infty$.

We may suppose that $\mathfrak{M}_r[F]>0$. Let I_G denote the integral on the right. By Hölder's inequality,

$$|I_G| \leqslant \mathfrak{M}_r[F]\,\mathfrak{M}_{r'}[G] \leqslant \mathfrak{M}_r[F],$$

a result true even if $\mathfrak{M}_r[F]=+\infty$. On the other hand, if $\mathfrak{M}_r[F]<+\infty$, we set

$$G_0(x)=|F(x)|^{r-1}\,\overline{\operatorname{sign} F(x)}/\mathfrak{M}_r^{r-1}[F] \quad \text{for} \quad r\geqslant 1.$$

We verify that $\mathfrak{M}_{r'}[G_0]=1$, $I_{G_0}=\mathfrak{M}_r[F]$. This proves (9·15) if $\mathfrak{M}_r[F]$ is finite.

If $\mathfrak{M}_r[F]=+\infty$, we have to show that there exist functions G with $\mathfrak{M}_{r'}[G]\leqslant 1$ and such that I_G exists and is arbitrarily large. Suppose first that $b-a<+\infty$, and let $F^n(x)$ denote the function equal to $F(x)$ where $|F(x)|\leqslant n$, and to 0 otherwise. Let $G^n(x)$ be derived from $F^n(x)$ in the same way as G_0 was derived from F. If n is large enough, $\mathfrak{M}_r[F^n]$ is positive (and finite), so that $\mathfrak{M}_{r'}[G^n]=1$ but

$$I_G=\int_a^b FG^n dx=\int_a^b F^nG^n dx=\mathfrak{M}_r[F_n]$$

is arbitrarily large with n.

If (a,b) is infinite, $(0,+\infty)$ for instance, we define $F^n(x)$ as previously, but only in the interval $0\leqslant x\leqslant n$, with $F^n(x)=0$ outside. This ensures the finiteness of $\mathfrak{M}_r[F^n]$ for every n, and the rest of the argument is unchanged.

Theorem (9·14) *also holds for* $r=\infty$. The proof of $|I_G|\leqslant\mathfrak{M}_r[F]$ remains unchanged in this case. On the other hand, if M is the essential upper bound of $|F|$, and if $0<M'<M$, the set E of points where $|F|\geqslant M'$ is of positive measure. If we choose a subset E_1 of E with $0<|E_1|<\infty$, then the function $G(x)$, equal to $\overline{\operatorname{sign} F(x)}/|E_1|$ in E_1 and to 0 elsewhere, has the property

$$\mathfrak{M}_{r'}[G]=\int_a^b |G|\,dx=1, \quad I_G=\frac{1}{|E_1|}\int_{E_1}|F|\,dx\geqslant M',$$

so that $\sup|I_G|\geqslant M$.

We conclude with the following theorem:

(9·16) THEOREM. *Let $f(x)$ be a non-negative function defined for $x\geqslant 0$, and let $r>1$, $s<r-1$. Then if $f^r(x)\,x^s$ is integrable over $(0,\infty)$ so is $\{x^{-1}F(x)\}^r x^s$, where $F(x)=\int_0^x f\,dt$. Moreover,*

$$\int_0^\infty \left\{\frac{F(x)}{x}\right\}^r x^s\,dx \leqslant \left(\frac{r}{r-s-1}\right)^r \int_0^\infty f^r(x)\,x^s dx. \tag{9·17}$$

We may suppose that $f\not\equiv 0$. Hölder's inequality

$$\int_0^x f t^{s/r} t^{-s/r}\,dt \leqslant \left(\int_0^x f^r t^s\,dt\right)^{1/r}\left(\int_0^x t^{-s/(r-1)}\,dt\right)^{1/r'}$$

shows that f is integrable over any finite interval and that $F(x)=o(x^{(r-1-s)/r})$ as $x\to 0$. The last estimate holds also as $x\to\infty$. For, applying the preceding argument to the integral defining $F(x)-F(\xi)$, we have

$$F(x)-F(\xi)<\epsilon x^{(r-1-s)/r}$$

if $x>\xi$ and $\xi=\xi(\epsilon)$ is large enough. Hence $F(x)<2\epsilon x^{(r-1-s)/r}$ for large x; that is,

$$F(x)=o(x^{(r-1-s)/r}) \quad \text{as} \quad x\to\infty.$$

Now let $0<a<b<\infty$. Integrating $\int_a^b F^r x^{s-r}\,dx$ by parts and applying Hölder's inequality to $\int_a^b F^{r-1} f x^{s-r+1}\,dx = \int_a^b (f x^{s/r})(F^{r-1} x^{s-r+1-s/r})\,dx$, we obtain

$$\int_a^b \left(\frac{F}{x}\right)^r x^s\,dx \leqslant -\left[\frac{F^r x^{s-r+1}}{r-s-1}\right]_a^b + \frac{r}{r-s-1}\left\{\int_a^b f^r x^s\,dx\right\}^{1/r}\left\{\int_a^b \left(\frac{F}{x}\right)^r x^s\,dx\right\}^{1/r'}.$$

Divide both sides by the last factor on the right, which is positive if a and $1/b$ are small enough. Since the integrated term tends to 0 as $a\to 0$, $b\to\infty$, we are led to (9·17).

The cases $s=0$ and $s=r-2$ are the most interesting in application.

10. Convex functions

A function $\phi(x)$ defined in an open or closed interval (a, b) is said to be *convex* if for every pair of points P_1, P_2 on the curve $y=\phi(x)$ the points of the arc P_1P_2 are below, or on, the chord P_1P_2. For example, x^r, with $r\geqslant 1$, is convex in $(0, +\infty)$.

Jensen's inequality states that for any system of positive numbers $p_1, p_2, \ldots, p_n$, and for any system of points $x_1, x_2, \ldots, x_n$ in (a, b),

$$\phi\left(\frac{p_1x_1+p_2x_2+\ldots+p_nx_n}{p_1+p_2+\ldots+p_n}\right)\leqslant\frac{p_1\phi(x_1)+\ldots+p_n\phi(x_n)}{p_1+\ldots+p_n}. \tag{10·1}$$

For $n=2$ this is just the definition of convexity, and for $n>2$ it follows by induction. Zero values of the p's may be allowed provided that $\Sigma p_i \neq 0$. If $-\phi(x)$ is convex, $\phi(x)$ is called *concave*. Linear functions are the only ones which are both convex and concave. Concave functions satisfy the inequality opposite to (10·1).

Let P_1, P_2, P_3 be three points on the convex curve $y=\phi(x)$, in the order indicated. Since P_2 is below or on the chord P_1P_3, the slope of P_1P_2 does not exceed that of P_1P_3. Hence, if a point P approaches P_1 from the right the slope of P_1P is non-increasing. Thus the *right-hand side derivative $D^+\phi(x)$ exists for every $a\leqslant x<b$ and is less than $+\infty$.* Similarly, the left-hand side derivative $D^-\phi(x)$ exists for every $a<x\leqslant b$ and is greater than $-\infty$.

If P_1, P, P_2 are points on the curve, in this order, the slope of P_1P does not exceed that of PP_2. Making $P_1\to P$, $P_2\to P$, we have

$$-\infty<D^-\phi(x)\leqslant D^+\phi(x)<+\infty \quad (a<x<b). \tag{10·2}$$

In particular, $\phi(x)$ is continuous in the interior of (a, b). The function ϕ may, however, be discontinuous at the end-points a, b (take the example $\phi(x)=0$ for $0<x<1$, $\phi(0)=\phi(1)=1$).

From the proof of the existence of $D^+\phi(x_0)$ and $D^-\phi(x_0)$, and from (10·2), we see that every straight line l passing through the point $(x_0, \phi(x_0))$ and having a slope k satisfying $D^-\phi(x_0)\leqslant k\leqslant D^+\phi(x_0)$ has at least one point in common with the curve $y=\phi(x)$, and that the curve is above or on l. Such a straight line is called a *supporting line* for the curve $y=\phi(x)$.

Let $x_1<x<x_2$ be the abscissae of P_1, P, P_2. The slope of P_1P does not exceed that of PP_2. The former is at least $D^+\phi(x_1)$, the latter at most $D^-\phi(x_2)\leqslant D^+\phi(x_2)$; thus

$$D^+\phi(x_1)\leqslant D^-\phi(x_2), \quad D^+\phi(x_1)\leqslant D^+\phi(x_2) \quad (x_1<x_2). \tag{10·3}$$

The second inequality shows that *$D^+\phi(x)$ is a non-decreasing function of x*. The same holds for $D^-\phi(x)$. Since $D^+\phi(x)$ is non-decreasing, it is continuous except possibly at a denumerable set of points. Let x be a point of continuity of $D^+\phi$ and let $x_1 < x$. Then, by (10·3) and (10·2),

$$D^+\phi(x_1) \leqslant D^-\phi(x) \leqslant D^+\phi(x).$$

Since $D^+\phi(x_1) \to D^+\phi(x)$ as $x_1 \to x$, we see that $D^-\phi(x) = D^+\phi(x)$ and so $\phi'(x)$ exists and is finite. Summing up we have:

(10·4) THEOREM. *A function $\phi(x)$ convex in an interval (a, b) is continuous for $a < x < b$. The one-sided derivatives of ϕ exist, are non-decreasing and satisfy* (10·2). *The derivative $\phi'(x)$ exists except possibly at a denumerable set of points.*

We have seen that the continuity of ϕ is a consequence of convexity. If, however, we *assume* that $\phi(x)$ is continuous, we may modify the definition of convexity slightly. *A continuous function $\phi(x)$ is convex if and only if, given any arc P_1P_2 of the curve, there is a subarc $P'_1P'_2$ lying below or on the chord P_1P_2.* The condition is obviously necessary. Suppose that it is satisfied, but $\phi(x)$ is not convex. The curve would then contain an arc P_1P_2 for which a certain subarc $P'_1P'_2$ would be everywhere above the chord P_1P_2. Moving P'_1 to the left, P'_2 to the right, we may suppose that P'_1 and P'_2 are on the chord P_1P_2 and the rest of the arc $P'_1P'_2$ is above that chord. But then no subarc of $P'_1P'_2$ is below the chord $P'_1P'_2$, contrary to hypothesis.

A convex function has no proper maximum† *in the interior of the interval of definition.* For if x_0 were such a maximum, the arc $y = \phi(x)$, $|x - x_0| \leqslant \delta$ would, for δ small enough, be partly above the chord.

(10·5) THEOREM. *A necessary and sufficient condition that a function $\phi(x)$, continuous in (a, b), should be convex is that for no pair of values α, β should the sum $\phi(x) + \alpha x + \beta$ have a proper maximum in the interior of (a, b).*

A sum of two convex functions being convex, the necessity of the condition is evident. To prove its sufficiency, suppose that $\phi(x)$ is not convex. There is then an arc P_1P_2 of the curve with all points above or on the chord P_1P_2. Let x_1, x_2 be the abscissae of P_1, P_2, and let $y = -\alpha x - \beta$ be the equation of the chord. Then $\phi(x) + \alpha x + \beta$ vanishes at the end-points of (x_1, x_2) and takes some positive values inside; it has therefore a proper maximum inside (x_1, x_2) and so also inside (a, b).

The following generalizations of ordinary first and second derivatives are useful. Given an $F(x)$ defined in the neighbourhood of x_0, let us consider the ratios

$$\left.\begin{aligned} &\frac{F(x_0+h) - F(x_0-h)}{2h}, \\ &\frac{F(x_0+h) + F(x_0-h) - 2F(x_0)}{h^2}. \end{aligned}\right\} \qquad (10{\cdot}6)$$

The limits (if they exist) of these expressions, as $h \to 0$, will be called respectively the *first* and *second symmetric derivatives* of F at the point x_0, and will be denoted by $D_1F(x_0)$ and $D_2F(x_0)$. The limit superior and the limit inferior of the ratios (10·6) are called the

† We say that $\phi(x)$ has a *proper* maximum at x_0 if $\phi(x_0) \geqslant \phi(x)$ in a neighbourhood of x_0, but ϕ is not constant in any neighbourhood of x_0.

upper and the *lower* (first or second) symmetric derivatives and will be denoted by $\overline{D}_1 F(x_0)$, $\underline{D}_1 F(x_0)$, $\overline{D}_2 F(x_0)$, $\underline{D}_2 F(x_0)$. The second symmetric derivative is often called the *Schwarz* (*or Riemann*) *derivative*.

If $F'(x_0)$ exists, so does $D_1 F(x_0)$ and both have the same value. For the first ratio (10·6) is half the sum of the ratios $\{F(x_0+h)-F(x_0)\}/h$ and $\{F(x_0)-F(x_0-h)\}/h$, which tend to $F'(x_0)$.

If $F''(x_0)$ exists, so does $D_2 F(x_0)$ and both have the same value. For Cauchy's mean-value theorem applied to the second ratio (10·6), h being the variable, shows that it can be written as $\{F'(x_0+k)-F'(x_0-k)\}/2k$ for some $0<k<h$, and the last ratio tends to $F''(x_0)$ as $k\to 0$.

These proofs actually give slightly more, namely: (i) both $\overline{D}_1 F(x_0)$ and $\underline{D}_1 F(x_0)$ are contained between the least and the greatest of the four Dini numbers of F at x_0; (ii) if $F'(x)$ exists in the neighbourhood of x_0, then both $\overline{D}_2 F(x_0)$ and $\underline{D}_2 F(x_0)$ are contained between the least and the greatest of the four Dini numbers of F' at x_0.

(10·7) THEOREM. *A necessary and sufficient condition for a continuous $\phi(x)$ to be convex in the interior of (a, b) is that $\overline{D}_2\phi(x)\geqslant 0$ there.*

We may suppose that (a, b) is finite. Since

$$\phi(x+h)+\phi(x-h)-2\phi(x)=\{\phi(x+h)-\phi(x)\}-\{\phi(x)-\phi(x-h)\}\geqslant 0$$

for a convex ϕ, the necessity of the condition (even in the stronger form $\underline{D}_2\phi\geqslant 0$) follows. To prove the sufficiency, let us first assume slightly more, namely, that $\overline{D}_2\phi>0$ in (a, b). If ϕ were not convex, the function $\psi(x)=\phi(x)+\alpha x+\beta$ would, for suitable α, β, have a maximum at a point x_0 inside (a, b), so that $\psi(x_0+h)+\psi(x_0-h)-2\psi(x_0)$ would be non-positive for small h. Since this expression equals

$$\phi(x_0+h)+\phi(x_0-h)-2\phi(x_0),$$

it follows that $\overline{D}_2\phi(x_0)\leqslant 0$, contrary to hypothesis.

Returning to the general case, consider the functions $\phi_n(x)=\phi(x)+x^2/n$. We have

$$\overline{D}_2\phi_n(x)=\overline{D}_2\phi(x)+2/n>0,$$

so that ϕ_n is convex. The limit of a convergent sequence of convex functions is convex (applying (10·1) with $n=2$); and since $\phi_n\to\phi$, ϕ is convex.

A necessary and sufficient condition for a function ϕ twice differentiable to be convex is that $\phi''\geqslant 0$. This follows from (10·7).

Suppose that $\phi(u)$ is convex for $u\geqslant 0$, and that u_0 is a minimum of $\phi(u)$. *If $\phi(u)$ is not constant for $u\geqslant u_0$, then it must tend to $+\infty$ with u at least as rapidly as a fixed positive multiple of u.* For let $u_1>u_0$, $\phi(u_1)\neq\phi(u_0)$. Clearly $\phi(u_1)>\phi(u_0)$. If P_0, P_1, P are points of the curve with abscissae u_0, u_1, u, where $u>u_1$, the slope of P_0P is not less than that of P_0P_1. This proves the assertion.

If $\phi(u)$ is non-negative convex and non-decreasing in $(0, +\infty)$, but not constant, the relation $f\in L_\phi(a, b)$, $b-a<\infty$, implies $f\in L(a, b)$. For then there is a $k>0$ such that $\phi(|f(x)|)\geqslant k|f(x)|$, if $|f(x)|$ is large enough.

Jensen's inequality for integrals is

$$\phi\left\{\frac{\int_a^b f(x)\,p(x)\,dx}{\int_a^b p(x)\,dx}\right\} \leqslant \frac{\int_a^b \phi\{f(x)\}\,p(x)\,dx}{\int_a^b p(x)\,dx}, \tag{10·8}$$

the hypotheses being that $\phi(u)$ is convex in an interval $\alpha \leqslant u \leqslant \beta$, that $\alpha \leqslant f(x) \leqslant \beta$ in $a \leqslant x \leqslant b$, that $p(x)$ is non-negative and $\not\equiv 0$, and that all the integrals in question exist.

Let

$$\gamma = \int_a^b fp\,dx \Big/ \int_a^b p\,dx, \tag{10·9}$$

so that $\alpha \leqslant \gamma \leqslant \beta$. Suppose first that $\alpha < \gamma < \beta$, and let k be the slope of a supporting line of ϕ through the point $(\gamma, \phi(\gamma))$. Then

$$\phi(u) - \phi(\gamma) \geqslant k(u-\gamma) \quad (\alpha \leqslant u \leqslant \beta).$$

Replacing here u by $f(x)$, multiplying both sides by $p(x)$, and integrating over $a \leqslant x \leqslant b$, we get

$$\int_a^b \phi\{f(x)\}\,p(x)\,dx - \phi(\gamma)\int_a^b p(x)\,dx \geqslant k\left\{\int_a^b f(x)\,p(x)\,dx - \gamma\int_a^b p(x)\,dx\right\} = 0,$$

by (10·9). This gives (10·8). If $\gamma = \beta$, (10·9) can be written $\int_a^b (f-\beta)\,p\,dx = 0$, which shows that $f(x) = \beta$ at almost all points at which $p > 0$. But then both sides of (10·8) reduce to $\phi(\beta)$. Similarly if $\gamma = \alpha$.

Jensen's inequality for Stieltjes integrals is

$$\phi\left\{\frac{\int_a^b f(x)\,d\omega(x)}{\int_a^b d\omega(x)}\right\} \leqslant \frac{\int_a^b \phi\{f(x)\}\,d\omega(x)}{\int_a^b d\omega(x)}, \tag{10·10}$$

where $\omega(x)$ is non-decreasing but not constant. The proof is similar to the one above.

(10·11) Theorem. *A necessary and sufficient condition that $\phi(x)$ $(a < x < b)$ should be convex is that it should be the integral of a non-decreasing function.*

If $\phi(x)$ is convex, then, as is easily seen, the ratio $\{\phi(x+h) - \phi(x)\}/h$ is uniformly bounded for $x, x+h$ belonging to any interval (a', b') interior to (a, b). Thus $\phi(x)$ is absolutely continuous, and therefore is the integral of ϕ'. The latter exists outside a denumerable set and is non-decreasing on the set where it exists. Completing ϕ' at the exceptional points so that the new function is still non-decreasing, we see that ϕ is the integral of a non-decreasing function. Conversely, suppose that $\phi(x) = C + \int_c^x \psi(t)\,dt$, where $a < c < b$ and $\psi(t)$ is non-decreasing in (a, b). Let (a', b') be any subinterval of (a, b), and let $y = l(x)$ be the equation of the chord through $(a', \phi(a'))$ and $(b', \phi(b'))$. We have to show that $\phi(x) - \phi(a') \leqslant l(x) - l(a')$ for $a' < x < b'$, or, what is the same thing, that

$$\frac{1}{x-a'}\int_{a'}^x \psi\,dt \leqslant \frac{1}{b'-a'}\int_{a'}^{b'} \psi\,dt = \frac{\int_{a'}^x + \int_x^{b'}}{(x-a') + (b'-x)}.$$

Since the last expression is contained between $\int_{a'}^{x} \Big/ (x-a')$ and $\int_{x}^{b'} \Big/ (b'-x)$, of which the latter is not less than the former (since ψ does not decrease), the proof of (10·11) is completed.

Let now $\phi(x)$, $x \geq 0$, be an arbitrary function, non-negative, non-decreasing, vanishing at $x=0$ and tending to $+\infty$ with x. The curve $y=\phi(x)$ may possess discontinuities and stretches of constancy. If at each point x_0 of discontinuity of ϕ we adjoin to the curve $y=\phi(x)$ the vertical segment $x=x_0$, $\phi(x_0-0) \leq y \leq \phi(x_0+0)$, we obtain a continuous curve, and we may define a function $\psi(y)$ inverse to $\phi(x)$ by defining $\psi(y_0)$ $(0 \leq y_0 < \infty)$ to be any x_0 such that the point (x_0, y_0) is on the continuous curve. The stretches of constancy of ϕ then correspond to discontinuities of ψ, and conversely. The function $\psi(y)$ is defined uniquely except for the y's which correspond to the stretches of constancy of ϕ, but since the set of such stretches is denumerable, our choice of $\psi(y)$ has no influence upon the integral $\Psi(y)$ of $\psi(y)$, and it is easy to see that the inequality (9·1) is valid in this slightly more general case.

From (10·11) it follows that *every function $\Phi(x)$, $x \geq 0$, which is non-negative, convex, and satisfies the relation $\Phi(0)=0$ and $\Phi(x)/x \to \infty$*, may be considered as a Young's function (see p. 16). More precisely, to every such function corresponds another function $\Psi(x)$ with similar properties, such that

$$ab \leq \Phi(a) + \Psi(b)$$

for every $a \geq 0$, $b \geq 0$. It is sufficient to take for $\Psi(y)$ the integral over $(0, y)$ of the function $\psi(x)$ inverse to $\phi(x) = D^{+}\Phi(x)$. Since $\Phi(x)/x$ tends to $+\infty$ with x, it is easy to see that $\phi(x)$ and $\psi(x)$ also tend to $+\infty$ with x. We have $ab = \Phi(a) + \Psi(b)$ if and only if the point (a, b) is on the continuous curve obtained from the function $y=\phi(x)$.

A non-negative function $\psi(u)$ $(a \leq u \leq b)$ will be called *logarithmically convex* if

$$\psi(t_1 u_1 + t_2 u_2) \leq \psi^{t_1}(u_1)\, \psi^{t_2}(u_2)$$

for u_1 and u_2 in (a, b), t_1 and t_2 positive and of sum 1. It is immediate that then either ψ is identically zero, or else ψ is strictly positive and $\log \psi$ is convex.

(10·12) THEOREM. *For any given function f, and for $\alpha > 0$,*

(i) *$\mathfrak{A}_\alpha[f]$ is a non-decreasing function of α;*

(ii) *$\mathfrak{M}_\alpha^\alpha[f]$ and $\mathfrak{A}_\alpha^\alpha[f]$ are logarithmically convex functions of α;*

(iii) *$\mathfrak{M}_{1/\alpha}[f]$ and $\mathfrak{A}_{1/\alpha}[f]$ are logarithmically convex functions of α.*

If we substitute $|f|^\alpha$ for f and 1 for g in (9·5), and divide both sides by $b-a$, we have $\mathfrak{A}_\alpha[f] \leq \mathfrak{A}_{\alpha r}[f]$ for $r > 1$. This proves (i). The result is not true for $\mathfrak{M}_\alpha$, as may be seen from the example $a=0$, $b=2$, $f(x)=1$.

Let now $\alpha = \alpha_1 t_1 + \alpha_2 t_2$ with $\alpha_i, t_i > 0$, $t_1 + t_2 = 1$, and suppose that f belongs to both L^{α_1} and L^{α_2}. Replacing the integrand $|f|^\alpha$ in $\mathfrak{M}_\alpha^\alpha$ by $|f|^{\alpha_1 t_1} . |f|^{\alpha_2 t_2}$, and applying Hölder's inequality with $r = 1/t_1$, $r' = 1/t_2$, we find

$$\mathfrak{M}_\alpha^\alpha \leq \mathfrak{M}_{\alpha_1}^{\alpha_1 t_1} \mathfrak{M}_{\alpha_2}^{\alpha_2 t_2},$$

which expresses the logarithmic convexity of $\mathfrak{M}_\alpha^\alpha[f]$. Dividing both sides by $b-a$, we have the result for $\mathfrak{A}_\alpha^\alpha$. Thus (ii) is proved.

(iii) we apply Hölder's inequality with the exponents $r=\alpha/\alpha_1 t_1>1$, We get

$$\mathfrak{M}_{1/\alpha}[f]=\left\{\int_a^b |f|^{1/\alpha}\,dx\right\}^\alpha=\left\{\int_a^b |f|^{t_1/\alpha}\,|f|^{t_2/\alpha}\,dx\right\}^\alpha$$

$$\leqslant\left(\int_a^b |f|^{1/\alpha_1}\,dx\right)^{\alpha_1 t_1}\left(\int_a^b |f|^{1/\alpha_2}\,dx\right)^{\alpha_2 t_2}=\mathfrak{M}^{t_1}_{1/\alpha_1}\mathfrak{M}^{t_2}_{1/\alpha_2}.$$

Dividing both sides by $(b-a)^\alpha$ we have the result for $\mathfrak{A}_{1/\alpha}$.

11. Convergence in L^r

Let $f_1(x), f_2(x), \ldots$ be a sequence of functions belonging to $L^r(a,b)$, $r>0$. If there is a function $f \in L^r(a,b)$ such that $\mathfrak{M}_r[f-f_n;\,a,b]\to 0$ as $n\to\infty$, we say that $\{f_n(x)\}$ *converges in* $L^r(a,b)$ (or, simply, *in* L^r) *to* $f(x)$.

(11·1) THEOREM. *A necessary and sufficient condition that a sequence of functions $f_n(x)\in L^r(a,b)$, $r>0$, should converge in L^r to some $f(x)$ is that $\mathfrak{M}_r[f_m-f_n]$ should tend to 0 as $m,n\to\infty$.*

If $r\geqslant 1$, the necessity of the condition follows from Minkowski's inequality, since, if $\mathfrak{M}_r[f-f_m]\to 0$, $\mathfrak{M}_r[f-f_n]\to 0$, then

$$\mathfrak{M}_r[f_m-f_n]\leqslant \mathfrak{M}_r[f-f_m]+\mathfrak{M}_r[f-f_n]\to 0.$$

For $0<r<1$, we use instead the second inequality (9·13).

The proof of sufficiency depends on the following further theorems, themselves important.

(11·2) FATOU'S LEMMA. *Let $g_1(x), g_2(x), \ldots$ be non-negative functions, integrable over (a,b) and satisfying*

$$\int_a^b g_k\,dx\leqslant A<+\infty \quad (k=1,2,\ldots). \tag{11·3}$$

If $g(x)=\lim g_k(x)$ exists almost everywhere, then g is integrable and

$$\int_a^b g\,dx\leqslant A. \tag{11·4}$$

Let $h_k(x)=\inf\{g_k(x), g_{k+1}(x), \ldots\}$. The function h_k is measurable and majorized by g_k, and so integrable. Since $h_k\leqslant h_{k+1}$ and $h_k\to g$, (11·4) follows from (11·3) by Lebesgue's theorem on the integration of monotone sequences.

(11·5) THEOREM. *Let $\{u_n(x)\}$ be a sequence of non-negative functions, and write $I_n=\int_a^b u_n\,dx$. If $I_1+I_2+\ldots<\infty$, then $u_1(x)+u_2(x)+\ldots$ converges almost everywhere in (a,b) to a finite sum. In particular $u_n(x)\to 0$ almost everywhere in (a,b).*

For if $u_1+u_2+\ldots$ diverged to $+\infty$ in a set of positive measure, Lebesgue's theorem mentioned above would imply that $I_1+I_2+\ldots=\infty$.

(11·6) THEOREM. *If $\mathfrak{M}_r[f_m-f_n;\,a,b]\to 0$ as $m,n\to\infty$, we can find a subsequence $\{f_{n_k}\}$ of $\{f_n\}$ which converges almost everywhere in (a,b).*

Suppose first that $r \geqslant 1$. Let $\epsilon_i = \sup \mathfrak{M}_r[f_m - f_n]$ for $m, n \geqslant i$. Since $\epsilon_i \to 0$, we have $\epsilon_{n_1} + \epsilon_{n_2} + \ldots < +\infty$ if n_k increases sufficiently rapidly. By Hölder's inequality,

$$\int_a^b |f_{n_k} - f_{n_{k+1}}| \, dx \leqslant (b-a)^{1/r'} \mathfrak{M}_r[f_{n_k} - f_{n_{k+1}}] \leqslant (b-a)^{1/r'} \epsilon_{n_k}, \tag{11·7}$$

and so, by (11·5), the series $|f_{n_1}| + |f_{n_2} - f_{n_1}| + |f_{n_3} - f_{n_2}| + \ldots$ converges almost everywhere in (a, b). The function $f(x) = f_{n_1} + (f_{n_2} - f_{n_1}) + \ldots = \lim f_{n_k}$ thus exists almost everywhere.

To establish the existence of $f(x)$ for $0 < r < 1$, we note that

$$(|f_{n_1}| + |f_{n_2} - f_{n_1}| + \ldots)^r \leqslant |f_{n_1}|^r + |f_{n_2} - f_{n_1}|^r + \ldots;$$

when we integrate the right-hand side of this inequality we obtain a finite number, provided n_k increases so fast that $\epsilon_{n_1}^r + \epsilon_{n_2}^r + \ldots < \infty$.

In this proof we tacitly assumed that (a, b) was finite, but the argument holds even if $b - a = \infty$, since (11·7) remains valid if (a, b) is replaced by any of its finite subintervals.

Returning to (11·1), let $\{n_k\}$ and $\{\epsilon_i\}$ be the sequences of Theorem (11·6) and $f = \lim f_{n_k}$. We have $\mathfrak{M}_r[f_m - f_{n_k}] \leqslant \epsilon_m$ for $n_k \geqslant m$. By (11·2), $\mathfrak{M}_r[f_m - f] \leqslant \epsilon_m$, and this completes the proof.

It is important to observe that the function f satisfying $\mathfrak{M}_r[f - f_m] \to 0$ is unique. Suppose that $\mathfrak{M}_r[f - f_m] \to 0$, $\mathfrak{M}_r[g - f_m] \to 0$. If $r \geqslant 1$, by Minkowski's inequality,

$$\mathfrak{M}_r[f - g] \leqslant \mathfrak{M}_r[f - f_m] + \mathfrak{M}_r[g - f_m] \to 0, \quad \text{so that} \quad \mathfrak{M}_r[f - g] = 0, \quad f \equiv g.$$

If $0 < r < 1$, we use instead the inequality

$$\mathfrak{M}_r^r[f - g] \leqslant \mathfrak{M}_r^r[f - f_m] + \mathfrak{M}_r^r[g - f_m].$$

(11·8) THEOREM. *Suppose that* $f \in L^r(a, b)$, $0 < r < +\infty$. *Then, given any* $\epsilon > 0$, *there is a continuous function* $\phi(x)$ *such that* $\mathfrak{M}_r[f - \phi] < \epsilon$.

Suppose first that $r \geqslant 1$, $b - a < +\infty$. There is a bounded function $\psi(x)$ such that $\mathfrak{M}_r[f - \psi] < \frac{1}{2}\epsilon$; for, taking N large enough, we may define $\psi(x)$ as equal to f wherever $|f| \leqslant N$, and equal to 0 elsewhere. If we can find a continuous $\phi(x)$ such that $\mathfrak{M}_r[\psi - \phi] < \frac{1}{2}\epsilon$, the result will follow by Minkowski's inequality. Let us set $\psi(x) = 0$ outside (a, b), and let $\Psi(x)$ be the indefinite integral of $\psi(x)$. The functions

$$\psi_n(x) = n[\Psi(x + 1/n) - \Psi(x)]$$

are continuous and uniformly bounded, and, by Lebesgue's theorem on the differentiation of the indefinite integral, tend to $\psi(x)$ almost everywhere in (a, b). Thus $\mathfrak{M}_r[\psi - \psi_n] \to 0$, and it is enough to set $\phi = \psi_n$, with n large enough. The modifications in the case $0 < r < 1$ are obvious.

The above argument holds if $b - a = \infty$, provided that $f(x) = 0$ for $|x|$ large enough. The general case can be reduced to this one; for if we modify f by setting it equal to 0 outside a sufficiently large interval, we get a function f_1 with $\mathfrak{M}_r[f - f_1]$ arbitrarily small.

(11·9) THEOREM. *Suppose that a sequence of functions* $f_n(x)$ *converges almost everywhere in a finite interval* (a, b) *to a limit* $f(x)$, *and that* $\mathfrak{M}_r[f_n; a, b] \leqslant M < +\infty$ *for a fixed* $r > 0$ *and all* n. *Then* $\mathfrak{M}_s[f_n - f] \to 0$ *as* $n \to \infty$, *for* $0 < s < r$.

Obviously, $\mathfrak{M}_r[f] \leqslant M$. Let E be a set of points on which $\{f_n(x)\}$ converges uniformly to $f(x)$, and let $D=(a,b)-E$; $|D|$ can be arbitrarily small. Clearly

$$\int_a^b |f_n-f|^s dx = \int_E + \int_D \leqslant o(1) + \left(\int_D |f_n-f|^r dx\right)^{s/r} |D|^{1-s/r},$$

by Hölder's inequality. By Minkowski's inequality, if $r \geqslant 1$, the last term is not greater than $(2M)^s |D|^{1-s/r}$, and so is arbitrarily small with $|D|$. Hence $\mathfrak{M}_s[f_n-f] \to 0$. The proof is similar for $0<r<1$ except that we use (9·13) instead of Minkowski's inequality.

12. Sets of the first and second categories

Let A be a linear point-set. By a *portion* of A we shall mean any *non-empty* intersection AI of A with an *open* interval I.

Let B be a subset of A. B is said to be *dense* in A if every portion of A contains points of B. B is said to be *non-dense* in A if every portion of A contains a portion (*subportion* of A) without points in common with B. A set dense in $(-\infty, +\infty)$ will be called *everywhere dense.*

Let $B \subset A$. If B can be decomposed into a denumerable sum of subsets (not necessarily disjoint) non-dense in A, B will be said to be *of the first category* on A. Otherwise B will be called *of the second category* on A. When $B=A$, we say that A is of the first or second category (as the case may be) *on itself.*

If $A=(-\infty, +\infty)$, we shall simply say that B is of the first or second category, as the case may be.

The following fact is important:

(12·1) THEOREM. *A closed set A (in particular, an interval) is of the second category on itself.*

For suppose that $A=A_1+A_2+\dots$, where the A_i are non-dense on A. In particular, there is a portion I_1A of A containing no points of A_1. In that portion we choose a subportion I_2A containing no points of A_2. In I_2A we choose a subportion I_3A containing no points of A_3, and so on. We may suppose that I_{n+1} is strictly interior to I_n, and that $|I_n| \to 0$. The intervals $I_1, I_2, \dots$ have a point x in common, and since all of them contain points of the closed set A, x must belong to A. Since $x \in I_nA$, x cannot belong to any of $A_1, A_2, \dots, A_n$. This being true for all n, we obtain a contradiction with the relation $A=A_1+A_2+\dots$.

If $B_1, B_2, \dots$ are all of the first category on A, so is $B_1+B_2+\dots$; thus *a subset B of a closed set A and the complementary set A − B cannot both be of the first category on A.*

An everywhere dense set may be of the first category (for example, any denumerable dense set). However, *if a set E is both dense in an interval I, and a denumerable product of open sets, then E is of the second category on I.* For the complementary set $I-E$ is then a denumerable sum of closed sets. These closed sets cannot contain intervals, since that would contradict the assumption that E is dense in I; so they are non-dense in I. Hence $I-E$ is of the first category, and E is of the second category, on I.

(12·2) THEOREM. *Let $f_1(x), f_2(x), \dots$ be a sequence of functions continuous in $a \leqslant x \leqslant b$. If the set E of points x at which the sequence $\{f_n(x)\}$ is unbounded is dense in (a,b),*

E is of the second category on (a, b). (More precisely, the complement of E is of the first category on (a, b).)

It is enough to show that E is a denumerable product of open sets. But if E_N is the set of points at which at least one of the inequalities $|f_n(x)| > N$ is satisfied, then E_N is open, and $E = E_1 E_2 \ldots$.

A set $A \subset (0, 1)$ can be of measure 1 and of the first category, or of measure 0 and of the second category. Thus though we may think of the second category as 'richer' in points than the first category, the new classification cannot be compared with the one based on measure.

(12·3) THEOREM. *Let $f_1(x), f_2(x), \ldots$ be continuous on a closed set E; then*

(i) *if $\limsup_{n\to\infty} f_n(x) < +\infty$ at each point of E, then there is a portion P of E and a number M such that $f_n(x) \leqslant M$ for all n and all $x \in P$;*

(ii) *if $f_n(x)$ converges on E to $f(x)$, then for any $\epsilon > 0$ there is a portion P of E and a number n_0 such that*

$$|f(x) - f_n(x)| \leqslant \epsilon \quad \textit{for} \quad x \in P,\ n \geqslant n_0. \tag{12·4}$$

(iii) *If E is, in addition, non-denumerable (in particular, if E is perfect), then the conclusions of* (i) *and* (ii) *hold even if the hypotheses fail to be satisfied in a denumerable subset D of E.*

(i) Let E_M be the set of x such that $f_n(x) \leqslant M$ for all n. Each E_M is closed and $E = E_1 + E_2 + \ldots$. By (12·1), some E_M is not non-dense on E and so, being closed, must contain a portion P of E. This proves (i).

(ii) For every $k = 1, 2, \ldots$, let E_k be the set of points $x \in E$ such that $|f_m(x) - f_n(x)| \leqslant \epsilon$ for $m, n \geqslant k$. The sets E_k are closed and $E = E_1 + E_2 + \ldots$. As in (i), some E_{n_0} contains a portion P of E. We have $|f_m(x) - f_n(x)| \leqslant \epsilon$ for $x \in P$ and $m, n \geqslant n_0$; this implies (12·4).

(iii) We begin with the extension of (i). Let $x_1, x_2, \ldots$ be the elements of D, and let E'_n be the set E_n in the proof of (i) augmented by the points $x_1, x_2, \ldots, x_n$. E'_n is closed and $E = E'_1 + E'_2 + \ldots$. Hence a certain E'_{m_0} contains a portion of E. If we take m_0 so large that E'_{m_0} is infinite (observe that $E'_1 \subset E'_2 \subset \ldots$), E_{m_0} will also contain a portion of E.

The extension of (ii) is proved similarly.

13. Rearrangements of functions. Maximal theorems of Hardy and Littlewood

In this section, unless otherwise stated, we shall consider only functions $f(x)$, defined in a fixed finite interval, which are non-negative and almost everywhere finite. We may suppose that the interval is of the form $(0, a)$.

For any $f(x)$, we shall denote by $\mathrm{E}(f > y)$ the set of points x such that $f(x) > y$. The measure $|\mathrm{E}(f > y)| = m(y)$ of this set will be called the *distribution function* of f. Two functions f and g will be called *equidistributed* if they have the same distribution functions. It is then clear that if f is integrable over $(0, a)$, so is g, and the integrals are equal. If f and g are equidistributed, so are $\chi(f)$ and $\chi(g)$ for any non-negative and non-decreasing $\chi(u)$.

(13·1) THEOREM. *For any $f(x)$, there exist functions $f^*(x)$ and $f_*(x)$ $(0 < x < a)$ equidistributed with f and respectively non-increasing and non-decreasing.*

The function $m(y) = |E(f>y)|$ is non-increasing and continuous to the right. Clearly $m(y) = a$ for y negative, and $m(+\infty) = 0$. If $m(y)$ is continuous and strictly decreasing for $y \geqslant 0$, then its inverse function, which we shall denote by $f^*(x)$, is decreasing and equidistributed with $f(x)$.

The definition of f^* just given holds, suitably modified, in the general case. Let us consider the curve $x = m(y)$ and a point y_0 of discontinuity of it. We adjoin to the curve the whole segment of points (x, y_0) with $m(y_0+0) < x \leqslant m(y_0 - 0)$ (noting that the point $x = m(y_0) = m(y_0+0)$ belongs to the initial curve) and we do this for every y_0. Every line $x = x_0$, $0 < x_0 \leqslant a$, intersects the new curve in at least one point, whose ordinate we denote by $f^*(x_0)$. The function $f^*(x)$ is defined uniquely for $0 < x \leqslant a$, except at those points which correspond to the stretches of constancy of $m(y)$. Such x are denumerable and for them we take for $f^*(x)$ any value that preserves the monotonicity. Taking into account the discontinuities and the stretches of constancy of $m(y)$, we may verify geometrically that, for each y_0, the set of points x such that $f^*(x) > y_0$ is a segment, with or without end-points, of length $m(y_0)$. Thus $|E(f^* > y_0)| = |E(f > y_0)|$.

We define $f_*(x) = f^*(a-x)$; the properties of f_* then follow trivially from those of f^*.

Suppose that $f(x)$ is integrable over $(0, a)$. For every x, $0 < x \leqslant a$, we set

$$\theta(x) = \theta_f(x) = \sup_\xi \frac{1}{x-\xi}\int_\xi^x f(t)\,dt, \quad \text{where} \quad 0 \leqslant \xi < x. \tag{13·2}$$

Clearly $\theta(x)$ is finite at every point at which the integral of f is differentiable. If f is non-increasing, then

$$\theta_f(x) = \frac{1}{x}\int_0^x f\,dt. \tag{13·3}$$

In particular, this formula applies to the function $f^*(x)$ introduced above.

(13·4) Theorem of Hardy and Littlewood. *For any non-decreasing and non-negative function $\chi(t)$, $t \geqslant 0$, we have*

$$\int_0^a \chi\{\theta_f(x)\}\,dx \leqslant \int_0^a \chi\{\theta_{f^*}(x)\}\,dx = \int_0^a \chi\left\{\frac{1}{x}\int_0^x f^*\,dt\right\}dx. \tag{13·5}$$

First of all we observe that for any $g(x) \geqslant 0$ we have

$$\int_0^a g(x)\,dx = -\int_0^\infty y\,dm(y) = \int_0^\infty m(y)\,dy, \tag{13·6}$$

where $m(y) = |E(g > y)|$. For, if g is bounded, the first equation follows from the fact that the approximating Lebesgue sums for the first integral coincide with the approximating Riemann-Stieltjes sums for the second. In the general case, for $u > 0$,

$$-\int_0^u y\,dm(y) = \int_{E(g \leqslant u)} g(x)\,dx,$$

and the result follows by making $u \to \infty$. Finally, the second equality in (13·6) follows from integration by parts, if we observe that

$$ym(y) \to 0 \text{ as } y \to \infty \quad \left(\text{since } ym(y) \leqslant \int_{E(g>y)} g\,dx\right).$$

Comparing the extreme terms of (13·6) we see that if we have another function $g_1(x) \geqslant 0$ and the corresponding $m_1(y)$, then the inequality $m_1(y) \geqslant m(y)$ for all y

implies that the integral of g_1 is not less than that of g. Hence, $\chi(t)$ being monotone, the inequality (13·5) will follow if we show that

$$|\mathrm{E}(\theta_f > y_0)| \leqslant |\mathrm{E}(\theta_{f^*} > y_0)| \quad \text{for all } y_0. \qquad \textbf{(13·7)}$$

We break up the proof of this inequality into three stages.

(13·8) LEMMA. *Given a continuous $F(x)$, $0 \leqslant x \leqslant a$, let H denote the set of points x for which there is a point ξ in $0 \leqslant \xi < x$ such that $F(\xi) < F(x)$. The set H consists of a denumerable system of disjoint intervals (α_k, β_k) such that $F(\alpha_k) \leqslant F(\beta_k)$. All these intervals are open except possibly one terminating at a.*

Since small changes of x do not impair the inequality $F(\xi) < F(x)$, the set H is open, except possibly for the point a. Let (α_k, β_k) be any one of the disjoint intervals (open, except when $\beta_k = a$) constituting H. Assuming that $F(\alpha_k) > F(\beta_k)$, denote by x_0 the smallest number in (α_k, β_k) such that $F(x_0) = \frac{1}{2}\{F(\alpha_k) + F(\beta_k)\}$. Thus no ξ corresponding to x_0 can belong to (α_k, x_0), since the points of this interval satisfy the inequality $F(x) \geqslant F(x_0)$. Hence $\xi < \alpha_k$ and the inequalities $F(\xi) < F(x_0) < F(\alpha_k)$ imply that $\alpha_k \in H$, which is false. It follows that $F(\alpha_k) \leqslant F(\beta_k)$.

Remark. We actually have $F(\alpha_k) = F(\beta_k)$, unless $\beta_k = a$. For no $\beta_k < a$ belongs to H, so that $F(\alpha_k) \geqslant F(\beta_k)$.

(13·9) LEMMA. *If E is any set in $(0, a)$, then* $\displaystyle\int_E f\,dx \leqslant \int_0^{|E|} f^*\,dx.$

Let $g(x)$ be equal to $f(x)$ in E and to 0 elsewhere. Since $g \leqslant f$, we also have $g^* \leqslant f^*$ and

$$\int_E f\,dx = \int_0^a g\,dx = \int_0^a g^*\,dx = \int_0^{|E|} g^*\,dx \leqslant \int_0^{|E|} f^*\,dx,$$

which proves (13·9).

Let $E(y_0)$ and $E^*(y_0)$ denote the sets in (13·7), and let $E_1^*(y_0) = \mathrm{E}(\theta_{f^*} \geqslant y_0)$, with equality this time included. Having fixed y_0 we drop it as an argument and write E, E^*, E_1^*. If we set $F(x) = \displaystyle\int_0^x f\,dt - y_0 x$, the set E becomes the set H of Lemma (13·8). We show that

$$\int_0^{|E|} f^*\,dx \geqslant y_0 |E|. \qquad \textbf{(13·10)}$$

In fact, if (α_k, β_k) are the intervals making up E, then

$$\int_{\alpha_k}^{\beta_k} f\,dx \geqslant y_0(\beta_k - \alpha_k),$$

by (13·8), and summing over all k we get the inequality

$$\int_E f\,dx \geqslant y_0 |E|, \qquad \textbf{(13·11)}$$

from which (13·10) follows, by (13·9).

Return now to (13·3), with f replaced by f^*. Since the right-hand side is a continuous and non-increasing function, $|E_1^*|$ is the greatest number $x \leqslant a$ such that $x^{-1}\displaystyle\int_0^x f^*\,dt \geqslant y_0$. Hence, by (13·10), $|E| \leqslant |E_1^*|$; in full,

$$|\mathrm{E}(\theta_f > y_0)| \leqslant |\mathrm{E}(\theta_{f^*} \geqslant y_0)|.$$

If we replace here y_0 by $y_0+\epsilon$ and make ϵ decrease to 0, we get (13·7), and the proof of (13·4) is completed.

In addition to $\theta_f(x)$, we define the functions

$$\theta'_f(x)=\sup_{\xi}\frac{1}{\xi-x}\int_x^{\xi} f\,dt \quad (x<\xi\leqslant a),$$

$$\Theta_f(x)=\max\{\theta_f(x),\,\theta'_f(x)\}=\sup_{\xi}\frac{1}{\xi-x}\int_x^{\xi} f\,dt \quad (0\leqslant\xi\leqslant a). \tag{13·12}$$

The inequality in (13·5) holds if we replace θ_f by θ'_f and f^* by f_*. Since

$$\Theta=\max(\theta,\theta'),$$

we have $\chi(\Theta)\leqslant\chi(\theta)+\chi(\theta')$ and

$$\int_0^a\chi(\Theta_f)\,dx\leqslant\int_0^a\chi(\theta_{f^*})\,dx+\int_0^a\chi(\theta'_{f_*})\,dx=2\int_0^a\chi(\theta_{f^*})\,dx.$$

Hence:

(13·13) THEOREM. *If $f\in L(0,a)$ and $\Theta(x)=\Theta_f(x)$ is defined by* (13·12), *then for a non-negative and non-decreasing $\chi(u)$,*

$$\int_0^a\chi\{\Theta(x)\}\,dx\leqslant 2\int_0^a\chi\left(\frac{1}{x}\int_0^x f^*\,dt\right)dx. \tag{13·14}$$

From this we deduce, by specifying χ, the following corollaries:

(13·15) THEOREM. (i) *If $f\in L^r(0,a)$, $r>1$, then $\Theta(x)\in L^r$ and*

$$\int_0^a\Theta^r\,dx\leqslant 2\left(\frac{r}{r-1}\right)^r\int_0^a f^r\,dx.$$

(ii) *If $f\in L(0,a)$, then $\Theta(x)\in L^\alpha$ for every $0<\alpha<1$, and*

$$\int_0^a\Theta^\alpha\,dx\leqslant\frac{2a^{1-\alpha}}{1-\alpha}\left(\int_0^a f\,dx\right)^\alpha.$$

(iii) *If $f\log^+ f\in L(0,a)$, then $\Theta(x)\in L$ and*

$$\int_0^a\Theta\,dx\leqslant 4\int_0^a f\log^+ f\,dx+A,$$

with A depending on a only.

We have to estimate the right-hand side of (13·14) and we may suppose from the start that f is non-increasing, so that we may replace f^* by f there.

Case (i) then follows from Theorem (9·16), with $s=0$; it is enough to set $f(x)=0$ for $x>a$. Case (ii) follows by an application of Hölder's inequality. For, with $\chi(u)=u^\alpha$,

$$\int_0^a\chi\left(\frac{1}{x}\int_0^x f\,dt\right)dx=\int_0^a\frac{dx}{x^{\alpha(1-\alpha)}}\left\{\frac{1}{x^\alpha}\int_0^x f\,dt\right\}^\alpha\leqslant\left(\int_0^a\frac{dx}{x^\alpha}\right)^{1-\alpha}\left(\int_0^a\frac{dx}{x^\alpha}\int_0^x f\,dt\right)^\alpha$$

$$=\frac{a^{(1-\alpha)^2}}{(1-\alpha)^{1-\alpha}}\left\{\int_0^a f\,dt\int_t^a\frac{dx}{x^\alpha}\right\}^\alpha\leqslant\frac{a^{1-\alpha}}{1-\alpha}\left(\int_0^a f\,dt\right)^\alpha.$$

In case (iii), the right-hand side of (13·14), with $\chi(u)=u$, is

$$2\int_0^a\frac{dx}{x}\int_0^x f(t)\,dt=2\int_0^a f(t)\log\frac{a}{t}\,dt.$$

Let E_1 and E_2 be the sets of points at which, respectively, $f<(a/t)^{\frac{1}{2}}$ and $f \geqslant (a/t)^{\frac{1}{2}}$. Clearly the integral of $f \log (a/t)$ extended over E_1 does not exceed a finite constant depending only on a. In E_2 we have $1 \leqslant (a/t) \leqslant f^2$, so that

$$\int_{E_2} f \log (a/t)\, dt \leqslant 2\int_{E_2} f \log f dt \leqslant 2\int_0^a f \log^+ f dt,$$

and (iii) follows by collecting the estimates.

The example $f(x)=1/(x\log^2 x)$, considered in the interval $(0, \frac{1}{2})$, shows that if $f \epsilon \mathrm{L}$ the function Θ need not be integrable. In this case $\Theta(x)=1/(x\,|\log x\,|)$.

For applications to Fourier series a slight modification of the function $\Theta(x)$ is useful. Let $f(x)$ be periodic and integrable, but not necessarily non-negative (or even real-valued). We set

$$M(x)=M_f(x)=\sup_{0<|t|\leqslant\pi} \frac{1}{t}\int_0^t |f(x+u)|\, du = \sup_{0<|t|\leqslant\pi} \frac{1}{t}\int_x^{x+t} |f(u)|\, du \tag{13·16}$$

for $-\pi\leqslant x\leqslant\pi$. Clearly $M_f(x)$ does not exceed the function $\Theta_{|f|}(x)$ *formed for the interval* $(-2\pi, 2\pi)$, so that

$$\int_{-\pi}^{\pi} \chi\{M_f(x)\}\, dx \leqslant \int_{-\pi}^{\pi} \chi\{\Theta_{|f|}(x)\}\, dx.$$

From this and (13·15) we easily get the following analogues of (i), (ii), (iii):

$$\left.\begin{aligned}
&\int_{-\pi}^{\pi} M^r(x)\, dx \leqslant 4\left(\frac{r}{r-1}\right)^r \int_{-\pi}^{\pi} |f|^r\, dx \quad (r>1),\\
&\int_{-\pi}^{\pi} M^\alpha(x)\, dx \leqslant 4\frac{a^{1-\alpha}}{1-\alpha}\left(\int_{-\pi}^{\pi} |f|\, dx\right)^\alpha \quad (0<\alpha<1),\\
&\int_{-\pi}^{\pi} M(x)\, dx \leqslant 8\int_{-\pi}^{\pi} |f| \log^+|f|\, dx + A.
\end{aligned}\right\} \tag{13·17}$$

The following inequalities, implicitly contained in the preceding proofs, deserve separate mention. First, for f integrable and non-negative in $(0, a)$,

$$|\mathrm{E}(\theta_f>y)| \leqslant y^{-1}\int_0^a f dx, \quad |\mathrm{E}(\Theta_f>y)| \leqslant 2y^{-1}\int_0^a f dx. \tag{13·18}$$

The first of these inequalities is contained in (13·11), and the second follows from the first by (13·12). Finally, for an f periodic and integrable but not necessarily non-negative,

$$|\mathrm{E}\{M_f(x)>y;\ 0\leqslant x\leqslant 2\pi\}| \leqslant 4y^{-1}\int_0^{2\pi} |f(x)|\, dx. \tag{13·19}$$

Remark. While in parts (ii) and (iii) of (13·15) we must necessarily assume that a is finite, part (i) holds, for infinite intervals. Suppose, e.g., that $f \epsilon \mathrm{L}^r(-\infty, +\infty)$, $r>1$, and consider the analogue of (13·15) (i) for the interval $(-a, a)$ and the function f_a, which is f restricted to $(-a, a)$. The passage $a\rightarrow+\infty$ leads to

$$\int_{-\infty}^{+\infty} \Theta^r dx \leqslant 2\left(\frac{r}{r-1}\right)^r \int_{-\infty}^{+\infty} f^r dx.$$

MISCELLANEOUS THEOREMS AND EXAMPLES

1. A sequence $\{u_\nu\}$ is of bounded variation if and only if it is a difference of two non-negative and non-increasing sequences.

2. Of the two series,

(i) $\frac{1}{2}+\cos x+\cos 2x+\ldots+\cos nx+\ldots$,

(ii) $\sin x+\sin 2x+\ldots+\sin nx+\ldots$,

the first diverges for all x, the second for all $x\not\equiv 0 \bmod \pi$.

[(i) For no x_0 do we have $\cos nx_0\to 0$. For otherwise,

$$\sin^2 nx_0=1-\cos^2 nx_0\to 1,\quad \sin^2 nx_0=\tfrac{1}{2}(1-\cos 2nx_0)\to\tfrac{1}{2},$$

a contradiction.

(ii) If $\sin nx_0\to 0$, then

$$\sin(n+1)x_0-\sin(n-1)x_0=2\sin x_0\cos nx_0\to 0,$$

that is, $\sin x_0=0$ by (i).]

3. Using $S[|\sin x|]$, and (6·3), prove
$$|\sin x|=\frac{8}{\pi}\sum_{n=1}^{\infty}\frac{\sin^2 nx}{4n^2-1}.$$

4. Let $c_m=\frac{1}{2}(a_m-ib_m)$ for $m>0$, and let $c_{-m}=\bar{c}_m$. Show that a necessary and sufficient condition for the existence of $\lim\sum_{-M}^{N}c_m\,e^{imx_0}$ as M and N tend to $+\infty$ independently of each other, is the simultaneous convergence of both series

$$\sum_1^\infty(a_m\cos mx_0+b_m\sin mx_0)\quad\text{and}\quad\sum_1^\infty(a_m\sin mx_0-b_m\cos mx_0).$$

5. Each of the two systems

(i) $1, \cos x, \cos 2x, \ldots, \cos nx, \ldots$,

(ii) $\sin x, \sin 2x, \ldots, \sin nx, \ldots$

is orthogonal and complete over $(0,\pi)$.

6. Let $\{\phi_n\}$ denote Rademacher's system (see § 3) and let

$$\chi_0(t)=1,\quad \chi_N(t)=\phi_{n_1}(t)\,\phi_{n_2}(t)\ldots,$$

where $2^{n_1}+2^{n_2}+\ldots$, $n_1>n_2>\ldots$, is the dyadic development of the positive integer N. Show that the system $\{\chi_N\}$ is orthonormal over the interval $(0,1)$.

[The system χ_N, in a different form, was first considered by Walsh [1]. See also Paley [1], who gave the above definition, and Kaczmarz [1].]

7. Let $s_N(x)$ be the sum of the first N terms of the Fourier series $\Sigma a_\nu\chi_\nu(x)$ of $f(x)$, $0\leqslant x\leqslant 1$. Prove the formula

$$s_{2^n}(x_0)=\int_0^1 f(t)\prod_{k=0}^{n-1}(1+\phi_k(x_0)\,\phi_k(t))\,dt,$$

and show that $s_{2^n}(x)\to f(x)$ almost everywhere as $n\to\infty$. This implies, in particular, that *the system $\{\chi_N\}$ is complete over* $(0,1)$.

[If x_0 is not a binary rational, and I_{n-1} is the interval of constancy of ϕ_{n-1} containing x_0, then $|I_{n-1}|=2^{-n}$, all functions $\phi_0,\phi_1,\ldots,\phi_{n-1}$ are constant in I_{n-1}, and the integral above is $|I_{n-1}|^{-1}\int_{I_{n-1}}f(t)\,dt$. If, therefore, $F(x)=\int_0^x f\,dt$ is differentiable at x_0, we have $s_{2^n}(x_0)\to F'(x_0)$.]

8. Orthogonal systems can be defined in spaces of any dimension, intervals of integration being replaced by any fixed measurable set of positive measure. Show that if $\{\phi_m(x)\}$ and $\{\psi_n(y)\}$ are orthonormal and complete in intervals $a\leqslant x\leqslant b$, $c\leqslant y\leqslant d$ respectively, then the doubly infinite system $\{\phi_m(x)\,\psi_n(y)\}$ is orthonormal and complete in the rectangle

$$R:\ a\leqslant x\leqslant b,\quad c\leqslant y\leqslant d.$$

[If $\iint_R f(x,y)\,\bar\phi_m(x)\,\bar\psi_n(y)\,dx\,dy=0$ for all m, n, the functions $f_m(y)=\int_a^b f(x,y)\,\bar\phi_m(x)\,dx$ vanish for almost all y, and hence $f(x,y)$ vanishes almost everywhere on almost all lines $y=\text{const.}$]

CHAPTER II

FOURIER COEFFICIENTS. ELEMENTARY THEOREMS ON THE CONVERGENCE OF $S[f]$ AND $\tilde{S}[f]$

1. Formal operations on $S[f]$

(1·1) Theorem. *Let n be an integer, u a real number and*

$$f(x) \sim \sum_{\nu=-\infty}^{+\infty} c_\nu e^{i\nu x}. \tag{1·2}$$

Then

(i) $$\overline{f(x)} \sim \sum_{\nu=-\infty}^{+\infty} \bar{c}_\nu e^{-i\nu x} = \sum_{\nu=-\infty}^{+\infty} \bar{c}_{-\nu} e^{i\nu x},$$

(ii) $$f(nx) \sim \sum_{\nu=-\infty}^{+\infty} c_\nu e^{i\nu n x} \quad (n \neq 0),$$

(iii) $$e^{inx} f(x) \sim \sum_{\nu=-\infty}^{+\infty} c_\nu e^{i(\nu+n)x} = \sum_{\nu=-\infty}^{+\infty} c_{\nu-n} e^{i\nu x},$$

(iv) $$f(x+u) \sim \sum_{\nu=-\infty}^{+\infty} c_\nu e^{i\nu u} e^{i\nu x},$$

(v) $$\frac{1}{n}\sum_{k=0}^{n-1} f\left(\frac{x}{n} + \frac{2\pi k}{n}\right) \sim \sum_{\nu=-\infty}^{+\infty} c_{\nu n} e^{i\nu x} \quad (n > 0).$$

The proofs are simple:

(i) $$\frac{1}{2\pi}\int_0^{2\pi} \overline{f(t)}\, e^{-i\nu t}\, dt = \frac{1}{2\pi}\int_0^{2\pi} \overline{f(t)\, e^{i\nu t}}\, dt = \bar{c}_{-\nu}.$$

(ii) Suppose first that $n > 0$. We observe that

$$\frac{1}{n}\sum_{k=0}^{n-1} \exp(2\pi\mu k i/n) = \begin{cases}1\\0\end{cases} \quad (\mu = 0, \pm 1, \pm 2, \ldots), \tag{1·3}$$

according as μ is or is not a multiple of n. Now

$$\int_0^{2\pi} f(nt)\, e^{-i\mu t}\, dt = \frac{1}{n}\int_0^{2\pi n} f(t)\, e^{-i\mu t/n}\, dt = \int_0^{2\pi} f(t)\, e^{-i\mu t/n}\left\{\frac{1}{n}\sum_{k=0}^{n-1} e^{-2\pi i k\mu/n}\right\} dt,$$

and this is $2\pi c_\nu$ or 0, according as $\mu/n = \nu$ is an integer or not. The case $n < 0$ reduces to $n > 0$, since, as we easily see,

$$f(-x) \sim \Sigma c_\nu e^{-i\nu x}.$$

(iii) $$\int_0^{2\pi} f(t)\, e^{int}\, e^{-i\nu t}\, dt = 2\pi c_{\nu-n}.$$

(iv) $$\int_0^{2\pi} f(t+u)\, e^{-i\nu t}\, dt = e^{i\nu u}\int_0^{2\pi} f(t+u)\, e^{-i\nu(t+u)}\, dt = 2\pi\, e^{i\nu u}\, c_\nu.$$

(v) This follows from (iv) and (1·3).

If
$$f(x) \sim \tfrac{1}{2}a_0 + \sum_1^{\infty} (a_\nu \cos \nu x + b_\nu \sin \nu x), \tag{1·4}$$

en

$$f(x+u) \sim \sum_{\nu=0}^{\infty} \{A_\nu(u) \cos \nu x - B_\nu(u) \sin \nu x\}.$$

EM. *If $f(x)$ and $g(x)$ are integrable and periodic, so is the function*

$$h(x) = \frac{1}{2\pi} \int_0^{2\pi} f(x-t)\, g(t)\, dt. \tag{1·6}$$

If $f \sim \Sigma c_\nu e^{i\nu x}$, $g \sim \Sigma d_\nu e^{i\nu x}$, then

$$h(x) \sim \sum_{-\infty}^{+\infty} c_\nu d_\nu e^{i\nu x}. \tag{1·7}$$

We show first that the integral (1·6) exists for almost all x. We may assume that f and g are real-valued, and this case may, in turn, be reduced to $f, g \geqslant 0$. Then

$$\int_0^{2\pi} dx \int_0^{2\pi} f(x-t)\, g(t)\, dt = \int_0^{2\pi} g(t) \left\{ \int_0^{2\pi} f(x-t)\, dx \right\} dt = \int_0^{2\pi} g(t)\, dt \int_0^{2\pi} f(x)\, dx. \tag{1·8}$$

The operations performed here are justified since $f(x-t)\, g(t)$ is measurable in the (x,t) plane (being a product of measurable functions) and since (the integrand being non-negative) the order of integration is irrelevant. Thus $h(x)$ is integrable and, in particular, finite almost everywhere. It is clearly periodic.

The function $f(x-t)\, g(t)$ is integrable over the square $0 \leqslant x \leqslant 2\pi$, $0 \leqslant t \leqslant 2\pi$. Thus for general f and g, $|f(x-t)\, g(t)\, e^{-i\nu x}|$ is integrable over the square and the following argument is also legitimate:

$$\frac{1}{2\pi} \int_0^{2\pi} h(x)\, e^{-i\nu x}\, dx = \frac{1}{2\pi} \int_0^{2\pi} \left\{ \frac{1}{2\pi} \int_0^{2\pi} f(x-t)\, e^{-i\nu(x-t)} g(t)\, e^{-i\nu t}\, dt \right\} dx$$

$$= \frac{1}{2\pi} \int_0^{2\pi} g(t)\, e^{-i\nu t} \left\{ \frac{1}{2\pi} \int_0^{2\pi} f(x-t)\, e^{-i\nu(x-t)}\, dx \right\} dt = c_\nu d_\nu,$$

and the proof of (1·5) is completed.

It is useful to observe that (1·7) is obtained by the formal process of multiplying $S[f(x-t)] = \Sigma c_\nu e^{i\nu x} e^{-i\nu t}$ and $S[g(t)] = \Sigma d_\nu e^{i\nu t}$ termwise and integrating each product term over $0 \leqslant t \leqslant 2\pi$.

The function

$$h(x) = I(f,g) = \frac{1}{2\pi} \int_0^{2\pi} f(x-t)\, g(t)\, dt$$

of (1·6) is often called the *convolution*, or *composition*, of the functions f and g. Obviously $I(f,g) = I(g,f)$.

For f in (1·4) and $g \sim \tfrac{1}{2}a'_0 + \Sigma(a'_\nu \cos \nu x + b'_\nu \sin \nu x)$, (1·7) can also be written

$$\frac{1}{\pi} \int_0^{2\pi} f(x-t)\, g(t)\, dt \sim \tfrac{1}{2} a_0 a'_0 + \sum_1^{\infty} \{(a_\nu a'_\nu - b_\nu b'_\nu) \cos \nu x + (a_\nu b'_\nu + a'_\nu b_\nu) \sin \nu x\}. \tag{1·9}$$

Set $g(t) = \bar{f}(-t)$ in (1·6) and replace t by $-t$. We obtain the special but interesting case

$$\frac{1}{2\pi} \int_0^{2\pi} f(x+t) \bar{f}(t)\, dt \sim \sum_{-\infty}^{+\infty} |c_\nu|^2 e^{i\nu x}. \tag{1·10}$$

Suppose that the f and g in (1·6) are in L^2. Then $\Sigma |c_\nu|^2$ and $\Sigma |d_\nu|^2$ converge. If we can show that the integral (1·6), which by Schwarz's inequality exists for every x, is a continuous function of x, then we can replace the sign '$\sim$' in (1·7) by '$=$' (Chapter I, (6·3)). For this purpose we need the case $p=2$ of the following lemma:

(1·11) LEMMA. *If f is periodic and in* L^p, $1 \leqslant p < \infty$, *the expression*

$$J_p(t; f) = \mathfrak{M}_p[f(x+t) - f(x)] = \left\{ \int_0^{2\pi} |f(x+t) - f(x)|^p \, dx \right\}^{1/p}$$

tends to 0 *with* t.

This is immediate for f continuous. Using Theorem (11·8) of Chapter I and its notation, we get, applying Minkowski's inequality twice,

$$J_p(t; f) \leqslant J_p(t; \phi) + J_p(t; f - \phi) \leqslant J_p(t; \phi) + 2\mathfrak{M}_p[f - \phi] < o(1) + 2\epsilon.$$

Hence $J_p(t; f) < 3\epsilon$ for $|t|$ small enough, and (1·11) follows.

Return to (1·6). If f and g are in L^2, then

$$|h(x+u) - h(x)| \leqslant \int_0^{2\pi} |f(x+u-t) - f(x-t)| \, |g(t)| \, dt \leqslant J_2(u; f) \, \mathfrak{M}_2[g] \to 0$$

as $u \to 0$, which shows that h is continuous. Hence:

(1·12) THEOREM. *Suppose that f and g are in* L^2 *and have coefficients c_ν and d_ν respectively. Then*

$$\frac{1}{2\pi} \int_0^{2\pi} f(x-t) \, g(t) \, dt = \sum_{-\infty}^{+\infty} c_\nu d_\nu e^{i\nu x}$$

for all x, and the series on the right converges absolutely and uniformly. In particular

$$\left.\begin{aligned} \frac{1}{2\pi} \int_0^{2\pi} f(x+t) \bar{f}(t) \, dt &= \sum_{-\infty}^{+\infty} |c_\nu|^2 e^{i\nu x}, \\ \frac{1}{2\pi} \int_0^{2\pi} f(t) \, g(t) \, dt &= \sum_{-\infty}^{+\infty} c_\nu d_{-\nu}, \\ \frac{1}{2\pi} \int_0^{2\pi} |f(t)|^2 \, dt &= \sum_{-\infty}^{+\infty} |c_\nu|^2. \end{aligned}\right\} \tag{1·13}$$

The last equation is Parseval's formula for the trigonometric system. The name *Parseval's formula* is often given to the first two more general relations (1·13).

If f is real-valued and has coefficients a_ν, b_ν we may write the last equation in the form

$$\frac{1}{\pi} \int_0^{2\pi} f^2(t) \, dt = \tfrac{1}{2} a_0^2 + \sum_1^{\infty} (a_\nu^2 + b_\nu^2). \tag{1·14}$$

Return to (1·5). If f and g are integrable, so is h. The following generalization of this result is of importance.

(1·15) THEOREM. *Let f and g be periodic and in* L^p *and* L^q *respectively, where* $p \geqslant 1$, $q \geqslant 1$, $1/p + 1/q > 1$. *Let*

$$\frac{1}{r} = \frac{1}{p} + \frac{1}{q} - 1; \tag{1·16}$$

then the function $h(x) = I(f, g)$ *defined by* (1·6) *belongs to* L^r, *and*

$$\mathfrak{A}_r[h] \leqslant \mathfrak{A}_p[f] \, \mathfrak{A}_q[g],$$

where $\mathfrak{A}_r[h]$ *stands for* $\mathfrak{A}_r[h; 0, 2\pi]$, *and similarly for* $\mathfrak{A}_p[f]$, $\mathfrak{A}_q[g]$.

Since $|I(f, g)| \leqslant I(|f|, |g|)$, we may suppose that $f \geqslant 0$, $g \geqslant 0$. Let λ, μ, ν be positive numbers satisfying $1/\lambda + 1/\mu + 1/\nu = 1$. Writing

$$f(x-t) \, g(t) = f^{p/\lambda} g^{q/\lambda} \, . \, f^{p(1/p - 1/\lambda)} \, . \, g^{q(1/q - 1/\lambda)},$$

and applying Hölder's inequality with the exponents λ, μ, ν (Chapter I, (9·8)), we see that $h(x)$ does not exceed

$$\left[\frac{1}{2\pi}\int_0^{2\pi} f^p(x-t)\,g^q(t)\,dt\right]^{1/\lambda}\left[\frac{1}{2\pi}\int_0^{2\pi} f^{p\mu(1/p-1/\lambda)}(x-t)\,dt\right]^{1/\mu}\left[\frac{1}{2\pi}\int_0^{2\pi} g^{q\nu(1/q-1/\lambda)}(t)\,dt\right]^{1/\nu}.$$

We use this with $\lambda=r,\quad 1/p-1/\lambda=1/\mu,\quad 1/q-1/\lambda=1/\nu,$ **(1·17)**

so that λ, μ, ν are positive numbers and satisfy $1/\lambda+1/\mu+1/\nu=1$ by (1·16). The last two factors in the product are just $\mathfrak{A}_p^{p/\mu}[f]$ and $\mathfrak{A}_q^{q/\nu}[g]$. Hence

$$\mathfrak{A}_r[h]=\mathfrak{A}_\lambda[h]\leqslant \mathfrak{A}_p^{p/\mu}[f]\,\mathfrak{A}_q^{q/\nu}[g]\left\{\frac{1}{4\pi^2}\int_0^{2\pi}dx\int_0^{2\pi} f^p(x-t)\,g^q(t)\,dt\right\}^{1/r}.$$

The expression in curly brackets is $\mathfrak{A}[f^p]\,\mathfrak{A}[g^q]=\mathfrak{A}_p^p[f]\,\mathfrak{A}_q^q[g]$, and the right-hand side is therefore

$$\mathfrak{A}_p^{p(1/\mu+1/r)}[f]\,\mathfrak{A}_q^{q(1/\nu+1/r)}[g]=\mathfrak{A}_p[f]\,\mathfrak{A}_q[g],$$

by (1·17). This completes the proof.

The theorem holds when $1/p+1/q=1$. Moreover, by an argument similar to that preceding Theorem (1·12), $h(x)$ is then continuous.

Let $f_1, f_2, \dots, f_k$ be periodic integrable functions having respectively Fourier coefficients $\{c_n^{(1)}\}, \{c_n^{(2)}\}, \dots, \{c_n^{(k)}\}$. We define the convolution $h(x)$, or $I(f_1, f_2, \dots, f_k)$, of $f_1, \dots, f_k$ by the induction formula

$$I(f_1, f_2, \dots, f_k)=I(I(f_1, \dots, f_{k-1}), f_k).$$

Then $h(x)$ is a periodic integrable function, and obviously

$$h(x)=I(f_1, f_2, \dots, f_k)\sim \Sigma c_n^{(1)} c_n^{(2)} \dots c_n^{(k)}\, e^{inx}. \qquad \textbf{(1·18)}$$

It follows that the operation of convolution is commutative and associative. Commutativity is anyway an immediate consequence of the definition of convolution, while associativity can also be derived directly from the formula

$$h(x)=(2\pi)^{-k}\int_0^{2\pi}\dots\int_0^{2\pi} f_1(x-t_2-\dots-t_k)\,f_2(t_2)\dots f_k(t_k)\,dt_2\dots dt_k.$$

(1·19) THEOREM. *Let f_1, f_2, ..., f_k be periodic and of the classes L^{r_1}, L^{r_2}, ..., L^{r_k} respectively. Suppose that $r_j\geqslant 1$ for all j and that the number*

$$\frac{1}{r}=\frac{1}{r_1}+\frac{1}{r_2}+\dots+\frac{1}{r_k}-(k-1) \qquad \textbf{(1·20)}$$

is positive. Then the convolution $h(x)=I(f_1, \dots, f_k)$ is of class L^r. If the right-hand side of (1·20) *is zero, then h is continuous. Moreover*

$$\mathfrak{A}_r[h]\leqslant \mathfrak{A}_{r_1}[f_1]\dots \mathfrak{A}_{r_k}[f_k]. \qquad \textbf{(1·21)}$$

This follows by induction from (1·15).

Let $F(x)$ be a function satisfying the condition

$$F(x+2\pi)-F(x)=\text{const.}\quad(-\infty<x<+\infty), \qquad \textbf{(1·22)}$$

and of bounded variation in $(0, 2\pi)$. Let $G(x)$ be a similar function and write

$$H(x)=\frac{1}{2\pi}\int_0^{2\pi} F(x-t)\,dG(t), \qquad \textbf{(1·23)}$$

the *convolution of F and dG*. The integral here is taken in the Riemann-Stieltjes sense, and so exists for every x such that $F(x-t)$, *qua* function of t, and $G(t)$ have no discontinuity in common. In other words, it exists for every x which does not belong to the denumerable set D of numbers $\xi_j+\eta_k$, where $\{\xi_j\}$ and $\{\eta_k\}$ are the discontinuities of $F(t)$ and $G(t)$ respectively. Let E be the set complementary to D. The function H satisfies on E a condition analogous to (1·22). If $\{(a_i, b_i)\}$ is a finite system of non-overlapping intervals with end-points in E, then

$$\Sigma \,|\, H(b_i) - H(a_i) \,| \leqslant \frac{1}{2\pi}\int_0^{2\pi} \{\Sigma \,|\, F(b_i - t) - F(a_i - t)\,|\}\,|\, dG(t)\,|, \tag{1·24}$$

which shows that H is of bounded variation over any finite portion of E. It is therefore a difference of two functions monotone on E, and so can be extended to all x, e.g. by the condition

$$H(x) = \lim H(x') \quad (x \in D,\ x' \in E,\ x' \to x+0). \tag{1·25}$$

Such an extension does not increase the variation of monotone functions, nor, therefore, the total variation of H. Let V_F denote the total variation of F over $(0, 2\pi)$. Since, as we see from (1·24), the total variation of H over $(0, 2\pi)\,E$ does not exceed $V_F V_G/2\pi$, it follows that

$$V_H \leqslant V_F V_G/2\pi. \tag{1·26}$$

Summarizing, the integral (1·23) exists for all x outside a certain denumerable set D and can, by (1·25), be extended to all x as a function satisfying a condition similar to (1·22), of bounded variation over $(0, 2\pi)$, and also satisfying (1·26). Clearly if F and G are monotone so is H. We may add that if we used the Lebesgue-Stieltjes integral, $H(x)$ could be defined by (1·23) for all x, and would have the same properties as the $H(x)$ above.

Let c_n and c'_n be the Fourier coefficients of dF and dG. We shall show that

$$dH(x) \sim \Sigma c_n c'_n e^{inx}. \tag{1·27}$$

For let $x_0 < x_1 < \ldots < x_k = x_0 + 2\pi$ be points of E. The nth Fourier coefficient of dH is the limit of the sum

$$\frac{1}{2\pi}\sum_{j=1}^{k} e^{-ix_j n}[H(x_j) - H(x_{j-1})] = \frac{1}{4\pi^2}\sum_{j=1}^{k}\int_0^{2\pi}\left\{\int_{x_{j-1}-t}^{x_j - t} e^{-in(x_j - t)}\,dF(u)\right\} e^{-int}\,dG(t)$$

as $\rho = \max(x_j - x_{j-1}) \to 0$. Suppose ρ so small that the oscillation of e^{-inu} over every interval of length $\leqslant \rho$ is less than ϵ. Then on replacing $e^{-in(x_j - t)}$ in the last integral by e^{-inu} we introduce an error at most

$$\frac{\epsilon}{4\pi^2}\sum_{j=1}^{k}\int_0^{2\pi}\left\{\int_{x_{j-1}-t}^{x_j-t} |\, dF(u)\,|\right\} |\, dG(t)\,| = \frac{\epsilon}{4\pi^2}\int_0^{2\pi}\left\{\int_{x_0-t}^{x_k-t} |\, dF(u)\,|\right\} |\, dG(t)\,| = \frac{\epsilon}{4\pi^2} V_F V_G.$$

But $$\int_0^{2\pi}\left\{\int_{x_0-t}^{x_k-t} e^{-inu}\,dF(u)\right\} e^{-int}\,dG(t) = \int_0^{2\pi} e^{-inu}\,dF(u)\int_0^{2\pi} e^{-int}\,dG(t) = 4\pi^2 c_n c'_n.$$

This completes the proof. As we see from (1·27), by interchanging the roles of F and G in (1·23) we modify H only by an additive constant,† a result which can also be obtained directly from (1·23) by integration by parts.

If F (or G) is continuous, the integral (1·23) exists for all x, and since

$$2\pi \,|\, H(x+h) - H(x)\,| \leqslant \max_t |\, F(x+h-t) - F(x-t)\,| \int_0^{2\pi} |\, dG(t)\,|,$$

$H(x)$ is also continuous.

† See the last remark on p. 41.

A special case of $H(x)$ is the function

$$F^*(x) = \frac{1}{2\pi}\int_0^{2\pi} F(x+t)\, d\bar{F}(t). \tag{1·28}$$

$F^*(-x)$ is the convolution of $F(-t)$ and $d\bar{F}(t)$. By (1·27) (if $dF(x) \sim \Sigma c_n e^{inx}$, then $dF(-x) \sim -\Sigma c_{-n} e^{inx}$),

$$dF^*(x) \sim \Sigma \,|\, c_n \,|^2 e^{inx}. \tag{1·29}$$

We shall show that the absence of a jump of $F^*(x)$ at $x=0$ is equivalent to the continuity of F at every point. More precisely:

(1·30) THEOREM. *Let $x_1, x_2, \ldots$ be all the discontinuities of F in the interval $0 \leqslant x < 2\pi$, and let $d_j = F(x_j+0) - F(x_j-0)$. Then*

$$F^*(+0) - F^*(-0) = (2\pi)^{-1} \Sigma \,|\, d_j \,|^2.$$

For let $S_k(x)$ be a step function having jumps $d_1, d_2, \ldots, d_k$ at the points $x_1, x_2, \ldots, x_k$, continuous elsewhere, and satisfying a condition analogous to (1·22). The difference $F_k(x) = F(x) - S_k(x)$ is continuous at $x_1, x_2, \ldots, x_k$, and has jumps $d_{k+1}, d_{k+2}, \ldots$ at the points $x_{k+1}, x_{k+2}, \ldots$. The function $F^*(x)$ equals

$$\frac{1}{2\pi}\int_0^{2\pi} F_k(x+t)\, d\bar{F}(t) + \frac{1}{2\pi}\int_0^{2\pi} S_k(x+t)\, d\bar{F}(t) = H_1(x) + H_2(x).$$

For $\pm\delta \in E$, $$2\pi \,|\, H_1(+\delta) - H_1(-\delta) \,| \leqslant V_F \sup_t \,|\, F_k(t+\delta) - F_k(t-\delta) \,|,$$

$$2\pi[H_2(+\delta) - H_2(-\delta)] = \int_0^{2\pi} [S_k(t+\delta) - S_k(t-\delta)]\, d\bar{F}(t). \tag{1·31}$$

The first inequality shows that by taking k large enough (i.e. by removing the 'heavier' discontinuities from F) we can make $H_1(+0) - H_1(-0)$ arbitrarily small. For δ small enough, $S_k(t+\delta) - S_k(t-\delta)$ is d_j in the δ neighbourhood of x_j, $j = 1, 2, \ldots, k$, and is zero elsewhere. This shows that the integral in (1·31) tends to $|\, d_1 \,|^2 + \ldots + |\, d_k \,|^2$ as $\delta \to 0$. From these facts (1·30) follows.

2. Differentiation and integration of S[f]

Suppose that a periodic function $f(x)$ is an integral, i.e. is absolutely continuous. Integration by parts gives

$$c_\nu = \frac{1}{2\pi}\int_0^{2\pi} f e^{-i\nu x}\, dx = \frac{1}{2\pi i\nu}\int_0^{2\pi} f' e^{-i\nu x}\, dx = \frac{c'_\nu}{i\nu} \quad (\nu \neq 0),$$

so that $c'_\nu = i\nu c_\nu$, the c'_ν being the coefficients of f'. Since f is periodic, $c'_0 = 0$. Thus, if $S'[f]$ denotes the result of differentiating $S[f]$ term by term, we have $S'[f] = S[f']$, or

$$f' \sim i \sum_{\nu=-\infty}^{+\infty} \nu c_\nu e^{i\nu x} = \sum_{\nu=1}^{\infty} \nu(b_\nu \cos \nu x - a_\nu \sin \nu x).$$

From this follows the general result:

(2·1) THEOREM. *If $f(x)$ is a k-th integral ($k = 1, 2, \ldots$), then $S^{(k)}[f] = S[f^{(k)}]$.*

The following result shows what happens when f has discontinuities, for simplicity a finite number of them:

(2·2) THEOREM. *Suppose that* $f(x)$ *has discontinuities of the first kind (jumps) at the points* $x_1 < x_2 < \ldots < x_k < x_{k+1} = x_1 + 2\pi$, *and that* $f(x)$ *is absolutely continuous in each of the intervals* (x_i, x_{i+1}), *if completed by continuity at the end-points* x_i, x_{i+1}. *Let*

$$d_i = [f(x_i+0) - f(x_i-0)]/\pi, \quad D(x) = \tfrac{1}{2} + \sum_{\nu=1}^{\infty} \cos \nu x.$$

Then
$$S'[f] - S[f'] = d_1 D(x-x_1) + d_2 D(x-x_2) + \ldots + d_k D(x-x_k). \tag{2·3}$$

The series $D(x)$ diverges everywhere (see p. 34, Ex. 2), but is summable by various methods to 0 if $x \neq 0$ (see Chapter III, §§ 1, 2). The statement (2·3) is, of course, to be interpreted *formally*: corresponding coefficients of the series on the two sides are equal.

We may suppose that $f(x_i) = \tfrac{1}{2}[f(x_i+0) + f(x_i-0)]$ for all i. Let $\phi(x)$ be the function defined in Chapter I, (4·12). Then $S'[\phi] = D(x) - \tfrac{1}{2}$. The function

$$\Phi(x) = d_1 \phi(x-x_1) + \ldots + d_k \phi(x-x_k)$$

has the same points of discontinuity, and the same jumps, as f. The difference $g = f - \Phi$ is therefore continuous, indeed absolutely continuous. Moreover,

$$\Phi'(x) = -\tfrac{1}{2}(d_1 + \ldots + d_k) = C,$$

say, except at the points x_i, so that $g' = f' - C$ almost everywhere. Now

$$\begin{aligned} S'[f] &= S'[g+\Phi] = S'[g] + S'[\Phi] = S[g'] + S'[\Phi] \\ &= S[f'] - C + \sum_i d_i\{D(x-x_i) - \tfrac{1}{2}\} = S[f'] + \sum_i d_i D(x-x_i). \end{aligned}$$

This completes the proof of (2·3).

Let $F(x)$, $0 \leq x \leq 2\pi$, be a function of bounded variation, and let c_ν be the Fourier coefficients of dF. The difference $F(x) - c_0 x$ is periodic (Chapter I, § 5), and its Fourier coefficient C_ν, $\nu \neq 0$, is

$$\frac{1}{2\pi}\int_0^{2\pi} (F - c_0 x)\, e^{-i\nu x}\, dx = \frac{1}{2\pi i\nu}\int_0^{2\pi} e^{-i\nu x}\, d(F - c_0 x) = \frac{1}{2\pi i\nu}\int_0^{2\pi} e^{-i\nu x}\, dF = \frac{c_\nu}{i\nu}.$$

Let us agree to write
$$F(x) \sim c_0 x + C_0 + \sum_{\nu=-\infty}^{+\infty}{}' \frac{c_\nu}{i\nu} e^{i\nu x}$$

instead of
$$F(x) - c_0 x \sim C_0 + \sum_{\nu=-\infty}^{+\infty}{}' \frac{c_\nu}{i\nu} e^{i\nu x},$$

where the dash signifies that the term $\nu = 0$ is omitted in summation. Then $S[dF]$ is obtained by formal differentiation of the first of these series, and we have:

(2·4) THEOREM. *With the convention just stated, the class of Fourier-Stieltjes series coincides with the class of formally differentiated Fourier series of functions of bounded variation.*

If $S[dF]$ vanishes identically, $S[F]$ consists of a constant term C. Thus $F(x) \equiv C$, and F is equal to C at every point of continuity, that is, outside a certain denumerable set. Hence, *if two functions* F_1 *and* F_2 *with regular discontinuities have the same Fourier-Stieltjes coefficients, then* $F_1(x) - F_2(x) \equiv C$.

Let f be periodic and F the indefinite integral of f. Since $F(x+2\pi)-F(x)$ is equal to the integral of f over $(x, x+2\pi)$ or, what is the same thing, over $(0, 2\pi)$, *a necessary and sufficient condition for the periodicity of F is that the constant term c_0 of $\mathsf{S}[f]$ is zero.* Suppose this condition satisfied. Then by (2·1) $\mathsf{S}[f]=\mathsf{S}'[F]$, so that $\mathsf{S}[F]$ is obtained by formal integration of $\mathsf{S}[f]$. In other words,

$$F(x) \sim C + \sum_{\nu=-\infty}^{+\infty}{}' \frac{c_\nu}{i\nu} e^{i\nu x} = C + \sum_{\nu=1}^{\infty} (a_\nu \sin \nu x - b_\nu \cos \nu x)/\nu,$$

where C is the constant of integration.

If $c_0 \neq 0$, we apply the result to the function $f-c_0$, whose integral $F-c_0 x$ is periodic. Hence we have:

(2·5) Theorem. *If $f \sim \Sigma c_\nu e^{i\nu x}$, and F is the indefinite integral of f, then*

$$F(x) - c_0 x \sim C + \sum_{\nu=-\infty}^{+\infty}{}' c_\nu e^{i\nu x}/i\nu = C + \sum_{\nu=1}^{\infty} (a_\nu \sin \nu x - b_\nu \cos \nu x)/\nu.$$

Example. Let $B_0(x)$, $B_1(x)$, $B_2(x)$, ... be the periodic functions defined by the conditions

(i) $B_0(x) = -1$;

(ii) $B_k'(x) = B_{k-1}(x)$ for $k = 1, 2, \ldots$;

(iii) the integral of B_k over $(0, 2\pi)$ is zero for $k = 1, 2, \ldots$.

Using Chapter I, (4·12), one verifies by induction that

$$B_k(x) \sim \sum_{\nu=-\infty}^{+\infty}{}' \frac{e^{i\nu x}}{(i\nu)^k} \quad (k=1, 2, \ldots), \tag{2·6}$$

where the dash indicates that the term $\nu=0$ is omitted in summation. Inside $(0, 2\pi)$, $B_k(x)$ is a polynomial of degree k (*Bernoulli's polynomial*, except for a numerical factor). According as k is even or odd,

$$B_k(x) \sim 2(-1)^{\frac{1}{2}k} \sum_{\nu=1}^{\infty} \frac{\cos \nu x}{\nu^k}, \quad B_k(x) \sim 2(-1)^{\frac{1}{2}(k-1)} \sum_{\nu=1}^{\infty} \frac{\sin \nu x}{\nu^k}.$$

Suppose that f is a kth integral $(k=1, 2, \ldots)$. Replacing in (1·5) f by $f^{(k)}$, g by B_k, we have the useful formula

$$f(x) - c_0 = \frac{1}{2\pi} \int_0^{2\pi} f^{(k)}(x-t)\, B_k(t)\, dt. \tag{2·7}$$

3. Modulus of continuity. Smooth functions

Let $f(x)$ be defined in a closed interval I, and let

$$\omega(\delta) = \omega(\delta; f) = \sup |f(x_2) - f(x_1)| \quad \text{for} \quad x_1 \in I,\ x_2 \in I,\ |x_2 - x_1| \leqslant \delta.$$

The function $\omega(\delta)$ is called the *modulus of continuity* of f. If I is finite, then f is continuous in I if and only if $\omega(\delta) \to 0$ with δ. If for some $\alpha > 0$ we have $\omega(\delta) \leqslant C\delta^\alpha$, with C independent of δ, we shall say that f satisfies a *Lipschitz condition of order α in (a, b)*. We shall also say that f belongs to the class Λ_α; in symbols,

$$f \in \Lambda_\alpha.$$

Only the case $0 < \alpha \leqslant 1$ is interesting: if $\alpha > 1$, then $\omega(\delta)/\delta$ tends to zero with δ, $f'(x)$ exists and is zero everywhere, and f is a constant.

The function f belongs to Λ_1 if and only if f is the integral of a bounded function.

It is sometimes convenient to consider the classes λ_α defined for $0 \leqslant \alpha < 1$ by the condition $\omega(\delta) = o(\delta^\alpha)$, so that if I is finite λ_0 is the class of continuous functions. By λ_1 we mean the class of functions having a continuous derivative.

A function $F(x)$ is said to be *smooth* at the point x_0 if

$$\{F(x_0+h)+F(x_0-h)-2F(x_0)\}/h = o(1) \quad \text{as} \quad h \to 0. \tag{3·1}$$

This relation may also be written

$$\{F(x_0+h)-F(x_0)\}/h - \{F(x_0)-F(x_0-h)\}/h = o(1). \tag{3·2}$$

It follows immediately that if $F'(x_0)$ exists and is finite then F is smooth at x_0. The converse is obviously false, but (as we see from (3·2)) if F is smooth at x_0 and if a one-sided derivative of F at x_0 exists, the derivative on the other side also exists and both are equal. The curve $y = F(x)$ has then no angular points, and this is the reason for the terminology.

If F is smooth at every point of an interval I, we say that F is smooth in I. (If I is closed, this presupposes that F is defined in a larger interval containing I.) If F is continuous and satisfies (3·1) uniformly in $x_0 \in I$ we shall say that F is *uniformly smooth*, and also that F *belongs to the class* λ_*. The *class* Λ_* is defined by the condition that F is continuous and that the left-hand side of (3·1) is $O(1)$ uniformly in x_0.

If $F \in \lambda_1$, then $F \in \lambda_*$; similarly, if $F \in \Lambda_1$, then $F \in \Lambda_*$. Thus λ_* and Λ_* are respectively generalizations of λ_1 and Λ_1. They are sometimes important for trigonometric series as being more natural than λ_1 and Λ_1. On the other hand, basic properties of λ_1 and Λ_1 do not hold for λ_* and Λ_*. Thus there exist functions $F \in \Lambda_*$ which are nowhere differentiable and functions $F \in \lambda_*$ differentiable in a set of measure zero only (p. 48). However, we do have:

(3·3) THEOREM. *If $F(x)$ is real-valued, continuous and smooth in an interval I, the set E of points where $F'(x)$ exists and is finite is of the power of the continuum in every subinterval I' of I.*

We may suppose that $I' = I$. Let $L(x) = mx + n$ be the linear function coinciding with $F(x)$ at the end-points a, b of I. Then $G(x) = F(x) - L(x)$ is continuous and smooth, and vanishes for $x = a, b$. If x_0 is a point inside I where G attains its maximum or minimum, the two terms on the left in

$$\{G(x_0+h)-G(x_0)\}/h + \{G(x_0-h)-G(x_0)\}/h \to 0$$

are of the same sign for $|h|$ small. Thus the right- and left-hand derivatives of G at the point x_0 exist and are zero, so that $G'(x_0) = 0$, $F'(x_0) = m = [F(b)-F(a)]/(b-a)$.

Hence E is dense in I. Let now $a < c < b$. The above proof shows that there is a point x_1 inside (a, c) such that $F'(x_1)$ exists and equals the slope of the chord through $(a, F(a))$ and $(c, F(c))$. Hence, if the slopes corresponding to two different c's are different, the corresponding points x_1 must also be different. But unless $F(x)$ is a linear function, in which case (3·3) is obvious, the set of the different slopes, and so also of the points x_1, is of the power of the continuum.

It is well known that a function $f(x)$ may be non-measurable and yet satisfy the condition

$$f(x+h)+f(x-h)-2f(x) = 0$$

for all x and h. This is the reason why in the definition of classes λ_* and Λ_* we assumed the continuity of f. It turns out that the functions of λ_* and Λ_* have 'a considerable degree of continuity'.

(3·4) THEOREM. *Let $f(x)$ be defined in a finite interval (a, b). If $f \in \Lambda_*$, then*

$$\omega(\delta; f) = O(\delta \log \delta)$$

and in particular $f \in \Lambda_\alpha$ for every $\alpha < 1$. If $f \in \lambda_*$, then $\omega(\delta; f) = o(\delta \log \delta)$.

It is enough to prove the part concerning Λ_*. Let $M = \max |f(x)|$. The hypothesis $f \in \Lambda_*$ implies

$$|f(x+\tau) - 2f(x+\tfrac{1}{2}\tau) + f(x)| \leqslant A\tau,$$

for $x \in (a, b)$ and τ small enough, $0 < \tau \leqslant \gamma$. Let us fix x and set $f(x+\tau) - f(x) = g(\tau)$. The left-hand side of the inequality above is $|g(\tau) - 2g(\frac{1}{2}\tau)|$. Replacing here τ successively by $\tau/2, \tau/2^2, \ldots$ we get

$$|g(\tau) - 2g(\tau/2)| \leqslant A\tau, \quad |2g(\tau/2) - 2^2 g(\tau/2^2)| \leqslant A\tau, \quad \ldots,$$
$$|2^{n-1}g(\tau/2^{n-1}) - 2^n g(\tau/2^n)| \leqslant A\tau,$$

where n will be defined in a moment. By termwise addition,

$$|g(\tau) - 2^n g(\tau/2^n)| \leqslant An\tau. \tag{3·5}$$

Suppose now that h tends to 0 through positive values. Let $0 < h \leqslant \frac{1}{2}\gamma$ and let n be a positive integer such that $2^n h$ is in the interval $(\frac{1}{2}\gamma, \gamma)$. The inequality $2^n h \leqslant \gamma$ implies that $n = O(\log h)$. From (3·5), with $\tau = 2^n h$, we get

$$|g(h)| \leqslant \frac{2M}{2^n} + \frac{An\tau}{2^n} = \frac{2Mh}{2^n h} + Anh \leqslant \frac{2Mh}{\frac{1}{2}\gamma} + O(h \log h) = O(h \log h),$$

or $f(x+h) - f(x) = O(h \log h)$, which proves the theorem.†

A function $f(x)$ defined on a set E will be said to *have property* D if, given any two points α, β in E, the function f takes on the product set $(\alpha, \beta) E$ all values between $f(\alpha)$ and $f(\beta)$. Property D may be considered as a (rather weak) substitute for continuity. A classical result of Darboux asserts that any exact derivative has property D in an interval where it exists.

(3·6) THEOREM. *Under the hypothesis of* (3·3), *$F'(x)$ has property* D *on E.*

For let

$$\alpha < \beta, \quad \alpha \in E, \quad \beta \in E, \quad F'(\alpha) = A, \quad F'(\beta) = B.$$

Let C be any number between A and B, say $A < C < B$. We have to show the existence in (α, β) of a point γ such that $F'(\gamma) = C$. By subtracting Cx from F, we may suppose that $C = 0$. Then $A < 0 < B$. Consider the function $G(x) = \{F(x+h) - F(x)\}/h$, where $h < \beta - \alpha$ is fixed, positive, and so small that

$$G(\alpha) < 0, \quad G(\beta - h) = \{F(\beta) - F(\beta - h)\}/h > 0. \tag{3·7}$$

Since $G(x)$ is continuous, there is a point x_0 inside $(\alpha, \beta - h)$ such that $G(x_0) = 0$, that is, $F(x_0 + h) = F(x_0)$. If γ is a point inside $(x_0, x_0 + h)$ at which F attains its maximum or minimum, then $F'(\gamma) = 0 = C$. Since $(x_0, x_0 + h) \subset (\alpha, \beta)$, the theorem follows.

Remark. The argument even shows that if $A < C < B$, and

$$\liminf_{h \to 0} \{F(\alpha + h) - F(\alpha)\}/h \leqslant A, \quad \limsup_{h \to 0} \{F(\beta) - F(\beta - h)\}/h \geqslant B,$$

then there is a point γ between α and β such that $F'(\gamma) = C$.

† The same proof shows that if $f(x+h) + f(x-h) - 2f(x) = O(h^\alpha)$, $0 < \alpha < 1$, then $f \in \Lambda_\alpha$ (see also Remark (*d*) on p. 120).

Let us now confine our attention to periodic functions. Given an $f \in L^p$, $p \geq 1$, the expression

$$\omega_p(\delta) = \omega_p(\delta; f) = \sup_{0 \leq h \leq \delta} \left\{ \frac{1}{2\pi} \int_0^{2\pi} | f(x+h) - f(x) |^p \, dx \right\}^{1/p}$$

will be called the *integral modulus of continuity* (in L^p) of f. Theorem (1·11) implies that $\omega_p(\delta) \to 0$ with δ, for every $f \in L^p$. Obviously $\omega_p(\delta)$ is a non-decreasing function of δ and p. If f is continuous, then $\omega_p(\delta) \to \omega(\delta)$ as $p \to \infty$. Unlike $\omega(\delta)$, $\omega_p(\delta)$ is not affected by a change of f in a set of measure 0.

If $\omega_p(\delta) = O(\delta^\alpha)$, we write $f \in \Lambda_\alpha^p$; and if $\omega_p(\delta) = o(\delta^\alpha)$, then $f \in \lambda_\alpha^p$. Here again the case $\alpha > 1$ is of no interest: *if* $\omega_p(\delta) = o(\delta)$, *then* $f \equiv \text{const}$. Since $\omega_p(\delta) \geq \omega_1(\delta)$, it is enough to take $p = 1$. Let

$$F(x) = \int_0^x f \, dt, \quad 0 < x_2 - x_1 < 2\pi, \ \delta > 0.$$

Then
$$\left| \delta^{-1} \int_{x_1}^{x_2} \{ f(x+\delta) - f(x) \} \, dx \right| = \left| \delta^{-1} \int_{x_2}^{x_2+\delta} f(u) \, du - \delta^{-1} \int_{x_1}^{x_1+\delta} f(u) \, du \right|.$$

The left-hand side here is not greater than $2\pi \, \delta^{-1} \omega_1(\delta) = o(1)$, as $\delta \to 0$. The right-hand side tends to $| F'(x_2) - F'(x_1) |$, provided that $F'(x_1)$, $F'(x_2)$ exist. Hence $F'(x)$ is constant outside a set of measure 0, which means that $f \equiv \text{const}$.

We may also consider the class Λ_*^p of periodic functions $F \in L^p$, $p \geq 1$, such that

$$\left\{ \int_0^{2\pi} | F(x+h) + F(x-h) - 2F(x) |^p \, dx \right\}^{1/p} = O(h).$$

Replacing $O(h)$ by $o(h)$ we define the class λ_*^p; and for $p = \infty$ and F continuous we get the classes Λ_* and λ_* respectively.

4. Order of magnitude of Fourier coefficients

The Fourier coefficients c_ν of a function f satisfy the inequalities

$$| c_\nu | \leq \tfrac{1}{2} \omega(\pi / | \nu |), \quad | c_\nu | \leq \tfrac{1}{2} \omega_1(\pi / | \nu |) \quad (\nu \neq 0), \tag{4·1}$$

where ω and ω_1 denote the moduli of continuity of f (see § 3). For, replacing x by $x + \pi/\nu$ in the integral defining c_ν and taking the mean value of the new and old integrals, we find that $2\pi c_\nu$ is

$$\int_0^{2\pi} f(x) \, e^{-i\nu x} \, dx = -\int_0^{2\pi} f\left(x + \frac{\pi}{\nu}\right) e^{-i\nu x} \, dx = \frac{1}{2} \int_0^{2\pi} \left\{ f(x) - f\left(x + \frac{\pi}{\nu}\right) \right\} e^{-i\nu x} \, dx.$$

Hence
$$| c_\nu | \leq \frac{1}{4\pi} \int_0^{2\pi} \left| f(x) - f\left(x + \frac{\pi}{\nu}\right) \right| dx, \tag{4·2}$$

and the right-hand side here exceeds neither $\frac{1}{2}\omega(\pi/|\nu|)$ nor $\frac{1}{2}\omega_1(\pi/|\nu|)$. If $f \in L^p$, $p \geq 1$, (4·2) implies

$$| c_\nu | \leq \tfrac{1}{2} \omega_p(\pi / | \nu |) \tag{4·3}$$

(see Chapter I, (10·12) (i)).

From the second inequality (4·1), and the fact that $\omega_1(\delta) \to 0$ with δ, we obtain the following important theorem:

(4·4) Theorem of Riemann-Lebesgue. *The Fourier coefficients c_ν of an integrable f tend to 0 as $| \nu | \to \infty$.*

The same result holds of course for the coefficients a_ν, b_ν, since $c_\nu = \frac{1}{2}(a_\nu - ib_\nu)$ for $\nu > 0$.

A slightly different proof of (4·4) runs as follows. We set $f=f_1+f_2$, where f_1 is bounded and $\int_0^{2\pi}|f_2|\,dx<\epsilon$. Correspondingly $c_\nu=c'_\nu+c''_\nu$. Here $f_1\in L^2$, so that $c'_\nu\to 0$ (a consequence of $\Sigma|c'_\nu|^2<\infty$). Since

$$|c''_\nu|\leqslant\frac{1}{2\pi}\int_0^{2\pi}|f_2|\,dx<\epsilon/2\pi,$$

$|c_\nu|$ is less than ϵ for $|\nu|$ large enough. This concludes the proof. The reader will notice that it proves (4·4) for the general uniformly bounded orthonormal system.

The following corollary of (4·4) is useful:

(4·5) THEOREM. *Let E be a measurable set in $(0, 2\pi)$, and let $\xi_1, \xi_2, \ldots$ be any sequence of real numbers. Then*

$$\int_E \cos^2(nx+\xi_n)\,dx\to\tfrac{1}{2}|E|\quad(n\to\infty).$$

For the integrand here is $\frac{1}{2}+\frac{1}{2}\cos 2nx\cos 2\xi_n-\frac{1}{2}\sin 2nx\sin 2\xi_n$, and the integrals of $\cos 2nx$ and $\sin 2nx$ over E tend to 0 since they are the Fourier coefficients (with a factor π) of the characteristic function of the set E.

The following is a slightly more general form of (4·4):

(4·6) THEOREM. *Let $f\in L(a,b)$, where (a,b) is finite or infinite, and let λ be a real variable. Let $a\leqslant a'<b'\leqslant b$. The integral*

$$\gamma_\lambda=\gamma_\lambda(f)=\gamma_\lambda(f;a',b')=\int_{a'}^{b'}f(x)\,e^{-i\lambda x}\,dx$$

tends to 0 as $\lambda\to\pm\infty$, and the convergence is uniform in a' and b'.

Suppose first that $b-a<\infty$. If $f=C$ the result is obvious, since then $|\gamma_\lambda|\leqslant 2|C|/|\lambda|$. Hence the result holds if f is a step-function (that is, if (a,b) can be broken up into a finite number of subintervals in each of which f is constant). Since a continuous f may be uniformly approximated by step-functions, (4·6) is valid for continuous functions. Applying Theorem (11·8) of Chapter I with $r=1$ and writing f as $\phi+(f-\phi)$, we find that

$$|\gamma_\lambda(f)|\leqslant|\gamma_\lambda(\phi)|+|\gamma_\lambda(f-\phi)|\leqslant|\gamma_\lambda(\phi)|+\epsilon<2\epsilon$$

for $|\lambda|$ large enough.

If $b-a=\infty$, for example if $(a,b)=(-\infty,+\infty)$, we write $f=f_1+f_2$, where $f_1=f$ in the interval $(-N,+N)$ and $f_1=0$ elsewhere. If N is large enough, then

$$\int_a^b|f_2|\,dx<\epsilon,\quad|\gamma_\lambda(f)|\leqslant|\gamma_\lambda(f_1)|+|\gamma_\lambda(f_2)|\leqslant o(1)+\epsilon,$$

and the result follows.

(4·7) THEOREM. (i) *If $f\in\Lambda_\alpha$, $0<\alpha\leqslant 1$, or if only $f\in\Lambda_\alpha^p$, then $c_\nu=O(|\nu|^{-\alpha})$;*
(ii) *If $f\in\Lambda_*$, or if only $f\in\Lambda_*^p$, then $c_\nu=O(\nu^{-1})$.*

Case (i) follows from (4·1) and (4·3). *Here 'O' cannot be replaced by 'o'* (see below), *except in the extreme case $\alpha=1, f\in\Lambda_1$.* In this latter case f is an integral, $S'[f]$ is still a Fourier series, and $\nu c_\nu\to 0$.

To prove (ii) we replace x by $x \pm \pi/\nu$ in the integrals defining c_ν. Then

$$2\pi c_\nu = \int_0^{2\pi} f(x)\, e^{-i\nu x}\, dx = -\int_0^{2\pi} f\left(x \pm \frac{\pi}{\nu}\right) e^{-i\nu x}\, dx$$
$$= -\frac{1}{4}\int_0^{2\pi} \left[f\left(x+\frac{\pi}{\nu}\right) + f\left(x-\frac{\pi}{\nu}\right) - 2f(x) \right] e^{-i\nu x}\, dx,$$
$$8\pi \,|\, c_\nu \,| \leqslant \int_0^{2\pi} \left| f\left(x+\frac{\pi}{\nu}\right) + f\left(x-\frac{\pi}{\nu}\right) - 2f(x) \right| dx = O\left(\frac{1}{\nu}\right).$$

For $f \in \lambda_*^p$, we have $c_\nu = o(1/\nu)$.

A good illustration of (4·7) is the *Weierstrass function*

$$f(x) = f_\alpha(x) = \sum_1^\infty b^{-n\alpha} \cos b^n x, \tag{4·8}$$

where $b > 1$ is an integer and α is a positive number. The series here converges absolutely and uniformly. The results which follow hold also for $\Sigma b^{-n\alpha} \sin b^n x$.

(4·9) THEOREM. *If* $0 < \alpha < 1$, *then* $f_\alpha \in \Lambda_\alpha$. *The function* f_1 *belongs to* Λ_* *but not to* Λ_1.

Let $0 < \alpha < 1$, $h > 0$. Then

$$f(x+h) - f(x) = -\sum b^{-n\alpha} \sin b^n(x + \tfrac{1}{2}h)\, 2 \sin \tfrac{1}{2} b^n h$$
$$= -\sum_1^N - \sum_{N+1}^\infty = P + Q,$$

where $N = N(h)$ is the largest integer satisfying $b^N h \leqslant 1$, so that $b^{N+1} h > 1$. Now

$$|\, P \,| \leqslant \sum_1^N b^{-n\alpha} . 1 . b^n h = O\{h . (b^N)^{1-\alpha}\} = O(h . h^{\alpha-1}) = O(h^\alpha),$$
$$|\, Q \,| \leqslant \sum_{N+1}^\infty b^{-n\alpha} . 1 . 2 = O(b^{-(N+1)\alpha}) = O(h^\alpha).$$

Hence $P + Q = O(h^\alpha)$ uniformly in x, and $f \in \Lambda_\alpha$.

In order to show that $f_1 \in \Lambda_*$, we write

$$f_1(x+h) + f_1(x-h) - 2f_1(x) = -\sum b^{-n} \cos b^n x\, (2 \sin \tfrac{1}{2} b^n h)^2$$
$$= -\sum_1^N - \sum_{N+1}^\infty = R + T,$$

with the same N as before. Then

$$|\, R \,| \leqslant h^2 \sum_1^N b^n = h^2 O(b^N) = h^2 O(h^{-1}) = O(h),$$
$$|\, T \,| \leqslant \sum_{N+1}^\infty b^{-n} \leqslant b^{-N} = O(h),$$

so that $R + T = O(h)$ and $f_1 \in \Lambda_*$. That $f_1 \notin \Lambda_1$ follows from the fact that otherwise $S'[f_1]$ would be a Fourier series and the coefficients of $S'[f_1]$ would tend to zero, which is not the case.

Minor changes in the preceding argument give the following result:

(4·10) THEOREM. *Let* $\epsilon_n \to 0$ *and*

$$g(x) = g_\alpha(x) = \sum_1^\infty \epsilon_n b^{-n\alpha} \cos b^n x. \tag{4·11}$$

Then $g_\alpha \in \lambda_\alpha$ *for* $0 < \alpha < 1$, *and* $g_1 \in \lambda_*$.

Weierstrass showed that for α small enough the function $f_\alpha(x)$ is nowhere differentiable. The extension to $\alpha \leqslant 1$ was first proved by Hardy. (For $\alpha > 1, f'(x)$ clearly exists and is continuous, since $S'[f]$ then converges absolutely and uniformly.)

f_1 is an example of a function of class Λ_* which is nowhere differentiable. On account of (4·10) and (3·3), $g_1(x)$ is differentiable in a set of the power of the continuum in every interval. As we shall see in Chapter V, p. 206, if $\Sigma\epsilon_n^2 = \infty$ (for example if $\epsilon_n = n^{-\frac{1}{2}}$), then g_1 is differentiable in a set of measure zero only. Thus, *smooth functions may be non-differentiable almost everywhere.*

If we write (4·8) in the form $\Sigma a_k \cos kx$ then $a_k = O(k^{-\alpha})$, and for $k = b^n$ this is the exact estimate. This shows that the results of (4·7) (i) cannot be improved.

(4·12) THEOREM. *Let $F(x)$ be a function of bounded variation over $0 \leqslant x \leqslant 2\pi$, and let C_ν and c_ν be the coefficients of F and dF respectively. If V is the total variation of F over $(0, 2\pi)$, then*

$$|C_\nu| \leqslant \frac{V}{\pi|\nu|} \quad (\nu \neq 0), \quad |c_\nu| \leqslant \frac{V}{2\pi}. \tag{4·13}$$

The second inequality follows from the formula

$$2\pi|c_\nu| = \left|\int_0^{2\pi} e^{-i\nu x}\, dF(x)\right| \leqslant \int_0^{2\pi} |dF(x)| = V.$$

Integrating by parts, we see that

$$2\pi C_\nu = \int_0^{2\pi} e^{-i\nu x} F(x)\, dx = \frac{F(2\pi) - F(0)}{-i\nu} + \frac{2\pi c_\nu}{i\nu}$$

for $\nu \neq 0$, and the last sum is absolutely $\leqslant 2V/|\nu|$.

Thus the coefficients of a function of bounded variation are $O(1/\nu)$. The example of the series $\Sigma\nu^{-1}\sin\nu x$ (see Chapter I, (4·12)) shows that we cannot replace 'O' by 'o' here. The function in this example is, however, discontinuous; examples of *continuous* functions of bounded variation with coefficients not $o(1/\nu)$ are much less obvious and will be given later (see, for example, Chapter V, §§ 3 and 7).

Consider the Fourier sine series $\Sigma b_\nu \sin\nu x$ of a function $f(x)$ defined in $(0, \pi)$. For the existence of the coefficients

$$b_\nu = \frac{2}{\pi}\int_0^\pi f \sin\nu x\, dx,$$

it is not necessary to suppose that f is integrable over $(0, \pi)$; it is enough to assume the integrability of $f \sin x$, for then $f \sin\nu x$ is also integrable. In this case we shall call our series *a generalized Fourier sine series.* For example, we have, in this sense,

$$\tfrac{1}{2}\cot\tfrac{1}{2}x \sim \sin x + \sin 2x + \dots + \sin nx + \dots, \tag{4·14}$$

a relation suggested by making $r \to 1$ in the formula for $\Sigma r^\nu \sin\nu x$ (see Chapter I, § 1). We have only to verify that the numbers

$$\beta_\nu = \frac{2}{\pi}\int_0^\pi \tfrac{1}{2}\cot\tfrac{1}{2}x \sin\nu x\, dx$$

satisfy the relations $\beta_1 = 1$, $\beta_\nu - \beta_{\nu+1} = 0$ for $\nu = 1, 2, \dots$, so that $\beta_1 = \beta_2 = \dots = 1$.

This example shows that *the generalized Fourier sine coefficients need not tend to* 0.

They are, however, $o(\nu)$. For $b_{\nu+1}-b_{\nu-1}$ is the νth cosine coefficient of the integrable function $2f(x)\sin x$, and so tends to 0. Hence if, for example, ν is odd,

$$b_\nu = b_1+(b_3-b_1)+\ldots+(b_\nu-b_{\nu-2})=o(\nu);$$

and the same argument holds for ν even.

The following result both generalizes and illuminates the Riemann-Lebesgue theorem.

(4·15) THEOREM. *Let $\alpha(x)$ be integrable, $\beta(x)$ bounded, both periodic. Then*

$$\frac{1}{2\pi}\int_0^{2\pi}\alpha(x)\,\beta(nx)\,dx \to \frac{1}{2\pi}\int_0^{2\pi}\alpha(x)\,dx\;\frac{1}{2\pi}\int_0^{2\pi}\beta(x)\,dx \quad \text{as} \quad n\to\infty. \tag{4·16}$$

Observe that if, for every $\epsilon>0$, we have $\alpha=\alpha_1+\alpha_2$ with $\mathfrak{M}[\alpha_1]<\epsilon$ and with the relation (4·16) holding for α_2 and each bounded β, then (4·16) is true. Now (4·16) is certainly true if α is the characteristic function of an interval and so, more generally, a step function. If α is integrable, we set $\alpha=\alpha_1+\alpha_2$, where α_2 is a step function and $\mathfrak{M}[\alpha_1]$ small.

The Riemann-Lebesgue theorem is the special cases $\beta=e^{\pm ix}$. As the above proof shows, (4·16) holds if we replace $\beta(nx)$ by $\beta(nx+\theta_n)$, where θ_n are arbitrary numbers. In this, moreover, n may tend to infinity by continuous values.

5. Formulae for partial sums of $S[f]$ and $\tilde{S}[f]$

Given an integrable and periodic f, let

$$a_\nu=\frac{1}{\pi}\int_{-\pi}^{\pi}f(t)\cos\nu t\,dt,\quad b_\nu=\frac{1}{\pi}\int_{-\pi}^{\pi}f(t)\sin\nu t\,dt, \tag{5·1}$$

so that
$$\tfrac{1}{2}a_0+\sum_{\nu=1}^{\infty}(a_\nu\cos\nu x+b_\nu\sin\nu x),\qquad \sum_{\nu=1}^{\infty}(a_\nu\sin\nu x-b_\nu\cos\nu x)$$

are $S[f]$ and $\tilde{S}[f]$ respectively. The partial sums of $S[f]$ will be denoted by $S_n[f]$, or by $S_n(x;f)$, or simply by $S_n(x)$; those of $\tilde{S}[f]$ by $\tilde{S}_n[f]$, $\tilde{S}_n(x;f)$, or $\tilde{S}_n(x)$. Using (5·1), we have

$$S_n(x)=\frac{1}{\pi}\int_{-\pi}^{\pi}f(t)\left\{\frac{1}{2}+\sum_{\nu=1}^{n}\cos\nu(t-x)\right\}dt=\frac{1}{\pi}\int_{-\pi}^{\pi}f(t)\,D_n(t-x)\,dt,$$

$$\tilde{S}_n(x)=-\frac{1}{\pi}\int_{-\pi}^{\pi}f(t)\left\{\sum_{\nu=1}^{n}\sin\nu(t-x)\right\}dt=-\frac{1}{\pi}\int_{-\pi}^{\pi}f(t)\,\tilde{D}_n(t-x)\,dt,$$

where
$$D_n(v)=\frac{1}{2}+\sum_{\nu=1}^{n}\cos\nu v=\sin(n+\tfrac{1}{2})\,v/2\sin\tfrac{1}{2}v,$$

$$\tilde{D}_n(v)=\sum_{\nu=1}^{n}\sin\nu v=\{\cos\tfrac{1}{2}v-\cos(n+\tfrac{1}{2})\,v\}/2\sin\tfrac{1}{2}v$$

(cf. Chapter I, § 1). The polynomials $D_n(v)$ and $\tilde{D}_n(v)$ are called *Dirichlet's kernel* and *Dirichlet's conjugate kernel* respectively. The formulae for S_n and $\tilde{S}_n$ may also be written

$$S_n(x)=\frac{1}{\pi}\int_{-\pi}^{\pi}f(x+u)\,D_n(u)\,du,\quad \tilde{S}_n(x)=-\frac{1}{\pi}\int_{-\pi}^{\pi}f(x+u)\,\tilde{D}_n(u)\,du.$$

Sometimes there is a slight advantage in taking the last term in S_n or $\tilde{S}_n$ with a factor $\frac{1}{2}$. The new expressions will be called the *modified partial sums*, and will be denoted by S_n^* and $\tilde{S}_n^*$ respectively. Thus

$$S_n^*(x) = \tfrac{1}{2}a_0 + \sum_{\nu=1}^{n-1}(a_\nu \cos \nu x + b_\nu \sin \nu x) + \tfrac{1}{2}(a_n \cos nx + b_n \sin nx) = \tfrac{1}{2}\{S_n(x) + S_{n-1}(x)\},$$

and $\tilde{S}_n^*$ is defined similarly. If we set

$$\left.\begin{aligned} D_n^*(v) &= D_n(v) - \tfrac{1}{2}\cos nv = \sin nv/2\tan \tfrac{1}{2}v, \\ \tilde{D}_n^*(v) &= \tilde{D}_n(v) - \tfrac{1}{2}\sin nv = (1-\cos nv)/2\tan\tfrac{1}{2}v, \end{aligned}\right\} \tag{5·2}$$

and proceed as before, we get

$$S_n^*(x) = \frac{1}{\pi}\int_{-\pi}^{\pi} f(x+t)\, D_n^*(t)\, dt, \quad \tilde{S}_n^*(x) = -\frac{1}{\pi}\int_{-\pi}^{\pi} f(x+t)\, \tilde{D}_n^*(t)\, dt. \tag{5·3}$$

By (4·4), $S_n - S_n^*$ tends uniformly to 0; S_n^* and S_n are equivalent with regard to convergence, and S_n^* is slightly the simpler. Similarly for $\tilde{S}_n$, $\tilde{S}_n^*$. We call D_n^* the *modified Dirichlet kernel*, $\tilde{D}_n^*$ the *modified conjugate Dirichlet kernel*.

With a fixed f and a fixed point x we set

$$\phi(t) = \phi_x(t) = \phi_x(t; f) = \tfrac{1}{2}\{f(x+t) + f(x-t) - 2f(x)\},$$
$$\psi(t) = \psi_x(t) = \psi_x(t; f) = \tfrac{1}{2}\{f(x+t) - f(x-t)\},$$

and we shall adhere throughout the book to this notation.

The polynomial
$$D_n^*(u) = \tfrac{1}{2} + \cos u + \ldots + \tfrac{1}{2}\cos nu$$

is even, and integrating it term by term we see that

$$\int_{-\pi}^{\pi} D_n^*(t)\, dt = \pi.$$

Hence
$$\begin{aligned} S_n^*(x) - f(x) &= \frac{1}{\pi}\int_{-\pi}^{\pi} f(x+t)\, D_n^*(t)\, dt - \frac{f(x)}{\pi}\int_{-\pi}^{\pi} D_n^*(t)\, dt \\ &= \frac{2}{\pi}\int_0^{\pi} \phi_x(t)\, D_n^*(t)\, dt = \frac{2}{\pi}\int_0^{\pi} \frac{\phi_x(t)}{2\tan\frac{1}{2}t}\sin nt\, dt. \end{aligned} \tag{5·4}$$

$\tilde{D}_n^*(u)$ being odd, we similarly get

$$\tilde{S}_n^*(x) = -\frac{2}{\pi}\int_0^{\pi} \psi_x(t)\, \tilde{D}_n^*(t)\, dt = -\frac{2}{\pi}\int_0^{\pi} \frac{\psi_x(t)}{2\tan\frac{1}{2}t}(1-\cos nt)\, dt. \tag{5·5}$$

For future reference we also state the following formulae:

$$\left.\begin{aligned} S_n(x) &= \frac{1}{\pi}\int_{-\pi}^{\pi} f(x+t)\, D_n(t)\, dt = \frac{1}{\pi}\int_0^{\pi} [f(x+t) + f(x-t)]\, D_n(t)\, dt, \\ S_n(x) - f(x) &= \frac{2}{\pi}\int_0^{\pi} \phi_x(t)\, D_n(t)\, dt = \frac{2}{\pi}\int_0^{\pi} \phi_x(t)\frac{\sin(n+\frac{1}{2})t}{2\sin\frac{1}{2}t}\, dt, \\ \tilde{S}_n(x) &= -\frac{2}{\pi}\int_0^{\pi} \psi_x(t)\, \tilde{D}_n(t)\, dt = -\frac{2}{\pi}\int_0^{\pi} \psi_x(t)\frac{\cos\frac{1}{2}t - \cos(n+\frac{1}{2})t}{2\sin\frac{1}{2}t}\, dt. \end{aligned}\right\} \tag{5·6}$$

Our main task in this chapter will be to show that, subject to suitable conditions on f at the point x, $S_n(x; f)$, or, what amounts to the same thing, $S_n^*(x; f)$, tends to $f(x)$

as $n \to \infty$. The summation problem for the conjugate series $\tilde{\text{S}}[f]$ leads us to consider the expression

$$-\frac{2}{\pi}\int_0^\pi \frac{\psi_x(t)}{2\tan\frac{1}{2}t}\,dt = -\frac{1}{\pi}\int_0^\pi \frac{f(x+t)-f(x-t)}{2\tan\frac{1}{2}t}\,dt, \tag{5·7}$$

where the integral is meant as the limit (if it exists) of

$$-\frac{2}{\pi}\int_\epsilon^\pi \frac{\psi_x(t)}{2\tan\frac{1}{2}t}\,dt \tag{5·8}$$

for $\epsilon \to +0$. The value of the expression (5·7), wherever it exists, will be denoted by $\tilde{f}(x)$, and the function $\tilde{f}(x)$ will be said to be *conjugate* to $f(x)$. The expression (5·8) will be denoted by $\tilde{f}(x;\,\epsilon)$. We show later (see Chapter IV, § 3 and Chapter VII, § 1) that for any integrable f the function $\tilde{f}$ exists almost everywhere; but the proof of this is far from simple.

The expression (5·7) can also be written

$$-\frac{1}{\pi}\int_{-\pi}^{\pi} \frac{f(x+t)}{2\tan\frac{1}{2}t}\,dt \quad \text{or} \quad -\frac{1}{\pi}\int_{-\pi}^{\pi} \frac{f(t)}{2\tan\frac{1}{2}(t-x)}\,dt,$$

where the integrals are taken in the 'principal value' sense, that is are the limits, for $\epsilon \to +0$, of integrals taken over the complements of intervals of length 2ϵ around the point of non-integrability of the integrand ($t=0$ in the first case and $t=x$ in the second).

From (5·5) we get formally

$$\tilde{S}_n^*(x) - \tilde{f}(x) = \frac{2}{\pi}\int_0^\pi \frac{\psi_x(t)}{2\tan\frac{1}{2}t}\cos nt\,dt. \tag{5·9}$$

There is an analogy between this integral and the last integral in (5·4), though the latter always converges, even absolutely, whereas in (5·9) both $\tilde{f}(x)$ and the right-hand side may not exist at some points. We shall see later that to a theorem on the convergence (or summability) of S[f] there usually corresponds an analogous theorem for $\tilde{\text{S}}[f]$.

We record some inequalities useful in 'convergence theory':

$$|D_n^*(t)| \leqslant n, \quad |D_n^*(t)| \leqslant 1/t \quad (0 < t \leqslant \pi;\ n = 1, 2, \ldots). \tag{5·10}$$

For $|D_n^*| \leqslant \frac{1}{2} + 1 + \ldots + \frac{1}{2} = n$; and the second estimate follows from (5·2), since $2\tan\frac{1}{2}t \geqslant t$. The first inequality (5·10) is preferable for t not too large in comparison with $1/n$, for example for $0 \leqslant t \leqslant \pi/n$, the second for larger t's. Similarly

$$|\tilde{D}_n^*(t)| < n, \quad |\tilde{D}_n^*(t)| \leqslant 2/t. \tag{5·11}$$

Analogous inequalities hold for D_n and $\tilde{D}_n$.

With the notation of § 1 of Chapter I we have easily

$$\tfrac{1}{2}[f(x_0+t)+f(x_0-t)] \sim \sum_0^\infty A_n(x_0)\cos nt,$$

$$\tfrac{1}{2}[f(x_0+t)-f(x_0-t)] \sim -\sum_1^\infty B_n(x_0)\sin nt.$$

Thus S[f] at $x = x_0$ is the same as the Fourier series at $t=0$ of the even function $\frac{1}{2}[f(x_0+t)+f(x_0-t)]$; $\tilde{\text{S}}[f]$ at $x=x_0$ is the series conjugate to the Fourier series at $t=0$ of the odd function $\frac{1}{2}[f(x_0+t)-f(x_0-t)]$.

6. The Dini test and the principle of localization

(6·1) THEOREM. *If the first of the integrals*

$$\int_0^\pi \frac{|\phi_x(t)|}{2\tan\frac{1}{2}t}dt, \quad \int_0^\pi \frac{|\psi_x(t)|}{2\tan\frac{1}{2}t}dt \tag{6·2}$$

is finite, then $\mathsf{S}[f]$ *converges at the point* x *to sum* $f(x)$. *If the second integral is finite, then* $\tilde{f}(x)$ *exists and* $\tilde{\mathsf{S}}[f]$ *at* x *converges to* $\tilde{f}(x)$.

The formulae (5·4) and (5·9) display the fundamental fact that, formally at least, $S_n^*(x)-f(x)$ and $\tilde{S}_n^*(x)-\tilde{f}(x)$ are the sine and cosine coefficients of certain functions. In each of the cases the function concerned is, by hypothesis, integrable, and in the second case $\tilde{f}(x)$ exists. Thus, by (4·4), we have respectively

$$S_n^*(x)-f(x)\to 0, \quad \tilde{S}_n^*(x)-\tilde{f}(x)\to 0.$$

The first part of (6·1) is called the *Dini test* for the convergence of $\mathsf{S}[f]$. The second part is due to Pringsheim. Since $2\tan\frac{1}{2}t\simeq t$ as $t\to 0$, the finiteness of the integrals (6·2) is equivalent to that of

$$\int_0^\pi \frac{|\phi_x(t)|}{t}dt, \quad \int_0^\pi \frac{|\psi_x(t)|}{t}dt.$$

Both integrals are finite if, for example,

$$f(x+t)-f(x)=O(|t|^\alpha) \quad (\alpha>0)$$

as $t\to 0$, and in particular if $f'(x)$ exists and is finite. The first integral converges even if f is discontinuous at x, provided that $\phi_x(t)$ tends to 0 sufficiently rapidly. The second integral diverges if $f(x\pm 0)$ exist and are different, and we shall see later that $\tilde{\mathsf{S}}[f]$ always diverges at such points.

(6·3) THEOREM. *If* $f(x)$ *vanishes in an interval* I, *then* $\mathsf{S}[f]$ *and* $\tilde{\mathsf{S}}[f]$ *converge uniformly in every interval* I' *interior to* I, *and the sum of* $\mathsf{S}[f]$ *there is* 0.

If the word 'uniformly' is omitted, (6·3) is a corollary of (6·1). For if $x\in I'$, both $\phi_x(t)$ and $\psi_x(t)$ vanish for small $|t|$ and the integrals (6·2) converge. To prove the general result, we need the following lemma:

(6·4) LEMMA. *Let* f *be integrable,* g *bounded, and both periodic. Then the Fourier coefficients of the function* $\chi(t)=f(x+t)g(t)$ *tend to* 0 *uniformly in the parameter* x.

By the second inequality in (4·1) it is enough to show that $\omega_1(\delta;\chi)\to 0$ uniformly in x. Now

$$\int_{-\pi}^{\pi}|\chi(t+h)-\chi(t)|\,dt \leqslant \int_{-\pi}^{\pi}|f(x+t+h)-f(x+t)|\,|g(t+h)|\,dt + \int_{-\pi}^{\pi}|f(x+t)|\,|g(t+h)-g(t)|\,dt = P+Q,$$

say. Suppose that $|g|<M$, $|h|\leqslant\delta$. Then

$$P\leqslant M\omega_1(\delta;f)\to 0.$$

In order to show that $Q \to 0$, we set $f = f_1 + f_2$, where f_1 is bounded, say $|f_1| \leqslant B$, and $\int_{-\pi}^{\pi} |f_2|\,dt < \epsilon/4M$. Then

$$Q \leqslant \int_{-\pi}^{\pi} |f_1(x+t)|\,|g(t+h)-g(t)|\,dt + \int_{-\pi}^{\pi} |f_2(x+t)|\,|g(t+h)-g(t)|\,dt \leqslant B\omega_1(\delta; g) + \tfrac{1}{2}\epsilon,$$

and so is less than ϵ for δ small enough. This proves (6·4).

Returning to (6·3), let $x \in I'$. Then $f(x+t)=0$ for $|t|<\eta$, say. Let $\lambda(t)$ be the periodic function equal to 0 for $|t|<\eta$ and to 1 elsewhere. Using (5·3) and (5·2), we write

$$S_n^*(x) = \frac{1}{\pi}\int_{-\pi}^{\pi} f(x+t)\frac{\lambda(t)}{2\tan\frac{1}{2}t}\sin nt\,dt = \frac{1}{\pi}\int_{-\pi}^{\pi} f(x+t)\,g(t)\sin nt\,dt. \qquad \textbf{(6·5)}$$

Here $g(t) = \lambda(t)/2\tan\frac{1}{2}t$ is bounded. By (6·4), $S_n^*(x)$ tends uniformly to 0 for $x \in I'$.

Similarly $\tilde{S}_n^*(x)$ tends uniformly to $\tilde{f}(x)$ in I'; for the difference $\tilde{S}_n^*(x) - \tilde{f}(x)$ is represented by (6·5) with $\sin nt$ replaced by $\cos nt$.

The result may be stated differently. Let us call two series $u_0 + u_1 + \ldots$ and $v_0 + v_1 + \ldots$ (convergent or not) *equiconvergent* if the difference $(u_0 - v_0) + (u_1 - v_1) + \ldots$ converges to 0. If this difference converges, but not necessarily to 0, the two series will be called *equiconvergent in the wider sense*. It is clear what 'uniform equiconvergence' means. The theorem that follows is a consequence of (6·3) when we set $f = f_1 - f_2$.

(6·6) THEOREM. *If two functions f_1 and f_2 are equal in an interval I, then $\mathsf{S}[f_1]$ and $\mathsf{S}[f_2]$ are uniformly equiconvergent in any interval I' interior to I; $\tilde{\mathsf{S}}[f_1]$ and $\tilde{\mathsf{S}}[f_2]$ are uniformly equiconvergent in I' in the wider sense.*

Considering for simplicity convergence at a single point, we see that *the convergence of $\mathsf{S}[f]$ and $\tilde{\mathsf{S}}[f]$, and the sum of $\mathsf{S}[f]$ (but not that of $\tilde{\mathsf{S}}[f]$) at a point x, depend only on the behaviour of f in an arbitrarily small neighbourhood of x.*

Theorems (6·3) and (6·6) express the *Riemann-Lebesgue localization principle*.

(6·7) THEOREM. (i) *Let $f(x)$ be integrable, $\rho(x)$ bounded, both periodic. If at a point x_0 the Dini numbers of ρ are bounded, the series $\mathsf{S}[\rho f]$ and $\rho(x_0)\,\mathsf{S}[f]$ are equiconvergent for $x = x_0$. The series $\tilde{\mathsf{S}}[\rho f]$ and $\rho(x_0)\,\tilde{\mathsf{S}}[f]$ are equiconvergent at x_0 in the wider sense.*

(ii) *If $\rho(x) \in \Lambda_1$, the equiconvergence of $\mathsf{S}[\rho f]$ and $\rho(x_0)\,\mathsf{S}[f]$, and that (in the wider sense) of $\tilde{\mathsf{S}}[\rho f]$ and $\rho(x_0)\,\tilde{\mathsf{S}}[f]$, is uniform in x_0.*

If $\rho(x_0) = 1$, case (i) may be interpreted as follows: 'slight' modifications of f in the neighbourhood of x_0 which leave $f(x_0)$ unaltered have no influence either upon the convergence of $\mathsf{S}[f]$ and $\tilde{\mathsf{S}}[f]$ at x_0, or on the sum of $\mathsf{S}[f]$ at that point (though they can influence the sum of $\tilde{\mathsf{S}}[f]$).

To prove (i), we observe that

$$S_n^*(x_0; \rho f) - \rho(x_0)\,S_n^*(x_0; f) = \frac{1}{\pi}\int_{-\pi}^{\pi} f(x_0+t)\,g(t)\sin nt\,dt,$$

where

$$g(t) = g_{x_0}(t) = [\rho(x_0+t) - \rho(x_0)]/2\tan\tfrac{1}{2}t$$

is a bounded function. Hence the integral on the right, being the Fourier coefficient of an integrable function, tends to 0 with $1/n$. For $\tilde{S}_n^*(x_0; \rho f) - \rho(x_0)\tilde{S}_n^*(x_0; f)$ we have the value

$$-\frac{1}{\pi}\int_{-\pi}^{\pi}\frac{\rho(x_0+t)-\rho(x_0)}{2\tan\frac{1}{2}t}f(x_0+t)(1-\cos nt)\,dt$$
$$= -\frac{1}{\pi}\int_{-\pi}^{\pi}f(x_0+t)\,g(t)\,dt + \frac{1}{\pi}\int_{-\pi}^{\pi}f(x_0+t)\,g(t)\cos nt\,dt,$$

and the last integral tends to 0. This proves (i).

Let us set

$$\chi(t) = \chi_x(t) = f(x+t)\,g_x(t).$$

The above argument and (4·1) will give (ii), if we can prove that $\omega_1(\delta; \chi) \to 0$ uniformly in x as $\delta \to 0$. Arguing as in the proof of (6·4), and observing that $|g_x(t)| < M$, say, we have only to show that the integral $\int_{-\pi}^{\pi} |g_x(t+h) - g_x(t)|\,dt$ tends to 0 with h, uniformly in x. We break up the interval of integration into two parts: the interval $|t| \leqslant \epsilon/8M$, and the remainder of $(-\pi, \pi)$. The first integral does not exceed $2M \cdot 2\epsilon/8M = \frac{1}{2}\epsilon$. Outside the first interval the function $g_x(t)$ is continuous in t, uniformly in x, so that the second integral tends to 0 with h, uniformly in x. The whole is thus less than ϵ for small $|h|$, and this completes the proof. For the conjugate series we argue similarly.

Theorem (6·7) includes (6·3). For let $\rho(x)$ denote the continuous function which is equal to 0 in I', is equal to 1 outside I, and is linear elsewhere. For $x_0 \in I'$ we have $\rho(x_0)\,\mathsf{S}[f] = 0$, and since $\mathsf{S}[\rho f] = \mathsf{S}[f]$, (6·7) implies that $\mathsf{S}[f]$ and $\tilde{\mathsf{S}}[f]$ converge uniformly in I', the sum of $\mathsf{S}[f]$ being 0 there.

The analogue of (6·1) for uniform convergence is as follows.

(6·8) THEOREM. *Suppose that f is continuous in a closed interval $I = (a, b)$ and let $\omega(\delta)$ be its modulus of continuity there. If $\omega(\delta)/\delta$ is integrable near $\delta = 0$, and if the integrals*

$$\int_0^{\pi}\frac{|f(a)-f(a-t)|}{t}\,dt, \qquad \int_0^{\pi}\frac{|f(b+t)-f(b)|}{t}\,dt,$$

are finite, then both $\mathsf{S}[f]$ and $\tilde{\mathsf{S}}[f]$ converge uniformly in I, to f and $\tilde{f}$ respectively.

For let $\xi(t)$ be the sum of the numbers

$$\omega(t), \quad |f(a)-f(a-t)|, \quad |f(b+t)-f(b)| \quad \text{for} \quad 0 \leqslant t \leqslant h = b-a.$$

The function $\xi(t)/t$ is integrable. Write

$$S_n^*(x) - f(x) = \frac{1}{\pi}\left\{\int_{|t|\leqslant\sigma} + \int_{\sigma\leqslant|t|\leqslant\pi}\right\}[f(x+t)-f(x)]\,D_n^*(t)\,dt = P + Q, \qquad (6·9)$$

say, where $0 < \sigma \leqslant h$, and consider first the term P. Let $x \in I$. If $x+t$ is in I, then $|f(x+t) - f(x)| \leqslant \omega(|t|)$. If $x+t$ is not in I, say $x+t > b$, then

$$|f(x+t)-f(x)| \leqslant |f(b)-f(x)| + |f(t+x)-f(b)| \leqslant \omega(t) + |f(t+x)-f(b)|,$$

and since $|D_n^*(t)| \leqslant |t|^{-1}$ it is easy to see that

$$|P| \leqslant \frac{2}{\pi}\int_0^{\sigma}\frac{\xi(t)}{t}\,dt < \epsilon,$$

provided σ is small enough. Since Q is the Fourier coefficient of the function $\{f(x+t)-f(x)\}\,g(t)$, where $g(t)$ is 0 in $(-\sigma, \sigma)$ and $\frac{1}{2}\cot\frac{1}{2}t$ outside, we see from (6·4) that $Q \to 0$ uniformly in I. Hence $S_n^*(x) \to f(x)$ uniformly in I.

With the hypotheses of (6·8), the integral defining $\tilde{f}(x)$ converges absolutely and uniformly in I, since

$$\int_0^\sigma \frac{|\psi_x(t)|}{2\tan\frac{1}{2}t}dt \leqslant \int_0^\sigma \frac{\xi(t)}{t}dt \quad (0<\sigma\leqslant h).$$

In particular, $\tilde{f}(x)$ is continuous in I. An argument similar to that above shows that $\tilde{S}_n^*(x)-\tilde{f}(x)\to 0$ uniformly in I.

(6·10) THEOREM. *If $f\in \mathrm{L}$, $\rho\in\Lambda_1$, the integrals*

$$\int_{-\pi}^{\pi} \rho(t)f(t)\tfrac{1}{2}\cot\tfrac{1}{2}(t-x)\,dt, \quad \int_{-\pi}^{\pi} \rho(x)f(t)\tfrac{1}{2}\cot\tfrac{1}{2}(t-x)\,dt, \tag{6·11}$$

taken in the 'principal value' sense, are uniformly equiconvergent in the wider sense.

This is immediate, since $[\rho(t)-\rho(x)]\frac{1}{2}\cot\frac{1}{2}(t-x)$ is bounded in x, t.

7. Some more formulae for partial sums

Let ϵ be a fixed positive number less than π. It is sometimes convenient to use the formulae

$$\left.\begin{aligned} S_n(x) &= \frac{1}{\pi}\int_{-\epsilon}^{\epsilon} f(x+t)\frac{\sin nt}{t}dt + o(1),\\ S_n(x)-f(x) &= \frac{2}{\pi}\int_0^{\epsilon} \phi_x(t)\frac{\sin nt}{t}dt + o(1). \end{aligned}\right\} \tag{7·1}$$

In the former the $o(1)$ term tends to 0 uniformly in x; in the latter it tends to 0 for every x and uniformly in every interval where f is bounded.

To prove the first formula we note that the difference between the integral on the right and the integral defining $S_n^*(x)$ $(=S_n(x)+o(1))$ is the sine coefficient of the function $f(x+t)\,g(t)$, where $g(t)$ is the function equal to $1/t-\frac{1}{2}\cot\frac{1}{2}t=O(1)$ for $|t|<\epsilon$ and to $-\frac{1}{2}\cot\frac{1}{2}t$ at the remaining points of $(-\pi,\pi)$. Similarly, the difference between the second integral and the one defining $S_n^*(x)-f(x)$ is the sine coefficient of

$$\{f(x+t)-f(x)\}\,g(t),$$

and the second formula (7·1) follows.

We note also the formula

$$\tilde{S}_n(x) = -\frac{1}{\pi}\int_{-\epsilon}^{\epsilon} f(x+t)\frac{1-\cos nt}{t}dt + R_n(x),$$

where $R_n(x)$ tends uniformly to a continuous function of x.

It is instructive to compare this and the first formula (7·1) with the exact formulae

$$\frac{1}{\pi}\int_{-\infty}^{+\infty} f(x+t)\frac{\sin\omega t}{t}dt = \sum_{\nu\leqslant\omega}{}' A_\nu(x), \tag{7·2}$$

$$-\frac{1}{\pi}\int_{-\infty}^{+\infty} f(x+t)\frac{1-\cos\omega t}{t}dt = \sum_{\nu\leqslant\omega}{}' B_\nu(x), \tag{7·3}$$

where ω is positive but not necessarily an integer, the integrals are defined as $\lim\limits_{T\to+\infty}\int_{-T}^{T}$, and the dash indicates that if ω is an integer then the last term of the sum is taken with a factor $\frac{1}{2}$.

We take the first formula only, the proof of the second being analogous. The familiar equation
$$\frac{1}{\pi}\int_{-\infty}^{+\infty}\frac{\sin t}{t}\,dt=1$$
(see (8·4) below) shows that
$$\frac{1}{\pi}\int_{-\infty}^{+\infty}\frac{\sin \lambda t}{t}\,dt=\operatorname{sign}\lambda \quad (-\infty<\lambda<+\infty). \tag{7·4}$$
Hence, if $f(x)=e^{i\nu x}$, the left-hand side of (7·2) is
$$\pi^{-1}e^{i\nu x}\lim\int_{-T}^{T}e^{i\nu t}\frac{\sin\omega t}{t}\,dt=(2\pi)^{-1}e^{i\nu x}\int_{-\infty}^{+\infty}\left[\frac{\sin(\omega+\nu)t}{t}+\frac{\sin(\omega-\nu)t}{t}\right]dt,$$
and the last integral is 2π, π, 0 according as $|\nu|<\omega$, $|\nu|=\omega$, $|\nu|>\omega$. This proves the formula if f is a trigonometric polynomial. Hence we may assume that $c_\nu=0$ for $|\nu|\leqslant\omega$.

We now use a result which will be established in Chapter IV, p. 160, and which asserts that a Fourier series can be integrated termwise over any finite interval after having been multiplied by any function of bounded variation. Thus if $S[f]=\Sigma c_\nu e^{i\nu x}$ we have
$$\frac{1}{\pi}\int_{-T}^{T}f(x+t)\frac{\sin\omega t}{t}\,dt=\Sigma c_\nu e^{i\nu x}\frac{1}{\pi}\int_0^T\left[\frac{\sin(\nu+\omega)t}{t}-\frac{\sin(\nu-\omega)t}{t}\right]dt. \tag{7·5}$$
Integrating by parts twice we get
$$\int_u^\infty\frac{\sin t}{t}\,dt=\frac{\cos u}{u}+\frac{\sin u}{u^2}-2\int_u^\infty\frac{\sin t}{t^3}\,dt=\frac{\cos u}{u}+O\left(\frac{1}{u^2}\right)\quad(u>0).$$
Since $|\nu|>\omega$ and $\int_0^T=\int_0^\infty-\int_T^\infty$, the sum in (7·5) is
$$\frac{1}{\pi}\Sigma c_\nu e^{i\nu x}\left[\frac{\cos T(\nu-\omega)}{T(\nu-\omega)}-\frac{\cos T(\nu+\omega)}{T(\nu+\omega)}+O\left(\frac{1}{T^2\nu^2}\right)\right]$$
$$=\frac{1}{\pi T}\Sigma c_\nu e^{i\nu x}\left[\frac{\cos T(\nu-\omega)}{\nu}-\frac{\cos T(\nu+\omega)}{\nu}+O\left(\frac{1}{\nu^2}\right)\right]$$
$$=\frac{2\sin T\omega}{\pi T}\Sigma c_\nu\frac{e^{i\nu x}}{\nu}\sin\nu T+o(1)$$
as $T\to\infty$ (observe that $\Sigma|c_\nu|\nu^{-2}<\infty$). If F denotes the integral of f, F is bounded and periodic, and the penultimate term is $(\pi T)^{-1}\sin T\omega[F(x+T)-F(x-T)]=o(1)$. Hence (7·5) tends to 0 as $T\to\infty$ and this completes the proof of (7·2).

The integral (7·4) converges uniformly in λ outside an arbitrarily small neighbourhood of $\lambda=0$. From the preceding proof it follows that the integrals (7·2) and (7·3) converge uniformly over the set obtained by removing from any finite interval $|\omega|\leqslant\Omega$ arbitrarily small neighbourhoods of the points $0, \pm1, \pm2, \ldots$. (The neighbourhoods must be removed, since the right-hand sides of the formulae are, in general, discontinuous at the points ν.)

We have also the formula
$$\tilde f(x)=-\frac{1}{\pi}\int_{-\infty}^{+\infty}\frac{f(x+t)}{t}\,dt=-\frac{1}{\pi}\lim_{\epsilon\to0}\left\{\lim_{T\to+\infty}\left(\int_{-T}^{-\epsilon}+\int_\epsilon^T\right)\frac{f(x+t)}{t}\,dt\right\}, \tag{7·6}$$

valid at every point at which $\tilde{f}(x)$ exists. (The internal limit always exists.) For if we subtract from f a constant, which changes nothing in (7·6), we may assume that the integral of f over a period is zero, and then the application of the second mean-value theorem gives the existence of each of the integrals

$$-\frac{1}{\pi}\int_{\pi}^{\infty}\frac{f(x+t)}{t}dt, \quad -\frac{1}{\pi}\int_{-\infty}^{-\pi}\frac{f(x+t)}{t}dt \tag{7·7}$$

separately. Their sum is

$$-\frac{1}{\pi}\int_{-\pi}^{\pi} f(x+t)\left[\sum_{k=-\infty}^{+\infty}{}' \frac{1}{t+2k\pi}\right]dt = -\frac{1}{\pi}\int_{-\pi}^{\pi} f(x+t)\left[\tfrac{1}{2}\cot\tfrac{1}{2}t-\frac{1}{t}\right]dt,$$

where the dash $'$ indicates the omission of $k=0$ in the summation. This is

$$-\frac{1}{\pi}\left(\int_{-\pi}^{-\epsilon}+\int_{\epsilon}^{\pi}\right)\left[\tfrac{1}{2}\cot\tfrac{1}{2}t-\frac{1}{t}\right]f(x+t)\,dt+o(1)=\tilde{f}(x;\epsilon)+\frac{1}{\pi}\int_{\epsilon\leqslant|t|\leqslant\pi}\frac{f(x+t)}{t}dt+o(1)$$

as $\epsilon\to+0$, and (7·6) easily follows.

8. The Dirichlet-Jordan test

This name is usually given to the following theorem (see also (8·14) below).

(8·1) THEOREM. *Suppose that $f(x)$ is of bounded variation over $(0, 2\pi)$. Then*

(i) *at every point x_0, $S[f]$ converges to the value $\tfrac{1}{2}[f(x_0+0)+f(x_0-0)]$; in particular, $S[f]$ converges to $f(x)$ at every point of continuity of f;*

(ii) *if further f is continuous at every point of a closed interval I, then $S[f]$ converges uniformly in I.*

We prove first the following lemma, only part of which is needed here:

(8·2) LEMMA. *The integrals*

$$\frac{2}{\pi}\int_0^{\xi} D_n(t)\,dt, \quad \frac{2}{\pi}\int_0^{\xi} D_n^*(t)\,dt, \quad \frac{2}{\pi}\int_0^{\xi}\frac{\sin nt}{t}dt \quad (0\leqslant\xi\leqslant\pi)$$

are all uniformly bounded in n and ξ. The difference between any two of these integrals tends to 0 with $1/n$, uniformly in ξ.

Let us denote these integrals respectively by $\alpha_n(\xi)$, $\beta_n(\xi)$, $\gamma_n(\xi)$. Plainly, $\beta_n-\alpha_n$ is uniformly bounded and tends uniformly to 0 as $n\to\infty$. Furthermore,

$$\gamma_n-\beta_n=\frac{2}{\pi}\int_0^{\xi}\left\{\frac{1}{t}-\tfrac{1}{2}\cot\tfrac{1}{2}t\right\}\sin nt\,dt=\frac{2}{\pi}\int_0^{\pi} w_{\xi}(t)\sin nt\,dt,$$

where $w_{\xi}(t)$ is $1/t-\tfrac{1}{2}\cot\tfrac{1}{2}t$ in $(0,\xi)$ and 0 in (ξ,π). Since the total variation of $w_{\xi}(t)$ over $(0,\pi)$ is uniformly bounded, the last integral is uniformly bounded and tends uniformly to 0 (see (4·13)). It is thus enough to show the boundedness of

$$\gamma_n(\xi)=\frac{2}{\pi}\int_0^{n\xi}\frac{\sin u}{u}du=\frac{2}{\pi}G(n\xi),$$

where

$$G(v)=\int_0^{v}\frac{\sin u}{u}du; \tag{8·3}$$

and this will follow if we show that $G(v)$ tends to a limit as $v\to+\infty$. Since the integrand tends to 0 it is enough to prove the existence of $\lim G(n\pi)$. But $\alpha_n(\pi)=1$ and

$\alpha_n(\pi)-\gamma_n(\pi)\to 0$ together imply $G(n\pi)\to\frac{1}{2}\pi$, which proves (8·2). We have obtained incidentally the well-known formula

$$\int_0^\infty \frac{\sin t}{t}\,dt=\tfrac{1}{2}\pi \tag{8·4}$$

Return now to (8·1), and apply the last remark of § 5. Replacing $f(x)$ by

$$\tfrac{1}{2}[f(x_0+x)+f(x_0-x)],$$

we may assume that $x_0=0$ and that $f(x)$ is even. We have to show that $S_n(0)\to f(+0)$.

Suppose first that $f(x)$ is non-negative and non-decreasing in $(0,\pi)$. Let C be a number greater than $|\beta_n(\xi)|$ for all n and ξ. We write

$$S_n^*(0)-f(+0)=\frac{2}{\pi}\int_0^\pi [f(t)-f(+0)]\,D_n^*(t)\,dt=\frac{2}{\pi}\left(\int_0^\eta+\int_\eta^\pi\right)=A+B, \tag{8·5}$$

say, where η is so chosen that $|f(\eta)-f(+0)|<\epsilon/4C$. Since $f(t)-f(+0)$ is non-negative and non-decreasing the second mean-value theorem gives

$$|A|=\left|\{f(\eta)-f(+0)\}\frac{2}{\pi}\int_{\eta'}^{\eta} D_n^*(t)\,dt\right|\leqslant\frac{\epsilon}{4C}2C=\tfrac{1}{2}\epsilon\quad(0<\eta'<\eta).$$

For fixed η, B is a sine coefficient of the function $w_\eta(t)$ equal to 0 in $(0,\eta)$ and to $\{f(t)-f(+0)\}\,\frac{1}{2}\cot\frac{1}{2}t$ in (η,π). Thus, by (4·13),

$$B\to 0,\quad |A+B|<\epsilon\quad\text{for}\quad n>n_0,\quad S_n(0)\to f(+0).$$

In the general case f is, in $(0,\pi)$, the difference f_1-f_2 of two non-negative and non-decreasing functions (the positive and negative variations of f). If we define f_1 and f_2 in $(-\pi,0)$ by the condition of evenness, the formula $f=f_1-f_2$ becomes valid in $(-\pi,\pi)$ and the general result follows from the special case just proved.

Case (ii) follows from the argument just used if we note that the continuity of f in I implies the continuity of the positive and negative variations of f in that interval, and that all the estimates obtained above hold uniformly for $x_0\in I$.

In Chapter III, § 3, we give a different proof of (8·1) that does not require the theorem on the continuity of the positive and negative variations.

A sequence of functions $s_n(x)$ defined in the neighbourhood of $x=x_0$ and converging for $x=x_0$ (but not necessarily for $x\neq x_0$) is said to converge uniformly *at* x_0 to limit s, if to every $\epsilon>0$ there is a $\delta=\delta(\epsilon)$ and a $p=p(\epsilon)$ such that

$$|s_n(x)-s|<\epsilon\quad\text{for}\quad|x-x_0|<\delta\quad\text{and}\quad n>p.$$

An equivalent definition is that $s_n(x_n)\to s$ for each sequence $x_n\to x_0$.

(8·6) THEOREM. *If f is of bounded variation, $S[f]$ converges uniformly at every point of continuity of f.*

It is enough to consider the case when $x_0=0$ and f is even and non-decreasing in $(0,\pi)$. A proof similar to that of (8·1)(i) shows that if n is large enough and x small enough and, e.g., positive then $S_n(x)$ is arbitrarily close to $f(+0)=f(0)$. We omit the details since a simpler proof will be given in Chapter III, § 3.

(8·7) Theorem. *Let a_ν, b_ν be the coefficients of f, and let $F(x)$ be the indefinite integral of f. Then*

$$F(x)=\tfrac{1}{2}a_0x+C+\sum_{\nu=1}^{\infty}(a_\nu\sin\nu x-b_\nu\cos\nu x)/\nu, \tag{8·8}$$

the series on the right being uniformly convergent.

For the proof it is enough to observe that the right-hand side without the term $\tfrac{1}{2}a_0x$ is $\mathsf{S}[F-\tfrac{1}{2}a_0x]$ (see (2·5)), and that $F(x)-\tfrac{1}{2}a_0x$ is a continuous function of bounded variation.

It follows from (8·8) that

$$\int_\alpha^\beta f(x)\,dx=[\tfrac{1}{2}a_0x]_\alpha^\beta+\sum_{\nu=1}^{\infty}\left[\frac{a_\nu\sin\nu x-b_\nu\cos\nu x}{\nu}\right]_\alpha^\beta$$

for every α, β. Thus, *if $\mathsf{S}[f]$ is integrated term by term over any interval (α,β), the resulting series converges to $\int_\alpha^\beta f\,dx$.*

Putting $x=0$ in (8·8) we see that *the series $\Sigma b_\nu/\nu$ converges for any f.* This may be false for $\Sigma a_\nu/\nu$ (see Chapter V, (1·11)).

The following result is an analogue of (8·1)(i) for $\tilde{\mathsf{S}}[f]$:

(8·9) Theorem. *If $f(x)$ is of bounded variation, a necessary and sufficient condition for the convergence of $\tilde{\mathsf{S}}[f]$ at x is the existence of the integral*

$$\tilde{f}(x)=-\frac{2}{\pi}\int_0^\pi\frac{\psi_x(t)}{2\tan\frac{1}{2}t}dt=\lim_{h\to+0}\left\{-\frac{2}{\pi}\int_h^\pi\frac{\psi_x(t)}{2\tan\frac{1}{2}t}dt\right\}=\lim_{h\to+0}\tilde{f}(x;h), \tag{8·10}$$

which represents then the sum of $\tilde{\mathsf{S}}[f]$.

We first show that:

(8·11) Lemma. *If f is of bounded variation, $\tilde{S}_n^*(x)-\tilde{f}(x;\pi/n)$ tends to 0 at every point of continuity of f and is bounded at every point of discontinuity.*

Let $\psi(t)=\tfrac{1}{2}[f(x_0+t)-f(x_0-t)]$. Since $\tilde{\mathsf{S}}[f]$ at x_0 is the same thing as $\tilde{\mathsf{S}}[\psi]$ at $t=0$ we may suppose that $x_0=0$ and that $f(x)$ is odd. Hence $\psi_{x_0}(t)=f(t)$. Let us also temporarily assume that $f(x)$ is non-negative and non-decreasing in $(0,\pi)$. Then

$$\tilde{S}_n^*(0)-\tilde{f}(0;\pi/n)=-\frac{2}{\pi}\int_0^{\pi/n}f(t)\tilde{D}_n^*(t)\,dt+\frac{2}{\pi}\int_{\pi/n}^{\pi}\frac{f(t)}{2\tan\frac{1}{2}t}\cos nt\,dt=A+B. \tag{8·12}$$

Suppose first that f is continuous at $t=0$, i.e. that $f(t)\to 0$ with t. Since $|\tilde{D}_n^*|\leqslant n$,

$$|A|\leqslant\frac{2n}{\pi}\int_0^{\pi/n}f(t)\,dt\leqslant 2f(\pi/n)=o(1).$$

Given $\epsilon>0$, we choose η such that $f(\eta)<\epsilon$, and write

$$B=\frac{2}{\pi}\left(\int_{\pi/n}^{\eta}+\int_\eta^\pi\right)=B'+B'',$$

say. Applying the second mean-value theorem twice, we get

$$|B'|=\left|\frac{1}{\pi}\cot(\pi/2n)\int_{\pi/n}^{\eta'}f(t)\cos nt\,dt\right|=\left|\frac{1}{\pi}\cot(\pi/2n)f(\eta')\int_{\eta''}^{\eta'}\cos nt\,dt\right|<\frac{1}{\pi}\frac{2n}{\pi}\epsilon\frac{2}{n},$$

so that $|B'|<\epsilon$. Since B'' is a Fourier coefficient, it tends to 0. Thus (8·12) tends to 0. We prove similarly that when $f(+0)\neq 0$, (8·12) is bounded.

When f is no longer assumed to be non-negative and non-decreasing in $(0,\pi)$, we decompose it into its positive and negative variations. These are continuous at $t=0$ when f is, so that (8·11) is proved.

Now let $\pi/(n+1)\leqslant h\leqslant \pi/n$. In the general case of an f of bounded variation we have

$$\left|\tilde{f}\left(0;\frac{\pi}{n}\right)-\tilde{f}(0;h)\right|\leqslant\frac{2}{\pi}\int_{\pi/(n+1)}^{\pi/n}|f(t)|\frac{dt}{2\tan\frac{1}{2}t}\leqslant\frac{2(n+1)}{\pi^2}\left(\frac{\pi}{n}-\frac{\pi}{n+1}\right)\sup|f(t)|=o(1)$$

as $n\to\infty$, and this, together with (8·11), proves (8·9).

If f has a jump at a point x, then obviously $\tilde{f}(x;\pi/n)\to\pm\infty$. Thus $\tilde{S}[f]$ diverges at x to $\pm\infty$. This is also contained in the following more precise result, in which only the integrability of f is assumed.

(8·13) Theorem. *If $f(x_0\pm 0)$ exist, and if $f(x_0+0)-f(x_0-0)=l$, then*

$$\tilde{S}_n(x_0)/\log n\to -l/\pi.$$

We may suppose that $x_0=0$ and that f is odd. It is easy to verify the result for the function $\phi(x)$ defined in Chapter I, (4·12), using the fact that the partial sums of the harmonic series are asymptotically equal to $\log n$. Subtracting $(l/\pi)\,\phi(x)$ from f we obtain an odd function g continuous and vanishing at the origin, and it is enough to prove that $\tilde{S}_n(0;g)=o(\log n)$. For this purpose we write (cf. (5·11))

$$|\tilde{S}_n^*(0;g)|\leqslant\frac{2}{\pi}\int_0^{\pi}|g(t)|\,|\tilde{D}_n^*(t)|\,dt=\frac{2}{\pi}\int_0^{\pi/n}+\frac{2}{\pi}\int_{\pi/n}^{\pi}$$

$$\leqslant\frac{2n}{\pi}\int_0^{\pi/n}|g(t)|\,dt+\frac{4}{\pi}\int_{\pi/n}^{\pi}\frac{|g(t)|}{t}\,dt=o(1)+o(\log n)=o(\log n)$$

(cf. Chapter I, (8·1)).

A corollary of (8·13) is that, *if the Fourier coefficients a_n, b_n of f are $o(1/n)$, f cannot have discontinuities of the first kind.* For the hypothesis implies that

$$\tilde{S}_n(x)=o\left(1+\frac{1}{2}+\dots+\frac{1}{n}\right)=o(\log n).$$

In particular, *if the Fourier coefficients of a function f of bounded variation are $o(1/n)$, the function f has only removable discontinuities.* For $f(x+0)=f(x-0)$ for every x, and by changing the values of f at the at most denumerable set of points where f is discontinuous, we can make f everywhere continuous.

(8·14) Theorem. *Suppose that f is integrable and periodic, and of bounded variation in an interval I. Then $S[f]$ converges to $\frac{1}{2}[f(x+0)+f(x-0)]$ at every point x interior to I. If, in addition, f is continuous in I, the convergence is uniform in any interval interior to I. A necessary and sufficient condition for the convergence of $\tilde{S}[f]$ at an x interior to I is the existence of the integral $\tilde{f}(x)$, which represents then the sum of $\tilde{S}[f]$.*

For we can modify f by making it equal to 0 outside I. The new function is of bounded variation, and it is enough to combine (8·1) and (8·9) with (6·6).

9. Gibbs's phenomenon

We now study the partial sums $s_n(x)$ of the special series

$$\sum_{\nu=1}^{\infty} \frac{\sin \nu x}{\nu} = \tfrac{1}{2}(\pi - x) = \phi(x) \quad (0 < x < 2\pi) \tag{9·1}$$

in the neighbourhood of $x=0$. The series cannot converge uniformly there since $\phi(x)$ is discontinuous at $x=0$. Supposing that $x>0$, and using (8·2), we see that

$$s_n(x) + \tfrac{1}{2}x = \int_0^x D_n(t)\,dt = \int_0^{nx} \frac{\sin t}{t}\,dt + o(1),$$

uniformly in $0 \leqslant x \leqslant \pi$. Hence the $s_n(x)$ are uniformly bounded, and

$$s_n(x) = \int_0^{nx} \frac{\sin t}{t}\,dt + R_n(x), \tag{9·2}$$

where $|R_n(x)| < \epsilon$ if $x < \epsilon$, $n > n_0(\epsilon)$.

Consider the integral (8·3). The integrals of $(\sin t)/t$ over the intervals $(k\pi, (k+1)\pi)$ decrease in absolute value and are of alternating sign when $k = 0, 1, 2, \ldots$. This shows that the curve $y = G(x)$ has a wave-like shape with maxima $M_1 > M_3 > M_5 > \ldots$ at $\pi, 3\pi, 5\pi, \ldots$ and minima $m_2 < m_4 < m_6 < \ldots$ at $2\pi, 4\pi, \ldots$. Substituting $x = \pi/n$ in (9·2), we get

$$s_n(\pi/n) \to G(\pi) > G(\infty) = \tfrac{1}{2}\pi.$$

Thus, though $s_n(x)$ tends to $\phi(x)$ at every fixed x, $0 < x < 2\pi$, the curves $y = s_n(x)$, which pass through the point $(0, 0)$, condense to the interval $0 \leqslant y \leqslant G(\pi)$ of the y-axis (cf. also (9·4), below), the ratio of whose length to that of the interval $0 \leqslant y \leqslant \phi(+0) = \frac{1}{2}\pi$ is

$$\frac{2}{\pi}\int_0^{\pi} \frac{\sin t}{t}\,dt = 1{\cdot}179\ldots.$$

Similarly, to the left of $x=0$ the curves $y = s_n(x)$ condense to the interval $-G(\pi) \leqslant y \leqslant 0$. This behaviour is called *Gibbs's phenomenon*, and its generalized form may be described as follows. Suppose that a sequence $\{f_n(x)\}$ converges for $x_0 < x \leqslant x_0 + h$, say, to limit $f(x)$ and that $f(x_0+0)$ exists. Suppose that, when $n \to \infty$ and $x \to x_0$ independently, we have

$$\limsup f_n(x) > f(x_0+0) \quad \text{or} \quad \liminf f_n(x) < f(x_0+0);$$

then we say that $\{f_n(x)\}$ shows Gibbs's phenomenon in the right-hand neighbourhood of $x = x_0$. Similarly for the left-hand neighbourhood. If $f(x) = \lim f_n(x)$ is defined and continuous at x_0, the absence of the phenomenon at the point x_0 is equivalent to the uniform convergence of $\{f_n(x)\}$ at x_0.

(9·3) THEOREM. *If f is of bounded variation*† *and has no removable discontinuities, $S[f]$ shows Gibbs's phenomenon at every point of discontinuity of f and only there.*

We may suppose that f has only regular discontinuities, i.e.

$$f(x) = \tfrac{1}{2}\{f(x+0) + f(x-0)\}$$

for each x.

Suppose that $f(\xi+0) - f(\xi-0) = l \neq 0$. The function

$$\Delta(x) = f(x) - \frac{l}{\pi}\phi(x-\xi)$$

† It is enough to assume that the coefficients of f are $O(1/n)$; see Theorem (3·8) of Chapter III.

is continuous at ξ. Hence $\mathsf{S}[\Delta]$ converges uniformly at ξ (see (8·6)). The behaviour of $S_n(x;f)$ near x is then effectively dominated by that of $\mathsf{S}_n[l\pi^{-1}\phi(x-\xi)]$ so that $\mathsf{S}[f]$ will show Gibbs's phenomenon at $x=\xi$. If f is continuous at ξ, $\mathsf{S}[f]$ converges uniformly at ξ and the phenomenon is absent.

(9·4) THEOREM. *The partial sums $s_n(x)$ of the series (9·1) are strictly positive for $0<x<\pi$.*

The theorem is true for $n=1$. Suppose that it is established for $n-1$ and that $s_n(x)$ has a non-positive minimum at a point x_0, $0<x_0<\pi$. Since

$$s_n'(x_0)=D_n(x_0)-\tfrac{1}{2}=\{\sin(n+\tfrac{1}{2})x_0-\sin\tfrac{1}{2}x_0\}/2\sin\tfrac{1}{2}x_0=0,$$

we infer that $\sin(n+\frac{1}{2})x_0=\sin\frac{1}{2}x_0$ and so also that $|\cos(n+\frac{1}{2})x_0|=\cos\frac{1}{2}x_0$. Hence

$$\sin nx_0=\sin(n+\tfrac{1}{2})x_0\cos\tfrac{1}{2}x_0-\cos(n+\tfrac{1}{2})x_0\sin\tfrac{1}{2}x_0\geqslant 0.$$

It follows that $s_n(x_0)-s_{n-1}(x_0)\geqslant 0$ and so $s_{n-1}(x_0)\leqslant s_n(x_0)\leqslant 0$, contrary to hypothesis.

10. The Dini-Lipschitz test

We know that $S_n^*(x)-f(x)$ is formally a Fourier sine coefficient (see (5·4)). We may therefore apply to it the device which led to the estimate (4·2) for Fourier coefficients. We fix x and take

$$\phi(t)=\phi_x(t),\quad \chi(t)=\phi(t)\tfrac{1}{2}\cot\tfrac{1}{2}t,\quad \eta=\pi/n.$$

Then

$$\begin{aligned}\pi[S_n^*(x)-f(x)]&=2\int_0^\pi\chi(t)\sin nt\,dt=-2\int_{-\eta}^{\pi-\eta}\chi(t+\eta)\sin nt\,dt\\&=\int_0^\pi\chi(t)\sin nt\,dt-\int_{-\eta}^{\pi-\eta}\chi(t+\eta)\sin nt\,dt\\&=\int_\eta^{\pi-\eta}\{\chi(t)-\chi(t+\eta)\}\sin nt\,dt+\int_{\pi-\eta}^{\pi}\chi(t)\sin nt\,dt\\&\quad+\int_0^\eta\chi(t)\sin nt\,dt-\int_{-\eta}^{\eta}\chi(t+\eta)\sin nt\,dt.\end{aligned}$$

Denote the last four integrals by I_1, I_2, I_3, I_4 respectively. Since $|\sin nt\,\frac{1}{2}\cot\frac{1}{2}t|\leqslant n$, we have

$$|I_3|+|I_4|\leqslant n\int_0^\eta|\phi(t)|\,dt+n\int_{-\eta}^{\eta}|\phi(t+\eta)|\,dt\leqslant 2n\int_0^{2\eta}|\phi(t)|\,dt.$$

For $n\geqslant 2$ and $t\in(\pi-\eta,\pi)$,

$$|\chi(t)\sin nt|\leqslant|\phi(t)|\leqslant\tfrac{1}{2}\{|f(x+t)|+|f(x-t)|+2|f(x)|\},$$

and since the indefinite integral is a continuous function we have $I_2=o(1)$, and uniformly in every interval where f is bounded. Finally, $|I_1|$ does not exceed

$$\int_\eta^{\pi-\eta}|\phi(t)|\left\{\frac{1}{2\tan\frac{1}{2}t}-\frac{1}{2\tan\frac{1}{2}(t+\eta)}\right\}dt+\int_\eta^{\pi-\eta}\frac{|\phi(t)-\phi(t+\eta)|}{2\tan\frac{1}{2}t}\,dt.$$

The difference inside curly brackets here is $\frac{1}{2}\sin\frac{1}{2}\eta/\sin\frac{1}{2}t\sin\frac{1}{2}(t+\eta)<\pi^2\eta/4t^2$. Collecting results and observing that $2\tan\frac{1}{2}t\geqslant t$ in $(0,\pi)$ we have

(10·1) THEOREM. *Let $\eta=\pi/n$. For every x, $|S_n^*(x)-f(x)|$ is majorized by*

$$\frac{1}{\pi}\int_\eta^\pi\frac{|\phi(t)-\phi(t+\eta)|}{t}\,dt+\eta\int_\eta^\pi\frac{|\phi(t)|}{t^2}\,dt+2\eta^{-1}\int_0^{2\eta}|\phi(t)|\,dt+o(1),\qquad\textbf{(10·2)}$$

the $o(1)$ being uniform in every interval where f is bounded.

As a corollary we obtain the following:

(10·3) Theorem of Dini-Lipschitz. *If f is continuous and its modulus of continuity $\omega(\delta)$ satisfies the condition $\omega(\delta)\log\delta\to 0$ with δ, then $S[f]$ converges uniformly.*

For since

$$|\phi(t)-\phi(t+\eta)|\leqslant\tfrac{1}{2}|f(x+t)-f(x+t+\eta)|+\tfrac{1}{2}|f(x-t)-f(x-t-\eta)|\leqslant\omega(\eta), \quad (10\cdot 4)$$

the first term in (10·2) does not exceed $\omega(\eta)\log n=o(1)$. Similarly, since $\phi(t)\to 0$ uniformly in x, the remaining terms in (10·2) tend uniformly to 0 (Chapter I, (8·1)).

(10·5) Theorem. *If the modulus of continuity of f in an interval I is $o(|\log\delta|^{-1})$, then $S[f]$ converges uniformly in every interval interior to I.*

For the continuous function coinciding with f on I and, say, linear outside I satisfies the hypothesis of (10·3), and it is enough to apply (6·6).

We shall see in Chapter VIII, § 2, that the condition

$$f(x_0\pm t)-f(x_0)=o\left(\frac{1}{|\log t|}\right) \quad (t\to+0) \qquad (10\cdot 6)$$

does not ensure the convergence of $S[f]$ at x_0, so that (10·5) is primarily a result about uniform convergence. However:

(10·7) Theorem. *$S[f]$ converges at x_0 to sum $f(x_0)$ provided the condition (10·6) is satisfied and the coefficients of f are $O(n^{-\delta})$ for some $\delta>0$.*

Without loss of generality we may suppose that $x_0=0$, f is even, $f(0)=0$. It is also convenient to have $a_0=0$, which may be achieved by subtracting $\frac{1}{2}a_0(1-\cos x)$ from $S[f]$. Finally, suppose that $|a_n|\leqslant n^{-\delta}$ $(0<\delta<1)$ for $n=1, 2, \ldots$. We set $r=\frac{1}{2}\delta$ and write

$$S_n^*(0)=\frac{2}{\pi}\int_0^\pi f(t)\frac{\sin nt}{2\tan\frac{1}{2}t}\,dt=\int_0^{n^{-1}}+\int_{n^{-1}}^{n^{-r}}+\int_{n^{-r}}^{\pi}=P+Q+R.$$

Here $P\to 0$ as $n\to\infty$, since f is continuous at $t=0$ and $|D_n^*|\leqslant n$. If

$$\epsilon(t)=\sup|f(u)\log u| \quad \text{for} \quad 0<u\leqslant t,$$

then
$$Q\leqslant\epsilon(n^{-r})\int_{n^{-1}}^{n^{-r}}\frac{dt}{t\log(1/t)}=\epsilon(n^{-r})\log 1/r=o(1),$$

and it only remains to prove that $R\to 0$. To this end we shall take for granted a result which will be established later (Chapter IV, Theorem (8·16)), namely that Fourier series can be integrated term by term after multiplication by any function of bounded variation. Then

$$R=\sum_{\nu=1}^{\infty}a_\nu\frac{2}{\pi}\int_{n^{-r}}^{\pi}\frac{\sin nt\cos\nu t}{2\tan\frac{1}{2}t}\,dt.$$

We replace the products $\sin nt\cos\nu t$ by differences of sines and apply the second mean-value theorem to the factor $\frac{1}{2}\cot\frac{1}{2}t$. We find that for $\nu\neq n$ the factor of a_ν does not exceed $4n^r/\pi|\nu-n^r|$ in absolute value. The factor of a_n is bounded. Hence

$$|R|\leqslant o(1)+\frac{4}{\pi}\sum_{\nu=1}^{\infty}{}'\frac{\nu^{-\delta}n^r}{|\nu-n|}=o(1)+\sum_{\nu=1}^{n-1}+\sum_{\nu=n+1}^{\infty}=o(1)+R_1+R_2,$$

say, where the dash indicates the omission of the term $\nu=n$. Now

$$\frac{\pi}{4}R_1<\frac{n^r}{\frac{1}{2}n}\sum_{\nu=1}^{[\frac{1}{2}n]}\nu^{-\delta}+n^r(\tfrac{1}{2}n)^{-\delta}\sum_{[\frac{1}{2}n]+1}^{n-1}\frac{1}{n-\nu}=O(n^{-\frac{1}{2}\delta})+O(n^{-\frac{1}{2}\delta}\log n)=o(1),$$

$$\frac{\pi}{4}R_2<n^{r-\delta}\sum_{\nu=n+1}^{2n}\frac{1}{\nu-n}+n^r\sum_{2n+1}^{\infty}\frac{\nu^{-\delta}}{\frac{1}{2}\nu}=O(n^{-\frac{1}{2}\delta}\log n)+O(n^{-\frac{1}{2}\delta})=o(1),$$

so that $R\to 0$, and this completes the proof.

A similar argument shows that, *under the hypotheses of* (10·7),

$$\tilde{S}_n(x_0)-\tilde{f}(x_0;\,\pi/n)\to 0.$$

The proof of the following theorem is very similar to that of (10·3).

(10·8) THEOREM. *If* $f\in\Lambda_\alpha$, $0<\alpha\leqslant 1$, *then* $S_n(x;f)-f(x)=O(n^{-\alpha}\log n)$, *uniformly in* x.

For now $\phi(t)=O(t^\alpha)$, $\phi(t+\eta)-\phi(t)=O(\eta^\alpha)$ (cf. (10·4)). The first term in (10·2) is $O(n^{-\alpha}\log n)$; the second is $O(n^{-\alpha})$ or $O(n^{-1}\log n)$, according as $\alpha<1$ or $\alpha=1$; the third is $O(n^{-\alpha})$. A glance back at the source of the fourth term shows that it is $O(\eta)=O(n^{-\alpha})$.

It can be shown by examples that the factor $\log n$ in (10·8) cannot, in general, be omitted (see p. 315, Example 10). Suppose, however, that there is a constant C such that the function $f(x)+Cx$ is monotone for all x. (The function f itself, being periodic, cannot be monotone, unless it is constant.) Such functions f will be called *of monotonic type*. We have now:

(10·9) THEOREM. *If f is of monotonic type and of class* Λ_α, $0<\alpha<1$, *then*

$$S_n(x,f)-f(x)=O(n^{-\alpha}),\quad \tilde{S}_n(x,f)-\tilde{f}(x)=O(n^{-\alpha}), \qquad \textbf{(10·10)}$$

uniformly in x.

Suppose that $g(x)=f(x)+Cx$ is increasing. The difference $S_n(x)-f(x)$ is given by the integral of $\pi^{-1}\{f(x+t)-f(x)\}D_n(t)$ extended over $(-\pi,\pi)$. We can replace f by g in this, since the integral of $tD_n(t)$ over $(-\pi,\pi)$ is zero. It is enough to show that the integral over $(0,\pi)$ is $O(n^{-\alpha})$, the proof for the remaining integral being similar. Let $2^{k-1}\leqslant n<2^k$. Our integral is

$$\frac{1}{\pi}\int_0^\pi\{g(x+t)-g(x)\}\,D_n(t)\,dt=\int_0^{\pi 2^{-k}}+\sum_1^k\int_{\pi 2^{-j}}^{\pi 2^{-(j-1)}}=P+\sum_1^k Q_j.$$

Since $g(x+t)-g(x)=O(t^\alpha)$ and $D_n=O(n)$, it follows that $P=O(n2^{-(1+\alpha)k})=O(n^{-\alpha})$. Also $g(x+t)-g(x)$ is non-negative and increasing. Applying the second mean-value theorem twice, we get for Q_j the value

$$O(2^{-j\alpha})\int_\xi^{\pi 2^{-(j-1)}}\frac{\sin(n+\frac{1}{2})t}{2\sin\frac{1}{2}t}\,dt=O(2^{-j\alpha})\,O(2^j)\int_\xi^{\xi'}\sin(n+\tfrac{1}{2})t\,dt=O(2^{j(1-\alpha)}n^{-1}),$$

from which it follows that

$$\sum_1^k Q_j=\sum_1^k O(n^{-1})\,2^{j(1-\alpha)}=O(n^{-1}\,2^{k(1-\alpha)})=O(n^{-\alpha}).$$

Hence $P+\Sigma Q_j=O(n^{-\alpha})$, and the first estimate (10·10) follows. The proof of the second is similar.

11. Lebesgue's test

(11·1) THEOREM. *If f is integrable, then, for almost all x,*

$$\int_0^h |f(x \pm t) - f(x)|\, dt = o(h) \quad as \quad h \to +0. \tag{11·2}$$

This theorem is due to Lebesgue. It generalizes the familiar fact (to which it reduces if the sign of absolute value on the left of (11·2) is dropped) that the derivative of the indefinite integral of $f(x)$ exists and equals $f(x)$ for almost all x. The set of x for which we have (11·2) is sometimes called the *Lebesgue set* of f.

We shall prove the following slightly more general result.

(11·3) THEOREM. *Suppose that $f \in L^r$ $(r \geq 1)$. Then, for almost all x,*

$$\int_0^h |f(x \pm t) - f(x)|^r\, dt = o(h) \quad as \quad h \to +0. \tag{11·4}$$

Let α be any rational number. The function $|f(x) - \alpha|^r$ is integrable and so

$$h^{-1}\int_0^h |f(x \pm t) - \alpha|^r\, dt \to |f(x) - \alpha|^r$$

for almost all x. Let E_α be the set of the x for which this does not hold. Since $|E_\alpha| = 0$, the sum E of all the E_α is of measure 0. We shall prove (11·4) for x not in E. Suppose that x_0 is not in E and $\epsilon > 0$ is given, and let β be rational and such that $|f(x_0) - \beta| < \frac{1}{2}\epsilon$. In the inequality

$$\left\{\frac{1}{h}\int_0^h |f(x_0 \pm t) - f(x_0)|^r\, dt\right\}^{1/r} \leq \left\{\frac{1}{h}\int_0^h |f(x_0 \pm t) - \beta|^r\, dt\right\}^{1/r} + \left\{\frac{1}{h}\int_0^h |\beta - f(x_0)|^r\, dt\right\}^{1/r},$$

the first term on the right tends by hypothesis to $|f(x_0) - \beta| < \frac{1}{2}\epsilon$. The second term is $|\beta - f(x_0)|$. Thus the right-hand side is $< \epsilon$ for h small enough, and (11·3) follows.

We shall systematically use the notation

$$\Phi(h) = \Phi_{x_0}(h) = \int_0^h |\phi_{x_0}(t)|\, dt, \quad \Psi(h) = \Psi_{x_0}(h) = \int_0^h |\psi_{x_0}(t)|\, dt.$$

It follows from (11·1) *that* $\quad \Phi_x(h) = o(h), \quad \Psi_x(h) = o(h)$
for almost all x.

The following test for the convergence of $S[f]$ is due to Lebesgue:

(11·5) THEOREM. *$S[f]$ converges to $f(x)$ at every point x at which*

$$\Phi(h) = o(h), \quad \int_\eta^\pi \frac{|\phi(t) - \phi(t + \eta)|}{t}\, dt \to 0 \quad as \quad \eta = \pi/n \to 0, \tag{11·6}$$

and the convergence is uniform over any closed interval of continuity of f where the second condition (11·6) *is satisfied uniformly.*

We apply (10·1). The first term in (10·2) is $o(1)$ by hypothesis. The third term there is $2\eta^{-1}\Phi(2\eta) = o(1)$. Integration by parts gives for the second term the value

$$\eta\left\{[\Phi(t)\, t^{-2}]_\eta^\pi + 2\int_\eta^\pi \Phi(t)\, t^{-3}\, dt\right\} = o(1),$$

since $\Phi(t) = o(t)$ (see Chapter I, (8·1)). This completes the proof.

Using the analogue of (10·1) for conjugate series we find, by a similar argument, that *the conditions*

$$\Psi_x(h)=o(h), \quad \int_\eta^\pi \frac{|\psi(t)-\psi(t+\eta)|}{t}\,dt\to 0 \tag{11·7}$$

together imply the relation $\tilde S_n^*(x)-\tilde f(x;\pi/n)\to 0.$

If we also observe that, for $\pi/(n+1)\leqslant h\leqslant \pi/n$, the first condition (11·7) gives

$$|\tilde f(x;h)-\tilde f(x;\pi/(n+1))|\leqslant \frac{2}{\pi}\int_{\pi/(n+1)}^{\pi/n}\frac{|\psi(t)|}{t}\,dt\leqslant 2\pi^{-2}(n+1)\Psi(\pi/n)=o(1) \tag{11·8}$$

as $n\to\infty$, we deduce that *under the conditions* (11·7), $\tilde{\mathrm{S}}[f]$ *converges at the point* x *if and only if* $\tilde f(x)$ *exists.*

The conditions (11·7) are certainly satisfied if f satisfies the Dini-Lipschitz condition in an interval containing x.

In Chapter VIII, §4, we shall see that there exist integrable functions f such that $S_n(x;f)$ is unbounded at every x. We now show, in the opposite direction, that $S_n(x)$ and $\tilde S_n(x)$ are $o(\log n)$ at almost all x. More precisely,

(11·9) THEOREM. *If* $\Phi_{x_0}(h)=o(h)$, *then* $S_n(x_0;f)=o(\log n)$; *if* $\Psi_{x_0}(h)=o(h)$, *then* $\tilde S_n(x_0;f)=o(\log n)$.

By (5·4) and (5·10), $|S_n^*(x_0)-f(x_0)|$ does not exceed

$$n\int_0^{1/n}|\phi(t)|\,dt+\int_{1/n}^{\pi}t^{-1}|\phi(t)|\,dt=n\Phi(1/n)+[\Phi(t)t^{-1}]_{1/n}^{\pi}+\int_{1/n}^{\pi}t^{-2}\Phi(t)\,dt.$$

The sum of the first two terms on the right is $\Phi(\pi)/\pi=O(1)=o(\log n)$. Since $\Phi(t)=o(t)$, the remaining integral is $o(\log n)$. Thus $S_n^*(x_0)=o(\log n)$.

Similarly (cf. (5·5) and (5·11)),

$$|\tilde S_n^*(x_0)|\leqslant n\int_0^{1/n}|\psi(t)|\,dt+2\int_{1/n}^{\pi}\frac{|\psi(t)|}{t}\,dt=o(\log n).$$

By (11·9), $S_n(x)$ *and* $\tilde S_n(x)$ *are* $o(\log n)$ *at every point of continuity of* f. Moreover, *if* f *is continuous in an interval* I, *then* $S_n(x)$ *and* $\tilde S_n(x)$ *are* $o(\log n)$ *uniformly in every interval interior to* I. The proofs are slightly simpler than those of (11·9), no integration by parts being necessary.

The most important tests for the convergence of Fourier series are those of Dini, Dini-Lipschitz and Dirichlet-Jordan, each of which is based on a different idea. Lebesgue's test may be shown to include the other three, but in practice it is less convenient to use because the second condition (11·6) corresponds to no simple property of the function f. The following application of Lebesgue's test is, however, of interest.

(11·10) THEOREM. *Suppose that* $f\in\lambda_{1/p}^{p}$, $p>1$. *Then* $\mathrm{S}[f]$ *converges to* $f(x)$ *at each point* x *of the Lebesgue set of* f; *and the convergence is uniform over any closed arc of continuity of* f. *At each point of the Lebesgue set where* $\tilde f(x)$ *exists,* $\tilde{\mathrm{S}}[f]$ *converges to* $\tilde f(x)$.

It is enough to prove the part about $\mathrm{S}[f]$. This will follow if we show that the second condition (11·6) is satisfied everywhere and uniformly in x. By Hölder's inequality, with $p'=p/(p-1)$,

$$\int_\eta^\pi\frac{|\phi(t)-\phi(t+\eta)|}{t}\,dt\leqslant\left\{\int_\eta^\pi|\phi(t)-\phi(t+\eta)|^p\,dt\right\}^{1/p}\left\{\int_\eta^\pi\frac{dt}{t^{p'}}\right\}^{1/p'}=o(\eta^{1/p})\,O(\eta^{-1/p})=o(1),$$

uniformly in x.

12. Lebesgue constants

This name is given to the numbers

$$L_n = \frac{1}{\pi}\int_{-\pi}^{\pi} |D_n(t)|\,dt = \frac{2}{\pi}\int_0^{\pi}\left|\frac{\sin(n+\frac{1}{2})t}{2\sin\frac{1}{2}t}\right| dt.$$

It is clear that if $|f| \leqslant 1$ then

$$|S_n(x;f)| \leqslant \frac{1}{\pi}\int_{-\pi}^{\pi} |f(x+t)|\,|D_n(t)|\,dt \leqslant L_n$$

for all x; and for $f(t) = \operatorname{sign} D_n(t)$, we actually have $S_n(0;f) = L_n$. While the function sign $D_n(t)$ is discontinuous at a finite number of points, given any $\epsilon > 0$ we can, by smoothing this function slightly at the points of discontinuity, obtain a continuous f such that $S_n(0;f) > L_n - \epsilon$. Thus, for each n, L_n is

(i) the maximum of $|S_n(x;f)|$ for all x and f satisfying $|f| \leqslant 1$;

(ii) the upper bound† of $|S_n(x;f)|$ for all x and all continuous f satisfying $|f| \leqslant 1$.

We shall prove that

$$L_n = 4\pi^{-2}\log n + O(1) \simeq 4\pi^{-2}\log n \quad \text{as} \quad n \to \infty. \tag{12·1}$$

Since $|D_n - D_n^*| \leqslant \frac{1}{2}$ and the function $1/t - \frac{1}{2}\cot\frac{1}{2}t$ is bounded for $|t| \leqslant \pi$,

$$L_n = \frac{2}{\pi}\int_0^{\pi} |D_n^*(t)|\,dt + O(1) = \frac{2}{\pi}\int_0^{\pi}\frac{|\sin nt|}{t}\,dt + O(1)$$

$$= \frac{2}{\pi}\sum_{k=0}^{n-1}\int_{k\pi/n}^{(k+1)\pi/n}\frac{|\sin nt|}{t}\,dt + O(1) = \frac{2}{\pi}\int_0^{\pi/n}\sin nt\left\{\sum_{k=1}^{n-1}\frac{1}{t+k\pi/n}\right\}dt + O(1).$$

The sum in curly brackets lies between

$$n\pi^{-1}(1+\tfrac{1}{2}+\ldots+1/(n-1)) \quad \text{and} \quad n\pi^{-1}(\tfrac{1}{2}+\ldots+1/n),$$

and so is equal to $\pi^{-1}n[\log n + O(1)]$. Since the integral of $\sin nt$ over $(0, \pi/n)$ is $2/n$, we obtain (12·1).

We may add that, since $|D_n(t)|$ is uniformly bounded in any interval $\epsilon \leqslant t \leqslant \pi, 0 < \epsilon < \pi$, the formula (12·1) implies that

$$\frac{2}{\pi}\int_0^{\epsilon} |D_n(t)|\,dt = \frac{4}{\pi^2}\log n + O(1) \simeq \frac{4}{\pi^2}\log n \tag{12·2}$$

for any fixed ϵ $(0 < \epsilon \leqslant \pi)$.

The formulae

$$\tilde{L}_n = \frac{2}{\pi}\int_0^{\pi} |\tilde{D}_n(t)|\,dt \simeq \frac{2}{\pi}\log n, \quad \frac{2}{\pi}\int_0^{\pi}\tilde{D}_n^*(t)\,dt \simeq \frac{2}{\pi}\log n \tag{12·3}$$

are also useful. They are equivalent, since $|\tilde{D}_n - \tilde{D}_n^*| \leqslant \frac{1}{2}$ and $\tilde{D}_n^*(t) \geqslant 0$ in $(0, \pi)$. The left-hand member of the second formula represents $-\tilde{S}_n^*(0;f)$ where $f(t) = \operatorname{sign} t$ $(-\pi < t < \pi)$. Since f has jump 2 at $t = 0$, (8·13) gives $-S_n^*(0;f) \simeq 2\pi^{-1}\log n$ and (12·3) follows.

The first integral (12·3) is an analogue of the Lebesgue constant L_n, and is the maximum of $|\tilde{S}_n(0;f)|$ for all functions f with $|f| \leqslant 1$. This maximum is attained for $f(t) = \operatorname{sign}\tilde{D}_n(t)$.

† By *upper bound* we always mean the least upper bound. Similarly, for the lower bound.

13. Poisson's summation formula

The notion of the Fourier transform

$$\gamma(u) = \frac{1}{2\pi}\int_{-\infty}^{+\infty} g(x)\, e^{-iux}\, dx \tag{13·1}$$

of a function $g(x)$ defined in $(-\infty, +\infty)$ (see Chapter I, § 4) is useful in the theory of Fourier series in connexion with the following simple fact.

Suppose first that $g(x)$ is absolutely integrable over $(-\infty, +\infty)$. The series

$$\sum_{k=-\infty}^{+\infty} g(x + 2k\pi) \tag{13·2}$$

is then absolutely convergent at almost all x in $(0, 2\pi)$, as is seen from the inequality

$$\sum_{k=-\infty}^{+\infty} \int_0^{2\pi} |\, g(x+2k\pi)\, |\, dx = \int_{-\infty}^{+\infty} |\, g(x)\, |\, dx < \infty$$

(cf. Chapter I, (11·5)). Let $G_n(x)$ be the nth symmetric partial sum, and $G(x) = \lim G_n(x)$ the sum, of (13·2). The function $G(x)$ exists for almost all x and is periodic. Since the $G_n(x)$ are majorized by an integrable function, the Fourier coefficient c_ν of G is

$$\lim_{n\to+\infty} \frac{1}{2\pi}\int_0^{2\pi} G_n(x)\, e^{-i\nu x}\, dx = \lim_{n\to\infty} \frac{1}{2\pi}\int_{-2n\pi}^{2(n+1)\pi} g(x)\, e^{-i\nu x}\, dx = \gamma(\nu). \tag{13·3}$$

Thus, with our hypothesis, *the Fourier coefficient* c_ν *of the sum* $G(x)$ *of* (13·2) *is equal to the Fourier transform* $\gamma(\nu)$ *of* $g(x)$.

If, moreover, it happens that $\Sigma c_\nu e^{i\nu x} = \mathsf{S}[G]$ converges at $x = 0$, and to a value which is the sum of (13·2) at $x = 0$, we are led to the equation

$$\sum_{k=-\infty}^{+\infty} g(2k\pi) = \sum_{\nu=-\infty}^{+\infty} \frac{1}{2\pi}\int_{-\infty}^{+\infty} g(x)\, e^{-i\nu x}\, dx. \tag{13·4}$$

This is called the *Poisson summation formula*, and it has many applications. The sum on the right is defined as the limit of the symmetric partial sums.

Suppose, for example, that $g(x)$ is not only absolutely integrable over $(-\infty, +\infty)$ but also of bounded variation, and that $2g(x) = g(x+0) + g(x-0)$ for all x. We shall prove (13·4) under these hypotheses.

Let v_k be the total variation of g over the interval $I_k = (2k\pi, 2(k+1)\pi)$, $k = 0, \pm 1, \ldots$. Since the series (13·2) converges absolutely at some point x_0 in I_0, the inequalities $|\, g(x+2k\pi) - g(x_0 + 2k\pi)\, | \leqslant v_k$, for $x \in I_0$, and $\Sigma v_k < \infty$, prove that (13·2) converges absolutely and uniformly in I_0 to a sum $G(x)$, obviously of bounded variation and such that $2G(x) = G(x+0) + G(x-0)$. Formula (13·4) is then a consequence of Theorem (8·1).

The equation $c_\nu = \gamma(\nu)$, slightly modified, remains valid in cases when $g(x)$ is not absolutely integrable over $(-\infty, +\infty)$. Suppose, namely, that $g(x)$ is integrable over every finite interval and that

(i) $\displaystyle\int_{2k\pi}^{2(k+1)\pi} |\, g(x)\, |\, dx \to 0 \quad \text{as} \quad k \to \pm\infty;$

(ii) the function g^* defined by the formula

$$g^*(x)=g(x)-\frac{1}{2\pi}\int_{2k\pi}^{2(k+1)\pi} g(t)\,dt \quad \text{for} \quad 2k\pi\leqslant x<2(k+1)\pi \quad (k=0,\ \pm 1,\ \pm 2,\ \ldots),$$

is absolutely integrable over $(-\infty, +\infty)$.

Conditions (i) and (ii) are certainly satisfied if, for example, $g(x)$ tends to 0 monotonically in the neighbourhoods of $+\infty$ and of $-\infty$. We prove now the following theorem:

(13·5) THEOREM. *Under conditions* (i) *and* (ii) *the integral* (13·1), *defined as* $\lim_{\omega\to\infty}\int_{-\omega}^{+\omega}$, *exists for* $u=\pm 1, \pm 2, \ldots$, *and the Fourier coefficients* c_ν^* *of the function*

$$G^*(x)=\lim_{n\to+\infty}\left\{\sum_{k=-n}^{n} g(x+2k\pi)-\frac{1}{2\pi}\int_{-2n\pi}^{2n\pi} g(x)\,dx\right\} \tag{13·6}$$

satisfy the equations $c_0^*=0,\quad c_\nu^*=\gamma(\nu)$ *for* $\nu=\pm 1, \pm 2, \ldots$.

The function $G^*(x)$ can be written in the form $\Sigma g^*(x+2k\pi)$; thus, by the case already dealt with, it is integrable over $(0, 2\pi)$, and c_ν^* is given by (13·3), with G_n^* for G_n and g^* for g. Since the integral of $e^{-i\nu x}$ over a period is zero for $\nu=\pm 1, \pm 2, \ldots$, we have, by (i),

$$c_\nu^*=\lim_{\omega\to\infty}\frac{1}{2\pi}\int_{-\omega}^{\omega} g(x)\,e^{-i\nu x}\,dx=\gamma(\nu) \quad (\nu=\pm 1,\ \pm 2, \ldots).$$

The integral of $g^*(x)$ over any interval $(2k\pi, 2(k+1)\pi)$, $k=0, \pm 1, \ldots$, is evidently 0. Hence $\int_0^{2\pi} G^*(x)\,dx=0$ and $c_0^*=0$.

The following application of Theorem (13·5) will be useful in fractional integration (Chapter XII, § 8):

(13·7) THEOREM. *Let* $0<\alpha<1$, *and let* $\Psi_\alpha(x)$ *be the periodic function defined for* $0<x<2\pi$ *by the formula*

$$\Psi_\alpha(x)=\lim_{n\to\infty}\frac{2\pi}{\Gamma(\alpha)}\left\{x^{\alpha-1}+(x+2\pi)^{\alpha-1}+\ldots+(x+2n\pi)^{\alpha-1}-\frac{(2\pi)^{\alpha-1}}{\alpha}n^\alpha\right\}. \tag{13·8}$$

Then

$$\Psi_\alpha(x)\sim\sum_{\nu=-\infty}^{+\infty}{}' \frac{e^{-\frac{1}{2}\pi i\alpha\,\mathrm{sign}\,\nu}}{|\nu|^\alpha}e^{i\nu x}, \tag{13·9}$$

where the dash ′ indicates that the term $\nu=0$ *is omitted from the summation.*

The function $\Psi_\alpha(x)$ is the $G^*(x)$ corresponding to a $g(x)$ equal to $2\pi x^{\alpha-1}/\Gamma(\alpha)$ for $x>0$ and to 0 elsewhere. The coefficients of Ψ_α are

$$c_0^*=0,\quad c_\nu^*=\frac{1}{\Gamma(\alpha)}\int_0^\infty t^{\alpha-1}e^{-i\nu t}\,dt \quad \text{for} \quad \nu\neq 0.$$

We now observe that†

$$e^{-\frac{1}{2}\pi i\alpha}\,\Gamma(\alpha)=\int_0^\infty t^{\alpha-1}e^{-it}\,dt \quad \text{for} \quad 0<\alpha<1. \tag{13·10}$$

† This formula is easily obtainable from the classical definition $\Gamma(\alpha)=\int_0^\infty x^{\alpha-1}e^{-x}\,dx$ by applying Cauchy's theorem to the integral of the function $z^{\alpha-1}e^{-z}$, taken along the boundary of the domain limited by the arcs $0\leqslant\arg z\leqslant\frac{1}{2}\pi$ of the circles $|z|=\epsilon$ and $|z|=R$, and by the rectilinear segments (ϵ, R) and $(i\epsilon, iR)$ of the real and imaginary axes; if $0<\alpha<1$, the integrals taken along the circular arcs tend to 0 as $\epsilon\to 0$ and $R\to\infty$, and (13·10) follows.

Substituting $t=\nu u$, we see that $c_\nu^* = \nu^{-\alpha}\exp(-\frac{1}{2}\pi i\alpha)$ for $\nu>0$, and since $c_{-\nu}^* = \bar{c}_\nu^*$, we get (13·9).

By (2·6) of Chapter I, the series in (13·9) converges uniformly outside any neighbourhood of $x=0$. If we omit the term $x^{\alpha-1}$ in (13·8), the limit will exist uniformly in $0\leqslant x\leqslant 2\pi$ (this will be the function G^* corresponding to a $g(x)$ equal to $2\pi x^{\alpha-1}/\Gamma(\alpha)$ for $x>2\pi$ and to zero elsewhere). Hence, *with an error uniformly* $O(1)$, *the periodic function* $\Psi_\alpha(x)$ *is* 0 *for* $-\pi\leqslant x<0$ *and is* $2\pi x^{\alpha-1}/\Gamma(\alpha)$ *for* $0<x\leqslant\pi$.

By considering the Fourier series of $\Psi_\alpha(x)\pm\Psi_\alpha(-x)$, and using the relation $\Gamma(\alpha)\Gamma(1-\alpha)=\pi/\sin\pi\alpha$, we get the useful formulae

$$\left.\begin{aligned}\sum_{\nu=1}^{\infty}\frac{\cos\nu x}{\nu^\alpha} &= \Gamma(1-\alpha)\sin\tfrac{1}{2}\pi\alpha\,.\,x^{\alpha-1}+O(1)\simeq\Gamma(1-\alpha)\sin\tfrac{1}{2}\pi\alpha\,.\,x^{\alpha-1}\\ \sum_{\nu=1}^{\infty}\frac{\sin\nu x}{\nu^\alpha} &= \Gamma(1-\alpha)\cos\tfrac{1}{2}\pi\alpha\,.\,x^{\alpha-1}+O(1)\simeq\Gamma(1-\alpha)\cos\tfrac{1}{2}\pi\alpha\,.\,x^{\alpha-1}\end{aligned}\right\}\quad(0<x\leqslant\pi). \tag{13·11}$$

Poisson's summation formula may be written in a slightly different and more symmetric form. Let $a>0$ and let $g(x)=h(ax/2\pi)$. Then, by (13·4),

$$\sum_{k=-\infty}^{+\infty} h(ak) = \sum_{\nu=-\infty}^{+\infty}\frac{1}{a}\int_{-\infty}^{+\infty} h(y)\,e^{-2\pi i\nu y/a}\,dy. \tag{13·12}$$

If, as is often convenient, we modify the definition (13·1) of the Fourier transform by replacing the factor $1/2\pi$ there by $1/(2\pi)^{\frac{1}{2}}$, and accordingly set

$$\chi(u)=\frac{1}{(2\pi)^{\frac{1}{2}}}\int_{-\infty}^{+\infty} h(y)\,e^{-iuy}\,dy, \tag{13·13}$$

the terms on the right of (13·12) can be written $(2\pi)^{\frac{1}{2}}a^{-1}\chi(2\pi\nu/a)$ and Poisson's formula becomes

$$\sqrt{a}\sum_{k=-\infty}^{+\infty} h(ak)=\sqrt{b}\sum_{k=-\infty}^{+\infty}\chi(bk), \tag{13·14}$$

where a, b are any two positive numbers satisfying the condition $ab=2\pi$, and $\chi(u)$ is the Fourier transform (13·13) of h.

MISCELLANEOUS THEOREMS AND EXAMPLES

1. Given any sequence of positive numbers ϵ_n tending to 0, there is always a continuous function $f(x)$ whose Fourier coefficients satisfy the inequality $|a_n|+|b_n|\geqslant\epsilon_n$ for infinitely many n. (Lebesgue [1], Hardy [1].)

[Take, for example, $f(x)=\epsilon_{n_1}\cos n_1x+\epsilon_{n_2}\cos n_2x+\dots$, where $\{n_k\}$ increases so rapidly that $\Sigma\epsilon_{n_k}<\infty$.]

2. Suppose that $f\in\Lambda_{\alpha_1}^{p_1}$ and $f\in\Lambda_{\alpha_2}^{p_2}$, $p_1\geqslant 1$, $p_2\geqslant 1$. Show that $f\in\Lambda_\alpha^p$ if the point with Cartesian co-ordinates $(\alpha, 1/p)$ is on the segment joining $(\alpha_1, 1/p_1)$, $(\alpha_2, 1/p_2)$. Hardy and Littlewood [5].)

[Use Chapter I, (10·12) (iii).]

3. Using the equation

$$\sum_{\lambda}^{\mu}(a_n\sin nx-b_n\cos nx)/n=\frac{1}{\pi}\int_0^{2\pi}f(t)\sum_{\lambda}^{\mu}\frac{\sin n(x-t)}{n}\,dt,$$

prove (8·7) and the formula $\displaystyle\sum_1^\infty b_n/n=\frac{1}{\pi}\int_0^{2\pi}f(t)\,\tfrac{1}{2}(\pi-t)\,dt.$

4. The number $C_k=1+2^{-2k}+3^{-2k}+\dots$ $(k=1,2,\dots)$ is a rational multiple of π^{2k}.

[Integrate the series $\sin x+\frac{1}{2}\sin 2x+\dots$ an odd number of times and set $x=0$.]

5. If $f(x)$ is periodic and has k continuous derivatives, the Fourier coefficients of f satisfy the inequality

$$|c_n| \leqslant \frac{\omega(\pi/n, f^{(k)})}{2n^k} \quad (n>0).$$

If $f^{(k)}$ is of bounded variation, then

$$|c_n| \leqslant V/\pi n^{k+1} \quad (n>0),$$

where V is the total variation of $f^{(k)}$ over $0 \leqslant x \leqslant 2\pi$.

6. Suppose that $f(x)$, $0 \leqslant x \leqslant 2\pi$, has k continuous derivatives in the closed interval $(0, 2\pi)$, but is not necessarily continuous when continued periodically. Show that the Fourier coefficient c_n of f is

$$\frac{\alpha}{n} + \frac{\beta}{n^2} + \ldots + \frac{\lambda + o(1)}{n^k} \quad \text{as} \quad n \to \pm\infty.$$

7. Let $f(x) \sim \Sigma c_n e^{inx}$, $h>0$, $f_h(x) = \dfrac{1}{2h}\displaystyle\int_{x-h}^{x+h} f(t)\,dt$. Show that

$$f_h(x) \sim c_0 + \Sigma' c_n \left(\frac{\sin nh}{nh}\right) e^{inx}.$$

The dash indicates that the term $n=0$ is omitted in summation. The sign '$\sim$' can also be replaced by '$=$'.

[If $F(x)$ is the integral of f, then $f_h(x) = [F(x+h) - F(x-h)]/2h$. Apply (8·7).]

8. Let $0 \leqslant \alpha < 1$. The system $\{e^{i(n+\alpha)x}\}_{n=0,\pm1,\pm2,\ldots}$ is orthogonal and complete over any interval of length 2π.

Each of the systems $\cos(n+\frac{1}{2})x$ and $\sin(n+\frac{1}{2})x$, $n=0,1,2,\ldots$, is orthogonal and complete over $(0,\pi)$.

The joint system $\cos(n+\frac{1}{2})x$, $\sin(n+\frac{1}{2})x$ is orthogonal and complete over any interval of length 2π.

Show that
$$\tfrac{1}{2}(\pi - x) = \frac{2}{\pi}\sum_{-\infty}^{+\infty} \frac{e^{i(n+\frac{1}{2})x}}{(2n+1)^2} = \frac{4}{\pi}\sum_{0}^{\infty} \frac{\cos(n+\frac{1}{2})x}{(2n+1)^2} \quad (0 \leqslant x \leqslant 2\pi),$$

the series being absolutely and uniformly convergent.

9. Let $f(x)$ be defined for $0 \leqslant x \leqslant 2\pi$. No matter how well behaved the function $f(x)$, $S[f]$ cannot converge uniformly if $f(+0) \neq f(2\pi - 0)$. Suppose, however, that α is such that $g(x) = f(x)\,e^{-i\alpha x}$ takes the same values at the end-points 0 and 2π. (Such an α always exists, though it need not be real, provided $f(+0) \neq 0$ and $f(2\pi-0) \neq 0$.) If $S[g]$ converges uniformly to $g(x)$, we get the uniformly convergent representation

$$f(x) = \sum_{-\infty}^{+\infty} c_n e^{i(n+\alpha)x}, \quad \text{with} \quad c_n = \frac{1}{2\pi}\int_0^{2\pi} f(x)\,e^{-i(n+\alpha)x}\,dx.$$

The series here is a Fourier series if α is real; if $f(+0) = -f(2\pi - 0)$, we have $\alpha = \frac{1}{2}$ (compare Ex. 8).

10. If α is not a real integer, then

$$\sum_{n=-\infty}^{+\infty} \frac{\sin(n+\alpha)x}{n+\alpha} = \pi, \qquad \sum_{n=-\infty}^{+\infty} \frac{\cos(n+\alpha)x}{n+\alpha} = \pi\cot\pi\alpha$$

for $0 < x < 2\pi$.

[See Chapter I, (4·19).]

11. Let $f(x)$ be periodic, non-negative and not identically 0. Prove that the Fourier coefficients of f satisfy the inequalities

$$|a_m| < a_0, \quad |b_m| < a_0, \quad |c_m| < a_0 \quad (m \neq 0). \quad \text{(Carathéodory [1].)}$$

12. Let $g(x)$ be periodic, odd, non-negative in $(0,\pi)$ and not identically 0. The Fourier coefficients b_m of g satisfy the inequality

$$|b_m| < m b_1 \quad (m = 2, 3, \ldots). \quad \text{(Rogosinski [1], Dieudonné [1].)}$$

[Prove, by induction, that $|\sin mt| \leqslant m|\sin t|$, $m = 1, 2, \ldots$.]

13. Let $0<\alpha<1$. The (non-integrable) function $x^{-\alpha-1}$, $0<x\leqslant\pi$, has generalized sine coefficients $b_n\simeq Cn^\alpha$, where $C\neq 0$ is independent of n.

$$\left[(\tfrac{1}{2}\pi)\,b_n=\int_0^\pi x^{-\alpha-1}\sin nx\,dx=n^\alpha\int_0^{n\pi} x^{-\alpha-1}\sin x\,dx.\right]$$

14. Show that

$$r\sin x+r^3\sin 3x+\ldots=\frac{1}{2}\,\frac{k\sin x}{1-k^2\cos^2 x},\quad k=\frac{2r}{1+r^2}\quad (0\leqslant r<1),$$

$$\frac{1}{2\sin x}\sim\sin x+\sin 3x+\sin 5x+\ldots.$$

The second formula is, formally, a limiting case of the first when $r\to 1$.

15. Let $f(x)$, $0\leqslant x\leqslant\pi$, (not necessarily integrable) be such that $g(x)=f(x)\sin x$ is of bounded variation. Then $\lim b_{2k+1}=g(+0)+g(\pi-0)$, $\lim b_{2k}=g(0)-g(\pi-0)$. Hence $b_n=o(1)$ if, and only if, $g(+0)=g(\pi-0)=0$.

[Observe that
$$b_n=\frac{2}{\pi}\int_0^\pi g(x)\frac{\sin nx}{\sin x}\,dx=\frac{2}{\pi}\int_0^{2\pi} g(\tfrac{1}{2}x)\frac{\sin\frac{1}{2}nx}{2\sin\frac{1}{2}x}\,dx.$$

If $n=2k+1$, the last integral is $2S_k(0\,;\,g(\frac{1}{2}x))$.

16. Let
$$(*)\quad \Sigma c_n\,e^{i(n+\alpha)x}$$
be the Fourier series of any integrable $f(x)$, $0\leqslant x\leqslant 2\pi$, with respect to the orthogonal system $e^{i(n+\alpha)x}$, $0\leqslant\alpha<1$. (Thus $c_n=c_n^{(\alpha)}$.) Show that all the series $(*)$ are uniformly equiconvergent in every interval $(\epsilon, 2\pi-\epsilon)$. The equiconvergence cannot, in general, hold in the whole interval $(0,2\pi)$ (if, for instance, $\alpha=\frac{1}{2}$, the sum $S(x)$ of the series $(*)$ satisfies the condition $S(x+2\pi)=-S(x)$).

17. Let $f(x)$ be periodic and integrable, and consider the functions

$$f_1(t)=\frac{f(x_0+t)+f(x_0-t)-2f(x_0)}{4\tan\frac{1}{2}t},\quad f_2(t)=\frac{f(x_0+t)+f(x_0-t)-2f(x_0)}{4\sin\frac{1}{2}t},$$

which are periodic and odd. Show that the generalized sine coefficients of $f_1(t)$ are $S_n^*(x_0\,;\,f)-f(x_0)$, and that the coefficients of $f_2(t)$ with respect to the system $\sin(n+\frac{1}{2})\,t$ are $S_n(x_0)-f(x_0)$.

18. Let $f(x)$ be periodic and integrable, and suppose that the even functions

$$f_3(t)=\frac{f(x_0+t)-f(x_0-t)}{4\tan\frac{1}{2}t},\quad f_4(t)=\frac{f(x_0+t)-f(x_0-t)}{4\sin\frac{1}{2}t}$$

are integrable over $(0,\pi)$. (This amounts to the integrability of $|f(x_0+t)-f(x_0-t)|\,t^{-1}$.) Show that the cosine coefficients of $f_3(t)$ and the coefficients of $f_4(t)$ with respect to the system $\cos(n+\frac{1}{2})\,t$ are respectively $\tilde{S}_n^*(x_0)-\tilde{f}(x_0)$ and $\tilde{S}_n(x_0)-\tilde{f}(x_0)$.

19. If $f(x)$ is the characteristic function of the interval $(-h,h)$, $0<h\leqslant\pi$, then

$$\tilde{f}(x)=\frac{1}{\pi}\log\left|\frac{\sin\frac{1}{2}(x+h)}{\sin\frac{1}{2}(x-h)}\right|.$$

[Observe that $\int_a^b \frac{1}{2}\cot\frac{1}{2}t\,dt=\log|\sin\frac{1}{2}b/\sin\frac{1}{2}a|$ for any subinterval (a,b) of $(-2\pi,2\pi)$.]

20. Let $g(x)=0$ for $|x|\leqslant h$, and $g(x)=\frac{1}{2}\cot\frac{1}{2}x$ elsewhere in $(-\pi,\pi)$, $0<h\leqslant\pi$. Show that

$$\tilde{g}(x)=\frac{1}{2}\left(1-\frac{h}{\pi}\right)-\frac{1}{\pi}\,\tfrac{1}{2}\cot\tfrac{1}{2}x\log\left|\frac{\sin\frac{1}{2}(x+h)}{\sin\frac{1}{2}(x-h)}\right|.$$

In particular,
$$\frac{1}{\pi}\int_{-\pi}^{\pi}\tfrac{1}{2}\cot\tfrac{1}{2}x\log\left|\frac{\sin\frac{1}{2}(x+h)}{\sin\frac{1}{2}(x-h)}\right|dx=\pi-h,$$

the integrand on the left being non-negative. (M. Riesz [1].)

21. Considering $\mathsf{S}[\cos\alpha x]$ at the points $x=0,\pi$ show that

$$\frac{\pi}{\sin\alpha\pi}=\frac{1}{\alpha}+2\alpha\sum_{k=1}^{\infty}\frac{(-1)^k}{\alpha^2-k^2}=\lim_{N\to+\infty}\sum_{k=-N}^{N}\frac{(-1)^k}{\alpha+k},$$

$$\frac{\pi}{\tan\alpha\pi}=\frac{1}{\alpha}+2\alpha\sum_{k=1}^{\infty}\frac{1}{\alpha^2-k^2}=\lim_{N\to+\infty}\sum_{k=-N}^{N}\frac{1}{\alpha+k}.$$

22. If $f(x_0+t)+f(x_0-t)$ increases monotonically to $+\infty$ as $t\to 0$, $0<t<t_0$, then $\mathsf{S}[f]$ diverges to $+\infty$ at the point x_0.

[Let $[f(x_0+t)+f(x_0-t)]/t=\chi(t)$. Then

$$\int_0^{\pi/n}\chi(t)\sin nt\,dt+o(1)\geqslant\pi S_n^*(x_0)\geqslant\int_0^{\pi/n}[\chi(t)-\chi(t+\pi/n)]\sin nt\,dt+o(1)$$

$$\geqslant\frac{1}{2}\int_0^{\pi/n}\chi(t)\sin nt\,dt+o(1).]$$

23. Using for the Lebesgue constants (see § 12) the formula

$$L_n=\frac{1}{\pi}\int_0^{2\pi}\operatorname{sign}\sin(n+\tfrac{1}{2})x\left[\frac{1}{2}+\sum_{k=1}^{n}\cos kx\right]dx,$$

and integrating termwise, prove that

$$L_n=\frac{1}{2n+1}+\frac{2}{\pi}\sum_{k=1}^{n}\frac{1}{k}\tan\frac{\pi k}{2n+1}.\quad\text{(Fejér [8].)}$$

24. Using $\mathsf{S}[|\sin x|]$ (see p. 34, Example 3) and the formula

$$(\sin kx)^2/\sin x=\sin x+\sin 3x+\ldots+\sin(2k-1)x$$

prove

$$L_n=\frac{16}{\pi^2}\sum_{k=1}^{\infty}\frac{1+\frac{1}{3}+\frac{1}{5}+\ldots+\frac{1}{2k(2n+1)-1}}{4k^2-1}.$$

This equation shows, in particular, that $\{L_n\}$ is strictly increasing. (Szegö [1].)

25. Show that the conclusions of (6·7) (i) remain valid if the hypotheses are replaced by the following ones: (i) $f(x)$ is bounded; (ii) $\rho(x)$ is integrable and satisfies

$$\int_{-\pi}^{\pi}|\rho(x_0+t)-\rho(x_0)|\,|t|^{-1}dt<\infty.\quad\text{(W. H. Young [11].)}$$

26. Consider the periodic functions $f_p(x)$ defined by the formulae

$$f_p(x)=|x|^{1/p'}\quad(1<p<\infty),$$

$$f_1(x)=\log|x|,$$

for $|x|\leqslant\pi$, $p'=p/(p-1)$. Show that (i) $f_p(x)$ belongs to Λ_*^p, $1\leqslant p<\infty$, but not to Λ_*^q, $q>p$; (ii) $f_p(x)$ does not belong to Λ_1^p.

27. Let $0\leqslant\alpha<1$, $-\infty<\beta<+\infty$. The modulus of continuity of

$$f_{\alpha,\beta}(x)=\sum_{n=1}^{\infty}b^{-n\alpha}n^{-\beta}\cos b^n x\quad(b=2,3,\ldots)$$

is $O(\delta^\alpha\log^{-\beta}1/\delta)$ if $\alpha>0$, and is $O(\log^{-(\beta-1)}1/\delta)$ if $\alpha=0$, $\beta>1$.

CHAPTER III

SUMMABILITY OF FOURIER SERIES

1. Summability of numerical series

We consider a doubly infinite matrix

$$M=\begin{pmatrix} a_{00} & a_{01} & \dots & a_{0n} & \dots \\ a_{10} & a_{11} & \dots & a_{1n} & \dots \\ \dots & \dots & \dots & \dots & \dots \\ a_{n0} & a_{n1} & \dots & a_{nn} & \dots \\ \dots & \dots & \dots & \dots & \dots \end{pmatrix}$$

of numbers. With every sequence $s_0, s_1, s_2, \dots$ we associate the sequence $\{\sigma_n\}$ given by

$$\sigma_n = a_{n0}s_0 + a_{n1}s_1 + \dots + a_{n\nu}s_\nu + \dots \quad (n=0,1,2,\dots), \tag{1·1}$$

provided the series on the right converges for all n. If σ_n tends to a limit s we shall say that the sequence $\{s_\nu\}$, or the series whose partial sums are s_ν, is *summable* M to limit (sum) s. The σ_n are also called the *linear means* (determined by the matrix M) of the $\{s_\nu\}$. Matrices M such that $a_{n\nu}=0$ for $n<\nu$ are called *triangular*.

Let us suppose that the numbers

$$N_n = |a_{n0}| + |a_{n1}| + \dots, \quad A_n = a_{n0} + a_{n1} + \dots$$

exist (and are finite) for all n. The matrix will be called *regular* if the following conditions are satisfied:

(i) $\lim\limits_{n\to+\infty} a_{n\nu} = 0$ for $\nu = 0, 1, \dots$;

(ii) the N_n are bounded;

(iii) $\lim\limits_{n\to\infty} A_n = 1$.

The finiteness of N_n implies the existence of A_n. It also implies the convergence of the series (1·1) for every bounded (in particular, convergent) sequence $\{s_\nu\}$.

(1·2) THEOREM. *If* M *is a regular matrix, and if s_ν tends to a finite limit s, then $\sigma_n \to s$.*

For let $s_\nu = s + \epsilon_\nu$, $\epsilon_\nu \to 0$. Correspondingly, $\sigma_n = \sigma_n' + \sigma_n''$, where

$$\sigma_n' = sA_n, \quad \sigma_n'' = \epsilon_0 a_{n0} + \epsilon_1 a_{n1} + \dots.$$

Here $\sigma_n' \to s$ by condition (iii). Let N be the upper bound of the N_n, and let $|\epsilon_\nu| < \eta/2N$ for $\nu > \nu_0$, where η is any given positive number. Then

$$|\sigma_n''| \leqslant (|\epsilon_0||a_{n0}| + \dots + |\epsilon_{\nu_0}||a_{n\nu_0}|) + (|a_{n,\nu_0+1}| + |a_{n,\nu_0+2}| + \dots)\eta/2N.$$

The first sum on the right tends to 0 as $n\to\infty$ (condition (i)), and so is less than $\frac{1}{2}\eta$ for $n > n_0$. The remainder does not exceed $N\eta/2N = \frac{1}{2}\eta$. Hence $|\sigma_n''| < \eta$ for $n > n_0$, $\sigma_n'' \to 0$, $\sigma_n \to s$, as desired.

We note that if $s=0$ condition (iii) is not needed in the above argument.

Condition (ii) by itself shows that the boundedness of $\{s_\nu\}$ implies that of $\{\sigma_n\}$; for if $|s_\nu| \leqslant A$ for all ν, then $|\sigma_n| \leqslant AN$.

It is interesting to observe that conditions (i), (ii), (iii) are also *necessary*, if $\{\sigma_n\}$ is to tend to s for every $\{s_\nu\} \to s$. For consider the sequence $s_\nu = 1$ for all ν, and the sequence $s_\nu = 0$ for all $\nu \neq \mu$, $s_\mu = 1$. In the first case $s = 1$, in the second $s = 0$. Since $\sigma_n = A_n$ in the first case and $\sigma_n = a_{n\mu}$ in the second, the necessity of conditions (i) and (iii) is evident. The necessity of (ii) is less simple and will not be discussed here (see Chapter IV, p. 168).

Condition (ii) is a consequence of (iii) if the numbers $a_{n\nu}$ are non-negative. Such matrices M are called *positive*.

(1·3) THEOREM. *If* M *is a positive regular matrix, then*

$$\liminf s_\nu \leqslant \liminf \sigma_n \leqslant \limsup \sigma_n \leqslant \limsup s_\nu \tag{1·4}$$

for any sequence $\{s_\nu\}$ for which the σ_n are defined.

In particular, if M *is positive Theorem* (1·2) *holds for $s = +\infty$ and $s = -\infty$.*

Let $\liminf s_\nu = \underline{s}$, $\limsup s_\nu = \bar{s}$. To prove the last inequality in (1·4), we may suppose that $\bar{s} < +\infty$. Let α be any number greater than $\bar{s}$. Then $s_\nu < \alpha$ for $\nu > \nu_0$ and, by (i),

$$\sigma_n \leqslant o(1) + \alpha(a_{n,\nu_0+1} + a_{n,\nu_0+2} + \dots) = o(1) + \alpha(A_n + o(1)).$$

Hence, by (iii), $\limsup \sigma_n \leqslant \alpha$, and so $\limsup \sigma_n \leqslant \bar{s}$. The first inequality (1·4) is proved similarly.

If M is not positive, (1·4) need not be true. However, we have:

(1·5) THEOREM. *Let* M *be a regular matrix, and let $C = \limsup N_i$. For any $\{s_\nu\}$ for which $\underline{s}$ and $\bar{s}$ are finite, the numbers $\underline{\sigma} = \liminf \sigma_n$, $\bar{\sigma} = \limsup \sigma_n$ are both contained in the interval whose end-points are*

$$\tfrac{1}{2}(\underline{s} + \bar{s}) \pm C \,.\, \tfrac{1}{2}(\bar{s} - \underline{s}).$$

In other words, $\underline{\sigma}$ and $\bar{\sigma}$ lie in the interval concentric with $(\underline{s}, \bar{s})$ and C times as large

Let us set $s_\nu = s'_\nu + s''_\nu$, where $s'_\nu = \frac{1}{2}(\underline{s} + \bar{s})$ for all ν. Then $\limsup |s''_\nu| = \frac{1}{2}(\bar{s} - \underline{s})$. Correspondingly $\sigma_n = \sigma'_n + \sigma''_n$, where

$$\sigma'_n \to \tfrac{1}{2}(\underline{s} + \bar{s}), \quad \limsup |\sigma''_n| \leqslant C \,.\, \tfrac{1}{2}(\bar{s} - \underline{s}).$$

This completes the proof.

(1·6) THEOREM. *Let $p_0, p_1, p_2, \dots$ and $q_0, q_1, q_2, \dots$ be two sequences and let*

$$P_n = p_0 + p_1 + \dots + p_n, \quad Q_n = q_0 + q_1 + \dots + q_n, \quad q_n > 0 \textit{ for all } n, \quad Q_n \to \infty.$$

Under these conditions, if $p_n/q_n \to s$ then $P_n/Q_n \to s$.

Set $s_\nu = p_\nu/q_\nu$, $\sigma_n = P_n/Q_n$. Then

$$\sigma_n = (q_0 s_0 + q_1 s_1 + \dots + q_n s_n)/Q_n.$$

The σ's here are linear means of the s's, and we may verify that the matrix M is a positive regular matrix. It is therefore enough to apply (1·3).

In particular, taking $q_\nu = 1$ for all ν, we obtain the classical result of Cauchy: *if $s_\nu \to s$, then $(s_0 + s_1 + \dots + s_n)/(n+1) \to s$.*

Given a sequence $s_0, s_1, s_2, \dots$ we define, for every $k = 0, 1, \dots$, the sequence $S_0^k, S_1^k, S_2^k, \dots$ by the conditions

$$S_n^0 = s_n, \quad S_n^k = S_0^{k-1} + S_1^{k-1} + \dots + S_n^{k-1} \quad (k = 1, 2, \dots;\ n = 0, 1, \dots).$$

Similarly, for $k=0,1,2,\ldots$ we define the sequence of numbers A_0^k, A_1^k, A_2^k, ... by the conditions

$$A_n^0=1, \quad A_n^k=A_0^{k-1}+A_1^{k-1}+\ldots+A_n^{k-1} \quad (k=1,2,\ldots;\ n=0,1,\ldots).$$

We say that the sequence $s_0, s_1, s_2, \ldots$ (or the series whose partial sums are s_n) is *summable by the k-th arithmetic mean of Cesàro*, or, briefly, *summable* (C, k), to limit (sum) s, if

$$\lim_{n\to\infty} S_n^k/A_n^k=s.$$

Summability (C, 0) is ordinary convergence. Summability (C, k) of a sequence implies summability (C, $k+1$) to the same limit (take $p_n=S_n^k$, $q_k=A_n^k$ in (1·6)). To find the numerical values of the A_n^k we use the following proposition: *If*

$$A_n=a_0+a_1+\ldots+a_n$$

for all n, and if $|x|<1$, *then*

$$\sum_{n=0}^{\infty} a_n x^n=(1-x)\sum_{n=0}^{\infty} A_n x^n, \tag{1·7}$$

provided either series converges. For if $\Sigma A_n x^n$ converges and if we multiply out the right-hand side and collect similar terms, we obtain the series on the left. Conversely, if $|x|<1$ and $\Sigma a_n x^n$ converges, then

$$(1-x)^{-1}\sum_0^{\infty} a_n x^n=\sum_0^{\infty} x^n \sum_0^{\infty} a_n x^n=\sum_0^{\infty} A_n x^n,$$

by Cauchy's rule of multiplication of power series, and $\Sigma A_n x^n$ converges.

In particular,

$$\sum_{n=0}^{\infty} A_n^k x^n=(1-x)^{-1}\sum_{n=0}^{\infty} A_n^{k-1}x^n=(1-x)^{-2}\sum_{n=0}^{\infty} A_n^{k-2}x^n=\ldots=(1-x)^{-(k+1)},$$

$$\sum_{n=0}^{\infty} S_n^k x^n=(1-x)^{-1}\sum_{n=0}^{\infty} S_n^{k-1}x^n=(1-x)^{-2}\sum_{n=0}^{\infty} S_n^{k-2}x^n=\ldots=(1-x)^{-k}\sum_{n=0}^{\infty} S_n^0 x^n.$$

This permits us to restate our definition as follows: *A sequence $s_0, s_1, \ldots$ (or a series $u_0+u_1+u_2+\ldots$ with partial sums $s_0, s_1, s_2, \ldots$) is summable* (C, α) *to limit (sum) s, if*

$$\sigma_n^\alpha=S_n^\alpha/A_n^\alpha\to s \quad as \quad n\to\infty, \tag{1·8}$$

S_n^α and A_n^α being given by the formulae

$$\left.\begin{aligned} &\sum_{n=0}^{\infty} A_n^\alpha x^n=(1-x)^{-\alpha-1},\\ &\sum_{n=0}^{\infty} S_n^\alpha x^n=(1-x)^{-\alpha}\sum_{n=0}^{\infty} s_n x^n=(1-x)^{-\alpha-1}\sum_{n=0}^{\infty} u_n x^n.\end{aligned}\right\} \tag{1·9}$$

We may then also write (C, α) $\lim s_n=s$, or (C, α) $\sum_0^\infty u_n=s$, as the case may be.

If $\{\sigma_n^\alpha\}$ is bounded, we say that $\{s_n\}$ is *bounded* (C, α).

In these new definitions α need no longer be a non-negative integer. The only restriction is that $\alpha\neq-1,-2,-3,\ldots$ (otherwise, as may be seen from the first formula (1·9), A_n^α is zero for large enough n). It turns out, however, that only the case $\alpha>-1$ is of interest. The numbers S_n^α and σ_n^α will be called respectively the *Cesàro sums* and *Cesàro means* of order α of the sequence $\{s_\nu\}$ (series Σu_ν). The A_n^α are called the *Cesàro numbers* of order α. It is useful to remember that in the case of a series $u_0+u_1+u_2+\dots$ we have $S_n^{-1}=u_n$.

From the definitions of A_n^α and S_n^α it follows that

$$\text{(i)}\ A_n^{\alpha+\beta+1}=\sum_{\nu=0}^{n} A_\nu^\alpha A_{n-\nu}^\beta,\quad \text{(ii)}\ S_n^{\alpha+\beta+1}=\sum_{\nu=0}^{n} S_\nu^\alpha A_{n-\nu}^\beta, \tag{1·10}$$

for all α and β. In particular, replacing α by $\alpha-1$, β by 0,

$$A_n^\alpha=\sum_{\nu=0}^{n} A_\nu^{\alpha-1},\quad S_n^\alpha=\sum_{\nu=0}^{n} S_\nu^{\alpha-1}. \tag{1·11}$$

Hence
$$A_n^\alpha-A_{n-1}^\alpha=A_n^{\alpha-1},\quad S_n^\alpha-S_{n-1}^\alpha=S_n^{\alpha-1}. \tag{1·12}$$

From (1·10) (ii) we get the fundamental formula

$$S_n^\beta=\sum_{\nu=0}^{n} A_{n-\nu}^{\beta-1}s_\nu=\sum_{\nu=0}^{n} A_{n-\nu}^{\beta}u_\nu, \tag{1·13}$$

which shows that
$$\sigma_n^\beta=\sum_{\nu=0}^{n}\frac{A_{n-\nu}^{\beta-1}}{A_n^\beta}s_\nu=\sum_{\nu=0}^{n}\frac{A_{n-\nu}^{\beta}}{A_n^\beta}u_\nu. \tag{1·14}$$

The first formula in (1·9) implies, first,

$$A_n^\alpha=\frac{(\alpha+1)(\alpha+2)\dots(\alpha+n)}{n!}=\binom{n+\alpha}{n}\simeq\frac{n^\alpha}{\Gamma(\alpha+1)}\quad(\alpha\neq-1,-2,\dots), \tag{1·15}$$

so that
$$\sum_{\nu=0}^{\infty}|A_\nu^\alpha|<+\infty\quad\text{for}\quad\alpha<-1, \tag{1·16}$$

and secondly

(1·17) THEOREM. *A_n^α is positive for $\alpha>-1$, increasing (as a function of n) for $\alpha>0$ and decreasing for $-1<\alpha<0$; and $A_n^0=1$ for all n. If $\alpha<-1$, A_n^α is of constant sign for n large enough.*

The Γ in (1·15) is Euler's gamma function, and in fact the formula itself is just Gauss's definition of that function. Later (in Chapter V, § 2) we shall need a slightly more precise formula, namely

$$A_n^\alpha=\frac{n^\alpha}{\Gamma(\alpha+1)}\left\{1+O\left(\frac{1}{n}\right)\right\}. \tag{1·18}$$

To prove it we note that $\log(1+u)=u+O(u^2)$ for small $|u|$, and hence

$$\log A_n^\alpha=\sum_{\nu=1}^{n}\log\left(1+\frac{\alpha}{\nu}\right)=\alpha\sum_{\nu=1}^{n}\frac{1}{\nu}+\sum_{\nu=1}^{n}O\left(\frac{1}{\nu^2}\right)=\alpha\log n+\text{const.}+O\left(\frac{1}{n}\right), \tag{1·19}$$

by formula (8·9) of Chapter I and the fact that the remainders of a series with terms $O(1/\nu^2)$ are $O(1/n)$. Comparing the last formula with (1·15) we see that our constant is $\log\{1/\Gamma(\alpha+1)\}$, and (1·18) follows from (1·19).

It is useful to note that if α is a positive integer, then

$$A_n^\alpha=\binom{n+\alpha}{n}=\binom{n+\alpha}{\alpha}=\frac{(n+1)(n+2)\dots(n+\alpha)}{\alpha!}. \tag{1·20}$$

(1·21) THEOREM. *If a series is summable* (C, α), $\alpha>-1$, *to sum s, it is also summable* (C, $\alpha+h$) *to s, for every $h>0$.*

(1·22) Theorem. *If a series $u_0+u_1+\dots$ is summable* (C, α), $\alpha>-1$, *to a finite sum then $u_n=o(n^\alpha)$.*

Let σ_n^α be the Cesàro means of the series. Then

$$\sigma_n^{\alpha+h}=\left(\sum_{\nu=0}^{n}A_{n-\nu}^{h-1}S_\nu^\alpha\right)\Big/A_n^{\alpha+h}=\left(\sum_{\nu=0}^{n}A_{n-\nu}^{h-1}A_\nu^\alpha\sigma_\nu^\alpha\right)\Big/A_n^{\alpha+h}.$$

Hence the $\sigma_n^{\alpha+h}$ are linear means of the σ_n^α, and using (1·10) and (1·15) we verify that conditions (i), (ii) and (iii) of regularity are satisfied for $\alpha>-1$, $h>0$. This proves (1·21). Moreover, since the matrix here is positive, *the limits of indetermination of the sequence $\{\sigma_n^{\alpha+h}\}$ are contained between those of $\{\sigma_n^\alpha\}$.*

To prove (1·22) we write

$$u_n/A_n^\alpha=\left(\sum_{\nu=0}^{n}A_{n-\nu}^{-\alpha-2}S_\nu^\alpha\right)\Big/A_n^\alpha=\left(\sum_{\nu=0}^{n}A_{n-\nu}^{-\alpha-2}A_\nu^\alpha\sigma_\nu^\alpha\right)\Big/A_n^\alpha.$$

Suppose that $\sigma_\nu^\alpha\to0$ (by subtracting a constant from u_0, we may assume that $u_0+u_1+\dots$ is summable (C, α) to 0). We have to show that the coefficients of the σ_ν^α here satisfy conditions (i) and (ii) (condition (iii) is superfluous since $\sigma_\nu^\alpha\to0$). Condition (i) follows from (1·15), since $\alpha>-1$. To prove (ii) suppose first that $\alpha\geqslant0$. Then, since $A_n^\alpha\geqslant A_\nu^\alpha>0$, we have

$$N_n\leqslant\sum_{\nu=0}^{n}|A_{n-\nu}^{-\alpha-2}|=\sum_{\nu=0}^{n}|A_\nu^{-\alpha-2}|\leqslant\sum_{\nu=0}^{\infty}|A_\nu^{-\alpha-2}|<\infty$$

(cf. (1·16)). If $-1<\alpha<0$, then $A_0^{-\alpha-2}=1$, $A_\nu^{-\alpha-2}<0$ for $\nu>0$, and using (1·11) we verify that $N_n=2$ for all n. This proves (1·22).

(1·23) Theorem. *Under the hypotheses of* (1·22), *if $\gamma<\alpha$ then $S_n^\gamma=o(n^\alpha)$.*

For $\gamma=-1$, this is Theorem (1·22). (1·23) follows from the preceding argument applied to the formula

$$S_n^\gamma/A_n^\alpha=\left(\sum_{\nu=0}^{n}A_{n-\nu}^{\gamma-\alpha-1}A_\nu^\alpha\sigma_\nu^\alpha\right)\Big/A_n^\alpha.$$

For $\gamma\neq-1,-2,\dots$ the conclusion may be written $\sigma_n^\gamma=o(n^{\alpha-\gamma})$.

Consider a series $u_0+u_1+\dots$. Its partial sums will be denoted by s_n, its first arithmetic means by σ_n. Thus

$$\sigma_n=\frac{s_0+s_1+\dots+s_n}{n+1}=\sum_{\nu=0}^{n}\left(1-\frac{\nu}{n+1}\right)u_\nu. \tag{1·24}$$

It is often useful to consider the difference

$$\Delta_n=s_n-\sigma_n=\frac{u_1+2u_2+\dots+nu_n}{n+1} \tag{1·25}$$

If $\Delta_n\to0$, and if the series is summable (C, 1), *it is also convergent.* In particular, a series $u_0+u_1+\dots$ summable (C, 1) and having terms $u_\nu=o(1/\nu)$ is convergent†. If $u_\nu=O(1/\nu)$ and if the series is bounded (C, 1), the partial sums of the series are bounded. Less obvious is the following:

(1·26) Theorem of Hardy. *If $u_0+u_1+\dots$ is summable* (C, 1) *and if $u_\nu=O(1/\nu)$, the series is convergent.*

† Another useful corollary of (1·25) is: if Σu_n converges then $u_1+2u_2+\dots+nu_n=o(n)$.

We shall return to this presently. Meanwhile we observe that the condition $\Delta_n \to 0$ may be satisfied in cases other than $u_\nu = o(1/\nu)$. We note two such cases:

(*a*) $\Sigma \nu |u_\nu|^2 = M$ is finite;

(*b*) $u_\nu \to 0$, $u_\nu = 0$ if ν does not belong to a sequence $n_1 < n_2 < n_3 < \dots$ of integers satisfying $n_{k+1}/n_k > q$, where q is fixed and greater than 1.

Suppose (*a*) is satisfied. By Schwarz's inequality,

$$|\Delta_n| \leqslant \frac{1}{n+1} \sum_{\nu=1}^{n} \nu^{\frac{1}{2}} |u_\nu| \nu^{\frac{1}{2}} \leqslant \frac{1}{n+1} \left(\sum_{\nu=1}^{n} \nu |u_\nu|^2 \right)^{\frac{1}{2}} \left(\sum_{\nu=1}^{n} \nu \right)^{\frac{1}{2}} \leqslant \left(\sum_{1}^{\infty} \nu |u_\nu|^2 \right)^{\frac{1}{2}},$$

so that $\limsup |\Delta_n| \leqslant M^{\frac{1}{2}}$. But $\limsup |\Delta_n|$ is not affected if we replace each of $u_0, u_1, u_2, \dots, u_k$ by zeros. Taking k large enough we may make M arbitrarily small; thus $\lim \Delta_n = 0$.

We may replace (*a*) by (*a'*) $\qquad \Sigma \nu^p |u_\nu|^{p+1} < \infty$,

where $p > 0$. The conclusion and proof hold, if in the latter we use Hölder's inequality instead of Schwarz's.

In case (*b*), let $|u_{n_\nu}| = \epsilon_\nu$ and let $n_k \leqslant n < n_{k+1}$. Then

$$|\Delta_n| \leqslant (n_k+1)^{-1} \sum_{\nu=1}^{k} n_\nu \epsilon_\nu \leqslant \sum_{\nu=1}^{k} (n_\nu/n_k)\, \epsilon_\nu \leqslant \sum_{\nu=1}^{k} \epsilon_\nu q^{\nu-k},$$

and the sum on the right is a linear mean of $\{\epsilon_\nu\}$. Conditions (i) and (ii) of regularity are satisfied, so that $\epsilon_\nu \to 0$ implies $\Delta_n \to 0$.

A series Σu_ν will be said to possess a *gap* (p, q) if $c_i = 0$ for $p < \nu \leqslant q$. Case (*b*) may be generalized as follows.

(1·27) Theorem. *If a series Σu_ν, with partial sums s_n, has infinitely many gaps (m_k, m_k') such that $m_k'/m_k \geqslant q > 1$, and is summable* (C, 1) *to sum s, then $s_{m_k} \to s$ (and so also $s_{m_k'} \to s$).*

We may suppose that $s = 0$. Since $s_0 + s_1 + \dots + s_n = (n+1)\,\sigma_n$, we have

$$\begin{aligned}(m_k' - m_k)\, s_{m_k} &= s_{m_k} + s_{m_k+1} + \dots + s_{m_k'-1} \\ &= m_k' \sigma_{m_k'-1} - m_k \sigma_{m_k-1} = o(m_k') + o(m_k) = o(m_k') = o(m_k' - m_k), \qquad \textbf{(1·28)}\end{aligned}$$

whence $s_{m_k} = o(1)$ and (1·27) follows.

If we assume nothing about the summability of Σu_ν and set

$$s^* = \sup_k |s_{m_k}|, \quad \sigma^* = \sup_n |\sigma_n|,$$

the identity $(m_k' - m_k)\, s_{m_k} = m_k' \sigma_{m_k'-1} - m_k \sigma_{m_k-1}$ gives

$$|s_{m_k}| \leqslant \sigma^* \frac{m_k' + m_k}{m_k' - m_k} \leqslant \frac{q+1}{q-1} \sigma^*,$$

and so
$$s^* \leqslant A_q \sigma^*, \qquad \textbf{(1·29)}$$

where A_q depends on q only.

Let k be a positive integer. Consider the expression

$$\sigma_{n,k} = \frac{s_n + s_{n+1} + \dots + s_{n+k-1}}{k} = \frac{(n+k)\,\sigma_{n+k-1} - n\sigma_{n-1}}{k} = \left(1 + \frac{n}{k}\right) \sigma_{n+k-1} - \frac{n}{k} \sigma_{n-1}. \qquad \textbf{(1·30)}$$

We verify that
$$\sigma_{n,k}=s_n+\sum_{\nu=n+1}^{n+k-1}\left(1-\frac{\nu-n}{k}\right)u_\nu. \tag{1·31}$$

(If $k=1$, we interpret the sum Σ on the right as 0.) If k tends to ∞ with n in such a way that n/k is bounded, then $\sigma_{n,k}$ defines a method of summability which is at least as strong as the (C, 1) method: *if* $\sigma_n\to s$, *then* $\sigma_{n,k}\to s$. This follows from (1·30) if we set $\sigma_\nu=s+\epsilon_\nu$, $\epsilon_\nu\to 0$.

The peculiarity of the method is that $\sigma_{n,k}$ is obtained from s_n by adding to it a linear combination, with coefficients positive and less than 1, of the terms $u_{n+1}, u_{n+2}, \ldots, u_{n+k-1}$ (cf. (1·31)); it is useful in certain applications. The case
$$\sigma_{n,n}=2\sigma_{2n-1}-\sigma_{n-1}$$
is particularly simple. We may call the $\sigma_{n,k}$ the *delayed first arithmetic means*. We observe that $\sigma_{n,1}=s_n$, $\sigma_{0,n}=\sigma_{n-1}$.

Returning to (1·26), suppose that $\sigma_\nu\to s$, $|u_\nu|<A/\nu$ for $\nu=1,2,\ldots$. By the remark just made,
$$|\sigma_{n,k}-s_n|\leqslant\sum_{\nu=n+1}^{n+k-1}|u_\nu|\leqslant A\sum_{n+1}^{n+k-1}\frac{1}{\nu}<A\frac{k-1}{n}.$$

Let ϵ be any positive number, and let $k=[n\epsilon]+1$.† The last expression is then $A[n\epsilon]/n\leqslant A\epsilon$. Since $n/k<n/n\epsilon=1/\epsilon$ is bounded, $\sigma_{n,k}\to s$, so that $\limsup|s_n-s|\leqslant A\epsilon$. This gives, ϵ being arbitrary, $\lim s_n=s$.

The series $u_0+u_1+\ldots$ is said to be summable by *Abel's* method (sometimes called *Poisson's*), or *summable* A, to sum s, if the series $u_0+u_1x+u_2x^2+\ldots$ is convergent for $|x|<1$ and if
$$\lim_{x\to 1-0}\sum_{\nu=0}^{\infty}u_\nu x^\nu=s, \tag{1·32}$$
where x tends to 1 along the real axis. By (1·7), summability A of a sequence $\{s_\nu\}$ may be defined as the existence of
$$\lim_{x\to 1-0}(1-x)\sum_{\nu=0}^{\infty}s_\nu x^\nu.$$

(1·33) THEOREM. *If the series* $u_0+u_1+\ldots$ *is summable* (C, α), $\alpha>-1$, *to sum* s (*finite or not*), *it is also summable* A *to* s.

For let $f(x)=u_0+u_1x+u_2x^2+\ldots$, and let $\{x_n\}$ be any sequence of points on the real axis tending to 1 from the left. By (1·9),
$$f(x_n)=(1-x_n)^{\alpha+1}\sum_{\nu=0}^{\infty}S_\nu^\alpha x_n^\nu=(1-x_n)^{\alpha+1}\sum_{\nu=0}^{\infty}\sigma_\nu^\alpha A_\nu^\alpha x_n^\nu.$$

The expression on the right is a linear mean of the sequence $\{\sigma_\nu^\alpha\}$, and corresponds to the matrix M with $a_{n\nu}=A_\nu^\alpha(1-x_n)^{\alpha+1}x_n^\nu$. This matrix is positive and satisfies conditions (i) and (iii). Hence $f(x_n)\to s$, and (1·33) is proved.

The argument also shows that the *limits of indetermination by the method* A (i.e. $\liminf_{x\to 1}f(x)$ and $\limsup_{x\to 1}f(x)$) *are contained between those by the method* (C, α).

(1·34) THEOREM. *If* $u_0+u_1+\ldots$ *is summable* (C, α), $\alpha>-1$, *to a finite sum* s, *then* (1·32) *holds as* x *tends to* 1 *along any path* L *lying between two chords of the unit circle which pass through* $x=1$.

† By $[x]$ we denote the integral part of x.

Such paths L will hereafter be spoken of as *non-tangential*. In the neighbourhood of the point 1 they are characterized by an inequality

$$|1-x|/(1-|x|) \leqslant \text{const.} \tag{1·35}$$

For the proof of (1·34) we observe that, if $\{x_n\}$ is any sequence of points tending to 1 along L, then $f(x_n)$ is as before a linear mean of the σ_ν^α generated by a matrix M satisfying conditions (i) and (iii). M is no longer positive, but

$$\sum_{\nu=0}^{\infty} |a_{n\nu}| = \sum_{\nu=0}^{\infty} A_\nu^\alpha |1-x_n|^{\alpha+1} |x_n|^\nu = |1-x_n|^{\alpha+1}/(1-|x_n|)^{\alpha+1}$$

is bounded, by (1·35). This proves (1·34).

(1·36) THEOREM OF TAUBER. *Let s_n and $f(x)$ denote the partial sums and the Abel means of a series Σu_n with terms $o(1/n)$. Then, if $N = \left[\dfrac{1}{1-x}\right]$, we have*

$$f(x) - s_N \to 0 \quad \textit{as} \quad x \to 1-0. \tag{1·37}$$

In particular, the series is Abel summable if and only if it converges.

The relation (1·37) *still holds if the condition $u_n = o(1/n)$ is replaced by*

$$u_1 + 2u_2 + \ldots + nu_n = o(n).$$

Suppose first that $\eta_n = nu_n \to 0$. The left-hand side in (1·37) is

$$\sum_1^N u_n(x^n - 1) + \sum_{N+1}^{\infty} u_n x^n = P + Q,$$

say. Observing that $N \leqslant 1/(1-x) < N+1$ and $1 - x^n \leqslant n(1-x)$, we have

$$|P| \leqslant (1-x) \sum_1^N |\eta_n| \leqslant N^{-1} \sum_1^N |\eta_n| \to 0,$$

$$|Q| \leqslant \sum_{N+1}^{\infty} \frac{|\eta_n|}{n} x^n \leqslant (N+1)^{-1} \max_{n>N} |\eta_n| \sum_{N+1}^{\infty} x^n \leqslant \frac{1}{(N+1)(1-x)} \max_{n>N} |\eta_n| \to 0.$$

Hence $P + Q \to 0$, and (1·37) follows.

Let now $v_0 = 0$, $v_n = u_1 + 2u_2 + \ldots + nu_n$ for $n > 0$, and suppose that $v_n = o(n)$. Summation by parts gives

$$\sum_0^n u_k = u_0 + \sum_{k=1}^{n} \frac{v_k - v_{k-1}}{k} = u_0 + \sum_{k=1}^{n} \frac{v_k}{k(k+1)} + \frac{v_n}{n+1}.$$

Since $v_n = o(n)$, the series $\sum_0^\infty u_k$ and $u_0 + \sum_1^\infty v_k/k(k+1)$ are equiconvergent. If, therefore, t_n and $g(x)$ are the partial sums and the Abel means of the second series, we have $s_N - t_N \to 0$, $f(x) - g(x) \to 0$. But the terms of the second series are $o(1/n)$, so that, by the case already dealt with we have $g(x) - t_N \to 0$. This and the preceding two relations imply (1·37).

The following theorem partly generalizes (1·36).

(1·38) THEOREM OF LITTLEWOOD. *A series $\sum_0^\infty u_n$ summable* A *and with terms $O(1/n)$ converges.*

We may suppose that $|u_n| \leqslant 1/n$ for $n>0$ and that $f(x)=\Sigma u_n x^n \to 0$ as $x\to 1$. If we substitute here x^k for x $(k=1,2,\dots)$, we see that for every power polynomial $P(x)$ without constant term we have

$$\sum_0^\infty u_n P(x^n) \to 0$$

as $x\to 1$.

Suppose that given any two numbers $0<\xi'<\xi<1$ and a positive δ we can find a $P(x)$ such that

(i) $0\leqslant P(x)\leqslant 1$ in $(0,1)$,

(ii) $P(x)\leqslant \delta x$ in $(0,\xi')$,

(iii) $1-P(x)\leqslant \delta(1-x)$ in $(\xi,1)$.

We show that we can then prove the convergence of Σu_n. Given any $0<x<1$, let $N=N(x)$ be the greatest non-negative integer satisfying $x^N\geqslant\xi$, and let $N'=N'(x)$ be the least positive integer satisfying $x^{N'}\leqslant\xi'$. Both N and N' are non-decreasing functions of x taking successively all values $1,2,3,\dots$ as $x\to1$. Clearly $N<N'$ and

$$N\simeq\frac{\log(1/\xi)}{\log(1/x)},\quad N'\simeq\frac{\log(1/\xi')}{\log(1/x)}.$$

For a P satisfying (i), (ii), (iii) we have,

$$\sum_0^\infty u_n P(x^n)-s_N=\sum_1^N u_n\{P(x^n)-1\}+\sum_{N+1}^{N'} u_n P(x^n)+\sum_{N'+1}^\infty u_n P(x^n)$$
$$=A(x)+B(x)+C(x),$$

say, and

$$|A(x)|\leqslant\delta\sum_1^N\frac{1}{n}(1-x^n)\leqslant\delta\sum_1^N(1-x)=\delta N(1-x),$$

$$|C(x)|\leqslant\delta\sum_{N'+1}^\infty\frac{x^n}{n}<\frac{\delta}{N'(1-x)},$$

$$|B(x)|\leqslant\sum_{N+1}^{N'}\frac{1}{n}<\frac{N'-N}{N}.$$

Take any $\epsilon>0$. From the asymptotic expressions for N and N' we see that if ξ' and ξ are sufficiently close to each other and both away from 0 and 1 (we may take ξ' and ξ symmetrically with respect to $\frac{1}{2}$), we have $\limsup|B(x)|\leqslant\epsilon$. Having fixed ξ' and ξ, and observing that $N(1-x)$ and $N'(1-x)$ tend to finite non-zero limits, we obtain $\limsup|A(x)|<\epsilon$, $\limsup|C(x)|<\epsilon$, if δ is small enough. Since $\Sigma u_n P(x^n)\to0$, we get $\limsup|s_N|\leqslant3\epsilon$, that is $s_N\to0$.

It remains to construct the required P, and it is enough to assume that $\xi'=\frac{1}{2}-\eta$, $\xi=\frac{1}{2}+\eta$. Write

$$R_k(x)=\{4x(1-x)\}^k,\quad P(x)=\int_0^x R_k(t)\,dt\bigg/\int_0^1 R_k(t)\,dt,$$

where k is a positive integer. Clearly P satisfies (i). Since $R_k(x)\leqslant(1-4\eta^2)^k$ in $(0,1)$ but outside $(\frac{1}{2}-\eta,\frac{1}{2}+\eta)$, and

$$\int_0^1 R_k(t)\,dt\geqslant\int_{\frac{1}{2}-\frac{1}{2}k}^{\frac{1}{2}+\frac{1}{2}k}\left(1-\frac{1}{k^2}\right)^k dt\simeq\frac{1}{k},$$

we find that

$$P(x) \leqslant x \max_{0\leqslant x\leqslant \frac{1}{2}-\eta} R_k(x) \Big/ \int_0^1 R_k(t)\,dt \leqslant \delta x$$

for $0 \leqslant x \leqslant \frac{1}{2} - \eta$ and k large enough. This is (ii), and (iii) is proved similarly.

Return for a moment to (1·2). If the sequence σ_n in (1·1) is of bounded variation, we say that $\{s_n\}$ (or the series whose partial sums are s_n) is *absolutely summable* M. Only absolute summability A, however, is of interest for trigonometric series. The parameter n in this case is a continuous variable and the definition must be modified in an obvious way: a series Σu_n is absolutely summable A if the function $f(r) = \Sigma u_n r^n$ is of bounded variation over $0 \leqslant r < 1$. Every absolutely convergent series with, say, real terms is a difference of two convergent series with non-negative terms, and, correspondingly, $f(r)$ is a difference of two non-decreasing bounded functions. Hence *every absolutely convergent series is absolutely summable* A.

Parallel to the theory of divergent series, one can construct a theory of divergent integrals. We shall only consider the analogue of the method (C, 1). Suppose we have a function $A(u)$ defined for $u \geqslant 0$ and integrable over every finite interval $0 \leqslant u \leqslant u_0$. We say that *the function $A(u)$ tends to limit s, as $u \to \infty$, by the method of the first arithmetic mean*, if

$$u^{-1}\int_0^u A(v)\,dv \to s \quad \text{as} \quad u \to +\infty. \tag{1·39}$$

We shall then write $(\mathrm{C}, 1) \lim\limits_{u\to\infty} A(u) = s$. It is easily seen that if $A(u)$ tends to s as $u \to \infty$, then we also have (1·39).

Consider an integral

$$\int_0^\infty a(v)\,dv, \tag{1·40}$$

where $a(v)$ is integrable over every finite interval $0 \leqslant v \leqslant v_0$. We shall say that (1·40) is *summable* (C, 1) *to* s if the partial integrals $A(u) = \int_0^u a(v)\,dv$ satisfy (1·39), and we shall write

$$(\mathrm{C}, 1)\int_0^\infty a(v)\,dv = s.$$

The latter relation is satisfied if (1·40) converges to s, that is, if $A(u) \to s$ as $u \to +\infty$. As an example one easily verifies that

$$\int_0^\infty e^{ixv}\,dv$$

is summable (C, 1) to sum ix^{-1} or $+\infty$, according as $x \neq 0$ or $x = 0$.

Consider a series $a_0 + a_1 + \ldots$. Generalizing the notion of the partial sum $A_n = a_0 + \ldots + a_n$, we shall introduce that of *sum function* $A(u)$, defined by the formula

$$A(u) = \sum_{n\leqslant u} a_n \quad (u \geqslant 0).$$

Thus $A(u)$ is a step function such that $A(u) = A_n$ for $n \leqslant u < n+1$. At $u = n$, $A(u)$ has jump a_n.

(1·41) THEOREM. *A necessary and sufficient condition for $a_0 + a_1 + \ldots$ to be summable* (C, 1) *to finite sum s is that* $(\mathrm{C}, 1)\lim\limits_{u\to\infty} A(u) = s$.

For suppose that $n \leqslant u < n+1$. Then, denoting by s_n the partial sums of the series,

$$\frac{1}{u}\int_0^u A(t)\,dt = \frac{s_0+\ldots+s_{n-1}+(u-n)\,s_n}{u} = \frac{n}{u}\sigma_{n-1}+\frac{u-n}{u}s_n. \tag{1·42}$$

If $\sigma_n \to s$, then $s_n = o(n)$, and so the right-hand side of (1·42) tends to s. The converse follows by substituting $u=n$ in (1·42).

The definition of summability (C, 1) of $a_0+a_1+\ldots$ in terms of the sum function $A(u)$ is often useful. Expressing the second member of (1·42) in terms of the a's, we get

$$\frac{1}{u}\int_0^u A(t)\,dt = \sum_{\nu\leqslant u} a_\nu\left(1-\frac{\nu}{u}\right), \tag{1·43}$$

a formula analogous to (1·24). (1·43) may be called the *integral* (C, 1) *mean of* Σa_n.

Remark. In the foregoing discussion of summability we confined ourselves to series of constants. Analogous results hold for series of functions and various modes of convergence or summability, bounded, uniform, etc. For example, *if a series* $\Sigma u_\nu(t)$ *is uniformly summable* (C, α), $\alpha > -1$, *on a set E of points t, it is also uniformly summable* (C, β), $\beta > \alpha$, *and uniformly summable* A, *on E*. We shall use such results in the sequel without stating them explicitly. Changes of argument that the extensions call for can easily be supplied by the reader.

2. General remarks about the summability of $S[f]$ and $\tilde{S}[f]$

Given a sequence $s_0, s_1, \ldots$, consider the linear means

$$\sigma_n = a_{n0}s_0 + a_{n1}s_1 + \ldots + a_{nk}s_k + \ldots \tag{2·1}$$

generated by a matrix M satisfying conditions (i), (ii) and (iii) of regularity (see § 1). If the s_k are the partial sums of a series Σu_k, and we write $s_k = u_0+u_1+\ldots+u_k$ in (2·1) we get

$$\sigma_n = \alpha_{n0}u_0 + \alpha_{n1}u_1 + \ldots + \alpha_{nk}u_k + \ldots, \tag{2·2}$$

where $\alpha_{nk} = a_{nk} + a_{n,k+1} + \ldots$. On the right here we have linear means of the series Σu_k.

The passage from (2·1) to (2·2) can be justified under very general conditions (it is trivially true if the matrix M is row-finite). We shall not do so here since in all the cases which interest us either the justification is immediate or, as happens for instance in the case of Abel summability, the form (2·2) is simpler and more natural than (2·1). We shall take (2·2) as a fresh starting-point and apply the idea to Fourier series.

We recall that the case when the parameter n is a continuous variable may be reduced to the standard case by considering discrete sequences of the parameter, and in what follows we shall not dwell on this point.

In all cases which interest us the α_{nk} satisfy the condition

$$\sum_k |\alpha_{nk}| < \infty \tag{2·3}$$

for all n, and we assume this once for all for the sake of simplicity. It is automatically satisfied if the matrix is row-finite.

A very important role is played by the linear means of the two basic series

$$\tfrac{1}{2} + \cos t + \cos 2t + \ldots, \tag{2·4}$$

$$0 + \sin t + \sin 2t + \ldots. \tag{2·5}$$

These means will be denoted by $K_n(t)$ and $\tilde{K}_n(t)$:

$$K_n(t) = \tfrac{1}{2}\alpha_{n0} + \sum_{k=1}^{\infty} \alpha_{nk} \cos kt, \tag{2·6}$$

$$\tilde{K}_n(t) = \sum_{k=1}^{\infty} \alpha_{nk} \sin kt. \tag{2·7}$$

They are both continuous and are even and odd functions respectively of t. If the linear forms are given in the form (2·1) then, clearly,

$$K_n(t) = \sum_{k=0}^{\infty} a_{nk} D_k(t) = \frac{1}{2\sin\frac{1}{2}t} \sum_{k=0}^{\infty} a_{nk} \sin(k+\tfrac{1}{2})t, \tag{2·8}$$

$$\tilde{K}_n(t) = \sum_{k=0}^{\infty} a_{nk} \tilde{D}_k(t) = A_n \cdot \tfrac{1}{2}\cot\tfrac{1}{2}t - \frac{1}{2\sin\frac{1}{2}t} \sum_{k=0}^{\infty} a_{nk} \cos(k+\tfrac{1}{2})t, \tag{2·9}$$

where D_k and $\tilde{D}_k$ denote the Dirichlet and the conjugate Dirichlet kernel and $A_n = a_{n0} + a_{n1} + \dots$.

It is customary to call linear means of (2·4) *kernels* corresponding to method M. Linear means of (2·5) are called *conjugate kernels*.

Let a_k, b_k be the Fourier coefficients of an f. The linear means for $S[f]$ and $\tilde{S}[f]$ are

$$\sigma_n(x) = \sigma_n(x; f) = \tfrac{1}{2}a_0\alpha_{n0} + \sum_{k=1}^{\infty} (a_k \cos kx + b_k \sin kx)\,\alpha_{nk}, \tag{2·10}$$

$$\tilde{\sigma}_n(x) = \tilde{\sigma}_n(x; f) = \sum_{k=1}^{\infty} (a_k \sin kx - b_k \cos kx)\,\alpha_{nk}. \tag{2·11}$$

Under the hypothesis (2·3) we have

$$\sigma_n(x) = \frac{1}{\pi}\int_{-\pi}^{\pi} f(x+t)\,K_n(t)\,dt, \tag{2·12}$$

$$\tilde{\sigma}_n(x) = -\frac{1}{\pi}\int_{-\pi}^{\pi} f(x+t)\,\tilde{K}_n(t)\,dt. \tag{2·13}$$

For the left-hand side of (2·12) is

$$\alpha_{n0}\frac{1}{2\pi}\int_{-\pi}^{\pi} f(t)\,dt + \sum_{k=1}^{\infty} \alpha_{nk}\frac{1}{\pi}\int_{-\pi}^{\pi} f(t)\cos k(t-x)\,dt = \frac{1}{\pi}\int_{-\pi}^{\pi} f(t)\left[\tfrac{1}{2}\alpha_{n0} + \sum_{k=1}^{\infty} \alpha_{nk}\cos k(t-x)\right]dt$$

$$= \frac{1}{\pi}\int_{-\pi}^{\pi} f(t)\,K_n(t-x)\,dt = \frac{1}{\pi}\int_{-\pi}^{\pi} f(x+t)\,K_n(t)\,dt,$$

the interchange of the order of integration and summation being justified by (2·3); and similarly we prove (2·13).

We shall always assume that

$$\alpha_{n0} = 1 \quad \text{for} \quad n = 0, 1, \dots. \tag{2·14}$$

(Observe that in any case $\alpha_{n0} \to 1$, if the linear means (2·2) of the series $1+0+0+\dots$, converging to 1 are also to tend to 1.) We shall call (2·14) *condition* (A). It can also be written

$$\text{(A)} \qquad \frac{1}{\pi}\int_{-\pi}^{\pi} K_n(t)\,dt = 1.$$

If it is satisfied, then $\quad \sigma_n(x) - f(x) = \dfrac{2}{\pi}\displaystyle\int_{0}^{\pi} \phi_x(t)\,K_n(t)\,dt, \qquad$ (2·15) since the left-hand side is

$$\frac{1}{\pi}\int_{-\pi}^{\pi} f(x+t)\,K_n(t)\,dt-\frac{1}{\pi}\int_{-\pi}^{\pi} f(x)\,K_n(t)\,dt=\frac{1}{\pi}\int_{-\pi}^{\pi}\{f(x+t)-f(x)\}\,K_n(t)\,dt,$$

and the last integral equals the right-hand side of (2·15) since $K_n(t)$ is even.

If K_n satisfies conditions (A) and

$$\text{(B)}\quad K_n \geqslant 0$$

for all n, we shall call K_n a *positive kernel*. If we have merely

$$\text{(B}'\text{)}\quad \frac{1}{\pi}\int_{-\pi}^{\pi} |\,K_n(t)\,|\,dt \leqslant C,$$

with C independent of n, we shall call the kernel K_n *quasi-positive*. Every positive kernel is quasi-positive, as we see from condition (A).

The following theorem is an immediate corollary of (2·12):

(2·16) Theorem. *If K_n is a positive kernel, then for any f satisfying*

$$m \leqslant f \leqslant M$$

we have

$$m \leqslant \sigma_n(x; f) \leqslant M. \tag{2·17}$$

If K_n is quasi-positive, $|\,f\,| \leqslant M$ implies

$$|\,\sigma_n(x; f)\,| \leqslant CM, \tag{2·18}$$

with the same C as in condition (B′).

We now introduce a condition (C) in addition to (A) and (B), or (B′), so far considered. Let

$$\mu_n(\delta)=\max_{\delta\leqslant t\leqslant \pi} |\,K_n(t)\,| \quad (0<\delta\leqslant\pi); \tag{2·19}$$

we shall say that the kernels K_n satisfy *condition* (C), if

$$\text{(C)}\quad \mu_n(\delta)\to 0 \quad \textit{for each fixed } \delta \quad (0<\delta\leqslant\pi).$$

Condition (C) implies that (2·4) is uniformly summable M to zero outside an arbitrarily small neighbourhood of $t=0$.

If condition (C) is satisfied, then the decomposition (see (2·12))

$$\sigma_n(x)=\frac{1}{\pi}\int_{-\delta}^{\delta} f(x+t)\,K_n(t)\,dt+\frac{1}{\pi}\int_{\delta\leqslant|t|\leqslant\pi} f(x+t)\,K_n(t)\,dt, \tag{2·20}$$

in which the last term is numerically majorized by

$$\mu_n(\delta)\frac{1}{\pi}\int_{\delta\leqslant|t|\leqslant\pi} |\,f(x+t)\,|\,dt \leqslant \mu_n(\delta)\frac{1}{\pi}\int_{-\pi}^{\pi} |\,f(t)\,|\,dt,$$

shows that the behaviour of $\sigma_n(x)$ at the point x depends solely on the values of f in an arbitrarily small neighbourhood $(x-\delta, x+\delta)$. We know, of course, that this result holds also for the Dirichlet kernel, which does not satisfy condition (C), but in the new case the last integral in (2·20) is uniformly small for all functions f such that $\int_{-\pi}^{\pi} |f(t)\,|\,dt$ is bounded.

(2·21) Theorem. *Suppose that the kernel K_n satisfies conditions* (A), (B)*—or merely* (B′)*— and* (C). *For any integrable f, if the numbers $f(x_0 \pm 0)$ exist and are finite, then we have*

$$\sigma_n(x_0)\to \tfrac{1}{2}\{f(x_0+0)+f(x_0-0)\}. \tag{2·22}$$

In particular, if f is continuous at x_0, we have

$$\sigma_n(x_0)\to f(x_0). \tag{2·23}$$

If f is continuous at every point of a closed interval $I=(\alpha,\beta)$, the relation (2·23) *holds uniformly in $x_0\in I$. In particular, if f is everywhere continuous we have* (2·23) *uniformly in x_0.*

Suppose first that K_n is positive. By changing if necessary the value of $f(x_0)$ we may suppose that $f(x_0)=\frac{1}{2}[f(x_0+0)+f(x_0-0)]$, so that

$$|\phi_{x_0}(t)|<\tfrac{1}{2}\epsilon \quad\text{for}\quad 0\leqslant t\leqslant\delta=\delta(\epsilon). \tag{2·24}$$

By (2·15), $|\sigma_n(x_0;f)-f(x_0)|$ does not exceed

$$\frac{2}{\pi}\int_0^\pi|\phi(t)|\,K_n(t)\,dt=\frac{2}{\pi}\left(\int_0^\delta+\int_\delta^\pi\right)\leqslant\frac{\epsilon}{\pi}\int_0^\delta K_n(t)\,dt+\frac{2\mu_n(\delta)}{\pi}\int_\delta^\pi|\phi|\,dt$$
$$<\frac{\epsilon}{\pi}\int_0^\pi K_n(t)\,dt+\frac{2\mu_n(\delta)}{\pi}\int_0^\pi|\phi(t)|\,dt=P+Q,$$

say. Here
$$P=\tfrac{1}{2}\epsilon \tag{2·25}$$
by condition (A), and
$$Q\to 0 \tag{2·26}$$
by condition (C), so that
$$P+Q<\epsilon \quad\text{for}\quad n>n_0, \tag{2·27}$$
which proves the first part of (2·21).

If f is continuous at every point of I (which is understood to mean that f is also continuous to the left at $x=\alpha$ and to the right at $x=\beta$), we can find a δ independent of x_0 such that (2·24) holds for all $x_0\in I$. As before we have (2·25). The integral in Q does not exceed

$$\int_0^\pi(|f(x_0+t)|+|f(x_0-t)|+2|f(x_0)|)\,dt=\int_{-\pi}^{\pi}|f(t)|\,dt+2\pi|f(x_0)|,$$

(2·26) holds uniformly in I, and (2·27) holds for all $x_0\in I$.

Only minor modifications are needed in the preceding argument if the kernel K_n is quasi-positive. In the inequality for $|\sigma_n-f|$ we have to replace K_n by $|K_n|$, which gives, with the previous notation,

$$P\leqslant\tfrac{1}{2}C\epsilon,\quad Q\to 0,$$

and the conclusion follows as before.

(2·28) Theorem. *Suppose that a positive kernel K_n satisfies condition* (C), *and that $m\leqslant f(x)\leqslant M$ for $x\in I=(a,b)$. Then for any $\epsilon>0$ and $0<\delta<\frac{1}{2}(b-a)$ there is an n_0 such that*

$$m-\epsilon\leqslant\sigma_n(x)\leqslant M+\epsilon \quad\textit{for}\quad x\in I_\delta=(a+\delta,b-\delta),\ n>n_0. \tag{2·29}$$

It is enough to prove the second inequality. In view of the remark concerning the decomposition (2·20), we have

$$\sigma_n(x)=\frac{1}{\pi}\int_{-\delta}^{\delta}f(x+t)\,K_n(t)\,dt+o(1),$$

where the $o(1)$ is uniform in x. Suppose that $x\in I_\delta$. Then $x+t\in I$ for $|t|<\delta$, the integral last written does not exceed

$$M\frac{1}{\pi}\int_{-\delta}^{\delta}K_n(t)\,dt\leqslant M\frac{1}{\pi}\int_{-\pi}^{\pi}K_n(t)\,dt=M,$$

and the result follows.

Given a function $f(x)$, let $M(a,b)$ and $m(a,b)$ be the upper and lower bounds of f in the interval $a<x<b$. For every x we set

$$M(x)=\lim_{h\to+0} M(x-h,x+h), \quad m(x)=\lim_{h\to+0} m(x-h,\ x+h).$$

These limits exist, since the expressions under the limits $h\to+0$ are monotone functions of h. We shall call $M(x)$ and $m(x)$ the *maximum* and *minimum* of f *at* the point x.

(2·30) Theorem. *If a positive kernel K_n satisfies condition* (C), *then for any sequence $\{x_n\}\to x_0$ we have*

$$m(x_0)\leqslant \liminf \sigma_n(x_n)\leqslant \limsup \sigma_n(x_n)\leqslant M(x_0).$$

In particular, $\{\sigma_n(x)\}$ converges uniformly at every point of continuity of f.

For if h is small enough the values of f in the interval (x_0-h, x_0+h) will be contained between $m(x_0)-\epsilon$ and $M(x_0)+\epsilon$, and so for n large enough $\sigma_n(x_n)$ is, by (2·28), contained between $m(x_0)-2\epsilon$ and $M(x_0)+2\epsilon$.

A special case of (2·30) may be stated separately: *if $f(x)$ tends to $+\infty$ as $x\to x_0$* (that is, if $M(x_0)=m(x_0)=+\infty$) *then $\sigma_n(x_0)\to+\infty$.*

As regards the convergence of $\{\tilde\sigma_n(x)\}$, there are no results as simple as (2·21). Let us suppose that $A_n=1$ in (2·9). (This holds for all important methods of summability; in any case, $A_n\to 1$, by condition (iii) of regularity.) Then the difference

$$H_n(t)=\tilde K_n(t)-\tfrac12\cot\tfrac12 t$$

has some resemblance to $K_n(t)$ (see (2·8)), which suggests that to results about $\sigma_n(x)$ there correspond results about

$$\tilde\sigma_n(x)-\left\{-\frac{1}{\pi}\int_{1/n}^{\pi}\{f(x+t)-f(x-t)\}\tfrac12\cot\tfrac12 t\,dt\right\}=\tilde\sigma_n(x)-\tilde f(x;\,1/n).$$

Without aiming unduly at generality we shall find that this is actually so for the methods which are of fundamental importance for Fourier series, namely those of Cesàro and Abel.

3. Summability of $S[f]$ and $\tilde S[f]$ by the method of the first arithmetic mean

The kernel corresponding to the (C, 1) method for (2·4) is

$$K_n(t)=\frac{1}{n+1}\sum_{\nu=0}^{n} D_\nu(t)=\frac{1}{n+1}\sum_{\nu=0}^{n}\frac{\sin(\nu+\frac12)t}{2\sin\frac12 t}. \tag{3·1}$$

Multiplying the numerator and denominator of the right-hand side by $2\sin\frac12 t$ and replacing the products of sines in the numerator by differences of cosines we easily find

$$K_n(t)=\frac{1}{n+1}\,\frac{1-\cos(n+1)t}{(2\sin\frac12 t)^2}=\frac{2}{n+1}\left\{\frac{\sin\frac12(n+1)t}{2\sin\frac12 t}\right\}^2. \tag{3·2}$$

Thus *the* (C, 1) *kernel is positive.*

We shall from now on always use the symbol $K_n(t)$ for the (C, 1) kernel (and later $\tilde K_n$ for the corresponding conjugate kernel). K_n is also called the *Fejér* kernel. K_n has the properties:

$$\text{(A)}\quad \frac{1}{\pi}\int_{-\pi}^{\pi} K_n(t)\,dt=1; \qquad \text{(B)}\quad K_n(t)\geqslant 0;$$

$$\text{(C)}\quad \mu_n(\delta)\to 0 \quad \text{for each} \quad 0<\delta\leqslant\pi,$$

where
$$\mu_n(\delta) = \max_{\delta \leqslant t \leqslant \pi} K_n(t).$$

Condition (A) follows from the corresponding property of the D_ν (see (3·1)), and (C) from the inequality
$$\mu_n(\delta) \leqslant 1/2(n+1)\sin^2 \tfrac{1}{2}\delta.$$

Hence, with the terminology of the previous section, *the kernel* $K_n(t)$ *is positive and satisfies condition* (C). In the following theorem, which is a consequence of (2·16) and (2·21),
$$\sigma_n(x) = \sigma_n(x; f) = \frac{2}{\pi(n+1)} \int_{-\pi}^{\pi} f(x+t) \left\{ \frac{\sin \frac{1}{2}(n+1)t}{2 \sin \frac{1}{2}t} \right\}^2 dt \tag{3·3}$$
denotes the (C, 1) mean of $S[f]$, a notation we adhere to from now on.

(3·4) THEOREM OF FEJÉR. *At every point* x_0 *at which the limits* $f(x_0 \pm 0)$ *exist (and, if both are infinite, are of the same sign) we have*
$$\sigma_n(x_0) \to \tfrac{1}{2}\{f(x_0+0) + f(x_0-0)\}.$$

In particular, $\sigma_n(x_0) \to f(x_0)$ *at every point of continuity of* f. *The convergence of the* σ_n *is uniform over every closed interval of points of continuity. In particular,* $\sigma_n(x)$ *converges uniformly to* $f(x)$ *if* f *is everywhere continuous.*

If $m \leqslant f(x) \leqslant M$ *for all* x, *then*
$$m \leqslant \sigma_n(x) \leqslant M \quad (n = 0, 1, \ldots). \tag{3·5}$$

Since $K_n(t)$ is zero only at a finite number of points, we easily see that, if $f \not\equiv$ constant, (3·5) can be replaced by the stronger inequality
$$m < \sigma_n(x) < M. \quad (\text{If } f \equiv C, \text{ then } \sigma_n(x) = C \text{ for all } x \text{ and } n.)$$

The theorem of Fejér has a number of important applications, some of which we now give.

If $S[f]$ *converges at a point* x_0 *of continuity of* f, *then its sum must necessarily be* $f(x_0)$. More generally, *if* $S[f]$ *converges at a point* x_0 *of simple discontinuity of* f *then its sum is* $s = \frac{1}{2}\{f(x_0+0) + f(x_0-0)\}$.

For at x_0 the series is certainly summable (C, 1) to s and so, if it converges, its sum must be s.

A similar argument shows that *if at* x_0 *the function* f *is continuous or has a simple discontinuity, the number* $\frac{1}{2}\{f(x_0+0)+f(x_0-0)\}$ *is contained between the limits of indetermination of* $\{S_n(x_0; f)\}$.

If $\tilde{S}[f]$ *is a Fourier series,* $\tilde{S}[f] = S[g]$, *then* $f + ig$ *cannot have a simple discontinuity at any point.* For if, for example, $f(x_0 \pm 0)$ exist, are finite and different, and if, say, $f(x_0+0) - f(x_0-0) > 0$, then, by Chapter II, (8·13), $\tilde{S}_n(x_0; f) \to -\infty$, and so also $\tilde{\sigma}_n(x_0; f) = \sigma_n(x_0; g) \to -\infty$, which is impossible since g is bounded near x_0. In particular, *if both* f *and* g *are of bounded variation, they are continuous.*

The trigonometric system is complete (Chapter I, § 6). For if all coefficients of a continuous function f are zero, the $\sigma_n(x; f)$ vanish identically, and so does $f(x) = \lim \sigma_n(x; f)$. For f discontinuous, we apply the same argument as in Chapter I, § 6, or use theorem (3·9) below.

(3·6) THEOREM OF WEIERSTRASS. *If f is periodic and continuous, then for every $\epsilon>0$ there is a trigonometric polynomial $T(x)$ such that $|f(x)-T(x)|<\epsilon$ for all x.*

We may take $T(x)=\sigma_n(x;f)$, with n sufficiently large.

(3·7) THEOREM. *If $f(x)$ is bounded and has Fourier coefficients $O(1/n)$ (in particular, if f is of bounded variation) the partial sums of $S[f]$ are uniformly bounded.*

For the σ_n are uniformly bounded and the assumption about the coefficients implies (see (1·25)) that the $s_n-\sigma_n$ are uniformly bounded.

If we use Theorem (1·26) and the fact that the coefficients of a function of bounded variation are $O(1/n)$ (cf. also the Remark at the end of § 1), the Dirichlet-Jordan theorem (8·1) of Chapter II becomes a corollary of Fejér's.

Theorem (8·6) of Chapter II may be generalized as follows:

(3·8) THEOREM. *Suppose that the Fourier coefficients of f are $O(1/n)$ and that x_0 is a point of continuity of f. Then $S[f]$ converges uniformly at x_0.*

This is a consequence of the general fact that, if a series $\Sigma u_n(x)$ is uniformly summable (C, 1) at x_0 to sum $f(x_0)$ (that $S[f]$ is uniformly summable (C, 1) at every point of continuity of f follows from Theorem (2·30)), and if the $u_n(x)$ are uniformly $O(1/n)$, then the series converges uniformly at x_0. For with the notation of the proof of Theorem (1·26) we have $|\sigma_{n,k}(x)-s_n(x)|<A\epsilon$ for all x. Since $\sigma_n(x)$ converges uniformly at x_0 to limit $f(x_0)$, the same holds for $\sigma_{n,k}(x)$, after (1·30), and we have $|\sigma_{n,k}(x)-f(x_0)|<\epsilon$ for $|x-x_0|<\delta$ and $n>n_0$. For such n and x we have $|s_n(x)-f(x_0)|<(A+1)\epsilon$ and (3·8) is established.

(3·9) THEOREM OF LEBESGUE. *$S[f]$ is summable (C, 1) to $f(x)$ at every point x where $\Phi_x(h)=o(h)$ (in particular, almost everywhere).*

We note that

$$K_n(t)<n+1, \quad K_n(t)\leqslant\frac{A}{(n+1)t^2} \quad (0<t\leqslant\pi;\ A \text{ an absolute const.}), \tag{3·10}$$

the first inequality following from (3·1) and the estimate $|D_\nu|\leqslant\nu+\frac{1}{2}<n+1$, and the second from (3·2). The first inequality (3·10) is suitable for t not too large in comparison with $1/n$, the second for the remaining t. Applying this to the formula

$$\sigma_n(x)-f(x)=\frac{2}{\pi}\int_0^\pi\phi_x(t)\,K_n(t)\,dt \tag{3·11}$$

(cf. (2·15)), we see that $|\sigma_n(x)-f(x)|$ is majorized by

$$\frac{2}{\pi}\int_0^\pi|\phi_x(t)|\,K_n(t)\,dt\leqslant\frac{2(n+1)}{\pi}\int_0^{1/n}|\phi_x(t)|\,dt+\frac{2A}{\pi}\int_{1/n}^\pi\frac{|\phi_x(t)|}{(n+1)t^2}dt=P+Q. \tag{3·12}$$

Clearly,
$$P\leqslant(n+1)\,\Phi(1/n)\leqslant 2n\Phi(1/n)\to 0, \tag{3·13}$$

and integrating by parts we find that Q does not exceed

$$\frac{2A}{\pi(n+1)}[\Phi(t)t^{-2}]_{1/n}^{\pi}+\frac{4A}{\pi(n+1)}\int_{1/n}^\pi\frac{\Phi(t)}{t^3}dt\leqslant\frac{2A}{\pi^3(n+1)}\Phi(\pi)+\frac{1}{n}\int_{1/n}^\pi o\left(\frac{1}{t^2}\right)dt=o(1). \tag{3·14}$$

Thus $P+Q=o(1)$ and the theorem is proved.

As an application we obtain a new proof of Parseval's formula for the trigonometric system (Chapter II, § 1). For let $f \in L^2$. Then

$$\frac{1}{2\pi}\int_0^{2\pi} |\sigma_n(x;f)|^2\,dx = \sum_{\nu=-n}^{+n}\left(1-\frac{|\nu|}{n+1}\right)^2 |c_\nu|^2 \leqslant \sum_{\nu=-\infty}^{+\infty} |c_\nu|^2,$$

and since $\sigma_n \to f$ almost everywhere, Fatou's lemma (Chapter I, (11·2)) gives

$$\frac{1}{2\pi}\int_0^{2\pi} |f|^2\,dx \leqslant \sum_{\nu=-\infty}^{+\infty} |c_\nu|^2.$$

This combined with the opposite inequality of Bessel gives Parseval's formula.

The following theorem completes (3·4):

(3·15) THEOREM. *If* $f \in \Lambda_\alpha$, $0<\alpha<1$, *then* $\sigma_n(x)-f(x)=O(n^{-\alpha})$ *uniformly in* x. *If* $f\in\Lambda_*$ *(in particular, if* $f\in\Lambda_1$*), then* $\sigma_n(x)-f(x)=O(n^{-1}\log n)$.

For in the majorant (3·12) for $|\sigma_n - f|$ we have $\phi(t)=O(t^\alpha)$, which immediately gives $P=O(n^{-\alpha})$ or $P=O(n^{-1})$, and $Q=O(n^{-\alpha})$ or $Q=O(n^{-1}\log n)$, according as $f\in\Lambda_\alpha$ or $f\in\Lambda_*$.

A slight generalization of the first part of (3·15) will be needed later.

(3·16) THEOREM. *Let* $\omega^*(t)$ *be a non-negative and increasing function defined in a right-hand neighbourhood of* $t=0$. *Suppose that* $\omega^*(t)\,t^{-\alpha}$ *is decreasing for some* α, $0<\alpha<1$. *Let* $\omega(t)$ *be the modulus of continuity for a periodic* f. *Then, if* $\omega(t)=O(\omega^*(t))$ *as* $t\to+0$, *we have* $\sigma_n - f = O(\omega^*(1/n))$. *Similarly,* $\omega(t)=o(\omega^*(t))$ *implies*

$$\sigma_n - f = o(\omega^*(1/n)).$$

Consider, for example, the 'o' case. Without loss of generality we may suppose that $\omega^*(t)$ is defined and satisfies the required conditions in $(0,\pi)$. For if the initial interval of definition is $(0,\epsilon)$, with $0<\epsilon<\pi$, it is enough to set $\omega^*(t)=\omega^*(\epsilon)$ for $\epsilon\leqslant t\leqslant\pi$. As in (3·12), we consider the terms P and Q. Clearly,

$$\tfrac{1}{2}\pi P \leqslant (n+1)\int_0^{1/n}\omega\left(\frac{1}{n}\right)dt = o\left(\omega^*\left(\frac{1}{n}\right)\right),$$

$$\tfrac{1}{2}\pi Q \leqslant \frac{A}{n+1}\int_{1/n}^{\pi}\frac{\omega(t)}{t^2}\,dt = \frac{A}{n+1}\int_{1/n}^{\pi} o\left\{\frac{\omega^*(t)}{t^\alpha t^{2-\alpha}}\right\}dt$$

$$\leqslant \frac{A}{n+1}\,\frac{\omega^*(1/n)}{(1/n)^\alpha}\int_{1/n}^{\pi} o(t^{\alpha-2})\,dt = \frac{A}{n+1}\,\frac{\omega^*(1/n)}{(1/n)^\alpha}\,o(n^{1-\alpha}) = o\left(\omega^*\left(\frac{1}{n}\right)\right),$$

which completes the proof.

We turn to the (C, 1) summability of $\tilde{S}[f]$. To avoid repetition we state once for all that in taking arithmetic (or any linear) means of a trigonometric series we shall always take into account the constant term with which the series begins, even if that term (as in $\tilde{S}[f]$ or $S'[f]$) happens to be zero (cf. (2·5)).

The conjugate Fejér kernel is

$$\tilde{K}_n(t) = \frac{1}{n+1}\sum_{\nu=0}^{n}\tilde{D}_\nu(t) = \tfrac{1}{2}\cot\tfrac{1}{2}t - \frac{1}{n+1}\sum_{\nu=0}^{n}\frac{\cos(\nu+\frac{1}{2})t}{2\sin\frac{1}{2}t}$$

$$= \tfrac{1}{2}\cot\tfrac{1}{2}t - \frac{1}{n+1}\,\frac{\sin(n+1)t}{(2\sin\frac{1}{2}t)^2}, \qquad \textbf{(3·17)}$$

by an argument similar to the one used to prove (3·2). The inequality

$$\sin(n+1)t \leqslant (n+1)\sin t$$

applied to
$$\tilde{K}_n(t) = \frac{(n+1)\sin t - \sin(n+1)t}{(n+1)(2\sin\frac{1}{2}t)^2}$$
gives
$$\tilde{K}_n(t) > 0 \quad \text{for} \quad 0 < t < \pi,\ n = 1, 2, \ldots, \tag{3·18}$$
so that $\tilde{K}_n(t)\,\mathrm{sign}\,t \geq 0$ in $(-\pi, \pi)$.

The (C, 1) means of $\tilde{S}[f]$ are given by the formula
$$\tilde{\sigma}_n(x) = \tilde{\sigma}_n(x; f) = -\frac{1}{\pi}\int_{-\pi}^{\pi} f(x+t)\,\tilde{K}_n(t)\,dt = -\frac{2}{\pi}\int_0^{\pi} \psi_x(t)\,\tilde{K}_n(t)\,dt. \tag{3·19}$$

It is clear that if $\tilde{f}$ is integrable and $\tilde{S}[f] = S[\tilde{f}]$, then $\tilde{\sigma}_n(x; f) = \sigma_n(x; \tilde{f})$.

(3·20) THEOREM. *At every point at which* $\Psi_x(h) = o(h)$ *(in particular, almost everywhere) we have*
$$\tilde{\sigma}_n(x) - \tilde{f}(x; 1/n) \to 0. \tag{3·21}$$

For let $\tilde{K}_n(t) = \frac{1}{2}\cot\frac{1}{2}t - H_n(t)$. The inequalities
$$|\tilde{K}_n(t)| \leq \tfrac{1}{2}n, \quad |H_n(t)| \leq \frac{A}{(n+1)t^2} \quad (0 < t \leq \pi) \tag{3·22}$$
are immediate (see (3·17)). By (3·19) and (3·17), $|\tilde{\sigma}_n(x) - \tilde{f}(x; 1/n)|$ does not exceed
$$\frac{2}{\pi}\left|\int_0^{1/n} \psi(t)\,\tilde{K}_n(t)\,dt - \int_{1/n}^{\pi} \psi(t)\,H_n(t)\,dt\right|$$
$$\leq \frac{n}{\pi}\int_0^{1/n} |\psi(t)|\,dt + \frac{2A}{\pi(n+1)}\int_{1/n}^{\pi} \frac{|\psi(t)|}{t^2}\,dt = P^* + Q^*,$$
and an argument similar to (3·13) and (3·14) gives $P^* + Q^* = o(1)$. This proves the theorem.

Suppose that $1/(n+1) \leq h \leq 1/n$. Since $\tilde{f}(x; h) - \tilde{f}(x; 1/n) \to 0$ at every point at which $\Psi_x(h) = o(h)$ (see Chapter II, (11·8)), we conclude that *at such points the summability* (C, 1) *of* $\tilde{S}[f]$ *is equivalent to the existence of* $\tilde{f}(x)$. It will be shown later (see Chapter IV, §3, and Chapter VII, §1) that $\tilde{f}(x)$ exists almost everywhere for any integrable f. Assuming this we can state the following:

(3·23) THEOREM. $\tilde{S}[f]$ *is summable* (C, 1) *to sum* $\tilde{f}(x)$ *almost everywhere.*

It is of interest to consider the integral (C, 1) means of $S[f]$ and $\tilde{S}[f]$ (see p. 83). Returning to the formulae (7·2) and (7·3) of Chapter II, we find that, except when ω is an integer, the right-hand sides there are the sum functions for $S[f]$ and $\tilde{S}[f]$. The left-hand sides are uniformly bounded for ω confined to an arbitrary finite interval, and converge uniformly except in the neighbourhood of integral values of ω. Hence if we integrate the equations with respect to ω, we can interchange the order of integrations on the left, obtaining the formulae
$$\sum_{\nu \leq \omega}\left(1 - \frac{\nu}{\omega}\right)A_\nu(x) = \frac{1}{\pi}\int_{-\infty}^{+\infty} f(x+t)\,\frac{2\sin^2\frac{1}{2}\omega t}{\omega t^2}\,dt, \tag{3·24}$$
$$\sum_{\nu \leq \omega}\left(1 - \frac{\nu}{\omega}\right)B_\nu(x) = -\frac{1}{\pi}\int_{-\infty}^{+\infty} f(x+t)\left\{\frac{1}{t} - \frac{\sin\omega t}{\omega t^2}\right\}dt, \tag{3·25}$$
analogous to (3·3) and (3·19). The left-hand sides here are continuous functions of ω and the first integral converges absolutely.

4. Convergence factors

In Chapter VIII we shall see that a Fourier series may diverge almost everywhere. We may therefore ask about *convergence factors* of the Fourier series, that is to say sequences $\{\lambda_\nu\}$ such that, for each Fourier series $\Sigma(a_\nu \cos \nu x + b_\nu \sin \nu x)$, the series

$$\tfrac{1}{2}a_0\lambda_0 + \sum_{\nu=1}^{\infty} (a_\nu \cos \nu x + b_\nu \sin \nu x)\,\lambda_\nu$$

converges almost everywhere.

A sequence of numbers $\lambda_0, \lambda_1, \ldots$ is said to be *convex* if $\Delta^2\lambda_\nu \geqslant 0$ for all ν, where

$$\Delta\lambda_\nu = \lambda_\nu - \lambda_{\nu+1}, \quad \Delta^2\lambda_\nu = \Delta\lambda_\nu - \Delta\lambda_{\nu+1}.$$

Geometrically, this amounts to saying that the polygonal line with vertices at the points (ν, λ_ν) is convex. We shall show that if $\{\lambda_\nu\}$ is convex and bounded, then it is non-increasing. By hypothesis, $\Delta\lambda_\nu$ is non-increasing. It cannot be negative for any value of ν, for then it would be less than a negative constant for all subsequent values of ν, which would imply that $\lambda_\nu \to +\infty$, contrary to hypothesis. Thus $\Delta\lambda_n = \lambda_n - \lambda_{n+1} \geqslant 0$ for all n, so that

$$\lambda_n \geqslant \lambda_{n+1} \to \lambda > -\infty.$$

In the equation

$$\lambda_0 - \lambda = \Delta\lambda_0 + \Delta\lambda_1 + \ldots$$

the terms on the right are non-increasing, and so, by the classical theorem of Abel, $n\Delta\lambda_n \to 0$. Taking this into account and summing the series $1.\Delta\lambda_0 + 1.\Delta\lambda_1 + \ldots$ by parts we get:

(4·1) THEOREM. *If $\{\lambda_n\}$ is convex and bounded, then $\{\lambda_n\}$ is non-increasing, $n\Delta\lambda_n \to 0$, and the series*

$$\sum_{n=0}^{\infty} (n+1)\,\Delta^2\lambda_n \tag{4·2}$$

converges to sum $\lambda_0 - \lim \lambda_n$.

It is geometrically obvious that *if a function $\lambda(x)$ is convex, the sequence $\{\lambda_n\} = \{\lambda(n)\}$ is convex.*

In particular, if we take $\lambda_n = 1/\log n$ for $n = 2, 3, \ldots$ and choose λ_0, λ_1 suitably, $\{\lambda_n\}$ will be convex.

(4·3) THEOREM. *Let s_n and σ_n be the partial sums and the* (C, 1) *means of a series $u_0 + u_1 + \ldots$. If σ_n converges, and if $s_n = o(\mu_n)$, where $\{1/\mu_n\}$ is convex and tends to* 0, *then $u_0\mu_0^{-1} + u_1\mu_1^{-1} + \ldots$ converges.*

For applying summation by parts twice to the nth partial sum of the last series we find that it is equal to

$$\sum_{k=0}^{n-2} (k+1)\,\sigma_k \Delta^2 \frac{1}{\mu_k} + n\sigma_{n-1}\Delta\frac{1}{\mu_{n-1}} + s_n \frac{1}{\mu_n} \to \sum_{k=0}^{\infty} (k+1)\,\sigma_k \Delta^2 \frac{1}{\mu_k}.$$

Take $\mu_n = \log n$ for $n \geqslant 2$. From (4·3), (3·9), (3·23) and Theorem (11·9) of Chapter II, we deduce:

(4·4) THEOREM. *If a_k, b_k are the Fourier coefficients of a function, both series*

$$\sum_{k=2}^{\infty} \frac{a_k \cos kx + b_k \sin kx}{\log k}, \quad \sum_{k=2}^{\infty} \frac{a_k \sin kx - b_k \cos kx}{\log k} \tag{4·5}$$

converge almost everywhere.

The result obviously holds if $\log k$ here is replaced by k^α, $\alpha > 0$.

The first series (4·5) converges at every point where $\Phi_x(h) = o(h)$, in particular at every point of continuity of f. *If f is continuous in (a, b), the series is uniformly convergent in every $(a+\epsilon, b-\epsilon)$, $\epsilon > 0$.*

5. Summability (C, α)

(5·1) THEOREM OF M. RIESZ. *The theorems* (3·4) *of Fejér* (*except for the last sentence*) *and* (3·9) *of Lebesgue hold if summability* (C, 1) *is replaced by* (C, α), $\alpha > 0$.

Let $K_n^\alpha(t)$ denote the (C, α) kernel and $\sigma_n^\alpha(x) = \sigma_n^\alpha(x; f)$ the (C, α) means of $S[f]$. Then

$$K_n^\alpha(t) = \sum_{\nu=0}^{n} A_{n-\nu}^{\alpha-1} D_\nu(t) / A_n^\alpha, \tag{5·2}$$

$$\sigma_n^\alpha(x) = \frac{1}{\pi}\int_{-\pi}^{\pi} f(x+t)\, K_n^\alpha(t)\, dt, \tag{5·3}$$

$$\sigma_n^\alpha(x) - f(x) = \frac{2}{\pi}\int_0^{\pi} \phi_x(t)\, K_n^\alpha(t)\, dt, \tag{5·4}$$

the last equation being a consequence of the validity of condition (A) (p. 85) for $K_n^\alpha(t)$.

It is enough to consider the case $0 < \alpha < 1$. We shall show that then

$$|K_n^\alpha(t)| < n+1 \leqslant 2n, \quad |K_n^\alpha(t)| \leqslant A_\alpha n^{-\alpha} t^{-(\alpha+1)} \tag{5·5}$$

for $n = 1, 2, \ldots$, $0 < t \leqslant \pi$, with A_α depending on α only.

These inequalities are analogous to (3·10) and reduce to the latter for $\alpha = 1$. Once (5·5) is established, the proof of the extended (3·9) goes as before. Similarly, to extend (3·4) it is enough to show that the kernel $K_n^\alpha(t)$ is quasi-positive and satisfies condition (C) (p. 86). Both these facts are corollaries of (5·5): for

$$\int_0^\pi |K_n^\alpha(t)|\, dt < 2n\int_0^{1/n} dt + A_\alpha n^{-\alpha}\int_{1/n}^{\pi} t^{-\alpha-1}\, dt < 2 + A_\alpha/\alpha,$$

and
$$\max_{\delta \leqslant t \leqslant \pi} |K_n^\alpha(t)| \leqslant A_\alpha n^{-\alpha} \delta^{-(\alpha+1)} \to 0.$$

It remains to prove (5·5). The first part follows from (5·2) and the estimate

$$|D_\nu| \leqslant \nu + \tfrac{1}{2} < n+1.$$

For the second we have, from (5·2),

$$K_n^\alpha(t) = \frac{1}{2A_n^\alpha \sin \frac{1}{2}t}\, \mathscr{I} \sum_{\nu=0}^{n} A_{n-\nu}^{\alpha-1} e^{i(\nu+\frac{1}{2})t} = \mathscr{I}\, \frac{e^{i(n+\frac{1}{2})t}}{2A_n^\alpha \sin\frac{1}{2}t} \sum_{\nu=0}^{n} A_\nu^{\alpha-1} e^{-i\nu t}$$

$$= \mathscr{I}\left\{\frac{e^{i(n+\frac{1}{2})t}}{2A_n^\alpha \sin\frac{1}{2}t}\left[(1-e^{-it})^{-\alpha} - \sum_{\nu=n+1}^{\infty} A_\nu^{\alpha-1} e^{-i\nu t}\right]\right\}. \tag{5·6}$$

Since $A_\nu^{\alpha-1}$ decreases monotonically to 0, the last series converges for $0 < t \leqslant \pi$ and the modulus of its sum is $\leqslant 2A_{n+1}^{\alpha-1} |1 - e^{-it}|^{-1}$ (see Chapter I, (2·2)). So, since $|\mathscr{I} z| \leqslant |z|$, $|K_n^\alpha(t)|$ is majorized by

$$\left\{(2\sin\tfrac{1}{2}t)^{-\alpha-1}\frac{1}{A_n^\alpha} + \frac{2A_{n+1}^{\alpha-1}}{A_n^\alpha}(2\sin\tfrac{1}{2}t)^{-2}\right\} \leqslant A_\alpha\{n^{-\alpha}t^{-\alpha-1} + n^{-1}t^{-2}\} \tag{5·7}$$

for $0<t\leqslant\pi$. If $nt\geqslant 1$, then
$$nt^2=(nt)^{1-\alpha}\,n^{\alpha}t^{\alpha+1}\geqslant n^{\alpha}t^{\alpha+1},$$
the right-hand side of (5·7) does not exceed $2A_\alpha n^{-\alpha}t^{-\alpha-1}$, and the second part of (5·5) follows. For $0<t\leqslant 1/n$ the second part of (5·5) is a consequence of the first. This completes the proof of (5·1).

Let $\tilde{\sigma}_n^\alpha(x)=\tilde{\sigma}_n^\alpha(x;f)$ be the (C, α) means of $\tilde{S}[f]$. The theorem which follows is an extension of (3·20) and (3·23) to summability (C, α).

(5·8) THEOREM. *Let* $0<\alpha<1$. *At every point* x *at which* $\Psi_x(h)=o(h)$ *we have*
$$\tilde{\sigma}_n^\alpha(x)-\tilde{f}(x;\,1/n)\to 0.$$
In particular, $\tilde{S}[f]$ *is almost everywhere summable* (C, α) *to sum* $\tilde{f}(x)$.

The proof is analogous to that of (3·20). Let $\tilde{K}_n^\alpha(t)$ be the conjugate (C, α) kernel. Then
$$\tilde{\sigma}_n^\alpha(x)=-\frac{2}{\pi}\int_0^\pi \psi_x(t)\,\tilde{K}_n^\alpha(t)\,dt, \tag{5·9}$$
$$\tilde{K}_n^\alpha(t)=\sum_{\nu=0}^{n} A_{n-\nu}^{\alpha-1}\tilde{D}_\nu(t)/A_n^\alpha, \tag{5·10}$$
$$\tilde{K}_n^\alpha(t)=\tfrac{1}{2}\cot\tfrac{1}{2}t-\frac{1}{A_n^\alpha}\sum_{\nu=0}^{n}A_{n-\nu}^{\alpha-1}\frac{\cos(\nu+\frac{1}{2})t}{2\sin\frac{1}{2}t}=\tfrac{1}{2}\cot\tfrac{1}{2}t-H_n^\alpha(t), \tag{5·11}$$
say. We show that (for $0<\alpha<1$)
$$|\tilde{K}_n^\alpha(t)|\leqslant n,\quad |H_n^\alpha(t)|\leqslant A_\alpha n^{-\alpha}t^{-(\alpha+1)}\quad (n^{-1}\leqslant t\leqslant\pi). \tag{5·12}$$
The first inequality here follows from (5·10) and the estimate $|\tilde{D}_\nu|\leqslant\nu\leqslant n$. For the second inequality we note that for $H_n^\alpha(t)$ we have a formula analogous to (5·6) with $\mathscr{I}$ replaced by $\mathscr{R}$, and the previous argument is still applicable.

To prove (5·8) we write (5·9) in the form
$$\tilde{\sigma}_n^\alpha(x)-\tilde{f}(x;\,1/n)=-\frac{2}{\pi}\int_0^{1/n}\psi_x(t)\,\tilde{K}_n^\alpha(t)\,dt+\frac{2}{\pi}\int_{1/n}^{\pi}\psi_x(t)\,H_n^\alpha(t)\,dt, \tag{5·13}$$
so that, as in the proof of (3·20),
$$|\tilde{\sigma}_n^\alpha(x)-\tilde{f}(x;\,1/n)|\leqslant\frac{2}{\pi}n\int_0^{1/n}|\psi_x(t)|\,dt+\frac{2A_\alpha}{\pi n^\alpha}\int_{1/n}^{\pi}\frac{|\psi_x(t)|}{t^{\alpha+1}}\,dt\to 0,$$
and the theorem follows.

For later applications we shall need a refinement of (5·6), namely, *if* $-1<\alpha<1$ *we have*
$$K_n^\alpha(t)=\frac{1}{A_n^\alpha}\frac{\sin[(n+\frac{1}{2}+\frac{1}{2}\alpha)t-\frac{1}{2}\pi\alpha]}{(2\sin\frac{1}{2}t)^{\alpha+1}}+\frac{2\theta\alpha}{n(2\sin\frac{1}{2}t)^2}\quad(|\theta|\leqslant 1). \tag{5·14}$$

Applying repeated summation by parts to the last series in (5·6) we obtain more and more accurate approximations for $K_n^\alpha(t)$. For example,
$$K_n^\alpha(t)=\mathscr{I}\left\{\frac{e^{i(n+\frac{1}{2})t}}{A_n^\alpha(2\sin\frac{1}{2}t)}\left[\frac{1}{(1-e^{-it})^\alpha}-A_{n+1}^{\alpha-1}\frac{e^{-i(n+1)t}}{1-e^{-it}}-\sum_{\nu=n+1}^{\infty}A_{\nu+1}^{\alpha-2}\frac{e^{-i(\nu+1)t}}{1-e^{-it}}\right]\right\}$$
$$=\frac{1}{A_n^\alpha}\frac{\sin[(n+\frac{1}{2}+\frac{1}{2}\alpha)t-\frac{1}{2}\pi\alpha]}{(2\sin\frac{1}{2}t)^{\alpha+1}}+\frac{\alpha}{n+1}\frac{1}{(2\sin\frac{1}{2}t)^2}+\frac{2\theta\alpha(1-\alpha)}{(n+1)(n+2)(2\sin\frac{1}{2}t)^3}. \tag{5·15}$$

Similar formulae hold for $\tilde{K}_n^\alpha(t)$. We may add that the estimates (5·5) and (5·12) hold also for $-1<\alpha\leqslant 0$.

6. Abel summability

Let a_ν, b_ν be the Fourier coefficients of f. The Abel (or, simply, the A) means of $S[f]$ and $\tilde{S}[f]$ are the functions

$$\left.\begin{aligned} f(r,x) &= \tfrac{1}{2}a_0 + \sum_{\nu=1}^{\infty}(a_\nu \cos \nu x + b_\nu \sin \nu x)\, r^\nu \\ \tilde{f}(r,x) &= \sum_{\nu=1}^{\infty}(a_\nu \sin \nu x - b_\nu \cos \nu x)\, r^\nu \end{aligned}\right\} \quad (0 \leqslant r < 1), \tag{6·1}$$

and we wish to investigate their limits as $r \to 1$. Since $a_\nu, b_\nu \to 0$, the series converge absolutely and uniformly for $0 \leqslant r \leqslant 1-\delta$, $\delta > 0$. Thus $f(r,x)$ and $\tilde{f}(r,x)$ (a notation which we shall use systematically) are continuous functions of the point re^{ix} for $r<1$.

The Abel means of the even series $\frac{1}{2} + \Sigma \cos \nu t$ and the odd series $\Sigma \sin \nu t$ are

$$P(r,t) = \tfrac{1}{2} + \sum_{\nu=1}^{\infty} r^\nu \cos \nu t = \frac{1}{2}\,\frac{1-r^2}{1-2r\cos t + r^2}, \tag{6·2}$$

$$Q(r,t) = \sum_{\nu=1}^{\infty} r^\nu \sin \nu t = \frac{r \sin t}{1-2r\cos t + r^2} \tag{6·3}$$

(see Chapter I, § 1). They are called the *Poisson kernel* and the *Poisson conjugate kernel* respectively. The standard formulae (2·12) and (2·13) (where now the continuous variable r plays the role of the former n) will now be

$$f(r,x) = \frac{1}{\pi}\int_{-\pi}^{\pi} f(x+t)\, P(r,t)\, dt = \frac{1}{\pi}\int_{-\pi}^{\pi} f(t)\, P(r,t-x)\, dt, \tag{6·4}$$

$$\tilde{f}(r,x) = -\frac{1}{\pi}\int_{-\pi}^{\pi} f(x+t)\, Q(r,t)\, dt = -\frac{1}{\pi}\int_{-\pi}^{\pi} f(t)\, Q(r,t-x)\, dt. \tag{6·5}$$

The right-hand sides here are usually called the *Poisson integral* and the *conjugate Poisson integral* of f. Thus the expressions 'Abel mean of $S[f]$' and 'Poisson integral of f' are synonymous.

The denominator $\quad \Delta(r,t) = 1 - 2r\cos t + r^2 \quad (0 \leqslant r < 1)$

of P and Q is positive for all t. It follows that

$$\left.\begin{aligned} P(r,t) &> 0 \quad \text{for all } t, \\ Q(r,t) &> 0 \quad \text{for } 0 < t < \pi. \end{aligned}\right\} \tag{6·6}$$

Hence P is a positive kernel. For fixed r, its maximum and minimum are attained at $t=0$ and $t=\pi$ respectively. Thus

$$\frac{1}{2}\,\frac{1-r}{1+r} \leqslant P(r,t) \leqslant \frac{1}{2}\,\frac{1+r}{1-r}. \tag{6·7}$$

It is sometimes convenient to use the inequality

$$P(r,t) \leqslant A\,\frac{\delta}{\delta^2 + t^2} \quad (\delta = 1-r,\ |t| \leqslant \pi), \tag{6·8}$$

in which A is an absolute constant. For $0 \leqslant r \leqslant \frac{1}{2}$ it is immediate, since both $P(r,t)$ (see (6·7)) and $\delta/(\delta^2+t^2)$ are then contained between two positive constants. For $\frac{1}{2} \leqslant r < 1$,

$$P(r,t) = \frac{1}{2}\,\frac{(1+r)(1-r)}{(1-r)^2 + 4r\sin^2\frac{1}{2}t} < \frac{\delta}{\delta^2 + 4\,.\,\frac{1}{2}(\pi^{-1}t)^2} < \tfrac{1}{2}\pi^2\,\frac{\delta}{\delta^2+t^2},$$

and (6·8) holds again. In particular, (6·7) and (6·8) imply

$$P(r,t) \leqslant \frac{1}{\delta}, \quad P(r,t) \leqslant \frac{A\delta}{t^2} \quad (0 < t \leqslant \pi,\ 0 \leqslant r < 1), \tag{6·9}$$

inequalities similar to those satisfied by Fejér's kernel (see (3·10)) if we replace δ by $1/(n+1)$.

The kernel $P(r,t)$ is positive, satisfies condition (A), (p. 85), i.e.

$$\frac{1}{\pi}\int_{-\pi}^{\pi} P(r,t)\,dt = 1 \tag{6·10}$$

(as is seen by termwise integration of the series in (6·2)), and also condition (C) (as is seen from the second inequality (6·9)). Thus all the assumptions which led us to Theorem (3·4) from (2·21) are valid, and we have the following result:

(6·11) THEOREM. *Theorem* (3·4) *of Fejér holds if we replace the* (C, 1) *means of* $S[f]$ *by the Abel means.*

Of course, this result, like some of the results established below, is also a consequence of the theorem of Fejér and of the fact that summability (C, 1) implies summability A for *any* series. But a direct study of summability A of Fourier series is of interest for two reasons. First, summability A of $S[f]$ may hold under weaker conditions for f than summability (C, 1); secondly, summability A of $S[f]$ has special features which are absent in (C, 1). For example, we may consider not only the radial but also the non-tangential, and even unrestricted, limit of $f(r,x)$ as (r,x) tends to a point on the unit circle.

The functions $f(r,x)$ and $\tilde{f}(r,x)$ of (6·1) are the real and imaginary parts of the function

$$\Phi(z) = \tfrac{1}{2}a_0 + \sum_{\nu=1}^{\infty} (a_\nu - ib_\nu)\,z^\nu, \quad z = re^{ix},$$

regular for $|z| < 1$. Thus $f(r,x)$ and $\tilde{f}(r,x)$ are *harmonic*, that is as functions of Cartesian co-ordinates ξ, η they satisfy Laplace's equation

$$u_{\xi\xi} + u_{\eta\eta} = 0.$$

Each real-valued function harmonic in the interior of a circle is the real part of a regular function†. Hence, if $u(r,x)$ is harmonic in the circle $0 \leqslant r < R$, we have

$$u(r,x) = \tfrac{1}{2}a_0 + \sum_{\nu=1}^{\infty} (a_\nu \cos \nu x + b_\nu \sin \nu x)\,r^\nu \quad (0 \leqslant r < R).$$

Let $f(x)$ be a continuous and periodic function and let (r,x) be the polar co-ordinates of a point. Theorem (6·11) asserts that the Poisson integral $f(r,x)$ of $f(x)$ tends uniformly to $f(x)$ as $r \to 1$. In other words, Poisson's integral gives, for the case of a circle, a solution (or indeed, as is shown in the theory of harmonic functions, the unique solution) of the following *problem of Dirichlet*. Given

(i) a plane region D limited by a simple closed curve C,

(ii) a function $f(p)$ defined and continuous for $p \in C$,

find a function $F(p)$ harmonic in D, continuous in $D + C$, and coinciding with $f(p)$ on C. As we shall see below, in the case of the unit circle the Poisson integral solves a more general Dirichlet problem in which $f(p)$ is an arbitrary integrable function.

† See, e.g., Littlewood, *Lectures on the theory of functions*, p. 84.

(6·12) THEOREM. *If $m \leqslant f(x) \leqslant M$ for all x, then*

$$m \leqslant f(r,x) \leqslant M \quad (0 \leqslant r < 1,\ 0 \leqslant x \leqslant 2\pi).$$

In particular, $f(r,x) \geqslant 0$ if $f \geqslant 0$.

If $m \leqslant f(x) \leqslant M$ for $x \in (a,b)$, then for every $\epsilon, \eta > 0$ there is a number r_0 such that

$$m - \epsilon < f(r,x) < M + \epsilon \quad \text{for} \quad r_0 \leqslant r < 1,\ x \in (a+\eta, b-\eta).$$

This is a special case of (2·16) and (2·28).

Since $P(r,x)$ is strictly positive for $r < 1$, it follows that if $m \leqslant f(x) \leqslant M$ in $(0, 2\pi)$ and $f \not\equiv$ const., then we have the sharper estimate

$$m < f(r,x) < M \quad (0 \leqslant r < 1).$$

Let $m(x_0)$ and $M(x_0)$ be the minimum and maximum of f at x_0 (see p. 88).

(6·13) THEOREM. *If L is any path leading from the interior of the unit circle to the point $(1, x_0)$, then the limits of indetermination of $f(r,x)$ as the point (r,x) approaches $(1, x_0)$ along L are contained between $m(x_0)$ and $M(x_0)$.*

For if (r_n, x_n) is any sequence of points, with $r_n < 1$, approaching $(1, x_0)$, then (see (2·30))

$$m(x_0) \leqslant \liminf f(r_n, x_n) \leqslant \limsup f(r_n, x_n) \leqslant M(x_0).$$

A special case of (6·13) asserts that, *if f is continuous at x_0, then $f(r,x)$ tends to $f(x_0)$ along L.*

Suppose that f has at x_0 a discontinuity of the first kind and that

$$f(x_0) = \tfrac{1}{2}\{f(x_0+0) + f(x_0-0)\}.$$

Let

$$d = f(x_0+0) - f(x_0-0)$$

be the jump of f at x_0. Without loss of generality we may suppose that $x_0 = 0$. The function

$$\phi(x) \sim \sum_1^\infty \frac{\sin \nu x}{\nu}$$

has jump π at $x = 0$ (see p. 9). It follows that

$$g(x) = f(x) - \frac{d}{\pi}\phi(x)$$

is continuous at $x = 0$, and $g(0) = f(0)$. Moreover,

$$f(r,x) = g(r,x) + \frac{d}{\pi}\phi(r,x) = g(r,x) + \frac{d}{\pi}\arctan\frac{r \sin x}{1 - r\cos x} \tag{6·14}$$

(Chapter I, § 1). Along any path L leading to $(1,0)$ from the interior of the unit circle we have $\lim g(r,x) = g(0) = f(0)$ so that the behaviour of $f(r,x)$ along L depends on that of the last term in (6·14).

The arctan in (6·14) is the angle, numerically $< \tfrac{1}{2}\pi$ and reckoned clockwise, which the segment joining (r,x) to $(1,0)$ makes with the negatively directed real axis. Hence:

(6·15) THEOREM. *Suppose that $f(x)$ has at x_0 a discontinuity of the first kind and that $f(x_0) = \tfrac{1}{2}\{f(x_0+0) + f(x_0-0)\}$. Let L be any path approaching the point $A(1, x_0)$ from inside* *the unit circle and making at A an angle θ, $-\frac{1}{2}\pi \leqslant \theta \leqslant \frac{1}{2}\pi$, reckoned clockwise, with the radius at A directed inwards. Then* $\lim f(r,x)$ *along L exists and equals*

$$f(x_0)+\frac{d}{\pi}\theta, \quad \textit{where} \quad d=f(x_0+0)-f(x_0-0). \tag{6·16}$$

If L does not have a tangent at A, $f(r,x)$ oscillates finitely as $(r,x)\to(1,x_0)$ along L.

If L is tangent to the unit circle at A, then $\theta=\frac{1}{2}\pi$ or $\theta=-\frac{1}{2}\pi$, and correspondingly (6·16) reduces to $f(x_0+0)$ or $f(x_0-0)$.

Let $\chi(x)$ be the characteristic function of an interval (α,β). The Poisson integral

$$\chi(r,x)=\frac{1}{\pi}\int_\alpha^\beta P(r,t-x)\,dt \tag{6·17}$$

of χ has a simple and useful interpretation.

Let C be the unit circumference with centre 0, z and ζ points inside and on C respectively. The point ζ' at which the ray ζz intersects C will be called *opposite* to ζ, with respect to z. If s is an arc of C, the arc s' consisting of the points opposite to those of s will be called *opposite* to s (with respect to z).

(6·18) THEOREM. *With the notation of* (6·17), $2\pi\chi(r,x)$ *is the length of the arc* $s'=(e^{i\alpha'},e^{i\beta'})$ *opposite to* $s=(e^{i\alpha},e^{i\beta})$ *with respect to* $z=r\,e^{ix}$.

Let s, s' also denote the lengths of s, s'. The angle γ which s subtends at z is $\frac{1}{2}(s+s')$. Clearly it is also

$$\mathscr{I}\int_{e^{i\alpha}}^{e^{i\beta}} d\log(\zeta-z)=\mathscr{I}\int_{e^{i\alpha}}^{e^{i\beta}}\frac{d\zeta}{\zeta-z}=\mathscr{R}\int_\alpha^\beta\frac{dt}{1-r\,e^{i(x-t)}}=\int_\alpha^\beta\{\tfrac{1}{2}+P(r,t-x)\}\,dt.$$

Thus
$$\tfrac{1}{2}(s+s')=\tfrac{1}{2}s+\pi\chi(r,x),$$
from which the theorem follows.

The level curves of $\chi(r,x)$, being the curves on which γ is constant, are those arcs of the circles through $e^{i\alpha}$, $e^{i\beta}$ which are inside C.

7. Abel summability (*cont.*)

From Theorem (3·9) we deduce that $S[f]$ is summable A at every point x for which $\Phi_x(h)=o(h)$, in particular almost everywhere. This result will be superseded by a somewhat stronger one (see (7·9) below). We shall first prove results about summability A of formally differentiated Fourier series.

As in Chapter I, p. 22, if

$$\lim_{h\to+0}\frac{F(x_0+h)-F(x_0-h)}{2h}=D_1F(x_0) \tag{7·1}$$

exists it is called the first symmetric derivative of F at x_0. In the general case the upper and lower limits, as $h\to 0$, of the ratio in (7·1), are called the *upper* and *lower* first symmetric derivatives. We shall denote them by $\overline{D}_1F(x_0)$ and $\underline{D}_1F(x_0)$ respectively. If $F'(x_0)$ exists so does $D_1F(x_0)$, and their values are equal.

(7·2) FATOU'S THEOREM. *Suppose that*

$$F(x)\sim\tfrac{1}{2}A_0+\Sigma(A_\nu\cos\nu x+B_\nu\sin\nu x)$$

and that $D_1F(x_0)$ exists, finite or infinite. Then $S'[F]$ is summable A at x_0 to sum $D_1F(x_0)$, that is,

$$\sum_{\nu=1}^{\infty} \nu(B_\nu \cos \nu x_0 - A_\nu \sin \nu x_0)\, r^\nu \tag{7·3}$$

tends to $D_1F(x_0)$ as $r \to 1$.

More generally, the limits of indetermination of (7·3) as $r \to 1$ are contained between $\underline{D}_1F(x_0)$ and $\overline{D}_1F(x_0)$.

It is enough to prove the second part. If $F(r, x)$ is the Poisson integral of F, then (7·3) is $\{\partial F(r,x)/\partial x\}_{x=x_0}$. From

$$F(r,x) = \frac{1}{\pi}\int_{-\pi}^{\pi} F(t)\, P(r, t-x)\, dt$$

we get $$\frac{1}{r}\left(\frac{\partial F(r,x)}{\partial x}\right)_{x=x_0} = -\frac{1}{\pi r}\int_{-\pi}^{\pi} F(t)\, P'(r, t-x_0)\, dt = \frac{1}{\pi}\int_{-\pi}^{\pi} g(t)\, K(r,t)\, dt, \tag{7·4}$$

where the dash denotes differentiation with respect to t,

$$g(t) = \{F(x_0+t) - F(x_0-t)\}/2\sin t,$$

$$K(r,t) = -r^{-1}P'(r,t)\sin t = (1-r^2)\sin^2 t/\Delta^2(r,t).$$

We note that $K(r, t)$ has the properties (A), (B), (C) of kernels. Property (B) is immediate; so is (C), even for $P'(r,t)$. In order to show that

$$\frac{1}{\pi}\int_{-\pi}^{\pi} K(r,t)\, dt = 1, \tag{7·5}$$

we take $F(x) = \sin x$, $x_0 = 0$. Then $F(r,x) = r\sin x$, $g(t) = 1$ and a comparison of the extreme terms in (7·4) gives (7·5).

Since the maximum and minimum of $g(t)$ at $t=0$ are $\overline{D}_1F(x_0)$ and $\underline{D}_1F(x_0)$, (7·2) would follow from (2·30) (with $x_n = x_0$ for all n) if we knew that $g(t)$ was integrable. The latter is not necessarily true (except when, for example, $\underline{D}_1F(x_0)$ and $\overline{D}_1F(x_0)$ are both finite), but this does not affect the proof, as we see from the following argument. Let $0 < \delta < \pi$. Since $P'(r,t)$ satisfies condition (C), the right-hand side in (7·4) is

$$\frac{1}{\pi}\int_{-\delta}^{\delta} g(t)\, K(r,t)\, dt + o(1).$$

The integral here is contained between the upper and lower bounds of $g(t)$ in $(-\delta, \delta)$ times $\frac{1}{\pi}\int_{-\delta}^{\delta} K\,dt$. But the latter integral tends to 1 as $r \to 1$. Hence the limits of indetermination of (7·3), as $r \to 1$, are contained between the upper and lower bounds of $g(t)$ in $(-\delta, \delta)$. Taking δ arbitrarily small we may replace these bounds by $\overline{D}_1F(x_0)$ and $\underline{D}_1F(x_0)$, and (7·2) is established.

(7·6) THEOREM. *If $F'(x_0)$ exists and is finite, then*

$$\partial F(r,x)/\partial x \to F'(x_0)$$

as (r,x) approaches $(1, x_0)$ non-tangentially.

We may suppose that $x_0 = 0$, $F(0) = 0$. Let us also temporarily suppose that $F'(0) = 0$.

Given any $\epsilon > 0$, let σ be such that $|F(u)| \leqslant \epsilon |u|$ for $|u| \leqslant 2\sigma$. From (7·4) and using the property (c) of $P'(r,t)$ we have

$$\frac{\partial F(r,x)}{\partial x} = -\frac{1}{\pi}\int_{-\pi}^{\pi} F(x+t)\,P'(r,t)\,dt = -\frac{1}{\pi}\int_{-\sigma}^{\sigma} F(x+t)\,P'(r,t)\,dt + o(1) = A + o(1). \quad (7·7)$$

Suppose that $|x| \leqslant \sigma$. Then $|F(x+t)| \leqslant \epsilon |x+t|$ in A and

$$|A| \leqslant \frac{\epsilon}{\pi}\int_{-\sigma}^{\sigma} (|x|+|t|)\,|P'|\,dt$$

$$< -\frac{2\epsilon}{\pi}\int_0^{\pi} (|x|+t)\,P'\,dt < \frac{2\epsilon}{\pi}\left\{|x|\,P(r,0) + \int_0^{\pi} P\,dt\right\} < \frac{2\epsilon}{\pi}\left\{\frac{|x|}{1-r} + \frac{\pi}{2}\right\}. \quad (7·8)$$

The expression in curly brackets remains bounded in a non-tangential approach. Hence, taking ϵ arbitrarily small, we see that (7·7) tends to 0, under the hypothesis $F'(0)=0$.

In the general case, we write

$$F(x) = \{F(x) - F'(0)\sin x\} + F'(0)\sin x.$$

The derivative of the expression in curly brackets at $x=0$ is zero, and for the function $F'(0)\sin x$, whose Poisson integral is $F'(0)\,r\sin x$, the theorem is obvious.

(7·9) THEOREM. *Let $F(x)$ be the indefinite integral of an integrable and periodic f. Then*

(i) *$S[f]$ is summable A to sum $D_1 F(x_0)$ at every point x_0 at which $D_1 F(x_0)$ exists, finite or infinite;*

(ii) *at every point at which $F'(x_0) = f(x_0)$ exists and is finite (in particular almost everywhere) the Poisson integral $f(r,x)$ of f tends to $f(x_0)$ as $(r,x) \to (1,x_0)$ non-tangentially.*

For supposing, as we may, that the constant term of $S[f]$ is zero, we have $S[f] = S'[F]$, $f(r,x) = \partial F(r,x)/\partial x$, and the theorem follows from (7·2) and (7·6). Part (i) here is not a consequence of (3·9), since the condition $\Phi_x(h) = o(h)$ is more stringent than the existence of the symmetric derivative.

It is sometimes important to know the behaviour of the Poisson integral for a tangential approach. The following result, in which for simplicity we take $x_0 = 0$, will be useful later and indicates the type of estimate one can expect.

(7·10) THEOREM. *Suppose that*

$$\left|\frac{1}{h}\int_0^h f(t)\,dt\right| \leqslant M \quad \text{for} \quad |h| \leqslant \pi. \quad (7·11)$$

Then, with $\delta = 1-r$ and denoting by K an absolute constant,

$$|f(r,x)| \leqslant KM\left(1 + \frac{|x|}{\delta}\right), \quad (7·12)$$

$$\left|\int_0^x f(r,u)\,du\right| \leqslant KM\,|x|. \quad (7·13)$$

Suppose that the constant term of $S[f]$ is zero. Then $F(x) = \int_0^x f\,dt$ is periodic. For $f(r,x) = \partial F(r,x)/\partial x$ we have (7·7) with $\sigma = \pi$ and no term $o(1)$. (7·8) with $\epsilon = M$ then shows that we have (7·12) with $K=1$.

In the general case we put $f=f_1+f_2$, where

$$f_2=\frac{1}{2\pi}\int_{-\pi}^{\pi} f\,dt$$

is a constant. Clearly $|f_2|\leqslant M$, and $f_1=f-f_2$ satisfies (7·11) with $2M$ instead of M; also the constant term of $S[f_1]$ is zero. It follows that $f(r,x)=f_1(r,x)+f_2(r,x)$ satisfies (7·12) with $K=3$.

For (7·13) we first suppose as before that $F(x)=\int_0^x f\,dt$ is periodic. From the equation preceding (7·4) we obtain

$$\int_0^x f(r,u)\,du=F(r,x)-F(r,0)=\frac{1}{\pi}\int_{-\pi}^{\pi} F(t)\{P(r,t-x)-P(r,t)\}\,dt,$$

so that the left-hand side of (7·13) does not exceed

$$M\frac{1}{\pi}\int_{-\pi}^{\pi} |t|\,|P(r,t-x)-P(r,t)|\,dt=MI(r,x),$$

say, and it is enough to show that

$$I(r,x)\leqslant K|x| \quad (|x|\leqslant\pi). \tag{7·14}$$

The cases $x>0$ and $x<0$ in (7·13) are equivalent and we may suppose that $0<x\leqslant\pi$, or even $0<x\leqslant\frac{1}{2}\pi$, since otherwise (7·14) is obvious.

Now

$$\pi I(r,x)=\int_{-x}^{x}+\int_{x}^{\pi}+\int_{-\pi+x}^{-x}+\int_{-\pi}^{-\pi+x}=I_1+I_2+I_3+I_4.$$

Clearly,

$$I_1\leqslant x\int_{-\pi}^{\pi}\{P(r,t)+P(r,t-x)\}\,dt=2\pi x,$$

$$I_2=\int_0^{\pi-x}(t+x)\,P(r,t)\,dt-\int_x^{\pi} tP(r,t)\,dt$$

$$<x\int_x^{\pi-x}P(r,t)\,dt+\int_0^x(t+x)\,P(r,t)\,dt$$

$$<x\int_0^{\pi}P(r,t)\,dt+2x\int_0^{\pi}P(r,t)\,dt=\tfrac{3}{2}\pi x.$$

The same argument gives $I_3<\frac{3}{2}\pi x$. Since the integrand in I_4 is uniformly bounded we see that $I_4(x)=O(x)$. Collecting results we obtain (7·14). For general f, we apply the same decomposition $f=f_1+f_2$ as above.

The series

$$\sum_{\nu=1}^{\infty}\nu(A_\nu\cos\nu x+B_\nu\sin\nu x)$$

is both the conjugate of $S'[F]$ and the formal derivative of $\tilde{S}[F]$. We shall denote it by $\tilde{S}'[F]$. Obviously,

$$\sum_{\nu=1}^{\infty}\nu(A_\nu\cos\nu x+B_\nu\sin\nu x)\,r^\nu=\partial\tilde{F}(r,x)/\partial x.$$

(7·15) Theorem. *If F is periodic and integrable the difference*

$$\frac{\partial\tilde{F}(r,x)}{\partial x}-\left(-\frac{1}{\pi}\int_{1-r}^{\pi}\frac{F(x+t)+F(x-t)-2F(x)}{4\sin^2\frac{1}{2}t}\,dt\right) \tag{7·16}$$

tends to 0 *with* $1-r$ *at every* x *at which* F *is smooth, i.e. at which*

$$F(x+t)+F(x-t)-2F(x)=o(t). \tag{7·17}$$

The formula (6·5), with F for f, gives

$$\frac{\partial \tilde{F}(r,x)}{\partial x}=-\frac{1}{\pi}\int_{-\pi}^{\pi}F(t)\frac{\partial}{\partial x}Q(r,t-x)\,dt=\frac{1}{\pi}\int_0^{\pi}\{F(x+t)+F(x-t)-2F(x)\}\,Q'(r,t)\,dt,$$

since Q' is even and the integral of Q' over $(-\pi,\pi)$ is 0. We note that

$$\left.\begin{aligned}Q'(r,t)=\frac{r[(1+r^2)\cos t-2r]}{(1-2r\cos t+r^2)^2},\quad Q'(1,t)=-\frac{1}{2(1-\cos t)},\\ |Q'(r,t)|\leqslant r+2r^2+3r^3+\ldots=r/(1-r)^2.\end{aligned}\right\} \tag{7·18}$$

Let $$\xi(t)=F(x+t)+F(x-t)-2F(x),\quad \delta=1-r.$$

By (7·17), $\xi(t)=o(t)$. We split the integral last written into two, denoted by A, B with ranges respectively $0\leqslant t\leqslant\delta$, $\delta\leqslant t\leqslant\pi$. Then (see (7·18))

$$|A|\leqslant\frac{r}{\pi\delta^2}\int_0^{\delta}|\xi(t)|\,dt=\delta^{-2}\int_0^{\delta}o(t)\,dt=o(1),$$

$$B=\frac{1}{\pi}\int_{\delta}^{\pi}\xi(t)\,Q'(1,t)\,dt+\frac{1}{\pi}\int_{\delta}^{\pi}\xi(t)\,[Q'(r,t)-Q'(1,t)]\,dt=B_1+B_2,$$

say. Here B_1 equals the expression in parentheses in (7·16), and (7·15) will be proved if we show that $B_2\to 0$. Collecting separately the terms with $\cos t$ and $\cos^2 t$ in the numerator we find

$$Q'(r,t)-Q'(1,t)=\frac{(1-r)^2[(1+r)^2-2r\cos t-2r\cos^2 t]}{2(1-\cos t)\,\Delta^2(r,t)}=\frac{\delta^2[\Delta(r,t)+2r\sin^2 t]}{2(1-\cos t)\,\Delta^2(r,t)}.$$

Since $\Delta>4r\sin^2\frac{1}{2}t$, the last expression is $O(\delta^2 t^{-4})$. Hence

$$|B_2|\leqslant\frac{1}{\pi}\int_{\delta}^{\pi}o(t)\,O(\delta^2t^{-4})\,dt=\delta^2\int_{\delta}^{\pi}o(t^{-3})\,dt=o(1),$$

which completes the proof of Theorem (7·15).

Thus, under the hypothesis (7·17) (in particular, if $F'(x)$ exists and is finite) the summability A of $\tilde{S}'[F]$ at x is equivalent to the existence of the integral

$$F^*(x)=-\frac{1}{\pi}\int_0^{\pi}\frac{F(x+t)+F(x-t)-2F(x)}{4\sin^2\frac{1}{2}t}\,dt=\lim_{\delta\to+0}\left\{-\frac{1}{\pi}\int_{\delta}^{\pi}\right\}. \tag{7·19}$$

We show below (Chapter IV, § 3) that if $F'(x)$ exists at every point of a set E then the integral (7·19) exists almost everywhere in E; we infer that the series $\tilde{S}'[F]$ is then summable A *almost everywhere* in E.

(7·20) THEOREM. *If* f *is integrable and* F *the indefinite integral of* f, *then*

$$\tilde{f}(r,x)-\left(-\frac{1}{\pi}\int_{1-r}^{\pi}[f(x+t)-f(x-t)]\,\tfrac{1}{2}\cot\tfrac{1}{2}t\,dt\right)\to 0\quad(r\to 1) \tag{7·21}$$

at every point where F *is smooth, in particular where* f *is continuous. If* f *is everywhere continuous, the convergence is uniform.*

This is a strengthening of (3·20), since the condition of smoothness, which can be written $\int_0^h \psi_x(t)\,dt = o(h)$, is less stringent than $\Psi_x(h) = o(h)$.

Suppose that the constant term of $S[f]$ is 0. Then F is periodic and $\tilde{f}(r,x) = \partial\tilde{F}(r,x)/\partial x$. Integration by parts gives

$$\int_\delta^\pi \frac{\xi(t)}{4\sin^2\frac{1}{2}t}\,dt = \xi(\delta)\,\tfrac{1}{2}\cot\tfrac{1}{2}\delta + \int_\delta^\pi \frac{\psi(t)}{\tan\frac{1}{2}t}\,dt, \tag{7·22}$$

where $\xi(t) = F(x+t) + F(x-t) - 2F(x)$. The integrated term is $o(1)$ by hypothesis and (7·20) follows from (7·15).

Theorem (7·20) shows that under the smoothness condition (7·17) $\tilde{S}[f]$ is summable A at x if and only if $\tilde{f}(x)$ exists. We show now that *the mere existence of* $\tilde{f}(x)$ *implies* (7·17) *and consequently* (by (7·20)) *the summability* A *of* $\tilde{S}[f]$ *at* x.

Since the ratio $t/\tan\frac{1}{2}t$ and its reciprocal are both bounded and monotone near $t=0$, an application of the second mean-value theorem shows that the existence of $\tilde{f}(x)$ is equivalent to the existence of the integral $\int_0^\pi t^{-1}\psi(t)\,dt$. The relation $\xi(t) = o(t)$ will follow if we apply to the latter integral the following lemma, with $\alpha = 1$ and $h(u) = u^{-1}\psi(u)$:

(7·23) LEMMA. *Suppose that* $h(u)$, $0 < u \leqslant a$, *is integrable over each interval* (ϵ, a), $\epsilon > 0$, *and that the (improper) integral* $\int_0^a h\,du = \lim_{\epsilon\to+0}\int_\epsilon^a h\,du$ *exists. Then, if* $\alpha > 0$,

$$\int_0^t h(u)\,u^\alpha\,du = o(t^\alpha) \quad (t\to 0). \tag{7·24}$$

Let $H(u) = \int_0^u h(v)\,dv$. For $t > \epsilon$ integration by parts gives

$$\int_\epsilon^t u^\alpha h(u)\,du = [u^\alpha H(u)]_\epsilon^t - \alpha\int_\epsilon^t u^{\alpha-1}H(u)\,du.$$

If we make $\epsilon\to 0$ and observe that $H(u) = o(1)$, (7·24) follows.

Let $\Xi(t) = \int_0^t \xi(u)\,du$. A minor modification in the proof of (7·15) (integration by parts so as to have $\Xi(t)$ instead of $\xi(t)$) shows that (7·16) tends to 0 if (7·17) is replaced by $\Xi(t) = o(t^2)$. If (7·19) exists, so does $\int_0^\pi t^{-2}\xi(t)\,dt$. By (7·23), with $\alpha = 2$, we then have $\Xi(t) = o(t^2)$. Hence *the existence of the integral* (7·19) *implies summability* A *of* $\tilde{S}'[F]$ *at* x.

Let $\Sigma c_\nu z^\nu = \phi(z)$ be the power series whose real and imaginary parts on $|z| = 1$ are $S[f]$ and $\tilde{S}[f]$ respectively. If this power series is summable (C, 1) at a point e^{ix_0}, then $\phi(z)$ tends to a limit as z approaches e^{ix_0} non-tangentially (cf. (1·34)). Thus, considering the imaginary part of $\phi(z)$ and applying (3·9) and (3·23), we get:

(7·25) THEOREM. *If* $f \in \mathrm{L}$, *then for almost all* x *the harmonic function* $\tilde{f}(r,x)$ *tends to a limit as* (r,x) *approaches* $(1,x_0)$ *non-tangentially.*

The fact that $S[f]$ is almost everywhere summable A to f can be complemented as follows:

(7·26) THEOREM. *Given any set E in $(0, 2\pi)$ of measure zero, there is a periodic and integrable $f(x) \geqslant 0$ such that for every $x_0 \in E$ we have $f(r, x) \to +\infty$ as $r\,e^{ix}$ approaches e^{ix_0} from the interior of the unit circle.*

For let G_n be an open set containing E and such that $|G_n| < 1/n^4$ Let $f_n(x) = n^2$ in G_n, $f_n(x) = 0$ elsewhere. Let $f(x) = \Sigma f_n(x)$. Obviously, $f \geqslant 0$, $f_n \leqslant f$ for every n and

$$\int_0^{2\pi} f\,dx = \Sigma \int_0^{2\pi} f_n\,dx < \Sigma n^{-4}.\,n^2 < \infty,$$

so that $f \in L$. If $x_0 \in E$, then $x_0 \in G_n$ and so, for $r\,e^{ix} \to e^{ix_0}$,

$$\liminf f(r, x) \geqslant \liminf f_n(r, x) = n^2,$$

so that $\lim f(r, x) = +\infty$.

A similar argument shows that $S[f]$ *is summable* $(C, 1)$ *to* $+\infty$ *at every point of* E.

8. Summability of $S[dF]$ and $\tilde{S}[dF]$

Let $F(x)$, $0 \leqslant x \leqslant 2\pi$, be a function of bounded variation. From (7·6) we see that at every point where $F'(x)$ exists and is finite, $S[dF]$ is summable A to sum $F'(x)$. Similarly Theorem (7·15) implies that at every such point summability A of $\tilde{S}[dF]$ is equivalent to the existence of the integral (7·19).

(8·1) THEOREM. *Let $\sigma_n^\alpha(x)$ and $\tilde{\sigma}_n^\alpha(x)$ be the α-th Cesàro means of $S[dF]$ and $\tilde{S}[dF]$. If $0 < \alpha \leqslant 1$, then*

$$\sigma_n^\alpha(x) \to F'(x), \tag{8·2}$$

$$\tilde{\sigma}_n^\alpha(x) - \left\{-\frac{1}{\pi}\int_{1/n}^{\pi} \frac{F(x+t) + F(x-t) - 2F(x)}{(2\sin\frac{1}{2}t)^2}\,dt\right\} \to 0 \tag{8·3}$$

for almost all x.

We shall only sketch the proof, which is similar to those of (5·1) and (5·8). First we prove the following analogue of Theorem (11·1) of Chapter II:

(8·4) THEOREM. *Let $F(x)$ be of bounded variation,*

$$F_x(t) = F(x+t) - F(x-t) - 2tF'(x),$$

$$G_x(t) = F(x+t) + F(x-t) - 2F(x),$$

and let $\Phi_x(h)$, $\Psi_x(h)$ be the total variations of the functions $F_x(t)$, $G_x(t)$ over the interval $0 \leqslant t \leqslant h$. Then

$$\Phi_x(h) = o(h), \quad \Psi_x(h) = o(h)$$

for almost all x.

Let γ be any number, and let $V_\gamma(t)$ be the total variation of the function $F(t) - \gamma t$. For almost all x, we have $V'_\gamma(x) = |F'(x) - \gamma|$, that is,

$$h^{-1}\int_0^h |d_t\{F(x \pm t) - \gamma(\pm t)\}| \to |F'(x) - \gamma| \quad \text{as} \quad h \to +0,$$

where the suffix t indicates that the variation is taken with respect to the variable t.

Considering rational values of γ and arguing as in the proof of Theorem (11·1) in Chapter II, we prove that

$$\int_0^h |d_t\{F(x\pm t)-(\pm t)F'(x)\}| = o(h),$$

and hence
$$\int_0^h |d_t F_x(t)| = o(h), \quad \int_0^h |d_t G_x(t)| = o(h),$$
for almost all x.

It is now easy to prove (8·2) for all x with $\Phi_x(h)=o(h)$. For

$$\sigma_n^\alpha(x) = \frac{1}{\pi}\int_{-\pi}^{\pi} K_n^\alpha(x-t)\,dF(t) = \frac{1}{\pi}\int_0^\pi K_n^\alpha(t)\,d_t\{F(x+t)-F(x-t)\},$$

$$\sigma_n^\alpha(x) - F'(x) = \frac{1}{\pi}\int_0^\pi K_n^\alpha(t)\,d_tF_x(t),$$

$$|\sigma_n^\alpha(x)-F'(x)| \leqslant \frac{1}{\pi}\int_0^\pi |K_n^\alpha(t)|\,|d_tF_x(t)| = \frac{1}{\pi}\int_0^{1/n} + \frac{1}{\pi}\int_{1/n}^\pi = A+B,$$

say. Here $A \leqslant 2n\Phi_x(1/n)=o(1)$. Integration by parts, in view of (5·5), gives

$$B \leqslant C\pi^{-1}n^{-\alpha}[\Phi_x(t)\,t^{-\alpha-1}]_{1/n}^{\pi} + C(1+\alpha)\,\pi^{-1}n^{-\alpha}\int_{1/n}^\pi \Phi_x(t)\,t^{-\alpha-2}dt = o(1),$$

and this yields (8·2). To obtain (8·3) we note that

$$\tilde\sigma_n^\alpha(x) = -\frac{1}{\pi}\int_0^\pi \tilde K_n^\alpha(t)\,d_t[F(x+t)+F(x-t)] = -\frac{1}{\pi}\int_0^\pi \tilde K_n^\alpha(t)\,d_tG_x(t),$$

$$\tilde\sigma_n^\alpha(x) - \left(-\frac{1}{\pi}\int_{1/n}^\pi \frac{d_tG_x(t)}{2\tan\frac12 t}\right) = -\frac{1}{\pi}\int_0^{1/n}\tilde K_n^\alpha(t)\,d_tG_x(t) + \frac{1}{\pi}\int_{1/n}^\pi H_n^\alpha(t)\,d_tG_x(t)$$

(cf. (5·13)). The two terms on the right are $o(1)$, since $\Psi_x(h)=o(h)$. Integration by parts gives

$$\int_{1/n}^\pi \frac{d_tG_x(t)}{2\tan\frac12 t}dt - \int_{1/n}^\pi \frac{G_x(t)}{(2\sin\frac12 t)^2}dt = \left[\frac{G_x(t)}{2\tan\frac12 t}\right]_{1/n}^\pi = o(1) \tag{8·5}$$

for all x at which F is smooth, and this proves (8·3).

Arguing as in the proofs of Theorem (11·9), Chapter II, we find that the *the partial sums of* $\mathsf{S}[dF]$ *and* $\tilde{\mathsf{S}}[dF]$ *are* $o(\log n)$ *almost everywhere.*

Taking for granted (again) that the integral (7·19) exists almost everywhere (see Chapter VII, (1·6)) we see that $\tilde{\mathsf{S}}[dF]$ *is summable* (C, α), $\alpha>0$, *at almost all points.* This implies, in turn, that Theorems (4·4) and (7·25) are valid for Fourier-Stieltjes series.

9. Fourier series at simple discontinuities

Given a numerical sequence $s_0, s_1, s_2, \ldots$, consider the numbers

$$\tau_n = \left(\sum_{\nu=1}^n \frac{s_\nu}{\nu}\right)\Big/\log(n+1) \quad (n=1,2,\ldots). \tag{9·1}$$

We can verify that the matrix transforming $\{s_n\}$ into $\{\tau_n\}$ satisfies the conditions of regularity (§ 1). The method of summability defined by (9·1) is called the *logarithmic mean.*

Let $\sigma_\nu = (s_0 + s_1 + \ldots + s_\nu)/(\nu+1)$. Substituting $s_\nu = (\nu+1)\sigma_\nu - \nu\sigma_{\nu-1}$ into (9·1), we get

$$\tau_n = \left(-\sigma_0 + \sum_{\nu=1}^{n-1} \frac{\sigma_\nu}{\nu} + \frac{n+1}{n}\sigma_n\right) \Big/ \log(n+1). \tag{9·2}$$

The matrix transforming $\{\sigma_n\}$ into $\{\tau_n\}$ again satisfies the conditions of regularity, so that the logarithmic mean is at least as strong as (C, 1). If $\sigma_n = (-1)^n$, then $\tau_n \to 0$ though $\lim \sigma_n$ does not exist; thus the method of the logarithmic mean is actually stronger than (C, 1).

Theorem (8·13) of Chapter II can be restated as follows: at every point x where f has a jump d, the terms

$$\nu(b_\nu \cos \nu x - a_\nu \sin \nu x)$$

of $S'[f]$ are summable by the logarithmic mean to d/π. Thus the terms of the differentiated Fourier series determine the jumps of the function.

It can be shown by examples (see p. 314, Example 3) that in general we cannot here replace the logarithmic mean by (C, 1). This, however, can be done if the function is of bounded variation. In this case we assume, slightly more generally, that $F(x)$ is not necessarily periodic but satisfies the condition

$$F(x+2\pi) - F(x) = \text{const.} \quad (-\infty < x < +\infty),$$

and is of bounded variation in $(0, 2\pi)$.

(9·3) THEOREM. *Let* $dF(x) \sim \Sigma c_\nu e^{i\nu x}$. *Then*

$$\frac{1}{n} \sum_{\nu=-n}^{n} c_\nu e^{i\nu x_0} \to \pi^{-1}[F(x_0+0) - F(x_0-0)]. \tag{9·4}$$

If
$$F \sim \tfrac{1}{2}A_0 + \Sigma(A_\nu \cos \nu x + B_\nu \sin \nu x),$$

then $S[dF] = S'[F]$ and (9·4) can be written

$$\frac{1}{n} \sum_{\nu=1}^{n} \nu(B_\nu \cos \nu x_0 - A_\nu \sin \nu x_0) \to \pi^{-1}[F(x_0+0) - F(x_0-0)]. \tag{9·5}$$

In the proof of (9·4) we may suppose that $x_0 = 0$. We verify (9·4) (or (9·5)) for the function

$$\tfrac{1}{2}(\pi - x) = \sin x + \tfrac{1}{2}\sin 2x + \ldots \quad (0 < x < 2\pi).$$

Subtracting a multiple of it from F we may suppose that F is continuous at $x = 0$, and we have to show that

$$\sum_{\nu=-n}^{n} c_\nu = \frac{1}{\pi}\int_{-\pi}^{\pi} D_n(t)\, dF(t) = \frac{1}{\pi}\int_{-\eta}^{\eta} + \frac{1}{\pi}\int_{\eta<|t|\leqslant\pi} = A + B$$

is $o(n)$. We choose η so small that the total variation of F over $(-\eta, \eta)$ is $\leqslant \epsilon$. Then

$$|A| \leqslant \pi^{-1}(n+\tfrac{1}{2})\int_{-\eta}^{\eta} |dF(t)| < 2n\pi^{-1}\epsilon < \epsilon n,$$

$$|B| \leqslant \pi^{-1} \max_{\eta\leqslant t\leqslant\pi} |D_n(t)| \int_{\eta\leqslant|t|\leqslant\pi} |dF(t)| = O(1) = o(n).$$

Hence $A + B = o(n)$ and (9·3) follows.

We apply Theorem (9·3) to the function F^* associated with F by means of the formula (1·28) of Chapter II. Using the results (1·29) and (1·30) of Chapter II, we obtain:

(9·6) Theorem. *Let $dF(x) \sim \Sigma c_\nu e^{i\nu x}$, and let $d_1, d_2, \dots$ be all the jumps of $F(x)$ in $0 \leqslant x < 2\pi$. Then*

$$\lim_{n\to\infty} \frac{1}{2n+1} \sum_{\nu=-n}^{n} |c_\nu|^2 = (2\pi)^{-2} \Sigma |d_j|^2, \tag{9·7}$$

and F is everywhere continuous if and only if

$$R_n = \frac{1}{2n+1} \sum_{-n}^{+n} |c_\nu|^2 \to 0. \tag{9·8}$$

In particular, F is continuous if $c_\nu \to 0$.

The condition $R_n \to 0$ is equivalent to

$$R'_n = \frac{1}{2n+1} \sum_{-n}^{n} |c_\nu| \to 0. \tag{9·9}$$

For $R_n \leqslant R'_n \max |c_k|$, and so (9·9) implies (9·8). The converse follows from Schwarz's inequality:

$$R'_n = (2n+1)^{-1} \sum_{-n}^{n} |c_\nu| \,.\, 1 \leqslant (2n+1)^{-1} \left(\sum_{-n}^{n} |c_\nu|^2 \right)^{\frac{1}{2}} (2n+1)^{\frac{1}{2}} = R_n^{\frac{1}{2}}.$$

Return to the hypothesis $f \in L$. For applications it is of interest to investigate the Abel summability of the sequence $\{\nu(b_\nu \cos \nu x - a_\nu \sin \nu x)\}$, that is, the existence of the limit of

$$(1-r) \sum_{\nu=1}^{\infty} \nu(b_\nu \cos \nu x - a_\nu \sin \nu x) r^\nu = (1-r) \frac{\partial}{\partial x} f(r, x) \tag{9·10}$$

as $r \to 1$. Instead of $d(x_0) = \lim_{t\to+0} [f(x_0+t) - f(x_0-t)]$ we may consider the *generalized jump*

$$\delta(x_0) = \lim_{h\to 0} \frac{1}{h} \int_0^h [f(x_0+t) - f(x_0-t)]\, dt = \lim_{h\to 0} \frac{F(x_0+h) + F(x_0-h) - 2F(x_0)}{h},$$

where F is the integral of f. Thus $\delta(x_0) = 0$ is the same thing as the smoothness of F at x_0. The existence of $d(x_0)$ implies that of $\delta(x_0)$ and both numbers are equal.

(9·11) Theorem. *If $\delta(x_0)$ exists and is finite then $(1-r) f_x(r, x)$, with $x = x_0$, tends to $\delta(x_0)/\pi$ as $r \to 1$.*

We may assume that $a_0 = 0$, so that F is periodic and $f_x(r, x) = F_{xx}(r, x)$. We may also take $x_0 = 0$. If $f \sim \Sigma \nu^{-1} \sin \nu x$, then (9·11) is obviously valid and so we may confine ourselves to the case $\delta(x_0) = 0$. Since

$$P''(r, t) = (1-r^2)\, r\, \frac{[2r(1+\sin^2 t) - (1+r^2)\cos t]}{(1 - 2r\cos t + r^2)^3} \tag{9·12}$$

is even in t, and since $\int_0^\pi P''\, dt = P'(r, \pi) - P'(r, 0) = 0$, we have

$$\frac{\partial^2 F(r, x)}{\partial x^2} = \frac{1}{\pi} \int_{-\pi}^{\pi} F(t) \frac{\partial^2}{\partial x^2} P(r, t-x)\, dt = \frac{1}{\pi} \int_{-\pi}^{\pi} \tfrac{1}{2}\{F(x+t) + F(x-t) - 2F(x)\} P''(r, t)\, dt. \tag{9·13}$$

It is enough to show that the expression

$$K(r, t) = (1-r)\, t P''(r, t)$$

has the properties (B′) and (C) of kernels (see § 2). (Property (A) is not needed because $\delta(x_0) = 0$.)

Property (C) follows from (9·12). The latter formula also shows that $P''(r,t)$ changes sign in $(0,\pi)$ once only, namely, for $t=\tau=\tau(r)$ satisfying

$$\frac{\cos\tau}{1+\sin^2\tau}=\frac{2r}{1+r^2},$$

so that $\tau\to 0$ as $r\to 1$. Furthermore,

$$\frac{(1-r)^2}{1+r^2}=\frac{1-\cos\tau+\sin^2\tau}{1+\sin^2\tau}\simeq\frac{3}{2}\tau^2,$$

so that
$$\tau\simeq\sqrt{3}\,(1-r). \qquad \textbf{(9·14)}$$

For (B′) we have to show that

$$\int_0^\pi t\,|\,P''(r,t)\,|\,dt\leqslant\frac{C}{1-r}.$$

The left-hand side here is

$$\left(-\int_0^\tau+\int_\tau^\pi\right)tP''\,dt=-2\tau P'(r,\tau)+2P(r,\tau)-P(r,0)-P(r,\pi)<-2\tau P'(r,\tau)+2P(r,\tau).$$

Since $P(r,\tau)\leqslant 1/(1-r)$, it is enough to obtain a similar estimate for $-\tau P'(r,\tau)$, which is easy by means of (9·14). This completes the proof of (9·11).

It is to be observed that, in the argument beginning with (9·13), we did not make use of the fact that F was an integral. Thus the reasoning shows that for any integrable F which is smooth at x we have

$$F_{xx}(r,x)=o\{(1-r)^{-1}\}. \qquad \textbf{(9·15)}$$

With the obvious extension to uniformity we can state the following result:

(9·16) THEOREM. *If $F\in\lambda_*$, we have* (9·15), *uniformly in x. If $F\in\Lambda_*$, the conclusion holds with 'O' instead of 'o'.*

10. Fourier sine series

If $f(x)$ is odd, its Poisson integral may be written

$$f(r,x)=\frac{1}{\pi}\int_0^\pi f(t)\,[P(r,x-t)-P(r,x+t)]\,dt. \qquad \textbf{(10·1)}$$

Since $P(r,u)$ is even and decreases in $0<u<\pi$, the difference in square brackets is positive for $0<x<\pi$. Thus, if $f(t)$ is non-negative and $f(t)\not\equiv 0$ in $(0,\pi)$, $f(r,x)$ is strictly positive for $0<x<\pi$. (Of course, $f(r,0)=f(r,\pi)=0$.) If $m\leqslant f(t)\leqslant M$, and $f(t)\not\equiv$ const. for $0\leqslant t\leqslant\pi$, then (10·1) implies

$$\frac{m}{\pi}\int_0^\pi [P(r,x-t)-P(r,x+t)]\,dt<f(r,x)<\frac{M}{\pi}\int_0^\pi [P(r,x-t)-P(r,x+t)]\,dt$$

for $0<x<\pi$. These inequalities may be rewritten

$$m\mu(r,x)<f(r,x)<M\mu(r,x),$$

where $\mu(r,x)$ (positive for $0<x<\pi$) is the Poisson integral of the function

$$\mu(t)=\operatorname{sign} t \quad (|\,t\,|<\pi).$$

This results holds if summability A is replaced by summability (C, 3):

(10·2) Theorem. *If $f(x)\not\equiv 0$ is odd and non-negative in $(0,\pi)$, then the third arithmetic means of $S[f]$ are strictly positive for $0<x<\pi$. More generally, if $f(x)\not\equiv$ const. and $m\leqslant f(x)\leqslant M$ in $(0,\pi)$, then*

$$m\sigma_n^3(x;\mu)<\sigma_n^3(x;f)<M\sigma_n^3(x;\mu)\quad(0<x<\pi;\ n=1,2,\ldots).$$

For the proof it is enough (arguing as in the case of Poisson's kernel) to show that the kernel $K_n^3(t)$ is strictly decreasing in $(0,\pi)$, or, $K_n^3(t)$ being a polynomial, that $\{K_n^3(t)\}'\leqslant 0$ there. The expression $\{K_n^3(t)\}'$ is the Cesàro mean $S_n^3(t)/A_n^3$ of the series $\frac{1}{2}+\cos t+\cos 2t+\ldots$ differentiated term by term. Then (1·9) gives the identity

$$\sum_{n=0}^{\infty} S_n^3(t)\,r^n=(1-r)^{-4}P'(r,t)=-\left[\frac{1}{2}\frac{1-r^2}{(1-r)^2\Delta(r,t)}\right]^2\frac{4r\sin t}{1-r^2},\tag{10·3}$$

where $$\Delta(r,t)=1-2r\cos t+r^2.$$

Using (1·9) again we see that the expression in square brackets is the power series

$$K_0(t)+2K_1(t)\,r+3K_2(t)\,r^2+\ldots,$$

where $K_n(t)$ is Fejér's kernel and is non-negative. Since $r/(1-r^2)=r+r^3+\ldots$ also has non-negative coefficients, we see that $S_n^3(t)\leqslant 0$ in $(0,\pi)$, and (10·2) follows.

11. Gibbs's phenomenon for the method (C,α)

This phenomenon was defined in §9 of Chapter II. Let $M(x_0)$ and $m(x_0)$ be the maximum and minimum of f at x_0 (see §2). Since for every $\{x_n\}\to x_0$ the limits of indetermination of $\sigma_n(x_n)$ are contained between $m(x_0)$ and $M(x_0)$ (see (2·30)), the $(C,1)$ mean of $S[f]$ does not show the phenomenon. It is easy to see that if the phenomenon for (C,α) is not shown for $\alpha=\alpha_1$ then it is not shown for any $\alpha>\alpha_1$. For if

$$m(x_0)-\epsilon\leqslant\sigma_n^{\alpha_1}(x)\leqslant M(x_0)+\epsilon$$

for $|x-x_0|\leqslant\eta$, $n>n_0$, then

$$m(x_0)-2\epsilon\leqslant\sigma_n^{\alpha}(x)\leqslant M(x_0)+2\epsilon$$

for $|x-x_0|\leqslant\eta$, $n>n_1$ (see (1·5)). It is therefore enough to consider the range $0<\alpha<1$.

(11·1) Theorem. *There is an absolute constant α_0, $0<\alpha_0<1$, with the following property: if $f(x)$ has a simple discontinuity at a point ξ, the means $\sigma_n^\alpha(x;f)$ show Gibbs's phenomenon at ξ for $\alpha<\alpha_0$ but not for $\alpha\geqslant\alpha_0$.*

Since $S[f]$ is uniformly summable (C,α) at every point of continuity, it is enough (as in Chapter II, §9) to prove (11·1) for the function

$$f(x)\sim\sin x+\tfrac{1}{2}\sin 2x+\ldots$$

at $\xi=0$. Observing that $S'[f]=\cos x+\cos 2x+\ldots$, we find

$$\left.\begin{aligned}\sigma_n^\alpha(x)&=-\tfrac{1}{2}x+\int_0^x K_n^\alpha(t)\,dt,\\ \sigma_n^\alpha(x)&=\tfrac{1}{2}(\pi-x)-\int_x^\pi K_n^\alpha(t)\,dt.\end{aligned}\right\}\tag{11·2}$$

Let us first take $\alpha=1$. The first formula then gives

$$\sigma_n(x) = -\tfrac{1}{2}x + \frac{2}{n+1}\int_0^x \frac{\sin^2\frac{1}{2}(n+1)t}{t^2}\,dt + \frac{2}{n+1}\int_0^x \sin^2\tfrac{1}{2}(n+1)t\left[\frac{1}{(2\sin\frac{1}{2}t)^2} - \frac{1}{t^2}\right]dt$$

$$= -\tfrac{1}{2}x + \int_0^{\frac{1}{2}(n+1)x}\left(\frac{\sin u}{u}\right)^2 du + R_n(x), \tag{11·3}$$

where $R_n(x)=O(1/n)$ uniformly in x. Observing that $\sigma_n(x)\to\frac{1}{2}(\pi-x)$ for $0<x<2\pi$, we deduce the formula

$$\int_0^\infty \left(\frac{\sin u}{u}\right)^2 du = \tfrac{1}{2}\pi. \tag{11·4}$$

(This is an analogue of (8·4) in Chapter II and can be deduced from it by integration by parts.) From (11·3) and (11·4) we get

$$\sigma_n(x) = \tfrac{1}{2}(\pi-x) - \int_{\frac{1}{2}(n+1)x}^\infty \left(\frac{\sin u}{u}\right)^2 du + R_n(x) < \tfrac{1}{2}\pi - \int_{nx}^\infty \left(\frac{\sin u}{u}\right)^2 du + R_n(x),$$

for $n\geqslant 1$. Hence

(i) *given any* $l>0$, *there is a* $\delta=\delta(l)>0$ *and an* $n_0=n_0(l)$ *such that*

$$\sigma_n(x) < \tfrac{1}{2}\pi - \delta \quad \text{for} \quad 0\leqslant x\leqslant l/n,\ n>n_0.$$

We shall now use the approximate formulae (5·15) for $K_n^\alpha(t)$. Integrating the right-hand side there over (x,π), applying the second mean-value theorem to the first integral and using the second formula (11·2), we get for $\sigma_n^\alpha(x)$ the value

$$\tfrac{1}{2}(\pi-x) - \frac{\alpha}{n+1}\tfrac{1}{2}\cot\tfrac{1}{2}x + \frac{2\theta_1}{nA_n^\alpha(2\sin\frac{1}{2}x)^{\alpha+1}} + \frac{B}{n^2x^2}, \tag{11·5}$$

where $|\theta_1|\leqslant 1$ and $|B|$ is less than an absolute constant. Since $A_n^\alpha\geqslant Cn^\alpha$ for $n\geqslant 1$ and $0\leqslant\alpha\leqslant 1$, we see that, for nx large enough, of the three last terms in (11·5) the first is the largest in absolute value. Therefore

(ii) *if* $\frac{1}{2}\leqslant\alpha\leqslant 1$, *there is an* l_1 *such that* $|\sigma_n^\alpha(x)|\leqslant\frac{1}{2}\pi$ *for* $l_1/n\leqslant x\leqslant\pi$, $n\geqslant n_1$.

We shall now show that

(iii) *if* $1-\alpha$ *is small enough then* $|\sigma_n^\alpha(x)|\leqslant\frac{1}{2}\pi$ *for* $0\leqslant x\leqslant l_1/n$.

This, in conjunction with (ii), will prove that if α is close enough to 1 the $\sigma_n^\alpha(x)$ do not show Gibbs's phenomenon. First of all we verify the inequality

$$A_k^\alpha/A_n^\alpha \geqslant A_k^\beta/A_n^\beta \quad \textit{for} \quad -1<\alpha<\beta,\ 0\leqslant k\leqslant n.$$

From it we deduce that $|\sigma_n^\alpha(x)-\sigma_n^\beta(x)|$ is less than

$$\sum_{\nu=1}^n \left(\frac{A_{n-\nu}^\alpha}{A_n^\alpha} - \frac{A_{n-\nu}^\beta}{A_n^\beta}\right)\left|\frac{\sin\nu x}{\nu}\right| \leqslant x\sum_{\nu=1}^n\left(\frac{A_{n-\nu}^\alpha}{A_n^\alpha} - \frac{A_{n-\nu}^\beta}{A_n^\beta}\right) = x\left(\frac{A_n^{\alpha+1}}{A_n^\alpha} - \frac{A_n^{\beta+1}}{A_n^\beta}\right) = \frac{nx(\beta-\alpha)}{(\alpha+1)(\beta+1)}. \tag{11·6}$$

For $\beta=1$ the last term is less than $\frac{1}{2}nx(1-\alpha)$, and so, using (i), it is enough to take α such that $\frac{1}{2}(1-\alpha)\,l_1\leqslant\delta(l_1)$.

In order to show that for α positive and small enough the phenomenon does occur, we consider the difference $\sigma_n^\alpha-\sigma_n^0=\sigma_n^\alpha-s_n$, which, by (11·6), is numerically less than $nx\alpha/(\alpha+1)<nx\alpha$. Since $s_n(\pi/n)$ tends to a limit greater than $\frac{1}{2}\pi$ (Chapter II, § 9), it follows that $\liminf\sigma_n^\alpha(\pi/n)>\frac{1}{2}\pi$, and so the phenomenon does occur, for α small enough.

We have therefore shown the existence of α_0, $0<\alpha_0<1$, such that for $\alpha<\alpha_0$ we have the phenomenon, while for $\alpha>\alpha_0$ we do not. If we show that the set G of α for which the phenomenon occurs is open, it cannot occur for $\alpha=\alpha_0$, and the proof of (11·1) will be complete.

Let $0<\alpha'<\alpha_0$. As in (ii) we see that there is an l' such that

$$|\sigma_n^\alpha(x)|\leqslant\tfrac{1}{2}\pi \quad \text{for} \quad \alpha'\leqslant\alpha\leqslant 1, \quad l'/n\leqslant x\leqslant\pi,\ n\geqslant n'.$$

From the majorant (11·6) for $|\sigma_n^\alpha-\sigma_n^\beta|$ we deduce that $\sigma_n^\alpha(x)$ is a uniformly continuous function of α in the range $0\leqslant\alpha\leqslant 1$, $0\leqslant x\leqslant l'/n$, $n=1,2,\ldots$. If the Gibbs's phenomenon occurs for a value $\alpha>\alpha'$, that is, if there is an $x_n\to 0$ such that $\sigma_n^\alpha(x_n)>\tfrac{1}{2}\pi+\epsilon$ for all n large enough, then, first, $0<x_n\leqslant l'/n$, and secondly, if $\beta-\alpha$ is small enough, $\sigma_n^\beta(x_n)\geqslant\tfrac{1}{2}\pi+\tfrac{1}{2}\epsilon$. This shows that G is open and completes the proof of (11·1).

12. Theorems of Rogosinski

Let $\lambda(t)$ be a function with $\lambda(0)=1$. Any series $u_0+u_1+\ldots$ may be considered as the series

$$\sum_{\nu=0}^{\infty} u_\nu\lambda(\nu t) \tag{12·1}$$

at the point $t=0$. Let $s_n(t)$ be the nth partial sum of (12·1), and s_n the nth partial sum of Σu_ν. We shall investigate the behaviour of $s_n(t)$ as $n\to\infty$ and simultaneously $t\to 0$.

(12·2) THEOREM. *Suppose that $\alpha_n=O(1/n)$. (i) If $\lambda(t)$ is continuous at $t=0$ and is of bounded variation over every finite interval, then $s_n\to s$ implies $s_n(\alpha_n)\to s$. (ii) If $\lambda''(t)$ exists and is bounded over every finite interval, then the summability* (C, 1) *of $u_0+u_1+\ldots$ to sum s implies*

$$s_n(\alpha_n)-(s_n-s)\,\lambda(n\alpha_n)\to s. \tag{12·3}$$

(i) Summation by parts gives

$$s_n(\alpha_n)=\sum_{\nu=0}^{n-1} s_\nu[\lambda(\nu\alpha_n)-\lambda((\nu+1)\,\alpha_n)]+s_n\lambda(n\alpha_n). \tag{12·4}$$

This is a linear transformation of $\{s_\nu\}$ which satisfies the conditions of regularity, and (i) follows.

(ii) Let σ_n be the (C, 1) means of $u_0+u_1+\ldots$. Summation by parts gives

$$\begin{aligned}&s_n(\alpha_n)-s_n\lambda(n\alpha_n)+\sigma_n\lambda(n\alpha_n)\\&=\sum_{\nu=0}^{n-2}(\nu+1)\,\sigma_\nu[\lambda(\nu\alpha_n)-2\lambda((\nu+1)\,\alpha_n)+\lambda((\nu+2)\,\alpha_n)]+n\sigma_{n-1}[\lambda((n-1)\,\alpha_n)-\lambda(n\alpha_n)]\\&\qquad+\sigma_n\lambda(n\alpha_n),\end{aligned} \tag{12·5}$$

and the right-hand side is a linear transformation of $\{\sigma_\nu\}$. The example $u_0=1$, $u_1=u_2=\ldots=0$ gives $s_0=s_1=\ldots=1$, $\sigma_0=\sigma_1=\ldots=1$ and shows that the sum of the coefficients of the σ_ν on the right is 1. This proves condition (iii) of regularity.

We now observe that for any fixed x,

$$\lambda(x+u)-\lambda(x)=u\lambda'(x+\theta u),$$

$$\lambda(x+u)+\lambda(x-u)-2\lambda(x)=\tfrac{1}{2}u^2\lambda''(x+\theta_1 u),$$

where θ and $|\theta_1|$ are between 0 and 1. Let $|n\alpha_n| \leqslant h$ for all n, and denote by M the common bound of $|\lambda(t)|$, $|\lambda'(t)|$ and $|\lambda''(t)|$ in the interval $(-h, h)$; then we find that the sum of the moduli of the coefficients of the σ_ν on the right in (12·5) is at most

$$M \cdot \tfrac{1}{2}\alpha_n^2 \sum_0^{n-2} (\nu+1) + nM\,|\alpha_n| + M \leqslant M(\tfrac{1}{4}h^2 + h + 1).$$

This proves condition (ii) of regularity. Condition (i) follows from the continuity of λ at $t=0$. Hence the left-hand side of (12·5) tends to s and (12·3) follows.

We note that if $\alpha_n = \alpha/n$, where α is a zero of $\lambda(t)$, then (12·3) simplifies.

We also observe that if the terms of $u_0 + u_1 + \ldots$ depend on a parameter, and if the hypotheses concerning this series are satisfied uniformly, the conclusions also hold uniformly.

Suppose now that $\lambda(t)$ satisfies the same conditions as before, except for the condition $\lambda(0)=1$. The case $\lambda(0) \neq 0$ reduces to $\lambda(0)=1$ by considering $\lambda(t)/\lambda(0)$. If, however, $\lambda(0)=0$, condition (iii) of regularity is no longer satisfied and the matrices generating the transformations (12·4) and (12·5) will have sums 0 in each row.

The result for this case is:

(12·6) THEOREM. *If* $\lambda(0)=0$, *and if the other conditions of* (12·2) *are satisfied, we have*

$$s_n(\alpha_n) \to 0, \quad s_n(\alpha_n) - (s_n - s)\,\lambda(n\alpha_n) \to 0$$

respectively, according as $u_0 + u_1 + \ldots$ *is convergent or summable* (C, 1) *to sum* s.

The most important special cases are $\lambda(t) = \cos t$ and $\lambda(t) = \sin t$. The reason for this is that, if $S_n(x)$ denotes the partial sum of any series

$$\tfrac{1}{2}a_0 + \sum_{\nu=1}^{\infty} (a_\nu \cos \nu x + b_\nu \sin \nu x), \tag{12·7}$$

then

$$\left.\begin{aligned} \tfrac{1}{2}[S_n(x+\alpha_n) + S_n(x-\alpha_n)] &= \tfrac{1}{2}a_0 + \sum_{\nu=1}^{n} (a_\nu \cos \nu x + b_\nu \sin \nu x)\cos \nu\alpha_n \\ \tfrac{1}{2}[S_n(x+\alpha_n) - S_n(x-\alpha_n)] &= \qquad - \sum_{\nu=1}^{n} (a_\nu \sin \nu x - b_\nu \cos \nu x)\sin \nu\alpha_n \end{aligned}\right\} \tag{12·8}$$

are $s_n(\alpha_n)$ with $\lambda(t) = \cos t, \sin t$. Thus from the first formula we get:

(12·9) THEOREM. *Let* $\alpha_n = O(1/n)$, *and let* $S_n(x)$ *be the partial sums of* (12·7). *Then*

$$\tfrac{1}{2}[S_n(x+\alpha_n) + S_n(x-\alpha_n)] \to s$$

at every point where $S_n(x) \to s$; *and*

$$\tfrac{1}{2}[S_n(x+\alpha_n) + S_n(x-\alpha_n)] - (S_n(x) - s)\cos n\alpha_n \tag{12·10}$$

tends to s *at every point where* (12·7) *is summable* (C, 1) *to* s. *In particular, if* (12·7) *is* S[*f*], (12·10) *tends to* $f(x)$ *at every point of continuity of* f, *and the convergence is uniform over any closed interval of continuity.*

If $\alpha_n = \tfrac{1}{2}p\pi/n$, where p is any fixed odd integer, (12·10) becomes

$$\tfrac{1}{2}[S_n(x + \tfrac{1}{2}p\pi/n) + S_n(x - \tfrac{1}{2}p\pi/n)], \tag{12·11}$$

and *this expression gives a method of summability of trigonometric series not weaker than the method* (C, 1).

If $\alpha_n = \frac{1}{2}p\pi/n + O(n^{-2})$, the last term in (12·10) is $o(1)$, since $S_n(x) = o(n)$ wherever (12·7) is summable (C, 1). In particular, what was said about (12·11) also applies to

$$\frac{1}{2}\left[S_n\left(x+\frac{p\pi}{2n+1}\right)+S_n\left(x-\frac{p\pi}{2n+1}\right)\right] \quad (p=1,3,5,\ldots). \tag{12·12}$$

(12·13) Theorem. *Let (12·7) be $\mathfrak{S}[f]$ and let ξ be a point of continuity of f. Then, for any sequence $h_n \to 0$, the expression*

$$\tau_n(\xi) = \tfrac{1}{2}[S_n(\xi + h_n + \tfrac{1}{2}\pi/n) + S_n(\xi + h_n - \tfrac{1}{2}\pi/n)] \tag{12·14}$$

tends to $f(\xi)$.

For (12·5), with $\lambda(t) = \cos t$, $\alpha_n = \pi/2n$ shows that $\limsup |s_n(\alpha_n)| \leqslant A \limsup |\sigma_\nu|$, where A is an absolute constant. Similarly

$$\limsup |\tau_n(\xi)| \leqslant A \limsup_{\nu\to\infty,\ n\geqslant\nu} |\sigma_\nu(\xi + h_n; f)|.$$

We may suppose that $f(\xi) = 0$. Taking into account that $\sigma_\nu(\xi + h_n; f) \to 0$ (see (2·30)), we have $\tau_n(\xi) \to 0$.

In Chapter VIII we shall see that $S_n(x; f)$ may diverge at a point ξ of continuity of f. Theorem (12·13) indicates the existence of a certain symmetry in the behaviour of the curves $y = S_n(x)$ near ξ: *The mean of the values of $S_n(x)$ at the end-points of any interval of length π/n differs little from $f(\xi)$, if the distance of its midpoint from ξ is small.*

These are applications of $\lambda(t) = \cos t$. Now let $\lambda(t) = \sin t$. From the second formula (12·8), and from (12·6), we get

(12·15) Theorem. *Let $\alpha_n = O(1/n)$, and let $S_n(x)$ and $\tilde{S}_n(x)$ be the partial sums of (12·7) and of the conjugate series. Then*

$$S_n(x+\alpha_n) - S_n(x-\alpha_n) \to 0$$

at every point x where the conjugate series converges. At every point where it is summable (C, 1) to $\tilde{s}$, we have

$$\tfrac{1}{2}[S_n(x+\alpha_n) - S_n(x-\alpha_n)] + (\tilde{S}_n(x) - \tilde{s}) \sin n\alpha_n \to 0.$$

Hence, if q is any fixed integer,

$$S_n(x + q\pi/n) - S_n(x - q\pi/n) \to 0$$

at every point where $\{\tilde{S}_n(x)\}$ is summable (C, 1), and in particular almost everywhere if (12·7) is an $\mathfrak{S}[f]$.

(12·16) Theorem. *Let $S_n(x)$ be the partial sums of $\sum_0^\infty c_\nu e^{i\nu x}$ and let $\alpha_n = O(1/n)$. Then $S_n(x+\alpha_n) \to s$ if $S_n(x) \to s$; and*

$$S_n(x+\alpha_n) - (S_n(x) - s)\, e^{in\alpha_n} \to s \tag{12·17}$$

if $\{S_n(x)\}$ is summable (C, 1) to s.

Apply (12·2), with $\lambda(t) = e^{it}$, to $S_n(x+\alpha_n) = \sum_0^n c_\nu e^{i\nu x} . e^{i\nu\alpha_n}$.

13. Approximation to functions by trigonometric polynomials

Given a periodic and continuous function $f(x)$, the *deviation* $\delta(f, T)$ of a trigonometric polynomial $T(x)$ from f is defined by the formula

$$\delta(f, T) = \max |f(x) - T(x)|.$$

The lower bound of the numbers $\delta(f, T)$ for all polynomials

$$T(x) = \tfrac{1}{2}a_0 + \sum_1^n (a_\nu \cos \nu x + b_\nu \sin \nu x)$$

of given order n will be denoted by $E_n[f]$ and called the *best approximation* of f of order n. By (3·6), $E_n[f]$ tends (monotonically) to 0 as $n \to \infty$.

(13·1) THEOREM. *Given f and n, $E_n[f]$ is 'attained'; that is to say there is a polynomial $T^*(x) = T^*(x; f, n)$ of order n such that $\delta(f, T^*) = E_n[f]$.*

For let $T^1, T^2, \ldots,$ be a sequence of polynomials of order n such that

$$\delta(f, T^k) \leqslant E_n[f] + 1/k \quad (k = 1, 2, \ldots). \tag{13·2}$$

In particular, the T^k are uniformly bounded. Thus if $a_0^k, \ldots, b_n^k$ are the coefficients of T^k, these numbers are all bounded. By the theorem of Bolzano-Weierstrass, there is a subsequence of the points $(a_0^k, \ldots, b_n^k)$ of $(2n+1)$-dimensional space which tends to a limit, $(a_0^*, \ldots, b_n^*)$. The corresponding $T^k(x)$ then tend uniformly to a polynomial $T^*(x)$ of order n. From (13·2) we get $\delta(f, T^*) \leqslant E_n[f]$, and since the opposite inequality is obvious (13·1) follows.

Let us write

$$f(x) = T^*(x) + R(x),$$

so that $|R(x)| \leqslant E_n = E_n[f]$. Let $s_k(x)$ and $r_k(x)$ denote the partial sums, and $\sigma_k(x)$ and $\rho_k(x)$ the (C, 1) means, of $S[f]$ and $S[R]$ respectively. For $k \geqslant n$ we have $s_k = T^* + r_k$, so that

$$h^{-1} \sum_n^{n+h-1} s_k = T^* + h^{-1} \sum_n^{n+h-1} r_k, \tag{13·3}$$

$$\left(1 + \frac{n}{h}\right) \sigma_{n+h-1} - \frac{n}{h} \sigma_{n-1} = T^* + \left(1 + \frac{n}{h}\right) \rho_{n+h-1} - \frac{n}{h} \rho_{n-1}. \tag{13·4}$$

Since $|\rho_k| \leqslant E_n$ for all k, the right-hand side of (13·4) differs from T^* by not more than $(1 + 2n/h) E_n$, and so from f by not more than $2(1 + n/h) E_n$. The left-hand side of (13·4) is a delayed arithmetic mean of $S[f]$ (see p. 80). For $h = n$ we get:

(13·5) THEOREM. *Let $\sigma_n(x) = \sigma_n(x; f)$. Then the difference between f and*

$$\tau_n(x) = 2\sigma_{2n-1}(x) - \sigma_{n-1}(x)$$

never exceeds $4E_n[f]$.

We know that $\tau_n(x)$ is obtained by adding to $S_n(x; f)$ a simple linear combination of the next $n-1$ terms of $S[f]$. In this way we obtain a polynomial whose approximation to f is almost as good as the best approximation E_n. (One must not forget, however, that τ_n is of order $2n-1$.)

(13·6) THEOREM. *Let $f(x)$ be periodic and k times differentiable. If $|f^{(k)}(x)| \leqslant M$, then*

$$E_n[f] \leqslant A_k M n^{-k} \quad (n = 1, 2, \ldots). \tag{13·7}$$

If $f^{(k)}$ is continuous and has modulus of continuity $\omega(\delta)$, then

$$E_n[f] \leqslant B_k \omega\left(\frac{2\pi}{n}\right) n^{-k} \quad (n = 1, 2, \ldots). \tag{13·8}$$

The constants A_k and B_k here depend on k only.

Let $\tau_m = 2\sigma_{2m-1} - \sigma_{m-1}$. Using the formula (3·24) with $\omega = 2m$ and with $\omega = m$, we get

$$\tau_m(x) = \frac{2}{\pi m}\int_{-\infty}^{+\infty} f(x+t)\frac{h(mt)}{t^2}\,dt, \tag{13·9}$$

where $h(t) = \sin^2 t - \sin^2 \tfrac{1}{2}t = \tfrac{1}{2}(\cos t - \cos 2t)$,

and, by (11·4), $$\tau_m(x) - f(x) = \frac{2}{\pi}\int_0^\infty \left\{f\left(x+\frac{t}{m}\right) + f\left(x-\frac{t}{m}\right) - 2f(x)\right\}\frac{h(t)}{t^2}\,dt. \tag{13·10}$$

We introduce the functions

$$H_0(t) = h(t)/t^2, \quad H_i(t) = \int_t^\infty H_{i-1}(u)\,du \quad (i = 1, 2, \ldots),$$

and temporarily take for granted that

(i) the integral defining $H_i(t)$ is absolutely convergent for $i = 1, 2, \ldots$;

(ii) $H_3(0) = H_5(0) = H_7(0) = \ldots = 0$.

Then, integrating by parts as many times as the existence of derivatives of f permits, we find for $\tau_m - f$ the values

$$\begin{aligned}\frac{2}{\pi}\int_0^\infty \left\{f\left(x+\frac{t}{m}\right) + f\left(x-\frac{t}{m}\right) - 2f(x)\right\} H_0(t)\,dt &= \frac{2}{\pi m}\int_0^\infty \left\{f'\left(x+\frac{t}{m}\right) - f'\left(x-\frac{t}{m}\right)\right\} H_1(t)\,dt \\ &= \frac{2}{\pi m^2}\int_0^\infty \left\{f''\left(x+\frac{t}{m}\right) + f''\left(x-\frac{t}{m}\right)\right\} H_2(t)\,dt \\ &= \frac{2}{\pi m^3}\int_0^\infty \left\{f'''\left(x+\frac{t}{m}\right) - f'''\left(x-\frac{t}{m}\right)\right\} H_3(t)\,dt \\ &= \ldots.\end{aligned}$$

Hence $$|\tau_m(x) - f(x)| \leqslant C_k m^{-k} M \quad \text{with} \quad C_k = \frac{4}{\pi}\int_0^\infty |H_k(t)|\,dt.$$

so that $E_{2m-1} \leqslant C_k m^{-k} M$. The same inequality holds for E_{2m} ($\leqslant E_{2m-1}$), and it follows for any n (whether even or odd) that

$$E_n \leqslant C_k(\tfrac{1}{2}n)^{-k} M,$$

which is (13·7) with $A_k = 2^k C_k$.

We now prove (i). It is enough to show that each $H_i(t)$ is $O(t^{-2})$ near $t = \infty$. (Since $H_0(t)$ is bounded, this will also imply that each $H_i(t)$ is bounded for $0 \leqslant t < \infty$.) The fact is obvious for $H_0(t)$. Let now $h_i(t)$ denote that ith integral of $h(t)$ which is periodic and has no constant term; $h_i(t)$ is either

$$\pm\tfrac{1}{2}(\cos t - 2^{-i}\cos 2t) \quad \text{or} \quad \pm\tfrac{1}{2}(\sin t - 2^{-i}\sin 2t).$$

Integration by parts gives

$$H_1(t) = -\frac{h_1(t)}{t^2} - 2!\frac{h_2(t)}{t^3} - \ldots - p!\frac{h_p(t)}{t^{p+1}} + (p+1)!\int_t^\infty \frac{h_p(u)}{u^{p+2}}\,du. \tag{13·11}$$

Here p is any positive integer. Since $|h_p(u)| \leqslant 1$, the last term is numerically not greater than $p!\,t^{-(p+1)}$. Hence $H_1(t) = O(t^{-2})$. If we integrate (13·11) over $(t, +\infty)$, integrating the terms on the right by parts, we find that $H_2(t)$ is a sum of a linear combination of $h_2(t)t^{-2}$, $h_3(t)t^{-3}$, ... and a remainder $O(t^{-p})$. Similarly, $H_3(t)$ is a sum of a linear combination of $h_3(t)t^{-2}$, $h_4(t)t^{-3}$, ... and a remainder $O(t^{-(p-1)})$, and so on. This proves (i).

To prove (ii), we apply (13·9) to the functions $f(t)=1$ and $f(t)=\cos t$. Correspondingly, $\tau_m(x)=1$ and $\tau_m(x)=\cos x$, $m\geqslant 1$. This gives, at $x=0$,

$$\frac{4}{\pi}\int_0^\infty H_0(t)\,dt=1,\quad \frac{4}{\pi}\int_0^\infty \cos\frac{t}{m}H_0(t)\,dt=1.$$

Integrating the second integral by parts twice and using the first identity, we get

$$\int_0^\infty \cos\frac{t}{m}H_2(t)\,dt=0.$$

In this, $\cos(t/m)\to 1$ uniformly over any finite interval as $m\to\infty$, so that $\int_0^\infty H_2\,dt=0$. Similarly we prove that $\int_0^\infty H_4\,dt=0$, and so on. This gives (ii).

The second part of (13·6) is obtainable from the first by a simple device. Given any periodic and integrable $f(x)$, whose integral is $F(x)$, and a number $\delta>0$, let

$$f_\delta(x)=\frac{1}{2\delta}\int_{-\delta}^{\delta}f(x+t)\,dt=\frac{1}{2\delta}\int_{x-\delta}^{x+\delta}f(t)\,dt=\frac{F(x+\delta)-F(x-\delta)}{2\delta}. \tag{13·12}$$

The function $f_\delta(x)$, which is also periodic, is called the *moving average* of $f(x)$. If f has k continuous derivatives, f_δ has $k+1$ such derivatives. For f absolutely continuous, we have $(f_\delta)'=(f')_\delta$. Though we shall not need the fact here, we observe that the modulus of continuity of f_δ never exceeds that of f. Clearly,

$$|f_\delta(x)-f(x)|\leqslant\omega(\delta;f). \tag{13·13}$$

Returning to the second part of (13·6) we write

$$f(x)=f_\delta(x)+g(x).$$

Then $f_\delta(x)$ has $k+1$ derivatives, and, by (13·12) and (13·13),

$$|f_\delta^{(k+1)}(x)|=\frac{|f^{(k)}(x+\delta)-f^{(k)}(x-\delta)|}{2\delta}\leqslant(2\delta)^{-1}\omega(2\delta;f^{(k)})\leqslant\delta^{-1}\omega(\delta;f^{(k)}),$$

$$|g^{(k)}(x)|=|f^{(k)}(x)-f_\delta^{(k)}(x)|\leqslant\omega(\delta;f^{(k)}).$$

Hence, by (13·7),

$$E_n[f]\leqslant E_n[f_\delta]+E_n[g]=A_{k+1}n^{-k-1}\delta^{-1}\omega(\delta;f^{(k)})+A_k n^{-k}\omega(\delta;f^{(k)}),$$

and setting here $\delta=2\pi/n$, we get (13·8) with $B_k=A_k+A_{k+1}/2\pi$.

(13·14) THEOREM. *If f has a continuous k-th derivative ($k=0,1,\ldots$), and if $f^{(k)}\in\Lambda_\alpha$, $0<\alpha\leqslant 1$, then*

$$E_n[f]=O(n^{-k-\alpha}). \tag{13·15}$$

This inequality, with $\alpha=1$, holds if $f^{(k)}$ merely belongs to Λ_*.

It is only the last statement that requires a proof. Suppose that

$$|f^{(k)}(x+t)+f^{(k)}(x-t)-2f^{(k)}(x)|\leqslant Mt,$$

where M is independent of x and t. Let $f_{\delta\delta}(x)$ be the moving average of f_δ, and let $f(x)=f_{\delta\delta}(x)+g(x)$. Thus $f_{\delta\delta}$ has $k+2$ derivatives and

$$|f_{\delta\delta}^{(k+2)}(x)|=\frac{|f_\delta^{(k+1)}(x+\delta)-f_\delta^{(k+1)}(x-\delta)|}{2\delta}=\frac{|f^{(k)}(x+2\delta)+f^{(k)}(x-2\delta)-2f^{(k)}(x)|}{4\delta^2}\leqslant\frac{M}{2\delta}.$$

It follows from (13·12) that

$$f_{\delta\delta}(x)=\frac{1}{4\delta^2}\int_{-\delta}^{\delta}\int_{-\delta}^{\delta}f(x+u+v)\,du\,dv=\frac{1}{4\delta^2}\int_{-2\delta}^{2\delta}f(x+t)\,(2\delta-|t|)\,dt$$
$$=\frac{1}{4\delta^2}\int_0^{2\delta}\{f(x+t)+f(x-t)\}\,(2\delta-t)\,dt,$$

and since the operation δ commutes with differentiation,

$$|g^{(k)}(x)|=|f_{\delta\delta}^{(k)}(x)-f^{(k)}(x)|=\frac{1}{4\delta^2}\left|\int_0^{2\delta}\{f^{(k)}(x+t)+f^{(k)}(x-t)-2f^{(k)}(x)\}\,(2\delta-t)\,dt\right|.$$

The last integrand is numerically not greater than $Mt(2\delta-t)\leqslant M\delta^2$, so that $|g^{(k)}(x)|\leqslant\frac{1}{2}M\delta$. Hence, by (13·7),

$$E_n[f]\leqslant E_n[f_{\delta\delta}]+E_n[g]\leqslant A_{k+2}n^{-k-2}(2\delta)^{-1}M+\tfrac{1}{2}A_kn^{-k}M\delta.$$

On setting $\delta=2\pi/n$ here, we get

$$E_n[f]\leqslant BMn^{-k-1},\quad B=A_{k+2}/4\pi+\pi A_k.$$

Our next aim is the converse of (13·14).

(13·16) LEMMA. *Let $T(x)$ be a polynomial of order n, and $M=\max|T(x)|$. Then*

$$|T'(x)|\leqslant 2nM,\quad |\tilde{T}'(x)|\leqslant 2nM. \tag{13·17}$$

This lemma is purely utilitarian; in Chapter X, § 3, we shall show that the factors 2 on the right are superfluous. We write

$$T(x)=\frac{1}{\pi}\int_0^{2\pi}T(t)\,D_n(x-t)\,dt,\quad T'(x)=-\frac{1}{\pi}\int_0^{2\pi}T(x+t)\,D_n'(t)\,dt.$$

Since T is a polynomial of order n, in the last integral we can add to

$$D_n'(t)=-\sum_1^n k\sin kt$$

any polynomial Q all of whose terms have rank greater than n. If we choose

$$Q=-\sum_1^{n-1}k\sin(2n-k)\,t$$

and take together the terms $k\sin kt$ and $k\sin(2n-k)\,t$ we get

$$\left.\begin{aligned}T'(x)&=\frac{2n}{\pi}\int_0^{2\pi}T(x+t)\sin nt\,K_{n-1}(t)\,dt,\\ |T'|&\leqslant 2n\frac{1}{\pi}\int_0^{2\pi}MK_{n-1}(t)\,dt=2nM,\end{aligned}\right\} \tag{13·18}$$

as desired. For the second part we take the formula

$$\tilde{T}'(x)=\frac{1}{\pi}\int_0^{2\pi}T(x+t)\,\tilde{D}_n'(t)\,dt=\frac{1}{\pi}\int_0^{2\pi}T(x+t)\left\{\sum_1^n k\cos kt\right\}dt,$$

and add to the expression in curly brackets the polynomial $\sum_1^{n-1}k\cos(2n-k)\,t$, obtaining

$$\tilde{T}'(x)=\frac{2n}{\pi}\int_0^{2\pi}T(x+t)\cos nt K_{n-1}(t)\,dt,\quad |\tilde{T}'|\leqslant 2nM. \tag{13·19}$$

(13·20) THEOREM. *Suppose that f satisfies (13·15) for some $k=0,1,2,\ldots$, and $0<\alpha\leqslant 1$. Then f has k continuous derivatives. If $\alpha<1$, then $f^{(k)}\in\Lambda_\alpha$. If $\alpha=1$ then $f^{(k)}$ belongs to Λ_* (though not necessarily to Λ_1) and $\omega(\delta;f^{(k)})=O(\delta\log\delta)$.*

Suppose that $E_n[f]\leqslant Mn^{-(k+\alpha)}$, $n=1,2,\ldots$. Let T_n be a polynomial of best approximation of order n for f. Then $f(x)=\lim_n T_{2^n}(x)$ or

$$f(x)=T_2+(T_4-T_2)+\ldots+(T_{2^n}-T_{2^{n-1}})+\ldots, \tag{13·21}$$

the series converging uniformly. Since $u_n=T_{2^n}-T_{2^{n-1}}$ is a polynomial of order 2^n with absolute value not greater than

$$|T_{2^n}-f|+|f-T_{2^{n-1}}|\leqslant 2E_{2^{n-1}}[f]=O(2^{-n(k+\alpha)}),$$

the first inequality (13·17) applied j times shows that $u_n^{(j)}=O(2^{-n(k-j+\alpha)})$. Hence the series (13·21) differentiated termwise j times, $j\leqslant k$, converges absolutely and uniformly. In particular, $f^{(k)}$ exists and is continuous.

We set $f-T_2=g$. It is enough to show that the conclusions of (13·20) are satisfied by g. Let $0<h\leqslant\frac{1}{2}$, and let N be the positive integer satisfying $2^{-N}<h\leqslant 2^{-(N-1)}$. Since $g=u_2+u_3+\ldots$, where $u_n=T_{2^n}-T_{2^{n-1}}$, we have

$$g^{(k)}(x+h)-g^{(k)}(x)=\Sigma\{u_n^{(k)}(x+h)-u_n^{(k)}(x)\}=\sum_2^N+\sum_{N+1}^\infty=P+Q. \tag{13·22}$$

The polynomial u_n is of order 2^n, and $u_n=O(2^{-n(k+\alpha)})$. Hence, by the mean-value theorem and the first inequality (13·17),

$$\begin{aligned}|P|\leqslant h\sum_2^N\max|u_n^{(k+1)}(x)|&\leqslant h\sum_2^N(2.2^n)^{k+1}.O(2^{-n(k+\alpha)})\\&=h\sum_2^N O(2^{n(1-\alpha)})=hO(2^{N(1-\alpha)})=hO(h^{\alpha-1})=O(h^\alpha),\end{aligned}$$

provided $0<\alpha<1$. Next, and this is true for $0<\alpha\leqslant 1$, we have

$$\begin{aligned}|Q|\leqslant\sum_{N+1}^\infty 2|\max u_n^{(k)}(x)|&\leqslant\sum_{N+1}^\infty 2(2.2^n)^k O(2^{-n(k+\alpha)})\\&=\sum_{N+1}^\infty O(2^{-n\alpha})=O(2^{-N\alpha})=O(h^\alpha).\end{aligned}$$

Hence $g^{(k)}\in\Lambda_\alpha$ for $0<\alpha<1$.

If $\alpha=1$, we still have $Q=O(h)$, and the estimate above for P becomes

$$P=O(hN)=O(h\log h).$$

Hence $P+Q=O(h\log h)$ and $\omega(\delta;g^{(k)})=O(\delta\log\delta)$.

It remains to prove that $g^{(k)}\in\Lambda_*$ for $\alpha=1$. With the same relation between h and N we write

$$\begin{aligned}g^{(k)}(x+h)+g^{(k)}(x-h)-2g^{(k)}(x)&=\Sigma\{u_n^{(k)}(x+h)+u_n^{(k)}(x-h)-2u_n^{(k)}(x)\}\\&=\sum_2^N+\sum_{N+1}^\infty=P_1+Q_1.\end{aligned} \tag{13·23}$$

The terms of Q_1 are numerically not greater than $4\max|u_n^{(k)}(x)|$, so that automatically we get the same estimate as for Q, namely $Q_1 = O(h)$. By the mean-value theorem, and arguing as for P,

$$|P_1| \leqslant h^2 \sum_2^N \max |u_n^{(k+2)}| = h^2 \sum_2^N O(2^n) = O(h^2 2^N) = O(h).$$

Hence $P_1 + Q_1 = O(h)$ and $g^{(k)} \in \Lambda_*$.

Remarks. (*a*) From (13·6) and (13·20) we see that *a necessary and sufficient condition that $E_n[f] = O(n^{-(k+\alpha)})$, where k is a non-negative integer and $0 < \alpha \leqslant 1$, is that f should have a continuous k-th derivative belonging to Λ_α if $\alpha < 1$, and to Λ_* if $\alpha = 1$.*

In particular, a necessary and sufficient condition for $f \in \Lambda_\alpha$, $0 < \alpha < 1$, is $E_n[f] = O(n^{-\alpha})$, and for $f \in \Lambda_*$ is $E_n = O(n^{-1})$.

(*b*) Since the class Λ_1 is a proper subset of Λ_* (Chapter II, (4·9)), we cannot replace Λ_* by Λ_1 in (13·20). Taking for instance $k = 0$, we can also verify this by a simple example. Let $f(x) = \sum_1^\infty 2^{-m} \cos 2^m x$, $2^N \leqslant n < 2^{N+1}$. Then

$$|f(x) - S_n(x; f)| = \left| \sum_{N+1}^\infty 2^{-m} \cos 2^m x \right| \leqslant \sum_{N+1}^\infty 2^{-m} = 2^{-N} < 2/n.$$

In particular, $E_n[f] < 2/n$, so that, by (13·20), $f \in \Lambda_*$ (a fact which we have verified directly in Chapter II, p. 47). However, f is not in Λ_1, since $S'[f]$ is not a Fourier series. Thus there is no simple characterization of the class Λ_1 in terms of the order of best approximation.

(*c*) If $f \in \Lambda_*$, then $E_n[f] = O(1/n)$ and so $\omega(\delta; f) = O(\delta \log \delta)$ by (13·20). It follows that *every $f \in \Lambda_*$ has modulus of continuity $O(\delta \log \delta)$*, and, in particular, belongs to Λ_α, $0 < \alpha < 1$. This result is not new (see Chapter II, (3·4)).

(*d*) The proof of (3·15) shows that, if

$$f(x+t) + f(x-t) - 2f(x) = O(t^\alpha) \quad (t > 0) \tag{13·24}$$

for $0 < \alpha < 1$, then $\sigma_n[f] - f = O(n^{-\alpha})$. In particular $E_n[f] = O(n^{-\alpha})$, so that $f \in \Lambda_\alpha$, that is, $f(x+t) - f(x) = O(t^\alpha)$. Since the latter condition implies (13·24), we see that Λ_α, $0 < \alpha < 1$, can be defined as the class of continuous functions satisfying (13·24).† It is only for $\alpha = 1$ that condition (13·24) yields a new class, Λ_*, larger than Λ_1.

(*e*) *For every continuous f,*

$$|f(x) - S_n(x; f)| \leqslant (L_n + 1) E_n[f], \tag{13·25}$$

where L_n is the Lebesgue constant (Chapter II, § 12). For let T_n be a polynomial of best approximation of order n for f, and let $f = T_n + g$. Then

$$|f - S_n[f]| = |T_n - S_n[T_n] + g - S_n[g]| = |g - S_n[g]| \leqslant |g| + |S_n[g]|$$
$$\leqslant \max |g| + L_n \max |g| = (L_n + 1) E_n[f].$$

In particular, since $L_n = O(\log n)$,

$$f(x) - S_n(x; f) = O(n^{-k-\alpha} \log n) \quad (k = 0, 1, \ldots; 0 < \alpha \leqslant 1), \tag{13·26}$$

provided f has k derivatives, and $f^{(k)} \in \Lambda_\alpha$ for $\alpha < 1$, $f^{(k)} \in \Lambda_*$ for $\alpha = 1$.

The inequality (13·25) shows that the approximation of f by $S_n[f]$ is at most $L_n + 1 = O(\log n)$ times worse than the best approximation. For $k = 0$, (13·26) gives

† This can also be shown directly by the method which gave Theorem (3·6) of Chapter II.

Theorem (10·8) of Chapter II. If $\omega(\delta; f) = o(|\log \delta|^{-1})$, we have $E_n = o(1/\log n)$ by (13·6), and (13·25) shows that $S[f]$ converges uniformly to f. This is Theorem (10·3) of Chapter II.

(13·27) THEOREM. *Under the hypothesis of* (13·20), $\tilde{f}$ *satisfies the same conclusions as* f.

Denote the series (13·21) again by $u_1 + u_2 + \dots$. We shall show that

$$\tilde{f} = \tilde{u}_1 + \tilde{u}_2 + \dots \quad (\tilde{u}_n = \tilde{T}_{2^n} - \tilde{T}_{2^{n-1}} \text{ for } n > 1). \tag{13·28}$$

Let $\tilde{L}_n$ be the constant introduced in Chapter II, (12·3). Since $\tilde{L}_n = O(\log n)$,

$$\max_x |\tilde{u}_n(x)| \leqslant \tilde{L}_{2^n} \max |u_n(x)| = O(n)\, O(2^{-n(k+\alpha)}) = O(n2^{-n\alpha}),$$

which shows that the series $\Sigma \tilde{u}_n$ converges uniformly to a function $f^*(x)$. If f_n are the partial sums of the series $u_1 + u_2 + \dots$, the partial sums in (13·28) are $\tilde{f}_n$. Since $f_n \to f$ and $\tilde{f}_n \to f^*$, both uniformly, each Fourier coefficient of f_n tends to the corresponding coefficient of f, and the coefficients of $\tilde{f}_n$ tend to those of f^*. Hence $f^* = \tilde{f}$, and (13·28) follows.

The series (13·28) differentiated termwise j times, $j \leqslant k$, also converges uniformly. For

$$\max |\tilde{u}_n^{(j)} x| \leqslant (2 \,.\, 2^n)^j \,.\, \max |\tilde{u}_n(x)| = O(n 2^{-n(k-j+\alpha)}) = O(n2^{-n\alpha}).$$

It remains to show that $\tilde{f}^{(k)}$ satisfies the same conclusions as $f^{(k)}$ in (13·20). Suppose first that $k \geqslant 1$. Then, as in (13·22),

$$\tilde{g}^{(k)}(x+h) - \tilde{g}^{(k)}(x) = \Sigma\{\tilde{u}_n^{(k)}(x+h) - \tilde{u}_n^{(k)}(x)\};$$

and from this point on the proof remains exactly the same as before, since in terms of $\max |u_n|$ we have the same estimates for the derivatives of $\tilde{u}_n$ as we had for the derivatives of u_n. (We use the second inequality (13·17) instead of the first.) Similarly for the analogue of (13·23).

Suppose now that $k = 0$ and that $\alpha < 1$. Since subtracting a constant from f does not affect $E_n[f]$ we may suppose that the constant term of $S[f]$ is 0, so that the integral F of f is periodic. By (13·6), $E_n[F] = O(n^{-1-\alpha})$, and, by the case just disposed of, $\tilde{F}$ has a derivative, obviously $\tilde{f}$, belonging to Λ_α. The argument holds when $k = 0$, $\alpha = 1$.

On taking $k = 0$ in (13·14) and (13·27), we obtain the following result:

(13·29) THEOREM. *If* f *belongs to* Λ_α, $0 < \alpha < 1$, *so does* $\tilde{f}$; *if* f *belongs to* Λ_* *so does* $\tilde{f}$.

It must be added that if $f \in \Lambda_1$, the function $\tilde{f}$ need not belong to Λ_1, since the function conjugate to a bounded function need not be bounded.

Theorem (13·29) can also be proved directly; we shall confine our attention to the first part, slightly generalizing it.

(13·30) THEOREM. *If the modulus of continuity of* f *is* $\omega(h)$, *that of* $\tilde{f}$ *does not exceed*

$$A\left[\int_0^h t^{-1}\omega(t)\,dt + h\int_h^\pi t^{-2}\omega(t)\,dt\right] = A\int_0^h dt \int_t^\pi u^{-2}\omega(u)\,du \quad (h \leqslant \tfrac{1}{2}\pi), \tag{13·31}$$

where A *is an absolute constant.*

We may suppose that $\omega(t)/t$ is integrable near 0, so that the integral defining $\tilde{f}$ is absolutely convergent, since otherwise there is nothing to prove. Now the identity (13·31) follows by integration by parts. (We note that $\int_t^\pi u^{-2}\omega(u)\,du = o(t^{-1})$ as $t\to 0$.)

For the proof of (13·30) we consider the formulae

$$\tilde{f}(x) = -\frac{1}{\pi}\int_{-\pi}^{\pi}\frac{f(x+t)-f(x)}{2\tan\frac{1}{2}t}\,dt,\quad \tilde{f}(x+h) = -\frac{1}{\pi}\int_{-\pi}^{\pi}\frac{f(x+t)-f(x+h)}{2\tan\frac{1}{2}(t-h)}\,dt.$$

The first is obvious since $\cot\frac{1}{2}t$ is odd, and the second follows from the first on replacing x by $x+h$ and t by $t-h$. The integrands are respectively majorized by $\omega(|t|)\,|t|^{-1}$ and $\omega(|t-h|)\,|t-h|^{-1}$. Thus if we cut out the interval $(-2h, 2h)$ from the interval of integration $(-\pi,\pi)$ and write $I(h) = \frac{1}{\pi}\int_0^h t^{-1}\omega(t)\,dt$, we commit errors at most $2I(2h) \leqslant 4I(h)$ in the first integral and at most $I(h)+I(3h) \leqslant 4I(h)$ in the second. It follows that with an error not greater than $8I(h)$ we have

$$\tilde{f}(x+h)-\tilde{f}(x) = -\frac{1}{\pi}\left(\int_{-\pi}^{-2h}+\int_{2h}^{\pi}\right)[f(x+t)-f(x)]\,[\tfrac{1}{2}\cot\tfrac{1}{2}(t-h)-\tfrac{1}{2}\cot\tfrac{1}{2}t]\,dt$$
$$+\frac{1}{\pi}[f(x+h)-f(x)]\left(\int_{-\pi}^{-2h}+\int_{2h}^{\pi}\right)\tfrac{1}{2}\cot\tfrac{1}{2}(t-h)\,dt.$$

The first term on the right is absolutely less than

$$\frac{1}{2\pi}\left(\int_{-\pi}^{-2h}+\int_{2h}^{\pi}\right)\frac{\omega(|t|)\sin\frac{1}{2}h}{|\sin\frac{1}{2}(t-h)\sin\frac{1}{2}t|}\,dt = \int_{2h}^{\pi}O(ht^{-2})\,\omega(t)\,dt = O\left(h\int_h^\pi t^{-2}\omega(t)\,dt\right).$$

A simple integration shows that the coefficient of $f(x+h)-f(x)$ in the remaining term is $O(1)$, so that the total contribution from it is $O\{\omega(h)\}$. Since

$$\omega(h) \leqslant \omega(h)\,2h\int_h^\pi t^{-2}dt \leqslant 2h\int_h^\pi t^{-2}\omega(t)\,dt,$$

we see, collecting results, that $|\tilde{f}(x+h)-\tilde{f}(x)|$ does not exceed the first expression (13·31), and since the latter increases with h, the inequality for $\omega(h;\tilde{f})$ follows.

We know that the approximation of f by $S_n[f]$ is only $O(\log n)$ times worse than the best approximation $E_n[f]$ of f. The approximation of f by $\tau_n = 2\sigma_{2n-1}-\sigma_{n-1}$ is of the same order as $E_n[f]$. It is curious that the σ_n themselves give only mediocre approximations, though they converge uniformly to every continuous f.

(13·32) THEOREM. *If $\sigma_n(x;f)-f(x) = o(1/n)$ uniformly in x, then $f\equiv$ const.*

For if c_k are the complex coefficients of f, then

$$(2\pi)^{-1}\int_0^{2\pi}\{f(x)-\sigma_n(x)\}\,e^{-ikx}\,dx = |k|\,c_k/(n+1)\quad (|k|\leqslant n),$$

and the relation $f-\sigma_n = o(1/n)$ implies that the left-hand side here is $o(1/n)$, which means that $c_k = 0$ for $k\neq 0$, that is, $f\equiv c_0$.

Thus if $f \not\equiv$ const., the (C, 1) means of $S[f]$ never give approximation with error $o(1/n)$. A similar argument shows that *the Abel means $f(r,x)$ of $S[f]$ never give approximation with error $o(1-r)$, unless $f \equiv$ const.* Generally, let us consider a matrix (with finite or infinite rows)

$$\begin{pmatrix} \gamma_{00} & \gamma_{01} & \cdots & \gamma_{0m_0} & \cdots & \cdots \\ \gamma_{10} & \gamma_{11} & \cdots & \cdots & \gamma_{1m_1} & \cdots \\ \cdots & \cdots & \cdots & \cdots & \cdots & \cdots \\ \gamma_{n0} & \gamma_{n1} & \cdots & \cdots & \cdots & \gamma_{nm_n} \\ \cdots & \cdots & \cdots & \cdots & \cdots & \cdots \end{pmatrix}. \tag{13·33}$$

If $f \sim \Sigma A_k(x)$, and if $f_n = \sum_0^{m_n} \gamma_{nk} A_k(x)$ is to approach f with an error $o(\rho_n)$, or $O(\rho_n)$, where $\rho_n \to 0$, then in each column of (13·33) $1-\gamma_{nk}$ must be $o(\rho_n)$ or $O(\rho_n)$, as the case may be. If f_n is either $S_n(x;f)$ or $\tau_n(x;f)$, then $\gamma_{nk}=1$ for k fixed and n large enough, so that the condition is satisfied. This condition is only necessary, not sufficient, but it explains the fact that for trigonometric series with coefficients rapidly tending to 0 ordinary partial sums may give a better approximation than stronger methods of summability.

Let $0<\alpha<1$. By (3·15) and (13·20), we have $\sigma_n[f]-f=O(n^{-\alpha})$ if and only if $f \in \Lambda_\alpha$. We shall now prove the following theorem.

(13·34) THEOREM. *A necessary and sufficient condition for $\sigma_n[f]-f=O(n^{-1})$ is that $\tilde{f} \in \Lambda_1$.*

Necessity. The hypothesis $\sigma_n[f]-f=O(1/n)$ indicates that $\tilde{f}$ exists and is continuous. It will be convenient to write $T_n(x)$ for $\sigma_n(x;f)$. Let $f=T_n+g_n$. Then

$$\sigma_n[f]-f=\{\sigma_n[T_n]-T_n\}+\{\sigma_n[g_n]-g_n\}=O(1/n).$$

The hypothesis $g_n=O(1/n)$ leads to $\sigma_n[g_n]=O(1/n)$, and so the same estimate holds for $T_n-\sigma_n[T_n]$. From the general formula (1·25) we deduce that

$$T_n-\sigma_n[T_n]=\tilde{T}'_n/(n+1),$$

and since the left-hand side is $O(1/n)$, it follows that $\tilde{T}'_n(x)=O(1)$. Hence the $\tilde{T}_n(x)$ satisfy condition Λ_1 uniformly in n, and so $\tilde{f}=\lim \tilde{T}_n$ is in Λ_1.

Sufficiency. Interchanging the roles of f and $\tilde{f}$ it is enough to show that if $f \in \Lambda_1$, then $\tilde{\sigma}_n[f]-\tilde{f}=O(1/n)$. From (3·17) we see that $\tilde{\sigma}_n(x)-\tilde{f}(x)$ equals

$$\frac{1}{\pi(n+1)}\int_0^\pi \{f(x+t)-f(x-t)\}\frac{\sin (n+1)t}{(2\sin \frac{1}{2}t)^2}\,dt=\int_0^{1/n}+\int_{1/n}^\pi=P+Q,$$

say. Assuming that $|f(x+t)-f(x)| \leqslant M|t|$, we have

$$|P| \leqslant \pi^{-1}(n+1)^{-1}\int_0^{1/n} 2Mt(n+1)t\left(\frac{2}{\pi}t\right)^{-2} dt=O(1/n).$$

By the second mean-value theorem,

$$\left|\int_t^\pi \frac{\sin (n+1)u}{(2\sin \frac{1}{2}u)^2}\,du\,.\right| \leqslant \frac{2}{n+1}\frac{1}{(2\sin \frac{1}{2}t)^2} \leqslant \frac{A}{nt^2}.$$

Denote the integral on the left by $R_n(t)$. Since f' exists almost everywhere and $|f'|\leqslant M$, integration by parts gives

$$Q=\pi^{-1}(n+1)^{-1}\left\{[-R_n(t)\,(f(x+t)-f(x-t))]_{1/n}^{\pi}+\int_{1/n}^{\pi}[f'(x+t)+f'(x-t)]\,R_n(t)\,dt\right\},$$

$$|Q|\leqslant\pi^{-1}(n+1)^{-1}\left\{|R_n(1/n)|\,.\,2Mn^{-1}+\int_{1/n}^{\pi}2M\,.\,An^{-1}t^{-2}dt\right\}=O(n^{-1}).$$

Hence $P+Q=O(1/n)$, $\tilde{\sigma}_n-\tilde{f}=O(1/n)$.

(13·35) THEOREM. *If* $S[f]$ *is of power series type, then* $f-\sigma_n[f]=O(1/n)$ *if and only if* $f\in\Lambda_1$.

For in this case $\tilde{f}=-i(f-c_0)$.

MISCELLANEOUS THEOREMS AND EXAMPLES

1. Let a simple, closed, convex curve L be given by the equations $x=\phi(t)$, $y=\psi(t)$, $0\leqslant t\leqslant 2\pi$. Show that if $\phi_n(t)$ and $\psi_n(t)$ are the (C, 1) means of $S[\phi]$ and $S[\psi]$, then the curves $x=\phi_n(t)$, $y=\psi_n(t)$ are in the interior of L. (Fejér [5].)

[If A, B, C are constants and $A\phi(t)+B\psi(t)+C$ is non-negative, but not identically zero, then $A\phi_n(t)+B\psi_n(t)+C>0$. Here, as in Example 2 below, the result is immediately extensible to any non-negative kernel, in particular to Poisson's.]

2. Let $\phi(t)$ and $\psi(t)$ be periodic functions, $\phi_n(t)$ and $\psi_n(t)$ the (C, 1) means of $S[\phi]$ and $S[\psi]$, and L, L_n the lengths of the curves $x=\phi(t)$, $y=\psi(t)$ and $x=\phi_n(t)$, $y=\psi_n(t)$, $0\leqslant t\leqslant 2\pi$, respectively. Show that $L_n\leqslant L$ for all n. (Compare Pólya and Szegö, *Aufgaben und Lehrsätze*, I, p. 56, Problem 89.)

[Let $0=t_0<t_1<\dots<t_k=2\pi$. Let $\Delta_j\phi=\phi(t_j)-\phi(t_{j-1})$, $\Delta_j\phi(t)=\phi(t_j+t)-\phi(t_{j-1}+t)$, and similarly for ψ, ϕ_n, ψ_n. Then

$$|\Delta_j\phi_n\cos\alpha+\Delta_j\psi_n\sin\alpha|\leqslant\pi^{-1}\int_0^{2\pi}|\Delta_j\phi(t)\cos\alpha+\Delta_j\psi(t)\sin\alpha|\,K_n(t)\,dt$$

for all α and j. We integrate this with respect to α over $0\leqslant\alpha\leqslant 2\pi$, interchange the order of integration on the right, and use the equation $\int_0^{2\pi}|a\cos\alpha+b\sin\alpha|\,d\alpha=4(a^2+b^2)^{\frac{1}{2}}$. Summation with respect to j gives

$$\sum_j\{(\Delta_j\phi_n)^2+(\Delta_j\psi_n)^2\}^{\frac{1}{2}}\leqslant\pi^{-1}\int_0^{2\pi}\sum_j\{(\Delta_j\phi(t))^2+(\Delta_j\psi(t))^2\}^{\frac{1}{2}}K_n(t)\,dt\leqslant\pi^{-1}\int_0^{2\pi}LK_n(t)\,dt=L,$$

so that $L_n\leqslant L$.]

3. Let $f(x)$ be periodic, integrable and equal to 0 for $x_0<x<x_0+h$. Let Γ be any circle tangent internally to the unit circle Γ_1 at the point e^{ix_0}. Show that $f(r,x)$ tends to 0 as re^{ix} approaches e^{ix_0} through that part Δ of the cuspidal region between Γ and Γ_1 for which $x>x_0$. What localization theorem does this give? (Hardy and Rogosinski, *Fourier series*, p. 65.)

[Let $x_0=0$. For re^{ix} tending to e^{ix_0} through Δ we have

$$f(r,x)=\pi^{-1}\int_{-h}^{0}P(r,x-t)\,f(t)\,dt+o(1).$$

If re^{ix} belongs to Δ, so does $re^{i(x-t)}$ for $t<0$. It is now enough to observe that

$$P(r,u)=\tfrac{1}{2}\mathscr{R}\{(1+re^{iu})/(1-re^{iu})\}$$

is bounded for $z=re^{iu}$ situated between Γ and Γ_1 (the function $\zeta=\frac{1}{2}(1+z)/(1-z)$ maps this domain into a vertical strip of the ζ plane), and that $\int_{-h}^{0}|f|\,dt$ is small with h.]

4. The nth partial sum of $\frac{1}{2}+r\cos x+r^2\cos 2x+\dots$ is non-negative for $0\leqslant r\leqslant\frac{1}{2}$, though not necessarily for $r>\frac{1}{2}$. (Fejér [6].)

[The partial sum is

$$P_n(r,x)=\frac{1-r^2-2r^{n+1}[\cos(n+1)x-r\cos nx]}{2(1-2r\cos x+r^2)}.$$

The sum $\frac{1}{2}+r\cos x$ is negative for $x=\pi$ if $r>\frac{1}{2}$.]

5. Let $f_n(r,x)$ be the nth partial sum of the series $f(r,x)$ in (6·1). Show that if $m\leqslant f(x)\leqslant M$ for all x, then $m\leqslant f_n(r,x)\leqslant M$ for $r\leqslant\frac{1}{2}$, but not necessarily for $r>\frac{1}{2}$. (Fejér [6].)

[A corollary of Example 4.]

6. For any $N=1,2,\dots$ there is a number r_N with the following property. Under the hypotheses of Example 5, we have $m\leqslant f_n(r,x)\leqslant M$ for $0\leqslant r\leqslant r_N$, $n\geqslant N$. Moreover, $r_N\leqslant r_{N+1}$, $r_N\to 1$ as $N\to\infty$. (Schur and Szegö [1]. See Example 4.)

7. The (C, 1) means $\sigma_n(x)$ of the series $\Sigma n^{-1}\sin nx$ are positive and less than $\frac{1}{2}(\pi-x)$ in the interval $0<x<\pi$.

[See Chapter II, (9·4). Also

$$\sigma_n(x)=-\tfrac{1}{2}x+\int_0^x K_n(t)\,dt<-\tfrac{1}{2}x+\int_0^\pi K_n\,dt=\tfrac{1}{2}(\pi-x).]$$

8. If $u_0+u_1+u_2+\dots$ is summable (C, α), or A, to sum s, so is $0+u_0+u_1+\dots$, while $u_1+u_2+\dots$ is summable to $s-u_0$.

9. For any series $u_0+u_1+\dots$ with partial sums s_n, let

$$s_n^*=\tfrac{1}{2}(s_n+s_{n-1})=u_0+u_1+\dots+\tfrac{1}{2}u_n$$

be the modified partial sums. Let $\sigma_n^{*\alpha}=s_n^{*\alpha}/A_n^\alpha$ be the (C, α) means of the sequence $\{s_n^*\}$. Show that

$$\sum_0^\infty s_n^{*\alpha}r^n=\frac{1}{2}\frac{1+r}{(1-r)^{\alpha+1}}\sum_0^\infty u_n r^n.$$

10. Let $\sigma_n^{*\alpha}(x;f)$ be the (C, α) means of $S_n^*(x;f)$. Show that under the hypotheses of (10·2),

$$m\sigma_n^{*2}(x;\mu)\leqslant\sigma_n^{*2}(x;f)\leqslant M\sigma_n^{*2}(x;\mu)\quad(0\leqslant x\leqslant\pi).$$

[For the termwise differentiated series $\frac{1}{2}+\cos t+\cos 2t+\dots$ we have

$$-\sum_0^\infty S_n^{*2}(t)\,r^n=\frac{1}{2}\frac{1+r}{(1-r)^3}\frac{1-r^2}{\Delta^2(r,t)}r\sin t=\left[\frac{1}{2}\frac{1-r^2}{(1-r)^2\Delta}\right]^2 2r\sin t,$$

so that $S_n^{*2}(t)<0$ for $0<t<\pi$.]

11. The (C, 3) means in (10·2) cannot be replaced by (C, 2) means. (Fejér [4].)

[$\{K_n^2(t)\}'$ is positive if $\sin(n+\frac{3}{2})t=0$, $\cos(n+\frac{3}{2})t=-1$, $\cos\frac{1}{2}t<\frac{1}{2}$.]

12. If $F'(x_0)=\lim_{h\to 0}[F(x_0+h)-F(x_0-h)]/2h$ exists and is finite, then at the point x_0 $\mathfrak{S}'[F]$ is summable by the logarithmic mean to sum $F'(x_0)$.

13. Let $f(z)=c_0+c_1z+c_2z^2+\dots$ be regular for $|z|<1$, continuous for $|z|\leqslant 1$. Let a, b, α, β be numbers, real or complex, satisfying $\alpha+\beta=1$, $\alpha e^a+\beta e^b=0$. Show that then $\alpha s_n(z\,e^{a/n})+\beta s_n(z\,e^{b/n})$ converges uniformly to $f(z)$ for $|z|\leqslant 1$. Here $s_n(z)=c_0+c_1z+\dots+c_nz^n$. (Rogosinski and Szegö [1].)

[The argument closely resembles that of § 12.]

14. Let $S_n^*(x)$ be the modified partial sums of $\mathfrak{S}[f]$. At every point x at which $\Phi_x(t)=o(t)$, a necessary and sufficient condition for the convergence of the series $\Sigma(S_k^*-f)/k$ is the convergence of the integral $\displaystyle\int_0^\pi\frac{\phi_x(t)}{2\sin\frac{1}{2}t}dt$. (See Hardy and Littlewood [7].)

[Let $\qquad u_n(x)=\sin x+2^{-1}\sin 2x+\dots+n^{-1}\sin nx,\quad r_n(x)=\tfrac{1}{2}(\pi-x)-u_n(x).$

Plainly $|u_n(x)|\leqslant nx$, and applying summation by parts we get $r_n(x)=O(1/nx)$. Let $T_n(x)$ be the nth partial sum of the given series. Then

$$T_n(x)=\frac{2}{\pi}\int_0^{\pi}\frac{\phi_x(t)}{2\tan\frac{1}{2}t}u_n(t)\,dt=\frac{2}{\pi}\int_0^{1/n}+\frac{2}{\pi}\int_{1/n}^{\pi}=A+B.$$

Here $A\to 0$, and on account of the inequality for r_n,

$$T_n(x)-\frac{2}{\pi}\int_{1/n}^{\pi}\frac{\phi_x(t)}{2\tan\frac{1}{2}t}\,\frac{\pi-t}{2}\,dt\to 0.]$$

15. Suppose that $F_x(h)$ is the integral of $\phi_x(t)$ over $0\leqslant t\leqslant h$ and that $\Phi_x(h)$ has its usual meaning. Neither of the conditions

$$\text{(i)}\ \ F_x(h)=o(h),\quad \text{(ii)}\ \ \Phi_x(h)=O(h)$$

taken separately implies the summability (C, 1) of $S[f]$ at x. Show that if both of them are satisfied, then $S[f]$ is summable (C, 1) at x to sum $f(x)$. (Hardy and Littlewood [8].)

This generalization of Theorem (3·9) is typical and many other results can be generalized similarly.

[The proof is similar to that of (3·9) except that now we split the integral (3·11) into integrals extended over intervals $(0, k/n)$, $(k/n, \pi)$, where k is large but fixed. By (ii) and the second estimate (3·10), the second integral is small with $1/k$. The Fejér kernel has a bounded number of maxima and minima in $(0, k/n)$ and so, by the second mean-value theorem and (i), the first integral tends to 0.]

16. Let f be periodic and k times continuously differentiable, and let $T_n(x)$ be a polynomial of best approximation of order n to f. Then $T_n^{(k)}(x)$ tends uniformly to $f^{(k)}(x)$. (E. Stein [1].)

[If $\tau_n(x)=\tau_n(x;f)$ is the delayed (C, 1) mean of $S[f]$, then

$$T_n^{(k)}(x)-f^{(k)}(x)=[T_n(x)-\tau_n(x;f)]^{(k)}+[\tau_n(x;f^{(k)})-f^{(k)}(x)].$$

The second term on the right tends uniformly to 0. The preceding term is, by (13·16),

$$O(n^k)\max_x|T_n(x)-\tau_n(x;f)|,$$

and it is enough to observe that T_n-f and τ_n-f are both $o(n^{-k})$.]

CHAPTER IV

CLASSES OF FUNCTIONS AND FOURIER SERIES

1. The class L^2

Let $\phi_1(x)$, $\phi_2(x)$, ... be a system orthonormal in (a, b). If $c_1, c_2, \ldots$ are the Fourier coefficients of an $f \in L^2$, with respect to $\{\phi_n\}$, the series $\Sigma \mid c_\nu \mid^2$ converges. The converse is one of the most important results of the Lebesgue theory of integration.

(1·1) THEOREM OF RIESZ AND FISCHER. *Let $\phi_1, \phi_2, \ldots$ be an orthonormal set of functions in (a, b) and let $c_1, c_2, \ldots$ be any sequence of numbers such that $\Sigma \mid c_\nu \mid^2$ converges. Then there is a function $f \in L^2(a, b)$ such that the Fourier coefficient of f with respect to ϕ_ν is c_ν for all ν, and moreover*

$$\int_a^b |f|^2 dx = \sum_{\nu=1}^{\infty} |c_\nu|^2, \tag{1·2}$$

$$\int_a^b |f - s_n|^2 dx \to 0, \tag{1·3}$$

where s_n is the n-th partial sum of the series $c_1\phi_1 + c_2\phi_2 + \ldots$.

The equation

$$\int_a^b |s_{n+k} - s_n|^2 = \sum_{n+1}^{n+k} |c_\nu|^2$$

implies that $\mathfrak{M}_2[s_m - s_n] \to 0$ as $m, n \to \infty$. By Theorem (11·1) of Chapter I, there is a function $f \in L^2$ such that $\mathfrak{M}_2[f - s_n] \to 0$. If $n \geqslant j$,

$$c_j = \int_a^b s_n \bar{\phi}_j dx = \int_a^b f \bar{\phi}_j dx + \int_a^b (s_n - f) \bar{\phi}_j dx.$$

By Schwarz's inequality, the last integral does not exceed $\mathfrak{M}_2[s_n - f] = o(1)$ in absolute value, and making $n \to \infty$ we see that c_j is the Fourier coefficient of f with respect to ϕ_j. Since s_n is now the nth partial sum of the Fourier series of f, the left-hand side of (1·3) is

$$\int_a^b |f|^2 dx - (|c_1|^2 + \ldots + |c_n|^2)$$

(Chapter I, (7·4)), and (1·2) follows on making $n \to \infty$.

In Chapter I, §3, we defined complete orthonormal systems. A system $\{\phi_\nu\}$ orthonormal in (a, b) is said to be *closed* if for each $f \in L^2(a, b)$ we have the Parseval formula

$$\int_a^b |f|^2 dx = \sum_{\nu=1}^{\infty} |c_\nu|^2 \quad \left(c_\nu = \int_a^b f \bar{\phi}_\nu dx\right). \tag{1·4}$$

In the domain of functions of the class L^2 the notions of 'closed' and 'complete' systems are equivalent. Every closed system is obviously complete. To prove the converse, let $c_1, c_2, \ldots$ be the Fourier coefficients of an $f \in L^2(a, b)$ with respect to $\{\phi_\nu\}$. Since $\Sigma \mid c_\nu \mid^2$ converges, there is, by (1·1), a $g \in L^2$ with Fourier coefficients c_ν and such that $\mathfrak{M}_2^2[g] = |c_1|^2 + |c_2|^2 + \ldots$. Since f and g have the same Fourier coefficients and $\{\phi_\nu\}$ is complete, we have $f \equiv g$ and (1·4) follows.

If the system $\{\phi_\nu\}$ in (1·1) is complete, the function f there is uniquely determined. Suppose now that $\{\phi_\nu\}$ is not complete and let $\{\psi_n\}$ be one of its completions, so that the system $\phi_1, \phi_2, \ldots, \psi_1, \psi_2, \ldots$ is orthonormal and complete in (a, b). Let $d_n = \int_a^b f\bar{\psi}_n\,dx$. From the Parseval formula,

$$\int_a^b |f|^2\,dx = \Sigma\,|c_\nu|^2 + \Sigma\,|d_n|^2,$$

and from (1·2) we get $d_1 = d_2 = \ldots = 0$. Thus, *if $\{\phi_\nu\}$ is not complete, the function f of the Riesz-Fischer theorem is uniquely determined by the condition that its Fourier coefficients with respect to the ϕ_ν are c_ν and with respect to any system completing $\{\phi_\nu\}$ are zero.*

(1·5) THEOREM. *A system $\{\phi_\nu\}$ orthonormal in (a, b) is complete if and only if for any $f \in L^2(a, b)$ and any $\epsilon > 0$ there is a linear combination $S = \gamma_1\phi_1 + \ldots + \gamma_n\phi_n$ with constant coefficients such that $\mathfrak{M}_2[f - S] < \epsilon$.*

For the completeness is equivalent to (1·4), and this in turn is equivalent to $\mathfrak{M}_2[f - s_n] \to 0$, where s_n is the partial sum of the Fourier series $c_1\phi_1 + c_2\phi_2 + \ldots$ of f. Hence if $\{\phi_\nu\}$ is complete we can find an $S = s_n$ such that $\mathfrak{M}_2[f - S] < \epsilon$. Conversely, if $\mathfrak{M}_2[f - S] < \epsilon$ for some $S = \gamma_1\phi_1 + \ldots + \gamma_n\phi_n$ then $\mathfrak{M}_2[f - s_n] \leqslant \mathfrak{M}_2[f - S] < \epsilon$ (Chapter I, (7·3)), so that $\mathfrak{M}_2[f - s_n] \to 0$.

The trigonometric system is complete (Chapter I, § 6). It is thus closed, and we get one more proof of Parseval's formula for this system (see also § 1 of Chapter II and § 3 of Chapter III).

Let a_ν, b_ν be the trigonometric Fourier coefficients of an $f \in L^2$. By (1·1),

$$\tilde{S}[f] = \Sigma(a_\nu \sin \nu x - b_\nu \cos \nu x) \tag{1·6}$$

is the Fourier series of a function of class L^2. Thus $\tilde{S}[f]$ is summable (C, 1) almost everywhere and, by Theorem (3·20) of Chapter III, the conjugate function $\tilde{f}(x)$ exists and is equal to the (C, 1) sum of (1·6) almost everywhere. Hence $\tilde{S}[f] = S[\tilde{f}]$ and, by Parseval's formula,

$$\frac{1}{\pi}\int_0^{2\pi} f^2\,dx = \tfrac{1}{2}a_0^2 + \frac{1}{\pi}\int_0^{2\pi} \tilde{f}^2\,dx. \tag{1·7}$$

Though the problem of the convergence almost everywhere of Fourier series will be discussed only in a later chapter (see Chapter XIII), a special result may be mentioned here.

(1·8) THEOREM. *The series*

$$\tfrac{1}{2}a_0 + \sum_{n=1}^{\infty} (a_n \cos nx + b_n \sin nx) = \sum_0^{\infty} A_n(x) \tag{1·9}$$

converges almost everywhere if $\Sigma(a_n^2 + b_n^2)\log^2 n$ is finite.

For the condition implies that $\Sigma A_n(x) \log n$ is a Fourier series, and it is enough to apply Theorem (4·4) of Chapter III.

We shall prove in Chapter XIII, § 1, that the finiteness of $\Sigma(a_n^2 + b_n^2)\log n$ is sufficient for the convergence almost everywhere of (1·9); but the proof of that is much less simple. Whether the hypothesis $\Sigma(a_n^2 + b_n^2) < \infty$ is sufficient remains an open question.

2. A theorem of Marcinkiewicz

The fact that every integrable function is almost everywhere the derivative of its indefinite integral is fundamental in questions about the representation of functions by their Fourier series. But certain problems require more than even the strengthened forms (11·1) and (11·3) of Chapter II, and we need to go more deeply into the structure of functions and point sets. The following result is particularly useful:

(2·1) THEOREM OF MARCINKIEWICZ. *Let P be a closed set in a finite interval (a,b) and let $\chi(t)=\chi_P(t)$ be the distance of the point t from P. Then*

(i) *for every $\lambda>0$ the integral*

$$I_\lambda(x)=I_\lambda(x,P)=\int_a^b \frac{\chi^\lambda(t)}{|t-x|^{\lambda+1}}\,dt \tag{2·2}$$

is finite at almost all points of P; more generally, if f is an integrable function in (a,b), the integral

$$J_\lambda(x)=J_\lambda(x,f,P)=\int_a^b \frac{f(t)\,\chi^\lambda(t)}{|t-x|^{\lambda+1}}\,dt \tag{2·3}$$

converges absolutely at almost all points of P and

$$\int_P |J_\lambda(x)|\,dx \leqslant 2\lambda^{-1}\int_a^b |f(x)|\,dx. \tag{2·4}$$

(ii) *If all intervals contiguous to P are of length less than 1, the integrals*

$$I_0(x)=\int_a^b \frac{\{\log 1/\chi(t)\}^{-1}}{|t-x|}\,dt,\quad J_0(x)=\int_a^b \frac{f(t)\,\{\log 1/\chi(t)\}^{-1}}{|t-x|}\,dt \tag{2·5}$$

converge absolutely at almost all points of P and

$$\int_P |J_0(x)|\,dx \leqslant A\int_a^b |f(x)|\,dx, \tag{2·6}$$

where A is a positive constant independent of f.

(i) It is enough to consider J_λ. We may suppose that $f\geqslant 0$. Then $0\leqslant J_\lambda(x)\leqslant\infty$, and if we prove (2·4) the finiteness of $J_\lambda(x)$ at almost all points of P will follow. Observe that the function $\chi(t)$ vanishes on P and its graph over any interval d contiguous to P is an isosceles triangle of height $\frac{1}{2}|d|$; also $\chi(t)$ is linear to the right and left of P.

Integration in (2·3) may be confined to the set $Q=(a,b)-P$. We have

$$\int_P J_\lambda(x)\,dx=\int_Q f(t)\,\chi^\lambda(t)\left\{\int_P \frac{dx}{|t-x|^{\lambda+1}}\right\}dt, \tag{2·7}$$

the interchange of the order of integration being justified by the positiveness of the integrand. To estimate the inner integral, fix a point t interior to an interval (α,β) contiguous to P and suppose, say, that t is closer to α than to β. Then

$$\int_P \frac{dx}{|t-x|^{\lambda+1}}\leqslant 2\int_{t-\alpha}^\infty u^{-\lambda-1}\,du=2\lambda^{-1}(t-\alpha)^{-\lambda}=2\lambda^{-1}\chi^{-\lambda}(t), \tag{2·8}$$

an estimate which still holds if t is to the right or left of P. Substituting this in (2·7) we obtain (2·4).

(ii) We consider J_0 and suppose that $f \geqslant 0$. If $\lambda = 0$, $l = b - a$, the left-hand side of (2·8) does not exceed

$$2\int_{t-\alpha}^{l} u^{-1}du = 2\log l - 2\log(t-\alpha) = 2\log l + 2\log 1/\chi(t) < A\log 1/\chi(t),$$

and this, combined with

$$\int_P J_0(x)\,dx = \int_Q f(t)\{\log 1/\chi(t)\}^{-1}\left\{\int_P \frac{dx}{|t-x|}\right\}dt,$$

immediately gives (2·6).

A modification of the function $\chi(t)$ is sometimes useful. Denote by $\chi^*(t)$ $(=\chi_P^*(t))$ the function equal to 0 in P, equal to $|d|$ if t is in an interval d contiguous to P, and equal, say, to 0 to the left and right of P.

(2·9) THEOREM. *With the hypotheses of Theorem* (2·1) *the integrals*

$$J_\lambda^*(x) = \int_a^b \frac{f(t)\,\chi^{*\lambda}(t)}{|t-x|^{\lambda+1}}dt, \quad J_0^*(x) = \int_a^b \frac{f(t)\{\log 1/\chi^*(t)\}^{-1}}{|t-x|}dt$$

converge at almost all points of P.

It is enough to consider J_λ^*, the proof for J_0^* being similar. We may suppose that $f \geqslant 0$. Since $\chi(t) \leqslant \frac{1}{2}\chi^*(t)$ for t between the extreme points of P, the convergence of J_λ^* is a stronger result than the convergence of J_λ. We can, however, deduce the former from the latter. Given any $\epsilon > 0$, let Q_ϵ denote the union of the intervals making up Q, each expanded concentrically in the ratio $1+\epsilon$ (in this process some of the expanded intervals may become overlapping). Let P_ϵ be the closed set complementary to Q_ϵ with respect to (a, b). Since $Q_\epsilon \supset Q$, $|Q_\epsilon - Q| \leqslant (b-a)\epsilon \to 0$ with ϵ, we have

$$P_\epsilon \subset P, |P - P_\epsilon| \to 0.$$

We easily see that $$\chi_P^*(t) \leqslant 2\epsilon^{-1}\chi_{P_\epsilon}(t).$$

Since $J_\lambda(x, f, P_\epsilon)$ is finite almost everywhere in P_ϵ, the same holds for $J_\lambda^*(x, f, P)$. Making ϵ approach 0 we see that $J_\lambda^*(x, f, P)$ is finite almost everywhere in P.

Remarks. (*a*) In the case of sets P having period 2π it is sometimes more convenient to use the integral

$$J_\lambda'(x) = \int_0^{2\pi} \frac{f(t)\,\chi^\lambda(t)}{|2\sin\frac{1}{2}(x-t)|^{\lambda+1}}dt = \int_{-\pi}^{\pi} \frac{f(x+t)\,\chi^\lambda(x+t)}{|2\sin\frac{1}{2}t|^{\lambda+1}}dt \tag{2·10}$$

instead of $J_\lambda(x)$, and to make a corresponding modification of J_0. Theorem (2·1) changes little; the factor $2/\lambda$ in (2·4) must be replaced by another factor depending on λ:

$$\int_P |J_\lambda'(x)|\,dx \leqslant A_\lambda \int_0^{2\pi} |f(x)|\,dx. \tag{2·11}$$

The proof remains the same.

(*b*) Though we shall not use the fact here, it is of interest to observe that an analogue of (2·1) holds in Euclidean space of any number of dimensions. Suppose, for instance, that P is a closed set contained in a finite circle K and that $\chi(t)$ is the distance of the point t in the plane from P. If f is integrable over K, then the two integrals

$$\int_K \frac{f(t)\,\chi^\lambda(t)}{|t-x|^{\lambda+2}}d\sigma \quad (\lambda > 0), \qquad \int_K \frac{f(t)\{\log 1/\chi(t)\}^{-1}}{|t-x|^2}d\sigma,$$

where $|t-x|$ denotes the distance between t and x and $d\sigma$ is an element of area, converge absolutely almost everywhere in P. When the dimension of the space increases by 1, so do the exponents in the denominators of the integrals considered.

(c) The convergence of the integral $I_\lambda^*(x)$ has a simple geometric interpretation. Let $d_1, d_2, \ldots$ be the intervals contiguous to a bounded closed and non-dense set P. For any $x \in P$, denote by $\delta_i(x)$ the distance of x from $d_i = (a_i, b_i)$; thus $\delta_i(x) = \min(|x-a_i|, |x-b_i|)$. Almost every point $x \in P$ is a point of density of P, and at a point of density $|d_i| = o\{\delta_i(x)\}$ as d_i approaches x. The finiteness of $I_\lambda^*(x)$ almost everywhere in P may be interpreted as follows: *the series*

$$\Sigma\{|d_i|/\delta_i(x)\}^{\lambda+1} \tag{2·12}$$

converges for every $\lambda > 0$ and almost all $x \in P$.

For let x be a point of density of P, and $\eta > 0$ so small that $|d_i| \leqslant \delta_i(x)$ for all the d_i situated entirely in $(x-\eta, x+\eta)$. Let $d_1', d_2', \ldots$ be all the d's situated, say, in $(x, x+\eta)$. We may suppose that $x+\eta$ is not interior to any d'. Let $\delta_i'(x)$ be the distance of x from d_i'. Then

$$\Sigma |d_i'|^\lambda . |d_i'| . [2\delta_i'(x)]^{-\lambda-1} \leqslant \int_x^{x+\eta} \frac{\chi^{*\lambda}(t)}{|t-x|^{\lambda+1}} dt \leqslant \Sigma |d_i'|^\lambda . |d_i'| . [\delta_i'(x)]^{-\lambda-1}$$

and the convergence of $\Sigma\{|d_i'|/\delta_i'(x)\}^{\lambda+1}$ is equivalent to that of the integral. Similarly for $(x-\eta, x)$. Since the series (2·12) and the integral I_λ^* extended over the d's outside $(x-\eta, x+\eta)$ are finite in any case, the assertion follows. Similar interpretations can be given for $I_0^*(x)$ and $J_\lambda^*(x)$.

(d) If in (2·3) and (2·6) we replace $f(t)\,dt$ by $dF(t)$, where F is of bounded variation, the resulting integrals converge almost everywhere in P. More interesting for applications, however, is the following result in which, for simplicity, we consider integrals of the type (2·10).

Suppose that $\mu(t)$ is a positive measure on the circumference of the unit circle such that

$$\left| \int_0^t d\mu \right| \leqslant A\,|t| \tag{2·13}$$

for all t. Then, if $\lambda > 0$, the integral

$$\int_{-\pi}^{\pi} f(x+t)\,\chi^\lambda(x+t) \frac{d\mu(t)}{(2\sin\frac{1}{2}t)^{\lambda+1}} = \int_{-\pi}^{\pi} f(t)\,\chi^\lambda(t) \frac{d_t\mu(t-x)}{|2\sin\frac{1}{2}(t-x)|^{\lambda+1}}$$

converges almost everywhere in P. The proof remains unchanged if we note (using integration by parts) that (2·13) implies

$$\int_{\delta \leqslant |t| \leqslant \pi} |t|^{-\lambda-1}\, d\mu(t) = O(\delta^{-\lambda}).$$

3. Existence of the conjugate function

(3·1) THEOREM. *If $f \in L$, then*

$$\tilde{f}(x) = -\frac{1}{\pi}\int_0^\pi [f(x+t) - f(x-t)]\,\tfrac{1}{2}\cot\tfrac{1}{2}t\,dt = -\frac{1}{\pi}\lim_{\epsilon\to+0}\int_\epsilon^\pi \tag{3·2}$$

exists for almost all x.

This result was already stated and used in Chapter III, § 3, and we shall give two proofs of it, one now and the other in Chapter VII, § 1. The latter proof is much the shorter of the two, but it uses the theory of analytic functions. On the other hand, what we prove here is more general, and is not so easily accessible by complex methods.

(3·3) THEOREM. *Suppose that $F \in L$ is periodic and has a finite derivative at every point of a set E of positive measure. Then the integral*

$$F^*(x) = -\frac{1}{\pi}\int_0^\pi \frac{F(x+t) + F(x-t) - 2F(x)}{(2\sin\frac{1}{2}t)^2}\,dt = -\frac{1}{\pi}\lim_{\epsilon\to+0}\int_\epsilon^\pi \tag{3·4}$$

exists at almost all points of E.

To see that (3·1) follows from (3·3) we may suppose that $a_0 = 0$, which does not affect $\tilde{f}$. The indefinite integral F of f is then periodic, and

$$\int_\epsilon^\pi \frac{F(x+t)+F(x-t)-2F(x)}{(2\sin\frac{1}{2}t)^2}\,dt = \frac{F(x+\epsilon)+F(x-\epsilon)-2F(x)}{2\tan\frac{1}{2}\epsilon} + \int_\epsilon^\pi \frac{f(x+t)-f(x-t)}{2\tan\frac{1}{2}t}\,dt. \tag{3·5}$$

At every point x where F is differentiable the integrated term tends to zero with ϵ, and the existence of $F^*(x)$ is equivalent to that of $\tilde{f}(x)$. Moreover, $F^*(x) = \tilde{f}(x)$.

We turn now to (3·3). The special case $F = \int f\,dx$, $f \in L^2$, was given in § 1. It follows that $F^*(x)$ exists almost everywhere if F is the integral of a function in L^2, and in particular if $F \in \Lambda_1$.

We write
$$\rho(x,h) = [F(x+h)-F(x)]/h$$
and denote by E_k the set of $x \in E$ such that $|\rho(x,h)| \leqslant k$ for $|h| < 1/k$. Hence
$$E_1 \subset E_2 \subset \ldots \subset E_k \subset \ldots \subset E, \quad |E_k| \to |E|.$$

Fix k, say $k = M$, and consider any closed subset P of E_M. We shall prove that $F^*(x)$ exists almost everywhere in P. Since $|E-P|$ may be arbitrarily small, (3·3) will follow.

By hypothesis,
$$|F(x+h)-F(x)| \leqslant M|h| \quad \text{for} \quad x \in P,\ |h| \leqslant 1/M. \tag{3·6}$$

Let $G(x)$ be the function coinciding with F on P and linear in the closed intervals $d_1, d_2, \ldots$ contiguous to P. We prove that $G(x) \in \Lambda_1$, and for this it is enough to prove
$$|G(x+h)-G(x)| \leqslant A|h| \quad \text{for} \quad |h| \leqslant 1/M, \tag{3·7}$$
with A independent of x, h. Suppose, for example, that $h > 0$. We consider first two special cases: (i) both x and $x+h$ belong to P; (ii) the interior of $(x, x+h)$ contains no points of P. In case (i), (3·7), with $A = M$, follows from (3·6). In case (ii), $(x, x+h)$ is contained in an interval d contiguous to P, and G is linear there. Thus, if $|d| \leqslant 1/M$, (3·7), with $A = M$, again follows from (3·6). Since there are only a finite number (if any) of d's with $|d| > 1/M$, (3·7) is always true in case (ii), provided A is large enough.

If neither (i) nor (ii) holds, $(x, x+h)$ contains points of P in its interior. Let $x+h_1$ and $x+h_2$, $h_1 \leqslant h_2$, be the extreme points of P in $(x, x+h)$. The absolute increment of G over $(x, x+h)$ does not exceed the sum of the increments over the intervals $(x, x+h_1)$, $(x+h_1, x+h_2)$, $(x+h_2, x+h)$, and each of the latter is at most A times the length of the corresponding interval, in virtue of cases (i) and (ii). This leads again to (3·7). Hence $G \in \Lambda_1$.

We set $H(x) = F(x) - G(x)$, so that
$$F(x) = G(x) + H(x). \tag{3·8}$$

From (3·6) and (3·7) we see that $H(x)$ satisfies an inequality analogous to (3·6), with $M' = M + A$ for M. Since, however, $H(x)$ vanishes in P, this implies that, except in a finite number of intervals exterior to P,
$$|H(x)| \leqslant M'\chi(x), \tag{3·9}$$
where $\chi(x)$ is the distance of x from P.

The functions G and H being periodic, (3·3) will follow if we show that the integrals $G^*(x)$ and $H^*(x)$ exist almost everywhere in P. This has already been proved for G^*, since $G \in \Lambda_1$. Consider $H^*(x)$. If $x \in P$, then $H(x)=0$ and (3·9) gives

$$\int_0^{1/M} \frac{|H(x+t)+H(x-t)-2H(x)|}{(2\sin\frac{1}{2}t)^2} dt \leqslant M' \int_{-1/M}^{1/M} \frac{\chi(x+t)}{(2\sin\frac{1}{2}t)^2} dt.$$

The integral on the right is finite almost everywhere in P. (See (2·10) and (2·11) with $\lambda=1, f=1$; we could also use the finiteness of the I_λ in (2·2).) The same therefore holds for the integral on the left, and the finiteness is not affected if on the left we replace the interval of integration $(0, 1/M)$ by $(0, \pi)$. Hence the integral $H^*(x)$ converges, even absolutely, almost everywhere in P, which completes the proof of (3·3).

In this proof we tacitly assumed that the sets E_k were measurable (to ensure the existence of closed subsets P). To prove the measurability it is enough to show that for any fixed α the function

$$\rho(x)=\rho(x;\alpha)=\sup_{0<|t|<\alpha}\left|\frac{F(x+t)-F(x)}{t}\right|$$

is lower semi-continuous, that is that $\liminf_{x\to x_0} \rho(x) \geqslant \rho(x_0)$. The inequality is immediate if we interpret $[F(x+t)-F(x)]/t$ as the slope of a chord. For if, e.g., F is continuous at x_0 and if x_1, $|x_1-x_0|<\alpha$, is such that the absolute value of the slope of the chord joining the point $P_0(x_0, F(x_0))$ to $P_1(x_1, F(x_1))$ exceeds $\rho(x_0)-\epsilon$, then the absolute value of the slope of the chord PP_1 exceeds $\rho(x_0)-2\epsilon$, provided the abscissa x of P is close enough to x_0, and the inequality follows. If F is discontinuous at x_0 both sides of the inequality are $+\infty$.

Immediate consequences of (3·1) are Theorems (5·8) and (3·23) of Chapter III, initially stated without proof. From (3·3) and from Theorem (7·15) of Chapter III we also deduce

(3·10) THEOREM. *If $F(x)$, periodic and integrable, is differentiable in a set E of positive measure, the series $\tilde{S}'[F]$ is summable A to sum (3·4) almost everywhere in E.*

The existence of $\tilde{f}(x)$ is not trivial even if $f(x)$ is continuous. The existence of the $\tilde{f}$ is due not to the smallness of $f(x+t)-f(x-t)$ for small $|t|$ but to the interference of positive and negative values; in fact, as we will show, *there exist continuous functions f such that the integral*

$$\int_0^\pi \frac{|f(x+t)-f(x-t)|}{t} dt \qquad (3\cdot 11)$$

diverges at every x. It will slightly simplify the notation if we consider functions of period 1 and replace the upper limit of integration π in (3·11) by 1. We first need the following lemma:

(3·12) LEMMA. *Let $g(x)$ be a function of period 1 such that $|g(x)|\leqslant 1$, $|g'(x)|\leqslant 1$, and that for no value of x does the difference $g(x+u)-g(x-u)$ vanish identically in u.*† *Then*

$$\int_{1/n}^1 \frac{|g(nx+nt)-g(nx-nt)|}{t} dt \geqslant C\log n, \quad \int_0^1 \frac{|g(nx+nt)-g(nx-nt)|}{t} dt \leqslant C_1 \log n$$

for $n=2,3,\ldots$, C and C_1 being positive constants independent of n.

Let $nx=y$, $nt=u$. Since g is periodic, the first integral is

$$\int_0^1 |g(y+u)-g(y-u)| \sum_{\nu=1}^{n-1} \frac{1}{u+\nu} du \geqslant \left(\sum_{\nu=2}^{n} \frac{1}{\nu}\right) \int_0^1 |g(y+u)-g(y-u)|\, du.$$

The first factor on the right exceeds a multiple of $\log n$, and the second, as a periodic, continuous and nowhere vanishing function of y, is bounded below by a positive number. This gives the first part of the lemma. Similarly we obtain the second part, observing that

$$\int_0^1 |g(y+u)-g(y-u)|\, u^{-1} du < \infty.$$

† For $g(x)$, $0\leqslant x\leqslant 1$, we may take, for example, the polygonal line with vertices $(0,0)$, $(\frac{1}{3},\frac{1}{3})$, $(1,0)$.

We now set
$$f(x)=\sum_{n=1}^{\infty} a_n g(\lambda_n x), \tag{3·13}$$
where the numbers $a_n>0$ and the integers $0<\lambda_1<\lambda_2<\ldots$ will be determined in a moment. Then
$$\int_{1/\lambda_\nu}^{1} \frac{|f(x+t)-f(x-t)|}{t}\,dt \geqslant a_\nu \int_{1/\lambda_\nu}^{1} \frac{|g(\lambda_\nu x+\lambda_\nu t)-g(\lambda_\nu x-\lambda_\nu t)|}{t}\,dt$$
$$-\left(\sum_{n=1}^{\nu-1}+\sum_{n=\nu+1}^{\infty}\right) a_n \int_{1/\lambda_\nu}^{1} \frac{|g(\lambda_n x+\lambda_n t)-g(\lambda_n x-\lambda_n t)|}{t}\,dt$$
$$\geqslant Ca_\nu \log\lambda_\nu - C_1 \sum_{n=1}^{\nu-1} a_n \log\lambda_n - 2\log\lambda_\nu \sum_{n=\nu+1}^{\infty} a_n, \tag{3·14}$$
since $|g(\lambda_n x+\lambda_n t)-g(\lambda_n x-\lambda_n t)|\leqslant 2$. If we take $a_n=1/n!$, $\lambda_n=2^{(n!)^2}$, the right-hand side of (3·14) divided by $\nu!$ tends to $C\log 2>0$, and this shows that (3·11) diverges everywhere.

It is interesting to observe that the integrals
$$\int_0^\pi \frac{f(x+t)-f(x)}{t}\,dt \quad\text{and}\quad \int_0^\pi \frac{f(x+t)+f(x-t)-2f(x)}{t}\,dt, \tag{3·15}$$
though apparently similar to (3·2), can diverge everywhere for a continuous f. The proof is analogous to that given above, but slightly less simple.

The theorem which follows will find an application in Chapter XII. Its proof is similar to that of (3·1) but the details are somewhat more elaborate.

(3·16) THEOREM. *If $f\in L$, $y>0$, then the measure of the set $E_y=E_y(f)$ where $|\tilde f(x)|>y$ satisfies*
$$|E_y|\leqslant \frac{A}{y}\int_0^{2\pi} |f|\,dx, \tag{3·17}$$
where A is an absolute constant.

We may suppose that $f\geqslant 0$. For if $f=f_1+f_2$, then $E_{2y}(f)\subset E_y(f_1)+E_y(f_2)$,
$$|E_{2y}(f)|\leqslant |E_y(f_1)|+|E_y(f_2)|. \tag{3·18}$$
Hence if f_1 and f_2 are the positive and negative parts of f, and if the theorem holds for f_1 and f_2, it holds for f.

We may also suppose that
$$\int_0^{2\pi} f\,dx=1. \tag{3·19}$$
The function $F(x)=\int_0^x f\,dt$ is non-decreasing in $(-\infty,+\infty)$, and
$$\tilde f(x)=-\frac{1}{\pi}\int_0^\pi \frac{F(x+t)+F(x-t)-2F(x)}{(2\sin\frac12 t)^2}\,dt \tag{3·20}$$
almost everywhere.

Fix y and denote by Q the set of x for which
$$\frac{F(\xi)-F(x)}{\xi-x}>y \tag{3·21}$$
for some ξ in the interior of $(x, x+2\pi)$. Q is open (possibly empty) and periodic. The complement P of Q is closed. If P and Q are not empty, Q is the union of a family $\{(a_i, b_i)\}$ of disjoint open intervals such that
$$\frac{F(b_i)-F(a_i)}{b_i-a_i}=y. \tag{3·22}$$
This fact is proved in the same way as Lemma (13·8) of Chapter I and the Remark to it

Write now $F = G + H$, where G coincides with F on P and is linear in each interval (a_i, b_i); hence $H = 0$ on P. If Q is empty, we write $F = G$, $H = 0$, and disregard H in the argument which follows.

The function $G(x)$ *is in* Λ_1. More precisely,

$$0 \leqslant \frac{G(x+h) - G(x)}{h} \leqslant y \quad \text{for} \quad 0 < h < 2\pi.$$

The first inequality is obvious since G, like F, is non-decreasing. The second inequality is immediate if both x and $x + h$ are either in P or in the same interval contiguous to P; the general case follows from these two by the argument used on p. 132.

Hence G is the indefinite integral of a (periodic) function $g = G'$. We have $0 \leqslant g(x) \leqslant y$ for almost all x, and $g(x) = f(x)$ almost everywhere in P (since $G = F$ in P). Clearly H is the indefinite integral of $h = H'$, and

$$f = g + h, \quad \tilde{f} = \tilde{g} + \tilde{h}.$$

Since $H = 0$ in P, the integral of h over each interval (a_i, b_i) is $H(b_i) - H(a_i) = 0$, and

$$\int_{a_i}^{b_i} f\,dx = \int_{a_i}^{b_i} g\,dx \quad (i = 1, 2, \dots), \qquad \int_0^{2\pi} f\,dx = \int_0^{2\pi} g\,dx. \tag{3·23}$$

Since $|E_{2y}(f)| \leqslant |E_y(g)| + |E_y(h)|$ it is enough to show that each of the two terms on the right is majorized by

$$\frac{A}{y}\int_0^{2\pi} f\,dx = \frac{A}{y}.$$

For g we have

$$|E_y(g)| \leqslant y^{-2}\int_0^{2\pi} \tilde{g}^2\,dx \leqslant y^{-2}\int_0^{2\pi} g^2\,dx$$
$$\leqslant y^{-1}\int_0^{2\pi} g\,dx = y^{-1}\int_0^{2\pi} f\,dx,$$

by (3·23).

It remains to estimate $|E_y(h)|$. First, summing from (3·22) over all (a_i, b_i) in a period, we get

$$|Q| = \Sigma(b_i - a_i) \leqslant \frac{F(2\pi) - F(0)}{y} = \frac{1}{y}\int_0^{2\pi} f(x)\,dx = \frac{1}{y}. \tag{3·24}$$

Next, let $\chi^*(x)$ be the function equal to $b_i - a_i$ in each (a_i, b_i), and to 0 in P. We show that

$$H(x) \leqslant y\chi^*(x). \tag{3·25}$$

This is obvious for x in P, since both sides are 0. If $a_i < x < b_i$, then

$$0 \leqslant F(x) - F(a_i) \leqslant y(x - a_i), \quad 0 \leqslant G(x) - G(a_i) \leqslant y(x - a_i);$$

and the equations $H(a_i) = 0$, $H = F - G$ imply

$$|H(x)| = |H(x) - H(a_i)| \leqslant y(x - a_i) < y(b_i - a_i),$$

which gives (3·25).

Write

$$I(x) = \frac{1}{\pi}\int_{-\pi}^{\pi} \frac{\chi^*(t)}{\{2\sin\frac{1}{2}(x-t)\}^2}\,dt = \frac{1}{\pi}\int_{-\pi}^{\pi} \frac{\chi^*(x+t)}{(2\sin\frac{1}{2}t)^2}\,dt.$$

In view of (3·25), if we apply (3·20) to h we obtain

$$|\tilde{h}(x)| \leqslant yI(x) \quad \text{for} \quad x \in P. \tag{3·26}$$

Let Q^* be the set obtained by expanding each (a_i, b_i) concentrically three times, and let P^* be the complement of Q^*. We have

$$\int_{P^*} I(x)\,dx \leqslant B\,|\,Q\,|, \tag{3·27}$$

where B is an absolute constant; this is a result analogous to (2·11) and the proof is essentially the same.

It is now easy to estimate $|\,E_y(h)\,|$. The intersection of $E_y(h)$ with Q^* has measure not greater than $|\,Q^*\,| \leqslant 3\,|\,Q\,|$. In P, and *a fortiori* in P^*, we have (3·26), and so, if $|\,\tilde{h}(x)\,| > y$, then $I(x) > 1$. But, by (3·27), the subset of P^* where $I(x) > 1$ has measure not greater than $B\,|\,Q\,|$. Hence, collecting results and using (3·24), we find

$$|\,E_y(h)\,| \leqslant 3\,|\,Q\,| + B\,|\,Q\,| \leqslant (B+3)\,y^{-1}.$$

This completes the proof of (3·16).

4. Classes of functions and (C, 1) means of Fourier series

We know that the necessary and sufficient condition for the numbers c_ν ($\nu = 0, \pm 1, \pm 2, \ldots$) to be the Fourier coefficients of a function L^2 is that the sum $\Sigma\,|\,c_\nu\,|^2$ be finite. It is natural to ask whether anything so simple can be proved for the classes L^r with $r \neq 2$. The answer is no, and it is this fact which makes the Parseval formula and the Riesz-Fischer theorem such exceptionally powerful tools of investigation. We shall now consider criteria of a different kind involving the Cesàro or Abel means of the series considered.

One point must be made clear. What matters in the proofs that follow is that the (C, 1) kernel, and Abel's kernel, satisfy conditions (A), (B), (C) stated in § 2 of Chapter III (and in particular are positive), and also that $S[f]$ is summable by these methods. The arguments are therefore applicable without change to any other kernel with these properties. Logically the (C, 1) method is simpler than Abel's, but the latter is often more significant, especially in applications to harmonic and analytic functions.

We pursue the following course: in §§ 4 and 5 the results will be proved for the (C, 1) means, and in § 6 the analogues for Abel means will be stated without proof.

Besides the classes L_ϕ, L^r introduced in § 9 of Chapter I, we shall consider other classes of functions. We shall denote by B, C, A and V the classes of periodic functions which are respectively bounded, continuous, absolutely continuous, and of bounded variation. If

$$\sum_{\nu=-\infty}^{+\infty} c_\nu\, e^{i\nu x} \tag{4·1}$$

is the Fourier series of a function of a definite class, we say that the series itself belongs to that class. By S we denote the class of Fourier-Stieltjes series. The (C, 1) means of (4·1) will be denoted by $\sigma_n(x)$.

(4·2) THEOREM. (i) *A necessary and sufficient condition for* $\Sigma\, c_\nu e^{i\nu x}$ *to belong to class* C *is the uniform convergence of* $\{\sigma_n(x)\}$.

(ii) *A necessary and sufficient condition for* $\Sigma\, c_\nu e^{i\nu x}$ *to belong to class* B *is the uniform boundedness of the* $\sigma_n(x)$.

The necessity of the condition in (i) is Fejér's theorem (Chapter III, (3·4)). To prove the sufficiency, we note that, for $n \geqslant |k|$,

$$\left(1-\frac{|k|}{n+1}\right)c_k = \frac{1}{2\pi}\int_0^{2\pi} \sigma_n(x)\, e^{-ikx}\, dx.$$

As $n \to \infty$ the left-hand side tends to c_k and the right-hand side to the kth Fourier coefficient of the continuous function $f(x) = \lim \sigma_n(x)$.

The necessity of the condition in (ii) is contained in Theorem (2·30) of Chapter III: if K is the essential upper bound of $|f|$ then $|\sigma_n(x)| \leqslant K$. Conversely, if $|\sigma_n| \leqslant K$, then, for large n,

$$K^2 \geqslant \frac{1}{2\pi}\int_0^{2\pi} |\sigma_n|^2\, dx = \sum_{k=-n}^{n} |c_k|^2 \left(1-\frac{|k|}{n+1}\right)^2 \geqslant \sum_{k=-\nu}^{\nu} |c_k|^2 \left(1-\frac{|k|}{n+1}\right)^2,$$

where ν is any fixed positive integer not exceeding n. Making $n \to \infty$ we get

$$|c_{-\nu}|^2 + \ldots + |c_0|^2 + \ldots + |c_\nu|^2 \leqslant K^2,$$

and this is true for every ν. Hence $\Sigma |c_k|^2$ converges, and, by the Riesz-Fischer theorem, (4·1) is an $S[f]$ with $f \in L^2$. Therefore $\sigma_n(x) \to f(x)$ almost everywhere, and the inequalities $|\sigma_n(x)| \leqslant K$ imply that $|f(x)| \leqslant K$ almost everywhere.

(4·3) THEOREM. *The series* $\Sigma c_\nu e^{i\nu x}$ *belongs to class* S *if and only if* $\mathfrak{M}[\sigma_n] = O(1)$.

We first suppose that $\Sigma c_\nu e^{i\nu x}$ is an $S[dF]$. Then

$$\left.\begin{aligned} \sigma_n(x) &= \frac{1}{\pi}\int_0^{2\pi} K_n(t-x)\, dF(t), \\ |\sigma_n(x)| &\leqslant \frac{1}{\pi}\int_0^{2\pi} K_n(t-x)\, |dF(t)|, \end{aligned}\right\} \tag{4·4}$$

where $|dF(t)|$ stands for $dV(t)$, $V(t)$ denoting the total variation of F over $(0, t)$. Let $V = V(2\pi)$. The right-hand side of the last inequality is a trigonometric polynomial in x whose constant term is $V/2\pi$. Integration over $0 \leqslant x \leqslant 2\pi$ therefore gives

$$\mathfrak{M}[\sigma_n] \leqslant \int_0^{2\pi} |dF(t)| = V, \tag{4·5}$$

and one part of (4·3) is established. For the other we need the following classical result, which we take for granted here:

(4·6) THEOREM OF HELLY. *Let* $\{F_n(x)\}$ *be a sequence of functions uniformly bounded and of uniformly bounded variation in an interval* (a, b). *Then there is a subsequence* $\{F_{n_k}(x)\}$ *converging at every point of* (a, b) *to a function* $F(x)$ *of bounded variation.*

The hypothesis of uniform boundedness may be replaced by the boundedness of $\{F_n\}$ at a single point x, since the former is implied by the latter together with the uniformly bounded variation.

Returning to Theorem (4·3), suppose that $\mathfrak{M}[\sigma_n] \leqslant V$ for all n and let

$$F_n(x) = \int_0^x \sigma_n(t)\, dt.$$

The functions $F_n(x)$ are of uniformly bounded variation in $(0, 2\pi)$ and vanish at $x=0$. Hence there is a subsequence $\{F_{n_j}(x)\}$, uniformly bounded and everywhere convergent to a function $F(x)$ of bounded variation (total variation $\leqslant V$) in $(0, 2\pi)$. If $|k| \leqslant n_j$, integrating by parts and making $j \to \infty$ we get

$$\left(1-\frac{|k|}{n_j+1}\right)c_k = \frac{1}{2\pi}\int_0^{2\pi} \sigma_{n_j} e^{-ikx}\,dx = \frac{1}{2\pi}F_{n_j}(2\pi) + \frac{ik}{2\pi}\int_0^{2\pi} F_{n_j} e^{-ikx}\,dx,$$

$$c_k = \frac{1}{2\pi}F(2\pi) + \frac{ik}{2\pi}\int_0^{2\pi} F e^{-ikx}\,dx = \frac{1}{2\pi}\int_0^{2\pi} e^{-ikx}\,dF(x),$$

since $F(0)=0$. Hence (4·1) is $S[dF]$ and the proof of (4·3) is complete. The result can be stated in the following equivalent form.

(4·7) THEOREM. *A necessary and sufficient condition for $\Sigma c_\nu e^{i\nu x}$ to belong to class V is that $\mathfrak{M}[\sigma_n'] = O(1)$, that is, that the σ_n be of uniformly bounded variation.*

The following result completes (4·3):

(4·8) THEOREM. *A necessary and sufficient condition for $\Sigma c_\nu e^{i\nu x}$ to be an $S[dF]$ with F non-decreasing is that $\sigma_n \geqslant 0$ for all n.*

The necessity follows from (4·4) since $K_n(u) \geqslant 0$. Conversely, if $\sigma_n(x) \geqslant 0$, the $F_n(x)$ in the proof of (4·3) are non-decreasing and so is $F(x) = \lim F_{n_j}(x)$.

(4·9) THEOREM. *A necessary and sufficient condition for $\Sigma c_\nu e^{i\nu x}$ to be an $S[dF]$ with F non-decreasing is that*

$$\sum_{\mu,\nu=0}^{n} c_{\mu-\nu}\xi_\mu\bar{\xi}_\nu \geqslant 0 \tag{4·10}$$

for all $n \geqslant 0$ and for all (complex) $\xi_0, \xi_1, \ldots, \xi_n$.

If $c_\nu = (2\pi)^{-1}\int_0^{2\pi} e^{-i\nu x}\,dF$, with F non-decreasing, then

$$2\pi\sum_{\mu,\nu=0}^{n} c_{\mu-\nu}\xi_\mu\bar{\xi}_\nu = \int_0^{2\pi}\left(\sum_{\mu,\nu=0}^{n} e^{-i(\mu-\nu)x}\xi_\mu\bar{\xi}_\nu\right)dF = \int_0^{2\pi}\left|\sum_0^n e^{-i\mu x}\xi_\mu\right|^2 dF \geqslant 0.$$

Conversely, if we take $\xi_\nu = e^{i\nu x}$ for all ν, and denote by σ_n the (C, 1) means of the series $\Sigma c_\nu e^{i\nu x}$, the left-hand side of (4·10) becomes

$$(n+1)c_0 + n(c_1 e^{ix} + c_{-1}e^{-ix}) + \ldots + (c_n e^{inx} + c_{-n}e^{-inx}) = (n+1)\sigma_n(x),$$

and it is enough to apply (4·8).

Let u^+ and u^- denote respectively $\max(u, 0)$ and $\max(-u, 0)$, so that

$$u^+ = \tfrac{1}{2}(|u|+u), \quad u^- = \tfrac{1}{2}(|u|-u). \tag{4·11}$$

Since the integral of σ_n over $(0, 2\pi)$ is constant (being $2\pi c_0$), the first equation shows that *the conditions*

$$\mathfrak{M}[\sigma_n] = O(1), \quad \mathfrak{M}[\sigma_n^+] = O(1)$$

are equivalent.

(4·12) THEOREM. *Suppose that $\Sigma c_\nu e^{i\nu x}$ is an $S[dF]$ and that*

$$F(x) = \tfrac{1}{2}[F(x+0) + F(x-0)] \tag{4·13}$$

for all x. Let V, P, N denote respectively the total, positive and negative variations of F over an interval $\alpha \leqslant x \leqslant \beta$. Then

$$\int_\alpha^\beta |\sigma_n|\, dx \to V, \quad \int_\alpha^\beta \sigma_n^+\, dx \to P, \quad \int_\alpha^\beta \sigma_n^-\, dx \to N. \tag{4·14}$$

Since $S[dF] = S'[F]$ (Chapter II, (2·4)), (4·13) implies that

$$\int_\alpha^\beta \sigma_n dx \to F(\beta) - F(\alpha) \tag{4·15}$$

(Chapter III, (3·4)). It is enough to prove the first formula (4·14), since the other two relations in (4·14) follow from this, combined with (4·15), (4·11) and

$$2P = V + (F(\beta) - F(\alpha)), \quad 2N = V - (F(\beta) - F(\alpha)).$$

That $\liminf \mathfrak{M}[\sigma_n; \alpha, \beta] \geqslant V$ is clear. For by (4·15) the functions $F_n(x) = \int_\alpha^x \sigma_n dt$ converge to $F(x) - F(\alpha)$ on (α, β) and $\mathfrak{M}[\sigma_n; \alpha, \beta]$ is the total variation of F_n over (α, β). The total variation V of the limit cannot exceed the limit inferior of the total variations of the F_n. It remains therefore to show that $\limsup \mathfrak{M}[\sigma_n; \alpha, \beta] \leqslant V$. If (α, β) coincides with $(0, 2\pi)$, this follows from (4·5), and (4·12) is established in this particular case. If $\beta - \alpha < 2\pi$, suppose that the inequality we want to prove is false. If V' is the total variation of F over the closed interval $(\beta, \alpha + 2\pi)$ we therefore have

$$\int_\alpha^{\alpha+2\pi} |\sigma_n|\, dx \to V + V', \quad \limsup \int_\alpha^\beta |\sigma_n|\, dx > V.$$

This implies that $\liminf \mathfrak{M}[\sigma_n; \beta, \alpha + 2\pi] < V'$, contrary to the opposite inequality which we have already proved (with α, β for $\beta, \alpha + 2\pi$). This proves (4·12).

We know that $\sigma_n(x; dF) \to F'(x)$ for almost all x (Chapter III, (8·1)), and we shall sometimes write $\sigma(x)$ $(= \lim \sigma_n(x))$ instead of $F'(x)$.

Let $P(x)$ be the positive variation of $F(x)$ over (α, x) and let

$$P(x) = P_a(x) + P_s(x)$$

be the decomposition of P into its absolutely continuous and singular parts. Both $P_a(x)$ and $P_s(x)$ are non-negative and non-decreasing for $x \geqslant \alpha$. Moreover, as is well known, we have almost everywhere

$$P_s'(x) = 0, \quad P_a'(x) = P'(x) = (F'(x))^+ = \sigma^+(x).$$

By (4·12),

$$\int_\alpha^\beta \sigma_n^+\, dx \to \{P_a(\beta) - P_a(\alpha)\} + \{P_s(\beta) - P_s(\alpha)\} = \int_\alpha^\beta \sigma^+ dx + P_s(\beta). \tag{4·16}$$

A necessary and sufficient condition for $P(x)$ to be absolutely continuous over (α, β) is that $P_s(\beta) = 0$, or

$$\int_\alpha^\beta \sigma_n^+\, dx \to \int_\alpha^\beta \sigma^+ dx. \tag{4·17}$$

Similarly, a necessary and sufficient condition for the negative variation $N(x)$ of F to be absolutely continuous in (α, β) is

$$\int_\alpha^\beta \sigma_n^-\, dx \to \int_\alpha^\beta \sigma^- dx. \tag{4·18}$$

If both $P(x)$ and $N(x)$ are absolutely continuous over (α, β), adding (4·17) and (4·18) we get

$$\int_\alpha^\beta |\sigma_n|\, dx \to \int_\alpha^\beta |\sigma|\, dx, \tag{4·19}$$

and conversely this relation implies both (4·17) and (4·18). (For then, from (4·16) and from a similar formula for σ_n^-, we see that $P_s(\beta) = N_s(\beta) = 0$.) Thus (4·19) is both necessary and sufficient for $F(x)$ to be absolutely continuous over (α, β). Hence:

(4·20) THEOREM. *Let $\Sigma c_\nu e^{i\nu x}$ be an $S[dF]$, where F satisfies* (4·13). *The conditions* (4·19), (4·17) *and* (4·18) *are necessary and sufficient for the function F, its positive variation, and its negative variation, respectively, to be absolutely continuous over (α, β).*

Let $F_k(x)$, $0 < x < 2\pi$, be a sequence of uniformly bounded functions. If $F_k(x)$ tends almost everywhere to a limit $F(x)$, then $C_n^k \to C_n$ as $k \to \infty$, where C_n^k and C_n denote the nth coefficients of F_k and F respectively. The converse is obviously false. If, for instance, $I_1, I_2, \ldots$ is any sequence of intervals whose length tends to zero, such that every point in $(0, 2\pi)$ belongs to infinitely many I_k, then the sequence of the characteristic functions $F_k(x)$ of the I_k diverges at every x, though $|C_n^k| \leqslant |I_k|/2\pi \to 0$, as $k \to \infty$ (uniformly in n). The converse is, however, true if the functions F_k are monotone.

(4·21) THEOREM OF CARATHÉODORY. *Let $\{F_k(x)\}$, $0 < x < 2\pi$, be a sequence of uniformly bounded and non-decreasing functions, and let C_n^k be the (complex) Fourier coefficients of F_k. If $\lim_{k\to\infty} C_n^k = C_n$ exists for every n, then the numbers C_n are the Fourier coefficients of a bounded non-decreasing function $F(x)$, $0 < x < 2\pi$, and $F_k(x) \to F(x)$ at every point x at which F is continuous.*

By (4·6) there is a subsequence of $\{F_k\}$ converging to a non-decreasing $F(x)$, $0 < x < 2\pi$. Obviously the Fourier coefficients of F are the C_n, and we have only to show that $F_k(\xi) \to F(\xi)$ for any point ξ of continuity of F interior to $(0, 2\pi)$. Suppose that $F_k(\xi)$ does not tend to $F(\xi)$. We can then find a subsequence $\{F_{k_j}\}$ such that $\lim F_{k_j}(\xi)$ exists and differs from $F(\xi)$, e.g. is greater than $F(\xi)$. We can select a subsequence $\{F_{k_j'}(x)\}$ of $\{F_{k_j}(x)\}$ such that $\lim F_{k'}(x) = G(x)$ exists everywhere. The Fourier coefficients of $G(x)$ are again C_n, so that $F(x) \equiv G(x)$. On the other hand,

$$G(\xi) = \lim F_{k_j'}(\xi) = \lim F_{k_j}(\xi) > F(\xi),$$

and since $G(x)$ is non-decreasing and $F(x)$ is continuous at $x = \xi$, we have $G(x) > F(x)$ in some interval to the right of ξ, so that $G(x) \not\equiv F(x)$. This contradiction shows that $F_k(\xi) \to F(\xi)$.

We shall now extend (4·21) to Fourier-Stieltjes series. Except when otherwise stated, every non-decreasing function Φ considered below will be defined for all x and will satisfy the condition

$$\Phi(x + 2\pi) - \Phi(x) = \Phi(2\pi) - \Phi(0).$$

(4·22) THEOREM. *Let $F_1(x)$, $F_2(x)$, ... be a sequence of non-decreasing functions and let c_n^k be the Fourier coefficients of dF_k. Then*

(i) *If $\lim_{k\to\infty} c_n^k = c_n$ exists for every n, there is a non-decreasing function $F(x)$ such that* *the Fourier coefficients of dF are c_n. Moreover, there are constants B_k such that $\{F_k(x)-B_k\}$ converges to $F(x)$ at every point of continuity of F.*

(ii) *Conversely, if for a sequence of constants B_k the sequence $\{F_k(x)-B_k\}$ converges to a (non-decreasing) function $F(x)$ at every point of continuity of F, then, denoting by c_n the Fourier coefficients of dF, we have $c_n^k \to c_n$ for all n.*

(i) Let B_k be the constant term of $S[F_k]$. The $F_k - B_k$ being uniformly bounded, there is a subsequence $\{F_{k_j}(x)-B_{k_j}\}$ converging to a limit $F(x)$ everywhere in $0 \leqslant x \leqslant 2\pi$. Let C_n^k be the Fourier coefficients of $F_k - B_k$. Then $C_0^k = 0$ for all k, and for $n \neq 0$ integration by parts gives

$$C_n^k = \frac{1}{2\pi}\int_0^{2\pi} (F_k - B_k)\, e^{-inx}\, dx = (c_n^k - c_0^k)/in. \tag{4·23}$$

Hence $\lim C_n^k = C_n$ exists for every n. By Theorem (4·21), $F_k(x) - B_k$ converges to $F(x)$ at every point of continuity of F interior to $(0, 2\pi)$. Hence if in (4·23) we make $k \to \infty$, we get $C_n = (c_n - c_0)/in$ for $n \neq 0$. On the other hand, if γ_n are the Fourier coefficients of dF, integration by parts gives

$$C_n = \frac{1}{2\pi}\int_0^{2\pi} F\, e^{-inx}\, dx = (\gamma_n - \gamma_0)/in,$$

so that $c_n - c_0 = \gamma_n - \gamma_0$ for $n \neq 0$. Moreover,

$$2\pi\gamma_0 = F(2\pi) - F(0) = \lim \{F_{k_j}(2\pi) - F_{k_j}(0)\} = \lim 2\pi c_0^{k_j} = 2\pi c_0,$$

so that $\gamma_0 = c_0$. Hence $c_n = \gamma_n$ for all n.

Let us now continue $F(x)$ outside $(0, 2\pi)$ by the condition

$$F(x+2\pi) - F(x) = F(2\pi) - F(0).$$

This, together with

$$F_k(x+2\pi) - F_k(x) = 2\pi c_0^k, \quad F(x+2\pi) - F(x) = 2\pi c_0, \quad c_0^k \to c_0,$$

implies that $F_k(x) - B_k$ converges to $F(x)$ at every point of continuity of F distinct from 0 (mod 2π). This in turn implies convergence also at the points congruent to 0 (mod 2π), if F is continuous there.

(ii) Let us write F_k instead of $F_k - B_k$, which does not change the Fourier-Stieltjes coefficients. Let C_n^k, C_n be the Fourier coefficients of F_k, F considered in $(0, 2\pi)$. Obviously, $C_n^k \to C_n$. If F is continuous at x, then

$$2\pi c_0^k = F_k(x+2\pi) - F_k(x) \to F(x+2\pi) - F(x) = 2\pi c_0,$$

and so $c_0^k \to c_0$. For $n \neq 0$,

$$(c_n^k - c_0^k)/in = C_n^k \to C_n = (c_n - c_0)/in,$$

which gives $c_n^k \to c_n$.

We can apply Theorem (4·22) to the problem of the distribution mod 1 of sequences

$$x_1,\ x_2,\ \ldots,\ x_k,\ \ldots \tag{4·24}$$

of real numbers. We wind the real axis around the circle Γ of length 1, and consider (4·24) as points on Γ, not distinguishing points congruent mod 1. Given any semi-open arc $\alpha < x \leqslant \beta$ on Γ, $0 \leqslant \beta - \alpha \leqslant 1$, we denote by $\nu_k(\alpha, \beta)$ the number of points among $x_1, x_2, \ldots, x_k$ which fall in that arc. We shall say that a function $F(x)$ is a *distribution function* of (4·24), if $F(x)$ is non-decreasing over $(-\infty, +\infty)$ and satisfies the condition $F(x+1)-F(x)=1$, and if

$$\nu_k(\alpha,\beta)/k \to F(\beta)-F(\alpha)$$

for any arc (α,β) whose end-points are points of continuity of F. If (4·24) has a distribution function, the latter is determined except for an arbitrary additive constant. If $F(x)=x+C$, we say that (4·24) is equidistributed mod 1, or simply *equidistributed*.

(4·25) THEOREM. *A necessary and sufficient condition for* (4·24) *to have a distribution function is that the limits*

$$\lim_{k\to\infty} \frac{1}{k}\{e^{-2\pi i n x_1}+e^{-2\pi i n x_2}+\dots+e^{-2\pi i n x_k}\}=c_n \tag{4·26}$$

exist for $n=0, \pm 1, \pm 2, \dots$. *If these limits do exist, they are the Fourier-Stieltjes coefficients with respect to the interval* $(0,1)$ *of the distribution function of* (4·24).

Sufficiency. Let $F_k(x)$ be the non-decreasing function defined by the conditions $F_k(x)=\nu_k(0,x)/k$ for $0<x\leqslant 1$ and $F_k(x+1)-F_k(x)=1$ for all x. In particular, $F_k(0)=0$, $F_k(1)=1$. $F_k(x)$ is a step function having jumps at the points $x_1,\dots,x_k$ and the point congruent to them mod 1; the expression under the limit sign in (4·26) is $c_n^k=\int_0^1 e^{-2\pi i n x}\,dF_k$. If $c_n^k\to c_n$ for all n, (4·22) implies that the c_n are the Fourier-Stieltjes coefficients of a non-decreasing F satisfying $F(x+1)-F(x)=1$. (Since $c_0^k=1$ for all k, we also have $c_0=1$.) Moreover, there are constants B_k such that $F_k(x)-B_k\to F(x)$ at the points of continuity of F. It follows that for any arc (α,β) at whose end-points F is continuous,

$$\nu_k(\alpha,\beta)/k=F_k(\beta)-F_k(\alpha)\to F(\beta)-F(\alpha).$$

Necessity. Suppose (4·24) has a distribution function F. Let α be any point of continuity of F and let F_k be the functions defined above. If x is any point of continuity of F situated in $(\alpha,\alpha+1)$, the expression $F_k(x)-F_k(\alpha)=\nu_k(\alpha,x)/k$ tends to a limit. Since $F_k(x+1)-F_k(x)=1$, it must tend to a limit for every point of continuity of F. By (4·22), the Fourier-Stieltjes coefficients of $F_k(x)-F_k(\alpha)$ must tend to limits as $k\to\infty$. This proves (4·26), since the ratio there is $\int_0^1 e^{-2\pi i n x}\,dF_k(x)$.

(4·27) THEOREM. *A necessary and sufficient condition for* (4·24) *to be equidistributed is that the limits* (4·26) *exist for* $n=\pm 1, \pm 2, \dots$ *and are all equal to* 0.

This can be seen at once if we note that the Fourier-Stieltjes series of $x+C$ consists of the constant term 1 only and that the limit c_0 in (4·26) always exists and equals 1.

A corollary of (4·27) is that *for any irrational x the sequence $x, 2x, 3x, \dots, kx, \dots$ is equidistributed.* For if $x_s=sx$, and if $n\neq 0$, the absolute value of the expression under the limit sign in (4·26) is

$$k^{-1}\left|\sum_{s=1}^{k} e^{-2\pi i s n x}\right| \leqslant 2k^{-1}\,|\,1-e^{-2\pi i n x}\,|^{-1}=o(1).$$

Similarly we prove the fact (which will be used in Chapter VIII, § 4) that if x is irrational, the sequence $x, 3x, 5x, 7x, \dots$ is equidistributed.

(4·28) THEOREM. *Let $m_1, m_2, \dots$ be any sequence of distinct positive integers, and let $\alpha_1, \alpha_2, \dots$ be any sequence of real numbers. Then for almost all x the sequence $m_s(x-\alpha_s)$ is equidistributed.*

It is enough to prove that for almost all x we have

$$\frac{1}{k}\sum_{s=1}^{k} e^{2\pi i m_s n(x-\alpha_s)} = o(1) \quad (k\to\infty;\ n=\pm 1, \pm 2, \ldots),$$

and this will follow (cf. footnote on p. 78) if we show that all the series

$$\sum_{s=1}^{\infty} \frac{e^{2\pi i m_s n(x-\alpha_s')}}{s} \quad (n=\pm 1, \pm 2, \ldots)$$

converge almost everywhere. The latter, in turn, is a corollary of the following lemma.

(4·29) LEMMA. *Let $\phi_1(x)$, $\phi_2(x)$, ... be a system orthonormal and uniformly bounded in (a,b). Then the series*

$$\sum_{s=1}^{\infty} \frac{\phi_s(x)}{s} \tag{4·30}$$

converges almost everywhere in (a,b).

Let s_N be the partial sums of (4·30), and let $f(x)$ be the function such that $\mathfrak{M}_2[f-s_N]\to 0$ (§ 1). For $N=k^2$ we have

$$\int_a^b |f-s_N|^2 = \sum_{N+1}^{\infty} \frac{1}{s^2} < \frac{1}{N} = \frac{1}{k^2}.$$

Thus the series $\sum \int_a^b |f-s_{k^2}|^2\,dx$ converges, which implies that $s_{k^2}\to f$ almost everywhere (Chapter I, (11·5)). For general N we find a k such that $k^2 \leqslant N < (k+1)^2$. Then s_N is obtained by augmenting s_{k^2} by less than $(k+1)^2-k^2=O(k)$ terms, each of which is $O(1/k^2)$. Thus the contribution of the additional terms is $O(k)\,O(1/k^2)=o(1)$, and $s_N\to f$ almost everywhere.

5. Classes of functions and (C, 1) means of Fourier series (*cont.*)

Let $\mathfrak{F}$ be a family of functions $F(x)$, $\alpha\leqslant x\leqslant\beta$, having the following property: for every $\epsilon>0$ there is a $\delta>0$ such that

$$|\Sigma[F(b_k)-F(a_k)]| < \epsilon \tag{5·1}$$

for every $F\in\mathfrak{F}$ and every finite system S of non-overlapping subintervals (a_k, b_k) of (α,β) satisfying $\Sigma(b_k-a_k)<\delta$. We shall then say that the functions $F\in\mathfrak{F}$ are *uniformly absolutely continuous* in (α,β). Clearly, the limit of an everywhere convergent sequence of functions from $\mathfrak{F}$ is absolutely continuous.

Let $\phi(u)$ be non-negative and non-decreasing for $u\geqslant 0$, and such that $\phi(u)/u\to\infty$ with u. Let $\mathfrak{f}$ be a family of functions $f(x)$ defined on $(0, 2\pi)$ and such that

$$\int_0^{2\pi} \phi(|f(x)|)\,dx \leqslant C,$$

where C is independent of f. *The integrals F of the $f\in\mathfrak{f}$ are then uniformly absolutely continuous.*

We have to show that the sums in (5·1), which are $\int_S f\,dx$, are uniformly small with $|S|$. Given any $M>0$, let u_0 be such that $\phi(u)/u\geqslant M$ for $u\geqslant u_0$. We set $|f|=f_1+f_2$, where $f_1=|f|$ if $|f|\leqslant u_0$ and $f_1=0$ otherwise. Thus the values of f_2 are either 0 or else at least u_0. Now

$$\left|\int_S f\,dx\right| \leqslant \int_S f_1\,dx + \int_S f_2\,dx \leqslant u_0|S| + M^{-1}\int_S \phi(f_2)\,dx \leqslant u_0|S| + CM^{-1}.$$

The last sum is small if we first take M large, but fixed, and then $|S|$ small.

(5·2) THEOREM. *A necessary and sufficient condition for*

$$\tfrac{1}{2}a_0+\sum_1^\infty (a_\nu \cos \nu x+b_\nu \sin \nu x)=\sum_0^\infty A_n(x) \tag{5·3}$$

to belong to L *is that the functions*

$$F_n(x)=\int_0^x \sigma_n(t)\,dt \quad (n=1, 2, \ldots) \tag{5·4}$$

should be uniformly absolutely continuous in $(0, 2\pi)$.

If the F_n are uniformly absolutely continuous, then *a fortiori* they are of uniformly bounded variation, $\mathfrak{M}[\sigma_n]=O(1)$, and $\Sigma A_n(x)$ is an $\mathsf{S}[dF]$. Since F is the limit of an everywhere convergent subsequence of F_n, F is absolutely continuous, and $\mathsf{S}[dF]=\mathsf{S}[f]$, where $f=F'$.

Conversely, suppose that $\Sigma A_n(x)$ is an $\mathsf{S}[f]$. Suppose for simplicity that $a_0=0$. The functions F_n in (5·4) are then, except for an additive constant depending on n, the (C, 1) means of the Fourier series of the integral F of f. Thus

$$|\Sigma[F_n(b_k)-F_n(a_k)]|$$

$$=\left|\frac{1}{\pi}\int_{-\pi}^{\pi} \Sigma[F(b_k+t)-F(a_k+t)]\,K_n(t)\,dt\right| \leqslant \max_t |\Sigma[F(b_k+t)-F(a_k+t)]|,$$

which is small with S. This proves (5·2).

Obviously, $\Sigma A_n(x)$ *belongs to class* A *if and only if the* σ_n *are uniformly absolutely continuous.*

(5·5) THEOREM. (i) *A necessary and sufficient condition for* $\Sigma A_n(x)$ *to belong to* L *is that* $\mathfrak{M}[\sigma_m-\sigma_n]\to 0$ *as* $m, n\to\infty$.

(ii) *If* $\Sigma A_n(x)$ *is* $\mathsf{S}[f]$, *then* $\mathfrak{M}[\sigma_n-f]\to 0$.

Suppose that $\Sigma A_n(x)$ is $\mathsf{S}[f]$. Integrating the inequality

$$|\sigma_n(x)-f(x)| \leqslant \frac{1}{\pi}\int_{-\pi}^{\pi} |f(x+t)-f(x)|\,K_n(t)\,dt \tag{5·6}$$

over $0\leqslant x\leqslant 2\pi$, we find

$$\mathfrak{M}[\sigma_n-f]\leqslant \frac{1}{\pi}\int_{-\pi}^{\pi}\eta(t)\,K_n(t)\,dt, \quad \text{where} \quad \eta(t)=\int_0^{2\pi}|f(x+t)-f(x)|\,dx.$$

Since $\eta(t)$ is continuous and vanishes at $t=0$ (Chapter I, (11·8)), and since the right-hand side of the last inequality is the (C, 1) mean of $\mathsf{S}[\eta]$ at $t=0$, we find that $\mathfrak{M}[\sigma_n-f]\to 0$. This proves (ii) and also the necessity of the condition in (i), since

$$\mathfrak{M}[\sigma_m-\sigma_n]\leqslant \mathfrak{M}[\sigma_m-f]+\mathfrak{M}[\sigma_n-f]\to 0 \quad \text{as} \quad m, n\to\infty.$$

Conversely, if $\mathfrak{M}[\sigma_m-\sigma_n]\to 0$ there is an $f\in$ L such that $\mathfrak{M}[\sigma_n-f]\to 0$ (Chapter I, (11·1)). For $n>|k|$,

$$2\pi\left(1-\frac{|k|}{n+1}\right)c_k=\int_{-\pi}^{\pi}\sigma_n e^{-ikt}\,dt=\int_{-\pi}^{\pi} f e^{-ikt}\,dt+\int_{-\pi}^{\pi}(\sigma_n-f)\,e^{-ikt}\,dt.$$

Making $n\to\infty$ and observing that the absolute value of the last term does not exceed $\mathfrak{M}[\sigma_n-f]$, we see that c_k is the kth coefficient of f. This proves (5·5).

(5·7) THEOREM. *Let $\phi(u), u \geqslant 0$, be convex, non-negative, non-decreasing, and such that $\phi(u)/u \to \infty$ with u. A necessary and sufficient condition for $\Sigma A_n(x)$ to belong to L_ϕ is that*

$$\int_0^{2\pi} \phi(|\sigma_n(x)|)\, dx \leqslant C, \tag{5·8}$$

where C is finite and independent of n.

To prove the necessity of the condition, we consider

$$|\sigma_n(x)| \leqslant \frac{1}{\pi}\int_0^{2\pi} K_n(x-t)\,|f(t)|\,dt. \tag{5·9}$$

By Jensen's inequality, taking into account that the integral of the function $p(t) = K_n(x-t)$ over $(0, 2\pi)$ is π, we find

$$\phi(|\sigma_n(x)|) \leqslant \frac{1}{\pi}\int_0^{2\pi} K_n(x-t)\,\phi(|f(t)|)\,dt. \tag{5·10}$$

If we integrate this with respect to x and invert the order of integration on the right we get

$$\int_0^{2\pi} \phi(|\sigma_n|)\,dx \leqslant \int_0^{2\pi} \phi(|f|)\,dt, \tag{5·11}$$

which proves the necessity of the condition.

As regards the sufficiency, Jensen's inequality

$$\phi\left(\frac{1}{2\pi}\int_0^{2\pi} |\sigma_n|\,dx\right) \leqslant \frac{1}{2\pi}\int_0^{2\pi} \phi(|\sigma_n|)\,dx \leqslant \frac{C}{2\pi}$$

implies that $\mathfrak{M}[\sigma_n] = O(1)$, so that $\Sigma A_n(x)$ is an $S[dF]$. Moreover, the functions (5·4) are uniformly absolutely continuous. Hence F is absolutely continuous, and $\Sigma A_n(x)$ is $S[f]$, $f = F'$. Since $\sigma_n(x) \to f(x)$ almost everywhere, (5·8) implies that $\mathfrak{M}[\phi(|f|)] \leqslant C$, that is, $f \in L_\phi$.

In particular, *a necessary and sufficient condition for* (5·3) *to belong to* L^r, $r > 1$, *is* $\mathfrak{M}_r[\sigma_n] = O(1)$. As (4·3) shows, the result fails for $r = 1$.

(5·12) THEOREM. *Suppose that $\phi(u), u \geqslant 0$, is convex, non-negative and non-decreasing. If $f \in L_\phi$, then*

$$\int_0^{2\pi} \phi(|\sigma_n|)\,dx \to \int_0^{2\pi} \phi(|f|)\,dx.$$

In particular, if $f \in L^r$, $r \geqslant 1$, then $\mathfrak{M}_r[\sigma_n] \to \mathfrak{M}_r[f]$.

After (5·11), it is enough to show that $\liminf \int_0^{2\pi} \phi(|\sigma_n|)\,dx \geqslant \int_0^{2\pi} \phi(|f|)\,dx$. Let E be any set of points at which the σ_n are uniformly bounded. Since $\sigma_n \to f$ almost everywhere, we have $\int_E \phi(|\sigma_n|)\,dx \to \int_E \phi(|f|)\,dx$ and hence

$$\liminf \int_0^{2\pi} \phi(|\sigma_n|)\,dx \geqslant \int_E \phi(|f|)\,dx.$$

The right-hand side here can be made arbitrarily close to $\int_0^{2\pi} \phi(|f|)\,dx$, since $|E|$ can be made arbitrarily close to 2π. This completes the proof of (5·12).

Suppose that a convex and non-negative function $\phi(u)$, $u \geqslant 0$, has $\phi(0)=0$ and is non-decreasing. Supposing that $\Sigma A_n(x)$ belongs to L_ϕ, we may ask whether

$$\int_0^{2\pi} \phi(|\sigma_n - f|)\,dx \to 0. \tag{5·13}$$

Applying Jensen's inequality to (5·6) we see that (5·13) holds provided the function $\eta(t) = \int_{-\pi}^{\pi} \phi\{|f(x+t)-f(x)|\}\,dx$ is integrable and tends to 0 with t. This is not always true if $\phi(u)$ increases too rapidly with u, but we can save the situation by adding a harmless factor to the argument of ϕ in (5·13): *If $f \in L_\phi$, then*

$$\eta^*(t) = \int_{-\pi}^{\pi} \phi\{\tfrac{1}{4}|f(x+t)-f(x)|\}\,dx$$

is integrable and tends to 0 with t.

In fact, let $f = g+h$, where g is bounded and $\int_{-\pi}^{\pi} \phi(|h|)\,dx < \epsilon$. By Jensen's inequality,

$$\int_0^{2\pi} \phi\{\tfrac{1}{4}|f(x+t)-f(x)|\}\,dx \leqslant \frac{1}{2}\int_0^{2\pi} \phi\{\tfrac{1}{2}|g(x+t)-g(x)|\}\,dx + \frac{1}{2}\int_0^{2\pi} \phi\{\tfrac{1}{2}|h(x+t)-h(x)|\}\,dx,$$

where the last term does not exceed $\frac{1}{4}\int_0^{2\pi} \phi(|h(x+t)|)\,dx + \frac{1}{4}\int_0^{2\pi} \phi(|h(x)|)\,dx < \frac{1}{2}\epsilon$, and the preceding term is bounded and tends to 0 with t. (Our hypothesis about ϕ implies that in every interval $0 \leqslant u \leqslant a$ we have $\phi(u) \leqslant Mu$, where $M = \phi(a)/a$.) Hence the total is less than ϵ for $|t|$ small, which proves the assertion. We thus obtain the following:

(5·14) Theorem. *Suppose that $\phi(u)$, $u \geqslant 0$, is convex, non-negative, non-decreasing and that $\phi(0)=0$. If $\Sigma A_n(x)$ is an $S[f]$ with $f \in L_\phi$, then*

$$\int_0^{2\pi} \phi(\tfrac{1}{4}|f-\sigma_n|)\,dx \to 0.$$

In particular, if $f \in L^r$, $r \geqslant 1$, then $\mathfrak{M}_r[f-\sigma_n] \to 0$.

(5·15) Theorem. *Suppose that $\Sigma A_n(x)$ is an $S[dF]$ with $F(x) = \frac{1}{2}\{F(x+0)+F(x-0)\}$ for all x.*

(i) *Either of the following two conditions is both necessary and sufficient for F to be absolutely continuous over a closed interval (α, β):*

(*a*) *the functions $F_n(x) = \int_\alpha^x \sigma_n\,dt$ are uniformly absolutely continuous over (α,β);*

(*b*) $\mathfrak{M}[\sigma_m - \sigma_n; \alpha, \beta] \to 0$.

(ii) *If the functions $\sigma_n(t)$ in (a) and (b) are replaced by $\sigma_n^+(t)$, we obtain necessary and sufficient conditions* ((*a'*), (*b'*), say) *for the positive variation of F to be absolutely continuous in (α,β).*

It is easy to see that (*b*) implies (*a*), so that for (i) it is enough to prove the sufficiency of (*a*) and the necessity of (*b*). The former is immediate, since then the function F in the proof of (4·3) is absolutely continuous in (α,β).

Suppose then, that $\Sigma A_n(x)$ is an $S[dF]$ and that F is absolutely continuous in $\alpha \leqslant x \leqslant \beta$. If we show that $\mathfrak{M}[\sigma_m - \sigma_n; \alpha', \beta'] \to 0$ for any interval (α', β') interior to (α, β), the necessity of (*b*) will follow. For, by (4·12), $\mathfrak{M}[\sigma_\nu; \alpha, \alpha']$ and $\mathfrak{M}[\sigma_\nu; \beta', \beta]$ tend to the total variations of F over (α, α') and (β', β), and so are small with $\alpha'-\alpha$, $\beta-\beta'$. The same follows for the integrals of $|\sigma_m-\sigma_n|$ over (α, α') and (β', β).

Let $f(x)=F'(x)$. To show that $\mathfrak{M}[\sigma_m-\sigma_n; \alpha', \beta']\to 0$, it is enough to prove that $\mathfrak{M}[\sigma_m-f; \alpha', \beta']\to 0$. We observe that

$$\sigma_m(x)=\frac{1}{\pi}\int_0^{2\pi} K_m(x-t)\,dF(t)=\frac{1}{\pi}\int_\alpha^\beta+\frac{1}{\pi}\int_\beta^{\alpha+2\pi}=v_m+w_m, \tag{5·16}$$

say. For $x\in(\alpha', \beta')$ and $t\in(\beta, \alpha+2\pi)$ the integrand of w_m tends uniformly to 0, so that $\mathfrak{M}[w_m; \alpha', \beta']\to 0$. Since $v_m=\sigma_m(x; f^*)$, where $f^*=f$ in (α, β), $f^*=0$ elsewhere, we have $\mathfrak{M}[v_m-f^*; 0, 2\pi]\to 0$, and so also $\mathfrak{M}[v_m-f; \alpha', \beta']\to 0$. Hence

$$\mathfrak{M}[\sigma_m-f; \alpha', \beta']\leqslant\mathfrak{M}[v_m-f; \alpha', \beta']+\mathfrak{M}[w_m; \alpha', \beta']\to 0,$$

and (i) is proved.

Analogously, for (ii) we must show the sufficiency of (a') and necessity of (b'). If the functions $F_n^*(x)=\int_\alpha^x \sigma_n^+\,dt$ are uniformly absolutely continuous in (α, β), their limit, which represents the positive variation of F over (α, x) (cf. (4·12)), is absolutely continuous there.

To prove the necessity of condition (b'), we begin with the case $(\alpha, \beta)=(0, 2\pi)$. Let $V(x)$, $P(x)$ and $N(x)$ be the total, positive and negative variations of F over $(0, x)$. Then

$$\sigma_n=\sigma_n'-\sigma_n'', \quad \text{where} \quad \sigma_n'=\sigma_n[dP]\geqslant 0, \quad \sigma_n''=\sigma_n[dN]\geqslant 0. \tag{5·17}$$

The relations $P'+N'=V'=|F'|$, $P'-N'=F'$ (known to be true almost everywhere), show that $P'=F'^+$, $N'=F'^-$ almost everywhere.

The inequalities $\sigma_n\leqslant\sigma_n'$, $\sigma_n'\geqslant 0$ show that $0\leqslant\sigma_n^+\leqslant\sigma_n'$. If we define $\theta_n(x)$ by

$$\sigma_n^+(x)=\theta_n(x)\,\sigma_n'(x)$$

at the points where $\sigma_n'\neq 0$, and $\theta_n(x)=1$ elsewhere, then $0\leqslant\theta_n(x)\leqslant 1$ for all x and n. We observe that, almost everywhere, $\sigma_n\to F'$ (Chapter III, (8·1)), and so also $\sigma_n^+\to F'^+$. The same fact applied to $\sigma_n[dP]$ gives $\sigma_n'\to P'=F'^+$. Hence $\theta_n(x)$ tends to 1 at almost all points where $p(x)=P'(x)\neq 0$.

Using now (for the first time) the hypothesis that $P(x)$ is absolutely continuous, we show that $\mathfrak{M}[\sigma_n^+-p; 0, 2\pi]\to 0$. In fact,

$$\int_0^{2\pi}|\sigma_n^+-p|\,dx=\int_0^{2\pi}|\sigma_n'\theta_n-p|\,dx\leqslant\int_0^{2\pi}|\sigma_n'-p|\,\theta_n\,dx+\int_0^{2\pi}|\theta_n-1|\,p\,dx.$$

The first integral on the right is majorized by $\mathfrak{M}[\sigma_n'-p]\to 0$. The last integral on the right also tends to 0, since the integrand $|\theta_n(x)-1|\,p(x)$ is majorized by $p(x)\in L$ and tends to 0 almost everywhere. Thus

$$\mathfrak{M}[\sigma_n^+-p]\to 0, \quad \mathfrak{M}[\sigma_m^+-\sigma_n^+]\leqslant\mathfrak{M}[\sigma_m^+-p]+\mathfrak{M}[\sigma_n^+-p]\to 0,$$

and the necessity of condition (b') is proved when $(\alpha, \beta)=(0, 2\pi)$.

To remove this restriction, we proceed as in case (b). It is enough to show that $\mathfrak{M}[\sigma_m^+-p; \alpha', \beta']\to 0$ for any (α', β') interior to (α, β). Assume for simplicity that (α, β) is included in $(0, 2\pi)$, and return to (5·16). The v_m there is $\sigma_m(x; dF^*)$, where F^* equals $F(x)$ in (α, β), $F(\alpha)$ in $(0, \alpha)$ and $F(\beta)$ in $(\beta, 2\pi)$. The positive variation P^* of F^*

is absolutely continuous, so that, if $p^* = P^{*\prime}$, $\mathfrak{M}[v_m^+ - p; \alpha', \beta'] \leqslant \mathfrak{M}[v_m^+ - p^*; 0, 2\pi] \to 0$. Since w_m tends uniformly to 0 over (α', β') we get $\mathfrak{M}[\sigma_m^+ - p; \alpha', \beta'] \to 0$.

Condition (b) is satisfied if there is a non-negative non-decreasing convex function $\phi(u)$, $u \geqslant 0$, such that $\phi(u)/u \to \infty$ with u, and if $\mathfrak{M}[\phi(|\sigma_n|); \alpha, \beta] = O(1)$. Similarly condition (b') is satisfied if $\mathfrak{M}[\phi(\sigma_n^+); \alpha, \beta] = O(1)$.

Many results of this and the preceding section hold, though some inequalities become less precise, for the kernel (C, α), $0 < \alpha < 1$. Let

$$\lambda_n = \lambda_n^\alpha = \frac{1}{\pi}\int_0^{2\pi} |K_n^\alpha(t)|\, dt, \quad \lambda = \sup_n \lambda_n \quad (\alpha > 0).$$

The proofs of the following results for $0 < \alpha < 1$ are essentially the same as for $\alpha = 1$.

(5·18) THEOREM. *Let ϕ be the same as in* (5·7). *If $\mathfrak{M}[\phi(|\sigma_n^\alpha|)] = O(1)$, then* (5·3) *belongs to* L_ϕ. If (5·3) *is an* $S[f]$, $f \in L_\phi$, *then* $\mathfrak{M}[\phi(|\sigma_n^\alpha|/\lambda)] = O(1)$.

If, in addition, $\phi(0) = 0$, then $\mathfrak{M}[\phi(|f - \sigma_n^\alpha|/4\lambda)] \to 0$ as $n \to \infty$.

(5·19) THEOREM. *A necessary and sufficient condition for* (5·3) *to belong to* S *is* $\mathfrak{M}[\sigma_n^\alpha] = O(1)$. *A necessary and sufficient condition for* (5·3) *to belong to* L *is* $\mathfrak{M}[\sigma_m^\alpha - \sigma_n^\alpha] \to 0$ *as* $m, n \to \infty$.

If we replace the σ_n by the partial sums s_n in the theorems of this and the preceding section, the conditions we obtain remain sufficient, though no longer necessary. The proofs of sufficiency remain the same, except at one point; we cannot use the fact that $s_n(x; f) \to f(x)$ almost everywhere, for this is false (see Chapter VIII, § 3). But for this purpose it is enough to know that there is a subsequence $\{s_{n_k}(x; f)\}$ converging to $f(x)$ almost everywhere, and we shall see in Chapter VII, § 6, that this is true.

Another observation on the sufficiency conditions in the theorems of this and the preceding section is also useful. In showing that a certain behaviour of the σ_n (or s_n) implies that the series belongs to a definite class, it is not really necessary to consider all positive integers n; it is enough to suppose that the condition is satisfied for some sequence $\{n_k\}$ tending to $+\infty$. Thus, if $\{\sigma_{n_k}(x)\}$ or $\{s_{n_k}(x)\}$ converges uniformly, the series belongs to class C (see (4·2)); if $\mathfrak{M}[s_{n_k}] = O(1)$, it is an $S[dF]$ (see (4·3)); if the $s_{n_k}(x)$ are non-negative, the series is an $S[dF]$ with F non-decreasing (see (4·8)), etc.

This makes it possible to state in a slightly different form some of the theorems proved above. For example, *a necessary and sufficient condition for $\Sigma A_n(x)$ to belong to class* C *is that the $\sigma_n(x)$ are uniformly continuous*. The necessity follows from the inequality (5·9), which, applied to $f(x+h) - f(x)$, yields

$$\omega(\delta; \sigma_n) \leqslant \omega(\delta; f).$$

Conversely, if the $\sigma_n(x)$ are uniformly continuous there is, by Arzelà's well-known theorem, a subsequence $\{\sigma_{n_k}(x)\}$ converging uniformly to a continuous function $f(x)$, and so $\Sigma A_n(x)$ is an $S[f]$, $f \in C$.

(5·20) THEOREM. *If $\mathfrak{M}[s_{n_k}] = O(1)$ for a sequence of partial sums of $\Sigma A_n(x)$ (in particular, if the s_{n_k} are non-negative), the series is an $S[dF]$ with F continuous.*

We know already that $\Sigma A_n(x)$ is an $S[dF]$, and so need only prove the continuity of F. Suppose that $F(x_0 + 0) - F(x_0 - 0) = d \neq 0$ for some x_0, and suppose for simplicity that $x_0=0$ and that $2F(0)=F(+0)+F(-0)$. Let $\phi(x)\sim\Sigma\nu^{-1}\sin\nu x$ (see Chapter I, (4·12)). We may write

$$F(x)=\{F(x)-(d/\pi)\,\phi(x)\}+(d/\pi)\,\phi(x)=F_1(x)+F_2(x),$$

say, where F_1 is continuous at $x=0$. Correspondingly

$$\mathsf{S}[dF]=\mathsf{S}[dF_1]+\mathsf{S}[dF_2],\quad s_n=s_n^1+s_n^2.$$

Since $\phi(0)=\phi(2\pi)$, we have $\mathsf{S}[dF_2]=\mathsf{S}'[F_2]$ and the nth partial sum of $\mathsf{S}[dF_2]$ is $(d/\pi)\,[D_n(x)-\frac{1}{2}]$. Thus whatever the value of $\epsilon>0$, $\mathfrak{M}[s_n^2;\,-\epsilon,\epsilon]\simeq C\log n$, where C is a positive constant (Chapter II, (12·2)). If we can show that for ϵ small enough and $n>n_0$ we have $\mathfrak{M}[s_n^1;\,-\epsilon,\epsilon]<\frac{1}{2}C\log n$, it will follow that $\mathfrak{M}[s_n;\,-\epsilon,\epsilon]$, and so also $\mathfrak{M}[s_n]$, tends to ∞, contrary to hypothesis.

Let $I=(-\epsilon,\epsilon)$, $I'=(-2\epsilon,2\epsilon)$. If $x\in I$, then

$$|\,s_n^1(x)\,|=\left|\frac{1}{\pi}\int_{-\pi}^{\pi}D_n(x-t)\,dF_1(t)\right|\leqslant\frac{1}{\pi}\int_{I'}|\,D_n(x-t)\,|\,|\,dF_1(t)\,|+O(1),$$

$D_n(u)$ being uniformly bounded for $\epsilon\leqslant|\,u\,|\leqslant\pi$. Integrating this over I and writing L_n for Lebesgue's constant, we have

$$\int_I|\,s_n^1(x)\,|\,dx\leqslant\frac{1}{\pi}\int_{I'}|\,dF(t)\,|\int_I|\,D_n(x-t)\,|\,dx\leqslant L_n\int_{I'}|\,dF_1(t)\,|.$$

Since $L_n=O(\log n)$, and the variation of F_1 over I is small with ϵ, owing to the continuity of F_1 at 0, we have $\mathfrak{M}[s_n^1;\,-\epsilon,\epsilon]<\frac{1}{2}C\log n$ for ϵ small enough and $n>n_0$. This proves Theorem (5·20).

6. Classes of functions and Abel means of Fourier series

Let
$$f(\rho,x)=\tfrac{1}{2}a_0+\sum_{n=1}^{\infty}(a_n\cos nx+b_n\sin nx)\,\rho^n\quad(0\leqslant\rho<1)\tag{6·1}$$

be the harmonic function associated with the series

$$\tfrac{1}{2}a_0+\sum_{n=1}^{\infty}(a_n\cos nx+b_n\sin nx)=\sum_{n=0}^{\infty}A_n(x).\tag{6·2}$$

The analogues for Abel means of the results obtained in the preceding two sections may be stated as follows. (As was explained in § 4, we omit the proofs.)

(6·3) THEOREM. *A necessary and sufficient condition for* $\Sigma A_n(x)$ *to belong to class* C *or, what is the same thing, for*

$$f(\rho,x)=\frac{1}{2\pi}\int_0^{2\pi}\frac{1-\rho^2}{1-2\rho\cos(x-t)+\rho^2}f(t)\,dt\quad(0\leqslant\rho<1)\tag{6·4}$$

with $f(t)$ *continuous, is that* $f(\rho,x)$ *should converge uniformly as* $\rho\to1$. *A necessary and sufficient condition for* $\Sigma A_n(x)$ *to belong to class* B *is that* $f(\rho,\,x)$ *should be bounded for* $0\leqslant\rho<1$.

(6·5) THEOREM. *A necessary and sufficient condition for* $\Sigma A_n(x)$ *to belong to class* S, *or, what is the same thing, for*

$$f(\rho,x)=\frac{1}{2\pi}\int_0^{2\pi}\frac{1-\rho^2}{1-2\rho\cos(x-t)+\rho^2}dF(t)\quad(0\leqslant\rho<1),\tag{6·6}$$

where $F(t)$ is of bounded variation, is that the integral

$$\int_0^{2\pi} |f(\rho,x)|\,dx = \mathfrak{M}[f(\rho,x)] \tag{6·7}$$

should be bounded as $\rho\to 1$. The latter condition is equivalent to $\mathfrak{M}[f^+(\rho,x)] = O(1)$. If we have (6·6), *then*

$$\int_0^{2\pi} |f(\rho,x)|\,dx \leqslant \int_0^{2\pi} |dF(x)|.$$

(6·8) THEOREM. *A necessary and sufficient condition for $f(\rho,x)$ to be representable by* (6·6), *with $F(t)$ non-decreasing, is that $f(\rho,x)\geqslant 0$ for $0\leqslant\rho<1$.*

(6·9) THEOREM. *If $f(\rho,x)$ is given by* (6·6), *and if*

$$F(x) = \tfrac{1}{2}\{F(x+0)+F(x-0)\}, \tag{6·10}$$

then $$\int_\alpha^\beta |f(\rho,x)|\,dx \to V,\quad \int_\alpha^\beta f^+(\rho,x)\,dx\to P,\quad \int_\alpha^\beta f^-(\rho,x)\,dx\to N,$$

where V, P, N are the total, positive and negative variations of F over (α,β).

This result leads to the following:

(6·11) THEOREM. *Let $F(\rho,x)$ be the Poisson integral of a periodic F of bounded variation satisfying* (6·10). *Then the total (positive, negative) variation of $F(\rho,x)$ over an arc $\alpha\leqslant x\leqslant\beta$ tends to the total (positive, negative) variation of $F(x)$ over $\alpha\leqslant x\leqslant\beta$ as $\rho\to 1$.*

(6·12) THEOREM. *Each of the following conditions is both necessary and sufficient for $\Sigma A_n(x)$ to belong to L (that is, for* (6·4) *to hold with an $f\in$ L):*

(i) $\int_0^x f(\rho,u)\,du$ *is a uniformly absolutely continuous function of x for $0\leqslant\rho<1$;*

(ii) $\int_0^{2\pi} |f(\rho,x)-f(\rho',x)|\,dx \to 0$ *as $\rho,\rho'\to 1$.*

(6·13) THEOREM. *Let $\phi(u)$ be non-negative, convex, and non-decreasing for $u\geqslant 0$, and let $\phi(u)/u\to\infty$ with u. A necessary and sufficient condition for $\Sigma A_n(x)$ to belong to L_ϕ is*

$$\int_0^{2\pi} \phi(|f(\rho,x)|)\,dx = O(1)\quad (0\leqslant\rho<1). \tag{6·14}$$

(6·15) THEOREM. *If $\Sigma A_n(x)$ is an $S[f]$, $f\in L_\phi$, where $\phi(u)$ is convex non-negative and non-decreasing for $u\geqslant 0$, then*

$$\int_0^{2\pi} \phi(|f(\rho,x)|)\,dx \to \int_0^{2\pi} \phi(|f|)\,dx\quad (\rho\to 1). \tag{6·16}$$

If in addition $\phi(0)=0$, then

$$\int_0^{2\pi} \phi(\tfrac{1}{4}|f(\rho,x)-f(x)|)\,dx \to 0\quad (\rho\to 1).$$

(6·17) THEOREM. *A necessary and sufficient condition for $\Sigma A_n(x)$ to belong to L^r, $r>1$, is*

$$\int_0^{2\pi} |f(\rho,x)|^r\,dx = O(1)\quad (\rho\to 1).$$

If $\Sigma A_n(x)$ is an $\mathrm{S}[f]$ with $f\in \mathrm{L}^r$, $r\geqslant 1$, then

$$\int_0^{2\pi} |f(\rho,x)-f(x)|^r\,dx\to 0 \quad (\rho\to 1).$$

(6·18) THEOREM. *If $\Sigma A_n(x)$ is an $\mathrm{S}[dF]$ with F satisfying* (6·10), *each of the conditions* (i), (ii) *below is both necessary and sufficient for F to be absolutely continuous in (α,β):*

(i) *The functions $\int_\alpha^x f(\rho,u)\,du$ are uniformly absolutely continuous in (α,β);*

(ii) $\mathfrak{M}[f(\rho,x)-f(\rho',x);\,\alpha,\beta]\to 0$ *as* $\rho,\rho'\to 1$.

If $f(\rho,x)$ is replaced by $f^+(\rho,x)$, we obtain necessary and sufficient conditions for the absolute continuity in (α,β) of the positive variation of F.

(6·19) THEOREM. *Let $\Sigma A_n(x)$ be an $\mathrm{S}[dF]$, let F satisfy* (6·10), *and let*

$$f(x)=\lim f(\rho,x)=F'(x).$$

Of the two conditions

$$\int_\alpha^\beta |f(\rho,x)|\,dx\to\int_\alpha^\beta |f(x)|\,dx,\qquad \int_\alpha^\beta f^+(\rho,x)\,dx\to\int_\alpha^\beta f^+(x)\,dx,$$

the first is necessary and sufficient for F to be absolutely continuous in (α,β), the second for the positive variation of F to be absolutely continuous there.

The analogue of (5·11) for Abel means is

$$\int_0^{2\pi}\phi(|f(\rho,x)|)\,dx\leqslant\int_0^{2\pi}\phi(|f(x)|)\,dx. \tag{6·20}$$

Let $0\leqslant\rho<\rho'<1$, so that $\rho=\rho' R$, with $0<R<1$. From (6·1) we see that $f(\rho,x)$ is the Poisson integral of $f(\rho',x)$, and (6·20) implies that

$$\int_0^{2\pi}\phi(|f(\rho,x)|)\,dx\leqslant\int_0^{2\pi}\phi(|f(\rho',x)|)\,dx \quad (0\leqslant\rho<\rho'<1). \tag{6·21}$$

Thus

(6·22) THEOREM. *If $\phi(u)$ is non-negative, non-decreasing, and convex for $u\geqslant 0$, and $f(\rho,x)$ is harmonic for $\rho<1$, the integral $\int_0^{2\pi}\phi(|f(\rho,x)|)\,dx$ is a non-decreasing function of ρ.*

The case $\phi(u)=u^r$, $r\geqslant 1$, is particularly important.

If $f(\rho,x)$ is given by (6·6), then writing $F=F_1-F_2$, where F_1, F_2 are non-decreasing, we represent $f(\rho,x)$ as a difference of two non-negative harmonic functions. If $f(\rho,x)$ is non-negative, the integral (6·7) is bounded (being in fact πa_0). The same holds if $f(\rho,x)$ is a difference of two non-negative harmonic functions. Thus

(6·23) THEOREM. *A necessary and sufficient condition for a harmonic function $f(\rho,x)$, $0\leqslant\rho<1$, to be representable in the form* (6·6), *with F of bounded variation, is that $f(\rho,x)$ should be a difference of two non-negative harmonic functions.*

Let $z=\rho e^{ix}$. The Poisson kernel $P(\rho,x)$ is the real part of

$$\tfrac12+z+z^2+\ldots=\tfrac12(1+z)/(1-z).$$

Thus, *the harmonic function* (6·6) *is the real part of the function*

$$\Phi(z)=\frac{1}{2\pi}\int_0^{2\pi}\frac{e^{it}+z}{e^{it}-z}\,dF(t)\quad (z=\rho\, e^{ix}), \tag{6·24}$$

regular in $|z|<1$. The imaginary part of $\Phi(z)$ is

$$\tilde{f}(\rho,x)=\sum_{\nu=1}^{\infty}(a_\nu\sin\nu x-b_\nu\cos\nu x)\,\rho^\nu=\frac{1}{\pi}\int_0^{2\pi}\frac{\rho\sin(x-t)}{1-2\rho\cos(x-t)+\rho^2}\,dF(t), \tag{6·25}$$

the harmonic function conjugate to $f(\rho,x)$ and vanishing at the origin. Hence

(6·26) THEOREM. *A function* $\Phi(z)$, *with* $\mathscr{I}\Phi(0)=0$, *regular for* $|z|<1$, *has a non-negative real part there if and only if* $\Phi(z)$ *is given by the formula* (6·24) *with* $F(t)$ *non-decreasing and bounded.*

The boundedness of the integral (6·7) does not imply the boundedness of the integral with $\tilde{f}(\rho,x)$, as we see by the example

$$f(\rho,x)=P(\rho,x),\quad \tilde{f}(\rho,x)=Q(\rho,x).$$

(That $\mathfrak{M}[Q(\rho,x)]\neq O(1)$ may be verified either directly, or by observing that $\sin x+\sin 2x+\ldots$ is not an $S[dF]$.) However:

(6·27) THEOREM. *Suppose that the integral* (6·7) *does not exceed* C *for* $0\leqslant\rho<1$. *Then the integral of* $\rho^{-1}|\tilde{f}(\rho,x)|$ (and *a fortiori* that of $|\tilde{f}(\rho,x)|$) *along any diameter of the unit circle does not exceed* $\frac{1}{2}C$.

The result is quite elementary, and in order not to use the representation (6·6), whose proof is rather deep, let us suppose first that $f(\rho,x)$ is continuous for $\rho\leqslant 1$. Then $f(\rho,x)$ is the Poisson integral of $f(x)=f(1,x)$, and

$$\tilde{f}(\rho,x)=-\frac{1}{\pi}\int_{-\pi}^{\pi}f(t+x)\,Q(\rho,t)\,dt,$$

$$\int_0^1\rho^{-1}(|\tilde{f}(\rho,x)|+|\tilde{f}(\rho,x+\pi)|)\,d\rho\leqslant\frac{1}{\pi}\int_{-\pi}^{\pi}|f(t+x)|\times\left\{\int_0^1\rho^{-1}(|Q(\rho,t)|+|Q(\rho,t+\pi)|)\,d\rho\right\}dt.$$

In estimating the term in curly brackets we may suppose that $0<t<\pi$. Then $Q(\rho,t)>0$, $Q(\rho,t+\pi)<0$, the term in question is

$$\lim_{R\to1}\int_0^R\rho^{-1}[Q(\rho,t)-Q(\rho,t+\pi)]\,d\rho=\lim_{R\to1}\int_0^R\left[\sum_1^\infty\rho^{\nu-1}\sin\nu t-\sum_1^\infty(-1)^\nu\rho^{\nu-1}\sin\nu t\right]d\rho$$
$$=\lim_{R\to1}2\sum_{\nu=1}^\infty R^{2\nu-1}\frac{\sin(2\nu-1)t}{2\nu-1}=\tfrac{1}{2}\pi$$

(see Chapter I, (4·13)), and the whole expression on the right is

$$\frac{1}{\pi}\frac{\pi}{2}\int_{-\pi}^{\pi}|f(t+x)|\,dt=\frac{1}{2}\int_{-\pi}^{\pi}|f(t)|\,dt\leqslant\tfrac{1}{2}C.$$

In the general case we fix R, $0<R<1$, and apply the result obtained to the function

$$f_1(\rho,x)=f(\rho R,x)$$

harmonic and continuous for $\rho \leqslant 1$. Since $\mathfrak{M}[f_1(\rho, x); 0, 2\pi] \leqslant C$, the integral

$$\int_0^1 \rho^{-1}(|\tilde{f}_1(\rho, x)| + |\tilde{f}_1(\rho, x+\pi)|)\, d\rho = \int_0^R \rho^{-1}(|\tilde{f}(\rho, x)| + |\tilde{f}(\rho, x+\pi)|)\, d\rho$$

does not exceed $\frac{1}{2}C$. The proof is completed by letting R tend to 1.

Let $U(\rho, x)$ be any function harmonic for $\rho < 1$ and let $V(\rho, x)$ be the conjugate function. The harmonic function $v(\rho, x) = V_x(\rho, x)$ vanishes at the origin (observe that V is of the form $\Sigma A_n(x)\rho^n$) and is the conjugate of $u(\rho, x) = U_x(\rho, x)$. Suppose U satisfies

$$\mathfrak{M}[U_x(\rho, x); 0, 2\pi] \leqslant C \quad \text{for} \quad 0 \leqslant \rho < 1.$$

Then, by (6·27) and the Cauchy-Riemann equations,

$$\int_D |\rho^{-1}v|\, d\rho = \int_D |\rho^{-1}V_x|\, d\rho = \int_D |U_\rho|\, d\rho \leqslant \tfrac{1}{2}C,$$

the integration being along any diameter D of the unit circle. The last integral is the total variation of U over D. Thus (6·27) may be re-stated as follows:

(6·28) THEOREM. Let $U(\rho, x)$ *be harmonic for* $\rho < 1$. *If the total variation of* U *over any circle* $\rho = \rho_0 < 1$ *does not exceed* C, *the total variation of* U *over any diameter of the unit circle does not exceed* $\frac{1}{2}C$.

Consider the Poisson integral $f(\rho, x)$ of an f in L^p, $p \geqslant 1$ (Cf. Chapter III, (6·4)), and suppose that

$$\frac{1}{2\pi}\int_0^{2\pi} |f(\rho, x)|^p\, dx \leqslant M^p \quad (0 \leqslant \rho < 1). \tag{6·29}$$

We shall deduce from this an estimate for $\mathfrak{A}_r[f(\rho, x)]$ for $r > p$. Let us apply to (6·4) Theorem (1·15) of Chapter II. If q is defined by $1/r = 1/p + 1/q - 1$ (so that $q > 1$), then

$$\mathfrak{A}_r[f(\rho, x)] \leqslant \mathfrak{A}_p[f]\, \mathfrak{A}_q[2P(\rho, t)]. \tag{6·30}$$

In order to estimate $\mathfrak{A}_q[P(\rho, t)]$ we use the inequalities (6·9) of Chapter III, where we may suppose that $A > 1$, and find

$$\mathfrak{A}_q^q[P(\rho, t)] \leqslant \frac{2}{2\pi}\int_0^\delta \delta^{-q}\, dt + \frac{2}{2\pi}A^q\delta^q \int_\delta^\infty t^{-2q}\, dt \leqslant A^q\,\delta^{1-q},$$

$$\mathfrak{A}_q[P(\rho, t)] \leqslant A\delta^{-1/q'}. \tag{6·31}$$

Hence, observing that $1/q' = 1/p - 1/r$, we get

(6·32) THEOREM. *If* (6·29) *holds for some* $p \geqslant 1$, *then*

$$\mathfrak{A}_r[f(\rho, x)] \leqslant BM(1-\rho)^{1/r - 1/p}$$

for $r > p$, B *denoting an absolute constant.*

Subtracting from f a suitable polynomial, we may make M as small as we please. Thus,

(6·33) THEOREM. *If* $\Sigma A_n(x)$ *is in* L^p, $p \geqslant 1$, *then*

$$\mathfrak{A}_r[f(\rho, x)] = o\{(1-\rho)^{1/r - 1/p}\} \quad \textit{as} \quad \rho \to 1.$$

The following result generalizes (6·32):

(6·34) THEOREM. *If $\mathfrak{A}_p[f(\rho,x)] \leqslant M(1-\rho)^{-\beta}$ for some $p \geqslant 1$, $\beta > 0$, then*

$$\mathfrak{A}_r[f(\rho,x)] \leqslant MB_\beta(1-\rho)^{-\beta+1/r-1/p} \quad \text{for} \quad r > p,$$

with B_β depending on β only.

Let

$$0 < \rho < 1, \quad \rho_1 = \rho^{\frac{1}{2}}, \quad g(x) = f(\rho_1, x).$$

Since $\rho_1 > \rho$, $f(\rho,x)$ is the Poisson integral of $g(x)$: $f(\rho,x) = g(\rho_1,x)$. By hypothesis,

$$\mathfrak{A}_p[g] = \mathfrak{A}_p[f(\rho_1,x)] \leqslant M(1-\rho_1)^{-\beta},$$

and by (6·32) applied to g,

$$\mathfrak{A}_r[f(\rho,x)] = \mathfrak{A}_r[g(\rho_1,x)] \leqslant BM(1-\rho_1)^{-\beta}(1-\rho_1)^{1/r-1/p}$$
$$= BM(1-\rho_1)^{-\beta+1/r-1/p}.$$

Since $(1-\rho_1)/(1-\rho)$ is contained between $\frac{1}{2}$ and 1, (6·34) follows with $B_\beta = 2^{\beta+1}B$. The 'O' in the conclusion is not replaceable by 'o' here, as it is in (6·32); see Example 6 at the end of the chapter.

The theorem which follows is an analogue of (6·32) for trigonometric polynomials. It suggests that to estimates of harmonic functions $f(\rho,x)$ there should correspond estimates for polynomials of order $n \sim 1/(1-\rho)$.

(6·35) THEOREM. *If T is a polynomial of order n, then*

$$\mathfrak{A}_r[T] \leqslant Bn^{1/p-1/r}\mathfrak{A}_p[T] \tag{6·36}$$

for $r > p \geqslant 1$, with B an absolute constant.

The Fejér kernel $K_n(t)$ satisfies an inequality

$$\mathfrak{A}_q[K_n] \leqslant An^{1/q'} \tag{6·37}$$

analogous to (6·31), since, as we have already observed (p. 97), the estimates for $K_n(t)$ and $P(\rho,t)$ are similar if we identify n and $1/(1-\rho)$. If the σ_k are the (C, 1) means of T, we have the inequalities (compare (6·30))

$$\mathfrak{A}_r[\sigma_k] \leqslant \mathfrak{A}_p[T]\,\mathfrak{A}_q[2K_k] \leqslant A\mathfrak{A}_p[T]\,k^{1/q'}. \tag{6·38}$$

For the delayed means $\tau_n = 2\sigma_{2n-1} - \sigma_{n-1}$ (p. 80) we have therefore

$$\mathfrak{A}_r[\tau_n] \leqslant A\mathfrak{A}_p[T]\{2(2n)^{1/q'} + n^{1/q'}\} \leqslant 5A\mathfrak{A}_p[T]\,n^{1/q'},$$

and it is enough to observe that $\tau_n = T$.

7. Majorants for the Abel and Cesàro means of $S[f]$

These means have simple estimates in terms of the non-negative function

$$M(x) = M_f(x) = \sup_{|t| \leqslant \pi} \frac{1}{t}\int_0^t |f(x+u)|\,du$$

introduced in Chapter I, § 13. The proofs will be based on the following lemma:

(7·1) LEMMA. *Let $\chi(t,p)$, $-\pi \leqslant t \leqslant \pi$, be a non-negative function depending on a parameter p and satisfying the conditions*

$$\text{(i)} \int_{-\pi}^{\pi} \chi(t,p)\,dt \leqslant K, \qquad \text{(ii)} \int_{-\pi}^{\pi} \left| t\frac{\partial}{\partial t}\chi(t,p)\right| dt \leqslant K_1, \tag{7·2}$$

where K *and* K_1 *are independent of* p. *If we set*

$$h(x,p)=\int_{-\pi}^{\pi} f(x+t)\,\chi(t,p)\,dt, \tag{7·3}$$

then

$$\sup_p |h(x,p)| \leqslant AM(x), \tag{7·4}$$

where A *depends only on* K *and* K_1.

For, fixing x, let $F(t)=\int_0^t f(x+u)\,du$. Then integrating in (7·3) by parts and using the inequality $|F(t)|\leqslant|t|\,M(x)$, we get

$$|h(x,p)|\leqslant M(x)\left\{\int_{-\pi}^{\pi}\left|t\frac{\partial}{\partial t}\chi(t,p)\right|dt+[\pi\chi(\pi,p)+\pi\chi(-\pi,p)]\right\}.$$

The expression in square brackets does not exceed $K+K_1$, as we see by writing (7·2) (i) in the form

$$-\int_{-\pi}^{\pi} t\frac{\partial\chi}{\partial t}dt+\pi[\chi(\pi,p)+\chi(-\pi,p)]\leqslant K$$

and applying (7·2) (ii). Summing up,

$$|h(x,p)|\leqslant(2K_1+K)\,M(x),$$

and (7·4) is established.

It is useful to observe that if $t\,\partial\chi/\partial t$ is of constant sign and if $\chi(\pm\pi,p)$ are bounded functions of p, then (7·2) (ii) is a consequence of (7·2) (i). This follows at once if we drop the absolute value sign in (7·2) (ii) and integrate by parts.

Combining (7·4) with the inequalities (13·17) of Chapter I, we get the following:

(7·5) THEOREM. *Under the hypotheses of* (7·1), *the function*

$$N(x)=\sup_p|h(x,p)|$$

satisfies the inequalities

$$\left.\begin{aligned}&\int_{-\pi}^{\pi}N^r(x)\,dx\leqslant A_r\int_{-\pi}^{\pi}|f|^r dx\quad(r>1),\\&\int_{-\pi}^{\pi}N^\alpha(x)\,dx\leqslant A_\alpha\left(\int_{-\pi}^{\pi}|f|\,dx\right)^\alpha dx\quad(0<\alpha<1),\\&\int_{-\pi}^{\pi}N(x)\,dx\leqslant A\int_{-\pi}^{\pi}|f|\log^+|f|\,dx+A,\end{aligned}\right\} \tag{7·6}$$

where the constants depend only on the indices shown explicitly, and on K *and* K_1.

It is useful to note that A_r remains bounded as $r\to+\infty$.

We note some special functions χ. The Poisson kernel $P(\rho,t)$ is one; the first inequality (7·2) is familiar, and the second follows from it since $t\,dP/dt\leqslant 0$ and

$$P(\rho,\pm\pi)=O(1).$$

The Fejér kernel $K_n(t)$ satisfies the first inequality but not the second. The same holds for the kernel $K_n^\delta(t)$, $0<\delta\leqslant 1$, which, in addition, is not of constant sign if $\delta<1$. The kernel $K_n^\delta(t)$ can, however, be majorized by a function satisfying (7·2), namely,

$$|K_n^\delta(t)|\leqslant\frac{c(\delta)\,n}{(1+n\,|t|)^{\delta+1}}\quad\text{for}\quad n\geqslant 1,\ |t|\leqslant\pi, \tag{7·7}$$

where $c(\delta)$ depends on δ only $(0<\delta\leqslant 1)$. For let $H_n(t)$ be the expression on the right. It exceeds at least one of $2^{-\delta-1}c(\delta)\,n$ and $c(\delta)/2^{\delta+1}n^\delta\,|t|^{\delta+1}$. Hence, by Chapter III, (5·5), it exceeds $|K_n^\delta(t)|$, provided that $c(\delta)$ is large enough. It is easy to see that $H_n(t)$ satisfies the first inequality (7·2), from which the second follows since $|tH_n'(t)|\leqslant(1+\delta)H_n(t)$. Thus:

(7·8) THEOREM. *The inequalities* (7·6) *hold if* $N(x)$ *is one of the functions*

$$\sup_{\rho<1}|f(\rho,x)|,\quad \sup_{n\geqslant 1}|\sigma_n^\delta(x)|.\dagger$$

The constants here depend again only on the indices shown explicitly and, in the second case, also on δ.‡

Let $\zeta=\rho e^{i\theta}$. For any $0\leqslant\sigma<1$, let Ω_σ denote the open domain bounded by the two tangents from $\zeta=1$ to the circle $|\zeta|=\sigma$, and by the more distant arc of the circle between the points of contact. By $\Omega_\sigma(x)$ we mean the domain Ω_σ rotated around the origin by an angle x. If $f(\rho,\theta)$ is the Poisson integral of f, we set

$$N(x)=N_{\sigma,f}(x)=\sup_{\zeta\in\Omega_\sigma(x)}|f(\rho,\theta)|. \tag{7·9}$$

Clearly, N is an increasing function of σ.

(7·10) THEOREM. *The function* $N(x)$ *in* (7·9) *satisfies the inequalities* (7·6), *where the constants will also depend on* σ.

Fix x, and let $\zeta=\rho e^{i\theta}$, $p=\rho\, e^{i(\theta-x)}$. For $\zeta\in\Omega_\sigma(x)$ we have

$$f(\rho,\theta)=\int_{-\pi}^{\pi}f(x+t)\,\chi(t,p)\,dt,\quad\text{where}\quad \chi(t,p)=\frac{1}{\pi}P(\rho,t+x-\theta).$$

The expression $\chi(t,p)$ here depends on the variable t and on the parameter p which is a point of Ω_σ. That (7·2) (i) holds is obvious. The left-hand side of (7·2) (ii), with $\xi=x-\theta$, $P'=dP/dt$, is

$$\frac{1}{\pi}\int_{-\pi}^{\pi}|tP'(\rho,t+\xi)|\,dt\leqslant\frac{2}{\pi}\int_{-\pi}^{\pi}\tfrac{1}{2}\pi\,|\sin\tfrac{1}{2}t\,P'(\rho,t+\xi)|\,dt$$
$$=\int_{-\pi}^{\pi}|\sin\tfrac{1}{2}(t-\xi)\,P'(\rho,t)|\,dt\leqslant\tfrac{1}{2}\int_{-\pi}^{\pi}|tP'(\rho,t)|\,dt+\tfrac{1}{2}|\xi|\int_{-\pi}^{\pi}|P'(\rho,t)|\,dt.$$

The penultimate integral is, as we know, bounded. The last term is

$$-|\xi|\int_0^\pi\frac{d}{dt}P(\rho,t)\,dt\leqslant\frac{|\xi|}{1-\rho}=\frac{|x-\theta|}{1-\rho}.$$

Considering separately the cases $\rho\geqslant\sigma$ and $\rho<\sigma$ we see that the last expression does not exceed a constant depending on σ only. This proves (7·2) (ii) and so also the theorem.

The most important special case of (7·10) is $\sigma=0$, when Ω_σ degenerates into a radius of the unit circle and (7·10) reduces to (7·8).

Results of this section can be extended to Fourier–Stieltjes series, and the generalizations do not require new ideas. For simplicity we confine our attention to Theorem (7·8).

† The conclusion holds actually for $n\geqslant 0$. It is enough, for example, to replace n by $n+1$ on the right of (7·7) and the inequality will hold for $n\geqslant 0$. The point is clearly without importance.

‡ The result holds for $\delta>1$. This follows from the fact, easy to verify (see Chapter III, (1·10)(ii)) that $N(x)=N^\delta(x)$ is a non-increasing function of δ for $\delta>-1$.

(7·11) Theorem. *Let* σ_n *and* $f(r,x)$ *be the* (C, 1) *and Abel means of an* S[dF], *and let* $N(x)$ *be one of the functions of Theorem* (7·8). *Then*

$$\mathfrak{M}_\alpha[N] \leqslant C_\alpha \int_0^{2\pi} |dF(x)| \quad (0<\alpha<1), \tag{7·12}$$

and
$$\int_0^{2\pi} |\sigma_n(x) - F'(x)|^\alpha dx \to 0, \quad \int_0^{2\pi} |f(r,x) - F'(x)|^\alpha dx \to 0 \quad (0<\alpha<1) \tag{7·13}$$

Let $0 < R < 1$, $N_R(x) = \max_{r \leqslant R} |f(r,x)|$. By (7·8) and the last inequality in theorem (6·5),

$$\mathfrak{M}_\alpha[N_R(x)] \leqslant C_\alpha \int_0^{2\pi} |f(R,x)|\, dx \leqslant C_\alpha \int_0^{2\pi} |dF(x)|,$$

and making $R \to 1$ we obtain (7·12) for Abel means. By considering the (C, 1) means of $f(R,x)$ and making $R \to 1$ we prove (7·12) in the remaining case.

The relations (7·13) follow from the fact that $|\sigma_n(x) - F'(x)|^\alpha$ and $|f(r,x) - F'(x)|^\alpha$ tend to 0 almost everywhere (see Chapter III, (7·2) and § 8) and are majorized by integrable functions.

8. Parseval's formula

Let $f(x)$ and $g(x)$ be periodic and of class L². If their coefficients are respectively c_ν and c'_ν, we have the Parseval formula (Chapter II, (1·13))

$$\frac{1}{2\pi}\int_0^{2\pi} fg\, dx = \sum_{\nu=-\infty}^{+\infty} c_\nu c'_{-\nu}, \tag{8·1}$$

or, what is the same thing,

$$\frac{1}{2\pi}\int_0^{2\pi} f\bar{g}\, dx = \sum_{\nu=-\infty}^{+\infty} c_\nu \bar{c}'_\nu. \tag{8·2}$$

Both series on the right here converge absolutely. If f and g are real-valued, and if $f \sim \frac{1}{2}a_0 + \Sigma(a_\nu \cos \nu x + b_\nu \sin \nu x)$, $g \sim \frac{1}{2}a'_0 + \Sigma(a'_\nu \cos \nu x + b'_\nu \sin \nu x)$, we have

$$\frac{1}{\pi}\int_0^{2\pi} fg\, dx = \tfrac{1}{2}a_0 a'_0 + \sum_{\nu=1}^{\infty} (a_\nu a'_\nu + b_\nu b'_\nu). \tag{8·3}$$

The above formulae hold in other cases besides the one in which $f \in$ L², $g \in$ L². Two classes K and K$_1$ of functions will be called *complementary*, if (8·1) holds for every $f \in$ K, $g \in$ K$_1$ in the sense that the series on the right is summable by some method of summation. It will appear that the Fourier series of functions belonging to complementary classes have in many cases the same or analogous properties; and the Parseval formula (8·1), in which f and g enter symmetrically, is the means for discovering these related properties. The formula is obvious (by termwise integration) if f is a trigonometric polynomial and g any integrable function.

Let $\ldots, \mu_{-1}, \mu_0, \mu_1, \ldots$ be a two-way infinite sequence of numbers. Suppose that along with c_ν, c'_ν the numbers $c_\nu \mu_\nu$, $c'_\nu/\mu_{-\nu}$ are also Fourier coefficients, say of functions f^*, g_*, and that Parseval's formula for f^* and g_* is valid. Then (8·1) gives

$$\int_0^{2\pi} fg\, dx = \int_0^{2\pi} f^* g_*\, dx. \tag{8·4}$$

The number $\mu_{-\nu}$ is necessarily distinct from 0 if $c'_\nu \neq 0$, but if $c'_\nu = 0$ the value ascribed to

$c'_\nu/\mu_{-\nu}$ has no influence upon the result, and for the sake of simplicity we may take it equal to 0 even if $\mu_{-\nu}=0$.

Suppose, for example, that $c'_0=0$, and that $\mu_\nu = i\nu$ for each ν. Then $f^*=f'$ and $g_*=-G$, where G is the indefinite integral of g, and is in our case a periodic function. Thus

$$\int_0^{2\pi} fg\,dx = -\int_0^{2\pi} f'G\,dx, \tag{8·5}$$

a formula which, of course, may also be obtained by integration by parts. (Less trivial is the analogue of (8·5) for fractional derivatives and integrals; see Chapter XII, §8.) If $\mu_\nu = -i \operatorname{sign} \nu$, then $f^*=\tilde{f}$, $g_*=\tilde{g}$, and, formally,

$$\frac{1}{2\pi}\int_0^{2\pi} fg\,dx = c_0\,c'_0 + \frac{1}{2\pi}\int_0^{2\pi} \tilde{f}\tilde{g}\,dx. \tag{8·6}$$

(8·7) Theorem. *The following are pairs of complementary classes:* (i) L^r *and* $L^{r'}$ $(r>1)$; (ii) B *and* L. (iii) L_Φ *and* L_Ψ, *if* Φ *and* Ψ *are complementary functions in the sense of Young;* (iv) C *and* S. *In all these cases the series in* (8·1) *are summable* (C, 1).

Part (iv) here is to be understood in the sense that if c_ν are the coefficients of an $S[f]$, and c'_ν the coefficients of an $S[dG]$, we have (8·2) with $f\bar{g}$ replaced by $fd\bar{G}$. Part (ii) is a limiting case $(r=\infty)$ of (i).

Let $\sigma_n(x)$ be the (C, 1) means of $S[f]$, τ_n the (symmetric) (C, 1) means of the series in (8·1), and Δ_n the difference between the integral in (8·1) and τ_n. Then

$$\Delta_n = \frac{1}{2\pi}\int_0^{2\pi} (f-\sigma_n)\,g\,dx, \tag{8·8}$$

and, by Hölder's inequality,

$$2\pi\,|\,\Delta_n\,| \leqslant \mathfrak{M}_r[f-\sigma_n]\,\mathfrak{M}_{r'}[g].$$

Hence $\Delta_n \to 0$ as $n\to\infty$ (cf. (5·14)), and (i) follows. The argument holds for $r = 1$ (using (5·5)), which proves (ii). To prove (iii), which generalizes (i), we apply Young's inequality (Chapter I, (9·1)) to $\Delta_n/16$:

$$2\pi\,|\,\Delta_n\,|/16 \leqslant \mathfrak{M}[\Phi\{\tfrac{1}{4}\,|\,f-\sigma_n\,|\}] + \mathfrak{M}[\Psi\{\tfrac{1}{4}\,|\,g\,|\}].$$

By virtue of (5·14), we get $\limsup|\,\Delta_n\,| \leqslant 8\pi^{-1}\mathfrak{M}[\Psi\{\tfrac{1}{4}\,|\,g\,|\}]$. Let $g=g'+g''$, where g' is a trigonometric polynomial and $\mathfrak{M}[\Psi\{\tfrac{1}{4}\,|\,g''\,|\}]<\epsilon$. (By (5·14) we may take $g'=\sigma_m(x;\,g)$ with m sufficiently large.) Substituting g' and g'' for g in (8·8) we find expressions Δ'_n and Δ''_n such that $\Delta_n = \Delta'_n + \Delta''_n$. Since g' is only a polynomial, $\Delta'_n \to 0$. On the other hand,

$$\limsup|\,\Delta''_n\,| \leqslant 8\pi^{-1}\mathfrak{M}[\Psi(\tfrac{1}{4}\,|\,g''\,|)] < 8\pi^{-1}\epsilon.$$

Thus $\limsup|\,\Delta_n\,| \leqslant 8\pi^{-1}\epsilon$, so that $\Delta_n \to 0$.

In (iv), $g(x)$ is replaced by $dG(x)$ in (8·1) and f is continuous. Then $2\pi\,|\,\Delta_n\,|$ does not exceed $\max|\,f(x)-\sigma_n(x)\,|$ multiplied by the total variation of G over $(0, 2\pi)$. Thus $\Delta_n \to 0$.

Let $g(x)$ be the characteristic function of a set E and $f(x)$ an integrable function. Parseval's formulae (8·1) and (8·3) can then be written

$$\int_E f\,dx = \sum_{\nu=-\infty}^{\infty} c_\nu \int_E e^{i\nu x}\,dx = \tfrac{1}{2}a_0\,|\,E\,| + \sum_{\nu=1}^{\infty}\int_E (a_\nu \cos\nu x + b_\nu \sin\nu x)\,dx.$$

Hence

(8·9) THEOREM. *If* $S[f]$ *is integrated termwise over any measurable set* E, *the resulting series is summable* (C, 1) *to sum* $\int_E f\,dx$.

Applying (8·1) to the functions $f(x-t)$ and $g(t)$ of the variable t, we find that in each of the cases listed in (8·7) we have

$$\frac{1}{2\pi}\int_0^{2\pi} f(x-t)\,g(t)\,dt = \sum_{\nu=-\infty}^{+\infty} c_\nu c'_\nu e^{i\nu x}, \tag{8·10}$$

where the series on the right is uniformly summable (C, 1). Moreover:

(8·11) THEOREM. *Given any pair of integrable functions* f, g, *the formula* (8·10) *holds, in the* (C, 1) *sense, almost everywhere in* x.

The proof follows from the fact that the left-hand side $h(x)$ of (8·10) is an integrable function, and that the series on the right is $S[h]$ (Chapter II, (1·5)).

Let us substitute $g(x)\,e^{-inx}$ for $g(x)$ in (8·1), and let c''_ν be the coefficients of $g(x)\,e^{-inx}$. Since $c''_{-\nu} = c'_{n-\nu}$, we find

$$\frac{1}{2\pi}\int_0^{2\pi} fg\,e^{-inx}\,dx = \sum_{\nu=-\infty}^{+\infty} c_\nu c'_{n-\nu}. \tag{8·12}$$

Thus:

(8·13) THEOREM. *The Fourier series of the product of the functions* $f \in L_\Phi$, $g \in L_\Psi$ (Φ *and* Ψ *being complementary functions in the sense of Young*) *is obtained by the formal multiplication of* $S[f]$ *and* $S[g]$ *by Laurent's rule. The series* (8·12) *defining the coefficients of* fg *are summable* (C, 1). *The result holds if* $f \in B$, $g \in L$.

It is obvious that each of the inequalities $\Sigma\,|\,c_\nu\,| < \infty$, $\Sigma\,|\,c'_\nu\,| < \infty$ implies the absolute convergence of the series in (8·12). If both inequalities hold, $S[fg]$ converges absolutely.

Let $f(x)$ be continuous and $G(x)$ of bounded variation. If c_ν, c'_ν are the coefficients of $S[f]$, $S[dG]$, we have the following analogue of (8·12):

$$\frac{1}{2\pi}\int_0^{2\pi} e^{-inx} f\,dG = \sum_{\nu=-\infty}^{+\infty} c_\nu c'_{n-\nu}. \tag{8·14}$$

In the results above we may replace summability (C, 1) by (C, α), $\alpha > 0$. The problem of replacing summability (C, α) by ordinary convergence is more delicate. Going over the proofs of parts (i) and (ii) of (8·7), we see that we may replace summability (C, 1) there by convergence provided $\mathfrak{M}_r[f - s_n] \to 0$, where $s_n = S_n(x; f)$. In Chapter VII, § 6, we shall see that this in fact happens if $f \in L^r$, $r > 1$ (though not for $r = 1$; see Chapter V, (1·12)). Thus, at least in (8·7)(i), the Parseval series converges. In particular, *if* $f \in L^r$, $g \in L^{r'}$ $r > 1$, *we have convergence in* (8·12). The proof of the following theorem is much easier:

(8·15) THEOREM. *If f is integrable and g of bounded variation, the series in* (8·1) *converges.*

Let δ_n be the difference between the integral and the nth partial sum of the series in (8·1). Then

$$|\,\delta_n\,| = \left|\frac{1}{2\pi}\int_0^{2\pi} [g(x) - S_n(x;\,g)]\,f(x)\,dx\right| \leqslant \frac{1}{2\pi}\int_0^{2\pi} |\,g - S_n[g]\,|\,|\,f\,|\,dx.$$

Since the $S_n[g]$ are uniformly bounded and tend to g outside a denumerable set, the last integrand is majorized by an integrable function and tends to 0 almost everywhere. Thus $\delta_n \to 0$.

From (8·15) we obtain:

(8·16) Theorem. *If f is integrable and periodic, (α, β) is a finite interval, and $g(x)$ any function of bounded variation in (α, β), not necessarily periodic, then*

$$\int_\alpha^\beta fg\,dx = \sum_{\nu=-\infty}^{+\infty} c_\nu \int_\alpha^\beta g\,e^{i\nu x}\,dx. \tag{8·17}$$

Thus Fourier series can be integrated term by term after multiplication by any function of bounded variation. If $\beta - \alpha = 2\pi$ this is nothing but (8·1), and the case $\beta - \alpha < 2\pi$ may be included by setting $g(x) = 0$ in $\beta < x < \alpha + 2\pi$; in the general case we break up (α, β) into a finite number of intervals of length not exceeding 2π.

The last result can be extended to the case of an infinite interval. Without loss of generality we may take $(\alpha, \beta) = (-\infty, +\infty)$. We have in fact

(8·18) Theorem. *The formula*

$$\int_{-\infty}^{+\infty} fg\,dx = \sum_{\nu=-\infty}^{+\infty} c_\nu \int_{-\infty}^{\infty} g(x)\,e^{i\nu x}\,dx \tag{8·19}$$

holds, and the series on the right converges for any integrable and periodic f, provided that $g(x)$ is (i) *integrable, and* (ii) *of bounded variation, over $(-\infty, +\infty)$.*

Let $G(x) = \sum_{-\infty}^{+\infty} g(x + 2k\pi)$. If the series converges at some point, it converges uniformly and its sum is of bounded variation over $(0, 2\pi)$ (Chapter II, § 13). On the other hand, since

$$\sum_{k=-\infty}^{+\infty} \int_0^{2\pi} |\, g(x + 2k\pi)\,|\,dx = \int_{-\infty}^{+\infty} |\, g(x)\,|\,dx < \infty,$$

the series defining $G(x)$ certainly has points of convergence.

Let c'_ν be the Fourier coefficients of $G(x)$. We may replace g by G in (8·1). Since a uniformly convergent series can be integrated term by term over $(0, 2\pi)$ after multiplication by any integrable function, and since f is periodic, it follows from the definition of G that

$$\int_0^{2\pi} fG\,dx = \int_{-\infty}^{+\infty} fg\,dx, \quad \int_0^{2\pi} G(x)\,e^{-inx}\,dx = \int_{-\infty}^{+\infty} g(x)\,e^{-inx}\,dx,$$

and Parseval's formula for f and G takes the form (8·19).

The hypothesis that g is integrable over $(-\infty, +\infty)$ is of course essential for the validity of (8·19). However, *if $c_0 = 0$, condition* (i) *in* (8·18) *may be replaced by the condition* (i′) *$g(x) \to 0$ as $|\,x\,| \to \infty$.*

For let $g^*(x) = g(2k\pi)$ for $2k\pi \leqslant x < 2(k+1)\pi$, $k = 0, \pm 1, \ldots$, and let v_k be the total variation of $g(x)$ over $2k\pi \leqslant x \leqslant 2(k+1)\pi$. The function $g^*(x)$ is of bounded variation over $(-\infty, +\infty)$. Since $\gamma(x) = g(x) - g^*(x)$ does not exceed v_k in absolute value for $2k\pi \leqslant x \leqslant 2(k+1)\pi$, $\gamma(x)$ is both integrable and of bounded variation over $(-\infty, +\infty)$.

Apply (8·19) to f and γ. Since the integral of f over a period is zero and $g(x)\to 0$ as $|x|\to\infty$, it is easy to verify that

$$\int_{-\infty}^{+\infty} f\gamma\,dx=\int_{-\infty}^{+\infty} fg\,dx,\quad \int_{-\infty}^{+\infty}\gamma\, e^{-i\nu x}\,dx=\int_{-\infty}^{+\infty} g\, e^{-i\nu x}\,dx$$

for $\nu=\pm 1, \pm 2, \ldots$, and the formula (8·19) for f and γ reduces to that for f and g.

Equations (8·1) and (8·10) can be extended to the case of several factors. Consider a finite set of functions $f, f_1, f_2, \ldots$, with Fourier coefficients $c_\nu, c'_\nu, c''_\nu, \ldots$, respectively. Multiplying formally $S[f], S[f_1], S[f_2], \ldots$, and integrating the result over $(0, 2\pi)$, we get

$$\frac{1}{2\pi}\int_0^{2\pi} ff_1f_2\ldots dt=\sum_{\lambda+\mu+\nu+\ldots=0} c_\lambda c'_\mu c''_\nu\ldots,\tag{8·20}$$

a formula which in the case of two functions reduces to (8·1). The argument is valid if the series $\Sigma|c'_\mu|, \Sigma|c''_\nu|, \ldots$ all converge. (Nothing is assumed about $\Sigma|c_\lambda|$.) For let $F=f_1f_2\ldots$. Then $S[F]=S[f_1]S[f_2]\ldots=\Sigma\gamma_n e^{int}$, where $\gamma_n=\Sigma c'_\mu c''_\nu\ldots$ for $\mu+\nu+\ldots=n$. The series for γ_n converges absolutely, and $\Sigma|\gamma_n|<\infty$. The left-hand side of (8·20) is thus

$$\frac{1}{2\pi}\int_0^{2\pi} fF\,dt=\sum_{\lambda+n=0} c_\lambda\gamma_n=\sum_{\lambda+\mu+\nu+\ldots=0} c_\lambda c'_\mu c''_\nu\ldots,$$

and the series here are absolutely convergent. In particular, (8·20) holds if all the functions $f, f_1, f_2, \ldots$, except possibly one, are trigonometric polynomials.

Restricting ourselves to three functions we also have the following result:

(8·21) THEOREM. *Let* $x(t)\sim\Sigma x_\nu e^{i\nu t}$, $y(t)\sim\Sigma y_\nu e^{i\nu t}$, $h(t)\sim\Sigma h_\nu e^{i\nu t}$, *and let* $x\in L^2$, $y\in L^2$, $h\in B$. *Then*

$$\frac{1}{2\pi}\int_0^{2\pi} x(t)\,y(t)\,h(t)\,dt=\sum_{\lambda+\mu+\nu=0} x_\lambda y_\mu h_\nu,\tag{8·22}$$

provided the sum on the right is treated as $\lim_{L,M\to\infty}\sum_{\lambda=-L}^{L}\sum_{\mu=-M}^{M} x_\lambda y_\mu h_{-\lambda-\mu}$.

Denote the last sum by $S_{L,M}$, and let $X_L(t)$ and $Y_M(t)$ denote the partial sums of $S[x]$ and $S[y]$. The integral in (8·22) with x, y replaced by X_L, Y_M becomes $S_{L,M}$. Let $H=\sup|h(t)|$. Then

$$\left|\int_0^{2\pi} xyh\,dt-\int_0^{2\pi} X_LY_Mh\,dt\right|\leqslant H\left\{\int_0^{2\pi}|x-X_L|\,|y|\,dt+\int_0^{2\pi}|X_L|\,|y-Y_M|\,dt\right\}$$
$$\leqslant H\{\mathfrak{M}_2[x-X_L]\,\mathfrak{M}_2[y]+\mathfrak{M}_2[X_L]\,\mathfrak{M}_2[y-Y_M]\}\to 0$$

as $L, M\to\infty$. This proves (8·21).

If $y(t)=\bar{x}(t)\sim\Sigma\bar{x}_{-n}e^{int}$, (8·22) gives

$$\frac{1}{2\pi}\int_0^{2\pi}|x(t)|^2 h(t)\,dt=\sum_{\lambda,\mu=-\infty}^{+\infty} x_\lambda\bar{x}_\mu h_{\mu-\lambda},\tag{8·23}$$

a formula with applications in the theory of quadratic forms.

(8·24) THEOREM. *With the notation of* (8·21), *suppose that* $x(t)\in L^2$, $y(t)\in L^2$, $h(t)\in L$; *then*

$$\frac{1}{4\pi^2}\int_0^{2\pi}\int_0^{2\pi} x(u)\,y(v)\,h(-u-v)\,du\,dv=\sum_{n=-\infty}^{+\infty} x_n y_n h_n,$$

where the series on the right converges absolutely.

For the proof, we apply (8·1) to the product of $x(u)$ and $y_1(u) = \frac{1}{2\pi}\int_0^{2\pi} y(v)h(-u-v)\,dv$. The latter function belongs to L^2 and has coefficients $y_{-n}h_{-n}$ (Chapter II, (1·15), (1·5)).

9. Linear operations

We are now going to prove a number of results on linear operations which will later find application to trigonometric series.

We consider a set E of arbitrary elements $x, y, z, \ldots$. It is often convenient to call E a *space*, and its elements $x, y, z, \ldots$ *points*. E will be called a *metric space* if to every pair of points x, y of E corresponds a non-negative number $d(x, y)$, called the *distance* between the points x and y, satisfying the following conditions:

(i) $d(x,y) = d(y,x)$;

(ii) $d(x,z) \leqslant d(x,y) + d(y,z)$ (*triangle inequality*);

(iii) $d(x,y) = 0$ if and only if $x = y$.

We say that a sequence $\{x_n\}$ of points of E tends to limit x, $x \in E$, and write $\lim x_n = x$, or $x_n \to x$, if $d(x, x_n) \to 0$ as $n \to \infty$.

Once distance has been introduced, there are various associated notions familiar from the theory of Euclidean spaces. First, by the *sphere* with centre x_0 and radius ρ we mean the set of points $x \in E$ such that $d(x, x_0) \leqslant \rho$; this sphere will be denoted by $S(x_0, \rho)$. This notion enables us, in turn, to introduce various kinds of point sets, such as *open*, *closed*, *non-dense*, *dense*, *everywhere dense*, the definitions being the same as in Euclidean spaces. Furthermore, we may consider sets of the *first category*, i.e. denumerable sums of non-dense sets, and sets of the *second category*, i.e. sets which are not of the first category (cf. Chapter I, § 12).

A metric space E is said to be *complete*, if for any sequence of points x_n such that $d(x_m, x_n) \to 0$ as $m, n \to \infty$ there is a point x such that $d(x_m, x) \to 0$. The inequality $d(x, x') \leqslant d(x, x_m) + d(x_m, x')$ shows that such a point x must be unique. It is a very important fact that *a complete metric space E is of the second category*, i.e. is not a sum of a sequence of sets non-dense in E. The proof of this in the general case is essentially the same as in the case (discussed in Chapter I, § 12) when E is a one-dimensional Euclidean space.

E is called *separable* if there is a denumerable set dense in E.

A space E, not necessarily metric, will be called *linear* if the following conditions are satisfied:

(i) there is a commutative and associative operation called *addition*, denoted by $+$ and applicable to any two points x, y of E; whenever x and y belong to E, so does $x + y$;

(ii) there is a unique element o (the *null* element) such that $x + o = x$ for every $x \in E$;

(iii) there is an operation called *multiplication*, applicable to every $x \in E$ and every scalar† α, and denoted by '.'. Instead of $\alpha . x$ we often write αx. Multiplication is assumed to have the properties

$$1.x = x, \quad 0.x = o, \quad \alpha.x \in E \quad \text{if} \quad x \in E,$$

and further to be distributive in α and in x, and associative in α. The latter means that $\beta.(\alpha.x) = \beta\alpha.x$.

The formula $x - y = x + (-1)\,y$ defines *subtraction* of elements of E.

† The only fields of scalars we use are the complex numbers or the real numbers.

Suppose that to every element x of a linear space E corresponds a unique non-negative number $\|x\|$ called the *norm* of x satisfying the conditions

$$\|x+y\| \leqslant \|x\| + \|y\|, \quad \|\alpha x\| = |\alpha| \, \|x\|,$$
$$\|x\| = 0 \quad \text{if and only if } x = o.$$

If the distance between any two points x, y of our linear space E is defined by the formula

$$d(x, y) = \|x - y\|,$$

this distance satisfies conditions (i), (ii), (iii) imposed above, and E becomes a *normed linear space*. A complete normed linear space is usually called a *Banach space*.

We shall now give a few examples of spaces. In each case the points of E will be either numbers or functions, and addition and multiplication have their usual interpretation. No confusion will arise if the null point o is denoted by 0.

(*a*) Let E be the set of all complex (or only all real) numbers. If $\|x\| = |x|$, we have a Banach space.

(*b*) Let E be the set C of all continuous functions $x(t)$ defined in a fixed interval (a, b), and let $\|x\| = \sup |x(t)|$ for $t \in (a, b)$. Then E is a Banach space. The relation $x_n \to x$ means that $x_n(t)$ converges uniformly to $x(t)$.

(*c*) Let E be the set of all complex-valued functions $x(t)$ defined and essentially bounded in (a, b), and let $\|x\|$ be the essential upper bound of $|x(t)|$ in (a, b) (cf. Chapter I, § 9). E is again a Banach space, and $x_n \to x$ means that $x_n(t)$ converges to $x(t)$ uniformly outside a set of measure 0.

(*d*) Let $p \geqslant 1$, and let E be the set of all complex-valued functions $x(t) \in \mathrm{L}^p(a, b)$. Let

$$\|x\| = \|x\|_p = \mathfrak{M}_p[x; a, b].$$

The space is linear, normed and complete (cf. Chapter I (9·11), (11·1)). For $p = \infty$, we obtain case (*c*).

(*e*) Let $1 \leqslant p < \infty$, and let E be the set of all sequences $x = \{x_k\}$ of complex numbers such that $\Sigma |x_k|^p < \infty$. Let

$$\|x\| = \|x\|_p = (\Sigma |x_k|^p)^{1/p}.$$

The space is linear, normed, and (as is easily seen) complete. It is often denoted by l^p.

(*f*) Let E be the set of all bounded sequences $x = \{x_k\}$ of complex numbers. If we set $\|x\| = \sup_k |x_k|$, we get a Banach space. It is the limiting case $p = \infty$ of l^p.

(*g*) Let E be the set of all convergent sequences $x = \{x_k\}$ of complex numbers, and once again let $\|x\| = \sup |x_k|$. The set E (a subset of the preceding E) is a normed linear space. It is also complete. For suppose that $x^m = \{x_1^m, x_2^m, \ldots\} \in E$ for $m = 1, 2, \ldots$, and that $\|x^m - x^n\| \to 0$ as $m, n \to \infty$. This is equivalent to saying that $|x_k^m - x_k^n| \to 0$ as $m, n \to \infty$, uniformly in k. This implies the existence of an $x = \{x_k\}$ such that $|x_k^m - x_k| \to 0$ as $m \to \infty$, uniformly in k. We shall show that (i) $\{x_k\}$ is convergent, (ii) $\|x^m - x\| \to 0$.

To prove (i) observe that

$$|x_k - x_l| \leqslant |x_k - x_k^m| + |x_k^m - x_l^m| + |x_l^m - x_l|.$$

The first and third terms on the right are less than ϵ for m large enough, uniformly in k and l. Having fixed such a large m, we make the second term less than ϵ by taking k and l large. Hence $|x_k - x_l| < 3\epsilon$ for k, l large, and so $\{x_k\}$ is convergent. Assertion (ii) follows from the fact, established above, that $|x_k^m - x_k| \to 0$ as $m \to \infty$, uniformly in k.

(*h*) Let $p \geqslant 1$, and let H_p be the class of all functions $x(t)$ of period 2π, of class L^p, with $\mathfrak{S}[x]$ of power series type. If $\| x \| = \mathfrak{M}_p[x;\, 0, 2\pi]$, the space becomes a Banach space.

(*i*) Let X be the set of all characteristic functions in (a, b), that is, of functions $x(t)$ taking almost everywhere in (a, b) the values 0 and 1 only. The set X is not a linear space, since $2x$ need not be a characteristic function if x is one. However, if for $x \in X$, $y \in X$ we set $d(x, y) = \mathfrak{M}[x - y;\, a, b]$, X becomes a complete metric space.

Let us consider in addition to the space E another space U. If to every $x \in E$ corresponds a uniquely determined point $u = u(x)$ in U, we say that $u(x)$ is a functional *operation* (or *transformation*) defined in E. If the spaces E and U are linear, and if for any numbers λ_1 and λ_2 we have

$$u(\lambda_1 x_1 + \lambda_2 x_2) = \lambda_1 u(x_1) + \lambda_2 u(x_2),$$

the operation $u(x)$ is called *linear*. If both E and U are metric, and if whenever $x_n \to x$ we have $u(x_n) \to u(x)$, we say that u is *continuous* at the point x. If a linear operation is continuous at some point, it is continuous at any other point, i.e. is continuous everywhere.

(9·1) Theorem. *A necessary and sufficient condition for a linear operation $u(x)$ to be continuous in E is the existence of a finite number M such that*

$$\| u(x) \| \leqslant M \| x \| \quad \textit{for every } x \in E. \tag{9·2}$$

The sufficiency of the condition is obvious. To prove its necessity, suppose that the ratio $\| u(x) \| / \| x \|$ is unbounded. Then there is a sequence of points x_n, $x_n \neq 0$, such that $\| u(x_n) \| \geqslant n \| x_n \|$. Multiplying x_n by a suitable constant we may assume that $\| x_n \| = 1/n$. Thus $x_n \to 0$, while the preceding inequality gives $\| u(x_n) \| \geqslant 1$, contradicting the continuity of u at $x = 0$.

The norms on the two sides of (9·2) may have different meanings, since the spaces E and U may be different.

A linear operation which is continuous is usually called *bounded*.

If U is the space of all complex, or all real, numbers and $\| u(x) \| = | u(x) |$, the linear operation u is called a *functional*.

The smallest number M satisfying (9·2) for all $x \in E$ will be called the *norm of the operation* and often denoted by M_u.†

(9·3) Theorem. *Let E be a normed linear space and L a linear subspace of E dense in E. Let $u = u(x)$ be a linear operation defined for $x \in L$, taking values from a Banach space U and satisfying an inequality*

$$\| u(x) \| \leqslant M \| x \| \quad (x \in L). \tag{9·4}$$

Then $u(x)$ can be uniquely extended as a linear operation to the whole of E without increasing the M in (9·4).

For let x_* be a point of E, $\{x_n\}$ a sequence of points from L such that $\| x_* - x_n \| \to 0$,

† One often denotes the norm of the operation $u(x)$ by $\| u \|$. This terminology and notation—very natural in a systematic study of the subject—arise from the fact that the set of all linear operations $u(x)$ defined on E may be considered as a new space, to whose points u are assigned norms $\| u \|$. Of course, $\| u \|$ and $\| u(x) \|$ mean different things.

and $u_n = u(x_n)$, $n = 1, 2, \ldots$. We have $\| x_m - x_n \| \to 0$ and so, by (9·4), $\| u_m - u_n \| \to 0$. The completeness of U implies the existence of a $u_* = \lim u_n$ and we set $u_* = u(x_*)$.

The number u_* is independent of the choice of $\{x_n\} \to x_*$. For if we take another sequence $\{x_n'\} \to x_*$ and set $u_*' = \lim u(x_n')$, the joint sequence $\{x_n''\} = x_1, x_1', x_2, x_2', \ldots$ will also converge to x_* and the number $u_*'' = \lim u(x_n'')$ will be equal to both u_* and u_*'. Hence $u_* = u_*'$.

The validity of (9·4) at the point x_n, together with the relations $\| x_n \| \to \| x_* \|$, $\| u_n \| \to \| u_* \|$ (consequences of $x_n \to x_*$, $u_n \to u_*$), implies its validity at x_*. Similarly, the validity of

$$u(\alpha x + \beta y) = \alpha u(x) + \beta u(y)$$

at the points x, y of L is preserved in E. This shows that the extended operation satisfies (9·4) and is additive. The inequality (9·4) implies the continuity of u and there is at most one continuous extension of $u(x)$ from a set dense in E.

The following theorem is basic for the theory of linear operations:

(9·5) THEOREM OF BANACH-STEINHAUS. *Let $\{u_n(x)\}$ be a sequence of bounded linear operations defined in a Banach space E, and let M_{u_n} be the norm of the operation u_n. If* $\sup \| u_n(x) \|$ *is finite for every point x belonging to a set F of the second category in E (in particular if it is finite for every $x \in E$), then the sequence M_{u_n} is bounded. In other words, there is a constant M such that*

$$\| u_n(x) \| \leqslant M \| x \| \quad \textit{for} \quad x \in E \textit{ and } n = 1, 2, \ldots.$$

The proof is based on two lemmas.

(9·6) LEMMA. *Let $\{u_n(x)\}$ be a sequence of bounded linear operations defined in a normed linear space E. If F is the set of points x at which* $\sup \| u_n(x) \| < \infty$, *then $F = F_1 + F_2 + \ldots$, where the sets F_i are closed and the sequence $\{\| u_n(x) \|\}$ is uniformly bounded on each of them.*

For let F_{mn} be the set of points x such that $\| u_m(x) \| \leqslant n$. The operations u_m being continuous, the sets F_{mn} are closed. So are the products $F_n = F_{1n} F_{2n} F_{3n} \ldots$. We note that $\| u_m(x) \| \leqslant n$ for $x \in F_n$, $m = 1, 2, \ldots$, and that $F = F_1 + F_2 + \ldots$.

(9·7) LEMMA. *If the set F in Lemma (9·6) is of the second category, then there is a sphere $S(x_0, \rho)$, $\rho > 0$, and a number K such that $\| u_m(x) \| \leqslant K$ for $x \in S(x_0, \rho)$ and $m = 1, 2, \ldots$.*

For since $F = F_1 + F_2 + \ldots$ and F is of the second category, at least one of the sets $F_1, F_2, \ldots$, say F_K, is not non-dense. Thus there is a sphere $S(x_0, \rho)$ in which F_K is dense. Since F_K is closed, F_K contains $S(x_0, \rho)$ and consequently $\| u_m(x) \| \leqslant K$ for $x \in S(x_0, \rho)$ and $m = 1, 2, \ldots$.

Returning to the proof of (9·5), let $S(x_0, \rho)$ be the sphere of (9·7). Every $x \in S(0, \rho)$ can be written in the form $(x_0 + x) - x_0 = x_1 - x_0$, say, where $x_1 \in S(x_0, \rho)$. Hence $\| u_n(x) \| \leqslant \| u_n(x_1) \| + \| u_n(x_0) \| \leqslant 2K$ for $n = 1, 2, \ldots$. It follows that

$$\| u_n(x) \| \leqslant 2K/\rho = M$$

on the sphere $\| x \| = 1$, so that $\| u_n(x) \| \leqslant M \| x \|$ for all x and n.

The theorem may also be stated as follows: *If the sequence $\{\| u_n(x) \|\}$ is unbounded at some point, the set of points x at which the sequence is bounded is of the first category in E.*

We shall apply (9·5) to the functionals

$$u(x)=\int_a^b x(t)\,y(t)\,dt, \tag{9·8}$$

where the function $x=x(t)$ is the variable point of a Banach space E, and $y=y(t)$ is a fixed function such that the product $x(t)\,y(t)$ is integrable over (a,b) for any $x\in E$. For $x(t)\in \mathrm{L}^r$ we define $\|x\|$ as $\mathfrak{M}_r[x;\,a,b]$.

(9·9) LEMMA. (i) *If the integral* (9·8) *exists for every bounded, or even only for every bounded and continuous function* $x(t)$, *then* $y\in\mathrm{L}(a,b)$.

(ii) *Conversely, if* (9·8) *exists for every* $x\in\mathrm{L}(a,b)$, *then the function* y *is essentially bounded.*

(iii) *If* (9·8) *exists for every* $x\in\mathrm{L}^r(a,b)$, $r>1$, *then* $y\in\mathrm{L}^{r'}(a,b)$, *with* $r'=r/(r-1)$.

Part (i) is trivial (take $x(t)\equiv 1$). For (ii), let $y^n(t)$ be the function $y(t)$ truncated by n (that is, equal to $y(t)$ wherever the latter function is absolutely not greater than n, and equal to 0 elsewhere). The existence of (9·8) implies that of

$$u_n(x)=\int_a^b x(t)\,y^n(t)\,dt \quad (n=1,2,\ldots) \tag{9·10}$$

for every $x\in\mathrm{L}(a,b)$. These formulae define a sequence of functionals in $\mathrm{L}(a,b)$, and we easily find that M_{u_n} is the essential upper bound of $|\,y^n(t)\,|$. The existence of (9·8) implies that $u_n(x)$ converges for each $x\in\mathrm{L}(a,b)$. Thus, by (9·5), $M_{u_n}=O(1)$, that is, $y(t)$ is essentially bounded. For (iii), let y^n and $u_n(x)$ have the same meaning as before. Then $u_n(x)$ is a functional in $\mathrm{L}^r(a,b)$ with norm $M_{u_n}=\mathfrak{M}_{r'}[y^n;\,a,b]$ (cf. Chapter I, (9·14)). Thus, by (9·5), $\mathfrak{M}_{r'}[y^n]=O(1)$, that is, $\mathfrak{M}_{r'}[y]<\infty$.

(9·11) THEOREM. (i) *If the sequence*

$$u_n(x)=\int_a^b x(t)\,y_n(t)\,dt \tag{9·12}$$

is bounded for every bounded, or even only merely for every bounded and continuous, function $x(t)$, $t\in(a,b)$, *then* $\mathfrak{M}[y_n;\,a,b]=O(1)$.

(ii) *If* $u_n(x)$ *is bounded for every* $x\in\mathrm{L}(a,b)$, *then the essential upper bounds of the* $|\,y_n\,|$ *are bounded.*

(iii) *If* $u_n(x)$ *is bounded for every* $x\in\mathrm{L}^r(a,b)$, $1<r<\infty$, *then* $\mathfrak{M}_{r'}[y_n;\,a,b]=O(1)$.

For (i), we observe that by (9·9) (i) each of the functions y_n is integrable, so that $u_n(x)$ is a functional in the space B or the space C. Taking $x=\overline{\mathrm{sign}\,y_n}$, we see that the norm M_{u_n} in B equals $\mathfrak{M}[y_n]$, and an application of (9·5) gives $\mathfrak{M}[y_n]=O(1)$. The same proof holds in the case of $x\in\mathrm{C}$, provided we can show that $M_{u_n}=\mathfrak{M}[y_n]$ also in this case. The latter is, however, obvious, since the function $\mathrm{sign}\,y_n(t)$ is almost everywhere the limit of a convergent sequence of functions continuous and not exceeding 1 in absolute value.

In case (ii) we proceed similarly: each of the functions y_n is essentially bounded (cf. (9·9) (ii)), and M_{u_n} is the essential upper bound of $|\,y_n\,|$.

Finally, in case (iii) each y_n belongs to $\mathrm{L}^{r'}(a,b)$, u_n is a functional in L^r, and $M_{u_n}=\mathfrak{M}_{r'}[y_n]$ (Chapter I, (9·14)).

(9·13) THEOREM. (i) *Suppose that $y_n(t) \in L(a,b)$ for $n = 1, 2, \ldots$ and that the sequence (9·12) converges for every bounded $x(t)$. Then the functions $Y_n(t) = \int_a^t y_n(u)\,du$ are uniformly absolutely continuous.*

(ii) *The conclusion holds if (9·12) converges for those $x(t)$ which are characteristic functions of measurable sets.*

It is enough to prove (ii), and the proof will be similar to that of (9·5).

We show that $\int_E y_n\,dt$ is small with $|E|$, uniformly in n. Let X be the set of all characteristic functions $x(t)$ in (a,b), and consider the integral (9·12), which we denote by $I_n(x)$. By hypothesis, $I_n(x)$ converges for every $x \in X$. We have to show that $|I_n(x)| \leqslant \epsilon$ for all n, if

$$\|x\| = \mathfrak{M}[x;\, a, b] \leqslant \delta = \delta(\epsilon).$$

Let $I_{\mu,\nu}(x) = I_\mu(x) - I_\nu(x)$ and let X_n be the set of points $x \in X$ such that $|I_{\mu,\nu}(x)| \leqslant \frac{1}{4}\epsilon$ for all $\mu, \nu \geqslant n$. Then $X = \Sigma X_n$. The sets X_n are closed, and since X, being a complete metric space, is not of the first category, one of the X_n, say, X_{n_0}, contains a sphere $S(x_0, \delta')$. We now observe that X, though not a linear space, has a property which may be used instead of linearity: if x is any point of $S(0, \delta')$, we can find two points x_1 and x_2 in $S(x_0, \delta')$ such that $x = x_1 - x_2$. It is enough to set

$$x_1(t) = x(t) + x_0(t)\,[1 - x(t)], \quad x_2(t) = x_0(t)\,[1 - x(t)].$$

Clearly $|I_{\mu,\nu}(x)| = |I_{\mu,\nu}(x_1) - I_{\mu,\nu}(x_2)| \leqslant \frac{1}{2}\epsilon$ for $\|x\| \leqslant \delta'$, $\mu \geqslant n_0$, $\nu \geqslant n_0$.

It follows that $|I_\mu(x) - I_{n_0}(x)| = |I_{\mu,n_0}(x)| \leqslant \frac{1}{2}\epsilon$ for $\|x\| \leqslant \delta'$, $\mu \geqslant n_0$. Since $I_{n_0}(x)$ is small with $\|x\|$, there is a $\delta'' > 0$ such that $|I_\mu(x)| \leqslant \epsilon$ for $\|x\| \leqslant \delta''$ and $\mu \geqslant n_0$. Finally, since $I_1(x), I_2(x), \ldots, I_{n_0-1}(x)$ are small with $\|x\|$, there is a δ such that $|I_n(x)| \leqslant \epsilon$ for $\|x\| \leqslant \delta$ and all n.

We shall occasionally need the following analogue of (9·9) for series:

(9·14) THEOREM. *Let $a_1 + a_2 + \ldots$ be a fixed series. Then*

(i) *if*

$$\sum_1^\infty a_k b_k \qquad \textbf{(9·15)}$$

converges for every bounded $\{b_k\}$, or even for every $\{b_k\}$ tending to 0, we have $\Sigma |a_k| < \infty$;

(ii) *if (9·15) converges for every convergent Σb_k, the sequence $\{a_k\}$ is of bounded variation;*

(iii) *if (9·15) converges for every $\{b_k\} \in l^r$, $1 < r < \infty$, we have $\{a_k\} \in l^{r'}$.*

(i) If $\Sigma |a_k| = \infty$, then (9·15) diverges for the bounded sequence $b_n = \operatorname{sign} \bar{a}_n$, or even for the convergent sequence $b_n = \epsilon_n \operatorname{sign} \bar{a}_n$, where $\{\epsilon_n\}$ is a positive sequence tending to 0 sufficiently slowly.

(ii) It is not difficult to see that the hypotheses of (ii) imply that $a_n = O(1)$. Let Σb_k be any series convergent to 0 and let $t_k = b_1 + b_2 + \ldots + b_k$. Using summation by parts we can write the convergent series (9·15) in the form $\sum_1^\infty t_k(a_k - a_{k+1})$. The latter series converges for every sequence $\{t_k\}$ tending to 0. Hence, by (i), $\Sigma |a_k - a_{k+1}| < \infty$ (see also Chapter I, (2·4)).

(iii) For a fixed n, the sum $s_n = \sum_1^n a_k b_k$ is a linear functional in the space l^r of sequences $\{b_k\}$, and has norm $N_n = (|a_1|^{r'} + \ldots + |a_n|^{r'})^{1/r'}$. By (9·5), the hypothesis that $s_n = O(1)$ for every $\{b_k\} \in l^r$ implies $N_n = O(1)$, that is, $\{a_n\} \in l^{r'}$.

Consider a doubly infinite matrix $\mathfrak{A}=\{\alpha_{nm}\}_{n,m=0,1,\ldots}$ defining a method of summation of sequences (Chapter III, § 1). For every sequence $x=\{x_k\}$ of numbers we set

$$y_n=\alpha_{n0}x_0+\alpha_{n1}x_1+\ldots+\alpha_{nk}x_k+\ldots. \tag{9·16}$$

Suppose $x_k \to s$; then $y_n \to s$, provided $\mathfrak{A}$ satisfies the three conditions (i), (ii), (iii) of regularity. We have already proved that conditions (i) and (iii) are also necessary, and we are now going to prove the necessity of (ii).

We slightly generalize the notion of summability $\mathfrak{A}$ by assuming that the series (9·16) converge only for n sufficiently large, say for $n \geqslant n_0$, where n_0 depends on $\{x_k\}$. We shall show that, *if for every convergent sequence* $\{x_k\}$ *the series* (9·16) *converge for all n sufficiently large, and if* $y_n=O(1)$ (we do not require the existence of $\lim y_k$, still less the relation $\lim y_n=\lim x_k$), *then the numbers*

$$N_n=|\alpha_{n0}|+|\alpha_{n1}|+\ldots+|\alpha_{nk}|+\ldots$$

are finite and bounded for n sufficiently large. (If the matrix $\mathfrak{A}$ is row-finite, the finiteness of all the N_n is obvious.)

Let E be the set of all convergent sequences $x=\{x_k\}$. It is a Banach space (see example (*g*) on p. 163). Write $E=E_1+E_2+\ldots+E_m+\ldots$, where E_m is the set of all the convergent sequences x such that the series (9·16) converge for all $n \geqslant m$. (If $\mathfrak{A}$ is row-finite, then $E_1=E_2=\ldots=E$.) Since E is not of the first category, some E_{m_0} is not of the first category. For every $x=\{x_k\} \in E_{m_0}$, the series (9·16) converge if $n \geqslant m_0$. This means that

$$y_{n,j}=\alpha_{n0}x_0+\alpha_{n1}x_1+\ldots+\alpha_{nj}x_j$$

tends to a limit as $j \to \infty$, for fixed $n \geqslant m_0$. Here $y_{n,j}$ is a linear functional in E. By (9·5), the norms of $y_{n,j}$, that is, the numbers $|\alpha_{n0}|+|\alpha_{n1}|+\ldots+|\alpha_{nj}|$, are bounded as $j \to \infty$, which means that the numbers N_n are finite for $n \geqslant m_0$. It remains to show that the N_n are bounded. The finiteness of N_n implies that the y_n, $n \geqslant n_0$, in (9·16) are linear functionals in E, and, by hypothesis, $y_n=O(1)$ for each $x \in E$. It is therefore enough to apply (9·5) once more.

We conclude this section by considering an application of Fourier series to a class of linear transformations of the space l^2 of sequences $x=\{x_n\}$ two-way infinite and with $\|x\|=(\Sigma|x_n|^2)^{\frac{1}{2}}<\infty$. Series $\sum\limits_{-\infty}^{+\infty}$ will be understood here to mean $\sum\limits_{-\infty}^{-1}+\sum\limits_{0}^{\infty}$.

Let a_{mn} be a two-way infinite matrix of numbers, with m, n ranging from $-\infty$ to $+\infty$. With every $x \in l^2$ we associate the point $y=\{y_m\}$, where

$$y_m=\sum_n a_{mn}x_n \quad (m=0, \pm 1, \ldots).$$

By (9·14), y is defined for every $x \in l^2$ if and only if $\sum\limits_n |a_{mn}|^2<\infty$ for each m. In investigating whether or not $y \in l^2$, we shall confine our attention to a special case, namely that of $a_{mn}=a_{m-n}$, where $a=\{a_n\}$ is a two-way infinite sequence. This case is easily treated by means of Fourier series. Thus let

$$y_m=\sum_n a_{m-n}x_n \quad (m=0, \pm 1, \ldots). \tag{9·17}$$

If $\Sigma a_k e^{ikt}$ is the Fourier series of a function $a(t)$, we shall call $a(t)$ the *characteristic* of the transformation (9·17).

(9·18) THEOREM. (i) *A necessary and sufficient condition that the y defined by* (9·17) *should satisfy* $y \in l^2$ *for every* $x \in l^2$ *is that the transformation should have a characteristic which is an essentially bounded function. If M is the essential upper bound of* $|a(t)|$, *then* $\|y\| \leqslant M\|x\|$ *and M is the norm of the transformation.*

(ii) *Suppose that the transformation* (9·17) *has a characteristic* $a(t) \in L^2$. *A necessary and sufficient condition that for every* $y \in l^2$ *there is an* $x \in l^2$ *satisfying* (9·17) *is that* $1/a(t)$

should be essentially bounded. If $1/a(t)$ *is bounded, then the solution* x *is unique and is given by the formula*

$$x_n = \sum_m a'_{n-m} y_m \quad (n=0, \pm 1, \ldots), \tag{9·19}$$

where a'_k *are the Fourier coefficients of* $1/a(t)$.

We know that the condition $\| a \| < \infty$ is both necessary and sufficient for the transformation (9·17) to be defined for all $x \in l^2$. If $x(t) \sim \Sigma x_k e^{ikt}$, $a(t) \sim \Sigma a_k e^{ikt}$, then $y_m = \frac{1}{2\pi}\int_0^{2\pi} a(t)\, x(t)\, e^{-imt}\, dt$ and, by Parseval's formula,

$$\| y \|^2 = \frac{1}{2\pi}\int_0^{2\pi} | a(t)\, x(t) |^2\, dt. \tag{9·20}$$

If $\| y \|$ is to be finite for every $x \in l^2$, then $a(t)$ must be essentially bounded (see (9·9)). Assuming this, let M be the essential upper bound of $| a(t) |$. By (9·20), $\| y \| \leqslant M \| x \|$, and M cannot be replaced by any smaller number, that is, M is the norm of the transformation (9·17). This proves (i).

Passing to (ii), let $y(t) \sim \Sigma y_k e^{ikt} \in L^2$ be given. If (9·17) has a solution $x = \{x_k\}$ such that $x(t) \sim \Sigma x_k e^{ikt} \in L^2$, then

$$y(t) = a(t)\, x(t). \tag{9·21}$$

Since $y(t) \in L^2$ is arbitrary, $a(t) \neq 0$ almost everywhere, and there is at most one $x = \{x_k\}$ satisfying (9·17). If $x(t) = y(t)\, a^{-1}(t)$ is to belong to L^2 for every $y(t) \in L^2$, then $1/a(t)$ must be essentially bounded. If the latter condition is satisfied, the point $x = \{x_k\}$ defined by the formula $y(t)\, a^{-1}(t) \sim \Sigma x_k e^{ikt}$ is the (unique) solution of (9·17) and is given by (9·19). This proves (ii).

Linear transformations in l^2 which preserve the norm and have inverses are called *unitary*. As is seen from (9·21), *the transformation* (9·17) *is unitary if and only if* $| a(t) | = 1$ *almost everywhere*. In this case, $1/a(t) = \bar{a}(t)$, and (9·19) may be written

$$x_n = \sum_m \bar{a}_{m-n} y_m. \tag{9·22}$$

Consider a product (i.e. a successive application) of two transformations of type (9·17), having respectively characteristics $a(t)$ and $b(t)$. As is seen from (9·21), this product is of the same type, with characteristic $a(t)\, b(t)$.

If the characteristic $a(t)$ depends on a parameter α in accordance with the formula

$$a(t) = a_\alpha(t) = e^{\alpha\phi(t)},$$

where $\phi(t)$ is a measurable function, then the transformations (9·17), which we shall call T_α, form a group with the property

$$T_{\alpha+\beta} = T_\alpha T_\beta, \quad T_\alpha^{-1} = T_{-\alpha}.$$

If α is real and $\phi(t)$ purely imaginary, the transformations are unitary.

Examples. (*a*) Suppose that $a(t) = \Sigma' \frac{e^{int}}{n} = i(\pi - t)$ in $(0, 2\pi)$ (see Chapter I, (4·12)). Then (9·17) may be written

$$y_m = \sum_n{}' \frac{x_n}{m-n}. \tag{9·23}$$

This transformation, which has norm π, may be considered as a discrete analogue of the conjugate function. There exist $y \in l^2$ such that (9·17) has no solution $x \in l^2$.

(*b*) If $a_\alpha(t) = e^{i\alpha(n-t)}$, (9·17) reduces to

$$y_m = \frac{\sin \pi\alpha}{\pi} \sum_n \frac{x_n}{\alpha + m - n}$$

if α is non-integral, and to $$y_m = (-1)^\alpha x_{m+\alpha}$$

for α integral. We have here a group of unitary transformations.

10. Classes L^*_Φ

As before, let $L_\Phi(a, b)$ denote the class of all functions f such that $\Phi(|f|)$ is integrable over (a, b). The class L^r, corresponding to $\Phi(u) = u^r$, is the most important special case, but occasionally quite natural problems lead to other classes. For example, the class $L \log^+ L$ of functions f such that $|f| \log^+ |f|$ is integrable, is of importance in several problems. This leads us to the question whether a class L_Φ can be so modified as to give a normed linear space.

First of all we have to define a norm $\|x\| = \|x\|_\Phi$, and if the definition is to be useful, the finiteness of $\|x\|$ and the integrability of $\Phi(|x(t)|)$ should be to a reasonable extent equivalent. The idea (modelled on the case $\Phi(u) = u^r$) of setting

$$\|x\| = \Phi_{-1}\left[\int_a^b \Phi(|x|)\, dt\right],$$

where Φ_{-1} is the function inverse to Φ, must be discarded. First of all, the condition $\|\alpha x\| = |\alpha| \, \|x\|$ would not, in general, be satisfied. Further, if $\Phi(u)$ increases too rapidly the integrability of $\Phi(|x|)$ need not imply that of $\Phi(2|x|)$. For these reasons we must proceed differently, and it turns out that a simple solution is possible if simultaneously with Φ we consider another function Ψ such that the pair Φ, Ψ is complementary in the sense of Young (Chapter I, § 9). We have shown on p. 25 that for any function $\Phi(u)$, $u \geq 0$, which is non-negative, convex, vanishing at the origin and such that $\Phi(u)/u \to \infty$ with u, we can find such a Ψ. In the rest of this section we suppose that Φ has all the properties just stated.

Consider the functions $x(t)$, $a \leq t \leq b$, such that the product $x(t)\, y(t)$ is integrable over (a, b) for *every* $y(t) \in L_\Psi(a, b)$, and set

$$\|x\| = \|x\|_\Phi = \sup_y \left| \int_a^b x(t)\, y(t)\, dt \right|, \tag{10·1}$$

the sup being with respect to all y with

$$\rho_y = \int_a^b \Psi(|y|)\, dt \leq 1.$$

The class of such x will be denoted by L^*_Φ. It is not obvious that $\|x\|$ must be finite. We prove this a little later; in the meanwhile, to avoid difficulty, we denote by L^*_Φ the class of functions $x(t)$, $a \leq t \leq b$, such that the norm $\|x\|_\Phi$ just defined is finite. Using Young's inequality, we see that $L_\Phi \subset L^*_\Phi$. It is immediate that L^*_Φ is a normed linear space. Classes L^*_Φ are often called *Orlicz* spaces.

We shall prove that L^*_Φ *is a complete space*. Suppose that $\|x_m - x_n\| \to 0$ for $m, n \to \infty$, so that $\|x_m - x_n\| < \epsilon$ for $m, n > \nu = \nu(\epsilon)$. It follows that

$$\left| \int_a^b (x_m - x_n)\, y\, dt \right| \leq \epsilon, \tag{10·2}$$

and so also $$\int_a^b |x_m - x_n||y|\,dt \leqslant \epsilon, \tag{10·3}$$

if $\rho_y \leqslant 1$, and $m, n > \nu$. Let α be such that $(b-a)\Psi(\alpha) = 1$. Taking $y(t) = \alpha\,\overline{\text{sign}\,(x_m - x_n)}$, we get from (10·2) that $\mathfrak{M}[x_m - x_n;\, a, b] \leqslant \epsilon/\alpha$. Since ϵ is arbitrary, there is a sequence $\{x_{m_k}(t)\}$ converging almost everywhere to a function $x(t)$ (Chapter I, (11·1)), and (10·3) shows that $\mathfrak{M}[(x - x_n)y;\, a, b] \leqslant \epsilon$ if $\rho_y \leqslant 1$. Thus $\|x - x_n\| \leqslant \epsilon$ for $n > \nu$, and the completeness of the space L_Φ^* follows.

We have tacitly assumed here that $b - a < \infty$, but the result holds if $b - a = \infty$. In fact, proceeding as before, we show that $\mathfrak{M}[x_m - x_n;\, a', b'] \to 0$ for any interval (a', b') satisfying $a < a' < b' < b$. Thence we infer the existence of a subsequence of $\{x_m(t)\}$ converging almost everywhere in (a, b), and the rest of the proof is unchanged. In what follows, to shorten the exposition, we shall assume that $b - a < \infty$. The results, however, are also valid for $b - a = \infty$, as simple modifications of the proofs show.

We have already observed that if $x \in L_\Phi$, then $x \in L_\Phi^*$. More generally, if there is a number $\theta > 0$ such that $\theta x \in L_\Phi$, then $x \in L_\Phi^*$. Conversely, we show that, *if* $x \in L_\Phi^*$, *then there is a constant* $\theta > 0$ *such that* $\theta x \in L_\Phi$. More precisely,

(10·4) THEOREM. *If* $x \in L_\Phi^*$, $\|x\| \neq 0$, *then*
$$\int_a^b \Phi(|x|/\|x\|)\,dt \leqslant 1. \tag{10·5}$$

We begin by showing that
$$\left|\int_a^b xy\,dt\right| \begin{cases} \leqslant \|x\|, & \text{if } \rho_y \leqslant 1; \\ \leqslant \|x\|\,\rho_y, & \text{if } \rho_y > 1. \end{cases} \tag{10·6}$$

The first inequality is obvious. Suppose now that $\rho_y > 1$ and replace y by y/ρ_y in the integral on the left. Since Ψ is convex and $\Psi(0) = 0$, $\Psi(|y|/\rho_y) \leqslant \Psi(|y|)/\rho_y$, so that
$$\int_a^b \Psi(|y|/\rho_y)\,dt \leqslant 1, \qquad \left|\int_a^b x\frac{y}{\rho_y}\,dt\right| \leqslant \|x\|,$$
which is the second inequality (10·6). It follows that the integral in (10·6) does not exceed $\|x\|\,\rho_y'$, where
$$\rho_y' = \max(\rho_y, 1).$$

We shall now prove (10·5) for x bounded. Let $\phi = \Phi'$. For the special case $y = \phi(|x|/\|x\|)$ that of equality in Young's inequality (pp. 16, 25), we have
$$\rho_y' \geqslant \left|\int_a^b \frac{x}{\|x\|}\,y\,dt\right| = \int_a^b \Phi\left[\frac{|x|}{\|x\|}\right]dt + \rho_y.$$
We may suppose that the integral preceding ρ_y is not 0. From this it follows, first, that (for the special y) $\rho_y' > \rho_y$, and so $\rho_y' = 1$, and then that
$$\int_a^b \Phi(|x|/\|x\|)\,dt \leqslant 1.$$

In the general case, let $x_n(t)$ be equal to $x(t)$ wherever $|x(t)| \leqslant n$ and equal to 0 elsewhere. Then (10·5) holds if the integrand on the left is $\Phi(|x_n|/\|x_n\|)$, n being so large that $\|x_n\| \neq 0$, hence also for $\Phi(|x_n|/\|x\|)$, and finally, making $n \to \infty$, for $\Phi(|x|/\|x\|)$.

(10·7) Theorem. (i) *If the integral* (9·8) *exists for every* $x \in L^*_\Phi(a, b)$, *then* $y \in L^*_\Psi(a, b)$.

(ii) *If the sequence* (9·12) *is bounded for every* $x \in L^*_\Phi$, *then* $\| y_n \|_\Psi = O(1)$.

(iii) *If the sequence* (9·12) *is bounded for every* $x \in L_\Phi$ *then there is a constant* $\theta > 0$ *such that* $\mathfrak{M}[\Psi(\theta \mid y_n \mid)] = O(1)$.

(i) Let y^n be the function y truncated by n, and let us consider the integral $\int_a^b xy^n dt$. Since each y^n, as a bounded function, belongs to L_Ψ, and since

$$| u_n(x) | \leqslant \| x \|_\Phi \rho'_{y^n},$$

$u_n(x)$ is a functional in L^*_Φ. By hypothesis, $\{u_n(x)\}$ converges for every $x \in L^*_\Phi$, so that there is a constant M such that $\| u_n(x) \| \leqslant M \| x \|_\Phi$ for $n = 1, 2, \ldots$. Now take any x such that $\mathfrak{M}[\Phi(| x |)] \leqslant 1$. Such an x belongs to L^*_Φ and has norm $\| x \|_\Phi \leqslant 2$. But since the inequality $\left| \int_a^b xy^n dt \right| \leqslant 2M$ is valid for every x with $\mathfrak{M}[\Phi(| x |)] \leqslant 1$ we have $\| y^n \|_\Psi \leqslant 2M$ for $n = 1, 2, \ldots$, and so also

$$\| y \|_\Psi \leqslant 2M.$$

(ii) By virtue of (i), each y_n belongs to L^*_Ψ. It follows from the inequality

$$\lambda \mid u_n(x) \mid \leqslant \| x \|_\Phi \rho'_{\lambda y_n},$$

where $\lambda = \lambda_n$ is a positive constant so small that $\lambda y_n \in L_\Psi$, that $u_n(x)$ is a functional in L^*_Φ. Thus, by (9·5), $| u_n(x) | \leqslant M \| x \|_\Phi$, for $n = 1, 2, \ldots$. In particular, $\left| \int_a^b xy_n dt \right| \leqslant 2M$ for x satisfying $\mathfrak{M}[\Phi(| x |)] \leqslant 1$. Thus

$$\| y_n \|_\Psi \leqslant 2M \quad \text{for} \quad n = 1, 2, \ldots.$$

(iii) Suppose that $x \in L^*_\Phi$. Then $x/\| x \|_\Phi \in L_\Phi$, and the sequence $\{u_n(x)/\| x \|_\Phi\}$ is bounded for every $x \in L^*_\Phi$. By (ii), $\| y_n \|_\Psi \leqslant N$, say, for $n = 1, 2, \ldots$, so that

$$\int_a^b \Psi(\theta \mid y_n \mid) \, dt \leqslant 1 \quad \text{for} \quad \theta = 1/N.$$

This completes the proof of (10·7).

We can now dispose of the superfluous restriction in the definition of L^*_Φ, and show that if $\int_a^b xy \, dt$ exists for every y with $\mathfrak{M}[\Psi(| y |)] \leqslant 1$, then the norm $\| x \|$ defined by (10·1) is finite. Let $x^n(t)$ be the function x truncated by n. By hypothesis, the sequence of integrals $v_n(y) = \int_a^b x^n y \, dt$ converges for every $y \in L^*_\Phi$, for then

$$\int_a^b \Psi(| y | / \| y \|) \, dt \leqslant 1.$$

It follows, as in (i), that $\| x^n \|_\Phi = O(1)$, that is, $\| x \|_\Phi < \infty$.

It is useful to note that (10·7) holds if we consider the class $L^*_\Phi(E)$, where E is an arbitrary set.

We have already observed that a necessary and sufficient condition that $x(t) \in L^*_\Phi$ is the existence of a constant $\theta > 0$ such that $\theta x \in L_\Phi$. It follows that if there is a constant C such that

$$\Phi(2u) \leqslant C\Phi(u) \tag{10·8}$$

for u large enough, and if $b - a < \infty$, the classes L_Φ and L^*_Φ are identical.

A simple calculation shows that if $\Phi(u)=u^r$, $r>1$, then $\| x \|_\Phi = r'^{1/r'} r^{1/r} \mathfrak{M}_r[x]$, so that, apart from a numerical constant factor, we have the same norm as in Example (*d*) on p. 163.

We may define the norm of x in a somewhat different way. We fix a $\Phi(u)$ $(0 \leqslant u < \infty)$ convex, non-decreasing, and satisfying $\Phi(0)=0$ and $\Phi(u)/u \to \infty$ as $u \to \infty$. Let $\Psi(v)$ be complementary to $\Phi(u)$ in the sense of Young, and let $\phi = \Phi'$, $\psi = \Psi'$. It will be convenient to normalize Φ by the condition

$$\Phi(1)+\Psi(1)=1. \tag{10·9}$$

This can be done, for example, by replacing $\phi(u)$ by $\phi(ku)$, k being such that the area of the rectangle with opposite vertices at $(0, 0)$ and $(k, \phi(k))$ is 1 (this area is a continuous function of k, and increases from 0 to ∞ with k); the new $\Phi(u)$ is then the old $k^{-1}\Phi(ku)$.

Consider the class of all functions $x(t)$, $a \leqslant t \leqslant b$, such that $\Phi(\theta \,|\, x \,|)$ is integrable over (a, b) for some $\theta > 0$ (the normalization does not affect the class). We then have

$$\int_a^b \Phi(\lambda^{-1} \,|\, x \,|)\, dt \leqslant \Phi(1) \tag{10·10}$$

for all large enough positive λ. Let λ_0 be the lower bound of all λ's satisfying (10·10); λ_0 is non-negative, and is 0 if and only if $x \equiv 0$. We call λ_0 the *norm* of x, and denote it by $\mathrm{N}x$ or $\mathrm{N}_\Phi x$. $\mathrm{N}x$ is non-negative, is 0 if and only if $x \equiv 0$, and $\mathrm{N}(\alpha x) = |\, \alpha \,|\, \mathrm{N}x$ for every scalar α. We also have

$$\mathrm{N}(x+y) \leqslant \mathrm{N}x + \mathrm{N}y. \tag{10·11}$$

For if $\lambda > \mathrm{N}x$, $\mu > \mathrm{N}y$, then from (10·10) and the analogous inequality for $|\, y \,|/\mu$ we obtain, using Jensen's inequality,

$$\begin{aligned} \int_a^b \Phi\left(\frac{|\, x+y \,|}{\lambda+\mu}\right) dt &\leqslant \frac{\lambda}{\lambda+\mu} \int_a^b \Phi\left(\frac{|\, x \,|}{\lambda}\right) dt + \frac{\mu}{\lambda+\mu} \int_a^b \Phi\left(\frac{|\, y \,|}{\mu}\right) dt \\ &\leqslant \frac{\lambda}{\lambda+\mu} \Phi(1) + \frac{\mu}{\lambda+\mu} \Phi(1) = \Phi(1). \end{aligned}$$

This implies that $\mathrm{N}(x+y) \leqslant \lambda+\mu$, and making $\lambda \to \mathrm{N}x$, $\mu \to \mathrm{N}y$ we deduce (10·11). Thus $\mathrm{N}x$ has all the properties of a norm.

If $\Phi(u) = u^p/p$, where $1 < p < \infty$, then $\Psi(v) = v^{p'}/p'$, (10·9) holds, and we immediately see that $\mathrm{N}x = \| x \|_p$.

The class of all functions $x(t)$ such that θx is in $L_\Phi(a, b)$ for some $\theta > 0$ is, as we know, the class L_Φ^* for which we have already defined a norm $\| x \|_\Phi$. Write $\Phi(1) = \nu$. Since $0 < \nu < 1$ we deduce from (10·5) that

$$\int_a^b \Phi\left(\frac{\nu x}{\| x \|}\right) dt \leqslant \nu \int_a^b \Phi\left(\frac{x}{\| x \|}\right) dt \leqslant \nu = \Phi(1),$$

so that $\mathrm{N}x \leqslant \| x \|/\nu$. On the other hand, if $\lambda > \mathrm{N}x$ then for all y with $\mathfrak{M}[\Psi(|\, y \,|)] \leqslant 1$ we have

$$\left| \int_a^b xy\, dt \right| \leqslant \lambda \int_a^b \left| \frac{x}{\lambda} \right| |\, y \,|\, dt \leqslant \lambda \left[\int_a^b \Phi\left(\frac{|\, x \,|}{\lambda}\right) dt + \int_a^b \Psi(|\, y \,|)\, dt \right] \leqslant \lambda[\Phi(1)+1] < 2\lambda,$$

since, by (10·9), $\Phi(1)<1$. Taking the upper bound of the first integral with respect to all y, and making $\lambda\to \mathrm{N}x$, we see that $\|x\|_\Phi\leqslant 2\mathrm{N}x$. Thus

$$\Phi(1)\,\mathrm{N}x\leqslant\|x\|_\Phi\leqslant 2\mathrm{N}x. \tag{10·12}$$

These inequalities show that the norms $\|x\|_\Phi$ and $\mathrm{N}x$ are *equivalent*, in the sense that their ratio is contained between two positive absolute constants. They show that, since L_Φ^* is complete with respect to the norm $\|x\|_\Phi$, it is also complete (and so is a Banach space) with respect to the norm $\mathrm{N}_\Phi x$.

We apply the foregoing to homogenizing certain inequalities.

Suppose we have an operation $y=Tx$ transforming functions $x(s)\in\mathrm{L}_\Phi(a,b)$ into functions $y(t)\in\mathrm{L}_{\Phi_1}(a_1,b_1)$, and such that

$$\int_{a_1}^{b_1}\Phi_1(|y|)\,dt\leqslant A\int_a^b\Phi(|x|)\,ds+B, \tag{10·13}$$

where Φ and Φ_1 are Young's functions, and A and B are constants independent of x. The operation T need not be linear; we assume only that it is *positively homogeneous*, by which we mean that, for each scalar α, if Tx is defined so is $T(\alpha x)$, and

$$|T(\alpha x)|=|\alpha|\,|Tx|.$$

(10·14) THEOREM. *Under the hypotheses just stated, Tx is defined for all $x\in L_\Phi^*(a,b)$, and we have*

$$\|y\|_{\Phi_1}\leqslant C\|x\|_\Phi, \tag{10·15}$$

or, equivalently,

$$\mathrm{N}_{\Phi_1}y\leqslant C\mathrm{N}_\Phi x, \tag{10·16}$$

where the C's are constants independent of x.

The advantage of (10·15) and (10·16) over (10·13) is the homogeneity of the relations.

That Tx is defined in $\mathrm{L}_\Phi^*(a,b)$ follows from the positive homogeneity of T and the fact that $x\in\mathrm{L}_\Phi^*$ implies $x/\|x\|_\Phi\in\mathrm{L}_\Phi$ (see (10·5)). Write $\lambda=\mathrm{N}_\Phi x$, $\lambda_1=\mathrm{N}_{\Phi_1}y$. From (10·13) applied to $x/\|x\|_\Phi$ we see that $\Phi_1(y/\|x\|_\Phi)$ is integrable, so that $y\in\mathrm{L}_{\Phi_1}^*(a_1,b_1)$ and $\lambda_1=\mathrm{N}_{\Phi_1}y$ is finite. We have to show that $\lambda_1\leqslant C\lambda$.

If $\lambda=0$, then $x\equiv 0$ and (by the positive homogeneity of T) $y\equiv 0$, $\lambda_1=0$, so that (10·16) holds. We may therefore suppose that $\lambda>0$. We may also suppose that $\lambda_1>\lambda$, for if $\lambda_1\leqslant\lambda$ we have (10·16) with $C=1$.

Let $\lambda<\lambda'<\lambda_1'<\lambda_1$. Applying (10·13) to x/λ' we obtain

$$\int_{a_1}^{b_1}\Phi_1\left(\frac{|y|}{\lambda'}\right)dt\leqslant A\int_a^b\Phi\left(\frac{|x|}{\lambda'}\right)ds+B\leqslant A\Phi(1)+B=C_1,$$

say, and since Φ_1 is convex and $\Phi_1(0)=0$,

$$\int_{a_1}^{b_1}\Phi_1\left(\frac{|y|}{\lambda'}\right)dt=\int_{a_1}^{b_1}\Phi_1\left(\frac{\lambda_1'}{\lambda'}\frac{|y|}{\lambda_1'}\right)dt\geqslant\frac{\lambda_1'}{\lambda'}\int_{a_1}^{b_1}\Phi_1\left(\frac{|y|}{\lambda_1'}\right)dt\geqslant\frac{\lambda_1'}{\lambda'}\Phi_1(1)$$

which, combined with the preceding inequalities, gives $\lambda_1'\leqslant C\lambda'$, and so also (10·16), with $C=C_1/\Phi_1(1)$. This completes the proof of (10·14).

Remark. The argument holds in the extreme case when $\Phi(u)=u$ or $\Phi_1(u)=u$.

As an illustration consider functions $x(s)$ integrable over a finite interval (a,b) and the operation $\Theta=Tx$ defined by

$$\Theta(t)=\sup_h\frac{1}{h}\int_t^{t+h}|x(s)|\,ds.$$

Though not linear, the operation is positively homogeneous and satisfies (10·13) with $(a_1, b_1) = (a, b)$, $\Phi(u) = u \log^+ u$, $\Phi_1(u) = u$ (Chapter I, (13·15) (iii)). It therefore also satisfies (10·15) and (10·16). A similar remark applies to the operation $Tf = \tilde{f}$ (see Chapter VII, (2·9)).

The theorem which follows generalizes Hölder's inequality.

(10·17) Theorem. *If $x(t) \in L^*_\Phi(a, b)$, $y(t) \in L^*_\Psi(a, b)$, where Φ and Ψ are complementary in the sense of Young, then xy is integrable over (a, b) and*

$$\left| \int_a^b xy \, dt \right| \leqslant N_\Phi x . N_\Psi y. \tag{10·18}$$

This is immediate if either $N_\Phi x$ or $N_\Psi y$ is 0. If neither is 0, then for $\lambda > N_\Phi x$, $\mu > N_\Psi y$ we have

$$\left| \int_a^b \frac{x}{\lambda} \frac{y}{\mu} dt \right| \leqslant \int_a^b \left| \frac{x}{\lambda} \right| \left| \frac{y}{\mu} \right| dt \leqslant \int_a^b \Phi\left(\frac{|x|}{\lambda} \right) dt + \int_a^b \Psi\left(\frac{|y|}{\mu} \right) dt \leqslant \Phi(1) + \Psi(1) = 1, \tag{10·19}$$

and making $\lambda \to N_\Phi x$, $\mu \to N_\Psi y$ we obtain (10·18).

We now consider cases of equality in (10·18) and we may suppose $\lambda_0 = N_\Phi(x)$ and $\mu_0 = N_\Psi y$ are both positive, for otherwise we always have equality. Recalling the definition of λ_0, we observe that if (10·10) holds for all $\lambda > \lambda_0$ then, by Fatou's lemma, it holds for $\lambda = \lambda_0$. Examples show that we can have strict inequality in (10·10) even if $\lambda = \lambda_0$, and we are interested in cases when

$$\int_a^b \Phi\left(\frac{|x|}{\lambda_0} \right) dt = \Phi(1). \tag{10·20}$$

If the integral (10·10) is finite for some $\lambda < \lambda_0$, then it is a continuous and strictly decreasing function of λ near λ_0, and (10·20) holds. Hence, for example, (10·20) holds if $\phi(2u) \leqslant C\phi(u)$ for all u; or if the inequality holds for u large enough and (a, b) is finite.

Suppose now that we have (10·20), and that the integral of $\Psi(\mu_0^{-1} |y|)$ over (a, b) is also 1. If we substitute λ_0, μ_0 for λ, μ in (10·19), the extreme terms are equal and, considering the cases of equality in Young's inequality $\xi\eta \leqslant \Phi(\xi) + \Psi(\eta)$, we come to the following conclusion: if $\lambda_0 \neq 0$, $\mu_0 \neq 0$, we have equality in (10·18) if and only if

(i) arg (xy) is constant almost everywhere in the set where $xy \neq 0$;

(ii) the point $(x(t)\lambda_0^{-1}, y(t)\mu_0^{-1})$ is almost always on the continuous curve obtained from $\eta = \phi(\xi)$ by adjoining vertical segments at the points of discontinuity of ϕ.

The arguments of this section apply to Stieltjes integrals $\int_a^b \Phi(|x(t)|) \, d\mu(t)$, where $\mu(t)$ is non-decreasing, and in particular to sums $\Sigma \Phi(a_i)$. We may define norms $\| a \|_\Phi$ and $N_\Phi a$ for sequences $a = \{a_i\}$, we have Hölder's inequality $|\Sigma a_i b_i| \leqslant N_\Phi a . N_\Psi b$, an analogue of (10·14), etc.

11. Conversion factors for classes of Fourier series

Consider two trigonometric series

$$\sum_{-\infty}^{\infty} c_\nu e^{i\nu x}, \tag{11·1}$$

$$\sum_{-\infty}^{\infty} \lambda_\nu e^{i\nu x}, \tag{11·2}$$

and the associated series
$$\sum_{-\infty}^{\infty} c_\nu \lambda_\nu e^{i\nu x}. \tag{11·3}$$

By $\{\lambda_\nu\}$ we shall now understand the two-way infinite sequence $\dots, \lambda_{-1}, \lambda_0, \lambda_1, \dots$. Given two classes P and Q of trigonometric series we shall say that $\{\lambda_\nu\}$ is of *type* (P, Q) if whenever (11·1) belongs to P, (11·3) belongs to Q.

(11·4) THEOREM. *A necessary and sufficient condition for $\{\lambda_\nu\}$ to be of any one of the types* (B, B), (C, C), (L, L), (S, S) *is that $\Sigma \lambda_\nu e^{i\nu x}$ be a Fourier-Stieltjes series.*

Let (11·1) be an $S[f]$, and let $\sigma_n(x)$, $l_n(x)$, $\sigma_n^*(x)$ denote the (C, 1) means of the series (11·1), (11·2) and (11·3) respectively. Then

$$\sigma_n^*(x) = \sum_{\nu=-n}^{n} \left(1 - \frac{|\nu|}{n+1}\right) c_\nu \lambda_\nu e^{i\nu x} = \frac{1}{2\pi}\int_0^{2\pi} l_n(t) f(x-t)\, dt. \tag{11·5}$$

Let $x=0$. If $\{\lambda_n\}$ is of type (C, C) or (B, B), the sequence $\{\sigma_n^*(0)\}$ is bounded for every $f \in C$, and by (9·11) we have $\mathfrak{M}[l_n(t)] = O(1)$, whence (11·2) belongs to S. Conversely, if (11·2) is an $S[dL]$, we have

$$\sigma_n^*(x) = \frac{1}{2\pi}\int_0^{2\pi} \sigma_n(x-t)\, dL(t), \tag{11·6}$$

so that the uniform boundedness of $\{\sigma_n(x)\}$ implies that of $\{\sigma_n^*(x)\}$. Similarly, if $\sigma_m(x) - \sigma_n(x)$ tends uniformly to 0 as $m, n \to \infty$, so does $\sigma_m^*(x) - \sigma_n^*(x)$. This completes the proof of (11·4) for the types (B, B) and (C, C).

If $\{\lambda_n\}$ is of type (S, S), it transforms, in particular, the series $\sum_{-\infty}^{\infty} e^{i\nu x} \in S$ into the series (11·2), and the latter must therefore belong to S. Conversely, if (11·2) is an $S[dL]$, (11·6) gives

$$|\sigma_n^*(x)| \leqslant \frac{1}{2\pi}\int_0^{2\pi} |\sigma_n(x-t)|\, |dL(t)|. \tag{11·7}$$

Integrating this over $(0, 2\pi)$ and inverting the order of integration on the right, we get $\mathfrak{M}[\sigma_n^*] \leqslant (v/2\pi)\,\mathfrak{M}[\sigma_n]$, where v is the total variation of $L(t)$ over $(0, 2\pi)$. Hence (11·3) belongs to S if (11·1) does.

It remains only to consider the case (L, L). Since

$$|\sigma_m^*(x) - \sigma_n^*(x)| \leqslant \frac{1}{2\pi}\int_0^{2\pi} |\sigma_m(x-t) - \sigma_n(x-t)|\, |dL(t)|,$$

$$\mathfrak{M}[\sigma_m^* - \sigma_n^*] \leqslant (v/2\pi)\,\mathfrak{M}[\sigma_m - \sigma_n],$$

the sufficiency of the condition is obvious (see (5·5)). To prove the necessity, let us consider for each n a system $I_n = (\alpha_1^n, \beta_1^n)$, (α_2^n, β_2^n), ... of non-overlapping intervals. It follows from (11·5) that

$$\int_{I_n} \sigma_n^*(x)\, dx = \frac{1}{2\pi}\int_0^{2\pi} f(t) \left\{ \int_{I_n} l_n(x-t)\, dx \right\} dt. \tag{11·8}$$

Suppose that (11·2) does not belong to S, so that the indefinite integrals of the functions $l_n(x)$ are not of uniformly bounded variation. We can then find a sequence $I_1, I_2, \dots$ such that the coefficient of $f(t)$ in (11·8) is not bounded for $t=0$. Since it is continuous in t (for each n), its essential upper bound is unbounded; thus by (9·11) there is an

$f \in \mathrm{L}$ such that the right-hand side of (11·8) is unbounded, and *a fortiori* $\mathfrak{M}[\sigma_n^*] \neq O(1)$. Hence (11·3) does not belong to S, and $\{\lambda_n\}$ is not of type (L, L).

Let $\tilde{\mathrm{P}}$ denote the class of trigonometric series conjugate to those belonging to P.

(11·9) THEOREM. *A necessary and sufficient condition that* $\{\lambda_n\}$ *should be of any one of the types* $(\tilde{\mathrm{B}}, \mathrm{B})$, $(\tilde{\mathrm{C}}, \mathrm{C})$, $(\tilde{\mathrm{L}}, \mathrm{L})$, $(\tilde{\mathrm{S}}, \mathrm{S})$ *is that the series conjugate to* $\Sigma\lambda_n e^{inx}$ *should belong to* S.

This follows from (11·4). For let $\epsilon_\nu = -i \operatorname{sign} \nu$. Saying that $\{\lambda_n\}$ is of type $(\tilde{\mathrm{B}}, \mathrm{B})$ means that whenever $\Sigma c_\nu e^{i\nu x}$ belongs to B, so does $\Sigma \epsilon_\nu \lambda_\nu c_\nu e^{i\nu x}$. A necessary and sufficient condition for this is, by (11·4), that $\Sigma \epsilon_\nu \lambda_\nu e^{i\nu x} \in \mathrm{S}$. Similarly for the remaining types.

(11·10) THEOREM. (i) *A necessary and sufficient condition that* $\{\lambda_n\}$ *should be of either of the types* (B, C), (S, L) *is that* $\Sigma\lambda_n e^{inx}$ *should belong to* L.

(ii) *The types* $(\tilde{\mathrm{B}}, \mathrm{C})$ *and* $(\tilde{\mathrm{S}}, \mathrm{L})$ *are characterized by the fact that the series conjugate to* $\Sigma\lambda_n e^{inx}$ *belongs to* L.

It is enough to prove (i), the proof of (ii) being then analogous to that of (11·9). Considering the series $\sum_{-\infty}^{+\infty} e^{i\nu x} \in \mathrm{S}$, we see the necessity of the condition for type (S, L). The sufficiency follows from the sufficiency for (L, L) in (11·4) on interchanging the roles of c_ν and λ_ν so that (11·3) belongs to L.

Let now f be any element from B. If $\{\lambda_n\}$ is of type (B, C), we see, on taking $x=0$ in (11·5) and using (9·13) (i), that the indefinite integrals of the $l_n(x)$ must be uniformly absolutely continuous. Thus $\Sigma\lambda_\nu e^{i\nu x} \in \mathrm{L}$. Conversely, if the latter condition is satisfied (11·5) implies that

$$2\pi \,|\, \sigma_m^*(x) - \sigma_n^*(x) \,| \leqslant \mathfrak{M}[l_m - l_n] \sup |\, f \,|.$$

Thus $\{\sigma_m^*(x)\}$ converges uniformly and (11·3) belongs to C.

Let $\chi(u)$, $u \geqslant 0$, be a function non-negative, non-decreasing, convex and such that $\chi(u)/u \to \infty$ with u.

(11·11) THEOREM. *If* (11·1) *is an* $S[f]$ *such that* $\chi(|\, f\,|)$ *is integrable, and if* (11·2) *is an* $S[dL]$, *then* (11·3) *is an* $S[g]$ *such that* $\chi(2\pi \,|\, g\,|/v)$ *is integrable, v denoting the total variation of L over* $(0, 2\pi)$.

Jensen's inequality applied to (11·6) gives

$$\chi\{2\pi \,|\, \sigma_n^*(x) \,|/v\} \leqslant v^{-1} \int_0^{2\pi} \chi(|\, \sigma_n(x-t) \,|) \,|\, dL(t) \,|.$$

It is now sufficient to integrate this inequality over $0 \leqslant x \leqslant 2\pi$, invert the order of integration on the right, and apply (5·7). In particular, *if* (11·2) *belongs to* S, $\{\lambda_n\}$ *is of type* $(\mathrm{L}^r, \mathrm{L}^r)$ *for every* $r \geqslant 1$.

The condition imposed here on (11·2) is sufficient only. That it is not necessary is seen by the example $\chi(u) = u^2$, since by the Parseval formula and the Riesz-Fischer theorem $\{\lambda_n\}$ is of type $(\mathrm{L}^2, \mathrm{L}^2)$ if and only if $\lambda_n = O(1)$.

Let now Φ, Ψ and Φ_1, Ψ_1 be two pairs of Young's complementary functions.

(11·12) THEOREM. *The types* $(\mathrm{L}_\Phi^*, \mathrm{L}_{\Phi_1}^*)$ *and* $(\mathrm{L}_{\Psi_1}^*, \mathrm{L}_\Psi^*)$ *are identical.*

(11·13) Theorem. *A necessary and sufficient condition that $\Sigma c_\nu e^{i\nu x}$ should belong to L^*_Φ is that for every $g \in L^*_\Psi$ with Fourier coefficients c'_ν the series*

$$\sum_{-\infty}^{\infty} c_\nu c'_\nu \qquad \textbf{(11·14)}$$

should be bounded (C, 1) (*or, what is here equivalent, summable* (C, 1)).

If $f \in L^*_\Phi$, $g \in L^*_\Psi$, there exist positive constants λ, μ such that $\lambda f \in L_\Phi$, $\mu g \in L_\Psi$, and the necessity in (11·13) follows from (8·7). For the sufficiency, let $\sigma_n(x)$ and τ_n denote the (C, 1) means of $\Sigma c_\nu e^{i\nu x}$ and (11·14) respectively. Then

$$\tau_n = \frac{1}{2\pi}\int_0^{2\pi} g(-t)\,\sigma_n(t)\,dt.$$

Since $\{\tau_n\}$ is assumed bounded for every $g \in L^*_\Psi$, it follows that $\Sigma c_\nu e^{i\nu x}$ belongs to L^*_Φ (see (10·7)). This proves (11·13).

To prove (11·12), we now note that, by (11·13), if $\{\lambda_n\}$ is of type $(L^*_\Phi, L^*_{\Phi_1})$ then for every $f \in L^*_\Phi$ with Fourier coefficients c_ν and every $g \in L^*_{\Psi_1}$ with coefficients c'_ν, the series $\sum_{-\infty}^{\infty} \lambda_\nu c_\nu c'_\nu$ is finite (C, 1). By (11·13) this also means that $\Sigma \lambda_\nu c'_\nu e^{i\nu x} \in L^*_\Psi$, so that $\{\lambda_n\}$ is of type $(L^*_{\Psi_1}, L^*_\Psi)$.

As corollaries we get

(i) *if Φ and Ψ are complementary functions, the types* (L^*_Φ, L^*_Φ) *and* (L^*_Ψ, L^*_Ψ) *are identical;*

(ii) *if $r>1$, $s>1$, the types* (L^r, L^s) *and* $(L^{s'}, L^{r'})$ *are identical; in particular,*

$$(L^r, L^r) = (L^{r'}, L^{r'}).$$

Suppose that (11·2) is $S[dL]$ and (11·1) and (11·3) are respectively $S[f]$ and $S[g]$, with f continuous. The formula

$$g(x) = \frac{1}{2\pi}\int_0^{2\pi} f(x-t)\,dL(t)$$

leads immediately to the following inequality for the moduli of continuity of f and g:

$$\omega(\delta; g) \leqslant \omega(\delta; f)\frac{1}{2\pi}\int_0^{2\pi} |dL|. \qquad \textbf{(11·15)}$$

(11·16) Theorem. *Suppose that $\Sigma \lambda_n e^{inx}$ belongs to* S. *Then $\{\lambda_n\}$ is of type* $(\Lambda_\alpha, \Lambda_\alpha)$, $0<\alpha\leqslant 1$, *and of types* (Λ_*, Λ_*) *and* (λ_*, λ_*).

The assertion concerning the type $(\Lambda_\alpha, \Lambda_\alpha)$ follows from (11·15), and the remainder from the similar inequality

$$\max_x |g(x+h)+g(x-h)-2g(x)| \leqslant \max_x |f(x+h)+f(x-h)-2f(x)|\frac{1}{2\pi}\int_0^{2\pi} |dL|.$$

The results obtained above may be stated in terms of 'real Fourier series'. If $c_\nu = \frac{1}{2}(a_\nu - ib_\nu)$, $c_{-\nu} = \bar{c}_\nu$, (11·1) becomes

$$\tfrac{1}{2}a_0 + \sum_{\nu=1}^{\infty} (a_\nu \cos \nu x + b_\nu \sin \nu x). \qquad \textbf{(11·17)}$$

If the λ_ν are real and $\lambda_{-\nu}=\lambda_\nu$, (11·2) and (11·3) become

$$2\left(\tfrac{1}{2}\lambda_0+\sum_{\nu=1}^{\infty}\lambda_\nu\cos\nu x\right), \tag{11·18}$$

$$\tfrac{1}{2}a_0\lambda_0+\sum_{\nu=1}^{\infty}(a_\nu\cos\nu x+b_\nu\sin\nu x)\,\lambda_\nu, \tag{11·19}$$

respectively. In translating the former results we have merely to substitute the new names (11·17), (11·18), (11·19) for the old (11·1), (11·2), (11·3).

The following illustration is useful.

Let $\lambda_0, \lambda_1, \lambda_2, \ldots$ be an arbitrary positive sequence tending to 0 and convex from some place on. For example, we may take

$$\lambda_n=\frac{1}{n^\alpha}\ (\alpha>0),\quad \lambda_n=\frac{1}{\log n},\quad \lambda_n=\frac{1}{\log\log n},\quad \ldots$$

for n large enough. In Chapter V, § 1, we show that series $\Sigma\lambda_\nu\cos\nu x$ belong to L. It follows that such sequences are of types (B, C) and (S, L).

It will also be shown (Chapter V, § 1) that the series $2\Sigma\mu_\nu\sin\nu x$ conjugate to (11·18) belongs to L, if $\mu_1, \mu_2, \ldots$ are positive and monotonically decreasing and $\Sigma\mu_\nu/\nu<\infty$. Thus, in particular, the sequences

$$\mu_n=\frac{1}{(\log n)^{1+\epsilon}},\quad \mu_n=\frac{1}{\log n\,(\log\log n)^{1+\epsilon}},\quad \ldots$$

are both of type $(\tilde{\mathrm{B}}, \mathrm{C})$ and $(\tilde{\mathrm{S}}, \mathrm{L})$, provided $\epsilon>0$. For $\epsilon=0$ this is no longer true.

MISCELLANEOUS THEOREMS AND EXAMPLES

1. Let $\phi_1(x)$, $\phi_2(x)$, ... be a system orthonormal over $(0, 2\pi)$. Then the system $(2\pi)^{-\frac{1}{2}}$, $\tilde{\phi}_1(x)$, $\tilde{\phi}_2(x)$, ... is also orthonormal there. If the first system is complete so is the second.

[Use the formula $\int_0^{2\pi} f\tilde{g}\,dx=-\int_0^{2\pi}\tilde{f}g\,dx$ for $f\in \mathrm{L}^2$, $g\in\mathrm{L}^2$.]

2. Let ϕ_1, ϕ_2, ... be an orthonormal system of uniformly bounded functions in (a, b). Let s_k be the partial sums of the series

$$a_1\phi_1+a_2\phi_2+\ldots$$

and suppose that the functions $S_k(x)=\int_a^x s_k(t)\,dt$ are uniformly absolutely continuous in (a, b), which is certainly the case if all the $|s_k|$ are majorized by an integrable function. Then the series is a Fourier series.

[By (4·6), a subsequence $S_{n_k}(x)$ converges to an absolutely continuous function $F(x)=\int_a^x f\,dt$. We show that for any bounded g

$$\int_a^b g s_{n_k}\,dt\to\int_a^b gf\,dt.$$

This is obvious for any step function g, and any bounded g can be uniformly approximated by such step functions except in sets of arbitrarily small measure. For $g=\bar{\phi}_m$ we get $a_m=\int_a^b f\bar{\phi}_m\,dt$.]

3. The integrals $\displaystyle\int_0^\pi\frac{f(x\pm t)-f(x)}{t}\,dt,\quad \int_0^\pi\frac{f(x+t)+f(x-t)-2f(x)}{t}\,dt$

can diverge for all x, even if f is continuous (§ 3). Show that if $f\in\mathrm{L}$, and if one of the integrals exists for $x\in E$, the other two exist almost everywhere in E.

[The integral $\displaystyle\int_0^\pi\frac{f(x+t)-f(x-t)}{t}\,dt$ exists almost everywhere.]

4. If $f \sim \frac{1}{2}a_0 + \sum_1^\infty (a_n \cos nx + b_n \sin nx)$, then for almost all x we have the formula

$$\sum_{n=1}^{\infty} \frac{a_n \cos nx + b_n \sin nx}{n} = \frac{1}{\pi}\int_{-\pi}^{\pi} f(x+t) \log \frac{1}{|2 \sin \frac{1}{2}t|}\, dt,$$

the series on the left being convergent almost everywhere.

5. Let $f(x)$ be continuous and periodic. (*a*) A necessary and sufficient condition for $\tilde{f}(x)$ to be continuous is that

$$\tilde{f}(x; h) = -\frac{1}{\pi}\int_h^{\pi} \psi_x(t) \cot \tfrac{1}{2}t\, dt$$

converges uniformly as $h \to +0$. (*b*) If $f \in L$, then a necessary and sufficient condition for $\tilde{f} \in L$ is that $\tilde{f}(x; h)$ tends to a limit in L. (For (*a*) see Zamansky [3].)

[For (*a*) use uniformity in the relation (3·21) of Chapter III.]

6. Let $P(r, x)$ be the Poisson kernel and $q \geq 1$ be fixed. Then

$$\left\{\int_0^{2\pi} P^q(r, x)\, dx\right\}^{1/q} \sim (1-r)^{-1/q'},$$

which shows that, unlike (6·32), we cannot replace 'O' by 'o' in the conclusion of (6·34).

7. If the integral modulus of continuity $\omega_1(\delta; f)$ is $o\{(\log 1/\delta)^{-1}\}$, then $\mathfrak{M}[f - S_n[f]] \to 0$. This is an integral analogue of the Dini-Lipschitz test (Chapter II, § 10).

8. A necessary and sufficient condition for a periodic f to belong to the class Λ_1^1 is that f should coincide almost everywhere with a function of bounded variation. (Hardy and Littlewood [9_1].)

[*Necessity*: If $\sigma_n = \sigma_n(x; f)$ then $\mathfrak{M}[\sigma_n(x+h) - \sigma_n(x)] \leq \mathfrak{M}[f(x+h) - f(x)] \leq Ch$, $\mathfrak{M}[\sigma_n'] \leq C$ and we apply (4·7). *Sufficiency*: If f is non-decreasing over $0 < x < 2\pi$, $\mathfrak{M}[f(x+h) - f(x)]$ is

$$\int_0^{2\pi-h} [f(x+h) - f(x)]\, dx + \int_{2\pi-h}^{2\pi} O(1) dx = \left(\int_{2\pi-h}^{2\pi} - \int_0^h\right) f dx + O(h) = O(h).]$$

9. A necessary and sufficient condition for a periodic f to be in Λ_1^p, $p > 1$, is that f should be equivalent to the integral of a function of L^p. (Hardy and Littlewood [9_1].)

[*Necessity*: Arguing as in Example 8 we get $\mathfrak{M}_p[\sigma_n'] \leq C$ and apply (5·7). *Sufficiency*:

$$\mathfrak{M}_p^p[f(x+h) - f(x)] \leq \int_0^{2\pi}\left\{\int_x^{x+h} |f'(t)|^p dt\right\}^p dx \leq h^p \int_0^{2\pi} |f'(t)|^p dt.]$$

10. A trigonometric series with (C, 1) means σ_n is a Fourier series if and only if there is a function $\Phi(u)$, $u \geq 0$, non-negative, non-decreasing, satisfying the condition $\Phi(u)/u \to \infty$ with u, and such that $\int_0^{2\pi} \Phi(|\sigma_n|)\, dx = O(1)$.

[If f is in L, so is $\Phi(|f|)$ with a suitable Φ satisfying the above conditions. For any such Φ there is also a *convex* $\Phi_1 \leq \Phi$ satisfying the conditions.]

11. Let $\Phi(u)$, $u \geq 0$, be convex and non-negative, and suppose that $\Phi(u)/u \to \infty$ as $u \to \infty$. A necessary and sufficient condition for a trigonometric series to be an $S[dF]$, with F having an absolutely continuous positive variation $P(x)$ such that $P'(x) \in L_\Phi$, is that $\int_0^{2\pi} \Phi(f^+(r e^{ix}))\, dx = O(1)$ as $r \to 1$ (see (6·13)). Similarly for σ_n. The case $\Phi(u) = u^p$ is the most interesting one.

12. Given a periodic $f \in L^p$, $1 \leq p < \infty$, consider the set $U = U_f$ of all functions ϕ which are finite linear combinations, with constant coefficients, of functions $f(x+\lambda)$, where $-\infty < \lambda < +\infty$. Show that a necessary and sufficient condition for U to be dense in L^p is that no Fourier coefficient of f should vanish. (This is an elementary analogue of a deeper theorem of Wiener [3], concerning Fourier transforms.)

[*Necessity*: If e^{ikx} is absent in $S[f]$ it is absent in $S[\phi]$, and so

$$\mathfrak{A}_p[e^{ikx} - \phi] \geq \mathfrak{A}_1[e^{ikx} - \phi] = \mathfrak{A}_1[1 - \phi e^{-ikx}] \geq \left|\frac{1}{2\pi}\int_0^{2\pi} (1 - \phi e^{-ikx})\, dx\right| = 1$$

for all ϕ. *Sufficiency*: It is enough to show that if all terms in $S[f]$ are non-zero, then for any integer k and any $\epsilon>0$ there is a ϕ such that $\mathfrak{M}_p[\phi-e^{ikx}]<\epsilon$. We may suppose that $k=0$ and that the constant term of $S[f]$ is 1. Let $f_1=\sigma_{n-1}[f]$ with n so large that $\mathfrak{M}_p[f-f_1]<\epsilon$. By Chapter II, (1·1), $n^{-1}\sum_{j=1}^{n} f_1(x+2\pi j/n)=1$, so that

$$\mathfrak{M}_p[n^{-1}\sum_j f(x+2\pi j/n)-1]<\epsilon.$$

The argument (suggested by R. Salem) holds for $p=\infty$ if L^p is replaced by C.]

13. Let $g=Tf$ be a linear operation, defined for real-valued f, with g real-valued and such that

$$\text{(i)} \quad \| g \|_r \leqslant M \| f \|_r.$$

For every complex-valued $f=f_1+if_2 \in L^r$ let us set $T[f]=T[f_1]+iT[f_2]$. Then (i) still holds.

[Let $g_\nu=Tf_\nu$. Integrating the inequality

$$\| g_1 \cos\alpha+g_2\sin\alpha \|_r^r \leqslant M^r \| f_1\cos\alpha+f_2\sin\alpha \|_r^r$$

over $0\leqslant\alpha\leqslant 2\pi$, interchanging the order of integration and observing that

$$\int_0^{2\pi} | a\cos\alpha+b\sin\alpha |^r d\alpha=(a^2+b^2)^{\frac{1}{2}r} C_r,$$

we get $$\| (g_1^2+g_2^2)^{\frac{1}{2}} \|_r \leqslant M \| (f_1^2+f_2^2)^{\frac{1}{2}} \|_r.]$$

14. Let $1\leqslant p\leqslant\infty$, $n=0,1,\ldots$. Let L_n^p (the generalized Lebesgue constant) be defined as $\sup_f \mathfrak{M}_p[S_n]$ for all f with $\mathfrak{M}_p[f]\leqslant 1$, where $S_n=S_n[f]$. Show that $L_n^p=L_n^{p'}$.

15. Let L^+ denote the class of $S[f]$ with f integrable and non-negative. We define B^+, C^+ correspondingly, and we denote by S^+ the class of $S[dF]$ with F non-decreasing. Then $\{\lambda_n\}$ is of any one of the types (B^+,B^+), (C^+,C^+), (L^+,L^+), (S^+,S^+) if and only if $\Sigma\lambda_n e^{inx}\in S^+$; and $\{\lambda_n\}$ is of type (B^+,C^+) or of type (S^+,L^+) if and only if $\Sigma\lambda_n e^{inx}\in L^+$.

16. Let $\sigma_n(x)$ and $f(r,x)$ be the (C, 1) and Abel means of a trigonometric series T. Each of the conditions

$$\| \sigma_n \|_\Phi=O(1), \quad \| f(r,x) \|_\Phi=O(1)$$

(the norms being understood in the sense of § 10) is both necessary and sufficient for T to belong to L_Φ^*.

17. If $f\in L_\Phi^*$ and if Φ satisfies the condition

$$\text{(i)} \quad \Phi(2u)/\Phi(u)=O(1) \quad \text{as} \quad u\to\infty,$$

then $$\| f-\sigma_n \|_\Phi \to 0, \quad \| \sigma_n \|_\Phi \to \| f \|_\Phi.$$

Similarly for the Abel mean $f(r,x)$.

18. Condition (i) of Example 17 is both necessary and sufficient for the space L_Φ^* to be separable. (Orlicz [2].) [*Sufficiency* follows from Example 17.]

19. Given a matrix $\{a_{mn}\}$, $m,n=0,1,\ldots$, consider the linear means

$$\sigma_m=a_{m0}s_0+a_{m1}s_1+\ldots+a_{mn}s_n+\ldots$$

of any sequence $\{s_n\}$. Show that a necessary and sufficient condition that any convergent $\{s_n\}$ be transformed into a convergent $\{\sigma_m\}$ (not necessarily with the same limit) is that

(i) $\lim_m a_{mn}$ exist for each n;

(ii) $\sum_n |a_{mn}|\leqslant C$, with C independent of m;

(iii) $A_m=\sum_n a_{mn}$ tend to a limit as $m\to\infty$.

If these conditions are satisfied and if the limits in (i) and (iii) are α_n and A respectively, then $\Sigma|\alpha_n|\leqslant C$ and for any $\{s_n\}\to s$ we have

$$\sigma_m\to As+\alpha_0(s_0-s)+\alpha_1(s_1-s)+\ldots. \quad \text{(Schur [1].)}$$

[The proof is similar to the arguments on pp. 74–5 and 168.]

CHAPTER V

SPECIAL TRIGONOMETRIC SERIES

In this chapter we study some special trigonometric series which not only are interesting in themselves but also illustrate the general theory.

1. Series with coefficients tending monotonically to zero

In Chapter I, § 2, we proved that if $\{a_\nu\}$ is monotonically decreasing to zero, both the series

$$\tfrac{1}{2}a_0 + \sum_{\nu=1}^{\infty} a_\nu \cos \nu x, \tag{1·1}$$

$$\sum_{\nu=1}^{\infty} a_\nu \sin \nu x, \tag{1·2}$$

converge uniformly outside an arbitrarily small neighbourhood of $x=0$. If $a_\nu \geqslant 0$ for all ν then obviously a necessary and sufficient condition for the uniform convergence everywhere of (1·1) is the convergence of Σa_ν. For (1·2), the situation is less obvious.

(1·3) THEOREM. *Suppose that $a_\nu \geqslant a_{\nu+1}$ and $a_\nu \to 0$. Then a necessary and sufficient condition for the uniform convergence of* (1·2) *is $\nu a_\nu \to 0$.*

If (1·2) converges uniformly and if $x = \pi/2n$, then

$$\sum_{[\frac{1}{2}n]+1}^{n} a_\nu \sin \nu x \geqslant \sin \tfrac{1}{4}\pi \,.\, a_n \sum_{[\frac{1}{2}n]+1}^{n} 1 \geqslant \sin \tfrac{1}{4}\pi \,.\, a_n \tfrac{1}{2}n,$$

so that $na_n \to 0$ as $n \to \infty$. This proves the necessity.

Conversely, let $\nu a_\nu \to 0$, so that $\epsilon_k = \sup\limits_{\nu \geqslant k} \nu a_\nu \to 0$. Let $0 < x \leqslant \pi$, and let $N = N_x$ be the integer satisfying

$$\pi/(N+1) < x \leqslant \pi/N.$$

We split the remainder $R_m(x) = a_m \sin mx + \ldots$ of (1·2) into two parts, $R_m = R'_m + R''_m$, where R'_m consists of the terms with indices $\nu < m+N$, and R''_m of those with $\nu \geqslant m+N$. Then

$$|R'_m(x)| = \left| \sum_{m}^{m+N-1} a_\nu \sin \nu x \right| \leqslant x \sum_{m}^{m+N-1} \nu a_\nu \leqslant xN\epsilon_m \leqslant \pi\epsilon_m. \tag{1·4}$$

Summing by parts and using the inequality $|\tilde{D}_m(x)| \leqslant \pi/x$ (Chapter II, § 5), we find

$$|R''_m| = \left| \sum_{m+N}^{\infty} (a_\nu - a_{\nu+1}) \tilde{D}_\nu(x) - a_{m+N} \tilde{D}_{m+N-1}(x) \right| \leqslant 2a_{m+N}\pi/x \leqslant 2(N+1)\, a_{m+N} \leqslant 2\epsilon_m.$$

Hence $|R_m| < 6\epsilon_m$, and the uniform convergence of (1·2) follows.

Remarks. (*a*) The above proof of sufficiency could be simplified if we knew that $\{\nu a_\nu\}$ decreased monotonically to 0. For then we could write (1·2) in the form $\Sigma \nu a_\nu (\nu^{-1} \sin \nu x)$; and since the partial sums of the series $\Sigma \nu^{-1} \sin \nu x$ are uniformly bounded it would be enough to apply Theorem (2·4) of Chapter I.

(*b*) If $a_\nu \geqslant a_{\nu+1} \to 0$, the condition $\nu a_\nu = O(1)$ is both necessary and sufficient for the uniform boundedness of the partial sums of (1·2); the proof is an obvious adaptation of that just given. As the example $\Sigma \nu^{-1} \sin \nu x$ shows, this condition does not imply uniform convergence.

(*c*) *Under the hypotheses $a_\nu \geqslant a_{\nu+1}$, $a_\nu \to 0$, the condition $\nu a_\nu \to 0$ is both necessary and sufficient for $\Sigma a_\nu \sin \nu x$ to be the Fourier series of a continuous function.* It is enough to prove that the condition is necessary. Suppose then that the (C, 1) means $\sigma_n(x)$ of $\Sigma a_\nu \sin \nu x$ converge uniformly. In particular $\sigma_n(\pi/2n) \to 0$. Since $\sin u \geqslant (2/\pi)u$ in $(0, \frac{1}{2}\pi)$, we have

$$\sum_{\nu=1}^{n} a_\nu \left(1 - \frac{\nu}{n+1}\right) \frac{2}{\pi} \left(\frac{\pi\nu}{2n}\right) \to 0.$$

Keeping $m = [\frac{1}{2}n]$ terms on the left, we obtain successively

$$\frac{1}{n} \sum_{\nu=1}^{m} \nu a_\nu \to 0, \quad \frac{1}{n} a_m \sum_{\nu=1}^{m} \nu \to 0, \quad m a_m \to 0.$$

There is a corresponding modification of (*b*).

(1·5) THEOREM. *If $a_\nu \to 0$ and the sequence $a_0, a_1, \ldots$ is convex, the series* (1·1) *converges, save possibly at $x = 0$, to a non-negative and integrable sum $f(x)$, and is the Fourier series of f.*

Summing twice by parts we have

$$s_n(x) = \sum_{\nu=0}^{n-2} (\nu+1) \Delta^2 a_\nu K_\nu(x) + n K_{n-1}(x) \Delta a_{n-1} + D_n(x) a_n, \tag{1·6}$$

where s_n is the partial sum of (1·1) and D_ν and K_ν are Dirichlet's and Fejér's kernels. If $x \neq 0$, the last two terms tend to 0 as $n \to \infty$. Thus $s_n(x)$ tends to the limit

$$f(x) = \sum_{\nu=0}^{\infty} (\nu+1) \Delta^2 a_\nu K_\nu(x), \tag{1·7}$$

which is non-negative, $\{a_\nu\}$ being convex. Also

$$\int_{-\pi}^{\pi} f(x)\, dx = \sum_{\nu=0}^{\infty} (\nu+1) \Delta^2 a_\nu \int_{-\pi}^{\pi} K_\nu(x)\, dx = \pi \sum_{\nu=0}^{\infty} (\nu+1) \Delta^2 a_\nu < +\infty$$

(see Chapter III, § 4) so that f is integrable.

In proving that (1·1) is $S[f]$, we may suppose that $a_0 = 0$. Since $a_1, a_2, \ldots$ monotonically decreases to 0 an application of (1·3) (see also Remark (*a*) above) shows that the series $\Sigma \nu^{-1} a_\nu \sin \nu x$ obtained by termwise integration of (1·1) converges uniformly to a continuous function $F(x)$, and so is $S[F]$. Since $F'(x)$ exists and is continuous for $x \neq 0$, and since $F(x)$ is continuous everywhere, F is a primitive of $F' = f$. If we first integrate over the interval (ϵ, π) and then make $\epsilon \to 0$ we get, since $F(0) = F(\pi) = 0$,

$$\frac{a_\nu}{\nu} = \frac{2}{\pi} \int_0^{\pi} F(x) \sin \nu x\, dx = \frac{2}{\pi\nu} \int_0^{\pi} f(x) \cos \nu x\, dx,$$

$$a_\nu = \frac{2}{\pi} \int_0^{\pi} f(x) \cos \nu x\, dx \quad (\nu > 0).$$

This is also true for $\nu = 0$, since, F being periodic, $\int_{-\pi}^{\pi} f dt = 0 = \pi a_0$. So (1·1) is $S[f]$.

In this proof that (1·1) is a Fourier series we actually used only the hypothesis that $\{a_\nu\}$ decreases monotonically to 0. Hence we have:

(1·8) Theorem. *If $a_\nu \to 0$, $\Delta a_\nu \geqslant 0$, the sum $f(x)$ of* (1·1) *is continuous for $x \neq 0$, has a Riemann integral (in general improper) and is the Fourier-Riemann series of f.*

If $\{a_\nu\}$ is also convex, then, as we have just seen, f is non-negative and F is a Lebesgue integral of f. The mere fact that $\{a_\nu\}$ is monotone, however, does not ensure the L-integrability of f; we have, in fact,

(1·9) Theorem. *There is a series* (1·1) *with coefficients monotonically decreasing to* 0 *and sum $f(x)$ not integrable* L.

Suppose we have a sequence $0 = \lambda_1 < \lambda_2 < \dots$ such that a_k is constant for $\lambda_n < k \leqslant \lambda_{n+1}$, $n = 1, 2, \dots$. Summation by parts gives

$$f(x) = \sum_{\nu=0}^{\infty} \Delta a_\nu . D_\nu(x) = \sum_{n=1}^{\infty} \alpha_n D_{\lambda_n}(x), \quad \alpha_n = \Delta a_{\lambda_n}. \tag{1·10}$$

We now observe that

$$\int_0^\pi | D_n | \, dx \leqslant C \log n, \quad \int_{1/n}^\pi | D_n | \, dx \geqslant C_1 \log n.$$

These both follow from Chapter II, (12·1); for since $D_n(x) = O(n)$, the difference between the two integrals is $O(1)$. From (1·10), observing that $| D_n(x)| < 2/x$ for $0 < x \leqslant \pi$, we get

$$\int_{1/\lambda_m}^\pi |f| \, dx \geqslant C_1 \alpha_m \log \lambda_m - C \sum_{n=1}^{m-1} \alpha_n \log \lambda_n - 2 \log(\pi \lambda_m) \sum_{n=m+1}^{\infty} \alpha_n.$$

Taking
$$\alpha_n = 1/n!, \quad \lambda_n = 2^{(n!)^2}$$

and arguing as on p. 134, we find that the last integral is unbounded as $m \to \infty$.

Given an arbitrary sequence of positive numbers $\epsilon_n \to 0$, we can easily construct, e.g. geometrically, a convex sequence $\{a_n\}$ such that $a_n \geqslant \epsilon_n$ and $a_n \to 0$. Thus *there exist Fourier series with coefficients a_n tending to* 0 *arbitrarily slowly* (see also Ex. 1 on p. 70).

If a_n, b_n are the Fourier coefficients of an integrable function, the series $\Sigma b_n/n$ converges (Chapter II, § 8). The example of the Fourier series

$$\sum_{n=2}^{\infty} \frac{\cos nx}{\log n} \tag{1·11}$$

shows that $\Sigma a_n/n$ may be divergent.

Let s_n and σ_n be the partial sums and the (C, 1) means of (1·1). With the hypotheses of (1·5), $\mathfrak{M}[f - \sigma_n] \to 0$ (cf. Chapter IV, (5·5)).

(1·12) Theorem. *With the hypotheses of* (1·5), $\mathfrak{M}[f - s_n]$ *tends to* 0 *if and only if* $a_n = o\{(\log n)^{-1}\}$.

For, subtracting (1·7) from (1·6), we see that $| f(x) - s_n(x) |$ is contained between

$$a_n | D_n(x) | \pm \left\{ \sum_{\nu=n-1}^{\infty} (\nu+1) \Delta^2 a_\nu K_\nu(x) + \Delta a_{n-1} K_{n-1}(x) \, n \right\}.$$

If we integrate this over $(-\pi, \pi)$ (the terms in the curly brackets are non-negative) we find that

$$\mathfrak{M}[f - s_n] = \pi a_n L_n + o(1),$$

where L_n is the Lebesgue constant (Chapter II, § 12). Since L_n is exactly of order $\log n$, (1·12) follows.

If $a_n \log n \to \infty$, e.g. if $a_n = (\log n)^{-\frac{1}{2}}$ for $n > 1$, then $\mathfrak{M}[f - s_n] \to \infty$, and so also $\mathfrak{M}[s_n] \to \infty$. The series (1·11), which is important for some problems, is a limiting case, since here $\mathfrak{M}[f - s_n]$ is bounded and even tends to a limit, but the limit is not zero. (Also $\mathfrak{M}[s_n]$ tends to a finite limit.)

We now pass to the series $\Sigma a_\nu \sin \nu x$ with $a_1 \geqslant a_2 \geqslant \ldots \to 0$. Summation by parts shows that for its partial sum $t_n(x)$ we have

$$t_n(x) = \sum_{\nu=1}^{n-1} \tilde{D}_\nu(x)\, \Delta a_\nu + a_n \tilde{D}_n(x) \to \sum_{\nu=1}^{\infty} \tilde{D}_\nu(x)\, \Delta a_\nu = g(x), \tag{1·13}$$

say, as $n \to \infty$. If we substitute $\tilde{D}_\nu^*$ for $\tilde{D}_\nu$, we get a function $g^*(x)$, $0 \leqslant x \leqslant \pi$, differing from $g(x)$ by the continuous function $\frac{1}{2}\Sigma \Delta a_\nu \sin \nu x$. The series defining g^* has non-negative terms, and since the integral of $\tilde{D}_n^*(x)$ over $(0, \pi)$ is exactly of order $\log n$ (Chapter II, (12·3)), we conclude that g^*, and so also g, is integrable over $(0, \pi)$ if and only if $\Sigma \Delta a_\nu \log \nu$ converges. If we assume this convergence, and observe that

$$a_n \log n = \log n \sum_{\nu=n}^{\infty} \Delta a_\nu \leqslant \sum_{\nu=n}^{\infty} \Delta a_\nu \log \nu \to 0,$$

we see that $\mathfrak{M}[g - t_n] \to 0$. In particular, $\Sigma a_\nu \sin \nu x$ is $S[g]$. Thus:

(1·14) Theorem. *Suppose that $a_1 \geqslant a_2 \geqslant \ldots \to 0$. The sum $g(x)$ of $\Sigma a_\nu \sin \nu x$ is then integrable if and only if $\Sigma \Delta a_\nu \log \nu < \infty$. If this condition is satisfied, then $\Sigma a_\nu \sin \nu x$ is $S[g]$, and $\mathfrak{M}[g - t_n] \to 0$.*

Under the hypotheses of (1·14), g^* is non-negative in $(0, \pi)$, so that g is there bounded below. *If the sequence $a_1, a_2, \ldots$ is also convex, then $g(x)$ is positive in $(0, \pi)$, unless* $a_1 = a_2 = \ldots = 0$. To prove this we apply summation by parts to the series $\Sigma \tilde{D}_\nu \Delta a_\nu$, and use the fact that $\tilde{K}_\nu \geqslant 0$ in $(0, \pi)$ (Chapter III, (3·18)).

(1·15) Theorem. *If we assume only that $a_\nu \to 0$ and $\Delta a_\nu \geqslant 0$, then $\Sigma a_\nu \sin \nu x$ is a generalized Fourier sine series* (Chapter II, § 4).

For a simple calculation shows that then

$$2g(x) \sin x = a_1 + a_2 \cos x + \sum_{\nu=2}^{\infty} (a_{\nu+1} - a_{\nu-1}) \cos \nu x.$$

The series on the right converges uniformly, and so is $S[2g \sin x]$. Writing the Fourier formulae for the coefficients $a_1, a_2, a_3 - a_1, \ldots$, we get by addition the formula

$$a_\nu = \frac{2}{\pi} \int_0^\pi g(x) \sin \nu x\, dx \quad (\nu = 1, 2, \ldots).$$

The integrand here is continuous, since $g \sin x$ is continuous.

Let h_n be the partial sums of the harmonic series $1 + \frac{1}{2} + \frac{1}{3} + \ldots$, so that $h_n \simeq \log n$. Let $a_n \to 0$, $\Delta a_n \geqslant 0$. From the formula

$$\sum_{\nu=1}^{n} \frac{a_\nu}{\nu} = \sum_{\nu=1}^{n-1} h_\nu \Delta a_\nu + a_n h_n \tag{1·16}$$

it follows that if $\Sigma \nu^{-1} a_\nu$ is finite, so is $\Sigma \Delta a_\nu \log \nu$. Conversely, the finiteness of the latter series implies that $a_n \log n \leqslant \sum_{n}^{\infty} \Delta a_\nu \log \nu = o(1)$, so that, by (1·16), $\Sigma \nu^{-1} a_\nu$ is finite. Thus, *if* $a_1 \geqslant a_2 \geqslant \ldots \to 0$, *the conditions* $\Sigma \nu^{-1} a_\nu < \infty$ *and* $\Sigma \Delta a_\nu \log \nu < \infty$ *are equivalent.* Hence in (1·14) we can replace the convergence of $\Sigma \Delta a_n \log n$ by that of $\Sigma n^{-1} a_n$.

That the latter series is convergent if (1·2) is a Fourier series, no matter what the a_n, we know already. This implies in particular that the series

$$\sum_{n=2}^{\infty} \frac{\sin nx}{\log n}, \tag{1·17}$$

conjugate to the Fourier series (1·11), is not a Fourier series.

2. The order of magnitude of functions represented by series with monotone coefficients

We begin by giving one more proof of the formulae

$$\left.\begin{aligned} \sum_{n=1}^{\infty} n^{-\beta} \cos nx \simeq x^{\beta-1} \Gamma(1-\beta) \sin \tfrac{1}{2}\pi\beta \\ \sum_{n=1}^{\infty} n^{-\beta} \sin nx \simeq x^{\beta-1} \Gamma(1-\beta) \cos \tfrac{1}{2}\pi\beta \end{aligned}\right\} \quad (x \to +0,\ 0 < \beta < 1), \tag{2·1}$$

established in Chapter II, § 13.

By Chapter III, (1·9),

$$\sum_{n=0}^{\infty} A_n^{-\beta} r^n e^{inx} = (1 - r e^{ix})^{\beta - 1} \quad (0 \leqslant r < 1). \tag{2·2}$$

Since the $A_n^{-\beta}$ are positive and decreasing to 0, $\Sigma A_n^{-\beta} e^{inx}$ converges for $x \neq 0$; and making $r \to 1$ we deduce from (2·2) that

$$\sum_{n=0}^{\infty} A_n^{-\beta} e^{inx} = (2 \sin \tfrac{1}{2}x)^{\beta-1} \exp\{\tfrac{1}{2} i (\pi - x)(1-\beta)\} \quad (0 < x < 2\pi). \tag{2·3}$$

Using the formula (1·18) of Chapter III (with $\alpha = -\beta$), and the relation $2 \sin \frac{1}{2}x \simeq x$, we deduce (2·1) from (2·3) by separating the real and the imaginary parts there.

Remark. The argument shows that *the second formula* (2·1) *holds for* $0 < \beta < 2$.

We shall use the formulae (2·1) to obtain some more general ones. In dealing with series (1·1) it is often convenient to assume that $a_n = a(n)$, where $a(u)$ is a function defined for all real $u \geqslant 0$. Usually, indeed, a_n is given as $a(n)$.

A positive $b(u)$ defined for $u > u_0$ will be called a *slowly varying* function if, for any $\delta > 0$, $b(u)\, u^{\delta}$ is an increasing, and $b(u)\, u^{-\delta}$ a decreasing, function of u for u large enough. *If $b(u)$ is slowly varying, then*

$$b(ku) \simeq b(u) \quad as \quad u \to \infty, \tag{2·4}$$

for every fixed $k > 0$, and even uniformly in every interval $\eta \leqslant k \leqslant 1/\eta$, $0 < \eta < 1$.

For if, for example, $1 \leqslant k \leqslant 1/\eta$, then

$$b(ku)/(ku)^{\delta} \leqslant b(u)/u^{\delta}, \quad b(ku) \leqslant k^{\delta} b(u) \leqslant \eta^{-\delta}\, b(u)$$

for large u. Similarly, $b(ku) \geqslant \eta^{\delta}\, b(u)$. Making δ arbitrarily small, we prove (2·4) for $1 \leqslant k \leqslant 1/\eta$. The case $\eta \leqslant k \leqslant 1$ is proved similarly.

If $\beta_1, \beta_2, \ldots$ are real, each of the functions

$$\log^{\beta_1} x, \quad \log \log^{\beta_2} x, \quad \ldots, \tag{2·5}$$

and so also any product of any finite number of them, is slowly varying.

(2·6) THEOREM. *Let $a_n = n^{-\beta} b_n$, where $0<\beta<1$ and $b(u)$, with $b(n)=b_n$, is a slowly varying function. Let*

$$f_\beta(x)=\sum_{n=1}^{\infty} n^{-\beta} b_n \cos nx, \quad g_\beta(x)=\sum_{n=1}^{\infty} n^{-\beta} b_n \sin nx. \tag{2·7}$$

Then, for $x\to +0$,

$$f_\beta(x)\simeq x^{\beta-1} b(x^{-1})\,\Gamma(1-\beta)\sin\tfrac{1}{2}\pi\beta, \quad g_\beta(x)\simeq x^{\beta-1} b(x^{-1})\,\Gamma(1-\beta)\cos\tfrac{1}{2}\pi\beta. \tag{2·8}$$

It is enough to prove the formula for g_β, the proof for f_β being the same. Set

$$B=\Gamma(1-\beta)\cos\tfrac{1}{2}\pi\beta.$$

We note that, for *all* $M, N>0$,

$$\sum_{n<M} n^{-\beta} < C_\beta M^{1-\beta}, \quad \Big|\sum_{n>N} n^{-\beta}\sin nx\Big| < CN^{-\beta}x^{-1} \quad (0<x\leqslant\pi), \tag{2·9}$$

where C is an absolute constant and C_β depends on β only. The first inequality follows from

$$\sum_{n<M} n^{-\beta} < \int_0^M t^{-\beta}dt = \frac{M^{1-\beta}}{1-\beta}$$

and the second from Chapter I, (2·2).

Let $0<\omega<\Omega<+\infty$. Then

$$\sum n^{-\beta}\sin nx = \sum_{n<\omega/x} + \sum_{\omega/x\leqslant n\leqslant\Omega/x} + \sum_{n>\Omega/x} = S_1+S_2+S_3, \tag{2·10}$$

with a corresponding decomposition $T_1+T_2+T_3$ of the series $\Sigma b_n n^{-\beta}\sin nx$. By (2·9),

$$|S_1| < C_\beta \omega^{1-\beta}x^{\beta-1}, \quad |S_3| < C\Omega^{-\beta}x^{\beta-1}. \tag{2·11}$$

We fix an $\epsilon>0$. By virtue of (2·1) we have

$$(B-\epsilon)\,x^{\beta-1}\leqslant S_2\leqslant (B+\epsilon)\,x^{\beta-1}, \tag{2·12}$$

provided ω and $1/\Omega$ are small enough (but fixed) and x is near enough to 0.

We fix $\delta>0$ such that $\delta<\beta$ and $\beta+\delta<1$. Since $u^\delta b(u)$ tends monotonically to $+\infty$ for large u, we have $u^\delta b(u)\geqslant v^\delta b(v)$ for u large and *all* $v\leqslant u$. Hence, if x is small,

$$\begin{aligned}|T_1| &\leqslant \sum_{n<\omega/x} b_n n^\delta n^{-\beta-\delta} \leqslant b(\omega/x)\,(\omega/x)^\delta \sum_{n<\omega/x} n^{-\beta-\delta}\\ &\leqslant b(\omega/x)\,(\omega/x)^\delta\, C_{\beta+\delta}(\omega/x)^{1-\beta-\delta} = C_{\beta+\delta}\,\omega^{1-\beta}x^{\beta-1}b(\omega/x).\end{aligned}$$

Since $b(\omega/x)\simeq b(1/x)$, the last product is less than $\epsilon x^{\beta-1}b(1/x)$, provided ω is small enough and x is near 0.

By (2·11) with $\beta-\delta$ for β and Chapter I, (2·1),

$$|T_3| = \Big|\sum_{n>\Omega/x} b_n n^{-\delta} n^{-\beta+\delta}\sin nx\Big| \leqslant 2b(\Omega/x)\,(\Omega/x)^{-\delta}\,C(\Omega/x)^{-\beta+\delta}\,x^{-1} = 2C\Omega^{-\beta}x^{\beta-1}b(\Omega/x),$$

and so

$$|T_3| < \epsilon b(1/x)\,x^{\beta-1}$$

for Ω large enough and x near 0.

On the other hand,

$$T_2 = b(1/x)\sum_{\omega/x}^{\Omega/x} n^{-\beta}\sin nx + \sum_{\omega/x}^{\Omega/x}\{b(n)-b(1/x)\}\,n^{-\beta}\sin nx = T_2'+T_2''.$$

Here $T_2' = b(1/x)\,S_2$ and so, by (2·12), is contained between $(B\pm\epsilon)\,x^{\beta-1}b(1/x)$. By (2·4), $|T_2''|$ does not exceed

$$\max_{\omega/x\leqslant n\leqslant\Omega/x} |b(n)-b(1/x)|\sum_{\omega/x}^{\Omega/x} n^{-\beta} = o\{b(1/x)\}\,O(\Omega/x)^{1-\beta} = o\{x^{\beta-1}b(1/x)\}.$$

Collecting results we see that $g_\beta = T_1 + T_3 + T_2' + T_2''$ is contained between

$$(B \pm 3\epsilon)\, x^{\beta-1} b(1/x)$$

for x small, and so
$$g_\beta(x) \simeq B x^{\beta-1} b(1/x) \quad \text{as} \quad x \to +0,$$
which completes the proof of (2·6).

Remark. The second formula (2·8) *holds for* $0 < \beta < 2$. *In particular*

$$g_1(x) = \Sigma n^{-1} b_n \sin nx \simeq \tfrac{1}{2}\pi b(1/x) \quad (x \to +0). \tag{2·13}$$

It is enough to indicate modifications of the preceding proof. We easily verify that the second inequality (2·9) holds for any $\beta > 0$; hence the inequality (2·11) for S_3 and the estimate for T_3 hold for $\beta > 0$. On the other hand, if $\beta < 2$ we have

$$|S_1| \leqslant \sum_{n<\omega/x} n^{-\beta}.nx = x \sum_{n<\omega/x} n^{1-\beta} \leqslant C_\beta \omega^{2-\beta} x^{\beta-1},$$

and instead of the previous inequality for T_1 we have $|T_1| \leqslant C_{\beta+\delta}\, \omega^{2-\beta} x^{\beta-1} b(\omega/x)$. Using the fact that the second equation (2·1) holds for $0 < \beta < 2$, we see that the estimates for S_2 and T_2 remain unchanged, and the proof concludes as before.

If $b(u)$ is slowly varying, then $b(u)/u$ is ultimately decreasing and so the series $\Sigma n^{-1} b_n$ and the integral $\int_1^\infty u^{-1} b(u)\, du$ are simultaneously finite or infinite. Write

$$\left.\begin{aligned} B(u) &= \int_1^u t^{-1} b(t)\, dt, \quad & R(u) &= \int_u^\infty t^{-1} b(t)\, dt, \\ B^*(u) &= \sum_{n \leqslant u} n^{-1} b(n), \quad & R^*(u) &= \sum_{n \geqslant u} n^{-1} b_n. \end{aligned}\right\} \tag{2·14}$$

Then, as $u \to \infty$,

$$\text{(i)} \quad b(u) = o\{B(u)\}, \quad B(u) \simeq B^*(u), \quad \text{if} \quad B(u) \neq O(1);$$

$$\text{(ii)} \quad b(u) = o\{R(u)\}, \quad R(u) \simeq R^*(u), \quad \text{if} \quad B(u) = O(1).$$

For let $k > 1$. For large u

$$B(u) > \int_{u/k}^u t^{-1} b(t)\, dt \simeq b(u) \int_{u/k}^u t^{-1}\, dt = b(u) \log k.$$

Taking k large we obtain $b(u) = o\{B(u)\}$, whether $B(u)$ is bounded or not. Similarly, $b(u) = o\{R(u)\}$. Since $B(u) - B^*(u)$ tends to a finite limit, we have $B(u) \simeq B^*(u)$ if either side is unbounded. Let now $u \to \infty$ and let N be an integer satisfying $N < u \leqslant N+1$. Then

$$R(N+1) \leqslant R^*(N+1) = R^*(u) < R(N), \quad R(N+1) \leqslant R(u) < R(N).$$

Since $R(N+1) \simeq R(N)$ ($b(u)$ being a slowly varying function), the second formula (ii) follows.

(2·15) THEOREM. *According as* $\Sigma b_n/n$ *diverges or converges,*

$$\left.\begin{aligned} f_1(x) &= \Sigma n^{-1} b_n \cos nx \simeq B(1/x), \\ \textit{or} \qquad f_1(0) - f_1(x) &\simeq R(1/x), \end{aligned}\right\} \quad (x \to +0). \tag{2·16}$$

Considering the first formula (2·16), we write

$$B^*(1/x) - f_1(x) = \sum_{n \leqslant 1/x} b_n n^{-1}(1 - \cos nx) - \sum_{n > 1/x} b_n n^{-1} \cos nx = U_1 + U_2.$$

Then (writing $b_n n^{-1}$ as $b_n n^{-\delta} n^{-1+\delta}$ and arguing as for T_3 in Theorem (2·6), but with $\Omega = 1$) we have

$$U_2 = O\{b(1/x)\} = o\{B(1/x)\} = o\{B^*(1/x)\}.$$

Since $b_n n^{-1}(1-\cos nx) < \frac{1}{2} n b_n x^2$, the familiar argument shows that

$$U_1 = O\{b(1/x)\} = o\{B^*(1/x)\}.$$

Hence

$$f_1(x) \simeq B^*(1/x) \simeq B(1/x).$$

If $f_1(0) = \Sigma b_n/n < \infty$, then $f_1(0) - f_1(x)$ is

$$\sum_{n<1/x} b_n n^{-1}(1-\cos nx) + \sum_{n \geqslant 1/x} b_n n^{-1} - \sum_{n \geqslant 1/x} b_n n^{-1} \cos nx = V_1 + R^*(1/x) + V_2.$$

Here again both V_1 and V_2 are $O\{b(1/x)\} = o\{R^*(1/x)\}$, so that

$$f_1(0) - f_1(x) \simeq R^*(1/x) \simeq R(1/x).$$

We pass now to the limiting case $\beta = 0$ of (2·6).

(2·17) THEOREM. *If $b(u)$ decreases to 0 and if $-ub'(u)$ is slowly varying, then, for* $x \to +0$,

$$\left.\begin{aligned} f_0(x) &\simeq -\tfrac{1}{2}\pi x^{-2} b'(1/x), \\ g_0(x) &\simeq x^{-1} b(1/x). \end{aligned}\right\} \tag{2·18}$$

The hypotheses here are satisfied if, for instance, $b(u)$ is for large u one of the functions (2·5), and so also if it is a product of a finite number of them. In particular,

$$\left.\begin{aligned} \sum_{n=2}^{\infty} \frac{\cos nx}{\log n} &\simeq \tfrac{1}{2}\pi x^{-1} \log^{-2}(1/x), \\ \sum_{n=2}^{\infty} \frac{\sin nx}{\log n} &\simeq x^{-1} \log^{-1}(1/x). \end{aligned}\right\} \quad (x \to +0). \tag{2·19}$$

Since $x^{-1}\log^{-1}(1/x)$ is not integrable, the second formula shows that $\Sigma(\sin nx)/\log n$ is not a Fourier series, a fact we already know (see p. 186).

We begin with f_0, and set

$$c(u) = u[b(u) - b(u+1)], \quad c_n = c(n) = n\Delta b_n.$$

We have $c(u) = -ub'(u+\theta)$, $0 < \theta < 1$, and so $c(u) \simeq -ub'(u)$, since $-ub'(u)$ is slowly varying. Clearly,

$$\sum_1^{\infty} b_n \cos nx = \sum_1^{\infty} \Delta b_n \{D_n(x) - \tfrac{1}{2}\} = (2\tan \tfrac{1}{2}x)^{-1} \sum_1^{\infty} n^{-1} c_n \sin nx + O(1). \tag{2·20}$$

If the numbers c_n satisfy the conditions for the b_n in the proof of (2·13), that formula and (2·20) give

$$f_0(x) \simeq \tfrac{1}{2}\pi x^{-1} c(x^{-1}) + O(1) \simeq -\tfrac{1}{2}\pi x^{-2} b'(x^{-1}), \tag{2·21}$$

since $x^{-1}c(x^{-1}) \to \infty$.

On analysing the proof of (2·15) we see that it is sufficient that (ultimately)

(i) $c_n/n = \Delta b_n$ should decrease;

(ii) $c_n n^{\delta} = n^{1+\delta}\Delta b_n$ should increase for some $0 < \delta < 1$;

(iii) $c(ku) \simeq c(u)$ uniformly in every interval $\eta \leqslant k \leqslant 1/\eta$.

Since $c(u) \simeq -ub'(u)$, and since $-ub'(u)$ is slowly varying, condition (iii) holds.

Since $-ub'(u)$ is slowly varying, $-b'(u)$ is decreasing. Therefore $b(u)$ is convex, and $c_n/n = \Delta b_n$ is decreasing (condition (i)). Finally,

$$\frac{b(n)-b(n+1)}{b(n+1)-b(n+2)} = \frac{b'(n+\theta)}{b'(n+\theta+1)} \leqslant \left(\frac{n+1+\theta}{n+\theta}\right)^{1+\delta} \quad (0<\theta<1),$$

by Cauchy's mean-value theorem and by the fact that $-u^{1+\delta}b'(u)$ increases. Since the last ratio is less than $\{(n+1)/n\}^{1+\delta}$, we see that $n^{1+\delta}\Delta b_n$ increases if n is replaced by $n+1$, and (ii) follows.

The proof for g_0 is similar. Since

$$g_0(x) = (2\tan\tfrac{1}{2}x)^{-1}\Sigma n^{-1}c_n(1-\cos nx) + O(1),$$

we apply the second formula (2·16) with c_n for b_n, so that the term $R(1/x)$ is

$$\int_{x^{-1}}^{\infty} u^{-1}c(u)\,du = \int_{x^{-1}}^{\infty} \{b(u)-b(u+1)\}\,du$$
$$= \int_{x^{-1}}^{1+x^{-1}} b(u)\,du \simeq b(x^{-1}).$$

Hence $g_0(x) \simeq x^{-1}b(x^{-1})$.

Theorems of this section have analogues in which the roles of the function f and of its coefficients are reversed. In these we assume that $f(x)$, $0<x\leqslant\pi$, is sufficiently 'well-behaved' in every interval (ϵ, π), $\epsilon>0$, but tends to ∞ in a specific way as $x \to +0$, and we inquire about the behaviour of the cosine and sine coefficients a_n, b_n of f. The two results that follow are analogues of (2·1) and of (2·6).

(2·22) THEOREM. *Let $0<\beta<1$. For the coefficients of the function*

$$f(x) = x^{-\beta} \quad (0<x\leqslant\pi)$$

we have

$$\left.\begin{aligned} \tfrac{1}{2}\pi a_n &\simeq n^{\beta-1}\Gamma(1-\beta)\sin\tfrac{1}{2}\pi\beta, \\ \tfrac{1}{2}\pi b_n &\simeq n^{\beta-1}\Gamma(1-\beta)\cos\tfrac{1}{2}\pi\beta. \end{aligned}\right\} \tag{2·23}$$

(2·24) THEOREM. *Let $0<\beta<1$, and let $b(x)$ be a function of bounded variation in every interval (ϵ,π), slowly varying as $x\to+0$. Then for the coefficients of $x^{-\beta}b(x)$, $0<x\leqslant\pi$, we have the relations*

$$\left.\begin{aligned} \tfrac{1}{2}\pi a_n &\simeq n^{\beta-1}b(1/n)\,\Gamma(1-\beta)\sin\tfrac{1}{2}\pi\beta, \\ \tfrac{1}{2}\pi b_n &\simeq n^{\beta-1}b(1/n)\,\Gamma(1-\beta)\cos\tfrac{1}{2}\pi\beta. \end{aligned}\right\} \tag{2·25}$$

For the a_n in (2·22) we have

$$a_n = 2\pi^{-1}\int_0^{\pi} t^{-\beta}\cos nt\,dt = 2\pi^{-1}n^{\beta-1}\int_0^{n\pi} t^{-\beta}\cos t\,dt.$$

A similar formula holds for b_n, and the integrals on the right tend, for $n\to\infty$, to the real and imaginary parts respectively of the integral

$$\int_0^{\infty} t^{-\beta}e^{it}\,dt = \Gamma(1-\beta)\exp\{\tfrac{1}{2}\pi i(1-\beta)\}$$

(see Chapter II, (13·10)), whence (2·23) follows.

We omit the details of the proof of (2·24), which does not differ essentially from that of (2·6) (we split the interval $(0,\pi)$ into $(0,\omega/n)$, $(\omega/n, \Omega/n)$ and $(\Omega/n, \pi)$). The hypothesis on $b(x)$ means that for every $\delta > 0$ the functions $x^{\delta}b(x)$ and $x^{-\delta}b(x)$ are respectively increasing and decreasing in some right-hand neighbourhood of $x=0$. That $b(x)$ is of bounded variation in every interval (ϵ, π) guarantees that the contribution to a_n, b_n of the integrals extended over (ϵ, π) are $O(1/n)$ and so are small in comparison with the right-hand sides in (2·25).

It may be added that since periodic odd functions are usually discontinuous at the points $\pm\pi$, their Fourier coefficients cannot tend rapidly to 0; and thus we can often not obtain simple formulae for the Fourier coefficients but only for the Fourier transforms, i.e. for the integrals of $f\cos mx$ and of $f\sin mx$ extended over the interval $0 \leqslant x < \infty$ (f in general not being periodic).

For some problems it is of importance not only to estimate the sum of the series but also to find a common majorant for the partial sums (or the remainders) of it. For the series considered in this section such estimates are implicitly contained in the proofs given above. We shall be satisfied here with the following inequalities, in which $0 < \beta < 1$ and $0 < x \leqslant \pi$:

$$\left|\sum_{n=1}^{N} n^{-\beta}\cos nx\right| \leqslant C_\beta x^{\beta-1}, \tag{2·26}$$

$$\left|\sum_{n=1}^{N} n^{-\beta}\sin nx\right| \leqslant C_\beta x^{\beta-1}, \tag{2·27}$$

$$\left|\sum_{n=1}^{N} n^{-1}\cos nx\right| \leqslant \log(1/x) + C \tag{2·28}$$

Consider, for example, (2·27). If $N < 1/x$, the sum here is identical with the sum S_1 in (2·10) corresponding to an $\omega \leqslant 1$, and so the inequality follows from the estimate (2·11) for S_1. If $N > 1/x$, the sum in (2·27) differs from $\sum_1^\infty$ by the sum S_3 in (2·10) corresponding to an $\Omega \geqslant 1$, so that (2·27) follows from the second formula (2·8) and second inequality (2·11). A similar proof holds for (2·26). Finally, (2·28) follows by combining an analogous argument with the proof of (2·16) for $b_n = 1$.

Since $\tilde{D}_n(x)$ is bounded below on $(0, \pi)$, uniformly in n, summation by parts shows that, for each given $\alpha > 0$, the partial sums of $\Sigma n^{-\alpha}\sin nx$ are uniformly bounded below on $(0, \pi)$. For $\Sigma n^{-\alpha}\cos nx$ the situation is different.

(2·29) THEOREM. *There is an α_0, $0 < \alpha_0 < 1$, such that for each $\alpha \geqslant \alpha_0$ the partial sums s_n of $\Sigma n^{-\alpha}\cos nx$ are uniformly bounded below, and for each $\alpha < \alpha_0$ they are not; α_0 is the (unique) root of the equation*

$$\int_0^{\frac{3}{2}\pi} \frac{\cos u}{u^\alpha}\,du = 0 \quad (0 < \alpha < 1).$$

Summation by parts shows that if the s_n are uniformly bounded below for some α, the same holds for any larger α. It is therefore enough to consider the case $0 < \alpha < 1$; we may also suppose that $0 < x \leqslant \pi$.

First we reduce the problem to one for integrals; we show that

$$\left|\int_0^n \frac{\cos ux}{u^\alpha}\,du - \frac{2\sin\frac{1}{2}x}{x}\sum_{\nu=1}^{n}\frac{\cos\nu x}{\nu^\alpha}\right| \leqslant C_\alpha, \tag{2·30}$$

where C_α depends on α only. For

$$\int_{\nu-\frac{1}{2}}^{\nu+\frac{1}{2}} \cos ux\, du = \frac{2\sin\frac{1}{2}x}{x}\cos\nu x,$$

$$\sum_{\nu=1}^{n}\int_{\nu-\frac{1}{2}}^{\nu+\frac{1}{2}} \frac{\cos ux}{u^\alpha}\,du = \sum_{\nu=1}^{n}\int_{\nu-\frac{1}{2}}^{\nu+\frac{1}{2}} \cos ux\left\{\frac{1}{u^\alpha}-\frac{1}{\nu^\alpha}\right\}du + \frac{2\sin\frac{1}{2}x}{x}\sum_{\nu=1}^{n}\frac{\cos\nu x}{\nu^\alpha},$$

and since, by the mean-value theorem, $u^{-\alpha}-\nu^{-\alpha}=O(\nu^{-\alpha-1})$ here, the first sum on the right is $O(1)$ uniformly in n and x, and (2·30) follows:

It is therefore enough to consider the α, $0<\alpha<1$, for which

$$\int_0^n \frac{\cos ux}{u^\alpha}\,du = x^{\alpha-1}\int_0^v \frac{\cos u}{u^\alpha}\,du \quad (v=nx)$$

is bounded below. The expression is bounded below if and only if the last integral, *qua* function of v, has a non-negative minimum in $(0,\infty)$. This minimum is attained for $\alpha=\frac{3}{2}\pi$ and is

$$m(\alpha)=\int_0^{\frac{3}{2}\pi}\frac{\cos u}{u^\alpha}\,du.$$

Since $m(0)<0$, $m(1)=+\infty$, it is enough to show that $m(\alpha)$ increases with α, $0<\alpha<1$.

Now
$$m'(\alpha)=\int_0^{\frac{3}{2}\pi}\log\frac{1}{u}\frac{\cos u}{u^\alpha}\,du > \int_0^{\frac{1}{2}\pi}\log\frac{1}{u}\frac{\cos u}{u^\alpha}\,du,$$

and since $u^{-\alpha}\cos u$ decreases in $0\leqslant u\leqslant\frac{1}{2}\pi$, the last integral exceeds

$$\cos 1\left\{\int_0^1 \log\frac{1}{u}\,du - \int_1^{\frac{1}{2}\pi}\log u\,du\right\} > \cos 1\left\{1-\int_1^{\frac{1}{2}\pi} 1\,du\right\} > 0,$$

and the theorem follows.

Theorem (2·24) can be used to obtain asymptotic expressions for the coefficients of certain Taylor series.

(2·31) THEOREM. *Let* $F(z)=\dfrac{1}{(1-z)^{\alpha+1}}\left\{\log\dfrac{a}{1-z}\right\}^\beta = \sum_{n=0}^{\infty} A_n^{\alpha,\beta}z^n,$ (2·32)

where α, β *are real numbers and* $a\geqslant 2$.† *Then*

$$A_n^{\alpha,\beta}\simeq\frac{n^\alpha}{\Gamma(\alpha+1)}(\log n)^\beta \quad \text{if} \quad \alpha\neq -1,-2,\ldots, \tag{2·33}$$

$$A_n^{\alpha,\beta}\simeq(-1)^{\alpha-1}(|\alpha|-1)!\,\beta n^\alpha(\log n)^{\beta-1} \quad \textit{if} \quad \alpha=-1,-2,\ldots. \tag{2·34}$$

The $A_n^{\alpha,\beta}$ are generalizations of the Cesàro numbers A_n^α, and (2·33) generalizes Chapter III, (1·15).

We first consider (2·33). For the function $F(z)=F_{\alpha,\beta}(z)$ we have $F'_{\alpha-1,\beta}=\alpha F_{\alpha,\beta}+\beta F_{\alpha,\beta-1}$, so that

$$nA_n^{\alpha-1,\beta}=\alpha A_{n-1}^{\alpha,\beta}+\beta A_{n-1}^{\alpha,\beta-1}. \tag{2·35}$$

We deduce from this that if (2·33) is valid for some α, it holds for $\alpha-1$. Hence if (2·33) is proved for $-1<\alpha<0$, it holds for all negative non-integral α. On the other hand, suppose we have (2·33) for some $\alpha>-1$. Since $F_{\alpha+1,\beta}=(1-z)^{-1}F_{\alpha,\beta}$ we have (see Chapter III, (1·7))

$$A_n^{\alpha+1,\beta}=\sum_{\nu=0}^{n}A_\nu^{\alpha,\beta}\simeq\frac{1}{\Gamma(\alpha+1)}\sum_{\nu=2}^{n}\nu^\alpha(\log\nu)^\beta+O(1)$$

$$\simeq\frac{1}{\Gamma(\alpha+1)}\frac{n^{\alpha+1}}{\alpha+1}(\log n)^{\beta\ddagger}=\frac{n^{\alpha+1}}{\Gamma(\alpha+2)}(\log n)^\beta,$$

† If $a\geqslant 2$, $\log\{a/(1-z)\}$ has no zero for $|z|<1$.

‡ It is not difficult to see that if $\alpha>-1$ and $\phi(u)$ is slowly varying, then

$$\sum_{\nu=1}^{n}\nu^\alpha\phi(\nu)\simeq\frac{n^{\alpha+1}}{\alpha+1}\phi(n).$$

and (2·33) holds with $\alpha+1$ for α. It follows that if we prove (2·33) for $-1<\alpha<0$, we have it for all non-integral α.

We now need Theorem (2·24). First we observe that if $0\leqslant\eta<\eta'\leqslant\pi$, then under the hypotheses of (2·24) we have $\int_\eta^{\eta'} x^{-\beta}b(x)\cos nx\,dx=O(n^{\beta-1}b(1/n))$, uniformly in η and η'; this is implicit in the proof of the theorem.

It follows from this that if $\lambda(x)$ is of bounded variation in $(0,\pi)$, then under the hypotheses of (2·24) the Fourier coefficients of $x^{-\beta}b(x)\lambda(x)$ are given by (2·25) with factor $\lambda(+0)$ inserted on the right. It is enough to prove this for λ monotone. In computing the coefficients it is enough to integrate over an arbitrarily small interval $(0,\epsilon)$ since the contribution of the remainder of $(0,\pi)$ is $O(1/n)$. Writing $\lambda(x)=\lambda(+0)+\{\lambda(x)-\lambda(+0)\}$ and applying to the integral containing $\lambda(x)-\lambda(+0)$ the second mean-value theorem and the remark of the preceding paragraph, we arrive at the conclusion.

We can now prove (2·33) for $-1<\alpha<0$. Since near $x=0$ we have $F(re^{ix})=O\{|x|^{-\alpha-1}\log^\beta(1/|x|)\}$ uniformly in r, $0\leqslant r<1$, it follows that $|F(re^{ix})|$ is majorized over $(0,\pi)$ by an integrable function of x. Hence $\Sigma A_n^{\alpha,\beta}e^{inx}=\mathfrak{S}[F(e^{ix})]$ and $\Sigma A_n^{\alpha,\beta}\cos nx=\mathfrak{S}[\mathcal{R}F(e^{ix})]$. Now

$$\mathcal{R}F(e^{ix})=\frac{1}{(2\sin\frac{1}{2}x)^{\alpha+1}}\left\{\log^2\frac{a}{2\sin\frac{1}{2}x}+\tfrac{1}{4}(\pi-x)^2\right\}^{\frac{1}{2}\beta}\cos\Phi, \tag{2·36}$$

where
$$\Phi=\tfrac{1}{2}(\pi-x)(\alpha+1)+\beta\arctan\left\{\frac{\frac{1}{2}(\pi-x)}{\log\frac{1}{2}(a\operatorname{cosec}\frac{1}{2}x)}\right\}.$$

Since the factor $\{\dots\}^{\frac{1}{2}\beta}$ in (2·36) is slowly varying, $\cos\Phi$ and $\{x/(2\sin\frac{1}{2}x)\}^{\alpha+1}$ are of bounded variation and tend respectively to $\cos\frac{1}{2}\pi(\alpha+1)=-\sin\frac{1}{2}\pi\alpha$ and 1 as $x\to0$, we find from the previous remark and the first formula (2·25) that

$$A_n^{\alpha,\beta}\simeq-\frac{2}{\pi}\sin\tfrac{1}{2}\pi\alpha\cos\tfrac{1}{2}\pi\alpha\,\Gamma(-\alpha)\,n^\alpha\log^\beta n,$$

which is (2·33) in view of the equation $\Gamma(z)\,\Gamma(1-z)=\pi/\sin\pi z$.

Thus the theorem is proved for all non-integral α. If we now prove (2·33) for $\alpha=0$, we prove the theorem also for $\alpha=1,2,\dots$ and, using (2·35), for $\alpha=-1,-2,\dots$. It remains therefore to show that $A_n^{0,\beta}\simeq(\log n)^\beta$.

This can be deduced from (2·33) with, say, $\alpha=-\frac{1}{2}$. For $F_{0,\beta}=F_{-\frac{1}{2},\beta}F_{-\frac{1}{2},0}$ implies that

$$A_n^{0,\beta}=\sum_{\nu=0}^{n}A_\nu^{-\frac{1}{2},\beta}A_{n-\nu}^{-\frac{1}{2}}.$$

Let θ be a small positive number, say $0<\theta<\frac{1}{2}$, and let $n'=[n\theta]$. We split the last sum into two, P_n and Q_n, extended respectively over $\nu\leqslant n'$ and $\nu>n'$. Observing that $A_\nu^{-\frac{1}{2}}$ is positive, decreasing and $A_\nu^{-\frac{1}{2}}\leqslant C\nu^{-\frac{1}{2}}$ for $\nu>0$, and also that the $A_\nu^{-\frac{1}{2},\beta}$ are all positive from some place on, we have

$$\begin{aligned}|P_n|&\leqslant A_{n-n'}^{-\frac{1}{2}}\sum_{\nu=0}^{n'}|A_\nu^{-\frac{1}{2},\beta}|=A_{n-n'}^{-\frac{1}{2}}\left\{\sum_{\nu=0}^{n'}A_\nu^{-\frac{1}{2},\beta}+O(1)\right\}\\&=A_{n-n'}^{-\frac{1}{2}}\{A_{n'}^{\frac{1}{2},\beta}+O(1)\},\end{aligned} \tag{2·37}$$

and find that $|P_n|\leqslant\frac{1}{2}\epsilon(\log n)^\beta$ if θ is small enough (but fixed) and n large enough.

Since $A_\nu^{-\frac{1}{2},\beta}\simeq A_\nu^{-\frac{1}{2}}(\log\nu)^\beta$, we have

$$Q_n=\sum_{\nu=n'+1}^{n}A_\nu^{-\frac{1}{2},\beta}A_{n-\nu}^{-\frac{1}{2}}\simeq(\log n)^\beta\sum_{\nu=n'+1}^{n}A_\nu^{-\frac{1}{2}}A_{n-\nu}^{-\frac{1}{2}}.$$

Now the last sum is

$$\left(\sum_{\nu=0}^{n}-\sum_{\nu=0}^{n'}\right)A_\nu^{-\frac{1}{2}}A_{n-\nu}^{-\frac{1}{2}}=1-\sum_{\nu=0}^{n'}A_\nu^{-\frac{1}{2}}A_{n-\nu}^{-\frac{1}{2}}$$

(see Chapter III, (1·10) (i)), and so, arguing as in (2·37), is arbitrarily close to 1 if θ is sufficiently small. Collecting results we see that $A_\nu^{0,\beta}=P_n+Q_n$ is contained between $(1\pm\epsilon)\log^\beta n$ for n sufficiently large, so that $A_n^{0,\beta}\simeq\log^\beta n$. This completes the proof of (2·31).

A similar argument can be applied (though the details are a little awkward) to obtain the asymptotic values of the coefficients of

$$F(z)=\frac{1}{(1-z)^{\alpha+1}}\left\{\log\frac{a_1}{1-z}\right\}^{\beta_1}\left\{\log_2\frac{a_2}{1-z}\right\}^{\beta_2}\cdots\left\{\log_k\frac{a_k}{1-z}\right\}^{\beta_k},$$

where α and the β_j are real, the a_j positive and so large that F is regular for $|z|<1$.

3. A class of Fourier-Stieltjes series

We begin by constructing a class of perfect non-dense sets. Let OA be a segment of length l whose end-points have abscissae x and $l+x$. Let $\alpha(1), \alpha(2), \ldots, \alpha(d)$ be d numbers such that

$$0 \leqslant \alpha(1) < \alpha(2) < \ldots < \alpha(d) < 1.$$

We consider d *closed* intervals with end-points $l\alpha(j)+x$ and $l\alpha(j)+l\eta+x$, where η is a positive number so small that the intervals have no points in common and are all contained in OA. These intervals will be called 'white'. The complementary intervals with respect to the closed interval OA will be called 'black' intervals, and are removed. The dissection of OA thus obtained will be said to be *of the type*

$$[d;\ \alpha(1), \ldots, \alpha(d);\ \eta].$$

Starting with the interval $(0, 2\pi)$ we perform a dissection of the type

$$[d_1;\ \alpha_1(1), \ldots, \alpha_1(d_1);\ \eta_1],$$

and remove the black intervals. On each white interval remaining we perform a dissection of the type $[d_2;\ \alpha_2(1), \ldots, \alpha_2(d_2);\ \eta_2]$, and remove the black intervals—and so on. After p operations we have $d_1 d_2 \ldots d_p$ white intervals, each of length $2\pi\eta_1\eta_2\ldots\eta_p$. When $p\to\infty$ we obtain a closed set P of measure

$$2\pi \lim d_1 \ldots d_p \eta_1 \ldots \eta_p.$$

(The limit exists, since $\eta_p < 1/d_p$ for each p.) In subsequent applications we will have $d_p \geqslant 2$ for each p. The set P will then be perfect and non-dense.

The abscissae of the left-hand ends of the white intervals of rank p, i.e. after the pth step in construction, are, as induction shows, given by the formula

$$x = 2\pi[\alpha_1(\theta_1)+\eta_1\alpha_2(\theta_2)+\eta_1\eta_2\alpha_3(\theta_3)+\ldots+\eta_1\eta_2\ldots\eta_{p-1}\alpha_p(\theta_p)], \tag{3·1}$$

where θ_k takes the values $1, 2, \ldots, d_k$. The abscissae of the points of P are given by the same series continued to infinity.

The structure of the set P displays a certain homogeneity. Consider a neighbourhood of a point of P. This neighbourhood contains a white interval I of a certain rank p. If $I_1 = I, I_2, \ldots, I_k$ are all the white intervals of rank p, then the sets $I_1 P$, $I_2 P, \ldots, I_k P$ are congruent and contained in intervals without points in common, and their union is P.

We now construct a non-decreasing function $F(x)$ constant in the intervals contiguous to P and increasing at every point of P. For each k let $\lambda_k(1), \lambda_k(2), \ldots, \lambda_k(d_k)$ be d_k positive numbers of sum 1. We denote by μ_k the largest of the $\lambda_k(j)$. Let $F_p(x)$ be a continuous non-decreasing function defined by the conditions:

(*a*) $F_p(0)=0,\ F_p(2\pi)=1$;

(*b*) $F_p(x)$ increases linearly by $\lambda_1(\theta_1)\lambda_2(\theta_2)\ldots\lambda_p(\theta_p)$ in each of the white intervals with left-hand ends given by (3·1);

(c) $F_p(x)$ is constant in every black interval of the pth stage of dissection.

We see at once, by considering $|F_{p+1}-F_p|$, that if the series

$$\Sigma(\mu_1\mu_2\ldots\mu_p)$$

converges, which we shall always suppose, then F_p tends uniformly, as $p\to\infty$, to a function $F(x)$ having the properties stated above and satisfying $F(0)=0$, $F(2\pi)=1$.

The particular function F obtained by taking $\lambda_k(j)=1/d_k$ we will call the *Lebesgue function* constructed on the set.

We shall compute the Fourier-Stieltjes coefficient

$$c_n=\frac{1}{2\pi}\int_0^{2\pi} e^{-inx}\,dF(x).$$

Using the formula (3·1) for the abscissae of the left-hand ends of the white intervals, we find that c_n is the limit, for $p\to\infty$, of the sum

$$(2\pi)^{-1}\Sigma\lambda_1(\theta_1)\,\lambda_2(\theta_2)\ldots\lambda_p(\theta_p)\exp\{-2\pi in[\alpha_1(\theta_1)+\eta_1\alpha_2(\theta_2)+\ldots+\eta_1\ldots\eta_{p-1}\alpha_p(\theta_p)]\}$$

extended over all possible combinations of θ's, θ_k taking the values $1,2,\ldots,d_k$. Writing

$$Q_k(\phi)=\lambda_k(1)\,e^{i\alpha_k(1)\phi}+\lambda_k(2)\,e^{i\alpha_k(2)\phi}+\ldots+\lambda_k(d_k)\,e^{i\alpha_k(d_k)\phi},$$

we can write the preceding sum in the form $(2\pi)^{-1}\prod_1^p\bar{Q}_k(2\pi n\eta_1\ldots\eta_{k-1})$ (with $\eta_1\ldots\eta_{k-1}=1$ for $k=1$). Hence, passing to the limit,

$$c_n=\frac{1}{2\pi}\prod_{k=1}^{\infty}\bar{Q}_k(2\pi n\eta_1\ldots\eta_{k-1}). \tag{3·2}$$

We shall now consider a few examples.

(1) *Sets of the Cantor type.* These are obtained by successive dissections of the type

$$[2;\,0,\,1-\xi_k;\,\xi_k],$$

where $0<\xi_k<\frac{1}{2}$. The points of the set are of the form

$$x=2\pi[\epsilon_1(1-\xi_1)+\epsilon_2\xi_1(1-\xi_2)+\ldots+\epsilon_p\xi_1\ldots\xi_{p-1}(1-\xi_p)+\ldots],$$

where ϵ_p is 0 or 1.

The polynomial $Q_k(\phi)$ corresponding to the Lebesgue function is

$$Q_k(\phi)=\tfrac{1}{2}(1+e^{i(1-\xi_k)\phi})=e^{i(1-\xi_k)\frac{1}{2}\phi}\cos\tfrac{1}{2}(1-\xi_k)\,\phi,$$

and hence the Fourier-Stieltjes coefficient of the Lebesgue function F is

$$\begin{aligned}c_n&=\frac{1}{2\pi}\prod_{k=1}^{\infty}\exp\{-n\pi i(1-\xi_k)\,\xi_1\ldots\xi_{k-1}\}\cos\pi n\xi_1\ldots\xi_{k-1}(1-\xi_k)\\&=\frac{(-1)^n}{2\pi}\prod_{k=1}^{\infty}\cos\pi n\xi_1\ldots\xi_{k-1}(1-\xi_k).\end{aligned}$$

If we write $\xi_1\ldots\xi_{k-1}(1-\xi_k)=r_k$, then

$$\sum_1^{\infty}r_k=1,\quad r_p>\sum_{p+1}^{\infty}r_k. \tag{3·3}$$

Also

$$x=2\pi\sum_1^{\infty}\epsilon_kr_k,\quad c_n=(-1)^n(2\pi)^{-1}\prod_1^{\infty}\cos\pi nr_k. \tag{3·4}$$

A still more special case is obtained if all the ξ_k are the same number ξ, $0<\xi<\frac{1}{2}$. The set is then said to be of Cantor type with *constant ratio of dissection*, and

$$x=2\pi(1-\xi)\sum_1^\infty \epsilon_k\xi^{k-1},\quad c_n=(-1)^n(2\pi)^{-1}\prod_{k=1}^\infty \cos \pi n\xi^{k-1}(1-\xi). \tag{3·5}$$

The classical Cantor case is $\xi=\frac{1}{3}$, when

$$x=4\pi\sum_1^\infty \epsilon_k 3^{-k},\quad c_n=(-1)^n(2\pi)^{-1}\prod_{k=1}^\infty \cos(2\pi n3^{-k}),$$

the c_n again corresponding to the Lebesgue function.

Suppose that $\xi=1/q$, $q=3,4,5,\ldots$. For $n=q^m$, (3·5) gives

$$c_{q^m}=(-1)^{q^m}(2\pi)^{-1}\prod_{k=1}^\infty \cos\pi(q-1)q^{m-k}=-(2\pi)^{-1}\prod_{k=1}^\infty \cos\pi(q-1)q^{-k}.$$

The latter expression is independent of m and different from zero. Hence

(3·6) Theorem. *If $\xi=1/q$, $q=3,4,\ldots$, the Fourier-Stieltjes coefficients of the Lebesgue function $F(x)$ do not tend to zero.*

Thus the periodic function $F(x)-x/2\pi$ is of bounded variation and continuous, and yet its Fourier coefficients are not $o(1/n)$ (compare Chapter II, p. 48).

(2) *Symmetrical perfect sets of order d.* These are obtained, d being an integer not less than 1, by successive dissections of the type

$$[d+1;\,0,(1-\xi_k)d^{-1},\,2(1-\xi_k)d^{-1},\,\ldots,\,1-\xi_k;\,\xi_k],$$

where $0<\xi_k<1/(d+1)$. The Cantor set corresponds to $d=1$. The points $\alpha_k(j)$, $j=1,2,\ldots,d+1$, are in arithmetic progression, the first point being 0 and the last $1-\xi_k$. The points of the set are

$$x=2\pi d^{-1}[\epsilon_1(1-\xi_1)+\epsilon_2\xi_1(1-\xi_2)+\ldots+\epsilon_p\xi_1\ldots\xi_{p-1}(1-\xi_p)+\ldots],$$

where ϵ_p takes the values $0,1,\ldots,d$. On setting $\gamma_k=(1-\xi_k)/d$, we can write the polynomial Q_k corresponding to the Lebesgue function as

$$(d+1)^{-1}(1+e^{i\gamma_k\phi}+e^{2i\gamma_k\phi}+\ldots+e^{di\gamma_k\phi})=(d+1)^{-1}e^{\frac{1}{2}di\gamma_k\phi}\frac{\sin\{(d+1)\frac{1}{2}\gamma_k\phi\}}{\sin\frac{1}{2}\gamma_k\phi}.$$

Hence
$$2\pi c_n=(-1)^n\prod_{k=1}^\infty \frac{\sin\{(d+1)\pi nd^{-1}\xi_1\ldots\xi_{k-1}(1-\xi_k)\}}{(d+1)\sin\{\pi nd^{-1}\xi_1\ldots\xi_{k-1}(1-\xi_k)\}}.$$

The set being always a symmetrical perfect set, let $d=2g$ be even, and let us construct a function such that for each k the number $\lambda_k(j)$ is the coefficient $\lambda(j)$ of z^{j-1} in the expansion of

$$(g+1)^{-2}(1+z+z^2+\ldots+z^g)^2.$$

Then $\lambda(1)+\lambda(2)+\ldots+\lambda(2g+1)=1$, as required; and, with $\gamma_k=(1-\xi_k)/2g$,

$$Q_k(\phi)=\left(\frac{1+e^{i\gamma_k\phi}+\ldots+e^{gi\gamma_k\phi}}{g+1}\right)^2=\left[e^{\frac{1}{2}gi\gamma_k\phi}\frac{\sin\{(g+1)\frac{1}{2}\gamma_k\phi\}}{(g+1)\sin(\frac{1}{2}\gamma_k\phi)}\right]^2,$$

$$2\pi c_n=(-1)^n\prod_{k=1}^\infty\left[\frac{\sin\left\{(g+1)\dfrac{\pi n}{2g}\xi_1\ldots\xi_{k-1}(1-\xi_k)\right\}}{(g+1)\sin\left\{\dfrac{\pi n}{2g}\xi_1\ldots\xi_{k-1}(1-\xi_k)\right\}}\right]^2.$$

(3) Consider a symmetrical perfect set of order d, and let $\lambda_k(j)$ be equal for all k to the coefficient $\lambda(j)$ of z^{j-1} in the expansion of $2^{-d}(1+z)^d$. Then

$$Q_k(\phi)=[\tfrac{1}{2}(1+e^{i\gamma_k\phi})]^d\quad(\gamma_k=(1-\xi_k)d^{-1}),$$

$$2\pi c_n=(-1)^n\prod_{k=1}^\infty[\cos\pi nd^{-1}\xi_1\ldots\xi_{k-1}(1-\xi_k)]^d.$$

We shall now consider the modulus of continuity of the function F, confining our attention to the case where $d_p = d$, $\eta_p = \eta$ and $\mu_k = \max[\lambda_k(1), \lambda_k(2), \ldots, \lambda_k(d)] = \mu$ are all constant. We shall show that then *F satisfies a Lipschitz condition of order* $|\log \mu|/|\log \eta|$.

Let x and $x' > x$ be two points of P. If x and x' are end-points of the same interval contiguous to P, then

$$F(x') - F(x) = 0.$$

If not, let p be the order of the dissection when for the first time appear at least two black intervals in (x, x'). Thus there is at least one white interval of rank p included in (x, x'), and so

$$x' - x \geqslant 2\pi\eta^p.$$

On the other hand, at the dissection of order $p-1$ there is at most one black interval (β, β') in (x, x'). It follows that

$$F(x') - F(x) = F(x') - F(\beta') + F(\beta) - F(x) \leqslant 2\mu^{p-1}.$$

Thus $$F(x') - F(x) \leqslant A(x'-x)^{|\log \mu|/|\log \eta|},$$

A being independent of x.

The extension to the case in which x, or x', or both, are outside P is immediate, since we apply the preceding inequality to the interval (x_1, x_1'), where x_1 and x_1' are the first and last points of P in (x, x').

Example. The Lebesgue function constructed on a symmetrical perfect set of order d and of constant ratio of dissection ξ, belongs to Λ_α with $\alpha = \log(d+1)/|\log \xi|$.

4. The series $\Sigma n^{-\frac{1}{2}-\alpha} e^{icn\log n} e^{inx}$

The power series
$$\sum_{n=1}^{\infty} e^{icn\log n} \frac{e^{inx}}{n^{\frac{1}{2}+\alpha}}, \tag{4·1}$$

which was first studied by Hardy and Littlewood, possesses many interesting properties. We suppose that α is real and c positive.

(4·2) THEOREM. *If $0 < \alpha < 1$, the series (4·1) converges uniformly in the interval $0 \leqslant x \leqslant 2\pi$ to a function $\phi_\alpha(x) \in \Lambda_\alpha$.*

The theorem is a consequence of certain lemmas, due to van der Corput, of considerable interest in themselves.

Given a real-valued function $f(u)$ and numbers $a < b$, we set

$$F(u) = e^{2\pi i f(u)},$$

$$I(F; a, b) = \int_a^b F(u)\,du, \quad S(F; a, b) = \sum_{a<n\leqslant b} F(n),$$

$$D(F; a, b) = I(F; a, b) - S(F; a, b).$$

(4·3) LEMMA. (i) *If $f(u)$, $a \leqslant u \leqslant b$, has a monotone derivative $f'(u)$, and if there is a positive λ such that $f' \geqslant \lambda$ or $f' \leqslant -\lambda$ in (a, b), then $|I(F; a, b)| < 1/\lambda$.*

(ii) *If $f''(u) \geqslant \rho > 0$ or $f''(u) \leqslant -\rho < 0$, then $|I(F; a, b)| \leqslant 4\rho^{-\frac{1}{2}}$.*

(i) Since $I(F; a, b) = (2\pi i)^{-1}\int_a^b dF(u)/f'(u)$, the second mean-value theorem, applied to the real and imaginary parts of the last integral, shows that $|I| \leqslant 2/\pi\lambda < 1/\lambda$.

(ii) We may suppose that $f''\geqslant\rho$. (Otherwise replace f by $-f$ and I by $\bar{I}$.) Then f' is increasing. Suppose for the moment that f' is of constant sign in (a,b), say $f'\geqslant 0$. If $a<\gamma<b$, then $f'\geqslant(\gamma-a)\rho$ in (γ,b). Therefore

$$|I(F;a,b)|\leqslant|I(F;a,\gamma)|+|I(F;\gamma,b)|\leqslant(\gamma-a)+1/(\gamma-a)\rho,$$

and, choosing γ so as to make the last sum a minimum, we find that $|I(F;a,b)|\leqslant 2\rho^{-\frac{1}{2}}$. In the general case (a,b) is a sum of two intervals in each of which f' is of constant sign, and (ii) follows by adding the inequalities for these two intervals.

(4·4) LEMMA. *If $f'(u)$ is monotone and $|f'|\leqslant\frac{1}{2}$ in (a,b), then*

$$|D(F;a,b)|\leqslant A,$$

where A is an absolute constant.

Suppose first that a and b are not integers. The sum S is then $\int_a^b F(u)\,d\psi(u)$, where $\psi(u)$ is a function constant in the intervals $n<u<n+1$ and having jumps 1 at the points n. If we take $\psi(u)=[u]+\frac{1}{2}$ for u non-integral ($[u]$ being the integral part of u) and $\psi(n)=n$, then

$$D(F;a,b)=\int_a^b F(u)\,d\chi(u),\quad\text{where}\quad \chi(u)=u-[u]-\tfrac{1}{2}\quad(u\neq 0,\pm 1,\ldots).$$

The function χ has period 1, and integration by parts gives

$$D(F;a,b)=-I(F'\chi;a,b)+R,\quad |R|\leqslant 1.$$

The partial sums of $\mathsf{S}[\chi]=-\Sigma(\sin 2\pi nu)/\pi n$ are uniformly bounded. Multiplying $\mathsf{S}[\chi]$ by F' and integrating over (a,b) we find that $D-R$ is equal to the sum of the expressions

$$\frac{1}{2\pi in}\left\{\int_a^b\frac{f'(u)}{f'(u)+n}\,d\,e^{2\pi i\{f(u)+nu\}}-\int_a^b\frac{f'(u)}{f'(u)-n}\,d\,e^{2\pi i\{f(u)-nu\}}\right\}\tag{4·5}$$

for $n=1,2,\ldots$. The ratios $f'/(f'\pm n)$ being monotone, the second mean-value theorem shows that (4·5) numerically does not exceed $2/\pi n(n-\frac{1}{2})$, and so the series with terms (4·5) converges absolutely and uniformly. This completes the proof if a and b are not integers. If a or b is an integer, it is enough to observe that $D(F;a,b)$ differs from $\lim_{\epsilon\to 0}D(F;a+\epsilon;b-\epsilon)$ by 1 at most.

Remark. The condition $|f'|\leqslant\frac{1}{2}$ can be replaced by $|f'|\leqslant 1-\epsilon$, with $\epsilon>0$, if we simultaneously replace A by A_ϵ.

(4·6) LEMMA. *If $f''(u)\geqslant\rho>0$ or $f''(u)\leqslant-\rho<0$, then*

$$|S(F;a,b)|\leqslant[|f'(b)-f'(a)|+2](4\rho^{-\frac{1}{2}}+A).$$

We may suppose that $f''\geqslant\rho$. Let α_p be the point (if any) where $f'(\alpha_p)=p-\frac{1}{2}$, and let

$$F_p(u)=e^{2\pi i\{f(u)-pu\}},$$

for $p=0,\pm 1,\pm 2,\ldots$. Then $|f'(u)-p|\leqslant\frac{1}{2}$ in (α_p,α_{p+1}). Let $\alpha_r,\alpha_{r+1},\ldots,\alpha_{r+s}$ be the points α, if any such exist, belonging to the interval $a\leqslant u\leqslant b$. From (4·3) and (4·4) it follows that

$$S(F;\alpha_p,\alpha_{p+1})=S(F_p;\alpha_p,\alpha_{p+1})=I(F_p;\alpha_p,\alpha_{p+1})-D(F_p;\alpha_p,\alpha_{p+1})$$

does not exceed $4\rho^{-\frac{1}{2}}+A$ in absolute value. The same holds for $S(F; a, \alpha_r)$ and $S(F; \alpha_{r+s}, b)$. Since $S(F; a, b)$ is the sum of these expressions for the intervals (a, α_r), (α_r, α_{r+1}), ..., (α_{r+s}, b), whose number is $s+2=f'(\alpha_{r+s})-f'(\alpha_r)+2$, the lemma follows.

To complete the proof of (4·2) we need the following result:

(4·7) THEOREM. *The partial sums $s_N(x)$ of the series $\Sigma e^{icn\log n} e^{inx}$ are $O(N^{\frac{1}{2}})$, uniformly in x.*

The function $f(u)=(2\pi)^{-1}(cu\log u+ux)$ has an increasing derivative. If $\nu \geqslant 0$ is an integer and $a=2^\nu$, $b=2^{\nu+1}$, a simple application of Lemma (4·6) shows that $|S(F; a, b)| \leqslant C2^{\frac{1}{2}\nu}$, with C depending on c only. The same holds if $2^\nu=a<b<2^{\nu+1}$. If $2^n<N\leqslant 2^{n+1}$, then

$$|s_N(x)| \leqslant 1+|S(F; 1, 2)|+|S(F; 2, 4)|+\dots+|S(F; 2^n, N)|$$
$$\leqslant 1+C(1+2^{\frac{1}{2}}+\dots+2^{\frac{1}{2}n}) \leqslant C_1 2^{\frac{1}{2}n} \leqslant C_1 N^{\frac{1}{2}},$$

with C_1 depending on c only, and (4·7) is established.

Return to (4·2). Summation by parts gives for the Nth partial sum of (4·1) the value

$$\sum_{\nu=1}^{N-1} s_\nu(x)\,\Delta\nu^{-\frac{1}{2}-\alpha}+s_N(x)\,N^{-\frac{1}{2}-\alpha}. \tag{4·8}$$

Since $\Delta\nu^{-\frac{1}{2}-\alpha}=O(\nu^{-\frac{3}{2}-\alpha})$, we conclude from (4·8) and from the estimate $s_\nu(x)=O(\nu^{\frac{1}{2}})$ that the partial sums of (4·1) are

(i) uniformly convergent if $\alpha>0$;
(ii) uniformly $O(\log N)$ if $\alpha=0$;
(iii) uniformly $O(N^{-\alpha})$ if $\alpha<0$.

Let $0<\alpha<1$. Making $N\to\infty$ in (4·8) we obtain

$$\phi_\alpha(x+h)-\phi_\alpha(x)=\sum_{\nu=1}^{\infty}\{s_\nu(x+h)-s_\nu(x)\}\,\Delta\nu^{-\frac{1}{2}-\alpha}=\sum_{\nu=1}^{N}+\sum_{\nu=N+1}^{\infty}=P+Q,$$

say. Let $0<h\leqslant 1$, $N=[1/h]$. The terms of Q are $O(\nu^{\frac{1}{2}})\,\Delta\nu^{-\frac{1}{2}-\alpha}=O(\nu^{-1-\alpha})$ so that

$$Q=O(N^{-\alpha})=O(h^\alpha).$$

On the other hand, since $s'_\nu(x)$—apart from a numerical factor—is the partial sum of (4·1) with $\alpha=-\frac{3}{2}$, we have $s'_\nu(x)=O(\nu^{\frac{3}{2}})$, by case (iii) above. Therefore, applying the mean-value theorem to the real and imaginary parts of $s_\nu(x+h)-s_\nu(x)$, we get

$$|P| \leqslant \sum_{\nu=1}^{N} O(h\nu^{\frac{3}{2}})\,\Delta\nu^{-\frac{1}{2}-\alpha}=O(h)\sum_{1}^{N}\nu^{-\alpha}=O(hN^{1-\alpha})=O(h^\alpha).$$

Since P and Q are $O(h^\alpha)$, so is $\phi_\alpha(x+h)-\phi_\alpha(x)$; and thus $\phi_\alpha \in \Lambda_\alpha$.

Theorem (4·2) ceases to be true when $\alpha=0$. It may be shown that in this case (4·1) is nowhere summable A and so certainly is not a Fourier series. (Another interesting consequence is that the function

$$\sum_{1}^{\infty}\frac{e^{icn\log n}}{n^\alpha}z^n,$$

which is regular for $|z|<1$, cannot be continued across $|z|=1$ for any α.) However, we have

(4·9) THEOREM. *If $\beta>1$ and c is positive, the series*

$$\sum_{n=2}^{\infty}\frac{e^{icn\log n}}{n^{\frac{1}{2}}(\log n)^\beta}e^{inx} \tag{4·10}$$

converges uniformly for $0\leqslant x\leqslant 2\pi$.

We replace $\Delta\nu^{-\frac{1}{2}-\alpha}$ by $\Delta\nu^{-\frac{1}{2}}\log^{-\beta}\nu = O(\nu^{-\frac{3}{2}}\log^{-\beta}\nu)$, $N^{-\frac{1}{2}-\alpha}$ by $N^{-\frac{1}{2}}\log^{-\beta}N$ in (4·8), and observe that a series with terms $O(\nu^{-1}\log^{-\beta}\nu)$ converges.

(4·11) THEOREM. *There is a continuous function $f(x)$ such that, if a_n, b_n are the Fourier coefficients of f, the series $\Sigma(|a_n|^{2-\epsilon}+|b_n|^{2-\epsilon})$ diverges for every $\epsilon > 0$.*

For if $f(x)$ is either the real or imaginary part of the function (4·10), with $\beta > 1$, and if $\rho_n = (a_n^2 + b_n^2)^{\frac{1}{2}}$, then $\rho_n = n^{-\frac{1}{2}}\log^{-\beta} n$, $\Sigma\rho_n^{2-\epsilon}$ diverges, and this is equivalent to the divergence of $\Sigma(|a_n|^{2-\epsilon}+|b_n|^{2-\epsilon})$.

5. The series $\Sigma\nu^{-\beta}e^{i\nu^\alpha}e^{i\nu x}$

We shall now discuss the series

$$\sum_{\nu=1}^{\infty} \nu^{-\beta} e^{i\nu^\alpha} e^{i\nu x}. \tag{5·1}$$

Here, once for all, $0 < \alpha < 1$, $-\pi \leqslant x \leqslant \pi$.

(5·2) THEOREM. (i) *If $\beta > 1 - \frac{1}{2}\alpha (> \frac{1}{2})$, the series (5·1) converges uniformly to a continuous sum $\psi_{\alpha,\beta}(x)$.*

(ii) *If, in addition, $\frac{1}{2}\alpha + \beta < 2$, then $\psi_{\alpha,\beta}(x) \in \Lambda_{\frac{1}{2}\alpha+\beta-1}$.*

For fixed x and for $u > 0$, the function $f(u) = (2\pi)^{-1}(u^\alpha + ux)$ has a decreasing derivative

$$f'(u) = (2\pi)^{-1}(\alpha u^{\alpha-1} + x). \tag{5·3}$$

Hence, n_0 being any positive integer, (4·3) (ii) gives

$$\left|\int_{n_0}^{n} e^{2\pi i f(u)}\,du\right| \leqslant 4(2\pi)^{\frac{1}{2}}\{\alpha(1-\alpha)\}^{-\frac{1}{2}} n^{1-\frac{1}{2}\alpha} = A_\alpha n^{1-\frac{1}{2}\alpha} \quad (n \geqslant n_0). \tag{5·4}$$

Since $f'(u) \to x/2\pi$ as $u \to \infty$, we have $|f'(u)| \leqslant \frac{3}{4}$ for $u \geqslant n_0$ and n_0 large enough. By Lemma (4·4), and the Remark following it, $|D(F; n_0, n)| \leqslant A$. Combining this with (5·4), we get

$$\left|\sum_{\nu=1}^{n} e^{i\nu^\alpha} e^{i\nu x}\right| \leqslant \left|\sum_{1}^{n_0}\right| + \left|\sum_{n_0+1}^{n}\right| < n_0 + O(n^{1-\frac{1}{2}\alpha}) + A = O(n^{1-\frac{1}{2}\alpha}).$$

Let $s_n(x)$ be the sum on the left. Then, summing by parts, we get for the Nth partial sum of (5·1),

$$\sum_{n=1}^{N-1} s_n(x)\,\Delta n^{-\beta} + s_N(x)\,N^{-\beta}. \tag{5·5}$$

The terms of the sum here are $O(n^{1-\frac{1}{2}\alpha}) . O(n^{-\beta-1}) = O(n^{-\frac{1}{2}\alpha-\beta})$, and $s_N N^{-\beta} = O(N^{1-\frac{1}{2}\alpha-\beta})$. It follows that, under the hypotheses of (i), (5·5) tends uniformly to a limit as $N \to \infty$.

We also observe that if $\frac{1}{2}\alpha + \beta = 1$ then (5·5) is uniformly $O(\log N)$, and if $\frac{1}{2}\alpha + \beta < 1$ it is uniformly $O(N^{1-\frac{1}{2}\alpha-\beta})$.

To prove (ii), we make $N \to \infty$ in (5·5). We have

$$\psi_{\alpha,\beta}(x) = \sum_{1}^{\infty} s_n(x)\,\Delta n^{-\beta},$$

$$\psi_{\alpha,\beta}(x+h) - \psi_{\alpha,\beta}(x) = \sum_{1}^{\infty}\{s_n(x+h) - s_n(x)\}\,\Delta n^{-\beta} = \sum_{1}^{N} + \sum_{N+1}^{\infty} = P + Q,$$

say, where $0<h\leqslant 1$ and $N=[1/h]$. The terms of Q are $O(n^{1-\frac{1}{2}\alpha}).O(n^{-\beta-1})=O(n^{-\frac{1}{2}\alpha-\beta})$. Hence†

$$Q=O(N^{1-\frac{1}{2}\alpha-\beta})=O(h^{\frac{1}{2}\alpha+\beta-1}).$$

Applying the mean-value theorem to the real and imaginary parts of s_n, and using the remark just made (for $\beta=-1$), we find that the terms of P are $O(hn^{2-\frac{1}{2}\alpha}).O(n^{-\beta-1})$. Hence

$$P=O(hN^{2-\frac{1}{2}\alpha-\beta})=O(h^{\frac{1}{2}\alpha+\beta-1}).$$

It follows that $\psi_{\alpha,\beta}\in\Lambda_{\frac{1}{2}\alpha+\beta-1}$.

(5·6) THEOREM. *Let* $\beta>0$. *Then*

(i) *the series* $\Sigma\nu^{-\beta}e^{i\nu^\alpha}e^{i\nu x}$ *converges uniformly for* $\epsilon\leqslant|x|\leqslant\pi$, $\epsilon>0$, *and, in particular, converges for* $x\neq 0$;

(ii) *if* $\frac{1}{2}\alpha+\beta<1$, *the sum* $\psi_{\alpha,\beta}(x)$ *of the series is*

$$O(x^{-(1-\alpha-\beta)/(1-\alpha)}),\quad O(\log 1/x),\quad O(1)\quad \textit{for}\quad x\to+0, \textit{ according as } \alpha+\beta<1, =1, >1,$$

and is $$O(|x|^{-(1-\frac{1}{2}\alpha-\beta)/(1-\alpha)})\quad \textit{for}\quad x\to-0;$$

(iii) *if* $\frac{1}{2}\alpha+\beta=1$, *then*

$$\psi_{\alpha,\beta}(x)=O(1)\quad \textit{for}\quad x\to+0,\quad \psi_{\alpha,\beta}(x)=O(\log|x|)\quad \textit{for}\quad x\to-0;$$

(iv) *if* $\beta>\frac{1}{2}\alpha$, (5·1) *is* $\mathfrak{S}[\psi_{\alpha,\beta}]$.

(i) For $\epsilon\leqslant|x|\leqslant\pi$ and $u\geqslant n_0=n_0(\epsilon)$, $|f'(u)|$ has a positive lower bound. By Lemma (4·3) (i), the left-hand side of (5·4) is uniformly bounded. Using Lemma (4·4) and the Remark to it, we see that the partial sums $s_n(x)$ of $\Sigma e^{i(\nu^\alpha+\nu x)}$ are uniformly bounded for $\epsilon\leqslant|x|\leqslant\pi$. An application of partial summation completes the proof of (i).

(ii) By C we shall here denote a positive constant independent of x and n. First, let $0<x\leqslant\pi$. We shall show that

$$|s_n(x)|\leqslant Cn^{1-\alpha},\quad |s_n(x)|\leqslant C/x\quad (0<x\leqslant\pi). \tag{5·7}$$

In virtue of (4·4) and (i), it is enough to prove that, for x small enough, these inequalities are satisfied by the integrals $I_n(x)=\int_1^n e^{i\nu^\alpha}e^{i\nu x}d\nu$. The new inequalities follow immediately, if we observe (see (5·3)) that $f'(u)$ exceeds both $Cu^{\alpha-1}$ and Cx, and apply (4·3).

For fixed x, the first inequality (5·7) is more advantageous if n is small, the second if n is large. For $n\sim x^{1/(\alpha-1)}$, the right-hand sides in (5·7) are of the same order. Hence, setting $M=[x^{1/(\alpha-1)}]$, we have

$$\psi_{\alpha,\beta}(x)=\sum_{n=1}^{\infty}s_n(x)\,\Delta n^{-\beta}=\sum_1^M+\sum_{M+1}^{\infty}=A+B.$$

The terms of A are $O(n^{1-\alpha}).O(n^{-\beta-1})$, and the terms of B are $O(x^{-1}\Delta n^{-\beta})$. It follows that, if $\alpha+\beta<1$, then

$$\psi_{\alpha,\beta}(x)=O(M^{1-\alpha-\beta})+O(x^{-1}M^{-\beta})=O(x^{-(1-\alpha-\beta)/(1-\alpha)}).$$

Similarly we get the other estimates in (ii) for $x\to+0$.

The case $-\pi\leqslant x<0$ is slightly less simple, since then $f'(u)$ is not of constant sign. The single zero of f' is

$$u_0=u_0(x)=(|x|/\alpha)^{-1/(1-\alpha)}.$$

† The interval $(x, x+h)$ may be partly outside $(-\pi,\pi)$, but since it is interior to $(-2\pi, 2\pi)$ it is easy to see that the conclusion holds.

It tends to ∞ as $x \to -0$, and we need only consider small x. Clearly $|f'| \geqslant C|x|$ for $u \geqslant 2u_0$. Set $N = [2u_0]$ and split the series (5·1) into two parts, Σ_1 and Σ_2, corresponding to $n \leqslant N$ and $n > N$. Since in any case $s_n(x) = O(n^{1-\frac{1}{2}\alpha})$ uniformly in x, we have, by (5·5),

$$\Sigma_1 = O((2u_0)^{1-\frac{1}{2}\alpha-\beta}) = O(|x|^{-(1-\frac{1}{2}\alpha-\beta)/(1-\alpha)}).$$

If we can get the same estimate for Σ_2, the proof of (ii) in the case $x < 0$ will be complete.

Now, summing by parts,

$$\Sigma_2 = \sum_{N+1}^{\infty} \{s_n(x) - s_N(x)\} \Delta n^{-\beta}.$$

Using the fact that $|f'| \geqslant C|x|$ for $u \geqslant N+1$ and applying Lemmas (4·4) and (4·3) (i), we obtain $s_n(x) - s_N(x) = O(x^{-1})$ for $n > N$, which leads to

$$\Sigma_2 = O(x^{-1}N^{-\beta}) = O(|x|^{-(1-\alpha-\beta)/(1-\alpha)}) = O(|x|^{-(1-\frac{1}{2}\alpha-\beta)/(1-\alpha)}).$$

(iii) The proof is contained in that of (ii).

(iv) It follows from (i), (ii) and (iii) that the function $\psi_{\alpha,\beta}(x)$ is always integrable L over $(0, \pi)$. The estimates for $x \to -0$ involve a much larger order of magnitude, since the exponent $(1-\frac{1}{2}\alpha-\beta)/(1-\alpha)$ can be arbitrarily large if $1-\alpha$ and β are sufficiently small. However, $\psi_{\alpha,\beta}$ is L-integrable over $(-\pi, 0)$ if $\beta > \frac{1}{2}\alpha$.

On the other hand, it is easy to see that (5·1) *is a Fourier-Riemann series if merely* $0 < \alpha < 1$, $\beta > 0$. For, when it is integrated termwise with respect to x, it converges absolutely and uniformly to a continuous function $\Psi(x)$, such that $\Psi'(x) = \psi_{\alpha,\beta}(x)$ for $x \neq 0$ (see (i)). Thus $\Psi(x)$ is the Riemann integral of $\psi_{\alpha,\beta}(x)$, and (5·1) a Fourier-Riemann series of $\psi_{\alpha,\beta}$. Hence, for $\beta > \frac{1}{2}\alpha$, we have a Fourier-Lebesgue series.

The special case $\beta = \frac{1}{2}\alpha$ in (5·6) (ii) leads to the estimate $O(1/x)$ for $x \to -0$. For a later application we shall need the following result:

(5·8) THEOREM. *If* $0 < \alpha < 1$ *and* γ *is real, the function*

$$\chi_{\alpha,\gamma}(x) = \sum_{2}^{\infty} n^{-\frac{1}{2}\alpha} (\log n)^{-\gamma} e^{in^\alpha} e^{inx} \tag{5·9}$$

is of the form

$$O\left(x^{-(1-\frac{3}{2}\alpha)/(1-\alpha)} \log^{-\gamma} \frac{1}{x}\right) \quad \textit{and} \quad O\left(x^{-1} \log^{-\gamma} \frac{1}{|x|}\right) \quad \textit{for} \quad x \to +0 \textit{ and } x \to -0,$$

respectively. Moreover, if $\gamma > 1$, $\chi_{\alpha,\gamma}$ *is integrable and* (5·9) *is* $S[\chi_{\alpha,\gamma}]$.

The proof is essentially the same as that of parts (ii) and (iv) of (5·6).

6. Lacunary series

Lacunary trigonometric series are series in which the terms that differ from zero are 'very sparse'. Such series may be written in the form

$$\sum_{k=1}^{\infty} (a_k \cos n_k x + b_k \sin n_k x) = \sum_{k=1}^{\infty} A_{n_k}(x), \tag{6·1}$$

supposing for simplicity that the constant term also vanishes. We define a lacunary series more specifically as one for which the n_k satisfy for all k an inequality

$$n_{k+1}/n_k > q > 1,$$

that is, for which they increase at least as rapidly as a geometric progression with ratio greater than 1.

Given a lacunary series (6·1), consider the sum

$$\sum_{k=1}^{\infty}(a_k^2+b_k^2)=\sum_{k=1}^{\infty}\rho_k^2. \tag{6·2}$$

(6·3) THEOREM. *If $\Sigma\rho_k^2$ is finite, the series $\Sigma A_{n_k}(x)$ converges almost everywhere.*

Let $s_m(x)$ and $\sigma_m(x)$ denote the partial sums and the arithmetic means of $\Sigma A_{n_k}(x)$ with the vacant terms replaced by 0's. The sequence $\sigma_m(x)$ converges almost everywhere and (6·3) follows from the fact that $s_m(x)-\sigma_m(x)\to 0$ for every x (Chapter III, (1·27)).

The converse of (6·3) is also true and lies deeper. *If $\Sigma A_{n_k}(x)$ converges in a set of points of positive measure, then $\Sigma\rho_k^2$ is finite.* We shall prove an even more general theorem. Let T* be any linear method of summation satisfying the first and third conditions of regularity (Chapter III, § 1); the second need not be satisfied. All linear methods of summation used in analysis are T* methods.

(6·4) THEOREM. *If $\Sigma A_{n_k}(x)$ is summable T* in a set E of positive measure, then $\Sigma\rho_k^2$ converges.*

We need the following lemma:

(6·5) LEMMA. *Suppose we are given a set $\mathscr{E}\subset(0,2\pi)$ of positive measure, and numbers $\lambda>1$, $q>1$. Then there exists an integer $h_0=h_0(\mathscr{E},\lambda,q)$ such that for any trigonometric polynomial $P(x)=\Sigma(a_j\cos n_jx+b_j\sin n_jx)$ with $n_{j+1}/n_j>q>1$ and $n_1\geqslant h_0$ we have*

$$\lambda^{-1}|\mathscr{E}|\tfrac{1}{2}\Sigma(a_j^2+b_j^2)\leqslant\int_{\mathscr{E}}P^2(x)\,dx\leqslant\lambda|\mathscr{E}|\tfrac{1}{2}\Sigma(a_j^2+b_j^2). \tag{6·6}$$

The inequalities hold also if $P(x)$ is an infinite series with $\Sigma(a_j^2+b_j^2)<\infty$.

Write the polynomial P in the complex form $\Sigma c_\nu e^{in_\nu x}$, with $n_{-\nu}=-n_\nu$. Then

$$\Sigma|c_\nu|^2=\tfrac{1}{2}\Sigma(a_\nu^2+b_\nu^2).$$

We have

$$\int_{\mathscr{E}}P^2(x)\,dx=\int_{\mathscr{E}}(\Sigma c_\mu e^{in_\mu x})(\Sigma\bar{c}_\nu e^{-in_\nu x})\,dx$$

$$=|\mathscr{E}|\Sigma|c_\nu|^2+\sum_{\mu\neq\nu}c_\mu\bar{c}_\nu\int_{\mathscr{E}}e^{i(n_\mu-n_\nu)x}\,dx. \tag{6·7}$$

Let γ_m denote the complex coefficients of the characteristic function of $\mathscr{E}$. The last integral is then $2\pi\gamma_{n_\nu-n_\mu}$. By Schwarz's inequality, the modulus of the last sum does not exceed

$$2\pi(\sum_{\mu,\nu}|c_\mu c_\nu|^2)^{\frac{1}{2}}(\sum_{\mu\neq\nu}|\gamma_{n_\nu-n_\mu}|^2)^{\frac{1}{2}}=2\pi(\sum_\nu|c_\nu|^2)(\sum_{\mu\neq\nu}|\gamma_{n_\nu-n_\mu}|^2)^{\frac{1}{2}}. \tag{6·8}$$

We assert that there exists a number $\Delta=\Delta(q)$ such that no integer N can be represented more than Δ times in the form $n_\nu-n_\mu$ with $\mu\neq\nu$.

It is enough to assume that $0<\mu<\nu$ and consider the two cases:

(i) $N=n_\nu+n_\mu$,

(ii) $N=n_\nu-n_\mu$.

In case (i) we have $\frac{1}{2}N < n_\nu < N$, and since the n_ν increase at least as rapidly as q^ν, the number of n_ν satisfying this inequality does not exceed $y+1$, where $q^y=2$. In case (ii), since $n_\mu < n_\nu/q$, we have $n_\nu - n_\nu/q < N$, that is $n_\nu < Nq/(q-1)$. On the other hand, $n_\nu > N$. Since the number of the n_ν between N and $Nq/(q-1)$ is bounded (the bound depending on q), the existence of $\Delta(q)$ follows.

Thus the last factor on the right of (6·8) does not exceed $\{2\Delta(|\gamma_h|^2+|\gamma_{h+1}|^2+\dots)\}^{\frac{1}{2}}$, where h is the least integer representable in the form $n_\nu - n_\mu$ with $1 \leqslant \mu < \nu$. But

$$n_\nu - n_\mu \geqslant n_\nu - n_{\nu-1} \geqslant n_\nu(1-q^{-1}) \geqslant n_1(1-q^{-1}).$$

This shows that h is large with n_1.

The γ's depend only on $\mathscr{E}$, and $\Sigma|\gamma_\nu|^2 = |\mathscr{E}|(2\pi)^{-1} \leqslant 1$. Hence if n_1 is large enough, $n_1 \geqslant h_0(\mathscr{E}, \lambda, q)$, we can make the right-hand side of (6·8) less than

$$(1-\lambda^{-1})|\mathscr{E}| < (\lambda-1)|\mathscr{E}|,$$

and we obtain (6·6) in virtue of (6·7).

If P is an infinite series with $\Sigma(a_j^2+b_j^2)$ finite, we first apply (6·6) to the partial sums P_l of P. Then making $l \to \infty$ and observing that

$$\int_{\mathscr{E}} P_l^2\, dx \to \int_{\mathscr{E}} P^2\, dx,$$

we get the required result.

Passing to the proof of (6·4), we denote by β_{mn} the elements of the matrix T* considered. The hypothesis is that for every $x \subset E$ each of the series $\sum\limits_n \beta_{mn} s_n(x)$, $m=0,1,2,\dots$, converges to a sum $\tau_m(x)$, which tends to a finite limit as $m \to \infty$. We begin with the case when the matrix is row-finite. If we set $\beta_{mn}+\beta_{m,n+1}+\dots = R_{mn}$, then

$$\tau_m(x) = \sum_{k=1}^{\infty} A_{n_k}(x)\, R_{mn_k}, \tag{6·9}$$

where $A_{n_k}(x) = a_k \cos n_k x + b_k \sin n_k x$. The sum here has only a finite number of terms different from zero. Since $\tau_m(x)$ converges in E, we can find a subset $\mathscr{E}$ of E with $|\mathscr{E}|>0$ and a number M such that $|\tau_m(x)| \leqslant M$ for all $x \in \mathscr{E}$ and all m. (For $E = E_1 + E_2 + \dots$, where E_p is the set of points $x \in E$ such that $|\tau_m(x)| \leqslant p$ for all m. At least one set E_p, say E_M, is of positive measure and may be taken for $\mathscr{E}$.)

We now apply (6·5) with $\lambda = 2$. The set $\mathscr{E}$ and the numbers q, λ determine an integer h_0 such that (6·6) holds for $n_1 > h_0$. The latter condition may be assumed satisfied here, since we may always reject a finite number of terms from $\Sigma A_{n_k}(x)$ without influencing its summability T* (although this can affect the value of the constant M). Thus

$$\tfrac{1}{4}|\mathscr{E}| \sum_k (a_k^2+b_k^2) R_{mn_k}^2 \leqslant \int_{\mathscr{E}} \tau_m^2(x)\, dx \leqslant M^2 |\mathscr{E}|,$$

$$\Sigma(a_k^2+b_k^2) R_{mn_k}^2 \leqslant 4M^2.$$

Let now $K>0$ be any fixed integer. Since $\lim\limits_{m\to\infty} R_{mn_k} = 1$, $k=1,2,\dots$, the last inequality gives

$$\sum_{k=1}^{K} (a_k^2+b_k^2) R_{mn_k}^2 \leqslant 4M^2, \qquad \sum_{k=1}^{K} (a_k^2+b_k^2) \leqslant 4M^2,$$

and the convergence of (6·2) follows.

We can remove the restriction on $\{\beta_{pq}\}$ to be row-finite, as follows. Let $\tau_m^*(x)$ be an expression analogous to $\tau_m(x)$ (cf. (6·9)), except that the upper limit of summation is not $+\infty$ but a number $N=N(m)$. We take N so large that the following conditions are satisfied:

(*a*) $|\tau_m(x)-\tau_m^*(x)|<1/m$ for $x\in E-E^m$, $|E^m|<|E|\,2^{-m-1}$;

(*b*) $\lim_{m\to\infty}(\beta_{m0}+\beta_{m1}+\ldots+\beta_{mN})=1$.

If $E^*=E^1+E^2+\ldots$, then $|E^*|<|E|$, and in the set $E-E^*$, which is of positive measure, the linear means $\tau_m^*(x)$ tend to a finite limit. But condition (*b*) ensures that the $\tau_m^*(x)$ are T* means corresponding to a matrix with only a finite number of terms different from zero in each row. Thus the general case is reduced to the special one already dealt with.

Remarks. (*a*) If the $\Sigma\rho_k^2$ is infinite, (6·4) implies that $\Sigma A_{n_k}(x)$ is non-summable almost everywhere by any method of summation. Considering, in particular, the method (C, 1) we get: *If $\Sigma\rho_k^2$ diverges, $\Sigma A_{n_k}(x)$ is not a Fourier series.*

(*b*) If $\Sigma\rho_k^2$ is infinite, then not only does the sequence of the partial sums of $\Sigma A_{n_k}(x)$ diverge almost everywhere, but so does every subsequence of this sequence. For selecting such a subsequence amounts to an application of a linear method of summation, in whose matrix each row consists entirely of zeros except for a single element 1.

(*c*) The proof of (6·4) holds if we assume that $\Sigma A_{n_k}(x)$ is merely *bounded* T* at every point of E, $|E|>0$. For some problems it is desirable to have a similar result for one-sided boundedness.

(6·10) THEOREM. *Suppose that $\Sigma\rho_k^2$ diverges, and let $\tau_m(x)$ be the* T* *means of $\Sigma A_{n_k}(x)$. Then the set of points x at which*

$$\tau_m^+(x)=o\left\{\sum_{k=1}^{\infty}\rho_k^2R_{mn_k}^2\right\}^{\frac{1}{2}}\qquad(\rho_k^2=a_k^2+b_k^2)\tag{6·11}$$

is of measure zero.

Here $\tau_m^+(x)=\max\{0,\tau_m(x)\}$. The sum in curly brackets, which we shall denote by Γ_m^2, tends to $+\infty$ with m, since $R_{mn_k}\to 1$ for fixed k. Hence (6·10) implies that *if the* T* *means of $\Sigma A_{n_k}(x)$ are bounded above (or below) at every point of a set of positive measure, the series $\Sigma\rho_k^2$ converges.*

Suppose that we have (6·11) for every $x\in E$, $|E|>0$, and that $\Sigma\rho_k^2$ diverges. Given any $\epsilon>0$, there is a set $\mathscr{E}\subset E$ with $|\mathscr{E}|>\frac{1}{2}|E|$ such that $\tau_m(x)/\Gamma_m\leqslant\epsilon$ in $\mathscr{E}$, for $m>m_0$. By dropping the few first terms of $\Sigma A_{n_k}(x)$, we may, without changing $\mathscr{E}$, suppose n_1 as large as we please. Let α_n, β_n be the Fourier coefficients of the characteristic function of the set $\mathscr{E}$. Then

$$\int_{\mathscr{E}}|\tau_m(x)|\,dx\leqslant\int_{\mathscr{E}}\{|\tau_m(x)-\epsilon\Gamma_m|+\epsilon\Gamma_m\}\,dx=\int_{\mathscr{E}}\{2\epsilon\Gamma_m-\tau_m(x)\}\,dx$$

$$=2\epsilon\Gamma_m|\mathscr{E}|-\pi\sum_{k=1}^{\infty}(\alpha_{n_k}a_k+\beta_{n_k}b_k)\,R_{mn_k}\leqslant 2\epsilon\Gamma_m|\mathscr{E}|+\pi\Gamma_m\{\Sigma(\alpha_{n_k}^2+\beta_{n_k}^2)\}^{\frac{1}{2}}.$$

The right-hand side here is less than $\epsilon\Gamma_m(2|\mathscr{E}|+\pi)$ if n_1 is large enough. This shows that

$$\int_{\mathscr{E}}|\tau_m|\,dx=o(\Gamma_m).\tag{6·12}$$

By (6·5), the left-hand side of

$$\int_{\mathscr{E}} \tau_m^2 dx \leqslant \left(\int_{\mathscr{E}} |\tau_m| dx\right)^{\frac{2}{3}} \left(\int_{\mathscr{E}} \tau_m^4 dx\right)^{\frac{1}{3}}$$

(an immediate consequence of Hölder's inequality) exceeds some fixed multiple of Γ_m^2 for n_1 large enough. By Theorem (8·20), which will be proved below, the integral $\int_{\mathscr{E}} \tau_m^4 dx$ ($\leqslant \mathfrak{M}_4^4[\tau_m]$) does not exceed a fixed multiple of Γ_m^4. Thus, $\int_{\mathscr{E}} |\tau_m|$ exceeds some fixed multiple of Γ_m. This contradicts (6·12) and proves Theorem (6·10).

In this argument we tacitly assumed that the Γ_m were finite. This follows from the hypothesis that the series defining $\tau_m(x)$ converges in a set of positive measure.

Consider the two lacunary series

$$\Sigma b^{-n\alpha} \cos b^n x = f_\alpha(x), \quad \Sigma b^{-n\alpha} \epsilon_n \cos b^n x = g_\alpha(x)$$

(already discussed in Chapter II, § 4), where α is positive, b is an integer not less than 2, and $\epsilon_n \to 0$. From (6·4) we deduce that, if $0 < \alpha \leqslant 1$, then the continuous function $f_\alpha(x)$ is differentiable at most in a set of measure zero. For

$$\frac{f_\alpha(x+h) - f_\alpha(x-h)}{2h} = -\Sigma b^{n(1-\alpha)} \sin b^n x \left(\frac{\sin b^n h}{b^n h}\right).$$

At every point of differentiability of f_α the left-hand side tends to $f'_\alpha(x)$ as $h \to 0$, which means that $S'[f_\alpha] = -\Sigma b^{n(1-\alpha)} \sin b^n x$ is summable by a linear method of summation to a finite limit. Hence, if f'_α existed in a set of positive measure, we should have $\Sigma b^{2n(1-\alpha)} < \infty$, which is false.

This result asserts less than the classical result of Weierstrass-Hardy (see p. 48) that f_α is *nowhere* differentiable if $0 < \alpha \leqslant 1$. The proof of the latter result, however, uses the special structure of the coefficients and exponents in $S[f]$, while the proof given above is valid for general lacunary series for which no such results are possible (see Example 17, p. 230). For example, the above proof shows that $g_1(x)$ is almost nowhere differentiable if $\Sigma \epsilon_n^2 = \infty$. On the other hand, we know that $g_1(x)$ is smooth and so certainly differentiable in a set of points having the power of the continuum (Chapter II, § 3).

Theorem (6·4) shows that if a lacunary series 'behaves well' on a set E of positive measure, then it 'behaves well' in $(0, 2\pi)$. We shall now give another example of this principle.

(6·13) THEOREM. (i) *Suppose that* $\Sigma A_{n_k}(x)$ *converges on a set* E, $|E| > 0$, *to a function* $f(x)$ *which coincides on* E *with another function* $g(x)$ *defined over an interval* $I = (\alpha, \beta) \supset E$ *and analytic on* I. *Then the series*

$$\Sigma(a_k \cos n_k x + b_k \sin n_k x)\,\rho^{n_k}$$

converges in some circle $|z| < 1 + \epsilon$ ($z = \rho e^{ix}$, $\epsilon > 0$).

(ii) *If* $\Sigma A_{n_k}(x)$ *converges to zero on a set* E *of positive measure, then the series vanishes identically*†.

The hypothesis concerning g means that in the neighbourhood of every point $x \in (\alpha, \beta)$, g is represented by a power series.

† The result holds if the series contains a constant term.

(6·14) LEMMA. *If H is any measurable set in $(0, 2\pi)$, then we can find a sequence of numbers $h_m \to 0$ such that for almost all $x \in H$ and for $m > m_0(x)$ the points $x \pm h_m$ are in H.*

Let $\chi(x)$ be the characteristic function of H. Then

$$I(t) = \int_0^{2\pi} |\chi(x+t) - \chi(x)| \, dx \to 0$$

as $t \to 0$. By Theorem (11·6) of Chapter I there is a sequence $k_m \to 0$ such that $\chi(x+k_m) - \chi(x) \to 0$ almost everywhere, and so also almost everywhere in E. Since χ only takes the values 0 and 1, we have $\chi(x+k_m) = \chi(x)$ for almost all $x \in E$ and $m > m_0(x)$. Moreover, $\{k_m\}$ may be a subsequence of any sequence tending to 0. Therefore, repeating the argument with the integral $J(k_m) = \int_0^{2\pi} |\chi(x-k_m) - \chi(x)| \, dx$ we obtain an $\{h_m\}$ with the required properties.

We apply this to $H = E$ in (6·13) (i). For almost all $x \in E$ and sufficiently large m,

$$\frac{g(x+h_m) - g(x-h_m)}{2h_m} = \frac{f(x+h_m) - f(x-h_m)}{2h_m} = \Sigma n_k (b_k \cos n_k x - a_k \sin n_k x) \frac{\sin n_k h_m}{n_k h_m}.$$

As $m \to \infty$, the left-hand side here tends to $g'(x)$. It follows that the series

$$\Sigma n_k (b_k \cos n_k x - a_k \sin n_k x)$$

is summable by a linear method of summation almost everywhere in E. Let $S, S', S'', \ldots$ denote respectively the series $\Sigma A_{n_k}(x)$ and the series obtained from it by successive termwise differentiations. By (6·4), $\Sigma n_k^2 (a_k^2 + b_k^2)$ converges. Hence there is a subset $E_1 \subset E$, $|E_1| = |E|$, such that S' converges in E_1 to sum $g'(x)$. Similarly, repeating the argument, there is a set $E_2 \subset E_1$, $|E_2| = |E_1|$, such that S'' converges in E_2 to sum $g''(x)$, and so on.

All the $S^{(\nu)}$ converge in the set $E^* = E E_1 E_2 \ldots$. Clearly $|E^*| = |E|$. We apply Lemma (6·5) with $\lambda = 2$, to $P = S^{(\nu)}$, $\mathscr{E} = E^*$. We may suppose n_1 so large that (6·6) holds. Thus, with $\gamma_k^2 = a_k^2 + b_k^2$,

$$\Sigma \gamma_k^2 n_k^{2\nu} \leqslant \frac{4}{|E^*|} \int_{E^*} |g^{(\nu)}(x)|^2 \, dx. \tag{6·15}$$

We may further suppose that the interval (α, β) is closed. The classical inequality of Cauchy for the coefficients of power series then gives

$$|g^{(\nu)}(x)| \leqslant M \nu! \, \delta^{-\nu} \quad (\alpha \leqslant x \leqslant \beta, \ \nu = 1, 2, \ldots)$$

with suitable M and δ. Applying this to (6·15), and keeping only one term on the left, we get

$$\gamma_k^2 n_k^{2\nu} \leqslant (2M\nu! \, \delta^{-\nu})^2 \leqslant (2M\nu^\nu \delta^{-\nu})^2,$$

$$\gamma_k^{1/n_k} \leqslant (2M)^{1/n_k} \left(\frac{\nu}{\delta n_k}\right)^{\nu/n_k}$$

If we set $\nu = [\frac{1}{2}\delta n_k]$ = integral part of $\frac{1}{2}\delta n_k$, we obtain

$$\limsup \gamma_k^{1/n_k} \leqslant 2^{-\frac{1}{2}\delta} < 1,$$

and part (i) of (6·13) is established.

As a corollary of this we have the following classical theorem.

(6·15) Theorem of Hadamard. *If a power series*

$$\Sigma c_k z^{n_k}, \quad n_{k+1}/n_k \geqslant q > 1,$$

converges for $|z|<1$ *and is analytically continuable across an arc of* $|z|=1$, *then the radius of convergence of* (6·17) *exceeds* 1.

For $\Sigma c_k e^{in_k x}$ is, by hypothesis, Abel summable to a function $g(x)$ analytic on an arc (α, β) and is also, by (6·4) and (6·3), convergent almost everywhere in (α, β).

In case (ii), $g \equiv 0$. The rejection of the first few terms of (6·1), so as to make n_1 large enough and (6·6) applicable, amounts to making g a polynomial of order $m < n_1$. Clearly $|g^{(\nu)}|$ is majorized by Mm^ν, where M is now the sum of the moduli of the coefficients of g; and (6·15) leads successively to

$$\Sigma \gamma_k^2 n_k^{2\nu} \leqslant 4M^2 m^{2\nu}, \quad \gamma_1 n_1^\nu \leqslant 2Mm^\nu.$$

The last inequality is impossible, for ν large enough, unless $\gamma_1 = 0$. Similarly $\gamma_2 = \gamma_3 = \ldots = 0$, and case (ii) is established.

Remark. It follows from (6·13) (ii) that if two lacunary series S_1 and S_2 have the same exponents (or, what amounts to the same thing, if the joint sequence of the exponents in S_1 and S_2 is still lacunary), and if they converge to the same sum on a set of positive measure, then $S_1 \equiv S_2$. The result holds for any two lacunary series, but the proof is then more difficult.

7. Riesz products

Consider the infinite product

$$\prod_{\nu=1}^{\infty} (1 + \alpha_\nu \cos n_\nu x), \tag{7·1}$$

where the positive integers n_ν satisfy a condition

$$n_{\nu+1}/n_\nu \geqslant q > 1,$$

and $-1 \leqslant \alpha_\nu \leqslant 1$, $\alpha_\nu \neq 0$ for all ν. Let

$$\mu_k = n_k + n_{k-1} + \ldots + n_1, \quad \mu_k' = n_{k+1} - n_k - \ldots - n_1 \quad (k = 1, 2, \ldots).$$

Then
$$\mu_k < n_k(1 + q^{-1} + q^{-2} + \ldots) = n_k q/(q-1),$$
$$\mu_k' > n_{k+1}(1 - q^{-1} - q^{-2} - \ldots) \geqslant n_k q(q-2)/(q-1).$$

Thus $\mu_k' > \mu_k$ if $q - 2 \geqslant 1$, that is, if $q \geqslant 3$, which we assume henceforth.

The kth partial product of (7·1) is a non-negative trigonometric polynomial

$$p_k(x) = 1 + \sum_{\nu=1}^{\mu_k} \gamma_\nu \cos \nu x = \prod_{i=1}^{k} (1 + \alpha_i \cos n_i x), \tag{7·2}$$

where $\gamma_\nu = 0$ if ν is not of the form $n_{i'} \pm n_{i''} \pm \ldots$, with $k \geqslant i' > i'' > \ldots$. The difference

$$p_{k+1} - p_k = p_k \alpha_{k+1} \cos n_{k+1} x$$

is a polynomial whose lowest term is of rank $\mu_k' > \mu_k$. Hence the passage from p_k to p_{k+1} consists in adding to (7·2) a group of terms whose ranks all exceed μ_k. Making $k \to \infty$, we obtain from (7·2) an infinite series

$$1 + \sum_{\nu=1}^{\infty} \gamma_\nu \cos \nu x \tag{7·3}$$

in which $\gamma_\nu = 0$ if $\nu \neq n_i \pm n_{i'} \pm n_{i''} \pm \ldots$, $i > i' > \ldots$. We shall say that (7·3) *represents* the product (7·1). The partial sums $s_n(x)$ of (7·3) have the property $s_{\mu_k}(x) = p_k(x) \geqslant 0$. It follows from Theorem (5·20) of Chapter IV that (7·3) is the Fourier-Stieltjes series of a non-decreasing continuous function $F(x)$. This function is obtained by integrating (7·3) termwise. In particular

$$F(x) - F(0) = \lim_{k\to\infty} \int_0^x p_k(t)\,dt. \tag{7·4}$$

Thus

(7·5) Theorem. *The series* (7·3) *representing the product* (7·1), *with* $n_{k+1}/n_k \geqslant 3$, $-1 \leqslant \alpha_\nu \leqslant +1$, *is the Fourier-Stieltjes series of a non-decreasing continuous function* F *defined by* (7·4).

The series (7·3) is formally obtained by multiplying out (7·1) and replacing the products of cosines by linear combinations of cosines. *No two terms thus obtained are of the same rank*, since every integer N can be represented in the form $n_i \pm n_{i'} \pm n_{i''} \pm \ldots$, with $i > i' > i'' > \ldots$, at most once. (Such sums, being greater than μ'_{i-1}, must be positive.) For suppose we have another representation $N = n_k \pm n_{k'} \pm \ldots$, $k > k' > \ldots$, with $k \neq i$, say $k < i$. Then $n_i = an_{i-1} + bn_{i-2} + cn_{i-3} + \ldots$, where a, b, c, ... take only the values 0, ± 1 and ± 2. The right-hand side of this equation is less than

$$2n_{i-1}(1 + 3^{-1} + 3^{-2} + \ldots) = 3n_{i-1},$$

and so cannot be equal to n_i. Hence $k = i$, and we have $n_{i'} \pm n_{i''} \pm \ldots = n_{k'} \pm n_{k''} \pm \ldots$. This gives $i' = k'$, and so on.

In particular, $\gamma_{n_\nu} = \alpha_\nu$. If α_ν does not tend to 0 (e.g. if $\alpha_\nu = 1$, $n_\nu = 3^\nu$) we obtain, with F. Riesz, a new example (historically the first) of a continuous function of bounded variation whose Fourier coefficients are not $o(1/n)$.

The products (7·1) are called *Riesz products*.

(7·6) Theorem. *If* $-1 \leqslant \alpha_\nu \leqslant 1$, $n_{\nu+1}/n_\nu \geqslant q > 3$, *and* $\Sigma\alpha_\nu^2 = \infty$, *then the function* F *of* (7·4) *has a derivative* 0 *almost everywhere.*

By Chapter III, (8·1), the series (7·3) is almost everywhere summable (C, 1) to sum $F'(x)$. The series has infinitely many gaps (μ_k, μ'_k), and since

$$\mu'_k/\mu_k \geqslant n_{k+1}(1 - q^{-1} - q^{-2} - \ldots)/n_k(1 + q^{-1} + q^{-2} + \ldots) \geqslant q - 2 > 1,$$

Chapter III, (1·27) shows that the partial products $p_k(x)$ converge to $F'(x)$ almost everywhere. The inequality $1 + u \leqslant e^u$ gives

$$0 \leqslant p_k(x) \leqslant \exp\left(\sum_{\nu=1}^{k} \alpha_\nu \cos n_\nu x\right).$$

Since $\Sigma\alpha_\nu^2 = \infty$, the partial sums of $\Sigma\alpha_\nu \cos n_\nu x$ take arbitrarily large negative values at almost all points (see (6·10)). Thus $\liminf p_k(x) = 0$, that is, $F'(x) = 0$ almost everywhere. Incidentally we have proved that *the product* (7·1) *converges to* 0 *almost everywhere.*

Remark. Using Theorem (6·3) we easily prove that if in (7·6) we assume that $\Sigma\alpha_\nu^2 < \infty$, then (7·1) converges almost everywhere to finite values different from 0.

(7·7) Theorem. *If* $\alpha_\nu \to 0$, $\Sigma\alpha_\nu^2 = \infty$, *and* $n_{\nu+1}/n_\nu \geqslant q > 3$, *then both the series* (7·3) *representing* (7·1) *and its conjugate series converge almost everywhere, the former to zero.*

The partial sums $s_{\mu_k}(x)$ of (7·3) converge almost everywhere. The same holds for the partial sums $\tilde{s}_{\mu_k}(x)$ of the conjugate series since the latter is summable (C, 1) almost everywhere and has the same gaps as (7·3). If $t_n = s_n + i\tilde{s}_n$ are the partial sums of $1 + \sum_1^\infty \gamma_\nu e^{i\nu x}$, then $M(x) = \sup_k |t_{\mu_k}(x)|$ is finite almost everywhere.

Take any point x at which t_{μ_k} converges, so that $M = M(x)$ is finite, fix k and let A be so large that

$$\left|\sum_{\mu}^{\mu_{k-1}} \gamma_\nu e^{i\nu x}\right| \leqslant A \quad (1 \leqslant \mu \leqslant \mu_{k-1}), \tag{7·8$_{k-1}$}$$

$$\left|\sum_{\mu}^{\mu_i} \gamma_\nu e^{i\nu x}\right| \leqslant A - 2M \quad (\mu_{i-1} < \mu \leqslant \mu_i;\ i = 1, 2, \ldots, k-1;\ \mu_0 = 0). \tag{7·9$_{k-1}$}$$

The number A *prima facie* depends on k, but we give an inductive proof that the inequalities are true for all k and an A independent of k.

We have
$$s_{\mu_k} = \left(1 + \sum_{\nu=1}^{\mu_{k-1}} \gamma_\nu \cos \nu x\right)(1 + \alpha_k \cos n_k x)$$
$$= s_{\mu_{k-1}} + \alpha_k \cos n_k x + \tfrac{1}{2}\alpha_k \sum_{\nu=1}^{\mu_{k-1}} \gamma_\nu[\cos(n_k - \nu)x + \cos(n_k + \nu)x]. \tag{7·10}$$

Since $n_k \pm \nu > 0$, the passage from s_{μ_k} to t_{μ_k} consists of replacing cosines by exponentials. We shall now estimate t_λ for $\mu_{k-1} < \lambda \leqslant \mu_k$.

Consider separately the cases

$$(a)\quad \mu_{k-1} < \lambda < n_k, \qquad (b)\quad n_k \leqslant \lambda \leqslant \mu_k.$$

In case (a), as we see from (7·10),

$$t_\lambda = t_{\mu_{k-1}} \quad \text{or} \quad t_\lambda = t_{\mu_{k-1}} + \tfrac{1}{2}\alpha_k \sum_{\nu=n_k-\lambda}^{\mu_{k-1}} \gamma_\nu e^{i(n_k-\nu)x},$$

according as $\lambda < n_k - \mu_{k-1}$ or $\lambda \geqslant n_k - \mu_{k-1}$. In the latter case, the last term on the right is absolutely $\leqslant \frac{1}{2}|\alpha_k| \left|\sum_{n_k-\lambda}^{\mu_{k-1}} \gamma_\nu e^{i\nu x}\right| \leqslant \frac{1}{2}|\alpha_k| A$, by (7·8$_{k-1}$). In case (b),

$$t_{n_k} = t_{n_k-1} + \alpha_k e^{in_k x}, \quad t_\lambda = t_{\mu_k} - \tfrac{1}{2}\alpha_k \sum_{\lambda-n_k+1}^{\mu_{k-1}} \gamma_\nu e^{i(n_k+\nu)x} \quad \text{for} \quad \lambda > n_k,$$

and the last term is again absolutely $\leqslant \frac{1}{2}|\alpha_k| A$. Hence

$$|t_\lambda - t_{\mu_{k-1}}| \leqslant \tfrac{1}{2}|\alpha_k|(A+2) \quad \text{or} \quad |t_\lambda - t_{\mu_k}| \leqslant \tfrac{1}{2}|\alpha_k| A \tag{7·11}$$

for $\mu_{k-1} < \lambda \leqslant n_k$ or $n_k < \lambda \leqslant \mu_k$ respectively (the additional $|\alpha_k|$ on the right of the first inequality being actually needed for $\lambda = n_k$ only). In particular,

$$|t_\lambda| \leqslant M + \tfrac{1}{2}|\alpha_k|(A+2) \leqslant M + \tfrac{1}{2}A + 1$$

for all λ in the range $(\mu_{k-1}+1, \mu_k)$.

Suppose now that A is so large that

$$2M + (\tfrac{1}{2}A + 1) \leqslant A - 2M.$$

Then, if $\mu_{k-1} < \mu \leqslant \mu_k$,

$$\left|\sum_{\mu}^{\mu_k} \gamma_\nu e^{i\nu x}\right| \leqslant |t_{\mu_k}| + |t_{\mu-1}| \leqslant M + M + \tfrac{1}{2}A + 1 \leqslant A - 2M.$$

If $\mu_{j-1} < \mu \leqslant \mu_j$, $j < k$, we have

$$\left|\sum_{\mu}^{\mu_k} \gamma_\nu e^{i\nu x}\right| \leqslant \left|\sum_{\mu}^{\mu_j}\right| + \left|\sum_{\mu_j+1}^{\mu_k}\right| \leqslant A - 2M + 2M = A.$$

Thus ($7\cdot9_{k-1}$) and ($7\cdot8_{k-1}$) imply ($7\cdot9_k$) and ($7\cdot8_k$), which shows that these hold for all k. In particular, the $t_\lambda(x)$ are bounded, even if α_ν does not tend to 0. If $\alpha_\nu \to 0$ then, by (7·11), t_λ converges almost everywhere. Since s_{μ_k} converges to 0 almost everywhere, so does s_λ. This completes the proof of (7·7).

(7·12) THEOREM. *Let $n_{\nu+1}/n_\nu \geqslant 3$ for all ν, and let $\Sigma\alpha_\nu^2 < \infty$. Then (a) the (complex-valued) series*

$$1 + \sum_1^\infty \delta_\nu \cos \nu x \tag{7·13}$$

representing the product

$$\prod_1^\infty (1 + i\alpha_\nu \cos n_\nu x) \tag{7·14}$$

is the Fourier series of a bounded function; (b) if $n_{\nu+1}/n_\nu \geqslant q > 3$, the series (7·13) and its conjugate converge almost everywhere.

Here the δ_ν are obtained from the $i\alpha_\nu$ in the same way as the γ_ν were obtained from the α_ν. The δ_ν are either real or purely imaginary. Obviously

$$\left|\prod_{\nu=1}^{k} (1 + i\alpha_\nu \cos n_\nu x)\right| \leqslant \prod_1^k (1 + \alpha_\nu^2)^{\frac12} < \prod_1^\infty (1 + \alpha_\nu^2)^{\frac12} < \infty,$$

and so there is a subsequence of the partial sums of (7·13) which is uniformly bounded. This proves (*a*). (See Chapter IV, p. 148).

If t_n denotes the partial sums of $1 + \sum_1^\infty \delta_\nu e^{i\nu x}$, the proof of (*b*) is a repetition of that of Theorem (7·7) with minor modifications caused by the terms of (7·13) being imaginary. For clearly the partial sums of order μ_k of both (7·13) and its conjugate $\Sigma\delta_\nu \sin \nu x$ converge almost everywhere. Hence t_{μ_k} converges almost everywhere, and the proof analogous to that of (7·7) shows that t_λ converges almost everywhere. It is now enough to observe that at each x where both $t_\lambda(x)$ and $t_\lambda(-x)$ converge, we have the convergence of $\Sigma\delta_\nu \cos \nu x$ and $\Sigma\delta_\nu \sin \nu x$.

Remarks. (*a*) Theorems (7·6), (7·7) and (7·12) remain valid if in (7·1) and (7·14) $\alpha_\nu \cos n_\nu x$ is replaced by $\alpha_\nu \cos n_\nu x + \beta_\nu \sin n_\nu x = \rho_\nu \cos(n_\nu x + \theta_\nu)$, with obvious conditions on the ρ_ν. The proofs remain the same.

(*b*) The indices of the non-zero terms of (7·3) and (7·13) are confined to the intervals (μ'_{k-1}, μ_k). Since the latter interval contains n_k, and since

$$\mu_k/\mu'_{k-1} \leqslant n_k(1 + q^{-1} + \ldots)/n_k(1 - q^{-1} - \ldots) = q/(q-2),$$

we see that no matter how small ϵ is, $\epsilon > 0$, the indices of the non-zero terms of these series will lie in the intervals $(n_k(1-\epsilon), n_k(1+\epsilon))$, provided q is large enough, $q > q_0(\epsilon)$. We shall use this remark later (Chapter VI, § 6).

(*c*) Theorems (7·6), (7·7) and (7·12) (*b*) remain valid for $n_{\nu+1}/n_\nu \geqslant 3$.

For (7·6) this is proved by splitting (7·1) into two subproducts, corresponding respectively to even or odd ν. At least one subproduct satisfies the hypotheses of (7·6), with $q = 9$. Hence in virtue of the remark to Theorem (7·6) $p_k(x)$ converges to 0 almost everywhere. Using the fact (to be proved in Chapter XIII, (5·13)) that if a series

$\Sigma A_k(x)$ is summable (C, 1) almost everywhere to sum $\sigma(x)$ and if a sequence $\{s_{n_k}\}$ of its partial sums converges almost everywhere to limit $s(x)$, then $s(x)=\sigma(x)$ almost everywhere, we see that $F'(x)=0$ almost everywhere.

The extensions of (7·7) and (7·12)(*b*) are based on a theorem (see Chapter XIII, p. 176, Remark (i)) that if a sequence of the partial sums of an $S[f]$ or $S[dF]$ converges almost everywhere, so does the same sequence of the partial sums of the conjugate series. In our case, $s_{\mu_k}(x)$ converges almost everywhere, and so the same holds for $\tilde{s}_{\mu_k}(x)$. From this point on, the proofs remain unchanged.

8. Rademacher series and their applications

Several properties of lacunary trigonometric series are shared by *Rademacher series*

$$\sum_{\nu=0}^{\infty} c_\nu \phi_\nu(t), \tag{8·1}$$

the functions ϕ_ν being those defined in Chapter I, § 3. This is not entirely surprising in view of the definition

$$\phi_\nu(t)=\operatorname{sign}\sin 2^{\nu+1}\pi t.$$

Rademacher series have a close connexion with the calculus of probabilities and are typical of a very large class of series arising there. We shall need only simple properties of (8·1) which can be proved directly.

We suppose that the c_ν in (8·1) may be complex numbers.

(8·2) THEOREM. *The series* (8·1) *converges almost everywhere if* $\Sigma |c_\nu|^2<\infty$. *If* $\Sigma|c_\nu|^2=\infty$, *then, whatever the method* T* *of summation,* (8·1) *is almost everywhere non-summable* T*.

The proof of the second part of (8·2) follows the same line as that of (6·4), and may be left to the reader. We need only observe that the system of functions

$$\phi_{j,k}(t)=\phi_j(t)\,\phi_k(t) \quad (0\leqslant j<k<\infty),$$

is orthonormal over (0, 1). (Similarly we can prove an analogue of (6·10).)

If $\Sigma|c_\nu|^2\neq\infty$, the series (8·1), with partial sums $s_n(t)$, is the Fourier series of a function $f\in L^2$. Moreover (see Chapter IV, (1·1))

$$\int_0^1 |f-s_n|^2\,dt\to 0, \quad \int_0^1 |f-s_n|\,dt\to 0, \quad \int_a^b (s_n-f)\,dt\to 0,$$

where $0\leqslant a<b\leqslant 1$. The third relation, which holds uniformly in a and b, is a consequence of the second, and the second follows from the first by an application of Schwarz's inequality.

Let $F(t)$ be the indefinite integral of $f(t)$, and let E, $|E|=1$, be the set of points where $F'(t)$ exists and is finite. We have just proved that the integral of s_n over any interval I tends to the integral of f over I. Therefore the integral of s_n-s_{k-1} over I tends to the integral of $f-s_{k-1}$, as $n\to\infty$. Let I be of the form $(l2^{-k}, (l+1)\,2^{-k})$, $l=0, 1, \ldots, 2^k-1$. Since the integral of $\phi_j(t)$ over I is zero for $j\geqslant k$, the integral of s_n-s_{k-1} over I is zero. It follows that for the intervals I just mentioned the integral of f over I equals the integral of s_{k-1} over I.

Let now $t_0 \neq p/2^q$, $t_0 \in E$, $t_0 \in I_k = (l2^{-k}, (l+1)2^{-k})$. Since s_{k-1} is constant over I_k,

$$s_{k-1}(t_0) = \frac{1}{|I_k|}\int_{I_k} s_{k-1}(t)\,dt = \frac{1}{|I_k|}\int_{I_k} f(t)\,dt \to F'(t_0) \quad \text{as} \quad k\to\infty,$$

which completes the proof of (8·2).

The analogue of Lemma (6·5) will be needed later and so we state it separately.

(8·3) Lemma. *Given any set* $\mathscr{E}\subset(0,1)$ *and any number* $\lambda>1$, *there is an integer* $h_0 = h_0(\mathscr{E},\lambda)$ *such that for any finite sum* $P(t) = \sum_{h_0}^{N} c_k\phi_k(t)$,

$$\lambda^{-1}|\mathscr{E}|\Sigma|c_k|^2 \leqslant \int_{\mathscr{E}} |P(t)|^2\,dt \leqslant \lambda|\mathscr{E}|\Sigma|c_k|^2.$$

The result holds for $N=\infty$ *provided* $\Sigma|c_k|^2<\infty$.

The proof is similar to that of Lemma (6·5) and we leave it to the reader.

(8·4) Theorem. *If* $\Sigma|c_\nu|^2<\infty$, *the sum* $f(t)$ *of* (8·1) *belongs to* L^r *for all* $r>0$. *More precisely,*

$$A_r(\Sigma|c_\nu|^2)^{\frac12} \leqslant \left\{\int_0^1 |f|^r\,dt\right\}^{1/r} \leqslant B_r(\Sigma|c_\nu|^2)^{\frac12} \quad (r>0), \tag{8·5}$$

where A_r, B_r *are positive and finite, and depend only on* r. *Moreover* $B_r\leqslant 2k^{\frac12}$, *where* $2k$ *is the least even integer not less than* r.

We suppose first that the c_ν are real and that $r=2k$ is an even integer. Then

$$\int_0^1 s_n^{2k}(t)\,dt = \Sigma A_{\alpha_1\alpha_2\ldots\alpha_j} c_{m_1}^{\alpha_1} c_{m_2}^{\alpha_2}\ldots c_{m_j}^{\alpha_j}\int_0^1 \phi_{m_1}^{\alpha_1}\ldots\phi_{m_j}^{\alpha_j}\,dt, \tag{8·6}$$

where $A_{\alpha_1\alpha_2\ldots\alpha_j} = (\alpha_1+\alpha_2+\ldots+\alpha_j)!/\alpha_1!\,\alpha_2!\ldots\alpha_j!$, and $\alpha_1, \alpha_2, \ldots, \alpha_j$ are any positive integers whose sum is $2k$. The indices $m_1, m_2, \ldots, m_j$ vary between 0 and n. It is easily verified that the integrals on the right vanish unless $\alpha_1, \alpha_2, \ldots, \alpha_j$ are all even, in which case the integrals are equal to 1. Observing that

$$\Sigma A_{\beta_1\beta_2\ldots\beta_j} c_{m_1}^{2\beta_1} c_{m_2}^{2\beta_2}\ldots c_{m_j}^{2\beta_j} = (c_1^2+c_2^2+\ldots+c_n^2)^k \; (\beta_1+\beta_2+\ldots+\beta_j=k)$$

we obtain the second inequality (8·5) with $f=s_n$, $r=2k$, and B_{2k}^{2k} equal to the upper bound of the ratio $A_{2\beta_1\ldots2\beta_j}/A_{\beta_1\ldots\beta_j}$. Since $s_n(t)\to f(t)$ almost everywhere, the inequality for f follows.

If we observe that

$$\frac{A_{2\beta_1,\ldots,2\beta_j}}{A_{\beta_1\ldots\beta_j}} = \frac{(k+1)(k+2)\ldots2k}{\Pi(\beta_i+1)\ldots2\beta_i}$$

we see that

$$B_{2k}^{2k} \leqslant (k+1)(k+2)\ldots2k/2^k \leqslant k^k, \quad B_{2k}\leqslant k^{\frac12}.$$

The second inequality (8·5), being true for $r=2, 4, \ldots$, must hold for any $r>0$, since $\mathfrak{M}_r[f;\,0,1]$ is a non-decreasing function of r (Chapter I, (10·12) (i)). Clearly $B_r\leqslant k^{\frac12}$, where $2k$ is the least even integer not less than r.

The first inequality (8·5) is immediate for $r\geqslant2$, for then

$$\mathfrak{M}_r[f]\geqslant\mathfrak{M}_2[f] = (\Sigma c_\nu^2)^{\frac12} = \gamma,$$

say. If $0<r<2<4$, let t_1 and t_2 be positive and such that $t_1+t_2=1$, $2=rt_1+4t_2$. The function $\mathfrak{M}_\alpha^\alpha[f]$ being logarithmically convex in α (Chapter I, (10·12) (ii)),

$$\gamma^2 = \mathfrak{M}_2^2[f] \leqslant \mathfrak{M}_r^{rt_1}\mathfrak{M}_4^{4t_2} \leqslant \mathfrak{M}_r^{rt_1}[f]\,(2^{\frac12}\gamma)^{4t_2},$$

which gives $\mathfrak{M}_r[f]\geqslant\gamma2^{-(2-r)/r}$.

If $c_\nu = c'_\nu + ic''_\nu$ and $f = f' + if''$ are complex, then

$$\mathfrak{M}_r[f] \leqslant \mathfrak{M}_r[f'] + \mathfrak{M}_r[f''] \leqslant B_r\{(\Sigma c'^2_\nu)^{\frac{1}{2}} + (\Sigma c''^2_\nu)^{\frac{1}{2}}\} \leqslant 2B_r(\Sigma |c_\nu|^2)^{\frac{1}{2}},$$

and the second inequality (8·5) follows with B_r doubled. Also if, for example, $(\Sigma c'^2_\nu)^{\frac{1}{2}} \geqslant (\Sigma c''^2_\nu)^{\frac{1}{2}}$, then

$$\mathfrak{M}_r[f] \geqslant \mathfrak{M}_r[f'] \geqslant A_r(\Sigma c'^2_\nu)^{\frac{1}{2}} \geqslant \tfrac{1}{2}A_r(\Sigma |c_\nu|^2)^{\frac{1}{2}},$$

which gives the first inequality (8·5) with half the previous A_r.

The estimate $B_{2k} \leqslant 2k^{\frac{1}{2}}$ enables us to strengthen the second inequality (8·5).

(8·7) THEOREM. *If* $\Sigma |c_\nu|^2 < \infty$, $\exp\{\mu |f(t)|^2\}$ *is integrable for every* $\mu > 0$.

For

$$\int_0^1 \exp(\mu |f|^2)\, dt = \sum_{k=0}^{\infty} \frac{\mu^k}{k!} \int_0^1 |f|^{2k}\, dt \leqslant \sum_{k=0}^{\infty} \frac{k^k}{k!} (4\mu\gamma^2)^k. \tag{8·8}$$

Since $k^k/k! < \Sigma k^n/n! = e^k$, the series on the right converges if $4e\mu\gamma^2 < 1$, that is if γ is small enough. It follows that for every $\mu > 0$ the function $\exp(\mu |f - s_n|^2)$ is integrable if only n is large enough. Since $|f|^2 \leqslant 2[|f - s_n|^2 + |s_n|^2]$, and $s_n(t)$ is bounded, the integrability of $\exp(\mu |f|^2)$ follows.

Theorems on Rademacher series enable us to prove some results about the series

$$\pm \tfrac{1}{2}a_0 + \sum_{n=1}^{\infty} \pm (a_n \cos nx + b_n \sin nx), \tag{8·9}$$

which we obtain from the standard series

$$\tfrac{1}{2}a_0 + \sum_{n=1}^{\infty} (a_n \cos nx + b_n \sin nx) = \sum_0^{\infty} A_n(x) \tag{8·10}$$

by changing the signs of the terms of the latter in an arbitrary way. Neglecting the sequences ± 1 containing only a finite number of $+1$ or of -1, we may write (8·9) in the form

$$\sum_{n=0}^{\infty} A_n(x)\, \phi_n(t), \tag{8·11}$$

where the ϕ_n are the Rademacher functions and the parameter t, $t \neq p/2^q$, runs through the interval $(0, 1)$. If the values of t for which the series (8·11) has some property P form a set of measure 1, we shall say that *almost all* the series (8·9) possess the property P.

(8·12) THEOREM. *If*

$$\tfrac{1}{4}a_0^2 + \sum_{n=1}^{\infty} (a_n^2 + b_n^2) \tag{8·13}$$

is finite, then almost all series (8·9) *converge almost everywhere in the interval* $0 \leqslant x \leqslant 2\pi$. *If* (8·13) *is infinite, then, whatever method* T* *of summability we consider, almost all series* (8·9) *are almost everywhere non-summable* T*.

Let $S_t(x)$ denote the series (8·11), and if the latter converges let $S_t(x)$ also denote its sum. Let E be the set of points (x, t) of the rectangle $0 \leqslant x \leqslant 2\pi$, $0 \leqslant t \leqslant 1$ where the series converges. If (8·13) is finite then, by (8·2), the intersection of E with every line $x = x_0$ is of measure 1. Since E is measurable, its plane measure is 2π, and therefore the intersection of E with almost every line $t = t_0$ is of measure 2π. This is just the first part of (8·12). The second part follows by the same argument provided we can show that *the divergence of* (8·13) *implies the divergence of* $A_1^2(x) + A_2^2(x) + \ldots$ *for almost all* x.

To establish this, suppose that $A_1^2(x)+A_2^2(x)+\ldots$ converges in a set H, $|H|>0$. Then there is a subset H' of H, $|H'|>0$, and a constant M such that in H' the sum of our series does not exceed M. Let $A_n(x)=\rho_n\cos(nx+\xi_n)$, $\rho_n\geqslant 0$. Integrating the series over H' we get

$$\sum_{n=1}^{\infty}\rho_n^2\int_{H'}\cos^2(nx+\xi_n)\,dx\leqslant M\,|H'|.$$

Since the coefficient of ρ_n^2 tends to $\frac{1}{2}|H'|>0$ (see Chapter II, (4·5)), the convergence of $\Sigma\rho_n^2$ follows, contrary to hypothesis. Thus (8·12) is proved. As a corollary, taking for example $\mathrm{T}^*=(\mathrm{C},1)$, we get

(8·14) THEOREM. *If* (8·13) *diverges, almost all the series* (8·9) *are not Fourier series.*

The theorem of Riesz-Fischer asserts that if (8·13) is finite, (8·10) is a Fourier series. We now see that the Riesz-Fischer theorem is in a way the best possible, since:

(8·15) THEOREM. *No condition on the moduli of the numbers a_n, b_n which permits* (8·13) *to diverge can possibly be a sufficient condition for* (8·10) *to be a Fourier series.*

(8·16) THEOREM. *If* (8·13) *is finite, then for almost all t the sum $S_t(x)$ of* (8·9) *belongs to every* L^r, $r>0$. *More generally, for any μ, $\exp\{\mu S_t^2(x)\}$ is integrable over $0\leqslant x\leqslant 2\pi$ for almost all t.*

Let γ^2 denote the sum of (8·13), and let μ be so small that the right-hand side of (8·8) converges. If $K=K(\mu,\gamma)$ is the sum of the latter series, we have as in (8·7)

$$\int_0^1\exp\{\mu S_t^2(x)\}\,dt\leqslant K.$$

Integrate this over $0\leqslant x\leqslant 2\pi$ and interchange the order of integration; then

$$\int_0^1 dt\int_0^{2\pi}\exp\{\mu S_t^2(x)\}\,dx\leqslant 2\pi K. \tag{8·17}$$

The inner integral here is finite for almost all t. To remove the assumption that μ is small we argue as in the proof of (8·7).

We shall now consider lacunary series

$$\Sigma(a_k\cos n_kx+b_k\sin n_kx)\quad(n_{k+1}/n_k\geqslant q>1) \tag{8·18}$$

and the sums

$$\gamma^2=\Sigma(a_k^2+b_k^2). \tag{8·19}$$

(8·20) THEOREM. *Suppose that $n_{k+1}/n_k\geqslant q>1$ for all k and that* (8·19) *is finite, so that* (8·18) *is an* $\mathsf{S}[f]$. *Then*

$$A_{r,q}\{\Sigma(a_k^2+b_k^2)\}^{\frac{1}{2}}\leqslant\left\{\frac{1}{2\pi}\int_0^{2\pi}|f|^r\,dx\right\}^{1/r}\leqslant B_{r,q}\{\Sigma(a_k^2+b_k^2)\}^{\frac{1}{2}} \tag{8·21}$$

for every $r>0$, where $A_{r,q}$ and $B_{r,q}$ depend on r and q only. If $\gamma\leqslant 1$, then also

$$\int_0^{2\pi}\exp\mu f^2\,dx\leqslant C, \tag{8·22}$$

provided $\mu\leqslant\mu_0(q)$, with C an absolute constant.

It is enough to prove (8·22), since then the second inequality (8·21) follows. The first inequality (8·21) follows from the second by the convexity argument used in the proof of (8·4).

We first suppose that $q \geqslant 3$, and consider the series

$$S_t(x) = \sum_{\nu=1}^{\infty} (a_\nu \cos n_\nu x + b_\nu \sin n_\nu x)\, \phi_{n_\nu}(t).$$

Then (8·17) is valid, provided μ is small enough, with K an absolute constant. It follows that there is a $t_0 \neq p/2^q$ such that

$$\int_0^{2\pi} \exp\{\mu S_{t_0}^2(x)\}\, dx \leqslant 2\pi K. \tag{8·23}$$

Consider the Riesz product (§ 7)

$$p_k(x) = \prod_{\nu=1}^{k} (1 + \phi_{n_\nu}(t_0) \cos n_\nu x) = 1 + \Sigma \gamma_\nu \cos \nu x.$$

We have $\gamma_{n_\nu} = \phi_{n_\nu}(t_0)$ for $\nu = 1, 2, \ldots, k$, and

$$s_{n_k}(x, f) = \sum_{\nu=1}^{k} A_{n_\nu}(x) = \frac{1}{\pi} \int_0^{2\pi} S_{t_0}(x+u)\, p_k(u)\, du.$$

The function $\chi(v) = \exp(\mu v^2)$ is increasing and convex for $v \geqslant 0$. The function $(2\pi)^{-1} p_k(x)$ is non-negative, and its integral over $(0, 2\pi)$ is 1. Jensen's inequality therefore gives

$$\chi(\tfrac{1}{2} |s_{n_k}(x,f)|) \leqslant \chi\left(\frac{1}{2\pi} \int_0^{2\pi} |S_{t_0}(x+u)|\, p_k(u)\, du\right) \leqslant \frac{1}{2\pi} \int_0^{2\pi} \chi(|S_{t_0}(x+u)|)\, p_k(u)\, du,$$

$$\int_0^{2\pi} \chi(\tfrac{1}{2} |s_{n_k}(x,f)|)\, dx \leqslant \int_0^{2\pi} \chi(|S_{t_0}(x)|)\, dx \leqslant 2\pi K,$$

by (8·23). So, making $k \to \infty$,

$$\int_0^{2\pi} \exp(\tfrac{1}{4} \mu f^2)\, dx \leqslant 2\pi K. \tag{8·24}$$

This is just (8·22) except that μ is replaced by $\frac{1}{4}\mu$. The right-hand side here, $2\pi K$, is an absolute constant, since $\gamma \leqslant 1$ and μ is small enough.

For general $q > 1$ we decompose (8·18) into Q lacunary series for each of which $q \geqslant 3$. (For Q we may take the least integer y such that $q^y \geqslant 3$). Correspondingly, $f = f_1 + f_2 + \ldots + f_Q$. By Jensen's inequality, and by (8·22) in the case $q \geqslant 3$,

$$\int_0^{2\pi} \exp\{\mu (f/Q)^2\}\, dx \leqslant Q^{-1} \sum_{k=1}^{Q} \int_0^{2\pi} \exp\{\mu f_k^2\}\, dx \leqslant C,$$

since the γ's corresponding to the f_k are not greater than 1. This proves (8·22) in the general case.

The device used in the proof of (8·7) shows that, under the hypotheses of (8·20), the left-hand side of (8·22) is finite for every $\mu > 0$.

In what follows $\qquad f^+ = \max\{f, 0\}, \quad f^- = \max\{-f, 0\}.$

(8·25) THEOREM. *Suppose that* $\gamma^2 = \Sigma(a_k^2 + b_k^2) < \infty$ *and write* $f(x) = \Sigma A_{n_k}(x)$. *Then both* f^+ *and* f^-, *and so also* $|f|$, *are not less than* $\gamma \lambda_q$ *in sets of points of measure not less than* $2\pi\mu_q$, *where* λ_q *and* μ_q *are positive numbers depending on* q *only.*

The proof is based on the following lemma, useful also in other problems:

(8·26) LEMMA. *Suppose that* $g(x) \geqslant 0$ *is defined in a set* E, $|E| > 0$, *and that*

$$\text{(i)} \quad \frac{1}{|E|} \int_E g\, dx \geqslant A > 0, \qquad \text{(ii)} \quad \frac{1}{|E|} \int_E g^2\, dx \leqslant B.$$

Then for any $0<\delta<1$ *the subset* E_δ *of* E *in which* $g(x)\geqslant \delta A$ *is of measure not less than* $|E|(1-\delta)^2(A^2/B)$.

The integral of g over the set $E-E_\delta$, in which $g<\delta A$, is less than $\delta A\,|E|$. Hence the integral of g over E_δ exceeds $(1-\delta)A\,|E|$, by (i). On the other hand,

$$\int_{E_\delta} g\,dx \leqslant \left(\int_{E_\delta} g^2\,dx\right)^{\frac{1}{2}} |E_\delta|^{\frac{1}{2}} \leqslant (B\,|E|)^{\frac{1}{2}}\,|E_\delta|^{\frac{1}{2}},$$

by (ii). Hence
$$A\,|E|\,(1-\delta)\leqslant (B\,|E|)^{\frac{1}{2}}\,|E_\delta|^{\frac{1}{2}},$$
that is,
$$|E_\delta|\geqslant |E|\,(A^2/B)(1-\delta)^2.$$

Return to (8·25). It is enough to prove the result for f^+. Since the integral of f over $(0,2\pi)$ is 0,

$$\frac{1}{2\pi}\int_0^{2\pi} f^+\,dx=\frac{1}{4\pi}\int_0^{2\pi}|f|\,dx\geqslant \tfrac{1}{2}A_{1,q}\gamma$$

(see (8·21)). Since
$$\frac{1}{2\pi}\int_0^{2\pi}(f^+)^2\,dx\leqslant \frac{1}{2\pi}\int_0^{2\pi} f^2\,dx=\tfrac{1}{2}\gamma^2,$$
an application of (8·26) with $\delta=\frac{1}{2}$ shows that f^+ exceeds $\frac{1}{4}\gamma A_{1,q}=\gamma\lambda_q$ in a set of measure not less than $2\pi\,\frac{1}{8}A_{1,q}^2=2\pi\mu_q$.

The following analogue of (8·25) for Rademacher functions will be needed later:

(8·27) Theorem. *Let* $f(x)=\Sigma c_n\phi_n(x)$, $0\leqslant x\leqslant 1$, *where the* c_n *are real and* $\gamma^2=\Sigma c_n^2<\infty$. *There exist two positive absolute constants* ϵ, η *such that both* f^+ *and* f^- *(and so also* $|f|$*) are not less than* $\gamma\eta$ *in sets of measure not less than* ϵ.

This is a consequence of Lemma (8·26) and of the inequalities $\mathfrak{M}_2[f]\leqslant B_2\gamma$, $\mathfrak{M}_1[f]\geqslant A_1\gamma$ (see (8·5)).

(8·28) Theorem. *Let* $f(r,x)=\Sigma(a_k\cos n_kx+b_k\sin n_kx)\,r^{n_k}$ *be the function associated with* (8·18), *harmonic for* $r<1$. *If* $\Sigma(a_k^2+b_k^2)=\infty$, *and if* $\omega(u)$ *is any function defined for* $u\geqslant 0$ *and monotonically tending to* $+\infty$ *with* u, *then*

$$\int_0^{2\pi}\omega(|f(r,x)|)\,dx\to+\infty \quad as \quad r\to 1.$$

This follows from (8·25), since the integrand is not less than $\omega\{\lambda_q[\Sigma(a_k^2+b_k^2)\,r^{2n_k}]^{\frac{1}{2}}\}$ in a set of measure not less than $2\pi\mu_q$.

(8·29) Theorem. *Let* $f_t(r,x)$ *be the harmonic function* $\Sigma A_n(x)\phi_n(t)r^n$. *If* $\Sigma(a_k^2+b_k^2)=\infty$, *and if* $\omega(u)$ *is as in* (8·28), *then for almost all* t *the integral* $\int_0^{2\pi}\omega\{|f_t(r,x)|\}\,dx$ *is unbounded as* $r\to 1$.

It is enough to show the existence of sequences $\{r_k\}\to 1$ and $\{M_k\}\to+\infty$ with the following properties: for almost all t we have

$$|f_t(r_k,x)|\geqslant M_k \tag{8·30}$$

for infinitely many k and for $x\in X=X_{t,k}$, with $|X|\geqslant\sigma>0$, where σ is an absolute constant. For then our integral exceeds $\sigma\omega(M_k)$ for $r=r_k$ and infinitely many k.

Applying (8·27) to $f_t(r,x)$ we see that for every $r<1$ the set $E=E_r$ of those points of the rectangle $0\leqslant x\leqslant 2\pi$, $0\leqslant t\leqslant 1$ at which

$$|f_t(r,x)|\geqslant \eta\{\Sigma A_n^2(x)\,r^{2n}\}^{\frac{1}{2}} \tag{8·31}$$

has an intersection of measure not less than ϵ with every line $x=\text{const}$. Hence $|E| \geqslant 2\pi\epsilon$. For $\delta<1$, let H_δ denote the set of the numbers t_0 such that the intersection of E with the line $t=t_0$ is of measure not greater than $2\pi\delta$. Then

$$2\pi\delta\,|H_\delta| + 2\pi(1-|H_\delta|) \geqslant |E| \geqslant 2\pi\epsilon,$$

which gives $|H_\delta| \leqslant (1-\epsilon)/(1-\delta)$. For $\delta=\epsilon^2$ we get $|H_\delta| \leqslant 1/(1+\epsilon)$. Hence *the set of the numbers t_0 such that the intersection of E with the line $t=t_0$ is of measure greater than $2\pi\epsilon^2$ has measure* $\geqslant \epsilon/(1+\epsilon) > \frac{1}{2}\epsilon$ (supposing, as we may, that $\epsilon<1$).

Clearly we must show that $\eta\{\Sigma A_n^2(x)\,r^{2n}\}^{\frac{1}{2}}$ becomes large, as $r\to 1$, outside a set of x of small measure. More precisely, we will show that *there exist sequences $r_k\to 1$, $M'_k\to\infty$ such that*

$$\eta\{\Sigma A_n^2(x)\,r_k^{2n}\}^{\frac{1}{2}} \geqslant M'_k \tag{8·32}$$

outside a set of measure at most $\pi\epsilon^2$.

Suppose we have already proved the last statement. What we have then shown is the following: there exist sequences $r_k\to 1$ and $M'_k\to\infty$, and a sequence of sets T_k, $|T_k| > \frac{1}{2}\epsilon$, such that for each t in T_k,

$$|f_t(r_k, x)| \geqslant M'_k$$

for all x in a set $X_{t,k}$ of measure at least $\pi\epsilon^2$.

Let $$T_0 = \limsup T_k.$$

Clearly $|T_0| \geqslant \frac{1}{2}\epsilon$, and the assertion containing (8·30) is true in T_0 with $M_k = M'_k$. Since the replacement of t by $t+p2^{-q}$ affects only a finite number of terms of $f_t(r,x)$, (8·30) is valid in the union of all translations of T_0 by $p2^{-q}$, provided we set, e.g. $M_k=\frac{1}{2}M'_k$. This union is of measure 1 and the theorem follows.

We now prove the assertion containing (8·32). We set $A_n(x) = \rho_n \cos(nx+x_n)$ and distinguish two cases:

$$\text{(i) } \rho_n \neq O(1), \quad \text{(ii) } \rho_n = O(1).$$

In case (i) there is a sequence $n_1<n_2<\ldots$ such that $\rho_{n_k}\to\infty$. Let $r_k = 1-1/n_k$. We have

$$\eta\{\Sigma A_n^2(x)\,r_k^{2n}\}^{\frac{1}{2}} \geqslant \eta\,|A_{n_k}(x)|\,r_k^{n_k} \geqslant \eta\rho_{n_k}|\cos(n_k x + x_{n_k})|\,e^{-1}$$

for any k. The set of points where $|\cos(jx+x_j)| \geqslant \delta$ does not hold is of measure less than $\pi\epsilon^2$ provided δ is small enough (the limitation imposed on δ is independent of j). Taking $M'_k = \eta\delta e^{-1}\rho_{n_k}$, (8·32) follows for case (i).

Let us now pass to case (ii). We may assume that $\rho_n \leqslant 1$ for all n. Given any $M>0$, we shall show that the measure of the set $B=B_r$ of points x for which

$$\Sigma A_n^2(x)\,r^{2n} \leqslant M$$

tends to 0 as $r\to 1$. Set $h(r) = \Sigma\rho_n^2 r^{2n}$ and integrate the last inequality over B. We get

$$\tfrac{1}{2}|B|\,\Sigma\rho_n^2 r^{2n} + \tfrac{1}{2}\pi\Sigma\rho_n^2 r^{2n}(\alpha_{2n}\cos 2x_{2n} - \beta_{2n}\sin 2x_{2n}) \leqslant M\,|B|, \tag{8·33}$$

where α_j, β_j are the Fourier coefficients of the characteristic function of B. Schwarz's inequality shows that the second term on the left numerically does not exceed

$$\tfrac{1}{2}\pi\{(\Sigma(\alpha_{2n}^2+\beta_{2n}^2)\}^{\frac{1}{2}}\,(\Sigma\rho_n^4 r^{4n})^{\frac{1}{2}} \leqslant \tfrac{1}{2}\pi(|B|\,\pi^{-1})^{\frac{1}{2}}\,h^{\frac{1}{2}}(r)$$

by Bessel's inequality, since $\rho_n \leqslant 1$, $r<1$. If for a sequence of r's tending to 1 we had $|B|$ greater than a positive constant, then, since $h^{\frac{1}{2}}(r) = o\{h(r)\}$, (8·33) would give $\frac{1}{4}|B|\,h(r) \leqslant M\,|B|$ for such r's and $1-r$ small enough, which is false.

This shows that $|B_r| \to 0$ as $r \to 1$. Then taking, for example, $M'_k = k$ we find an $r_k < 1$ such that

$$\eta\{\Sigma A_n^2(x)\, r_k^{2n}\}^{\frac{1}{2}} > M'_k$$

outside a set of x of measure not exceeding $\pi\epsilon^2$. Hence $|f_t(r,x)| \geqslant M'_k$ for t in a set T_k of measure not less than $\frac{1}{2}\epsilon$ and for x in a set $X_{t,k}$ of measure not less than $\pi\epsilon^2$, the same conclusion (with a different M'_k) which we reached in the case (i). This completes the proof of (8·29).

The series $\Sigma \pm A_n(x)$ will be called *randomly continuous* if almost all of them are Fourier series of continuous functions.

Let $\Sigma(a_k^2 + b_k^2)$ be finite. Then for almost all t the sum $S_t(x)$ of (8·9) belongs to every L^r. It is natural to ask whether the $\Sigma \pm A_k(x)$ are randomly continuous. That this is not so follows from the fact (see Chapter VI, (6·1)) that if a lacunary series is the Fourier series of a bounded function, then the sum of the moduli of the terms of the series is finite. Thus for *no* sequence of signs is

$$\pm \sin 10x \pm 2^{-1} \sin 10^2 x \pm \dots \pm n^{-1} \sin 10^n x \pm \dots$$

the Fourier series of a bounded (still less of a continuous) function. We have, however the following theorem:

(8·34) THEOREM. *Let $s_{n,t}(x)$ denote the partial sums of the series* (8·9).

(i) *If $\gamma^2 = \Sigma(a_k^2 + b_k^2) < \infty$, then for almost all t we have $s_{n,t}(x) = o\{(\log n)^{\frac{1}{2}}\}$, uniformly in x.*

(ii) *If $\Sigma(a_k^2 + b_k^2)(\log k)^{1+\epsilon} < \infty$ for some $\epsilon > 0$, then almost all series* (8·9) *converge uniformly and so are Fourier series of continuous functions.*

As the lacunary series

$$\Sigma \pm (n \log n)^{-1} \sin 10^n x$$

shows, (ii) is false for $\epsilon = 0$.

(i) Consider the inequality (8·17). As its proof shows, it holds for arbitrarily large μ, provided γ is small enough. It then also holds for the partial sums $s_{n,t}(x)$ (for which γ is decreased):

$$\int_0^1 dt \int_0^{2\pi} \exp\{\mu s_{n,t}^2(x)\}\, dx \leqslant 2\pi K, \tag{8·35}$$

with K independent of n. Fix t and let $M_n(t)$ be the maximum of $|s_{n,t}(x)|$. Let x_0 be a point at which this maximum is attained. Since the derivative of $s_{n,t}$ does not exceed $2nM_n(t)$ (Chapter III, (13·17)), $s_{n,t}$ cannot change by more than $\frac{1}{2}M_n(t)$ over any interval of length $1/4n$. In particular, $|s_{n,t}|$ exceeds $\frac{1}{2}M_n(t)$ for $x_0 \leqslant x \leqslant x_0 + 1/4n$, and

$$\int_{x_0}^{x_0+1/4n} \exp\{\mu s_{n,t}^2(x)\}\, dx \geqslant \frac{1}{4n} \exp\{\tfrac{1}{4}\mu M_n^2(t)\}.$$

The integral on the left is increased if it is taken the whole interval $(0, 2\pi)$. By (8·35), we have

$$\int_0^1 \exp\{\tfrac{1}{4}\mu M_n^2(t)\}\, dt \leqslant 8K\pi n.$$

Hence

$$\int_0^1 \exp\{\tfrac{1}{4}\mu(M_n^2(t) - \alpha \log n)\}\, dt \leqslant 8K\pi n^{1-\frac{1}{4}\mu\alpha}$$

for any $\alpha > 0$. Take $\alpha\mu = 12$. Then the right-hand sides, being $O(n^{-2})$, form a convergent series. By Chapter I, (11·5), the series $\Sigma \exp\{\frac{1}{4}\mu M_n^2 - 3 \log n\}$ converges almost everywhere, and so $M_n^2(t) \leqslant 12\mu^{-1}\log n$, for almost all t and large enough n. Since by dropping the first few terms we can make γ arbitrarily small, and so μ arbitrarily large, we have

$$M_n(t) = o\{(\log n)\}^{\frac{1}{2}}$$

for almost all t, and (i) follows.

(ii) Let
$$S_{n,t}(x) = \sum_1^n A_k(x)\,\phi_k(t)\,(\log k)^{\frac{1}{2}+\frac{1}{2}\epsilon}.$$

By (i), $S_{n,t}(x) = o\{(\log n)^{\frac{1}{2}}\}$ for almost all t, uniformly in x. We fix such a t and suppose for simplicity that $a_1 = b_1 = 0$. Then summation by parts gives

$$s_{n,t}(x) = \sum_2^{n-1} S_{k,t}(x)\,\Delta\frac{1}{(\log k)^{\frac{1}{2}+\frac{1}{2}\epsilon}} + S_{n,t}(x)\frac{1}{(\log n)^{\frac{1}{2}+\frac{1}{2}\epsilon}} = \sum_2^n o\{(\log k)^{\frac{1}{2}}\}\,O\left\{\frac{1}{k(\log k)^{\frac{3}{2}+\frac{1}{2}\epsilon}}\right\} + o(1).$$

The terms of the last series being $o\{k^{-1}(\log k)^{-1-\frac{1}{2}\epsilon}\}$, $s_{n,t}(x)$ converges uniformly as $n \to \infty$.

We may ask if the random continuity of the series $\Sigma \pm A_n(x)$ implies that almost all the series converge uniformly. This is an open problem, but we can prove the following result:

(8·36) Theorem. *Let $\{n_k\}$ be any lacunary sequence of indices $(n_{k+1}/n_k > q > 1)$. If $\Sigma \pm A_n(x)$ is randomly continuous, the sequence $\{s_{n_k,t}(x)\}$ converges uniformly in x for almost all t.*

Let t_0 be any fixed number, not a diadic fraction. We first note that almost all the series
$$\Sigma\phi_m(t_0)\,\phi_m(t)\,A_m(x)$$

are of the class C (i.e. are Fourier series of continuous functions). For let $E \subset (0,1)$, $|E| = 1$, be such that $\Sigma A_m(x)\,\phi_m(t)$ is in C for $t \in E$. For each $t \in E$ we define t' by

$$\phi_m(t')\,\phi_m(t_0) = \phi_m(t) \quad (m = 0, 1, 2, \ldots).$$

This transformation merely interchanges diadic intervals of the same rank. Since any open set can be covered by non-overlapping diadic intervals, it follows that the transformation preserves the measure of any open set, and so also (passing to the complements) of any closed set and, finally, of any measurable set. In particular, the set of the t' is also of measure 1.

We now split the series $\Sigma A_m(x)\,\phi_m(t)$ into blocks $P_k = \sum_{n_k+1}^{n_{k+1}} A_m(x)\,\phi_m(t)$. By the remark just made, the two series

$$P_0 + P_1 + P_2 + P_3 + \ldots, \quad P_0 - P_1 + P_2 - P_3 + \ldots$$

are in C for almost all t (there is a t_0 such that $\{\phi_m(t_0)\}$ takes the necessary sequence of values ± 1). Hence the series $P_0 + P_2 + P_4 + \ldots$ and $P_1 + P_3 + P_5 + \ldots$ are in C for almost every t. But both series have gaps and Theorem (1·27) of Chapter III shows that $\{s_{n_k,t}(x)\}$ converges uniformly in x for almost all t.

We now prove a theorem of a slightly different character.

(8·37) Theorem. *If the power series $\Sigma a_n z^n$ has radius of convergence 1, then almost all the functions*

$$f_t(z) = \sum_0^\infty a_n z^n \phi_n(t), \quad z = r\,e^{i\theta}$$

are not continuable across $|z| = 1$.

Suppose that for every t in a set E of positive outer measure there is an arc (α, β) on $|z|=1$ and two positive numbers δ, M such that in the domain

$$\Delta: 1-2\delta \leqslant r \leqslant 1+2\delta, \quad \alpha-\delta \leqslant \theta \leqslant \beta+\delta$$

$f_t(z)$ is regular and numerically not greater than M. The numbers α, β, M, δ depend on t, but taking them rational we can select a subset of E—call it E again—also of positive outer measure such that they are independent of $t \in E$.

Let Δ^* denote the domain $1-\delta \leqslant r \leqslant 1$, $\alpha \leqslant \theta \leqslant \beta$, and let $\epsilon > 0$ be so small that every circle with centre $z \in \Delta^*$ and radius ϵ is contained in Δ. By Cauchy's theorem,

$$|f_t^{(p)}(z)| \leqslant \frac{Mp!}{\epsilon^p} \leqslant C^p p^p \quad \text{for} \quad p=1, 2, \ldots, t \in E, z \in \Delta^*.$$

Let $\mathscr{E} \supset E$ be the set of *all* t for which these inequalities are satisfied. Clearly, $\mathscr{E}$ is measurable (even measurable B) and $|\mathscr{E}| > 0$. But, for $|z| < 1$,

$$f_t^{(p)}(z) = \sum_0^\infty b_n \phi_n(t), \quad \text{with} \quad b_n = a_n n(n-1) \ldots (n-p+1) r^{n-p} e^{i(n-p)\theta},$$

and so, applying (8·3) with $\lambda=2$ and supposing p large enough, we get

$$\Sigma |b_n|^2 \leqslant 2C^{2p} p^{2p}.$$

In particular (making $r \to 1$)

$$a_n n(n-1) \ldots (n-p+1) \leqslant 2^{\frac{1}{2}} C^p p^p.$$

Set $p = [\eta n] + 1$, where $0 < \eta < 1$. For n large enough,

$$|a_n| \{n(1-\eta)\}^{n\eta} \leqslant C^{2n\eta} (2n\eta)^{n\eta+1},$$

and so,

$$\limsup |a_n|^{1/n} \leqslant \{2C^2\eta/(1-\eta)\}^\eta < 1,$$

for η fixed and sufficiently small. Hence the radius of convergence of $\Sigma a_n z^n$ is greater than 1, contrary to hypothesis, and this contradiction proves the theorem.

Random insertion of the signs ± 1 into a trigonometric series has a close connexion with the random insertion of the factors 0, 1, that is, with the random suppression of terms. It is enough to replace the $\phi_n(t)$ in (8·11) by

$$\phi_n^*(t) = \tfrac{1}{2}(1+\phi_n(t)). \tag{8·38}$$

The functions $\phi_n^*(t)$ take the values 0, 1, each in sets of measure $\frac{1}{2}$.

In the two theorems that follow, T is any linear method of summation which satisfies conditions (i), (ii), (iii) of regularity (Ch. III, §1), and T* is any linear method which satisfies conditions (i) and (iii).

(8·39) Theorem. (i) *If Σc_n is summable by a method* T, *and if $\Sigma |c_n|^2 < \infty$, then*

$$\Sigma c_n \phi_n^*(t) \quad (0 \leqslant t \leqslant 1) \tag{8·40}$$

is summable T *almost everywhere.*

(ii) *Conversely, if* (8·40) *is summable in a set E of positive measure by a method* T*, *then Σc_n is summable* T* *and $\Sigma |c_n|^2 < \infty$.*

Case (i) is immediate since, by (8·2), $\Sigma c_n \phi_n(t)$ is convergent, and so also summable T, almost everywhere, and the same holds for (8·40). In case (ii), if E is the set of all points where $\Sigma c_n \phi_n^*$ is summable T*, the measure of E must be 1. For the replacement of t by $t + p/2^q$ changes only a finite number of terms in (8·40), and so E is invariant under translations by $p/2^q$. This means that the average density of E in each of the intervals

$$I_{p,q} = (p/2^q, (p+1)/2^q)$$

is the same, and so equal to $|E|$. Since $|E| > 0$, the density theorem for measurable sets asserts that the relative density of E in some of the intervals $I_{p,q}$ must be arbitrarily close to 1. Hence $|E| = 1$. Let E^* be the reflexion of E in the point $t = \frac{1}{2}$. Since $|E^*| = 1$, there is a $t_0 \in EE^*$. Adding the series (8·40) for $t = t_0$ and $t = 1 - t_0$, and observing that $\phi_n(1-t_0) = -\phi_n(t_0)$ for all n, we obtain the summability T* of Σc_n. This shows that $\Sigma c_n \phi_n(t)$ is summable T* in E, and the finiteness of $\Sigma |c_n|^2$ follows from (8·2).

(8·41) THEOREM. *Suppose that* $\Sigma(a_n^2+b_n^2)=\infty$. *Then for every method of summation* T*, *and for almost all t, the series*

$$\Sigma(a_n \cos nx + b_n \sin nx)\, \phi_n^*(t) = \Sigma A_n(x)\, \phi_n^*(t) \tag{8·42}$$

is summable T* *for almost no x. In particular, almost no series* (8·42) *is a Fourier series.*

If the first conclusion were false, there would exist a set X, $|X|>0$, such that for each $x_0 \in X$ the series $\Sigma A_n(x_0)\,\phi_n^*(t)$ is summable T* for all t in a set of positive measure. That, by (8·39)(ii), would imply that $\Sigma A_n^2(x_0)<\infty$, and so also $\Sigma(a_n^2+b_n^2)<\infty$, contrary to hypothesis.

In the case $\Sigma(a_n^2+b_n^2)<\infty$, there seems to be less analogy between the series (8·11) and (8·42). Thus, by (8·34), almost all series $\Sigma\phi_n(t)\, n^{-1}\sin nx$ are Fourier series of continuous functions, whereas, because of the function $\Sigma n^{-1}\sin nx$, almost all series $\Sigma\phi_n^*(t)\, n^{-1}\sin nx$ are Fourier series of discontinuous functions.

9. Series with 'small' gaps

This name will be given to the series

$$\Sigma(a_k \cos n_k\theta + b_k \sin n_k\theta),$$

where the indices $n_1<n_2<\dots$ satisfy an inequality

$$n_{k+1}-n_k \geqslant q > 0 \quad \text{for} \quad k=1,2,\dots,$$

that is, increase at least as rapidly as an arithmetic progression with difference q. Only the case $q>1$ need be considered. Every lacunary series (see § 6) is a member of this class (at least after the rejection of the first few terms), but not conversely.

Theorems about lacunary series proved in § 6 show that if they 'behave well' on a set E of positive measure, they 'behave well' in $(0, 2\pi)$. It will now be shown that if E is a large enough interval, somewhat similar conclusions hold for series with 'small' gaps. It will be convenient to write the series in the complex form.

(9·1) THEOREM. *Let*

$$P(\theta)=\sum_{k=-N}^{N} c_k e^{in_k\theta} \quad (n_{-k}=-n_k) \tag{9·2}$$

be a finite sum with

$$n_{k+1}-n_k \geqslant q>0 \quad (k=0,1\dots), \tag{9·3}$$

and let I be any interval of length greater than $2\pi/q$, *so that*

$$|I| = 2\pi(1+\delta)/q \quad (\delta>0).$$

Then

$$\Sigma |c_k|^2 \leqslant A_\delta \frac{1}{|I|}\int_I |P(\theta)|^2\, d\theta, \tag{9·4}$$

$$|c_k| \leqslant A_\delta \frac{1}{|I|}\int_I |P(\theta)|\, d\theta, \tag{9·5}$$

where A_δ *depends only on* δ.

The results hold for infinite sums if the series (9·2) *converges uniformly.*

The inequality (9·4) somewhat resembles the first inequality (6·6), and there is also a resemblance between the proofs. The proof of (9·4) consists in showing that, for a suitable function χ, the integral $\int_I |P|^2 \chi\, d\theta$ majorizes a fixed multiple of $\Sigma|c_k|^2$. In the lacunary case we had $\chi\equiv 1$ on I. The sparsity of terms in a lacunary series made it possible to base the proof only on the most obvious properties of the coefficients γ_k of the function χ (completed by 0 outside I), namely, on $\Sigma|\gamma_k|^2<\infty$. In the present case we need, as we shall see below, at least $\Sigma|\gamma_k|<\infty$, a condition which does not hold for a discontinuous characteristic function and which therefore requires a different choice of χ.

Though we are not interested in the generalization for its own sake, the proof of (9·1) runs more smoothly if we do not require the n_k to be integers. The simultaneous transformations $\theta\to c\theta$, $n_k\to n_k/c$ change neither P nor the right-hand sides of (9·4) and (9·5). Since we may also assume that I is symmetric with respect to $\theta=0$ (if θ_0 is the midpoint of I, transformation to $\theta-\theta_0$ does not alter the $|c_n|$), it is enough to take

$$I=(-\pi,\pi),\quad q=1+\delta.$$

Let χ be any real-valued function vanishing outside I and $\gamma(u)$ its Fourier transform (p. 8). Then

$$\begin{aligned}\frac{1}{2\pi}\int_{-\pi}^{\pi}\chi(x)\,|P(x)|^2\,dx&=\frac{1}{2\pi}\int_{-\infty}^{+\infty}(\Sigma c_k\,e^{in_kx})\,(\Sigma\bar{c}_l\,e^{-in_lx})\,\chi(x)\,dx=\Sigma c_k\bar{c}_l\gamma(n_l-n_k)\\&\geqslant\gamma(0)\,\Sigma\,|c_k|^2-\sum_{k\neq l}\tfrac{1}{2}(|c_k|^2+|c_l|^2)\,|\gamma(n_l-n_k)|\\&=\sum_k|c_k|^2\{\gamma(0)-\sum_l{}'\,|\gamma(n_l-n_k)|\},\end{aligned}\tag{9·6}$$

where the dash indicates that $l\neq k$ in the summation. If χ is bounded, say not greater than M, and the expression in curly brackets exceeds a positive number Γ depending on δ only, a comparison of the extreme terms of (9·6) gives

$$\Sigma\,|c_k|^2\leqslant\frac{M}{\Gamma}\frac{1}{2\pi}\int_{-\pi}^{\pi}|P|^2\,dx,$$

which is (9·4) with the simplifications adopted.

To show that this hypothetical situation can be realized, let

$$\chi(x)=2\pi\cos\tfrac{1}{2}x\quad\text{for}\quad|x|\leqslant\pi,\quad\chi(x)=0\text{ elsewhere.}\tag{9·7}$$

Then
$$\gamma(u)=\frac{4\cos\pi u}{1-4u^2},$$
and since $|n_k-n_l|\geqslant|k-l|\,q$,

$$\begin{aligned}\sum_l{}'\,|\gamma(n_l-n_k)|&\leqslant\sum_l{}'\frac{4}{4(k-l)^2q^2-1}<\frac{8}{q^2}\sum_{l=1}^{\infty}\frac{1}{4l^2-1}\\&=\frac{4}{q^2}\sum_{l=1}^{\infty}\left(\frac{1}{2l-1}-\frac{1}{2l+1}\right)=\frac{4}{q^2}=\frac{\gamma(0)}{(1+\delta)^2}.\end{aligned}\tag{9·8}$$

This gives (9·4) with $A_\delta=2\pi(1+\delta)^2/4\delta(2+\delta)<A\,(1+\delta^{-1})$.

To prove (9·5), let $|c_j|$ be the largest of the $|c_k|$. Then

$$\begin{aligned}\left|\frac{1}{2\pi}\int_{-\infty}^{+\infty}\chi(x)\,P(x)\,e^{-in_jx}\,dx\right|&=|\sum_k c_k\gamma(n_j-n_k)|\\&\geqslant|c_j|\,\gamma(0)-\sum_{k\neq j}|c_k|\,|\gamma(n_j-n_k)|\geqslant|c_j|\left(4-\frac{4}{(1+\delta)^2}\right),\end{aligned}$$

using (9·8). Since the left-hand side here does not exceed $\int_{-\pi}^{\pi}|P|\,dx$, (9·5) follows, with the same A_δ as before.

The inequality opposite to (9·4) is also true. It is easier and is valid under more general conditions.

(9·9) THEOREM. *Let P and $\{n_k\}$ be the same as in* (9·1) *and let J be any interval of length $2\pi\eta/q$, where $\eta>0$. Then*

$$\frac{1}{|J|}\int_J |P(x)|^2\,dx \leqslant B_\eta \Sigma |c_k|^2. \tag{9·10}$$

We may suppose that $q=1$. The inequality (9·10) follows from Parseval's formula if the n_k are integers and, say, $|J|\leqslant 2\pi$. For the left-hand side is then not greater than

$$\frac{2\pi}{|J|}\frac{1}{2\pi}\int_0^{2\pi}|P|^2\,dx=\eta^{-1}\Sigma|c_k|^2.$$

To prove (9·10) in the general case, we note that in the last term of (9·6) we have $\Sigma'\leqslant\gamma(0)$. Hence,

$$\frac{1}{2\pi}\int_{-\pi}^{\pi}\chi|P|^2\,dx\leqslant\Sigma|c_k|^2\{\gamma(0)+\Sigma'\}\leqslant 2\gamma(0)\,\Sigma|c_k|^2,$$

$$2^{-\frac{1}{2}}\int_{-\frac{1}{2}\pi}^{\frac{1}{2}\pi}|P|^2\,dx\leqslant 8\Sigma|c_k|^2.$$

This estimate holds for the integral of $|P|^2$ over any interval of length π and leads to (9·10) if $\eta=1$, and so also if $\eta<1$ (with $B_\eta=\eta^{-1}B_1$).

If $\eta>1$, we split J into a finite number of intervals J_k of lengths contained between π and 2π and observe that the left-hand side of (9·10) does not exceed the largest of the ratios $|J_k|^{-1}\int_{J_k}|P|^2\,dx$. Thus (9·9) is established, with $B_\eta<A(1+\eta^{-1})$.

The following generalization of (6·16) follows easily from (9·1).

(9·11) THEOREM. *If the radius of convergence R of the power series*

$$\Sigma c_k z^{n_k}=f(z),\quad n_{k+1}-n_k\to\infty, \tag{9·12}$$

is 1, *the function f is not continuable across $|z|=1$.*

For suppose that a closed arc I on $|z|=1$ is one of regularity for f. There are then constants C, δ such that (compare a similar argument on p. 221)

$$|f^{(p)}(z)|\leqslant C^p p!\leqslant C^p p^p \quad\text{for}\quad p=1,2,\ldots,\ \theta\in I,\ 1-\delta\leqslant r<1.$$

We reject the first few terms of the series (9·12), possibly altering the value of C, so as to make (9·5) applicable to $P(\theta)=f^{(p)}(re^{i\theta})$. Making $r\to1$ we get

$$|c_k|\,n_k(n_k-1)\ldots(n_k-p+1)\leqslant A\limsup_{r\to1}\frac{1}{|I|}\int_I|f^{(p)}(re^{i\theta})|\,d\theta\leqslant AC^p p^p.$$

For $p=[n_k\epsilon]+1$ and k large enough we therefore have

$$|c_k|\{n_k(1-\epsilon)\}^{n_k\epsilon}\leqslant(2Cn_k\epsilon)^{n_k\epsilon},$$

$$\limsup|c_k|^{1/n_k}\leqslant\left(\frac{2C\epsilon}{1-\epsilon}\right)^{\epsilon}<1,$$

provided ϵ is small enough. This gives $R>1$, and the contradiction proves (9·11).

The proof actually gives more than is explicitly stated, namely, *if*

$$\liminf(n_{k+1}-n_k)\geqslant\gamma,$$

then every arc of length greater than $2\pi/\gamma$, and so also every arc of length $2\pi/\gamma$, on $|z|=1$ contains at least one singular point of f.

10. A power series of Salem

We return to Theorem (8·34). It holds of course for power series. Following Salem we complete it as follows:

(10·1) THEOREM. *Let $r_1, r_2, \ldots$ be a sequence of positive numbers tending monotonically to 0, such that Σr_n^2 converges and that $\{1/r_n\}$ is concave. Then there is a sequence of numbers ϵ_n, $|\epsilon_n| = 1$, such that $\Sigma \epsilon_n r_n e^{inx}$ converges uniformly.*†

Examples of sequences $\{r_n\}$ satisfying the hypothesis are

$$n^{-\frac{1}{2}}(\log n)^{-\frac{1}{2}-\epsilon}, \quad n^{-\frac{1}{2}}(\log n)^{-\frac{1}{2}}(\log\log n)^{-\frac{1}{2}-\epsilon}, \quad \ldots,$$

etc., for $\epsilon > 0$ and n large enough. The factors ϵ_n are not ± 1 and we know nothing about the set of admissible $\{\epsilon_n\}$, except for the obvious fact that it is of the power of the continuum.

We may suppose that $r_n = r(n)$, where $r(u)$ is monotonically decreasing and differentiable, and $1/r(u)$ is concave. The convergence of Σr_n^2 is equivalent to that of $\int_1^\infty r^2(u)\,du$.

The proof of (10·1) is based on certain extensions of the lemmas of van der Corput proved in § 4 to the expressions

$$I(F; a, b) = \int_a^b r(u)\,F(u)\,du, \quad S(F; a, b) = \sum_{a<n\leqslant b} r(n)\,F(n),$$

where $r(u)$ is a positive decreasing function and $F(u) = \exp 2\pi i f(u)$. Some of these extensions are immediate consequences of the case $r \equiv 1$, others are less so.

We take for the variable of integration the primitive $R = R(u)$ of $r(u)$. It is an increasing function of u, so that $u = u(R)$ and $I = \int e^{2\pi i f}\,dR$, with

$$f_R = f'(u)/r(u), \quad f_{RR} = f''(u)/r^2(u) - f'(u)\,r'(u)/r^3(u).$$

Here $r' \leqslant 0$. Hence $f_{RR} \geqslant f''(u)/r^2(u)$, if only $f'(u) \geqslant 0$. Applying Lemma (4·3), we get the following result:

(10·2) LEMMA. *If $f''(u) > 0$ and $f'(u) \geqslant 0$, then*

$$|I(F; a, b)| \leqslant 4 \max\,(r(u)/\{f''(u)\}^{\frac{1}{2}}).$$

The case $f'' < 0$, $f' \geqslant 0$ is slightly less simple, and we shall have to introduce the additional hypothesis that r'/f'' is monotone.

(10·3) LEMMA. *If $f'' < 0$, $f' \geqslant 0$, and r'/f'' is monotone, then*

$$|I| \leqslant 4 \max \frac{r}{|f''|^{\frac{1}{2}}} + \max \frac{|r'|}{|f''|}.$$

Write
$$I = \int_a^b [r(u) - r(b)]\,F(u)\,du + r(b) \int_a^b F(u)\,du = P + Q. \tag{10·4}$$

† For r_n monotonically decreasing to 0 and $\{1/r_n\}$ convex, the theorem is an immediate consequence of (8·34). For then $1/r_n$ exceeds a fixed positive multiple of n, $r_n = O(1/n)$, the hypotheses of (8·34) are satisfied and we can take for $\{\epsilon_n\}$ almost any sequence of ± 1.

Then $$|Q| \leqslant 4r(b)\max\{1/|f''(u)|^{\frac{1}{2}}\} \leqslant 4\max\{r(u)/|f''(u)|^{\frac{1}{2}}\},$$

$$P = \frac{1}{2\pi i}\int_a^b \frac{r(u)-r(b)}{f'(u)-f'(b)}\,\frac{f'(u)-f'(b)}{f'(u)}\,d\,e^{2\pi i f(u)}.$$

The factor $(f'(u)-f'(b))/f'(u)$ is decreasing and contained between 0 and 1. The derivative of the preceding factor can be written

$$\frac{f''(u)}{f'(u)-f'(b)}\left[\frac{r'(u)}{f''(u)} - \frac{r(u)-r(b)}{f'(u)-f'(b)}\right] = \frac{f''(u)}{f'(u)-f'(b)}\left[\frac{r'(u)}{f''(u)} - \frac{r'(v)}{f''(v)}\right] \quad (u<v<b),$$

and so is of constant sign. Hence, applying the second mean-value theorem to the two monotone factors, we obtain

$$|P| \leqslant \frac{2}{\pi}\max\frac{r(u)-r(b)}{f'(u)-f'(b)}\,.\max\frac{f'(u)-f'(b)}{f'(u)} \leqslant \max\frac{r'}{f''}\,.\,1,$$

and collecting results we get (10·4).

Changing f into $-f$ (which does not affect $|I|$), we may replace the hypotheses of (10·3) by $f''>0, f'\leqslant 0$. If $f''>0$, but nothing is assumed about the sign of f', we split (a, b) into subintervals in which the sign of f' is constant, and deduce from (10·2) and (10·3):

(10·5) LEMMA. *If $f''(u)$ is of constant sign and r'/f'' is monotone, then*

$$|I(F; a, b)| \leqslant 8\max(r/|f''|^{\frac{1}{2}}) + \max|r'/f''|.$$

Remark. The term $\max|r'/f''|$ is necessary here. Take, for instance,

$$r=f', \quad I=(2\pi i)^{-1}(e^{2\pi i f(b)} - e^{2\pi i f(a)}),$$

and supposing that $f(u)$ increases indefinitely with u, take $f(b)=f(a)+\frac{1}{2}$. Then $|I|=1/\pi$. But choosing, for example, $f=\log\log u$, we see that $\max(r/|f''|^{\frac{1}{2}})$ in (a, b) tends to 0 as $a\to\infty$.

(10·6) LEMMA. *If $f'(u)$ is monotone and $|f'|\leqslant\frac{1}{2}$, then*

$$|I(F; a, b) - S(F; a, b)| \leqslant A\max r(u).$$

Here A is an absolute constant. For $r(u)=1$ this is (4·4). The general case is reduced to this by applying the second mean-value theorem to the equation

$$I - S = \int_a^b r(u)\,F(u)\,d\chi(u),$$

with χ defined as in § 4.

(10·7) LEMMA. *Suppose that $f''(u)$ is of constant sign, and r'/f'' and $r/|f''|^{\frac{1}{2}}$ are monotone. Then*

$$|S| \leqslant 16\max(r/|f''|^{\frac{1}{2}}) + 2\max|r'/f''| + 2A\max r + \int_a^b (8r|f''|^{\frac{1}{2}} + |r'| + Ar|f''|)\,du, \tag{10·8}$$

where A is the constant in (10·6).

The proof is similar to that of (4·6). By hypothesis, f' is monotone, say increasing. Let α_k be defined by the condition $f'(\alpha_k)=k-\frac{1}{2}$, for k integral, and let $\alpha_r, \alpha_{r+1}, \ldots, \alpha_{r+s}$

be the points α, if such exist, in the interval $a \leqslant u \leqslant b$. We consider the values a, α_r, $\alpha_{r+1}, \ldots, \alpha_{r+s}$, b of u, to which correspond the values

$$f'(a), \quad r-\tfrac{1}{2}, \quad r+\tfrac{1}{2}, \quad \ldots, \quad r+s-\tfrac{1}{2}, \quad f'(b)$$

of $v=f'(u)$. In the interval (α_k, α_{k+1}) we have $|f'-k| \leqslant \frac{1}{2}$. Let

$$S_k = \sum_{\alpha_k < n \leqslant \alpha_{k+1}} r(n)\, e^{2\pi i f(n)} = \sum_{\alpha_k < n \leqslant \alpha_{k+1}} r(n)\, e^{2\pi i \{f(n)-kn\}}.$$

Since $(f-uk)'=f'-k$, $(f-uk)''=f''$, we get from (10·5) and (10·6)

$$|S_k| \leqslant 8 \max (r/|f''|^{\frac{1}{2}}) + \max |r'/f''| + A \max r, \tag{10·9}$$

where the max are taken over $\alpha_k \leqslant u \leqslant \alpha_{k+1}$. (10·9) holds also for the incomplete intervals (a, α_r) and (α_{r+s}, b).

Let now $\phi(u)$ be any positive monotone function in (a, b), say decreasing, and consider the sum

$$\sigma(a,b) = \Sigma \max_{\alpha_k \leqslant u \leqslant \alpha_{k+1}} \phi(u) = \phi(a) + \phi(\alpha_r) + \ldots + \phi(\alpha_{r+s}),$$

which takes into account also the intervals (a, α_r) and (α_{r+s}, b). If we introduce the new variable $v=f'(u)$, $\phi(u)$ becomes a decreasing function $\Phi(v)$, and σ is

$$\begin{aligned}
\Phi[f'(a)] + \Phi(r-\tfrac{1}{2}) + \ldots + \Phi(r+s-\tfrac{1}{2}) &\leqslant \Phi[f'(a)] + \Phi(r-\tfrac{1}{2}) + \int_{r-\frac{1}{2}}^{r+s-\frac{1}{2}} \Phi(v)\, dv \\
&\leqslant 2\Phi[f'(a)] + \int_{f'(a)}^{f'(b)} \Phi(v)\, dv \\
&= 2\phi(a) + \int_a^b \phi(u) f''(u)\, du.
\end{aligned}$$

Since $\phi(u)$ is positive and monotone,

$$\sigma \leqslant 2 \max \phi + \int_a^b \phi f''\, du.$$

This and the inequalities (10·9) give (10·8) (of course, we can omit the term $|r'|$ on the right and increase the value of A by 1).

We now pass to the proof of Theorem (10·1). We consider the series

$$\sum_1^\infty r(n)\, e^{2\pi i [g(n)+nx]}, \tag{10·10}$$

where
$$g''(u) = r^2(u) \Big/ \int_u^\infty r^2(t)\, dt, \quad g(u) = \int_1^u \log \left(\int_v^\infty r^2 dt \right)^{-1} dv \quad (u \geqslant 1),$$

and apply to it the estimate (10·8) with $f(u) = g(u) + ux$. Since $f''(u) = g''(u)$, the estimate will be valid uniformly in x. We shall show that S is small for a large and $b > a$.

Now $r(u)\{f''(u)\}^{-\frac{1}{2}} = \left(\int_u^\infty r^2 dt \right)^{\frac{1}{2}}$ is decreasing and tends to 0. Also

$$|r'|/f'' = |r'|\, r^{-2} \int_u^\infty r^2 dt$$

decreases monotonically to 0, since $|r'|\, r^{-2} = (1/r)'$ is positive and decreases, $1/r$ being concave. Thus the first three terms on the right in (10·8) are small for large a. Finally, $1/r$ being concave,

$$\int_u^\infty r^2 dt = \int_u^\infty \left| \frac{r^2}{r'} \right| |r'|\, dt \geqslant \frac{r^2(u)}{|r'(u)|} \int_u^\infty |r'|\, dt = \frac{r^3(u)}{|r'(u)|}.$$

Hence $rf''=r^3\Big/\int_u^\infty r^2\,dt \leqslant |\,r'\,|$, and the integral in (10·8) does not exceed

$$8\int_a^b r(f'')^{\frac{1}{2}}\,du+(A+1)\int_a^b |\,r'\,|\,du \leqslant 8\int_a^\infty \frac{r^2(u)\,du}{\left(\int_u^\infty r^2 dt\right)^{\frac{1}{2}}}+(A+1)\,r(a)$$

$$=16\left(\int_a^\infty r^2(u)\,du\right)^{\frac{1}{2}}+(A+1)\,r(a),$$

which is small for large a. Hence (10·10) converges uniformly, and (10·1) follows, with $\epsilon_n=e^{2\pi i g(n)}$.

MISCELLANEOUS THEOREMS AND EXAMPLES

1. If $a_n=\epsilon_n/n$ and $\{\epsilon_n\}$ is a positive decreasing sequence, the partial sums t_n of the series $\Sigma a_n \sin nx$ are positive for $0<x<\pi$.

[Sum by parts and use Chapter II, (9·4).]

2. If $\Sigma(a_k \cos kx+b_k \sin kx)=A_k(x)$ is a Fourier series, the series $\Sigma A_k(x)/\log k$ converges in the metric L. So does $\Sigma B_k(x)/(\log k)^{1+\epsilon}$ for $\epsilon>0$, though not for $\epsilon=0$.

[Consider the series $\Sigma \cos nx/\log n$ and $\Sigma \sin nx/(\log n)^{1+\epsilon}$.]

3. Suppose that $\{a_n\}$ is positive, convex and monotonically decreasing. Then the modified partial sums $t_n^*=\frac{1}{2}(t_n+t_{n-1})$ of $\Sigma a_n \sin nx$ are positive for $0<x<\pi$. This need not be true for the t_n.

[Sum by parts twice and use the fact that $\tilde{K}_m>0$, $\tilde{D}_m^*>0$ inside $(0,\pi)$. For the negative assertion, consider $\Sigma r^m \sin mx$, $n=2$, $r>\frac{1}{2}$.]

4. Let
$$R_k(x)=\frac{1}{2}+\sum_{n=1}^\infty \left(\frac{\sin nh}{nh}\right)^k \cos nx,$$
where $k=1,2,\ldots$ and h is a constant such that $0<kh\leqslant\pi$. Show that R_k vanishes in (kh,π) and is a polynomial of degree $k-1$ in each of the intervals $((k-2)\,h,kh)$, $((k-4)\,h,(k-2)\,h)$, ... (For $k=1,2$, see p. 10.)

[Consider the function $B_k(x)$ from p. 42 and the kth difference
$$B_k(x+kh)-\binom{k}{1}B_k(x+(k-2)\,h)+\ldots\pm B_k(x-kh).$$
The result can also be obtained by repeated application of Theorem (1·5) of Chapter II to $R_1(x)$.]

5. Let $h_1, h_2, \ldots, h_s, \ldots$ be positive numbers with $\Sigma h_s<\infty$. Let
$$R(x)=\frac{1}{2}+\sum_{n=1}^\infty a_n \cos nx, \quad \text{where} \quad a_n=\prod_{s=1}^\infty \left(\frac{\sin nh_s}{nh_s}\right).$$
The function $R(x)$ has derivatives of all orders and is not constant. If $d=h_1-(h_2+h_3+\ldots)$ is positive, $R(x)$ is constant in $(0,d)$. If $h=h_1+h_2+\ldots<\pi$, then $R(x)=0$ in (h,π). Cf. Mandelbrojt [2].

[Observe that $a_1\neq 0$, $a_n=O(n^{-k})$ for each fixed k.]

6. If a_n decreases monotonically to 0 and if $\Sigma a_n \sin nx \in$ L, then $\Sigma a_n \cos nx \in$ L.

[$\Sigma a_n/n<\infty$.]

7. If $\{a_n\}$ is positive, convex, and tends to 0, then the sums of both the series $\Sigma a_n \cos nx$ and $\Sigma a_n \sin nx$ have continuous derivatives in the interior of $(0,2\pi)$.

[The series $\Sigma a_n e^{inx}$ differentiated termwise is uniformly summable (C, 1) in every closed interval interior to $(0,2\pi)$.]

8. Let a_n be the cosine coefficients of a function $f(x)$ such that $f(x)\log 1/|\,x\,|$ is integrable over $(-\pi,\pi)$. Then the series $\Sigma a_n/n$ converges and
$$\sum_{n=1}^\infty a_n/n=-\frac{1}{\pi}\int_0^{2\pi} f(x)\log(2\sin\tfrac{1}{2}x)\,dx.$$
In particular, this holds for functions f such that $f\log^+|\,f\,|$ is integrable. (Hardy and Littlewood [12].)

[Multiply both sides of the first formula (2·8) of Chapter I by $f(x)$ and integrate over $(-\pi, \pi)$. The argument is justified because the partial sums of $\Sigma(\cos nx)/n$ are uniformly $O(\log 1/|x|)$ near $x=0$ (see (2·28)).]

9. Let $a_1 \geqslant a_2 \geqslant \ldots \to 0$, $a_1 > 0$, and let $t_n(x)$ be the partial sums of $\Sigma a_\nu \sin \nu x$, $g(x) = \lim t_n(x)$. Then

(i) $$\lim g(x)/x = \Sigma \nu a_\nu$$

(even if the series on the right diverges). In particular $g(x)$ is strictly positive in some interval $0 < x \leqslant \delta$.

(ii) There is an interval $0 < x \leqslant \delta$ in which all the t_n are strictly positive. Hartman and Wintner [1].

[(i) Using (1·13) we have

$$(*) \quad g(x) = -\tfrac{1}{2}a_1 \tan \tfrac{1}{4}x + \sum_{\nu=1}^{\infty} \Delta a_\nu \cdot \frac{1-\cos(\nu+\frac{1}{2})x}{2\sin\frac{1}{2}x}.$$

Since the terms of the last series are non-negative and the cofactor of Δa_ν is asymptotically equal to $\frac{1}{2}(\nu+\frac{1}{2})^2 x$ as $x \to +0$, we obtain $\lim g(x)/x = -\frac{1}{8}a_1 + \frac{1}{2}\Sigma(\nu+\frac{1}{2})^2 \Delta a_\nu$, and summation by parts shows that this expression is $\Sigma \nu a_\nu$.]

10. If $a_1 \geqslant a_2 \geqslant \ldots \to 0$, $a_1 > 0$, $g(x) \sim \Sigma a_\nu \sin \nu x$, $0 < \gamma < 2$, then $x^{-\gamma} g(x) \in L(0, \pi)$ if and only if $\Sigma \nu^{\gamma-1} a_\nu$ converges. (Boas [1_{III}], Heywood [2]; for generalizations see Aljančić, Bojanić and Tomić [1].)

[Consider the formula (*) in Example 9. The integral of $x^{-\gamma}[1-\cos(\nu+\frac{1}{2})x](2\sin\frac{1}{2}x)^{-1}$ over $(0, \pi)$ being exactly of the order ν^γ (the proof is the same as that of Chapter II, (12·1)), $x^{-\gamma}g(x)$ is in $L(0, \pi)$ if and only if $\Sigma \nu^\gamma \Delta a_\nu < \infty$, or, what is the same, $\Sigma \nu^{\gamma-1} a_\nu < \infty$.]

11. Let $b(u)$ be a slowly varying function. Then, with the notation (2·7), and assuming $1 < \gamma < 2$, we have the formulae

$$f_\gamma(x) - f_\gamma(0) \simeq x^{\gamma-1} b(x^{-1})\, \Gamma(1-\gamma) \sin \tfrac{1}{2}\pi\gamma,$$

$$g_\gamma(x) \simeq x^{\gamma-1} b(x^{-1})\, \Gamma(1-\gamma) \cos \tfrac{1}{2}\pi\gamma,$$

analogous to (2·8) (see also the remark on p. 188).

[Set $\gamma = \beta + 1$ and integrate the relations (2·8) near $x = 0$. By repeating this procedure we obtain analogous relations for $k < \gamma < k+1$, $k = 2, 3, \ldots$.]

12. If $b(u)$ is slowly varying then, with the same notation as above,

$$\left.\begin{aligned} f_2(x) - f_2(0) &\simeq -\tfrac{1}{2}\pi x b(1/x), \\ g_2(x) \simeq x B(1/x), \quad g_2(x) - x f_1(0) &\simeq x R(1/x), \end{aligned}\right\} \quad \text{(i)}$$

the last two relations being respectively valid according as $\Sigma b_n/n$ diverges or converges.

[Integrate (2·13) and (2·16). In the proof of (i) we need the fact—easily obtainable by differentiation—that if $b(u)$ is slowly varying so are $B(u)$ and $R(u)$, for $\Sigma b_n/n$ divergent and convergent respectively. A similar argument gives formulae for f_k and g_k if $k = 3, 4, \ldots$.]

13. For $\alpha = 1$, the sum $\phi_\alpha(x)$ of (4·1) belongs to Λ_*. (It has been already pointed out that it does not belong to Λ_1.) For $\frac{1}{2}\alpha + \beta = 2$ the sum $\psi_{\alpha,\beta}(x)$ of (5·1) belongs to Λ_*.

14. If $\gamma > 0$, $\delta > \frac{1}{2}(1+\gamma)$, the series $\Sigma \dfrac{\exp[2\pi i n(\log n)^\gamma]}{n^{\frac{1}{2}}(\log n)^\delta} z^n$ converges uniformly on $|z| = 1$. (Ingham [2].)

[The proof follows the same line as in § 4. We have $\sum_2^n \nu^{-\frac{1}{2}} \exp 2\pi i(\nu(\log\nu)^\gamma + i\nu x)] \leqslant A(\log n)^{\frac{1}{2}(1+\gamma)}$.]

A periodic and integrable function $f(x)$ is said to be of *bounded deviation* (Hadamard) if for each fixed interval (a, b) we have $\int_a^b f e^{-inx} dx = O(1/n)$. In particular, the Fourier coefficients of such functions are $O(1/n)$. Every function of bounded variation is of bounded deviation. Examples of functions of bounded deviation but not of bounded variation have been given by Alexits [2], Bray (see Mandelbrojt [1]) and Hille [3].

15. The function $x \sin (1/x)$ ($|x| \leqslant \pi$) is of bounded deviation. (Bray; see Mandelbrojt [1].)

[It is enough to show that the integral of $x \exp (nx \pm 1/x)$ over any subinterval of $(0, \pi)$ is uniformly $O(1/n)$. The substitution $x^2 = u$ reduces the problem to the function $\exp (nu^{\frac{1}{2}} \pm u^{-\frac{1}{2}})$ and any subinterval (c, d) of $(0, \pi^2)$. We may suppose that $c \geqslant C/n$, since the integral over a subinterval of $(0, C/n)$ is $O(1/n)$. For a suitable C the derivative of $nu^{\frac{1}{2}} \pm u^{-\frac{1}{2}}$ in $(C/n, \pi^2)$ is monotone and greater than $C_1 n$. Apply (4·3) (i).]

16. Show that if in the lacunary series $\Sigma(a_k \cos n_k x + b_k \sin n_k x)$ we have $|a_k| + |b_k| = O(1/n_k)$, then the sum $f(x)$ of the series is of bounded deviation. It need not be of bounded variation, as the example of the function $f(x) = \Sigma 2^{-k} \cos 2^k x$, differentiable almost nowhere, shows. (Hille [3].)

17. Suppose that for the lacunary series $\Sigma(a_k \cos n_k x + b_k \sin n_k x) = f(x)$ the sum $\Sigma(|a_k| + |b_k|)$ is finite but $\Sigma n_k^2(a_k^2 + b_k^2)$ infinite. Then $f(x)$ is differentiable almost nowhere. [Compare p. 206.]

18. Suppose that $\Sigma(|a_n| + |b_n|)/n$ converges but $\Sigma(a_n^2 + b_n^2)$ diverges. (Take, for example, $a_n = n^{-\frac{1}{2}}$, $b_n = 0$.) Then almost all continuous functions

$$\Sigma \pm n^{-1}(a_n \cos nx + b_n \sin nx)$$

are differentiable almost nowhere.

19. The conclusion in Example 7 is false if $\{a_n\}$ is merely monotonically decreasing to zero. The sums of the series $\Sigma a_n \cos nx$ and $\Sigma a_n \sin nx$ may be then differentiable almost nowhere.

[Summation by parts gives

$$\Sigma a_n \cos nx = (2 \sin y)^{-1} \Sigma \Delta a_n \sin (2n+1) y,$$

with $y = \frac{1}{2}x$. Consider the case when the last series is lacunary and apply Example 17.]

20. If $n_{k+1}/n_k \geqslant 3$ and $\Sigma \alpha_k^2 < \infty$, then the Riesz product $\Pi(1 + \alpha_k \cos n_k x) = 1 + \Sigma \gamma_n \cos nx$ is the Fourier series of a function f such that $\exp (\lambda(\log |f|)^2)$ is integrable for every $\lambda > 0$. In particular, $f \in L^p$ for every $p > 0$.

21. Consider the product $\Pi(1 + \alpha_k \cos n_k x)$, where $n_{k+1}/n_k > q > 2, |d_k| \leqslant 1$. It is then no longer true that p_k is a partial sum of p_{k+1}, but the rank of the lowest term in $p_{k+1} - p_k$ tends to infinity with k. In particular, p_k tends termwise to a trigonometric series. If $|\alpha_k| \leqslant 1$, the latter is the Fourier–Stieltjes series of a continuous, non-decreasing function. (Wiener and Wintner [1].)

22. Let K_m denote Fejér's kernel. The Nth partial product of the Riesz product $\prod_0^\infty (1 + \cos 2^n x)$ is $2K_{2^N-1}(x)$ and so tends termwise to $1 + 2\cos x + 2\cos 2x + \dots$, the Fourier-Stieltjes series of a discontinuous function. (Wiener and Wintner [1].)

23. Suppose that $n_{k+1}/n_k \geqslant q > 2$, $|\alpha_k| \leqslant 1$, $dF \sim \Pi(1 + \alpha_k \cos n_k x)$.
Show that $F \in \Lambda_\alpha$, where $\alpha = 1 - \log 2/\log q$.

24. Let $-1 \leqslant \alpha_k \leqslant 1$, $\Sigma \alpha_k^2 = \infty$. Show that for almost all sequences of signs ± 1 the product $\Pi(1 \pm \alpha_k \cos kx)$ diverges to 0 almost everywhere in x.

25. Suppose that $-1 \leqslant \alpha_k \leqslant +1$, $\Sigma \alpha_k^2 = \infty$, $n_{k+1}/n_k \geqslant 3$. Show that the trigonometric series representing $\Pi(1 + i\alpha_k \cos n_k x)$ diverges almost everywhere (cf. (7·12)).

26. Let M be any (not necessarily linear) method of summation of series $u_0 + u_1 + \dots$ which has the following properties: (*a*) if Σu_n converges to sum s, it is also summable M to s; (*b*) if Σu_n and Σv_n are summable M to sums s and t respectively, then $\Sigma(au_n + bv_n)$ is summable M to sum $as + bt$; (*c*) if $u_0 + u_1 + u_2 + \dots$ is summable M to s, the series $u_1 + u_2 + \dots$ is summable M to $s - u_0$. Show that the series

$$(*) \quad \cos x + \cos 2x + \dots + \cos 2^n x + \dots$$

cannot be summable M on a measurable set $E \subset (0, 2\pi)$, $|E| > 0$, to a finite measurable sum. (Kolmogorov [4].)

[Suppose (*) is summable M on E, $|E| > 0$, to a measurable sum $f(x)$. Since summability of (*) at x implies summability at $2^k x$, we have $|E| = 2\pi$. In particular, $f(x)$ is bounded on a set $H \subset E$, of measure arbitrarily close to 2π. Let $\chi(x)$ be the characteristic function of H, and let H_N be the set whose characteristic function is $\chi(2^N x)$. By Chapter II, (4·15), $|HH_N| \to |H|^2/2\pi$, and so, for suitable H, and for N large enough, $|HH_N|$ is arbitrarily close to 2π. For $x \in HH_N$ we have

$$f(x) = \cos x + \ldots + \cos 2^{N-1}x + f(2^N x),$$

and we arrive at a contradiction since $f(x)$ and $f(2^N x)$ are bounded on HH_N, while, by (8·25), the sum $\cos x + \ldots + \cos 2^{N-1}x$ is large with N on a substantial subset of $(0, 2\pi)$, and so also at some points of HH_N.]

27. Let $E \subset (0, 2\pi)$, $|E| > 0$, $n_{k+1}/n_k \geqslant q > 1$,

$$\gamma^2 = \Sigma(a_k^2 + b_k^2) < \infty, \quad f = \Sigma(a_k \cos n_k x + b_k \sin n_k x).$$

Show that there are two positive constants λ_q and μ_q, depending on q only, such that

$$f^+ \geqslant \lambda_q \gamma, \quad f^- \geqslant \lambda_q \gamma$$

in subsets of E of measure not less than $\mu_q |E|$, provided n_1 is large enough.

[The proof is analogous to that of (8·25). We show first that

$$\int_E f^4 dx \leqslant A_q \gamma^4 |E|,$$

provided n_1 is large enough. This, together with the first inequality (6·6) (valid with P replaced by f), gives $\int_E |f| \, dx \geqslant B_q \gamma |E|$. Since

$$\int_E f^+ dx + \int_E f^- dx = \int_E |f| \, dx, \quad \int_E f^+ dx - \int_E f^- dx = \int_0^{2\pi} f\chi \, dx,$$

where χ is the characteristic function of E, and since the last integral is small in comparison with γ if n_1 is large enough, we see that $\int_E f^+ dx \geqslant \frac{1}{3} B_q \gamma |E|$. Combining this with

$$\int_E (f^+)^2 dx \leqslant \int_E f^2 dx \leqslant \tfrac{1}{4} |E| \gamma^2$$

(see (6·6)) and applying (8·26), we obtain the conclusion for f^+.]

CHAPTER VI

THE ABSOLUTE CONVERGENCE OF TRIGONOMETRIC SERIES

1. General series

The absolute convergence of the complex trigonometric series $\Sigma c_k e^{ikx}$ (in particular, of a power series in e^{ix}) at a single point x_0 implies the convergence of $\Sigma |c_k|$, and so the absolute (and uniform) convergence of $\Sigma c_k e^{ikx}$ for all x. For the series

$$\tfrac{1}{2}a_0 + \sum_{k=1}^{\infty} (a_k \cos kx + b_k \sin kx) = \sum_{k=0}^{\infty} A_k(x) \tag{1·1}$$

the situation is less simple. The convergence of

$$\sum_{k=1}^{\infty} (|a_k| + |b_k|) \tag{1·2}$$

naturally implies the absolute (and uniform) convergence of (1·1) for all x; on the other hand, (1·1) may converge absolutely at an infinite set of points without (1·2) converging. An example is given by

$$\Sigma \sin n!\, x,$$

whose terms vanish from some point onwards for every x commensurable with π.

(1·3) THEOREM OF DENJOY-LUSIN. *If* (1·1) *converges absolutely for x belonging to a set A of positive measure,* (1·2) *converges.*

Suppose for simplicity that $a_0 = 0$, and let the kth term of (1·1) be $\rho_k \sin(kx + x_k)$, with $\rho_k = (a_k^2 + b_k^2)^{\frac{1}{2}} \geq 0$. The function

$$\alpha(x) = \sum_{k=1}^{\infty} \rho_k |\sin(kx + x_k)| \tag{1·4}$$

is finite at every $x \in A$. Hence there is a set $E \subset A$, $|E| > 0$, such that $\alpha(x)$ is bounded on E, $\alpha(x) < M$ say. Since the partial sums of (1·4) are uniformly bounded on E, the series may be termwise integrated over E:

$$\sum_{k=1}^{\infty} \rho_k \int_E |\sin(kx + x_k)|\, dx = \int_E \alpha(x)\, dx \leq M|E|. \tag{1·5}$$

To prove the convergence of $\rho_1 + \rho_2 + \ldots$, which is equivalent to that of (1·2), it is enough to show that the coefficients of ρ_k in (1·5) all exceed an $\epsilon > 0$. This is immediate, if we observe that by replacing the integrand by $\sin^2(kx + x_k)$ we do not increase the integral and that the new integral tends to $\tfrac{1}{2}|E|$ (Chapter II, (4·5)).

(1·6) THEOREM. *Suppose that* $|a_1| \geq |a_2| \geq \ldots$.

(i) *If $\Sigma a_n \cos nx$ converges absolutely at a point x_0, then $\Sigma |a_n| < \infty$.*

(ii) *The same holds for the series $\Sigma a_n \sin nx$, if $x_0 \not\equiv 0 \pmod{\pi}$.*

In (i) we may suppose that $0 < x_0 < \pi$. The hypothesis implies that $\Sigma |a_n| \cos^2 nx_0$ is finite. Since $2\cos^2 nx_0 = 1 + \cos ny_0$, with $y_0 = 2x_0$, and since the partial sums of $\Sigma |a_n| \cos ny_0$ are bounded, the result follows. Part (ii) is proved similarly.

(1·7) Theorem. *Let $\rho_n=(a_n^2+b_n^2)^{\frac{1}{2}}$. If $\rho_1\geqslant\rho_2\geqslant\dots$, and if $\Sigma A_n(x)$ converges absolutely at two points x', x'' with $|x'-x''|<\pi$, then $\Sigma\rho_n<\infty$.*

Write (1·1) in the form $\Sigma\rho_n\sin(nx+x_n)$, and let $t=x'-x''$. Since

$$nt=(nx'+x_n)-(nx''+x_n),$$

it follows that
$$|\sin nt|\leqslant|\sin(nx'+x_n)|+|\sin(nx''+x_n)|; \tag{1·8}$$

thus the series $\Sigma\rho_n|\sin nt|$ converges, and (1·7) follows from (1·6) (ii).

It is obvious that (1·6) and (1·7) hold if we suppose only that $\{|a_n|\}$ and $\{\rho_n\}$ respectively are of bounded variation.

The set A in Theorem (1·3) is of positive measure. This, while sufficient to ensure the convergence of (1·2), is not necessary. The problem of characterizing those sets A for which the absolute convergence of (1·1) in A implies the finiteness of (1·2) is still unsolved. The results that follow, however, throw some light on the situation.

Suppose that $\rho_1+\rho_2+\dots=\infty$ for (1·1) and let A be the set of points at which $\alpha(x)$ is finite. The complementary set is a product of a sequence of open sets (see the proof of Theorem (12·2) of Chapter I). Hence A is the sum of a sequence of closed sets. None of these closed sets contains an interval, for otherwise we should have $|A|>0$, and $\rho_1+\rho_2+\dots<\infty$. It follows that all of them are non-dense, and A is of the first category. Thus:

(1·9) Theorem. *If $\Sigma A_n(x)$ converges absolutely in a set of the second category* (*even one of measure* 0), *the series* (1·2) *converges.*

The set A of points where $\Sigma A_n(x)$ converges absolutely has curious properties. Let $\bar{A}$ be the set of points of absolute convergence of the series $\Sigma B_n(x)$ conjugate to $\Sigma A_n(x)$ and let C and $\bar{C}$ respectively be the sets of points where $\Sigma A_n(x)$ and $\Sigma B_n(x)$ converge, not necessarily absolutely. It will be convenient to place all these sets on the circumference of the unit circle.

(1·10) Theorem. *Every point of A is a point of symmetry of the sets A, $\bar{A}$, C, $\bar{C}$.*

We have
$$A_n(x+h)+A_n(x-h)=2A_n(x)\cos nh,$$

$$B_n(x+h)-B_n(x-h)=2A_n(x)\sin nh.$$

The first formula implies that if $\Sigma|A_n(x)|<\infty$ and if $\Sigma A_n(x+h)$ converges, or converges absolutely, so does $\Sigma A_n(x-h)$. This proves the symmetry property for A and C. The second formula gives the proof for $\bar{A}$, $\bar{C}$.

By (1·3), the set A (and similarly $\bar{A}$) has measure either 0 or 2π.

(1·11) Theorem. *If A is infinite, then C, and similarly $\bar{C}$, has measure either* 0 *or* 2π.

By (1·10), if $x\in A$, and $x+h\in A$, then all the points $x+h$, $x+2h$, $x+3h$, ... belong to A. Since A is infinite, h may be arbitrarily small, so that A is everywhere dense. Suppose that C and its complement C' are both of positive measure, and let c, c' be points of density for C and C' respectively. There is an $\epsilon>0$ such that if any interval I of length $\leqslant 2\epsilon$ contains c, then $|IC|>\frac{1}{2}|I|$, and if any interval J of length $\leqslant 2\epsilon$ contains c', then $|JC'|>\frac{1}{2}|J|$. Let $I=(c-\epsilon,c+\epsilon)$, and take an x_0 belonging to A

and distant less than $\frac{1}{2}\epsilon$ from the midpoint of the arc (c, c'). Reflexion in x_0 takes C into itself and I into an interval J, $|J| = 2\epsilon$, containing c'. The inequalities

$$|JC| > \tfrac{1}{2}|J|, \quad |JC'| > \tfrac{1}{2}|J|$$

being incompatible, we have a contradiction. The argument for $\bar{C}$ is identical.

There exist trigonometric series absolutely convergent in a perfect set but not everywhere (see Example 1 at the end of the chapter). On the other hand, we shall prove the existence of perfect sets P of measure zero which as regards the absolute convergence of trigonometric series resemble the sets of positive measure: if (1·1) is absolutely convergent in P, (1·2) is finite.

A point set S will be called a *basis*, if every real x can be represented in the form $\alpha_1 x_1 + \alpha_2 x_2 + \ldots + \alpha_m x_m$, where $\alpha_1, \alpha_2, \ldots$ are integers, $x_1, x_2, \ldots$ belong to S, and m depends on x. We may also write

$$x = \epsilon_1 x_1 + \ldots + \epsilon_n x_n,$$

where $\epsilon_j = \pm 1$ and the x_j are not necessarily different.

(1·12) THEOREM. *If S is a basis, and if $\Sigma A_n(x)$ is absolutely convergent in S, then $\Sigma \rho_n$ is finite.*

We shall reduce the general case to that of a purely sine series, which is immediate. In fact the inequality

$$|\sin n(\epsilon_1 x_1 + \epsilon_2 x_2 + \ldots + \epsilon_m x_m)| \leqslant |\sin nx_1| + \ldots + |\sin nx_m| \quad (\epsilon_j = \pm 1), \quad \textbf{(1·13)}$$

which it is easy to prove by induction, shows that if $\Sigma |b_n \sin nx|$ converges in S, it converges everywhere, and so $\Sigma |b_n| < \infty$. In the general case, we need the following lemma:

(1·14) LEMMA. *Let S be a basis, and let $S^* = S_u$ be the set S translated by u. There is then a set T of the second category such that for every $y \in T$ we have*

$$y = \alpha_1 x_1^* + \alpha_2 x_2^* + \ldots + \alpha_m x_m^*,$$

with α_j integral and $x_j^* \in S^*$ for all j.

By hypothesis, for every x we have

$$x = \alpha_1(x_1^* - u) + \alpha_2(x_2^* - u) + \ldots,$$

that is,

$$x + ku = \alpha_1 x_1^* + \ldots + \alpha_m x_m^*,$$

where $k = k(x)$ is an integer. Let E_n be the set of x for which $k(x) = n$. At least one of the E_n, say E_{n_0}, is of the second category, and we may take for T the set E_{n_0} translated by $n_0 u$. We say that S^* is a *basis for* T. (Incidentally it is not difficult to deduce that S^* is a basis (cf. Example 2 at the end of the chapter), but we do not need this.)

Passing to the proof of (1·12), let v be any point of S, and let $x = y + v$. Then

$$A_n(x) = A_n(v) \cos ny - B_n(v) \sin ny.$$

By hypothesis $\Sigma |A_n(v)| < \infty$, and therefore $\Sigma B_n(v) \sin ny$ converges absolutely in a set S^* obtained from S by a translation $-v$. By (1·14), S^* is a basis for a set T of the second category. The argument which we applied to sine series shows that $\Sigma B_n(v) \sin ny$ is absolutely convergent in T, and so for all y (see (1·9)). The same holds for the series $\Sigma\{A_n(v)\cos ny - B_n(v)\sin ny\} = \Sigma A_n(x)$. Thus (1·2) is finite.

The above proof may be modified slightly, by basing it on Theorem (1·3) instead of (1·9). It is enough to observe that (1·14) and its proof remain valid if we replace there the words 'second category' by 'positive outer measure'. Since the set of points y for which $\Sigma \mid B_n(v) \sin ny \mid$ converges is measurable, it is of positive measure, and therefore is the whole interval $(0, 2\pi)$.

To give an example of (1·12), we shall show that *Cantor's ternary set C constructed on* $(0, 1)$ (or on any other interval) *is a basis*. More precisely, we shall show that *the set of all sums $x+y$, with $x \in C$, $y \in C$, fills the whole interval* $(0, 2)$.

Consider the set K of all points (x, y) of the plane such that $x \in C$, $y \in C$. The set K may be also obtained as follows. Divide the square $0 \leqslant x \leqslant 1$, $0 \leqslant y \leqslant 1$, which we call Q_0, into nine equal parts, and let Q_1 be the sum of the four closed corner squares. Repeat the procedure for each of these corner squares, denoting the sum of the new corner squares by Q_2, and so on. Plainly, $K = Q_0 Q_1 Q_2 \ldots$. The projection of any Q_j on the diagonal $y = x$ of Q_0 fills up that diagonal. In other words, any straight line L_h with equation $x + y = h$, $0 \leqslant h \leqslant 2$, meets every Q_j at one point at least. Since the Q_j are closed and form a decreasing sequence, it follows that $KL_h \neq 0$, and this is just what we wanted to prove. Similarly we can show that *the set of all differences $x - y$ ($x \in C, y \in C$) fills the interval* $(-1, 1)$.

2. Sets N

In what follows we shall repeatedly use the following classical result of Dirichlet:

(2·1) LEMMA. *Let $\alpha_1, \alpha_2, \ldots, \alpha_k$ be any k real numbers, and let Q be any positive integer. Then we can find an integer q with $1 \leqslant q \leqslant Q^k$ and integers $p_1, p_2, \ldots, p_k$ such that*

$$\left|\alpha_j - \frac{p_j}{q}\right| < \frac{1}{Qq} \leqslant \frac{1}{q^{1+1/k}} \quad (j = 1, 2, \ldots, k). \tag{2·2}$$

Let $\langle x \rangle = x - [x]$ be the fractional part of x. Consider the k-dimensional half-closed unit cube I defined by the inequalities $0 \leqslant x_j < 1$ for $j = 1, \ldots, k$, and divide I into Q^k congruent half-closed sub-cubes by drawing hyperplanes parallel to the faces of I at distances $1/Q$. Of the $Q^k + 1$ points with co-ordinates

$$\langle n\alpha_1 \rangle, \quad \langle n\alpha_2 \rangle, \quad \ldots, \quad \langle n\alpha_k \rangle \quad (n = 0, 1, \ldots, Q^k),$$

at least two, say those corresponding to $n = q_1$ and $n = q_2 > q_1$, are in the same sub-cube of I. Hence $\mid \langle q_2\alpha_j \rangle - \langle q_1\alpha_j \rangle \mid < 1/Q$ for all j, or, setting $[q_1\alpha_j] = p'_j$, $[q_2\alpha_j] = p''_j$,

$$\mid (q_2 - q_1)\alpha_j - (p''_j - p'_j) \mid < 1/Q,$$

which is the first inequality (2·2) with $q = q_2 - q_1 \geqslant 1$, $p_j = p''_j - p'_j$.

Remarks. (*a*) The fractions p_j/q may not all be irreducible.

(*b*) The first inequality (2·2) shows that given any $\epsilon > 0$ we can find an integer $q > 0$ such that all the products $q\alpha_1$, $q\alpha_2$, $\ldots$, $q\alpha_k$ differ from integers by less than ϵ. It is enough to take $Q \geqslant 1/\epsilon$.

(*c*) For any system $\alpha_1, \alpha_2, \ldots, \alpha_k$ there exist fractions p_j/q *with* q *arbitrarily large*, satisfying

$$|\alpha_j - p_j/q| < q^{-1-1/k} \quad (j=1, \ldots, k). \tag{2·3}$$

For we may take, successively, $Q=1, 2, \ldots$. If the corresponding q's are bounded, or even have a bounded subsequence only, we may suppose that q is the same for infinitely many Q's. Making a further selection, we may suppose (since the fractions p_j/q are bounded) that the p_j also stay constant. For $Q \to \infty$, the first inequality (2·2) then gives $\alpha_j = p_j/q$. Thus the q in (2·2) can be arbitrarily large, unless all the α_j are rational. In the latter case, however, we even have $\alpha_j - p_j/q = 0$ for infinitely many q's.

In the preceding section we investigated properties of sets E such that whenever the series $\Sigma A_k(x)$ converges absolutely in E it converges everywhere. We now prove a number of results about those sets which do not have this property.

A set E will be called of *type* N if there is a series $\Sigma A_k(x)$ which converges absolutely in E but not everywhere (and so has $\Sigma\rho_k = \infty$). If the series in question is a sine series, we shall call E of *type* N_s. Every set of type N_s is also of type N. The converse will be proved below.

(2·4) Lemma. *Every denumerable set is of type* N.

Let E consist of the numbers $\pi\alpha_1, \pi\alpha_2, \ldots$. By Remark (*b*), for each k there exists an integer $q = n_k$ such that

$$|\sin \pi n_k \alpha_j| < 1/k^2 \quad \text{for} \quad j = 1, 2, \ldots, k.$$

We may suppose that $n_{k+1} > n_k$. Then the terms of the series $\Sigma |\sin n_k x|$ are less than $1/k^2$ for $x = \pi\alpha_j$ and $k \geqslant j$, and the series converges in E, though not everywhere.

(2·5) Theorem. *Every set* E *of type* N *is also of type* N_s.

The proof is in two parts.

(2·6) Lemma. *Let* E *be of type* N, *and let* x_0 *be a fixed point of* E. *Let* E_{x_0} *denote the set* E *translated by* $-x_0$, *that is, the set of all points* $x - x_0$ *with* $x \in E$. *Then* E_{x_0} *is of type* N_s.

For using the inequality (1·8) with $x' = x$, $x'' = x_0$, we see that if the series

$$\Sigma\rho_n \sin(nx + x_n), \quad \text{with} \quad \Sigma\rho_n = \infty,$$

converges absolutely in E, so does the series $\Sigma\rho_n \sin n(x - x_0)$; that is, $\Sigma\rho_n \sin nx$ converges absolutely in E_{x_0}.

(2·7) Lemma. *If a set* E *is of type* N_s, *the set* E_0 *obtained by the addition to* E *of any point* x_0 *outside* E *is also of type* N_s.

For suppose that $\Sigma\rho_n |\sin nx|$ converges for $x \in E$ and that $\Sigma\rho_n = \infty$. Let ω_n be numbers not less than 1, monotonically increasing to $+\infty$, and such that

$$\Sigma\rho_n/\omega_n = \infty, \quad \Sigma\rho_n/\omega_n^2 < \infty.$$

(We may take, for example, $\omega_n = \sum_1^n \rho_k$ for n large.) By (2·1), with $k=1$, there exist for each n integers $q_n > 0$ and p_n such that

$$1 \leqslant q_n \leqslant \omega_n, \quad \left| q_n \frac{n x_0}{\pi} - p_n \right| < \frac{1}{[\omega_n]}.$$

Hence $$|q_n n x_0 - p_n \pi| < \frac{\pi}{[\omega_n]} < \frac{2\pi}{\omega_n}, \quad |\sin nq_n x_0| < \frac{2\pi}{\omega_n},$$

$$\Sigma \frac{\rho_n}{\omega_n} |\sin nq_n x_0| < \Sigma \frac{2\pi\rho_n}{\omega_n^2} < \infty.$$

For $x \in E$ we have

$$\Sigma \frac{\rho_n}{\omega_n} |\sin nq_n x| \leqslant \Sigma \frac{\rho_n q_n}{\omega_n} |\sin nx| \leqslant \Sigma \rho_n |\sin nx| < \infty.$$

Thus $\Sigma(\rho_n/\omega_n) \sin nq_n x$ converges absolutely for $x = x_0$ and for $x \in E$, though $\Sigma(\rho_n/\omega_n)$ diverges.

The integers nq_n are not necessarily increasing, but obviously only a finite number of them can be equal to any given integer. Hence, rearranging $\Sigma(\rho_n/\omega_n) \sin nq_n x$, we obtain a trigonometric sine series converging absolutely in E_0 but not everywhere.

Return to (2·5) and let $x_0 \in E$. By (2·6), there is a series $\Sigma\rho_n |\sin nx|$ with $\Sigma\rho_n = \infty$ converging in E_{x_0}. By (2·7) there is a series $\Sigma r_n |\sin nx|$ with $\Sigma r_n = \infty$ converging in E_{x_0} and at x_0. Thus

$$\Sigma r_n |\sin n(x - x_0)| < \infty \quad (x \in E); \qquad \Sigma r_n |\sin nx_0| < \infty;$$

and so $$\Sigma r_n |\sin nx| < \infty \quad \text{for} \quad x \in E,$$

which proves (2·5).

Theorem (2·5) gives a new proof of (1·12), since the latter, as we observed, is immediate for sine series.

The properties N and N_s being equivalent, it follows from (2·7) that property N is not affected if we add to the set one point, and so if we add any finite number of points.

More generally

(2·8) Theorem. *If E is of type* N *and D is denumerable, then $E + D$ is of type* N.

We know that E is of type N_s. Let the points of D be $x_1, x_2, \ldots$, and let E_k be the set E augmented by the points $x_1, x_2, \ldots, x_k$.

As the proof of (2·7) shows, given any series $\Sigma r_n |\sin nx|$ convergent in E, with $\Sigma r_n = \infty$, we can construct a series $\Sigma r'_n |\sin nx|$ convergent in E_k. Moreover,

$$\Sigma r'_n |\sin nx| \leqslant \Sigma r_n |\sin nx| \quad \text{for all } x.$$

More than this is true; for a moment's consideration shows that, given any $N \geqslant 1$, we can find an $\bar{N} \geqslant N$ such that

$$\sum_1^N r'_n |\sin nx| \leqslant \sum_1^{\bar{N}} r_n |\sin nx|.$$

Multiplying $\Sigma r'_n |\sin nx|$ by a sufficiently small number, we may also suppose that the sum of the series at the points $x_1, x_2, \ldots, x_k$ is less than a given $\epsilon > 0$.

Consider now a series $\Sigma\rho'_n \sin nx$ with $\Sigma\rho'_n = \infty$ and

$$\Sigma\rho'_n |\sin nx| < \infty \quad \text{for} \quad x \in E_1. \tag{2·9}$$

Take N_1 so large that

$$\sum_1^{N_1} \rho'_n \geqslant 1, \quad \sum_1^{N_1} \rho'_n |\sin nx| \leqslant \tfrac{1}{2} \quad \text{for} \quad x = x_1.$$

Starting with the remainder $\sum_{N_1+1}^{\infty} \rho'_n \sin nx$, we construct a series $\Sigma\rho''_n |\sin nx|$ convergent in E_2, with $\Sigma\rho''_n = \infty$. Let $N_2 > N_1$ be such that

$$\sum_{N_1+1}^{N_2} \rho''_n \geqslant 1, \quad \sum_{N_1+1}^{N_2} \rho''_n |\sin nx| < \frac{1}{2^2} \quad \text{for} \quad x = x_1, x_2,$$

and $\bar{N}_2 \geq N_2$ such that

$$\sum_{N_1+1}^{N_2} \rho''_n |\sin nx| \leq \sum_{N_1+1}^{\bar{N}_2} \rho'_n |\sin nx|.$$

Generally, having defined N_{k-1} and $\bar{N}_{k-1} \geq N_{k-1}$, we consider $\sum_{\bar{N}_{k-1}+1}^{\infty} \rho'_n \sin nx$ and construct a series $\sum_{\bar{N}_{k-1}+1}^{\infty} \rho_n^{(k)} \sin nx$, with $\Sigma \rho_n^{(k)} = \infty$. We take numbers $N_k > \bar{N}_{k-1}$ and $\bar{N}_k \geq N_k$ such that

$$\sum_{\bar{N}_{k-1}+1}^{N_k} \rho_n^{(k)} \geq 1, \quad \sum_{\bar{N}_{k-1}+1}^{N_k} \rho_n^{(k)} |\sin nx| \leq \frac{1}{2^k} \quad \text{for} \quad x = x_1, x_2, \ldots, x'_k, \tag{2·10}$$

$$\sum_{\bar{N}_{k-1}+1}^{N_k} \rho_n^{(k)} |\sin nx| \leq \sum_{\bar{N}_{k-1}+1}^{\bar{N}_k} \rho'_n |\sin nx| \quad \text{for all } x, \tag{2·11}$$

and so on.

The series

$$\Sigma \rho_n \sin nx = \sum_{k=1}^{\infty} \sum_{n=\bar{N}_{k-1}+1}^{N_k} \rho_n^{(k)} \sin nx,$$

where $N_0 = 0$, $\bar{N}_1 = N_1$, is the required series. For

$$\sum_{k=1}^{\infty} \sum_{n=\bar{N}_{k-1}+1}^{N_k} \rho_n^{(k)} |\sin nx|$$

converges in E, as may be seen from (2·11) and (2·9). It converges in D by virtue of the second inequality (2·10). Finally, the first inequality (2·10) shows that $\Sigma \rho_n = \infty$.

The following theorem shows that we cannot replace D in (2·8) by an arbitrary set of type N:

(2·12) THEOREM. *There exist two sets A and B of type* N *such that $A+B$ is not of type* N.

Every x, $0 \leq x \leq 2\pi$, may be written in the form

$$2\pi \sum_{k=1}^{\infty} \epsilon_k 2^{-k} \quad (\epsilon_k = 0, 1).$$

Let $k_1 = 1 < k_2 < \ldots < k_p < \ldots$ be given, and let

$$\Delta_p = 2\pi \sum_{k_p}^{k_{p+1}-1} \epsilon_k 2^{-k} \quad (\epsilon_k = 0, 1).$$

The expansions

$$\Delta_1 + \Delta_3 + \Delta_5 + \ldots, \qquad \Delta_2 + \Delta_4 + \Delta_6 + \ldots$$

represent two perfect and non-dense sets A and B in $(0, 2\pi)$. Since all points in $(0, 2\pi)$ can be written in the form $x+y$, with $x \in A$, $y \in B$, it follows that $A+B$ is not of type N. But for a suitable choice of the k_p, both A and B will be of type N.

For

$$2^{k_{2p}}(\Delta_1 + \Delta_3 + \ldots + \Delta_{2p-1}) \equiv 0 \pmod{2\pi},$$

$$2^{k_{2p}}(\Delta_{2p+1} + \Delta_{2p+3} + \ldots) < 4\pi(2^{-(k_{2p+1}-k_{2p})} + \ldots) < 8\pi\, 2^{-(k_{2p+1}-k_{2p})}.$$

Take, for example, $k_p = p^2$, and consider the series $\Sigma |\sin 2^{k_{2p}} x|$. The preceding inequalities immediately show that it converges in A. Thus A is of type N. Likewise B is of type N.

It is obvious that any translation of a set of type N is of type N. We have the following deeper result, in which E_λ denotes the set of points $x\lambda$ with $x \in E$.

(2·13) THEOREM. *If E is of type* N, *so is E_λ for every λ.*

The case $\lambda = 0$ is trivial. We are considering E as a periodic set of period 2π. Hence E_λ has period $2\pi|\lambda|$, and the set E_λ reduced mod 2π is, in general, 'richer' in points than a portion of E_λ situated in an interval of length 2π.

Since E is of type N_s, there is a series $\Sigma\rho_n|\sin(nx/\lambda)|$ converging in E_λ, with $\Sigma\rho_n=\infty$. Let $\{\omega_n\}$ be, as in the proof of (2·7), an increasing sequence of numbers satisfying $\Sigma\rho_n\omega_n^{-1}=\infty$, $\Sigma\rho_n\omega_n^{-2}<\infty$. By means of (2·1), we choose integers q_n, p_n such that

$$1\leqslant q_n\leqslant\omega_n,\quad \left|q_n\frac{n}{\lambda}-p_n\right|<\frac{1}{[\omega_n]}<\frac{2}{\omega_n}.$$

Then
$$|\sin p_n x|<\left|\sin\frac{2x}{\omega_n}\right|+\left|\sin q_n\frac{n}{\lambda}x\right|<\frac{2|x|}{\omega_n}+q_n\left|\sin\frac{n}{\lambda}x\right|,$$

$$\frac{\rho_n}{\omega_n}|\sin p_n x|<2|x|\frac{\rho_n}{\omega_n^2}+\rho_n\left|\sin\frac{n}{\lambda}x\right|.$$

Hence the series $\Sigma\rho_n\omega_n^{-1}|\sin p_n x|$ converges in E_λ, and since $\Sigma\rho_n\omega_n^{-1}=\infty$, E_λ is of type N (only a finite number of the p_n can take a given value).

We shall now obtain a necessary condition for a set $E\subset(0,2\pi)$ to be of type N. Let $F(x)$, $0\leqslant x\leqslant 2\pi$, represent a mass distribution of total mass 1, concentrated on E; that is,

$$\int_0^{2\pi}dF=\int_E dF=1. \tag{2·14}$$

If E is closed, F is a non-decreasing function constant in the intervals contiguous to E.

By hypothesis, there is a series $\Sigma\rho_n|\sin nx|$ convergent in E, with $\Sigma\rho_n=\infty$. In particular, $\Sigma\rho_n\sin^2 nx<\infty$ in E, and so also

$$\sum_1^N\rho_n\sin^2 nx\Big/\sum_1^N\rho_n\to 0\quad(x\in E).$$

The ratio here does not exceed 1. Multiplying it by dF and integrating over $(0,2\pi)$, we get

$$\sum_1^N\rho_n\int_0^{2\pi}\sin^2 nx\,dF\Big/\sum_1^N\rho_n\to 0.$$

Hence
$$\liminf_{n\to\infty}\int_0^{2\pi}\sin^2 nx\,dF=0,$$

or
$$\limsup_{n\to\infty}\int_0^{2\pi}\cos 2nx\,dF=1. \tag{2·15}$$

Thus

(2·16) Theorem. *If E is of type* N, *then for any positive mass distribution dF satisfying* (2·14) *we have* (2·15).

This means that not only do the Fourier-Stieltjes coefficients of dF not tend to zero, but their upper limit is as large as possible. The fact that for any distribution of mass 1 over E we have (2·15), might be interpreted as indicating the 'smallness' of E (see also Example 10, p. 251).

Remarks. (*a*) It seems not to be known whether the condition (2·15), which is fulfilled for *all* functions of the type described above, is sufficient for E to be of type N. It is almost immediate, however, that if (2·15) is satisfied for a given F, then there is a subset E' of E of type N such that $\int_{E-E'}dF=0$.

For (2·15) implies the existence of a sequence $\{n_k\}$ such that $\int_0^{2\pi} \sin^2 n_k x\, dF \leqslant 2^{-k}$. Hence

$$\int_0^{2\pi} \left\{ \sum_1^{\infty} |\sin n_k x| \right\} dF \leqslant \Sigma\, 2^{-k/2} < \infty,$$

so that $\Sigma |\sin n_k x| < \infty$ for $x \in E' \subset E$, where E' is such that the variation of F over $E - E'$ is zero.

(*b*) We shall apply (2·16) to symmetrical perfect sets of the Cantor type (Chapter V, § 3). We know that the points of such a set P are:

$$x = 2\pi(\epsilon_1 r_1 + \epsilon_2 r_2 + \ldots) \quad (\epsilon_k = 0, 1), \tag{2·17}$$

where $r_1 + r_2 + \ldots = 1$ and $r_p > r_{p+1} + r_{p+2} + \ldots$, or alternatively, putting in evidence the successive ratios of dissection, $r_k = \xi_1 \ldots \xi_{k-1}(1 - \xi_k)$.

Let F be the Lebesgue function for P. Then (Chapter V, (3·4))

$$\int_0^{2\pi} \cos 2nx\, dF = \prod_{k=1}^{\infty} \cos 2\pi n r_k.$$

If P is of type N, we have the *necessary* condition

$$\limsup_{n \to \infty} \prod_{k=1}^{\infty} \cos^2 2\pi n r_k = 1,$$

or

$$\limsup_{n \to \infty} \prod_{k=1}^{\infty} (1 - \sin^2 2\pi n r_k) = 1. \tag{2·18}$$

The latter condition is equivalent to

$$\liminf_{n \to \infty} \sum_{k=1}^{\infty} \sin^2 2\pi n r_k = 0. \tag{2·19}$$

That (2·18) implies (2·19) follows from the inequality $1 + x \leqslant e^x$ and the fact that the product in (2·18) does not exceed 1. The converse is a consequence of the inequality

$$(1 - \epsilon_1)(1 - \epsilon_2) \ldots (1 - \epsilon_N) \geqslant 1 - (\epsilon_1 + \ldots + \epsilon_N),$$

valid for $\epsilon_j \leqslant 1$ and easily verifiable by induction.

We now show that, *if the sequence* r_k (which is decreasing) *is such that* $r_k / r_{k+1} = O(1)$, *then* P *is not of type* N.

Suppose that $r_k / r_{k+1} \leqslant M$, and take any α, $0 < \alpha < 1/M$. Then, if n is large enough, there is at least one r_k in the interval $(\alpha/4n, 1/4n)$, and for this r_k

$$\sin^2 2\pi n r_k \geqslant \sin^2 \left(2\pi n \frac{\alpha}{4n} \right) = \sin^2 \tfrac{1}{2}\pi\alpha,$$

which makes (2·19) impossible.

In particular, if $\xi_k \geqslant \delta > 0$ for all k, P is not of type N.

(*c*) The condition $\liminf_{n \to \infty} \sum_{k=1}^{\infty} |\sin 2\pi n r_k| = 0$, very similar to (2·19), is *sufficient* for P to be of type N. For let $\{n_p\}$ be such that

$$\sum_{k=1}^{\infty} |\sin 2\pi n_p r_k| < \eta_p,$$

with $\Sigma \eta_p < \infty$. The relation (2·17) implies

$$|\sin n_p x| \leqslant \sum_{k=1}^{\infty} |\sin 2\pi n_p r_k| \leqslant \eta_p$$

and gives $\Sigma |\sin n_p x| < \infty$, for $x \in P$.

3. The absolute convergence of Fourier series

(3·1) THEOREM OF S. BERNSTEIN. *If* $f \in \Lambda_\alpha$, $\alpha > \frac{1}{2}$, *then* S[f] *converges absolutely. For* $\alpha = \frac{1}{2}$ *this is not necessarily true.*

Suppose that $\Sigma A_n(x)$ is $S[f]$. Then

$$f(x+h)-f(x-h) \sim -2\sum_{n=1}^{\infty} B_n(x)\sin nh,$$

$$\frac{1}{\pi}\int_0^{2\pi}[f(x+h)-f(x-h)]^2\,dx = 4\sum_{n=1}^{\infty}\rho_n^2\sin^2 nh. \tag{3·2}$$

If $\omega(\delta)$ is the modulus of continuity of f, the left-hand side of (3·2) does not exceed $2\omega^2(2h)$. On setting $h=\pi/2^{\nu+1}$, $\nu=1,2,\ldots$, we obtain

$$\sum_{n=1}^{\infty}\rho_n^2\sin^2\frac{\pi n}{2^{\nu+1}} \leqslant \tfrac{1}{2}\omega^2\left(\frac{\pi}{2^\nu}\right),$$

$$\sum_{n=2^{\nu-1}+1}^{2^\nu}\rho_n^2\sin^2\frac{\pi n}{2^{\nu+1}} \leqslant \tfrac{1}{2}\omega^2\left(\frac{\pi}{2^\nu}\right),$$

$$\sum_{n=2^{\nu-1}+1}^{2^\nu}\rho_n^2 \leqslant \omega^2\left(\frac{\pi}{2^\nu}\right), \tag{3·3}$$

since in the preceding sum the co-factors of the ρ_n^2 all exceed $\frac{1}{2}$. By Schwarz's inequality,

$$\sum_{2^{\nu-1}+1}^{2^\nu}\rho_n \leqslant \left(\sum_{2^{\nu-1}+1}^{2^\nu}\rho_n^2\right)^{\frac{1}{2}}\left(\sum_{2^{\nu-1}+1}^{2^\nu}1^2\right)^{\frac{1}{2}} \leqslant 2^{\frac{1}{2}\nu}\omega\left(\frac{\pi}{2^\nu}\right), \tag{3·4}$$

and finally,

$$\sum_{n=2}^{\infty}\rho_n = \sum_{\nu=1}^{\infty}\sum_{n=2^{\nu-1}+1}^{2^\nu}\rho_n \leqslant \sum_{\nu=1}^{\infty}2^{\frac{1}{2}\nu}\omega\left(\frac{\pi}{2^\nu}\right). \tag{3·5}$$

If $\omega(\delta)\leqslant C\delta^\alpha$, $\alpha>\frac{1}{2}$, the last series converges and the first part of (3·1) is established. The proof of the second part is contained in that of (3·10).

Remark. The above proof gives slightly more than is actually stated in (3·1). For $S[f]$ to converge absolutely it is enough that $\Sigma 2^{\frac{1}{2}\nu}\omega(\pi 2^{-\nu})$ converges. This condition is readily seen to be equivalent to the convergence of the series $\Sigma n^{-\frac{1}{2}}\omega(\pi/n)$, or of the integral

$$\int_0^1 \delta^{-\frac{3}{2}}\omega(\delta)\,d\delta.$$

(3·6) THEOREM. *If $f(x)$ is of bounded variation and of the class Λ_α, for some $\alpha>0$, $S[f]$ converges absolutely.*

The second condition imposed on f is not superfluous, as the example

$$f(x)\sim\sum_{n=2}^{\infty}\frac{\sin nx}{n\log n} \tag{3·7}$$

shows. Here $f(x)$ is of bounded variation, indeed absolutely continuous (Chapter V, (1·5)), but $S[f]$ is not absolutely convergent.

Let $\omega(\delta)$ be the modulus of continuity of f, and V the total variation of f over $(0,2\pi)$. Obviously

$$\sum_{k=1}^{2N}\left[f\left(x+\frac{k\pi}{N}\right)-f\left(x+\frac{(k-1)\pi}{N}\right)\right]^2 \leqslant \omega\left(\frac{\pi}{N}\right)\sum_{k=1}^{2N}\left|f\left(x+\frac{k\pi}{N}\right)-f\left(x+\frac{(k-1)\pi}{N}\right)\right| \leqslant \omega\left(\frac{\pi}{N}\right)V.$$

We integrate this over $(0, 2\pi)$ and observe that all the integrals on the left are equal. This gives successively, with $N = 2^\nu$,

$$2N\int_0^{2\pi}\left[f\left(x+\frac{\pi}{2N}\right)-f\left(x-\frac{\pi}{2N}\right)\right]^2 dx \leqslant 2\pi V\omega(\pi/N),$$

$$\sum_{n=1}^{\infty}\rho_n^2\sin^2\frac{\pi n}{2N}\leqslant \tfrac{1}{4}N^{-1}V\omega\left(\frac{\pi}{N}\right),$$

$$\sum_{n=2^{\nu-1}+1}^{2^\nu}\rho_n^2\leqslant \tfrac{1}{2}2^{-\nu}V\omega(\pi 2^{-\nu}),$$

$$\sum_{n=2^{\nu-1}+1}^{2^\nu}\rho_n\leqslant \tfrac{1}{2}V^{\frac{1}{2}}\omega^{\frac{1}{2}}(\pi 2^{-\nu}),$$

$$\sum_{n=2}^{\infty}\rho_n\leqslant \tfrac{1}{2}V^{\frac{1}{2}}\sum_{\nu=1}^{\infty}\omega^{\frac{1}{2}}(\pi 2^{-\nu}).$$

If $\omega(\delta)\leqslant C\delta^\alpha$, $\alpha>0$, the series on the right converges and (3·6) follows. The convergence of $\Sigma\omega^{\frac{1}{2}}(\pi 2^{-\nu})$ is equivalent to that of $\Sigma n^{-1}\omega^{\frac{1}{2}}(\pi/n)$, or of

$$\int_0^1 \delta^{-1}\omega^{\frac{1}{2}}(\delta)\,d\delta.$$

(3·8) THEOREM. *If f is absolutely continuous and if $f'\in L^p$, $p>1$, then $S[f]$ converges absolutely.*

This follows from (3·6); for if $f'\in L^p$, $p>1$, then $f\in\Lambda_{1/p'}$, since, if $0<h\leqslant 2\pi$,

$$|f(x+h)-f(x)|\leqslant\int_x^{x+h}|f'(t)|\,dt\leqslant\left(\int_x^{x+h}|f'|^p\,dt\right)^{1/p}h^{1/p'}\leqslant\left(\int_0^{2\pi}|f'|^p\,dt\right)^{1/p}h^{1/p'}.$$

For $p=2$ (and so $p\geqslant 2$), (3·8) is immediate. For if a_n, b_n are the coefficients of f', those of f are $-b_n/n$, a_n/n, and the inequalities

$$|a_n|\,n^{-1}\leqslant\tfrac{1}{2}(a_n^2+n^{-2}),\quad |b_n|\,n^{-1}\leqslant\tfrac{1}{2}(b_n^2+n^{-2}),$$

coupled with Bessel's inequality $\Sigma(a_n^2+b_n^2)<\infty$, imply the absolute convergence of $S[f]$. (The argument can be extended to general $p>1$ if instead of Bessel's inequality we use its extension for $1<p\leqslant 2$, the Hausdorff-Young theorem (2·3) of Chapter XII.)

(3·9) THEOREM. *The conclusion of* (3·8) *remains valid if instead of the integrability of $|f'|^p$ we assume that of $|f'|\log^+|f'|$.*

The proof is postponed to Chapter VII, p. 287, where the result will be obtained as a corollary of the following theorem:

If $S[f]$ and $\tilde{S}[f]$ are both Fourier series of functions of bounded variation, $S[f]$ converges absolutely.

Here we only observe that the integrability of $|f'|(\log^+|f'|)^{1-\epsilon}$, $\epsilon>0$, would not be enough. For if f is given by (3·7), then $S[f]$ converges absolutely only for $x=0$ and $x=\pi$. But $f'(x)\sim 1/x\log^2 x$ as $x\to+0$ (Chapter V, (2·19)), so that $|f'|(\log^+|f'|)^{1-\epsilon}$ is integrable for every $0<\epsilon<1$.

The problem of absolute convergence of Fourier series may be generalized as follows. Given a series $\Sigma A_n(x)$, we ask about the values of the exponent β which make

$$\Sigma(|a_n|^\beta + |b_n|^\beta)$$

convergent. Theorem (3·1) is a special case of the following result:

(3·10) THEOREM. *If* $f \in \Lambda_\alpha$, $0<\alpha\leqslant 1$, *then* $\Sigma(|a_n|^\beta+|b_n|^\beta)$ *converges for* $\beta > 2/(2\alpha+1)$, *but not necessarily for* $\beta = 2/(2\alpha+1)$.

The proof of the first part is like that of the first part of (3·1). Let $\gamma = 2/(2\alpha+1)$. Since $0<\gamma<2$, we may suppose further that $0<\beta<2$. By Hölder's inequality and (3·3), we have

$$\sum_{2^{\nu-1}+1}^{2^\nu} \rho_n^\beta \leqslant \left(\sum_{2^{\nu-1}+1}^{2^\nu} \rho_n^2\right)^{\frac{1}{2}\beta} \left(\sum_{2^{\nu-1}+1}^{2^\nu} 1\right)^{1-\frac{1}{2}\beta} \leqslant 2^{\nu(1-\frac{1}{2}\beta)}\,\omega^\beta(\pi\, 2^{-\nu}),$$

$$\sum_{n=2}^{\infty} \rho_n^\beta \leqslant \sum_{\nu=1}^{\infty} 2^{\nu(1-\frac{1}{2}\beta)}\,\omega^\beta(\pi\, 2^{-\nu}),$$

and the last series converges if $\omega(\delta) \leqslant C\delta^\alpha$, since $1-\beta(\frac{1}{2}+\alpha)<0$. This gives the finiteness of $\Sigma\rho_n^\beta$, and so also of $\Sigma(|a_n|^\beta+|b_n|^\beta)$.†

The second parts of (3·1) and of (3·10) are corollaries of the results obtained in Chapter V, § 4. It was proved there that the real and imaginary parts of the series

$$\sum_{n=1}^{\infty} \frac{e^{in\log n}}{n^{\frac{1}{2}+\alpha}} e^{inx} \quad (0<\alpha<1) \tag{3·11}$$

belong to Λ_α, and it is easy to see that, for both, $\Sigma\rho_n^{2/(2\alpha+1)}=\infty$. The series

$$\sum_{n=1}^{\infty} \frac{e^{in\log n}}{(n\log n)^{\frac{3}{2}}} e^{inx} \tag{3·12}$$

belongs to Λ_1 (Chapter V, (4·9)), but $\Sigma\rho_n^{\frac{2}{3}}=\infty$.

(3·13) THEOREM. *If f is of bounded variation, and if $f\in\Lambda_\alpha$, $0<\alpha\leqslant 1$, then* $\Sigma(|\alpha_n|^\beta+|b_n|^\beta)$ *converges for* $\beta>2/(\alpha+2)$, *though not necessarily for* $\beta=2/(\alpha+2)$.

The proof of the convergence for $\beta>2/(2+\alpha)$ is analogous to the proofs of Theorems (3·6) and (3·10), and is left to the reader.

In proving that $\Sigma(|a_n|^\beta+|b_n|^\beta)$ can diverge for $\beta=2/(2+\alpha)$, we may suppose that $0<\alpha<1$, since the case $\alpha=1$ is taken care of by (3·12). We start from the series

$$\sum_{n=2}^{\infty} n^{-\frac{1}{2}\alpha}(\log n)^{-\gamma} e^{in^\alpha} e^{inx} \quad (0<\alpha<1).$$

It was proved in Chapter V, (5·8), that this is a Fourier series if $\gamma>1$. So the integrated series

$$-i\Sigma n^{-\frac{1}{2}\alpha-1}(\log n)^{-\gamma} e^{in^\alpha} e^{inx} \tag{3·14}$$

is the Fourier series of a function of bounded variation (indeed absolutely continuous). By Chapter V, (5·2), the sum of (3·14) without the logarithmic factor is a function of the class Λ_α, and by Chapter IV, (11·16), the sum of (3·14) is in Λ_α. On the other hand,

$$\Sigma|c_n|^{2/(2+\alpha)} = \Sigma n^{-1}(\log n)^{-\gamma[2/(2+\alpha)]}=\infty,$$

if γ is close enough to 1. It follows that for both the real and the imaginary parts of (3·14) the series $\Sigma(|a_n|^\beta+|b_n|^\beta)$ diverges if $\beta=2/(2+\alpha)$.

† If $A_n(x)=\rho_n\cos(nx+\phi_n)$, then $(|a_n|^\beta+|b_n|^\beta)/\rho_n^\beta=|\cos\phi_n|^\beta+|\sin\phi_n|^\beta$ is contained between $2^{-\frac{1}{2}\beta}$ and 2, so that the series $\Sigma(|a_n|^\beta+|b_n|^\beta)$ and $\Sigma\rho_n^\beta$ both converge or both diverge.

4. Inequalities for polynomials

The problem of the absolute convergence of Fourier series has close connexion with the following one, for trigonometric polynomials.

Consider all polynomials

$$T(x) = \tfrac{1}{2}a_0 + \sum_{k=1}^{n} (a_k \cos kx + b_k \sin kx), \tag{4·1}$$

of fixed order $n \geqslant 1$, such that $|T(x)| \leqslant 1$. How large can the sum

$$\Gamma(T) = \tfrac{1}{2}|a_0| + \sum_{k=1}^{n} (|a_k| + |b_k|)$$

be? The answer is given by the following theorem, in which A and B denote positive absolute constants.

(4·2) Theorem. (i) *If* $|T(x)| \leqslant 1$, *then* $\Gamma(T) \leqslant An^{\frac{1}{2}}$;

(ii) *Conversely, for every n there is a polynomial* (4·1) *such that* $|T(x)| \leqslant 1$, $\Gamma(T) \geqslant Bn^{\frac{1}{2}}$.

Since $|a_k| + |b_k|$ is contained between ρ_k and $2^{\frac{1}{2}}\rho_k$, (4·2) is not affected if we replace $\Gamma(T)$ by $\Gamma^*(T) = \tfrac{1}{2}\rho_0 + \sum_1^n \rho_k$.

Part (i) is immediate, since

$$\Gamma^*(T) = \tfrac{1}{2}\rho_0 + \sum_1^n \rho_k \leqslant \left(\tfrac{1}{2}\rho_0^2 + \sum_1^n \rho_k^2\right)^{\frac{1}{2}} \left(\tfrac{1}{2} + \sum_1^n 1^2\right)^{\frac{1}{2}}$$
$$= \left(\frac{1}{\pi}\int_0^{2\pi} T^2\,dx\right)^{\frac{1}{2}} (n + \tfrac{1}{2})^{\frac{1}{2}} \leqslant 2^{\frac{1}{2}}(n+\tfrac{1}{2})^{\frac{1}{2}} = (2n+1)^{\frac{1}{2}} \leqslant 3^{\frac{1}{2}}n^{\frac{1}{2}}. \tag{4·3}$$

To prove (ii), consider the polynomial

$$g(x) = g_t(x) = \sum_{k=1}^{N} \{\phi_{2k-1}(t)\cos kx + \phi_{2k}(t)\sin kx\},$$

where $\phi_1, \phi_2, \ldots$ are Rademacher's functions (Chapter I, § 3). By Chapter V, (8·4),

$$\int_0^1 dt \int_0^{2\pi} |g_t(x)|\,dx = \int_0^{2\pi} dx \int_0^1 |g_t(x)|\,dt \geqslant A_1\left\{\int_0^{2\pi} (\cos^2 x + \sin^2 x + \ldots + \sin^2 Nx)^{\frac{1}{2}}\,dx\right\}$$
$$= 2\pi A_1 N^{\frac{1}{2}}.$$

It follows that there is a $t = t_0$ such that $\displaystyle\int_0^{2\pi} |g_{t_0}(x)|\,dx \geqslant 2\pi A_1 N^{\frac{1}{2}}$.

Let a_k, b_k be the Fourier coefficients of the function $f(x) = \operatorname{sign} g_{t_0}(x)$. Let σ_ν be the (C, 1) means of $S[f]$, and let

$$\tau_N = 2\sigma_{2N-1} - \sigma_{N-1} = \tfrac{1}{2}a_0 + \sum_{k=1}^{N} A_k(x) + \sum_{k=N+1}^{2N-1} \left(1 - \frac{k-N}{N}\right) A_k(x)$$

be the delayed means of $S[f]$ (Chapter III, § 1); for $n = 2N-1$ or $n = 2N$, $\tfrac{1}{3}\tau_N$ will be the required polynomial T.

Clearly, $|\tfrac{1}{3}\tau_N| \leqslant 1$, since $|f| \leqslant 1$ and so $|\sigma_\nu| \leqslant 1$. It is therefore enough to show that $\Gamma(\tau_N) \geqslant CN^{\frac{1}{2}}$ for some fixed C. Now

$$\Gamma(\tau_N) \geqslant \sum_1^N (|a_k| + |b_k|) \geqslant \left|\sum_1^N a_k\phi_{2k-1}(t_0) + b_k\phi_{2k}(t_0)\right| = \left|\frac{1}{\pi}\int_0^{2\pi} g_{t_0}\tau_N\,dx\right|.$$

Since $g_{l_0}(x)$ is a polynomial of order N, and the Nth partial sums of τ_N and of $S[f]$ coincide, this last expression is

$$\left|\frac{1}{\pi}\int_0^{2\pi} g_{l_0} f\,dx\right| = \frac{1}{\pi}\int_0^{2\pi} |g_{l_0}|\,dx = 2A_1 N^{\frac{1}{2}}.$$

This completes the proof of (ii).

(4·4) THEOREM. *For any $r>0$ and any polynomial* (4·1), *let*

$$\Gamma_r(T) = \{\tfrac{1}{2}|a_0|^r + \Sigma(|a_k|^r + |b_k|^r)\}^{1/r}.$$

Then, (i) *for any $1 \leqslant r \leqslant 2$ and any $T(x)$ with $|T| \leqslant 1$, we have $\Gamma_r(T) \leqslant An^{-\frac{1}{2}+1/r}$, and* (ii) *for any $1 \leqslant r \leqslant 2$ and any n, there is a polynomial* (4·1) *satisfying $|T| \leqslant 1$ with $\Gamma_r(T) \geqslant Bn^{-\frac{1}{2}+1/r}$.*

The restriction on r is essential: for $r > 2$, part (i) is false and part (ii) trivial.

Part (i) is proved here as in (4·2), except that we use Hölder's inequality instead of Schwarz's. For (ii) we have, by Hölder's inequality,

$$\Gamma(T) \leqslant \Gamma_r(T)(2n+1)^{(r-1)/r} \leqslant 3\Gamma_r(T)\,n^{(r-1)/r}$$

for any T of order n; and the T for which we have $\Gamma(T) \geqslant Bn^{\frac{1}{2}}$ satisfies also the inequality $\Gamma_r(T) \geqslant \frac{1}{3}Bn^{-\frac{1}{2}+1/r}$.

5. Theorems of Wiener and Lévy

It is obvious that the absolute convergence of $S[f]$ at a point x_0 is not a local property, but depends on the behaviour of f in the whole interval $(0, 2\pi)$. However,

(5·1) THEOREM. *If to every point x_0 there corresponds a neighbourhood I_{x_0} of x_0 and a function $g(x) = g_{x_0}(x)$ such that* (i) *$S[g]$ converges absolutely and* (ii) *$g(x) = f(x)$ in I_{x_0}, then $S[f]$ converges absolutely.*

By the Heine-Borel theorem we can find a finite number of points $x_1 < x_2 < \ldots < x_m$ such that the intervals $I_{x_1}, I_{x_2}, \ldots, I_{x_m}$ cover $(0, 2\pi)$. Let $I_{x_k} = (u_k, v_k)$. We may suppose that the successive intervals overlap and even that $u_k < v_{k-1} < u_{k+1} < v_k$, $k = 1, 2, \ldots, m$, where $u_{m+1} = u_1 + 2\pi$, $v_0 = v_m - 2\pi$. Let $\lambda_k(x)$ be the periodic and continuous function equal to 1 in (v_{k-1}, u_{k+1}), vanishing outside (u_k, v_k), and linear in (u_k, v_{k-1}) and (u_{k+1}, v_k). It is readily seen that $\lambda_1(x) + \lambda_2(x) + \ldots + \lambda_m(x) = 1$. Since $\lambda_k'(x)$ is of bounded variation, the Fourier coefficients of λ_k are $O(n^{-2})$, and $S[\lambda_k]$ converges absolutely.

Since $S[f\lambda_k] = S[g_{x_k}\lambda_k] = S[g_{x_k}]\,S[\lambda_k]$, it follows that $S[f\lambda_k]$ converges absolutely (Chapter IV, §8). To complete the proof of (5·1), we observe that

$$S[f] = S[f.(\lambda_1 + \ldots + \lambda_m)] = S[f\lambda_1] + \ldots + S[f\lambda_m].$$

(5·2) THEOREM. (i) *Suppose that $S[f]$ converges absolutely, that the values (in general, complex) of $f(t)$ lie on a curve C, and that $\phi(z)$ is an analytic (not necessarily single-valued) function of a complex variable regular at every point of C. Then $S[\phi(f)]$ converges absolutely.*

(ii) *In particular, if $f(t) \neq 0$, and if $S[f]$ converges absolutely, so does $S[1/f]$.*

For every $g(x) = \Sigma a_k e^{ikx}$ we write

$$\|g\| = \Sigma|a_k|, \quad \mathrm{M}g = \max_x |g(x)|. \tag{5·3}$$

Clearly,

$$\|\Sigma g_i\| \leqslant \Sigma\|g_i\|, \quad \|g_1 g_2\| \leqslant \|g_1\|\,\|g_2\|. \tag{5·4}$$

(5·5) LEMMA. *(a) If* $g(x)=\Sigma a_k e^{ikx}$ *is twice continuously differentiable, then*

$$\| g \| \leqslant 4(\mathrm{M}g + \mathrm{M}g''). \tag{5·6}$$

(b) If $g(x,\theta)$ *is periodic in* x, *and if for each value of the parameter* θ, $0 \leqslant \theta \leqslant 2\pi$, *we have* $\| g(x,\theta) \| \leqslant A$, *then*

$$\left\| \int_0^{2\pi} g(x,\theta)\, d\theta \right\| \leqslant 2\pi A.$$

(a) We have $|a_0| \leqslant \mathrm{M}g$, and if $n \neq 0$ integration by parts gives $|a_n| \leqslant n^{-2}\mathrm{M}g''$. Since $\Sigma' n^{-2} = \frac{1}{3}\pi^2 < 4$, (5·6) follows. Part (b) is immediate.

Return to (5·2) (i). Since ϕ is analytic for $z=f(x)$, there is a $\rho>0$ such that $\phi(z)$ is regular in each circle $|z-f(x)| \leqslant 2\rho$. Let $s(x)$ be a partial sum of $S[f]$ such that

$$\mathrm{M}(s-f) \leqslant \| s-f \| \leqslant \tfrac{1}{2}\rho.$$

Then $\phi[s(x)+\rho e^{i\theta}]$ is twice continuously differentiable in x and θ. Hence, for each θ, $\| \phi[s(x)+\rho e^{i\theta}] \|$ is finite, and this norm is a bounded function of θ.

On the other hand, we have

$$(s+\rho e^{i\theta}-f)^{-1} = \rho^{-1} e^{-i\theta} \left\{ 1 + \sum_1^\infty (f-s)^n \rho^{-n} e^{-in\theta} \right\}$$

and therefore, by (5·4),

$$\| (s+\rho e^{i\theta}-f)^{-1} \| \leqslant \rho^{-1} \left\{ 1 + \sum_1^\infty (\tfrac{1}{2}\rho)^n \rho^{-n} \right\} = 2\rho^{-1}.$$

Since, by Cauchy's formula,

$$\phi[f(x)] = \frac{1}{2\pi} \int_0^{2\pi} \frac{\phi[s(x)+\rho e^{i\theta}]}{s(x)+\rho e^{i\theta}-f(x)} \rho e^{i\theta}\, d\theta,$$

part (b) of Lemma (5·5) implies that $\| \phi[f(x)] \|$ is finite. This completes the proof of (5·2).

Theorems (5·1) and (5·2) (ii) are due to Wiener, (5·2) (i) to Lévy. A corollary of (5·2) (ii) is that, *if* $F(z)$ *is regular for* $|z|<1$, *continuous and distinct from* 0 *in* $|z| \leqslant 1$, *and if the Taylor series of* F *converges absolutely on* $|z|=1$, *then the Taylor series of* $1/F$ *also converges absolutely on* $|z|=1$.

We verify without difficulty that the argument of the preceding proof also yields the following result:

(5·7) THEOREM. *Let* $f(x)$ *have an absolutely convergent Fourier series, and let* $F_n(x)$ *be a sequence of analytic functions converging uniformly to* 0 *in a neighbourhood of the range of* $f(x)$. *Then* $\| F_n[f(x)] \| \to 0$.

From this we deduce the following theorem:

(5·8) THEOREM. *If* $f(x)$ *has an absolutely convergent Fourier series, then*

$$\lim_{n\to\infty} \| f^n \|^{1/n} = \mathrm{M}f.$$

Let $R > \mathrm{M}f$, $F_n(z) = (z/R)^n$. By (5·7),

$$\lim_{n\to\infty} \| F_n[f(x)] \| = \lim_{n\to\infty} \frac{\| f^n \|}{R^n} = 0,$$

which implies that
$$\limsup_{n\to\infty} \frac{\| f^n \|^{1/n}}{R} \leqslant 1.$$

Since this holds for every $R > \mathrm{M}f$, it follows that

$$\limsup \| f^n \|^{1/n} \leqslant \mathrm{M}f.$$

On the other hand, we have

$$\| f^n \| \geqslant \mathrm{M}f^n = (\mathrm{M}f)^n,$$

whence $\liminf \| f^n \|^{1/n} \geqslant \mathrm{M}f$. This completes the proof of (5·8).

6. The absolute convergence of lacunary series

(6·1) THEOREM. *If a lacunary series*

$$\Sigma(a_j \cos n_j x + b_j \sin n_j x) = \Sigma \rho_j \cos (n_j x + x_j) \quad (n_{j+1}/n_j > q > 1) \tag{6·2}$$

is an $\mathrm{S}[f]$ *with* f *bounded above (or below), then* $\Sigma \rho_j < \infty$.

(6·3) THEOREM. *If the partial sums* s_m *of any lacunary series* (6·2) *satisfy the condition*

$$\limsup_{m\to\infty} s_m(x) < +\infty \tag{6·4}$$

(or $\liminf s_m(x) > -\infty$*) at every point of an interval* (α, β)*, then* $\Sigma \rho_j < \infty$.

Theorem (6·1) is a corollary of (6·3). For the hypothesis of (6·1) implies that the (C, 1) means $\sigma_m(x)$ of (6·2) are bounded above. Since $s_m(x) - \sigma_m(x) \to 0$ (owing to the lacunary character of the series and the relations $a_k, b_k \to 0$; see p. 79), the $s_m(x)$ are bounded above and $\Sigma \rho_j < \infty$ by (6·3).

Nevertheless, it is more convenient to begin by proving (6·1), which is the easier of the two; the more so since part of the argument will be used later in the proof of (6·3).

We first recall a few properties of the Riesz products, defined in § 7 of Chapter V.

Let $m_1, m_2, \ldots, m_k$ be a sequence of positive integers satisfying $m_{j+1}/m_j \geqslant Q \geqslant 3$ for all j, and let

$$R(x) = \prod_{j=1}^{k} \{1 + \cos (m_j x + \xi_j)\}. \tag{6·5}$$

$R(x)$ is a non-negative polynomial with constant term 1,

$$R(x) = 1 + \Sigma \gamma_\nu \cos (\nu x + \eta_\nu), \tag{6·6}$$

so that

$$\frac{1}{2\pi}\int_0^{2\pi} R(x)\, dx = 1. \tag{6·7}$$

Moreover, $0 \leqslant \gamma_\nu \leqslant 1$ for all ν, and $\gamma_\nu = 0$ unless

$$\nu = m_j \pm m_{j'} \pm m_{j''} \pm \ldots \quad (k \geqslant j > j' > j'' > \ldots). \tag{6·8}$$

There is at most one such representation of any given ν, and so in particular the m_jth term of (6·6) is $\cos (m_j x + \xi_j)$.

Let $q > 1$ be given. Since the sum in (6·8) is contained between

$$m_j(1 - Q^{-1} - Q^{-2} - \ldots) = m_j(Q-2)/(Q-1) \quad \text{and} \quad m_j(1 + Q^{-1} + Q^{-2} + \ldots) = m_j Q/(Q-1),$$

the ν with $\gamma_\nu \neq 0$ are confined to the intervals $(m_j/q, m_j q)$ provided Q is large enough, say $Q > Q_0(q)$. We shall call the interval $(m_j/q, m_j q)$ the q-neighbourhood of m_j.

Return now to (6·1), and split $\{n_j\}$ into r subsequences

$$\{n_{jr+s}\}_{j=0,1,\ldots}, \quad \text{where} \quad s = 1, 2, \ldots, r.$$

We take r so large that $Q=q^r$ exceeds both 3 and the number $Q_0(q)$. Fix s, and set

$$R^s(x)=\prod_{j=0}^{k}\{1+\cos(n_{jr+s}x+x_{jr+s})\}. \tag{6·9}$$

Since the ranks of the (non-zero) terms of R^s are in the q-neighbourhoods of the n_{jr+s} $(j=0, 1, \dots, k)$, and the only non-zero term of $\mathsf{S}[f]$ in such a neighbourhood has rank n_{jr+s}, it follows that

$$\frac{1}{\pi}\int_0^{2\pi} f(x)\,R^s(x)\,dx=\frac{1}{\pi}\int_0^{2\pi} f(x)\left\{\sum_{j=0}^{k}\cos(n_{jr+s}x+x_{jr+s})\right\}dx=\sum_{j=0}^{k}\rho_{jr+s}. \tag{6·10}$$

But if M is the upper bound of f the left-hand side here does not exceed

$$M\frac{1}{\pi}\int_0^{2\pi} R^s(x)\,dx=2M,$$

by (6·7). It follows that the right-hand side of (6·10) does not exceed $2M$. Making $k\to\infty$ and summing the inequalities over $s=1, 2, \dots, r$, we see that $\Sigma\rho_j\leqslant 2Mr$, and (6·1) is established.

Under the hypothesis of (6·3) there is a subinterval I of (α,β) such that the s_m are uniformly bounded above, say $s_m\leqslant M$, in I (see Chapter I, (12·3)). Hence (6·3) is a corollary of the following lemma about polynomials.

(6·11) LEMMA. *Given any interval I and any $q>1$, we can find an integer $n_0=n_0(q, |I|)$ and a constant $A=A(q, |I|)$ such that for every lacunary polynomial*

$$T(x)=\Sigma(a_j\cos n_jx+b_j\sin n_jx)=\Sigma\rho_j\cos(n_jx+x_j)\quad(n_{j+1}/n_j>q>1) \tag{6·12}$$

for which $n_1\geqslant n_0$ and $T(x)\leqslant M$ in I, we have

$$\Sigma\rho_j\leqslant AM.$$

We need more information about the γ_ν in (6·6).

Suppose ν is given by (6·8) but is not equal to m_j. Then

$$|\nu-m_j|\geqslant m_{j'}(1-Q^{-1}-Q^{-2}-\dots)=m_{j'}(Q-2)/(Q-1)\geqslant\tfrac{1}{2}m_1.$$

We have already observed that ν is contained between $m_j(Q-2)/(Q-1)$ and $m_jQ/(Q-1)$, and so between $\frac{1}{2}m_j$ and $\frac{3}{2}m_j$. It follows that ν must differ from m_{j+1} (if such an m exists) by at least $\frac{3}{2}m_j\geqslant\frac{3}{2}m_1$. Similarly it must differ from m_{j-1} (if such an m exists) by at least

$$\tfrac{1}{2}m_j-\tfrac{1}{3}m_j=\tfrac{1}{6}m_j\geqslant\tfrac{1}{2}m_{j-1}\geqslant\tfrac{1}{2}m_1.$$

Collecting these estimates we see that *each ν with $\gamma_\nu\neq 0$ in* (6·6) *either is an m_j or differs from all m_j by at least $\frac{1}{2}m_1$.*

Return to $T(x)$. As before, we split $\{n_j\}$ into r subsequences $\{n_{jr+s}\}$, where $s=1, 2, \dots, r$, and r satisfies the conditions imposed above. Let T^s consist of the terms of T which have rank n_{jr+s} for some j (so that T is the sum of all T^s) and let R^s be defined by (6·9). We set

$$U=R^1+R^2+\dots+R^r.$$

An argument similar to (6·10) gives

$$\frac{1}{\pi}\int_0^{2\pi} TU=\sum_s\frac{1}{\pi}\int_0^{2\pi} TR^s\,dx=\sum_s\frac{1}{\pi}\int_0^{2\pi} T^sR^s\,dx=\Sigma\rho_j. \tag{6·13}$$

To estimate the left-hand side here we write

$$T=\Sigma c_j e^{in_jx}, \quad U=\Sigma \delta_\nu e^{i\nu x}, \quad \mu=\Sigma\rho_j,$$

where $n_{-j}=-n_j$. Since the only overlapping terms in different R^s are the constant ones, we have $\delta_0=r$, $|\delta_\nu|\leqslant\frac{1}{2}$ for $\nu\neq 0$. Hence

$$|c_j|=\tfrac{1}{2}\rho_{|j|}, \quad |\delta_\nu|\leqslant r \quad \text{for all } j \text{ and } \nu. \tag{6·14}$$

Let
$$TU=\Sigma C_m e^{imx}, \quad \text{where} \quad C_m=\sum_{n_j-\nu=m} c_j\bar{\delta}_\nu, \tag{6·15}$$

and let $N=[\frac{1}{2}n_1]$. We observe that

(i) $C_0=\frac{1}{2}\mu$ (since the left-hand side of (6·13) is $2C_0$);

(ii) $|C_m|\leqslant r\Sigma|c_j|=r\mu$ (by (6·15) and (6·14));

(iii) $C_m=0$ for $0<|m|<N$ (since $\delta_\nu=0$ in R^s if ν differs from any n_j by less than N).

Let now $\lambda(x)$ be the periodic and continuous function equal to 0 outside I, equal to 1 at the midpoint of I, and linear in each half of I. The series $\mathsf{S}[\lambda]=\Sigma\lambda_\nu e^{i\nu x}$ converges absolutely. The actual values of the λ_ν (see Chapter I, (4·16)) are not needed. Consider the formula

$$\frac{1}{2\pi}\int_0^{2\pi}\lambda T U\,dx=\Sigma\lambda_m\bar{C}_m.$$

On the one hand, by (i), (ii) and (iii), the sum here exceeds

$$\lambda_0C_0-\sum_{|m|\geqslant N}|\lambda_m C_m|\geqslant\tfrac{1}{2}\mu(\lambda_0-2r\sum_{|m|\geqslant N}|\lambda_m|); \tag{6·16}$$

on the other hand, the integral is

$$\frac{1}{2\pi}\int_I\lambda TU\,dx\leqslant\frac{M}{2\pi}\int_I\lambda U\,dx\leqslant\frac{M}{2\pi}\int_0^{2\pi}U\,dx=Mr. \tag{6·17}$$

If $N=[\frac{1}{2}n_1]$ is so large that the expression in brackets on the right of (6·16) exceeds $\frac{1}{2}\lambda_0$, a comparison of (6·17) and (6·16) shows that $\frac{1}{4}\lambda_0\mu\leqslant Mr$, so that $\mu\leqslant AM$ where $A=4r/\lambda_0$, and (6·11) is proved.

Remarks. (*a*) Theorem (6·3) holds if instead of the partial sums s_m of (6·2) we consider linear means. It is easier to deal with means applicable to the terms rather than to the partial sums of series (see p. 84).

Suppose we have a matrix $\{\alpha_{mn}\}_{m,\,n=0,1,\dots}$ and a series $u_0+u_1+\dots$. Consider the expression

$$\sigma_m=\sum_n\alpha_{mn}u_n.$$

The only thing we assume about the α_{mn} is that $\lim\limits_m\alpha_{mn}=1$ for each n. Suppose now that the means

$$\sigma_m(x)=\Sigma\alpha_{mn_j}\rho_j\cos(n_jx+x_j) \tag{6·18}$$

of (6·2) satisfy the following conditions for $x\in(\alpha,\beta)$:

(i) they exist (i.e. the series (6·18) converge);

(ii) they are continuous;

(iii) $\limsup\limits_m\sigma_m(x)<+\infty$.

Then, as before, there exists a subinterval I of (α,β) in which the σ_m are uniformly bounded above, say by M. Observing that (6·11) holds not only for polynomials but also for infinite series (as is seen by first considering the (C, 1) means of T and then making a passage to the limit), we have

$$\sum_j|\alpha_{mn_j}|\rho_j\leqslant AM.$$

If we retain only a fixed finite number of terms on the left and then make $m \to \infty$, we see that each partial sum of $\Sigma\rho_j$ is $\leqslant AM$, and so also $\Sigma\rho_j \leqslant AM$.

Condition (ii) is certainly satisfied if the matrix is row-finite. It is also satisfied in some other cases, for example for summability A.

(*b*) The conclusion of (6·3) holds if instead of (6·4) we suppose that for each $x \in (\alpha, \beta)$ we have at least one of the inequalities

$$\limsup s_m(x) < +\infty, \quad \liminf s_m(x) > -\infty. \tag{6·19}$$

It is enough to show that the new hypothesis implies that at least one of the inequalities (6·19) is satisfied in a subinterval of (α, β). For suppose this is not the case, and let E^+ and E^- denote respectively the subsets of (α, β) in which $\limsup s_m = +\infty$ and $\liminf s_m = -\infty$. By Chapter I, (12·2), the set E^+ (being a set of points at which the sequence of continuous functions

$$s_m^+(x) = \max\{0, s_m(x)\}$$

is unbounded) is the complement of a set of the first category in (α, β). Similarly E^- is the complement of a set of the first category in (α, β). It follows that E^+E^- is the complement of a set of the first category in (α, β) and, in particular, is not empty, contrary to hypothesis.

The conclusion may be stated slightly differently: if $\Sigma\rho_j = +\infty$, then in a set of points which is dense and of the second category in every interval we have simultaneously

$$\liminf s_m(x) = -\infty, \quad \limsup s_m(x) = +\infty. \tag{6·20}$$

There is a corresponding result for the linear means discussed in (*a*).

(*c*) If q is large, Theorem (6·3) can be obtained by the following simple geometrical argument. Let I be a subinterval of (α, β) in which $s_m(x) \leqslant M$ for all m. Consider the curves $y = \cos(n_j x + x_j)$ for $j = 1, 2, \ldots$, and denote, generally, by I_j any of the intervals in which $\cos(n_j x + x_j) \geqslant \frac{1}{2}$. Assuming, as we may, that n_1 is large enough, we select an interval I_1 totally included in I. Since n_2/n_1 is large we can find an interval I_2 totally included in the I_1 just considered, and so on. Let x^* be the point common to $I, I_1, I_2, \ldots$. Since

$$\sum_{j=1}^{m} \rho_j \cos(n_j x^* + x_j) \leqslant M$$

for all m and since $\cos(n_j x^* + x_j) \geqslant \frac{1}{2}$, it follows that $\Sigma\rho_j$ converges.

This argument is valid for $q \geqslant 4$. For let d_j and δ_j denote respectively the length of I_j and the distance between two consecutive I_j. We can find an $I_{j+1} \subset I_j$ if

$$2d_{j+1} + \delta_{j+1} \leqslant d_j. \tag{6·21}$$

Observing that $d_j = 2\pi/3n_j$, $\delta_j = 4\pi/3n_j$, we find that (6·21) is equivalent to $n_{j+1}/n_j \geqslant 4$.

By considering the inequalities $\cos(n_j x + x_j) \geqslant \epsilon > 0$, where ϵ is arbitrarily small but fixed, we can extend the preceding argument to the case $q > 3$. The argument does not work, however, for general $q > 1$.

MISCELLANEOUS THEOREMS AND EXAMPLES

1. The set of points where $\Sigma n^{-1} \sin n!x$ converges absolutely contains a perfect subset. [Consider the graphs of the curves $y = \sin n!x$.]

2. (i) Every measurable set of positive measure is a basis. (Steinhaus [5].)

(ii) Every set whose complement is of the first category is a basis. (Niemytski [1].)

[Suppose $|E| > 0$, $x \in E$, $y \in E$. To prove (i), it is enough to show that the set of differences $x - y$ contains an interval. Let E_h denote the set E translated by h. Considering the neighbourhood of a point of density of E, we see without difficulty that $EE_h \neq 0$ for all small enough h. The proof of (ii) is similar.]

3. Let C be Cantor's ternary set constructed on $(0, 2\pi)$. Every $x \in C$ can be written in the form $2\pi(\alpha_1 3^{-1} + \alpha_2 3^{-2} + \ldots)$, where α_i is either 0 or 2. Using this, show that every $z \in (0, 4\pi)$ can be written in the form $x + y$ with $x \in C$, $y \in C$ (which shows that C is a basis).

4. Let A be the set of points of absolute convergence of $\Sigma A_n(x)$. Then A is invariant under

(i) the translation, (ii) the symmetry,

making any two points a, b of A correspond. (Arbault [1].)

[If [1·7] is written $\Sigma\rho_n \sin(nx+x_n)$, and if $a, b, c \in A$, an argument similar to the proof of (1·7) shows that the series converges absolutely at the points (i) $a-b+c$, (ii) $a+b-c$.]

5. If a trigonometric series converges absolutely on a perfect set E of Cantor type (Chapter V, § 3), then the series converges uniformly on E. (Arbault [1], Malliavin [1].)

[A consequence of Chapter I, (12·3) (ii), of the homogeneous character of E, and of the proof of (1·10).]

6. A necessary and sufficient condition for $S[h]$ to converge absolutely is that h be the convolution (Chapter II, § 1) of two functions f and g of the class L^2. (M. Riesz; see Hardy and Littlewood [12].)

[The sufficiency of the condition follows from Chapter II, (1·7); if $\Sigma c_n e^{inx}$ converges absolutely, consider the functions with Fourier coefficients $|c_n|^{\frac{1}{2}}$ and $|c_n|^{\frac{1}{2}} \operatorname{sign} c_n$.]

7. The hypotheses of theorems (3·1) and (3·10) are unnecessarily stringent. The conclusions hold if the condition $f \in \Lambda_\alpha$ is replaced by $f \in \Lambda_\alpha^2$.

8. Let $0<\alpha\leqslant 1$, $1\leqslant p\leqslant 2$. If a_n, b_n are the coefficients of an $f\in\Lambda_\alpha^p$, then $\Sigma(|a_n|^\beta+|b_n|^\beta)$ converges if $\beta>p/\{p(1+\alpha)-1\}$. (Szász [2].)

[The proof is similar to that of (3·10) if instead of Parseval's formula we use the inequality of Hausdorff-Young, which will be established in Chapter XII, § 2.]

9. (i) If $f\in\Lambda_\alpha$, $0<\alpha\leqslant 1$, then $\Sigma n^{\beta-\frac{1}{2}}(|a_n|+|b_n|)$ converges for $\beta<\alpha$. (Weyl [1], Hardy [1].)

(ii) If f is, in addition, of bounded variation, then $\Sigma n^{\beta/2}(|a_n|+|b_n|)<\infty$.

(iii) If $f\in\Lambda_\alpha^p$ for $0<\alpha\leqslant 1$, $1\leqslant p\leqslant 2$, then $\Sigma n^\gamma(|a_n|+|b_n|)<\infty$ for $\gamma<\alpha-1/p$.

10. Let $F(x)$, $0\leqslant x\leqslant 2\pi$, be a non-decreasing function of jumps; that is, the increment of F over any interval is equal to the sum of all the jumps of F in that interval. Let $d_1, d_2, \dots$ be all the jumps of F. Show that

$$\limsup_{n\to\infty}\int_0^{2\pi}\cos 2nx\, dF(x)=\Sigma d_i.$$

[Compare (2·16).]

11. If a lacunary series $\Sigma(a_k\cos n_kx+b_k\sin n_kx)$, $n_{k+1}/n_k>q>1$, is absolutely summable A (see p. 83) in a set E of positive measure, then $\Sigma(|a_k|+|b_k|)$ converges.

[Supposing for simplicity that the series is purely cosine, write $f(r,x)=\Sigma a_k r^{n_k}\cos n_kx$. By reducing E we may suppose that $\int_0^1\left|\frac{\partial}{\partial r}f(r,x)\right|dr$ is uniformly bounded for $x\in E$. Integrating the integral with respect to x, changing the order of integration, and applying Chapter V, (6·5), we deduce that $\int_0^1(\Sigma a_k^2n_k^2r^{2n_k})^{\frac{1}{2}}\,dr$ is finite. If we consider the integral over $1-q/n_k\leqslant r\leqslant 1-1/n_k$ and in the sum Σ keep only the kth term we deduce that $\Sigma|a_k|$ converges.]

CHAPTER VII

COMPLEX METHODS IN FOURIER SERIES

1. Existence of conjugate functions

So far, though using complex numbers, we have not systematically applied complex variable theory; in particular, we have not applied Cauchy's theorem. Our nearest approach was the use of harmonic functions, which are the real parts of analytic functions.

In operating, however, with harmonic functions there are certain inconveniences; thus even the square of a harmonic function need not be harmonic. The situation is different for analytic functions; elementary operations performed on such functions lead again to analytic functions. So does the operation of taking a function of a function. Dealing directly with analytic functions instead of with their real parts accordingly offers distinct advantages. The significance and value of complex methods in the theory of Fourier series is well seen in the problem of the summability of conjugate series and the existence of conjugate functions.

In Chapter III we proved a number of results on the summability of Fourier series. The corresponding results for conjugate series were much less complete. The obstacle was that at that time we knew nothing about the *existence* of the integral

$$\tilde{f}(x) = -\frac{1}{\pi}\int_0^{\pi}\frac{f(x+t)-f(x-t)}{2\tan\frac{1}{2}t}\,dt = \lim_{\epsilon\to+0}\left\{-\frac{1}{\pi}\int_{\epsilon}^{\pi}\right\}, \tag{1·1}$$

all that we proved being that, almost everywhere, the existence of $\tilde{f}(x)$ was equivalent to the summability A of $\tilde{S}[f]$. Later, in Chapter IV, § 3, we gave a proof of the existence of (1·1), and so, by the equivalence, of the summability A of $\tilde{S}[f]$, almost everywhere. We shall now prove the latter fact by complex methods, without appealing to the existence of (1·1).

(1·2) THEOREM. *For any* $f \in L$, $\tilde{S}[f]$ *is summable* A *almost everywhere. More generally, the harmonic function conjugate to the Poisson integral of* f *has a non-tangential limit at almost all points of the unit circle.*

We may suppose that $f \geqslant 0$, $f \not\equiv 0$. Let $z = r\,e^{ix}$, and let $u(r,x)$ be the Poisson integral and $v(r,x)$ the conjugate harmonic function. The function $u(r,x)+iv(r,x)$ is regular for $|z|<1$ and its values there belong to the right half-plane. Hence

$$G(z) = e^{-u(r,x)-iv(r,x)} \tag{1·3}$$

is regular and absolutely less than 1 for $|z|<1$. It follows that $G(z)$ is, *qua* harmonic function, the Poisson integral of a bounded function. Thus the non-tangential limit of $G(r\,e^{ix})$ exists almost everywhere. Since the non-tangential limit of $u(r,x)$ also exists almost everywhere and is finite $(=f)$, the limit of G must be distinct from 0 almost everywhere. Hence a finite non-tangential limit of $v(r,x)$ exists almost everywhere, and (1·2) is established. As corollaries (see Chapter III, (7·20), (5·8)) we have:

(1·4) THEOREM. *For an integrable* f, *the integral* (1·1) *exists almost everywhere.*

(1·5) THEOREM. *For almost all x, $\tilde{S}[f]$ is summable* (C, α), $\alpha > 0$, (*and so also summable* A) *to sum* $\tilde{f}(x)$.

Extensions to Fourier-Stieltjes series are easy. Thus we have:

(1·6) THEOREM. *Let $F(x)$ be of bounded variation in $0 \leqslant x \leqslant 2\pi$. The conclusion of* (1·2) *holds if $\tilde{S}[f]$ there is replaced by $\tilde{S}[dF]$. The integral*

$$F^*(x) = -\frac{1}{\pi}\int_0^\pi \frac{F(x+t)+F(x-t)-2F(x)}{4\sin^2\frac{1}{2}t}\,dt = \lim_{\epsilon\to 0}\left\{-\frac{1}{\pi}\int_\epsilon^\pi\right\} \tag{1·7}$$

exists almost everywhere and represents for almost all x the (C, α), $\alpha > 0$, (*and so also the* A) *sum of $\tilde{S}[dF]$.*

Suppose, as we may, that F is non-decreasing. The Poisson integral $u(r, x)$ of dF is then non-negative. Let $v(r, x)$ be the conjugate of $u(r, x)$ and let $G(z)$ be given by (1·3). Arguing as before, we prove that the non-tangential limit of $v(r, x)$ exists almost everywhere. In particular $\tilde{S}[dF]$ is summable A almost everywhere. By Chapter III, (7·15), the integral (1·7) exists almost everywhere, and by Chapter III, (8·3), $\tilde{S}[dF]$ is almost everywhere summable (C, α), $\alpha > 0$, with sum (1·7).

The integral (1·7) may be written

$$-\frac{1}{\pi}\int_{-\pi}^{\pi}\frac{dF(t)}{2\tan\frac{1}{2}(t-x)} = -\frac{1}{\pi}\int_{-\pi}^{\pi}\frac{dF(t+x)}{2\tan\frac{1}{2}t}, \tag{1·8}$$

where the integrals are to be taken in the 'principal value' sense (see p. 51).

2. The Fourier character of conjugate series

Of the series
$$\sum_2^\infty \frac{\cos nx}{\log n}, \quad \sum_2^\infty \frac{\sin nx}{\log n}, \tag{2·1}$$

the first is a Fourier series, and the second is not (Chapter V, §1). Thus $\tilde{S}[f]$ need not be a Fourier series.

Of the series
$$\sum_1^\infty \frac{\sin nx}{n}, \quad \sum_1^\infty \frac{\cos nx}{n}, \tag{2·2}$$

the first is the Fourier series of the bounded function $\frac{1}{2}(\pi - x)$, $0 < x < 2\pi$, the second of the unbounded function $-\log|2\sin\frac{1}{2}x|$ (Chapter I, (2·8)). Of the series

$$\sum_2^\infty \frac{\sin nx}{n\log n}, \qquad \sum_2^\infty \frac{\cos nx}{n\log n}, \tag{2·3}$$

the first is the Fourier series of a continuous function, since it converges uniformly (by Theorem (1·3) of Chapter V), the second is the Fourier series of a discontinuous one. (The latter series diverges to $+\infty$ at $x = 0$, and so is also (C, 1) summable to $+\infty$ there; hence it is the Fourier series of an unbounded, and so also discontinuous, function.) However,

(2·4) THEOREM OF M. RIESZ. *If $f \in L^p$, $1 < p < +\infty$, then $\tilde{f} \in L^p$, and*

$$\mathfrak{M}_p[\tilde{f}] \leqslant A_p \mathfrak{M}_p[f], \tag{2·5}$$

with A_p depending on p only. Moreover, $\tilde{S}[f] = S[\tilde{f}]$.

As we have just seen, (2·4) fails for $p=1$ and for $p=\infty$. For $p=1$, its place is taken by the following two theorems:

(2·6) THEOREM. *If $f\in L$, then $\tilde f\in L^{\mu}$ for every $0<\mu<1$, and there is a constant B_μ, depending on μ only, such that*

$$\mathfrak{M}_\mu[\tilde f]\leqslant B_\mu \mathfrak{M}[f]. \tag{2·7}$$

(2·8) THEOREM. *If $f\log^+|f|$ is integrable, then $\tilde f\in L$, and $\tilde{S}[f]=S[\tilde f]$. Moreover, there are two absolute constants A, B such that*

$$\int_0^{2\pi}|\tilde f|\,dx\leqslant A\int_0^{2\pi}|f|\log^+|f|\,dx+B. \tag{2·9}$$

This result admits of a partial converse, namely,

(2·10) THEOREM. *If $f\in L$ and is bounded below, and if $\tilde{S}[f]$ is a Fourier series, then $f\log^+|f|\in L$.*

It follows from (2·4) that if f is bounded, then $\tilde f$ belongs to every L^p. More generally, we have

(2·11) THEOREM. (i) *If $|f|\leqslant 1$, then*

$$\int_0^{2\pi}\exp(\lambda|\tilde f|)\,dx\leqslant C_\lambda \tag{2·12}$$

for $0<\lambda<\frac{1}{2}\pi$.

(ii) *If f is continuous, then $\exp\lambda|\tilde f|$ is integrable for all $\lambda>0$.*

The proof of all these results can be based on Cauchy's formula

$$\frac{1}{2\pi i}\int_{|z|=r}z^{-1}G(z)\,dz=\frac{1}{2\pi}\int_0^{2\pi}G(r\,e^{ix})\,dx=G(0)\quad(0<r<1), \tag{2·13}$$

valid for any $G(z)$ regular in $|z|<1$.

Let $f(r,x)$ be the Poisson integral of f, and $\tilde f(r,x)$ the conjugate harmonic function. The inequalities (2·5), (2·7), (2·9) and (2·12) imply the corresponding relations for $f(r,x)$ and $\tilde f(r,x)$, and conversely are implied by the latter, on making $r\to 1$ and applying the relation $\mathfrak{M}_p[f(r,x)]\to\mathfrak{M}_p[f]$ (or Theorem (6·15) of Chapter IV) and Fatou's lemma. Thus it is enough to prove the inequalities for the functions $f(r,x)$ and $\tilde f(r,x)$, which we shall also denote by $u(z)$ and $v(z)$.

Begin with (2·7), and suppose at first that $f\geqslant 0$ and that $f\not\equiv 0$, since otherwise the theorem is obvious. The regular function

$$F(z)=u(z)+iv(z)\quad(z=r\,e^{ix},\ 0\leqslant r<1) \tag{2·14}$$

has positive real part, and so $F(z)\neq 0$. We may also write

$$F(z)=R\,e^{i\Phi},\quad\text{where}\quad R>0,\ |\Phi|\leqslant\tfrac{1}{2}\pi. \tag{2·15}$$

The function

$$G(z)=F^{\mu}(z)=R^{\mu}e^{i\mu\Phi}$$

is regular for $|z|<1$ and real at $z=0$. On taking real parts, the second equation (2·13) gives

$$\frac{1}{2\pi}\int_0^{2\pi}R^{\mu}\cos\mu\Phi\,dx=F^{\mu}(0)=\left(\frac{1}{2\pi}\int_0^{2\pi}f\,dx\right)^{\mu}, \tag{2·16}$$

$F(0)$ being the constant term of $S[f]$. Since $R \geqslant |v|$ and $\cos \mu\Phi \geqslant \cos \frac{1}{2}\mu\pi$, we have

$$\left(\int_0^{2\pi} |v|^\mu dx\right)^{1/\mu} \leqslant B_\mu \int_0^{2\pi} f dx,$$

where
$$B_\mu^\mu = (2\pi)^{1-\mu}/\cos \tfrac{1}{2}\mu\pi < 2\pi/(1-\mu).$$

On making $r \to 1$, we get (2·7).

If f is of variable sign we write

$$f_1 = f^+, \quad f_2 = f^-, \quad f = f_1 - f_2, \quad \tilde{f} = \tilde{f}_1 - \tilde{f}_2, \tag{2·17}$$

$$\mathfrak{M}_\mu^\mu[\tilde{f}] \leqslant \mathfrak{M}_\mu^\mu[\tilde{f}_1] + \mathfrak{M}_\mu^\mu[\tilde{f}_2] \leqslant B_\mu^\mu \left\{ \left(\int_0^{2\pi} f_1 dx\right)^\mu + \left(\int_0^{2\pi} f_2 dx\right)^\mu \right\} \leqslant 2B_\mu^\mu \left(\int_0^{2\pi} |f|\, dx\right)^\mu,$$

since $f_1 + f_2 = |f|$. This proves (2·6) with B_μ^μ increased by the factor 2.

For (2·5) we need the inequality

$$|\sin \phi|^p \leqslant \alpha \cos p\phi + \beta \cos^p \phi \quad (|\phi| \leqslant \tfrac{1}{2}\pi), \tag{2·18}$$

where p is positive and *not an odd integer*, and α, β are constants depending on p. We may suppose that $\phi \geqslant 0$. Since $\cos \frac{1}{2}p\pi \neq 0$, we have for a suitable α (possibly negative) $\alpha \cos p\phi \geqslant 1$ in a certain neighbourhood of $\phi = \frac{1}{2}\pi$. (2·18) is valid *a fortiori* in this neighbourhood, provided $\beta \geqslant 0$. Having fixed α, we take β so large that $\beta \cos^p \phi \geqslant |\alpha| + 1$, and so

$$\alpha \cos p\phi + \beta \cos^p \phi \geqslant 1 \geqslant \sin^p \phi,$$

in the remainder of $(0, \frac{1}{2}\pi)$. This completes the proof of (2·18).

Suppose once more that $f \geqslant 0$, and let $F(z)$ be given by (2·15). If we substitute Φ for ϕ in (2·18), multiply both sides by R^p and integrate over $0 \leqslant x \leqslant 2\pi$, we have

$$\int_0^{2\pi} |v|^p dx \leqslant \beta \int_0^{2\pi} u^p dx + \alpha \int_0^{2\pi} R^p \cos p\Phi\, dx = \beta \int_0^{2\pi} u^p dx + 2\pi\alpha \left(\frac{1}{2\pi}\int_0^{2\pi} f dx\right)^p, \tag{2·19}$$

using (2·16) with μ replaced by p. Hence, making $r \to 1$,

$$\int_0^{2\pi} |\tilde{f}|^p dx \leqslant \beta \int_0^{2\pi} f^p dx + 2\pi |\alpha| \left(\frac{1}{2\pi}\int_0^{2\pi} f dx\right)^p \leqslant (\beta + |\alpha|) \int_0^{2\pi} |f|^p dx,$$

and this is (2·5) with $A_p^p = \beta + |\alpha|$. (Incidentally, if $\alpha < 0$, for example if $1 < p \leqslant 2$, we may drop the last term in (2·19), and take $A_p^p = \beta$.)

For f of variable sign we use (2·17). By Minkowski's inequality,

$$\mathfrak{M}_p[\tilde{f}_1] \leqslant \mathfrak{M}_p[\tilde{f}_1] + \mathfrak{M}_p[\tilde{f}_2] \leqslant A_p\{\mathfrak{M}_p[f_1] + \mathfrak{M}_p[f_2]\} \leqslant 2A_p \mathfrak{M}_p[f]. \tag{2·20}$$

We have now proved (2·5) for $p > 1$, $p \neq 3, 5, 7, \ldots$, and in particular for $1 < p \leqslant 2$. The proof of (2·5) will be completed if we show that

(2·21) THEOREM. *If the inequality* (2·5) *is valid for a certain* $p > 1$ (*that is, for all* f *in* L^p) *it is also valid for the conjugate exponent* $p' = p/(p-1)$. *Moreover, we can take* $A_{p'} = A_p$.

Let g be any trigonometric polynomial with $\mathfrak{M}_p[g] \leqslant 1$. Parseval's relation (Chapter IV, (8·6)) gives

$$\int_0^{2\pi} \tilde{f}(r, x)\, g(x)\, dx = -\int_0^{2\pi} f(r, x)\, \tilde{g}(x)\, dx.$$

By Theorem (9·14) of Chapter I, and using the fact that polynomials are dense in $L^{p'}$, the upper bound of the left-hand side for all g is $\mathfrak{M}_{p'}[\tilde{f}(r,x)]$, and since

$$\left|\int_0^{2\pi} f(r,x)\,\tilde{g}\,dx\right| \leqslant \mathfrak{M}_{p'}[f(r,x)]\,\mathfrak{M}_p[\tilde{g}] \leqslant \mathfrak{M}_{p'}[f(r,x)]\,A_p\,\mathfrak{M}_p[g] \leqslant A_p\,\mathfrak{M}_{p'}[f(r,x)],$$

by Hölder's inequality and the hypothesis that (2·5) holds for p, it follows that $\mathfrak{M}_{p'}[\tilde{f}(r,x)] \leqslant A_p \mathfrak{M}_{p'}[f(r,x)]$ and, on making $r\to 1$,

$$\mathfrak{M}_{p'}[\tilde{f}] \leqslant A_p \mathfrak{M}_{p'}[f].$$

This completes the proof of (2·21).

To complete the proof of (2·4) we have to show that $\tilde{S}[f] = S[\tilde{f}]$. This follows from the fact that $\mathfrak{M}_p[f(r,x)] = O(1)$ implies $\mathfrak{M}_p[\tilde{f}(r,x)] = O(1)$ (see Chapter IV, (6·17)). Thus $\tilde{S}[f]$ is the Fourier series of the function $\lim_{r\to 1} \tilde{f}(r,x) = \tilde{f}(x)$ (see also Theorem (4·4) below).

We base the proof of (2·8) on the inequality

$$|\sin\phi| \leqslant \gamma(\cos\phi\log\cos\phi + \phi\sin\phi) + \delta\cos\phi \quad (|\phi| \leqslant \tfrac{1}{2}\pi), \tag{2·22}$$

analogous to (2·18) and similarly proved (γ, δ are positive absolute constants). First suppose that $f \geqslant 0$, $f \not\equiv 0$. We again set $F(z) = u(z) + iv(z) = Re^{i\Phi}$ and apply (2·13) to the function $G(z) = F(z)\log F(z)$, regular for $|z| < 1$. Taking real parts we get

$$\frac{1}{2\pi}\int_0^{2\pi} \{R\cos\Phi\log R - R\Phi\sin\Phi\}\,dx = u(0)\log u(0), \tag{2·23}$$

or $\displaystyle\int_0^{2\pi} R(\cos\Phi\log\cos\Phi + \Phi\sin\Phi)\,dx = \int R\cos\Phi\log(R\cos\Phi)\,dx - 2\pi u(0)\log u(0).$

Multiplying this by γ and adding $\delta\int_0^{2\pi} R\cos\Phi\,dx$ to both sides of the resulting equation we get, by (2·22),

$$\int_0^{2\pi} |v(re^{ix})|\,dx \leqslant \gamma\int_0^{2\pi} u(re^{ix})\log u(re^{ix})\,dx + \delta\int_0^{2\pi} u(re^{ix})\,dx - 2\pi\gamma u(0)\log u(0). \tag{2·24}$$

We now make a slightly stronger assumption about f, namely, that $f \geqslant e = 2{\cdot}718\ldots$. Then $u \geqslant e$ and (2·24) gives

$$\int_0^{2\pi} |v(re^{ix})|\,dx \leqslant (\gamma+\delta)\int_0^{2\pi} u(re^{ix})\log u(re^{ix})\,dx.$$

Making here $r\to 1$ we get (2·9), with $A = \gamma+\delta$, $B = 0$.

In the general case, let E_1, E_2, be respectively the sets of points at which $f \geqslant e$, $f \leqslant -e$. Let $f = f_1 + f_2 + f_3$, where $f_1 = \max(f, e)$, $f_2 = \min(f, -e)$, so that $|f_3| \leqslant e$. Then

$$\mathfrak{M}[\tilde{f}] \leqslant \mathfrak{M}[\tilde{f}_1] + \mathfrak{M}[\tilde{f}_2] + \mathfrak{M}[\tilde{f}_3].$$

Since $\displaystyle\mathfrak{M}[\tilde{f}_k] \leqslant A\int_0^{2\pi} |f_k|\log|f_k|\,dx \leqslant A\int_{E_k} |f|\log^+|f|\,dx + 2\pi e \;(k=1,2),$

$$\mathfrak{M}[\tilde{f}_3] \leqslant (2\pi)^{\frac{1}{2}}\,\mathfrak{M}_2[\tilde{f}_3] \leqslant (2\pi)^{\frac{1}{2}}\,\mathfrak{M}_2[f_3] \leqslant 2\pi e,$$

we have (2·9) with $A = \gamma+\delta$, $B = 6\pi e$.

In proving (2·10), we may add a constant to f, which does not affect $\tilde{f}$, and thus suppose that $f \geqslant 1$. Then $u \geqslant 1$. Since $\Phi\sin\Phi \geqslant 0$, $\cos\Phi\log\cos\Phi \leqslant 0$, (2·23) gives

$$\int_0^{2\pi} u\log u\,dx \leqslant \int_0^{2\pi} u\log R\,dx \leqslant \tfrac{1}{2}\pi\int_0^{2\pi} |v|\,dx + 2\pi u(0)\log u(0), \tag{2·25}$$

and the integral on the right is bounded since $\tilde{S}[f]$ is a Fourier series. Making $r \to 1$ and using Fatou's lemma (Chapter I, (11·2)), we see that $f \log f$ is integrable.

It remains to show that, under the hypotheses of (2·8), $\tilde{S}[f] = S[\tilde{f}]$. Instead, however, of giving a direct proof, which is somewhat less simple than in the case of Theorem (2·4), we prefer to appeal to Theorem (4·4) below according to which $\tilde{S}[f] = S[\tilde{f}]$ whenever $f \in L$.

We pass to the proof of (2·11). Let u, v, F have the usual meaning. Applying (2·13) to the functions $G(z) = \exp\{\pm i\lambda F(z)\}$ and taking the real parts, we get

$$\int_0^{2\pi} \cos \lambda u \exp\{\mp \lambda v\}\, dx = 2\pi \cos\{\lambda u(0)\} \leqslant 2\pi.$$

Adding these inequalities and observing that

$$\exp(\lambda\,|\,v\,|) \leqslant \exp(\lambda v) + \exp(-\lambda v), \quad \cos \lambda u \geqslant \cos \lambda \quad (|\,u\,| \leqslant 1,\ 0 \leqslant \lambda < \tfrac{1}{2}\pi),$$

we obtain
$$\int_0^{2\pi} \exp(\lambda\,|\,v\,|)\, dx \leqslant \frac{4\pi}{\cos \lambda} \quad (0 \leqslant \lambda < \tfrac{1}{2}\pi),$$

which reduces to (2·12) when $r \to 1$.

If f is continuous, let T be a polynomial such that $|\,f - T\,| < \epsilon$. By the foregoing, $\exp \lambda\,|\,\tilde{f} - \tilde{T}\,|$ is integrable provided $\lambda\epsilon < \frac{1}{2}\pi$. Since $\exp \lambda\,|\,\tilde{T}\,|$ is bounded, the integrability of $\exp \lambda\,|\,\tilde{f}\,|$ follows for λ arbitrarily large ($\lambda < \pi/2\epsilon$, with ϵ arbitrarily small).

Remarks. (a) Let $\omega(t)$, $t \geqslant 0$, be a function which is non-negative, increasing, and $o(t \log t)$ for $t \to \infty$. Let $f(x)$ be periodic, not less than 1, and such that $\omega\{f(x)\}$ is integrable while $f \log f$ is not. By (2·10), $\tilde{S}[f]$ is not a Fourier series, and, by Theorem (4·4) below, $\tilde{f}$ is not integrable. This shows that the integrability of $f \log f$ in Theorem (2·8) cannot be replaced by any similar but weaker condition.

(b) *There is an* $f \in L$ *such that* $\tilde{f}$ *is not integrable over any finite interval.* Let $g(x)$ be the periodic function equal to

$$1 + \frac{1}{|\,x\,| \log^2 |x/2\pi|}$$

for $-\pi \leqslant x < \pi$. Thus g exceeds 1 and is integrable while $g \log g$ is not. Let $\{r_k\}$ be dense in $(0, 2\pi)$, Σa_k a convergent series of positive terms whose sum exceeds 1, and

$$f(x) = \Sigma a_k g(x - r_k). \tag{2·26}$$

The function f is periodic, exceeds 1 and is integrable over $(0, 2\pi)$ (Chapter I, (11·5)); but $f \log f$ is not integrable over any interval. We shall show that $\tilde{f}$ is not integrable over any interval.

For suppose that $\tilde{f}$ is integrable in some interval (a, b). Let (a', b') be interior to (a, b) and let $\lambda(x)$ be periodic, continuous, equal to 1 in (a', b'), equal to 0 outside (a, b) and linear in (a, a') and (b', b). By Theorem (6·7) of Chapter II, the difference $\lambda\tilde{f} - (\widetilde{\lambda f})$ is bounded, and since, by assumption, $\lambda\tilde{f}$ is in L, so is $(\widetilde{\lambda f})$. Hence, by Theorem (2·10) and Remark (a), $f(x)\lambda(x)$ is in the class $L \log^+ L$ and *a fortiori* each $\lambda(x)\, g(x - r_k)$ is in $L \log^+ L$, which is false for some k.

(c) Theorem (2·4) has useful variants. Let

$$F(z) = u(z) + iv(z) \quad (v(0) = 0)$$

be regular for $|z|<1$. Then (2·5) with $f(x)=u(re^{ix})$, $\tilde{f}(x)=v(re^{ix})$, and Minkowski's inequality, immediately give

$$\mathfrak{M}_p[F(re^{ix})] \leqslant A'_p \mathfrak{M}_p[u(re^{ix})] \quad (0 \leqslant r < 1) \tag{2·27}$$

with A'_p not exceeding the A_p in (2·5) by more than 1.

Another variant will be needed later. Let $u(re^{ix}) = \sum_{-\infty}^{+\infty} c_n r^{|n|} e^{inx}$ be harmonic (not necessarily real-valued) for $r<1$, and let $\Phi(z) = \sum_0^\infty c_n z^n$. Then

$$\mathfrak{M}_p[\Phi(re^{ix})] \leqslant A''_p \mathfrak{M}_p[u(re^{ix})]. \tag{2·28}$$

Suppose first that u is real-valued, and let $v(z)$ be the harmonic function conjugate to $u(z)$ with $v=0$ at the origin. Then $2\Phi = u+iv+c_0 = F+c_0$, and (2·28) follows from (2·27) and the inequality

$$|c_0| = \left|\frac{1}{2\pi}\int_0^{2\pi} u(re^{ix})\,dx\right| \leqslant \mathfrak{A}_p[u(re^{ix})] \leqslant \mathfrak{M}_p[u(re^{ix})].$$

Suppose now that u is complex-valued, $u=u_1+iu_2$, and that, correspondingly, $\Phi=\Phi_1+i\Phi_2$. Then, writing $\mathfrak{M}_p[\Phi]$ for $\mathfrak{M}_p[\Phi(re^{ix})]$ with a similar notation for u,

$$\mathfrak{M}_p[\Phi] \leqslant \mathfrak{M}_p[\Phi_1] + \mathfrak{M}_p[\Phi_2] \leqslant A''_p(\mathfrak{M}_p[u_1] + \mathfrak{M}_p[u_2]) \leqslant 2A''_p \mathfrak{M}_p[u].$$

Similar variants can be stated for Theorems (2·6), (2·8) and (2·11).

We shall now consider a few more substitutes for Theorem (2·4) in the cases $p=1,\infty$. One of them is Theorem (6·27) of Chapter IV, which may be stated as follows:

(2·29) Theorem. *Let $f \in L$ and let $v(z) = \tilde{f}(r,x)$. Then*

$$\int_D |v(z)|\,|dz| \leqslant \int_D |v(z)z^{-1}|\,|dz| \leqslant \frac{1}{2}\int_0^{2\pi} |f(x)|\,dx$$

for any diameter D of the circle $|z|<1$.

Still another substitute for (2·4) at $p=1,\infty$ is as follows:

(2·30) Theorem. *Let $u(r,x)$, $v(r,x)$ be conjugate harmonic functions for $0 \leqslant r < 1$, and let $\alpha>0$, $\delta = 1-r$.*

(i) *If $u(r,x) = O(\delta^{-\alpha})$ as $r \to 1$ (uniformly in x), then $v(r,x) = O(\delta^{-\alpha})$.*

(ii) *If $\mathfrak{M}[u(r,x)] = O(\delta^{-\alpha})$, then $\mathfrak{M}[v(r,x)] = O(\delta^{-\alpha})$.*

We first prove the following lemma:

(2·31) Lemma. *If $\Phi(z) = U(r,x) + iV(r,x)$ is regular for $|z|<1$, and if $U(r,x)$ is the Poisson integral of a function $U(x)$, then*

$$|\Phi'(z)| \leqslant \frac{2\delta^{-1}}{1+r} U^*(r,x) \leqslant 2\delta^{-1} U^*(r,x), \tag{2·32}$$

where $U^*(r,x)$ is the Poisson integral of $|U(x)|$.

For if P, Q are the Poisson kernel and its conjugate, and if

$$S(r,t) = P + iQ = \tfrac{1}{2}(1+re^{it})/(1-re^{it}),$$

we verify that

$$|dS(r,t)/dt| = |re^{it}(1-re^{it})^{-2}| = 2r\delta^{-1}(1+r)^{-1}P(r,t), \tag{2·33}$$

so that

$$|\Phi'(z)| = r^{-1}\left|\frac{d}{dx}\frac{1}{\pi}\int_0^{2\pi} U(t)\,S(r,t-x)\,dt\right| \leqslant \frac{2}{\delta(1+r)}\frac{1}{\pi}\int_0^{2\pi} |U(t)|\,P(r,t-x)\,dt,$$

which is (2·32).

By a change of variable, for any function $F(z)=u(r,x)+iv(r,x)$ regular for $|z|\leqslant R$ we obtain

$$|F'(z)| \leqslant \frac{2R}{R+r}\frac{u_R^*(r,x)}{R-r} \leqslant \frac{2}{R-r}u_R^*(r,x), \tag{2·34}$$

where $u_R^*(r,x)$ is the function harmonic for $r<R$ and coinciding with $|u(R,x)|$ for $r=R$.

Returning to (2·30), let $r=1-\delta$, $R=1-\frac{1}{2}\delta$, and

$$\phi(\delta)=\max_x |u(r,x)|, \quad \psi(\delta)=\mathfrak{M}[u(r,x)], \quad J(r)=\mathfrak{M}[F(r\,e^{ix})],$$

where $F=u+iv$. By (2·34),

$$|F'(r\,e^{ix})| \leqslant 4\delta^{-1}u_R^*(r,x) = 4\delta^{-1}\phi(\tfrac{1}{2}\delta),$$

$$\mathfrak{M}[F'(r\,e^{ix})] \leqslant 4\delta^{-1}\mathfrak{M}[u_R^*(r,x)] \leqslant 4\delta^{-1}\mathfrak{M}[u_R^*(R,x)] = 4\delta^{-1}\psi(\tfrac{1}{2}\delta),$$

$$0\leqslant J'(r) = \int_0^{2\pi}\frac{d}{dr}(u^2+v^2)^{\frac{1}{2}}\,dx = \int_0^{2\pi}(uu_r+vv_r)(u^2+v^2)^{-\frac{1}{2}}\,dx$$

$$\leqslant \int_0^{2\pi}(u_r^2+v_r^2)^{\frac{1}{2}}\,dx = \mathfrak{M}[F'(r\,e^{ix})] \leqslant 4\delta^{-1}\psi(\tfrac{1}{2}\delta).$$

Integrating the inequalities for $|F'|$, J' with respect to r we have

$$|F(r\,e^{ix})-F(0)| \leqslant 4\int_\delta^1 t^{-1}\phi(\tfrac{1}{2}t)\,dt, \quad J(r)-J(0)\leqslant 4\int_\delta^1 t^{-1}\psi(\tfrac{1}{2}t)\,dt.$$

If $\phi(\delta)$ or $\psi(\delta)$ is $O(\delta^{-\alpha})$, the corresponding integral is also $O(\delta^{-\alpha})$, and (2·30) follows.

(2·35) Theorem. *Let $u(r,x)$ be harmonic for $r<1$ and let $1\leqslant p\leqslant\infty$, $\alpha>0$. Then each of the relations*

$$\left.\begin{array}{c}\text{(i)}\quad \mathfrak{M}_p[u(r,x)]=O(\delta^{-\alpha}),\\ \text{(ii)}\quad \mathfrak{M}_p[u_x(r,x)]=O(\delta^{-\alpha-1}),\qquad \text{(iii)}\quad \mathfrak{M}_p[u_r(r,x)]=O(\delta^{-\alpha-1})\end{array}\right\} \tag{2·36}$$

implies the other two.

That (ii) and (iii) are equivalent follows from the Cauchy-Riemann equation $ru_r=v_x$ together with (2·4) if $1<p<\infty$, or (2·30) if $p=1,\infty$. The equivalence of (i) and (ii) follows from (2·4), (2·30) and the following lemma:

(2·37) Lemma. *If $F(z)$ is regular for $|z|<1$, the two relations*

$$\text{(i)}\quad \mathfrak{M}_p[F(r\,e^{ix})]=O(\delta^{-\alpha}),\qquad \text{(ii)}\quad \mathfrak{M}_p[F'(r\,e^{ix})]=O(\delta^{-\alpha-1}) \tag{2·38}$$

are equivalent.

Suppose that $\mathfrak{M}_p[F(\rho\,e^{ix})]\leqslant A(1-\rho)^{-\alpha}$. Let $z=r\,e^{ix}$. If C denotes the circumference with centre z and radius $\frac{1}{2}\delta$ we have

$$F'(z)=\frac{1}{2\pi i}\int_C\frac{F(\zeta)}{(\zeta-z)^2}d\zeta, \quad |F'(z)|\leqslant\frac{1}{\pi\delta}\int_0^{2\pi}|F(z+\tfrac{1}{2}\delta\,e^{i\theta})|\,d\theta.$$

Raising both sides of the last inequality to the power p (if $p<\infty$) and integrating with respect to x we obtain, by Minkowski's inequality (Chapter I, (9·11)),

$$\mathfrak{M}_p[F'(r\,e^{ix})]\leqslant\frac{1}{\pi\delta}\int_0^{2\pi}\mathfrak{M}_p[F(r\,e^{ix}+\tfrac{1}{2}\delta\,e^{i\theta})]\,d\theta\leqslant\frac{1}{\pi\delta}\int_0^{2\pi}A(\tfrac{1}{2}\delta)^{-\alpha}\,d\theta=\frac{2^{\alpha+1}A}{\delta^{\alpha+1}}.$$

The first inequality still holds for $p=\infty$. Hence (i) implies (ii).

Conversely, suppose that $\mathfrak{M}_p[F'(\rho\,e^{ix})]\leqslant B(1-\rho)^{-\alpha-1}$. Since

$$|F(r\,e^{ix})-F(0)|\leqslant\int_0^r|F'(\rho\,e^{ix})|\,d\rho,$$

Minkowski's inequality gives

$$\mathfrak{M}_p[F(r\,e^{ix})-F(0)]\leqslant\int_0^r\mathfrak{M}_p[F'(\rho\,e^{ix})]\,d\rho\leqslant\int_0^r B(1-\rho)^{-\alpha-1}\,d\rho<B\alpha^{-1}\delta^{-\alpha},$$

and so $$\mathfrak{M}_p[F(r\,e^{ix})]\leqslant B\alpha^{-1}\delta^{-\alpha}+|F(0)|\,(2\pi)^{1/p}=O(\delta^{-\alpha}).$$

The following result is an analogue of (2·6) for Fourier–Stieltjes series.

(2·39) Theorem. *Let $F(x)$ be of bounded variation over $0\leqslant x\leqslant 2\pi$, and let $F^*(x)$ be defined by* (1·7). *Then for each $0<\mu<1$ we have*

$$\mathfrak{M}_\mu[F^*]\leqslant B_\mu\int_0^{2\pi}|dF(x)|. \tag{2·40}$$

Let $u(r,x)$ be the Poisson–Stieltjes integral of F and $v(r,x)$ the conjugate harmonic function. By (2·7),

$$\left\{\int_0^{2\pi}|v(r,x)|^\mu\,dx\right\}^{1/\mu}\leqslant B_\mu\int_0^{2\pi}|u(r,x)|\,dx\leqslant B_\mu\int_0^{2\pi}|dF(x)|. \tag{2·41}$$

Since $v(r,x)\to F^*(x)$ for almost all x as $r\to 1$, (2·41) implies (2·40).

3. Applications of Green's formula

The main tool in the preceding section was Cauchy's formula (2·13). Some of the results can also be obtained by means of Green's formula. This latter method gives good estimates for the constants occurring in the inequalities.

Let $w=w(\xi,\eta)$ be a function which, with its derivatives of the first two orders, is continuous in $\xi^2+\eta^2<a^2$. Let $I(r)$ denote the integral of w round the circle with centre 0 and radius $r<a$. In investigating the behaviour of $I(r)$ as $r\to a$, it is natural to consider the derivative $dI(r)/dr$, and this in turn suggests an application of Green's formula

$$\int_{C_r}\frac{\partial w}{\partial r}\,ds=\iint_{S_r}\Delta w\,d\sigma. \tag{3·1}$$

Here S_r is the circle $\xi^2+\eta^2\leqslant r^2$, C_r its circumference,

$$\Delta w=w_{\xi\xi}+w_{\eta\eta}$$

is the Laplacian, and $\partial/\partial r$ means differentiation in the direction of the radius vector. Incidentally, (3·1) is immediate if we introduce polar co-ordinates ρ, x and use the formula

$$\Delta w=\rho^{-1}(\rho w_\rho)_\rho+\rho^{-2}w_{xx}. \tag{3·2}$$

For owing to the periodicity of w in x, the integral over $0 \leqslant x \leqslant 2\pi$ of the last term of (3·2) is 0, so that the right-hand side of (3·1) is

$$\int_0^r \int_0^{2\pi} \rho^{-1}(\rho w_\rho)_\rho \rho \, d\rho \, dx = \int_0^{2\pi} w_r r \, dx = \int_{C_r} w_r \, ds.$$

Let $\zeta = \xi + i\eta$ and let $\qquad F(\zeta) = u(\xi, \eta) + iv(\xi, \eta)$

be regular in the neighbourhood of a point $\zeta_0 = \xi_0 + i\eta_0$. Let

$$w = w(u, v),$$

with its first and second derivatives, be real-valued and continuous in the neighbourhood of the point $u_0 = u(\xi_0, \eta_0)$, $v_0 = v(\xi_0, \eta_0)$. Then

$$w_{\xi\xi} + w_{\eta\eta} = (w_{uu} + w_{vv}) \, | F'(\zeta) |^2 \tag{3·3}$$

at the point $\zeta = \zeta_0$.

For $w_\xi = w_u u_\xi + w_v v_\xi$,

$$w_{\xi\xi} = (w_{uu} u_\xi + w_{uv} v_\xi) u_\xi + (w_{vu} u_\xi + w_{vv} v_\xi) v_\xi + w_u u_{\xi\xi} + w_v v_{\xi\xi}.$$

If to this equation we add the corresponding one for $w_{\eta\eta}$, and observe that u, v satisfy the Laplace and Cauchy-Riemann equations, we get (3·3).

Two cases of (3·3) are of interest. First, if $R = | F(\zeta) |$ and if $w = w(R)$, then by (3·2),

$$\Delta w = R^{-1}(R w_R)_R \, | F' |^2. \tag{3·4}$$

If $w = w(u)$ is independent of v then, by (3·3),

$$\Delta w = w_{uu} \, | F' |^2. \tag{3·5}$$

Let $F(\zeta) = u + iv$ be regular for $| \zeta | < 1$, and let p be any real number. Then, by (3·4) and (3·5),

$$\Delta \, | F |^p = p^2 \, | F |^{p-2} \, | F' |^2, \quad \Delta u^p = p(p-1) \, u^{p-2} \, | F' |^2. \tag{3·6}$$

The first formula is valid at every point where $F \neq 0$ (otherwise we have to suppose that $p \geqslant 2$), the second where $u \neq 0$. We apply them to the proof of Theorem (2·4). We know that it is enough to prove (2·4) in the case when $1 < p \leqslant 2$ and when u, the Poisson integral of f, is positive. Since $| F | \geqslant u$, (3·6) then gives

$$\Delta \, | F |^p \leqslant p' \Delta u^p, \quad p' = p/(p-1). \tag{3·7}$$

We now set $\quad I(r) = \int_0^{2\pi} u^p(r, x) \, dx, \quad J(r) = \int_0^{2\pi} | F(r e^{ix}) |^p \, dx, \qquad (3·8)$

and apply (3·1) to $w = u^p$ and $w = | F |^p$. The left-hand sides of (3·1) will then be $rI'(r)$ and $rJ'(r)$ respectively. Thus (3·7) implies,

$$J'(r) \leqslant p' I'(r).$$

Integrating this inequality with respect to r and observing that $I(0) = J(0) \geqslant 0$, $p' > 1$, we get $J(r) \leqslant p' I(r)$. Since $| F | \geqslant | v |$, this gives

$$\mathfrak{M}_p[v(r e^{ix})] \leqslant A_p \mathfrak{M}_p[u(r e^{ix})],$$

where $A_p = p'^{1/p}$ for $1 < p \leqslant 2$. If u is of variable sign, the value of A_p must be multiplied by 2. An application of (2·21) shows that

$$A_p \leqslant 2p^{1/p'} < 2p \quad \text{for} \quad p \geqslant 2. \tag{3·9}$$

A similar argument gives (2·6) and (2·8).

In (2·8) we may begin (as in the preceding section) by supposing that $f \geqslant e$; then $u \geqslant e$. Let $J(r)$ and $I(r)$ denote the integrals of $|F(re^{ix})|$ and $u \log u$ over $0 \leqslant x \leqslant 2\pi$. By (3·4) and (3·5),

$$\Delta |F| = |F'|^2 |F|^{-1}, \quad \Delta(u \log u) = |F'|^2 u^{-1},$$

so that

$$\Delta |F| \leqslant \Delta(u \log u), \quad J'(r) \leqslant I'(r), \quad J(r) \leqslant I(r)$$

(since $J(0) \leqslant I(0)$). This gives (2·9) with $A = 1$, $B = 0$. In the general case, arguing as on p. 256, we have $A = 1$, $B = 6\pi e$.

To prove (2·6) for $f \geqslant 0$, we replace p by μ in (3·6), obtaining $\Delta |F|^\mu \leqslant -\mu(1-\mu)^{-1} \Delta u^\mu$. Hence, if $J(r)$ and $I(r)$ are given by (3·8) with μ for p, we have successively

$$J'(r) \leqslant -\mu(1-\mu)^{-1} I'(r),$$

$$J(r) - J(0) \leqslant \mu(1-\mu)^{-1}\{I(0) - I(r)\} \leqslant \mu(1-\mu)^{-1} I(0),$$

$$J(r) \leqslant (1-\mu)^{-1} I(0),$$

which leads to (2·7) with $B_\mu^\mu = (2\pi)^{1-\mu}(1-\mu)^{-1}$. The argument on p. 255 shows that B_μ must be multiplied by $2^{1/\mu}$ for f of variable sign.

4. Integrability B

There exist functions $f \in L$ such that $\tilde{f}$ is not integrable. It is interesting to observe that, with a suitable definition, more general than that of Lebesgue, of an integral, the function $\tilde{f}$ is integrable and $\tilde{S}[f]$ is the Fourier series of $\tilde{f}$.

Given any function $f(x)$, $a \leqslant x < b$, we define the function f for all x by periodicity: $f(x+h) = f(x)$, where $h = b - a$. Let $a = x_0 < x_1 < x_2 < \ldots < x_n = b$ be any subdivision of (a, b), and let $\rho = \max(x_j - x_{j-1})$. Let ξ_j be an arbitrary point in (x_{j-1}, x_j). Consider the expressions

$$I(t) = \sum_{j=1}^{n} f(\xi_j + t)(x_j - x_{j-1}), \tag{4·1}$$

which are Riemann sums for the family of functions $f_t(x) = f(x+t)$. If $f(x)$ is R-integrable, so is $f(t+x)$, and $I(t)$ tends to a limit as $\rho \to 0$, no matter what the choice of x_j and ξ_j. Owing to the periodicity of f, this limit is independent of t. But even if f is not integrable R the sum $I(t)$ may tend to a limit J *in measure* as $\rho \to 0$. By this we mean that given any $\epsilon > 0$ we can find a $\delta = \delta(\epsilon)$ such that if $\rho < \delta$, then $|I(t) - J| < \epsilon$, except for a set T of t's of measure $< \epsilon$. (T may depend on the x_j and ξ_j.) If this is so, we say that f is *integrable* B over (a, b), and that J is the integral of f over (a, b). We may also say that f is integrable R 'in measure'.

(4·2) THEOREM. *If f is integrable* L *over (a, b), then it is also integrable* B, *and both integrals have the same value.*

For let $f = f_1 + f_2$, and correspondingly $I(t) = I_1(t) + I_2(t)$, where f_1 is continuous and $\mathfrak{M}[f_2; a, b] < \epsilon^2/3(b-a)$. Then

$$\int_a^b |I_2(t)|\, dt \leqslant \Sigma(x_j - x_{j-1}) \int_a^b |f_2(\xi_j + t)|\, dt < \Sigma(x_j - x_{j-1})\, \epsilon^2/3(b-a) = \tfrac{1}{3}\epsilon^2,$$

so that the set T of t's for which $|I_2(t)| > \frac{1}{3}\epsilon$ is of measure less than ϵ. If J, J_1, J_2 are the integrals of f, f_1, f_2 over (a, b), then

$$|I(t) - J| = |I_1(t) + I_2(t) - J_1 - J_2| \leqslant |I_1(t) - J_1| + |I_2(t)| + |J_2|.$$

Here $|I_1(t)-J_1|<\frac{1}{3}\epsilon$, provided $\rho<\delta=\delta(\epsilon)$. Again, $|I_2(t)|\leqslant\frac{1}{3}\epsilon$ for $t\notin T$. By hypothesis $|J_2|<\frac{1}{3}\epsilon^2/(b-a)<\frac{1}{3}\epsilon$, assuming as we may that $\epsilon<b-a$. Hence

$$|I(t)-J|<\epsilon \quad \text{for} \quad t\notin T, \quad |T|<\epsilon,$$

provided $\rho\leqslant\delta$; and (4·2) follows.

(4·3) Theorem. *For every periodic $f\in$ L, $\tilde{f}$ is integrable* B *over* $(0,2\pi)$. *Moreover $\tilde{f}(x)\,e^{-ikx}$ is integrable* B *over* $(0,2\pi)$ *for $k=0,\pm 1,\ldots$ and* $\tilde{\mathsf{S}}[f]=\mathsf{S}[\tilde{f}]$.

Fix $k=0,\pm 1,\pm 2,\ldots$, and let $\tilde{I}^k(t)$ be the sum (4·1) with f replaced by $\tilde{f}(x)\,e^{-ikx}$. Then

$$|\tilde{I}^k(t)|=|\sum_j \tilde{f}(t+\xi_j)\,e^{-ik(t+\xi_j)}\,\delta_j|=|\sum_j \tilde{f}(t+\xi_j)\,e^{-ik\xi_j}\,\delta_j|,$$

where $\delta_j=x_j-x_{j-1}$. The last sum is conjugate to $\Sigma f(t+\xi_j)\,e^{-ik\xi_j}\delta_j$. Thus, by (2·7),

$$\mathfrak{M}_{\frac{1}{2}}[\tilde{I}^k(t)]\leqslant B_{\frac{1}{2}}\int_0^{2\pi}|\Sigma f(t+\xi_j)\,e^{-ik\xi_j}\delta_j|\,dt\leqslant 2\pi B_{\frac{1}{2}}\int_0^{2\pi}|f(t)|\,dt.$$

This shows that $|\tilde{I}^k(t)|<\epsilon$ outside a set of measure less than ϵ, provided that $\mathfrak{M}[f]$ is small enough, say less than $\eta=\eta(\epsilon)$.

We now set $f=f_1+f_2$, where f_1 is a polynomial and $\mathfrak{M}[f_2]<\eta$. Correspondingly, $I(t)=I_1(t)+I_2(t)$ and (for the coefficients) $c_k=c'_k+c''_k$. Obviously $\tilde{I}^k_1(t)$ tends (as $\rho\to 0$, and uniformly in t) to the kth coefficient of $2\pi\tilde{f}_1$, that is to say, to $-2\pi i(\operatorname{sign} k)\,c'_k$. On the other hand, $|\tilde{I}^k_2(t)|<\epsilon$ for $t\notin T$, $|T|<\epsilon$; and $|c''_k|<\eta/2\pi$.

Thus for ρ small enough and $t\notin T$, the sum $\tilde{I}^k(t)$ differs from $-2\pi i(\operatorname{sign} k)\,c'_k$ by less than 2ϵ, or from $-2\pi i(\operatorname{sign} k)\,c_k$ by less than $2\epsilon+\eta$. Since ϵ, η are arbitrarily small, $\tilde{I}^k(t)$ tends in measure to $-2\pi i(\operatorname{sign} k)\,c_k$ as $\rho\to 0$. This shows that $\tilde{f}e^{-ikx}$ is integrable B over $(0,2\pi)$ and that $\mathsf{S}[\tilde{f}]=\tilde{\mathsf{S}}[f]$. In particular (for $k=0$), the function $\tilde{f}$ is integrable B over $(0,2\pi)$ and the value of the integral is zero. This completes the proof of (4·3).

(4·4) Theorem. *If f and $\tilde{f}$ are both integrable* L, *then* $\tilde{\mathsf{S}}[f]=\mathsf{S}[\tilde{f}]$.

This is a corollary of (4·3). A different proof will be found on p. 285.

5. Lipschitz conditions

(5·1) Theorem. (i) *A necessary and sufficient condition for a function $u(r,x)$ harmonic for $r<1$ to be the Poisson integral of an $f(x)\in\Lambda_\alpha$, $0<\alpha\leqslant 1$, is that*

$$u_x(r,x)=O(\delta^{\alpha-1}), \tag{5·2}$$

where $\delta=1-r$, uniformly in x as $r\to 1$.

(ii) *A necessary and sufficient condition for u to be the Poisson integral of an f in Λ_*, or in λ_*, is that $u_{xx}=O(\delta^{-1})$, or $o(\delta^{-1})$, respectively.*

Since Λ_1 is the class of the integrals of bounded functions, the case $\alpha=1$ of (i) follows from Theorem (6·3) of Chapter IV. We may suppose then that $0<\alpha<1$. We need the following estimates for the derivatives of the Poisson kernel:

$$|P_t(r,t)|\leqslant\delta^{-2}, \quad |P_t(r,t)|\leqslant At^{-2} \quad (|t|\leqslant\pi;\ \delta=1-r). \tag{5·3}$$

These follow from (2·33) and the inequalities (6·9) of Chapter III. Since the integral of $P_t(r,t)$ over $0 \leqslant t \leqslant 2\pi$ is zero,

$$u_x(r,x) = -\frac{1}{\pi}\int_{-\pi}^{\pi} f(x+t)\,P_t(r,t)\,dt = \frac{1}{\pi}\int_{-\pi}^{\pi}[f(x)-f(x+t)]\,P_t(r,t)\,dt$$

$$= \frac{1}{\pi}\int_{|t|\leqslant\delta} + \frac{1}{\pi}\int_{\delta\leqslant|t|\leqslant\pi} = A+B,$$

$$|A| \leqslant \int_{|t|\leqslant\delta} O(|t|^\alpha)\,\delta^{-2}\,dt = O(\delta^{\alpha-1}), \quad B = \int_\delta^\pi O(|t|^\alpha)\,O(t^{-2})\,dt = O(\delta^{\alpha-1}),$$

by (5·3), and the necessity of (5·2) is established.

Conversely, suppose (5·2) is satisfied. Let $v(r,x)$ be the function conjugate to u and let $F(z) = u+iv$. By (5·2) and (2·30) (i), $F'(z) = O(\delta^{\alpha-1})$ as $r \to 1$. From

$$F(r\,e^{ix}) - F(0) = \int_0^r F'(\rho\,e^{ix})\,e^{ix}\,d\rho = \int_0^r O\{(1-\rho)^{\alpha-1}\}\,d\rho,$$

we deduce that $F(e^{ix}) = \lim\limits_{r\to 1} F(r\,e^{ix})$ exists, uniformly in x. Hence $F(r\,e^{ix})$ is the Poisson integral of $F(e^{ix})$ and it is enough to show that $F(e^{ix}) \in \Lambda_\alpha$.

For $0<h<1$ we have

$$F(e^{i(x+h)}) - F(e^{ix}) = \left(\int_{I_1} + \int_{I_2} + \int_{I_3}\right) F'(z)\,dz = J_1+J_2+J_3, \tag{5·4}$$

where I_1 is the segment $1 \geqslant r \geqslant 1-h$ of the radius $\arg z = x$; I_2 is the arc $z = (1-h)\,e^{it}$, $x \leqslant t \leqslant x+h$; and I_3 is the segment $1-h \leqslant r \leqslant 1$ of the radius $\arg z = x+h$. Since $F'((1-h)\,e^{it}) = O(h^{\alpha-1})$, it follows that $J_2 = h.O(h^{\alpha-1}) = O(h^\alpha)$. Also

$$J_1 = \int_0^h O(\delta^{\alpha-1})\,d\delta = O(h^\alpha),$$

and similarly $J_3 = O(h^\alpha)$. Thus $J_1+J_2+J_3 = O(h^\alpha)$ and $F(e^{ix}) \in \Lambda_\alpha$.

That the condition in (ii) is necessary follows from Theorem (9·16) of Chapter III. In the proof of sufficiency we may restrict ourselves to the case $u_{xx} = O(\delta^{-1})$, the argument in the 'o' case being similar.

The hypothesis implies that $v_{xx} = O(\delta^{-1})$ (by (2·30)), so that $F_{xx}(r\,e^{ix}) = O(\delta^{-1})$. Since $d/dx = iz\,d/dz$, we have $(zF'(z))' = O(\delta^{-1})$. Integrating this along radii we see that $F'(z) = O(\log 1/\delta)$ as $r \to 1$, and one more radial integration shows that $F(z)$ is continuous in the closed circle $|z| \leqslant 1$.

Let now $0<h<1$, $\zeta = e^{ix}$, $\zeta_h = (1-h)\,e^{ix}$ We need the formula

$$F(\zeta) = F(\zeta_h) + (\zeta-\zeta_h)\,F'(\zeta_h) + \int_{\zeta_h}^{\zeta} (\zeta - z)\,F''(z)\,dz, \tag{5·5}$$

where the integration is along the radius. For the proof we consider the formula

$$F(\zeta_\epsilon) = F(\zeta_h) + [(z-\zeta)\,F'(z)]_{z=\zeta_h}^{z=\zeta_\epsilon} + \int_{\zeta_h}^{\zeta_\epsilon} (\zeta-z)\,F''(z)\,dz,$$

easily obtained by integration by parts. If we let $\epsilon \to 0$ and observe that

$$F(\zeta_\epsilon) \to F(\zeta), \quad \epsilon F'(\zeta_\epsilon) = O(\epsilon\log\epsilon) = o(1),$$

we obtain (5·5).

The relations $(zF')' = O(\delta^{-1})$, $F'(z) = O(\log 1/\delta)$ imply that $F''(z) = O(\delta^{-1})$. Thus the integrand in the last term of (5·5) is $O(1)$, the term itself is $O(h)$, and we have

$$F(e^{ix}) = F(r_h e^{ix}) + h r_h^{-1} g(r_h e^{ix}) + O(h),$$

where $r_h = 1 - h$, $g(z) = zF'$ (so that $g' = O(\delta^{-1})$). Subtracting this from the similar equation with x replaced by $x + h$, we obtain

$$F(e^{i(x+h)}) - F(e^{ix}) = \int_{r_h e^{ix}}^{r_h e^{i(x+h)}} F'(z)\, dz + O(h),$$

since the difference of the values of g at the points $(1-h)\, e^{ix}$ and $(1-h)\, e^{i(x+h)}$ is equal to the integral of g' along the arc joining the points and so is $h \,.\, O(h^{-1}) = O(1)$. From the similar equation with $x + h$ replaced by $x - h$, observing that the sum of the integrals on the right is

$$i\int_0^h \{g(r_h e^{i(x+t)}) - g(r_h e^{i(x-t)})\}\, dt = i\int_0^h dt \int_{r_h e^{i(x-t)}}^{r_h e^{i(x+t)}} g'(z)\, dz = \int_0^h dt \,.\, O(th^{-1}) = O(h),$$

we have $F(e^{i(x+h)}) + F(e^{i(x-h)}) - 2F(e^{ix}) = O(h)$, that is, $F \in \Lambda_*$.

Remark. Combining (5·1) and (2·30) we obtain a new proof of the result that if f is in one of the classes Λ_α, with $0 < \alpha < 1$, Λ_*, λ_*, then $\tilde{f}$ is in the same class (see Chapter III, (13·29)).

Though the function conjugate to a continuous function need not be continuous (see (2·3)), it still retains some traces of continuity, as is shown by the following theorem (for the definition of property D see Chapter II, §3):

(5·6) THEOREM. *Let $f(x)$ be periodic and continuous. The conjugate function*

$$\tilde{f}(x) = \lim_{\epsilon \to +0} \left\{ -\frac{1}{\pi} \int_\epsilon^\pi \frac{f(x+t) - f(x-t)}{2 \tan \frac{1}{2} t}\, dt \right\} \tag{5·7}$$

has property D *on the set E of points where $\tilde{f}(x)$ exists.*

We may suppose that the constant term of $S[f]$ is zero. The integral F of f is then periodic and has a continuous derivative, and so is in the class λ_*. Thus the conjugate function $\tilde{F}$ is also in λ_*. It is the integral of an $\tilde{f} \in L$. If $\tilde{f}(r, x)$ is the Poisson integral of $\tilde{f}$ and $\tilde{f}(x; \epsilon)$ is the expression under the limit sign in (5·7), then

$$\tilde{f}(r, x) - \tilde{f}(x; 1 - r) \to 0 \quad \text{as} \quad r \to 1, \tag{5·8}$$

by Theorem (7·20) of Chapter III. Suppose that $a < b$, $\tilde{f}(a) = A$, $\tilde{f}(b) = B$, and that C is any number between A and B. We have to show that there is a c between a and b such that $\tilde{f}(c) = C$. Suppose, for example, that $A < C < B$. By (5·8),

$$\lim_{r\to 1} \tilde{f}(r, a) = A, \quad \lim_{r \to 1} \tilde{f}(r, b) = B.$$

By Theorem (7·2) of Chapter III, using the smoothness of $\tilde{F}$,

$$\liminf_{h \to 0} \{\tilde{F}(a+h) - \tilde{F}(a)\}/h \leqslant A, \quad \limsup_{h \to C} \{\tilde{F}(b+h) - \tilde{F}(b)\}/h \geqslant B.$$

By the remark on p. 44, there is then a point c between a and b such that $\tilde{F}'(c) = C$ Hence $\tilde{f}(r, c) \to C$, $\tilde{f}(c; 1 - r) \to C$ (by (5·8)), and (5·6) follows.

6. Mean convergence of $S[f]$ and $\tilde{S}[f]$

Theorems on conjugate functions lead to results about the partial sums of $S[f]$ and $\tilde{S}[f]$.

Let $s_n(x)=S_n(x;f)$, $\tilde{s}_n(x)=\tilde{S}_n(x;f)$, and let $s_n^*(x)$ and $\tilde{s}_n^*(x)$ be the modified partial sums of $S[f]$ and $\tilde{S}[f]$ (Chapter II, § 5). We have

$$\left.\begin{aligned} s_n^*(x) &= \frac{1}{\pi}\int_{-\pi}^{\pi} f(x+t)\frac{\sin nt}{2\tan\frac{1}{2}t}\,dt,\\ \tilde{s}_n^*(x)-\tilde{f}(x) &= \frac{1}{\pi}\int_{-\pi}^{\pi} f(x+t)\frac{\cos nt}{2\tan\frac{1}{2}t}\,dt.\end{aligned}\right\} \tag{6·1}$$

Replacing here $\sin nt$ by $\sin n(t+x)\cos nx - \cos n(t+x)\sin nx$, and similarly for $\cos nt$, we obtain formulae expressing s_n^* and $\tilde{s}_n^*$ in terms of conjugate functions:

$$\left.\begin{aligned} s_n^*(x) &= \tilde{g}_n(x)\sin nx - \tilde{h}_n(x)\cos nx,\\ \tilde{s}_n^*(x)-\tilde{f}(x) &= -\tilde{g}_n(x)\cos nx - \tilde{h}_n(x)\sin nx,\end{aligned}\right\} \tag{6·2}$$

where g_n and h_n are respectively $f(x)\cos nx$ and $f(x)\sin nx$. The right-hand sides here are defined almost everywhere.

It follows that

$$\left.\begin{aligned} |s_n^*| &\leqslant |\tilde{g}_n| + |\tilde{h}_n|,\\ |\tilde{s}_n^*| &\leqslant |\tilde{g}_n| + |\tilde{h}_n| + |\tilde{f}|.\end{aligned}\right\} \tag{6·3}$$

(6·4) THEOREM. *If $f\in L^p$, $1<p<\infty$, then*

$$\mathfrak{M}_p[s_n]\leqslant C_p\mathfrak{M}_p[f], \quad \mathfrak{M}_p[\tilde{s}_n]\leqslant C_p\mathfrak{M}_p[f], \tag{6·5}$$

$$\mathfrak{M}_p[f-s_n]\to 0, \quad \mathfrak{M}_p[\tilde{f}-\tilde{s}_n]\to 0 \quad (n\to\infty), \tag{6·6}$$

where C_p depends only on p.

Since $s_n - s_n^*\to 0$, $\tilde{s}_n-\tilde{s}_n^*\to 0$ uniformly in x, it is sufficient to prove (6·6) with s_n, $\tilde{s}_n$ replaced by s_n^*, $\tilde{s}_n^*$. The inequalities

$$|s_n - s_n^*| \leqslant \frac{1}{2\pi}\int_0^{2\pi}|f|\,dx = \mathfrak{A}[f]\leqslant \mathfrak{A}_p[f]\leqslant\mathfrak{M}_p[f],$$

and the corresponding inequalities for $\tilde{s}_n-\tilde{s}_n^*$, show that it is enough to prove (6·5) for s_n^*, $\tilde{s}_n^*$.

Using Minkowski's inequality, (6·3) and (2·5), we have

$$\left.\begin{aligned} \mathfrak{M}_p[s_n^*]\leqslant\mathfrak{M}_p[\tilde{g}_n]+\mathfrak{M}_p[\tilde{h}_n]\leqslant A_p\{\mathfrak{M}_p[g_n]+\mathfrak{M}_p[h_n]\}\leqslant 2A_p\mathfrak{M}_p[f],\\ \mathfrak{M}_p[\tilde{s}_n^*]\leqslant 3A_p\mathfrak{M}_p[f],\end{aligned}\right\} \tag{6·7}$$

which proves (6·5). Now let $f=f'+f''$, where f' is a polynomial and $\mathfrak{M}_p[f'']<\epsilon$. Then, with an obvious notation,

$$s_n^* = s_n^{*\prime}+s_n^{*\prime\prime}, \quad f-s_n^* = (f'-s_n^{*\prime})+(f''-s_n^{*\prime\prime}),$$

$$\mathfrak{M}_p[f-s_n^*]\leqslant\mathfrak{M}_p[f'-s_n^{*\prime}]+\mathfrak{M}_p[f'']+\mathfrak{M}_p[s_n^{*\prime\prime}]=\mathfrak{M}_p[f'']+\mathfrak{M}_p[s_n^{*\prime\prime}],$$

the last equation being valid for n large enough. By (6·7) the last sum does not exceed $(2A_p+1)\mathfrak{M}_p[f''] < (2A_p+1)\epsilon$, so that $\mathfrak{M}_p[f-s_n^*]\to 0$ and, by (2·4),

$$\mathfrak{M}_p[\tilde{f}-\tilde{s}_n^*] \leqslant A_p \mathfrak{M}_p[f-s_n^*]\to 0.$$

(6·8) Theorem. *If* $f\in L$, $0<\mu<1$, *then*

$$\mathfrak{M}_\mu[s_n] \leqslant C_\mu \mathfrak{M}[f], \quad \mathfrak{M}_\mu[\tilde{s}_n] \leqslant C_\mu \mathfrak{M}[f],$$

$$\mathfrak{M}_\mu[f-s_n]\to 0, \quad \mathfrak{M}_\mu[\tilde{f}-\tilde{s}_n]\to 0.$$

(6·9) Theorem. *If* $|f|\log^+|f|$ *is integrable, then*

$$\mathfrak{M}[s_n] \leqslant C\int_0^{2\pi} |f|\log^+|f|\,dx + C, \quad \mathfrak{M}[\tilde{s}_n] \leqslant C\int_0^{2\pi} |f|\log^+|f|\,dx + C,$$

$$\mathfrak{M}[f-s_n]\to 0, \quad \mathfrak{M}[\tilde{f}-\tilde{s}_n]\to 0.$$

We confine the proofs to the case of the s_n, the argument for the $\tilde{s}_n$ being similar. It is again enough to consider s_n^* instead of s_n. The proof of (6·8) is analogous to that of (6·4) provided we use Theorem (2·6) instead of (2·4) and the inequality

$$\mathfrak{M}_\mu^\mu[\phi+\psi] \leqslant \mathfrak{M}_\mu^\mu[\phi] + \mathfrak{M}_\mu^\mu[\psi]$$

instead of Minkowski's.

If $f\log^+|f|\in L$, then

$$\mathfrak{M}[s_n^*] \leqslant 2A\int_0^{2\pi} |f|\log^+|f|\,dx + 2B \tag{6·10}$$

by (6·3), (2·9) and an argument analogous to (6·7). We apply this to kf, where k is a positive constant (clearly if $f\log^+|f|$ is integrable so is $kf\log^+|kf|$). We have

$$\mathfrak{M}[s_n^*] \leqslant 2A\int_0^{2\pi} |f|\log^+|kf|\,dx + 2B/k.$$

We fix k so that $2B/k<\epsilon$ and again put $f=f'+f''$, where f' is a polynomial and the integrals of both $2A|f''|\log^+|kf''|$ and $|f''|$ over $(0, 2\pi)$ are less than ϵ. (By Theorem (5·14) of Chapter IV, we may take $f'=\sigma_m(x;f)$, with m large enough.) Using the same argument as before, we have, for n large enough,

$$\mathfrak{M}[f-s_n^*] \leqslant \mathfrak{M}[f''] + \mathfrak{M}[s_n^{*\prime\prime}] \leqslant \mathfrak{M}[f''] + 2A\int_0^{2\pi} |f''|\log^+|kf''|\,dx + 2B/k < 3\epsilon,$$

so that $\mathfrak{M}[f-s_n^*]\to 0$.

As corollaries of (6·4) and (6·9) we have

(6·11) Theorem. (i) *If* $f\sim\Sigma c_n e^{inx}\in L^p$ *and* $g\sim\Sigma c_n' e^{inx}\in L^{p'}$, *where* $1<p<\infty$, *the series in Parseval's formula*

$$\frac{1}{2\pi}\int_0^{2\pi} fg\,dx = \sum_{-\infty}^{+\infty} c_n c_{-n}' \tag{6·12}$$

is convergent.

(ii) *The conclusion holds if* $f\log^+|f|$ *is integrable, and* g *is bounded.*

This is a generalization of Theorem (8·7) of Chapter IV. That (i) follows from (6·4) was already indicated on p. 159. Part (ii) follows similarly from the relation $\mathfrak{M}[f-s_n]\to 0$ (see (6·9)). We also see that if $f\log^+|f|$ is integrable then in Theorem (8·9) of Chapter IV we can replace summability (C, 1) by convergence.

(6·13) Theorem. *For any $f \in L$ there is a sequence $\{n_k\}$ such that $s_{n_k}(x) \to f(x)$, $\tilde{s}_{n_k}(x) \to \tilde{f}(x)$ almost everywhere.*

This follows from (6·8) and Theorem (11·6) of Chapter I. First we select a sequence $\{m_k\}$ such that $s_{m_k} \to f$ almost everywhere, and then from $\{m_k\}$ a subsequence $\{n_k\}$ such that $\tilde{s}_{n_k} \to \tilde{f}$ almost everywhere.

The sequence $\{n_k\}$ in (6·13) depends essentially on f. It will be shown in Chapter XV—and this is deeper—that for $f \in L^p$, $p > 1$, we can take a fixed sequence independent of f.

Theorem (6·4) ceases to be true for $p = 1$ and $p = \infty$ since $\mathfrak{M}[f - s_n]$ need not tend to zero for f integrable (Chapter V, (1·12)), nor need s_n converge (let alone converge uniformly) to f for f continuous (see Chapter VIII, § 1). However, if f and $\tilde{f}$ are both integrable or both continuous, then $S[f]$ and $\tilde{S}[f]$ behave in much the same way. We have in fact:

(6·14) Theorem. (i) *If f and $\tilde{f}$ are both continuous and $S[f]$ converges uniformly, so does $\tilde{S}[f]$. If f and $\tilde{f}$ are both bounded and $S[f]$ has bounded partial sums, so has $\tilde{S}[f]$.*

(ii) *Suppose that $\tilde{S}[f]$ is a Fourier series. Then, if $\mathfrak{M}[s_n]$ is bounded so is $\mathfrak{M}[\tilde{s}_n]$; and if $\mathfrak{M}[f - s_n]$ tends to 0, so does $\mathfrak{M}[\tilde{f} - \tilde{s}_n]$.*

We first prove the following lemma, only part of which is needed now:

(6·15) Lemma. *Let $T(x)$ be a trigonometric polynomial of order n and let $p \geqslant 1$. Then*

$$\mathfrak{M}_p[T'] \leqslant 2n\mathfrak{M}_p[T], \quad \mathfrak{M}_p[\tilde{T}'] \leqslant 2n\mathfrak{M}_p[T]. \tag{6·16}$$

For $p = \infty$, this reduces to Theorem (13·16) of Chapter III, and the general case can be proved in the same way. For the first formula (13·18) of Chapter III gives

$$|T'(x)| \leqslant 2n \frac{1}{\pi} \int_0^{2\pi} |T(x+t)| K_{n-1}(t)\, dt,$$

from which, applying Jensen's inequality, we immediately obtain the first inequality (6·16). The second is obtained similarly, starting with the first formula (13·19) of Chapter III.

Return to (6·14) and let σ_n and $\tilde{\sigma}_n$ be the (C, 1) means of $S[f]$ and $\tilde{S}[f]$. Suppose that $S[f]$ converges uniformly. By Chapter III, (1·25),

$$\tilde{\sigma}_n - \tilde{s}_n = \frac{s'_n}{n+1} = \frac{s'_n - s'_{n_0}}{n+1} + \frac{s'_{n_0}}{n+1} = P_n + Q_n, \tag{6·17}$$

where n_0 is fixed and so large that $|s_n - s_{n_0}| < \epsilon$ for $n > n_0$. The first inequality (6·16) for $p = \infty$ shows that $|P_n| < 2\epsilon$. Clearly $|Q_n| < \epsilon$ for n large enough, so that $|\tilde{\sigma}_n - \tilde{s}_n| < 3\epsilon$ and $\tilde{\sigma}_n - \tilde{s}_n \to 0$ uniformly in x. If $\tilde{f}$ is continuous, then $\tilde{\sigma}_n \to \tilde{f}$ uniformly in x. Thus $\tilde{s}_n \to \tilde{f}$ uniformly in x and the first part of (6·14) (i) follows. The part about bounded f is still simpler and needs no further explanation.

Part (ii) is proved in the same way by using the first inequality (6·16) for $p = 1$. Suppose, for example, that $\mathfrak{M}[f - s_n] \to 0$. Then $\mathfrak{M}[s_n - s_{n_0}] < \epsilon$ for n_0 large enough and $n > n_0$. By (6·17), $\mathfrak{M}[\tilde{\sigma}_n - \tilde{s}_n] < 3\epsilon$ for n large enough, which means that $\mathfrak{M}[\tilde{\sigma}_n - \tilde{s}_n] \to 0$. This and the relation $\mathfrak{M}[\tilde{f} - \tilde{\sigma}_n] \to 0$ (valid if $\tilde{S}[f]$ is a Fourier series) lead to $\mathfrak{M}[\tilde{f} - \tilde{s}_n] \to 0$.

Remark. The relation $\tilde{\sigma}_n - \tilde{s}_n \to 0$ in the proof of the first part of (6·14) (i) was established under the sole condition that $S[f]$ converges uniformly. Then $\tilde{S}[f] = S[\tilde{f}]$, so that $\{\tilde{\sigma}_n\}$ converges almost everywhere. We have, therefore:

(6·18) THEOREM. *If* $S[f]$ *converges uniformly,* $\tilde{S}[f]$ *converges at every point at which it is summable* (C, 1), *in particular almost everywhere.*

(6·19) THEOREM. *There is an absolute constant* $\lambda_0 > 0$ *such that if* $|f(x)| \leqslant 1$, *then*

$$\int_0^{2\pi} \exp(\lambda |s_n|)\, dx \leqslant A_\lambda, \quad \int_0^{2\pi} \exp(\lambda |\tilde{s}_n|)\, dx \leqslant A_\lambda, \tag{6·20}$$

$$\frac{1}{2\pi}\int_0^{2\pi} \exp(\lambda |s_n - f|)\, dx \to 1, \quad \frac{1}{2\pi}\int_0^{2\pi} \exp(\lambda |\tilde{s}_n - \tilde{f}|)\, dx \to 1 \quad (n \to \infty), \tag{6·21}$$

for $0 < \lambda < \lambda_0$. *If f is also continuous, the relations hold for any* $\lambda > 0$.

If $|f| \leqslant 1$, the functions g_n and h_n in the first formula (6·3) are numerically $\leqslant 1$. Hence, using Schwarz's inequality and Theorem (2·11), we have

$$\int_0^{2\pi} \exp(\lambda |s_n^*|)\, dx \leqslant \left(\int_0^{2\pi} \exp(2\lambda |\tilde{g}_n|)\, dx\right)^{\frac{1}{2}} \left(\int_0^{2\pi} \exp(2\lambda |\tilde{h}_n|)\, dx\right)^{\frac{1}{2}} \leqslant C_{2\lambda}$$

for $\lambda < \frac{1}{4}\pi$. Since $|s_n - s_n^*| \leqslant 1$, the first inequality (6·20) follows (for $\lambda < \frac{1}{4}\pi$).

We now prove the first formula (6·21). From $0 \leqslant e^u - 1 \leqslant u e^u$, $u \geqslant 0$, and Hölder's inequality, it follows that the difference between the two sides of the formula does not exceed

$$\frac{\lambda}{2\pi}\int_0^{2\pi} |f - s_n| \exp(\lambda |f - s_n|)\, dx \leqslant \frac{\lambda}{2\pi}\left(\int_0^{2\pi} |f - s_n|^p\, dx\right)^{1/p} \left(\int_0^{2\pi} \exp(\lambda p' |f - s_n|)\, dx\right)^{1/p'}.$$

If $\lambda < \frac{1}{4}\pi$ and p is so large that $p'\lambda < \frac{1}{4}\pi$, the last factor is bounded; and since the preceding factor is $o(1)$ (by (6·4)), the first formula (6·21) follows.

If f is continuous, then $f = f' + f''$, where f' is a polynomial and $|f''| < \epsilon$. Correspondingly, $s_n = s_n' + s_n''$ and

$$f - s_n = (f' - s_n') + (f'' - s_n'') = f'' - s_n''$$

for n large enough. Thus the first relations (6·20) and (6·21) are valid for $\lambda\epsilon < \lambda_0$, that is, for any $\lambda > 0$.

The results for $\tilde{s}_n$ are proved similarly.

(6·22) THEOREM. *Let* $\alpha > 0$, $\beta > -1$ *and let* $\sigma_n^\beta(x)$, $\tilde{\sigma}_n^\beta(x)$ *be respectively the* (C, β) *means of a trigonometric series* $\Sigma A_k(x)$ *and of its conjugate series* $\Sigma B_k(x)$.

If $\sigma_n^\beta(x) = O(n^\alpha)$ *uniformly in* x, *then* $\tilde{\sigma}_n^\beta(x) = O(n^\alpha)$ *uniformly in* x. *If* $\mathfrak{M}[\sigma_n^\beta] = O(n^\alpha)$, *then* $\mathfrak{M}[\tilde{\sigma}_n^\beta] = O(n^\alpha)$.

This is an analogue of Theorem (2·30). It is easy to verify that the (C, β) means $\sigma_n^\beta = \sum_{\nu=0}^{n} A_{n-\nu}^\beta u_\nu / A_n^\beta$ (see Chapter III, § 1) of any series Σu_ν satisfy the relations

$$\sigma_n^\beta - \sigma_n^{\beta+1} = \frac{1}{n+\beta+1}\frac{1}{A_n^\beta}\sum_{\nu=0}^{n} A_{n-\nu}^\beta \nu u_\nu,$$

$$\sigma_n^\beta - \sigma_{n-1}^\beta = \frac{\beta}{n(n+\beta)}\frac{1}{A_n^{\beta-1}}\sum_{\nu=0}^{n} A_{n-\nu}^{\beta-1} \nu u_\nu.$$

In particular, $$\tilde{\sigma}_n^\beta(x) - \tilde{\sigma}_n^{\beta+1}(x) = -(n+\beta+1)^{-1}\frac{d}{dx}\sigma_n^\beta(x), \tag{6·23}$$

$$\tilde{\sigma}_n^\beta(x) - \tilde{\sigma}_{n-1}^\beta(x) = -\beta n^{-1}(n+\beta)^{-1}\frac{d}{dx}\sigma_n^{\beta-1}(x). \tag{6·24}$$

Suppose that $|\sigma_n^\beta(x)| \leqslant \Phi(n)$ for all x and n. Then $|\{\sigma_n^\beta(x)\}'| \leqslant 2n\Phi(n)$ by (6·15), and

$$|\tilde{\sigma}_n^{\beta+1}(x) - \tilde{\sigma}_n^\beta(x)| \leqslant 2n\Phi(n)/(n+\beta+1) \leqslant 2\Phi(n), \tag{6·25}$$

by (6·23). By (6·24), with β replaced by $\beta+1$,

$$|\tilde{\sigma}_n^{\beta+1} - \tilde{\sigma}_{n-1}^{\beta+1}| \leqslant (\beta+1)n^{-2}.2n\Phi(n),$$

$$|\tilde{\sigma}_n^{\beta+1}| \leqslant |\tilde{\sigma}_1^{\beta+1}| + |\tilde{\sigma}_2^{\beta+1} - \tilde{\sigma}_1^{\beta+1}| + \ldots + |\tilde{\sigma}_n^{\beta+1} - \tilde{\sigma}_{n-1}^{\beta+1}| \leqslant 2(\beta+1)\sum_1^n \frac{\Phi(\nu)}{\nu} = O(n^\alpha),$$

provided $\Phi(\nu) = O(\nu^\alpha)$. This and (6·25) give $\tilde{\sigma}_n^\beta(x) = O(n^\alpha)$.

The proof for $\mathfrak{M}[\tilde{\sigma}_n^\beta]$ runs parallel.

The following theorem is an analogue of (6·8) for Fourier–Stieltjes series. The function F^* here is defined by (1·7).

(6·26) Theorem. (i) *Let s_n and $\tilde{s}_n$ be the partial sums of* $\mathsf{S}[dF]$ *and* $\tilde{\mathsf{S}}[dF]$. *Then, for each* $0<\mu<1$ *we have*

$$\mathfrak{M}_\mu[s_n] \leqslant B_\mu \int_0^{2\pi} |dF|, \quad \mathfrak{M}_\mu[\tilde{s}_n] \leqslant B_\mu \int_0^{2\pi} |dF|. \tag{6·27}$$

(ii) *If, in addition, the coefficients of dF tend to* 0, *then*

$$\mathfrak{M}_\mu[s_n - F'] \to 0, \quad \mathfrak{M}_\mu[\tilde{s}_n - F^*] \to 0. \tag{6·28}$$

(i) Let $dF \sim \Sigma c_\nu e^{i\nu x}$ and let $u(r,x)$ be the Abel mean of $\mathsf{S}[dF]$. By (6·8),

$$\left(\int_0^{2\pi}\left|\sum_{-n}^{n} c_\nu e^{i\nu x} r^{|\nu|}\right|^\mu dx\right)^{1/\mu} \leqslant B_\mu \int_0^{2\pi} |u(r,x)|\,dx \leqslant B_\mu \int_0^{2\pi} |dF(x)|,$$

and making $r\to 1$ we obtain the first inequality (6·27). The result for $\tilde{s}_n$ is proved similarly.

(ii) This lies deeper and the proof is based on results that will be proved later. The additional hypothesis $c_n \to 0$ is indispensable since, for example, the first relation (6·28) implies

$$\mathfrak{M}_\mu[s_n - s_{n-1}] \to 0,$$

and so also $c_n \to 0$; in particular, F is continuous (Chapter III, (9·6)). In view of (6·8) it is enough to prove the result for F singular. The first relation (6·28) then is $\mathfrak{M}_\mu[s_n] \to 0$.

We need the fact that if the coefficients c_n of dF tend to 0, and if $G(x) = \int_0^x \chi(t)\,dF(t)$, where χ is the characteristic function of an interval, then the coefficients of dG also tend to 0. This is a special case of the more general Theorem (10·9) of Chapter XII, but we give a proof here.

First of all, since F is continuous so is its total variation. It follows that given an $\eta>0$ we can find a polynomial T such that

$$\int_0^{2\pi} |\chi - T|\,|dF| < \eta.$$

(We first approximate to χ by a trapezoidal function and then to the latter by a polynomial.) Since the hypothesis $c_n \to 0$ implies that $\int_0^{2\pi} T(x)\,e^{-inx}\,dF(x) \to 0$, we immediately obtain that $\int_0^{2\pi} \chi(x)\,e^{-inx}\,dF(x) \to 0$, as asserted.

If F is singular, then given any $\epsilon>0$ we can find a finite system σ' of non-overlapping intervals such that if σ'' is the complement of σ', then

$$|\sigma'| < \epsilon, \quad \int_{\sigma''} |dF| < \epsilon. \tag{6·29}$$

Let χ' and χ'' be the characteristic functions of σ' and σ'', let

$$F_1(x) = \int_0^x \chi'\,dF, \quad F_2(x) = \int_0^x \chi''\,dF,$$

and let s_n' and s_n'' be the partial sums of $S[dF_1]$ and $S[dF_2]$. Since $s_n = s_n' + s_n''$, it is enough to show that $\mathfrak{M}_\mu[s_n']$ and $\mathfrak{M}_\mu[s_n'']$ are small with ϵ, provided n is large enough.

By (6·27) and (6·29),

$$\mathfrak{M}_\mu[s_n''] \leqslant B_\mu \int_0^{2\pi} |dF_2| = B_\mu \int_{\sigma''} |dF| < B_\mu \epsilon.$$

Passing to s_n', let σ_*' be the set obtained by expanding each of the intervals constituting σ' concentrically twice, and let σ_*'' be the complement of σ_*'. Since the coefficients of dF_1 tend to 0 and F_1 is constant on each interval of σ_*'', s_n' converges uniformly to 0 on σ_*'' and the integral of $|s_n'|^\mu$ over σ_*'' tends to 0†. If now $\mu < \mu_1 < 1$, Hölder's inequality gives

$$\int_{\sigma_*'} |s_n'|^\mu dx \leqslant \left(\int_{\sigma_*'} |s_n'|^{\mu_1} dx\right)^{\mu/\mu_1} |\sigma_*'|^{1-\mu/\mu_1} \leqslant B_{\mu_1}^\mu \left| \int_0^{2\pi} dF \right|^\mu (2\epsilon)^{1-\mu/\mu_1}.$$

It follows that $\mathfrak{M}_\mu[s_n']$ is small with ϵ if n is large enough. This completes the proof of the first relation (6·28), and the second is proved similarly.

7. Classes H^p and N

Let $p > 0$. A function $F(z)$, regular for $|z| < 1$, is said to *belong to the class* H^p, if

$$\mu_p(r) = \mu_p(r; F) = \frac{1}{2\pi} \int_0^{2\pi} |F(re^{ix})|^p dx \tag{7·1}$$

is bounded for $0 \leqslant r < 1$. We shall write H instead of H^1. If $p > 1$, H^p coincides with the class of power series whose real parts are Poisson integrals of functions $f(x) \in L^p$ (cf. (2·27) and Chapter IV, (6·17)). Thus a necessary and sufficient condition for $F(z)$ to belong to H^p, $p > 1$, is that

$$F(z) = \frac{1}{2\pi} \int_0^{2\pi} \frac{e^{it} + z}{e^{it} - z} f(t)\, dt + i\gamma, \tag{7·2}$$

where $f(t)$ is real-valued and of the class L^p, and $\gamma = \mathscr{I} F(0)$. If $\gamma = 0$, then, by (2·27) and $\mathfrak{M}_p[u] \leqslant \mathfrak{M}_p[f]$,

$$\mu_p(r; F) \leqslant A_p^p \int_0^{2\pi} |f|^p dt,$$

where A_p depends on p only.

The case $p = 2$ is particularly simple, since if

$$F(z) = \sum_0^\infty c_n z^n, \tag{7·3}$$

the Parseval formula

$$\mu_2(r; F) = \frac{1}{2\pi} \int_0^{2\pi} |F(re^{ix})|^2 dx = \Sigma |c_n|^2 r^{2n} \tag{7·4}$$

shows that $F \in \mathrm{H}^2$ *if and only if* $\Sigma |c_n|^2 < \infty$.

Clearly, *if* $F \in \mathrm{H}^\alpha$, *then* $F \in \mathrm{H}^\beta$ *for* $0 < \beta < \alpha$. For $\mu_\beta^{1/\beta} \leqslant \mu_\alpha^{1/\alpha}$, by Chapter I, (10·12).

A function $F(z)$, regular for $|z| < 1$, will be said to *belong to the class* N if the integral

$$\mu_0(r) = \mu_0(r; F) = \frac{1}{2\pi} \int_0^{2\pi} \log^+ |F(re^{ix})|\, dx$$

is bounded for $r < 1$‡.

† We use here the fact that if a trigonometric series S has coefficients tending to 0 and if the series obtained by integrating S termwise twice converges in an interval (a, b) to a linear function, then S converges uniformly to 0 in every closed interval interior to (a, b). See Chapter IX, (4·23).

‡ The most natural study of the classes H^p and N would be through the theory of *subharmonic functions* (see, for example, T. Rado, *Subharmonic functions*). Since, however, we are here mostly interested in certain applications, a direct study based on Jensen's formula seems preferable. In order to avoid trivial complications, we shall always tacitly assume that $F(z)$ does not vanish identically.

If $F(z)\in H^p$, *then* $F(z)\in N$. This follows from the inequality

$$u^p \geqslant p\log^+ u \quad (u\geqslant 0).$$

The inequality is obvious for $0\leqslant u\leqslant 1$. For $u\geqslant 1$ it is enough to observe that the derivative of the left-hand side exceeds that of the right, and that for $u=1$ the left-hand side exceeds the right.

We show later (see Theorem (7·25) below) that each $F(z)\in N$ has a non-tangential limit at almost all points of $|z|=1$. Applying Fatou's lemma we therefore see that, *if* $F\in H^p$, *then the radial limit* $F(e^{ix})=\lim F(re^{ix})$ *is in* L^p; *if* $F(z)\in N$, *then* $\log^+|F(e^{ix})|$ *is in* L.

Let $\zeta\neq 0$ be a fixed point in $|z|<1$, and let

$$\zeta^*=1/\bar{\zeta}$$

be the point conjugate to ζ with respect to the circumference $|z|=1$. The function

$$w=b(z)=b(z;\,\zeta)=\frac{z-\zeta}{z-\zeta^*}\frac{1}{|\zeta|} \tag{7·5}$$

is regular in the circle $|z|\leqslant 1$ and maps it in a one-one manner onto itself. In particular

$$|b(z)|=1 \quad\text{for}\quad |z|=1; \qquad |b(z)|<1 \quad\text{for}\quad |z|<1.$$

The point $z=\zeta$ corresponds to $w=0$. If $0<|\zeta|<R$, the function

$$w=b(z/R;\,\zeta/R)$$

has a zero at $z=\zeta$ and maps $|z|\leqslant R$ onto $|w|\leqslant 1$.

Let $F(z)$ be regular for $|z|\leqslant R$ and have no zero there. Then $\log|F(z)|=\mathscr{R}\log F(z)$ is harmonic for $|z|\leqslant R$, and so

$$\frac{1}{2\pi}\int_0^{2\pi}\log|F(r\,e^{ix})|\,dx=\log|F(0)| \quad (0\leqslant r\leqslant R). \tag{7·6}$$

If $F(z)$ has no zeros on $|z|=R$, but does have some inside, say a zero of order $k\geqslant 0$ at $z=0$ and zeros $\zeta_1, \zeta_2, \ldots, \zeta_m$ distinct from the origin, then the function

$$F_1(z)=\frac{F(z)}{z^k\prod_{\nu=1}^{m} b(z/R,\,\zeta_\nu/R)} \tag{7·7}$$

is regular for $|z|\leqslant R$ and distinct from zero there. If we apply (7·6) to F_1 and note that $|\Pi|=1$ on $|z|=R$, we get the *Jensen formula*

$$\frac{1}{2\pi}\int_0^{2\pi}\log|F(R\,e^{ix})|\,dx=\log\left\{|F(z)\,z^{-k}|_{z=0}\prod_{\nu=1}^{m}R/|\zeta_\nu|\right\}+k\log R. \tag{7·8}$$

This formula holds also if $F(z)$ does have zeros on $|z|=R$. It is enough to show that both sides of (7·8) are continuous functions of R. For the right-hand side this is obvious. The same will follow for the left-hand side if we show that the integral

$$I(r)=\int_0^{2\pi}\log|r\,e^{ix}-\zeta|\,dx, \quad |\zeta|=R, \tag{7·9}$$

is a continuous function of r at $r=R$; for in the neighbourhood of $|z|=R$, the function F is a product of a non-vanishing regular function and of a finite number of linear factors $z-\zeta$, with $|\zeta|=R$. The value of $I(r)$ being independent of $\arg\zeta$, we may suppose that $\zeta=R$. Now, for $z=r\,e^{ix}$ with r near R,

$$\log(R+r)\geqslant\log|z-R|=\tfrac{1}{2}\log(r^2+R^2-2rR\cos x)\geqslant\tfrac{1}{2}\log[2rR(1-\cos x)],$$

and so the absolute value of the integrand in (7·9) is majorized by an integrable function. We can therefore proceed to the limit $r\to R$ under the integral sign; in other words $I(r)$ is continuous at $r=R$. Thus (7·8) holds for all functions regular in $|z|\leqslant R$.

Suppose now for simplicity, that $R=1$.

(7·10) Theorem. *If $F(z)$ is regular for $|z|\leqslant 1$, then $\log|F(z)|$ is majorized in $|z|<1$ by the Poisson integral of the function $\log|F(e^{ix})|$, that is,*

$$\log|F(\zeta)|\leqslant\frac{1}{2\pi}\int_0^{2\pi}\log|F(e^{ix})|\frac{1-\rho^2}{1-2\rho\cos(x-\xi)+\rho^2}dx\quad(\zeta=\rho\,e^{i\xi}).\qquad(7{\cdot}11)$$

The proof is similar to that of (7·8). If F has no zeros in $|z|\leqslant 1$, then $\log|F(r\,e^{ix})|$ is the Poisson integral of $\log|F(e^{ix})|$. If F does have zeros in $|z|<1$ but none on $|z|=1$, we apply the result to the function F_1 of (7·7); since $|F_1|=|F|$ on $|z|=1$, and $|F_1|>|F|$ in $|z|<1$, (7·11) follows. Finally, if F has zeros on $|z|=1$, we apply (7·11) to the function $F(Rz)$, where R is less than 1 and $F\neq 0$ on $|z|=R$, and then make $R\to 1$. This completes the proof of (7·10).

Let $\phi(u)$ be a function non-decreasing and convex for $-\infty<u<+\infty$. Jensen's inequality (Chapter I, (10·8)) applied to (7·11) gives

$$\phi(\log|F(\rho\,e^{i\xi})|)\leqslant\frac{1}{\pi}\int_0^{2\pi}\phi(\log|F(e^{ix})|)\,P(\rho,x-\xi)\,dx.$$

If we integrate both sides over $0\leqslant\xi\leqslant 2\pi$ and interchange the order of integration on the right, we get

$$\int_0^{2\pi}\phi(\log|F(\rho\,e^{i\xi})|)\,d\xi\leqslant\int_0^{2\pi}\phi(\log|F(e^{ix})|)\,dx.$$

Suppose now that $F(z)$ is regular for $|z|<1$, and let $R<1$. Applying the result to the function $F(zR)$, regular for $|z|\leqslant 1$, we obtain the following theorem.

(7·12) Theorem. *If $\phi(u)$ is non-decreasing and convex in $(-\infty,\infty)$, and $F(z)$ is regular in $|z|<1$, then the integral*

$$\frac{1}{2\pi}\int_0^{2\pi}\phi(\log|F(r\,e^{ix})|)\,dx$$

is a non-decreasing function of r.

In particular (taking $\phi(u)=\exp pu$, or $\phi(u)=u^+$), *$\mu_p(r)$ and $\mu_0(r)$ are non-decreasing functions of r.*

That $\mu_2(r)$ is a non-decreasing function of r follows already from (7·4).

(7·13) Theorem. *Let $F(z)$ be regular for $|z|<1$, and let $\zeta_1,\zeta_2,\ldots$ be all the zeros of F distinct from the origin (counted according to multiplicity). If the integral*

$$\int_0^{2\pi}\log|F(r\,e^{ix})|\,dx$$

is bounded above as $r\to 1$, and in particular if $F\in$ N, then the product $\Pi|\zeta_\nu|$ converges.

Consider (7·8) with $R<1$. We have

$$\log|F(z)z^{-k}|_{z=0}+k\log R+\sum_{n=1}^{m}\log(R/|\zeta_n|)=\frac{1}{2\pi}\int_0^{2\pi}\log|F(Re^{ix})|\,dx\leqslant M,$$

say, where $m=m(R)$ is the number of the zeros in $|z|\leqslant R$, and $|\zeta_1|\leqslant|\zeta_2|\leqslant\ldots$. Since, however, the terms $\log(R/|\zeta_n|)$ are non-negative, the inequality holds also for $m<m(R)$. Thus, making $R\to 1$, we see that for fixed m,

$$\log|F(z)z^{-k}|_{z=0}+\sum_{n=1}^{m}\log(1/|\zeta_n|)\leqslant M,$$

which proves (7·13).

The convergence of $\Pi|\zeta_n|$ is equivalent to that of

$$\Sigma(1-|\zeta_n|). \tag{7·14}$$

(7·15) Theorem. *Let $\zeta_1,\zeta_2,\ldots$ be a sequence of points such that $0<|\zeta_n|<1$, and that $\Pi|\zeta_n|$ converges. Then the product*

$$\Pi b(z;\zeta_n)=\Pi\frac{z-\zeta_n}{z-\zeta_n^*}\frac{1}{|\zeta_n|} \tag{7·16}$$

converges absolutely and uniformly in every circle $|z|\leqslant r<1$ to a function $\beta(z)$, regular and absolutely less than 1 in $|z|<1$, which has $\zeta_1,\zeta_2,\ldots$ as its only zeros there.

We first prove the convergence of $\Pi((z-\zeta_n)/(z-\zeta_n^*))$. Since, for $|z|\leqslant r$,

$$\left|1-\frac{z-\zeta_n}{z-\zeta_n^*}\right|=\left|\frac{\zeta_n-\zeta_n^*}{z-\zeta_n^*}\right|\leqslant\frac{1-|\zeta_n|^2}{1-r}\leqslant 2\frac{1-|\zeta_n|}{1-r}, \tag{7·17}$$

and since $\Sigma(1-|\zeta_n|)$ converges, the product (7·16) converges absolutely and uniformly for $|z|\leqslant r$. That the function $\beta(z)$ is regular for $|z|<1$ and that $\zeta_1,\zeta_2,\ldots$ are its only zeros there is clear. Each factor in (7·16) is absolutely less than 1 for $|z|<1$, so that $|\beta(z)|<1$ for $|z|<1$.

Given an $F(z)\in\mathrm{N}$, let $\zeta_1,\zeta_2,\ldots$ be all of its zeros situated in $|z|<1$ and distinct from the origin. If $F(z)$ has a zero of order $k\geqslant 0$ at $z=0$ the expression

$$B(z)=e^{i\gamma}z^k\prod_n\frac{z-\zeta_n}{z-\zeta_n^*}\frac{1}{|\zeta_n|}, \tag{7·18}$$

where γ is any real number, is called the *Blaschke product* of F. If $F(z)$ has no zero for $0<|z|<1$, the product $\prod_n$ is to be replaced by 1, and thus $B(z)=e^{i\gamma}z^k$ for such F. We have $|B(z)|\leqslant 1$ for $|z|<1$ and the ratio

$$G(z)=F(z)/B(z) \tag{7·19}$$

is regular and without zeros in $|z|<1$.

(7·20) Theorem. (i) *Suppose that $\mu_0(r;F)\leqslant\mu<\infty$ for $0\leqslant r<1$, and let $B(z)$ be the Blaschke product of F. Then, for $G(z)$ defined by (7·19), we have $\mu_0(r;G)\leqslant\mu$.*

(ii) *If $\mu_p(r;F)\leqslant\mu<\infty$ for $0\leqslant r<1$, then $\mu_p(r;G)\leqslant\mu$.*

Take case (ii) first, and let $B_n(z)$ be the nth partial product of (7·18). Since $|B_n(z)|$ tends uniformly to 1 as $r\to 1$, we have

$$\lim_{r\to 1}\mu_p(r;F/B_n)=\lim_{r\to 1}\mu_p(r;F)\leqslant\mu.$$

Hence $\mu_p(r; F/B_n) \leqslant \mu$ for every $r<1$, and this gives $\mu_p(r; G) \leqslant \mu$, since $B_n(z)$ tends uniformly to $B(z)$ on $|z|=r$. The same argument gives (i).

To sum up, if F belongs to H^p or to N, we have the *decomposition formula*

$$F(z)=B(z)\,G(z), \tag{7·21}$$

where G has no zeros and belongs to the same class as F, and $|B(z)| \leqslant 1$.

We shall always suppose the γ *in* (7·18) *selected so that* $G(0)$ *is real and positive.*

The importance of (7·21) is due to the fact that every branch of $\log G$, and of $G^\alpha = \exp(\alpha \log G)$, is regular for $|z|<1$. Thus G is more 'flexible' than F under certain operations. For example, if $G \in \mathrm{H}^\alpha$, then $G^{\alpha/\beta} \in \mathrm{H}^\beta$, and this makes it possible to extend the properties of certain especially simple classes H^α (for example, when $\alpha=2$) to other classes H^β. A very special but typical application of (7·21) deserves a separate statement.

(7·22) THEOREM. *A necessary and sufficient condition that* $F \in \mathrm{H}$ *is that* $F=F_1F_2$, *where* $F_1 \in \mathrm{H}^2$, $F_2 \in \mathrm{H}^2$.

The sufficiency of the condition follows from the inequality $2|F_1F_2| \leqslant |F_1|^2+|F_2|^2$, which we integrate over $0 \leqslant x \leqslant 2\pi$. To prove the necessity, suppose that $\mu_1(r; F) \leqslant \mu$. Let $F_1(z)$ be any branch of $G^{\frac{1}{2}}(z)$ (see (7·21)) and let $F_2 = BF_1$. Then

$$F=F_1F_2, \quad \mu_2(r; F_k) \leqslant \mu_1(r; G) \leqslant \mu$$

for $k=1, 2$.

A variant of (7·21) is also useful. Let

$$B(z)-1=B^*(z),$$

so that $B^*(z)$ has no zeros in $|z|<1$ (unless $B(z) \equiv 1$, a case which we exclude from consideration) and $|B^*(z)|<2$. Then (7·21) may be written

$$F(z)=G(z)+G^*(z), \tag{7·23}$$

where $G(z)$ and $G^*(z)=B^*(z)\,G(z)$ have no zeros for $|z|<1$, and $|G^*(z)| \leqslant 2|G(z)|$.

Since $|B(z)| \leqslant 1$ for $|z|<1$, the non-tangential limit

$$B(e^{ix_0}) = \lim_{z \to e^{ix_0}} B(z)$$

exists for almost all x_0. Moreover, where it exists, $|B(e^{ix_0})| \leqslant 1$.

(7·24) THEOREM. *For almost all x we have* $|B(e^{ix})|=1$.

We may suppose that $B(z)$ has infinitely many factors, for otherwise the result is immediate. Since $|B(e^{ix})| \leqslant 1$ almost everywhere, we have

$$\lim_{r \to 1} \mu_2(r; B) = \frac{1}{2\pi}\int_0^{2\pi} |B(e^{ix})|^2\,dx \leqslant 1$$

and (7·24) will follow if we show $\mu_2(r; B) \to 1$ as $r \to 1$.

Let $B_n(z)$ be the nth partial product of (7·18), and let $R_n(z)=B(z)/B_n(z)$. The function R_n is regular and numerically not greater than 1 for $|z|<1$. Moreover,

$$R_n(0)=|\zeta_{n+1}\zeta_{n+2}\cdots|.$$

The expression $\mu_2(r; R_n)$ is an increasing function of r, taking the value $|R_n(0)|^2$ at $r=0$. Making $r\to 1$ and observing that $|B_n(e^{ix})|=1$, we get

$$1\geqslant \lim_{r\to 1}\mu_2(r; B)=\lim_{r\to 1}\mu_2(r; R_n)\geqslant |R_n(0)|^2=|\zeta_{n+1}\zeta_{n+2}\cdots|^2.$$

Since the last number may be arbitrarily close to 1, we have $\mu_2(r; B)\to 1$.

(7·25) THEOREM. *If $F(z)\in \mathrm{N}$ (in particular, if $F\in \mathrm{H}^p$), the non-tangential limit $F(e^{ix})=\lim_{z\to e^{ix}} F(z)$ exists for almost all x. Moreover, $\log|F(e^{ix})|$ is integrable. In particular, $F(e^{ix})\neq 0$ almost everywhere.*

The function $G(z)$ in (7·21) belongs to N, and $\log|G(z)|$ is harmonic for $|z|<1$. Thus

$$\int_0^{2\pi} |\log|G(re^{ix})||\,dx \tag{7·26}$$

is also bounded as $r\to 1$. Hence, if we assume the existence of $F(e^{ix})$, and so by (7·24) the existence of $G(e^{ix})$, Fatou's lemma applied to (7·26) shows that $\log|G(e^{ix})|\in \mathrm{L}$, and hence that $\log|F(e^{ix})|\in \mathrm{L}$.

By Chapter IV, (6·5), $\log|G(z)|$ is a Poisson-Stieltjes integral. Hence, observing that $\mathscr{I}\log G(0)=0$, we can represent $G(z)$ by formula (6·24) of Chapter IV. Thus

$$G(z)=\exp\left\{\frac{1}{2\pi}\int_0^{2\pi}\frac{e^{it}+z}{e^{it}-z}\,d\lambda(t)\right\}, \tag{7·27}$$

where $\lambda(t)$ is real-valued and of bounded variation. Conversely, any $G(z)$ of the form (7·27) is of class N, without zeros in $|z|<1$ and with $G(0)>0$.

Let $\lambda(t)=\lambda_1(t)-\lambda_2(t)$, where $\lambda_1(t)$ and $\lambda_2(t)$ are bounded and non-increasing. Then

$$G(z)=G_1(z)/G_2(z), \tag{7·28}$$

where

$$G_k(z)=\exp\left\{\frac{1}{2\pi}\int_0^{2\pi}\frac{e^{it}+z}{e^{it}-z}\,d\lambda_k(t)\right\}\quad (k=1,2). \tag{7·29}$$

The functions G_k have no zeros, and $|G_k|\leqslant 1$ for $|z|<1$ since the real part of the exponent is non-positive. It follows that $G_k(e^{ix})$ exists almost everywhere. It is also almost everywhere different from zero, the real part of the exponent having a finite non-tangential limit almost everywhere. Hence $G(e^{ix})=G_1(e^{ix})/G_2(e^{ix})$, and so also $F(e^{ix})=B(e^{ix})\,G(e^{ix})$, exists (and is not 0) almost everywhere. This completes the proof of (7·25).

It follows from (7·25) that, *if $F_1\in \mathrm{N}$, $F_2\in \mathrm{N}$, and if $F_1(e^{ix})=F_2(e^{ix})$ in a set of x of positive measure, then $F_1(z)\equiv F_2(z)$.*

We have also, by Fatou's lemma, that *if $F(z)\in \mathrm{H}^p$, then $F(e^{ix})\in \mathrm{L}^p$ $(p>0)$.*

It is of interest to observe that, *for any set E situated on $|z|=1$ and of measure zero there is an $F(z)\not\equiv 0$, regular for $|z|<1$, bounded* (in particular, $F\in \mathrm{N}$) *and such that $F(z)$ tends to 0 as z approaches, in an arbitrary manner, any point of E.* It is enough to set

$$F(z)=\exp\{-f(r,x)-i\tilde{f}(r,x)\},$$

where $f(r,x)$ is the Poisson integral of the function f from Theorem (7·26) of Chapter III.

If E is, in addition, closed, then there exists an $F(z)$ regular for $|z|<1$, continuous for $|z|\leqslant 1$, vanishing on E and only there. Let $f(x)$ be integrable over $(0,2\pi)$, differentiable outside E, and tending to $+\infty$ as x tends to any point of E. Then $f(r, x) \to \infty$ as (r, x) approaches any point of E, and it is easy to see that the $F(z)$ just defined has the required properties. The function f may be constructed as follows.

Let (a_1, b_1), (a_2, b_2), ..., be the intervals contiguous to E, $d_i = b_i - a_i$, so that $\Sigma d_i = 2\pi$. Let $\{\epsilon_i\}$ be a positive sequence such that

$$\text{(i) } \epsilon_i/d_i \to \infty, \quad \text{(ii) } \Sigma\epsilon_i < \infty.$$

Let $f(x) = \epsilon_i(x - a_i)^{-\frac{1}{2}}(b_i - x)^{-\frac{1}{2}}$ in (a_i, b_i) $(i = 1, 2, \ldots)$, $f = +\infty$ in E. The integral of f over (a_i, b_i) being a fixed multiple of ϵ_i, f is integrable. The minimum of f in (a_i, b_i) is $2\epsilon_i/d_i$, and tends to $+\infty$ with i. Hence f has the needed properties.

From (7·21) and (7·27) we see that

(7·30) THEOREM. *The general function of class* N *is*

$$F(z) = B(z) \exp\left\{\frac{1}{2\pi}\int_0^{2\pi} \frac{e^{it} + z}{e^{it} - z}\, d\lambda(t)\right\}, \tag{7·31}$$

where $\lambda(t)$ *is any real-valued function of bounded variation, and* $B(z)$ *is any product* (7·18) *with* $0 < |\zeta_n| < 1$, $\Sigma(1 - |\zeta_n|) < +\infty$.

Closely related to (7·30) is the following result:

(7·32) THEOREM. *A necessary and sufficient condition for a function* $F(z)$ *regular in* $|z| < 1$ *to be of the class* N *is that* $F(z) = F_1(z)/F_2(z)$, *where* F_1 *and* F_2 *are both regular and bounded for* $|z| < 1$, *and* F_2 *has no zeros there.*

The necessity follows from (7·31), if we observe that, with the notation of (7·29), we have $G = G_1/G_2$, where G_1 and G_2 are numerically not greater than 1 and without zeros. Thus $F = F_1/F_2$, where $F_1 = BG_1$, $F_2 = G_2$.

Conversely, if F_1 and F_2 are both bounded, say less than 1 in absolute value, and $F_2 \neq 0$, then

$$\frac{1}{2\pi}\int_0^{2\pi} \log^+ \left|\frac{F_1(re^{ix})}{F_2(re^{ix})}\right| dx \leqslant \frac{1}{2\pi}\int_0^{2\pi} \log \frac{1}{|F_2(re^{ix})|}\, dx = \log \frac{1}{|F_2(0)|}.$$

We now show that the moduli of the boundary values of functions from N and H^p can be prescribed, roughly speaking, arbitrarily.

(7·33) THEOREM. *Let* $f(x)$, $0 \leqslant x \leqslant 2\pi$, *be non-negative and such that* $\log f(x) \in \mathrm{L}$ (*in particular* $f > 0$ *almost everywhere*). *Then*

(i) *There is an* $F \in \mathrm{N}$ *such that* $|F(e^{ix})| = f(x)$ *almost everywhere*;

(ii) *if, in addition,* $f \in \mathrm{L}^p$, $p > 0$, *then there is an* $F \in \mathrm{H}^p$ *such that* $|F(e^{ix})| = f(x)$ *almost everywhere.*

(i) Consider the F in (7·31) with $\lambda(x)$ the indefinite integral of $\log f(x)$ and $B \equiv 1$. Then $F \in \mathrm{N}$ and

$$|F(e^{ix})| = e^{\log f(x)} = f(x).$$

(ii) Let F be the same as in (i). Since $\phi(u) = e^{pu}$ is convex, we have, by Jensen's inequality,

$$|F(re^{ix})|^p = \exp\left\{p\frac{1}{\pi}\int_0^{2\pi} P(r, x-t) \log f(t)\, dt\right\}$$

$$\leqslant \frac{1}{\pi}\int_0^{2\pi} P(r, x-t)\, e^{p\log f(t)}\, dt = \frac{1}{\pi}\int_0^{2\pi} P(r, x-t)\, |f(t)|^p\, dt.$$

Hence $|F(re^{ix})|^p$ is majorized by the Poisson integral of $|f|^p$,

$$\int_0^{2\pi} |F(re^{ix})|^p\,dx \leqslant \int_0^{2\pi} |f(x)|^p\,dx, \quad F \in H^p.$$

(7·34) THEOREM. *If $F(z) \in H^p$, and r and ρ tend to 1, then*

$$\mathfrak{M}_p[F(re^{ix}) - F(\rho e^{ix})] \to 0, \quad \mathfrak{M}_p[F(re^{ix}) - F(e^{ix})] \to 0.$$

It is enough to prove the first relation, since the two are equivalent. Both relations have been proved in Chapter IV, (6·17) for $p > 1$. On account of the decomposition formula (7·23), it is enough to consider the case when $F(z)$ is without zeros. Let $F_1(z) = F^{\frac{1}{2}}(z)$, so that $F_1 \in H^{2p}$. Then

$$\int_0^{2\pi} |F(re^{ix}) - F(\rho e^{ix})|^p\,dx$$
$$\leqslant \left\{\int_0^{2\pi} |F_1(re^{ix}) - F_1(\rho e^{ix})|^{2p}\,dx\right\}^{\frac{1}{2}} \left\{\int_0^{2\pi} |F_1(re^{ix}) + F_1(\rho e^{ix})|^{2p}\,dx\right\}^{\frac{1}{2}},$$

by the inequality of Schwarz. If $2p > 1$, the first factor on the right tends to 0, and the second is bounded, so that the result follows for $p > \frac{1}{2}$. From this we similarly get the result for $p > \frac{1}{4}$, and so on.

(7·35) THEOREM. *If $F(z) \in H^\alpha$, and if $F(e^{ix}) \in L^\beta$ for some $\beta > \alpha$, then $F(z) \in H^\beta$.*

This is immediate if $\alpha = 2$; for if $F(z) = \Sigma c_n z^n$ is in H^2, it follows that $\Sigma |c_n|^2 < \infty$, $\Sigma c_n e^{inx}$ is the Fourier series of $F(e^{ix})$, $F(re^{ix})$ is the Poisson integral of $F(e^{ix})$, and since $F(e^{ix})$ is L^β, $F(re^{ix})$ is in H^β. It is also immediate for any $\alpha > 0$, if $F(z) \neq 0$ in $|z| < 1$. For if we set $F_1(z) = F^{\frac{1}{2}\alpha}(z)$, then

$$F_1(z) \in H^2, \quad F_1(e^{ix}) \in L^{2\beta/\alpha},$$

so that
$$F_1(z) \in H^{2\beta/\alpha}, \quad F(z) \in H^\beta.$$

In the general case, $F = BG$ with $G(z) \neq 0$, $G(z) \in H^\alpha$. Since $|F(e^{ix})| = |G(e^{ix})|$ almost everywhere, it follows that

$$G(e^{ix}) \in L^\beta, \quad G(z) \in H^\beta, \quad F = BG \in H^\beta.$$

For any $0 \leqslant \sigma < 1$, $0 \leqslant x \leqslant 2\pi$, let $\Omega_\sigma(x)$ denote the domain bounded by the two tangents from the point e^{ix} to the circle $|z| = \sigma$ and by the larger of the two arcs of that circle between the points of contact. For $\sigma = 0$, the domain reduces to the radius through e^{ix}.

(7·36) THEOREM. *For any $F \in H^p$, $p > 0$, let*

$$N(x) = N_{\sigma, F}(x) = \sup_{z \in \Omega_\sigma(x)} |F(z)|.$$

Then $N(x) \in L^p$ and

$$\left\{\int_0^{2\pi} N^p\,dx\right\}^{1/p} \leqslant A_p \left\{\int_0^{2\pi} |F(e^{ix})|^p\,dx\right\}^{1/p}. \tag{7·37}$$

This is contained in Chapter IV, (7·6), if, say, $p=2$. In the general case we make the usual decomposition $F=GB$ and have, since $G_1=G^{\frac{1}{2}p}$ is in H^2,

$$\int_0^{2\pi} N^p_{\sigma,F}\,dx \leqslant \int_0^{2\pi} N^p_{\sigma,G}\,dx = \int_0^{2\pi} N^2_{\sigma,G_1}\,dx \leqslant A_2^2\int_0^{2\pi} |G_1(e^{ix})|^2\,dx$$
$$= A_2^2\int_0^{2\pi} |G(e^{ix})|^p\,dx = A_2^2\int_0^{2\pi} |F(e^{ix})|^p\,dx.$$

This gives (7·37) with $A_p=A_2^{2/p}$.

The following result strengthens Theorems (2·4), (2·6), (2·8):

(7·38) THEOREM. *The function*

$$\tilde{f}^*(x)=\sup_{0<h\leqslant 1}\left|-\frac{1}{\pi}\int_h^{\pi}[f(x+t)-f(x-t)]\,\tfrac{1}{2}\cot\tfrac{1}{2}t\,dt\right| \tag{7·39}$$

satisfies the inequalities

$$\left.\begin{aligned}&\text{(i)}\quad \mathfrak{M}_p[\tilde{f}^*]\leqslant A_p\mathfrak{M}_p[f]\quad (p>1),\\ &\text{(ii)}\quad \mathfrak{M}_\mu[\tilde{f}^*]\leqslant A_\mu\mathfrak{M}[f]\quad (0<\mu<1),\\ &\text{(iii)}\quad \mathfrak{M}[\tilde{f}^*]\leqslant B\int_0^{2\pi}|f|\log^+|f|\,dx+C,\end{aligned}\right\} \tag{7·40}$$

where the coefficients on the right depend only on the parameters indicated.†

The expression inside the modulus signs in (7·39) is $\tilde{f}(x;h)$. Let

$$\psi_x(t)=\tfrac{1}{2}[f(x+t)-f(x-t)],$$

$$M(x)=M_f(x)=\sup_{0<|u|\leqslant\pi}\frac{1}{u}\int_0^u|f(x+t)|\,dt.$$

The difference $\tilde{f}(x;1-\rho)-\tilde{f}(\rho,x)$ may be written

$$\frac{2}{\pi}\int_0^{1-\rho}\psi_x(t)\,Q(\rho,t)\,dt-\frac{2}{\pi}\int_{1-\rho}^{\pi}\psi_x(t)\,\frac{(1-\rho)^2\,\frac{1}{2}\cot\frac{1}{2}t}{1-2\rho\cos t+\rho^2}\,dt=G_\rho(x)+H_\rho(x).$$

Since $|Q(\rho,t)|$ does not exceed $1/(1-\rho)$, we see that $|G_\rho(x)|\leqslant M(x)$. Let $R_\rho(t)$ be the co-factor of $\psi_x(t)$ in $H_\rho(x)$. We easily verify that

$$tR_\rho(t)=O(1),\quad \int_{1-\rho}^{\pi}\left|t\frac{d}{dt}R_\rho(t)\right|dt=O(1),$$

whence, integrating by parts, we deduce that $|H_\rho(x)|$ does not exceed a multiple of $M(x)$. Therefore

$$|\tilde{f}(x;1-\rho)-\tilde{f}(\rho,x)|\leqslant CM(x), \tag{7·41}$$

where C is an absolute constant.

Let $F(\rho e^{ix})=f(\rho,x)+i\tilde{f}(\rho,x)$, and let $N(x)$ be the upper bound of $|F(\rho e^{ix})|$ for $0\leqslant\rho<1$. Then

$$|\tilde{f}(x;1-\rho)|\leqslant|\tilde{f}(x;1-\rho)-\tilde{f}(\rho,x)|+|\tilde{f}(\rho,x)|\leqslant CM(x)+N(x),$$

and the right-hand side majorizes $\tilde{f}^*(x)$. If $f\in\mathrm{L}^p$, Minkowski's inequality gives

$$\mathfrak{M}_p[\tilde{f}^*]\leqslant C\mathfrak{M}_p[M]+\mathfrak{M}_p[N]\leqslant A_p\mathfrak{M}_p[f],$$

in virtue of (7·36), (2·4) and Chapter I, (13·17). This gives (i), and the proofs of (ii) and (iii) are similar.

† Though the point is not important, it is not difficult to see that (7·40) holds if in (7·39) we replace the condition on h by $0<h\leqslant\pi$. For if $1\leqslant h\leqslant\pi$, $|\tilde{f}(x;h)|$ does not exceed a fixed multiple of $\mathfrak{M}[f]$.

(7·42) Theorem. *Let $\tilde{\sigma}_n(x)$ be the* (C, 1) *means of* $\tilde{S}[f]$, *and let*

$$\tilde{\sigma}_*(x) = \sup_{n\geq 1} |\tilde{\sigma}_n(x)|.$$

We have the following inequalities analogous to (7·40):

$$\left.\begin{aligned} &\text{(i)} \quad \mathfrak{M}_p[\tilde{\sigma}_*] \leq A_p \mathfrak{M}_p[f] \quad (p>1),\\ &\text{(ii)} \quad \mathfrak{M}_\mu[\tilde{\sigma}_*] \leq A_\mu \mathfrak{M}[f] \quad (0<\mu<1),\\ &\text{(iii)} \quad \mathfrak{M}[\tilde{\sigma}_*] \leq B\int_0^{2\pi} |f| \log^+ |f|\, dx + C. \end{aligned}\right\} \tag{7·43}$$

The inequalities hold if we replace $\tilde{\sigma}_*$ by

$$\tilde{\sigma}_*^\alpha(x) = \sup_{n\geq 1} |\tilde{\sigma}_n^\alpha(x)| \quad (0<\alpha\leq 1),$$

where $\tilde{\sigma}_n^\alpha$ denote the (C, α) *means of* $\tilde{S}[f]$; *but the constants on the right then depend also on* α.

We shall only consider the case $\alpha = 1$, the proof for $0<\alpha\leq 1$ being essentially the same.

Part (i) of (7·43) follows immediately from the inequalities

$$\mathfrak{M}_p[\tilde{\sigma}_*] \leq A_p \mathfrak{M}_p[\tilde{f}], \quad \mathfrak{M}_p[\tilde{f}] \leq A_p \mathfrak{M}_p[f],$$

the first of which is an application of Theorem (7·8) of Chapter IV to $S[\tilde{f}] = \tilde{S}[f]$. The argument, however, does not work in cases (ii) and (iii) (in which the integrability of $\tilde{f}$ is not as good as that of f) and a different proof is needed.

We have (see the proof of Theorem (3·20) of Chapter III)

$$\tilde{\sigma}_n(x) - \tilde{f}(x;\, 1/n) = -\frac{2}{\pi}\int_0^{1/n} \psi_x(t)\, \tilde{K}_n(t)\, dt + \frac{2}{\pi}\int_{1/n}^{\pi} \psi_x(t)\, H_n(t)\, dt,$$

$$|\tilde{\sigma}_n(x) - \tilde{f}(x;\, 1/n)| \leq n\int_0^{1/n} |\psi_x(t)|\, dt + \frac{A}{n}\int_{1/n}^{\pi} t^{-2} |\psi_x(t)|\, dt.$$

It is easily seen that the right-hand side of the last inequality is majorized by a constant multiple of $M(x)$, and that Theorem (7·42) follows from (7·38) and the inequalities (13·17) of Chapter I for $M(x)$.

We shall now prove some results about Blaschke products.

By Chapter III, (7·9), the Poisson integral of any integrable function tends to a limit as the variable point approaches almost any point e^{ix_0} of the unit circumference, provided it remains within a fixed angle formed by two chords through e^{ix_0}. The result holds, in particular, for any harmonic function bounded inside the unit circle. Even for bounded functions the result fails if the angle is replaced by any fixed domain tangent to the unit circle; we now show this by means of Blaschke products.

(7·44) Theorem. *Let C_0 be any simple closed curve passing through $z=1$, situated, except for that point, totally inside the circle $|z|=1$, and tangent to the circle at that point. Let C_θ be the curve C_0 rotated around $z=0$ by an angle θ. There is a Blaschke product $B(z)$ which, for almost all θ_0, does not tend to any limit as $z\to e^{i\theta_0}$ inside C_{θ_0}.*

We may suppose that for r close to 1 the circle $|z|=r<1$ meets C_0 at exactly two points. (Otherwise we replace the region bounded by C_0 by a smaller region having the required property.) Let l_n denote the length of the arc of $|z|=1-1/n$ situated inside C_0, and let $m_n=[2\pi/l_n]+1$. Let S_n be any system of m_n equally spaced points situated on $|z|=1-1/n$. The circular distance between any two consecutive points is less than l_n, so that every C_θ contains in its interior a point of S_n. The sum σ_n of the distances of the points of S_n from the circumference $|z|=1$ is

$$m_n/n \leqslant (1+2\pi/l_n)/n = o(1),$$

since the tangency of C_0 to $|z|=1$ implies that $nl_n \to \infty$. Let us take n_k increasing so rapidly that $\Sigma\sigma_{n_k}<\infty$, and let $B(z)$ be the Blaschke product with zeros at the points of $S_{n_1}+S_{n_2}+\ldots$. Since $B(z)$ has infinitely many zeros inside every C_{θ_0}, the limit of $B(z)$ as $z\to e^{i\theta_0}$ in the interior of C_{θ_0} must be zero if it exists at all. By (7·24) such a limit exists for almost no θ_0.

(7·45) THEOREM. *A necessary and sufficient condition for a function $F(z)$ regular for $|z|<1$ to be a Blaschke product (that is, to be of the form* (7·18)) *is*

$$\int_0^{2\pi} |\log|F(re^{ix})||\,dx \to 0 \quad as \quad r\to 1. \tag{7·46}$$

Necessity. We may suppose that $F(0)\neq 0$ so that F is, except for a factor $e^{i\gamma}$, of the form (7·16). We may also suppose that F has infinitely many factors; otherwise (7·46) is obvious. Consider Jensen's formula (7·8) with $k=0$ and $F(0)=\Pi|\zeta_\nu|$. The right-hand side increases with R and so, if $m=m_0$ corresponds to a fixed $R_0<1$,

$$\frac{1}{2\pi}\int_0^{2\pi}\log|F(Re^{ix})|\,dx \geqslant \log\left\{|F(0)|\prod_1^{m_0}\frac{R}{|\zeta_\nu|}\right\} = \log\left\{R^{m_0}\prod_{m_0+1}^{\infty}|\zeta_\nu|\right\}$$

for $R>R_0$. Hence

$$\lim_{R\to 1}\frac{1}{2\pi}\int_0^{2\pi}\log|F(Re^{ix})|\,dx \geqslant \log(|\zeta_{m_0+1}||\zeta_{m_0+2}|\ldots),$$

a negative number whose modulus is arbitrarily small if m_0 is large enough. Since $|F|\leqslant 1$, this leads to (7·46).

Sufficiency. The hypothesis (7·46) implies that $F\in \mathrm{N}$. Let $B(z)$ be the Blaschke product of F, and let $G(z)=F(z)/B(z)$. Since (7·46) holds for $B(z)$, it holds for $G(z)$. Thus $G(z)$ is of the form (7·27), and so $\log|G(z)|$ is the Poisson-Stieltjes integral of λ. By Chapter IV, (6·9), the total variation of $\lambda(t)$ over $(0, 2\pi)$ is $\lim\int_0^{2\pi}|\log|G(re^{ix})||\,dx$, and so is zero. Hence $G(z)=1$, $F(z)=B(z)$.

The following result completes (7·24):

(7·47) THEOREM. *If a Blaschke product $B(z)$ contains infinitely many factors, the set of the radial limits $w=B(e^{ix})=\lim_{r\to 1}B(re^{ix})$ covers the whole circumference $|w|=1$ infinitely many times.*

It is clear that if $B(z)$ contains n linear factors, the numbers $w=B(e^{ix})$ cover $|w|=1$ exactly n times.

We shall deduce (7·47) from the following slightly more general result:

(7·48) THEOREM. *Suppose that $F(z)$ is regular and absolutely less than 1 for $|z|<1$. Suppose also that $|F(e^{ix})|=\lim_{r\to 1}|F(re^{ix})|=1$ at almost all points of an arc $a<x<b$.*

Then either $F(z)$ is analytically continuable across this arc or the values of $F(e^{ix})$, $a<x<b$, cover the circumference $|w|=1$ infinitely many times.

Since $B(z)$ in (7·47) has at least one singular point on $|z|=1$ (because any limit point of the zeros of $B(z)$ is singular), (7·47) is a consequence of (7·48).

To prove (7·48), fix any α, $|\alpha|=1$, and let $\zeta=L(w)$ be a linear function mapping $|w|\leqslant 1$ onto $\mathscr{R}\zeta\leqslant 0$ and making $w=\alpha$ and $\zeta=\infty$ correspond. The function $L\{F(z)\}$ is regular for $|z|<1$ and has a negative real part there whose boundary values are 0 almost everywhere on the arc $a<x<b$. By Chapter IV, (6·26),

$$L\{F(z)\}=-\frac{1}{2\pi}\int_0^{2\pi}\frac{e^{it}+z}{e^{it}-z}d\lambda(t)+i\gamma,\quad \gamma=\mathscr{I}L\{F(0)\}, \tag{7·49}$$

where $\lambda(t)$ is bounded and non-decreasing. The real part of $L\{F(z)\}$ tends radially to $\lambda'(t)$ at every point of differentiability of λ, so that $\lambda'(t)=0$ almost everywhere in (a,b).

Let E be the set of *points of increase* of λ (that is, of the points in the neighbourhood of which λ is not constant) situated in (a,b). There are two possibilities: (i) E is infinite, (ii) E is finite.

In case (i), there are infinitely many points in (a,b) at which the symmetric derivative of $\lambda(t)$ is infinite.This is obvious if $\lambda(t)$ is discontinuous at infinitely many points of (a,b). If the discontinuities are finite in number, (a,b) contains an interval (t',t'') where λ is continuous and in which there are infinitely many points of E. The derivative $\lambda'(t)$ must be infinite at infinitely many points of (t',t''), for otherwise $\lambda(t)$ would be absolutely continuous inside (t',t''), and so constant there (since $\lambda'(t)=0$ almost everywhere in (t',t'')), contrary to hypothesis. At every point at which $\lambda'(t)=\infty$, the symmetric derivative of λ is also ∞.

Thus, in case (i), $\mathscr{R}L\{F(z)\}$ tends to $-\infty$, and so $L\{F(z)\}$ tends to ∞, along infinitely many radii terminating on (a,b). From the definition of $L(w)$ it follows that $F(z)$ tends to α along these radii.

In case (ii), $\lambda(t)$ is a step function in (a,b). Hence, by (7·49), $\Phi(z)=L\{F(z)\}$ is regular on the arc $a<x<b$ of $|z|=1$, except for a finite number of points at which it has poles. It is easily seen, taking into account the boundedness of F, that $F(z)=L^{-1}\{\Phi(z)\}$ is regular at all points of the arc (a,b).

Return to Theorem (7·35). Its conclusion, that $F\in \mathrm{H}^{\beta}$, is no longer valid if the hypothesis $F(z)\in \mathrm{H}^{\alpha}$ is replaced by $F(z)\in \mathrm{N}$. For example, the function

$$F(z)=\exp\left(\frac{1}{2}\frac{1+z}{1-z}\right)=\exp\{P(r,x)+iQ(r,x)\}$$

is of the class N and its boundary values belong to L^{β} ($|F(e^{ix})|=1$ for $x\neq 0$), but $F(z)$ is not in H^{β} for any $\beta>0$. (Observe that $P(r,x)$ exceeds a constant multiple of $1/(1-r)$ for $|x|\leqslant 1-r$, so that $\mu_{\beta}(r;F)\to\infty$.)

To generalize (7·35), we introduce a subclass of the class N. An $F(z)\in \mathrm{N}$ will be said to belong to the class N′, if the function $\lambda(t)$ in (7·31) has its *positive variation absolutely continuous*.

(7·50) THEOREM. *Suppose that $F(z)\in \mathrm{N}'$ and that $\phi(u)$, $u\geqslant 0$, is non-negative, non-decreasing and convex. Then*

$$\int_0^{2\pi}\phi\{\log^+|F(re^{ix})|\}\,dx\leqslant\int_0^{2\pi}\phi\{\log^+|F(e^{ix})|\}\,dx.$$

We may suppose that the right-hand side is finite. Consider (7·31), and let $\lambda_1(t)$ and $\lambda_2(t)$ be the positive and negative variations of $\lambda(t)$. Let $u_1(r,x)$ and $u_2(r,x)$ be the Abel means of $S[d\lambda_1]=S[\lambda_1']$ and of $S[d\lambda_2]$; then

$$\log^+|F(r,x)|\leqslant\log^+|G(r,x)|=\{u_1(r,x)-u_2(r,x)\}^+\leqslant u_1(r,x),$$

and so
$$\int_0^{2\pi}\phi\{\log^+|F(r,x)|\}\,dx\leqslant\int_0^{2\pi}\phi\{u_1(r,x)\}\,dx\leqslant\int_0^{2\pi}\phi\{\lambda_1'(x)\}\,dx,$$

by Chapter IV, (6·20). It is enough to show that $\lambda_1'(x)=\log^+|F(e^{ix})|$ almost everywhere. This follows from the equations

$$\log|F(e^{ix})|=\lambda'(x),\quad \lambda_1'(x)=\{\lambda'(x)\}^+,$$

valid almost everywhere. Thus the proof of (7·50) is completed.

Let us again consider (7·31). By Chapter IV, (6·19), $F(z)$ is in N′ if and only if the integrals

$$\int_0^x\log^+|G(re^{it})|\,dt\quad(0\leqslant r<1),$$

are uniformly absolutely continuous. This is equivalent to the uniform absolute continuity of the integrals

$$\int_0^x\log^+|F(re^{it})|\,dt\quad(0\leqslant r<1),\tag{7·51}$$

in view of the inequalities

$$\log^+|G|+\log|B|\leqslant\log^+|F|\leqslant\log^+|G|\tag{7·52}$$

(see (7·21)) and the uniform absolute continuity of the integrals $\int_0^x\log|B(re^{it})|\,dt$ (see (7·45)).

The integrals (7·51) are uniformly absolutely continuous, if there exists a non-negative function $\psi(u)$, $u\geqslant 0$, such that $\psi(u)/u\to\infty$ with u and that

$$\int_0^{2\pi}\psi\{\log^+|F(re^{ix})|\}\,dx\leqslant C,$$

with C independent of r (Chapter IV, § 6). In particular, taking $\psi(u)=\exp\alpha u$, we see that $\mathrm{H}^\alpha\subset\mathrm{N}'$ and that (7·35) is a corollary of (7·50).

(7·53) THEOREM. *An* $F\in\mathrm{N}$ *is in* N′ *if and only if*

$$\lim_{r\to1}\int_0^{2\pi}\log^+|F(re^{ix})|\,dx=\int_0^{2\pi}\log^+|F(e^{ix})|\,dx.\tag{7·54}$$

To see this we observe that, by Chapter IV, (6·19), $F(z)$ is in N′ if and only if

$$\lim_{r\to1}\int_0^{2\pi}\log^+|G(re^{ix})|\,dx=\int_0^{2\pi}\log^+|G(e^{ix})|\,dx.\tag{7·55}$$

The right-hand sides in these two equations are the same. Thus (7·54) is equivalent to (7·55), in view of (7·52) and the fact that the integral of $\log|B|$ over $(0,2\pi)$ tends to 0 (see (7·45)).

Replacing, in (7·54), $\log^+$ by $\log^-$ we obtain a necessary and sufficient condition for the negative variation of λ to be absolutely continuous. Adding this to (7·54) we conclude that the relation

$$\lim_{r\to 1}\int_0^{2\pi} |\log|F(re^{ix})||\,dx = \int_0^{2\pi} |\log|F(e^{ix})||\,dx, \tag{7·56}$$

is valid if and only if λ is absolutely continuous.

We conclude this section by a few remarks about the classes H^p as abstract spaces. For $F(z)\in \mathrm{H}^p$ we set

$$\| F\|_p = \lim_{r\to 1}\mu_p^{1/p}(r;F) = \left\{\frac{1}{2\pi}\int_0^{2\pi} |F(e^{ix})|^p\,dx\right\}^{1/p}. \tag{7·57}$$

Then, except for the irrelevant factor $(2\pi)^{-1/p}$, $\| F\|_p$ is the usual norm, in L^p, of the boundary function $F(e^{ix})$. Obviously $\| F\|_p \geqslant 0$, and $\| F\|_p = 0$ if and only if $F\equiv 0$ (see (7·25)). Also $\| kF\|_p = |k|\,\| F\|_p$. The triangle inequality

$$\| F+G\|_p \leqslant \| F\|_p + \| G\|_p$$

is satisfied for $p\geqslant 1$ but not necessarily for $0<p<1$. In the latter case, if we define the distance $d(F,G)$ of two points F and G in H^p by the formula

$$d(F,G) = \| F-G\|_p^p = \frac{1}{2\pi}\int_0^{2\pi} |F(e^{ix})-G(e^{ix})|^p\,dx,$$

this distance does satisfy the triangle inequality, and H^p is a metric space. It is convenient, however, to keep the definition (7·57) of norm even in the case $0<p<1$, though the triangle property is then lacking.

We know (see (7·34)), that if $F\in \mathrm{H}^p$, then

$$\int_0^{2\pi} |F(e^{ix})-F(Re^{ix})|^p\,dx \to 0 \quad \text{as} \quad R\to 1. \tag{7·58}$$

Let
$$F(z) = c_0 + c_1 z + \dots + c_n z^n + \dots \tag{7·59}$$

and let $\epsilon>0$ be given. If in (7·58) we fix R sufficiently close to 1, and then fix N sufficiently large, the polynomial $P(z) = \sum_0^N c_n R^n z^n$ will satisfy the inequality $\| F-P\|_p < \epsilon$. Thus, with the metric (7·57), the set of all polynomials $d_0 + d_1 z + \dots + d_n z^n$ is dense in H^p. Since these polynomials belong to H^p, this class can be defined as the closure under the metric (7·57) of the set of all polynomials. Since we could have required that the coefficients d_k be rational, *the space* H^p *is separable.*

(7·60) THEOREM. *The space* H^p *is complete.*

We have to show that if $F_n \in \mathrm{H}^p$ for $n=1,2,\dots$ and if $\| F_m - F_n\|_p \to 0$ as $m,n\to\infty$, then there is a $\Phi\in \mathrm{H}^p$ such that $\| F_n - \Phi\|_p \to 0$.

We show first that, *if* $F\in \mathrm{H}^p$ *and if* $\| F\|_p \leqslant M$, *then*

$$|F(z)| \leqslant M(1-R)^{-1/p} \quad \textit{for} \quad |z|\leqslant R<1. \tag{7·61}$$

If $p=1$, (7·59) gives

$$|c_n|\,r^n \leqslant \frac{1}{2\pi}\int_0^{2\pi} |F(re^{ix})|\,dx \leqslant M \quad (n=0,1,\dots),$$

and making $r\to 1$ we have $|c_n|\leqslant M$. Hence

$$|F(z)| \leqslant M(1+R+\dots) = M/(1-R) \quad \text{for} \quad |z|\leqslant R.$$

If $p \neq 1$, we set $F = GB$, where B is the Blaschke product of F and $\| G \|_p \leqslant M$. The latter inequality can be written $\| G^p \|_1 \leqslant M^p$, and so, by (7·61) with $p = 1$, we have $| G^p(z) | \leqslant M^p/(1-R)$ for $|z| \leqslant R$. Since $|B| \leqslant 1$, the function $F = GB$ satisfies (7·61).

The hypotheses $F_n \in \mathrm{H}^p$ and $\| F_m - F_n \|_p \to 0$ give $\| F_m \|_p \leqslant M$ for all m and some M. By (7·61), the functions F_m are uniformly bounded in each circle $|z| \leqslant R < 1$. Hence† we can select from $\{F_m\}$ a subsequence converging uniformly in $|z| \leqslant R' < R$. Applying the diagonal procedure, we may suppose that this subsequence $\{F_{m_k}\}$ converges uni formly in each circle $|z| \leqslant R$, $R < 1$, to a function $\Phi(z)$ regular in $|z| < 1$. Let

$$\epsilon_N = \sup \| F_m - F_n \|_p \quad \text{for} \quad m, n \geqslant N;$$

hence $\epsilon_N \to 0$. If $R < 1$,

$$\frac{1}{2\pi}\int_0^{2\pi} | \Phi(R\,e^{ix}) - F_n(R\,e^{ix}) |^p \,dx = \lim_{k\to\infty} \frac{1}{2\pi}\int_0^{2\pi} | F_{m_k}(R\,e^{ix}) - F_n(R\,e^{ix}) |^p \,dx \leqslant \epsilon_n^p,$$

which shows that $\| \Phi - F_n \|_p \leqslant \epsilon_n$ and completes the proof of (7·60).

8. Power series of bounded variation

We shall now show that if $f(x)$ and $\tilde{f}(x)$ are both of bounded variation then $S[f]$ has a number of interesting properties. It will be convenient to state the results in a form bearing on the power series

$$F(z) = a_0 + a_1 z + a_2 z^2 + \dots \quad (|z| < 1). \tag{8·1}$$

We shall say that (8·1) is *of bounded variation* if its real and imaginary parts, for $z = e^{ix}$, are Fourier series of functions of bounded variation. We know (see p. 89) that $F(e^{ix}) = \lim_{r\to 1} F(r\,e^{ix})$ is then continuous. Consequently, (8·1) converges uniformly for $|z| = 1$, and so also for $|z| \leqslant 1$.

(8·2) THEOREM. *If $F(z)$ is of bounded variation, then $F(e^{ix})$ is absolutely continuous.*

An equivalent form of this is as follows:

(8·3) THEOREM. *If a trigonometric series S and its conjugate $\tilde{S}$ are both Fourier-Stieltjes series, then S and $\tilde{S}$ are ordinary Fourier series.*

To prove (8·3), let (8·1) be the power series which for $z = e^{ix}$ reduces to $S + i\tilde{S}$. By Chapter IV, (6·5), $F(z)$ is in H and so satisfies the first equation (7·34) with $p = 1$. Hence $S + i\tilde{S}$ is a Fourier series (Chapter IV, (6·12)) and (8·3) follows.

Part of this argument deserves a special statement: *a power series is of class* H *if and only if its real and imaginary parts for $z = e^{ix}$ are Fourier series.*

This remark, coupled with (7·35), gives a new proof of Theorem (4·4). Let $F(z)$ be the analytic function whose real part is the Poisson integral of an $f \in \mathrm{L}$. Then $F \in \mathrm{H}^\mu$, $0 < \mu < 1$, by (2·6). If we suppose that $\tilde{f} \in \mathrm{L}$, then $F(e^{ix}) = \lim F(re^{ix}) \in \mathrm{L}$, $F(z) \in \mathrm{H}$, and $\tilde{S}[f] = S[\tilde{f}]$.

† We use here the familiar fact that if the F_m are uniformly bounded in a circle their derivatives are uniformly bounded in each smaller concentric circle, so that the F_m are equicontinuous in the smaller circle.

From (8·2) we deduce the following:

(8·4) Theorem. *If the partial sums s_n of $\sum_{-\infty}^{+\infty} c_k e^{ikx}$ satisfy*

$$\int_0^{2\pi} |s_n| \, dx = O(1), \tag{8·5}$$

in particular if $s_n \geqslant 0$ for all n, then $c_k \to 0$.

The hypothesis implies that the series is an $\mathsf{S}[dF]$. Suppose that $|c_{n_\nu}| \geqslant \epsilon > 0$ for $n_1 < n_2 < \ldots \to \infty$. The series

$$\ldots + c_{n_\nu} + c_{n_\nu - 1} e^{-ix} + \ldots + c_{-n_\nu} e^{-2in_\nu x} + \ldots = e^{-in_\nu x} \mathsf{S}[dF],$$

$$c_{n_\nu} + c_{n_\nu - 1} e^{-ix} + \ldots + c_{-n_\nu} e^{-2in_\nu x} \qquad = e^{-in_\nu x} s_{n_\nu}(x),$$

are respectively $\mathsf{S}[dG_\nu]$ and $\mathsf{S}[dH_\nu]$, with

$$G_\nu(x) = \int_0^x e^{-in_\nu t} dF(t), \quad H_\nu(x) = \int_0^x e^{-in_\nu t} s_{n_\nu}(t) \, dt.$$

The total variations of G_ν and H_ν are uniformly bounded (see (8·5)). Taking subsequences, we may suppose that $\{G_\nu\}$ and $\{H_\nu\}$ converge to functions of bounded variation G and H. The coefficients of dG and dH are the limits of the corresponding coefficients of dG_ν and dH_ν. It follows that

(*a*) the constant term of $\mathsf{S}[dG]$ is numerically not less than ϵ;

(*b*) $\mathsf{S}[dH]$ has no terms with positive indices;

(*c*) the coefficients of $\mathsf{S}[dG]$ and $\mathsf{S}[dH]$ with non-positive indices are the same.

By (*b*) and (*c*), H and $G - H$ are absolutely continuous. Thus G is absolutely continuous. If we show that $G'(x) = 0$ almost everywhere, it will follow that $G(x) \equiv 0$, contradicting (*a*) and completing the proof.

Let $F = F_0 + F_1$, where F_0 and F_1 are the absolutely continuous and singular parts of F. By the Riemann-Lebesgue theorem,

$$G(x) = \lim_\nu \int_0^x e^{-in_\nu t} d(F_0 + F_1) = \lim_\nu \int_0^x e^{-in_\nu t} dF_1,$$

$$|G(x+h) - G(x)| = \left| \lim_\nu \int_x^{x+h} e^{-in_\nu t} dF_1 \right| \leqslant \int_x^{x+h} |dF_1| \quad (h > 0).$$

Since the last term is $o(h)$ for almost all x, $G'(x) = 0$ almost everywhere.

(8·6) Theorem. *A power series of bounded variation converges absolutely on $|z| = 1$.*

We begin by proving

(8·7) Theorem. *If $\Phi(z) = b_0 + b_1 z + \ldots \in \mathrm{H}$, then*

$$\Sigma \frac{|b_n|}{n+1} \leqslant \frac{1}{2} \int_0^{2\pi} |\Phi(e^{ix})| \, dx. \tag{8·8}$$

It is enough to prove this for functions regular in $|z| \leqslant 1$ and without zeros on $|z| = 1$. For then, applying the result to $\Phi(Rz)$, with $0 < R < 1$. and making $R \to 1$ we get (8·8) for general $\Phi \in \mathrm{H}$.

The inequality (8·8) is immediate if the b_n are all real and non-negative. For then, multiplying both sides of the equation

$$b_0 \sin x + b_1 \sin 2x + \ldots = \mathscr{I}\{e^{ix}\Phi(e^{ix})\} \tag{8·9}$$

by $\frac{1}{2}(\pi - x)$, integrating the result over $(0, 2\pi)$ (the left-hand side of (8·9) converges uniformly), and noting that the nth sine coefficient of $\frac{1}{2}(\pi - x)$ is $1/n$, we get

$$\Sigma \frac{b_n}{n+1} = \frac{1}{\pi}\int_0^{2\pi} \mathscr{I}\{e^{ix}\Phi(e^{ix})\}\,\tfrac{1}{2}(\pi - x)\,dx \leqslant \frac{1}{2}\int_0^{2\pi} |\Phi(e^{ix})|\,dx.$$

When the b_n are not all non-negative it is enough to construct a function $\Phi^*(z) = \Sigma b_n^* z^n$ such that $|b_n| \leqslant b_n^*$ and $\mu(1; \Phi^*) \leqslant \mu(1; \Phi)$ (see (7·1)).

Let $B(z)$ be the Blaschke product for $\Phi(z)$. In our case $B(z)$ has only a finite number of factors, the function $\Psi = \Phi/B$ is regular and non-zero for $|z| \leqslant 1$, and $|\Psi(e^{ix})| = |\Phi(e^{ix})|$. Set

$$\Phi_1 = B\Psi^{\frac{1}{2}} = \Sigma c_n' z^n, \quad \Phi_2 = \Psi^{\frac{1}{2}} = \Sigma c_n'' z^n \quad (\text{so that } \Phi = \Phi_1\Phi_2),$$

$$\Phi_1^* = \Sigma |c_n'| z^n, \quad \Phi_2^* = \Sigma |c_n''| z^n,$$

$$\Phi^* = \Phi_1^*\Phi_2^* = \Sigma b_n^* z^n.$$

The functions Φ_k, and so also the Φ_k^*, are regular for $|z| \leqslant 1$, $k = 1, 2$. Obviously

$$b_n^* = \Sigma |c_\nu'|\,|c_{n-\nu}''| \geqslant |\Sigma c_\nu' c_{n-\nu}''| = |b_n|.$$

Moreover, using Schwarz's inequality and the equation $\mu_2(1; \Phi_k) = \mu_2(1; \Phi_k^*)$ (a consequence of Parseval's formula),

$$\mu(1; \Phi^*) \leqslant \mu_2^{\frac{1}{2}}(1; \Phi_1^*)\,\mu_2^{\frac{1}{2}}(1; \Phi_2^*) = \mu_2^{\frac{1}{2}}(1; \Phi_1)\,\mu_2^{\frac{1}{2}}(1; \Phi_2) = \mu(1; \Psi) = \mu(1; \Phi).$$

This proves (8·7).

Returning to (8·6), let us suppose that $F(z)$ is of bounded variation. We apply (8·8) to $\Phi(z) = F'(Rz) = a_1 + 2a_2 Rz + \ldots$, where $0 < R < 1$. (Thus we use (8·8) only in the case when Φ is regular for $|z| \leqslant 1$.) We get

$$|a_1|R + |a_2|R^2 + \ldots \leqslant \frac{1}{2}\int_0^{2\pi} R\,|F'(Re^{ix})|\,dx.$$

The integral on the right is the total variation of $F(z)$ over $|z| = R$ and tends to the total variation V of $F(z)$ over $|z| = 1$ as $R \to 1$ (Chapter IV, (6·11)). Hence

$$|a_1| + |a_2| + \ldots \leqslant \tfrac{1}{2}V, \tag{8·10}$$

which is a more precise formulation of Theorem (8·6).

A corollary of (8·6) and of (2·8) is Theorem (3·9) of Chapter VI, stated there without proof: *If F is absolutely continuous and $F' \in \mathrm{L}\log^+\mathrm{L}$, then $S[F]$ converges absolutely.*

(8·11) THEOREM. *If s_n are the partial sums of $\sum_{-\infty}^{+\infty} c_k e^{ikx}$, and if*

$$\int_0^{2\pi} |s_n(x)|\,dx = O(1), \tag{8·12}$$

then

$$\frac{1}{n}\sum_{k=1}^{n} |c_k| = O\left(\frac{1}{\log n}\right). \tag{8·13}$$

In particular, if $|c_1| \geqslant |c_2| \geqslant \ldots$, then $c_n = O(1/\log n)$.

Let M be the upper bound of the integrals (8·12). Applying (8·8) to

$$e^{-i\nu x}s_\nu(x) = c_\nu + c_{\nu-1}e^{-ix} + \ldots + c_{-\nu}e^{-2i\nu x} \quad (\nu > 0),$$

we obtain

$$\frac{|c_\nu|}{1} + \frac{|c_{\nu-1}|}{2} + \ldots + \frac{|c_1|}{\nu} \leqslant \tfrac{1}{2}M.$$

Replace here ν by $2n, 2n-1, 2n-2, \ldots, 1$ and add the inequalities. Observing that $\sum_1^\mu k^{-1} \geqslant A\log(\mu+1)$ $(\mu = 1, 2, \ldots)$, we have

$$A\sum_{k=1}^{2n}|c_k|\log(2n-k+2) \leqslant \tfrac{1}{2}M\,.\,2n,$$

whence

$$A\log n\sum_{k=1}^{n}|c_k| \leqslant Mn,$$

and (8·13) follows.

If (8·12) is replaced by the stronger condition $\mathfrak{M}[s_m - s_n] \to 0$, we have (8·13) with '$o$' instead of '$O$'.

9. Cauchy's integral

Let $g(z)$ be any integrable function defined on the circumference $|z| = 1$. The integral

$$\frac{1}{2\pi i}\int_{|\zeta|=1}\frac{g(\zeta)}{\zeta - z}\,d\zeta, \tag{9·1}$$

exists for $|z| \neq 1$; it defines one function, $F(z)$, regular for $|z| < 1$, and another, $F_1(z)$, regular for $|z| > 1$. Clearly,

$$F(z) = \sum_{n=0}^{\infty}\frac{z^n}{2\pi i}\int_{|\zeta|=1}\zeta^{-n-1}g(\zeta)\,d\zeta \quad (|z| < 1), \tag{9·2}$$

$$F_1(z) = -\sum_{n=1}^{\infty}\frac{z^{-n}}{2\pi i}\int_{|\zeta|=1}\zeta^{n-1}g(\zeta)\,d\zeta \quad (|z| > 1). \tag{9·3}$$

In general, F and F_1 are not analytic continuations of each other.

Let

$$z = r\,e^{ix}, \quad z^* = 1/\bar{z} = r^{-1}\,e^{ix}.$$

We verify that

$$F(z) - F_1(z^*) = \frac{1}{2\pi i}\int_{|\zeta|=1}g(\zeta)\left\{\frac{1}{\zeta - z} - \frac{1}{\zeta - z^*}\right\}d\zeta = \frac{1}{\pi}\int_0^{2\pi}g(e^{it})\,P(r, t-x)\,dt. \tag{9·4}$$

Hence

$$F(z) - F_1(z^*) \to g(e^{ix_0}) \tag{9·5}$$

for almost all x_0, as $z \to e^{ix_0}$ along any non-tangential path.

Formula (9·1) may be written

$$F(r\,e^{ix}) = \frac{1}{2}\left[\frac{1}{\pi}\int_0^{2\pi}g(e^{it})\{P(r, t-x) + \tfrac{1}{2}\}\,dt + \frac{i}{\pi}\int_0^{2\pi}g(e^{it})\,Q(r, t-x)\,dt\right]. \tag{9·6}$$

Thus

$$F(e^{ix_0}) = \lim_{z\to e^{ix_0}}F(z),$$

where $z \to e^{ix_0}$ along a non-tangential path, exists for almost all x_0. By (9·5), a similar result holds for F_1 and non-tangential paths in $|z| > 1$.

From (9·6) and (2·6) we see that *the function $F(z)$ belongs to H^μ for every $0 < \mu < 1$.*

(9·7) Theorem. *Let $F(z)$ be defined for $|z|<1$ as the expression (9·1). A necessary and sufficient condition for $F(e^{ix_0}) = \lim_{z\to e^{ix_0}} F(z)$ to coincide with $g(e^{ix_0})$ almost everywhere is that $S[g(e^{ix})]$ is of power series type; that is,*

$$\int_{|\zeta|=1} \zeta^{n-1} g(\zeta)\, d\zeta = 0 \quad \textit{for} \quad n = 1, 2, \ldots. \tag{9·8}$$

By (9·3), condition (9·8) is equivalent to $F_1(z)\equiv 0$. But if $F_1(z)\equiv 0$, (9·4) shows that $F(z)$ is the Poisson integral of $g(e^{ix})$, and so $F(e^{ix}) = g(e^{ix})$ almost everywhere.

Conversely, suppose that $\lim_{r\to 1} F(re^{ix}) = g(e^{ix})$ almost everywhere. Since $g\in L$ and $F\in H^\mu$, $0<\mu<1$, it follows from (7·35) that $F\in H$ and so $\mathfrak{M}[F(re^{ix}) - g(e^{ix})]\to 0$ as $r\to 1$. Hence the Fourier coefficients of $F(re^{ix})$ tend to the corresponding coefficients of $g(e^{ix})$ as $r\to 1$. Since the coefficients of $F(re^{ix})$ with negative indices are zero, the same holds for $g(e^{ix})$. This completes the proof of (9·7).

We shall say that a function $\Phi(z)$, regular for $|z|<1$, is *representable by the Cauchy integral*, if

(i) $\Phi(e^{ix}) = \lim_{r\to 1}\Phi(re^{ix})$ exists for almost all x and is integrable;

(ii) $\Phi(z)$ is given by the integral (9·1) with $g(\zeta)$ replaced by $\Phi(\zeta)$.

(9·9) Theorem. *A function $F(z)$, regular for $|z|<1$, is representable by the Cauchy integral if and only if $F(z)\in H$.*

If $F(z)$ is given by (9·1), then $F\in H^\mu$, $0<\mu<1$, and the hypothesis that $\lim F(re^{ix})\in L$ implies that $F\in H$ (see (7·35)).

Conversely, suppose that $F(z) = \Sigma c_n z^n \in H$. Since $\mathfrak{M}[F(re^{ix}) - F(e^{ix})]\to 0$, we have

$$\int_0^{2\pi} F(e^{ix})\, e^{-inx}\, dx = \lim_{r\to 1}\int_0^{2\pi} F(re^{ix})\, e^{-inx}\, dx = \lim_{r\to 1} 2\pi c_n r^n = 2\pi c_n,$$

so that $F(z)$ is given by (9·2)—and so also by (9·1)—with $g(\zeta)$ replaced by $F(\zeta)$.

(9·10) Theorem. *Let $F(z)$ be regular for $|z|<1$. Then the following two conditions are equivalent:*

(i) *$F(z)$ is representable by the Cauchy integral;*

(ii) *$F(z)$ is the Poisson integral of $F(e^{ix})$.*

Both conditions imply that $F(e^{ix}) = \lim_{r\to 1} F(re^{ix})$ exists almost everywhere and is integrable. If $F(z)$ is the Poisson integral of $F(e^{ix})$ then $\mathfrak{M}[F(re^{ix})] = O(1)$. Hence $F\in H$ and, by (9·9), $F(z)$ is representable by the Cauchy integral. Conversely, if $F(z)$ is given by (9·1) with $g(e^{it}) = F(e^{it})$ then, by (9·8), the function F_1 formed with $g(e^{it}) = F(e^{it})$ vanishes for $|z|>1$. Substituting $F(e^{it})$ for $g(e^{it})$ in (9·4), we see that $F(z)$ is then the Poisson integral of $F(e^{ix})$.

10. Conformal mapping

The method of conformal mapping is useful for some problems of trigonometric series. We give here some basic facts which we shall need later.

Let

$$w = F(z) = c_0 + c_1 z + c_2 z^2 + \ldots + c_n z^n + \ldots \tag{10·1}$$

be a function regular and univalent for $|z| \leqslant R$. (*Univalent* means that to distinct z's there correspond distinct values of $F(z)$.) Then $F(z)$ maps the circle $|z| < R$ in a one-one way onto an open domain D bounded by a simple closed curve. If $w = u + iv$, the Jacobian $\partial(u, v)/\partial(x, y)$ is

$$\begin{vmatrix} u_x & u_y \\ v_x & v_y \end{vmatrix} = u_x^2 + u_y^2 = |F'(z)|^2, \qquad (10\cdot 2)$$

by the Cauchy-Riemann equations, and the area $|D|$ of D is

$$\int_0^R \int_0^{2\pi} |F'(r e^{ix})|^2 r\, dr\, dx = \int_0^R r\, dr \int_0^{2\pi} |\Sigma n c_n r^{n-1} e^{i(n-1)x}|^2 dx$$
$$= 2\pi \int_0^R (\Sigma n^2 |c_n|^2 r^{2n-1})\, dr = \pi \Sigma n |c_n|^2 R^{2n}. \qquad (10\cdot 3)$$

If F is regular and univalent for $|z| < R$ only, then the area of D is the limit as $R' \to R$ of the area of $D_{R'}$, the image of $|z| < R'$ for $R' < R$. Replacing R by R' in (10·3), and making $R' \to R$, we have

$$|D| = \iint_{|z|<R} |F'(r e^{ix})|^2 r\, dr\, dx = \pi \Sigma n |c_n|^2 R^{2n}, \qquad (10\cdot 4)$$

the right-hand side being finite or infinite. If $F(z)$ is regular but not necessarily univalent for $|z| < R$, the integral in (10·4) is, *by definition*, the area of the image of $|z| < R$ by the mapping $w = F(z)$. For $R = 1$, we get

$$|D| = \pi \sum_1^\infty n |c_n|^2. \qquad (10\cdot 5)$$

(10·6) THEOREM. *If the function* (10·1) *is regular for* $|z| < 1$, *and if* $\Sigma n |c_n|^2$ *is finite, then the series*

$$c_0 + c_1 e^{ix} + \dots + c_n e^{inx} + \dots \qquad (10\cdot 7)$$

converges for almost all x. *If in addition* $F(z)$ *can be extended so as to be continuous for* $|z| \leqslant 1$, *then* (10·7) *converges uniformly in* $0 \leqslant x \leqslant 2\pi$.

For the finiteness of (10·5) implies that of $\Sigma |c_n|^2$, and so (10·7) is a Fourier series. Thus the latter series is summable (C, 1) almost everywhere and, by the remark on p. 79, converges at every point of summability. If $F(z)$ is continuous for $|z| \leqslant 1$, then (10·7) is uniformly summable A to $F(e^{ix})$, and so is $S[F(e^{ix})]$. It is therefore uniformly summable (C, 1) and hence, owing to the finiteness of (10·5), uniformly convergent.

The following result is classical:

(10·8) THEOREM OF RIEMANN. *Let* D *be any simply connected domain whose boundary contains at least two points. Then given any point* $w_0 \in D$ *there is always a function* $F(z)$, *regular and univalent for* $|z| < 1$, *which maps* $|z| < 1$ *onto* D *in such a way that* $F(0) = w_0$.

We take this result for granted, since there are several proofs in the existing literature. By *simply connected* we mean a domain whose complement is a continuum. In our applications, we shall be interested only in the special case when D is the interior of a simple closed curve. In that case (10·8) can be completed as follows:

(10·9) THEOREM. *Let* D *be the interior of a simple closed curve* C. *Then the function* $F(z)$ *of* (10·8) *can be extended as a continuous function to* $|z| \leqslant 1$, *and the extended function gives a one-one bicontinuous correspondence between* $|z| \leqslant 1$ *and the closure* $\overline{D}$ *of* D.

We give the proof of (10·9) here. It is a simple consequence of the following facts:

(*a*) There is a dense sequence $\{x_n\}$ of numbers, all distinct, such that $\lim_{r\to 1} F(r e^{ix_n})$, which we shall denote by $F(e^{ix_n})$, exists for each n.

(*b*) The numbers $F(e^{ix_n})$ are all different and all situated on C.

(*c*) The set of the numbers $F(e^{ix_n})$ is dense on C.

(*d*) If $|z^{(k)}|\to 1$, $|z^{(k)}|<1$, then all the limit points of the sequence $\{F(z^{(k)})\}$ are on C.

We begin with the deduction of (10·9) from (*a*), (*b*), (*c*) and (*d*).

Let

$$w=\omega(e^{it}) \quad (0\leqslant t\leqslant 2\pi) \tag{10·10}$$

be the equation of C, and suppose that C is described in the positive direction as t increases from 0 to 2π. By (*b*), we can write $F(e^{ix_n})=\omega(e^{it_n})$. Since the mapping $w=F(z)$ is conformal at $z=0$, it follows that if e^{ix_m}, e^{ix_n}, e^{ix_p} are in a given order on $|z|=1$, then $F(e^{ix_m})$, $F(e^{ix_n})$, $F(e^{ix_p})$ are in the same order on C.

Let $z^*=e^{ix^*}$, and let $x_m<x^*<x_n$. The image of the sector $x_m<x<x_n$, $0<r<1$, is the interior of the curvilinear triangle limited by the curves $w=F(r e^{ix_m})$, $w=F(r e^{ix_n})$, $0<r<1$, and by the arc $t_m<t<t_n$ of C. It follows, in view of (*c*), that we can find x_m, x_n with $x_m<x^*<x_n$ and such that t_m, t_n are arbitrarily close. From this it follows that to a nested sequence of intervals (x_m, x_n) converging to x^* there corresponds a nested sequence of intervals (t_m, t_n) converging to a point t^*.

In view of (*d*) it now follows that if $\{z^{(k)}\}$ is any sequence of points with $|z^{(k)}|<1$, converging to z^*, then $F(z^{(k)})$ tends to $\omega(e^{it^*})$. Thus F can be extended as a continuous function to $|z|\leqslant 1$, and the values of $F(e^{ix})$ are all on C.

By (*c*), the values of $F(e^{ix})$ cover the whole of C. To show that to different x correspond different $F(e^{ix})$, suppose that $x'<x''$, $F(e^{ix'})=F(e^{ix''})$. If $x'<x_m<x_n<x''$, then *a fortiori* $F(e^{ix_m})=F(e^{ix_n})$, contrary to (*b*).

It remains to show that the function inverse to $F(z)$ is continuous in $\bar{D}$. This follows from the fact (an immediate consequence of the Bolzano–Weierstrass theorem) that a one-one mapping of a closed bounded set which is one way continuous is bicontinuous.

We shall now prove (*a*), (*b*), (*c*) and (*d*), beginning with (*d*). Suppose it is false. By selecting a subsequence of $z^{(k)}$, we may assume that $F(z^{(k)})$ tends to a point w_* in the interior of C. Since the inverse of the function $F(z)$ maps the neighbourhood of w_* onto a neighbourhood of a point z_*, $|z_*|<1$, it follows that all the points $z^{(k)}$, from some point onwards, are in an arbitrarily small neighbourhood of z_*. This contradicts the hypothesis that $|z^{(k)}|\to 1$.

To prove (*a*) it is enough to observe that the finiteness of (10·5) implies that of $\Sigma|c_n|^2$, so that (10·7), *qua* Fourier series, is summable A almost everywhere. A different argument, which avoids the use of the Riesz-Fischer theorem, runs as follows. The integral (10·4), with $R=1$, is finite. By Fubini's theorem, $|F'|^2$ is integrable on almost every radius, and so, by Schwarz's inequality, this holds for $|F'|$. For every such radius the curve $w=F(r e^{ix})$, $0\leqslant r<1$, is of finite length, and so $\lim_{r\to 1} F(r e^{ix})=F(e^{ix})$ exists.

That $F(e^{ix})$ is on C follows from (*d*). Suppose that $x'<x''$, and that $F(e^{ix'})=F(e^{ix''})=w$. Then, by (*d*), the function $F(z)-w$ tends uniformly to 0 as z approaches any arc $\alpha\leqslant x\leqslant\beta$ interior to (x', x''). The proof of (*b*) will therefore be complete if we prove the following lemma.

(10·11) Lemma. *Let $G(z)$ be any function regular and bounded for $|z|<1$. If $G(z)$ tends to 0 as z approaches an arc of the unit circle, then $G(z)\equiv 0$.*

This is a very special case of (7·25), but an elementary proof is immediate. Dividing G if necessary by z^k, we may suppose that $G(0)\neq 0$. Suppose that an arc of $|z|=1$ on which G vanishes has length greater than $2\pi/n$, and let $\epsilon=\exp(2\pi i/n)$. Then

$$H(z)=G(z)\,G(\epsilon z)\ldots G(\epsilon^{n-1}z)$$

is regular for $|z|<1$ and tends uniformly to 0 as $|z|\to 1$. Hence $H(z)\equiv 0$, which contradicts the inequality $H(0)=G^n(0)\neq 0$. This proves the lemma and so also (*b*).

It remains to prove (*c*). Let X, $|X|=2\pi$, be the set of angles x such that

$$F(e^{ix})=\lim F(r\,e^{ix})$$

exists. As the proofs of (*a*) show, (10·7) is $\mathfrak{S}[F(e^{ix})]$. (The second proof of (*a*) shows that $c_0+c_1r\,e^{ix}+\ldots$ tends almost everywhere to $F(e^{ix})$ as $r\to 1$, and is majorized by an integrable function.) The correspondence between X and points on C gives a function $t=t(x)$, defined and strictly increasing on X, if X is repeated periodically and $t(x)$ is correspondingly extended by the formula $t(x+2\pi)=t(x)+2\pi$. If (*c*) were false, the function $t(x)$, $x\in X$, would have jumps, and the same would hold for the bounded periodic function $w(t(x))=F(e^{ix})$, $x\in X$. This is impossible (see p. 89), and the proof of (10·9) is complete.

(10·12) Theorem. *If $F(z)=\Sigma c_n z^n$ maps $|z|<1$ conformally onto the interior of a simple closed curve C, then $\Sigma c_n e^{inx}$ converges uniformly in $0\leqslant x\leqslant 2\pi$.*

This is a corollary of (10·6).

(10·13) Theorem. *If the curve C limiting the domain D in* (10·9) *has a tangent at a point $w_1\in C$, then the mapping $w=F(z)$ is angle preserving at the point z_1 which corresponds to w_1.*

This means that if Γ' and Γ'' are any two curves approaching z_1 from $|z|<1$, having tangents at z_1 and intersecting at an angle α there, then the images C', C'' of Γ', Γ'' have tangents at w_1 and intersect at an angle α.

The existence of the tangent to C at w_1 means that when w tends to w_1 along C, the ray $\overline{w_1w}$ tends to a limiting position. In other words, as $z=e^{ix}$ approaches $z_1=e^{ix_1}$ from either side, the expression

$$g(z)=\arg\{F(z)-F(z_1)\}$$

tends to a limit, and the two limits differ by π.

The function $g(z)$, being the imaginary part of the function $\log\{F(z)-F(z_1)\}$ regular in $|z|<1$, is harmonic for $|z|<1$. It is continuous on $|z|=1$, for $z\neq z_1$. It is obviously bounded in the part of $|z|<1$ outside any given neighbourhood of z_1. It is also bounded in a neighbourhood of z_1, owing to the existence of the tangent to C at w_1. Thus $g(z)$ is harmonic and bounded for $|z|<1$, and so is the Poisson integral of $g(e^{ix})$.

The function $g(e^{ix})$ has jump $d=\pi$ at $x=x_1$. By Chapter III, (6·15), $\lim g(z)$ exists as z tends to z_1 along Γ' or Γ'', and by the same theorem the two limits differ by $\alpha d/\pi=\alpha$. The existence of the limits of $g(z)$ along Γ' and Γ'' implies the existence of tangents to C' and C'' at w_1. The angle between these tangents is α, and (10·13) follows.

Suppose now that the curve C in (10·12) is rectifiable. Then the function $F(e^{ix})$ is of bounded variation over $0 \leqslant x \leqslant 2\pi$. Using (8·6), we can therefore complete (10·12) as follows.

(10·14) THEOREM. *If the boundary of D is rectifiable, the power series for $F(z)$ converges absolutely on $|z| = 1$.*

Similarly, using (8·2), we have:

(10·15) THEOREM. *If the boundary of D is rectifiable, the function $F(e^{ix})$ is absolutely continuous.*

Incidentally, in the case which interests us the proofs of (8·2) and (8·6) simplify, since $F'(z) \neq 0$ and so the Blaschke products do not occur.

Let C be rectifiable, and let s be the arc length of C measured from a fixed point on C. We may take s as parameter and write (10·10) in the form

$$w = w(s) = u(s) + iv(s), \quad 0 \leqslant s \leqslant l, \; w(0) = w(l),$$

l being the length of C. To any point set A on C corresponds a set S on $(0, l)$. We shall call A measurable if S is, and the measure of A will be defined as that of S.

Let now $E = (\alpha, \beta)$ be any arc on $|z| = 1$, and let A be the image of E on C. The length of A is the total variation of $F(e^{ix})$ over E, and since $F(e^{ix})$ is absolutely continuous,

$$|A| = \int_E \left| \frac{dF(e^{ix})}{dx} \right| dx. \tag{10·16}$$

This formula holds if E is a sum of a finite number of non-overlapping open intervals, and so also if E is any open set on $|z| = 1$. Passing to complementary sets, we get (10·16) when E is closed. Since every measurable set E contains a closed set, and is contained in an open set, of measures arbitrarily close to that of E, (10·16) is valid for a general measurable set E on $|z| = 1$.

In particular, $|E| = 0$ implies $|A| = 0$.

Conversely, suppose that $|A| = 0$. Then $dF(e^{ix})/dx = 0$ almost everywhere in E. By Chapter III, (7·6), if $dF(e^{ix})/dx = 0$ for $x = x_0$ then $F'(z) \to 0$ as z tends to e^{ix_0} along a non-tangential path. By (7·25), this can happen only in a set of measure zero, since $F' \in \mathrm{H}$. Hence $|E| = 0$. Thus we have:

(10·17) THEOREM. *If C is rectifiable, then to sets of measure zero on $|z| = 1$ correspond sets of measure zero on C, and conversely.*

At every point $w_0 = F(e^{ix_0})$ such that $dF(e^{ix})/dx$ exists and is different from zero, the curve C has a tangent. We have just proved that $dF(e^{ix})/dx \neq 0$ almost everywhere. Hence C has a tangent almost everywhere (a result which, of course, also follows from the classical theorem about the existence of the tangent almost everywhere on a rectifiable curve). By (10·13), the mapping $w = F(z)$ is angle preserving almost everywhere on $|z| = 1$.

Let $z_0 = e^{ix_0}$ be such that $dF(e^{ix})/dx$ exists and has a finite value μ there. Then $izF'(z)$ tends to μ as $z \to z_0$ non-tangentially. Let z_1 be any point of the segment $\overline{z_0 z}$. Then

$$\frac{F(z) - F(z_0)}{z - z_0} = \lim_{z_1 \to z_0} \frac{F(z) - F(z_1)}{z - z_0} = \frac{1}{z - z_0} \lim_{z_1 \to z_0} \int_{z_1}^{z} F'(\zeta)\, d\zeta = \frac{1}{z - z_0} \int_{z_0}^{z} F'(\zeta)\, d\zeta,$$

where the last integral is taken along the segment $\overline{z_0 z}$. Clearly, the last ratio tends to μ/iz_0 as $z \to z_0$ non-tangentially. Hence, *at almost every point* z_0, $|z_0| = 1$, *the function* $F(z)$ *has an angular derivative*

$$\lim \frac{F(z) - F(z_0)}{z - z_0}, \quad \textit{where} \quad z \to z_0 \textit{ non-tangentially},$$

and this derivative is distinct from zero. In particular, the lengths $|z - z_0|$ and $|w - w_0|$ are asymptotically proportional as $z \to z_0$ non-tangentially.

We conclude by a simple application of conformal mapping to the problem of convergence of Fourier series.

It will be shown in the next chapter that for a continuous f, $S[f]$ need not converge everywhere, still less converge uniformly. However:

(10·18) THEOREM. *If $f(t)$ is continuous and periodic, there is a strictly increasing function $t = t(\theta)$, $0 \leqslant \theta \leqslant 2\pi$, mapping the interval $(0, 2\pi)$ onto itself and such that the Fourier series of $g(\theta) = f(t(\theta))$ converges uniformly.*

If there is a continuous and periodic function $\phi(t)$ such that

$$u = f(t), \quad v = \phi(t), \quad 0 \leqslant t \leqslant 2\pi,$$

represents a simple closed curve C, then the conclusion of (10·18) is a consequence of (10·12). For C can then also be given by an equation $w = F(e^{i\theta})$ where $F(z)$ is a function regular for $|z| < 1$, continuous for $|z| \leqslant 1$, with $S[F(e^{i\theta})]$ uniformly convergent. The mapping $t = t(\theta)$ is one-one and continuous. If we select F so that $\theta = 0$ for $t = 0$, then $t(\theta)$ has the required properties.

The function $\phi(t)$ need not always exist, but it does exist if, for example,

$$f(0) = f(a) = f(2\pi) \quad \text{for some} \quad 0 < a < 2\pi$$

and if f is strictly less than $f(0)$ inside $(0, a)$ and strictly greater than $f(2\pi)$ inside $(a, 2\pi)$, or conversely. For then we may take for ϕ any function strictly increasing in $(0, a)$ and strictly decreasing in $(a, 2\pi)$.

It is therefore enough to show that in the general case there exists a function $\omega(t)$, $0 \leqslant t \leqslant 2\pi$, with $\omega(0) = \omega(2\pi)$, continuous and of bounded variation, such that $f_1 = f + \omega$ is of the type just mentioned. For, clearly, $S[\omega(t(\theta))]$ is uniformly convergent for any increasing $t(\theta)$ mapping $(0, 2\pi)$ onto itself.

By subtracting a constant from f, we may suppose that $\int_0^{2\pi} f dt = 0$. We may also suppose that $f(0) = f(2\pi) = 0$. (Otherwise subtract from f the function $f(0) \cos t$, whose integral over $(0, 2\pi)$ is zero and which may be incorporated in $\omega(t)$.) It follows that f must vanish at some point a in the interior of $(0, 2\pi)$. Let M_1 and M_2 respectively be $\max |f(t)|$ in the intervals $(0, a)$ and $(a, 2\pi)$, and let M_1, M_2 be attained at points t_1 and t_2 of these intervals. Define $\omega_1(t)$ as

$$\max_{0 \leqslant \xi \leqslant t} |f(\xi)| \quad \text{for} \quad 0 \leqslant t \leqslant t_1; \qquad \max_{t \leqslant \xi \leqslant a} |f(\xi)| \quad \text{for} \quad t_1 \leqslant t \leqslant a;$$

$$-\max_{a \leqslant \xi \leqslant t} |f(\xi)| \quad \text{for} \quad a \leqslant t \leqslant t_2; \qquad -\max_{t \leqslant \xi \leqslant 2\pi} |f(\xi)| \quad \text{for} \quad t_2 \leqslant t \leqslant 2\pi.$$

The function $\omega_1(t)$ is continuous and of bounded variation, with $\omega_1(0) = \omega_1(2\pi) = 0$. Moreover, $|f| \leqslant |\omega_1|$, so that $f + \omega_1$ is non-negative in $(0, a)$ and non-positive in $(a, 2\pi)$.

If $\omega_2(t)$ is any continuous function of bounded variation vanishing at the points $0, a, 2\pi$, strictly positive inside $(0, a)$, and strictly negative inside $(a, 2\pi)$, then the sum $\omega = \omega_1 + \omega_2$ has the required properties.

MISCELLANEOUS THEOREMS AND EXAMPLES

1. Let $F(z) = u + iv$ be regular for $|z| < 1$, and suppose that u and v are non-negative. Show that then

$$\int_0^{2\pi} |F(re^{ix})|^{2-\epsilon} dx \leqslant C_\epsilon |F(0)|^{2-\epsilon} \quad (0 < r < 1;\ 0 < \epsilon < 2).$$

If $\epsilon = 0$, the integral on the left need not be bounded.

It follows that if the real and imaginary parts u, v of $F(z), |z| < 1$, are both bounded below, then both $u(e^{ix})$ and $v(e^{ix})$ belong to $L^{2-\epsilon}$.

[Let $F_1(z) = e^{-\pi i/4} F(z) = u_1 + iv_1$, so that $|v_1| \leqslant u_1$. Apply to F_1 the argument used in the proof of (2·6). The function $F(z) = e^{\frac{1}{4}\pi i}\{(1+z)/(1-z)\}^{\frac{1}{2}}$, mapping $|z| < 1$ onto the first quadrant, is a counter-example for $\epsilon = 0$.]

2. Show that the constant A_p in (2·4) satisfies an inequality $A_p \geqslant Ap$ for $p \geqslant 2$. This, coupled with (3·9), shows that $A_p \sim Ap$ for $p \to \infty$. (Titchmarsh [4].)

[Consider $f(x) = \frac{1}{2}(\pi - x)$, $0 < x < 2\pi$, and observe that $\tilde{f}(x) = \log(2\sin\frac{1}{2}x) \sim \log x$ for $x \to +0$.]

3. For any $f \in \mathrm{L}$ and $0 < \epsilon < \pi$, let $\tilde{f}(x;\epsilon) = -\dfrac{1}{\pi}\displaystyle\int_\epsilon^\pi \dfrac{f(x+t) - f(x-t)}{2\tan\frac{1}{2}t}\, dt$. Then

$$(a) \qquad \tilde{f}(x;\epsilon) = \frac{1}{\pi^2}\int_0^{2\pi} \tilde{f}(x+t)\, \tfrac{1}{2}\cot\tfrac{1}{2}t \log\left|\frac{\sin\frac{1}{2}(t+\epsilon)}{\sin\frac{1}{2}(t-\epsilon)}\right| dt,$$

$$(b) \qquad \mathfrak{M}_p[\tilde{f}(x;\epsilon)] \leqslant (1 - \epsilon/\pi)\, \mathfrak{M}_p[\tilde{f}\,], \quad p \geqslant 1. \quad \text{(M. Riesz [1].)}$$

[Let $g(t)$ be the function equal to 0 in $(-\epsilon, \epsilon)$ and to $\frac{1}{2}\cot\frac{1}{2}t$ elsewhere in $(-\pi, \pi)$. Observe that, by Chapter IV, (8·6),

$$\pi\tilde{f}(x;\epsilon) = -\int_{-\pi}^{\pi} f(x+t)\, g(t)\, dt = -\int_{-\pi}^{\pi} \tilde{f}(x+t)\, \tilde{g}(t)\, dt,$$

and use Chapter II, p. 72, Example 20.]

4. Let Φ, Ψ and Φ_1, Ψ_1 be two pairs of Young's complementary functions. If for every $f \in \mathrm{L}^*_\Phi$, $\tilde{f}$ belongs to $\mathrm{L}^*_{\Phi_1}$ and if A is a constant independent of f such that $\|\tilde{f}\|_{\Phi_1} \leqslant A\|f\|_\Phi$ (such an A always exists), then for every $g \in \mathrm{L}^*_{\Psi_1}$ we have $\tilde{g} \in \mathrm{L}^*_\Psi$. Moreover,

$$\|\tilde{g}\|_\Psi \leqslant 2A\|g\|_{\Psi_1}.$$

[It is sufficient to show that if $\|v\|_{\Phi_1} \leqslant A\|u\|_\Phi$ for every $u + iv$ regular for $|z| < 1$ and satisfying $v(0) = 0$, then $\|v\|_\Psi \leqslant 2A\|u\|_{\Psi_1}$. If h is any trigonometric polynomial such that $\mathfrak{M}[\Phi(|h|)] \leqslant 1$, we have

$$\|v\|_\Psi = \sup_h \left|\int_0^{2\pi} vh\, dx\right| = \sup_h \left|\int_0^{2\pi} u\tilde{h}\, dx\right| \leqslant 2A\sigma\|u\|_{\Psi_1},$$

where $\sigma = \max\{1, \sup\mathfrak{M}[\Phi_1(|\tilde{h}|/2A)]\}$ (Chapter IV, (10·6)). On the other hand, since $\mathfrak{M}[\Phi(|h|)] \leqslant 1$, we have $\|h\|_\Phi \leqslant 2$, and so $\|\tilde{h}\|_{\Phi_1} \leqslant A\|h\|_\Phi \leqslant 2A$. Hence (Chapter IV, (10·4)) $\mathfrak{M}[\Phi_1(|\tilde{h}|/2A)] \leqslant 1$, $\sigma = 1$, and $\|v\|_\Psi \leqslant 2A\|u\|_{\Psi_1}$.]

5. Let $s(x)$ be concave and non-negative for $x \geqslant 0$, have a continuous derivative for $x > 0$, and tend to $+\infty$ with x. Let $S(x) = \displaystyle\int_0^x s(t)\, dt$. Let $R(x)$ be non-negative and convex for $x \geqslant 0$, have its first two derivatives continuous for $x > 0$, and tend to $+\infty$ with x. Suppose, in addition, that there is a constant C such that

$$S''(x) + S'(x)/x \leqslant CR''(x).$$

Then $f \in \mathrm{L}_R$ implies $\tilde{f} \in \mathrm{L}_S$.

[The proof is substantially the same as that in §3. Observe that $S(2x) \leqslant C_1 S(x)$, with C_1 independent of x. If $S(x) \leqslant R(x)$, then we have $\mathfrak{M}[S(|\tilde{f}|)] \leqslant C_2\mathfrak{M}[R(|f|)]$, with C_2 independent of f.]

6. If $\alpha>0$ and if $|f|(\log^+|f|)^\alpha\in L$, then $|\tilde f|\log^{\alpha-1}(2+|\tilde f|)\in L$, and

$$\int_0^{2\pi}|\tilde f|\log^{\alpha-1}(2+|\tilde f|)\,dx\leqslant A_\alpha\int_0^{2\pi}|f|(\log^+|f|)^\alpha\,dx+B_\alpha.$$

7. The preceding result is false for $\alpha=0$. (Titchmarsh [1].)
[Take $f(x)=\Sigma\cos nx/\log\log n$. By Chapter V, (2·17), $|\tilde f|/\log|\tilde f|$ is of the order

$$(x\log 1/x\log\log 1/x)^{-1}\quad\text{for}\quad x\to+0.$$

The true form of Example 6 for $\alpha=0$ is theorem (2·6).]

8. Example 6 has a converse analogous to (2·10), namely: if both $|f|(\log^+|f|)^{\alpha-1}$ and $|\tilde f|(\log^+|\tilde f|)^{\alpha-1}$ are integrable for an $\alpha\geqslant1$, and if f is bounded below, then $|f|(\log^+|f|)^\alpha\in L$. The result can also be stated in the form of an inequality analogous to (2·25).

9. If $\int_0^{2\pi}\exp|f|^\alpha\,dx<\infty$, where $\alpha>0$, then $\exp\lambda|\tilde f|^\beta$ is integrable for $\beta=\alpha/(\alpha+1)$ and $\lambda<\lambda_0(\alpha)$.

10. Let $f(r,x)$ be the Poisson-Stieltjes integral of a function $F(x)$ of bounded variation over $0\leqslant x\leqslant 2\pi$, F^* given by (1·7). Then, for $0<\mu<1$,

$$\text{(i)}\quad \mathfrak{M}_\mu[\tilde f(r,x)]\leqslant B_\mu\int_0^{2\pi}|dF|,$$

$$\text{(ii)}\quad \mathfrak{M}_\mu[\tilde f(r,x)-F^*(x)]\to0\quad(r\to1)$$

(see (2·39)). Similarly, if σ_n is the (C, 1) mean of $S[dF]$, then

$$\text{(iii)}\quad \mathfrak{M}_\mu[\tilde\sigma_n]\leqslant B_\mu\int_0^{2\pi}|dF|,$$

$$\text{(iv)}\quad \mathfrak{M}_\mu[\tilde\sigma_n-F^*]\to0\quad(n\to\infty).$$

[For (ii) and (iv) use Chapter I, (11·9).]

11. Let $0<\alpha<1$, $1\leqslant p\leqslant\infty$. Then (i) $f\in\Lambda^p_\alpha$ if and only if its Poisson integral $u(r,x)$ satisfies $\mathfrak{M}_p[u_x(r,x)]=O(\delta^{\alpha-1})$ (Hardy and Littlewood [10], [19]); (ii) $f\in\Lambda^p_*$ if and only if

$$\mathfrak{M}_p[u_{xx}(r,x)]=O(\delta^{-1}).$$

12. Let $f\in\Lambda^p_\alpha$, $q>p\geqslant1$. Then $f\in\Lambda^q_{\alpha-p^{-1}+q^{-1}}$.
[A corollary of (5·1) and of Chapter IV, (6·34).]

13. Let $f\in\Lambda^p_*$, $q>p\geqslant1$. Then $f\in\Lambda^q_{1-p^{-1}+q^{-1}}$.

14. A trigonometric series T is of the class Λ_α, $0<\alpha\leqslant1$, if and only if its (C, 1) mean $\sigma_n(x)$ satisfies the condition $\sigma_n'(x)=O(n^{1-\alpha})$ uniformly in x. The condition $\mathfrak{M}_p[\sigma_n']=O(n^{1-\alpha})$ is necessary and sufficient for T to belong to Λ^p_α, $1<p<\infty$.

15. Let $f(x)$ and $\phi(x)$ be defined almost everywhere in the intervals $a\leqslant x<b$ and $a\leqslant x\leqslant b$ respectively, and extended outside (a,b) by the conditions

$$f(x+h)=f(x),\quad \phi(x+h)-\phi(x)=\phi(b)-\phi(a),$$

where $h=b-a$. Modifying slightly the notation of § 4, we may say that the integral (*) $\int_a^b f(x)\,d\phi(x)$ exists in the B-sense, and has value J, if the sums

$$I(t)=\Sigma f(t+\xi_i)\,[\phi(t+x_i)-\phi(t+x_{i-1})]$$

tend in measure to J as $\max(x_j-x_{j-1})\to0$. Show that

(i) If ϕ is of bounded variation and the integral (*) exists as a Lebesgue–Stieltjes integral, then it exists also in the B-sense, and both integrals have the same value;

(ii) The conjugate series of an $S[dF]$ is $S[d\tilde F]$, the coefficients of the latter series being defined in the B-sense.

16. Under the hypothesis of (7·15) the product (7·16) converges absolutely and uniformly outside every circle $|z| = r > 1$, provided we suppress in the product the finite number of terms having poles there. If an arc of $|z| = 1$ does not contain a limit point of the sequence $\{\xi_n\}$, the product converges absolutely and uniformly in the two-sided neighbourhood of that arc, and so the inner and outer functions are analytic continuations of each other.

17. Under the hypothesis of (7·15) the function (7·16), regular for $|z| < 1$, has a radial limit at the point e^{ix}, if

$$\Sigma \frac{1 - |\zeta_\nu|}{|e^{ix} - \zeta_\nu|} < \infty. \quad \text{(Frostman [1].)}$$

[We verify, for instance geometrically, that for $z = re^{ix}$ the second term in (7·17) is majorized by a constant multiple of $(1 - |\zeta_\nu|)/|e^{ix} - \zeta_\nu^*|$ for $0 < r < 1$. Observe that

$$|e^{ix} - \zeta_\nu^*| = |e^{ix} - \zeta_\nu|\,|\zeta_\nu|^{-1}.]$$

CHAPTER VIII

DIVERGENCE OF FOURIER SERIES

1. Divergence of Fourier series of continuous functions

In Chapter II we gave sufficient conditions for the convergence of Fourier series. We shall now investigate how far these tests are best possible. It will appear that (apart from possible minor improvements) the problem of the convergence of Fourier series *at an individual point* has reached a stage where we can hardly hope for essentially new positive results, at least if we use only the classical devices of Chapter II. Such tests as Dini's or Dini-Lipschitz's represent limits beyond which we encounter actual divergence.

(1·1) THEOREM. *There exists a continuous function whose Fourier series diverges at a point.*

This was first shown by P. du Bois Reymond. Since then several other such constructions have been found, and we reproduce here two of them. One is due to Fejér and is remarkable for its elegance and simplicity. The other, that of Lebesgue, lies nearer the root of the matter and can be used in many similar problems.

(i) *Lebesgue's proof.* We know that if n is large enough, there is a continuous function $f(x)=f_n(x)$ not exceeding 1 in absolute value, such that $S_n(0;f_n)$ is as large as we please; we may take for $f(x)$ the function sign $D_n(x)$ smoothed out at the points of discontinuity (Chapter II, § 12). This function f depends on n. To obtain a fixed continuous f such that $S_n(0;f)$ is unbounded as $n\to+\infty$, we appeal to Theorem (9·11) of Chapter IV. If we replace there $y_n(t)$ by $D_n(t)$, $x(t)$ by $f(t)$, and use the fact that the Lebesgue constant L_n tends to $+\infty$, we see† that there is a continuous f such that $S_n(0;f)\neq O(1)$.

Let $\{\lambda_n\}$ be any sequence of positive numbers tending to $+\infty$ more slowly than $\log n$. Since the integral of $|D_n(t)|\lambda_n^{-1}$ over $(0,\pi)$ tends to $+\infty$, by applying Chapter IV, (9·11) again, we obtain the following result:

(1·2) THEOREM. *Given any sequence $\lambda_n=o(\log n)$, there is a continuous f such that $S_n(0;f)\geqslant\lambda_n$ for infinitely many n.*

We know that $S_n(x;f)=o(\log n)$, uniformly in x, if f is continuous (Chapter II, § 11). We now see that this result cannot be improved.

Applying Theorem (9·5) of Chapter IV *in its most general form* to the proof of (9·11) in Chapter IV, we obtain a result from which we may conclude that the set of continuous functions f with $S[f]$ convergent at the point 0, or at any other fixed point, forms a set of the first category in the space C of all continuous and periodic functions. Thus the set of continuous functions with Fourier series

† Theorem (9·11) of Chapter IV (which is due to Lebesgue) lies rather deep, but in the special case $y_n(t)=D_n(t)$ it is not difficult to prove it directly; see Lebesgue's *Leçons sur les séries trigonométriques*.

convergent at some rational point or other is again a set of the first category. Hence, we have the following theorem:

(1·3) THEOREM. *If we reject from the space C a certain set of the first category, the Fourier series of the remaining functions have points of divergence in every interval.*

(ii) *Fejér's proof.* Let $N > n > 0$, and let us consider the two polynomials

$$Q(x, N, n) = 2\sin Nx \sum_1^n \frac{\sin kx}{k}, \tag{1·4}$$

$$R(x, N, n) = 2\cos Nx \sum_1^n \frac{\sin kx}{k}. \tag{1·5}$$

Clearly,

$$Q = \frac{\cos(N-n)x}{n} + \frac{\cos(N-n+1)x}{n-1} + \dots$$
$$+ \frac{\cos(N-1)x}{1} - \frac{\cos(N+1)x}{1} - \dots - \frac{\cos(N+n)x}{n} \tag{1·6}$$

is a purely cosine polynomial with terms of rank varying from $N-n$ to $N+n$. Similarly, $R = -\tilde{Q}$ is a purely sine polynomial.

Since the partial sums of the series $\sin x + \frac{1}{2}\sin 2x + \dots$ are uniformly bounded, the polynomials Q and $\tilde{Q}$ are uniformly bounded in x, N, n, say

$$|Q| \leqslant C, \quad |\tilde{Q}| \leqslant C. \tag{1·7}$$

On the other hand, at $x = 0$ the sum of the first (or last) n terms of $Q(x, N, n)$ in (1·6) is $1/n + \dots + 1/2 + 1 > \log n$, and so is large with n.

Let $\{N_k\}$ and $\{n_k\}$ be any two sequences of positive integers, with $n_k < N_k$, and let $\alpha_k > 0$, $\alpha_1 + \alpha_2 + \dots < \infty$. The series

$$\Sigma \alpha_k Q(x, N_k, n_k), \tag{1·8}$$

$$\Sigma \alpha_k \tilde{Q}(x, N_k, n_k) \tag{1·9}$$

converge to continuous functions, which we denote by $f(x)$, $g(x)$ respectively. If $N_k + n_k < N_{k+1} - n_{k+1}$ for all k, the ranks of non-zero terms in different Q's do not overlap. Similarly for the $\tilde{Q}$'s. Therefore, unbracketing the terms, we may represent (1·8) and (1·9) as trigonometric series

$$\Sigma a_\nu \cos \nu x, \tag{1·10}$$

$$\Sigma a_\nu \sin \nu x. \tag{1·11}$$

Denoting the partial sums of these series by $s_n(x)$ and $t_n(x)$, we see that $s_{N_k+n_k}(x)$ and $t_{N_k+n_k}(x)$ converge uniformly, so that the series (1·10) and (1·11) are $S[f]$ and $S[g]$, respectively. Moreover, $S[g] = \tilde{S}[f]$. Since

$$|s_{N_k+n_k}(0) - s_{N_k}(0)| > \alpha_k \log n_k,$$

$S[f]$ will certainly be divergent at $x = 0$ if $\alpha_k \log n_k$ does not tend to zero. Thus, *if, for example,*

$$\alpha_k = k^{-2}, \quad \tfrac{1}{2}N_k = n_k = 2^{k^3}, \tag{1·12}$$

the Fourier series of the continuous function f defined by (1·8) *diverges at the point* 0.

It is easy to see that both (1·10) and (1·11) converge uniformly for $\delta \leqslant |x| \leqslant \pi$ for any $\delta > 0$. This follows from the fact that the partial sums of $Q(x, N_k, n_k)$ and $\tilde{Q}(x, N_k, n_k)$

are bounded in that interval uniformly in x and k. (Observe that the partial sums of the series $\frac{1}{2}+\cos x+\cos 2x+\ldots$ are uniformly bounded in (δ,π) and use (2·4) of Chapter I.) Since $\mathsf{S}[g]$ contains sines only, it converges for $x=0$, and so everywhere.

(1·13) THEOREM. *There is a continuous function whose Fourier series converges everywhere, but not uniformly.*

Consider the sum $g(x)$ of (1·9) satisfying (1·12). The corresponding series (1·11) converges everywhere. But for $x=\pi/4n$, $N=2n$, the sum of the first n terms of $\tilde{Q}(x,N,n)$ exceeds

$$(1+2^{-1}+\ldots+n^{-1})\sin\tfrac{1}{4}\pi > 2^{-\frac{1}{2}}\log n,$$

so that

$$|\, t_{3n_k}(x_k)-t_{2n_k}(x_k)\,| \geqslant 2^{-\frac{1}{2}}\alpha_k \log n_k$$

for some x_k, which proves (1·13).

(1·14) THEOREM. *There is a power series $c_0+c_1z+\ldots$ regular for $|z|<1$, continuous for $|z|\leqslant 1$, and divergent for $z=1$.*

With the previous notation, $\mathsf{S}[g]=\tilde{\mathsf{S}}[f]$, and so the power series $c_0+c_1z+\ldots$ which reduces to $\mathsf{S}[f]+i\mathsf{S}[g]$ for $z=e^{ix}$ has the required properties.

If we set

$$\alpha_k=2^{-k},\quad n_k=\tfrac{1}{2}N_k=2^{2^k} \tag{1·15}$$

in (1·8) and (1·9), the partial sums $s_n(x)$ and $t_n(x)$ are uniformly bounded. (Under the hypothesis (1·12) they were not.) As before, $\{s_n(0)\}$ diverges, and $\{t_n(x)\}$ converges everywhere but not uniformly.

(1·16) THEOREM. *There exist continuous functions $F(x)$ and $G(x)=\tilde{F}(x)$ such that $\mathsf{S}[F]$ diverges at a dense set of points, while $\mathsf{S}[G]$ converges everywhere, though in no interval uniformly.*

Let $f(x)$ and $g(x)$ be the functions (1·8) and (1·9) satisfying (1·15). Let E be a denumerable set of points $r_1, r_2, \ldots$ dense in $(0,2\pi)$, and let $\epsilon_1+\epsilon_2+\ldots$ be any convergent series with positive terms. We set

$$F(x)=\Sigma\epsilon_i f(x-r_i),\quad G(x)=\Sigma\epsilon_i g(x-r_i),$$

and denote by $F_k(x)$ and $G_k(x)$ the partial sums of these series. The series defining F converges uniformly, and we obtain a partial sum of $\mathsf{S}[F]$ by adding the corresponding partial sums of $\mathsf{S}[\epsilon_i f(x-r_i)]$ for all i.

Write

$$F=F_k+R_k,\quad G=G_k+R_k^*,$$

$$\mathsf{S}[F]=\mathsf{S}[F_k]+\mathsf{S}[R_k],\quad \mathsf{S}[G]=\mathsf{S}[G_k]+\mathsf{S}[R_k^*],$$

and suppose that an $\eta>0$ is given. We know that the partial sums of $\mathsf{S}[f]$ and $\mathsf{S}[g]$ are uniformly bounded, say are less than A in absolute value. Thus the partial sums of $\mathsf{S}[R_k]$ and $\mathsf{S}[R_k^*]$ are less than

$$A(\epsilon_{k+1}+\epsilon_{k+2}+\ldots)<\eta$$

in absolute value, provided $k=k(\eta)$ is large enough. We conclude that

(i) $\mathsf{S}[F]$ diverges at every point r_i, $1\leqslant i\leqslant k$, at which the oscillation of the partial sums of $\mathsf{S}[\epsilon_i f(x-r_i)]$ exceeds η;

(ii) if x is not in E, the oscillation of the partial sums of $\mathsf{S}[F]$ at x is less than η;

(iii) the oscillation of $\mathsf{S}[G]$ is less than η at every x.

Since η and $1/k$ may be arbitrarily small, we see from (i) and (ii) that $S[F]$ diverges in E and converges outside E. From (iii) we see that $S[G]$ converges everywhere. It remains only to show that this convergence is non-uniform in the neighbourhood of every r_h.

Since $S[g(x-r_h)]$ converges non-uniformly in the neighbourhood of r_h, so does the Fourier series of $\epsilon_h g(x-r_h)+R_k^*$, provided k is large enough. This sum differs from G by a sum of a finite number of functions whose Fourier series all converge uniformly in a sufficiently small neighbourhood of r_h. Thus $S[G]$ converges non-uniformly in the neighbourhood of r_h.

The preceding argument gives more than we set out to prove, since it shows that, *given any denumerable set E in $(0, 2\pi)$ there is a continuous f such that $S[f]$ diverges in E and converges outside E.*

The problem of the existence of a continuous f with $S[f]$ divergent everywhere (or even almost everywhere) is still open, and seems to be very difficult. It a simple matter, however, to construct a continuous f with $S[f]$ divergent in a non-denumerable set of points. For let $r_1, r_2, \ldots$ be a sequence containing every rational point of $(0, 2\pi)$ infinitely many times, and let

$$f(x)=\Sigma k^{-2}Q(x-r_k, 2.\,2^{k^3}, 2^{k^3}).$$

The function f is continuous, and $S[f]$ is obtained by replacing each Q by the expression (1·6). At every rational point in $(0, 2\pi)$, $S[f]$ will contain infinitely many blocks of terms with sums exceeding $k^{-2}\log 2^{k^3}$ for some arbitrarily large values of k. It follows that the $S_n(x;f)$ are unbounded in a dense set of points, and it is enough to apply Chapter I, (12·2). Our $S[f]$ even turns out to be divergent in a set of the second category.

(1·17) THEOREM. *There is a power series $\Sigma c_n z^n=\Phi(z)$ regular for $|z|<1$, continuous for $|z|\leqslant 1$, convergent on $|z|=1$ but non-uniformly on every arc of $|z|=1$.*

We first construct a series with a single point of non-uniform convergence. Let

$$P(z)=P(z,N,n)=\sum_{k=1}^{n}\frac{z^{N-k}}{k}-\sum_{k=1}^{n}\frac{z^{N+k}}{k}$$

be the power polynomial whose real part on $z=e^{ix}$ is Q (see (1·6)), and let $P^*(x)=P(e^{ix})$. By (1·7), $|P^*(x)|\leqslant C_1$, and summation by parts easily gives

$$|S_m(x;P^*)|\leqslant C_2|x|^{-1} \quad (0\leqslant|x|\leqslant\pi), \tag{1·18}$$

where C_1 and C_2 are absolute constants. In particular,

$$\left|S_m\left(\frac{1}{k}; P^*\right)\right|\leqslant C_2 k \quad (k=1,2,\ldots). \tag{1·19}$$

Consider the series

$$\sum_{k=1}^{\infty}\alpha_k P(z\,e^{i/k}, N_k, n_k) \quad (\alpha_k=k^{-3},\ \tfrac{1}{2}N_k=n_k=2^{k^3}). \tag{1·20}$$

It converges uniformly in $|z|\leqslant 1$ to a continuous $\Phi(z)$. Since any partial sum of the Taylor series $\Sigma c_n z^n$ of Φ is a partial sum of (1·20) augmented by a partial sum of some $\alpha_k P(z\,e^{i/k}, N_k, n_k)$, we immediately deduce from (1·18) that $\Sigma c_n\,e^{inx}$ converges uniformly outside every neighbourhood of $x=0$, and from (1·19) that $\Sigma c_n\,e^{inx}$ converges at $x=0$.

On the other hand, $\Sigma c_n e^{inx}$ does not converge uniformly on $|z|=1$ since the sum of the first half of the terms of $\alpha_k P(z e^{i/k}, N_k, n_k)$ exceeds at $z=e^{-i/k}$ a fixed positive multiple of $\alpha_k \log n_k \geqslant C_3 > 0$.

To obtain non-uniform convergence on every arc of $|z|=1$ we proceed as in the proof of (1·16).

Divergent Fourier series of *bounded* functions can also be obtained by means of Riesz products (Chapter V, § 7).

(1·21) THEOREM. *The Riesz product*

$$\prod_{k=1}^{\infty}\left(1+i\frac{\cos 10^k x}{k}\right) \tag{1·22}$$

is a Fourier series whose partial sums are uniformly bounded and which is divergent in a set of points of the power of the continuum in every interval.

In Chapter V, § 7, we proved that the partial products $p_N(x)$ of (1·22) are uniformly bounded, which implies that (1·22) is an $S[f]$ with f bounded. We note that every partial sum of $S[f]$ is, for some N, of the form $p_N +$ (a partial sum of $p_{N+1}-p_N$). Now

$$|p_{N+1}-p_N| = (N+1)^{-1}|\cos 10^{N+1}x|\,|p_N| = O(N^{-1}).$$

Since $p_{N+1}-p_N$ is a polynomial of order $10+10^2+\ldots+10^{N+1}<10^{N+2}$, and since the Lebesgue constant L_n is $O(\log n)$, it follows that all the partial sums of $p_{N+1}-p_N$ are $O(N).O(N^{-1})=O(1)$, uniformly in N. Since the p_N are uniformly bounded, the same holds for the partial sums of $S[f]$.

Using the formula $1+z=\exp\{z+O(|z|^2)\}$, for $|z|$ small, we have

$$p_N(x)=\exp\left\{i\sum_{1}^{N}k^{-1}\cos 10^k x\right\}.\exp F_N(x), \tag{1·23}$$

where $\{F_n\}$ converges uniformly to a finite limit. Since the series $\Sigma k^{-1}\cos 10^k x$ diverges in a set E which is of the power of the continuum (even of the second category) in every interval (see Chapter VI, (6·3), especially Remarks (*b*) and (*c*), to it), and since the terms of the series tend to 0, the divergence of p_N—and so also of $S[f]$—in E follows.

Our function is complex-valued, $f=f_1+if_2$, but as may be seen from (1·23) both $S[f_1]$ and $S[f_2]$ diverge in E.

2. Further examples of divergent Fourier series

The Dini-Lipschitz theorem (Chapter II, (10·3)) asserts that if the modulus of continuity $\omega(\delta)$ of a function f is $o\{(\log\delta)^{-1}\}$, then $S[f]$ converges uniformly. The same argument shows that if $\omega(\delta)=O\{(\log\delta)^{-1}\}$ the $S_n(x;f)$ are uniformly bounded. However,

(2·1) THEOREM. *There exist two continuous functions $f(x)$ and $g(x)=\tilde{f}(x)$, both having modulus of continuity $O\{(\log\delta)^{-1}\}$, such that $S[f]$ diverges at some point, while $S[g]$ converges everywhere but not uniformly.*

As before, we define f and g by the series (1·8) and (1·9), and assume (1·15). We know that $S[f]$ oscillates finitely at $x=0$, and that $S[g]$ converges everywhere, but not uniformly. To prove the inequalities for $\omega(\delta;f)$ and $\omega(\delta;g)$, for example for the former, we take any $0<h\leqslant\frac{1}{2}$ and define $\nu=\nu(h)$ as the largest integer k satisfying $2^{2^k}\leqslant 1/h$. We break up the sum defining f into two parts, $f_1(x)$ and $f_2(x)$, the latter consisting of the terms with indices greater than ν. Then by (1·7),

$$|f_2(x+h)-f_2(x)|\leqslant 2C\sum_{\nu+1}^{\infty}2^{-k}=4C.2^{-\nu-1}\leqslant 4C/|\log h|. \tag{2·2}$$

Now, by (1·4),

$$Q'(x,N,n)=-N\tilde{Q}(x,N,n)+2\sin Nx\sum_{1}^{n}\cos kx,$$

$$|Q'|\leqslant NC+2n=nC' \quad \text{for} \quad N=2n,\ C'=2C+2.$$

(We could also use Chapter III, (13·16).) By the mean-value theorem.

$$\begin{aligned}|f_1(x+h)-f_1(x)|&\leqslant C'h[2^{-1}2^{2^1}+2^{-2}2^{2^2}+\ldots+2^{-\nu}2^{2^\nu}]\\&=O(h2^{-\nu}2^{2^\nu})=O(2^{-\nu})=O\{(\log h)^{-1}\}.\end{aligned} \tag{2·3}$$

Since $|f(x+h)-f(x)|$ does not exceed the sum of (2·2) and (2·3), (2·1) is established. Arguing as in the preceding section, we can construct f and $g=\tilde{f}$, with moduli of continuity $O\{(\log\delta)^{-1}\}$, such that $S[f]$ diverges in a set of points everywhere dense, and $S[g]$ converges everywhere but non-uniformly in every interval. Also the function Φ of Theorem (1·17) may be made to have modulus of continuity $O\{(\log\delta)^{-1}\}$ on $|z|=1$.

We shall now show that, in a sense, the Dini condition (Chapter II, (6·1)) cannot be improved.

(2·4) THEOREM. *Given any continuous $\mu(t)\geqslant 0$ such that $\mu(t)/t$ is not integrable in the neighbourhood of $t=0$, we can find a continuous function f such that $|f(t)-f(0)|\leqslant\mu(t)$ for small t and $S[f]$ diverges at $t=0$.*

Let

$$\chi_n(t)=\mu(t)\frac{\sin nt}{2\tan\frac{1}{2}t}.$$

If $\mathfrak{M}[\chi_n]\neq O(1)$ as $n\to\infty$, we can find a continuous $g(x)$, $|g|\leqslant 1$, such that the integral of $\chi_n(t)g(t)$ over $(-\pi,\pi)$ is unbounded for $n\to\infty$ (Chapter IV, (9·11)). This means that $S[f]$, with $f=g\mu$, diverges at the point 0. Since we may suppose that $\mu(0)=0$, we have $|f(t)-f(0)|=|f(t)|\leqslant\mu(t)$.

It remains to show that our hypotheses imply $\mathfrak{M}[\chi_n]\neq O(1)$. This we shall deduce from Theorem (4·15) of Chapter II. Let us there set $\beta(t)=|\sin t|$, and $\alpha(t)=|\mu(t)\,\frac{1}{2}\cot\frac{1}{2}t|$ if $0<\epsilon\leqslant|t|\leqslant\pi$, $\alpha(t)=0$ elsewhere. Denote the corresponding integral by $I_n(\epsilon)$. Since $\mathfrak{M}[\chi_n]\geqslant I_n(\epsilon)$, we have

$$\liminf\mathfrak{M}[\chi_n]\geqslant\liminf I_n(\epsilon)=\lim I_n(\epsilon).$$

Since the function $\mu(t)\,\frac{1}{2}\cot\frac{1}{2}t$ is not integrable, we may make $\lim I_n(\epsilon)$ as large as we please by taking ϵ small enough. This shows that $\mathfrak{M}[\chi_n]\to\infty$ and the proof of (2·4) is completed.

The case $\mu(t)=o(\log 1/|t|)^{-1}$ (taking, for example, $\mu(t)=(\log 1/|t|)^{-1}(\log\log 1/|t|)^{-1}$ for small $|t|$) is of interest in connexion with the Dini-Lipschitz test. It shows that *the latter is primarily a test for uniform convergence, and the relation*

$$f(x_0+t)-f(x_0)=o\{(\log 1/|t|)^{-1}\} \quad \textit{as} \quad t\to 0$$

does not ensure the convergence of $S[f]$ at x_0.

This observation can be completed as follows. Let f be a continuous function with $S[f]$ divergent at the point 0, such that $f(0)=0$, $f(t)=o\{(\log|t|)^{-1}\}$. Let $f_1(t)=f(t)$, $f_2(t)=0$ for $0\leqslant t\leqslant\pi$, and $f_1(t)=0$, $f_2(t)=f(t)$ for $-\pi\leqslant t<0$. Since $f=f_1+f_2$, it follows that either $S[f_1]$ or $S[f_2]$, say the former, diverges at $t=0$. Let $(a,b)=(-\frac{1}{4}\pi,0)$. The function f_1 is zero in (a,b), and so *a fortiori* its modulus of continuity there is $o\{(\log\delta)^{-1}\}$. Moreover,

$$f_1(a-t)-f_1(a)=o\{(\log 1/t)^{-1}\},\quad f_1(b+t)-f_1(b)=o\{(\log 1/t)^{-1}\}\quad\text{as}\quad t\to+0,$$

so that $f_1(x+t)-f_1(x)=o\{(\log|t|)^{-1}\}$ uniformly in $a\leqslant x\leqslant b$. None the less $S[f_1]$ diverges at an end-point of (a,b). Hence in Theorem (10·5) of Chapter II we do not have uniform convergence in (a,b), even if we assume additionally that $f(x+t)-f(x)=o\{(\log|t|)^{-1}\}$ at the endpoints of (a,b).

We know that Fourier series with coefficients $O(1/n)$ converge at every point at which they are summable (C, 1). Fourier series of continuous functions with such coefficients converge uniformly. We shall now show that the condition $O(1/n)$ cannot be replaced by anything weaker.

(2·5) THEOREM. *Given any function* $\chi(u)$, $0\leqslant u<+\infty$, *tending to* $+\infty$ *with* u, *we can find two continuous functions* $f(x)$ *and* $g(x)=\tilde{f}(x)$ *having coefficients* $O\{\chi(n)/n\}$ *and such that*

(i) $S[f]$ *diverges at a point;*

(ii) $S[g]=\tilde{S}[f]$ *converges everywhere, but not uniformly.*

Part (i) shows, in particular, that Theorem (1·26) of Chapter III does not hold if we weaken the condition that the terms of the series are $O(1/n)$. Replacing, if necessary, $\chi(u)$ by $\chi_1(u)=\inf_{v\geqslant u}\chi(v)$, we may assume that $\chi(u)$ is increasing.

In the proof of (i) we shall slightly modify the polynomials Q, using instead the polynomials

$$Q(x;\,N,n,m)=2\sin Nx\sum_{m}^{n}\frac{\sin\nu x}{\nu}\quad(0<m<n<N),\tag{2·6}$$

which are uniformly bounded and have only terms of ranks between $N-n$ and $N+n$. The sum of the terms of rank not exceeding N has at $x=0$ the value

$$\sum_{m}^{n}\frac{1}{\nu}>\int_{m}^{n}\frac{du}{u}=\log(n/m).$$

Thus if we set $$f(x)=\Sigma k^{-2}Q(x;\,N_k,n_k,m_k),$$

where $$N_{k-1}+n_{k-1}<N_k-n_k,\quad \log(n_k/m_k)>k^2,\tag{2·7}$$

then, as in the examples considered previously, $S[f]$ will diverge at $x=0$. We have only to show that by a proper selection of the N_k, n_k, m_k we can attain the estimate

$$a_n=O\{\chi(n)/n\}$$

for the (cosine) coefficients of f.

Since the numerically largest coefficient in $Q(x;\,N_k,n_k,m_k)$ is $1/m_k$, and the rank of the terms is between $N_k\pm n_k$, our requirement will be satisfied if

$$k^{-2}(N_k+n_k)/m_k=O\{\chi(N_k-n_k)\},$$

or, setting $N_k-n_k=\delta_k$, if

$$k^{-2}(2n_k+\delta_k)/m_k=O\{\chi(\delta_k)\}.\tag{2·8}$$

Suppose that the numbers $N_1, n_1, m_1, \ldots, N_{k-1}, n_{k-1}, m_{k-1}$ have already been determined. We take a number $\rho_k > 0$ such that $\log \rho_k > k^2$, then an integer $\delta_k > 0$ such that

$$k^{-2}\rho_k \leqslant \chi(\delta_k), \quad N_{k-1} + n_{k-1} < \delta_k, \tag{2·9}$$

and finally an m_k such that

$$k^{-2}\delta_k / m_k \leqslant \chi(\delta_k). \tag{2·10}$$

We set

$$n_k = [\rho_k m_k] + 1, \; N_k = n_k + \delta_k,$$

and shall show that (2·7) and (2·8) hold. The inequality $\log \rho_k > k^2$ and the second condition (2·9) imply (2·7). The definition of n_k, the first inequality (2·9) and the inequality (2·10) lead to (2·8). This proves part (i) of (2·5).

Part (ii) is obtained similarly, by using instead of Q the conjugate polynomial $\tilde{Q}$, obtained from (2·6) by replacing there $\sin Nx$ by $-\cos Nx$. The same device as in the proof of (1·16) leads to continuous functions with coefficients $O\{\chi(n)/n\}$ whose Fourier series diverge in a dense set, or converge everywhere but in no interval uniformly.

3. Examples of Fourier series divergent almost everywhere

(3·1) THEOREM OF KOLMOGOROV. *There exists an $f \in L$ such that $S[f]$ diverges almost everywhere.*

In § 4 we shall show that there is an f such that $S[f]$ diverges at every point. The proof of this result is, however, more difficult and it is preferable to deal with it separately.

(3·2) LEMMA. *There exists a sequence of non-negative trigonometric polynomials f_n with constant term $\frac{1}{2}$, having the following properties:*

For each n there is a number A_n and a set $E_n \subset (0, 2\pi)$ such that

(i) $A_n \to \infty$;

(ii) $|E_n| \to 2\pi$;

(iii) *if $x \in E_n$, then for a suitable $l = l_x$ we have*

$$|S_l(x; f_n)| > A_n. \tag{3·3}$$

This lemma is the main part of the proof of (3·1). We take it temporarily for granted, only observing meanwhile that if (3·3) is to hold in a 'large' set of points x then l must necessarily depend on x. For the hypotheses of (3·2) imply that $\int_0^{2\pi} f_n \, dx = \pi$ and so, by Chapter VII, (6·8), that the integral $\int_0^{2\pi} |S_m(x; f_n)|^{\frac{1}{2}} dx$ is uniformly bounded in n, m. This shows, by (i), that the measure of the set of points where we have (3·3) for a fixed l tends to 0 as $n \to \infty$.

Assuming (3·2) we prove (3·1) as follows. Let the order of f_n be ν_n. We may suppose that the l in (iii) satisfies $1 \leqslant l \leqslant \nu_n$. Let $\{n_k\}$ increase so rapidly that

$$\Sigma A_{n_k}^{-\frac{1}{2}} < \infty.$$

The polynomial $f_{n_k}(x) - \frac{1}{2}$ has constant term 0. Hence, if q_k increases rapidly enough. the order of the polynomial $f_{n_k}(q_k x) - \frac{1}{2}$ is less than the lowest rank of the non-zero terms in $f_{n_{k+1}}(q_{k+1}x) - \frac{1}{2}$; in particular the polynomials $f_{n_k}(q_k x) - \frac{1}{2}$ do not overlap.

We write
$$f(x)=\sum_{k=1}^{\infty}\frac{f_{n_k}(q_k x)-\frac{1}{2}}{A_{n_k}^{\frac{1}{2}}}. \tag{3·4}$$

Since $\Sigma A_{n_k}^{-\frac{1}{2}}\int_0^{2\pi}|f_{n_k}(q_k x)-\frac{1}{2}|\,dx\leqslant\Sigma A_{n_k}^{-\frac{1}{2}}\int_0^{2\pi}\{f_{n_k}(x)+\frac{1}{2}\}dx=2\pi\Sigma A_{n_k}^{-\frac{1}{2}}<\infty,$

the series in (3·4) converges (absolutely) for almost all x, and its partial sums are absolutely majorized by an integrable function; in particular, f is integrable. The majorized convergence implies that $S[f]$ is obtained by adding formally the Fourier series of the individual terms on the right of (3·4), that is, by writing out in full the successive polynomials $\phi_k=A_{n_k}^{-\frac{1}{2}}\{f_{n_k}(q_k x)-\frac{1}{2}\}$.

Let $\mathscr{E}_k$ be the set of points x such that $q_k x\in E_{n_k}$. Clearly $|\mathscr{E}_k|=|E_{n_k}|$. By (iii), at every point $x\subset\mathscr{E}_k$ some partial sum of $\phi_k(x)$, that is, some connected block of terms of $S[f]$, exceeds $(A_{n_k}-\frac{1}{2})A_{n_k}^{-\frac{1}{2}}$ in absolute value. Since this tends to $+\infty$ with k, it follows that $S[f]$ diverges at every x which belongs to infinitely many $\mathscr{E}_k$. Since $|\mathscr{E}_k|\to 2\pi$, $S[f]$ diverges almost everywhere. More precisely, $\{S_n(x;f)\}$ is unbounded at almost all x.

In the foregoing argument, starting out with the given polynomials $f_{n_k}(x)$ we formed non-overlapping polynomials $\phi_k(x)$. The non-overlapping can also be achieved by multiplying $f_{n_k}(x)$ by exponentials $e^{i\mu_k x}$. Write
$$g(x)=\Sigma\, e^{i\mu_k x}f_{n_k}(x)\,A_{n_k}^{-\frac{1}{2}}. \tag{3·5}$$

Having chosen the n_k as before, we take for μ_k positive integers increasing so rapidly that $\mu_k+\nu_{n_k}<\mu_{k+1}-\nu_{n_{k+1}}$ for each k. Then the terms on the right of (3·5) are (complex-valued) non-overlapping polynomials which when written out in full in the complex form give $S[g]$. The complex form of $S[g]$ contains only exponentials e^{inx} with $n>0$. It follows that $S[g]$ is of power-series type. At each point in E_{n_k} a connected block of terms of $S[g]$ exceeds $A_{n_k}/A_{n_k}^{\frac{1}{2}}=A_{n_k}^{\frac{1}{2}}$ in absolute value. Hence $S[g]$ diverges at every point which belongs to infinitely many E_{n_k}, that is, diverges almost everywhere, and we obtain the following result:

(3·6) THEOREM. *There exists an* $S[g]$ *of power-series type which diverges almost everywhere.*

The real and imaginary parts of $S[g]$ are Fourier series. It will be shown in Chapter XIII, (5·1), that for each $f\in L$ the sets of the points of convergence of $S[f]$ and $\tilde{S}[f]$ are the same, except for a set of measure 0. Therefore (3·6) shows that there exists a real-valued $f\in L$ such that both $S[f]$ and $\tilde{S}[f]$ are Fourier series and both diverge almost everywhere. It is curious that, as we shall see presently, this result cannot be achieved by the construction (3·4).

Return to the proof of (3·2). We write
$$a_j=\frac{4\pi j}{2n+1}\quad (j=0,1,\ldots,n).$$

The polynomial f_n is defined as an average of Fejér kernels,
$$f_n(x)=\frac{1}{n}\{K_{m_1}(x-a_1)+K_{m_2}(x-a_2)+\ldots+K_{m_n}(x-a_n)\},$$

where the integers m_j are such that

$$m_1 \geqslant n^4, \quad m_{j+1} > 2m_j, \quad 2m_j + 1 \text{ is a multiple of } 2n+1.$$

Clearly $f_n \geqslant 0$ and the constant term of f_n is $\frac{1}{2}$.

We have

$$S_{m_j}(x; f_n) = \frac{1}{n}\sum_{i=1}^{j} K_{m_i}(x-a_i) + \frac{1}{n}\sum_{i=j+1}^{n}\left\{\frac{1}{2} + \sum_{l=1}^{m_j}\frac{m_i - l + 1}{m_i + 1}\cos l(x-a_i)\right\}, \tag{3·7}$$

and since
$$m_i - l + 1 = (m_i - m_j) + (m_j - l + 1),$$
we also have

$$S_{m_j}(x; f_n) = \frac{1}{n}\sum_{i=1}^{j} K_{m_i}(x-a_i) + \frac{1}{n}\sum_{i=j+1}^{n}\frac{m_j+1}{m_i+1}K_{m_j}(x-a_i) + \frac{1}{n}\sum_{i=j+1}^{n}\frac{m_i-m_j}{m_i+1}D_{m_j}(x-a_i).$$

Let
$$\Delta_i = (a_{i-1}, a_i), \quad \Delta_i' = (a_i - n^{-2}, a_i + n^{-2}) \quad (i = 1, 2, \ldots, n).$$

The estimate $K_m(t) = O(m^{-1}t^{-2})$, together with $m_i \geqslant n^4$, shows that $K_{m_i}(x-a_i)$ is uniformly bounded outside Δ_i', and so the contribution of the first two terms on the right in the last formula for $S_{m_j}(x; f_n)$ is less than an absolute constant A outside $\Sigma\Delta_i'$:

$$S_{m_j}(x; f_n) \geqslant \frac{1}{n}\left|\sum_{i=j+1}^{n}\frac{m_i - m_j}{m_i+1}D_{m_j}(x-a_i)\right| - A \quad (j = 1, 2, \ldots, n;\, x \notin \Sigma\Delta_i'). \tag{3·8}$$

We observe that since

$$(2m_j+1)\tfrac{1}{2}a_i = \frac{2m_j+1}{2n+1}2i\pi \equiv 0 \pmod{2\pi},$$

we have
$$D_{m_j}(x-a_i) = \frac{\sin(m_j+\frac{1}{2})(x-a_i)}{2\sin\frac{1}{2}(x-a_i)} = \frac{\sin(m_j+\frac{1}{2})x}{2\sin\frac{1}{2}(x-a_i)}.$$

Suppose now that $x \in \Delta_j$, $1 \leqslant j \leqslant n - \sqrt{n}$. Owing to the condition $m_{i+1} \geqslant 2m_i + 1$, the multipliers of the D_{m_j} in (3·8) are not less than $\frac{1}{2}$, and since the denominators $2\sin\frac{1}{2}(x-a_i)$ are all of constant sign for $i > j$ we get

$$\begin{aligned}
\frac{1}{n}\left|\sum_{i=j+1}^{n}\frac{m_i-m_j}{m_i+1}D_{m_j}(x-a_i)\right| &\geqslant \frac{1}{2n}|\sin(m_j+\tfrac{1}{2})x|\sum_{i=j+1}^{n}\frac{1}{a_i - a_{j-1}}\\
&= \frac{1}{2n}|\sin(m_j+\tfrac{1}{2})x|\frac{2n+1}{4\pi}\sum_{k=2}^{n-j+1}\frac{1}{k}\\
&\geqslant \frac{1}{4\pi}|\sin(m_j+\tfrac{1}{2})x|(-1+\log(n-j))\\
&\geqslant \frac{1}{4\pi}|\sin(m_j+\tfrac{1}{2})x|(-1+\tfrac{1}{2}\log n)\\
&\geqslant \frac{1}{9\pi}|\sin(m_j+\tfrac{1}{2})x|\log n,
\end{aligned} \tag{3·9}$$

for n large enough; thus at the points not in $\Sigma\Delta_i'$ for which

$$|\sin(m_j+\tfrac{1}{2})x| \geqslant \frac{9\pi}{(\log n)^{\frac{1}{2}}}, \tag{3·10}$$

we have
$$|S_{m_j}(x; f_n)| \geqslant (\log n)^{\frac{1}{2}} - A = A_n. \tag{3·11}$$

The set of x in $(0, 2\pi)$ where (3·10) fails has measure $O\{(\log n)^{-\frac{1}{2}}\}$. Therefore, if from

$(0, a_{[n-n^{\frac{1}{2}}]})$ we remove the points where (3·10) fails, and those which are in $\Sigma\Delta_i'$, and denote the remainder by E_n, then

$$2\pi - |E_n| = O\{(\log n)^{-\frac{1}{2}}\} + O(n \cdot n^{-2}) + O(n^{\frac{1}{2}} \cdot n^{-1}) = o(1).$$

For each $x \in E_n$ and a suitable $j = j(x)$ we have (3·11), and (3·2) follows.

This completes the proofs of (3·1) and (3·6).

We know (Chapter II, (11·9)) that for any $f \in \mathrm{L}$ we have $S_n(x; f) = o(\log n)$ for almost all x. It is conceivable that this result is best possible, that is, for any sequence of positive numbers $\lambda_n = o(\log n)$ there is an $f \in \mathrm{L}$ such that at almost every point x we have $S_n(x; f) > \lambda_n$ for infinitely many n (compare (1·2)). The problem is open.

(3·12) THEOREM. *There is an $f \in \mathrm{L}$ such that $\{S_{2^n}[f]\}$ diverges almost everywhere.*

Consider the f in (3·4) and write m_j^n for the m_j in the proof of (3·2). The proofs of (3·1) and (3·2) show that $\{S_{N_k}[f]\}$ diverges almost everywhere, if $\{N_k\}$ consists of the numbers

$$q_k m_j^{n_k} \quad (k = 1, 2, \ldots; \ 1 \leqslant j \leqslant n_k - \sqrt{n_k}), \tag{3·13}$$

and it is enough to show that, with a suitable choice of the n_k and q_k, all the N_k are powers of 2.

In the construction (3·4), $\{n_k\}$ is any sequence increasing sufficiently rapidly, and we may select for the n_k powers of 2; after the n_k have been chosen, q_k is any sequence increasing rapidly enough. For each n_k we may select the $m_j = m_j^{n_k}$ arbitrarily, provided they increase sufficiently rapidly and $2m_j + 1$ is divisible by $2n_k + 1$. Now if $n_k = 2^\nu$, $2m_j + 1$ is divisible by $2^{\nu+1} + 1$ provided that $m_j = 2^\mu$ and $\mu + 1$ is an odd multiple of $\nu + 1$. Hence all the m_j may be taken as powers of 2, and selecting the q_k likewise we can make all the numbers (3·13) powers of 2. This completes the proof of (3·12).

We shall show in Chapter XV, § 4, that if $f \in \mathrm{L}^p$, $p > 1$, then $\{S_{\lambda_k}[f]\}$ converges almost everywhere, provided that the λ_k increase at least as rapidly as a geometric progression, $\lambda_{k+1}/\lambda_k > \lambda > 1$; and the result holds for $p = 1$, provided $S[f]$ is of power-series type. Theorem (3·12) indicates that the requirement that $S[f]$ be of power-series type is indispensable. It also indicates that the construction (3·4) cannot give an $f \in \mathrm{L}^p$, $p > 1$, with $S[f]$ diverging almost everywhere (the fact that the sum (3·4) is not in L^p can also be verified directly by showing that the integral over $(0, 2\pi)$ of the pth power of the general term of the series (3·4) tends to $+\infty$). An even stronger statement holds: for f in (3·4), $\tilde{S}[f]$ is not a Fourier series. For otherwise $S[f]$ would be the real part of an $S[g]$ of power-series type, and so $S_{2^n}[f]$ would converge almost everywhere which, with a suitable selection of the n_k and q_k, is not the case.

From what has just been said we see that, for the g in (3·6), $S_{2^n}[g]$ converges almost everywhere. But again g is not in any L^p, $p > 1$, and the problem whether there is an $f \in \mathrm{L}^p$, $p > 1$, with $S[f]$ diverging almost everywhere, is open.

(3·14) THEOREM OF MARCINKIEWICZ.† *There is an $f \in \mathrm{L}$ such that $\{S_n(x; f)\}$ oscillates finitely at almost every point x.*

† The theorem shows that the behaviour of the S_n is different, for example, from the behaviour of the difference quotient of a function. For a well-known theorem of Denjoy (see Saks, *Theory of the Integral*, p. 270) asserts that if the Dini numbers of a function are finite at each point of a set E, then the function is differentiable almost everywhere in E.

The proof consists in a refinement of the construction (3·4). We consider the same polynomials f_n as before, select increasing n_k so that $\Sigma 1/\log n_k$ converges, and write

$$f(x)=\sum_{k=1}^{\infty}\frac{f_{n_k}(q_k x)-\frac{1}{2}}{\log n_k}, \tag{3·15}$$

where the q_k are so chosen that the polynomials on the right do not overlap. It will be convenient to assume that q_{k+1} divides q_k. The series converges almost everywhere and $f\in \mathrm{L}$.

(i) *The partial sums of* S$[f]$ *are bounded almost everywhere.* In view of the convergence of (3·15) almost everywhere, it is enough to show that for almost all x the partial sums of the individual polynomials on the right of (3·15) are bounded.

Considering n large enough, write

$$f_n^*(x)=f_n(x)/\log n,\quad \Delta_j^*=(a_{j-1}+1/n\log n, a_j-1/n\log n),\quad U_n=\sum_1^n \Delta_j^*,$$

and denote by C various positive absolute constants. We first show that, for all p and n,

$$|S_p(x;f_n^*)|\leqslant C \quad \text{if} \quad x\in U_n. \tag{3·16}$$

The estimate $D_p(x)=O(1/x)$, valid for $|x|\leqslant\pi$ uniformly in p, and the equation $K_m=(m+1)^{-1}\{D_0+D_1+\ldots+D_m\}$ show that the partial sums of $K_m(x)$ are also $O(1/x)$ in $|x|\leqslant\pi$, uniformly in m. Applying this to f_n, we find that

$$|S_p(x;f_n^*)|\leqslant\frac{C}{n\log n}\left[\frac{1}{|x-a_1|}+\frac{1}{|x-a_2|}+\ldots+\frac{1}{|x-a_n|}\right].$$

Now if $x\in\Delta_j^*$ then both $1/|x-a_{j-1}|$ and $1/|x-a_j|$ are less than $n\log n$, and the sum of the remaining terms in square brackets is majorized by $Cn\left(1+\frac{1}{2}+\ldots+\frac{1}{n}\right)\simeq Cn\log n$. This proves (3·16).

Denote the complement of U_n by V_n, and the set of points x such that $q_k x\in V_{n_k}$ by W_{n_k}. Since

$$|W_{n_k}|=|V_{n_k}|\leqslant n_k\frac{2}{n_k\log n_k}=\frac{2}{\log n_k},$$

we have $\Sigma|W_{n_k}|<\infty$. This shows that almost every point x_0 is outside all the W_{n_k} with large enough k, that is, $q_k x_0$ is in all U_{n_k} with large enough k. In view of (3·16), $|S_p(x_0;f_{n_k}^*(q_k x))|\leqslant C$ for almost all x_0 and k large enough. This shows that the partial sums of the polynomials on the right of (3·15) are bounded at almost all points x, and proves (i).

(ii) S$[f]$ *diverges almost everywhere.* We suppose that the m_j satisfy the same conditions as before, except that on m_1 we impose the milder condition $m_1\geqslant n^3$.

The estimate $K_m(x)=O(m^{-1}x^{-2})$ shows that for $x\in U_n$ we have

$$K_{m_i}(x-a_j)=O\{m_i^{-1}(x-a_j)^{-2}\}=O(n^{-1}\log^2 n).$$

Hence the A in (3·8) now becomes $O(n^{-1}\log^2 n)$ and from (3·9) we see that

$$|S_{m_j}(x;f_n^*)|>\frac{1}{9\pi}|\sin(m_j+\tfrac{1}{2})x|-C\frac{\log n}{n} \tag{3·17}$$

for $x\in\Delta_j^*$, $1\leqslant j\leqslant n-\sqrt{n}$.

Denote by H_j the set of points in Δ_j^* for which

$$|\sin(m_j+\tfrac{1}{2})x|\geqslant\tfrac{1}{2}.$$

Since $m_j|\Delta_j^*|$ is large, we have $|H_j| > \theta\,|\Delta_j^*|$, θ being a positive absolute constant, and so for $E_n = \sum_{j \leqslant n-\sqrt{n}} H_j$ we obtain

$$|E_n| \geqslant (n-\sqrt{n}-1)\left[\frac{4\pi}{2n+1} - \frac{2}{n\log n}\right]\theta = 2\pi\theta[1-o(1)].$$

Let now $\mathscr{E}_k$ be the set of x such that $q_k x \in E_{n_k}$. Hence

$$|\mathscr{E}_k| = |E_{n_k}| \geqslant 2\pi\theta[1-o(1)]. \tag{3·18}$$

If $x_0 \in \mathscr{E}_k$, then $q_k x_0 \in E_{n_k}$, and from (3·17) we deduce that, with a suitable $m_j = m_j(x_0, k)$, we have

$$|S_{q_k m_j}(x_0; f^*_{n_k}(q_k x))| = |S_{m_j}(q_k x_0; f^*_{n_k}(x))| > \frac{1}{18\pi} - C\frac{\log n_k}{n_k} > C > 0 \tag{3·19}$$

for k large enough.

In view of (3·18), the set $\mathscr{E}$ of the points which belong to infinitely many $\mathscr{E}_k$ has measure at least $2\pi\theta$. At every point of $\mathscr{E}$ infinitely many terms of the series (3·15) have, by (3·19), partial sums exceeding C in absolute value. It follows that $S[f]$ diverges everywhere in $\mathscr{E}$. Let D be the set of the points of divergence of $S[f]$. We have just seen that $|D| \geqslant 2\pi\theta$. But since q_{k+1} is a multiple of q_k, D has arbitrarily small periods and so, being measurable, must be of measure either 0 or 2π. It follows that $|D| = 2\pi$ and (ii) is proved.

Theorem (3·14) is a corollary of (i) and (ii).

(3·20) THEOREM. *There is an $S[g]$ of power-series type which diverges boundedly at almost all points.*

We may be brief. Write

$$g(x) = \sum e^{i\mu_k x}\frac{f_{n_k}(x)}{\log n_k},$$

where $\sum 1/\log n_k < \infty$ and the μ_k make the polynomials $e^{i\mu_k x} f_{n_k}$ non-overlapping. The proof that the $S_p(x; g)$ are bounded is similar to that of (i); since not only the partial sums of $K_m(x)$ but also those of $\tilde{K}_m(x)$ are uniformly $O(1/x)$, it follows that the partial sums of the polynomials $e^{i\mu x} K_m(x)$ (written out in the complex form) are uniformly $O(1/x)$, after which we proceed as before. To show that $S[g]$ diverges almost everywhere, we can no longer assert that the set D of its points of divergence has arbitrarily small periods. It is, however, clear that, for any interval I,

$$|IE_{n_k}| = \frac{|I|}{2\pi}|E_{n_k}| + o(1) \geqslant |I|\,\theta + o(1).$$

Hence, if E is the set of points which belong to infinitely many E_{n_k}, we have $|IE| \geqslant |I|\,\theta$. It follows that $|E| = 2\pi$, and so also $|D| = 2\pi$, since $D \supset E$.

4. An everywhere divergent Fourier series

(4·1) THEOREM OF KOLMOGOROV. *There is an $f \in L$ such that $S[f]$ diverges everywhere.*

We need the following lemma:

(4·2) LEMMA. *There is a sequence of non-negative trigonometric polynomials $F_1, F_2, \ldots, F_n, \ldots$, of orders $\nu_1 < \nu_2 < \ldots$, with constant terms* 1 *and having the following properties:*

With each n we can associate a number A_n, a set $E_n \subset (0, 2\pi)$, and an integer λ_n such that

(i) $A_n \to \infty$;

(ii) $E_1 \subset E_2 \subset \ldots$, $\Sigma E_n = (0, 2\pi)$;

(iii) $\lambda_n \to \infty$;

(iv) *for each $x \in E_n$ there is a $k = k(x)$ satisfying $\lambda_n \leqslant k \leqslant \nu_n$ such that*

$$S_k(x; F_n) > A_n.$$

Assuming temporarily the validity of the lemma we prove (4·1) as follows.

Let $n_1 = 1$ and suppose that $n_1, n_2, \ldots, n_{i-1}$ have been defined. We select n_i so that

$$(a)\quad \lambda_{n_i} > \nu_{n_{i-1}}, \qquad (b)\quad A_{n_i} > 4A_{n_{i-1}}, \qquad (c)\quad A_{n_i}^{\frac{1}{4}} > \nu_{n_{i-1}}.$$

Having defined $\{n_i\}$ we write

$$f(x) = \sum_{k=1}^{\infty} A_{n_k}^{-\frac{1}{2}} F_{n_k}(x). \tag{4·3}$$

Condition (b) implies that $\Sigma A_{n_k}^{-\frac{1}{2}} < \infty$, so that $f(x) \in L$. Let $x \in E_{n_i}$. We have $f = u + v + w$, where $v = A_{n_i}^{-\frac{1}{2}} F_{n_i}$, u is the sum of the terms preceding v in (4·3), and w the sum of the terms following v. Hence

$$S_p(x; f) = S_p(x; u) + S_p(x; v) + S_p(x; w). \tag{4·4}$$

By (iv), there is a $k = k(x, i)$ satisfying $\lambda_{n_i} \leqslant k \leqslant \nu_{n_i}$ such that

$$\left.\begin{aligned} &S_k(x; v) \geqslant A_{n_i}^{\frac{1}{2}}. \\ \text{By } (a), \qquad &S_k(x; u) = u(x_0) \geqslant 0. \end{aligned}\right\} \tag{4·5}$$

Since the coefficients of any integrable g are absolutely not greater than $\pi^{-1}\mathfrak{M}[g]$, and so $|S_k[g]| \leqslant (2k+1)\pi^{-1}\mathfrak{M}[g]$, ($b$) and ($c$) give

$$\begin{aligned} |S_k(x; w)| &\leqslant 2\pi \frac{2k+1}{\pi} \sum_{j=i+1}^{\infty} A_{n_j}^{-\frac{1}{2}} \leqslant 2(2k+1) A_{n_{i+1}}^{-\frac{1}{2}} (1 + \tfrac{1}{2} + \tfrac{1}{4} + \ldots) \\ &< 12k A_{n_{i+1}}^{-\frac{1}{2}} \leqslant 12\nu_{n_i} A_{n_{i+1}}^{-\frac{1}{2}} < 12. \end{aligned}$$

This, together with (4·4) and (4·5), leads to

$$S_k(x; f) \geqslant A_{n_i}^{\frac{1}{2}} - 12.$$

Since each x in $(0, 2\pi)$ belongs to all E_{n_i} with i large enough, $S[f]$ diverges everywhere.

We pass to the construction of the F_n.

We fix n, write

$$x_j = \frac{2\pi j}{2n+1} \qquad (j = 0, 1, \ldots, 2n)$$

(so that x_{2j} is our previous a_j) and denote by I'_j the interval $(x_j - \delta, x_j + \delta)$, where δ is a fixed number, less than $\pi/(2n+1)$, to be determined later. The interval $(x_j + \delta, x_{j+1} - \delta)$ is denoted by I_j.

F_n is defined as a sum of two polynomials

$$F_n = f_n + \phi_n,$$

where f_n is of the type considered in the preceding section,

$$f_n(x) = \frac{1}{n} \sum_{i=1}^{n} K_{m_i}(x - x_{2i}), \tag{4·6}$$

though the m_i are not the same as before; f_n is non-negative and has constant term $\frac{1}{2}$. It will be shown that appropriate partial sums of f_n can be made large at each point of ΣI_j. This is no longer true for $x \in \Sigma I'_j$, and it is to overcome this difficulty that we consider $F_n = f_n + \phi_n$.

We deal with the polynomial ϕ_n first. We want it to be non-negative, to have constant term $\frac{1}{2}$ (so that $F_n = f_n + \phi_n$ will be non-negative with constant term 1) and to be large, say $\phi_n(x) \geqslant n$, in all intervals I'_j. This is possible, if δ is small enough, by taking, for example,

$$\phi_n(x) = K_m((2n+1)x), \tag{4·7}$$

where m is large enough. Denote the order of ϕ_n by m_0.

By a further reduction of δ we may suppose that

$$D_{m_0}(x) \geqslant 0 \quad \text{for} \quad x \in I'_0. \tag{4·8}$$

Having fixed δ and ϕ_n we proceed to the construction of f_n in (4·6), where $m_0 \leqslant m_1 < m_2 < \dots$. The m_j will be defined presently. If $m_j \leqslant k < m_{j+1}$, then

$$S_k(x; f_n) = \frac{1}{n}\sum_{i=1}^{j} K_{m_i}(x - x_{2i}) + \frac{1}{n}\sum_{i=j+1}^{n}\left\{\frac{1}{2} + \sum_{l=1}^{k}\frac{m_i - l + 1}{m_i + 1}\cos l(x - x_{2i})\right\},$$

and since $m_i - l + 1 = (m_i - k) + (k - l + 1)$ and $K_m \geqslant 0$, we have

$$S_k(x; f_n) \geqslant \frac{1}{n}\sum_{i=j+1}^{n}\frac{m_i - k}{m_i + 1}D_k(x - x_{2i}) \quad (m_j \leqslant k < m_{j+1}), \tag{4·9}$$

and so also

$$S_k(x; f_n) \geqslant -\frac{\sin(k+\frac{1}{2})x}{n}\sum_{i=j+1}^{n}\frac{m_i - k}{m_i + 1}\,\frac{1}{2\sin\frac{1}{2}(x_{2i} - x)}, \tag{4·10}$$

provided $2k+1$ is a multiple of $2n+1$.

We shall show in a moment that if $m_1 < m_2 < \dots < m_n$ increase fast enough, then to every $x \in I_{2j} + I_{2j+1}$ there corresponds an integer $k = k_x$ such that

(*a*) $2k+1$ is a multiple of $2n+1$,

(*b*) $m_j \leqslant k < \frac{1}{2}m_{j+1}$,

(*c*) $\sin(k+\frac{1}{2})x < -\frac{1}{2}$.

Take this result temporarily for granted. For such a k, (4·10) gives

$$S_k(x; f_n) \geqslant \frac{1}{4n}\sum_{i=j+1}^{n}\frac{1}{x_{2i} - x} > \frac{2n+1}{16\pi n}\sum_{i=j+1}^{n}\frac{1}{i-j} > C\log(n-j),$$

and if $j \leqslant n - \sqrt{n}$ we have

$$S_k(x; f_n) > C\log n. \tag{4·11}$$

Collecting results we come to the following conclusion about $F_n = f_n + \phi_n$: *if $x \in I_{2j} + I_{2j+1}$ and $0 \leqslant j \leqslant n - \sqrt{n}$, then there is an integer k satisfying $\frac{1}{2}m_{j+1} > k \geqslant m_j \geqslant m_0$ such that*

$$S_k(x; F_n) > C\log n. \tag{4·12}$$

For, with the previous k,

$$S_k(x; F_n) = S_k(x; f_n) + S_k(x; \phi_n) = S_k(x; f_n) + \phi_n \geqslant C\log n.$$

We now consider $S_p(x; F_n)$ in the intervals I'_j and show that

$$S_{m_0}(x; F_n) > \tfrac{1}{2}n \quad \text{for} \quad x \in I'_j \quad (j = 0, 1, \dots, 2n), \tag{4·13}$$

if n is sufficiently large.

In the equation
$$S_{m_0}(x; F_n) = S_{m_0}(x; f_n) + S_{m_0}(x; \phi_n),$$
the second term on the right exceeds n for $x \in \Sigma I'_j$. If we show that
$$S_{m_0}(x; f_n) > -C \log n, \quad \text{for} \quad x \in \Sigma I'_j \quad (n > 1), \tag{4·14}$$
we shall have
$$S_{m_0}(x; F_n) \geqslant n - C \log n > \tfrac{1}{2}n,$$
and (4·13) will be established.

Let therefore $x \in I'_l$ for some fixed l. We return to (4·9) and write $k = m_0$, $j = 0$. It is easy to see that if $x \in I'_l$, then the distance between x and x_{2i} is at least $\pi \,|\, l - 2i \,|/(2n+1)$. For l odd this gives
$$S_{m_0}(x; f_n) \geqslant -\frac{1}{n} \sum_{i=1}^{n} \frac{1}{2} \frac{\pi}{|\, x - x_{2i} \,|}$$
$$\geqslant -\frac{2n+1}{2n} \sum_{i=1}^{n} \frac{1}{|\, l - 2i \,|} \geqslant -C \log n.$$
If l is even, then $x - x_l \in I'_0$ and, by (4·8), $D_{m_0}(x - x_l) \geqslant 0$; hence
$$S_{m_0}(x; f_n) \geqslant -\frac{2n+1}{2n} \sum_{i=1}^{n}{}' \frac{1}{|\, l - 2i \,|} \geqslant -C \log n.$$
Thus we have proved (4·14) and so also (4·13).

Let E_n denote the interval $0 \leqslant x \leqslant 4\pi(n - \sqrt{n})/(2n+1)$. From (4·12) and (4·13) we see that if n is large enough then for each $x \in E_n$ we have $S_k(x; F_n) \geqslant C \log n$, where $k \geqslant m_0$. The sets E_n, the numbers $A_n = C \log n$ and the indices $\lambda_n = m_0$ $(= m_0^n)$ satisfy the conditions of Lemma (4·2).

It remains to show that if the m_i increase fast enough, then for each $x \in I_{2j} + I_{2j+1}$, $j = 0, 1, \ldots, n$, we can find an integer k satisfying conditions (*a*), (*b*) and (*c*) above. It is enough to show that if $m_0, m_1, \ldots, m_j$ have already been defined, we can find an integer m'_j such that (*a*) and (*c*) hold for some $k = k_x$ between m_j and m'_j; for then we may take for m_{j+1} any integer greater than $2m'_j$.

We fix j and write $2k + 1 = \rho(2n+1)$, where ρ is an (odd) integer, and
$$x_{2j+2} - x = 4\pi\theta/(2n+1), \quad \text{where} \quad 0 < \theta < 1.$$
Then
$$-\sin(k + \tfrac{1}{2})\, x = \sin(k + \tfrac{1}{2})(x_{2j+2} - x) = \sin 2\pi\rho\theta,$$
and x is in $I_{2j} + I_{2j+1}$ if and only if θ is in $S = (\eta, \tfrac{1}{2} - \eta) + (\tfrac{1}{2} + \eta, 1 - \eta)$, where η is positive, less than $\tfrac{1}{4}$, and depends only on δ and n (more precisely, on the ratio $\delta/(2n+1)$). For our purposes it is enough to show that if ρ_0 is odd and fixed then for every $\theta \in S$ there is an odd $\rho = \rho(\theta) \geqslant \rho_0$ such that
$$\sin 2\pi\rho\theta > \tfrac{1}{2}.$$
For then, by continuity, the last inequality is satisfied in an interval around θ, and using the Heine-Borel theorem we establish the existence of a finite upper bound for such $\rho(\theta)$.

Let $J = (\tfrac{1}{12}, \tfrac{5}{12})$. We have to show that if $\theta \in S$, then infinitely many of the numbers $\theta, 3\theta, 5\theta, \ldots$ are in J. For θ irrational, this follows from the fact that the latter sequence is equidistributed (see p. 142). Suppose now that $\theta = p/q$ is rational and irreducible and consider the cases (*a*) q odd, (*b*) q even.

In case (*a*), the q numbers $\rho_0 p, (\rho_0+2)p, \dots, (\rho_0+2q-2)p$ are all distinct mod q, so that when we divide them by q we obtain the ratios $0, 1/q, \dots, (q-1)/q$ as fractional parts. If $q>3$ at least one of the fractional parts must be in the interior of J since $|J|=\frac{1}{3}$; if $q=3$, J contains the fraction $\frac{1}{3}$. In case (*b*), q is even, hence p is odd. The $\frac{1}{2}q$ numbers $\rho_0 p, (\rho_0+2)p, (\rho_0+4)p, \dots, (\rho_0+q-2)p$ are odd and distinct mod q. Hence dividing them by q we get as fractional parts all the numbers $1/q, 3/q, \dots, (q-1)/q$. If $2/q<\frac{1}{3}$, that is, if $q>6$, at least one of these fractions is in J; for $q=4$ and $q=6$ we verify that $\frac{1}{4}$ and $\frac{1}{6}$ are in J (the case $q=2$ must be excluded since $\frac{1}{2}$ is not in S).

This completes the proof of (4·1).

In view of Theorem (3·14) we may ask if there exists an $S[f]$ oscillating finitely at *each* point. If true, this result lies a good deal deeper than Theorem (4·1) and requires completely different methods. For if $S_n(x;f)=O(1)$ at each x, then there is an interval (a,b) in which the $S_n(x;f)$ are uniformly bounded (Chapter I, (12·3)), and so also f is bounded. Hence, replacing f by 0 outside (a,b) we obtain a function f_1 bounded and such that $S[f_1]$ diverges (boundedly) at each point interior to (a,b). (It would then be easy to obtain a bounded function f_2 with $S[f_2]$ diverging everywhere.) However, the construction above gives divergence of $\{S_n(x;f)\}$ for indices $n=n_k$ tending to $+\infty$ as rapidly as we please, which is certainly impossible for bounded functions (Chapter XV, § 4).

MISCELLANEOUS THEOREMS AND EXAMPLES

1. In Theorem (8·9) of Chapter IV, summability (C, 1) cannot be replaced by ordinary convergence. More precisely, there exists an $f\in L$ and a set E such that if $S[f]$ is integrated termwise over E the resulting series diverges.

[A consequence of Chapter IV, (9·13).]

2. In Theorem (7·2) of Chapter III, summability A cannot be replaced by summability (C, 1). More precisely, there exists a function $F(x)$ having a finite (and so also a finite symmetric) derivative at the point $x=0$, but such that $S'[F]$ is not summable (C, 1) at $x=0$.

[Verify that $\int_0^\pi |K_n'(t)|\sin t\,dt \neq O(1)$. By Theorem (9·11) of Chapter IV, there exists a continuous function $g(t)$ with $\int_0^\pi g(t)\,K_n'(t)\sin t\,dt\neq O(1)$. Set $F(t)=g(t)\sin t$.]

3. Theorem (8·13) of Chapter II asserts that at every point x at which f has a jump d, the terms $\nu(b_\nu\cos\nu x - a_\nu \sin \nu x)$ of $S'[f]$ are summable to d/π by the method of the first logarithmic mean (Chapter III, § 9). In this result, logarithmic summability cannot be replaced by summability (C, 1). More precisely, there exists a continuous $f(x)$ such that $\sum_1^n \nu b_\nu \neq O(n)$.

[Observe that $\int_0^\pi |D_n'(t)|\,dt\neq O(n)$.]

4. A series $u_0+u_1+\dots+u_n+\dots$ is said to be summable by Borel's method, or summable B, to sum s, if

$$B(\lambda)=e^{-\lambda}\sum_{n=0}^{\infty} s_n\lambda^n/n! \to s \quad \text{as} \quad \lambda\to\infty,$$

where $s_n=u_0+u_1+\dots+u_n$. Show that

(i) if a series converges, it is summable B to the same sum;

(ii) a power series may be summable B outside its circle of convergence, so that the method B is rather strong; nevertheless

(iii) there exists a continuous $f(x)$ with $S[f]$ not summable B at some points. (Moore [1].)

[(i) Apply Chapter III, (1·2); (ii) the series $1+z+z^2+\dots$ is summable B for $\mathscr{R}z<1$; (iii) it is sufficient to show that the Lebesgue constants corresponding to the method B form an unbounded function. These constants are equal to

$$\frac{2}{\pi}\int_0^\pi e^{-\lambda(1-\cos t)}\frac{|\sin(\lambda\sin t+\frac{1}{2}t)|}{2\sin\frac{1}{2}t}\,dt,$$

and are of the order $\log\lambda$. Propositions (ii) and (iii) show that of the two methods B and (C, k), $k>0$, neither is stronger than the other.]

5. Let $B_\lambda(x,f)=e^{-\lambda}\sum_{n=0}^{\infty}S_n(x;f)\lambda^n/n!$ be the Borel means of $S[f]$. Show that at every point x_0 such that $\Phi_{x_0}(t)=o(t)$ (Chapter II, § 11), and in particular at every point of continuity,

$$B_\lambda(x_0,f)-\frac{1}{\pi}\int_{-\lambda^{-\frac{1}{2}}}^{\lambda^{-\frac{1}{2}}}f(x_0+t)\frac{\sin\lambda t}{t}\,dt\to 0$$

as $\lambda\to\infty$.

6. If $[f(x_0+t)-f(x_0)]\log 1/|t|\to 0$ as $t\to 0$, then $S[f]$ is summable B at x_0 to the value $f(x_0)$. (Hardy and Littlewood [2].)

[Apply the preceding result and observe that $\displaystyle\int_{1/n}^{1/n^{\frac{1}{2}}}\frac{dt}{t\log t}=O(1)$.]

7. Consider a sequence $p_0, p_1, p_2, \dots$ of positive numbers with the properties

$$P_n=p_0+p_1+\dots+p_n\to\infty,\quad p_n/P_n\to 0.$$

A series $u_0+u_1+\dots$, with partial sums s_n, is said to be summable by Nörlund's method corresponding to the sequence $\{p_\nu\}$, or summable N$\{p_\nu\}$, to sum s, if

$$\sigma_n=(s_0p_n+s_1p_{n-1}+\dots+s_np_0)/P_n=(u_0P_n+u_1P_{n-1}+\dots+u_nP_0)/P_n$$

tends to s as $n\to\infty$.

If $P_n=A_n^\alpha$, $\alpha>0$, we obtain as a special case Cesàro's method of summation. Show that

(i) if Σu_n converges, it is summable N$\{p_\nu\}$ to the same sum;

(ii) if $0<p_0\leqslant p_1\leqslant\dots$, and if Σu_ν is summable (C, 1), it is also summable N$\{p_\nu\}$ to the same sum.

[For more facts, and literature, concerning the methods N$\{p_\nu\}$, see, for example, Hardy's *Divergent series*.]

8. Let $p_\nu>p_{\nu+1}\to 0$, $P_\nu\to\infty$. A necessary and sufficient condition that the method N$\{p_\nu\}$ should sum $S[f]$ to sum $f(x_0)$ at every point x_0 of continuity of f, is that the sequence

$$\lambda_n=P_n^{-1}\sum_1^n P_\nu/\nu$$

should be bounded.

[For details, see Hille and Tamarkin [1_{II}] (or the first edition of this book, p. 186); also Karamata [6]

9. Using the polynomials (1·4) show that for any positive sequence $\epsilon_n\to 0$ there exists a continuous function f such that $|S_n(0;f)|\geqslant\epsilon_n\log n$ for infinitely many n (compare (1·2)).

10. Let $0<\alpha<1$. Show that the function

$$f(x)=\sum_1^\infty 4^{-k\alpha}Q(x,2.4^k,4^k)$$

(the Q's being defined by (1·4)) belongs to Λ_α and that $|S_m(0;f)-f(0)|\geqslant Cm^{-\alpha}\log m$ for infinitely many m. Thus the factor $\log n$ in Theorem (10·8) of Chapter II cannot be omitted. (Lebesgue [1].)

[The proof is similar to that of (2·1). For $\alpha=1$, f is of the class Λ_* but not of the class Λ_1. To obtain an example from the latter class a different argument is needed.]

11. There is an $f\in L$ with the following property: for almost every x we can find a sequence $q_1<q_2<\dots$ (depending in general on x) such that

$$\text{(i) } |S_{q_k}(x;f)|\to\infty,\quad \text{(ii) } \Sigma 1/q_k=\infty.$$

(This is not possible if $S[f]$ is of power series type; cf. Chapter XV, (4·3), (5·10).) [The proof is similar to that of (3·1).]

CHAPTER IX

RIEMANN'S THEORY OF TRIGONOMETRIC SERIES

1. General remarks. The Cantor-Lebesgue theorem

In the previous chapters we have almost exclusively been concerned with Fourier series. We shall now prove a number of properties of trigonometric series

$$\tfrac{1}{2}a_0 + \sum_{n=1}^{\infty} (a_n \cos nx + b_n \sin nx), \tag{1·1}$$

with coefficients tending to 0, but otherwise quite arbitrary. Riemann found the first fundamental results in the subject, and these with their subsequent extensions constitute what is now called the *Riemann theory of trigonometric series*. The chief points of the theory are the *problem of uniqueness* and the *problem of localization*, which we now proceed to discuss.

In this chapter we suppose, unless otherwise stated, that the coefficients of the trigonometric series considered tend to 0. We recall the notation

$$\tfrac{1}{2}a_0 + \sum_{n=1}^{\infty} (a_n \cos nx + b_n \sin nx) = \sum_{n=0}^{\infty} A_n(x),$$

$$\sum_{n=1}^{\infty} (a_n \sin nx - b_n \cos nx) = \sum_{n=1}^{\infty} B_n(x),$$

$$A_n(x) = \rho_n \cos(nx + \alpha_n) \quad (\rho_n^2 = a_n^2 + b_n^2,\ \rho_n \geq 0).$$

(1·2) THE CANTOR-LEBESGUE THEOREM. *If $A_n(x) \to 0$ for x belonging to a set E of positive measure, then $a_n, b_n \to 0$.*

If ρ_n does not tend to 0, there exists a sequence $n_1 < n_2 < \dots$ of indices and an $\epsilon > 0$ such that $\rho_{n_k} > \epsilon$ for all k. From this and the relation $\rho_n \cos(nx + \alpha_n) \to 0$ in E we see that $\cos(n_k x + \alpha_{n_k}) \to 0$ in E. *A fortiori*, $\cos^2(n_k x + \alpha_{n_k}) \to 0$ in E. The terms of the last sequence do not exceed 1, so that, by Lebesgue's theorem on the integration of bounded sequences, the integral

$$\int_E \cos^2(n_k x + \alpha_{n_k})\, dx$$

tends to 0 whereas, after Theorem (4·5) of Chapter II, it tends to $\tfrac{1}{2}|E|$. The contradiction proves (1·2).

(1·3) THEOREM. *If $\Sigma A_n(x)$ converges in a set E of positive measure, then $a_n \to 0$, $b_n \to 0$. More generally, if $\Sigma A_n(x)$ is summable* (C, k), $k > -1$, *in a set E of positive measure, then $a_n = o(n^k)$, $b_n = o(n^k)$.*

We prove the second statement. A consequence of the hypothesis is that

$$a_n n^{-k} \cos nx + b_n n^{-k} \sin nx \to 0 \quad (x \in E),$$

(Chapter III, (1·22)) so that, by (1·2), $a_n n^{-k} \to 0$, $b_n n^{-k} \to 0$.

(1·4) THEOREM. *If $A_n(x)=O(1)$ for each $x\in E$, $|E|>0$, then $a_n=O(1)$, $b_n=O(1)$. If $\Sigma A_n(x)$ is finite* (C, k), $k>-1$, *in E, $|E|>0$, then $a_n=O(n^k)$, $b_n=O(n^k)$.*

The proof is left to the reader.

(1·5) THEOREM. *Let $\chi(n)$ be a positive sequence tending monotonically to* 0. *Suppose that $\Sigma A_n(x)$ converges in E, $|E|>0$, to sum $f(x)$ and that*

$$s_{n-1}(x)-f(x)=o\{\chi(n)\}$$

for each $x\in E$. Then $a_n=o\{\chi(n)\}$, $b_n=o\{\chi(n)\}$.

The result holds if the 'o' is replaced by 'O' throughout.

Considering the 'o' case, set $r_n(x)=f(x)-s_n(x)$. Then

$$A_n(x)=r_{n-1}(x)-r_n(x)=o\{\chi(n)\}+o\{\chi(n+1)\}=o\{\chi(n)\}$$

in E, and the conclusion follows from (1·2).

(1·2) asserts that if $|a_n|+|b_n|$ does not tend to 0, then $A_n(x)$ cannot tend to 0 except possibly in a set of measure 0. We shall investigate the nature of this set.

A set E is said to be of *type* H, if there is a sequence of positive integers $n_1<n_2<n_3<\dots$ and an interval Δ such that for each $x\in E$ no point

$$n_kx \quad (k=1, 2, 3, \dots)$$

is in $\Delta(\text{mod } 2\pi)$.

The closure of a set of type H is also of type H. A set of type H is non-dense.

Recalling the definition of an equidistributed sequence of numbers (Chapter IV, § 4) we see that if E is of type H, then for a suitable $\{n_k\}$ and each $x\in E$ the sequence n_kx, reduced mod 2π, is non-equidistributed over $(0, 2\pi)$ in a rather strong sense; not only do some intervals get less than their proper share of points n_kx, but there even exist intervals totally devoid of such points.

The following is an equivalent definition of sets of type H: a set $E\subset(0, 2\pi)$ is of type H if there exist integers $0<n_1<n_2<\dots$, a number α and a number $\delta<1$ such that

$$\cos(n_kx+\alpha)<\delta \quad \text{for} \quad x\in E \quad (k=1, 2, \dots). \tag{1·6}$$

We can construct a set of type H as follows. Denote by $\langle x\rangle$ the fractional part of x,

$$\langle x\rangle=x-[x],$$

fix a number d, $0<d<1$, and a number a, and consider for each integer $n>0$ the set E_n of points x such that

$$\left\langle\frac{nx-a}{2\pi}\right\rangle\leqslant d. \tag{1·7}$$

Then for any sequence $0<n_1<n_2<\dots$, the set

$$E=E_{n_1}E_{n_2}\dots E_{n_k}\dots$$

(possibly empty) is of type H. For if $x\in E$, then $x\in E_{n_k}$; thus n_kx is situated, mod 2π, in the interval $(a, a+2\pi d)$, and so never enters $\Delta=(a+2\pi d, a+2\pi)$.

The set E_n of points x satisfying (1·7) consists of intervals of length $2\pi d/n$ separated by intervals of length $2\pi(1-d)/n$. If we identify points congruent mod 2π (that is, if we consider E_n on the circumference of the unit circle), then E_n consists of n equal and equally distributed intervals, and $|E_n|=2\pi d$, a quantity independent of n and a.

The Cantor ternary set C constructed on $(0, 2\pi)$ is of type H with

$$n_k = 3^{k-1}, \quad \Delta = (\tfrac{2}{3}\pi, \tfrac{4}{3}\pi) \quad (k = 1, 2, \ldots).$$

For, as we easily see geometrically or deduce from the parametric representation of the points of C (Chapter V, §3), the set of the points $3^{k-1}x$, $x \in C$, is, when reduced mod 2π, identical with C.

The same argument shows generally that a perfect set of constant ratio of dissection $(\xi, 1-2\xi, \xi)$ is an H set if $\xi = 1/q$, $q = 3, 4, 5, \ldots$.

The following observation is useful. Suppose that in the definition of a set of type H we make Δ depend on k; that is, we assume that for $x \in E$ the point $n_k x$, reduced mod 2π, does not enter an interval $\Delta_k = (a_k, b_k)$. Then the set E is still of type H, provided the lengths of the Δ_k stay above a positive number η. For selecting a sequence k_j such that a_{k_j} and b_{k_j} converge, say to limits a and b, we see that $b - a \geq \eta$ and that any interval Δ' interior to (a, b) has the following property: for j large enough and for each $x \in E$, $n_{k_j} x$ never enters Δ' (mod 2π). Hence E is of type H.

Correspondingly, in (1·6) we may make α depend on k.

We may present the result in a slightly different form. Given any periodic set $\mathscr{E}$, denote by $\mathscr{E}^n$ the set of points nx, where $x \in \mathscr{E}$, and let l_n be the upper bound of the lengths of intervals without points in common with $\mathscr{E}^n$ (if $\mathscr{E}$ is closed, l_n is the length of the largest interval contiguous to $\mathscr{E}^n$). *$\mathscr{E}$ is of type* H *if and only if* $\limsup l_n > 0$.

A denumerable sum of sets H is called an H_σ *set*. H_σ sets are of the first category. They are also of measure 0. This is a corollary of the following theorem:

(1·8) THEOREM. *A set of type* H *is of measure* 0.

Let E be a set of type H associated with a sequence $\{n_k\}$ and an interval Δ. Let E_n be the set of points x such that nx is not in Δ (mod 2π). Clearly $E \subset \mathscr{E} = E_{n_1} E_{n_2} \ldots$, and it is enough to show that $|\mathscr{E}| = 0$.

Let $d = (2\pi - |\Delta|)/2\pi$. If we omit factors in $E_{n_1} E_{n_2} \ldots$ the product can only increase. Also, if S is any finite system of intervals, then $|SE_n| \to d|S|$ as $n \to \infty$. Let now $d < d_1 < 1$, $m_1 = n_1$, and suppose that $m_1, m_2, \ldots, m_{k-1}$ have been defined. If

$$S_{k-1} = E_{m_1} E_{m_2} \ldots E_{m_{k-1}},$$

we find an $m_k > m_{k-1}$ in $\{n_i\}$ such that

$$|S_{k-1} E_{m_k}| \leq d_1 |S_{k-1}|.$$

It follows that
$$|E_{m_1} E_{m_2} \ldots E_{m_k}| \leq 2\pi d_1^k.$$

Hence $|E_{m_1} E_{m_2} \ldots| = 0$ and, *a fortiori*, $|\mathscr{E}| = |E_{n_1} E_{n_2} \ldots| = 0$.

(1·9) THEOREM. *Except perhaps in a set of type* H_σ *(and measure* 0*) we have, as* $n \to \infty$,

$$\limsup A_n(x) = \limsup \rho_n, \quad \liminf A_n(x) = -\limsup \rho_n, \tag{1·10}$$

and, in particular, $\limsup |A_n(x)| = \limsup \rho_n$.

We first show that if $\alpha_1, \alpha_2, \ldots$ are real, and $n_1 < n_2 < \ldots$ positive integers, then, except perhaps in a set of type H_σ, we have

$$\limsup \cos(n_k x + \alpha_k) = +1, \quad \liminf \cos(n_k x + \alpha_k) = -1. \tag{1·11}$$

It is enough to prove that the first of these holds outside a set E of type H_σ. Let $0<\delta<1$, and denote by Z_i^δ the set of x such that

$$\cos(n_k x+\alpha_k)\leqslant\delta \quad \text{for} \quad k\geqslant i. \tag{1·12}$$

In view of the observation made above, Z_i^δ is an H set. Outside the H_σ set

$$Z^\delta = Z_1^\delta + Z_2^\delta + \dots$$

we have
$$\limsup\cos(n_k x+\alpha_k)\geqslant\delta,$$

and outside the H_σ set
$$E=Z^{\frac{1}{2}}+Z^{\frac{2}{3}}+Z^{\frac{3}{4}}+\dots$$

we have the first equation (1·11).

Return to (1·9). It is enough to consider the first equality (1·10). Let $\{n_k\}$ be such that $\limsup\rho_n=\lim\rho_{n_k}$. Outside the set E just considered (with α_{n_k} for α_k) we have

$$\limsup A_n(x)\geqslant\limsup A_{n_k}(x)=\lim\rho_{n_k}=\limsup\rho_n,$$

that is, $\limsup A_n(x)\geqslant\limsup\rho_n$. Since the opposite inequality is valid for all x, the first equation (1·10) holds in the complement of E. This completes the proof of (1·9).

A corollary of (1·9) is that if $|a_n|+|b_n|$ does not tend to 0, $A_n(x)$ can tend to 0 at most in a set H_σ. Hence (H_σ being of the first category), *if $A_n(x)$ tends to 0 in a set of the second category, then a_n and b_n tend to* 0.

2. Formal integration of series

Given a series

$$\tfrac{1}{2}a_0+\sum_{n=1}^{\infty}(a_n\cos nx+b_n\sin nx), \tag{2·1}$$

with $a_n, b_n\to 0$, consider the function

$$F(x)=\tfrac{1}{4}a_0x^2-\sum_{n=1}^{\infty}\frac{a_n\cos nx+b_n\sin nx}{n^2}, \tag{2·2}$$

obtained by integrating the series (2·1) formally twice. $F(x)$ is continuous since its series converges absolutely and uniformly. It is readily seen that

$$\frac{F(x+2h)+F(x-2h)-2F(x)}{4h^2}=A_0+\sum_{n=1}^{\infty}A_n(x)\left(\frac{\sin nh}{nh}\right)^2. \tag{2·3}$$

The numerator of the ratio on the left will be denoted by

$$\Delta^2F(x,2h).$$

We denote the upper and lower limits of $\Delta^2F(x,h)/h^2$, as $h\to0$, by $\overline{D}^2F(x)$ and $\underline{D}^2F(x)$ respectively; if $\overline{D}^2F(x)=\underline{D}^2F(x)$, the common value is denoted by $D^2F(x)$ and is called the *second symmetric derivative* of F at the point x. If $D^2F(x_0)$ exists we say that (2·1) is, at the point x_0, summable by the *Riemann method of summation*, or summable R, to sum $D^2F(x_0)$.

(2·4) Riemann's First Theorem. *If $\Sigma A_n(x)$, with $a_n, b_n\to0$, converges at a point x to a finite sum s, it is also summable* R *to s.*

We have to show that $\Delta^2F(x,2h_i)/4h_i^2\to s$ for every positive $\{h_i\}\to0$. Set

$$s_n=A_0+A_1+\dots+A_n, \quad (\sin^2h)/h^2=u(h).$$

Applying summation by parts we find that the right-hand side of (2·3) is, for $h=h_i$, equal to

$$\sum_{n=0}^{\infty} s_n\{u(nh_i)-u((n+1)h_i)\}. \tag{2·5}$$

This is a linear transformation of the sequence $\{s_n\}\to s$, and it is enough to show that the transformation satisfies the conditions of regularity (Chapter III, § 1). This is obvious for conditions (i) and (iii). To verify (ii) we observe that

$$\sum_{n=0}^{\infty} |u(nh_i)-u((n+1)h_i)| = \sum_{n=0}^{\infty}\left|\int_{nh_i}^{(n+1)h_i} u'(t)\,dt\right| \leqslant \int_0^{\infty} |u'(t)|\,dt, \tag{2·6}$$

and the last integral is finite since the integrand is $O(t^{-2})$ for $t\to\infty$.

Theorem (2·4) can be generalized as follows:

(2·7) THEOREM. *If $\Sigma A_n(x)$ has partial sums $s_n(x)$ bounded at x, and if*

$$\underline{s}(x)=\liminf s_n(x), \quad \bar{s}(x)=\limsup s_n(x),$$

then the numbers $\underline{D}^2F(x)$ and $\bar{D}^2F(x)$ are both contained in the interval $(s-k\delta, s+k\delta)$, where $s=\frac{1}{2}\{\bar{s}(x)+\underline{s}(x)\}$, $\delta=\frac{1}{2}\{\bar{s}(x)-\underline{s}(x)\}$, and k is an absolute constant.

This follows from Theorem (1·5) of Chapter III; k is the upper bound, for all $\{h_i\}$, of the sums on the left in (2·6).

(2·8) RIEMANN'S SECOND THEOREM. *If $a_n, b_n\to 0$, then*

$$\frac{F(x+2h)+F(x-2h)-2F(x)}{4h} = A_0h+\sum_{n=1}^{\infty} A_n\frac{\sin^2 nh}{n^2h}\to 0, \tag{2·9}$$

as $h\to 0$, uniformly in x.

It is again enough to prove that (2·9) holds for each positive $\{h_i\}\to 0$. The series in (2·9) is a linear transformation of the sequence $A_n\to 0$, so it is enough to verify conditions (i) and (ii) of regularity. (Condition (iii) is irrelevant here.) Condition (i) is obviously satisfied. To verify (ii) we observe that

$$h_i+\sum_{n=1}^{\infty}\frac{\sin^2 nh_i}{n^2h_i} \leqslant h_i+\sum_{n=1}^{N}\frac{n^2h_i^2}{n^2h_i}+\sum_{n=N+1}^{\infty}\frac{1}{n^2h_i} < (N+1)h_i+1/Nh_i. \tag{2·10}$$

If we set $N=[1/h_i]+1$, then $1/h_i<N\leqslant 1+1/h_i$ and the right-hand side† of (2·10) is less than 4 for $|h_i|\leqslant 1$. This completes the proof of (2·8).

It is clear that if a_n and b_n are $O(1)$, the ratio in (2·9) is uniformly bounded as $h\to 0$.

The relation (2·9) is satisfied at every point x, irrespective of the convergence or divergence of $\Sigma A_n(x)$. Following the terminology of Chapter II, § 3, we may say that $F(x)$ is *uniformly smooth*, or *belongs to the class* λ_*, or *satisfies condition* λ_*. The sum of any trigonometric series with coefficients $o(1/n^2)$ satisfies condition λ_*; for it can be obtained by twice formally integrating a series with coefficients tending to 0. If the coefficients of $\Sigma A_n(x)$ are $O(1/n^2)$, the sum of the series satisfies condition Λ_*.

The following theorem is a generalization of these results:

(2·11) THEOREM. *If the $\rho_n=(a_n^2+b_n^2)^{\frac{1}{2}}$ satisfy the condition*

$$\sigma_n=\sum_{k=1}^{n} k^2\rho_k=o(n), \tag{2·12}$$

† The left-hand side of (2·10) can also be summed exactly. For, substituting $x=0$ in the development (4·16) of Chapter I, we find

$$\tfrac{1}{2}h+\sum_{1}^{\infty}\frac{\sin^2 nh}{n^2h}=\tfrac{1}{2}\pi \quad \text{for} \quad 0<h\leqslant\tfrac{1}{2}\pi.$$

then $\Sigma A_n(x)$ *converges absolutely and uniformly, and its sum* $f(x)$ *satisfies condition* λ_*. *If we have* 'O' *in* (2·12) *instead of* 'o', *then* $f \in \Lambda_*$.

It is enough to consider the 'o' case. If $\rho_n = o(1/n^2)$, (2·11) is immediate. To prove the absolute and uniform convergence of (2·1) in the general case we observe that

$$\sum_1^\infty \rho_k = \lim_{N\to\infty} \sum_1^N (\sigma_k - \sigma_{k-1}) k^{-2} = \sum_1^\infty \sigma_k (k^{-2} - (k+1)^{-2}) = \sum_1^\infty o(k)\, O(k^{-3}) < \infty. \quad \textbf{(2·13)}$$

Now, supposing for simplicity that $a_0 = 0$,

$$\left|\frac{f(x+2h)+f(x-2h)-2f(x)}{4h}\right| \leqslant h^{-1} \sum_1^\infty \rho_k \sin^2 kh$$
$$\leqslant h \sum_1^N k^2 \rho_k + h^{-1} \sum_{N+1}^\infty \rho_k = P + Q,$$

say. Set $N = [1/h]$. Then $P \leqslant \sigma_N / N = o(1)$ as $h \to 0$, and arguing as in (2·13) we have

$$Q \leqslant h^{-1} \sum_{N+1}^\infty \sigma_k (k^{-2} - (k+1)^{-2}) = h^{-1} \sum_{N+1}^\infty o(k^{-2}) = h^{-1} o(N^{-1}) = o(1).$$

Hence $P + Q = o(1)$ and (2·11) follows.

Set

$$\rho_n = \eta_n / n \quad (n = 1, 2, \ldots).$$

The condition $\sum_1^n k\eta_k = o(n)$, identical with (2·12), is satisfied in the following two special cases (see Chapter III, p. 79).

(i) $\Sigma n \eta_n^2 < \infty$;

(ii) $\eta_n \to 0$; $\eta_n = 0$ except possibly for a sequence n_k such that $n_{k+1}/n_k > q > 1$.

The idea of associating with $\Sigma A_n(x)$ the function $F(x)$ of (2·2) is due to Riemann. In some cases we may also consider the function

$$L(x) = \tfrac{1}{2} a_0 x + \sum_1^\infty (a_n \sin nx - b_n \cos nx)/n, \qquad \textbf{(2·14)}$$

obtained from (2·1) by a single integration. Then

$$\frac{L(x+h) - L(x-h)}{2h} = A_0 + \sum_1^\infty A_n(x) \frac{\sin nh}{nh}. \qquad \textbf{(2·15)}$$

The difficulty in using $L(x)$ is that the series (2·14) need not converge everywhere even if $\Sigma A_n(x)$ does. (A simple example is provided by the series $\Sigma(\sin nx)/\log n$.) Furthermore, even if (2·14) does converge everywhere, $L(x)$ need not be a continuous function.

On the other hand, the series (2·14) converges almost everywhere. For its periodic part is the Fourier series of a function in L² (Chapter IV, § 1), and since this has terms $o(1/n)$ it must be convergent wherever it is summable (C, 1).

If $L(x)$ exists in the neighbourhood of a point x_0 and if

$$\{L(x_0 + h) - L(x_0 - h)\}/2h \to s \quad \text{as} \quad h \to 0,$$

we say that $\Sigma A_n(x)$ is summable at the point x_0 by the *Lebesgue method* of summation, or *summable* L, to sum s.

(2·16) THEOREM. *Suppose that a_n and b_n are $o(1/n)$ or, more generally, that*

$$\sum_1^n k\rho_k = o(n). \tag{2·17}$$

Then, if $s_n(x)$ are the partial sums of $\Sigma A_n(x)$ and $N = N(h) = [1/h]$, we have, uniformly in x,

$$\frac{L(x+h)-L(x-h)}{2h} - s_N(x) \to 0$$

as $h \to 0$. In particular, a necessary and sufficient condition for $\Sigma A_n(x)$ to converge at x to sum s (finite or infinite) is that it should be summable L *at x to sum s.*

From (2·17) and Theorem (2·11) we deduce that the function $L(x)$ exists everywhere and satisfies condition λ_*. The difference last written is

$$\sum_1^N A_n\left(\frac{\sin nh}{nh} - 1\right) + \sum_{N+1}^\infty A_n \frac{\sin nh}{nh} = P + Q.$$

Denote by τ_n the left-hand side of (2·17). Since $(\sin u)/u - 1 = O(u^2) = O(u)$ for $|u| \leqslant 1$, we have

$$|P| = O\left\{h\sum_1^N n\rho_n\right\} = O\{N^{-1}\tau_N\} = o(1),$$

$$|Q| \leqslant h^{-1}\sum_{N+1}^\infty \frac{\rho_n}{n} = h^{-1}\sum_{N+1}^\infty \frac{\tau_n - \tau_{n-1}}{n^2} \leqslant h^{-1}\sum_{N+1}^\infty \tau_n\left(\frac{1}{n^2} - \frac{1}{(n+1)^2}\right) = O(N)\sum_{N+1}^\infty o\left(\frac{1}{n^2}\right) = o(1).$$

Hence $P + Q = o(1)$ and (2·16) follows.

Part of Theorem (2·16) can be generalized as follows:

(2·18) THEOREM. *If a_n and b_n are $O(1/n)$, then $\Sigma A_n(x)$ is summable* L *at x_0 to a finite sum s if and only if it converges at x_0 to s.*

Let $\Sigma A_n(x)$ converge at x_0 to s. We may suppose that $s = 0$. Write

$$\frac{L(x_0+h)-L(x_0-h)}{2h} = \left\{\tfrac{1}{2}a_0 + \sum_1^{kN} A_n(x_0)\frac{\sin nh}{nh}\right\} + \left\{\sum_{kN+1}^\infty A_n(x_0)\frac{\sin nh}{nh}\right\} = P + Q,$$

where $N = [1/h]$ and k is a fixed integer. From

$$|Q| \leqslant h^{-1}\sum_{kN+1}^\infty |A_n(x_0)|\, n^{-1} \leqslant O(N)\sum_{kN+1}^\infty O(n^{-2}) = O(N)\,O(1/(kN)),$$

we see that $|Q|$ is arbitrarily small if k is large enough. By Theorem (12·2) of Chapter III, P tends to 0 as $h \to 0$. It follows that $\{L(x_0+h) - L(x_0-h)/2h\} \to 0$ as $h \to 0$.

Conversely, let $\Sigma A_n(x)$ be summable L at x_0 to sum s. We may suppose that $a_0 = 0$. The linear term in (2·14) then disappears and $\Sigma A_n(x)$ is $S'[L]$. By Theorem (7·9) of Chapter III, $\Sigma A_n(x)$ is, at x_0, summable A to sum s, and so also converges to s since its terms are $O(1/n)$ (Chapter III, (1·38)).

While a series $\Sigma A_n(x)$ with coefficients $o(1/n)$ is a Fourier series, this no longer holds if we impose only the condition (2·17); the lacunary series $\Sigma n^{-\frac{1}{2}} \cos 2^n x$ is an instance in point (see Chapter V, §6).

(2·19) THEOREM. *Suppose that $\Sigma A_n(x)$ satisfies (2·17) and is $S[f]$. Then*

(i) *if $f(x) \to s$, $-\infty \leqslant s \leqslant +\infty$, as $x \to x_0 + 0$, the series converges at x_0 to sum s;*

(ii) *if $f(x)$ is continuous† in an interval $a \leqslant x \leqslant b$, it converges uniformly in (a, b).*

† At a and b only one-sided continuity is required.

(i) $L(x)$ has at x_0 a right-hand derivative equal to s. Since $L(x)$ is smooth, the left-hand derivative at x also exists and equals s. Hence $\{L(x_0+h)-L(x_0-h)\}/2h \to s$ and, by (2·16), $s_N(x) \to s$.

(ii) If $h \to +0$, then $\{L(x+h)-L(x)\}/h \to f(x)$ uniformly in the interval

$$I: \; a \leqslant x \leqslant a+\tfrac{1}{2}(b-a).$$

Since $L \in \lambda_*$, we have $\{L(x)-L(x-h)\}/h \to f(x)$, and so also

$$\{L(x+h)-L(x-h)\}/2h \to f(x),$$

uniformly in I. The last relation holds uniformly also in the remaining part of (a,b), by an analogous argument, and it only remains to apply (2·16).

If we were to assume the two-sided continuity of f at a and b, the uniform convergence of $\Sigma A_n(x)$ in (a,b) would follow at once from the uniform summability (C, 1) there since, owing to (2·17), the s_n and the (C, 1) means of $\Sigma A_n(x)$ are uniformly equiconvergent.

We recall the following definition (Chapter II, § 3). A function $f(x)$ defined on a set E is said to have property D, if $f(x)$ takes all intermediate values; i.e. for any two points x_1 and x_2 in E and any number η between $f(x_1)$ and $f(x_2)$, there is a point $\xi \in E$ between x_1 and x_2 such that $f(\xi)=\eta$.

(2·20) THEOREM. *Suppose that* $\Sigma A_n(x)$ *satisfies* (2·17). *Then the set E of the points of convergence of the series is of the power of the continuum in every interval, and the sum* $f(x)$ *has property* D *on E.*

After (2·11), the function $L(x)$ in (2·14) *is in the class* λ_*. By Chapter II, (3·3), (3·6), a finite derivative $L'(x)$ exists, and *has property* D, in a set E which is of the power of the continuum in every interval. Since for smooth functions the existence of $L'(x)$ is equivalent to the existence of the first symmetric derivative, it is enough to apply (2·16).

Example. As the series

$$\Sigma \nu^{-\frac{1}{2}} \cos 10^{\nu} x \qquad \textbf{(2·21)}$$

shows, the set E in (2·20) may be of measure 0 (Chapter V, (6·4)). In every interval we can find points x_1 and x_2 such that $\limsup s_n(x_1) = +\infty$, $\liminf s_n(x_2) = -\infty$, s_n being the partial sums of (2·21) (Chapter VI, p. 250). It follows from (2·16) that the ratio $\{L(x+h)-L(x-h)\}/2h$ has at x_1 the limit superior $+\infty$ and at x_2 the limit inferior $-\infty$. By the Remark in Chapter II, p. 44, we deduce that given any finite η there is a point ξ between x_1 and x_2 such that (2·21) converges at ξ to sum η. Hence though the series (2·21) diverges almost everywhere it converges to any given sum at some point of any interval.

Given a function $f(x)$ on a (measurable) set E and a point x_0, we say that $f(x)$ has an *approximate limit as* $x \to x_0$, in symbols

$$\lim_{x \to x_0} \operatorname{ap} f(x) = s,$$

if $f(x)$ tends to s as x tends to x_0 through a set $\mathscr{E}$, subset of E, having x_0 as a point of density. Since two sets having x_0 as a point of density must have points in common in each neighbourhood of x_0, there is at most one approximate limit.

Correspondingly, the two limits

$$\lim_{h\to 0}\text{ap}\,\{f(x_0+h)-f(x_0)\}/h,\quad \lim_{h\to 0}\text{ap}\,\{f(x_0+h)-f(x_0-h)\}/2h$$

are called the *approximate* and the *symmetric approximate* derivative of f at x_0 and denoted by $f'_{\text{ap}}(x_0)$ and $f'_{\text{sap}}(x_0)$.

The following theorem shows the effect of a single integration on convergent trigonometric series:

(2·22) Theorem. *If $a_n, b_n \to 0$ and if $\Sigma A_n(x)$ converges at x_0 to a (finite) sum s, then $L'_{\text{sap}}(x_0)$ exists and equals s.*

We may suppose that $a_0=0$ and $s=0$. $L(x)$ exists almost everywhere. If $L(x_0\pm h)$ exist, then

$$\frac{L(x_0+h)-L(x_0-h)}{2h}=\sum_1^{2^N} A_n(x_0)\frac{\sin nh}{nh}+\sum_{2^N+1}^{\infty} A_n(x_0)\frac{\sin nh}{nh}=P(h)+Q(h),$$

where the integer N is defined by the condition $2^{-(N+1)}<h\leqslant 2^{-N}$. A simple summation by parts (or an application of Theorem (12·2) of Chapter III) shows that

$$P(h)\to 0\quad (h\to 0).$$

Applying summation by parts to $Q(h)$, and writing $s_n(x_0)=s_n$, we have

$$\begin{aligned}Q(h)&=o(1)+\sum_{2^N+1}^{\infty} s_n\left(\frac{\sin nh}{nh}-\frac{\sin (n+1)h}{(n+1)h}\right)\\ &=o(1)+\sum_{2^N+1}^{\infty} s_n\frac{\sin nh}{n(n+1)h}-\frac{2\sin\frac{1}{2}h}{h}\sum_{2^N+1}^{\infty}\frac{s_n\cos(n+\frac{1}{2})h}{n+1}\\ &=o(1)+Q_1(h)+Q_2(h),\end{aligned}$$

say. Clearly $\quad |Q_1|\leqslant h^{-1}\max_{n>2^N}|s_n|\sum_{2^N+1}^{\infty}\frac{1}{n(n+1)}=o(h^{-1}2^{-N})=o(1).$

Hence, collecting results, we see that (2·22) will be proved if we show that

$$\lim_{h\to 0}\text{ap}\, Q_2(h)=0.$$

Since

$$\begin{aligned}\int_{2^{-N-1}}^{2^{-N}} Q_2^2(h)\,dh&\leqslant\int_0^{\pi}\left(\sum_{2^N+1}^{\infty} s_n\frac{\cos(n+\frac{1}{2})h}{n+1}\right)^2 dh=2\int_0^{\frac{1}{2}\pi}\left(\sum_{2^N+1}^{\infty} s_n\frac{\cos(2n+1)t}{n+1}\right)^2 dt\\ &\leqslant\pi\sum_{2^N+1}^{\infty}\frac{s_n^2}{(n+1)^2}=o(2^{-N}),\end{aligned}$$

the integral on the left is $\epsilon_N 2^{-N-1}$, where $\epsilon_N\to 0$. Let E^N be the set of those h in $(2^{-N-1}, 2^{-N})$ at which $Q_2(h)$ exists and $|Q_2(h)|\geqslant\epsilon_N^{\frac{1}{4}}$. Then

$$\epsilon_N^{\frac{1}{2}}|E^N|\leqslant\epsilon_N 2^{-N-1},\quad |E^N|/2^{-(N+1)}\leqslant\epsilon_N^{\frac{1}{2}},$$

that is, the average density of E^N in $(2^{-N-1}, 2^{-N})$ does not exceed $\epsilon_N^{\frac{1}{2}}$. It follows immediately that the set $E=E^1+E^2+\dots$ has density 0 at $h=0$, and that $Q_2(h)\to 0$ as h tends to 0 outside E. Hence $\lim\text{ap}\, Q_2(h)=0$, and the theorem is proved.

Remark. Even without the hypothesis $a_n, b_n\to 0$ the convergence of $\Sigma A_n(x_0)$ to s implies that $A_n(x_0)\to 0$, so that $\Sigma A_n(x_0)\dfrac{\sin nh}{nh}$ converges for almost all h, and, as the foregoing proof shows, the sum of the last series has approximate limit s as $h \to 0$. The hypothesis $a_n, b_n = o(1)$ was used only to establish the existence of the function $L(x)$. (It would be enough to suppose that $a_n, b_n = O(n^{\frac{1}{2}-\epsilon})$, $\epsilon > 0$; see Chapter IV, (1·1).)

We now consider the special case when $\Sigma A_n(x)$ is of power-series type

$$\sum_0^\infty c_n e^{inx}, \tag{2·23}$$

with $c_n \to 0$. For such series

$$F(x) = \tfrac{1}{2}c_0 x^2 - \sum_1^\infty \frac{c_n}{n^2} e^{inx}. \tag{2·24}$$

If, for a fixed x_0, as $h \to 0$ we have

$$F(x_0 + h) = \alpha_0 + \alpha_1 h + \tfrac{1}{2}\alpha_2 h^2 + o(h^2), \tag{2·25}$$

we say that F has at x_0 a *second generalized derivative* equal to α_2. We observe that (2·25) implies not only

$$\frac{F(x_0+h) + F(x_0-h) - 2F(x_0)}{h^2} \to s \quad \text{but also} \quad \frac{F(x_0+2h) - 2F(x_0+h) + F(x_0)}{h^2} \to s.$$

(2·26) THEOREM. *If* (2·23) *converges at* x_0 *to (finite) sum* s, *then the function* $F(x)$ *of* (2·24) *has at* x_0 *a second generalized derivative* s.

We omit the proof here; a much more general result will be proved in Chapter XI, § 2.

(2·27) THEOREM. *If* (2·23) *converges at* x_0 *to (a finite) sum* s, *then the function*

$$L(x) = c_0 x + \sum_1^\infty \frac{c_n}{in} e^{inx}$$

has an approximate derivative at x_0 *equal to* s: $L'_{\text{ap}}(x_0) = s$.

This is an analogue of (2·22) and it is enough to sketch the proof. We may suppose that $c_0 = 0$, $x_0 = 0$, $s = 0$. The convergence of Σc_n implies that of $\Sigma c_n/n$, and defining N as before we have

$$\frac{L(h) - L(0)}{h} = \sum_1^{2^N} c_n \frac{e^{inh} - 1}{inh} + \sum_{2^N+1}^\infty c_n \frac{e^{inh} - 1}{inh} = P(h) + Q(h).$$

Here again $P(h) \to 0$. Summation by parts shows that $h^{-1} \sum_{2^N+1}^\infty c_n/n = o(1)$. Hence

$$Q(h) = o(1) + \sum_{2^N+1}^\infty \frac{c_n e^{inh}}{inh},$$

and the last term tends approximately to 0 as $h \to 0$.

It will be shown in Chapter XIV, (4·1), that if $\Sigma A_n(x)$ converges in a set E, the conjugate series $\Sigma B_n(x)$ converges almost everywhere in E. From this and (2·27) we deduce the following consequence:

(2·28) THEOREM. *If* $\Sigma A_n(x)$ *converges in a set* E, *then* $L(x) = \frac{1}{2}a_0 x + \Sigma B_n(x)/n$, *has an approximate derivative almost everywhere in* E.

3. Uniqueness of the representation by trigonometric series

The theory of Fourier series associates with every integrable and periodic function $f(x)$ a special trigonometric series—the Fourier series of $f(x)$—which, as we have shown, represents $f(x)$ in various ways. It is natural to inquire whether functions can be represented by trigonometric series other than Fourier series. The problem has a number of answers, since the word 'represent' may have various meanings. The problem of summability, and so of convergence, in mean was discussed in Chapter IV. In this section we consider the representation of functions by everywhere convergent trigonometric series. The following result is fundamental:

(3·1) THEOREM. (i) *If*

$$\tfrac{1}{2}a_0 + \sum_{n=1}^{\infty} (a_n \cos nx + b_n \sin nx) \tag{3·2}$$

converges everywhere to 0, *the series vanishes identically; that is, all its coefficients are* 0.

(ii) *More generally, if* (3·2) *converges*† *everywhere to an integrable function* $f(x)$, *then* (3·2) *is* $S[f]$.

We begin with the proof of (i), though it is a special case of (ii).

By (1·3), $a_n \to 0$, $b_n \to 0$. It follows that the function

$$F(x) = \tfrac{1}{4}a_0 x^2 - \sum_{n=1}^{\infty} \frac{a_n \cos nx + b_n \sin nx}{n^2} \tag{3·3}$$

is continuous. By (2·4), $D^2F(x) = 0$ for all x and so (Chapter I, (10·7)), $F(x)$ is linear, $F(x) = \alpha x + \beta$. Comparing this with (3·3), making $x \to \infty$, and observing that the periodic part of the series (3·3) is bounded, we find that $a_0 = 0$, $\alpha = 0$. Hence

$$\beta + \sum_{1}^{\infty} (a_n \cos nx + b_n \sin nx)/n^2 = 0. \tag{3·4}$$

The series on the left converges uniformly and so is the Fourier series of the right-hand side. Hence all the coefficients in (3·4) are 0, and this completes the proof of (i).

A corollary of (i) is that *if two trigonometric series converge everywhere to the same sum, then the two series are identical, that is, the corresponding coefficients of both series are the same.*

We precede the proof of (ii) by the following remark: *an arbitrary series* (3·2), *convergent or not, is* $S[f]$ *if the* $F(x)$ *of* (3·3) *satisfies an equation*

$$F(x) = \int_a^x dy \int_a^y f(t)\,dt + Ax + B, \tag{3·5}$$

where A and B are constants. For let $F_1(x) = F(x) - \tfrac{1}{4}a_0 x^2$. The periodic part of the series (3·3) is then $S[F_1]$, and (3·2) without the constant term is $S''[F_1]$. Since, by (3·5), F_1 is a second integral, we have

$$S''[F_1] = S[F_1''] = S[f - \tfrac{1}{2}a_0]$$

(Chapter II, (2·1)), and (3·2) is $S[f]$.

Hence (ii) will follow if we show that under the hypothesis of the theorem we have (3·5). The proof of the latter will be based on a number of lemmas which give more than we actually require but which will be useful in generalizations of (ii) (see especially § 8 below).

† By 'convergence' we mean convergence to a finite sum.

(3·6) LEMMA. *If a function $F(x)$, $a<x<b$, has a finite derivative $F'(x)$, and if at a point x_0 all the Dini numbers of $F'(x)$ are contained between m and M, then*

$$m \leqslant \underline{D}^2 F(x_0) \leqslant \overline{D}^2 F(x_0) \leqslant M.$$

This has already been proved in Chapter I, p. 23.

(3·7) LEMMA. *Suppose that $F(x)$ is continuous for $a<x<b$ and satisfies $\overline{D}^2 F \geqslant 0$ for all x in (a,b). Then F is convex.*

This is Theorem (10·7) of Chapter I.

Given a function $f(x)$, $a \leqslant x \leqslant b$, not necessarily finite-valued, we call a continuous $\psi(x)$ a *major function* for f, if for each x the Dini numbers of ψ are not less than $f(x)$ and are distinct from $-\infty$; a continuous $\phi(x)$ is a *minor function* for f, if the Dini numbers of $\phi(x)$ are not greater than $f(x)$ and are distinct from $+\infty$.

(3·8) LEMMA. *Let $f(x)$, $a \leqslant x \leqslant b$, be integrable, $f_1(x) = \int_a^x f\,dt$, and let ϵ be positive. Then there exist major and minor functions, ψ and ϕ, for f such that*

$$|f_1(x) - \psi(x)| < \epsilon, \quad |f_1(x) - \phi(x)| < \epsilon$$

in (a,b). The functions ψ and ϕ can even be absolutely continuous.

This is a well-known result from the theory of the Lebesgue integral and we take it for granted here.†

(3·9) LEMMA. *Let $f(x)$, $a<x<b$, be integrable over (a,b) and greater than $-\infty$. Let $F(x)$, $a<x<b$, be continuous and such that*

$$\overline{D}^2 F(x) \geqslant f(x). \tag{3·10}$$

Then

$$G(x) = F(x) - \int_a^x dy \int_a^y f(t)\,dt \tag{3·11}$$

is convex.

Let ϕ_n be a minor function of f such that $\left|\phi_n(x) - \int_a^x f(t)\,dt\right| \leqslant 1/n$, $n = 1, 2, \ldots$. Write

$$f_1(x) = \int_a^x f\,dt, \quad f_2(x) = \int_a^x f_1\,dt, \quad \Phi_n(x) = \int_a^x \phi_n\,dt.$$

By (3·10) and (3·6),

$$\overline{D}^2 F(x) \geqslant f(x) \geqslant \underline{D}^2 \Phi_n(x). \tag{3·12}$$

Since the extreme terms here cannot be infinities of the same sign, we have $\overline{D}^2(F - \Phi_n) \geqslant 0$, and $F - \Phi_n$ is convex in (a,b). As $n \to \infty$, $F - \Phi_n$ tends to $F - f_2$, which is therefore also convex, and the lemma follows.

(3·13) LEMMA. *Let $f(x)$, $a<x<b$, be integrable and finite-valued. Let $F(x)$ be continuous and such that*

$$\overline{D}^2 F(x) \geqslant f(x) \geqslant \underline{D}^2 F(x). \tag{3·14}$$

Then

$$F(x) = \int_a^x dy \int_a^y f(t)\,dt + Ax + B, \tag{3·15}$$

where A and B are constants.

† See Ch. J. de la Vallée-Poussin, *Intégrales de Lebesgue*, or S. Saks, *Theory of the Integral*.

Since the G in (3·11) is both convex and concave, it is linear.

Part (ii) of (3·1) now follows easily. For the F in (3·3) satisfies $D^2F=f$ for all x. Hence F satisfies (3·15) which, as we know, proves that $\Sigma A_n(x)$ is $S[f]$.

The hypothesis that $\Sigma A_n(x)$ converges was used only to show that $a_n, b_n \to 0$ and that $\Sigma A_n(x)$ is summable R. Theorem (3·1) would still hold, therefore, under these weaker assumptions.

(3·16) Lemma. *If $F(x)$ is convex in (a,b), then $D^2F(x)$ exists for almost all x in (a,b) and is integrable over any interval (a',b') totally interior to (a,b).*

Since $F(x)$ is the integral of a non-decreasing function $\xi(x)$ (Chapter I, (10·11)), we have

$$\frac{F(x+h)+F(x-h)-2F(x)}{h^2}=\frac{1}{h^2}\int_0^h\{\xi(x+t)-\xi(x-t)\}dt; \tag{3·17}$$

and $\xi'(x)$ exists almost everywhere in (a,b) and is integrable over (a',b'). At each point where a finite $\xi'(x)$ exists we have

$$\xi(x+t)-\xi(x-t)=2t\xi'(x)+o(t),$$

the right-hand side of (3·17) tends to $\xi'(x)$, and $D^2F(x)=\xi'(x)$.

(3·18) Theorem. *Suppose that $\Sigma A_n(x)$ converges everywhere to a sum $f(x)$ such that $f(x)\geqslant\chi(x)$, where χ is integrable. Then f is integrable and the series is $S[f]$.*

We may suppose, by changing χ if necessary in a set of measure 0, that χ is finite-valued. We have $D^2F\geqslant\chi$. By Lemma (3·9),

$$H(x)=F(x)-\int_a^x dy\int_a^y\chi(t)\,dt$$

is convex. Hence D^2H exists almost everywhere and is integrable over $(0,2\pi)$. It follows that $D^2F=f$ is integrable. Hence the series is $S[f]$.

In particular, if $\Sigma A_n(x)$ converges everywhere to a non-negative sum, the sum is integrable.

(3·19) Theorem. *Suppose that $a_n, b_n\to 0$ and denote by $s^*(x)$ and $s_*(x)$ the upper and lower sums of $\Sigma A_n(x)$:*

$$s^*(x)=\limsup_{n\to\infty}s_n(x),\quad s_*(x)=\liminf_{n\to\infty}s_n(x).$$

If $s^*(x)$ and $s_*(x)$ are finite outside a denumerable set E, and are both integrable, then $\Sigma A_n(x)$ is $S[f]$, where $f=\underline{D}^2F$, $F(x)=\frac{1}{4}a_0x^2-\Sigma A_n(x)/n^2$.

We need a few lemmas.

(3·20) Lemma. *Suppose that $F(x)$ is continuous for $a<x<b$, and that $\overline{D}^2F\geqslant 0$ for each x, except possibly in a denumerable set E. Suppose also that*

$$\limsup_{h\to+0}\Delta^2F(x,h)/h\geqslant 0\quad\textit{for}\quad x\in E, \tag{3·21}$$

where $\Delta^2F(x,h)=F(x+h)+F(x-h)-2F(x)$. Then F is convex in (a,b).

This is a generalization of (3·7). As in the proof of the latter we may suppose that $\overline{D}^2F>0$ outside E, for otherwise we apply the argument to $F(x)+x^2/n$ and then make $n\to\infty$.

Suppose that F is not convex. There is then an arc of the curve $y=F(x)$ lying partly above the corresponding chord. Let α, β be the terminal abscissae of the arc, and let $y=l_\mu(x)$ be the equation of the straight line through $(\alpha, F(\alpha))$ with slope μ. If μ exceeds the slope of the chord but is close to it, the function $G_\mu(x)=F(x)-l_\mu(x)$ takes positive values at some points of (α,β). Let $x_0=x_0(\mu)$ be a point where G_μ attains its absolute maximum in (α,β). We have $\alpha<x_0<\beta$, since $G_\mu(\alpha)=0$, $G_\mu(\beta)<0$.

Clearly $\Delta^2 G_\mu(x_0,h)\leqslant 0$ for small $h>0$. Hence $\Delta^2 F(x_0,h)\leqslant 0$ for such h and, in particular, $\overline{D}^2F(x_0)\leqslant 0$. Therefore $x_0\in E$.

Since (3·21) holds with G_μ for F, and since $\Delta^2 G_\mu(x_0,h)\leqslant 0$ for small h, we have

$$\limsup_{h\to+0} \Delta^2 G_\mu(x_0,h)/h=0.$$

But $G_\mu(x_0\pm h)-G_\mu(x_0)\leqslant 0$ for small h. Hence $\limsup\{G_\mu(x_0+h)-G_\mu(x_0)\}/h=0$ and

$$\limsup_{h\to+0}\{F(x_0+h)-F(x_0)\}/h=\mu.$$

This shows that to different admissible μ there correspond different $x_0=x_0(\mu)$, which is impossible since $x_0\in E$ and the set of the μ is of the power of the continuum.

(3·22) LEMMA. *Let $f(x)$, $a<x<b$, be integrable over (a,b) and greater than $-\infty$, except possibly in a denumerable set E. Let $F(x)$, $a<x<b$, be continuous and such that $\overline{D}^2F(x)\geqslant f(x)$ outside E. Suppose that F satisfies* (3·21). *Then* (3·11) *is convex.*

(3·23) LEMMA. *Let $f(x)$, $a<x<b$, be integrable over (a,b) and finite outside a denumerable set E. Let $F(x)$, $a<x<b$, be continuous and such that $\overline{D}^2F\geqslant f\geqslant \underline{D}^2F$ outside E. Suppose that*

$$\limsup_{h\to+0}\Delta^2F(x,h)/h\geqslant 0\geqslant \liminf_{h\to+0}\Delta^2F(x,h)/h \tag{3·24}$$

in E. Then F satisfies (3·15).

If we use Lemma (3·20) instead of (3·7), the proofs of (3·22) and (3·23) run parallel to those of (3·9) and (3·13).

We pass to the proof of (3·19). By Lemma (2·7), $\overline{D}^2F(x)$ and $\underline{D}^2F(x)$ are contained between

$$\tfrac{1}{2}\{s^*(x)+s_*(x)\}\pm\tfrac{1}{2}k\{s^*(x)-s_*(x)\}.$$

In particular, $\overline{D}^2F$ and $\underline{D}^2F$ are integrable. Write $f=\underline{D}^2F$. Then $\overline{D}^2F\geqslant f\geqslant \underline{D}^2F$, and since F is smooth an application of (3·23) shows that F differs from a second integral of f by a linear function, and in particular that D^2F exists almost everywhere. It follows that $\Sigma A_n(x)$ is $\mathsf{S}[f]$, with $f=\underline{D}^2F=D^2F$ almost everywhere.

The theorem below is a generalization of (3·19). Its proof is less simple and we shall obtain it as a corollary of results proved in § 8 below.

(3·25) THEOREM. *Suppose that $a_n, b_n\to 0$. If the upper and lower sums of $\Sigma A_n(x)$ are both finite, except possibly in a denumerable set E, and if $s_*(x)\geqslant\chi(x)$ where χ is integrable (in particular, if s_* is integrable), then the series is a Fourier series.*

Return to Lemma (3·20). The continuity of F there was used merely to guarantee that G_μ attains its maximum, and for that purpose the upper semi-continuity of F

is sufficient. The same applies to Lemma (3·22). Hence we have the following result, which will find application later:

(3·26) LEMMA. *Lemmas* (3·20) *and* (3·22) *hold if the condition that F is continuous is replaced by the condition that F is upper semi-continuous.*

4. The principle of localization. Formal multiplication of trigonometric series

We proved in Chapter II, § 6, that the behaviour of $S[f]$ at a point x_0 depends only on the values of f in an arbitrarily small neighbourhood of x_0. This is a special case of the following *localization principle* of Riemann for general trigonometric series with coefficients tending to 0:

The behaviour of a series

$$\tfrac{1}{2}a_0 + \sum_1^\infty (a_n \cos nx + b_n \sin nx) = \sum_0^\infty A_n(x) \tag{4·1}$$

at a point x_0 depends only on the values of the function

$$F(x) = \tfrac{1}{4}a_0 x^2 - \sum_1^\infty (a_n \cos nx + b_n \sin nx)/n^2 \tag{4·2}$$

in an arbitrarily small neighbourhood of x_0.

More precisely, and somewhat more generally, we have the following:

(4·3) THEOREM. *Let S_1 and S_2 be two trigonometric series with coefficients tending to* 0, *and let F_1 and F_2 be the functions F corresponding to S_1 and S_2. If $F_1(x) = F_2(x)$ in an interval (a, b), or more generally if $F_1(x) - F_2(x)$ is linear in (a, b), then in every interval (a', b') interior to (a, b)*

(i) *S_1 and S_2 are uniformly equiconvergent;*

(ii) *$\tilde{S}_1$ and $\tilde{S}_2$ are uniformly equiconvergent in the wider sense.*

If two integrable functions f_1 and f_2 coincide in an interval (a, b), then, since Fourier series may be integrated termwise, the F_1 and F_2 corresponding to $S[f_1]$ and $S[f_2]$ differ in (a, b) by a linear function. Hence the principle of localization for Fourier series—as well as for conjugate series (see Chapter II, (6·6))—is a special case of (4·3).

We base the proof of (4·3) on Rajchman's theory of formal multiplication of trigonometric series, which is of independent interest and has a number of other applications. It will be convenient to write our series in the complex form.

Given two series

$$\sum_{-\infty}^{+\infty} c_n e^{inx}, \tag{4·4}$$

$$\sum_{-\infty}^{+\infty} \gamma_n e^{inx}, \tag{4·5}$$

we call

$$\sum_{-\infty}^{+\infty} C_n e^{inx} \tag{4·6}$$

their *formal product* if

$$C_n = \sum_{p=-\infty}^{+\infty} c_p \gamma_{n-p} \tag{4·7}$$

for all n. We assume that the series defining the C_n converge absolutely. This is certainly the case if, for example, the c_n are bounded (in particular, if $c_n \to 0$) and $\Sigma |\gamma_n| < \infty$.

(4·8) LEMMA. *If $c_n \to 0$ as $n \to \pm\infty$, and if $\Sigma |\gamma_n| < \infty$, then $C_n \to 0$.*

Let $M = \max |c_n|$. As $n \to +\infty$, we have

$$|C_n| \leqslant M \sum_{p \leqslant \frac{1}{2}n} |\gamma_{n-p}| + \max_{p > \frac{1}{2}n} |c_p| \sum_{p > \frac{1}{2}n} |\gamma_{n-p}|$$

$$\leqslant M \sum_{q \geqslant \frac{1}{2}n} |\gamma_q| + \max_{p > \frac{1}{2}n} |c_p| \sum_{-\infty}^{+\infty} |\gamma_q| \to 0.$$

As regards the case $n \to -\infty$, it is enough to observe that

$$C_{-m} = \sum_{p=-\infty}^{+\infty} c'_p \gamma'_{m-p}, \quad \text{where} \quad c'_p = c_{-p},\ \gamma'_p = \gamma_{-p}.$$

Remark. If c_n and γ_n depend on a parameter, and the conditions imposed on c_n and γ_n are satisfied uniformly, then $C_n \to 0$ uniformly.

We shall be occupied a good deal with series $\Sigma \gamma_n e^{inx}$ which satisfy the condition

$$\Sigma |n\gamma_n| < \infty.$$

The condition is satisfied if, for instance, $\Sigma \gamma_n e^{inx}$ is the development of a function having three continuous derivatives.

Let

$$\Gamma_n = \sum_{n}^{\infty} |\gamma_p| \quad (n \geqslant 0), \qquad \Gamma_n = \sum_{-\infty}^{n} |\gamma_p| \quad (n < 0).$$

$\Sigma |n\gamma_n| < \infty$ is equivalent to $\Sigma \Gamma_n < \infty$. For

$$\sum_{n=1}^{\infty} (\Gamma_n + \Gamma_{-n}) = \sum_{n=1}^{\infty} \sum_{\nu=n}^{\infty} (|\gamma_\nu| + |\gamma_{-\nu}|) = \sum_{\nu=1}^{\infty} (|\gamma_\nu| + |\gamma_{-\nu}|)\, \nu.$$

(4·9) THEOREM OF RAJCHMAN. *Suppose that $c_n \to 0$, $\Sigma |n\gamma_n| < \infty$, and that $\Sigma \gamma_n e^{inx}$ converges to sum $\lambda(x)$. Then the two series*

$$\sum_{-\infty}^{+\infty} C_n e^{inx}, \qquad \sum_{-\infty}^{+\infty} \lambda(x)\, c_n e^{inx}$$

are uniformly equiconvergent. In particular, if $\lambda(x)$ is 0 in a set E, $\Sigma C_n e^{inx}$ converges uniformly to 0 in E.

We first prove the special case. Let

$$R_k(x) = \sum_{n \geqslant k} \gamma_n e^{inx}.$$

If $x_0 \in E$, then

$$|R_k(x_0)| \leqslant \Gamma_k, \quad |R_{-k}(x_0)| \leqslant \Gamma_{-k-1}$$

for $k \geqslant 0$, and so $\Sigma |R_k(x_0)|$ converges uniformly in E. Now

$$S_m(x_0) = \sum_{n=-m}^{m} C_n e^{inx_0} = \sum_{n=-m}^{m} e^{inx_0} \sum_{p=-\infty}^{+\infty} c_p \gamma_{n-p}$$

$$= \sum_{p=-\infty}^{+\infty} c_p e^{ipx_0} \sum_{n=-m}^{m} \gamma_{n-p} e^{i(n-p)x_0}$$

$$= \sum_{p=-\infty}^{+\infty} c_p e^{ipx_0} R_{-m-p}(x_0) - \sum_{p=-\infty}^{+\infty} c_p e^{ipx_0} R_{m-p+1}(x_0).$$

Applying Lemma (4·8) (with $c_p e^{ipx_0}$ and $R_p(x_0)$ for c_p and γ_p) and the Remark to it, we see that $S_m(x_0)$ tends uniformly to 0 in E.

The result just obtained remains valid even if the c_n and γ_n in (4·4) and (4·5) themselves depend on the variable x, provided that the formal product of the two series is defined by (4·6) and (4·7); for the result is nothing but a theorem on the Laurent multiplication of arbitrary two-way infinite series. Using this observation we can easily prove the general statement in (4·9).

Write

$$\gamma_0^* = \gamma_0 - \lambda(x), \quad \gamma_n^* = \gamma_n \quad \text{for} \quad n \neq 0,$$

and consider the formal product $\Sigma C_n^* e^{inx}$ of $\Sigma c_n e^{inx}$ and $\Sigma \gamma_n^* e^{inx}$. The remainders $R_k^*(x)$ of $\Sigma \gamma_n^* e^{inx}$ satisfy the inequalities

$$|R_k^*| \leqslant \Gamma_k, \quad |R_{-k}^*| \leqslant \Gamma_{-k-1}$$

for $k>0$. Hence $\Sigma C_n^* e^{inx}$ converges uniformly to 0 for all x and (4·9) follows if we observe that

$$C_n^* = C_n - \lambda(x)\, c_n. \qquad \textbf{(4·10)}$$

We state separately a number of immediate corollaries of (4·9).

(4·11) THEOREM. *If $\lambda(x_0) \neq 0$, a necessary and sufficient condition that $\Sigma C_n e^{inx_0}$ should converge is that $\Sigma c_n e^{inx_0}$ should converge.*

Let T be any linear method of summation satisfying the conditions of regularity (Chapter III, § 1). Observing that since $\Sigma C_n^* e^{inx}$ converges to 0 it is summable T to 0, we obtain the following:

(4·12) THEOREM. *If $\lambda(x_0) \neq 0$, a necessary and sufficient condition that $\Sigma C_n e^{inx_0}$ should be summable* T *is that $\Sigma c_n e^{inx_0}$ should be summable* T. *If the sum of the latter series is s, the sum of the former is $\lambda(x_0)\, s$.*

(4·13) THEOREM. *If $\Sigma c_n e^{inx}$ is uniformly convergent, or summable* T, *over a set $\mathscr{E}$, so is $\Sigma C_n e^{inx}$. The converse is also true if $|\lambda(x)| \geqslant \epsilon > 0$ in $\mathscr{E}$.*

Theorem (4·11) can be completed by considering limits of indetermination of the partial sums. Restricting ourselves to the case of ordinary convergence (there is no difficulty in stating a result for summability T) we have:

(4·14) THEOREM. *If the upper and lower sums of $\Sigma c_n e^{inx_0}$ are $\bar{s}$, $\underline{s}$, then the upper and lower sums of $\Sigma C_n e^{inx_0}$ are $\lambda(x_0)\bar{s}$, $\lambda(x_0)\underline{s}$ if $\lambda(x_0)>0$, and $\lambda(x_0)\underline{s}$, $\lambda(x_0)\bar{s}$ if $\lambda(x_0)<0$.*

We now prove an analogue of (4·9) for the series

$$\Sigma c_n \epsilon_n e^{inx} \quad (\epsilon_n = -i \operatorname{sign} n) \qquad \textbf{(4·15)}$$

conjugate to $\Sigma c_n e^{inx}$.

(4·16) THEOREM. *Under the hypotheses of* (4·9), *the series*

$$\Sigma C_n \epsilon_n e^{inx}, \quad \Sigma \lambda(x)\, c_n \epsilon_n e^{inx} \qquad \textbf{(4·17)}$$

are uniformly equiconvergent in the wider sense. In particular, if $\lambda(x)=0$ in a set E the series

$$\Sigma C_n \epsilon_n e^{inx_0} \qquad \textbf{(4·18)}$$

conjugate to the formal product converges uniformly in E.

Let $\tilde{S}_m(x_0)$ denote the partial sum of (4·18). Writing $c'_n = c_n e^{inx_0}$ and defining C'_n and γ'_n similarly, we have

$$\tilde{S}_m(x_0) = \sum_{-m}^{m} \epsilon_n C'_n = \sum_{-m}^{m} \epsilon_n \sum_{p=-\infty}^{+\infty} c'_p \gamma'_{n-p} = \sum_{p=-\infty}^{+\infty} c'_p \sum_{n=-m}^{m} \gamma'_{n-p} \epsilon_n$$

$$= -i \sum_{p=-\infty}^{+\infty} c'_p \sum_{n=1}^{m} (\gamma'_{n-p} - \gamma'_{-n-p})$$

$$= -i \sum_{p=-\infty}^{+\infty} c'_p \{R_{1-p}(x_0) - R_{m-p+1}(x_0) - R_{-m-p}(x_0) + R_{-p}(x_0)\}.$$

It follows, by the same proof as for (4·9), that if λ vanishes in E, and $x_0 \in E$, then

$$\tilde{S}_m(x_0) \to -i \sum_{-\infty}^{+\infty} c'_p \{R_{1-p}(x_0) + R_{-p}(x_0)\},$$

uniformly in x_0. This gives the second part of (4·16).

To prove the general statement, we use the same device as before and consider the formal product $\Sigma C^*_n e^{inx}$ of $\Sigma c_n e^{inx}$ and $\Sigma \gamma^*_n e^{inx}$. The coefficients C^*_n depend on x, but if we define the series 'conjugate' to $\Sigma C^*_n e^{inx}$ as $\Sigma C^*_n \epsilon_n e^{inx}$, the latter series will be uniformly convergent, as the proof just given shows, and it is enough to apply (4·10).

The following is one of the corollaries of (4·16).

(4·19) THEOREM. *If* $\Sigma c_n \epsilon_n e^{inx}$ *is uniformly summable* T *over a set* $\mathscr{E}$, *so is* $\Sigma C_n \epsilon_n e^{inx}$. *The converse is also true if* $|\lambda(x)| \geq \epsilon > 0$ *in* $\mathscr{E}$.

A feature of Theorems (4·9) and (4·16) is that we suppose next to nothing about one of the factors of the formal product, while we impose rather stringent conditions on the other. However, if the first series is a Fourier series, the conditions imposed upon the second can be relaxed. It is not difficult to see that Theorems (6·7) and (6·10) of Chapter II may be considered as theorems on the formal multiplication of trigonometric series in the case when the 'bad' series is a Fourier series.

We now pass to the proof of (4·3).

Let T be a linear method of summation satisfying conditions (i), (ii) and (iii) of regularity (Chapter III, § 1). We say that T is of *type* U if every trigonometric series with coefficients tending to 0 and summable T to a finite and integrable function $f(x)$ is S$[f]$. In § 3 we proved that ordinary convergence and Riemann summability are both of type U.

In what follows we frequently consider formal products of trigonometric series by the Fourier series of a function λ. We suppose always that λ has a continuous third derivative. Then the coefficients of λ are $O(n^{-3})$ and we can apply Theorems (4·9) and (4·16). It will also be convenient to suppose that if of two functions $\phi(x)$ and $\psi(x)$ one is equal to 0 in an interval (α, β), the product $\phi\psi$ is defined and equal to 0 in (α, β), even if the other factor is not defined in the interval.

We first prove the following result:

(4·20) THEOREM. *Let* T *be any method of summation of type* U. *If* $\Sigma c_n e^{inx}$, *with* $c_n \to 0$, *is summable* T *for* $a < x < b$ *to a finite and integrable function* $f(x)$, *then in any interval* (a', b') *totally interior to* (a, b) *the series is uniformly equiconvergent with* S$[\lambda f]$, *where* $\lambda(x)$ *is equal to* 1 *in* (a', b') *and equal to* 0 *outside* (a, b). *The series* $\Sigma c_n \epsilon_n e^{inx}$ *and* $\tilde{\text{S}}[\lambda f]$ *are uniformly equiconvergent in the wider sense in* (a', b').

To prove the first part of the theorem we observe that the formal product of $\Sigma c_n e^{inx}$ and $\mathsf{S}[\lambda]$ converges to 0 outside (a, b) and is summable T to λf in (a, b). Hence the product is summable T in the whole interval $(0, 2\pi)$ to sum λf. This sum is integrable. Hence the product is $\mathsf{S}[\lambda f]$, and we apply (4·9). To obtain the second part of the theorem we apply (4·16).

We are now in a position to prove (4·3).

Let $S = S_1 - S_2$. We have to show that both S and its conjugate converge uniformly over (a', b'), and that the sum of the former series is 0. Integrating S termwise twice, we obtain a function $F(x) = F_1(x) - F_2(x)$ which is linear in (a, b). Since $D^2F(x) = 0$ for each x interior to (a, b), S is summable R to 0 for $a < x < b$ and it is enough to apply (4·20) with $f = 0$.

As a special case of (4·3) we have the following theorem:

(4·21) THEOREM. *Let S be a trigonometric series with coefficients tending to* 0, *and let $F(x)$ be the sum of the series obtained by integrating S termwise twice. Suppose that*

$$F(x) = Ax + B + \int_a^x dy \int_a^y f(t)\, dt \quad (a \leqslant x \leqslant b), \tag{4·22}$$

where A and B are constants and f is integrable over (a, b). Let $f^*(x)$ be equal to $f(x)$ in (a, b) and to 0 *elsewhere. Then S and $\mathsf{S}[f^*]$ are uniformly equiconvergent in every interval (a', b') totally interior to (a, b); and $\tilde{S}$ and $\tilde{\mathsf{S}}[f^*]$ are uniformly equiconvergent in the wider sense in (a', b').*

It is enough to observe that Fourier series may be integrated termwise; hence if $F_1(x)$ is the sum of $\mathsf{S}[f^*]$ integrated termwise twice, $F_1(x)$ satisfies an equation similar to (4·22), and so $F(x) - F_1(x)$ is linear in (a, b).

A special case of (4·3), which was actually used in the proof of the theorem, deserves a separate statement:

(4·23) THEOREM. *If the sum $F(x)$ of a trigonometric series S integrated termwise twice is linear in (a, b), then S and $\tilde{S}$ are uniformly convergent in every interval interior to (a, b), the sum of S being* 0.

The following theorem gives an equivalent form of the condition (4·22):

(4·24) THEOREM. *Let $S = \Sigma A_n(x)$, $F(x) = \frac{1}{4}a_0 x^2 - \Sigma' A_n(x)/n^2$, and consider the series*

$$\tfrac{1}{2}a_0 x + \sum_1^\infty (a_n \sin nx - b_n \cos nx)/n. \tag{4·25}$$

We have (4·22) *for $a \leqslant x \leqslant b$ if and only if* (4·25) *converges uniformly in the closed interval $a \leqslant x \leqslant b$ to an indefinite integral of f.*

We may suppose that $a_0 = 0$, so that both F and (4·25) are periodic. Let F_1 be the F corresponding to (4·25). If F satisfies (4·22) for $a \leqslant x \leqslant b$, then F_1, which is obtained by integrating (4·2), is equal in (a, b) to a third (repeated) integral of f and is a second integral of the absolutely continuous function

$$L(x) = A + \int_a^x f dt. \tag{4·26}$$

Since Fourier series of absolutely continuous functions converge uniformly, an application of (4·20) to the series (4·25) shows that (4·25) converges uniformly to sum (4·26) in every (a', b') interior to (a, b). By (2·19), (4·25) converges to $L(x)$ uniformly in (a, b). Conversely, if (4·25) converges uniformly in (a, b) to (4·26), termwise integration of (4·25) immediately gives (4·22) in (a, b). The same conclusion holds if we only assume that (4·25) converges to (4·26) at every x interior to (a, b); for, by (4·20), the series (4·25) must converge uniformly in every (a', b') interior to (a, b) and so, by (2·19), uniformly in (a, b).

(4·27) THEOREM. *If $\Sigma A_n(x)$ converges in (a, b), except possibly for a denumerable set of points, to a finite and integrable function $f(x)$, then $\frac{1}{2}a_0 x + \Sigma B_n(x)/n$ converges uniformly in (a, b) to an indefinite integral of f.*

By (3·23), F satisfies (4·22) in (a, b) and it is enough to apply (4·24).

Suppose that $\Sigma A_n(x)$ is $S'[\Phi]$, where Φ is a function which in an interval (a, b) is the indefinite integral of an $f \in L(a, b)$. We then call the series a *restricted Fourier* series associated with the interval (a, b) and function f. Of course, the a_n, b_n need not tend to 0 (though they must be $o(n)$).

(4·28) THEOREM. *Let S, with coefficients tending to 0, be a restricted Fourier series associated with (a, b) and f, and let f^* be equal to f in (a, b) and to 0 elsewhere. Then S and $S[f^*]$ are uniformly equiconvergent in every interval (a', b') interior to (a, b); $\tilde{S}$ and $\tilde{S}[f^*]$ are, in (a', b'), uniformly equiconvergent in the wider sense.*

This is a corollary of (4·21), since the $F(x)$ corresponding to S satisfies the condition (4·22) in (a, b).

Return to the principle of localization. Riemann deduced it from an important formula which we are now going to prove in a somewhat more general form.

Let $a < a' < b' < b$, and let $\lambda(x)$ be a function equal to 1 in (a', b'), equal to 0 outside (a, b) and having coefficients $O(n^{-5})$. The latter condition could be relaxed but the point is without importance.

(4·29) THEOREM. *Let $\Sigma A_k(x)$ have coefficients tending to 0 and let*

$$F(x) = \tfrac{1}{4}a_0 x^2 - \Sigma' A_n(x)/n^2.$$

Then the sequences

$$\tfrac{1}{2}a_0 + \sum_{k=1}^{n} (a_k \cos kx + b_k \sin kx) - \frac{1}{\pi}\int_a^b F(t)\,\lambda(t)\frac{d^2}{dt^2} D_n(x-t)\,dt, \tag{4·30}$$

$$\sum_{k=1}^{n} (a_k \sin kx - b_k \cos kx) - \frac{1}{\pi}\int_a^b F(t)\,\lambda(t)\frac{d^2}{dt^2} \tilde{D}_n(x-t)\,dt \tag{4·31}$$

converge uniformly in (a', b'). In the case of (4·30) *the limit is* 0.

D_n and $\tilde{D}_n$ denote the Dirichlet kernel and the conjugate Dirichlet kernel. Since the integrals (4·30) and (4·31) depend only on the values of $F(x)$ in the interval (a, b), the theorem includes the principle of localization.

To grasp the meaning of the theorem, suppose for simplicity that $a_0 = 0$ and denote $\Sigma A_k(x)$ by S; F is then periodic and has coefficients $o(n^{-2})$. Assume for the moment that the formal product $S[F]\,S[\lambda] = \Sigma C_n e^{inx}$ has coefficients $o(n^{-2})$ (this is easy to show, but is not needed for the proof of (4·29)). Then $F\lambda - C_0$ may be considered as the function $F_1(x)$

corresponding to a trigonometric series S_1 with coefficients tending to 0. Since $\lambda=1$, and so $F=F_1+C_0$, in (a', b'), the difference $S-S_1$ converges uniformly to 0 in every interval (a'', b'') interior to (a', b') (see (4·3)). But (4·30) is the difference of the nth partial sums of the series S and $\mathsf{S}''[F_1]=S_1$. Hence (4·30) converges to 0 uniformly in every interval (a'', b'') interior to (a', b'); and a similar argument proves the uniform convergence of (4·31) in (a'', b''). In other words, Riemann's formulae are, very nearly, consequences of the principle of localization. The only drawback of this argument is that it gives convergence in (a'', b'') and not in (a', b'). Though this point is of minor importance, we shall prove the theorem in its complete form partly for aesthetic reasons and partly because the above argument cannot be applied to the case $a'=b'$ (considered by Riemann).

(4·32) LEMMA. *If V and W are trigonometric series, then*

$$(VW)''=V''W+2V'W'+VW'',$$

where products are formal products and dashes denote termwise differentiation.

If c_n, γ_n, C_n are the coefficients of V, W, VW respectively, then the nth coefficient of $(VW)''$ is

$$-\sum_{p=-\infty}^{+\infty} c_p \gamma_{n-p} n^2,$$

and it is enough to observe that

$$-n^2=-(n-p)^2+2i(n-p)\,ip-p^2.$$

The argument presupposes that the formal products in (4·32) exist.

Returning to (4·29), suppose that $a_0=0$ and denote $\Sigma A_k(x)$ by S. The difference (4·30) is the nth partial sum of the series

$$\begin{aligned} S-\mathsf{S}''[F\lambda] &= S-\{\mathsf{S}[F]\,\mathsf{S}[\lambda]\}'' \\ &= (S-\mathsf{S}''[F]\,\mathsf{S}[\lambda])-2\mathsf{S}'[F]\,\mathsf{S}'[\lambda]-\mathsf{S}[F]\,\mathsf{S}''[\lambda]. \end{aligned}$$

Since $\mathsf{S}''[F]=S$, we have

$$S-\mathsf{S}''[F\lambda]=S\mathsf{S}[1-\lambda]-2\mathsf{S}'[F]\,\mathsf{S}'[\lambda]-\mathsf{S}[F]\,\mathsf{S}''[\lambda]. \tag{4·33}$$

Observing that S, $\mathsf{S}'[F]$, $\mathsf{S}[F]$ have coefficients tending to 0, and $\mathsf{S}[1-\lambda]$, $\mathsf{S}'[\lambda]$, $\mathsf{S}''[\lambda]$ have coefficients $O(n^{-3})$ and converge to 0 in (a', b'), we deduce from (4·9) that $S-\mathsf{S}''[F\lambda]$ converges uniformly to 0 in (a', b'). This gives the first half of the theorem. To prove the second half, we note that, by (4·16), the series conjugate to each of the products on the right of (4·33) converge uniformly over (a', b'), and that (4·31) is the nth partial sum of the series conjugate to $S-\mathsf{S}''[F\lambda]$.

Since the general series $\Sigma A_n(x)$ can be represented as a sum of two series, one consisting of the single term $\frac{1}{2}a_0$ and the other of the remaining terms, it remains to prove (4·29) in the case $S=\frac{1}{2}a_0$. Integrating by parts twice we see that (4·30) and (4·31) are then equal to

$$\tfrac{1}{2}a_0-\frac{1}{\pi}\int_0^{2\pi}\{F(t)\,\lambda(t)\}''\,D_n(x-t)\,dt, \tag{4·34}$$

$$-\frac{1}{\pi}\int_0^{2\pi}\{F(t)\,\lambda(t)\}''\,\tilde{D}_n(x-t)\,dt, \tag{4·35}$$

respectively. Since $F(t)=\frac{1}{4}a_0t^2$ and $\{F(t)\,\lambda(t)\}''=\frac{1}{2}a_0$ in (a', b'), the simplest criteria for the convergence of Fourier series and conjugate series show that, for $a'\leqslant x\leqslant b'$, the expressions (4·34) and (4·35) tend uniformly to limits, the limits of the first being 0. This completes the proof of (4·29).

Remarks. (*a*) We supposed that $a' < b'$, but the theorem and its proof remain valid if $a' = b'$, provided $\lambda' = \lambda'' = 0$ at that point. The conditions $\lambda' = \lambda'' = 0$ are automatically satisfied in the whole interval (a', b') if $a' < b'$ and the coefficients of λ are $O(n^{-5})$.

(*b*) The proof which gave (4·29) in a somewhat weaker form, and which elucidated the meaning of Riemann's formulae, also shows that Rajchman's method of formal multiplication is sometimes more advantageous than the original method of Riemann. Following Rajchman, to show that the behaviour of F outside (a, b) has no effect upon the behaviour of S in the interior of (a, b), we multiply S by $\mathsf{S}[\lambda]$, where λ is a function vanishing outside (a, b); the behaviour of $S\mathsf{S}[\lambda]$ is known at every point. Riemann's method consists in integrating S twice, multiplying the resulting function $F(x)$ by $\lambda(x)$ and differentiating $\mathsf{S}[F\lambda]$ twice. That the resulting series S_1 is equiconvergent with S in (a', b') is simply Riemann's theorem; and it can easily be shown that S_1 converges to 0 outside (a, b). Riemann's theorem tells us nothing about the behaviour of S_1 in the remaining intervals (a, a') and (b', b). Using the theorems on formal multiplication we can deduce this behaviour from the formula (4·33), and we see that it involves not only the series S but also the series $\mathsf{S}'[F]$ obtained by a single integration of S.

However, it must be stressed that Riemann's idea of introducing the function F into problems of localization is of fundamental importance. The method of formal multiplication complements it but can in no way replace it.

5. Formal multiplication of trigonometric series (*cont.*)

We give a few more applications of the theory.

(5·1) Theorem. *Given an arbitrary closed set E on the real axis, of period 2π, there is a trigonometric series with coefficients tending to zero which converges in E and diverges elsewhere.*

We know that there is a trigonometric series S with coefficients tending to zero which diverges everywhere (Chapter VIII, §4). Let $\lambda(x)$ be a periodic function with coefficients $O(n^{-3})$ which is 0 in E and different from 0 elsewhere. (See below for the construction of λ.) The formal product of S by $\mathsf{S}[\lambda]$ has the required properties since, by (4·9), it converges in E and diverges elsewhere.

Since, by (4·16), the series conjugate to the product converges in E, the power series $\Sigma \alpha_n e^{inx}$, whose real part is that product, converges in E and diverges elsewhere. Consequently, we have the following:

(5·2) Theorem. *For any closed set E situated on the circumference of the unit circle there is a power series $\sum\limits_0^\infty \alpha_n z^n$ with coefficients tending to 0 which converges in E and diverges at the remaining points of the circumference.*

Return to the function λ. Consider all the intervals (α_n, β_n) contiguous to E and situated in a period. Let $\lambda_n(x)$ be the periodic function which is

$$\{\sin \tfrac{1}{2}(x - \alpha_n) \sin \tfrac{1}{2}(x - \beta_n)\}^4 \quad \text{in} \quad (\alpha_n, \beta_n)$$

and is 0 elsewhere. Set
$$\lambda(x) = \Sigma \mu_n \lambda_n(x),$$
where the numbers η_n are positive and $\Sigma \eta_n < \infty$. Clearly λ is 0 in E and only there, and termwise differentiation of the last series shows that $\lambda'''(x)$ exists and is continuous.

Remark. If the series S used in the last two proofs is a Fourier series, then we need not appeal to the theory of formal multiplication but can use instead Theorem (6·7) of Chapter II.

The only everywhere divergent trigonometric series with coefficients tending to 0 which we so far know is that obtained in Chapter VIII, § 4—which is in fact a Fourier series. We give here a simpler construction (for a series that is not a Fourier series).

(5·3) THEOREM. *The series*
$$\sum_{k=2}^{\infty} \frac{\cos k(x - \log\log k)}{\log k}$$
diverges for every x.

Write
$$l_n = [\log n], \quad L_n = \log\log n,$$
and set
$$I_n = (L_n, L_{n+1}),$$
$$G_n = \sum_{n+1}^{n+l_n} \frac{1}{\log k}, \quad G_n(x) = \sum_{n+1}^{n+l_n} \frac{\cos k(x - L_k)}{\log k}.$$
Clearly,
$$G_n \geqslant l_n / \log(n + l_n) \to 1, \tag{5·4}$$
and, since $|\sin u| \leqslant |u|$,
$$G_n - G_n(x) \leqslant \frac{1}{2\log n} \sum_{n+1}^{n+l_n} k^2 (x - L_k)^2. \tag{5·5}$$

Suppose that $x \in I_n$ and k has the same range as in (5·5). Then
$$|x - L_k| \leqslant L_{n+l_n} - L_n \leqslant l_n / n \log n \leqslant 1/n,$$
by an application of the mean-value theorem to the function $\log\log n$. Hence
$$G_n - G_n(x) \leqslant \frac{(n + l_n)^2 l_n}{2n^2 \log n}.$$

The right-hand side here tends to $\frac{1}{2}$ as $n \to \infty$. Applying this and (5·4) to the equation
$$G_n(x) = G_n - (G_n - G_n(x)),$$
we see that $G_n(x)$ stays above a fixed positive quantity, provided n is large enough. Since each x belongs (mod 2π) to infinitely many of the intervals I_n, the divergence of the series follows.

We shall now apply localization and formal multiplication to power series on the circle of convergence.

If the series
$$\sum_0^{\infty} \alpha_n z^n \tag{5·6}$$
converges at some point of the unit circle $|z| = 1$, then $\alpha_n \to 0$. The converse is false—the power series whose real part for $z = e^{ix}$ is the series in (5·3) diverges at every point on the unit circle. We have, however, the following result:

(5·7) THEOREM OF FATOU-RIESZ. *Suppose that $\alpha_n \to 0$, so that* (5·6) *converges in $|z| < 1$ to a regular function $\Phi(z)$. Then* (5·6) *converges at every point of the unit circle where $\Phi(z)$ is regular, and the convergence is uniform over every closed arc of regularity.*

It is enough to prove the part about uniform convergence. We may suppose, for simplicity, that $\alpha_0 = 0$. The continuous function

$$F(x) = -\sum_1^{\infty} \alpha_n n^{-2} e^{inx},$$

obtained by integrating $\Sigma\alpha_n e^{inx}$ termwise twice, is the boundary value, for $z = e^{ix}$, of the function

$$\Psi(z) = -\int_0^z \frac{d\zeta}{\zeta} \int_0^{\zeta} \frac{\Phi(w)}{w} dw.$$

Let (a, b) be a closed arc of regularity of Φ on the unit circle. For $\epsilon > 0$ sufficiently small, Φ is regular on the closed arc $(a-\epsilon, b+\epsilon)$. The same holds of Ψ; hence F has infinitely many derivatives on $(a-\epsilon, b+\epsilon)$. Let $\lambda(x)$ be periodic, equal to 1 in $(a-\frac{1}{2}\epsilon, b+\frac{1}{2}\epsilon)$ and to 0 outside $(a-\epsilon, b+\epsilon)$. The function $F_1(x) = \lambda(x) F(x)$ is periodic and has as many derivatives as we wish (as many as λ has), so that we may suppose that the coefficients of F_1 are, say, $O(n^{-4})$. But then $F_1(x)$ is obtained by integrating termwise twice a trigonometric series S_1 with coefficients $O(n^{-2})$. The latter series converges uniformly, and since $F = F_1$ on $(a-\frac{1}{2}\epsilon, b+\frac{1}{2}\epsilon)$, it follows, by (4·3), that $\Sigma\alpha_n e^{inx}$ and S_1 are uniformly equiconvergent on (a, b), that is, $\Sigma\alpha_n e^{inx}$ converges uniformly on (a, b). This proves the theorem.

If e^{ix_0} is a point of regularity of Φ and $\Sigma\alpha_n e^{inx_0}$ converges to s, then

$$s = \lim_{r\to 1} \Sigma\alpha_n r^n e^{inx_0} = \lim_{r\to 1} \Phi(r e^{ix_0}) = \Phi(e^{ix_0}).$$

Hence (5·6) converges to Φ at every point e^{ix} of regularity.

The condition imposed upon the behaviour of Φ on the circle of convergence can be considerably relaxed. If we only know that in (a, b) $F(x)$ is a second integral of a function f (which is certainly the case if, for instance, $\Sigma\alpha_n e^{inx}$ is uniformly summable A on (a, b), f being then continuous), it is enough to assume that f satisfies one of the tests for the convergence of Fourier series.

The proof of the theorem that follows, which contains (5·7) as a special case, employs this idea.

(5·8) THEOREM. *Let $\alpha_n \to 0$, and let $\Phi(z) = \alpha_0 + \alpha_1 z + \alpha_2 z^2 + \dots$, $|z| < 1$. Let Γ be a sector of $|z| < 1$ having the circular arc I ($z = e^{ix}$, $a \leq x \leq b$) on its boundary. Suppose that the integral*

$$J = \iint_{\Gamma} |\Phi'(re^{ix})|^2 r\,dr\,dx \tag{5·9}$$

is finite. (This is certainly the case if, for example, Φ is univalent and bounded in Γ.) Then $\Sigma\alpha_n z^n$ converges at almost all points of I; and converges uniformly on every arc I' totally interior to I, if $\Phi(z)$ admits of a continuous extension to I.

If Γ coincides with $|z| < 1$, (5·8) reduces to Theorem (10·6) of Chapter VII.

The finiteness of J implies the finiteness of $\int_0^1 |\Phi'(re^{ix})|^2 dr$, and so also of $\int_0^1 |\Phi'(re^{ix})|\, dr$, for almost all x in (a, b). In particular, the radial limit

$$\Phi(e^{ix}) = \lim_{r\to 1} \Phi(re^{ix}) = \lim_{r\to 1} \left\{\Phi(0) + \int_0^{re^{ix}} \Phi'(\zeta) d\zeta\right\}$$

exists for almost all x in (a, b).

Let $0<r<r_1<1$. Making first $r_1 \to 1$ and then $r \to 1$ in

$$\int_a^b |\Phi(r_1 e^{ix}) - \Phi(re^{ix})|^2 dx$$
$$= \int_a^b \left| \int_r^{r_1} \Phi'(\rho e^{ix})\, d\rho \right|^2 dx \leqslant (r_1 - r) \int_a^b \int_r^{r_1} |\Phi'(\rho e^{ix})|^2 dx\, d\rho \leqslant \frac{1-r}{r} J,$$

we see that $\int_a^b |\Phi(e^{ix}) - \Phi(re^{ix})|^2 dx \to 0$ as $r \to 1$. In particular, $\Phi(e^{ix}) \in L^2(a,b)$. It follows immediately that

$$\int_a^\xi \{\Phi(e^{iu}) - \Phi(re^{iu})\}\, du = \int_a^\xi \Phi(e^{iu})\, du - \Sigma \alpha_n (in)^{-1} r^n \{e^{inu}\}_a^\xi \to 0 \tag{5·10}$$

uniformly in $a \leqslant \xi \leqslant b$.

This implies that $\Sigma \alpha_n (in)^{-1} \{e^{inu}\}_a^\xi$ is uniformly summable A, and so, by Tauber's theorem (Chapter III, (1·36)), uniformly convergent in $a \leqslant \xi \leqslant b$. Thus

$$\int_a^\xi \Phi(e^{iu})\, du = \Sigma \alpha_n (in)^{-1} (e^{in\xi} - e^{-ina}) \quad (a \leqslant \xi \leqslant b).$$

Integrating this with respect to ξ we deduce that $\Sigma \alpha_n (in)^{-1} e^{ina}$ converges, and that $F(x)$ differs from the second integral of $\Phi(e^{ix})$ by a linear function. We show that $\Phi(e^{ix})$ satisfies condition $\lambda_{\frac{1}{2}}^2$ in every interval (a', b') totally interior to (a, b), i.e. that

$$\int_{a'}^{b'} |\Phi(e^{i(x+h)}) - \Phi(e^{ix})|^2 dx = o(h). \tag{5·11}$$

Since functions in $\lambda_{\frac{1}{2}}^2$ satisfy Lebesgue's test for the convergence of Fourier series (Chapter II, (11·10)†, (5·8) will follow by (4·21).

Suppose, for example, that $h>0$. If x and $x+h$ are in (a,b), and $\Phi(e^{ix})$ and $\Phi(e^{i(x+h)})$ exist, we have

$$|\Phi(e^{i(x+h)}) - \Phi(e^{ix})| \leqslant \left| \int_{C_1} \Phi'(z)\, dz \right| + \left| \int_{C_2} \Phi'(z)\, dz \right| + \left| \int_{C_3} \Phi'(z)\, dz \right| = P + Q + R, \tag{5·12}$$

where C_1 is the segment $z = \rho e^{ix}$, $1-h \leqslant \rho \leqslant 1$; C_2 is the arc $z = (1-h) e^{iu}$, $x \leqslant u \leqslant x+h$; and C_3 is the segment $z = \rho e^{i(x+h)}$, $1-h \leqslant \rho \leqslant 1$.

We easily see that $\int_{a'}^{b'} P^2 dx \leqslant h \int_{a'}^{b'} dx \int_{1-h}^1 |\Phi'(\rho e^{ix})|^2 d\rho = o(h)$,

and similarly $\int_{a'}^{b'} R^2 dx = o(h)$. If we also show that $\int_{a'}^{b'} Q^2 dx = o(h)$, (5·11) will follow.

Let (a'', b'') be an interval containing (a', b') and contained in (a,b). Write $r_h = 1-h$. For h small enough, we have

$$\int_{a'}^{b'} Q^2 dx \leqslant h \int_{a'}^{b'} dx \int_x^{x+h} |\Phi'(r_h e^{iu})|^2 du \leqslant h^2 \int_{a''}^{b''} |\Phi'(r_h e^{iu})|^2 du. \tag{5·13}$$

For any z_0 on the arc $z = r_h e^{iu}$, $a'' \leqslant u \leqslant b''$, with h small enough and $0 < \sigma < h$, we also have

$$|\Phi'(z_0)|^2 \leqslant \frac{1}{2\pi} \int_0^{2\pi} |\Phi'(z_0 + \sigma e^{i\phi})|^2 d\phi.$$

Multiplying this by σ and integrating over $0 < \sigma < h$, we obtain

$$|\Phi'(z_0)|^2 \leqslant \frac{1}{\pi h^2} \iint_{|z - z_0| < h} |\Phi'(z)|^2 d\omega,$$

† The argument there is given for functions in $\lambda_{\frac{1}{2}}^2(0, 2\pi)$, but remains valid in our case.

where $d\omega$ is an element of area. The last integral can also be written

$$\int_{|z|<1} |F'(z)|^2 \chi(z; z_0, h)\, d\omega,$$

where $\chi(z; z_0, h)$ is the characteristic function of the circle with centre z_0 and radius h. Hence, by (5·13),

$$\int_{a'}^{b'} Q^2 dx \leqslant \pi^{-1} \int_{a''}^{b''} \left\{ \int_{|z|<1} |F'(z)|^2 \chi(z; r_h e^{iu}, h)\, d\omega \right\} du$$

$$= \pi^{-1} \int_{|z|<1} |F'(z)|^2 \left\{ \int_{a''}^{b''} \chi(z; r_h e^{iu}, h)\, du \right\} d\omega.$$

The inner integral on the right is 0 if $|z| \leqslant 1-2h$; it is also 0 if z is outside Γ and h is small enough; for z in Γ and satisfying $|z| \geqslant 1-2h$ the integral is $O(h)$. Hence

$$\int_{a'}^{b'} Q^2 dx = O(h) \int_a^b du \int_{1-2h}^{1} |\Phi'(\rho e^{iu})|^2 d\rho = o(h).$$

This completes the proof of (5·8).

In the remainder of this section we give a few more theorems on the multiplication of trigonometric series.

Consider two series $\qquad S = \Sigma c_n e^{inx}, \quad T = \Sigma \gamma_n e^{inx},$

and their product $\qquad ST = \Sigma C_n e^{inx} \quad (C_n = \sum_{p+q=n} c_p \gamma_q).$

The numbers C_n are defined and tend to 0 as $n \to \infty$, provided

$$c_p \to 0, \quad \Sigma |\gamma_q| < \infty. \tag{5·14}$$

Simple examples show that, if $\Sigma|\gamma_q| < \infty$, the mere vanishing of

$$\phi(x) = \Sigma \gamma_n e^{inx} \tag{5·15}$$

at a point x_0 does not guarantee the convergence of ST at x_0 (see example 19 at the end of the chapter). Part (i) of the theorem that follows shows that the situation is different if ϕ vanishes in a neighbourhood of x_0:

(5·16) Theorem. *Suppose* (5·14) *is satisfied and write* $\phi(x) = \Sigma \gamma_n e^{inx}$. *Then*

(i) *if* $\phi = 0$ *in an interval* (a, b), *both the product* ST *and its conjugate* $(\widetilde{ST})$ *converge uniformly in every subinterval* $(a+\epsilon, b-\epsilon)$, *the sum of* ST *being* 0;

(ii) *if there is a function* $\phi_1(x)$ *which coincides in* (a, b) *with* $\phi(x)$ *and is such that* $S[\phi_1] = \Sigma \gamma'_n e^{inx}$ *satisfies* $\Sigma |n\gamma'_n| < \infty$ *then in every* $(a+\epsilon, b-\epsilon)$ *the two series*

$$ST - \phi(x) S, \quad (\widetilde{ST}) - \phi(x) \tilde{S}$$

converge uniformly, and the sum of the first is 0.

The proof of (i) is based on the following lemma:

(5·17) Lemma. *Under the hypothesis* (5·14), *and denoting by* $U = \Sigma \delta_n e^{inx}$ *any absolutely convergent trigonometric series, we have the associativity relation*

$$(ST)\, U = S(TU). \tag{5·18}$$

Since the coefficients of ST tend to 0, the left-hand side of (5·18) is defined, and its nth coefficient is

$$\sum_{q+t=n} \Big(\sum_{r+s=q} c_r \gamma_s \Big) \delta_t = \sum_{r+s+t=n} c_r \gamma_s \delta_t,$$

the series on the right being absolutely convergent. Since TU converges absolutely, the right-hand side of (5·18) is also defined and its nth coefficient is

$$\sum_{r+q=n}\Big(\sum_{s+t=q}\gamma_s\delta_t\Big)c_r=\sum_{r+s+t=n}c_r\gamma_s\delta_t.$$

This proves the lemma. (Of course, the assumption $c_n=O(1)$ would have been sufficient.)

Let $U=\mathsf{S}[\lambda]$, where λ equals 1 in $(a+\epsilon, b-\epsilon)$, equals 0 outside (a,b), and has coefficients $O(n^{-3})$. Since $TU\equiv 0$, (5·18) shows that $(ST)\,U$ converges uniformly to 0, and, using (4·9), we see that ST converges uniformly to 0 in $(a+\epsilon, b-\epsilon)$. This implies that $(\widetilde{ST})$ converges uniformly in $(a+2\epsilon, b-2\epsilon)$, and so also in $(a+\epsilon, b-\epsilon)$.

To prove (ii) we write $\quad ST=S\mathsf{S}[\phi]=S\mathsf{S}[\phi-\phi_1]+S\mathsf{S}[\phi_1].$

By (i), $S\mathsf{S}[\phi-\phi_1]$ converges uniformly to 0 in $(a+\epsilon, b-\epsilon)$ and, by (4·9), $S\mathsf{S}[\phi_1]$ is uniformly equiconvergent with $\phi_1(x)\,S=\phi(x)\,S$ in (a,b). Hence ST is uniformly equiconvergent with $\phi(x)\,S$ in $(a+\epsilon, b-\epsilon)$. Similarly we prove the result about $(\widetilde{ST})$.

(5·19) THEOREM. *Let $S=\Sigma c_\nu e^{i\nu x}$, where $c_\nu\to 0$, and let $T=\Sigma\gamma_\nu e^{i\nu x}$, where $\Sigma|\gamma_\nu|<\infty$. Then:*

(i) *if S converges to 0 in $a<x<b$, so does ST;*

(ii) *if S converges in (a,b) to a finite $f\in L(a,b)$ or, more generally, if the function F obtained by integrating S termwise twice is, in (a,b), the second integral of an $f\in L(a,b)$ then in every interval (a',b') interior to (a,b) ST is uniformly equiconvergent with $\mathsf{S}[\phi f^*]$, where $\phi(x)$ is the sum of $\Sigma\gamma_\nu e^{i\nu x}$, and $f^*=f$ in (a,b) and $f^*=0$ elsewhere.*

(i) If $c_\nu\to 0$, $\Sigma|\gamma_\nu|<\infty$, and if U is any absolutely convergent series, we have a formula

$$(ST)\,U=(SU)\,T,$$

which is proved in the same way as (5·18). Let now $U=\mathsf{S}[\lambda]$, where λ is the same as above. Since SU converges to 0 everywhere, and so is identically 0, the same holds for $(SU)\,T=(ST)\,U$, which implies that ST converges uniformly to 0 in (a',b').

(ii) By (4·3), $S=\mathsf{S}[f^*]+S_1$, where S_1 converges to 0 in $a<x<b$, and it is enough to observe that $\mathsf{S}[f^*]\,T=\mathsf{S}[\phi f^*]$ and that, by (i), ST converges to 0 in $a<x<b$.

(5·20) THEOREM. *If two trigonometric series S and T have coefficients $o(1/n)$ and $O(1/n)$ respectively and converge at x_0 to sums s and t, then the product ST converges at x_0 to sum st.*

The conclusion of the theorem is false if both S and T have coefficients $O(1/n)$. For if $S=T=\Sigma n^{-1}\sin nx$, both series converge at $x=0$ to sum 0, whilst ST, which is the Fourier series of $\{\tfrac{1}{2}(\pi-x)\}^2$ in $|x|\leqslant\pi$, converges at $x=0$ to $\tfrac{1}{4}\pi^2$ (cf. Example 21 on p. 371).

We may suppose that $x_0=0$. By subtracting s from the constant term of S, we may also suppose that $s=0$. Let

$$s_n=\sum_{-n}^{n}c_\nu,\quad t_n=\sum_{-n}^{n}\gamma_\nu,\quad S_n=\sum_{-n}^{n}C_\nu.$$

The series defining the C_ν converge absolutely, and

$$S_n=\sum_{\nu=-n}^{n}\sum_{p=-\infty}^{+\infty}c_p\gamma_{\nu-p}=\sum_{p=-\infty}^{+\infty}c_p\sum_{\nu=-n-p}^{n-p}\gamma_\nu,$$

$$=\sum_{|p|\leqslant\frac{1}{2}n}+\sum_{\frac{1}{2}n<|p|\leqslant n}+\sum_{n<|p|\leqslant 2n}+\sum_{|p|>2n}=P_1+P_2+P_3+P_4.$$

We have to show that $S_n \to 0$. (We shall see that in our special case, $s=0$, it is enough to suppose that $t_n = O(1)$.)

Let $m=[\frac{1}{2}n]$. For $|p| \leqslant m$, considering separately p positive and negative, we have

$$\sum_{\nu=-n-p}^{n-p} \gamma_\nu = \sum_{-(n-|p|)}^{n-|p|} \gamma_\nu + O\left\{\sum_{n-|p|+1}^{n+|p|} \frac{1}{\nu}\right\}, \tag{5·21}$$

so that
$$\sum_{\nu=-n-p}^{n-p} \gamma_\nu = t_{n-|p|} + O\left\{\frac{|p|}{n-|p|}\right\} = t_{n-|p|} + O\left(\frac{|p|}{n}\right),$$

and
$$P_1 = \sum_{|p|\leqslant m} c_p t_{n-|p|} + \frac{1}{n}\sum_{|p|\leqslant m} o(1) = \sum_{p=0}^{m} (s_p - s_{p-1})\, t_{n-p} + o(1)$$
$$= \sum_{p=0}^{m} s_p (t_{n-p} - t_{n-p-1}) + o(1)$$
$$= \sum_{p=0}^{m} s_p O\left(\frac{1}{n-p}\right) + o(1) \qquad = O\left(\frac{1}{n}\right) \sum_{p=0}^{m} o(1) + o(1) = o(1).$$

Using (5·21) also for $m < |p| < n$, we have

$$P_2 = \sum_{m<|p|<n} o\left(\frac{1}{p}\right) t_{n-|p|} + \sum_{m<|p|<n} o\left(\frac{1}{|p|}\right) \log\frac{n+|p|}{n-|p|} + o(1)$$
$$= o\left(\frac{1}{n}\right) \sum_{m<|p|<n} O(1) \;+ o\left(\frac{1}{n}\right) \sum_{m<|p|<n} \log\frac{n+|p|}{n-|p|} + o(1)$$
$$= o(1) \qquad\qquad + o(1)\int_{\frac{1}{2}}^{1} \log\frac{1+x}{1-x}\,dx + o(1) = o(1).$$

Similarly, observing that for $|p| > n$ we have

$$\sum_{\nu=-n-p}^{n-p} \gamma_\nu = O\left\{\sum_{|p|-n}^{|p|+n} \frac{1}{\nu}\right\} = O\left\{\log\frac{|p|+n}{|p|-n}\right\},$$

we find that P_3 tends to 0.

Finally, using the last equation we obtain

$$P_4 = \sum_{|p|>2n} o\left(\frac{1}{|p|}\right) O\left(\frac{n}{|p|}\right) = o(n) \sum_{p>2n} \frac{1}{p^2} = o(1);$$

and collecting results, $S_n = P_1 + P_2 + P_3 + P_4 = o(1)$. This proves (5·19).

Consider the sine and the cosine developments of an $f \in L(0, \pi)$. While the two developments are uniformly equiconvergent in every interval totally interior to $(0, \pi)$, their behaviour at $x=0$ (or at $x=\pi$) may be entirely different.

(5·22) THEOREM. *Suppose that*

$$\sum_{n=1}^{\infty} a_n \sin nx, \tag{5·23}$$

$$\tfrac{1}{2}a_0' + \sum_{n=1}^{\infty} a_n' \cos nx \tag{5·24}$$

are the sine and cosine developments of an $f \in L(0, \pi)$. *Then*

(i) *if* $a_n = o(1/n)$, (5·24) *converges at* $x=0$ *to sum* 0;

(ii) *if* $a_n = O(1/n)$ *and* f *is continuous at* 0, (5·24) *converges uniformly at* $x=0$;

(iii) *if* $a_n = O(1/n)$ *and* f *is continuous in a neighbourhood* $0 \leqslant x \leqslant \delta$, (5·24) *converges uniformly in* $(0, \delta)$.

(i) Let $\chi(x) = \operatorname{sign} x$ $(|x| < \pi)$. By Chapter I, (4·13),

$$S[\chi] = \frac{4}{\pi}\sum_{0}^{\infty}\frac{\sin(2\nu+1)x}{2\nu+1} = \frac{2}{\pi}\sum_{\nu=-\infty}^{+\infty}\frac{e^{(2\nu+1)ix}}{i(2\nu+1)}.$$

Since (5·24) is the product of (5·23) by $S[\chi]$, (i) follows from (5·20).

(ii) An analysis of the proof of (5·19) shows that if (with the previous notation)

$$|s_n| \leqslant \epsilon, \quad |t_n| \leqslant 1, \quad |c_n| \leqslant 1/|n|, \quad |\gamma_n| \leqslant 1/|n|$$

for all n, then $|S_n| \leqslant A\epsilon$, where A is an absolute constant.

Suppose first that $f(+0) = 0$; then the partial sums s_n of (5·23) converge uniformly at $x = 0$ (Chapter III, (3·8)). Hence, omitting if necessary the first few terms in (5·23), we have $|s_n(x)| \leqslant \epsilon$ for $|x| \leqslant \eta$. Since the partial sums of $S[\chi]$ are uniformly bounded, it follows that the partial sums $S_n(x)$ of (5·24) satisfy $|S_n(x)| \leqslant A\epsilon$ for $|x| \leqslant \eta$, and since the contribution to $S_n(x)$ of the omitted terms converges uniformly to a limit, (ii) is established. The case $f(+0) \neq 0$ is reduced to the preceding one by subtracting $f(+0)\, S[\chi]$ from (5·23).

(iii) At every x interior to $(0, \pi)$ the uniform convergence of (5·23) implies that of (5·24). Hence, using (ii), (5·24) converges uniformly at every point of the closed interval $(0, \delta)$, and so converges uniformly in that interval.

We may interchange the roles of the sine and cosine series in (5·22). While (i) has no interesting counterpart, (ii) has the following analogue; if $a_n' = O(1/n)$, f is continuous at $x = 0$ and $f(0) = 0$, then (5·23) converges uniformly at $x = 0$. A similar extension holds for (iii). The proofs remain unchanged.

6. Sets of uniqueness and sets of multiplicity

A set E is called a set of *uniqueness*, or U-set, if every trigonometric series converging to 0 outside E vanishes identically. In § 3 we showed that every denumerable set is a U-set. If E is not denumerable but does not contain any perfect subset (such sets exist if we assume Zermelo's axiom), E is also a U-set. This follows from the fact that the set of points where a trigonometric series does not converge to 0 is a Borel set, and so, if it does not contain a perfect subset, must be denumerable; this implies that the series vanishes identically.

If E is a U-set, so is every subset of E.

A set E which is not a U-set is called a set of *multiplicity*, or M-set. If E is an M-set, then there is a trigonometric series which converges to 0 outside E but does not vanish identically.

Every set E of positive measure is an M-*set*. For let E_1 be a subset of E which is perfect and of positive measure, and let $f(x)$ be the characteristic function of E_1. The Fourier series of f converges to 0 outside E_1, and so also outside E, but does not vanish identically since its constant term is $|E_1|/2\pi > 0$.

It follows that it is only the case of sets of measure 0 which requires study, and it is a very curious fact that among perfect sets of measure 0 we find both U-sets and M-sets. Whether a given set E of measure 0 is of type U or M seems to depend rather on the arithmetic than on the metric properties of E, and the problem of characterizing, in structural terms, sets U and M is still open.

We shall now construct a family of perfect U-sets.

(6·1) THEOREM. *A set E is a* U-*set if there exists a sequence of periodic functions $\lambda_1(x), \lambda_2(x), \dots$ having the following properties:*

(i) *all λ_k vanish on E;*

(ii) *each $\mathsf{S}[\lambda_k] = \Sigma\gamma_n^k e^{inx}$ satisfies $\sum_n |n\gamma_n^k| < \infty$;*

(iii) $\sum_n |\gamma_n^k| < A$ $(k = 1, 2, \dots;$ *A independent of k);*

(iv) $\lim_k \gamma_n^k = 0$ *for $n \neq 0$;*

(v) $\limsup |\gamma_0^k| > 0$.†

We may assume that $|E| = 0$. Suppose that

$$S = \sum_{-\infty}^{+\infty} c_n e^{inx} \tag{6·2}$$

converges to 0 outside E; by (1·2), $c_n \to 0$. By (i) and (ii), and (4·9), the product

$$S\mathsf{S}[\lambda_k] = \Sigma C_n^k e^{inx}$$

converges to 0 both in E (since $\lambda_k = 0$ in E) and outside E (since S converges to 0 outside E). It follows that the product vanishes identically. In particular

$$C_0^k = \sum_n c_{-n}\gamma_n^k = c_0\gamma_0^k + \Sigma' c_{-n}\gamma_n^k = 0.$$

But (iii) and (iv) imply that, for fixed N,

$$\sum_{|n|<N}{}' c_{-n}\gamma_n^k \to 0 \quad \text{as} \quad k \to \infty,$$

$$\Big|\sum_{|n|\geq N} c_{-n}\gamma_n^k\Big| \leq A \max_{n \geq N} |c_n|.$$

It follows that

$$\Sigma' c_{-n}\gamma_n^k \to 0 \quad \text{as} \quad k \to \infty,$$

and so $c_0\gamma_0^k \to 0$, which, by (v), gives $c_0 = 0$.

Hence the constant term of every series S converging to 0 outside E is 0. This shows that all the coefficients of S must be 0; for if $c_k \neq 0$ for some k, the formal product of S by e^{-ikx} is a series converging to 0 outside E with a non-zero constant term, which is impossible. This completes the proof of (6·1).

(6·3) THEOREM. *Sets of type* H *are sets of uniqueness.*

Let E be of type H. This means that there exists a sequence of integers $n_1 < n_2 < \dots$ and an interval Δ such that for $x \in E$ the numbers $n_k x$ are all outside Δ, mod 2π. Let $\lambda(x)$ be a periodic function having three continuous derivatives, equal to 0 outside Δ and positive in the interior of Δ. We write

$$\lambda(x) = \Sigma\gamma_n e^{inx},$$

$$\lambda_k(x) = \lambda(n_k x) = \Sigma\gamma_n e^{inn_k x} = \Sigma\gamma_n^k e^{inx},$$

and verify that the λ_k satisfy conditions (i)–(v) of Theorem (6·1). Hence E is a U set.

In particular, Cantor's ternary set constructed on $(0, 2\pi)$ is a U-set.

† Selecting a subsequence of $\{\lambda_k\}$ and dividing it by suitable constants we may replace condition (v) by (v′) $\gamma_0^k = 1$. Conditions (iv) and (v′) mean that $\mathsf{S}[\lambda_k]$ tends termwise to $\mathsf{S}[1]$.

We now consider a generalization of H-sets. Denote points, or vectors, in the Euclidean m-dimensional space R^m by $(x', x'', \ldots, x^{(m)})$. Suppose we have a sequence of vectors $V_1, V_2, \ldots$ with positive integral components,

$$V_k = (n_k', n_k'', \ldots, n_k^{(m)}). \tag{6·4}$$

We call the sequence $\{V_k\}$ *normal*, if for each fixed non-zero vector $(a', a'', \ldots, a^{(m)})$ with integral components we have

$$|a'n_k' + a''n_k'' + \ldots + a^{(m)}n_k^{(m)}| \to \infty \quad \text{as} \quad k \to \infty.$$

By taking all the $a^{(j)}$ except one equal to 0, we see that each $n_k^{(j)}$ tends to ∞ with k. The sequence $\{V_k\}$ is normal if, for example, n_k' and all the ratios n_k''/n_k', n_k'''/n_k'', $\ldots$, $n_k^{(m)}/n_k^{(m-1)}$ tend to $+\infty$.

A set E is said to be of type $H^{(m)}$, if there is a normal sequence $V_1, V_2, \ldots$ and a domain Δ in R^m such that for each $x \in E$ the points

$$(n_k'x, n_k''x, \ldots, n_k^{(m)}x) \tag{6·5}$$

are outside Δ (mod 2π) (that is, if Δ has no points with co-ordinates congruent mod 2π to the co-ordinates in (6·5)). For $m=1$ we get H sets. It is obvious that the closure of an $H^{(m)}$ set is also $H^{(m)}$.

(6·6) Theorem. *Sets* $H^{(m)}$ *are sets of uniqueness.*

It is enough to consider the case $m=2$, which is typical. Let E be of type $H^{(2)}$. We may suppose that Δ is a rectangle $a' \leqslant x' \leqslant b'$, $a'' \leqslant x'' \leqslant b''$. Denote by (n_k', n_k'') a normal sequence such that for each $x \in E$ the point $(n_k'x, n_k''x)$ never enters Δ. Let $\mu_1(x) = \Sigma\delta_\nu' e^{i\nu x}$ be a periodic function having three continuous derivatives which vanishes outside (a', b') and is positive in the interior of (a', b'); by $\mu_2(x) = \Sigma\delta_\nu'' e^{i\nu x}$ we denote an analogous function for (a'', b''). The product $\mu_1(x')\,\mu_2(x'')$ is zero outside Δ, and positive in the interior of Δ (mod 2π). Let

$$\lambda_k(x) = \mu_1(n_k'x)\,\mu_2(n_k''x) = \Sigma\delta_\nu' e^{i\nu n_k' x}\,\Sigma\delta_\nu'' e^{i\nu n_k'' x} = \Sigma\gamma_n^k e^{inx}.$$

Clearly $\lambda_k(x)$ vanishes in E. Also, since the formal product $\Sigma D_n e^{inx}$ of any two series $\Sigma d_\nu' e^{i\nu x}$ and $\Sigma d_\nu'' e^{i\nu x}$ satisfies the inequality $\Sigma|D_n| \leqslant \Sigma|d_\nu'|\,\Sigma|d_\nu''|$, we have

$$\sum_n |\gamma_n^k| \leqslant (\Sigma|\delta_\nu'|)(\Sigma|\delta_\nu''|) = A,$$

and on differentiating $S[\lambda_k]$,

$$\sum_n |n\gamma_n^k| \leqslant |n_k'|(\Sigma|\nu\delta_\nu'|)(\Sigma|\delta_\nu''|) + |n_k''|(\Sigma|\delta_\nu'|)(\Sigma|\nu\delta_\nu''|) < \infty.$$

Hence the λ_k satisfy conditions (i), (ii) and (iii) of (6·1). We shall now show that conditions (iv) and (v) also hold.

We write
$$\gamma_n^k = \sum_{\nu' n_k' + \nu'' n_k'' = n} \delta_{\nu'}'\delta_{\nu''}'', \tag{6·7}$$

and first consider the case $n=0$. Then

$$\gamma_0^k = \delta_0'\delta_0'' + \sum_{\nu' n_k' + \nu'' n_k'' = 0}{}' \delta_{\nu'}'\delta_{\nu''}'',$$

and we show that the last sum (in which $|\nu'| + |\nu''| > 0$) tends to 0 as $k \to \infty$. We fix an N and split the sum into two, $\Sigma^{(1)}$ and $\Sigma^{(2)}$, collecting in $\Sigma^{(1)}$ all the terms with $|\nu'| \leqslant N$, $|\nu''| \leqslant N$. Since the sequence $\{(n_k', n_k'')\}$ is normal, $\nu' n_k' + \nu'' n_k''$ cannot be 0 if

$|\nu'| \leqslant N$, $|\nu''| < N$, ν' and ν'' are not simultaneously 0, and k is large enough. Hence $\Sigma^{(1)} = 0$ for k large enough. In $\Sigma^{(2)}$ either $|\nu'|$ or $|\nu''|$ must be greater than N and so

$$|\Sigma^{(2)}| \leqslant \Big(\sum_{|\nu|>N} |\delta'_\nu|\Big)\Big(\sum_{-\infty}^{+\infty} |\delta''_\nu|\Big) + \Big(\sum_{|\nu|>N} |\delta''_\nu|\Big)\Big(\sum_{-\infty}^{+\infty} |\delta'_\nu|\Big).$$

Since the right-hand side is arbitrarily small for N large enough, we see that $\Sigma^{(1)} + \Sigma^{(2)} \to 0$ as $k \to \infty$. This proves that $\gamma_0^k \to \delta'_0 \delta''_0 > 0$ as $k \to \infty$, and so condition (v) is verified.

Suppose now that $n \neq 0$. In the formula (6·7) the sum on the right does not contain the pair $(\nu', \nu'') = (0, 0)$. Hence, arguing as before, we see that $\gamma_n^k \to 0$ for each $n \neq 0$ and condition (iv) is established. Thus the λ_k satisfy all the conditions of (6·1) and E is a U-set.

(6·8) THEOREM. *A necessary and sufficient condition for a closed set E to be a set of multiplicity is that there exists a function $\Phi(x)$, $-\infty < x < +\infty$, which is constant in each interval contiguous to E but not constant identically, and which after subtraction of a suitable linear function is periodic and has coefficients $o(1/n)$.*

Suppose that $S = \Sigma c_n e^{inx}$ converges to 0 outside E but not everywhere. By the principle of localization, S converges uniformly in every closed interval without points in common with E. Therefore, integrating S termwise, we see that in the equation

$$\Phi(x) = c_0 x + \Sigma' c_n (in)^{-1} e^{inx}, \tag{6·9}$$

the series on the right converges outside E, that the sum $\Phi(x)$ is constant in each interval contiguous to E, that the periodic part of the series is $\mathfrak{S}[\Phi - c_0 x]$, and that its coefficients are $o(1/n)$. The function Φ cannot be constant identically, since otherwise the periodic part of Φ would be equal to a linear function; and this is only possible when $c_0 = 0$ and $c_1 = c_{-1} = c_2 = \ldots = 0$.

Conversely, suppose that there is a function $\Phi(x)$ which is constant in each interval contiguous to E but not constant identically, and for which, with a suitable c_0, $\Phi - c_0 x$ is periodic and has coefficients $o(1/n)$. We write

$$\Phi(x) - c_0 x \sim C + \Sigma' c_n (in)^{-1} e^{inx}, \tag{6·10}$$

where $c_n \to 0$. Let $S_1 = \Sigma' c_n e^{inx}$ be the series obtained by termwise differentiation of the right-hand side. Since Fourier series can be integrated termwise, the Riemann function $F_1 = -\Sigma' c_n n^{-2} e^{inx}$ associated with S_1 and obtained by integrating the series Σ' in (6·10) differs from $-\frac{1}{2} c_0 x^2$ by a linear function, in each interval contiguous to E. By the principle of localization, (4·3), S_1 converges to $-c_0$ outside E, that is,

$$S = c_0 + S_1 = \Sigma c_n e^{inx}$$

converges to 0 outside E. The series S cannot vanish identically, since this, together with (6·10), would imply that Φ is identically constant. Hence E is an M-set.

Remark. If E is a set of multiplicity then there is a trigonometric series converging to 0 *outside E but not everywhere and having constant term* 0. Suppose that S converges to 0 outside E but not everywhere; S must have at least one non-zero coefficient c_k. If $l \neq k$,

$$T = c_k e^{-ilx} S - c_l e^{-ikx} S$$

is the required example: it has constant term 0, converges to 0 outside E but not everywhere, since the set E_1 of points at which S does not converge to 0 is certainly infinite and, by (4·9), T converges to 0 only in a finite subset of E_1, namely at those points where $c_k e^{-ilx} - c_l e^{-ikx} = 0$.

An immediate consequence of this is that, if E is a closed set of multiplicity, then there exists a *periodic* function $\Phi(x)$ constant in the intervals contiguous to E, but not everywhere, and having coefficients $o(1/n)$.

We call a set E *a set of multiplicity in the restricted sense*, or M_0-set, if there is a Fourier-Stieltjes series $S[d\Phi]$ converging to 0 outside E but not everywhere. If $|E| = 0$ (the only interesting case), then $\Phi'(x) = 0$ almost everywhere (see Chapter III, (8·1)) and Φ is singular.

(6·11) THEOREM. *A necessary and sufficient condition for a closed set E to be an* M_0-*set is that there exists a function* $\Phi(x)$, *of bounded variation over every finite interval, constant in each interval contiguous to E but not in* $(-\infty, +\infty)$, *satisfying the condition*

$$\Phi(x+2\pi) - \Phi(x) = \Phi(2\pi) - \Phi(0), \tag{6·12}$$

and having Fourier-Stieltjes coefficients tending to zero.

The proof is similar to that of (6·8). Suppose that $S = S[d\Phi] = \Sigma c_n e^{inx}$ converges to 0 outside E. Then the Fourier coefficients of $d\Phi$ tend to 0, the function Φ is, up to an additive constant, given by (6·9) and so, as the proof of the necessity part in (6·8) shows, is constant in each interval contiguous to E but not in $(-\infty, +\infty)$. Moreover, (6·9) implies (6·12). Conversely, suppose that there is a function Φ having the properties stated in Theorem (6·11). The proof of the sufficiency part of (6·8) shows that $S = S[d\Phi]$ converges to 0 outside E but not everywhere.

As in the case of M-sets, we can replace the condition (6·12) by that of periodicity of Φ.

In connexion with (6·11), we recall the fact that if the coefficients of $d\Phi$ tend to 0, then Φ is necessarily continuous (Chapter III, (9·6)).

(6·13) THEOREM. *Let E be a closed* U-*set and* $\Phi(x)$, $-\infty < x < +\infty$, *a function which is of bounded variation in* $(0, 2\pi)$, *satisfies* (6·12), *and is constant in each interval contiguous to E but not constant identically. Then the coefficients of* $d\Phi$ *are not* $o(1)$, *and the coefficients of* $\Phi(x)$, $0 \leqslant x < 2\pi$, *are not* $o(1/n)$.

That the coefficients of $d\Phi$ do not tend to 0 follows from (6·11). If $\Phi(2\pi - 0) \neq \Phi(+0)$, the function $\Phi(x)$ continued periodically from $0 \leqslant x < 2\pi$ has jumps, and so its coefficients are (trivially) not $o(1/n)$. If $\Phi(2\pi - 0) = \Phi(0)$, then $S[d\Phi] = S'[\Phi]$, and since the coefficients of $d\Phi$ do not tend to 0, those of $\Phi(x)$, $0 \leqslant x < 2\pi$, are not $o(1/n)$.

(6·14) THEOREM OF MENŠOV. *There exist perfect* M-*sets of measure* 0.

We do not give the original construction of Menchov, since the theorem is a consequence of results which will be proved later, and which we assume here.

Fix a number ξ, $0 < \xi < \frac{1}{2}$, and denote by $E(\xi)$ the set of 'constant ratio of dissection' (Chapter V, § 3) which we construct on $(0, 2\pi)$ by subdividing at each stage the 'white' intervals in the ratio ξ, $1 - 2\xi$, ξ and removing the central part. The measure of $E(\xi)$ is 0. Let $\Phi(x)$ be the Lebesgue singular function associated with $E(\xi)$. In Chapter V,

(3·5) we have obtained explicit formulae for the coefficients c_n of $d\Phi$. The problem of the behaviour of the c_n as $n\to\infty$ will be solved in Chapter XII, § 11, where we show that $c_n \neq o(1)$ *if and only if* $\theta = 1/\xi$ *is an algebraic integer all whose conjugates are of modulus less than* 1. It follows that except for such ξ's, which form a denumerable set, we always have $c_n \to 0$, and $E(\xi)$ is, by (6·11), an M- (indeed M_0-) set.

Algebraic integers whose conjugates are all in the interior of the unit circle will be called S *numbers*.

Remarks. (*a*) The result just stated shows that, among the sets $E(\xi)$, sets M are more numerous than sets U.

(*b*) The result does not say anything about the U or M character of $E(\xi)$ if θ is an S number, since it merely asserts that the coefficients of a *special* function Φ do not tend to 0. In the particular case, however, when θ is a rational integer ($\theta = 3, 4, \ldots$), $E(\xi) = E(1/\theta)$ is an H-set (see § 1) and so a U-set.

(*c*) Though all the sets $E(\xi)$ are of measure 0, it is natural to think that the 'thickness' of $E(\xi)$ increases with ξ. In particular, $E(\frac{1}{3})$ is 'thicker' than $E(\frac{4}{13})$. None the less $E(\frac{1}{3})$, as an H-set, is a set of uniqueness, while $E(\frac{4}{13})$ is a set of multiplicity. This shows that it is not so much metric as number-theoretic properties that determine the U or M character of a set.

If E_1 and E_2 are sets of uniqueness, their sum $E_1 + E_2$ may be a set of multiplicity. We obtain an example by breaking a perfect M-set of measure 0 into two subsets E_1 and E_2 without perfect subsets. Then E_1 and E_2 are U-sets (p. 344), but their sum is not. The sets E_1 and E_2 are of measure 0, but are not Borel sets. Whether the sum of two U-sets measurable B is a U-set is not known. In the case of closed sets we have the following theorem:

(6·15) THEOREM OF N. BARY. *If* $E_1, E_2, \ldots, E_n, \ldots$ *are closed* U-*sets, their sum* $E = E_1 + E_2 + \ldots + E_n + \ldots$ *is a* U-*set.*

The proof is based on two lemmas.

(6·16) LEMMA. *Let* $\mathscr{E}$ *be a closed* U-*set contained in an open interval* J. *Suppose that a trigonometric series* S

(i) *has partial sums bounded at each point of* $J - \mathscr{E}$;

(ii) *converges to* 0 *almost everywhere in* J:

Then it converges to 0 *everywhere in* J.

By (ii), the coefficients of S tend to 0. Let J_1 be an open subinterval of J without points in common with $\mathscr{E}$, and let $\lambda_1(x)$ be a function having three continuous derivatives, equal to 0 outside J_1, and positive in J_1. Since the partial sums of S are bounded at each point of J_1 and converge to 0 almost everywhere in J_1, the formal product $S_1 = \mathsf{S}[\lambda_1] S$ has partial sums bounded at each point of $(0, 2\pi)$ and converges to 0 for almost all x. By (3·19), S_1 is identically 0, and so S converges to 0 in J_1. It follows that S converges to 0 in $J - \mathscr{E}$.

If now $\lambda(x)$ vanishes outside J and is positive in J, the product $S\mathsf{S}[\lambda]$ converges to 0 at each point not in $\mathscr{E}$. Hence it vanishes identically. It follows that S converges to 0 in J.

(6·17) LEMMA. *A set* N *which is a denumerable product of open sets cannot be of the first category on itself; that is, it cannot be of the form* ΣN_i, *where the* N_i *are non-dense on* N.

When N is a closed set, this is Theorem (12·1) of Chapter I. The proof in the general case is similar. We denote by $J_1, J_2, \ldots$ open intervals and by $\bar{J}_1, \bar{J}_2, \ldots$ their closures. Write $N = G_1 G_2 \ldots$, where the G_i are open sets. By hypothesis, there is a J_1 such that $J_1 N$ is not empty and $\bar{J}_1$ has no point in common with N_1; we may also suppose that $\bar{J}_1 \subset G_1$. Similarly, there is a J_2 such that $\bar{J}_2 \subset J_1$, $J_2 N$ is not empty and $\bar{J}_2$ has no point in common with N_2; moreover $\bar{J}_2 \subset G_2$. Generally there is a J_n such that $\bar{J}_n \subset J_{n-1}$, $J_n N$ is not empty, $\bar{J}_n$ has no point in common with N_n and $\bar{J}_n \subset G_n$, $n = 2, 3, \ldots$. Since $\bar{J}_1 \supset \bar{J}_2 \supset \ldots$ and $\bar{J}_n \subset G_n$, there is a point x_0 common to all $\bar{J}_n$ and this point also belongs to $N = G_1 G_2 \ldots$. But since $x_0 \in \bar{J}_n$, x_0 cannot be in N_n for any n. This contradicts $N = \Sigma N_n$ and proves the lemma.

Return to (6·15) and suppose that there is a trigonometric series S converging to 0 outside E but not everywhere. Since $|E| \leqslant \Sigma |E_n| = 0$, S converges to 0 for almost all x. Let N be the set of points at which the partial sums of S are unbounded. N is not empty; for otherwise, by (3·19), S would be identically zero.

It was shown in Chapter I, § 12, that N is a denumerable product of open sets. Write $N_i = NE_i$. We have

$$N = NE = N\Sigma E_i = \Sigma N_i.$$

By (6·17), some N_i, say N_{i_0}, is not non-dense on N; thus there is an open interval J such that NJ is not empty and $N_{i_0} J = E_{i_0} NJ$ is dense on NJ. Since E_{i_0} is closed, we have $E_{i_0} NJ = NJ$ and, in particular,

$$NJ \subset E_{i_0} J,$$

which means that a certain portion of N consists entirely of points of E_{i_0}.

By reducing J if necessary, we may suppose that the end-points of J are not in E_{i_0}. Hence the set $\mathscr{E} = E_{i_0} \bar{J} = E_{i_0} J$ is a closed set of uniqueness situated in the open interval J. Since S converges to 0 almost everywhere in J and since at every point of the set

$$J - \mathscr{E} = J - E_{i_0} J \subset J - NJ$$

the partial sums of S are bounded, Lemma (6·16) shows that S converges to 0 in J. This contradicts the fact that $NJ \neq 0$ and proves (6·15).

Given any set E, denote by E_λ the set of points λx, where $x \in E$. The proof of the following theorem must be postponed to Chapter XVI, §10:

(6·18) THEOREM. *Suppose that E and E_λ are both subsets of $(0, 2\pi)$. Then, if E is a U-set, so is E_λ.*

The remaining two theorems of the section deal with slightly different aspects of the theory of sets of uniqueness.

We call any set E of measure 0 a U*-*set* if it has the following property: if a trigonometric series S converges outside E to a finite and integrable function f, then $S = \mathsf{S}[f]$. Every U*-set is also a U-set. Whether the converse is true is not known, except in the case of a closed E:

(6·19) THEOREM. *Let E be any closed set of measure 0 situated in $(0, 2\pi)$ and S any trigonometric series which converges outside E to a finite and integrable function f. Then the difference $S_1 = S - \mathsf{S}[f]$ is a series which converges to 0 outside E. In particular. $S = \mathsf{S}[f]$ if E is a U-set.*

This is immediate. For in each interval contiguous to E the Riemann function obtained by integrating S twice differs from the second integral of f by a linear function. Since the same holds for $\mathsf{S}[f]$, the principle of localization shows that S_1 converges to 0 outside E.

The hypothesis of (6·19) can be somewhat relaxed by assuming, instead of the convergence of S, that the upper and lower sums f^* and f_* of S are finite at each point outside E (except for a denumerable set) and are integrable. Then $f^* = f_*$ almost everywhere and $S - \mathsf{S}[f^*]$ converges to 0 outside E.

Return to U-sets. Let $\epsilon = (\epsilon_0, \epsilon_1, \epsilon_2, \ldots)$ be a sequence of positive numbers tending monotonically to 0. We call E a U(ϵ)*-set* if every series $S = \Sigma c_n e^{inx}$ converging to 0 outside E and satisfying the condition

$$|c_n| \leqslant \epsilon_{|n|} \tag{6·20}$$

for all n necessarily vanishes identically. Of course if two sequences ϵ and ϵ' are multiples of each other, the U(ϵ) and U(ϵ') sets are the same.

(6·21) THEOREM. *For each sequence $\epsilon = (\epsilon_0, \epsilon_1, \ldots)$ decreasing monotonically to* 0, *no matter how slowly, there exist* U(ϵ)*-sets of positive measure.*

The proof is similar to the proof that H-sets are U-sets.

Denote the distance of x from the nearest integer by $\{x\}$; thus $0 \leqslant \{x\} \leqslant \frac{1}{2}$. Let $\delta_1, \delta_2, \ldots$ be a sequence of positive numbers, less than $\frac{1}{2}$ and tending to 0, to be determined later; and let E_n be the set of points in $(-\pi, \pi)$ at which

$$\{nx/2\pi\} \geqslant \delta_n. \tag{6·22}$$

The intervals constituting E_n are separated by intervals of length $4\pi\delta_n/n$. We fix a sequence $n_1 < n_2 < \ldots$ and set

$$E = E_{n_1} E_{n_2} \ldots. \tag{6·23}$$

We first show that if $\{n_k\}$ increases fast enough, then $|E| > 0$. For write $S_k = E_{n_1} \ldots E_{n_k}$. We take $n_1 = 1$, and since $|E_n| \to 2\pi$, we can determine $n_2, n_3, \ldots$ successively so that $|S_{k+1}| \geqslant (1 - 2^{-k}) |S_k|$ for $k = 1, 2, \ldots$. Then

$$|E| = \lim |S_k| \geqslant |E_1| \prod_1^\infty (1 - 2^{-k}) > 0.$$

Next we consider the function

$$\lambda(x, h) = 1 + \sum_{-\infty}^{+\infty}{}' \left(\frac{\sin \nu h}{\nu h}\right)^3 e^{i\nu x} = \sum_{-\infty}^{+\infty} \gamma_\nu(h)\, e^{i\nu x},$$

where $0 < h \leqslant \frac{1}{3}\pi$. This function is 0 for $3h \leqslant |x| \leqslant \pi$ (see Chapter V, Example 4).

Finally, we determine a sequence of positive h_n tending to 0 such that

$$h_n^{-3} \epsilon_{[\frac{1}{2}n]} \to 0, \tag{6·24}$$

and define the δ_n in (6·22) by the condition

$$3h_n = 2\delta_n \pi. \tag{6·25}$$

Having thus defined the E_n, we shall show that each (6·23) is a U(ϵ)-set. Suppose that $S = \Sigma c_n e^{inx}$ satisfies (6·20) and converges to 0 outside E. In particular, S converges to 0 outside each E_{n_k}. Since, in view of (6·25), $\lambda(n_k x, h_{n_k})$ is 0 in E_{n_k}, the product $S\mathsf{S}[\lambda(n_k x, h_{n_k})]$ is identically 0. The mth coefficient of the product is

$$\sum_{\nu=-\infty}^{+\infty} c_{m-\nu n_k} \gamma_\nu(h_{n_k}) = c_m + \Sigma' = 0,$$

and, since $|c_n| \leqslant \epsilon_{|n|}$, for $n_k \geqslant 2|m|$ we get

$$|c_m| \leqslant \epsilon_{[\frac{1}{2}n_k]} \Sigma' |\gamma_\nu(h_{n_k})| \leqslant \epsilon_{[\frac{1}{2}n_k]} h_{n_k}^{-3} \Sigma' |\nu|^{-3}.$$

By (6·24), with $n = n_k$, the right-hand side here tends to 0 for $k \to \infty$. Hence $c_m = 0$ for each m. Since, with a suitable $\{n_k\}$, we have $|E| > 0$, (6·21) is established.

U(ϵ)-sets may have measure arbitrarily close to 2π; whether or not there are U(ϵ)-sets of measure 2π is an open problem.

7. Uniqueness of summable trigonometric series

In the preceding section we proved a number of theorems about the uniqueness of the representation of functions by means of convergent trigonometric series. Since, however, there exist Fourier series which diverge everywhere (Chapter VIII, §4), it is natural to ask for theorems of uniqueness for summable trigonometric series. We shall restrict ourselves to Abel's method of summation in view of its significance for the theory of functions. Since Abel's method also applies to series with terms not tending to zero, we may first inquire about the conditions which we must impose upon the coefficients of the series considered.

Let

$$P(r, x) = \tfrac{1}{2} + \sum_1^\infty r^n \cos nx = \frac{1}{2}\,\frac{1-r^2}{1-2r\cos x + r^2}.$$

The Abel means of the series

$$\sum_1^\infty n \sin nx \tag{7·1}$$

are

$$\sum_1^\infty n r^n \sin nx = -\frac{\partial}{\partial x} P(r, x) = \frac{r(1-r^2)\sin x}{(1-2r\cos x + r^2)^2},$$

and we see at once that (7·1) is summable A to 0 for all x. Hence there is no uniqueness for trigonometric series summable A and having coefficients $O(n)$.

Also, since

$$\tfrac{1}{2} + \sum_1^\infty \cos nx \tag{7·2}$$

is summable A to 0 for each $x \neq 0$, there are no (non-empty) sets of uniqueness for trigonometric series summable A and having bounded coefficients.

Given a trigonometric series

$$(S) \qquad \tfrac{1}{2}a_0 + \sum_1^\infty (a_n \cos nx + b_n \sin nx) = \sum_0^\infty A_n(x),$$

we shall write

$$\left.\begin{aligned} f(r, x) &= \Sigma A_n(x)\, r^n, \\ f^*(x) = \limsup_{r\to 1} \Sigma A_n(x)\, r^n, &\quad f_*(x) = \liminf_{r\to 1} \Sigma A_n(x)\, r^n, \end{aligned}\right\} \tag{7·3}$$

and call f^* and f_* the *upper* and *lower Abel sums* of S.

(7·4) THEOREM. *Suppose that S satisfies the conditions*

$$a_n = o(n), \quad b_n = o(n), \tag{7·5}$$

and is summable A *for each x to a function $f(x)$ finite and integrable. Then $S = \mathsf{S}[f]$. In particular, if $f \equiv 0$, then $a_0 = a_1 = b_1 = \ldots = 0$.*

This theorem can be generalized in several directions. The proofs of these generalizations are, however, far from simple, and it seems desirable to treat them separately (see § 8). Without loss of generality *we may suppose that* $a_0 = 0$.

The following lemma is basic for the whole argument.

(7·6) RAJCHMAN'S LEMMA. *Suppose that*

$$C - \sum_{1}^{\infty} \frac{a_n \cos nx + b_n \sin nx}{n^2} \tag{7·7}$$

is an $\mathsf{S}[F]$, *and that* (7·7) *is summable* A *at* x_0 *to* $F(x_0)$. *Then the intervals*

$$(\underline{D}^2 F(x_0), \overline{D}^2 F(x_0)) \quad \textit{and} \quad (f_*(x_0), f^*(x_0))$$

have points in common; that is,

$$\underline{D}^2 F(x_0) \leqslant f^*(x_0), \quad f_*(x_0) \leqslant \overline{D}^2 F(x_0). \tag{7·8}$$

It is enough to prove the first inequality (7·8), since, when applied to $-F(x)$, it gives the second.

Let $F(r, x)$ be the harmonic function associated with (7·7). We may suppose that $x_0 = 0$, $F(-x) = F(x)$, and $F(0) = \lim F(r, 0) = 0$. We write $F(r, 0) = G(r)$. The lemma will be established if we show that, for any finite m,

$$\underline{D}^2 F(0) > m \quad \text{implies} \quad f^*(0) \geqslant m. \tag{7·9}$$

We may even suppose that $m = 0$, for otherwise consider $F(x) - m(1 - \cos x)$ instead of $F(x)$.

Suppose then that $\underline{D}^2 F(0) > 0$ and that, contrary to what we want to prove, $f^*(0) < 0$. In the Laplace equation

$$\frac{1}{r^2} \frac{\partial^2 F(r, x)}{\partial x^2} + \frac{1}{r} \frac{\partial}{\partial r} \left(r \frac{\partial F(r, x)}{\partial r} \right) = 0, \tag{7·10}$$

the first term on the left is $r^{-2} f(r, x)$. It follows that $rG'(r)$ is an increasing function of r in an interval $r_0 \leqslant r < 1$. Since $G(r) \to 0$ as $r \to 1$, the Cauchy mean-value theorem gives

$$\frac{G(r)}{\log r} = \frac{G(r) - G(1)}{\log r - \log 1} = \rho G'(\rho) \quad (r_0 \leqslant r < \rho < 1),$$

and so, for some σ between ρ and 1,

$$\frac{G(r)}{\log r} - \frac{G(\rho)}{\log \rho} = \rho G'(\rho) - \sigma G'(\sigma) < 0.$$

If we show that this is impossible, by proving that

$$\limsup \left\{ \frac{G(r)}{\log r} \right\}' < 0, \tag{7·11}$$

the lemma will be established.

Write

$$\Delta = \Delta(r, t) = 1 - 2r \cos t + r^2,$$

$$\phi(t) = \{F(t) + F(-t) - 2F(0)\}/\sin^2 t = 2F(t)/\sin^2 t.$$

Representing $F(r,x)$ as the Poisson integral of $F(x)$ we have

$$r\frac{G(r)}{1-r^2}=\frac{1}{\pi}\int_0^\pi F(t)\frac{r}{\Delta}dt,$$

$$\left\{r\frac{G(r)}{1-r^2}\right\}'=\frac{1}{\pi}\int_0^\pi F(t)\frac{1-r^2}{\Delta^2}dt=\frac{1}{2\pi}\int_0^\eta \phi(t)\sin^2 t\frac{1-r^2}{\Delta^2}dt+o(1)$$

$$=-\frac{1}{2\pi r}\int_0^\eta \phi(t)\sin t\frac{\partial}{\partial t}P(r,t)\,dt+o(1),$$

where η is a fixed positive number less than $\frac{1}{3}\pi$.

By hypothesis, $\liminf \phi(t)>0$. Hence, choosing η small enough, we have $\phi(t)>h>0$ for t in $(0,\eta)$. We have also $\cos t\geqslant\frac{1}{2}$ in $(0,\eta)$. Noting that $\partial P/\partial t<0$ in $(0,\pi)$ we obtain

$$\liminf_{r\to1}\left\{r\frac{G(r)}{1-r^2}\right\}'\geqslant\liminf\left\{-\frac{h}{2\pi}\int_0^\eta \sin t\frac{\partial}{\partial t}P(r,t)\,dt\right\}$$

$$=\liminf\left\{\frac{h}{2\pi}\int_0^\eta \cos t\,P(r,t)\,dt\right\}$$

$$\geqslant\liminf\left\{\frac{h}{4\pi}\int_0^\eta P(r,t)\,dt\right\}$$

$$=\lim\left\{\frac{h}{4\pi}\int_0^\pi P(r,t)\,dt\right\}=\tfrac{1}{8}h>0.$$

Thus $\{rG(r)/(1-r^2)\}'$ stays above a positive quantity as $r\to1$. Now, with $c(r)=(1-r^2)/(r\log r)$ we have

$$\left\{\frac{G(r)}{\log r}\right\}'=c(r)\left\{\frac{rG(r)}{1-r^2}\right\}'+c'(r)\frac{rG(r)}{1-r^2},$$

and since $c(r)\to-2$, $c'(r)=O(1-r)$, we obtain (7·11), and so complete the proof of the lemma.

Return to (7·4), and suppose first that the series (7·7), obtained by integrating S twice, is an $\mathfrak{S}[F]$, with F continuous. By (7·6), $\underline{D}^2F\leqslant f\leqslant\overline{D}^2F$ for all x. By (3·13), F is a second integral of f, and this shows that $S=\mathfrak{S}[f]$ (see p. 326).

The hypothesis (7·5) and the Riesz-Fischer theorem imply that (7·7) is a Fourier series, and in view of what has just been proved, Theorem (7·4) will be established if we show that the hypotheses of (7·4) imply that the function

$$F(x)=\lim_{r\to1}F(r,x) \tag{7·12}$$

exists and is continuous.

The existence of $F(x)$ follows by twice applying to S the following lemma:

(7·13) LEMMA. *If the series $u_0+u_1+\dots$ is summable* A *(indeed, if only the upper and lower Abel sums of the series are finite), the series $\sum_1^\infty u_n/n$ is summable* A.

For if $g(r)=u_0+u_1r+\dots$, then

$$G(r)=\sum_1^\infty\frac{u_n}{n}r^n=\int_0^r\frac{g(\rho)-u_0}{\rho}d\rho,$$

and since the integrand is bounded, we have $|G(r)-G(r')|\to0$ as $r\to1$, $r'\to1$, which proves the lemma.

The following observation will be useful later. *Suppose that the u_n are functions of a parameter x, and that $g(r)$ is uniformly bounded for $0 \leqslant r < 1$ and x belonging to a set E; then $u_1 + \frac{1}{2}u_2 + \ldots$ is uniformly summable A for $x \in E$.*

Return to the function $F(x)$ in (7·12) and denote by D the set of discontinuities of F. We have to show that D is empty. We first show that *D has no isolated points.*

Suppose that x_0 is an isolated point of D. There is then an interval (a, b), containing x_0 in its interior, such that F is continuous in the interiors of (a, x_0) and (x_0, b). By (3·13) and the inequalities $\underline{D}^2F \leqslant f \leqslant \overline{D}^2F$, we have $F(x) = \int_a^x dy \int_a^y f(t)\,dt + Ax + B$ in the interior of (a, x_0), since Lemma (3·13) holds also for open intervals. In particular, $F(x_0 - 0)$ exists and is finite. Similarly, $F(x_0 + 0)$ exists. Since the coefficients of (7·7) are $o(1/n)$, we have, by (2·19) (i)†, $F(x_0 - 0) = F(x_0) = F(x_0 + 0)$ and F is continuous at x_0, contrary to hypothesis.

Next, let P be any perfect set. Consider $F(x)$ on P and denote by D_P the set of points $x \in P$ at which F is discontinuous *with respect to P*. We show that *D_P is non-dense on P.*

Let $r_1 < r_2 < \ldots \to 1$ be such that

$$\max_{r_n \leqslant r \leqslant r_{n+1}} |f(r, x) - f(r_n, x)| \leqslant 1 \quad (n = 1, 2, \ldots;\ 0 \leqslant x \leqslant 2\pi). \tag{7·14}$$

By hypothesis, $\{f(r_n, x)\}$ tends to a finite limit for each x. By Theorem (12·3) (i) of Chapter I, on each portion of P there is a sub-portion Π on which all the $f(r_n, x)$ are uniformly bounded; by (7·14), $f(r, x)$ is uniformly bounded for $x \in \Pi$, $0 \leqslant r < 1$. Applying twice to S the observation following Lemma (7·13), we see that $\lim_{r \to 1} F(r, x)$ exists uniformly on Π; thus D_P has no point in common with Π, and so is non-dense on P.

It is now easy to complete the proof of (7·4). Suppose that $D \neq 0$. D has no isolated points and is non-dense. Hence the closure $\overline{D}$ of D is perfect. Take a portion Π of $\overline{D}$ in which $F(x)$ is continuous with respect to $\overline{D}$; we may suppose that Π is perfect. Denote by $d_i = (a_i, b_i)$, $i = 1, 2, \ldots$, the open intervals contiguous to Π. $F(x)$ is continuous in each d_i and, as the proof of the absence of isolated points in D shows, $F(a_i + 0)$ and $F(b_i - 0)$ exist and equal $F(a_i)$ and $F(b_i)$ respectively.

Consider the function

$$F_1(x) = \int_0^x dy \int_0^y f(t)\,dt.$$

The difference $F_2(x) = F(x) - F_1(x)$ is linear in the closure of each d_i, and is also continuous at each point of Π with respect to $\overline{D}$. Hence at each point of Π $F(x)$ is continuous with respect to the whole neighbourhood, which is impossible since Π contains points of D. Hence $D = 0$, F is continuous for all x, and (7·4) is proved.

The proof above gives more than was actually stated. Suppose that S satisfies (7·5) and that both $f^*(x)$ and $f_*(x)$ are finite and integrable. $F(x)$ then exists everywhere and is continuous. Let $f(x) = f^*(x)$ wherever $\overline{D}^2F(x) \geqslant f^*(x)$, and $f(x) = \overline{D}^2F(x)$ elsewhere; f is integrable and finite, we have $\overline{D}^2F \geqslant f \geqslant \underline{D}^2F$ for all x, and the proof above shows that $S = \mathsf{S}[f]$. In particular, $f^* = f_*$ almost everywhere. All this, however, is included in the theorem which follows.

† By Tauber's theorem (Chapter III, (1·36)), Abel means and partial sums of any series with terms $o(1/n)$ are equi-convergent.

8. Uniqueness of summable trigonometric series (*cont.*)

We keep the notation of the preceding section.

(8·1) THEOREM. *Suppose that* (i) $S=\Sigma A_n(x)$ *has coefficients* $o(n)$, *and that*
(ii) $f^*(x)\geqslant\chi(x)$, *where* χ *is integrable;*
(iii) $f_*(x)$ *is finite except possibly in a denumerable set* E;
(iv) *for each* $x\in E$ *and* $r\to 1$ *we have*

$$(1-r)f(r,x)\to 0.$$

Then S *is a Fourier series.*

Observe that we do not assume the finiteness of f^*. Condition (iv) is automatically satisfied if instead of (i) we assume the much stronger condition $|a_n|+|b_n|\to 0$. Conditions (ii) and (iii) are certainly satisfied if both f^* and f_* are finite everywhere, and *one of them* is integrable.

Since the limits of indetermination of the partial sums of a series contain the upper and lower Abel sums of the series, Theorem (3·25), stated without proof, is a corollary of (8·1).

We begin with two lemmas on Fourier series.

(8·2) LEMMA. *If* F *is integrable and* $D^2F(x_0)$ *exists and is finite, then* $\mathfrak{S}''[F]$ *is summable* A *at* x_0 *to sum* $D^2F(x_0)$.

This lemma, though more elementary than (7·6), is not a consequence of it.

We may suppose that $x_0=0$, $F(0)=0$, F is even, and $D^2F(0)=0$. The Abel means of $\mathfrak{S}''[F]$ at $x=0$ are

$$\frac{2}{\pi}\int_0^\pi F(t)\,P''(r,t)\,dt, \tag{8·3}$$

where differentiation is with respect to t; and we have to show that this integral tends to 0 as $r\to 1$.

Since the part of (8·3) extended over any interval (ϵ,π), where $\epsilon>0$, tends to zero, and since $F(t)=o(t^2)$ for $t\to 0$, the lemma will follow if we show that

$$\int_0^\pi t^2\,|P''(r,t)|\,dt=O(1). \tag{8·4}$$

A simple computation shows that $P''(r,t)=0$ means that $y=\cos t$ is given by

$$2ry^2+(1+r^2)\,y-4r=0.$$

The product of the roots is -2, and so there is at most one value of t in $(0,\pi)$ satisfying $P''=0$; there is at least one such t since $P'=0$ for $t=0$ and $t=\pi$. Hence P'' changes sign exactly once in $(0,\pi)$, at a point $t=\eta=\eta(r)$. The integral (8·4) is

$$-\int_0^\eta t^2P''\,dt+\int_\eta^\pi t^2P''\,dt=-2\eta^2P'(r,\eta)+2\int_0^\eta tP'\,dt-2\int_\eta^\pi tP'\,dt$$

$$<-2\eta^2P'(r,\eta)-2\int_0^\pi tP'\,dt$$

$$<-2\eta^2P'(r,\eta)+2\int_0^\pi P\,dt=-2\eta^2P'(r,\eta)+\pi,$$

and since, as is easily verified, $P' = O(t^{-2})$ uniformly in r, (8·4) follows and the lemma is established.

(8·5) Lemma.† *Suppose that $C - \Sigma' A_n(x)/n^2$ is an* S[F] *and write*

$$\Delta^2 F(x,t) = F(x+t) + F(x-t) - 2F(x).$$

At each point x_0 at which the series is summable A *to $F(x_0)$, we have*

$$\liminf_{h\to+0} \Delta^2 F(x_0, h)/h \leqslant \pi \limsup_{r\to 1} (1-r) f(r, x_0), \tag{8·6}$$

$$\limsup_{h\to+0} \Delta^2 F(x_0, h)/h \geqslant \pi \liminf_{r\to 1} (1-r) f(r, x_0). \tag{8·7}$$

It is enough to prove (8·6). We may suppose that $x_0 = 0$, $F(0) = 0$ and F is even. Write $f(r,0) = g(r)$, $F(r,0) = G(r)$. It is enough to show that, for any finite m,

$$\liminf 2F(h)/h > m \quad \text{implies} \quad \pi \limsup (1-r) g(r) \geqslant m. \tag{8·8}$$

Consider for a moment the special case $F_1(x) = \frac{1}{6}\pi^2 - \sum_1^\infty n^{-2} \cos nx$; F_1 is an integral of $\Phi_1 = \Sigma n^{-1} \sin nx$, and Φ_1 has jump π at $x = 0$. We verify that both $2F_1(h)/h$ and $\pi(1-r) g_1(r)$ (where g_1 is the g corresponding to F_1) have limits π. Hence by subtracting $m\pi^{-1} F_1(x)$ from $F(x)$ we reduce the case of general m in (8·8) to that of $m = 0$.

The Laplace equation (7·10) gives

$$g(r) + r\frac{d}{dr}\left(r\frac{dG(r)}{dr}\right) = 0. \tag{8·9}$$

If the second inequality (8·8), with $m = 0$, is false, then $g(r) < -Ar/(1-r)$ for some $A > 0$ and r sufficiently close to 1. Combining this with (8·9) we get

$$\{rG'(r)\}' > A/(1-r).$$

This shows that $G'(r) \to +\infty$ as $r \to 1$, and since simultaneously $G(r) \to 0$, we see that $G(r)$ is strictly negative for r sufficiently close to 1.

Consider now the formulae

$$G(r) = \frac{2}{\pi}\int_0^\pi F(t) P(r,t)\, dt, \quad g(r) = \frac{2}{\pi}\int_0^\pi F(t) P''(r,t)\, dt. \tag{8·10}$$

The first inequality (8·8), with $m = 0$, shows that $F > 0$ near $x = 0$. Since, by the second formula (8·10), $\limsup (1-r) g(r)$ depends only on the values of F in an arbitrarily small neighbourhood of $x = 0$, we may suppose that $F > 0$ in $0 < x \leqslant \pi$ without impairing the truth or falsehood of (8·8). But with this new F the first formula (8·10) indicates that $G(r) > 0$ as $r \to 1$, contrary to what we have just shown. This contradiction proves (8·5).

We now pass to the proof of (8·1) and temporarily replace (i) by the hypothesis that $C - \Sigma' A_n(x)/n^2$ is an S[F], where F is continuous. Our first aim is to show that then

$$f^*(x) = f_*(x) \quad \textit{almost everywhere}. \tag{8·11}$$

† To understand the meaning of the lemma, suppose, for example, that $S'[F] = S[\Phi]$ and that Φ has a jump d at x_0, $d = \Phi(x_0+0) - \Phi(x_0-0)$. Then, as one easily sees, $\Delta^2 F(x_0, t)/t \to d$ as $t \to +0$. On the other hand, d is a (generalized) limit of the nth term of $\pi S'[\Phi] = \pi S''[F]$ at $x = x_0$ (Chapter III, (9·5)). If we take here the Abel limit, we have $d = \pi \lim (1-r) f(r, x_0)$.

Since $f_*(x) > -\infty$ outside a denumerable set, Theorem (12·3) (iii) of Chapter I shows, as in the proof of (7·4), that every interval J contains a subinterval I such that, with a suitable $A = A_I$,

$$f(r,x) \geqslant A \quad \text{for} \quad x \in I \quad (0 \leqslant r < 1).$$

In particular, $f_* \geqslant A$ in I. By (7·6), $\bar{D}^2 F \geqslant A$ in I; hence $F(x) - \frac{1}{2}Ax^2$ is convex in I. In particular, by (3·16), a finite $D^2 F$ exists almost everywhere in I. By (8·2), $f^* = f_*$ almost everywhere in I.

Denote by Q the union of all open intervals I such that $f^* = f_*$ almost everywhere in I. Q is open and $f^* = f_*$ at almost all $x \in Q$. The complement P of Q is a closed non-dense set. If we show that $P = 0$, (8·11) will follow.

Suppose that $P \neq 0$; this implies that $|P| > 0$ and, in particular, that P is non-denumerable. We have $f_* > -\infty$ in P, except possibly in EP. Hence there is a portion Π of P and a constant A such that

$$f_*(x) \geqslant A \quad \text{for} \quad x \in \Pi. \tag{8·12}$$

Let $\delta_1, \delta_2, \ldots$ be the intervals contiguous to Π. We have $f^* = f_*$ at almost all points of $\Sigma\delta_i$. By modifying χ in a set of measure 0 we may suppose that χ is finite in $\Sigma\delta_i$, except possibly in $E\Sigma\delta_i$, and that

$$f_*(x) \geqslant \chi(x) \quad \text{for} \quad x \in \Sigma\delta_i. \tag{8·13}$$

Let J be an interval such that $\Pi = JP$. Since $\bar{D}^2 F \geqslant f_*$, (8·12) and (8·13) imply that, in J, $\bar{D}^2 F$ majorizes an integrable function g which can be $-\infty$ only in a subset of E. But, by (8·5),

$$\limsup_{h \to 0} \Delta^2 F(x,h)/h \geqslant 0 \tag{8·14}$$

in E. Hence, by Lemma (3·22), the difference between F and a second integral of g is convex in J. It follows that $D^2 F$ exists and is finite almost everywhere in J, and so also $f^* = f_*$ almost everywhere in J. Hence $PJ = 0$, contrary to the hypothesis that $PJ \neq 0$. This contradiction proves (8·11).

Redefining χ in a set of measure 0 we may suppose that χ is finite outside E and that

$$f_* \geqslant \chi$$

for all x. In particular, $\bar{D}^2 F \geqslant \chi$ for all x, and $\bar{D}^2 F > -\infty$ outside E. Let χ_2 be a second integral of χ. Since we have (8·14) in E,

$$\Delta = F - \chi_2 \tag{8·15}$$

is convex in $(-\infty, +\infty)$. Hence, by (3·16),

$$D^2 F = D^2 \Delta + D^2 \chi_2 = D^2 \Delta + \chi$$

exists almost everywhere and is integrable over a period.

It follows that $f_*(x)$, which is equal to $D^2 F(x)$ almost everywhere, is integrable, and *we may take f_* for the function χ of Theorem* (8·1).

Since $\Delta(x)$ is convex, it has for each x a right-hand and left-hand derivative $D^+\Delta(x)$ and $D^-\Delta(x)$, both non-decreasing. Therefore $D^+F(x)$ and $D^-F(x)$ exist everywhere, and F is the integral of either, say of $D^+F(x) = \Phi(x)$. We shall show that Φ is continuous.

The only possible discontinuities of Φ are jumps. Since $S''[F] = S'[\Phi]$, and since, by the theorem of Fatou (Chapter III, (7·2)), the Abel sums of $S'[\Phi]$ at x_0 are contained between the upper and lower limits for $t\to 0$ of $\{\Phi(x_0+t)-\Phi(x_0-t)\}/2t$, we see that

$$\limsup_{t\to 0}\{\Phi(x_0+t)-\Phi(x_0-t)\}/2t \geqslant f_*(x_0) \geqslant \liminf_{t\to 0}\{\Phi(x_0+t)-\Phi(x_0-t)\}/2t.$$

If $x_0 \notin E$, and so $f_*(x_0)$ is finite, this implies that $\Phi(x_0+0)=\Phi(x_0-0)$. If $x_0\in E$, then the Abel means $f(r,x)$ of $S'[\Phi]=S''[F]$ satisfy $(1-r)f(r,x_0)\to 0$, so that, by Lemma (8·5),

$$\limsup_{h\to+0}\Delta^2F(x_0,h)/h \geqslant 0 \geqslant \liminf_{h\to+0}\Delta^2F(x_0,h)/h.$$

But the extreme terms here are both equal to $\Phi(x_0+0)-\Phi(x_0-0)$. Hence again $\Phi(x_0+0)=\Phi(x_0-0)$. It follows that $\Phi(x)=D^+F(x)$ is continuous for all x. This implies that $F'(x)$ exists everywhere and is continuous.

Thus $F'(x)=\Phi(x)$. From (8·15), with $\chi=f_*$, we deduce that

$$\Phi(x)=\int_0^x f_*(t)\,dt+\lambda(x), \tag{8·16}$$

where λ is non-decreasing and continuous. If we show that λ is constant, it will follow that F is a second integral, and so that S is a Fourier series.

By the theorem of Fatou,

$$\liminf_{h\to 0}\{\Phi(x+h)-\Phi(x-h)\}/2h \leqslant f_*(x). \tag{8·17}$$

Let $\psi(x)$ be a major function for f_*; we may suppose that ψ is absolutely continuous. All the Dini numbers of ψ are not less than $f_*(x)$ and, in particular,

$$\liminf\{\psi(x+h)-\psi(x-h)\}/2h \geqslant f_*(x).$$

Write $G(x)=\Phi(x)-\psi(x)$. G is continuous and of bounded variation, and in view of (8·17) satisfies

$$\liminf\{G(x+h)-G(x-h)\}/2h \leqslant 0 \tag{8·18}$$

for $x\notin E$. Let us temporarily take for granted the following lemma:

(8·19) LEMMA. *If $G(x)$ is continuous and of bounded variation, and outside a denumerable set E satisfies* (8·18), *then G is non-increasing.*

Hence $G(x)=\Phi(x)-\psi(x)$ is non-increasing. But $\psi(x)$ can be arbitrarily close to $\int_0^x f_*\,dt$. Hence $\Phi(x)-\int_0^x f_*\,dt$ is non-increasing, which is compatible with (8·16) only if λ is constant.

This proves Theorem (8·1) in the case when the hypothesis $a_n=o(n)$, $b_n=o(n)$ is replaced by the condition that S twice integrated is the Fourier series of a continuous function. This result is of independent interest. It covers the important special case of coefficients tending to 0 and, more generally, that of $|a_n|+|b_n|=O(n^\eta)$, $\eta<1$.

Remark. Suppose that conditions (ii), (iii) and (iv) of (8·1) are satisfied not in $(0,2\pi)$ but in an interval (a,b), and that $-\Sigma' A_n(x)/n^2$ is the Fourier series of a function $F(x)$ which is continuous in (a,b). Then f_* is integrable over each interval totally interior to (a,b) and

$$F(x)=\int_c^x dy\int_c^y f_*(t)\,dt+Ax+B \quad (a<x<b,\ c=\tfrac{1}{2}(a+b)).$$

The proof runs parallel to the preceding argument. In view of Lemma (3·26), the result holds if we assume that F is merely upper semi-continuous in (a, b). This remark will be used below.

We now complete the proof of (8·1) by showing that, under the hypotheses of (8·1),

$$F(x) = \lim F(r, x) \tag{8·20}$$

exists for each x and is continuous, and $-\Sigma A_n(x)/n^2 = \mathfrak{S}[F]$. We split the proof into several stages.

(i) *$F(x)$ exists for each x and satisfies $-\infty \leqslant F(x) < +\infty$.* If $x_0 \notin E$, then $f_*(x_0) > -\infty$ and the argument of Lemma (7·13) shows that $\lim \Sigma A_n(x_0) r^n/n$ exists and is finite or $+\infty$; repeating the argument we find that $-F(x) = \lim \Sigma A_n(x_0) r^n/n^2$ is finite or $+\infty$. If $x_0 \in E$, then $f(r, x_0) = o\{(1-r)^{-1}\}$, whence $\Sigma A_n(x_0) r^n/n = o\{\log 1/(1-r)\}$, and $\Sigma A_n(x_0) r^n/n^2$ tends to a finite limit.

(ii) *$F(x)$ is finite for almost all x.* By the Riesz-Fischer theorem, $\Sigma A_n(x)/n^2$ is a Fourier series.

(iii) *On each perfect set P there is a portion on which F is upper semi-continuous.* There is a portion Π of P and a number A such that

$$f(r, x) > A \quad (x \in \Pi,\ 0 \leqslant r < 1).$$

Without loss of generality we may suppose that $A = 0$, so that $f(r, x) > 0$ for $x \in \Pi$ (this we can do without impairing the hypothesis $a_0 = 0$; assuming, as we may, that the diameter of Π is less than π, we subtract from S a suitable monomial $B \cos(x - x_0)$). Since

$$F(r, x) = -\int_0^r \frac{d\rho}{\rho} \int_0^\rho \frac{f(\rho_1, x)}{\rho_1} d\rho_1$$

is a decreasing function of r for each $x \in \Pi$, $F(x)$ as a limit of a decreasing sequence of continuous functions is upper semi-continuous (but not necessarily finite-valued) on Π.

(iv) *Each interval J contains a subinterval I such that $F(x)$ is finite-valued and continuous on I.* We have $f(r, x) > A$ in a subinterval I of J. Suppose, as before, that $A = 0$. The function

$$F(r, x) = -\Sigma A_n(x) r^n/n^2 \tag{8·21}$$

has a non-negative second derivative with respect to x, and so is convex, in I. For each x in I, $F(r, x)$ tends to a limit which is finite or $-\infty$. It cannot tend to $-\infty$ at any point $x_0 \in I$. For otherwise, owing to the convexity of $F(r, x)$, the limit would be $-\infty$ in the whole interval I, contradicting (ii). It follows that $F(x)$, as a limit of convex functions $F(r, x)$, is convex, and so also finite-valued and continuous, in the interior of I.

(v) *$F(x)$ is finite-valued and upper semi-continuous.* Suppose that this is not so, and denote by D the set of points where $F(x)$ is either not finite or not upper semi-continuous. By (iv), the closure $\bar{D}$ of D is non-dense. We show that $\bar{D}$ has no isolated points.

Suppose that an $x_0 \in \bar{D}$ is isolated. Then F is finite-valued and upper semi-continuous in the interior of $(x_0 - \epsilon, x_0)$ and $(x_0, x_0 + \epsilon)$, for some $\epsilon > 0$. By the Remark above, the difference between F and a second integral of χ is convex in the interior of either interval. Hence $F(x_0 \pm 0)$ exist and are either finite or $+\infty$. Since the coefficients of $\mathsf{S}[F]$ are $o(1/n)$, we have $F(x_0+0)=F(x_0-0)=F(x_0)$ (cf. (2·19)). By (i), $F(x_0)<+\infty$. It follows that $F(x_0)$ is finite and F continuous at x_0, and so also in the interior of $(x_0-\epsilon, x_0+\epsilon)$, contradicting the supposition that the interval contains points of $\bar{D}$.

It follows that $\bar{D}$ is perfect. Take a portion Π of $\bar{D}$ on which F is, by (iii), upper semi-continuous. In each of the intervals (a_i, b_i) contiguous to Π, F is continuous and the numbers $F(a_i+0)$ and $F(b_i-0)$, which are equal to $F(a_i)$ and $F(b_i)$ respectively, are finite.

The set of points of Π where $F(x)=-\infty$ is non-dense on Π. For otherwise it would be dense on a portion Π_1 of Π and, owing to the upper semi-continuity of F on Π, would contain Π_1, and so also some of the points a_i, b_i; this is impossible since $F(a_i)=F(a_i+0)$ and $F(b_i)=F(b_i-0)$ are finite.

Collecting results we see that there is a portion of $\bar{D}$ (call it Π again) and an integrable χ such that

$$F(x)-\int_0^x dy\int_0^y \chi(t)\,dt \tag{8·22}$$

is finite-valued and upper semi-continuous on Π, and is convex in the closure of each interval contiguous to Π. It follows that (8·22), and so also $F(x)$, is finite-valued and upper semi-continuous on an interval J such that $\Pi=J\bar{D}$. This contradicts the fact that J contains points of $\bar{D}$. Hence (v) is proved.

(vi) $F(x)$ *is continuous*. Suppose that F is not continuous, and denote by D the set of points of discontinuity of F. The closure $\bar{D}$ of D is non-dense. We show, as in the proof of (v), that $\bar{D}$ has no isolated points, i.e. is perfect. Consider a portion Π of $\bar{D}$ such that

$$f_*(x)\geqslant A \quad \text{on } \Pi \tag{8·23}$$

for some constant A and denote by $\delta_1, \delta_2, \ldots$ the intervals contiguous to Π, and by J an interval such that $\Pi=J\bar{D}$. In each δ_i, (8·22) is convex, D^2F exists almost everywhere, and $f^*=f_*=D^2F$ almost everywhere. Hence, modifying χ in a set of measure 0, we may take χ finite outside $E\Sigma\delta_i$ and satisfying

$$f_*(x)\geqslant \chi(x) \quad \text{in } \Sigma\delta_i. \tag{8·24}$$

By (8·23) and (8·24), f_*, and so also $\underline{D}^2F$, exceeds in J an integrable function g finite outside a denumerable set in which we have (8·14). By Lemma (3·22), the difference between F and a second integral of g is convex. Hence F is continuous on J, contradicting the hypothesis that $J\bar{D}\neq 0$. Hence F is continuous everywhere and the proof of (8·1) is completed.

We have still, however, to prove Lemma (8·19). We may suppose that we have strict inequality in (8·18), for otherwise we argue with $G(x)-x/n$ and afterwards make $n\to\infty$. Suppose that G is not non-increasing. Then $G(\alpha)<G(\beta)$ for some $\alpha<\beta$. By a well-known result from the theory of integration†, the total variation of G on the set N of points where G has no derivative, finite or infinite, is 0. This means that we can cover N by a sequence of intervals $I_1, I_2, \ldots$ such that, if V_i is the total variation of G over I_i, then ΣV_i is arbitrarily small.

We may suppose that $\Sigma V_i<G(\beta)-G(\alpha)$. Observing that the projection of the arc $y=G(x)$, $x\in I_i$, on the y-axis does not exceed V_i, we see that there is a C, $G(\alpha)<C<G(\beta)$, such that in no I_i does G take the value C; because E is denumerable, we may also

† See de la Vallée-Poussin, *Intégrales de Lebesgue*, p. 93, or Saks, *Theory of the Integral*, p. 125.

suppose that $G(x) \neq C$ in E. Let X be the set of points where $G(x) = C$ and let x_0 be the last of these points; such an x_0 exists since X is closed. Since $G'(x_0)$ exists, $G(x_0) = C$, and $G(x) > C$ for $x > x_0$, we see that $G'(x_0) \geqslant 0$. Since $x_0 \notin E$ and (8·18), with strict inequality, is false at x_0, we come to a contradiction and the lemma is established.†

(8·25) THEOREM. *Suppose that a series S satisfies conditions* (i), (ii) *and* (iii) *of Theorem* (8·1), *and that condition* (iv) *is satisfied at all points of E, except possibly for a finite number, $x_1, x_2, \ldots, x_n$, of them. Then S differs from a Fourier series by*

$$\Sigma \alpha_i D(x - x_i),$$

where the α_i are constants and $D(x) = \frac{1}{2} + \cos x + \cos 2x + \ldots$.

We may again suppose that $a_0 = 0$. The same proof which gives (8·1) shows that the $F(x)$ in (8·20) is, in each interval (x_{i-1}, x_i), of the form

$$\int_0^x dy \int_0^y f(t)\,dt + A_i x + B_i. \tag{8·26}$$

F is continuous at the x_i but may have angular points there. Let

$$D_1(x) = \cos x + \cos 2x + \ldots.$$

Integrating D_1 twice we obtain a function having an angular point at $x = 0$ and only there. Therefore, if we subtract from S a linear combination, with constant coefficients, of the series $D_1(x - x_i)$, $i = 1, 2, \ldots, n$, the function F for that difference is smooth at the x_i and is of the form (8·26) with A_i, B_i independent of i. It follows that the difference considered is a Fourier series and the theorem is established.

If we confine our attention to series S with coefficients tending to 0, we may consider *sets of uniqueness for the method of Abel*. A set E will be said to be *of type* $\mathrm{U_A}$, if every series S with coefficients tending to 0, and summable A to 0 outside E, is necessarily identically 0. Every set $\mathrm{U_A}$ is also U; whether the converse is true is an open problem, except when E is closed, in which case an affirmative answer is a corollary of the following theorem:

(8·27) THEOREM. *Let E be a closed set of measure* 0 *and S a series with coefficients tending to* 0 *having its upper and lower Abel sums finite outside E, and one of them, say f_*, integrable. Then $S - \mathsf{S}[f_*]$ converges to* 0 *outside E; in particular $S = \mathsf{S}[f_*]$, if E is a* U *set.*

The function F obtained by integrating S twice is, in each interval contiguous to E, a second integral of f_*. Since the same holds for $\mathsf{S}[f_*]$, the principle of localization shows that $S - \mathsf{S}[f_*]$ converges to 0 outside E.

† It may be observed, though this is irrelevant for us, that if $\limsup_{h \to 0} \{G(x+h) - G(x-h)\}/2h \leqslant 0$ outside a denumerable set E and G is merely continuous, then G is non-increasing. For let $F(x)$ be an integral of $G(x)$. From

$$\frac{F(x+h) + F(x-h) - 2F(x)}{h^2} = \frac{1}{h^2}\int_0^h 2t\,\frac{G(x+t) - G(x-t)}{2t}\,dt$$

we easily deduce that $\underline{D}^2 F \leqslant 0$ outside E, and, since F is smooth, $-F$ is convex by Lemma (3·20), and hence G is non-increasing.

9. Localization for series with coefficients not tending to zero

This is a continuation of §§ 4, 5. We first consider formal multiplication of series with coefficients not necessarily tending to 0.

Consider the series

$$(S):\quad \Sigma c_n e^{inx},$$

$$(T):\quad \Sigma \gamma_n e^{inx}$$

and their formal product

$$(ST):\quad \Sigma C_n e^{inx},$$

where

$$C_n = \sum_p c_p \gamma_{n-p}.$$

Suppose that $c_n = o(|n|^k)$. The series defining C_n converge absolutely if $\Sigma |n|^k |\gamma_n| < \infty$. Since we are going to discuss the summability (C, k) of S, only the case $k > -1$ is of interest. The subcases $-1 < k < 0$ and $k \geq 0$ usually require slightly different arguments.

(9·1) LEMMA. *Let $k \geq 0$. If $c_n = o(|n|^k)$ and $\Sigma |n|^k |\gamma_n| < \infty$, then $C_n = o(|n|^k)$.*

The case $k=0$ is Lemma (4·8). Suppose that $n \to +\infty$, and write $|c_\nu| = \epsilon_\nu |\nu|^k$, $|\gamma_\nu| |\nu|^k = \eta_\nu$ for $\nu \neq 0$. Then

$$|C_n| \leq o(n^k) + \sum_{\nu=-\infty}^{+\infty}{}' \epsilon_\nu \eta_{n-\nu} \frac{|\nu|^k}{|n-\nu|^k},$$

where the dash signifies that the terms $\nu = 0$ and $\nu = n$ are omitted. We split the last sum in two, corresponding to $|\nu| \leq 2n$ and $|\nu| > 2n$. Then

$$|\Sigma'| \leq (2n)^k \sum_{|\nu| \leq 2n}{}' \epsilon_\nu \eta_{n-\nu} + 2^k \sum_{|\nu|>2n} \epsilon_\nu \eta_{n-\nu}$$

$$\leq \{O(n^k) + O(1)\} \sum_{-\infty}^{+\infty} \epsilon_\nu \eta_{n-\nu} = O(n^k)\, o(1) = o(n^k),$$

using (4·8), and so $C_n = o(n^k)$.

(9·2) LEMMA. *Let $-1 < k < 0$. If $c_n = o(|n|^k)$ and $\gamma_n = O(1/n)$, $\Sigma |\gamma_n| < \infty$, then $C_n = o(|n|^k)$.*

Let $n \to +\infty$, $|c_\nu| = \epsilon_\nu |\nu|^k$ and suppose that $|\gamma_\nu| \leq 1/|\nu|$ for $\nu \neq 0$. Then

$$|C_n| \leq o(n^k) + \sum_{\nu=-\infty}^{+\infty}{}' \epsilon_\nu |\nu|^k |\gamma_{n-\nu}|.$$

Split the last sum into two, corresponding to $|\nu| \leq \frac{1}{2}n$ and $|\nu| > \frac{1}{2}n$. Then

$$|\Sigma'| \leq \frac{1}{(\frac{1}{2}n)} \sum_{|\nu| \leq \frac{1}{2}n}{}' \epsilon_\nu |\nu|^k + o(n^k) \sum_{|\nu| > \frac{1}{2}n}{}' |\gamma_{n-\nu}|$$

$$= O(1/n)\, o(n^{1+k}) + o(n^k) \sum_{-\infty}^{+\infty} |\gamma_\nu| = o(n^k),$$

which gives $C_n = o(n^k)$.

(9·3) LEMMA. *Let $k \geq 0$, $S = \Sigma c_n e^{inx}$, $T = \Sigma \gamma_n e^{inx}$, $U = \Sigma \delta_n e^{inx}$, where*

$$c_n = o(|n|^k), \quad \Sigma |\gamma_n| |n|^k < \infty, \quad \Sigma |\delta_n| |n|^k < \infty.$$

Then

$$S(TU) = (ST)\, U.$$

For $k = 0$ this is Lemma (5·17), and the general case is proved in the same way.

(9·4) LEMMA. *Let $h>0$, $\alpha-h>-1$. If for the* (C, α) *means $\sigma_n^\alpha = S_n^\alpha/A_n^\alpha$ of a series $\sum_0^\infty u_n$ we have*

$$\sigma_n^\alpha = s+o(n^{-h}),$$

then Σu_n is summable (C, $\alpha-h$) *to s.*

We may suppose that $s=0$; hence $S_\nu^\alpha = o(\nu^{\alpha-h})$. By Chapter III, (1·10),

$$S_n^{\alpha-h} = \sum_{\nu=0}^{n} S_\nu^\alpha A_{n-\nu}^{-h-1} = \sum_{\nu\leqslant \frac12 n} + \sum_{\frac12 n<\nu\leqslant n} = P_n+Q_n,$$

$$|P_n| \leqslant O(n^{-h-1}) \sum_{\nu\leqslant\frac12 n} o(\nu^{\alpha-h}) = O(n^{-h-1})\, o(n^{\alpha-h+1}) = o(n^{\alpha-2h}) = o(n^{\alpha-h}),$$

$$|Q_n| \leqslant o(n^{\alpha-h}) \sum_{\frac12 n<\nu\leqslant n} |A_{n-\nu}^{-h-1}| \leqslant o(n^{\alpha-h}) \sum_{\nu=0}^{\infty} |A_\nu^{-h-1}| = o(n^{\alpha-h}).$$

Hence $S_n^{\alpha-h} = o(n^{\alpha-h})$, $\sigma_n^{\alpha-h}\to 0$, and the lemma is established.

Before we proceed further, consider the formal product of the two series

$$\sum_1^\infty \sin nx, \quad \sin x. \tag{9·5}$$

Simple computation shows that the product is $\frac12(1+\cos x)$ (which is to be expected since the first series (9·5) represents $\frac12\cot\frac12 x$). The two series (9·5) converge at $x=0$ to sum 0, but the product converges to sum 1. Since the second series (9·5) is a polynomial, we see that the mere vanishing at a point of the sum of the 'good' factor—no matter how rapidly its coefficients tend to 0—cannot guarantee that the product is 0 at that point.

In the remainder of this section we denote by k' the least integer $\geqslant k$: $k'=k$ if k is an integer, $k'=[k]+1$ otherwise.

(9·6) THEOREM. *Suppose that $c_n = o(|n|^k)$, $k\geqslant 0$, and that $T=\mathsf{S}[\lambda]$ satisfies the condition*

$$\Sigma |\gamma_n| |n|^{k+k'+1} < \infty$$

(*so that λ has at least $k'+1$ continuous derivatives*). *Suppose also that at every point of a set E we have $\lambda'=\lambda''=\ldots=\lambda^{(k')}=0$. Then at every point x in E the two series $\Sigma C_n e^{inx}$ and $\Sigma\lambda(x)c_n e^{inx}$ are uniformly equisummable* (C, k); *that is, the series*

$$\Sigma\{C_n - \lambda(x)c_n\}e^{inx} \tag{9·7}$$

is uniformly summable (C, k) *to 0 in E. In particular, if also $\lambda=0$ in E, then $\Sigma C_n e^{inx}$ is uniformly summable* (C, k) *to 0 in E.*

The result holds for $-1<k<0$, provided the condition on T is replaced by

$$\Sigma|n\gamma_n| < \infty, \quad \gamma_n = O(n^{-2}). \tag{9·8}$$

The proof for any particular point x_0 in E shows that the conclusion is uniform in E. Without loss of generality we may suppose that $x_0=0$. Hence

$$\lambda'(0)=\lambda''(0)=\ldots=\lambda^{(k')}(0)=0.$$

It is enough to consider the case $\lambda(0)=0$, since otherwise (as in the proof of (4·9)) we subtract $\lambda(0)$ from γ_0.

We begin with the easier case $-1<k<0$ and return to the proof of (4·9), where now $x_0=0$. The second condition (9·8) implies that $R_n(x_0)=O(1/n)$. The first condition

(9·8) is equivalent to $\Sigma\Gamma_n < \infty$. Hence, applying Lemma (9·2) to the formula for $S_m(x_0)$ on p. 331, we see that $S_m(x_0) = o(m^k)$ and, by Lemma (9·4), $\Sigma C_n e^{inx_0}$ is summable (C, k) to 0.

We split the proof for $k \geq 0$ into several stages.

(i) *k is an integer* ($k' = k$), $\lambda(x) = (1 - e^{ix})^{k+1}$. Write

$$\Delta c_\nu = c_\nu - c_{\nu-1}, \quad \Delta^l c_\nu = \Delta^{l-1} c_\nu - \Delta^{l-1} c_{\nu-1} \quad (l = 2, 3, \ldots).$$

We easily see that

$$(1 - e^{ix})^{k+1} \sum_{-\infty}^{+\infty} c_\nu e^{i\nu x} = \sum_{-\infty}^{+\infty} \Delta^{k+1} c_\nu e^{i\nu x},$$

the left-hand side here meaning the formal product. Hence, if S_n^l is the lth Cesàro sum of ST at $x = 0$, we have

$$\left.\begin{aligned} S_n^0 &= \sum_{-n}^{n} \Delta^{k+1} c_\nu = \Delta^k c_n - \Delta^k c_{-n-1}, \\ S_n^1 &= \sum_{\nu=0}^{n} S_\nu^0 = \Delta^{k-1} c_n + \Delta^{k-1} c_{-n-2} - 2\Delta^{k-1} c_{-1}, \\ S_n^2 &= \sum_{\nu=0}^{n} S_\nu^1 = \Delta^{k-2} c_n - \Delta^{k-2} c_{-n-3} + O(n), \\ &\cdots\cdots\cdots\cdots\cdots\cdots\cdots\cdots\cdots\cdots\cdots \\ S_n^k &= \sum_{\nu=0}^{n} S_\nu^{k-1} = c_n + (-1)^{k-1} c_{-n-k-1} + O(n^{k-1}). \end{aligned}\right\} \tag{9·9}$$

Hence $S_n^k / A_n^k = o(1)$ and ST is summable (C, k) to 0 at $x = 0$.

(ii) *k fractional*, $\lambda(x) = (1 - e^{ix})^{k'+1}$. Hence $k' - 1 < k < k'$. Substituting k' for k in (9·9) we obtain

$$S_n^{k'} / A_n^{k'} = o(n^{k-k'}),$$

and it is enough to apply Lemma (9·4) with $\alpha = k'$, $h = k' - k$.

(iii) *k and $\lambda(x)$ general.* By hypothesis, $\lambda(0) = \lambda'(0) = \ldots = \lambda^{(k')}(0) = 0$. Hence, if $h = 0, 1, \ldots, k' + 1$, the function $\lambda_h(x) = \lambda(x)(1 - e^{ix})^{-h}$, completed by continuity at $x = 0$, is continuous. Write

$$\lambda_h(x) = \lambda(x)(1 - e^{ix})^{-h} \sim \Sigma \gamma_\nu^h e^{i\nu x}.$$

Since $\lambda_{h-1} = (1 - e^{ix})\lambda_h$, we have $\gamma_n^{h-1} = \gamma_n^h - \gamma_{n-1}^h$, whence, using the fact that $\gamma_n^h \to 0$ as $n \to \pm\infty$,

$$\gamma_n^h = \sum_{\nu=-\infty}^{n} \gamma_\nu^{h-1}, \tag{9·10}$$

$$\gamma_n^h = -\sum_{\nu=n+1}^{\infty} \gamma_\nu^{h-1}. \tag{9·11}$$

for $h = 1, 2, \ldots, k' + 1$. We shall use (9·10) for $n < 0$ and (9·11) for $n > 0$. Observing that, for any $\alpha \geq 0$,

$$\sum_{n=1}^{\infty} |\gamma_n^h| n^\alpha \leq \sum_{n=1}^{\infty} n^\alpha \sum_{\nu=n+1}^{\infty} |\gamma_\nu^{h-1}| \leq \sum_{n=1}^{\infty} \sum_{\nu=n}^{\infty} = \sum_{\nu=1}^{\infty} |\gamma_\nu^{h-1}| \sum_{n=1}^{\nu} n^\alpha \leq \sum_{\nu=1}^{\infty} |\gamma_\nu^{h-1}| \nu^{\alpha+1},$$

and that, similarly, by (9·10)

$$\sum_{n=1}^{\infty} |\gamma_{-n}^h| n^\alpha \leq \sum_{\nu=1}^{\infty} |\gamma_{-\nu}^{h-1}| \nu^{\alpha+1},$$

we have
$$\sum_{n\neq 0} |\gamma_n^{k'+1}|\,|n|^k = \sum_{n=1}^{\infty} (|\gamma_n^{k'+1}| + |\gamma_{-n}^{k'+1}|)\, n^k$$
$$\leqslant \sum_{n=1}^{\infty} (|\gamma_n^{k'}| + |\gamma_{-n}^{k'}|)\, n^{k+1}$$
$$\dots\dots\dots\dots\dots\dots\dots\dots$$
$$\leqslant \sum_{n=1}^{\infty} (|\gamma_n| + |\gamma_{-n}|)\, n^{k+k'+1} < \infty, \qquad (9\cdot12)$$
by hypothesis.

We have
$$\lambda(x) = \Sigma\gamma_n e^{inx} = (1-e^{ix})^{k'+1} \Sigma\gamma_n^{k'+1} e^{inx}.$$

In forming ST we may first multiply S by $T_1 = \Sigma\gamma_n^{k'+1} e^{inx}$ and then the result by $U = (1-e^{ix})^{k'+1}$ (Lemma (9·3)). By (9·12) and Lemma (9·1), ST_1 has coefficients $o(|n|^k)$. Cases (i) and (ii) show that $ST = (ST_1)\,U$ is summable (C, k) to 0 at $x=0$. This completes the proof of the theorem. For $k=0$ we have a new proof of Theorem (4·9).

(9·13) THEOREM. *Under the hypotheses of* (9·6), *the two series* $(\widetilde{ST})$ *and* $\lambda(x)\tilde{S}$ *are uniformly equisummable* (C, k) *in the wider sense in* E.

The proof is similar to that of (9·6). We may again suppose that E reduces to the point 0 and that $\lambda(0) = \lambda'(0) = \dots = \lambda^{(k')}(0) = 0$; we have to show that $(\widetilde{ST})$ is summable (C, k) at $x=0$. We denote the lth Cesàro sums of $(\widetilde{ST})$ at 0 by $\tilde{S}_n^l$.

For $k \geqslant 0$ we consider the stages (i), (ii) and (iii) as before.

(i) Let $\epsilon_\nu = -i \operatorname{sign} \nu$. Then

$$\left.\begin{aligned}
\tilde{S}_n^0 &= \sum_{-n}^{n} \epsilon_\nu \Delta^{k+1} c_\nu = -i(\Delta^k c_n + \Delta^k c_{-n-1}) + i(\Delta^k c_0 + \Delta^k c_{-1}),\\
\tilde{S}_n^1 &= \sum_{\nu=1}^{n} \tilde{S}_\nu^0 = -i(\Delta^{k-1} c_n - \Delta^{k-1} c_{-n-2}) + i(\Delta^k c_0 + \Delta^k c_{-1})\, A_{n-1}^1 + O(1),\\
&\dots\dots\dots\dots\dots\dots\dots\dots\dots\dots\dots\dots\\
\tilde{S}_n^k &= \sum_{\nu=1}^{n} \tilde{S}_\nu^{k-1} = -i(c_n + (-1)^k c_{-n-k-1}) + i(\Delta^k c_0 + \Delta^k c_{-1})\, A_{n-1}^k + O(n^{k-1}).
\end{aligned}\right\} \qquad (9\cdot14)$$

Hence
$$\frac{\tilde{S}_n^k}{A_n^k} = o(1) + i(\Delta^k c_0 + \Delta^k c_{-1}) \frac{A_{n-1}^k}{A_n^k} \to i(\Delta^k c_0 + \Delta^k c_{-1}).$$

(ii) Substituting k' for k in the last formula (9·14) we have

$$\frac{\tilde{S}_n^{k'}}{A_n^{k'}} = i(\Delta^k c_0 + \Delta^k c_{-1}) + o(n^{k-k'}),$$

and we again apply Lemma (9·4).

(iii) Using the previous notation, $(\widetilde{ST})$ is the series conjugate to the product of ST_1 by $U = (1-e^{ix})^{k'+1}$. Since ST_1 has coefficients $o(|n|^k)$, $(\widetilde{ST})$ is, by (i) and (ii), summable (C, k) at $x=0$.

The case $-1 < k < 0$ is treated in the same way as in (9·6).

Theorems (9·6) and (9·13) lead to a principle of localization for series with coefficients not necessarily tending to 0.

In § 4 we associated with every trigonometric series S whose coefficients c_n tend to

0, a continuous function $F(x)$ obtained by integrating S twice. More generally, denote by $F(x)$ the function obtained by integrating S termwise p times:

$$F(x) = c_0 \frac{x^p}{p!} + \Sigma' \frac{c_n}{(in)^p} e^{inx}. \tag{9·15}$$

If $c_n = o(|n|^k)$, the series on the right converges absolutely and uniformly provided $p - k > 1$. We assume, however, only the weaker hypothesis that the periodic series in (9·15) is a Fourier series.

The theorem which follows is a generalization of (4·29):

(9·16) THEOREM. *Suppose that the coefficients c_n of S are $o(|n|^k)$, $k > -1$, and that the periodic part of the function $F(x)$ in* (9·15) *is a Fourier series. Let $s_n(x)$ and $\tilde{s}_n(x)$ be the partial sums of S and $\tilde{S}$. Let $\lambda(x)$ be a function differentiable sufficiently often, equal to 0 outside an interval (a, b) of length $< 2\pi$, and to 1 in a subinterval (a', b') of (a, b). Then the two sequences*

$$s_n(x) - \frac{(-1)^p}{\pi} \int_a^b F(t)\, \lambda(t) \frac{d^p}{dt^p} D_n(x-t)\, dt, \tag{9·17}$$

$$\tilde{s}_n(x) - \frac{(-1)^p}{\pi} \int_a^b F(t)\, \lambda(t) \frac{d^p}{dt^p} \tilde{D}_n(x-t)\, dt \tag{9·18}$$

are uniformly summable (C, k) *on* (a', b'), *the limit of the first being* 0.

The proof is similar to that of (4·29). Suppose first that $c_0 = 0$, so that F is periodic; we may also suppose that λ is periodic. If $T = \mathfrak{S}[\lambda]$, (9·17) is the nth partial sum of

$$S - (\mathfrak{S}[F]\, T)^{(p)} = (S - ST) - \sum_{q=1}^{p} \binom{p}{q} \mathfrak{S}^{(p-q)}[F]\, T^{(q)},$$

by an extension of (4·32) to the pth derivative. The series $\mathfrak{S}^{(p-q)}[F]$ have coefficients $o(|n|^k)$, and if we assume that $T^{(p)}$ (and so also $T^{(q)}$, $q < p$) satisfies the conditions imposed on the coefficients of T in (9·6), all the products $\mathfrak{S}^{(p-q)}[F]\, T^{(q)}$ are, by (9·6), uniformly summable (C, k) to 0 on (a', b'), and the same holds for $S - ST = S\mathfrak{S}[1-\lambda]$. Hence (9·17) is uniformly summable (C, k) to 0 on (a', b').

The case $c_0 \neq 0$ is dealt with as in the proof of (4·29), provided $k \geq 0$. If $-1 < k < 0$ a modification is needed. We may suppose that $S = c_0$, $F = c_0 x^p/p!$. Then (9·17) is

$$c_0 - c_0 \frac{1}{\pi} \int_0^{2\pi} \left\{\lambda(t) \frac{t^p}{p!}\right\}^{(p)} D_n(x-t)\, dt. \tag{9·19}$$

This difference tends to 0 in (a', b'). If it is, say, $O(1/n)$, then, by (9·4), it is summable (C, k) to 0 and the proof is completed. Suppose that $\lambda^{(p+2)}$ exists and is continuous; then $\{t^p \lambda(t)\}^{(p)}$ has two continuous derivatives, the terms of $\mathfrak{S}[(t^p\lambda)^{(p)}]$ are $O(1/n^2)$, and the remainders are $O(1/n)$, that is, (9·19) is $O(1/n)$, as desired.

The proof of the remaining part of (9·16) runs parallel.

A corollary of (9·16) is the following *principle of localization*:

(9·20) THEOREM. *Let S_1 and S_2 be two trigonometric series with coefficients $o(|n|^k)$, $k > -1$, and let F_1 and F_2 be the functions F corresponding to S_1 and S_2. If $F_1 = F_2$ in (a, b) or, more generally, if $F_1 - F_2$ is a polynomial of degree less than p there, then in every interval (a', b') interior to (a, b)*

(i) *S_1 and S_2 are uniformly equisummable* (C, k);

(ii) *$\tilde{S}_1$ and $\tilde{S}_2$ are uniformly equisummable* (C, k) *in the wider sense.*

To prove (i), write $S_1 - S_2 = S$, $F_1 - F_2 = F$. We have to show that if F is a polynomial of degree less than p on (a, b), then S is uniformly summable (C, k) to 0 on (a', b'). In view of the summability of (9·17) it is enough to show that

$$\frac{(-1)^p}{\pi}\int_{-\pi}^{\pi} F(t)\,\lambda(t)\frac{d^p}{dt^p}D_n(x-t)\,dt = \frac{1}{\pi}\int_{-\pi}^{\pi}\{F(t)\,\lambda(t)\}^{(p)}\,D_n(x-t)\,dt$$

is uniformly summable (C, k) to 0 on (a', b'). Since $(F\lambda)^{(p)} = 0$ in (a', b'), this is immediate. (If $-1 < k < 0$, this follows if we assume that $\lambda^{(p+2)}$ exists and is continuous.)

The proof of (ii) is similar.

(9·21) Theorem of M. Riesz. *Suppose that* $\Phi(z) = \sum_0^\infty \alpha_n z^n$ *is regular for* $|z| < 1$ *and that* $\alpha_n = o(n^k)$, $k > -1$. *Then* $\Sigma\alpha_n e^{inx}$ *is summable* (C, k) *at every point of regularity of* Φ, *and the summability is uniform over every closed arc of regularity.*

This is a generalization of (5·7) and a corollary of (9·20). For assuming, as we may, that $\alpha_0 = 0$ we see (as in the proof of (5·7)) that $F(x) = \Sigma\alpha_n(in)^{-p}e^{inx}$ has infinitely many derivatives on $(a-\epsilon, b+\epsilon)$ and so coincides on $(a-\frac{1}{2}\epsilon, b+\frac{1}{2}\epsilon)$ with a function F_1 corresponding to a Fourier series S_1 which has coefficients $O(n^{-2})$. Since S_1 is uniformly summable (C, k) (even if $-1 < k < 0$), $\Sigma\alpha_n e^{inx}$ is uniformly summable (C, k) on (a, b).

(9·22) Theorem. *If a trigonometric series* S *with coefficients* $o(n^k)$, $k > -1$, *is uniformly summable* A *to* 0 *over an arc* (a, b), *then* S *is uniformly summable* (C, k) *on every arc interior to* (a, b).

Since

$$u(r, x) = \Sigma c_n e^{inx} r^{|n|} \to 0$$

uniformly over (a, b), we find, on integrating this relation p times over (a, x) that $F(x)$ is in (a, b) a polynomial of degree $p-1$, and the theorem follows from (9·20) with $S_2 \equiv 0$.†

(9·23) Theorem. *If a trigonometric series* S *with coefficients* $o(n^k)$, $k > 0$, *is uniformly summable* A *over an arc* (a, b), *then* S *is uniformly summable* (C, k) *over each arc* (a', b') *interior to* (a, b).

Let $f(x)$ be any continuous and periodic function equal in (a, b) to the Abel sum of S. Clearly $S - \mathsf{S}[f]$ is uniformly summable A to 0 on (a, b) and so summable (C, k) on (a', b'). Since $\mathsf{S}[f]$ is uniformly summable (C, k) (Chapter III, (5·1)), the conclusions follows.

(9·24) Theorem. *Suppose that the condition* $\alpha_n \to 0$ *in Theorem* (5·8) *is replaced by* $\alpha_n = o(n^k)$, $k \geq 0$. *Then the conclusion of the theorem holds, provided we replace 'convergence' by 'summability* (C, k)*'.*

The proof is a minor modification of the proof of (5·8). We start from the formula (5·10) which was obtained without using the order of magnitude of the α_n. Integrating $p-1$ times the relation

$$\int_a^\xi \Phi(e^{iu})\,du - \Sigma\alpha_n(in)^{-1}r^n\{e^{inu}\}_a^\xi \to 0 \quad (r \to 1),$$

† Of course, (9·22) is also a corollary of (9·21), since by the *symmetry principle* of Schwarz (see e.g. J. E. Littlewood, *Lectures on the theory of functions*, p. 129) the harmonic function $u(r, x)$ is continuable across the arc (a, b), and so the same holds for the regular function $\Phi(z)$, $|z| < 1$, whose real part is u.

which holds uniformly in $a \leqslant \xi \leqslant b$, and arguing as below (5·10), we find that $F(x) = \Sigma\alpha_n(in)^{-p} e^{inx}$ differs from the pth integral of $\Phi(e^{iu})$ by a polynomial of degree $p-1$. Let S_1 be the Fourier series of the function equal to $\Phi(e^{iu})$ in (a, b) and, say, to 0 outside (a, b). The function F_1 obtained by integrating S_1 termwise p times differs from F by a polynomial of degree $p-1$. Since, as in the proof of (5·8), $\Phi(e^{iu})$ satisfies condition $\lambda^2_{\frac{1}{2}}$ in every subinterval (a', b') of (a, b), and since $k \geqslant 0$, S_1 is summable (C, k) at every point x interior to (a, b) at which Lebesgue's test is applicable, and by (9·20) the same holds for $\Sigma\alpha_n e^{inx}$. Hence the latter series is summable (C, k) at almost all points of (a, b). The part concerning uniform summability is proved similarly.

(9·25) THEOREM. *Let*

$$c_n = o(|n|^k), \quad \Sigma|\gamma_n||n|^k < \infty \quad (k > 0), \tag{9·26}$$

and write $S = \Sigma c_n e^{inx}$, $T = \Sigma\gamma_n e^{inx} = \mathfrak{S}[\phi]$. *If* ϕ *takes a constant value* ϕ_0 *in* (a, b), *the two series*

$$ST - \phi_0 S, \quad (\widetilde{ST}) - \phi_0 \tilde{S}$$

are uniformly summable (C, k) *in every interval* (a', b') *interior to* (a, b), *the sum of the first being* 0.

This is an analogue of (5·16), the condition on $\phi_1(x)$ $(=\phi_0)$ being much more stringent than before, since we are multiplying series with coefficients not tending to 0. It is enough to consider the case $\phi_0 = 0$. Let $\lambda(x)$ be periodic, equal to 1 in (a', b') and to 0 outside (a, b), and such that $\mathfrak{S}[\lambda] = U = \Sigma\delta_n e^{inx}$ satisfies $\Sigma|\delta_n||n|^{k+k'+1} < \infty$. In particular, $\Sigma|\delta_n||n|^k < \infty$. The last condition, together with (9·26), implies

$$(ST)\,U = S(TU)$$

(the proof being identical with that of Lemma (5·17); see Lemma (9·3)), and since TU is identically 0, so is $(ST)\,U$. In view of the conditions imposed on U, ST is uniformly summable (C, k) on (a', b') to 0. We prove the result about $(\widetilde{ST})$ similarly.

We conclude with a few remarks about the formal products of $S = \Sigma c_n e^{inx}$ and $T = \Sigma\gamma_n e^{inx} = \mathfrak{S}[\lambda]$ in the case when c_n does not necessarily tend to 0.

We impose on T two conditions. One requires that the γ_n tends to 0 sufficiently rapidly (which amounts to the requirement that λ have sufficiently many derivatives); the other demands that a sufficient number of the derivatives of λ vanish at the point x_0 at which we consider ST. While the first requirement is harmless and can be easily satisfied, it is not so with the second; and this somewhat restricts the use of formal multiplication in the case of coefficients not tending to 0.

Suppose, however, $P(x)$ is a trigonometric polynomial which has at x_0 a sufficient number of derivatives in common with λ. Since the behaviour of $S\mathfrak{S}[\lambda - P]$ at x_0 is governed by Theorem (9·6), the problem reduces to the study of the formal product SP or, ultimately, of the product

$$S\,e^{imx},$$

where m is an integer different from 0.

Suppose, for example, that $m > 0$ and write $T = e^{imx}$. The difference of the nth partial sums of ST and $e^{imx_0} S$ at $x = x_0$ is

$$\sum_{-n}^{n} c_{\nu-m} e^{i\nu x_0} - e^{imx_0} \sum_{-n}^{n} c_\nu e^{i\nu x_0} = -e^{imx_0}\left\{\sum_{n-m+1}^{n} c_\nu e^{i\nu x_0} - \sum_{-n-m}^{-n-1} c_\nu e^{i\nu x_0}\right\}.$$

It follows immediately that *the formal product* $S\,e^{imx}$ *and the series* $e^{imx_0}S$ *are equisummable* (C, k) *at* $x=x_0$ *if both sequences*

$$c_n\,e^{inx_0},\quad c_{-n}\,e^{-inx_0}\quad (n=0,1,2,\ldots)$$

are summable (C, k) *to* 0†. Under the same condition, $(\widetilde{Se^{imx}})$ and $e^{imx_0}\tilde{S}$ are equisummable (C, k) in the wider sense at $x=x_0$.

If

$$\sum_{-\infty}^{\infty} c_n\,e^{inx}=\tfrac{1}{2}a_0+\sum_1^{\infty}(a_n\cos nx+b_n\sin nx)=\sum_0^{\infty}A_n(x)$$

is real-valued, the condition reduces to the requirement that the two sequences $A_n(x_0)$ and $B_n(x_0)$ should be summable (C, k) to 0.

Suppose our function λ is 0 outside an interval (a,b) and 1 in a subinterval (a',b') of (a,b). If we want the polynomial P to be independent of x_0, it is convenient to assume that λ itself is a polynomial in each of the intervals (a,a') and (b',b). Then in each case we may take that polynomial for P, and the remarks just made, together with Theorem (9·6), may give us information about the behaviour (even uniformly) of ST.

MISCELLANEOUS THEOREMS AND EXAMPLES

1. Prove that

$$\limsup|a_n\cos nx+b_n\sin nx|=\limsup(a_n^2+b_n^2)^{\frac{1}{2}}$$

almost everywhere, by the same method which gave (1·2).

[Observe that if m is a positive integer, E an arbitrary set of positive measure and $n_k\to\infty$, then

$$\int_E \cos^{2m}(n_k x+\alpha_{n_k})\,dx\to|E|\binom{2m}{m}2^{-2m};$$

and that, for m large, the right-hand side is of order $m^{-\frac{1}{2}}$.]

2. Let $s_n(x)$ be the partial sums of $\Sigma A_n(x)$. The convergence of $\Sigma A_n(x)$ at a single point x_0 does not imply that $a_n\to 0$, $b_n\to 0$. Show that

(i) if

$$s_n(x_0+h)\to g,$$

for $|h|\leqslant A/n$, where g is finite, then a_n, $b_n\to 0$;

(ii) more generally, if $\alpha>-1$ and if the (C, α) means $\sigma_n^{\alpha}(x)$ of $\Sigma A_n(x)$ satisfy

$$\sigma_n^{\alpha}(x_0+h)\to g,$$

then $|a_n|+|b_n|=o(n^{\alpha})$.

[(i) The hypotheses imply that $A_n(x_0+h)\to 0$; consider the graph of the curve $y=A_n(x)$.]

3. Given any set E of positive measure and any integer $m\geqslant 1$, there is a positive number $\delta=\delta(E,m)$ such that for every sum $c_1\,e^{ip_1x}+c_2\,e^{ip_2x}+\ldots+c_m\,e^{ip_mx}$ with integral $p_1<p_2<\ldots<p_m$ we have

$$\int_E|\Sigma c_s\,e^{ip_sx}|^2\,dx\geqslant\delta\Sigma|c_s|^2.$$

[Apply induction; we may suppose that $p_1=0$.]

4. Every perfect set P contains a perfect subset of type H.

[Consider, for example, the intersection of P with the set E_n of points where $\cos nx\geqslant 0$; take n large.]

5. Suppose that $S=\Sigma A_n(x)$ has coefficients tending to 0, or even only bounded, and let $F(x)$ be the Riemann function for S. Show that if S converges at x_0 to sum s, then

$$\frac{F(x_0+\alpha+\beta)-F(x_0-\alpha+\beta)-F(x_0+\alpha-\beta)+F(x_0-\alpha-\beta)}{4\alpha\beta}=\Sigma A_n(x_0)\frac{\sin n\alpha}{n\alpha}\,\frac{\sin n\beta}{n\beta}\to s,$$

as α and β tend to 0 in such a way that α/β and β/α remain bounded. (Riemann [1].)

[The proof is similar to that of (2·4).]

† We use the fact, which is immediate, that if a sequence $u_0, u_1, u_2,\ldots$ is summable (C, k) to s so are $0, u_0, u_1, u_2,\ldots$ and $u_1, u_2, u_3,\ldots$

6. If $a_n, b_n \to 0$, then $F(x) = \frac{1}{4}a_0 x^2 - \Sigma A_n(x)/n^2$ satisfies

$$\{F(x_0+\alpha+\beta) - F(x_0-\alpha+\beta) - F(x_0+\alpha-\beta) + F(x_0-\alpha-\beta)\}/\alpha \to 0,$$

uniformly in x, as α and β tend to 0 in the same way as in Example 5. (Riemann [1].)

7. A sequence $\{a_n\}$ is said to be *summable* R′ to limit s if

$$\frac{2}{\pi}\sum_{n=1}^{\infty} a_n \frac{\sin^2 nh}{n^2 h}$$

converges near $h = 0$ and tends to s as $h \to 0$. Show that if $\{a_n\}$ converges to s, it is also summable R′ to s.

[See footnote on p. 320; the theorem is essentially the same as (2·8), where we have $s = 0$.]

8. The methods R and R′ are not comparable. (See Marcinkiewicz [4], Kuttner [1].)

9. If $S[F]$ has coefficients $o(1/n)$, then

$$\lim_{h\to 0} \text{ap}\, \{F(x+h) - F(x-h)\} = 0$$

for each x. (Rajchman and Zygmund [2].)

[The proof is similar to that of (2·22).]

10. Suppose that $S[f] = \Sigma A_n(x)$ has coefficients $O(1/n)$. Then a necessary and sufficient condition for the convergence of $\tilde{S}[f] = \Sigma B_n(x)$ at x_0 is the existence of the integral

$$-\frac{1}{\pi}\int_0^{\pi} \{f(x_0+t) - f(x_0-t)\}\, \tfrac{1}{2}\cot \tfrac{1}{2}t\, dt,$$

the value of the integral being the same as the sum of $\tilde{S}[f]$ at x_0. (Hardy and Littlewood [16].)

[This is an analogue of (2·18). Observe that

$$-\frac{1}{\pi}\int_h^{\pi} [f(x_0+t) - f(x_0-t)]\, \tfrac{1}{2}\cot \tfrac{1}{2}t\, dt = \frac{2}{\pi}\sum_1^{\infty} B_n(x_0) \int_h^{\pi} \sin nt\, \tfrac{1}{2}\cot \tfrac{1}{2}t\, dt,$$

and that making $h \to 0$ we apply to $\Sigma B_n(x_0)$ a method of summation somewhat resembling that of Lebesgue.]

11. If $\Sigma B_n(x) = \tilde{S}[f]$, and if $\Sigma B_n(x_0)$ converges to s, then

$$\lim_{h\to 0} \text{ap} \left\{ -\frac{1}{\pi}\int_h^{\pi} [f(x_0+t) - f(x_0-t)]\, \tfrac{1}{2}\cot \tfrac{1}{2}t\, dt \right\} = s.$$

[This is an analogue of (2·22).]

12. Suppose that $S = \Sigma A_n(x)$ converges everywhere to sum $f(x)$. If $f(x_0) > \alpha$, then the set, E, of points where $f(x) > \alpha$ is of positive measure. (Steinhaus [2].)

[Suppose that $|E| = 0$. Then, by (3·18), f is integrable and $S = S[f]$. Since $f \leqslant \alpha$ almost everywhere, the (C, 1) means of S are not greater than α, which contradicts the convergence of S at x_0 to $f(x_0) > \alpha$. Using formal multiplication we can prove that E is of positive measure in every neighbourhood of x_0.]

13. If $S = \Sigma A_n(x)$ converges everywhere to sum $f(x) \geqslant 0$, then $f \in L$, by (3·18). If S converges to an $f \geqslant 0$ in (a, b), then f need not be integrable in (a, b). (Consider, for example, $\Sigma(\sin nx)/\log n$ in $(0, \pi)$.) However, $f \in L^{1-\epsilon}(a, b)$ for every $\epsilon > 0$.

[It is enough to prove that $f^{1-\epsilon}$ is integrable near $x = b$. Suppose that $b = 0$, and let

$$\Phi(x) = \tfrac{1}{2}a_0 x + \Sigma B_n(x)/n, \quad F(x) = \tfrac{1}{4}a_0 x^2 - \Sigma A_n(x)/n^2.$$

By (2·8), F is in λ_*, and by Chapter II, (3·4), $\omega(\delta; F) = o(\delta \log \delta)$. Since Φ is monotonically increasing in $(a, 0)$, and F is an integral of Φ, we obtain successively, as $h \to +0$,

$$\int_{-h}^{-\frac{1}{2}h} \Phi\, dt = o(h \log h), \quad \Phi(-h) = o(\log h), \quad \int_{-h}^{-\frac{1}{2}h} f\, dt = o(\log h).$$

Hence

$$\int_{-h}^{-\frac{1}{2}h} f^{1-\epsilon}\, dt \leqslant \left(\int_{-h}^{-\frac{1}{2}h} f\, dt\right)^{1-\epsilon} (\tfrac{1}{2}h)^{\epsilon} = O(h^{\epsilon} \log h).$$

and the result follows on setting $h = 2^{-\nu}$ and summing over all ν large enough.]

14. Suppose that S converges for $a<x<b$ to a non-negative sum $f(x)$. A necessary and sufficient condition for f to be integrable over (a, b) is that

$$(*)\quad \tfrac{1}{2}a_0 x + \sum_{n=1}^{\infty} (a_n \sin nx - b_n \cos nx)/n$$

should converge for $x=a$ and $x=b$. (Verblunsky [2].)

[Let $F(x)$ be the sum of $(*)$. F is monotonically increasing in the interior of (a, b), and $f \in L(a, b)$ if and only if $F(a+0)$ and $F(b-0)$ are finite. Since the coefficients of $(*)$ are $o(1/n)$, it is enough to apply (2·19) (i).]

15. Let $E=E(\xi)$ be a perfect set of constant ratio of dissection ξ, $1-2\xi$, ξ $(\xi<\frac{1}{2})$ on $(0, 2\pi)$, and $\Phi(x)$ the Lebesgue singular function associated with it (see Chapter V, § 3). Show that $S[d\Phi]$ diverges, or more precisely is unbounded, at each point of E.

[Suppose that $0<\xi<\frac{1}{3}$; let $x_0 \in E$, $F(x)=\int_0^x \Phi\, dt$. If $S[d\Phi]$ converges, or is merely bounded, at x_0, then

$$(*)\quad \{F(x_0+\alpha+\beta)-F(x_0-\alpha+\beta)-F(x_0+\alpha-\beta)+F(x_0-\alpha-\beta)\}/4\alpha\beta=O(1)$$

as α and β tend to 0 in such a way that α/β and β/α are bounded (see Example 5 above). Let h_m be the length of a 'white' interval, and k_m the length of a 'black' interval, of rank m. (We recall that 'black' intervals of rank m are the central parts which we remove at the mth stage of construction, leaving the two adjoining 'white' intervals.) We have

$$h_m = 2\pi\xi^m, \quad k_m = 2\pi\xi^{m-1}(1-2\xi).$$

Consider a 'white' interval of rank m containing x_0 and the two adjoining 'black' intervals of ranks m and $m-l$, $l>0$. Clearly $h_m<k_m$, since $\xi<\frac{1}{3}$. It follows that x_0+h_m and x_0+k_m are in the same 'black' interval; similarly for x_0-h_m and x_0-k_m. Write $\alpha_m=\frac{1}{2}(h_m+k_m)$, $\beta_m=\frac{1}{2}(k_m-h_m)$ and apply $(*)$ with $\alpha=\alpha_m$, $\beta=\beta_m$. Observing that the increment of Φ over a 'white' interval of rank m is 2^{-m} (which is the property which characterizes the Lebesgue function among all singular functions corresponding to E), we see that the left side of $(*)$ is

$$\frac{1}{2\alpha_m}\left[\frac{1}{2\beta_m}\int_{\alpha_m-\beta_m}^{\alpha_m+\beta_m} \Phi(x_0+t)\,dt - \frac{1}{2\beta_m}\int_{-\alpha_m-\beta_m}^{-\alpha_m+\beta_m} \Phi(x_0+t)\,dt\right] = \frac{2^{-m}}{2\alpha_m} \to +\infty,$$

and so is not $O(1)$. This proves the theorem for $\xi<\frac{1}{3}$.]

16. Let $0<\xi<\frac{1}{3}$, and let E be the perfect set of constant ratio of dissection ξ, $1-2\xi$, ξ constructed on $(0, 2\pi)$. Let Φ be the Lebesgue singular function associated with E. Suppose that the coefficients of $S[d\Phi]=\Sigma c_n e^{inx}$ tend to 0, so that $S[d\Phi]$ converges to 0 outside E without being identically 0. There is then a linear method of summation M satisfying conditions (i), (ii) and (iii) of regularity (Chapter III, § 1) which sums $S[d\Phi]$ to 0 at every point. In other words, the empty set is a set of multiplicity for the method M. (Marcinkiewicz and Zygmund [3].)

[Let h_m, k_m, α_m, β_m have the same meaning as in Example 15. If $x_0 \in E$, the points $x_0+\alpha_m$, $x_0+\alpha_m\pm\beta_m$ are all in the same black interval, where Φ is constant and so F, obtained by integrating $S[d\Phi]$ twice, linear. Hence

$$\frac{F(x_0+\alpha_m+\beta_m)+F(x_0+\alpha_m-\beta_m)-2F(x_0+\alpha_m)}{\beta_m^2} = \Sigma c_n e^{inx_0}\left(\frac{\sin\frac{1}{2}n\beta_m}{\frac{1}{2}n\beta_m}\right)^2 e^{in\alpha m}=0.$$

This relation holds with $-\alpha_m$ for α_m. Taking half the sum of the two expressions we see that

$$(*)\quad \sum_{n=-\infty}^{+\infty} c_n e^{inx}\left(\frac{\sin\frac{1}{2}n\beta_m}{\frac{1}{2}n\beta_m}\right)^2 \cos n\alpha_m$$

is equal to 0 for each $x \in E$, and clearly tends to 0 as $m\to\infty$ if $x \notin E$. The expression $(*)$ defines a method of summation satisfying conditions (i), (ii) and (iii) of regularity.]

17. Given two sequences $u_0, u_1, \ldots$ and $v_0, v_1, \ldots$, let $w_n=u_0v_n+u_1v_{n-1}+\ldots+u_nv_0$. Fix $\{v_n\}$. A necessary and sufficient condition that $w_n\to 0$ for each $\{u_n\}\to 0$ is $\Sigma|v_n|<\infty$.

[Compare Chapter III, (1·2) and Chapter IV, p. 168.]

18. Given two two-way infinite sequences $\dots, u_{-1}, u_0, u_1, \dots$ and $\dots, v_{-1}, v_0, v_1, \dots$ let

$$w_n = \sum_{k=-\infty}^{+\infty} u_k v_{n-k} = \sum_{0}^{+\infty} + \sum_{-\infty}^{-1}.$$

Fix $\{v_n\}$. A necessary and sufficient condition that the w_n should exist for each $\{u_n\}$ tending to 0 as $n \to \pm\infty$, is that $\Sigma |v_n| < +\infty$. If $\Sigma |v_n| < +\infty$, then $w_n \to 0$ for each $\{u_n\} \to 0$.

19. Consider two series $(U) \sum_0^\infty u_n$ and $(V) \sum_0^\infty v_n$, and their Cauchy product $(W) \sum_0^\infty w_n$, where $w_n = u_0 v_n + u_1 v_{n-1} + \dots + u_n v_0$. Fix V. A necessary and sufficient condition that W converge for every U whose terms merely tend to 0 is that

(i) $\sum_0^\infty v_n$ converge to 0,

(ii) $\sum_0^\infty \left| \sum_{\nu=n}^\infty v_\nu \right| < +\infty.$

If (i) and (ii) hold, then W converges to 0. (Condition (ii) implies that $\Sigma |v_n| < \infty$.) [If U_n, V_n, W_n are the partial sums of U, V, W, then $W_n = u_0 V_n + u_1 V_{n-1} + \dots + u_n V_0$.]

20. Consider two series $(U) \sum_{-\infty}^{+\infty} u_n$ and $(V) \sum_{-\infty}^{+\infty} v_n$, and their Laurent product $(W) \sum_{-\infty}^{+\infty} w_n$, where $w_n = \sum_{-\infty}^{+\infty} u_k v_{n-k} = \sum_0^{+\infty} + \sum_{-\infty}^{-1}$. Fix V. Necessary and sufficient conditions that W be defined and converge symmetrically for every U with $u_n \to 0$ are

$$\text{(i)} \quad \Sigma |v_n| < \infty, \qquad \text{(ii)} \quad \Sigma v_n = 0, \qquad \text{(iii)} \quad \sum_{p=-\infty}^{+\infty} \left| \sum_{\nu=-n-p}^{n-p} v_\nu \right| < A,$$

with A independent of n. If (i), (ii) and (iii) hold then W converges to 0. (Condition (iii) is satisfied if $\sum_{n=1}^{+\infty} \left| \sum_{\nu=n}^{+\infty} v_\nu \right|$ and $\sum_{n=-\infty}^{-1} \left| \sum_{\nu=-\infty}^{n} v_\nu \right|$ are finite.)

21. If the series $u_0 + u_1 + u_2 + \dots$ and $v_0 + v_1 + v_2 + \dots$ have coefficients $O(1/n)$ and converge to sums s and t respectively, then the Cauchy product $w_0 + w_1 + w_2 + \dots$ of the two series converges to sum st. (Hardy [10]. Compare Theorem (5·20).)

[We may suppose that $s = t = 0$. If U_n, V_n, W_n are the partial sums of Σu_n, Σv_n, Σw_n, then

$$W_n = \sum_{k=0}^n U_k v_{n-k} = \sum_{k=0}^m + \sum_{k=m+1}^n = W'_n + W''_n,$$

say, where $m = [\tfrac{1}{2}n]$. Now

$$W'_n = \sum_{k=0}^m o(1)\, O\left(\frac{1}{n-k}\right) = O\left(\frac{1}{n}\right) \sum_{k=0}^m o(1) = o(1),$$

and after summation by parts we obtain a similar estimate for W''_n.]

22. Given a series $(U) \sum_0^\infty u_n$, write

$$u_n^{(0)} = u_n, \quad u_n^{(k)} = \sum_{\nu=n+1}^\infty u_\nu^{(k-1)} \quad (k = 1, 2, \dots),$$

provided the series for $u_n^{(k)}$ is defined and converges. U will be said to *converge k-tuply*, or *have convergence of order k*, if $\sum_n u_n^{(k)}$ converges. Ordinary convergence of U is convergence of order 0.

Suppose that $c_n \to 0$ and that $\Phi(z) = \sum_0^\infty c_n z^n$ is regular on a closed arc $z = e^{ix}$, $a \leqslant x \leqslant b$, of the unit circle. Then $\Sigma c_n e^{inx}$ has convergence of each finite order k in (a, b), and the convergence is uniform over (a, b). (Rajchman [5].)

[Except for the uniformity, the theorem is a consequence of the theorem (5·7) of Fatou. Suppose, for example, that $x=0$ is in (a,b). Then

$$\frac{\Phi(1)-\Phi(z)}{1-z}=\sum_0^\infty c_n^{(1)}z^n \quad \left(c_n^{(1)}=\sum_{n+1}^\infty c_\nu\right)$$

is regular at $z=1$, and, by the theorem of Fatou, $\Sigma c_n^{(1)}$ converges. This argument can be repeated. The uniformity of the k-tuple convergence can be proved by considering power series with coefficients depending on a parameter.]

23. Suppose that $c_n=O(n^{-\alpha})$, $\alpha>0$, and that $\Phi(z)=\sum_0^\infty c_n z^n$ is regular on a closed arc $z=e^{ix}$, $a\leqslant x\leqslant b$; then

$$\sum_0^n c_n e^{inx}-\Phi(e^{ix})=O(n^{-\alpha}) \quad (a\leqslant x\leqslant b),$$

uniformly in x. The result holds if $n^{-\alpha}$ is replaced by a positive sequence χ_n tending monotonically to 0 and such that χ_n/χ_{2n} is bounded. Also 'O' may be replaced by 'o' throughout.

24. (i) Let I_1 and I_2 be closed arcs of the unit circumference without points in common, and suppose that on each I_k $(k=1,2)$ we consider a trigonometric series S_k with coefficients $o(1)$. Then there is a trigonometric series S with coefficients $o(1)$ which is equiconvergent with S_1 on I_1 and with S_2 on I_2. (ii) The result holds if I_1 and I_2 intersect in one or two disjoint intervals (not points) provided S_1 and S_2 are equiconvergent in the interior of $I_1 I_2$. (Phragmén [1].)

[(i) Let λ be periodic, equal to 1 in I_1, equal to 0 in I_2, and let $T=\mathsf{S}[\lambda]$ have coefficients $O(n^{-3})$. Then $S=(S_1-S_2)T+S_2$ has the required property.

(ii) We define S as above, with λ this time equal to 1 in I_1-I_2, and to 0 in I_2-I_1.]

25. Let $\{S_k\}$ be a sequence of trigonometric series with coefficients uniformly tending to 0. If each $S_k=\Sigma c_n^k e^{inx}$ converges to 0 outside a closed set P, and if $c_n=\lim c_n^k$ exists for each n, then $S=\Sigma c_n e^{inx}$ converges to 0 outside P.

[The function $F=\frac{1}{2}c_0x^2-\Sigma' n^{-2}c_n e^{inx}$ is linear in each interval contiguous to P.]

26. Let $f\sim\Sigma(a_n\cos nx+b_n\sin nx)$, $0<\gamma<1$. If $|a_n|+|b_n|=o(n^{-1-\gamma})$, then $f\in\lambda_\gamma$. The result is false for $\gamma=1$; the true form of the theorem in this case is (2·8). If 'o' is replaced by 'O', then $f\in\Lambda_\gamma$.

NOTES

In these notes we give additional comments about the results in the text and bibliographical references; 'Miscellaneous Theorems and Examples' at the ends of the chapters also contain such references. Numbers in square brackets refer to the Bibliography at the end of the second volume. We do not try to be complete, especially as regards older literature. The reader interested in bibliographic details may consult

Burkchardt, 'Trigonometrische Reihen und Integrale', *Enzyklopädie der Math. Wiss.* II, i, 2, 1904–16, Art. II, A. 12, 1325–1396.

Hilb and Riesz, 'Neuere Untersuchungen über trigonometrische Reihen und Integrale', *Enzyklopädie der Math. Wiss.* II, 3, 1924, Art. II, C. 10, vol. I_3, pp. 1191–1224.

Plessner, 'Trigonometrische Reihen', in Pascal's *Repertorium der Höheren Mathematik*, I_3, 1325–1396.

Plancherel [1],

and the relevant sections in the periodicals *Jahrbuch über die Fortschritte der Mathematik*, *Zentralblatt für Mathematik*, *Mathematical Reviews* and *Referativnyĭ Zhurnal Matematika* (in Russian).

CHAPTER I

§3. A presentation of the general theory of orthogonal series is given in Kaczmarz and Steinhaus, *Theorie der Orthogonalreihen*. For the theory of orthogonal polynomials see Szegö, *Orthogonal Polynomials*, and for bibliography—Hille, Walsh and Shohat, *A Bibliography in Orthogonal Polynomials*.

Rademacher functions as an orthogonal system were first considered in Rademacher [1]. For various aspects of the theory see the references in Example 6 on p. 34, and also Šneider [1], [2]; Fine [1], [2]; Morgenthaler [1].

§4. The theory of Fourier–Lebesgue series was started by Lebesgue. Although his work in the theory of trigonometric series is basic, we do not attempt to give detailed references to him and work prior to his, and refer the reader to his *Leçons sur les séries trigonométriques*, which gives an adequate picture of the period.

For a discussion of the notion of integral in connexion with the theory of trigonometric series see Lusin [1], Denjoy, *Leçons sur le calcul des coefficients d'une série trigonométrique*, and Jeffery, *Trigonometric series*. Jeffery's book has further bibliographic references. See also Chapter XI, §§6–7 of this book.

§§9, 10. An exhaustive treatment of the subject, and bibliography, will be found in Hardy Littlewood and Pólya, *Inequalities*. Theorem (9·16) is due to Hardy [2].

§12. Sets of the first and second category were introduced by R. Baire. A detailed discussion of the notion may be found in Denjoy's book quoted above.

§13. The main results of the section are due to Hardy and Littlewood [1]; see also Hardy, Littlewood and Pólya, *Inequalities*, Chapter X. Flett [1] gives a new and somewhat simpler proof of (13·15) (but not of (13·13)).

CHAPTER II

§1. Theorems (1·5) and (1·15) are due to W. H. Young [2], [3], [4]; see also Hardy, Littlewood and Pólya, *Inequalities*, pp. 198 sqq. For (1·30) see Wiener [1], [3].

§3. Class Λ_α is often denoted by Lip α; classes λ_α, Λ_α^p and λ_α^p by lip α, Lip (α, p) and lip (α, p) respectively. The theorem that if $\omega_1(\delta; f) = o(\delta)$, then $f \equiv$ const., is due to Titchmarsh [5].

Riemann [1] was the first to consider smooth functions; examples indicating the importance of the notion will be found in Zygmund [1]. Theorem (3·3) is an unpublished result of Z. Zalcwasser; it generalizes an earlier result of Rajchman [1] that E is dense in I. (3·4) is proved in Zygmund [1]; the proof in the text was communicated to the author in 1952 by Vijayaraghavan.

§ 4. Estimates (4·1) and (4·12) are due to Lebesgue [1]. The Lipschitz character and the non-differentiability of the Weierstrass functions are studied in Hardy [1]. For (4·7), (4·9), (4·10) see Zygmund [1]. For Theorem (4·15) see Fejér [1]; it holds if $\alpha \in L^p$, $\beta \in L^q$, $1/p + 1/q = 1$.

§ 6. The proof of (6·3) is a good illustration of the fact that proving the uniform convergence of Fourier series may require more subtle devices than proving pointwise convergence. For (6·3) and (6·8) see Hobson [2]; Lemma (6·4) is taken from Plessner [1]; the first part of (6·7) (i) from Steinhaus [1].

§ 7. See Hardy [4].

§ 8. Theorem (8·9) is due to W. H. Young [5]. (8·13) was proved by Lukács [1], and completes an earlier result of Fejér [2] (see Chapter III, (9·3)).

§ 9. For Gibbs's phenomenon see Gronwall [1], Zalcwasser [1], Hardy and Rogosinski [1], Hyltén–Cavallius [1]. (9·4) is proved in Jackson [1], Landau [1], Turán [1].

§ 10. (10·7) is due to Hardy and Littlewood [2], [3]; (10·8) to Lebesgue [1]; for (10·9) see Salem and Zygmund [3].

§ 11. For (11·1) see Lebesgue [2]; for (11·3), Hardy and Littlewood [4]; generalizations of (11·5) will be found in Gergen [1]; (11·10) is proved in Hardy and Littlewood [10].

§ 12. The estimate (12·1) is due to Fejér [1] (see also Examples 23 and 24 on p. 73). The result may be considered as an extreme case ($r=0$) of the following theorem of Kolmogorov [1]: *If* C_r *is the class of all periodic functions* f *satisfying* $|f^{(r)}(x)| \leqslant 1$ *for all* x $(r = 1, 2, \ldots)$, *then*

$$\sup_{x, f \in C_r} |f(x) - S_n(x; f)| = \frac{4}{\pi^2} \frac{\log n}{n^r} + O(n^{-r}) \simeq \frac{4}{\pi^2} \frac{\log n}{n^r}.$$

Interesting expressions for Lebesgue constants in the case of power series were obtained by Landau [2].

§ 13. Theorem (13·7) is due to Weyl [1].

CHAPTER III

§ 1. A detailed exposition of the theory of divergent series will be found in Hardy's *Divergent Series*.

Theorem (1·2) is due to Toeplitz [1], and conditions (i), (ii), (iii) of regularity are sometimes called *Toeplitz conditions*.

Delayed arithmetic means were first considered by de la Vallée-Poussin; see his *Leçons sur l'approximation des fonctions*, p. 33. Delayed means can be defined for any method (C, α) as the (C, α) means of the sequence $s_n, s_{n+1}, \ldots$.

The proof of (1·38) follows the ideas of Karamata [1] and Ingham [1]; see also Wielandt [1] and Izumi [1] (the original proof is in Littlewood [3]). *Theorem* (1·38) *holds if the condition* $u_n = O(1/n)$ *is replaced by the one-sided condition* $u_n \leqslant A/n$ (Hardy and Littlewood [18]). To prove this by the method of the text, suppose that P satisfies condition (i), and that

$$\text{(ii}'\text{)}\ P(x) \leqslant \delta x(1-x) \text{ in } (0, \xi'), \quad \text{(iii}'\text{)}\ 1 - P(x) \leqslant \delta x(1-x) \text{ in } (\xi, 1);$$

(ii′) and (iii′) follow from (ii) and (iii) by a change of δ. It follows that the polynomials $P^*(x) = P(x) + \delta x(1-x)$ and $P_*(x) = P(x) - \delta x(1-x)$ are, like P, without constant terms, and that

$$1 \leqslant P^*(x) \leqslant 1 + \delta x(1-x) \text{ in } (\xi, 1); \quad -\delta x(1-x) \leqslant P_*(x) \leqslant 0 \text{ in } (0, \xi').$$

Considering $\Sigma u_n P^*(x^n) - s_N$, and arguing as in the proof of (1·38), we obtain that $\liminf s_N \geqslant 0$. Similarly, considering $\Sigma u_n P_*(x^n) - s_{N'}$ we deduce that $\limsup s_{N'} \leqslant 0$. Hence Σu_n converges to 0.

§ 2. The general remarks of this section are merely elaborations of the proof of the fundamental theorem (3·4) of the next section.

§ 3. For (3·4) see Fejér [3]; for (3·9), Lebesgue [2]; for (3·15) Bernstein [1] (also Nikolsky [2], Sz. Nagy [2]). Theorems (3·20) and (3·23) will be found in Privalov [1], Plessner [2].

§ 4. Theorems about the convergence factors for Fourier series are due to Hardy [3]; for conjugate series see Plessner [2].

§ 5. For (5·1), see M. Riesz [2]; for (5·8), Zygmund [2]; for (5·15), Kogbetliantz [2].

§ 6. Theorem (6·18) is due to H. A. Schwarz.

§ 7. Theorems (7·2), (7·6) and (7·9) are due to Fatou [1] (see also Grosz [1]). For (7·10), see Hardy and Littlewood [6]; Theorems (7·15) and (7·20) will be found in Privalov [1], Plessner [2]. For (7·26), in the case when E is closed, see Fatou [1].

§ 8. W. H. Young [6], M. Riesz [2], Plessner [2].

§ 9. For (9·3), see Fejér [2]; for (9·6), Wiener [1].

§ 10. A proof of (10·2) will be found in Fejér [4], [11]; see also the literature indicated in the latter paper.

§ 11. Theorem (11·1) was proved by Cramér [1]; see also Gronwall [2].

§ 12. The main results of the section are due to Rogosinski [2], [3], [4]; see also Bernstein [4]. Additional results will be found in Karamata [3], [4], Agnew [1].

§ 13. The literature on best approximation is very extensive. General presentations are given in de la Vallée Poussin's *Leçons sur l'approximation des fonctions d'une variable réelle*, Jackson's *The Theory of Approximation* and Achieser's *Lectures on the Theory of Approximation*. Theorems (13·6) and (13·14) are due to Jackson [2]; (13·20) to S. Bernstein [1]; the significance of the classes Λ_* and λ_* for best approximation is pointed out in Zygmund [1]. The proof of (13·16) given in the text is due to F. Riesz [1]. For the first part of (13·29), see Korn [1], Privalov [2]; for the second, Zygmund [1], the results hold if the classes Λ_α, Λ_* are replaced by λ_α, λ_* respectively.

For (13·32) see Hille [1]; the argument also shows that if $f \not\equiv$ const., then $\mathfrak{M}[\sigma_n - f] \neq o(1/n)$ The sufficiency part of (13·34) was proved by Alexits [1] (see also Zygmund [3]), the necessity by Zamansky [1].

The proof of (13·6) gives no information about the constants A_k and B_k. Favard [1] shows that if $f^{(r-1)}(x)$ is absolutely continuous and $|f^{(r)}(x)| \leqslant M$ almost everywhere ($r = 1, 2, \ldots$), then for each n there is a polynomial $T_n(x)$ of order n such that

$$|f(x) - T_n(x)| \leqslant \frac{4}{\pi}\frac{K_r}{(n+1)^r} M,$$

where $K_r = \sum_{k=0}^{\infty} \frac{(-1)^{k(r-1)}}{(2k+1)^{r+1}}$, and the result is best possible in the sense that for suitable f and x the last inequality becomes an equality. It follows in particular that the A_k and B_k in (13·16) can be replaced by absolute constants. Corresponding results for $\tilde{f}$ will be found in Achieser and Krein [1] (or in Achieser, *Lectures on the Theory of Approximation*).

For some other aspects of best approximation, see Zamansky [2], Sz. Nagy [1], N. Bary and Stečkin [1]. Best approximation in L^p is studied in Quade [1].

CHAPTER IV

§ 1. Theorem (1·1) was obtained independently by F. Riesz [1] and Fischer [1]. Several alternative proofs will be found in G. C. and W. H. Young [1]. In considering orthogonal systems we tacitly assumed that such systems must be denumerable. That this is actually so follows from the fact that the distance $\mathfrak{M}_2[\phi - \psi]$ of any two functions ϕ, ψ of an orthonormal system is $\sqrt{2}$, and that the space L^2, being separable, cannot contain a non-denumerable system of spheres exterior to each other.

The existence of $\tilde{f}$ for $f \in L^2$ was first proved by Lusin; see his paper [1].

§ 2. Marcinkiewicz was the first to recognize the importance of the integrals of the type I_λ (see, for example, his papers [1], [2], [3]); but instead of the function χ he uses only χ^* (and its modifications) which is somewhat more awkward to apply. Remark (*d*) is due to Ostrow and E. Stein [1].

§ 3. The existence of $\tilde{f}$ was initially proved by complex methods (for the literature see the notes to Chapter VII, § 1), and the proof is one of the rare instances of applications of analytic functions to the theory of the real variable. The first purely real variable proof of the existence of $\tilde{f}$ is due to Besicovitch [1], [2]; in [1] he treats the case of f in L^2, and in [2] the general case. See also Titchmarsh [1], Loomis [1], Stein and Weiss [3]. The proof of the text is, with slight modifications, that of Marcinkiewicz [2].

Theorem (3·16) was first proved, by complex methods, by Kolmogorov [2]; see also Titchmarsh [1]. The proof of the text follows the argument of Calderón and Zygmund [5].

Divergence almost everywhere of the integrals (3·11) and (3·15) was studied by Lusin [1], Titchmarsh [2], Hardy and Littlewood [7], Marcinkiewicz [3]. For the divergence everywhere see Kaczmarz [2], [3], Mazurkiewicz [1].

§ 4. For (4·3), (4·7) and (4·8), see W. H. Young [6]. For (4·9) see Toeplitz [2], F. Riesz [5], and the literature indicated there.

For (4·21) and (4·22) see, respectively, Carathéodory [2] and Lévy [1]; for (4·25) and (4·27), Weyl [2], Schoenberg [1].

§ 5. Theorem (5·2) will be found in G. C. and W. H. Young [1]; for (5·5), see Steinhaus [2], Grosz [1]; for (5·7), W. H. Young [7], Zygmund [4] (the case $\Phi(u)=u^r$, $r>1$, is also discussed in G. C. and W. H. Young [1]). Theorem (5·20) will be found in Sidon [1].

§ 6. For theorems which are analogues of results of the two preceding sections we give only occasional citations, and refer the reader to Fichtenholz [1]. In connexion with the representation (6·6) see F. Riesz [3], Herglotz [1]. For (6·27) see Zygmund [5], F. Riesz [4] and an earlier paper of Fejér and Riesz [1].

Theorems (6·32), (6·33) and (6·34) are from Hardy and Littlewood [9_{II}]. Theorem (6·35) is due to Nikolsky [1]; see also Szegö and Zygmund [1], N. Bary [1].

§ 7. Hardy and Littlewood [1]. The second relation (7·13) had been proved in Evans, *The Logarithmic Potential*, p. 144.

§ 8. For (8·7) see G. C. and W. H. Young [1], Steinhaus [2], Zygmund [4]. The validity of Parseval's formula in some other cases is considered in Edmonds [1], Hardy and Littlewood [20]. Theorems (8·15) and (8·18) are due to W. H. Young [1] and Hardy [5], respectively. An interesting application of (8·18) is given in Hardy [6].

§ 9. A general presentation of the theory of linear operations may be found in Banach's *Opérations linéaires* or in F. Riesz and Sz. Nagy, *Leçons d'Analyse fonctionnelle.*

Theorem (9·5) is taken from Banach and Steinhaus [1]; the idea of basing the proof on the notion of category of sets is due to Saks, and has proved very fruitful. For (9·13) see Saks [1]; (9·18) will be found in Toeplitz [3]; extensions to l^p in M. Riesz [1], Titchmarsh [4]. Forms (9·17) are usually called *Toeplitz forms;* for their theory see Grenander and Szegö, *Toeplitz forms and their applications.*

§ 10. Classes L^*_Φ were first introduced by Orlicz [1], initially under the hypothesis that $\Phi(2u)=O\{\Phi(u)\}$ for $u\to+\infty$ (that this restriction is not necessary and that L^*_Φ can be defined as the class of f such that $\Phi(k\,|f\,|)$ is integrable for some $k>0$, was shown in the first edition of this book). There exists considerable literature about Orlicz spaces; we mention here only Birnbaum and Orlicz [1], Orlicz [2], Zaanen [1], Morse and Transue [1], Luxemburg [1], G. Weiss [1]. The last three papers contain a definition of a norm analogous to (10·10) but with 1 instead of $\Phi(1)$ on the right. The definition (10·10) and subsequent developments (except for (10·14)) are taken from Billik [1] (the fact that in general we do not have equality in (10·20) was already pointed out by Morse and Transue, *loc. cit.*).

§ 11. A general point of view about classes (P, Q) was first formulated in Fekete [1]. For individual results see W. H. Young [8], Steinhaus [2], Sidon [2], M. Riesz [4], Bochner [1], Zygmund [6], Kaczmarz [4], Kaczmarz and Marcinkiewicz [1], Hille [3], Hille and Tamarkin [1_{II}], Karamata and Tomić [1], Karamata [5].

Salem [4], shows that for any Fourier series $\Sigma A_n(x)$ there is a sequence $\{\lambda_n\}$ monotonically increasing to $+\infty$ and such that $\Sigma A_n(x)\lambda_n$ is still a Fourier series (a similar result for functions in L^p, $p>1$, is proved in Littlewood and Paley [1_{III}]). The only condition on $\{\lambda_n\}$ being that it must increase sufficiently slowly, we may select it so that $\{1/\lambda_n\}$ is convex, in which case $\Sigma\lambda_n^{-1}\cos nx$ is a Fourier series (Chapter V, (1·5)). It follows that *every Fourier series $\Sigma A_n(x)$ is a convolution of two Fourier series* (in our case, a convolution of $\Sigma A_n(x)\lambda_n$ and $\frac{1}{2}\lambda_0^{-1}+\Sigma\lambda_n^{-1}\cos nx$).

CHAPTER V

§ 1. Theorem (1·3) is due to Chaundy and Jolliffe [1]. For (1·5) and (1·12) see W. H. Young [8], Kolmogorov [3]. In connexion with (1·14) see W. H. Young [8], Sidon [2], Hille and Tamarkin [1_{II}]. Other results about series with monotone coefficients will be found in Boas [1], Sz. Nagy [3], Hyltén-Cavallius [1].

§2. Connexion between the asymptotic behaviour of a function and its Fourier coefficients is a classical topic and has considerable literature. Results have been obtained under various hypotheses, and it is not always easy to compare them. In this section we give a few fundamental results, aiming at simplicity rather than generality. The definition of a slowly varying function, as we introduce it here, occurs in Hardy and Rogosinski [2], though the authors do not use it systematically. It seems to be most convenient for our purposes, though it differs from the generally adopted definition of Faber [1] and Karamata [2].

The main result of the section is Theorem (2·6). It is essentially contained in Hardy and Rogosinski [2], who, however, do not settle the limiting cases $\beta=0$ and $\beta=1$ completely. From the paper of Aljančić, Bojanić and Tomić [2], which appeared recently, we borrowed the remark that the part of Theorem (2·6) about sine series is valid for $1<\beta<2$. Estimates less precise than in Theorem (2·6) but valid under more general conditions will be found in Salem [2]; they are reproduced in the first edition of this book. Finally, there are a number of results converse to (2·6); we refer the reader to Hardy and Rogosinski [3], Heywood [1], Aljančić, Bojanić and Tomić [2], and the literature quoted there.

The main part of Theorem (2·29) is an unpublished result of J. E. Littlewood and R. Salem. They showed that there exists an α_0, $0<\alpha_0<1$, such that the partial sums s_n of $\Sigma n^{-\alpha}\cos nx$ are uniformly bounded from below for $\alpha>\alpha_0$ but not for $\alpha<\alpha_0$. The fact that the s_n are uniformly bounded from below for $\alpha=\alpha_0$, and that α_0 is the root of the equation given in the theorem, I owe to S. Izumi.

Theorem (2·31) is due to Faber [1], who bases the proof on Cauchy's formula. Another proof will be found in Littlewood's *Lectures*, p. 93.

§3. The formulae for the Fourier–Stieltjes coefficients of Cantor–Lebesgue functions were first obtained in Carleman [1]. Generalizations are due to Salem [9]. See also Hille and Tamarkin [2].

§4. The series (4·1) was first considered by Hardy and Littlewood [11], who showed that it satisfies a certain functional equation relating it to Weierstrass's functions. The proof is reproduced in Littlewood's *Lectures on the Theory of Functions*, pp. 100 sqq. (it could be shortened and made more straightforward by basing it on Poisson's summation formula rather than on Cauchy's formula). A new and different proof of the functional equation was given by Paley [2]. Generalizations will be found in Wilton [1], Randels [1], Ingham [2]. Hardy and Littlewood (*loc. cit.*) show that the series $\Sigma n^{-\frac{1}{2}}e^{icn\log n}e^{inx}$ diverges everywhere. M. Weiss [2] showed that its partial sums $s_n(x)$ satisfy almost everywhere the curious relation

$$\limsup_{n\to\infty}\frac{|s_n(x)|}{(\log n)^{\frac{1}{2}}(\log\log\log n)^{\frac{1}{2}}}=1,$$

and obtained a similar estimate for the Abel means of the series.

The proof of (4·2) follows, in the main, Hille [2]. For Lemmas (4·3), (4·4), (4·6) see van der Corput [1]. Theorem (4·9) is due to Carleman [2]; see also Gronwall [3].

§5. An asymptotic formula for the function (5·1) will be found in Hardy [7].

§6. Theorem (6·3) is due to Kolmogorov [5]; a generalization will be found in Erdös [1]. For (6·4), (6·5), (6·10) see Zygmund [8], [9]. Theorem (6·15) of Hadamard is classical. For (6·13) see Zygmund [7]; the paper also contains a proof of the remark at the end of the section.

The proof of Theorem (6·4) uses the lacunarity of $\{n_k\}$ only in a limited degree. The condition we actually need is that the equations $n_\mu\pm n_\nu=N$ have a bounded number of solutions, and this can occur for $\{n_k\}$ satisfying $n_{k+1}/n_k\to 1$. The situation is rather typical, and many results about lacunary series are actually proved for more general series. We do not consider these generalizations since the condition of lacunarity is the simplest and the case the most interesting.

We mention without proof a few other results about lacunary series. Hardy [1] proves the non-differentiability of the Weierstrass function $\Sigma a^n\cos b^n x$ under the most general condition $ab\geqslant 1$. Paley [2] gives (without proof) a result about the distribution of the values of a lacunary power series on the circle of convergence. Salem and Zygmund [4] show that the values of the Weierstrass function $\Sigma b^{-\alpha n}e^{ib^n x}$ $(0\leqslant x\leqslant 2\pi)$ cover a full square provided α is small enough. M. Weiss [1], completing earlier results of Salem and Zygmund [5] and Erdös and Gál [1], shows that for the lacunary series $\Sigma r_k\cos(n_k x+\alpha_k)$ we have the *Law of the Iterated Logarithm*:

$$\limsup\frac{|s_k(x)|}{(2R_k\log\log R_k)^{\frac{1}{2}}}=1\quad\left(R_k=\tfrac{1}{2}\sum_{\nu=1}^{k}r_\nu^2\right)$$

almost everywhere, provided $r_k = o\{(R_k/\log\log R_k)^{\frac{1}{2}}\}$. In Chapter XV, § 4, we prove the Central Limit Theorem for lacunary series.

An important result of Hardy and Littlewood [17] asserts that for numerical lacunary series (by which we mean a series of constants whose terms are all zero except at lacunary places) Abel summability implies convergence; a simplified proof is given in Ingham [1]. A corresponding result for absolute Abel summability will be found in Zygmund [20].

Properties of lacunary series with different definitions of lacunarity are studied in Mandelbrojt, *Séries de Fourier et Classes Quasi-analytiques de Fonctions*, and Levinson, *Gap and Density Theorems*.

§ 7. Riesz products (7·1) were introduced in F. Riesz [6]; the complex products (7·14) in Salem and Zygmund [1]. For the remainder of the section see Zygmund [9]; also Schaeffer [1].

§ 8. That a Rademacher series $\Sigma c_\nu \phi_\nu$ converges almost everywhere if $\Sigma |c_\nu|^2 < \infty$, was first proved by Rademacher [1] (the proof in the text is from Paley and Zygmund [1_{I}]); the converse was proved by Khintchin and Kolmogorov [1]. Lemma (8·3) and the non-summability of $\Sigma c_\nu \phi_\nu$ if $\Sigma |c_\nu|^2 = \infty$ will be found in Zygmund [8] (cf. also Marcinkiewicz and Zygmund [1]). The second inequality (8·4) is valid for a large class of independent random variables and is a classical result of the Calculus of Probability. The idea of deducing the first inequality (8·5) by means of convexity is from Littlewood [1].

Theorems about almost all series $\Sigma \pm A_n(x)$ are considered in Paley and Zygmund [1] (see, however, earlier papers of Steinhaus [3] and Littlewood [6], [7]). An interesting argument of a different type will be found in Salem [2, Chapter III]. For (8·22) see Zygmund [10]. The second inequality (8·21) can also be proved directly by the same method as the second inequality (8·5) (the argument is given in the first edition of this book, p. 216) but the values of $B_{r,q}$ obtained in this way are too large to give (8·22). For Lemma (8·26) see Paley and Zygmund [1_{III}, Lemma 19] and Salem and Zygmund [2]. (8·34) is taken from Paley and Zygmund [1_{III}], p. 192, the proof in the text from Salem and Zygmund [2]. The latter paper also contains (8·36). Theorem (8·37), from Paley and Zygmund [1_{III}], complements an earlier result of Steinhaus [3]. For series $\Sigma c_\nu \phi_\nu^*$ see Paley and Zygmund [1_{II}].

§ 9. See Wiener [2], Ingham [3]. Nothing seems to be known about possible extensions to classes L^p, $p \neq 2$.

§ 10. Salem [2, Chapter IV].

CHAPTER VI

§ 1. (1·3) was proved by Denjoy [1] and Lusin [2]; (1·6) by Fatou [2] (the proof in the text is due to S. Saks); (1·7) by Salem [1]. For (1·9) and (1·11) see Lusin [1]; for (1·10), Fatou [2]. (1·12) was proved by Niemytski [1] (for sine series). That Cantor's ternary set is a basis was proved by Steinhaus [4]; more general examples of bases were found later by Denjoy [2] and Mirimanoff [1].

§ 2. For Theorems (2·5), (2·7), (2·13), (2·16) and remarks to (2·16), see Salem [1], [6]; (2·8) is an unpublished result of P. Erdös. Theorem (2·12) was proved by Marcinkiewicz [3].

§ 3. Theorem (3·1) is proved in Bernstein [2], [3]; for (3·6) and (3·9) see Zygmund [11] (the quadratic variation used in the proof appears earlier in Wiener [1]); Salem [7] shows that in the condition $\int_0^1 \delta^{-1}\omega^{\frac{1}{2}}(\delta)\,d\delta < \infty$ we cannot replace $\omega^{\frac{1}{2}}$ by $\omega^{\frac{1}{2}+\epsilon}$. Theorem (3·13) is due to Szász [1]; for generalizations see Salem [2, Chapter V], Stečkin [1]. For (3·13) see Waraszkiewicz [1], Zygmund [5].

Given a closed periodic set P and a continuous periodic $f(x)$ defined on P, we may ask whether we can define f outside P in such a way that the Fourier series of the extended function converges absolutely. There are perfect sets P for which this is possible no matter what f; see Carleson [1], Reiter [1], Helson [2], Kahane and Salem [1]. See also Kahane [1].

§ 4. Theorem (4·2) is due to Bernstein [2], [3]; in the proof of (ii) he uses a different construction. A certain simplification in the proof of (ii) (see the first edition of this book, p. 144) we owe to R. Salem. For (4·3), see Szász [1].

§ 5. For Theorems (5·1) and (5·2) (ii) see Wiener [3]; for (5·3), Lévy [2]. For (5·7) see Beurling [1]. The proofs of (5·2) (ii) and (5·7) given in the text were communicated to us by A. P. Calderón.

§ 6. (6·1) is due to Sidon [3]; for (6·3) see Zygmund [12]. See also Stečkin [1_{III}].

CHAPTER VII

§ 1. Theorems (1·2), (1·4), (1·5) (for $\alpha = 1$) and (1·6) were first proved by Privalov [1] and, later and independently, by Plessner [2]; see also the literature indicated in the notes to Chapter IV, § 3. For (1·3) in the case $\alpha > 0$, see Zygmund [2].

§ 2. Theorems (2·4) and (2·21) are due to M. Riesz [1]. His proof was reproduced in the first edition of this book; the present proof is from Calderón [1]; generalizations may be found in Hardy and Littlewood [14]. Theorem (2·6) was proved by Kolmogorov [2]; see also Hardy [8], Tamarkin [1]. Theorems (2·8) and (2·11) (i) are taken from Zygmund [4]; for (2·11) see also Warschawski [1], and for (2·8), Calderón [1], Titchmarsh [3]. Theorem (2·10) was communicated to us orally by M. Riesz. Remark (*b*) on p. 257 is due to Lusin [1]. For (2·29) see Zygmund [5], F. Riesz [4], but, except for the best possible factor $\frac{1}{2}$, the result is essentially contained in an earlier paper by Prasad [1], who showed that if $f(x)$ is of bounded variation in an interval, then $S[f]$ is absolutely summable A at every point interior to the interval. Theorems (2·30) and (2·35) are taken from Hardy and Littlewood [13]; Lemma (2·31) from Zygmund [1].

§ 3. The idea of applying Green's formula is due to P. Stein [1], though some of the formulae used, like (3·3) and (3·4), will be found in earlier literature, for example, in Hardy [9]; see also Spencer [1]. The proof of (2·6) by Stein's method was communicated to us by Z. Zalcwasser.

§ 4. Integral B is one of the several generalizations of Lebesgue's integral suggested by Denjoy [3]; more details will be found in Boks [1] (the proof of (4·2) given in the text is due to S. Saks). Jessen [1] showed that, if we consider only partitions of (a, b) into 2^n equal parts and set $\xi_j = x_j$, then $I(t)$ tends to I for almost all t. Theorems (4·3) and (4·4) were obtained by Kolmogorov [2]; for (4·4) see also Titchmarsh [1], Smirnov [1]. For applications to conjugate functions of a somewhat different generalization of Lebesgue's integral, see Titchmarsh [1], Ulyanov [1].

§ 5. For classes Λ_α see Hardy and Littlewood [9_{II}]; for Λ_* and λ_*, Zygmund [1].

§ 6. The idea of expressing $S_n(x; f)$ in terms of conjugate functions seems to have appeared first in Kolmogorov [2]. For (6·4) and (6·11) (i), see M. Riesz [1]; for (6·8) and (6·13), Kolmogorov [2]; for (6·9), (6·11) (ii) and (6·19), Zygmund [4]. For (6·14) and (6·18) see Fejér [9], Zygmund [10]. (6·22) is taken from Salem and Zygmund [3].

§ 7. Classes H^r were first considered by Hardy [9]; classes N by Ostrowski [1] and Nevanlinna [1]. A simplified and systematic approach to classes H^r is due to F. Riesz [7], who proved the basic facts of the theory, in particular the decomposition theorem (7·20) (ii), the existence of boundary values, Theorems (7·24) and (7·34). The form (7·23) of the decomposition theorem was systematically used in their work by Hardy and Littlewood. Krylov [1] extends the theory of classes H^r to functions regular in a half-plane, but the paper also contains some novel facts.

It must be observed that the theory of classes H^r does not extend to harmonic functions: there exist functions $u(\rho, x)$ harmonic in the unit circle, satisfying $\mathfrak{M}_r[u(\rho, x)] = O(1)$ for all $r < 1$, and without radial limits; see Hardy and Littlewood [13].

The fact that a function regular and bounded in $|z| < 1$ can have non-tangential limit 0 in a prescribed set of measure 0 on $|z| = 1$ is proved in Privalov [1]. The structure of the set of zeros on $|z| = 1$ of functions which are regular in $|z| < 1$ and satisfy a Lipschitz condition of positive order in $|z| \leqslant 1$, is studied in Carleson [1].

For (7·33) see Szegö [2].

The theory of classes N was mainly developed by Nevanlinna; in particular, Theorem (7·32) is due to him. We prove here only results with applications to trigonometric series, and refer the reader interested in the general theory to Nevanlinna's book *Theorie der eindeutigen analytischer Funktionen*. It may be added that though in Theorem (7·13) we assume only that the integral of $\log |F(r e^{ix})|$ is bounded above, the class of functions regular in $|z| < 1$ and having this property does not seem to be of special interest, in particular the functions need not have boundary values (observe for example, that if $f(z)$ is any function regular in $|z| < 1$, then $F(z) = \exp f(z)$ is in the class).

Theorems (7·30) and (7·35) will be found in Smirnov [1], [2]; (7·36) in Hardy and Littlewood [1]; for (7·44) see Littlewood [2] and Zygmund [13]; for (7·45), Frostman [1]; for (7·47) and (7·48) Seidel [1] and Calderón, González-Domínguez and Zygmund [1], where also earlier literature is indicated. In connexion with (7·50) see Doob [1], Zygmund [12].

(7·61) was proved by Hardy and Littlewood, [10], [9].

Theorems (8·2) and (8·3) are due to F. and M. Riesz [1]. (8·4) was proved by Helson [1], settling an old conjecture of Steinhaus. M. Weiss [3] has shown that the hypothesis $\mathfrak{M}[s_n]=O(1)$ does not imply that $\Sigma c_k e^{ikx}$ is a Fourier series. Theorem (8·6) is due to Hardy and Littlewood [15]; see also Hardy, Littlewood and Pólya, *Inequalities*, p. 236, where a different proof is given, and Fejér [10]. Theorem (8·11) was proved by R. Salem and the author.

§ 9. Privalov [1], Fichtenholz [1].

§ 10. Literature about conformal mapping will be found in Gattegno and Ostrowski [1]. Theorem (10·6) is due to Fejér [12]; (10·14) to Hardy and Littlewood [15]; for (10·15) and (10·17) see F. and M. Riesz [1], Privalov [1]. For (10·18) see Bohr [1], Salem [3]; in a somewhat similar direction goes the following theorem of Menšov [1]: *given any periodic* $f \in L$ *and any* $\epsilon>0$ *we can modify* f *in a set of measure less than* ϵ *in such a way that the Fourier series of the new function converges uniformly.*

CHAPTER VIII

§ 1. The two proofs of (8·1) are, of course, not basically different. While Lebesgue's proof appeals directly to a general theorem on linear operations, the same principle is implicit in the more concrete construction of Fejér. The polynomials Q were introduced in Fejér [8]. In connexion with (1·13), (1·14), (1·16) see Fejér [8], Steinhaus [6]; for (1·17), see Fejér [7]; for (1·21), Zygmund [16].

We also mention the following two results.

(*a*) *There exists a continuous* f *such that* $S[f]=\Sigma a_k \cos kx$ *diverges at* $x=0$, *and* $|a_1| \geqslant |a_2| \geqslant \dots$. (Salem [10]).

(*b*) *There exists a continuous* f *with* $S[f]$ *uniformly convergent and such that* $S[f^2]$ *is divergent in a dense set* (Salem [5]).

§ 2. For (2·1) see Faber [2], Lebesgue [1]; for (2·5), Hardy and Littlewood [3].

§§ 3, 4. Theorems (3·1) and (4·1) are proved in Kolmogorov [6], [7]. Theorem (3·6) is proved in Hardy and Rogosinski, *Fourier Series*, pp. 70–2; the almost everywhere divergent Fourier series which they construct to prove (3·1) is of power series type, a fact which they note in passing, but do not stress, though the result is of definite interest. Zeller [1] uses Kolmogorov's construction to obtain a Fourier series whose set of points of convergence is an arbitrary denumerable sum of closed sets. Theorem (3·14) is taken from Marcinkiewicz [2], where also other examples of divergent Fourier series will be found.

CHAPTER IX

§ 1. Sets of type H were introduced by Rajchman [2]. For (1·9) and (1·10), see Steinhaus [7], Rajchman [2].

§ 2. Theorems (2·7) and (2·8) are proved in Riemann [1]; for (2·11), see Zygmund [1]. Theorem (2·16), under the hypothesis $k\rho_k \to 0$, is proved in Fatou [1]. For (2·20) see Zygmund [1]; for (2·22), Rajchman and Zygmund [2]; for (2·27), Verblunsky [1_{II}]. Theorem (2·28) was stated, under the hypothesis $|E|=2\pi$, by Lusin [1], but it seems that he never published the proof.

§ 3. Part (i) of (3·1) is due to G. Cantor; part (ii) and (3·19) to de la Vallée-Poussin [1]. Theorem (3·18) is proved in Steinhaus [2] (and generalizes an earlier unpublished result of Banach for $\chi\equiv 0$). N. Bary [5], generalizing earlier results of Lusin and Menšov, shows that *for any measurable and periodic* f *there is continuous and periodic* F *such that* $F'=f$ *almost everywhere and* $S'[F]$ *converges to* f *almost everywhere.*

§ 4. The essence of (4·3) is due to Riemann [1]; the part about the uniformity of convergence was first proved in Phragmén [1] and Neder [1]; the conjugate series are considered in Zygmund [14].

For (4·9) see Rajchman [3], [4]; a number of generalizations will be found in Zygmund [14]. For (4·29) see Riemann [1]; also Rajchman [3], Neder [1], Zygmund [14]. In connexion with (4·27), see Lusin [1], Hobson [1]. The notion of a restricted Fourier series was introduced by W. H. Young, [9], [10].

§ 5. Theorem (5·1) is proved in Rajchman [3]; for (5·2) see Zygmund [14]; the first example of a power series converging in a given interval and diverging in another given interval was constructed by Steinhaus [10]. It is well known (see Sierpiński [1]) that the set of points of convergence of a series of continuous functions is the most general set of type $F_{\sigma\delta}$ (that is a set of the form $\prod_k \sum_i F_{ik}$, where the F_{ik} are closed sets). The problem whether for any periodic set S of type $F_{\sigma\delta}$ we can find a trigonometric or power series having S as the set of points of convergence is open. For results in this direction see Mazurkiewicz [1], Herzog and Piranian [1], Zeller [1] (see the reference to Zeller in the Notes to Chapter VIII, §§ 3, 4).

The first example of a power series $\Sigma c_n z^n$ having coefficients tending to 0 and diverging everywhere on $|z|=1$ was constructed by Lusin [3]; the corresponding example for trigonometric series was obtained by Steinhaus [8]; for (5·3) see Steinhaus [9]. Mazurkiewicz [2] showed that for every linear method of summation there is a power series with coefficients tending to 0, non-summable at any point of $|z|=1$.

Theorem (5·7), without the uniformity of convergence, was proved by Fatou [1]; the complete theorem by M. Riesz [5], [6]. For (5·8), see Zygmund [15], Lusin [4].

Theorem (5·16) was proved, in part, by Phragmén [1]; the paper seems to have escaped attention, and the theorem was rediscovered (with a different proof) in Zygmund [17]. Phragmén seems to have been the first to consider formal products of trigonometric series, though in applications his theorem is not as satisfactory as Rajchman's Theorem (4·9). For (5·20) and (5·22) see Zygmund [18_{II}]. A number of results about formal multiplication will be found in Schmetterer [1], [2].

§ 6. That there exist perfect M-sets of measure 0 was first proved in Menšov [1], and the result marks the beginning of the modern theory of uniqueness. That M-sets are, in a certain sense, more numerous than U-sets follows also from the results of Salem [8]. The existence of perfect U-sets was proved independently by Rajchman [2] and N. Bary [4], [2]; the theorem that H-sets are sets of uniqueness was proved by Rajchman [2], [4], [5]; the proof of the second of these papers uses (8·1) implicitly, but an explicit formulation of (8·1) first occurs in Pyatetski-Shapiro [1]. The definition of sets $H^{(m)}$ and Theorem (6·6) are also due to Pyatetski-Shapiro [1]; he proves (*loc. cit*) that for each $m=2, 3, \ldots$ there are sets $H^{(m)}$ which are not denumerable sums of sets $H^{(m-1)}$. He also shows (*loc. cit.*) that there actually exist perfect sets of multiplicity which are not sets of multiplicity in the restricted sense.

Kahane and Salem [1], [2], have shown that if P is any perfect M-set of constant ratio of dissection, then there are trigonometric series other than Fourier–Stieltjes series converging to 0 outside P but not everywhere; on the other hand, there are perfect sets P such that every trigonometric series with coefficients $O(1/n)$ and converging to constants in the intervals contiguous to P, is necessarily the Fourier series of a function of bounded variation.

For more literature about sets of uniqueness and multiplicity the reader is referred to the monograph of N. Bary [3].

§ 7, 8. M. Riesz [7] was the first to consider uniqueness of summable trigonometric series. Uniqueness for Abel summable series was first studied by Rajchman [1]; the paper contains a proof of (7·6) (it can be shown that neither of the intervals in (7·6) need include the other; see Rajchman and Zygmund [1]). Rajchman considered the case of coefficients tending to 0, but his method can be applied to series which after two termwise integrations become Fourier series of continuous functions (see Zygmund [19]). Theorems for series with coefficients $o(n)$ are due to Verblunsky [$1_{\text{I, II}}$]. Considerable generalizations of Theorem (8·2) for the method (C, α) will be found in Wolf [1]. The uniqueness of series summable A to 0 is studied, by means of complex methods, in Wolf [2].

§ 9. See Zygmund [14] and, for (9·21), M. Riesz [7], Rajchman [4].

A. ZYGMUND

TRIGONOMETRIC SERIES

VOLUME II

The right of the
University of Cambridge
to print and sell
all manner of books
was granted by
Henry VIII in 1534.
The University has printed
and published continuously
since 1584.

CAMBRIDGE UNIVERSITY PRESS
Cambridge
New York Port Chester
Melbourne Sydney

CONTENTS

CHAPTER X

TRIGONOMETRIC INTERPOLATION

CHAPTER XI

DIFFERENTIATION OF SERIES. GENERALIZED DERIVATIVES

CHAPTER XII

INTERPOLATION OF LINEAR OPERATIONS. MORE ABOUT FOURIER COEFFICIENTS

CHAPTER XIII

CONVERGENCE AND SUMMABILITY ALMOST EVERYWHERE

CHAPTER XIV

MORE ABOUT COMPLEX METHODS

CHAPTER XV

APPLICATIONS OF THE LITTLEWOOD-PALEY FUNCTION TO FOURIER SERIES

CHAPTER XVI

FOURIER INTEGRALS

CHAPTER XVII

A TOPIC IN MULTIPLE FOURIER SERIES

CHAPTER X

TRIGONOMETRIC INTERPOLATION

1. General remarks

In this chapter trigonometric polynomials will be systematically referred to simply as *polynomials*. We shall refer to ordinary polynomials, when we have occasion to speak of them, as *power polynomials*.

A polynomial

$$T(x) = \tfrac{1}{2}a_0 + \sum_{k=1}^{n} (a_k \cos kx + b_k \sin kx) = \sum_{k=-n}^{n} c_k e^{ikx} \tag{1·1}$$

of order n has $2n+1$ coefficients, so that one would, in principle, expect that $2n+1$ constraints would be sufficient to determine T. A purely cosine polynomial of order n has $n+1$ coefficients; a sine one, n coefficients.

Let us fix $2n+1$ points

$$x_0, x_1, \ldots, x_{2n}$$

on the x-axis, distinct modulo 2π. (In what follows we shall speak simply of *distinct* points.) If desirable, we can always assume that these points are situated in any fixed interval of length 2π.

(1·2) THEOREM. *Given $2n+1$ distinct points $x_0, x_1, \ldots, x_{2n}$ and arbitrary numbers $y_0, y_1, \ldots, y_{2n}$, real or complex, there is always a unique polynomial* (1·1) *such that*

$$T(x_k) = y_k \quad (k = 0, 1, \ldots, 2n). \tag{1·3}$$

If we treat (1·3) as a system of linear equations in the c_k, the determinant of the system is

$$e^{-in(x_0+x_1+\ldots+x_{2n})} \begin{vmatrix} 1 & e^{ix_0} & \ldots & e^{2nix_0} \\ 1 & e^{ix_1} & \ldots & e^{2nix_1} \\ \ldots & \ldots & \ldots & \ldots \\ 1 & e^{ix_{2n}} & \ldots & e^{2nix_{2n}} \end{vmatrix} = e^{-in(x_0+x_1+\ldots+x_{2n})} \prod_{\mu>\nu} (e^{ix_\mu} - e^{ix_\nu}),$$

and this is different from 0.

The polynomial $T(x)$ just defined is called the (*trigonometric*) *interpolating polynomial* corresponding to the points (abscissae) x_k and the values (ordinates) y_k. The points $x_0, x_1, \ldots, x_{2n}$ are often called the *fundamental*, or *nodal*, points of interpolation.

Let $t_j(x)$ be the polynomial of order n which takes the value 1 when $x = x_j$ and the value 0 at the remaining points x_k. Then

$$T(x) = \sum_{j=0}^{2n} y_j t_j(x), \tag{1·4}$$

since the right-hand side is a polynomial of order n taking the value y_k for $x = x_k$, for all k. The polynomials $t_j(x)$, $j = 0, 1, \ldots, 2n$, are called the *fundamental polynomials* corresponding to the fundamental points $x_0, x_1, \ldots, x_{2n}$.

Clearly

$$t_j(x) = \prod_{k \neq j} 2 \sin \tfrac{1}{2}(x - x_k) \Big/ \prod_{k \neq j} 2 \sin \tfrac{1}{2}(x_j - x_k). \tag{1·5}$$

For the expression on the right is equal to 1 when $x = x_j$ and to 0 at the remaining x_k; and it is a polynomial of order n, since the numerator consists of $2n$ factors each of the form $\alpha e^{\frac{1}{2}ix} + \beta e^{-\frac{1}{2}ix}$.

It is also easy to see that if

$$\Delta(x) = \prod_{k=0}^{2n} 2 \sin \tfrac{1}{2}(x - x_k),$$

then

$$t_j(x) = \Delta(x)/\{2\Delta'(x_j) \sin \tfrac{1}{2}(x - x_j)\}. \tag{1·6}$$

By the *number of roots* of a polynomial $T(x)$ we shall mean the sum of the multiplicities of its distinct real roots (distinct mod 2π, that is). We have now:

(1·7) THEOREM. *The number of roots of any $T(x) \not\equiv 0$ of order n does not exceed $2n$.*

From (1·1) we see that

$$e^{-inx}\, T(x) = P(z), \tag{1·8}$$

where $z = e^{ix}$ and $P(z)$ is a power polynomial of degree $2n$ in z. If $x = \xi$ is a root of order k of $T(x)$, that is, if

$$T(\xi) = T'(\xi) = \dots = T^{(k-1)}(\xi) = 0, \quad T^{(k)}(\xi) \neq 0,$$

successive differentiation of (1·8) with respect to x shows that $\zeta = e^{i\xi}$ is a root of order k for $P(z)$, and conversely. Hence if the number of roots of $T(x)$ exceeds $2n$, the number of roots of $P(z)$, multiplicity being taken into account, also exceeds $2n$. Thus $P(z) \equiv 0$, that is, $T(x) \equiv 0$, contrary to the hypothesis.

As a corollary, we obtain that *if two polynomials $S(x)$ and $T(x)$ of order n vanish at the same $2n$ points $\xi_1, \xi_2, \dots, \xi_{2n}$ of the interval $0 \leqslant x < 2\pi$, then one of S and T is a multiple of the other.* (If k of the points ξ coincide, we mean that S and T have roots of multiplicity at least k there.) For suppose that $S \not\equiv 0$ (otherwise, $S = 0 . T$), and let $C = T(\xi)/S(\xi)$, where ξ is distinct from $\xi_1, \xi_2, \dots, \xi_{2n}$ and is such that $S(\xi) \neq 0$. The polynomial $T(x) - CS(x)$ of order n vanishes not only at the points $\xi_1, \xi_2, \dots, \xi_{2n}$ but also at ξ. Hence $T - CS \equiv 0$, $T \equiv CS$.

In particular, if T vanishes at the roots of $\cos(nx + \alpha)$, then $T \equiv C \cos(nx + \alpha)$.

(1·9) THEOREM. *If a cosine polynomial $C(x)$ of order n vanishes at $n+1$ points $\xi_0 < \xi_1 < \dots < \xi_n$ in $0 \leqslant x \leqslant \pi$, then $C(x) \equiv 0$.*

If $\xi_0 > 0$ or $\xi_n < \pi$, $C(x)$ vanishes at $2n+1$ points and so is identically zero. For $C(x)$ is even, and if, for example, $\xi_0 > 0$, $C(x)$ vanishes at $\pm\xi_0, \pm\xi_1, \dots, \pm\xi_{n-1}, \xi_n$ (ξ_n and $-\xi_n$ are not distinct if $\xi_n = \pi$). If simultaneously $\xi_0 = 0$ and $\xi_n = \pi$, then $C(x)$, being even, must have at least double roots at $x = 0, \pi$, so that the number of roots of $C(x)$ is at least $4 + 2(n-1) = 2n+2$, and again $C(x) \equiv 0$.

(1·10) THEOREM. *If a sine polynomial $S(x)$ of order n vanishes at n points $\xi_1 < \xi_2 < \dots < \xi_n$ interior to $(0, \pi)$, then $S(x) \equiv 0$.*

It is enough to observe that $S(x)$ vanishes at $2n+2$ distinct points $0, \pm\xi_1, \dots, \pm\xi_n, \pi$.

It is sometimes important to interpolate by means of purely cosine or purely sine polynomials.

(1·11) THEOREM. *Given any $n+1$ distinct points $\xi_0, \xi_1, \dots, \xi_n$ in $0 \leqslant x \leqslant \pi$, and any numbers $\eta_0, \eta_1, \dots, \eta_n$, there is a unique cosine polynomial $C(x)$ of order n such that $C(\xi_k) = \eta_k$ for all k.*

Observe that

$$\tau_j(x) = \prod_{k \neq j} (\cos x - \cos \xi_k) / \prod_{k \neq j} (\cos \xi_j - \cos \xi_k)$$

is a cosine polynomial of order n which is equal to 1 when $x=\xi_j$ and vanishes at the remaining ξ_k, so that

$$C(x)=\sum_{j=0}^{n}\eta_j\tau_j(x)$$

is a cosine polynomial having the required properties. Its uniqueness is a consequence of (1·9).

If the points ξ_j are all in the interior of $(0,\pi)$, the roots of $\cos x-\cos\xi_j$ are all simple, and so

$$\tau_j(x)=-\frac{\delta(x)\sin\xi_j}{\delta'(\xi_j)(\cos x-\cos\xi_j)},$$

where

$$\delta(x)=\prod_k(\cos x-\cos\xi_k).$$

The case in which $C(x)$ is of order $n-1$ and

$$\xi_0^{(n-1)}=\frac{\pi}{2n},\quad \xi_1^{(n-1)}=\frac{3\pi}{2n},\quad \ldots,\quad \xi_{n-1}^{(n-1)}=\frac{(2n-1)\pi}{2n}$$

is particularly interesting. Here $\delta(x)$ has the same roots as $\cos nx$, so that

$$\delta(x)=C\cos nx,$$

and it is easy to verify that now

$$C(x)=\frac{\cos nx}{n}\sum_{j=0}^{n-1}\frac{(-1)^j\sin\xi_j}{\cos\xi_j-\cos x}\eta_j\quad\left(\xi_j=(2j+1)\frac{\pi}{2n}\right). \tag{1·12}$$

(1·13) THEOREM. *Given any distinct points $\xi_1,\xi_2,\ldots,\xi_n$ interior to $(0,\pi)$ and any n numbers $\eta_1,\eta_2,\ldots,\eta_n$, there is a unique sine polynomial $S(x)$ of order n such that $S(\xi_k)=\eta_k$ for all k.*

It is enough to set

$$S(x)=\sum_{j=1}^{n}\eta_j\sigma_j(x),$$

where

$$\sigma_j(x)=\frac{\sin x\prod_{k\neq j}(\cos x-\cos\xi_k)}{\sin\xi_j\prod_{k\neq j}(\cos\xi_j-\cos\xi_k)}.$$

Clearly σ_j is a sine polynomial of order n which is equal to 1 when $x=\xi_j$ and vanishes at the remaining ξ_k.

Return to the general formula (1·4). Given any function $f(x)$ of period 2π, the interpolating polynomial which coincides with $f(x)$ at the points x_k (and so also at the points congruent to x_k mod 2π) is equal to

$$\sum_{j=0}^{2n}f(x_j)\,t_j(x). \tag{1·14}$$

Suppose now that for each n we have a system

$$x_0^{(n)},\quad x_1^{(n)},\quad \ldots,\quad x_{2n}^{(n)} \tag{1·15}$$

of $2n+1$ fundamental points. It is natural to ask for conditions under which the sum (1·14) will tend to $f(x)$ as $n\to\infty$. This problem of the representation of functions by interpolating polynomials has something in common with the problem of the representation of functions by their Fourier series. It is natural to expect that the geometric structure of the fundamental sets (1·15) is of great importance here. Little is known about the behaviour of the interpolating polynomials for the general system (1·15), and in what follows we shall be concerned almost exclusively with the case of *equidistant* nodal points. By this we mean that

$$x_j^{(n)} = x_0^{(n)} + \frac{2\pi j}{2n+1} \quad (j=0, 1, \ldots, 2n). \tag{1·16}$$

Thus the points $\exp(ix_j^{(n)})$, $j=0, 1, \ldots, 2n$, are equally spaced over the circumference of the unit circle. This case has been particularly well investigated, and is the most important in applications. Moreover, the analogy with Fourier series is here particularly striking.

If no confusion arises, we shall write x_j for $x_j^{(n)}$.

The polynomial coinciding with the periodic function $f(x)$ at the points (1·16) will be denoted by $I_n(x, f)$ or by $I_n[f]$, or simply by $I_n(x)$, and will be called the *n-th interpolating polynomial* of f.

Consider the Dirichlet kernel

$$D_n(u) = \tfrac{1}{2} + \sum_{k=1}^{n} \cos ku = \frac{\sin(n+\frac{1}{2})u}{2\sin\frac{1}{2}u}.$$

It is a polynomial of order n vanishing at the points $2\pi j/(2n+1)$, $j=1, 2, \ldots, 2n$, and equal to $n+\frac{1}{2}$ for $u=0$. Thus the polynomial $D_n(x-x_j)/(n+\frac{1}{2})$, which is equal to 1 when $x=x_j$ and to 0 at the remaining points x_k, is a fundamental polynomial for the system (1·16) and, by (1·14),

$$I_n(x, f) = \frac{2}{2n+1} \sum_{j=0}^{2n} f(x_j)\, D_n(x-x_j). \tag{1·17}$$

This expression can be written as a Stieltjes integral. Let ξ_0 be any real number, and for every positive integral p let $\omega_p(x)$, $-\infty < x < +\infty$, be any step function which has jumps $2\pi/p$ at the points

$$\xi_\nu = \xi_0 + 2\nu\pi/p \quad (\nu = 0, \pm 1, \pm 2, \ldots), \tag{1·18}$$

is constant in the interior of each interval $(\xi_\nu, \xi_{\nu+1})$ and has regular discontinuities at the ξ_ν. The function $\omega_p(x)$ is determined uniquely, except for an irrelevant additive constant, by the suffix p and by the position of any point ξ_ν; so no misunderstanding will occur if we denote the function simply by $\omega_p(x)$. If the set (1·18) contains a point ξ, or a point set E, we shall say that the function $\omega_p(x)$ is *associated* with ξ, or with E.

The formula (1·17) can now be written

$$I_n(x, f) = \frac{1}{\pi} \int_0^{2\pi} f(t)\, D_n(x-t)\, d\omega_{2n+1}(t), \tag{1·19}$$

where ω_{2n+1} is associated with the points (1·16). If $S(x)$ is a polynomial of order n, then $I_n(x, S) = S(x)$, since both sides are equal at the points (1·16). Thus

$$S(x) = \frac{1}{\pi} \int_0^{2\pi} S(t)\, D_n(x-t)\, d\omega_{2n+1}(t). \tag{1·20}$$

If $g(x)$ is periodic, then $\int_\alpha^{\alpha+2\pi} g\, d\omega_p$ is independent of α. In particular, the integral in (1·19) may be taken over any interval of length 2π.

If $f(x)$ is continuous, the integral in (1·19) certainly exists as a Riemann-Stieltjes integral. If f is discontinuous at some of the points (1·16), the integral does not exist in the Riemann-Stieltjes sense. We might here use the more general Lebesgue-Stieltjes definition, but it is much simpler to treat the integral in (1·19) merely as a different notation for the sum in (1·17), and we shall always do so. The advantage of the integral notation is that it brings to light the formal similarity between the nth interpolating polynomial of f and the nth partial sum

$$S_n(x;f)=\frac{1}{\pi}\int_0^{2\pi} f(t)\,D_n(x-t)\,dt \tag{1·21}$$

of $S[f]$. If we add a suitable constant to $\omega_{2n+1}(t)$, it will tend uniformly to t as n tends to infinity, and this might suggest that the behaviour of $I_n(x,f)$ as $n\to\infty$ should be similar to that of $S_n(x;f)$. We shall see later that within certain limits this is actually the case, though the parallelism does not go so far as might be expected from the formal resemblance of the integrals in (1·19) and (1·21).

In this chapter, unless otherwise stated, we shall consider only functions integrable in the classical Riemann sense and of period 2π. In particular, our functions will be bounded. The most interesting special case, and that in which the most important problems arise, is that of continuous functions. Usually the extension of results from continuous to R-integrable functions (if possible at all) does not require essentially new ideas; but R-integrability is as natural for the theory of interpolation as L-integrability is for the theory of Fourier series. That L-integrability is not of much use for interpolation is clear from the fact that the $I_n(x,f)$ are defined by the values of f at a denumerable set of points. By modifying f there, we can change the behaviour of the $I_n[f]$, while $S[f]$ remains unchanged.

The polynomial $\tilde{I}_n(x,f)$ conjugate to $I_n(x,f)$ is obtained from (1·17) by replacing each $D_n(x-x_j)$ by the conjugate Dirichlet kernel $\tilde{D}_n(x-x_j)$, where

$$\tilde{D}_n(x)=\sum_{k=1}^{n}\sin kx=\frac{\cos\frac{1}{2}x-\cos(n+\frac{1}{2})x}{2\sin\frac{1}{2}x}.$$

Thus

$$\tilde{I}_n(x,f)=\frac{1}{\pi}\int_0^{2\pi} f(t)\,\tilde{D}_n(x-t)\,d\omega_{2n+1}(t). \tag{1·22}$$

In particular, for any polynomial $S(x)$ of order n,

$$\tilde{S}(x)=\frac{1}{\pi}\int_0^{2\pi} S(t)\,\tilde{D}_n(x-t)\,d\omega_{2n+1}(t). \tag{1·23}$$

Trigonometric interpolation is analogous to interpolation by means of power polynomials. Given any $n+1$ distinct points $\zeta_0, \zeta_1, \ldots, \zeta_n$ of the complex plane, and any numbers $\eta_0, \eta_1, \ldots, \eta_n$, there is always a uniquely determined (interpolating) polynomial $P(\zeta)=c_0+c_1\zeta+\ldots+c_n\zeta^n$ of degree n satisfying

$$P(\zeta_k)=\eta_k \quad (k=0, 1, \ldots, n). \tag{1·24}$$

The uniqueness follows from the fact that the difference of two such polynomials would be a polynomial of degree n having at least $n+1$ zeros, and so would vanish identically. If we set

$$\left.\begin{aligned} w(\zeta)&=(\zeta-\zeta_0)(\zeta-\zeta_1)\ldots(\zeta-\zeta_n),\\ l_j(\zeta)&=w(\zeta)/w'(\zeta_j)(\zeta-\zeta_j), \end{aligned}\right\} \tag{1·25}$$

then $l_j(\zeta)$ is a polynomial of degree n equal to 1 at ζ_j and vanishing at the remaining ζ_k. Thus, if $F(\zeta)$ is any function such that $F(\zeta_j)=\eta_j$ for all j, we get the classical Lagrange interpolating formula

$$P(\zeta)=\sum_{j=0}^{n}\eta_j l_j(\zeta)=\sum_{j=0}^{n}\eta_j\frac{w(\zeta)}{w'(\zeta_j)(\zeta-\zeta_j)}$$
$$=\sum_{j=0}^{n}F(\zeta_j)\frac{w(\zeta)}{w'(\zeta_j)(\zeta-\zeta_j)}.$$

Assume now that all the points ζ_j are real and situated in the interval $-1\leqslant\zeta\leqslant 1$, and consider the (standard) mapping

$$\zeta=\cos x \tag{1·26}$$

of the interval $-1\leqslant\zeta\leqslant 1$ on to the interval $0\leqslant x\leqslant\pi$. It transforms any function $F(\zeta)$, defined in $-1\leqslant\zeta\leqslant 1$, into $F(\cos x)=f(x)$, say, and the points $\zeta_0, \zeta_1, \dots, \zeta_n$ into points $x_0, x_1, \dots, x_n$. The power polynomial $P_n(\zeta)$ coinciding with $F(\zeta)$ at the points ζ_k becomes $P_n(\cos x)$, a purely cosine polynomial coinciding with $f(x)$ at the points $x_0, x_1, \dots, x_n$.

Conversely, we suppose that $f(x)$ is any function defined for $0\leqslant x\leqslant\pi$, that $0\leqslant x_0<x_1<\dots<x_n\leqslant\pi$, and that $C_n(x)$ is the cosine polynomial of order n coinciding with f at the points $x_0, x_1, \dots, x_n$. We observe that $\cos kx$ is a power polynomial of degree k in $\cos x$. (This is obvious for $k=0, 1$, and for general k it follows by induction from the formula $\cos kx+\cos(k-2)x=2\cos x\cos(k-1)x$.) Thus the transformation (1·26), which carries the function $f(x)$ into an $F(\zeta)$ defined in $-1\leqslant\zeta\leqslant 1$, also carries $C_n(x)$ into a power polynomial $P_n(\zeta)$ coinciding with F at the points $\zeta_j=\cos x_j$.

The problem of interpolating by means of power polynomials $P_n(\zeta)$ on the interval $-1\leqslant\zeta\leqslant 1$ is thus equivalent to that of interpolating by means of cosine polynomials on $0\leqslant x\leqslant\pi$. The case of the so-called *Tchebyshev abscissae*

$$\zeta_0^{(n-1)}=\cos\frac{\pi}{2n},\quad \zeta_1^{(n-1)}=\cos\frac{3\pi}{2n},\quad \dots,\quad \zeta_{n-1}^{(n-1)}=\cos\frac{(2n-1)\pi}{2n}$$

is equivalent to cosine interpolation with equidistant fundamental points $\pi/2n$, $3\pi/2n, \dots, (2n-1)\pi/2n$.

2. Interpolating polynomials as Fourier series

Write

$$I_n(x,f)=\tfrac{1}{2}a_0^{(n)}+\sum_{\nu=1}^{n}(a_\nu^{(n)}\cos\nu x+b_\nu^{(n)}\sin\nu x)$$
$$=\sum_{\nu=-n}^{+n}c_\nu^{(n)}e^{i\nu x}.$$

If we replace $D_n(u)$ in (1·19) by $\frac{1}{2}+\cos u+\dots+\cos nu$ and compare the terms on both sides, we get

$$a_\nu^{(n)}=\frac{1}{\pi}\int_0^{2\pi}f(t)\cos\nu t\,d\omega_{2n+1}(t),\quad b_\nu^{(n)}=\frac{1}{\pi}\int_0^{2\pi}f(t)\sin\nu t\,d\omega_{2n+1}(t) \tag{2·1}$$

for $\nu=0, 1, 2, \dots, n$. Similarly we have

$$c_\nu^{(n)}=\frac{1}{2\pi}\int_0^{2\pi}f(t)\,e^{-i\nu t}\,d\omega_{2n+1}(t) \tag{2·2}$$

for $|\nu|\leqslant n$. The numbers $a_\nu^{(n)}$, $b_\nu^{(n)}$ will be called the *Fourier-Lagrange coefficients* of f (corresponding to the fundamental points (1·16)). The $c_\nu^{(n)}$ are the *complex* Fourier-

Lagrange coefficients of f. Where no ambiguity arises we shall write a_ν, b_ν, c_ν for $a_\nu^{(n)}$, $b_\nu^{(n)}$, $c_\nu^{(n)}$. For a fixed ν, the integral defining $a_\nu^{(n)}$ is an approximate Riemann sum for the integral

$$\frac{1}{\pi}\int_0^{2\pi} f(t)\cos\nu t\,dt,$$

and similarly for $b_\nu^{(n)}$, $c_\nu^{(n)}$. Thus *as* $n\to\infty$ *the* ν*-th Fourier-Lagrange coefficient of* f *tends to the* ν*-th Fourier coefficient of* f.

We recall a definition from Chapter I, § 3. Let $\phi_1(x), \phi_2(x), \ldots$ be defined in an interval (a, b), and let $w(x)$ be a non-decreasing function in (a, b). We say that the system of functions ϕ_ν is orthogonal over (a, b) with respect to the weight dw if

$$\int_a^b \phi_j(x)\,\overline{\phi}_k(x)\,dw(x) = \begin{cases} 0 & \text{for } j\neq k, \\ \lambda_k>0 & \text{for } j=k. \end{cases}$$

Given any function $f(x)$ defined in (a, b), we call the numbers

$$c_j = \frac{1}{\lambda_j}\int_a^b f(x)\,\overline{\phi}_j(x)\,dw(x)$$

the *Fourier coefficients of* f, and the series

$$c_1\phi_1 + c_2\phi_2 + \ldots$$

the *Fourier series of* f, all with respect to the system $\{\phi_\nu\}$ and the weight dw. The system is called *complete* if the vanishing of all the c_j implies that f vanishes almost everywhere with respect to dw; that is, that the variation of $w(x)$ over the set of points at which f does not vanish is 0.

Return to (2·1). Taking for $f(x)$ one of the functions

$$\tfrac{1}{2},\quad \cos x,\quad \sin x,\quad \ldots,\quad \cos nx,\quad \sin nx, \tag{2·3}$$

we immediately deduce that *this system is orthogonal over* $(0, 2\pi)$ (or any interval of length 2π) *with respect to the weight* $d\omega_{2n+1}$. The numbers λ here are equal to $\tfrac{1}{2}\pi, \pi, \pi, \ldots, \pi$. The formulae (2·1) imply that $I_n(x, f)$ *is the Fourier series of* f *with respect to the system* (2·3) *and the weight* $d\omega_{2n+1}$.

Similarly, we show that the system $e^{i\nu x}$, $\nu=0, \pm 1, \ldots, \pm n$, is orthogonal over $(0, 2\pi)$ with respect to $d\omega_{2n+1}$, and $I_n(x, f)$ is the Fourier series of f with respect to this system.

If for a given f the numbers $a_0, a_1, b_1, \ldots, a_n, b_n$ are all 0, then $I_n(x, f)\equiv 0$. This means that $f=0$ at the discontinuities of ω_{2n+1} (since $I_n(x, f)=f$ there), that is, that the total variation of ω_{2n+1} over the set where f does not vanish is 0. We may therefore say that *the system* (2·3) *is complete with respect to* $d\omega_{2n+1}$.

The orthogonality of the system (2·3) with respect to $d\omega_{2n+1}$ can also be proved directly; we have only to observe that

$$\int_0^{2\pi}\cos kx\,d\omega_N(x) + i\int_0^{2\pi}\sin kx\,d\omega_N(x) = \int_0^{2\pi} e^{ikx}\,d\omega_N(x), \tag{2·4}$$

and that the last integral is 0 if k is any integer not divisible by N (see Chapter II, (1·3)). Under this hypothesis both integrals on the left must vanish, and from this we easily infer that *the system* (2·3) *is orthogonal over any interval of length* 2π *with respect to* $d\omega_m$, *where* m *is any integer* (odd or even) *greater than* $2n$.

It follows from (2·4) that if $T(x)$ is any polynomial of order less than N, then

$$(2\pi)^{-1}\int_0^{2\pi} T\,d\omega_N$$

is equal to the constant term of $T(x)$. Thus

$$\int_0^{2\pi} T(x)\,d\omega_N(x)=\int_0^{2\pi} T(x)\,dx \tag{2·5}$$

for any polynomial T of order strictly less than N. In particular,

(2·6) THEOREM. *If $S(x)=\Sigma\gamma_\nu e^{i\nu x}$ and $T(x)=\Sigma\gamma'_\nu e^{i\nu x}$ are polynomials of order $n<\frac{1}{2}N$, we have the Parseval formulae*

$$\frac{1}{2\pi}\int_0^{2\pi} S(x)\,T(x)\,d\omega_N=\sum_{\nu=-n}^{n}\gamma_\nu\gamma'_{-\nu}, \tag{2·7}$$

$$\frac{1}{2\pi}\int_0^{2\pi} |S(x)|^2\,d\omega_N=\sum_{\nu=-n}^{+n}|\gamma_\nu|^2. \tag{2·8}$$

The case $N=2n+1$ is particularly important.

If a system of functions $\{\phi_n(x)\}$ is orthogonal in (a,b) with respect to the weight $dw(x)$, and if S_k is a linear combination of the functions $\phi_1, \phi_2, \ldots, \phi_k$ with arbitrary constant coefficients, then the quadratic approximation

$$\int_a^b |f(x)-S_k(x)|^2\,dw(x)$$

of f by S_k is a minimum for fixed k if S_k is the kth partial sum of the Fourier series of f with respect to the system $\{\phi_n\}$ and weight $dw(x)$ (Chapter I, § 7). This and the fact that $I_n(x,f)$ is a Fourier series give significance to the kth partial sum of $I_n(x,f)$,

$$\begin{aligned} I_{n,k}(x,f)&=\tfrac{1}{2}a_0^{(n)}+\sum_{\nu=1}^{k}(a_\nu^{(n)}\cos\nu x+b_\nu^{(n)}\cos\nu x)\\ &=\frac{1}{\pi}\int_0^{2\pi} f(t)\,D_k(t-x)\,d\omega_{2n+1}(t)\quad(k=0,1,\ldots,n), \end{aligned} \tag{2·9}$$

and we have the following theorem:

(2·10) THEOREM. *The polynomial $I_{n,k}[f]$ minimizes the integral*

$$\int_0^{2\pi}|f(x)-S(x)|^2\,d\omega_{2n+1}(x)$$

among polynomials S of order k.

Hence $I_{n,k}[f]$ is the unique solution of the following problem: *among all polynomials $S(x)$ of order $k\leqslant n$ find the one which would approximate best—in the sense of least squares —to the function f at the points $x_0, x_1, \ldots, x_{2n}$.* For $k<n$ we cannot in general expect that the minimizing S would coincide with f at those points.

3. The case of an even number of fundamental points

In principle, any $2n+1$ conditions will suffice to determine a polynomial $T(x)$ of order n. In the previous section we assigned the value of T at $2n+1$ pre-assigned points. Now we shall choose a set of $2n$ equidistant points depending on T, more precisely on the phase of the highest term of T; and we show that T is uniquely determined by the conditions of having given values at these $2n$ points.

For write T in the form

$$T(x)=\tfrac{1}{2}a_0+\sum_{\nu=1}^{n-1}(a_\nu\cos\nu x+b_\nu\sin\nu x)+\rho\cos(nx+\alpha) \tag{3·1}$$

(ρ not necessarily positive), and consider the function ω_{2n} associated with the roots of $\sin(nx+\alpha)$. Since the product of any two of the functions

$$\tfrac{1}{2},\quad \cos x,\quad \sin x,\quad \ldots,\quad \cos(n-1)x,\quad \sin(n-1)x,\quad \cos(nx+\alpha) \tag{3·2}$$

is a polynomial of order less than $2n$, it follows from (2·5) and the ordinary orthogonality of the system of functions (3·2) that this system is orthogonal with respect to $d\omega_{2n}$ over any interval of length 2π. (This holds for *any* ω_{2n}.) Thus the coefficients a_ν, b_ν, ρ of T can be determined in the usual Fourier fashion. However, while

$$\int_0^{2\pi}(\tfrac{1}{2})^2\,d\omega_{2n}=\tfrac{1}{2}\pi,$$

and

$$\int_0^{2\pi}\cos^2\nu t\,d\omega_{2n}=\int_0^{2\pi}\sin^2\nu t\,d\omega_{2n}=\pi$$

for $\nu=1, 2, \ldots, n-1$ (by (2·5)), we have

$$\int_0^{2\pi}\cos^2(nt+\alpha)\,d\omega_{2n}(t)=\int_0^{2\pi}1\,.\,d\omega_{2n}(t)=2\pi,$$

by virtue of the hypothesis on ω_{2n}. Thus

$$T(x)=\frac{1}{\pi}\int_0^{2\pi}T(t)\left\{\tfrac{1}{2}+\sum_{\nu=1}^{n-1}\cos\nu(t-x)+\tfrac{1}{2}\cos(nt+\alpha)\cos(nx+\alpha)\right\}d\omega_{2n}(t).$$

To the last term in curly brackets we may add $\tfrac{1}{2}\sin(nt+\alpha)\sin(nx+\alpha)$, which is 0 at the discontinuities of ω_{2n}. The expression in brackets then becomes $D_n^*(t-x)$, where

$$D_n^*(u)=\tfrac{1}{2}+\sum_{\nu=1}^{n-1}\cos\nu u+\tfrac{1}{2}\cos nu=\frac{\sin nu}{2\tan\frac{1}{2}u}$$

is the modified Dirichlet kernel (Chapter II, § 5), and we obtain the following result:

(3·3) THEOREM. *For any polynomial* (3·1) *we have*

$$T(x)=\frac{1}{\pi}\int_0^{2\pi}T(t)\,D_n^*(t-x)\,d\omega_{2n}(t), \tag{3·4}$$

provided ω_{2n} *is associated with the roots of* $\sin(nt+\alpha)$.

The right-hand side here depends solely on the values of T at these roots.

Let now α be any real number. *Any* polynomial $S(x)$ of order n can be written

$$S(x)=T(x)+\sigma\sin(nx+\alpha),$$

where T is of the form (3·1). Thus in (3·4) we may replace $T(x)$ by $S(x)-\sigma\sin(nx+\alpha)$ and $T(t)$ by $S(t)$, since $\sin(nt+\alpha)$ vanishes at the discontinuities of $\omega_{2n}(t)$. This gives

$$S(x)=\sigma\sin(nx+\alpha)+\frac{1}{\pi}\int_0^{2\pi}S(t)\,D_n^*(t-x)\,d\omega_{2n}(t). \tag{3·5}$$

In particular, let $\phi_{2n}(t)$ and $\psi_{2n}(t)$ be the functions ω_{2n} associated respectively with the zeros of $\cos nt$ and $\sin nt$. Let $u_1, u_2, \ldots, u_{2n}$ and $v_1, v_2, \ldots, v_{2n}$ be the discontinuities of ϕ_{2n} and ψ_{2n}. We may suppose that

$$u_k = (2k-1)\pi/2n, \quad v_k = k\pi/n \quad (k=1, 2, \ldots, 2n).$$

Then,

(3·6) THEOREM. *For every polynomial*

$$S(x) = \tfrac{1}{2}a_0 + \sum_{\nu=1}^{n} (a_\nu \cos \nu x + b_\nu \sin \nu x),$$

we have

$$S(x) = a_n \cos nx + \frac{1}{\pi}\int_0^{2\pi} S(t)\, D_n^*(t-x)\, d\phi_{2n}(t), \tag{3·7}$$

$$S(x) = b_n \sin nx + \frac{1}{\pi}\int_0^{2\pi} S(t)\, D_n^*(t-x)\, d\psi_{2n}(t). \tag{3·8}$$

These formulae are particularly useful for obtaining expressions for the polynomials $\tilde{S}$, S', $\tilde{S}'$, which, unlike S, contain $2n$ coefficients only. For example, differentiating (3·7), where the integral is actually a finite sum, we get

$$S'(x) = -na_n \sin nx + \frac{1}{\pi}\int_0^{2\pi} S(t)\left\{\frac{n\cos n(x-t)}{2\tan\frac{1}{2}(x-t)} - \frac{\sin n(x-t)}{4\sin^2\frac{1}{2}(x-t)}\right\} d\phi_{2n}(t). \tag{3·9}$$

In this put $x=0$ and recall that ϕ_{2n} is associated with the zeros of $\cos nt$; then we obtain

$$S'(0) = \frac{1}{n}\sum_{k=1}^{2n} S(u_k)\frac{(-1)^{k+1}}{(2\sin\frac{1}{2}u_k)^2}, \tag{3·10}$$

and applying the result to the polynomial $S(\theta+x)$ we have

$$S'(\theta) = \frac{1}{n}\sum_{k=1}^{2n} S(\theta+u_k)\frac{(-1)^{k+1}}{(2\sin\frac{1}{2}u_k)^2} \quad \left(u_k = \frac{(k-\frac{1}{2})\pi}{n}\right). \tag{3·11}$$

This formula for the derivative of a trigonometric polynomial has interesting applications. If we write

$$\alpha_k = n^{-1}(2\sin\tfrac{1}{2}u_k)^{-2},$$

it gives

$$|S'(\theta)| \leqslant \sum_{k=1}^{2n} \alpha_k |S(\theta+u_k)|. \tag{3·12}$$

Now $\alpha_1+\alpha_2+\ldots+\alpha_{2n}=n$, as we may easily verify by taking $S(x)=\sin nx$ in (3·10). Hence, if $|S(x)| \leqslant M$ for all x, we have

$$|S'(\theta)| \leqslant M(\alpha_1+\alpha_2+\ldots+\alpha_{2n}) = Mn.$$

More precisely, we have $|S'(\theta)| < Mn$, unless $S(\theta+u_k)$ is alternately $\pm M$ for $k=1, 2, \ldots, 2n$, that is, unless $S(\theta+x)$ coincides either with $M\sin nx$ or $-M\sin nx$ at the points u_k. To fix our ideas, consider the first case and let

$$\Delta(x) = S(\theta+x) - M\sin nx.$$

Then $\Delta(x)$ has roots $u_1, u_2, \ldots, u_{2n}$; but since $|S| \leqslant M$, both $S(\theta+x)$ and $M\sin nx$ attain their maxima and minima simultaneously at the points u_k and the roots must be at least double. It follows that $\Delta(x)$ has at least $4n > 2n+1$ roots. Hence

$$S(\theta+x) \equiv M\sin nx, \quad S(x) \equiv M\sin n(x-\theta),$$

and we get the following theorem:

(3·13) Theorem of S. Bernstein. *If a polynomial $S(x)$ of order n satisfies $|S(x)| \leqslant M$ for all x, then $|S'(x)| \leqslant nM$, with equality if and only if S is of the form $M\cos(nx+\alpha)$.*

Let now $\chi(u)$ be non-decreasing, non-negative and convex in $u \geqslant 0$. Dividing (3·12) by n and applying Jensen's inequality (Chapter I, (10·1)) we get

$$\chi(n^{-1}|S'(\theta)|) \leqslant \chi(n^{-1}\Sigma\alpha_k|S(\theta+u_k)|) \leqslant n^{-1}\Sigma\alpha_k\chi(|S(\theta+u_k)|), \tag{3·14}$$

and integrating over $0 \leqslant \theta \leqslant 2\pi$ we have

$$\int_0^{2\pi}\chi(n^{-1}|S'(\theta)|)\,d\theta \leqslant n^{-1}\Sigma\alpha_k\int_0^{2\pi}\chi(|S(\theta+u_k)|)\,d\theta$$

$$= n^{-1}\Sigma\alpha_k\int_0^{2\pi}\chi(|S(\theta)|)\,d\theta,$$

that is

$$\int_0^{2\pi}\chi(n^{-1}|S'(\theta)|)\,d\theta \leqslant \int_0^{2\pi}\chi(|S(\theta)|)\,d\theta. \tag{3·15}$$

Suppose now that χ is strictly increasing and that we have equality in (3·15). The two members here are the integrals of the extreme terms in (3·14), so that these terms must therefore be equal for all θ. Since χ is strictly increasing, this is only possible if we have equality for all θ in (3·12). The latter condition implies that for every θ the numbers $S(\theta+u_k)$ are of alternating sign. This in turn implies that the distance between two consecutive zeros of S never exceeds π/n. If $S \not\equiv 0$, none of these distances can be less than π/n, for otherwise S would have more than $2n$ zeros. Thus either $S \equiv 0$ or S has $2n$ equidistant zeros. In either case (see p. 2) $S = M\cos(nx+\xi)$. Hence

(3·16) Theorem. *For every function $\chi(u)$ non-negative, non-decreasing and convex in $u \geqslant 0$, we have*

$$\int_0^{2\pi}\chi(n^{-1}|S'(\theta)|)\,d\theta \leqslant \int_0^{2\pi}\chi(|S(\theta)|)\,d\theta. \tag{3·17}$$

If χ is strictly increasing, equality occurs if and only if $S = M\cos(nx+\xi)$. In particular,

$$\left(\int_0^{2\pi}|S'|^p\,d\theta\right)^{1/p} \leqslant n\left(\int_0^{2\pi}|S|^p\,d\theta\right)^{1/p} \quad \textit{for} \quad p \geqslant 1. \tag{3·18}$$

When $p \to \infty$, the last inequality reduces to $\max|S'| \leqslant n\max|S|$.

If $|S| \leqslant M$, then not only $|S'| \leqslant nM$ but also $|\tilde{S}'| \leqslant nM$. This is a corollary of the following result:

(3·19) Theorem. *If a polynomial $S(x)$ of order n satisfies $|S| \leqslant M$, then*

$$\{S'^2(x)+\tilde{S}'^2(x)\}^{\frac{1}{2}} \leqslant nM, \tag{3·20}$$

the sign of equality holding if and only if $S = M\cos(nx+\xi)$.

Since the integral in (3·7) is a linear combination of the $2n$ expressions $D_n^*(x-u_k)$, the conjugate polynomial $\tilde{S}(x)$ is given by

$$\tilde{S}(x) = a_n\sin nx + \frac{1}{\pi}\int_0^{2\pi}S(t)\,\tilde{D}_n^*(x-t)\,d\phi_{2n}(t), \tag{3·21}$$

where

$$\tilde{D}_n^*(u) = \sum_{k=1}^{n-1}\sin ku + \tfrac{1}{2}\sin nu = (1-\cos nu)\,\tfrac{1}{2}\cot\tfrac{1}{2}u$$

is the modified conjugate kernel of Dirichlet (Chapter II, (5·2)).

Let α be any real number. Multiplying (3·7) by $\cos\alpha$ and (3·21) by $\sin\alpha$ and subtracting we get

$$S(x)\cos\alpha-\tilde{S}(x)\sin\alpha=a_n\cos(nx+\alpha)+\frac{1}{\pi}\int_0^{2\pi}S(t)\frac{\sin[n(x-t)+\alpha]-\sin\alpha}{2\tan\frac{1}{2}(x-t)}\,d\phi_{2n}(t).$$

Suppose now that x_0 is any root of $\sin(nx+\alpha)$ such that $\cos(nx_0+\alpha)=1$. Differentiating the equation above with respect to x and putting $x=x_0$ we have (exactly as in the deduction of (3·10) from (3·9))

$$S'(x_0)\cos\alpha-\tilde{S}'(x_0)\sin\alpha=\frac{1}{n}\sum_{k=1}^{2n}\frac{(-1)^{k+1}+\sin\alpha}{4\sin^2\frac{1}{2}(x_0-u_k)}S(u_k). \qquad (3\cdot22)$$

The coefficients of $S(u_k)$ in the sum on the right are of alternating sign. Denoting their absolute values by β_k and applying the result to $T(x)=S(\theta+x-x_0)$, we obtain

$$|S'(\theta)\cos\alpha-\tilde{S}'(\theta)\sin\alpha|\leqslant\frac{1}{n}\sum_{k=1}^{2n}\beta_k|S(\theta+u_k-x_0)|. \qquad (3\cdot23)$$

Here we have $\beta_1+\beta_2+\ldots+\beta_{2n}=n^2$, by applying (3·22) to $S(x)=\sin nx$. Thus if $|S|\leqslant M$,

$$|S'(\theta)\cos\alpha-\tilde{S}'(\theta)\sin\alpha|\leqslant nM, \qquad (3\cdot24)$$

and the sign of equality occurs if and only if $S(\theta+u_k-x_0)$ is alternately $\pm M$, that is, if $S(x)$ is of the form $M\cos(nx+\xi)$. Since α is arbitrary, (3·24), with x for θ, implies (3·20). If we have equality in (3·20) for $x=\theta$, then for a suitable α we have equality in (3·24), and so $S(x)=M\cos(nx+\xi)$.

If $\chi(u)$ is non-negative, non-decreasing and convex, it follows from (3·23) that

$$\int_0^{2\pi}\chi(n^{-1}|S'(\theta)\cos\alpha-\tilde{S}'(\theta)\sin\alpha|)\,d\theta\leqslant\int_0^{2\pi}\chi(|S(\theta)|)\,d\theta. \qquad (3\cdot25)$$

If χ is strictly increasing, there is equality if and only if $S(x)=M\cos(nx+\xi)$. As a special case of (3·25) we have

$$\int_0^{2\pi}\chi(n^{-1}|\tilde{S}'(\theta)|)\,d\theta\leqslant\int_0^{2\pi}\chi(|S(\theta)|)\,d\theta,$$

and, in particular,

$$\left(\int_0^{2\pi}|\tilde{S}'|^p\,d\theta\right)^{1/p}\leqslant n\left(\int_0^{2\pi}|S|^p\,d\theta\right)^{1/p}\quad(p\geqslant1).$$

Another formula is obtained by taking $x=0$ in (3·21). We get

$$\tilde{S}(0)=-\frac{1}{n}\sum_{k=1}^{2n}S(u_k)\tilde{D}_n^*(u_k)=-\frac{1}{n}\sum_{k=1}^{n}\{S(u_k)-S(-u_k)\}\tfrac{1}{2}\cot\tfrac{1}{2}u_k,$$

whence

$$\tilde{S}(\theta)=-\frac{1}{n}\sum_{k=1}^{n}\{S(\theta+u_k)-S(\theta-u_k)\}\tfrac{1}{2}\cot\tfrac{1}{2}u_k$$

$$=-\frac{1}{\pi}\int_0^{\pi}\frac{S(\theta+t)-S(\theta-t)}{2\tan\frac{1}{2}t}\,d\phi_{2n}(t). \qquad (3\cdot26)$$

(3·27) THEOREM. *The formula* (3·26) *is valid for any polynomial* $S(x)$ *of order* $2n-1$.

Let $S(x)$ be a polynomial of order n. Treating it as a polynomial of order $n+1$ and applying (3·4) we see that

$$\tilde{S}(x)=\frac{1}{\pi}\int_0^{2\pi}S(t)\,\tilde{D}_{n+1}^*(x-t)\,d\omega_{2n+2}(t)$$

for *any* ω_{2n+2}. If the t_m denote successive discontinuities of ω_{2n+2} we have

$$\tilde{S}(0) = -\frac{1}{n+1}\sum_{m=-n}^{n+1} S(t_m)\frac{1-\cos(n+1)t_m}{2\tan\frac{1}{2}t_m}.$$

Take for the t_m the roots of $\sin(n+1)t=0$. Then

$$t_m = m\pi/(n+1), \quad \cos(n+1)t_m = (-1)^m;$$

and bracketing together the terms corresponding to opposite m we get (since the coefficient of $S(t_{n+1})$ is 0)

$$\tilde{S}(0) = -\frac{1}{n+1}\sum_{\substack{1\leqslant m\leqslant n\\ m \text{ odd}}}\{S(t_m)-S(-t_m)\}\cot\tfrac{1}{2}t_m,$$

or, applying this to $S(x+\theta)$,

$$\tilde{S}(\theta) = -\frac{1}{n+1}\sum_{\substack{1\leqslant m\leqslant n\\ m \text{ odd}}}\left\{S\left(\theta+\frac{m\pi}{n+1}\right)-S\left(\theta-\frac{m\pi}{n+1}\right)\right\}\cot\frac{m\pi}{2(n+1)}. \tag{3·28}$$

Suppose now that $S(x)$ is of order $2n-1$ and write $2n-1$ for n in (3·28). We obtain (3·26) and the theorem is proved.

Let $f(x)$ be periodic; let

$$E_n(x) = E_n(x,f) \tag{3·29}$$

be the Fourier series of f with respect to the system (3·2) and weight $d\omega_{2n}$, and assume that ω_{2n} is associated with the roots of $\sin(nx+\alpha)$. The argument which led to (3·4) gives

$$E_n(x) = \frac{1}{\pi}\int_0^{2\pi} f(t)\,D_n^*(x-t)\,d\omega_{2n}(t) = \frac{1}{n}\sum_{k=1}^{2n} f(x_k)\,D_n^*(x-x_k), \tag{3·30}$$

where $x_1, x_2, \ldots, x_{2n}$ are the distinct roots of $\sin(nx+\alpha)$. Since

$$D_n^*(x-x_k) = \sin n(x-x_k)\,\tfrac{1}{2}\cot\tfrac{1}{2}(x-x_k)$$

is a polynomial of order n, equal to n at x_k and to 0 at the remaining x_j, *$E_n(x,f)$ is a polynomial of order n coinciding with f at the points x_k.* The most general such polynomial is

$$E_n(x) + C\sin(nx+\alpha),$$

since $C\sin(nx+\alpha)$ is the most general polynomial of order n vanishing at the points x_k. *Among the polynomials of this family, E_n is characterized by the fact that the integral of its square over $(0, 2\pi)$ is a minimum.* For E_n, being a linear combination of the functions (3·2), is of the form (3·1). Hence E_n is orthogonal to $\sin(nx+\alpha)$, so that

$$\int_0^{2\pi}\{E_n + C\sin(nx+\alpha)\}^2\,dx = \int_0^{2\pi} E_n^2\,dx + C^2\int_0^{2\pi}\sin^2(nx+\alpha)\,dx > \int_0^{2\pi} E_n^2\,dx,$$

if $C \neq 0$.

The system $\frac{1}{2}, \cos x, \sin x, \ldots, \cos nx, \sin nx$ is complete (though not orthogonal) with respect to any $d\omega_{2n}$ that is, if the integrals of $f\cos kx\,d\omega_{2n}$ and $f\sin kx\,d\omega_{2n}$ over a period are 0 for $k = 0, 1, \ldots, n$, then $f=0$ at the discontinuities of ω_{2n}. For the hypothesis implies that the polynomial (3·30), which coincides with f at the discontinuities of ω_{2n}, vanishes identically.

It follows that *the system* (3·2) *is complete with respect to any $d\omega_{2n}$, provided ω_{2n} is not associated with the zeros of* $\cos(nx+\alpha)$ (in which case $\cos(nx+\alpha)$ vanishes identically, and (3·2) is no longer an orthogonal system, with respect to $d\omega_{2n}$). For suppose that ω_{2n} is associated with the zeros of $\cos(nx+\alpha')$, where, by hypothesis, $\alpha \not\equiv \alpha' \pmod{\pi}$, and that the Fourier coefficients of f with respect to the functions (3·2), and weight $d\omega_{2n}$, are all 0. Since $\cos nx$ and $\sin nx$ are linear combinations, with constant coefficients, of $\cos(nx+\alpha)$ and $\cos(nx+\alpha')$, the vanishing of the integrals of $f\cos(nx+\alpha)\,d\omega_{2n}$ and $f\cos(nx+\alpha')\,d\omega_{2n}$ over a period implies the vanishing of the integrals of $f\cos nx\,d\omega_{2n}$ and $f\sin nx\,d\omega_{2n}$, and the desired conclusion follows from the preceding result.

4. Fourier-Lagrange coefficients

(4·1) Theorem. *The Fourier-Lagrange coefficients*

$$c_\nu^{(n)} = \tfrac{1}{2}(a_\nu^{(n)} - ib_\nu^{(n)}) = \frac{1}{2\pi}\int_0^{2\pi} f(t)\,e^{-i\nu t}\,d\omega_{2n+1}(t) \tag{4·2}$$

tend to zero as $|\nu| \to \infty$, uniformly in $n \geqslant |\nu|$.

In other words, given an $\epsilon > 0$ we can find a $\nu_0 = \nu_0(\epsilon)$ such that $|c_\nu^{(n)}| < \epsilon$ for $n \geqslant |\nu| \geqslant \nu_0$. We give two proofs.

(i) Let
$$h = h_n = 2\pi/(2n+1).$$

Suppose first that f is the characteristic function of an interval $a \leqslant x \leqslant b$. If Σ' denotes summation over those k for which the discontinuities t_k of ω_{2n+1} are in (a, b), $|c_\nu^{(n)}|$ is equal to

$$\frac{1}{2n+1}\left|\Sigma' e^{-i\nu t_k}\right| \leqslant \frac{1}{2n+1}\frac{2}{|1-e^{-i\nu h}|} = \frac{1}{(2n+1)\sin\dfrac{\pi|\nu|}{2n+1}} \leqslant \frac{1}{2|\nu|}.$$

Hence $c_\nu^{(n)} \to 0$. The result holds if (a, b) is open or half-open, and so also if f is any step function.

For any f integrable R and for any $\epsilon > 0$ we can find a set S of non-overlapping intervals $i_1, i_2, \ldots, i_p$ such that the oscillation of f over each i_k is less than ϵ, and the set $R = (0, 2\pi) - S$ is of measure less than ϵ. Let $M(x)$ be the step function which in i_k is equal to the upper bound of f in i_k and in R is equal to 0. Then $0 \leqslant M(x) - f(x) \leqslant \epsilon$ in S, $|M(x) - f(x)| \leqslant M$ in R, where M is the upper bound of $|f(x)|$. Now

$$2\pi c_\nu = \int_0^{2\pi} M(t)\,e^{-i\nu t}\,d\omega_{2n+1} + \int_0^{2\pi}\{f(t) - M(t)\}\,e^{-i\nu t}\,d\omega_{2n+1} = c_\nu' + c_\nu'',$$

say. Obviously, $c_\nu' \to 0$; and

$$|c_\nu''| \leqslant \int_0^{2\pi} |f(t) - M(t)|\,d\omega_{2n+1} = \int_S + \int_R = \xi_\nu + \eta_\nu,$$

say. Since
$$\xi_\nu \leqslant \epsilon\int_S d\omega_{2n+1} \leqslant 2\pi\epsilon,$$

$$\eta_\nu \leqslant M\int_R d\omega_{2n+1} \to M|R| \quad \text{as} \quad n \to \infty,$$

it follows that c_ν'' is small with ϵ, provided n is large enough. Hence $c_\nu \to 0$.

(ii) Let γ_ν be the Fourier coefficients of f. Consider the Parseval formulae

$$\sum_{\nu=-n}^{n} |c_\nu^{(n)}|^2 = \frac{1}{2\pi}\int_0^{2\pi} |f|^2\, d\omega_{2n+1}, \tag{4·3}$$

$$\sum_{\nu=-\infty}^{+\infty} |\gamma_\nu|^2 = \frac{1}{2\pi}\int_0^{2\pi} |f|^2\, dx. \tag{4·4}$$

It is enough to show that for any $\epsilon > 0$ we can find a $\nu_0 = \nu_0(\epsilon)$ such that

$$\sum_{\nu_0 \leqslant |\nu| \leqslant n} |c_\nu^{(n)}|^2 < \epsilon^2. \tag{4·5}$$

This property could be described as *uniform convergence*, in the parameter n, of the series in (4·3). For this purpose choose $\nu_0 = \nu_0(\epsilon)$ such that

$$0 \leqslant \frac{1}{2\pi}\int_0^{2\pi} |f|^2\, dx - \sum_{|\nu|<\nu_0} |\gamma_\nu|^2 < \tfrac{1}{2}\epsilon^2.$$

As $n \to \infty$ the right-hand side of (4·3) tends to the right-hand side of (4·4). Moreover, $c_\nu^{(n)} \to \gamma_\nu$ for any fixed ν. Hence the last inequality gives

$$0 \leqslant \frac{1}{2\pi}\int_0^{2\pi} |f|^2\, d\omega_{2n+1} - \sum_{|\nu|<\nu_0} |c_\nu^{(n)}|^2 < \epsilon^2, \tag{4·6}$$

provided n is large enough, say $n \geqslant n_0 = n_0(\epsilon)$. (The difference here must be non-negative, by virtue of (4·3).) We have $n_0 \geqslant \nu_0$. Since increasing ν_0 can only make the difference in (4·6) smaller, we may suppose that $\nu_0 = n_0$. But, by (4·3), if $n \geqslant \nu_0$, the inequality (4·6) is identical with (4·5), which completes the proof.

Let Φ be a family of functions of period 2π. We call the functions f of Φ *uniformly integrable* R if

(*a*) the functions f of Φ are uniformly bounded;

(*b*) for every $\epsilon > 0$ there is a $p_0 = p_0(\epsilon)$ with the following property: for each f of Φ we can find $p \leqslant p_0$ intervals $i_1, i_2, \ldots, i_p$ in $(0, 2\pi)$ such that the oscillation of f over each i_k is less than ϵ and the set complementary to the i_k is of measure less than ϵ.

(4·7) THEOREM. *The Fourier-Lagrange coefficients* $c_\nu^{(n)}$ *tend uniformly to* 0, *as* $|\nu| \to \infty$, *for any set of functions* f *uniformly integrable* R.

The proof is identical with the first proof of (4·1).

(4·8) THEOREM. *If* f *is of bounded variation and* V *is the total variation of* f *over* $0 \leqslant x \leqslant 2\pi$, *then*

$$|c_\nu^{(n)}| \leqslant \frac{V}{2|\nu|}.$$

Write

$$S_{-1} = 0,$$

$$S_k = e^{-i\nu t_0} + e^{-i\nu t_1} + \ldots + e^{-i\nu t_k}$$

for $k \geqslant 0$. Then

$$c_\nu^{(n)} = (2n+1)^{-1} \sum_{k=0}^{2n} f(t_k)\,(S_k - S_{k-1})$$

$$= (2n+1)^{-1} \sum_{k=0}^{2n-1} [f(t_k) - f(t_{k+1})]\, S_k,$$

since $S_{2n}=0$. Thus if $h=2\pi/(2n+1)$ we have

$$2\pi\,|\,c_\nu^{(n)}\,|\leqslant h\max|\,S_k\,|\sum_{k=0}^{2n-1}|\,f(t_k)-f(t_{k+1})\,|$$

$$\leqslant\frac{2hV}{|\,1-e^{-i\nu h}\,|}\leqslant\frac{\pi V}{|\,\nu\,|},$$

and the proof is completed.

Theorems (4·1), (4·7) and (4·8) hold if in the definition of $c_\nu^{(n)}$ we replace ω_{2n+1} by ω_{2n}.

Let γ_ν and $c_\nu^{(n)}$ denote respectively the Fourier and the Fourier-Lagrange coefficients of f. Suppose that ω_{2n+1} is associated with the point u. Then

$$c_\nu^{(n)}=\sum_{\mu=-\infty}^{+\infty}\gamma_{\nu+\mu(2n+1)}\,e^{i(2n+1)\mu u}\qquad(|\,\nu\,|\leqslant n), \tag{4·9}$$

provided $S[f]$ converges to $f(t_k)$ at the points t_k and the series on the right is interpreted as the limit of the symmetric partial sums. For $(2n+1)\,c_\nu$ is equal to

$$\sum_{k=0}^{2n}f(t_k)\,e^{-i\nu t_k}=\sum_{k=0}^{2n}\sum_{m=-\infty}^{+\infty}\gamma_m\,e^{i(m-\nu)\,t_k}$$

$$=\sum_{m=-\infty}^{+\infty}\gamma_m\sum_{k=0}^{2n}e^{i(m-\nu)\,t_k},$$

and the coefficient of γ_m is either $(2n+1)\,e^{i(m-\nu)u}$ or 0, according as $m-\nu$ is or is not divisible by $2n+1$. In particular, if $S[f]$ is absolutely convergent, then

$$\sum_{\nu=-n}^{+n}|\,c_\nu^{(n)}\,|\leqslant\sum_{\nu=-\infty}^{+\infty}|\,\gamma_\nu\,|. \tag{4·10}$$

5. Convergence of interpolating polynomials

We now show that certain tests for the convergence of $S[f]$ (see Chapter II) remain valid for the polynomials $I_n[f]$, the proofs being essentially the same. There are, however, certain differences in the behaviour of S_n and I_n.

First of all, we cannot expect that the integral tests for the convergence of $S[f]$, such as Dini's or Lebesgue's tests (Chapter II, §§ 6, 11), will hold for interpolating polynomials. The reason is that the behaviour of the $I_n[f]$ depends on the values of f at a denumerable set of points only, and, roughly speaking, the behaviour of f at these points can be 'bad', though integral conditions may be satisfied.

Secondly, we cannot in general expect the $I_n[f]$ to have any definite behaviour at a point ξ of discontinuity, even if f has only a jump at ξ. For let $x_j^{(n)}$ and $x_{j+1}^{(n)}$ be two successive fundamental points straddling ξ. Since $I_n(x_k^{(n)},f)=f(x_k^{(n)})$, it follows that, if x_j or x_{j+1} comes close to ξ, then $I_n(\xi)$ comes close to $f(\xi-0)$ or $f(\xi+0)$, as the case may be. Thus to obtain information about the behaviour of $I_n(\xi)$ we would have to associate each ω_{2n+1} with a definite point, varying with n, and this we do not do.

In what follows we consider the partial sums $I_{n,\nu}$ of the I_n rather than the I_n themselves, and discuss the behaviour of $I_{n,\nu}$ as $\nu\to\infty$, n remaining always $\geqslant\nu$. As before, all functions mentioned will be supposed integrable R.

(5·1) THEOREM. *If $f=0$ in (a,b), then $I_{n,\nu}[f]$ tends uniformly to 0 in every interval $(a+\epsilon, b-\epsilon)$, where $\epsilon>0$.*

This is a corollary of the following result (compare Chapter II, (6·7)).

(5·2) THEOREM. *If $\lambda(x)$ is integrable R and has all its Dini numbers finite at $x=x_0$ then*

$$I_{n,\nu}(x_0, \lambda f)-\lambda(x_0)\, I_{n,\nu}(x_0, f)\to 0.$$

If $\lambda\in\Lambda_1$, the convergence is uniform in x_0.

By (4·1), it is enough to prove (5·2) for the polynomials $I^*_{n,\nu}$ obtained by inserting a factor $\frac{1}{2}$ in the last term of $I_{n,\nu}$. We have

$$\left.\begin{aligned} I^*_{n,\nu}(x_0,f)&=\frac{1}{\pi}\int_0^{2\pi} f(t)\, D^*_\nu(x_0-t)\, d\omega_{2n+1}(t),\\ I^*_{n,\nu}(x_0,\lambda f)-\lambda(x_0)\, I^*_{n,\nu}(x_0,f)&=\frac{1}{\pi}\int_0^{2\pi} f(t)\frac{\lambda(t)-\lambda(x_0)}{2\tan\frac{1}{2}(t-x_0)}\sin\nu(t-x_0)\, d\omega_{2n+1}(t).\end{aligned}\right\} \quad (5\cdot3)$$

By (4·1), the last integral tends to 0 if the ratio $[\lambda(t)-\lambda(x_0)]/(t-x_0)$ is bounded. For the discontinuities of the function

$$h_{x_0}(t)=[\lambda(t)-\lambda(x_0)]\,\tfrac{1}{2}\cot\tfrac{1}{2}(t-x_0)$$

are those of λ and, possibly, $t=x_0$, so that $f(t)\, h_{x_0}(t)$ is integrable R. If $\lambda\in\Lambda_1$, the $h_{x_0}(t)$ are uniformly integrable R, since they are uniformly continuous for $|t-x_0|\geq\epsilon>0$. Hence $f(t)\, h_{x_0}(t)$ is uniformly integrable R, and it is enough to apply (4·7).

To deduce (5·1) from (5·2), let $\lambda(t)$ be periodic, continuous, equal to 0 in $(a+\epsilon, b-\epsilon)$, equal to 1 outside (a,b), and linear elsewhere. Then $\lambda f=f$ and, for x_0 in $(a+\epsilon, b-\epsilon)$,

$$I_{n,\nu}(x_0,\lambda f)-\lambda(x_0)\, I_{n,\nu}(x_0,f)=I_{n,\nu}(x_0,f),$$

from which (5·1) follows.

Theorem (5·1) expresses the *principle of localization* for interpolating polynomials. If $f_1=f_2$ in (a,b), the sequences $I_{n,\nu}[f_1]$ and $I_{n,\nu}[f_2]$ are uniformly equiconvergent in $(a+\epsilon, b-\epsilon)$.

(5·4) THEOREM. *If f is of bounded variation, then $I_{n,\nu}(x,f)$ tends to $f(x)$ at every point of continuity of f. The convergence is uniform over every closed interval of continuity of f.*

We postpone the proof to the next section.

We mentioned above that the integral test of Dini cannot be expected to hold for the $I_{n,\nu}$. The following result is a substitute for Dini's test:

(5·5) THEOREM. *If $|f(\xi\pm t)-f(\xi)|\leq\mu(t)$ when $0<t\leq\eta$, where $\mu(t)$ is a non-decreasing function of t such that $\int_0^\eta t^{-1}\mu(t)\,dt<+\infty$, then $I_{n,\nu}(\xi,f)\to f(\xi)$.*

The integrability of $t^{-1}\mu(t)$ implies that $\mu(+0)=0$, so that f is continuous at ξ. Write

$$\begin{aligned} I^*_{n,\nu}(\xi)-f(\xi)&=\frac{1}{\pi}\int_{\xi-\pi}^{\xi+\pi}[f(t)-f(\xi)]\, D^*_\nu(\xi-t)\, d\omega_{2n+1}\\ &=\frac{1}{\pi}\int_{\xi-\delta}^{\xi+\delta}+\frac{1}{\pi}\int_R=P+Q,\end{aligned}$$

where R is the complement of $(\xi-\delta, \xi+\delta)$. For every fixed $\delta>0$, Q tends to 0. It is therefore enough to show that P is small with δ, if n is large enough.

For this purpose, given any $\epsilon>0$ choose δ so that $\int_0^{2\delta} t^{-1}\mu(t)\,dt<\epsilon$. The integral P is a finite sum extended over the fundamental points t_j belonging to $(\xi-\delta,\xi+\delta)$. We write $P=P_1+P_2$, where P_1 is the contribution of the two points t_{k-1} and t_k nearest to ξ (so that $t_{k-1}\leqslant\xi\leqslant t_k$). Evidently $P_1\to 0$. Let $h=2\pi/(2n+1)$. The contribution of the points $t_{k+1}, t_{k+2}, \ldots, t_{k+l}$ belonging to $(\xi,\xi+\delta)$ does not exceed

$$\frac{2}{2n+1}\sum_{j=k+1}^{k+l}\frac{|f(t_j)-f(\xi)|}{t_j-\xi}\leqslant\frac{h}{\pi}\sum_{i=1}^{l}\frac{\mu((i+1)h)}{ih}\leqslant\frac{3h}{\pi}\sum_{i=1}^{l}\frac{\mu((i+1)h)}{(i+2)h}$$
$$\leqslant\frac{3}{\pi}\int_{2h}^{(l+2)h}t^{-1}\mu(t)\,dt\leqslant\int_0^{2\delta}t^{-1}\mu(t)\,dt<\epsilon,$$

the penultimate inequality being valid for n large enough. A similar result holds for the fundamental points in $(\xi-\delta,\xi)$. Thus

$$|P|\leqslant o(1)+2\epsilon$$

for n large enough, and (5·5) follows.

The case in which $\mu(t)=(\log 1/t)^{-1-\epsilon}$ for small t is particularly interesting. Here ϵ must be positive so long as we consider only one point ξ. If the conditions are satisfied at all points of an interval, however, we may relax them slightly, and we have, in fact, the following theorem:

(5·6) THEOREM. *If f is continuous and its modulus of continuity is $o\{(\log 1/\delta)^{-1}\}$, then $I_{n,\nu}(x,f)$ tends uniformly to $f(x)$.*

Let ω_{2n+1} be associated with a point u, and let

$$\lambda_{n,\nu}=\frac{1}{\pi}\int_0^{2\pi}|D_\nu(t-x)|\,d\omega_{2n+1}. \tag{5·7}$$

This is an analogue of the Lebesgue constant (Chapter II, § 12), and depends on ν, n, x and u; more precisely, it depends on ν, n and $x-u$. We require the inequality

$$\lambda_{n,\nu}\leqslant A\log\nu\quad(1<\nu\leqslant n), \tag{5·8}$$

where A is an absolute constant, and this we take temporarily for granted. Let σ_ν be the (C, 1) means of S[f]. Since the modulus of continuity of f is $o\{(\log 1/\delta)^{-1}\}$, we have, by Chapter III, (3·16),

$$f-\sigma_\nu=o\{(\log\nu)^{-1}\}. \tag{5·9}$$

Now write

$$\left.\begin{aligned}f&=(f-\sigma_\nu)+\sigma_\nu,\\ I_{n,\nu}[f]&=I_{n,\nu}[f-\sigma_\nu]+I_{n,\nu}[\sigma_\nu].\end{aligned}\right\} \tag{5·10}$$

The last term is $\sigma_\nu(x)$ and tends uniformly to $f(x)$. It is therefore enough to show that $I_{n,\nu}[f-\sigma_\nu]$ tends uniformly to 0. But

$$\begin{aligned}|I_{n,\nu}[f-\sigma_\nu]|&=\left|\frac{1}{\pi}\int_0^{2\pi}\{f(t)-\sigma_\nu(t)\}\,D_\nu(t-x)\,d\omega_{2n+1}(t)\right|\\ &\leqslant\lambda_{n,\nu}\max_t|f(t)-\sigma_\nu(t)|\\ &=O(\log\nu)\,o(1/\log\nu)=o(1),\end{aligned} \tag{5·11}$$

by (5·8) and (5·9).

Returning now to (5·8), we have

$$\lambda_{n,\nu}=\frac{1}{\pi}\int_{-\pi}^{\pi}|D_\nu(t)|\,d\omega'_{2n+1}(t), \tag{5·12}$$

where ω'_{2n+1} is associated with the point $u-x$, and it is enough to show that

$$\int_0^\pi |D_\nu(t)|\,d\omega'_{2n+1}(t)=O(\log\nu).$$

Let t_k be the first discontinuity of ω'_{2n+1} satisfying $t_k\geqslant 1/\nu$, and let $h=2\pi/(2n+1)$. Using the fact that $D_\nu(t)$ is both $O(\nu)$ and $O(1/t)$, we write

$$\int_0^\pi |D_\nu(t)|\,d\omega'_{2n+1}\leqslant A\int_0^{t_k}\nu\,d\omega'_{2n+1}(t)+A\int_{t_{k+1}}^{\pi}t^{-1}d\omega'_{2n+1}(t)$$
$$=P+Q.$$

Since the number of discontinuities of ω'_{2n+1} in $(0, 1/\nu)$ is $O(n/\nu)$, we immediately see that

$$P=O(1).$$

Also, if t_{k+j} is the last discontinuity of ω'_{2n+1} in $(0,\pi)$, we have

$$Q\leqslant Ah[t_{k+1}^{-1}+t_{k+2}^{-1}+\dots+t_{k+j}^{-1}]$$
$$\leqslant A\int_{t_k}^{\pi}t^{-1}dt=O(\log\nu).$$

Hence $P+Q=O(\log\nu)$ and (5·8) is established.

Remark. The inequality (5·8) can also be deduced from the estimate for the Lebesgue constant L_ν (Chapter II, §12), and the following result, which will be proved in §7 below: *If $S(t)$ is a polynomial of order ν and $N>\nu$, then*

$$\int_0^{2\pi}|S(t)|\,d\omega_N(t)\leqslant A\int_0^{2\pi}|S(t)|\,dt,$$

where A is an absolute constant. If $S(t)=D_\nu(x-t)$, then

$$\int_0^{2\pi}|D_\nu(x-t)|\,d\omega_N(t)\leqslant A\int_0^{2\pi}|D_\nu(x-t)|\,dt$$
$$=A\int_0^{2\pi}|D_\nu(t)|\,dt=A\pi L_\nu.$$

Since $L_\nu=O(\log\nu)$, the result follows.

The same argument shows that

$$\int_0^{2\pi}|\tilde{D}_\nu(x-t)|\,d\omega_N\leqslant A\int_0^{2\pi}|\tilde{D}_\nu(t)|\,dt=O(\log\nu),$$

by Chapter II, (12·3).

(5·13) THEOREM. (i) *For any f,*

$$I_{n,\nu}(x,f)=O(\log\nu), \tag{5·14}$$

uniformly in x and n.

(ii) *If f is continuous at x_0, then*

$$I_{n,\nu}(x_0,f)=o(\log\nu). \tag{5·15}$$

(iii) *The latter relation holds uniformly in x_0 if f is everywhere continuous.*

(i) If $|f|\leqslant M$, $\quad |I_{n,\nu}[f]|\leqslant M\lambda_{n,\nu}=O(\log\nu)$.

(ii) Given an $\epsilon>0$, let δ be so small that the oscillation of f in $(x_0-\delta, x_0+\delta)$ is less than ϵ. Let $f_1(x)$ be the function equal to $f(x_0)$ in this interval and equal to $f(x)$ elsewhere. If we set $f=f_1+f_2$, then $|f_2(x)|\leqslant\epsilon$ for all x, whence

$$|I_{n,\nu}[f_2]|\leqslant\epsilon\lambda_{n,\nu}\leqslant A\epsilon\log\nu.$$

Since f_1 is constant in $(x_0-\epsilon, x_0+\epsilon)$, $I_{n,\nu}(x_0,f_1)$ converges to $f_1(x_0)$ and so is $o(\log\nu)$. Since $I_{n,\nu}[f]=I_{n,\nu}[f_1]+I_{n,\nu}[f_2]$, the result now follows.

(iii) As in (5·10) we have

$$I_{n,\nu}[f]=I_{n,\nu}[f-\sigma_\nu]+I_{n,\nu}[\sigma_\nu].$$

The last term is σ_ν, and so tends to $f(x)$ uniformly in x, while as in (i) the first term is $O(\log\nu).o(1)=o(\log\nu)$, again uniformly in x.

(5·16) Theorem. *If $S[f]$ converges absolutely, $I_{n,k}(x,f)$ converges uniformly to $f(x)$.*

Multiply both sides of (4·9) by $e^{i\nu x}$, sum the results for $\nu=0,\pm1,\pm2,\dots,\pm k$, and subtract from $f=\Sigma\gamma_j e^{ijx}$; we get

$$|f(x)-I_{n,k}|\leqslant 2\sum_{|j|>k}|\gamma_j|,$$

and the expression on the right tends to 0 as $k\to\infty$ (n always remaining $\geqslant k$).

The results of this section remain valid for the interpolating polynomials $E_n(x,f)$ (see § 3) and their partial sums $E_{n,k}$.

If at a point x_0 of continuity of f the expressions $I_{n,\nu}(x_0,f)$ tend to a limit, that limit must be $f(x_0)$. This follows immediately from a result which will be established in the next section, namely that

$$\frac{1}{n+1}\{I_{n,0}+I_{n,1}+\dots+I_{n,n}\}$$

tends to f at every point of continuity of f. Whether the existence of $\lim I_n(x_0,f)$ at a point of continuity implies that this limit must be $f(x_0)$, is a deeper problem (and the answer is not always affirmative).

(5·17) Theorem. *Suppose that the fundamental points are associated with the point* 0, *and let x_0 be a point of continuity of f. Then*

(i) *if $E_n(x_0,f)\to s$, $s=f(x_0)$;*

(ii) *if $I_n(x_0,f)\to s$, and x_0 is not of the form $2\pi p/q$, where p/q is an irreducible fraction with even denominator, $s=f(x_0)$.*

The proof of (i) is based on the result of Kronecker that *given any real x there are infinitely many fractions p/q such that*

$$|x-p/q|\leqslant q^{-2} \tag{5·18}$$

(see Chapter VI, § 2). The fundamental points in (i) are $t_k^n=\pi k/n$. By Kronecker's theorem applied to x_0/π, there are infinitely many pairs k, n with

$$|x_0-t_k^n|\leqslant\pi n^{-2}.$$

Consider only such t_k^n. The polynomials $E_n(x,f)$ are uniformly $O(\log n)$, and so their derivatives are uniformly $O(n\log n)$. By the mean-value theorem,

$$E_n(x_0)-E_n(t_k^n)=(x_0-t_k^n)E_n'(\theta)=O(n^{-2})O(n\log n)=o(1),$$

where θ is between x_0 and t_k^n. Since

$$E_n(t_k^n) = f(t_k^n) \to f(x_0), \quad E_n(x_0) \to s,$$

it follows that $s = f(x_0)$.

The proof of (ii) is the same as that of (i) if $x_0' = x_0/2\pi$ can be approximated with error $O(q^{-2})$ by infinitely many fractions p/q *with odd denominators*. This is clearly impossible if x_0' is an irreducible fraction p_0/q_0 with an even denominator, for if p/q is in its lowest terms,

$$|x_0' - p/q| = |p_0/q_0 - p/q| \geqslant 1/q_0 q.$$

We shall show that *if x is not of the form p_0/q_0 just described then there are infinitely many fractions p/q with odd denominators satisfying* (5·18).

We may assume that x is irrational, for if x is rational with odd denominator the result is obvious. We know that there are infinitely many fractions p/q satisfying (5·18). We may assume that the q here are even, for otherwise there is nothing to prove. Let ξ, η be a pair of integers such that $p\eta - q\xi = 1$. Clearly η is odd. We may assume also that $|\eta| \leqslant \frac{1}{2}q$, for if ξ_0, η_0 satisfy this relation, so also do $\xi_0 + tp$, $\eta_0 + tq$ for every integer t. We observe now that

$$\left|x - \frac{\xi}{\eta}\right| \leqslant \left|x - \frac{p}{q}\right| + \left|\frac{p}{q} - \frac{\xi}{\eta}\right| \leqslant \frac{1}{q^2} + \frac{1}{|q\eta|}$$

$$\leqslant \frac{1}{4\eta^2} + \frac{1}{2\eta^2} < \frac{1}{\eta^2}, \tag{5·19}$$

and it remains to prove the existence of infinitely many fractions ξ/η with the required property. If we had only a finite number of them, the η's would be bounded and infinitely many of them would have the same value η_0. By making q tend to $+\infty$ through the values giving $\eta = \eta_0$, we would obtain from the second inequality of (5·19) the result that $x = \xi_0/\eta_0$, contrary to the hypothesis that x is irrational.

6. Jackson polynomials and related topics

We note that the Fejér kernel

$$K_n(u) = \frac{2}{n+1}\left\{\frac{\sin \frac{1}{2}(n+1)u}{2\sin \frac{1}{2}u}\right\}^2$$

vanishes at the points $2\pi k/(n+1)$ for $k = 1, 2, \ldots, n$ and equals $\frac{1}{2}(n+1)$ at $u = 0$. Thus if $t_0, t_1, \ldots, t_n$ are any $n+1$ points equally spaced over $(0, 2\pi)$, for example

$$t_k = t_0 + \frac{2\pi k}{n+1} \quad (k = 0, 1, \ldots, n), \tag{6·1}$$

and if ω_{n+1} is associated with these points, then

$$J_n(x) = J_n(x, f) = \frac{2}{n+1}\sum_{k=0}^{n} f(t_k)\, K_n(x - t_k)$$

$$= \frac{1}{\pi}\int_0^{2\pi} f(t)\, K_n(t - x)\, d\omega_{n+1}(t)$$

is a polynomial of order n coinciding with f at these points.

Since $K_n'(u) = 0$ for $u = 2\pi k/(n+1)$, the derivative $J_n'(x, f)$ vanishes at the points t_k. Thus $J_n(x, f)$ *is a polynomial of order n coinciding with f at the points t_k and having a* *vanishing derivative there*. This means that J_n satisfies $2n+2>2n+1$ conditions; but, as we shall see later (see (6·7) below), the conditions governing the derivative J'_n are not independent.

The J_n are called *Jackson polynomials*.

The vanishing of J'_n at the points t_k ensures a certain smoothness of the graph of J_n and suggests that $f(x)$ might be represented better by $J_n(x,f)$ than by $I_n(x,f)$. We show that this is in fact true. It is convenient to consider more general polynomials of order n, namely,

$$J_{n,m}(x)=J_{n,m}(x,f)=\frac{1}{\pi}\int_0^{2\pi} f(t)\,K_n(x-t)\,d\omega_m(t), \tag{6·2}$$

where m is any integer greater than n. Thus $J_n=J_{n,n+1}$. The case in which $m=2n+1$ is also of particular interest.

(6·3) Theorem. *Let $f(x)$ be bounded and periodic. Then*

(i) *$J_{n,m}[f]$ remains within the same bounds as f;*

(ii) *$J_{n,m}(x,f)$ converges to $f(x)$ at every point x of continuity of f as $n\to\infty$, m remaining always greater than n. The convergence is uniform in every closed interval (α,β) of continuity of f.*

The proof is similar to that of Fejér's theorem (Chapter III, § 3) and is based on the fact that K_n is a positive kernel. By (2·5),

$$\frac{1}{\pi}\int_0^{2\pi} K_n(t-x)\,d\omega_m(t)=\frac{1}{\pi}\int_0^{2\pi} K_n(t-x)\,dt=1, \tag{6·4}$$

so that if $A\leqslant f(x)\leqslant B$ for all x, then $A\leqslant J_{n,m}(x)\leqslant B$, a property of $J_{n,m}$ similar to that of σ_n.

Consider now (ii), and suppose that $|f|\leqslant M$. Given $\epsilon>0$, we can find a $\delta>0$ such that $|f(t)-f(x)|\leqslant\epsilon$ for $x\in(\alpha,\beta)$ and $|t-x|\leqslant\delta$; we include here the case when (α,β) reduces to a single point. By (6·2) and (6·4),

$$\begin{aligned} J_{n,m}(x)-f(x)&=\frac{1}{\pi}\int_0^{2\pi}[f(t)-f(x)]\,K_n(t-x)\,d\omega_m(t)\\ &=\frac{1}{\pi}\int_{x-\pi}^{x+\pi}=\frac{1}{\pi}\int_{x-\delta}^{x+\delta}+\frac{1}{\pi}\int_R\\ &=P+Q,\end{aligned}$$

say, and

$$|P|\leqslant\frac{1}{\pi}\int_{x-\delta}^{x+\delta}\epsilon K_n(t-x)\,d\omega_m\leqslant\frac{\epsilon}{\pi}\int_0^{2\pi}K_n(t)\,d\omega_m=\epsilon,$$

$$|Q|\leqslant\frac{2M}{\pi}\int_R K_n(t-x)\,d\omega_m\leqslant\frac{2M}{\pi}\max_{\delta\leqslant u\leqslant\pi}K_n(u)\int_R d\omega_m.$$

Thus $Q\to 0$ as $n\to\infty$, $|P+Q|<2\epsilon$ for $n>n_0$, and (6·3) follows.

Let $B_{n,\nu}(x)=B_{n,\nu}(x,f)$ denote the arithmetic mean of the partial sums $I_{n,0}, I_{n,1}, \dots, I_{n,\nu}$ of $I_n(x,f)$. Thus

$$\begin{aligned} B_{n,\nu}&=\frac{1}{\nu+1}\sum_{k=0}^{\nu}I_{n,k}=\frac{1}{\pi}\int_0^{2\pi}f(t)\,K_\nu(t-x)\,d\omega_{2n+1}\\ &=J_{\nu,2n+1}\end{aligned} \tag{6·5}$$

for $0\leqslant\nu\leqslant n$, and, in particular, $B_{n,n}=J_{n,2n+1}$. This leads to the following interpretation of (6·3):

(6·6) THEOREM. *Let $f(x)$ be bounded and periodic, and let $B_{n,\nu}$ be the arithmetic mean of the polynomials $I_{n,0}, I_{n,1}, \ldots, I_{n,\nu}$. Then*

(i) *$B_{n,\nu}(x)$ remains within the same bounds as f;*

(ii) *$B_{n,\nu}(x,f)$ converges to $f(x)$ at every point of continuity of f as $\nu\to\infty$. The convergence is uniform in every closed interval of continuity of f.*

At the points of simple discontinuity of f the behaviour of the $J_{n,m}$ is in general indeterminate, as is that of the $I_{n,\nu}$.

Consider now any polynomial $S(x)$ of order n. The values which the derivative $S'(x)$ takes at the points $t_0, t_1, \ldots, t_n$ are not arbitrary, since

$$\frac{2\pi}{n+1}\sum_0^n S'(t_k)=\int_0^{2\pi} S'(x)\,d\omega_{n+1}(x)=\int_0^{2\pi} S'(x)\,dx=0. \tag{6·7}$$

However, *given any two sets of numbers*

$$y_0,\quad y_1,\quad \ldots,\quad y_n;\quad y_0',\quad y_1',\quad \ldots,\quad y_n',$$

such that

$$y_0'+y_1'+\ldots+y_n'=0$$

(so that the number of conditions is $2n+1$), *there is a unique polynomial $S(x)$ of order n such that*

$$S(t_k)=y_k,\quad S'(t_k)=y_k'\quad (k=0,1,\ldots,n).$$

The uniqueness follows from the fact that the difference of two such $S(x)$ would have at least $n+1$ double roots, and so would vanish identically. To prove the existence of S observe that

$$\tilde{D}^*_{n+1}(u)=\sum_1^n \sin\nu u+\tfrac{1}{2}\sin(n+1)u=\{1-\cos(n+1)u\}\tfrac{1}{2}\cot\tfrac{1}{2}u \tag{6·8}$$

is a polynomial of order $n+1$ which vanishes at the points $u=2\pi\nu/(n+1)$, $\nu=0,1,\ldots,n$, and has derivative zero at each of these points with the exception of $u=0$ $(\nu=0)$, where its derivative is $\frac{1}{2}(n+1)^2$. Thus

$$H_n(x)=2(n+1)^{-2}\sum_{k=0}^n y_k'\tilde{D}^*_{n+1}(x-t_k) \tag{6·9}$$

is a polynomial vanishing at the points t_k and having derivatives y_k' there. Since

$$\sum_0^n y_k'\sin(n+1)(x-t_k)=\sin(n+1)(x-t_0)\,.\sum_0^n y_k'=0,$$

we may replace $\tilde{D}^*_{n+1}$ by $\tilde{D}_n$ in (6·9), so that H_n is a polynomial of order n.

We can now define $S(x)$ as $H_n(x)+J_n(x)$, where

$$H_n(x)=2(n+1)^{-2}\sum_{k=0}^n y_k'\tilde{D}_n(x-t_k) \tag{6·10}$$

and $J_n(x)$ is the Jackson polynomial taking the values y_k at the points t_k (and having zero derivative there). Clearly $S(x)$ has the required properties.

The polynomials $H_n(x)$, which vanish at the points t_k and whose derivatives have given values there, are called the *interpolating polynomials of the second kind*.

The determination of a polynomial by means of the values which it and some of its derivatives take at fixed points is called *Hermite interpolation* (as distinct from *Lagrange interpolation*, where the derivatives are not considered). Thus the polynomials J_n and

H_n give the solution of the problem when the $n+1$ fundamental points $t_k^{(n)}$ are equally spaced and only the first derivative is taken into account.

Return to (6·9) and let
$$Y' = \max(|y_0'|, |y_1'|, \ldots, |y_n'|).$$

Then, by (6·10) and the Remark on p. 19,

$$|H_n(x)| \leqslant \frac{Y'}{(n+1)\pi} \int_0^{2\pi} |\tilde{D}_n(t-x)|\, d\omega_{n+1}(t)$$
$$\leqslant Y'.O(n^{-1}\log n), \qquad (6·11)$$

and we have now:

(6·12) THEOREM. *Let $f(x)$ be any bounded periodic function and let $S_n(x)$ denote the polynomial of order n such that*
$$S(t_k^{(n)}) = f(t_k^{(n)}), \quad S'(t_k^{(n)}) = y_{n,k}'$$
for $k=0, 1, \ldots, n$, where (for each n) $y_{n,0}'\, y_{n,1}' \ldots, y_{n,n}'$ are arbitrary numbers with sum equal to 0. If
$$Y_n' = \max(|y_{n,0}'|, |y_{n,1}'|, \ldots, |y_{n,n}'|) = o(n/\log n),$$
then, as $n \to \infty$, $S_n(x)$ tends to $f(x)$ at every point of continuity of f, and the convergence is uniform over each closed interval of continuity.

For $S_n = J_n + H_n$, where by (6·11), H_n tends uniformly to 0 and $J_n = J_n[f]$ satisfies the conclusions of (6·3).

The case $Y_n' = O(1)$ is of special interest.

We now apply (6·6) to the proof of (5·4). In Chapter III, § 1, we proved the theorem of Hardy, that if a series $u_0 + u_1 + \ldots$ is summable (C, 1) to s, and if $u_\nu = O(1/\nu)$, then the series converges to s. Suppose now that we have a family of finite sums
$$u_0^{(n)} + u_1^{(n)} + \ldots + u_n^{(n)} \qquad (6·13)$$
depending on the parameter n, and that
$$|u_\nu^{(n)}| \leqslant A/\nu,$$
where A is independent of n and ν. Let $s_{n,\nu}$ and $\sigma_{n,\nu}$ denote respectively the partial sums and the first arithmetic means of the series (6·13). Then, *if $\sigma_{n,\nu}$ tends to s as $\nu \to \infty$, n remaining always $\geqslant \nu$, so does $s_{n,\nu}$; if the $\sigma_{n,\nu}$ are bounded so are the $s_{n,\nu}$.*

The second part is an immediate consequence of the equation
$$s_{n,\nu} - \sigma_{n,\nu} = \frac{1}{\nu+1} \sum_1^\nu k u_k^{(n)}.$$

The proof of the first part is actually implicit in the proof of Hardy's theorem, as the reader will easily verify. Thus (5·4) follows from (6·6), if we use (4·8).

It is important to observe that the analogue of Fejér's theorem about the summability (C, 1) of $S[f]$ is obtained by considering the means (6·5), and not the means of the polynomials $I_0, I_1, \ldots$ (see, however, theorem (7·32) below). Although $I_n[f]$ represents a complete Fourier series with respect to a certain orthogonal system, there is little connexion between I_n and I_{n+1}.

(6·14) THEOREM. *Let $x_0^n < x_1^n < \ldots < x_{2n}^n$ be any $2n+1$ equidistributed points, and let $h = h_n = 2\pi/(2n+1)$. Let $U_n(x, f)$ be the polynomial of order n taking at the points x_ν^n the values*
$$\tfrac{1}{2}\{f(x_\nu^n + \tfrac{1}{2}h) + f(x_\nu^n - \tfrac{1}{2}h)\}.$$

Then $U_n(x, f)$ tends to f at every point of continuity of f, and the convergence is uniform over every closed arc of continuity.

Observing that $\omega_{2n+1}(x+\frac{1}{2}h)-\omega_{2n+1}(x-\frac{1}{2}h)$ is constant, we have

$$U_n(x, f) = \frac{1}{\pi}\int_0^{2\pi} \tfrac{1}{2}\{f(t+\tfrac{1}{2}h)+f(t-\tfrac{1}{2}h)\}\, D_n(t-x)\, d\omega_{2n+1}(t)$$

$$= \frac{1}{\pi}\int_0^{2\pi} f(t)\, \tfrac{1}{2}\{D_n(t-\tfrac{1}{2}h-x)+D_n(t+\tfrac{1}{2}h-x)\}\, d\omega_{2n+1}(t+\tfrac{1}{2}h)$$

$$= \tfrac{1}{2}a_0^{(n)} + \sum_{\nu=1}^{n} (a_\nu^{(n)} \cos \nu x + b_\nu^{(n)} \sin \nu x) \cos \tfrac{1}{2}\nu h,$$

where $a_\nu^{(n)}$, $b_\nu^{(n)}$ are the Fourier-Lagrange coefficients of f corresponding to the fundamental points $x_\nu^n+\frac{1}{2}h$. The last expression closely resembles the expression

$$\tfrac{1}{2}\{S_n(x+\tfrac{1}{2}h)+S_n(x-\tfrac{1}{2}h)\}$$

for the partial sums of Fourier series (Chapter III, § 12), and the two have similar properties. Summing by parts we find $U_n(x, f)$ equal to

$$\sum_{\nu=0}^{n-1} I_{n,\nu}\Delta \cos \tfrac{1}{2}\nu h + I_n \cos\frac{\pi n}{2n+1}$$

$$= \sum_{\nu=0}^{n-2} B_{n,\nu}(\nu+1)\,\Delta^2 \cos \tfrac{1}{2}\nu h + B_{n,n-1}\, n\Delta \cos \tfrac{1}{2}(n-1)\, h + I_n \cos\frac{\pi n}{2n+1}.$$

Disregarding the last term, which is $O(\log n)\, O(1/n) = o(1)$, we have here a transformation of the double sequence $B_{n,\nu}$ by means of a regular matrix (Chapter III, § 1), and (6·14) follows from (6·6).

It is also easy to verify that for any fixed f the U_n are uniformly bounded.

For the polynomial $I_n[f]$ the ratio of the order of the polynomial to the number of fundamental points is $n/(2n+1) \simeq \frac{1}{2}$. $I_n[f]$ need not tend to f, even if f is continuous; we may, however, remedy this defect if we allow the asymptotic value of the ratio of the order to the number of points to increase, no matter how slightly. (For the polynomials $J_n[f]$, which tend uniformly to f when f is continuous, the ratio is $n/(n+1) \simeq 1$.) This can be achieved either by keeping fixed the number of fundamental points and increasing the order of the polynomial, or, conversely, by keeping the order fixed and dropping a number of fundamental points. Both procedures will be considered here.

(6·15) THEOREM. *Given any $\epsilon > 0$ and any bounded periodic f, we can define a polynomial of order not exceeding $n(1+\epsilon)$ interpolating f at any equidistant points $x_0, x_1, \ldots, x_{2n}$ and converging to f at every point of continuity of f, and uniformly in every closed interval of continuity.*

Let $0 < h \leq n$. The required polynomial will be defined as

$$L_{n,h} = \frac{1}{2h+1} \sum_{\nu=n-h}^{n+h} I_{n,\nu}(x),$$

where
$$I_{n,\nu}(x) = \frac{1}{\pi}\int_0^{2\pi} f(t)\, D_\nu(t-x)\, d\omega_{2n+1}.$$

The $L_{n,h}$ are analogous to the delayed (C, 1) means introduced in Chapter III, p. 80, and

$$L_{n,h}=\frac{1}{2h+1}\left(\sum_{\nu=0}^{n+h}-\sum_{\nu=0}^{n-h-1}\right)I_{n,\nu}$$
$$=\frac{1}{2h+1}\{(n+h+1)\,J_{n+h,\,2n+1}-(n-h)\,J_{n-h-1,\,2n+1}\}.$$

If $h=[n\epsilon]$, it follows from (6·3) that $J_{n+h,\,2n+1}\to f$, $J_{n-h-1,\,2n+1}\to f$, and so also

$$L_{n,h}\to f.$$

The order of $L_{n,h}$ is $n+h\leqslant n(1+\epsilon)$. Finally,

$$L_{n,h}=\frac{1}{2h+1}\,\frac{1}{\pi}\int_0^{2\pi}\left\{\frac{1-\cos(n+h+1)(x-t)}{4\sin^2\frac{1}{2}(x-t)}-\frac{1-\cos(n-h)(x-t)}{4\sin^2\frac{1}{2}(x-t)}\right\}f(t)\,d\omega_{2n+1}(t)$$
$$=\frac{2}{2h+1}\,\frac{1}{\pi}\int_0^{2\pi}D_n(t-x)\,D_h(t-x)f(t)\,d\omega_{2n+1}(t)$$
$$=\frac{2}{2h+1}\,\frac{2}{2n+1}\sum_{k=0}^{2n}D_n(x-x_k)\,D_h(x-x_k)f(x_k),$$

so that
$$L_{n,h}(x_k)=f(x_k)$$
for all k, and (6·15) follows.

The polynomial of order n taking the values $y_0, y_1, \ldots, y_{2n}$ at the equidistant points $x_0<x_1<\ldots<x_{2n}$ may be written

$$\frac{\sin(n+\frac{1}{2})(x-x_0)}{2n+1}\left\{\sum_{j=0}^{2n}\frac{(-1)^j y_j}{\sin\frac{1}{2}(x-x_j)}\right\} \tag{6·16}$$

(see (1·17)). Let $2l$ be a fixed even integer and split the points $x_0, x_1, \ldots, x_{2n}$ into consecutive blocks of $2l$ elements and a terminal block of less than $2l$ elements. The number of full blocks is $[(2n+1)/2l]$. Given a periodic $f(x)$ we consider the polynomial (6·16), where the y_j are defined as follows. If x_j is in the terminal block we set $y_j=f(x_j)$. For all the other x_j, except for one in each block, we also set $y_j=f(x_j)$. Finally, if x_{i+1} is the first element of any block, we define the y_j corresponding to the exceptional x_j of that block by the relation

$$y_{i+1}-y_{i+2}+y_{i+3}-\ldots-y_{i+2l}=0. \tag{6·17}$$

We shall call the polynomial (6·16) defined in this manner a *2l-adjusted* interpolating polynomial. It interpolates f at least at

$$m_n=2n+1-[(2n+1)/2l]\geqslant(2n+1)(1-1/2l)$$

points. Hence by taking l large enough we may make the limit of n/m_n as close to $\frac{1}{2}$ as we please.

(6·18) THEOREM. *For a fixed l, the $2l$-adjusted interpolating polynomial of any bounded and periodic f converges to f at the points of continuity of f. The convergence is uniform over every closed interval of continuity.*

The difference between the expression in (6·16) and $f(x)$ is

$$\frac{\sin(n+\frac{1}{2})(x-x_0)}{2n+1}\left\{\sum_{j=0}^{2n}\frac{(-1)^j[y_j-f(x)]}{\sin\frac{1}{2}(x-x_j)}\right\}=\Sigma'+\Sigma'',$$

say, where Σ' corresponds to the x_j in $(x, x+\pi)$ and Σ'' to the other x_j. Let p be a fixed integer, and let $\Sigma' = \sigma' + \sigma''$, where σ' corresponds to the p points x_j nearest to x. If f is continuous at x, each $y_j - f(x)$ in σ' tends to 0 as $n \to \infty$. Since $|D_n(u)| \leqslant n + \frac{1}{2}$, $\sigma' \to 0$. Suppose that $|f| \leqslant M$. By virtue of (6·17), the absolute value of the partial sums of $y_k - y_{k+1} + y_{k+2} - \dots$ does not exceed $4lM$, whence (cf. Chapter I, (2·3))

$$|\sigma''| \leqslant \frac{4l+1}{2n+1} M \frac{1}{\sin \dfrac{p\pi}{2n+1}} \leqslant \frac{4l+1}{2p} M,$$

which is arbitrarily small if p is sufficiently large. Thus $\Sigma' \to 0$. Similarly $\Sigma'' \to 0$, and (6·18) follows.

7. Mean convergence of interpolating polynomials

In the study of mean convergence we shall consider not only the interpolating polynomials I_n and their partial sums $I_{n,\nu}$, but also the conjugate polynomials $\tilde{I}_n$, $\tilde{I}_{n,\nu}$. It will appear in the sequel that the latter tend in the mean to the conjugate function

$$\tilde{f}(x) = -\frac{1}{\pi} \int_0^\pi [f(x+t) - f(x-t)] \tfrac{1}{2} \cot \tfrac{1}{2} t \, dt = -\frac{1}{\pi} \lim_{\epsilon \to +0} \int_\epsilon^\pi,$$

although, in general, $\tilde{I}_n$ is not an interpolating polynomial for $\tilde{f}$.

Since every f integrable R is bounded, and so is in L^2, the proof of the existence of $\tilde{f}$ almost everywhere is particularly simple (see Chapter IV, § 1). However, $\tilde{f}$ can be unbounded even if f is continuous, and so need not be integrable R. We know that $\tilde{f}$ belongs to every L^p, where $p < \infty$; more generally, $\exp(\lambda |\tilde{f}|)$ is integrable if λ is positive and small enough (Chapter VII, (2·4), (2·11)).

In what follows we confine our attention to polynomials I_n. The argument is applicable without change to the polynomials E_n.

(7·1) Theorem. *For every f,*

$$\left.\begin{aligned} \int_0^{2\pi} I_{n,\nu}^2 \, dx &\leqslant \int_0^{2\pi} f^2 \, d\omega_{2n+1}, \\ \int_0^{2\pi} \tilde{I}_{n,\nu}^2 \, dx &\leqslant \int_0^{2\pi} f^2 \, d\omega_{2n+1}. \end{aligned}\right\} \tag{7·2}$$

Further,

$$\left.\begin{aligned} \lim_{\nu \to \infty} \int_0^{2\pi} (f - I_{n,\nu}[f])^2 \, dx &= 0, \\ \lim_{\nu \to \infty} \int_0^{2\pi} (\tilde{f} - \tilde{I}_{n,\nu}[f])^2 \, dx &= 0. \end{aligned}\right\} \tag{7·3}$$

By Parseval's formula,

$$\mathfrak{M}_2[I_{n,\nu}] \leqslant \mathfrak{M}_2[I_n], \quad \mathfrak{M}_2[\tilde{I}_{n,\nu}] \leqslant \mathfrak{M}_2[I_n].$$

Since

$$\int_0^{2\pi} I_n^2 \, dx = \int_0^{2\pi} I_n^2 \, d\omega_{2n+1} = \int_0^{2\pi} f^2 \, d\omega_{2n+1},$$

(7·2) follows.

If f is integrable R,

$$\mathfrak{M}_2[f-I_{n,\nu}] \leqslant \mathfrak{M}_2[f]+\mathfrak{M}_2[I_{n,\nu}],$$

$$\leqslant \mathfrak{M}_2[f]+\left(\int_0^{2\pi} f^2\,d\omega_{2n+1}\right)^{\frac{1}{2}} \to 2\mathfrak{M}_2[f],$$

since f^2, along with f, is integrable R. Thus

$$\limsup_{\nu\to\infty} \mathfrak{M}_2[f-I_{n,\nu}] \leqslant 2\mathfrak{M}_2[f].$$

Let now T be a polynomial such that $\mathfrak{M}_2[f-T]<\epsilon$, and let $f=f_1+T$. If ν exceeds the order of T,

$$I_{n,\nu}[f]=I_{n,\nu}[f_1]+I_{n,\nu}[T]=I_{n,\nu}[f_1]+T,$$

whence

$$\mathfrak{M}_2[f-I_{n,\nu}[f]]=\mathfrak{M}_2[f_1-I_{n,\nu}[f_1]],$$

$$\limsup \mathfrak{M}_2[f-I_{n,\nu}[f]] \leqslant 2\mathfrak{M}_2[f_1]<2\epsilon,$$

and the first equation (7·3) follows. The second equation is a consequence of the inequality

$$\mathfrak{M}_2[\tilde{f}-\tilde{I}_{n,\nu}] \leqslant \mathfrak{M}_2[f-I_{n,\nu}].$$

(7·4) THEOREM. *For any f and any finite interval (α,β),*

$$\lim_{\nu\to\infty}\int_\alpha^\beta I_{n,\nu}[f]\,dx=\int_\alpha^\beta f\,dx,$$

$$\lim_{\nu\to\infty}\int_\alpha^\beta \tilde{I}_{n,\nu}[f]\,dx=\int_\alpha^\beta \tilde{f}\,dx.$$

For the absolute value of the difference of the first two integrals does not exceed

$$\mathfrak{M}[f-I_{n,\nu};\,\alpha,\beta] \leqslant \mathfrak{M}[f-I_{n,\nu}] \leqslant (2\pi)^{\frac{1}{2}}\,\mathfrak{M}_2[f-I_{n,\nu}]\to 0,$$

assuming, as we may, that (α,β) is of length at most 2π. The proof of the second formula is similar. Both are analogues of the fact that $\mathsf{S}[f]$ can be integrated termwise over any interval (α,β), the result being $\int_\alpha^\beta f\,dx$.

The exponent 2 in (7·1) can be replaced by any number $p>0$, at the cost of a numerical factor. The proof is based on the following theorem, interesting in itself:

(7·5) THEOREM. *Let $S(x)$ be a polynomial of order n. Then*

$$\left\{\int_0^{2\pi} |S|^p\,d\omega_{2n+1}\right\}^{1/p} \leqslant A\left\{\int_0^{2\pi} |S|^p\,dx\right\}^{1/p} \quad (1\leqslant p\leqslant +\infty), \tag{7·6}$$

$$\left\{\int_0^{2\pi} |S|^p\,dx\right\}^{1/p} \leqslant A_p\left\{\int_0^{2\pi} |S|^p\,d\omega_{2n+1}\right\}^{1/p} \quad (1<p<+\infty), \tag{7·7}$$

where A is an absolute constant, and A_p depends on p only.

That (7·7) is false for $p=\infty$ (unlike (7·6), which is then obvious) follows from the fact, to be proved in § 8, that there is a continuous f such that the $S(x)=I_n(x,f)$ are unbounded although, of course, they are bounded at the fundamental points. For $p=1$, if $S(x)=D_n(x)$ and ω_{2n+1} is associated with the roots of $\sin(n+\frac{1}{2})x$, the integral on the right of (7·7) is $2\pi D_n(0)/(2n+1)=\pi$, while $\mathfrak{M}[S]=\pi L_n$ is unbounded as $n\to\infty$, L_n being the Lebesgue constant. Thus (7·7) fails also for $p=1$.

We prove first that for any function $\Phi(u)$, non-negative, non-decreasing and convex in $u \geqslant 0$,

$$\int_0^{2\pi} \Phi(\tfrac{1}{3} | S |) \, d\omega_{2n+1} \leqslant \int_0^{2\pi} \Phi(| S |) \, dx. \tag{7·8}$$

The inequality (7·6) with $A = 3$ is an immediate consequence of (7·8) with $\Phi(u) = u^p$.

In the formula

$$S(x) = \frac{1}{\pi} \int_0^{2\pi} S(t) \, D_n(t-x) \, dt$$

we may replace D_n by any polynomial whose nth partial sum is D_n. In particular, we may replace D_n by

$$2K_{2n-1} - K_{n-1},$$

K_ν being the Fejér kernel (see Chapter III, (1·31)). Hence we have

$$\tfrac{1}{3} | S(x) | \leqslant \tfrac{2}{3} \frac{1}{\pi} \int_0^{2\pi} | S(t) | \, K_{2n-1}(t-x) \, dt + \tfrac{1}{3} \frac{1}{\pi} \int_0^{2\pi} | S(t) | \, K_{n-1}(t-x) \, dt,$$

$$\Phi(\tfrac{1}{3} | S(x) |) \leqslant \tfrac{2}{3} \Phi\left\{\frac{1}{\pi} \int_0^{2\pi} | S(t) | \, K_{2n-1}(t-x) \, dt\right\} + \tfrac{1}{3} \Phi\left\{\frac{1}{\pi} \int_0^{2\pi} | S(t) | \, K_{n-1}(t-x) \, dt\right\},$$

$$\Phi(\tfrac{1}{3} | S(x) |) \leqslant \tfrac{2}{3} \frac{1}{\pi} \int_0^{2\pi} \Phi(| S(t) |) \, K_{2n-1}(t-x) \, dt + \tfrac{1}{3} \frac{1}{\pi} \int_0^{2\pi} \Phi(| S(t) |) \, K_{n-1}(t-x) \, dt,$$

using Jensen's inequality twice. If, after integrating the last inequality with respect to $\omega_{2n+1}(x)$ over $0 \leqslant x \leqslant 2\pi$, we interchange the order of integration (or rather of summation and integration) on the right and use the fact that

$$\frac{1}{\pi} \int_0^{2\pi} K_m(t-x) \, d\omega_{2n+1}(x) = \frac{1}{\pi} \int_0^{2\pi} K_m(t-x) \, dx = 1$$

for $m \leqslant 2n$, we get (7·8).

For the proof of (7·7) we take a function g such that

$$\mathfrak{M}_p[S] = \int_0^{2\pi} S g \, dx, \quad \mathfrak{M}_{p'}[g] = 1,$$

where $p' = p/(p-1)$ (see Chapter I, (9·14)). Then, by Hölder's inequality and (7·6),

$$\begin{aligned} \mathfrak{M}_p[S] = \int_0^{2\pi} S g \, dx = \int_0^{2\pi} S S_n[g] \, dx &= \int_0^{2\pi} S S_n[g] \, d\omega_{2n+1} \\ &\leqslant \left(\int_0^{2\pi} | S |^p \, d\omega_{2n+1}\right)^{1/p} \left(\int_0^{2\pi} | S_n[g] |^{p'} \, d\omega_{2n+1}\right)^{1/p'} \\ &\leqslant A\left(\int_0^{2\pi} | S |^p \, d\omega_{2n+1}\right)^{1/p} \mathfrak{M}_{p'}[S_n[g]]. \end{aligned} \tag{7·9}$$

By theorem (6·4) of Chapter VII, there is a constant $R_{p'}$ independent of n and g, such that the last factor does not exceed $R_{p'} \mathfrak{M}_{p'}[g] = R_{p'}$†. Hence the last product in (7·9) does not exceed $A R_{p'} \left(\int_0^{2\pi} | S |^p \, d\omega_{2n+1}\right)^{1/p}$. This gives (7·7) with

$$A_p = A R_{p'}.$$

† The constant R_p is denoted there by C_p. Arguing as in the proof of Theorem (2·21) of Chapter VII, we can show that the least value of R_p satisfies $R_p = R_{p'}$.

Remark. Let $\epsilon>0$, and let N be any integer, even or odd, not less than $(1+\epsilon)n$. *The inequality* (7·6) *holds if we replace* $d\omega_{2n+1}$ *by* $d\omega_N$, *but the constant* A *on the right will now depend on* ϵ. This is a consequence of the following generalization of (7·8):

$$\int_0^{2\pi}\Phi(\lambda^{-1}|S|)\,d\omega_N\leqslant\int_0^{2\pi}\Phi(|S|)\,dx\quad(N\geqslant(1+\epsilon)n,\ \lambda=2+\epsilon^{-1}).$$

To prove this inequality we proceed as in the proof of (7·8) but replace D_n by

$$K_{n,h}=\left(1+\frac{n}{h}\right)K_{n+h-1}-\frac{n}{h}K_{n-1}.$$

It is enough to suppose that $h\geqslant\epsilon n$ and set $N=n+h$.

The following is an analogue of (7·5) for power polynomials:

(7·10) THEOREM. *Let*

$$P(z)=c_0+c_1z+\ldots+c_nz^n.$$

Then

$$\left\{\int_0^{2\pi}|P(e^{it})|^p\,d\omega_{n+1}\right\}^{1/p}\leqslant A'\left\{\int_0^{2\pi}|P(e^{it})|^p\,dt\right\}^{1/p}\quad(1\leqslant p\leqslant+\infty),\tag{7·11}$$

$$\left\{\int_0^{2\pi}|P(e^{it})|^p\,dt\right\}^{1/p}\leqslant A_p'\left\{\int_0^{2\pi}|P(e^{it})|^p\,d\omega_{n+1}\right\}^{1/p}\quad(1<p<\infty).\tag{7·12}$$

This follows immediately from (7·5) if $n=2k$ is even. For then $|P(e^{it})|=|S(t)|$, where

$$S(t)=c_0e^{-ikt}+c_1e^{-i(k-1)t}+\ldots+c_ne^{ikt}$$

is a (complex-valued) trigonometric polynomial of order k.

If n is odd, write $P(z)=c_0+zQ(z)$ and denote $P(e^{it})$ and $Q(e^{it})$ by P and Q respectively. Observing that

$$|P|\leqslant|c_0|+|Q|,\quad|Q|\leqslant|c_0|+|P|,$$

and that Q is a power polynomial of even degree $n-1$, we get

$$\left(\frac{1}{2\pi}\int_0^{2\pi}|P|^p\,d\omega_{n+1}\right)^{1/p}\leqslant|c_0|+\left(\frac{1}{2\pi}\int_0^{2\pi}|Q|^p\,d\omega_{n+1}\right)^{1/p}$$

$$\leqslant|c_0|+A\mathfrak{A}_p[Q]\leqslant|c_0|+A|c_0|+A\mathfrak{A}_p[P].\tag{7·13}$$

Since

$$|c_0|=(2\pi)^{-1}\left|\int_0^{2\pi}P\,dt\right|\leqslant\mathfrak{A}_p[P],$$

(7·11) with $A'=2A+1$ follows immediately. An argument parallel to this, using the fact that $c_0=(2\pi)^{-1}\int_0^{2\pi}P\,d\omega_{n+1}$, gives (7·12) with $A_p'=2A_p+1$.

(7·14) THEOREM. *For every* f, *and for* $p>1$,

$$\mathfrak{M}_p[I_{n,\nu}]\leqslant A_p\left(\int_0^{2\pi}|f|^p\,d\omega_{2n+1}\right)^{1/p},\tag{7·15}$$

$$\mathfrak{M}_p[\tilde{I}_{n,\nu}]\leqslant A_p\left(\int_0^{2\pi}|f|^p\,d\omega_{2n+1}\right)^{1/p},\tag{7·16}$$

$$\mathfrak{M}_p[f-I_{n,\nu}]\to0,\quad\mathfrak{M}_p[\tilde{f}-\tilde{I}_{n,\nu}]\to0\quad(n\geqslant\nu\to\infty).\tag{7·17}$$

The inequalities $$\mathfrak{M}_p[I_{n,\nu}] \leqslant A_p \mathfrak{M}_p[I_n],$$

$$\mathfrak{M}_p[I_n] \leqslant A_p \left\{ \int_0^{2\pi} |I_n|^p \, d\omega_{2n+1} \right\}^{1/p} = A_p \left\{ \int_0^{2\pi} |f|^p \, d\omega_{2n+1} \right\}^{1/p}$$

give (7·15). The proof of (7·16) is similar. Once (7·15) and (7·16) are established, the proof of (7·17) follows the pattern of that of (7·3).

For a generalization of (7·14) we need the following result:

(7·18) THEOREM. *Let $S(x)$ be a polynomial of order n such that $|S(x)| \leqslant 1$ at the discontinuities of ω_{2n+1}. There are two positive absolute constants λ_0, μ_0 such that*

$$\int_0^{2\pi} \exp(\lambda_0 |S_\nu(x)|) \, dx \leqslant \mu_0, \tag{7·19}$$

$$\int_0^{2\pi} \exp(\lambda_0 |\tilde{S}_\nu(x)|) \, dx \leqslant \mu_0, \tag{7·20}$$

for the partial sums S_ν of S.

We first strengthen (7·7) by showing that

$$\mathfrak{M}_p[S_\nu] \leqslant A_p \left(\int_0^{2\pi} |S|^p \, d\omega_{2n+1} \right)^{1/p}, \tag{7·21}$$

$$\mathfrak{M}_p[\tilde{S}_\nu] \leqslant A_p \left(\int_0^{2\pi} |S|^p \, d\omega_{2n+1} \right)^{1/p}, \tag{7·22}$$

for $p > 1$, with A_p satisfying the same inequality $A_p \leqslant AR_{p'}$ as in (7·7). The proofs are similar to that of (7·7). We take a $g(x)$ satisfying $\mathfrak{M}_{p'}[g] = 1$ and such that

$$\mathfrak{M}_p[S_\nu] = \int_0^{2\pi} S_\nu g \, dx.$$

Then
$$\int_0^{2\pi} S_\nu g \, dx = \int_0^{2\pi} S_\nu S_\nu[g] \, dx = \int_0^{2\pi} S_\nu S_\nu[g] \, d\omega_{2n+1} = \int S S_\nu[g] \, d\omega_{2n+1}$$
$$\leqslant \left(\int_0^{2\pi} |S|^p \, d\omega_{2n+1} \right)^{1/p} \left\{ \int_0^{2\pi} |S_\nu[g]|^{p'} \, d\omega_{2n+1} \right\}^{1/p'}$$
$$\leqslant \left(\int_0^{2\pi} |S|^p \, d\omega_{2n+1} \right)^{1/p} A \left\{ \int_0^{2\pi} |S_\nu[g]|^{p'} \, dx \right\}^{1/p'},$$

and, as before, the last factor does not exceed $R_{p'}$. This proves (7·21).

Similarly, we choose an $h(x)$ with $\mathfrak{M}_{p'}[h] = 1$ and $\mathfrak{M}_p[\tilde{S}_\nu] = \int_0^{2\pi} \tilde{S}_\nu h \, dx$. This last integral is equal to $-\int_0^{2\pi} S_\nu \tilde{h} \, dx$, and arguing as before we get (7·22).

The constants A_p in (7·21) and (7·22) do not exceed Ap when $p \geqslant 2$. (Going through the proof of theorem (6·7) of Chapter VII, we see that our A_p does not exceed a fixed multiple, independent of p, of the constant A_p in Chapter VII, (2·5), and it is enough to apply Chapter VII, (2·21) and (3·8)). Thus

$$\int_0^{2\pi} \cosh \lambda |S_\nu| \, dx = 2\pi + \sum_{j=1}^{\infty} \frac{\lambda^{2j}}{(2j)!} \int_0^{2\pi} |S_\nu|^{2j} \, dx$$
$$\leqslant 2\pi + \sum_{j=1}^{\infty} \frac{\lambda^{2j}}{(2j)!} (A \,.\, 2j)^{2j} \int_0^{2\pi} |S|^{2j} \, d\omega_{2n+1}$$
$$\leqslant 2\pi + \sum_{j=1}^{\infty} \frac{\lambda^{2j}}{(2j)!} (2Aj)^{2j} \, 2\pi < \infty,$$

provided that $\lambda A < e^{-1}$ (using the inequality $n^n \leqslant n!\, e^n$). Since $e^{|u|} \leqslant 2\cosh u$, this proves (7·19). The proof of (7·20) is similar.

(7·23) THEOREM. *There are positive absolute constants λ_0, μ_0, λ such that for every f with $|f| \leqslant 1$,*

$$\int_0^{2\pi} \exp \lambda_0 |I_{n,\nu}|\, dx \leqslant \mu_0, \quad \int_0^{2\pi} \exp \lambda_0 |\tilde{I}_{n,\nu}|\, dx \leqslant \mu_0, \tag{7·24}$$

$$\frac{1}{2\pi}\int_0^{2\pi} \exp \lambda |f - I_{n,\nu}|\, dx \to 1, \quad \frac{1}{2\pi}\int_0^{2\pi} \exp \lambda |\tilde{f} - \tilde{I}_{n,\nu}|\, dx \to 1. \tag{7·25}$$

If f is continuous, (7·25) *holds for every $\lambda > 0$.*

The inequalities (7·24) are consequences of (7·18). Since $0 \leqslant e^u - 1 \leqslant u e^u$ for $u \geqslant 0$,

$$0 \leqslant \frac{1}{2\pi}\int_0^{2\pi} \{e^{\lambda|f - I_{n,\nu}|} - 1\}\, dx \leqslant \frac{\lambda}{2\pi}\int_0^{2\pi} |f - I_{n,\nu}|\, e^{\lambda|f - I_{n,\nu}|}\, dx$$
$$\leqslant \frac{\lambda}{2\pi}\mathfrak{M}_2[f - I_{n,\nu}]\left\{\int_0^{2\pi} e^{2\lambda|f - I_{n,\nu}|}\, dx\right\}^{\frac{1}{2}},$$

by Schwarz's inequality. Since $\mathfrak{M}_2[f - I_{n,\nu}] \to 0$, by (7·1), and since

$$\int_0^{2\pi} e^{2\lambda|f - I_{n,\nu}|}\, dx \leqslant e^{2\lambda}\int_0^{2\pi} e^{2\lambda|I_{n,\nu}|}\, dx = O(1)$$

for $2\lambda \leqslant \lambda_0$, we obtain the first formula (7·25). The proof of the second formula (7·25) is similar except that here we have (Chapter VII, (2·11))

$$\int_0^{2\pi} e^{2\lambda|\tilde{f} - \tilde{I}_{n,\nu}|}\, dx \leqslant \left(\int_0^{2\pi} e^{4\lambda|\tilde{f}|}\, dx\right)^{\frac{1}{2}} \left(\int_0^{2\pi} e^{4\lambda|\tilde{I}_{n,\nu}|}\, dx\right)^{\frac{1}{2}} = O(1),$$

if λ is small enough.

If f is continuous, let T be a polynomial such that $|f - T| < \epsilon$ and let $f = f_1 + T$. If ν exceeds the order of T and $2\lambda\epsilon \leqslant \lambda_0$, then

$$\int_0^{2\pi} e^{2\lambda|f - I_{n,\nu}[f]|}\, dx = \int_0^{2\pi} e^{2\lambda|f_1 - I_{n,\nu}[f_1]|}\, dx$$
$$\leqslant e^{2\lambda\epsilon}\int_0^{2\pi} e^{2\lambda|I_{n,\nu}[f_1]|}\, dx = O(1).$$

Since ϵ is arbitrarily small, λ may be as large as we please. Similarly for the $\tilde{I}_{n,\nu}$.

The inequalities (7·24) may be considered as substitutes for (7·7) when $p = \infty$. The following two theorems, which we state without proof, are substitutes for (7·7), and for its analogue for $\tilde{S}$, when $p = 1$.

(7·26) THEOREM. *For any polynomial S of order n,*

$$\int_0^{2\pi} |S|\, dx \leqslant A\int_0^{2\pi} |S| \log^+ |S|\, d\omega_{2n+1} + A,$$

$$\int_0^{2\pi} |\tilde{S}|\, dx \leqslant A\int_0^{2\pi} |S| \log^+ |S|\, d\omega_{2n+1} + A,$$

where A is a positive absolute constant.

(7·27) THEOREM. *For any S of order n and for $0<\mu<1$,*

$$\left(\int_0^{2\pi} |S|^\mu dx\right)^{1/\mu} \leqslant A_\mu \int_0^{2\pi} |S|\, d\omega_{2n+1},$$

$$\left(\int_0^{2\pi} |\tilde{S}|^\mu dx\right)^{1/\mu} \leqslant A_\mu \int_0^{2\pi} |S|\, d\omega_{2n+1},$$

where A_μ depends on μ only.

The proofs are based on Theorems (2·8) and (2·6) of Chapter VII.

(7·28) THEOREM. *Let $\delta>0$, $m\geqslant(1+\delta)\,2n$. Then for any polynomial S of order n we have*

$$\int_0^{2\pi} |S|\, dx \leqslant A_\delta \int_0^{2\pi} |S|\, d\omega_m, \tag{7·29}$$

$$\max |S(x)| \leqslant A_\delta \max |S(x_k)|, \tag{7·30}$$

where x_k are the discontinuities of ω_m, and A_δ depends on δ only.

Let

$$K_{n,h} = \frac{D_n + D_{n+1} + \ldots + D_{n+h-1}}{h} = \frac{(n+h)\,K_{n+h-1} - nK_{n-1}}{h}. \tag{7·31}$$

Then
$$S(x) = \frac{1}{\pi}\int_0^{2\pi} S(t)\, D_n(x-t)\, dt = \frac{1}{\pi}\int_0^{2\pi} S(t)\, K_{n,h}(x-t)\, dt$$
$$= \frac{1}{\pi}\int_0^{2\pi} S(t)\, K_{n,h}(x-t)\, d\omega_m(t),$$

provided $2n+h-1<m$. We take $h=m-2n$. Then

$$\int_0^{2\pi} |S(x)|\, dx \leqslant \int_0^{2\pi} |S(t)|\, d\omega_m(t)\,.\,\max_t \frac{1}{\pi}\int_0^{2\pi} |K_{n,h}(x-t)|\, dx,$$

$$\max |S(x)| \leqslant \max |S(x_k)| \max_x \frac{1}{\pi}\int_0^{2\pi} |K_{n,h}(x-t)|\, d\omega_m(t).$$

By (7·31) and (2·5), we have

$$\frac{1}{\pi}\int_0^{2\pi} |K_{n,h}(x-t)|\, d\omega_m(t) \leqslant \frac{1}{\pi}\int_0^{2\pi} \left\{\left(1+\frac{n}{h}\right) K_{n+h-1}(x-t) + \frac{n}{h} K_{n-1}(x-t)\right\} d\omega_m(t)$$
$$= \frac{1}{\pi}\left\{\left(1+\frac{n}{h}\right)\int_0^{2\pi} K_{n+h-1}(t)\, dt + \frac{n}{h}\int_0^{2\pi} K_{n-1}(t)\, dt\right\}$$
$$= 1 + \frac{2n}{h} \leqslant A(1+\delta^{-1}).$$

The same estimate (with $A=1$) holds if we replace $d\omega_m(t)$ by dt. This gives (7·29) and (7·30) with $A_\delta = A(1+\delta^{-1})$.

This estimate for A_δ is rather crude. To improve it, consider

$$\frac{1}{\pi}\int_0^{2\pi} |K_{n,h}(t)|\, dt = \frac{1}{\pi}\int_0^{\pi} \frac{|\sin(n+\frac{1}{2}h)t \sin \frac{1}{2}ht|}{2h \sin^2 \frac{1}{2}t}\, dt.$$

Since only small δ are of interest, we may suppose that $m\leqslant 3n$, $\delta\leqslant\frac{1}{2}$; thus $2\delta n\leqslant h\leqslant n$. We split the last integral into three, say I_1, I_2, I_3, extended respectively over the intervals $(0, 1/n)$, $(1/n, 1/h)$, $(1/h, \pi)$. Since $|\sin ku| \leqslant |k \sin u|$ for integral k, the integrand of I_1 does not exceed $\frac{1}{2}(2n+h)$ and $I_1=O(1)$. If we replace the numerator in the integrand of I_3 by 1, we easily find that $I_3=O(1)$. Finally, since the numerator of I_2 does not exceed $h\sin\frac{1}{2}t$,

$$I_2 \leqslant \frac{1}{\pi}\int_{1/n}^{1/h}\frac{dt}{2\sin\frac{1}{2}t} < \frac{1}{2}\int_{1/n}^{1/h}\frac{dt}{t} < \log\frac{n}{h} \leqslant \log\frac{1}{2\delta} < \log\frac{1}{\delta}.$$

Collecting results we see that for the A_δ in (7·29) we have

$$A_\delta = O(\log 1/\delta).$$

The same inequality holds for the A_δ in (7·30), since

$$\int_0^{2\pi} |K_{n,h}(x-t)|\,d\omega_m(t) \leqslant A\int_0^{2\pi} |K_{n,h}(t)|\,dt,$$

by the remark to Theorem (7·5).

We conclude this section by applying (7·18) to the problem of the behaviour of the $I_n[f]$.

(7·32) THEOREM. *Let $I_k = I_k[f]$ and let*

$$\xi_n = \frac{1}{n+1}\sum_{k=0}^{n} |I_k|, \quad \eta_n = \frac{1}{n+1}\sum_{k=0}^{n} |I_k - f|.$$

Then

(i) *if f is bounded, $\xi_n = O(\log\log n)$ for almost all x;*
(ii) *if f is continuous, $\xi_n = o(\log\log n)$ for almost all x;*
(iii) *if the modulus of continuity of f is $o\{(\log\log 1/\delta)^{-1}\}$; $\eta_n \to 0$ for almost all x.*

We say that a numerical sequence $\{s_n\}$ is *strongly summable* (C, 1) to the limit s if

$$\{|s_0 - s| + |s_1 - s| + \ldots + |s_n - s|\}/(n+1) \to 0$$

as $n\to\infty$. Strong summability (C, 1) implies ordinary summability (C, 1), but not conversely. Thus the conclusion of (iii) implies that $\{I_n[f]\}$ is summable (C, 1) almost everywhere. (For applications of strong summability to Fourier series see Chapter XIII, §§ 7,8.)

We easily verify that if $2^{m-1} \leqslant n \leqslant 2^m$, then

$$\xi_n \leqslant 2\xi_{2^m}, \quad \eta_n \leqslant 2\eta_{2^m}, \quad \log\log n \simeq \log\log 2^m.$$

It is therefore enough to prove (7·32) when n runs through the values 2^m.

(i) Let $|f| \leqslant 1$. Hölder's inequality and (7·19) give

$$\int_0^{2\pi} \exp\lambda_0\xi_n\,dx \leqslant \mu_0. \tag{7·33}$$

Let E_n denote the set of points x of $(0, 2\pi)$ at which $\xi_n \geqslant (2/\lambda_0)\log\log n$. Then

$$\int_{E_n} \exp\lambda_0\{(2/\lambda_0)\log\log n\}\,dx \leqslant \int_0^{2\pi}\exp\lambda_0\xi_n\,dx \leqslant \mu_0,$$

so that $|E_n| \leqslant \mu_0(\log n)^{-2}$. Hence $\Sigma|E_{2^m}| < \infty$, so that we have $\xi_{2^m} \leqslant (2/\lambda_0)\log\log 2^m$ for almost all x, provided that m is large enough.

(ii) By subtracting from f a polynomial (which adds $O(1)$ to ξ_n), we may assume that $|f|<\epsilon$. The inequality (7·33) is then valid with λ_0/ϵ for λ_0, so that

$$\xi_{2^m} \leqslant (2\epsilon/\lambda_0)\log\log 2^m$$

almost everywhere for m large enough. Hence $\xi_{2^m}=o(\log\log 2^m)$ almost everywhere.

(iii) Let $l_k=\log\log k$ when $k\geqslant 3$, and $l_k=\log\log 3$ otherwise. Let $f_k=f-\sigma_k[f]$. By Chapter III, (3·16), $|f_k|\leqslant \epsilon_k/l_k$, where $\epsilon_k\to 0$. Writing $I_k^*=I_k[f_k]$, we have

$$\int_0^{2\pi} \exp\{\lambda_0\epsilon_k^{-1}l_k\,|I_k^*|\}\,dx\leqslant \mu_0,$$

$$\int_0^{2\pi} \exp\left\{\frac{\lambda_0}{n+1}\sum_0^n \epsilon_k^{-1}l_k\,|I_k^*|\right\}dx\leqslant \mu_0.$$

As before, we find that

$$(n+1)^{-1}\sum_0^n \epsilon_k^{-1}l_k\,|I_k^*|=O(l_n)$$

almost everywhere, and so

$$(n+1)^{-1}\sum_0^n l_k\,|I_k^*|=o(l_n)$$

almost everywhere. Using the fact that $I_k^*=O(\log k)$ (see the proof of (5·13) (i)) and that $l_{\sqrt{n}}\simeq l_n$, we have

$$\sum_0^n |I_k^*|\leqslant \sum_{k\leqslant\sqrt{n}}|I_k^*|+O(l_n^{-1})\sum_{\sqrt{n}<k\leqslant n} l_k\,|I_k^*|$$

$$=O(\sqrt{n}\log n)+O(l_n^{-1})\,o(nl_n)=o(n)$$

almost everywhere. Since

$$I_k[f]-f=I_k[f_k]+\sigma_k-f=I_k^*+o(1),$$

we have

$$\eta_n\leqslant\frac{1}{n+1}\sum_{k=0}^n |I_k^*|+o(1)=o(1)$$

almost everywhere, which is the desired result.

8. Divergence of interpolating polynomials

We return for a while to the interpolating polynomials corresponding to *general* systems

$$x_0^n,\ x_1^n,\ \ldots,\ x_{2n}^n\quad (n=0,1,2,\ldots) \tag{8·1}$$

of fundamental points of order n. Let $t_j^n(x)$ be the corresponding fundamental polynomials, so that the interpolating polynomials at a given point ξ are

$$U_n(\xi,f)=\sum_{j=0}^{2n} t_j^n(\xi)f(x_j^n). \tag{8·2}$$

Fixing ξ and n, we have here a functional (Chapter IV, § 9) which we shall consider in the space C of all continuous and periodic functions. The norm of this functional, that is, the upper bound of the absolute value of (8·2) for all f with $|f|\leqslant 1$, is

$$\Lambda_n(\xi)=\sum_{j=0}^{2n}|t_j^n(\xi)|, \tag{8·3}$$

which is also the largest value at the point ξ of the interpolating polynomials of all such functions. It is equal to $U_n(\xi, f)$ whenever f satisfies the conditions

$$f(x_j^n) = \operatorname{sign} t_j^n(\xi) \tag{8·4}$$

for all j. It is an analogue of the Lebesgue constant for Fourier series, but depends on the point ξ.

(8·5) THEOREM. *For a fixed ξ,*

(i) *if $\Lambda_n(\xi) = O(1)$, then $U_n(\xi, f) \to f(\xi)$ for every continuous f;*

(ii) *if $\Lambda_n(\xi) \neq O(1)$, then $U_n(\xi, f) \neq O(1)$ for a certain continuous f.*

The proof is simple.

(i) Suppose that $\Lambda_n(\xi) \leqslant M$ for all n. Let $f = f_1 + f_2$, where $|f_1| < \epsilon$ and f_2 is a polynomial of order m. Then

$$U_n(\xi, f) - f(\xi) = U_n(\xi, f_1) - f_1(\xi) + U_n(\xi, f_2) - f_2(\xi),$$

and

$$|U_n(\xi, f_1) - f_1(\xi)| \leqslant |U_n(\xi, f_1)| + |f_1(\xi)| \leqslant (M+1)\epsilon,$$

while $U_n(\xi, f_2) = f_2(\xi)$ when $n \geqslant m$. Hence $U_n(\xi, f) \to f(\xi)$ as $n \to \infty$.

(ii) If $\Lambda_n(\xi) \neq O(1)$, the theorem of Banach-Steinhaus (Chapter IV, (9·5)) implies the existence of a continuous f such that $U_n(\xi, f) \neq O(1)$.

Suppose now that $\xi = \xi_n$ depends on n. The above arguments show that if

$$\Lambda_n(\xi_n) = O(1), \quad \text{then} \quad U_n(\xi_n, f) - f(\xi_n) \to 0$$

for each continuous f; and if

$$\Lambda_n(\xi_n) \neq O(1), \quad \text{then} \quad U_n(\xi_n, f) \neq O(1)$$

for some continuous f. Hence, if $\{\Lambda_n(\xi)\}$ is not uniformly bounded, there is a continuous f such that $\{U_n(x, f)\}$ is not uniformly bounded.

(8·6) THEOREM. *For any sequence of systems of fundamental points there is a continuous function f such that $\{U_n(x, f)\}$ is not uniformly bounded, and so, in particular, does not converge uniformly.*

We first prove the following result showing a connexion between the interpolating polynomials and the partial sums of Fourier series:

(8·7) THEOREM. *If $U_n(x, f, u)$ denotes the interpolating polynomial of order n corresponding to the system* (8·1) *translated by u, then*

$$\frac{1}{2\pi}\int_0^{2\pi} U_n(\xi, f, u)\, du = S_n(\xi; f). \tag{8·8}$$

This is immediate if f is a polynomial of order n, since then $U_n(x, f, u) = f(x)$ for all u. It is therefore enough to prove the formula for an f whose Fourier series begins with terms of rank greater than n, so that the right-hand side is 0.

We obtain the polynomial $U_n(\xi, f, u)$ if on the right-hand side of (8·2) we replace x_j^n by $x_j^n + u$, for all j. From (1·5) we see that $t_j^n(\xi)$ becomes a polynomial $t_j^n(\xi, u)$ of order n in u, and so

$$\int_0^{2\pi} t_j^n(\xi, u)\, f(x_j^n + u)\, du = 0$$

for all j. Therefore in our case the left-hand side of (8·8) is 0, the desired result.

Return to (8·6). It follows from (8·8) that given ξ, n, f we can find a $u_0 = u_0(\xi, n, f)$ such that

$$|U_n(\xi, f, u_0)| \geqslant |S_n(\xi; f)|.$$

There is a continuous f with $|f| \leqslant 1$ such that the right-hand side is arbitrarily close to the Lebesgue constant L_n, say exceeds $L_n - 1$. On the other hand,

$$t_j^n(x, u) = t_j^n(x-u),$$

and so
$$U_n(\xi, f(x), u) = U_n(\xi - u, f(x+u), 0). \tag{8·9}$$

Hence $\Lambda_n(\xi - u_0) \geqslant L_n - 1$, $\Lambda_n(x)$ is not bounded in n and x, and there is a continuous f such that $\{U_n(x, f)\}$ is not uniformly bounded.

The argument just used shows that the functions $\Lambda_n(\xi)$ are not uniformly bounded, but it does not prove the existence of a ξ_0 such that $\Lambda_n(\xi_0) \neq O(1)$, and so does not prove the existence of a continuous f such that $U_n(x, f)$ diverges at some point ξ_0.

The behaviour of $\Lambda_n(\xi)$ as a function of ξ may be very irregular, since $\Lambda_n(\xi)$ may be very large for some ξ's, but is equal to 1 whenever ξ is a fundamental point.

From now on we shall only consider the case of equidistant fundamental points. Then $t_j^n(\xi) = 2D_n(\xi - x_j)/(2n+1)$, and

$$\Lambda_n(\xi) = \frac{1}{\pi}\int_0^{2\pi} |D_n(\xi - t)|\, d\omega_{2n+1}(t)$$

is of period $h = 2\pi/(2n+1)$ *qua* function of ξ.

We compute the asymptotic behaviour of $\Lambda_n(\xi)$ as $n \to \infty$ for ξ situated midway between two consecutive fundamental points. Supposing, for example, that $\xi = 0$ is such a point, we find

$$\begin{aligned}\Lambda_n(0) &= \frac{h}{\pi}\left\{2\sum_{\nu=0}^{n-1} |D_n(\tfrac{1}{2}h + \nu h)| + |D_n(\tfrac{1}{2}h + nh)|\right\} \\ &= \frac{1}{\pi}\left\{\sum_{\nu=0}^{n-1} \frac{h}{\sin\frac{1}{2}(\frac{1}{2}+\nu)h} + \tfrac{1}{2}h\right\}.\end{aligned} \tag{8·10}$$

Since $1/\sin(\frac{1}{2}u)$ decreases in $0 < u \leqslant \pi$, we easily see that

$$\sum_0^{n-1} \frac{h}{\sin\frac{1}{2}(\frac{1}{2}+\nu)h} > \int_{\frac{1}{2}h}^{\pi} \frac{du}{\sin\frac{1}{2}u},$$

and that
$$\sum_1^{n-1} \frac{h}{\sin\frac{1}{2}(\frac{1}{2}+\nu)h} < \int_{\frac{1}{2}h}^{\pi-h} \frac{du}{\sin\frac{1}{2}u}.$$

These and (8·10) show that

$$\begin{aligned}\Lambda_n(0) &= \frac{1}{\pi}\int_{\frac{1}{2}h}^{\pi} \frac{du}{\sin\frac{1}{2}u} + O(1) \\ &= -\frac{2}{\pi}\log\tan\tfrac{1}{8}h + O(1) \\ &= \frac{2}{\pi}\log n + O(1).\end{aligned} \tag{8·11}$$

In particular, $\Lambda_n(0)$ is unbounded. It follows that:

(8·12) THEOREM. *If the fundamental points of interpolation are equidistant and $\xi = 0$ lies midway between two consecutive fundamental points, there is a continuous f such that $\{I_n(0, f)\} \neq O(1)$. (In particular, $\{I_n(0, f)\}$ diverges.)*

An argument very similar to that which led to the estimate (8·11) shows that if the fundamental points are equidistant, then

$$\Lambda_n(\xi) = \frac{2}{\pi} | \sin (n+\tfrac{1}{2})(\xi - x_0^n) | \log n + O(1). \tag{8·13}$$

Theorem (8·12) is an analogue of a result for Fourier series asserting that there is a continuous f with $S[f]$ diverging at a given point. While, however, the problem of the existence of a continuous f with $S[f]$ diverging almost everywhere is still open and seems to be difficult, the corresponding result for interpolating polynomials can be obtained comparatively easily.

(8·14) THEOREM. *For each n let the fundamental points of interpolation be equidistant and associated with the point 0. Then there is a continuous f such that $\{I_n(x, f)\}$ diverges almost everywhere.*

The proof is based on the following lemma, analogous to Lemma (3·2) of Chapter VIII:

(8·15) LEMMA. *For every positive integer n there is a periodic and continuous $f_n(x)$ satisfying the following conditions:*

(i) $|f_n| \leqslant 1$;

(ii) *$I_s(x, f_n)$ converges uniformly to $f_n(x)$ as $s \to \infty$;*

(iii) *there is a set $E_n \subset (0, 2\pi)$, a number M_n, and an integer q_n, such that*

(*a*) $\lim |E_n| = 2\pi$,

(*b*) $\lim M_n = +\infty$,

(*c*) *for each $x \in E_n$ there is a number $m(x)$, not less than n and not exceeding q_n, such that*

$$| I_{m(x)}(x, f_n) | \geqslant M_n.$$

Denote by N_p the set of the fundamental points

$$2\pi j/(2p+1) \quad (j = 0, 1, \ldots, 2p),$$

and select $(n-1)$ positive integers $p_1 < p_2 < \ldots < p_{n-1}$ such that $2p_1+1, 2p_2+1, \ldots, 2p_{n-1}+1$ are successive primes, and $p_1 \geqslant n$. It is easily seen that no two of the sets $N_{p_1}, N_{p_2}, \ldots, N_{p_{n-1}}$ have points in common other than 0.

We first assign the values of f_n at the points of the sets N_{p_k}, $k = 1, 2, \ldots, n-1$. Let P_k denote the interval $2\pi k/n \leqslant t \leqslant 2\pi$. We define

$$f_n(t) = \operatorname{sign} \cos (p_k + \tfrac{1}{2}) t$$

at those points of N_{p_k} which belong to P_k. (Thus $f_n = \pm 1$ at those points, since $\sin (p_k + \frac{1}{2}) t = 0$ at the points of N_{p_k}.) At the remaining points of N_{p_k} we set $f_n = 0$. This definition gives $f_n = 0$ at $t = 0$, and, since no other points of the sets N_{p_k} coincide, the function f_n is determined uniquely at the points of $N_{p_1} + N_{p_2} + \ldots + N_{p_{n-1}}$.

At the remaining points of the interval $0 \leqslant t \leqslant 2\pi$, we determine f_n by the conditions of linearity and periodicity. Thus f_n satisfies a Lipschitz condition, and conditions (i) and (ii) of (8·15) are satisfied.

In order to show that f_n satisfies condition (iii), let us denote by δ_k the set of points t such that

$$\frac{2\pi}{n}(k-1) \leqslant t < \frac{2\pi}{n} k, \quad | \sin (p_k + \tfrac{1}{2}) t | \geqslant (\log n)^{-\frac{1}{2}}. \tag{8·16}$$

For $x \in \delta_k$,

$$|I_{p_k}(x, f_n)| = \left|\frac{1}{\pi}\int_0^{2\pi} f_n(t)\, D_{p_k}(x-t)\, d\omega_{2p_k+1}(t)\right|$$
$$= \left|\frac{1}{\pi}\sin(p_k+\tfrac{1}{2})x\int_0^{2\pi} f_n(t)\frac{\cos(p_k+\frac{1}{2})t}{2\sin\frac{1}{2}(x-t)}d\omega_{2p_k+1}(t)\right|$$
$$\geqslant \frac{1}{\pi}(\log n)^{-\frac{1}{2}}\int_{P_k}\tfrac{1}{2}\operatorname{cosec}\tfrac{1}{2}(t-x)\, d\omega_{2p_k+1}(t), \tag{8·17}$$

since at the points of N_{p_k} which belong to P_k we have $f_n(t)\cos(p_k+\frac{1}{2})t = 1$, while $f_n = 0$ at the remaining points of N_{p_k}.

We easily see that the last integral in (8·17) exceeds

$$\int_{P_k}\frac{1}{t-x}d\omega_{2p_k+1}(t) \geqslant \int_{2\pi k/n}^{2\pi}\frac{1}{t-2\pi(k-1)/n}d\omega_{2p_k+1}(t) \tag{8·18}$$

(compare the first inequality (8·16)). If we replace $d\omega_{2p_k+1}(t)$ in the second integral in (8·18) by dt, the error introduced is $O(1)$; for the difference between the integral $\int_a^b g(t)\,dt$ of a positive decreasing function $g(t)$ and the Riemann sum obtained by subdividing (a, b) into m equal parts does not exceed $g(a)(b-a)/m$, and p_k is not small in comparison with n (since $p_1 \geqslant n$). Hence the second integral (8·18) is

$$\int_{2\pi k/n}^{2\pi}\frac{dt}{t-2\pi(k-1)/n} + O(1) = \log(n-k+1) + O(1).$$

If now $n-k \geqslant \sqrt{n}$, we have

$$\log(n-k+1) + O(1) \geqslant \tfrac{1}{2}\log n + O(1) > \tfrac{1}{3}\log n$$

for sufficiently large n.

Collecting these results, we see that for n sufficiently large, for $k \leqslant n - \sqrt{n}$, and for x belonging to the set δ_k defined by (8·16), we have

$$|I_{p_k}(x, f_n)| \geqslant \frac{1}{\pi}(\log n)^{-\frac{1}{2}}.\tfrac{1}{3}\log n = \frac{1}{3\pi}(\log n)^{\frac{1}{2}}. \tag{8·19}$$

Denote the last quantity by M_n. The inequality $|I_{p_k}(x, f_n)| \geqslant M_n$ is satisfied at every point of the set

$$E_n = \sum_{k\leqslant n-\sqrt{n}} \delta_k.$$

Since $M_n \to \infty$ and $|E_n| \to 2\pi$ as $n \to \infty$, and the p_k in (8·19) do not exceed p_{n-1} (which may therefore be taken for the q_n in condition (iii)), conditions (*a*), (*b*), (*c*) are satisfied, and the lemma is established.

We show now that if the f_n and the M_n are those of the lemma, then for suitably chosen $n_1 < n_2 < \ldots < n_i < \ldots$ the function

$$f(x) = \sum_{i=1}^{\infty} f_{n_i}(x)/M_{n_i}^{\frac{1}{2}} \tag{8·20}$$

has the properties enunciated in (8·14). We enumerate the conditions to be imposed on $\{n_i\}$.

Denote the complement of E_n by E'_n. Let $\{n_i\}$ be such that

$$\text{(i) } M_{n_{i+1}} > 4M_{n_i}, \quad \text{(ii) } \Sigma M_{n_i}^{-\frac{1}{2}} < 1, \quad \text{(iii) } \Sigma|E'_{n_i}| < \infty.$$

By (iii), almost all x belong to at most a finite number of the E'_{n_i} and so belong to infinitely many E_{n_i}.

Given $n_1, n_2, \ldots, n_{k-1}$, and so also $q_{n_1}, q_{n_2}, \ldots, q_{n_{k-1}}$, we require n_k to be so large that

$$\text{(iv)}\quad |I_s[F_{k-1}]| < 1 \quad \text{for} \quad s \geqslant n_k,$$

where F_{k-1} is the $(k-1)$th partial sum of the series (8·20). This is feasible since, by (ii), $|F_{k-1}| < 1$ and since $I_s[F_{k-1}]$ converges uniformly to F_{k-1}. Further, let

$$\text{(v)}\quad M^{\frac{1}{2}}_{n_k} > q_{n_{k-1}}.$$

It is immediately obvious that if n_i increases sufficiently rapidly it satisfies the conditions (i)–(v) just stated.

Consider now a point x_0 in E_{n_k} and write

$$f = F_{k-1} + f_{n_k} M^{-\frac{1}{2}}_{n_k} + R_k,$$

so that $R_k = f - F_k$. Then

$$I_s(x_0, f) = I_s(x_0, F_{k-1}) + I_s(x_0, f_{n_k})\, M^{-\frac{1}{2}}_{n_k} + I_s(x_0, R_k). \tag{8·21}$$

For a certain s not less than n_k and not exceeding q_{n_k} we have

$$|I_s(x_0, f_{n_k})\, M^{-\frac{1}{2}}_{n_k}| \geqslant M_{n_k} M^{-\frac{1}{2}}_{n_k} = M^{\frac{1}{2}}_{n_k}. \tag{8·22}$$

Since $s \geqslant n_k$, condition (iv) gives

$$|I_s(x_0, F_{k-1})| < 1. \tag{8·23}$$

Finally, since $s \leqslant q_{n_k}$,

$$|I_s(x_0, R_k)| \leqslant (2s+1) \max |R_k(x)| \leqslant 3q_{n_k} \sum_{j=k+1}^{\infty} M^{-\frac{1}{2}}_{n_j} < 6q_{n_k} M^{-\frac{1}{2}}_{n_{k+1}} < 6, \tag{8·24}$$

by virtue of conditions (i) and (v); the first inequality (8·24) uses the fact that the complex Fourier-Lagrange coefficients of R_k do not exceed $\max |R_k|$ in absolute value. From (8·21), (8·22), (8·23) and (8·24) we obtain

$$|I_s(x_0, f)| \geqslant M^{\frac{1}{2}}_{n_k} - 7.$$

Hence for each x_0 which belongs to infinitely many E_{n_k} we have

$$\limsup_{s\to\infty} |I_s(x_0, f)| = +\infty,$$

and (8·14) follows by virtue of condition (iii).

(8·25) THEOREM. *With equidistant fundamental points associated with* 0, *there is a continuous f such that $\{I_n(x, f)\}$ diverges almost everywhere, while* S[f] *converges uniformly.*

This shows that for a given continuous f the behaviour of the sequences $\{S_m[f]\}$ and $\{I_m[f]\}$ may be totally different. To prove (8·25) it is enough to show that the f_n in (8·15) can be found so that each $S[f_n]$ converges uniformly, and that $|S_m[f_n]| < A$, where A is independent of m and n. For using the decomposition $f = F_k + R_k$ above, we have $S_m[f] = S_m[F_k] + S_m[R_k]$. For k fixed, $S_m[F_k]$ converges uniformly to F_k. Also

$$|S_m[R_k]| = \left| \sum_{j=k+1}^{\infty} S_m[f_{n_j}]\, M^{-\frac{1}{2}}_{n_j} \right| \leqslant A \sum_{j=k+1}^{\infty} M^{-\frac{1}{2}}_{n_j}.$$

Since the right-hand side here is arbitrarily small for k large enough, the uniform convergence of S[f] follows.

Consider now the function $\lambda_\delta(x)$ which is continuous and even, equal to 1 when $x=0$ and to 0 when $2\delta \leqslant x \leqslant \pi$, and linear in $(0, 2\delta)$; $\lambda_\delta(x)$ is the 'roof function', and, by Chapter I, (4·16),

$$\lambda_\delta(x) = \frac{2\delta}{\pi}\left\{\tfrac{1}{2} + \sum_{n=1}^{\infty}\left(\frac{\sin n\delta}{n\delta}\right)^2 \cos nx\right\}.$$

We observe that

(i) $|S_m(x; \lambda_\delta)| < 1$ for all x, m, δ $(0 < \delta \leqslant \frac{1}{2}\pi)$;

(ii) given any $\epsilon > 0$ and $\eta > 0$, we can find a δ_0 such that $|S_m(x; \lambda_\delta)| < \epsilon$ for $\eta \leqslant |x| \leqslant \pi$, $0 < \delta \leqslant \delta_0$, and all m.

Property (i) is obvious since $|S_m(x; \lambda_\delta)|$ takes its largest value at $x=0$, where it is less than $\lambda_\delta(0)=1$.

To prove (ii) we write

$$|S_m(x; \lambda_\delta)| = \left|\frac{1}{\pi}\int_{-\pi}^{\pi} D_m(x-t)\,\lambda_\delta(t)\,dt\right| \leqslant \frac{1}{\pi}\int_{-2\delta}^{2\delta} |D_m(x-t)|\,dt. \qquad \textbf{(8·26)}$$

If, for example, $\delta \leqslant \frac{1}{4}\pi$, $0 \leqslant x \leqslant \pi$, then $|x-t| \leqslant \frac{3}{2}\pi$ and

$$|D_m(x-t)| \leqslant B/|x-t|, \qquad \textbf{(8·27)}$$

where B is an absolute constant. Now suppose that $\eta \leqslant x \leqslant \pi$ (since $\lambda_\delta(x)$ is even we need consider only positive x) and that $2\delta < \frac{1}{2}\eta$. Then (8·26) and (8·27) give

$$|S_m(x; \lambda_\delta)| \leqslant \frac{1}{\pi}\int_{-2\delta}^{2\delta} B(x-t)^{-1}\,dt \leqslant \frac{1}{\pi}B(\eta-2\delta)^{-1}.4\delta < \epsilon,$$

provided δ is small enough. This proves (ii).

Return now to the f_n in (8·15). The only values of f_n relevant for our purpose were those at the points of $N_{p_1}+N_{p_2}+\ldots+N_{p_{n-1}}$. Instead of using linear interpolation, we make f_n a 'roof function' in the neighbourhood of each point ξ of $N_{p_1}+\ldots+N_{p_{n-1}}$. In other words, we define f_n as a sum of a finite number of functions $\pm\lambda_\delta(x-\xi)$, where the ξ's belong to $N_{p_1}+\ldots+N_{p_{n-1}}$, and δ is so small that the 'roofs' do not overlap.

Let r be the number of points ξ in $N_{p_1}+\ldots+N_{p_{n-1}}$ at which $f_n = \pm 1$, let η be so small that the intervals $(\xi-\eta, \xi+\eta)$ do not overlap, and let $\epsilon = 1/r$. If now δ is sufficiently small, and in particular less than $\frac{1}{2}\eta$, it follows from (ii) that $|S_m(x; \lambda_\delta(x-\xi'))| < 1/r$ whenever $\xi \neq \xi'$ and x belongs to $(\xi-\eta, \xi+\eta)$. Further, by (i) $|S_m(x; \lambda_\delta(x-\xi))| < 1$, whence $|S_m(x; f_n)| < 2$ for x in each $(\xi-\eta, \xi+\eta)$. Since $|S_m(x; f_n)| < 1$ when x lies outside all the intervals $(\xi-\eta, \xi+\eta)$, we have $|S_m(x; f_n)| < 2$ for all x, m, n. Since $S[f_n]$ converges uniformly, this establishes (8·25).

In the two theorems which follow, we assume that 0 is a fundamental point.

(8·28) Theorem. *Given any sequence of positive numbers $\epsilon_n \to 0$ there is a continuous f such that for almost all x we have*

$$I_n(x, f) \neq O(\epsilon_n \log n). \qquad \textbf{(8·29)}$$

(8·30) Theorem. *There is a continuous f with modulus of continuity*

$$\omega(\delta) = O(1/\log(1/\delta))$$

such that the $I_n[f]$ (i) *are uniformly bounded,* (ii) *diverge almost everywhere.*

Theorem (8·28) shows that the estimate $I_n(x,f)=o(\log n)$ contained in (5·13) is best possible for continuous f not only at individual points (which is an easy consequence of (8·13)), but even almost everywhere. The situation here is different from that in the theory of Fourier series; if f is continuous the estimate $S_n[f]=o(\log n)$ cannot be improved at individual points (Chapter VIII, (1·2)) but $S_n[f]=o\{(\log n)^{\frac{1}{2}}\}$ almost everywhere (Chapter XIII, (1·2)).

That the $I_n[f]$, and even the $I_{n,\nu}[f]$, are uniformly bounded when

$$\omega(\delta)=O(|\log\delta|)^{-1}$$

is an analogue of (5·6), and is proved similarly. However, under the same hypothesis the $S_n[f]$ are not only uniformly bounded, but converge almost everywhere (Chapter XIII, (1·16)). Thus (8·30) gives another example of the difference in the behaviour of $\{I_n[f]\}$ and $\{S_n[f]\}$.

Return to (8·28). Its proof resembles that of (8·14), and we may be brief. It is enough to show the existence of an f such that at almost all x we have $|I_s(x,f)|\geqslant A\epsilon_s\log s$ for infinitely many s, A denoting a positive absolute constant. For applying this to the sequence $\{\epsilon_s^{\frac{1}{2}}\}$ we obtain Theorem (8·28).

We may suppose that $\epsilon_1>\epsilon_2>\dots$ and that $\epsilon_n\log n\to\infty$. We modify the definition of the set δ_k by substituting $\epsilon_n^{\frac{1}{2}}$ for $(\log n)^{-\frac{1}{2}}$ in the second inequality (8·16). The complement E_n' of the set $E_n=\sum_{k\leqslant n-\sqrt{n}}\delta_k$ has now measure $O(n^{-\frac{1}{2}})+O(\epsilon_n^{\frac{1}{2}})$. If $x\in\delta_k$, $k\leqslant n-\sqrt{n}$, we have

$$|I_{p_k}(x,f_n)|\geqslant\frac{1}{3\pi}\epsilon_n^{\frac{1}{2}}\log n,$$

instead of (8·19). We show that this leads to

$$|I_{p_k}(x,f_n)|\geqslant A\epsilon_{p_k}^{\frac{1}{2}}\log p_k. \tag{8·31}$$

Since $p_k\geqslant n$, $\epsilon_{p_k}\leqslant\epsilon_n$, it is enough to show that $\log n\geqslant A\log p_k$.

Let $\rho_1,\rho_2,\dots$ be the sequence of all primes. It is familiar that

$$\rho_m\leqslant Am\log m,$$

but for our purposes it is enough to assume the weaker inequality

$$\rho_m\leqslant Am^2; \tag{8·32}$$

indeed, $\rho_m\leqslant Am^r$, for any fixed r, would do. Let p_1 be the least number not less than n such that $2p_1+1$ is a prime. Hence $n\leqslant p_1<p_2<\dots<p_{n-1}$, and, say,

$$2p_1+1=\rho_{j+1},\quad 2p_2+1=\rho_{j+2},\quad \dots,\quad 2p_{n-1}+1=\rho_{j+n-1},$$

while $\rho_j<2n+1$. Obviously $j\leqslant n$, so that $j+n-1<2n$. It follows from this and (8·32) that $\rho_{j+n-1}<4An^2$, and so $\log n>A\log p_k$. This proves (8·31).

To complete the proof of (8·28) we select $\{n_i\}$ so that

$$\Sigma n_i^{-\frac{1}{2}}<\infty,\quad \Sigma\epsilon_{n_i}^{\frac{1}{2}}<\infty, \tag{8·33}$$

and write

$$f(x)=\Sigma\epsilon_{n_i}^{\frac{1}{2}}f_{n_i}(x).$$

If the n_i increase fast enough, then for every x which belongs to infinitely many E_{n_i} (and, by (8·33), almost every x belongs to all E_{n_i} with i large enough), there exist infinitely many s such that

$$I_s(x,f)=\epsilon_{n_i}^{\frac{1}{2}}I_s(x,f_{n_i})+O(1). \tag{8·34}$$

These s are the p_k of (8·31) with $n=n_i$. Hence the absolute value of the right-hand side of (8·34) exceeds $A\epsilon_s \log s - O(1)$ for such s, and so also exceeds $A\epsilon_s \log s$ for s large enough, which completes the proof of (8·28).

For the proof of (8·30) we modify the definition of f_n slightly by replacing the condition $p_1 \geqslant n$ by $p_1 \geqslant n^2$; the purpose of this is to make $p_1, p_2, \ldots, p_n$ large in comparison with n. Arguing as before we find that $p_{n-1} \leqslant An^4$.

Since the distance of any two points of $N_{p_1}+N_{p_2}+\ldots+N_{p_{n-1}}$ exceeds $2\pi/(2p_{n-1}+1)^2 > An^{-8}$, the slope of f_n is less than An^8. Hence the functions

$$g_n(t)=f_n(t)/\log n$$

satisfy the inequalities

$$|g_n(x+t)-g_n(x)| \leqslant \frac{2}{\log n}, \tag{8·35}$$

$$|g_n(x+t)-g(x)| \leqslant An^8 |t|. \tag{8·36}$$

Let
$$f(x)=\sum_{i=1}^{\infty}(-1)^i g_{n_i}(x), \tag{8·37}$$

where n_i increases so fast that
$$\log n_{i+1}/\log n_i \geqslant 2.$$

This implies, in particular, that $n_{i+1}/n_i > 2$ and $\Sigma 1/\log n_i < \infty$, so that f is continuous. If
$$n_{j+1}^{-16} \leqslant t < n_j^{-16} \quad (j=1,2,\ldots),$$

we have, by (8·35) and (8·36),

$$\begin{aligned} |f(x+t)-f(x)| &\leqslant \left(\sum_{i=1}^{j}+\sum_{i=j+1}^{\infty}\right)|g_{n_i}(x+t)-g_{n_i}(x)| \\ &\leqslant At\sum_{i=1}^{j} n_i^8 + 2\sum_{i=j+1}^{\infty}\frac{1}{\log n_i} \\ &\leqslant Atn_j^8(1+\tfrac{1}{2}+(\tfrac{1}{2})^2+\ldots)+\frac{2}{\log n_{j+1}}(1+\tfrac{1}{2}+\ldots) \\ &\leqslant Atn_j^8+\frac{4}{\log n_{j+1}} \\ &\leqslant At.t^{-\frac{1}{2}}+\frac{64}{\log 1/t}=O\left\{\left(\log\frac{1}{t}\right)^{-1}\right\}, \end{aligned}$$

so that the modulus of continuity $\omega(\delta)$ of f is $O\{(\log 1/\delta)^{-1}\}$.

For a fixed n we have (see (8·17))

$$I_{p_k}(x,f_n)=-\frac{1}{\pi}\sin(p_k+\tfrac{1}{2})x\int_{P_k}\tfrac{1}{2}\operatorname{cosec}\tfrac{1}{2}(t-x)\,d\omega_{2p_k+1}(t).$$

Denoting by δ_k the set of points in $2\pi(k-1)/n \leqslant x \leqslant 2\pi k/n$ at which

$$\sin(p_k+\tfrac{1}{2})x \leqslant -\tfrac{1}{2},$$

we see that
$$I_{p_k}(x,f_n) \geqslant \frac{1}{6\pi}\log n \quad \text{for} \quad x\in E_n=\sum_{k\leqslant n-\sqrt{n}}\delta_k. \tag{8·38}$$

Since the p_k are large in comparison with n, δ_k is highly periodic in $(2\pi(k-1)/n, 2\pi k/n)$, and if n_i increases fast enough almost all points of $(0,2\pi)$ belong to infinitely many

E_{n_i} with i even and to infinitely many E_{n_i} with i odd. Denote the jth partial sum of $\Sigma(-1)^i g_{n_i}(x)$ by $F_j(x)$, and write

$$f=F_{j-1}+(-1)^j g_{n_j}+R_j.$$

We deduce from (8·38) that at almost all x for infinitely many j and suitable p_k we have $I_{p_k}(x,(-1)^j g_{n_j})>1/6\pi$ as well as $I_{p_k}(x,(-1)^j g_{n_j})<-1/6\pi$. On the other hand, if n_i increases fast enough we have

$$I_{p_k}(x,F_{j-1})=F_{j-1}(x)+o(1)=f(x)+o(1), \quad I_{p_k}(x,R_j)=o(1),$$

uniformly in x. This shows that $\{I_n(x,f)\}$ diverges almost everywhere.

We conclude this section by pointing out one more difference in the behaviour of $\{I_n[f]\}$ and $\{S_n[f]\}$. It will be shown in Chapter XIII, (1·17), that if $n_1<n_2<\dots$ is any lacunary sequence of positive integers (that is, $n_{k+1}/n_k>q>1$), then $S_{n_k}(x;f)\to f(x)$ almost everywhere when f is continuous, and even when $f\in L^2$. Going through the proof of (8·15) and (8·14), and selecting for p_k integers satisfying the condition $p_{k+1}/p_k>q>1$, we verify immediately that we obtain a continuous f and that $\{I_{n_k}(x,f)\}$ diverges almost everywhere for some lacunary sequence $\{n_k\}$.

9. Divergence of interpolating polynomials (*cont.*)

(9·1) THEOREM. *Suppose that* 0 *is a fundamental point of interpolation for each* n. *There is then a continuous* g *such that* $I_n(x,g)$ *diverges for all* $x\neq 0$.

Since $I_n(0,g)=g(0)$ for each n, we cannot have divergence at $x=0$.

The proof of (9·1) is based on the fact that, for any real α, $\sin\alpha x$ and $\sin(\alpha+1)x$ cannot be small simultaneously, except in the neighbourhood of $x=0$ or $x=\pi$. More precisely, we have

$$\max\{|\sin\alpha x|,|\sin(\alpha+1)x|\}\geqslant\tfrac{1}{2}|\sin x|, \tag{9·2}$$

as we see from the equation $\sin x=\sin(\alpha+1)x\cos\alpha x-\sin\alpha x\cos(\alpha+1)x$.

In this section we systematically write α' for $\alpha+1$. As before, N_p denotes the set of points $2\pi j/(2p+1)$, where $j=0,1,\dots,2p$. We observe that N_p and $N_{p'}$ $(=N_{p+1})$ have only the point 0 in common. All the ω_j we consider in the proof of (9·1) have 0 as a point of discontinuity.

(9·3) LEMMA. *For each* $m=1,2,\dots$ *there is a continuous function* $f=f_m$ *such that* $|f|\leqslant 1$, *with the following properties:*

(i) $\{I_s(x,f)\}$ *converges uniformly;*

(ii) *for each* x *in the intervals* $(1/m,\pi-1/m)$ *and* $(\pi+1/m,2\pi-1/m)$ *there is an* $n(x)$ *such that* $|I_{n(x)}(x,f)|>m$.

We fix m, take an arbitrarily large number $M>0$ and select n so large that

$$\int_{2\pi/n}^{2\pi/n+1/m}\frac{d\omega_{2p+1}(t)}{2\sin\frac{1}{2}t}>M \tag{9·4}$$

for all $p\geqslant n$.

Next we fix numbers $p_1,p_2,\dots,p_{n-1}$ such that

$$n\leqslant p_1<p_2<\dots<p_{n-1}, \tag{9·5}$$

$$n(2p_i'+1)^3\leqslant 2p_{i+1}+1 \quad (i=1,2,\dots,n-2). \tag{9·6}$$

Consider now the sets N_{p_1} and $N_{p_1'}$. At the points of N_{p_1} we set

$$f(t) = \operatorname{sign}\cos(p_1+\tfrac{1}{2})\,t \quad \text{or} \quad f(t)=0,$$

according as $t \geqslant 2\pi/n$ or $t < 2\pi/n$. In particular, $f(0)=0$. Similarly at the points of $N_{p_1'}$ we set

$$f(t) = \operatorname{sign}\cos(p_1'+\tfrac{1}{2})\,t \quad \text{or} \quad f(t)=0,$$

according as $t \geqslant 2\pi/n$ or $t < 2\pi/n$.

Suppose we have already defined f at the points of

$$S_{k-1} = \sum_{1}^{k-1} (N_{p_i} + N_{p_i'}),$$

where $k < n-1$. At the points of N_{p_k} which are not in S_{k-1} we set

$$f(t) = \operatorname{sign}\cos(p_k+\tfrac{1}{2})\,t \quad \text{or} \quad f(t)=0, \tag{9·7}$$

according as $t \geqslant 2\pi k/n$ or $t < 2\pi k/n$. Similarly at the points of $N_{p_k'}$ which are not in S_{k-1} we set

$$f(t) = \operatorname{sign}\cos(p_k'+\tfrac{1}{2})\,t \quad \text{or} \quad f(t)=0$$

according as $t \geqslant 2\pi k/n$ or $t < 2\pi k/n$.

In this way we define f by induction in S_{n-1}. At the remaining points of $(0, 2\pi)$ we define f arbitrarily, provided $|f| \leqslant 1$ and $\{I_s(x,f)\}$ converges uniformly.

Suppose now that

$$2\pi(k-1)/n \leqslant x < 2\pi k/n, \quad x \leqslant 2\pi - 1/m, \tag{9·8}$$

and consider

$$I_{p_k}(x,f) = -\frac{1}{\pi}\sin(p_k+\tfrac{1}{2})\,x \int_0^{2\pi} f(t)\frac{\cos(p_k+\tfrac{1}{2})\,t}{2\sin\frac{1}{2}(t-x)}\,d\omega_{2p_k+1}(t)$$

$$= -\frac{1}{\pi}\sin(p_k+\tfrac{1}{2})\,x \left\{\int_{N_{p_k}-N_{p_k}S_{k-1}} + \int_{N_{p_k}S_{k-1}}\right\}.$$

We first show that

$$\sin(p_k+\tfrac{1}{2})\,x \int_{N_{p_k}S_{k-1}} = O(1). \tag{9·9}$$

The distance between any two points of S_{k-1} exceeds $l_{k-1} = 2\pi/(2p_{k-1}'+1)^2$. Hence there are at most two points of $N_{p_{k-1}}S_{k-1}$ whose distance from x is less than l_{k-1}. Since $|D_{p_k}| \leqslant p_k + \frac{1}{2}$, the contribution of these two points to the left side of (9·9) is $O(1)$. For the remaining t's in $N_{p_k}S_{k-1}$ we have $|t-x| > l_{k-1}$, and since the number of points in S_{k-1} does not exceed $2\sum_{1}^{k-1}(2p_i'+1) < 2n(2p_{k-1}'+1)$, the contribution of these t's to the left-hand side of (9·9) is

$$O\left\{\frac{n(2p_{k-1}'+1)}{l_{k-1}(2p_k+1)}\right\} = O\left\{\frac{n(2p_{k-1}'+1)^3}{2p_k+1}\right\} = O(1),$$

by (9·6). This proves (9·9).

Let $N^*_{p_k}$ denote the part of N_{p_k} situated in $2\pi k/n \leqslant t \leqslant 2\pi$. In view of (9·8) and (9·7), using the argument which led to (9·9) we have

$$\left| \sin(p_k+\tfrac{1}{2})x \int_{N_{p_k}-N_{p_k}S_{k-1}} f(t)\frac{\cos(p_k+\tfrac{1}{2})t}{2\sin\tfrac{1}{2}(t-x)}\,d\omega_{2p_k+1}\right|$$

$$= |\sin(p_k+\tfrac{1}{2})x| \int_{N^*_{p_k}-N^*_{p_k}S_{k-1}} \tfrac{1}{2}\operatorname{cosec}\tfrac{1}{2}(t-x)\,d\omega_{2p_k+1}$$

$$= |\sin(p_k+\tfrac{1}{2})x| \int_{N^*_{p_k}} \tfrac{1}{2}\operatorname{cosec}\tfrac{1}{2}(t-x)\,d\omega_{2p_k+1}+O(1)$$

$$\geqslant |\sin(p_k+\tfrac{1}{2})x| \int_{2\pi k/n}^{2\pi} \tfrac{1}{2}\operatorname{cosec}\tfrac{1}{2}\left(t-\frac{2\pi}{n}(k-1)\right)d\omega_{2p_k+1}(t)+O(1)$$

$$\geqslant |\sin(p_k+\tfrac{1}{2})x| \int_{2\pi/n}^{2\pi/n+1/m} \tfrac{1}{2}\operatorname{cosec}\tfrac{1}{2}t\,d\omega_{2p_k+1}\left(t+\frac{2\pi}{n}(k-1)\right)+O(1),$$

Replacing here $d\omega_{2p_k+1}(t+2\pi(k-1)/n)$ by $d\omega_{2p_k+1}(t)$ we commit an error $O(n/p_k)=O(1)$. Hence, collecting results,

$$|I_{p_k}(x,f)| \geqslant \frac{1}{\pi}|\sin(p_k+\tfrac{1}{2})x|\,M+O(1).$$

An identical argument gives, for x satisfying (9·8),

$$|I_{p'_k}(x,f)| \geqslant \frac{1}{\pi}|\sin(p'_k+\tfrac{1}{2})x|\,M+O(1).$$

By (9·2), these two estimates lead to

$$\max\{|I_{p_k}(x,f)|,|I_{p'_k}(x,f)|\} \geqslant \frac{1}{2\pi}M|\sin x|+O(1). \qquad \textbf{(9·10)}$$

Suppose now that x is in $(1/m, \pi-1/m)$ or $(\pi+1/m, 2\pi-1/m)$. Then

$$|\sin x| \geqslant \sin(1/m),$$

and the right-hand side of (9·10) exceeds $(2\pi)^{-1}M|\sin(1/m)|+O(1)$, and so also exceeds m if M is chosen large enough. This proves condition (ii) of (9·3).

The rest of the proof of (9·1) is simple. Using Lemma (9·3) we see that if m_i increases fast enough the function

$$g(x)=\Sigma m_i^{-\frac{1}{2}} f_{m_i}(x) \qquad \textbf{(9·11)}$$

is continuous and $\{I_s(x,g)\}$ diverges everywhere except at $x=0$, and possibly $x=\pi$. If $I_s(\pi,g)$ converges, it is enough to add to g a continuous h such that $\{I_s(x,h)\}$ diverges at π but converges elsewhere.

Remark. We might set $h(x)=f(x+\pi)$, where f is the function of Theorem (8·12), except that the proof of (8·12) does not guarantee that $\{I_s(x,f)\}$ converges for $x\neq 0$. To construct a continuous f such that $\{I_s(x,f)\}$ diverges at $x=0$ and converges elsewhere (the fundamental points being the same as in (8·12)), we may proceed directly. By (8·11), $\Lambda_n(0)\to\infty$. There is a continuous f_n, $|f_n|\leqslant 1$, such that $I_n(0,f_n)=\Lambda_n(0)$. If ϵ_n tends to 0 slowly enough and if we modify f_n by making it 0 outside $(-\epsilon_n,\epsilon_n)$ the modified f_n will satisfy $I_n(0,f_n)>\frac{1}{2}\Lambda_n(0)$. Since only the values of f_n at the fundamental points are relevant we may suppose that $\{I_s(x,f_n)\}$ converges uniformly for each n.

It follows that outside any interval $(-\delta, \delta)$ the $I_s(x, f_n)$ are bounded by a number which depends on δ but not on s and n. Hence, if n_i increases fast enough and

$$f(x) = \Sigma \Lambda_{n_i}^{-\frac{1}{2}}(0) f_{n_i}(x),$$

$\{I_s(x, f)\}$ diverges at $x=0$ and converges uniformly outside every $(-\delta, \delta)$.

The proof of the following theorem is identical with the proof of (8·25):

(9·12) THEOREM. *There is a g for which* (9·1) *holds, such that* $S[g]$ *converges uniformly.*

Suppose now that the fundamental points of interpolation are the roots of

$$\sin\{(n+\tfrac{1}{2})x+\beta\}=0 \quad (n=0,1,2,\ldots), \tag{9·13}$$

where β is fixed. Since the inequality (9·2) holds if αx on the left is replaced by $\alpha x+\beta$, the proof of (9·1) shows that there is a continuous g such that $I_s(x, g)$ diverges for all x, except possibly for $x=0$ and $x=\pi$. In view of (8·5)(ii), (8·13), and the remark above, if β is not an integral multiple of $\frac{1}{2}\pi$ there is a continuous h such that $\{I_s(x, h)\}$ diverges at the points of convergence of $\{I_s(x, g)\}$ and converges elsewhere. Hence:

(9·14) THEOREM. *If the fundamental points of interpolation are the roots of* (9·13), *where β is not an integral multiple of $\frac{1}{2}\pi$, then there is a continuous g such that $\{I_s[g]\}$ diverges everywhere.*

We may also consider the problem of the divergence of

$$E_n(x, f) = \frac{1}{\pi}\int_{-\pi}^{\pi} f(t)\, D_n^*(x-t)\, d\omega_{2n}(t)$$

(see § 3). The case when ω_{2n} is associated with the roots of

$$\cos nx = 0$$

is of particular interest. Arguing as before, but operating with $(0, \pi)$ instead of $(0, 2\pi)$, we find then that there is a continuous g which is 0 in $(-\pi, 0)$, such that $\{E_n(x, g)\}$ diverges in $0 \leqslant x \leqslant \pi$. It follows that if $g_1(x) = \frac{1}{2}\{g(x)+g(-x)\}$, then $E_n(x, g_1)$ diverges everywhere. Since g_1 is even, $E_n[g_1]$ is a cosine polynomial. It is of order $n-1$ since $\cos nx$ is 0 at the fundamental points. Hence, making the transformation $t=\cos x$ we obtain the following result:

(9·15) THEOREM. *There is a continuous function $G(t)$, $-1 \leqslant t \leqslant +1$, such that if $P_{n-1}(t)$ denotes the power polynomial of degree $n-1$ coinciding with $G(t)$ at the Tchebyshev abscissae*

$$\cos\frac{\pi}{2n}, \quad \cos\frac{3\pi}{2n}, \quad \ldots, \quad \cos\frac{(2n-1)\pi}{2n},$$

then $\{P_{n-1}(t)\}$ diverges for $-1 \leqslant t \leqslant +1$.

We conclude with a theorem which shows that the behaviour of the I_n depends not only on the properties of f but also on the selection of the fundamental points.

(9·16) THEOREM. *Let β be incommensurable with π, and let $I'_n(x, g)$ be the interpolating polynomials for g associated with the roots of* (9·13). *There is then a continuous g for which* (9·1) *holds such that $\{I'_s(x, g)\}$ converges uniformly.*

The proof is similar to that of (8·25) and we shall be brief.

As we have observed there, we may define f_n as a sum of a finite number of 'roof functions' $\pm\lambda_\delta(x-\xi)$, where the ξ's are in $S_{n-1}=\sum_1^{n-1}(N_{p_i}+N_{p_i'})$. Since β is incommensurable with π, the roots of (9·13) have no point in common with S_{n-1}. We may take δ arbitrarily small, and in the first instance so small that the 'roofs' corresponding to various ξ's do not overlap. The number of 'roofs' does not then exceed the number of points of S_{n-1} and, in particular, is independent of δ. If we can show the existence of an absolute constant A such that

$$|I_s'(x,f_n)|<A \tag{9·17}$$

for all x, s, n, then, since $\{I_s'(x,f_n)\}$ converges uniformly for each n, the polynomials $I_s'(x,g)$ for the g in (9·11) will converge uniformly. We can also select the h we add to g so that $\{I_s'(x,h)\}$ converges uniformly.

The inequality (9·17) will follow if we show that

(i) $|I_s'(x,\lambda_\delta(x-\xi))|\leqslant 1$ for all x, ξ, s;

(ii) given any $\epsilon>0$ and $\eta>0$ we can find δ_0 such that $|I_s'(x,\lambda_\delta(x-\xi))|<\epsilon$ for $\eta\leqslant|x-\xi|\leqslant\pi$, $0<\delta\leqslant\delta_0$, and all s.

Condition (i) follows immediately from condition (i) on p. 41, since, by (4·10), the sum of the moduli of the complex coefficients of $I_s'(x,\lambda_\delta(x-\xi))$ does not exceed the sum of the moduli of the coefficients of $\mathsf{S}[\lambda_\delta(x-\xi)]$, which is 1.

Condition (ii) is analogous to condition (ii) on p. 41, and is proved similarly; if ω_{2n+1}' is associated with the roots of (9·13) we have

$$|I_s'(x,\lambda_\delta(x-\xi)|\leqslant\frac{1}{\pi}\int_{-2\delta}^{2\delta}|D_s(x-\xi-t)|\,d\omega_{2s+1}'(t)<\epsilon,$$

provided $\eta\leqslant|x-\xi|\leqslant\pi$ and δ is small enough.

10. Polynomials conjugate to interpolating polynomials

Let $\tilde{D}_\nu(u)$ denote the Dirichlet conjugate kernel:

$$\tilde{D}_\nu(u)=\sum_1^\nu \sin ju=[\cos\tfrac{1}{2}u-\cos(\nu+\tfrac{1}{2})u]\,\tfrac{1}{2}\operatorname{cosec}\tfrac{1}{2}u.$$

The polynomial $\tilde{I}_{n,\nu}(x,f)$ conjugate to $I_{n,\nu}(x,f)$ is given by the formula

$$\begin{aligned}\tilde{I}_{n,\nu}(x)&=\frac{1}{\pi}\int_0^{2\pi}f(t)\,\tilde{D}_\nu(x-t)\,d\omega_{2n+1}(t)\\&=-\frac{1}{\pi}\int_0^{2\pi}[f(t)-f(x)]\frac{\cos\frac{1}{2}(t-x)-\cos(\nu+\frac{1}{2})(t-x)}{2\sin\frac{1}{2}(t-x)}\,d\omega_{2n+1}(t);\end{aligned} \tag{10·1}$$

for subtracting a constant from f does not change $\tilde{I}_{n,\nu}$. Let

$$h_\nu=2\pi/(2\nu+1).$$

Since $|\tilde{D}_\nu(u)|<\nu$, we see immediately that, if f is continuous at x, the interval

$(x-h_\nu, x+h_\nu)$ contributes only $o(1)$ to the last integral in (10·1) as $\nu\to\infty$. This suggests that we should consider the expression

$$\begin{aligned}\hat{f}_{n,\nu}(x) &= -\frac{1}{\pi}\int_0^{2\pi(h_\nu)} [f(t)-f(x)]\,\tfrac{1}{2}\cot\tfrac{1}{2}(t-x)\,d\omega_{2n+1}(t)\\ &= -\frac{2}{2n+1}\sum_{|x-x_j^n|\geqslant h_\nu} [f(x_j^n)-f(x)]\,\tfrac{1}{2}\cot\tfrac{1}{2}(x_j^n-x),\end{aligned} \tag{10·2}$$

where the symbol $\int_0^{2\pi(h_\nu)}$ means that we integrate only over the part of the interval $0\leqslant t\leqslant 2\pi$ for which $|t-x|\geqslant h_\nu$. The case $\nu=n$ is the most interesting, and for brevity we shall write $\hat{f}_n$ for $\hat{f}_{n,n}$. The $\hat{f}_n(x)$ is obtained by dropping no more than two terms from

$$-\frac{2}{2n+1}\sum_0^{2n}[f(x_j^n)-f(x)]\,\tfrac{1}{2}\cot\tfrac{1}{2}(x_j^n-x).$$

It is natural to expect that the limit

$$\hat{f}(x)=\lim_{n\to\infty}\hat{f}_n(x), \tag{10·3}$$

if it exists, plays in interpolation the same role as the conjugate function

$$\tilde{f}(x) = -\frac{1}{\pi}\int_0^{2\pi} f(t)\,\tfrac{1}{2}\cot\tfrac{1}{2}(t-x)\,dt = -\frac{1}{\pi}\lim_{\epsilon\to+0}\left\{\int_{x+\epsilon}^{x+\pi}+\int_{x-\pi}^{x-\epsilon}\right\} \tag{10·4}$$

does in the theory of Fourier series. It is easy to see that at the points where f is continuous the existence of the limit (10·4) when ϵ tends to 0 through the sequence of values h_ν, $\nu=1,2,\ldots$, implies the existence of the limit when ϵ tends continuously to 0. We shall write

$$\tilde{f}_\nu(x) = -\frac{1}{\pi}\int_{h_\nu}^{\pi}[f(x+t)-f(x-t)]\,\tfrac{1}{2}\cot\tfrac{1}{2}t\,dt. \tag{10·5}$$

We know that $\tilde{f}(x)$ exists almost everywhere for every $f\in L$, and, in particular, for every continuous f (see Chapter IV, (3·1) and Chapter VII, (1·4)). As a consequence of this, one has theorems for series conjugate to Fourier series, analogous to theorems for Fourier series. In interpolation the situation is different; even for continuous f the limit (10·3) may not exist anywhere. The same thing applies *a fortiori* to

$$\lim \hat{f}_{n,\nu}(x) \quad (n\geqslant\nu). \tag{10·6}$$

For this reason the behaviour of the polynomials $\tilde{I}_{n,\nu}(x,f)$ may be totally different from that of the $\tilde{S}_\nu(x;f)$. The convergence of $\tilde{I}_{n,\nu}$ in norm was discussed in § 7. Here we shall consider pointwise convergence. Use will be made of the fact that at every point x of continuity of f

$$\tilde{I}_{n,\nu}(x,f)-\hat{f}_{n,\nu}(x) = \frac{1}{\pi}\int_{-\pi}^{\pi(h_\nu)}[f(t)-f(x)]\frac{\cos(\nu+\frac{1}{2})(t-x)}{2\sin\frac{1}{2}(t-x)}\,d\omega_{2n+1}(t)+o(1) \tag{10·7}$$

as $\nu\to\infty$. This follows from (10·1), if we observe that the contribution of the interval $(x-h_\nu, x+h_\nu)$ is $o(1)$. It is also useful to observe that the $o(1)$ is uniform for any family of functions f which are equicontinuous.

(10·8) THEOREM. *If $\lambda(x)$ has all its Dini numbers finite at ξ, then*

$$\tilde{I}_{n,\nu}(\xi,\lambda f)-\lambda(\xi)\,\tilde{I}_{n,\nu}(\xi,f)$$

tends to a finite limit as $\nu\to\infty$. If $\lambda\in\Lambda_1$, the convergence is uniform in ξ.

The proof is similar to that of (5·2), and the limit in question is

$$-\frac{1}{\pi}\int_0^{2\pi} f(t)\frac{\lambda(t)-\lambda(\xi)}{2\tan\frac{1}{2}(t-\xi)}\,dt.$$

(10·9) THEOREM. *If f is 0 in (a,b), then $\tilde{I}_{n,\nu}(x,f)$ tends uniformly to a limit in every interval $(a+\epsilon, b-\epsilon)$ interior to (a,b).*

This is an analogue of (5·1) and is proved in the same way. It shows that if $f_1=f_2$ in (a,b), then $\tilde{I}_{n,\nu}[f_1]$ and $\tilde{I}_{n,\nu}[f_2]$ are uniformly equiconvergent in the wider sense in $(a+\epsilon, b-\epsilon)$.

(10·10) THEOREM. *Suppose that $|f(\xi\pm t)-f(\xi)|\leqslant\mu(t)$ for $0<t\leqslant\eta$, where $\mu(t)$ is a non-decreasing function of t such that $\int_0^\eta t^{-1}\mu(t)\,dt<\infty$. Then as $\nu\to\infty$,*

$$\hat{f}_{n,\nu}(\xi)\to\tilde{f}(\xi) \quad \textit{and} \quad \tilde{I}_{n,\nu}(\xi)\to\tilde{f}(\xi).$$

This is an analogue of (5·5). To prove that $\hat{f}_{n,\nu}(\xi)\to\tilde{f}(\xi)$, we may suppose that $\xi=0$, $f(0)=0$. For any fixed $0<\delta\leqslant\pi$ the part of the integral

$$\int_{-\pi}^{\pi(h_\nu)} f(t)\tfrac{1}{2}\cot\tfrac{1}{2}t\,d\omega_{2n+1} \tag{10·11}$$

which is taken over $\delta\leqslant|t|\leqslant\pi$ tends to $\int_{\delta\leqslant|t|\leqslant\pi} f(t)\tfrac{1}{2}\cot\tfrac{1}{2}t\,dt$, and it is enough to show that the part of (10·11) extended over $|t|\leqslant\delta$ is small with δ, provided that ν is sufficiently large. The proof of this is similar to the proof of (5·5). Hence $\hat{f}_{n,\nu}(\xi)\to\tilde{f}(\xi)$. The proof of $\tilde{I}_{n,\nu}(\xi,f)\to\tilde{f}(\xi)$ is also similar to that of (5·5).

(10·12) THEOREM. *If $S[f]$ converges absolutely, $\tilde{I}_{n,\nu}(x,f)$ converges uniformly to $\tilde{f}(x)$.*

The proof is similar to that of (5·16).

(10·13) THEOREM. *Suppose that f has a modulus of continuity $\omega(\delta)=o\{(\log 1/\delta)^{-1}\}$. Then as $\nu\to\infty$,*

$$\hat{f}_{n,\nu}(x)-\hat{f}_\nu(x)\to 0, \tag{10·14}$$

so that if one of the three quantities

$$\tilde{f}(x),\quad \lim_{n\to\infty}\hat{f}_n(x),\quad \lim_{\nu\to\infty}\hat{f}_{n,\nu}(x)$$

exists, so do the remaining two, and all three have the same value.† *In particular, the limit* (10·6) *exists almost everywhere.*

We may suppose that $x=0$, $f(0)=0$. We shall also suppose that $x_{2n}<0<x_0$. (The case when 0 is a fundamental point can be treated similarly.) Let $x_p=x_p^n$ and $x_q=x_q^n$ be the extreme fundamental points in the interval $h_\nu\leqslant t\leqslant\pi$. Then, since $\cot\frac{1}{2}\pi=0$,

$$\begin{aligned}&\int_{h_\nu}^{\pi} f(t)\tfrac{1}{2}\cot\tfrac{1}{2}t\,dt-\int_{h_\nu}^{\pi} f(t)\tfrac{1}{2}\cot\tfrac{1}{2}t\,d\omega_{2n+1}\\ &=\int_{x_p}^{x_q} f(t)\tfrac{1}{2}\cot\tfrac{1}{2}t\,dt-\int_{x_p}^{x_q} f(t)\tfrac{1}{2}\cot\tfrac{1}{2}t\,d\omega_{2n+1}+o(1)\\ &=\sum_p^{q-1}\int_{x_j}^{x_{j+1}}[f(t)-f(x_j)]\tfrac{1}{2}\cot\tfrac{1}{2}t\,dt+\sum_p^{q-1}f(x_j)\int_{x_j}^{x_{j+1}}\tfrac{1}{2}[\cot\tfrac{1}{2}t-\cot\tfrac{1}{2}x_j]\,dt+o(1).\end{aligned} \tag{10·15}$$

† Making first ν very large but fixed, and then making $n\to\infty$, we easily deduce that, at every x where f is continuous, if $\lim_{\nu\to\infty}\hat{f}_{n,\nu}(x)$ exists so does $\tilde{f}(x)$, and both are equal.

Numerically, the first sum on the right does not exceed

$$\omega(h_n)\int_{x_p}^{x_q} t^{-1}dt = o\{(\log n)^{-1}\}\log(\pi/x_1) = o\{(\log n)^{-1}\}O(\log n) = o(1).$$

The expression in square brackets in the second sum on the right of (10·15) is $O(1/ntx_j) = O(1/nx_j^2)$, and so the whole sum is numerically not greater than

$$O(1)\sum_{p}^{q-1} j^{-2}|f(x_j)| \leqslant O(1)\sum_{1}^{q-1} j^{-2}|f(x_j)| = o(1), \tag{10·16}$$

f being continuous and vanishing at the origin. Hence the expression on the left of (10·15) tends to 0 as $\nu\to\infty$. A similar conclusion holds if in (10·15) we replace $\int_{h_\nu}^{\pi}$ by $\int_{-\pi}^{-h_\nu}$, and the result (with uniformity in x) follows by addition.

(10·17) THEOREM. *The relation* (10·14) *holds if f is of bounded variation in the neighbourhood of x and continuous at x.*

We may again suppose that $x=0$, $f(0)=0$. We may also suppose that f is non-decreasing near 0. Since

$$\int_\alpha^\beta f(t)\tfrac{1}{2}\cot\tfrac{1}{2}t\,d\omega_{2n+1}(t)\to\int_\alpha^\beta f(t)\tfrac{1}{2}\cot\tfrac{1}{2}t\,dt$$

for any interval (α,β) interior to $(0,\pi)$ or to $(-\pi,0)$, we may assume that f is non-decreasing and bounded in $(-\pi,\pi)$. Consider the equation (10·15). We show exactly as before that the second term on the right is $o(1)$. The first term on the right is non-negative and does not exceed

$$h_n\sum_{p}^{q-1}\frac{f(x_{j+1})-f(x_j)}{x_j}\leqslant\sum_{1}^{q-1}\frac{f(x_{j+1})-f(x_j)}{j}, \tag{10·18}$$

since $x_1 > x_0 > 0$, $x_j \geqslant h_n j$. This last sum we split into two parts $\sum_{j\leqslant j_0} + \sum_{j>j_0} = P+Q$. Here $Q < f(\pi)/j_0$, and so is small when j_0 is large. If j_0 is fixed, each term of P tends to 0 as $n\to\infty$ (since $f(x_j^n)\to f(0)=0$ for each j), so that $P\to 0$. Hence the expression on the left of (10·15) tends to 0 and (10·14) follows as before. Since any function of bounded variation is differentiable almost everywhere, the existence of $\lim \tilde{f}_{n,\nu}(x)$ almost everywhere also follows from (10·10).

(10·19) THEOREM. *Suppose that f has a modulus of continuity $\omega(\delta) = o\{(\log 1/\delta)^{-1}\}$. Then as $\nu\to\infty$,*

$$\tilde{I}_{n,\nu}(x,f) - \tilde{f}_{n,\nu}(x)\to 0 \tag{10·20}$$

uniformly in x, so that a necessary and sufficient condition for the existence of $\lim \tilde{I}_{n,\nu}(x,f)$ *(or of* $\lim \tilde{I}_n(x,f)$*) at a point x is the existence of $\tilde{f}$ there. In particular,* $\lim \tilde{I}_{n,\nu}(x,f)$ *exists and equals $\tilde{f}(x)$ almost everywhere.*

Let $T(t)$ be a polynomial of order n. Then

$$T(x) = \frac{1}{\pi}\int_{-\pi}^{\pi} T(x+t)\frac{\sin(n+\frac{1}{2})t}{2\sin\frac{1}{2}t}\,d\omega_{2n+1}, \tag{10·21}$$

$$\tilde{T}(x) = -\frac{1}{\pi}\int_{-\pi}^{\pi} T(x+t)\left\{\tfrac{1}{2}\cot\tfrac{1}{2}t - \frac{\cos(n+\frac{1}{2})t}{2\sin\frac{1}{2}t}\right\}d\omega_{2n+1}. \tag{10·22}$$

Suppose that ω_{2n+1} is associated with the roots of

$$\sin[(n+\tfrac{1}{2})t+\alpha]=0, \tag{10·23}$$

and assume temporarily that $\alpha \not\equiv 0 \pmod{\pi}$. Multiplying (10·21) and (10·22) by $\cos\alpha$ and $\sin\alpha$ respectively and adding the results we get†

$$-\frac{1}{\pi}\int_{-\pi}^{\pi} T(x+t)\,\tfrac{1}{2}\cot\tfrac{1}{2}t\,d\omega_{2n+1}=T(x)\cot\alpha+\tilde{T}(x). \tag{10·24}$$

If the order of T is $\nu\leqslant n$, then we also have

$$\tilde{T}(x)=-\frac{1}{\pi}\int_{-\pi}^{\pi} T(x+t)\left\{\tfrac{1}{2}\cot\tfrac{1}{2}t-\frac{\cos(\nu+\frac{1}{2})t}{2\sin\frac{1}{2}t}\right\}d\omega_{2n+1},$$

which together with (10·24) implies that

$$\frac{1}{\pi}\int_{-\pi}^{\pi} T(x+t)\frac{\cos(\nu+\frac{1}{2})t}{2\sin\frac{1}{2}t}\,d\omega_{2n+1}=-T(x)\cot\alpha. \tag{10·25}$$

In proving (10·19) we may suppose that $x=0, f(0)=0$. We have $f-\sigma_\nu[f]=o(1/\log\nu)$ since the modulus of continuity of f is $o\{(\log 1/\delta)^{-1}\}$. Denote $\sigma_\nu[f]$ by T and write $f=g+T$, so that g (which depends on ν) is $o(1/\log\nu)$ uniformly in x. We write

$$\tilde{I}_{n,\nu}(0,f)-\hat{f}_{n,\nu}(0)=\{\tilde{I}_{n,\nu}(0,g)-\hat{g}_{n,\nu}(0)\}+\{\tilde{I}_{n,\nu}(0,T)-\hat{T}_{n,\nu}(0)\}, \tag{10·26}$$

and apply (10·7). Since the functions g are equicontinuous (indeed, tend uniformly to 0), we obtain

$$\begin{aligned}|\tilde{I}_{n,\nu}(0,g)-\hat{g}_{n,\nu}(0)| &\leqslant 2\max|g(t)|\frac{1}{\pi}\int_{-\pi}^{\pi(h_\nu)}\left|\frac{\cos(\nu+\frac{1}{2})t}{2\sin\frac{1}{2}t}\right|d\omega_{2n+1}+o(1)\\ &\leqslant \max|g(t)|\int_{-\pi}^{\pi(h_\nu)}\frac{d\omega_{2n+1}}{|t|}+o(1)=o(1),\end{aligned} \tag{10·27}$$

since the last integral is $O(\log\nu)$ and $g(t)=o(1/\log\nu)$.

Now $T(0)=f(0)-g(0)=-g(0)=o(1/\log\nu)$. Hence, by subtracting a constant $o(1/\log\nu)$ from $T(t)$ and adding the same constant to $g(t)$, which does not impair the estimate $g(t)=o(1/\log\nu)$, we may suppose that $T(0)=0$. Applying (10·7) to the polynomial T (which remains equicontinuous as $\nu\to\infty$), and applying (10·25) at $x=0$, we find that

$$\begin{aligned}\tilde{I}_{n,\nu}(0,T)-\hat{T}_{n,\nu}(0) &=\frac{1}{\pi}\int_{-\pi}^{\pi(h_\nu)} T(t)\frac{\cos(\nu+\frac{1}{2})t}{2\sin\frac{1}{2}t}\,d\omega_{2n+1}+o(1)\\ &=-\frac{1}{\pi}\int_{-h_\nu}^{h_\nu} T(t)\frac{\cos(\nu+\frac{1}{2})t}{2\sin\frac{1}{2}t}\,d\omega_{2n+1}+o(1).\end{aligned} \tag{10·28}$$

Suppose that $|f|\leqslant M$, so that $|T|\leqslant M$ also. By the mean-value theorem and Bernstein's inequality,

$$|T(t)|=|T(t)-T(0)|=|t|\,|T'(t_1)|\leqslant \nu M|t|,$$

† By making $\alpha\to 0$ we easily see that when $\alpha=0$ in (10·23) the finite part of the left-hand side of (10·24) is $\pi^{-1}h_n T'(x)+\tilde{T}(x)$, but we do not need this.

so that (10·28) gives

$$|\tilde{I}_{n,\nu}(0,T)-\hat{T}_{n,\nu}(0)| \leqslant \frac{\nu M}{\pi}\int_{-h_\nu}^{h_\nu} \frac{t}{2\sin\frac{1}{2}t}\,d\omega_{2n+1}+o(1)$$

$$\leqslant \tfrac{1}{2}\nu M\int_{-h_\nu}^{h_\nu} d\omega_{2n+1}+o(1) < AM+o(1), \tag{10·29}$$

where A is an absolute constant. From (10·26), (10·27) and (10·29) we get

$$\limsup_{\nu\to\infty} |\tilde{I}_{n,\nu}(0,f)-\hat{f}_{n,\nu}(0)| \leqslant AM. \tag{10·30}$$

To pass from this to (10·20) (for $x=0$), we write $f=f_1+f_2$, where f_1 is a polynomial and $|f_2|<\epsilon$. For f_1 we have (10·20) (if only on account of (10·10)), and for f_2 the right-hand side of (10·30) does not exceed $A\epsilon$.

This completes the proof of (10·19), except for one remark. In the argument we used the formula (10·25), which presupposes that ω_{2n+1} is not associated with the roots of $\sin(n+\frac{1}{2})t=0$. For n and ν fixed, however, both $\tilde{I}_{n,\nu}[f]$ and $\hat{f}_{n,\nu}$ are continuous functions of the α in (10·23), and since our estimates are uniform in α they hold also for $\alpha=0$.

(10·31) THEOREM. *We have $\tilde{I}_{n,\nu}(x,f)-\hat{f}_{n,\nu}(x)\to 0$ as $\nu\to\infty$ if f is of bounded variation in the neighbourhood of x and continuous at x.*

We again suppose that $x=0$, $f(0)=0$, and that f is non-decreasing. We have to show that the integral on the right of (10·7) is $o(1)$ for $x=0$, and it is enough to prove this for the part extended over (h_ν,π). We fix $\epsilon>0$ and write

$$\frac{1}{\pi}\int_{h_\nu}^{\pi} f(t)\frac{\cos(\nu+\frac{1}{2})t}{2\sin\frac{1}{2}t}\,d\omega_{2n+1}=\frac{1}{\pi}\int_{h_\nu}^{\delta}+\frac{1}{\pi}\int_{\delta}^{\pi}=P+Q,$$

where δ is so small that $f(\delta)<\epsilon$. We take ν so large that $h_\nu<\delta$. We know (Chapter I, §2) that if the a_j are positive and monotone

$$\left|\sum_M^N a_jb_j\right| \leqslant \begin{cases} a_M\max|B_j| & \text{for } \{a_j\} \text{ decreasing,}\\ 2a_N\max|B_j| & \text{for } \{a_j\} \text{ increasing,}\end{cases} \tag{10·32}$$

where $B_j=b_M+b_{M+1}+\ldots+b_j$. Now let $x_p=x_p^n$ and $x_q=x_q^n$ $(p<q)$ be the first and last fundamental points in (h_ν,δ). Applying (10·32) twice we get

$$|P|=\left|\frac{2}{2n+1}\sum_p^q f(x_j)\frac{\cos(\nu+\frac{1}{2})x_j}{2\sin\frac{1}{2}x_j}\right| \leqslant \frac{2}{2n+1}2f(x_q)\max_r\left|\sum_p^r\frac{\cos(\nu+\frac{1}{2})x_j}{2\sin\frac{1}{2}x_j}\right|$$

$$\leqslant \frac{4}{2n+1}\epsilon\frac{1}{2\sin\frac{1}{2}x_p}\max_r\left|\sum_p^r\cos(\nu+\tfrac{1}{2})x_j\right|$$

$$\leqslant \frac{4\epsilon}{2n+1}\,\frac{1}{2\sin\frac{1}{2}h_\nu}\,\frac{2}{2\sin\frac{1}{2}(\nu+\frac{1}{2})h_n}\leqslant\epsilon.$$

Finally, for fixed δ we have $Q\to 0$, by (4·1), and this completes the proof of (10·31).

That $\tilde{I}_{n,\nu}[f]\to\tilde{f}$ almost everywhere when f is of bounded variation also follows from (10·10) and the fact that f is differentiable almost everywhere.

(10·33) THEOREM. *If f is integrable* R, *then*

$$\mathfrak{M}_2[\tilde{f}-\hat{\tilde{f}}_{n,\nu}]\to 0 \tag{10·34}$$

as $\nu\to\infty$. In particular, if $\lim \hat{\tilde{f}}_{n,\nu}(x)$, or $\lim \hat{\tilde{f}}_n(x)$, exists when x belongs to a set E, this limit is $\tilde{f}(x)$ almost everywhere in E.

Consider the Jackson polynomial $J_{\nu,2n+1}(x,f)$ (see (6·2)) and its conjugate

$$\begin{aligned}\tilde{J}_{\nu,2n+1}(x) &= -\frac{1}{\pi}\int_{-\pi}^{\pi} f(t)\,\tilde{K}_\nu(t-x)\,d\omega_{2n+1}(t)\\ &= -\frac{1}{\pi}\int_{-\pi}^{\pi}[f(t)-f(x)]\left[\tfrac{1}{2}\cot\tfrac{1}{2}(t-x)-\frac{\sin(\nu+1)(t-x)}{(\nu+1)\,4\sin^2\frac{1}{2}(t-x)}\right]d\omega_{2n+1}.\end{aligned} \tag{10·35}$$

If we show that as $\nu\to\infty$ (with $n\geqslant\nu$),

$$\mathfrak{M}_2[\tilde{J}_{\nu,2n+1}-\tilde{f}]\to 0, \tag{10·36}$$

$$\mathfrak{M}_2[\tilde{J}_{\nu,2n+1}-\hat{\tilde{f}}_{n,\nu}]\to 0, \tag{10·37}$$

then (10·34) will follow by an application of Minkowski's inequality. But $J_{\nu,2n+1}$ is uniformly bounded and tends to f almost everywhere (indeed, at every point of continuity of f). Hence $\mathfrak{M}_2[J_{\nu,2n+1}-f]$ tends to 0. Since the last expression majorizes $\mathfrak{M}_2[\tilde{J}_{\nu,2n+1}-\tilde{f}]$, (10·36) follows.

Since $|\tilde{K}_\nu(u)|<\nu$, we have at every point of continuity of f

$$|\tilde{J}_{\nu,2n+1}(x)-\hat{\tilde{f}}_{n,\nu}(x)|\leqslant o(1)+\frac{1}{\pi(\nu+1)}\int_{-\pi}^{\pi(h_\nu)}\frac{|f(t)-f(x)|}{4\sin^2\frac{1}{2}(t-x)}\,d\omega_{2n+1}. \tag{10·38}$$

The second term on the right is $o(1)$ at every point of continuity of f. Moreover, it is easy to see that the right-hand side of (10·38) is uniformly bounded. Hence we have (10·37), and (10·33) follows.

(10·39) THEOREM. *There is a continuous $f(x)$ such that $\lim \hat{\tilde{f}}_n(x)$ (and,* a fortiori, *$\lim \hat{\tilde{f}}_{n,\nu}(x)$) exists at no point.*

It is enough to sketch the proof since it resembles those of (8·14) and (8·15). The set N_p is defined as before, and P_k is the shorter of the two intervals $(2\pi k/n, 2\pi)$ and $(2\pi k/n, 2\pi k/n+\pi)$. The function f_n, which we shall now write as f^n to avoid confusion, is defined to be equal to 1 at those points of N_{p_k} which belong to P_k, equal to 0 at the remaining points of N_{p_k}, and linear between any two consecutive points of $N_{p_1}+\ldots+N_{p_{n-1}}$. The second condition (8·16) is now dropped; then, for $2\pi(k-1)/n\leqslant x<2\pi k/n$, $k\geqslant n-\sqrt{n}$,

$$\begin{aligned}|\hat{\tilde{f}}^n_{p_k}(x)| &= \left|\frac{1}{\pi}\int_0^{2\pi(h_{p_k})}\frac{f^n(t)}{2\tan\frac{1}{2}(t-x)}\,d\omega_{2p_k+1}\right|\\ &\geqslant \frac{1}{\pi}\int_{P_k}\frac{1}{t-x}\,d\omega_{2p_k+1}+O(1)\geqslant\frac{1}{\pi}\int_{2\pi k/n}^{2\pi}\frac{dt}{t-2\pi(k-1)/n}+O(1)\\ &= \pi^{-1}\log(n-k+1)+O(1)\geqslant(3\pi)^{-1}\log n,\end{aligned} \tag{10·40}$$

when n is sufficiently large. If we define E_n as $\sum_{k\leqslant n-\sqrt{n}}\delta_k$, E_n is an interval $(0,a_n)$, where $a_n\to 2\pi$. Writing $\frac{1}{3}\log n=M_n$, defining $f(x)$ by (8·20), and repeating the argument on pp. 39–40, we find that $\lim \hat{\tilde{f}}_m(x)$ exists at no point of the interval $0\leqslant x<2\pi$.

The f in (10·39) may be such that $S[f]$ and $\tilde{S}[f]$ both converge uniformly.

(10·41) Theorem. *For the f satisfying* (10·39) *the polynomials*

$$\tilde{B}_n(x) = \tilde{J}_{n,2n+1}(x)$$

(*see* (6·5)) *diverge at every x as $n \to \infty$. This is,* a fortiori, *true for the polynomials $\tilde{B}_{n,\nu}$.*

Consider (10·38) with $\nu = n$; if f is continuous, we have

$$\tilde{B}_n(x) - \hat{f}_n(x) = o(1) \tag{10·42}$$

uniformly in x, and the assertion of (10·41) follows from the divergence of $\{\hat{f}_n(x)\}$.

The proof of the following result resembles that of (8·30) and we omit it:

(10·43) Theorem. *The f in* (10·39) *and* (10·41) *can have modulus of continuity $O\{(\log 1/\delta)^{-1}\}$.*

MISCELLANEOUS THEOREMS AND EXAMPLES

1. Let x_0 be a fundamental point of interpolation. Then

$$I_n(x,f) = \frac{2}{2n+1}\sin(n+\tfrac{1}{2})(x-x_0)\sum_{k=0}^{2n}\frac{(-1)^k f(x_k)}{2\sin\frac{1}{2}(x-x_k)}$$

$$= \frac{2}{2n+1}\sin(n+\tfrac{1}{2})(x-x_0)\sum_{k=-\infty}^{+\infty}\frac{(-1)^k f(x_k)}{x-x_k},$$

where $x_k = x_0 + 2\pi k/(2n+1)$;

$$E_n(x,f) = \frac{1}{n}\sin n(x-x_0)\sum_{k=0}^{2n-1}\frac{(-1)^k f(x_k)}{2\tan\frac{1}{2}(x-x_k)}$$

$$= \frac{1}{n}\sin n(x-x_0)\sum_{k=-\infty}^{+\infty}\frac{(-1)^k f(x_k)}{x-x_k},$$

where $x_k = x_0 + \pi k/n$;

$$J_n(x,f) = \frac{1}{(n+1)^2}\sin^2\tfrac{1}{2}(n+1)(x-x_0)\sum_{k=0}^{n}\frac{f(x_k)}{\sin^2\frac{1}{2}(x-x_k)}$$

$$= \frac{4}{(n+1)^2}\sin^2\tfrac{1}{2}(n+1)(x-x_0)\sum_{k=-\infty}^{+\infty}\frac{f(x_k)}{(x-x_k)^2},$$

where $x_k = x_0 + 2\pi k/(n+1)$. (See (1·19), (3·30) and (6·2)). (de la Vallée-Poussin [3].)

2. Let $\psi(t)$, $0 \leqslant t \leqslant 2\pi$, be of bounded variation. Necessary and sufficient conditions that

$$c_\nu = \frac{1}{2\pi}\int_0^{2\pi} S(t)\, e^{-i\nu t}\, d\psi(t) \quad (\nu = 0, \pm 1, \pm 2, \ldots, \pm n),$$

or, what is the same thing, that

$$S(x) = \frac{1}{\pi}\int_0^{2\pi} S(t)\, D_n(x-t)\, d\psi(t)$$

for every polynomial $S(x) = \sum_{-n}^{n} c_\nu e^{i\nu x}$ of order n, are (i) that $\mathsf{S}[d\psi]$ has constant term 1 and (ii) that all the other terms of $\mathsf{S}[d\psi]$ of rank at most $2n$ vanish.

Conditions (i) and (ii) are satisfied by any ω_{2n+1}, since

$$d\omega_{2n+1}(t) \sim \sum_{\nu=-\infty}^{+\infty} e^{i\nu(2n+1)(t-x_0)},$$

where x_0 is one of the discontinuities of ω_{2n+1}.

3. Consider $2n+1$ distinct points $t_0, t_1, \dots, t_{2n}$ of the interval $(0, 2\pi)$, and a step function $\psi(t)$ continuous outside these points. A necessary and sufficient condition that

$$c_\nu = \frac{1}{2\pi}\int_0^{2\pi} S(t)\, e^{-i\nu t}\, d\psi(t) \quad (\nu = 0, \pm 1, \dots, \pm n)$$

for any polynomial $S(t) = \sum_{-n}^{n} c_\nu e^{i\nu t}$ of order n, is that $\psi(t)$ be an ω_{2n+1}.

[Only the necessity requires proof. Consider the function $L(t) = t + \sum_{-\infty}^{+\infty}{}' e^{i\nu t}/i\nu$, which has jump 2π at $t=0$. Every step function ψ which has jump α_j at t_j for $j = 0, 1, \dots, 2n$ and is continuous elsewhere, is, except for an additive constant, of the form $(2\pi)^{-1}\sum \alpha_j L(t-t_j)$. Using the result of Example 2 we obtain a number of relations between the α_j and $z_j = e^{it_j}$, from which we conclude that $z_0^{2n+1} = z_1^{2n+1} = \dots = z_{2n}^{2n+1}$, which shows that the z_j are equally spaced on $|z| = 1$. Finally, we prove that all the α_j are equal.]

4. Let m be positive, not necessarily integral, and $-m < w < m$. Show that

$$\text{(i)} \quad \frac{1}{\pi}\int_{-\infty}^{+\infty} \frac{\sin wt}{t}\, d\omega_m(t) = \operatorname{sign} w,$$

$$\text{(ii)} \quad \frac{1}{\pi}\int_{-\infty}^{+\infty} \frac{\sin mt}{t}\, d\omega_m(t) = 2\cos^2 \tfrac{1}{2} m t_0,$$

where t_0 is a discontinuity of ω_m.

[(i) Using the formula $\sum_{\nu=-\infty}^{+\infty} (\alpha+\nu)^{-1} e^{i\lambda\nu} = \pi(i + \cot \alpha\pi)\, e^{-i\lambda\alpha}$, valid for $0 < \lambda < 2\pi$ and a non-integral α (Chapter I, § 4), show that

$$\text{(iii)} \quad \frac{1}{\pi}\int_{-\infty}^{+\infty} \frac{e^{iwt}}{t}\, d\omega_m(t) = \cot \tfrac{1}{2} m t_0 + i \operatorname{sign} w$$

for $|w| < m$.]

5. Let $T(x) = \sum_{-n}^{n} c_\nu e^{i\nu x}$ be a polynomial of order n and let $w > 0$, $w + n < m$. Show that

$$\frac{1}{\pi}\int_{-\infty}^{+\infty} T(t)\, \frac{\sin w(t-x)}{t-x}\, d\omega_m(t) = \frac{1}{\pi}\int_{-\infty}^{+\infty} T(x+t)\, \frac{\sin wt}{t}\, d\omega_m(t) = \sum_{|\nu| \leq w}{}^{*}\, c_\nu e^{i\nu x},$$

where the asterisk indicates that if w is integral the extreme terms of Σ are multiplied by $\frac{1}{2}$.
[Use formula (iii) of the preceding example.]

6. Let $I_n(f, x) = \sum_{-n}^{n} c_\nu e^{i\nu x}$, $0 < w < n+1$. Show that

$$\frac{1}{\pi}\int_{-\infty}^{+\infty} f(t)\, \frac{\sin w(t-x)}{t-x}\, d\omega_{2n+1}(t) = \sum_{|\nu| \leq w}{}^{*}\, c_\nu e^{i\nu x}.$$

[We have $f(x) = I_n(x, f)$ at the discontinuities of ω_{2n+1}.]

7. Let $T(x)$ be a polynomial of order n. Show that

$$\frac{1}{\pi}\int_{-\infty}^{+\infty} T(x+t)\, \frac{\sin nt}{t}\, d\omega_{2n}(t) = T(x),$$

provided that ω_{2n} is associated with the point 0.
[Verify when $T(x) = \alpha e^{inx} + \beta e^{-inx}$; apply Example 4.]

8. Let $I_n(x, f) = \sum_{-n}^{n} c_\nu e^{i\nu x}$, $0 < y \leq n+1$. Show that

$$\sum_{|\nu| \leq y} \left(1 - \frac{|\nu|}{y}\right) c_\nu e^{i\nu x} = \frac{1}{\pi}\int_{-\infty}^{+\infty} f(t)\, \frac{1 - \cos w(t-x)}{w(t-x)^2}\, d\omega_{2n+1}(t).$$

[Integrate the formula in Example 6 over $0 < w \leq y$.]

9. For any polynomial $P(z)=c_0+c_1z+\ldots+c_nz^n$ of degree n and for $1\leqslant p<\infty$ we have

$$\text{(i)}\qquad \left\{\int_0^{2\pi}|P'(e^{ix})|^p\,dx\right\}^{1/p}\leqslant A_p n\left\{\int_0^{2\pi}|\mathscr{R}P(e^{ix})|^p\,dx\right\}^{1/p},$$

where
$$\text{(ii)}\qquad A_p^p=\pi^{\frac{1}{2}}\frac{\Gamma(\frac{1}{2}p+1)}{\Gamma(\frac{1}{2}p+\frac{1}{2})},$$

and equality occurs in (i) if and only if $P(z)=Az^n$, where A is an absolute constant. Since the A_p tend to 1 as $p\to\infty$, (i) contains (3·20) as a limiting case.

[Suppose that c_0 is real, and denote by $S(x)$ and $\tilde{S}(x)$ the real and imaginary part of $P(e^{ix})$. By (3·25),

$$\text{(iii)}\qquad \int_0^{2\pi}|S'(x)\cos\alpha+\tilde{S}'(x)\sin\alpha|^p\,dx\leqslant n^p\int_0^{2\pi}|S(x)|^p\,dx$$

for any real α. Integrating this over $0\leqslant\alpha\leqslant 2\pi$, interchanging the order of integration on the left and observing that, for any real a, b,

$$\int_0^{2\pi}|a\cos\alpha+b\sin\alpha|^p\,d\alpha=(a^2+b^2)^{\frac{1}{2}p}\int_0^{2\pi}|\sin\alpha|^p\,d\alpha,$$

we obtain (i) with
$$A_p^p=2\pi\Big/\int_0^{2\pi}|\sin\alpha|^p\,d\alpha.$$

Using Euler's function $B(x,y)$ and the formula $\Gamma(2x)=\pi^{-\frac{1}{2}}\,2^{2x-1}\Gamma(x)\,\Gamma(x+\frac{1}{2})$, we easily obtain

$$\int_0^{2\pi}|\sin\alpha|^p\,d\alpha=2\int_0^{\pi}|\sin 2\alpha|^p\,d\alpha=2^{p+1}\int_0^{\pi}|\sin\alpha\,\cos\alpha|^p\,d\alpha$$
$$=2^{p+2}\int_0^{\frac{1}{2}\pi}(\sin\alpha\,\cos\alpha)^p\,d\alpha=2^{p+1}\int_0^1u^{\frac{1}{2}(p-1)}(1-u)^{\frac{1}{2}(p-1)}\,du$$
$$=2^{p+1}B(\tfrac{1}{2}p+\tfrac{1}{2},\tfrac{1}{2}p+\tfrac{1}{2})=2^{p+1}\Gamma^2(\tfrac{1}{2}p+\tfrac{1}{2})/\Gamma(p+1)=2\pi^{\frac{1}{2}}\,\Gamma(\tfrac{1}{2}p+\tfrac{1}{2})/\Gamma(\tfrac{1}{2}p+1),$$

which gives (ii). If we have equality in (i) we have equality in (iii), for some α. Hence (p. 11) $S(x)=M\cos(nx+\xi)$, $P(z)=Az^n$.]

10. Let
$$\psi(\theta)=\frac{\sin\pi\theta}{\pi}\left\{\frac{1}{\theta}-\frac{1}{\theta+1}+\frac{1}{\theta+2}-\ldots\right\}.$$

Then

(i) $\psi(\theta)+\psi(1-\theta)=1$;

(ii) $0<\psi(\theta)<1$ for $0<\theta<1$;

(iii) $\psi(\frac{1}{2})=\frac{1}{2}$.

[(i) Use the formula $\Sigma\dfrac{(-1)^\nu}{\theta+\nu}=\dfrac{\pi}{\sin\pi\theta}$.

(ii) We have $\psi(\theta)>0$ for $0<\theta<1$; use (i).]

11. Suppose that f is periodic and of bounded variation, and has a jump at ξ. Let $x_\nu=x_\nu^n$ be the fundamental points for $I_n[f]$. Suppose that $x_m<\xi<x_{m+1}$ and write

$$\theta=(\xi-x_m)\frac{2n+1}{2\pi},$$

so that $\theta=\theta_n$ is contained between 0 and 1. Then $I_n(\xi,f)$ is equiconvergent (as $n\to\infty$) with

$$\psi(\theta_n)f(\xi-0)+\psi(1-\theta_n)f(\xi+0),$$

where ψ is the function of Example 10. (de la Vallée-Poussin [3].)

[Since $I_n[f]$ converges to f at the points of continuity of f, it is enough to verify the result for a function f which is equal to 1 in a left-hand and to 0 in a right-hand neighbourhood of ξ. Apply the second formula for I_n in Example 1.

A similar result holds for $E_n[f]$ (with the same ψ and $\theta=(\xi-x_m)\,n/\pi$) and for $J_n[f]$.]

12. Let $\{n_k\}$ be lacunary, that is, let $n_{k+1}/n_k>q>1$ for all k. Then, almost everywhere,

(i) $I_{n_k}(x,f)=O(\log\log n_k)$ in the general case;

(ii) $I_{n_k}(x,f)=o(\log\log n_k)$ if f is continuous;

(iii) $I_{n_k}(x, f) = O(1)$ if f has modulus of continuity $O\left\{1/\log\log\frac{1}{\delta}\right\}$;

(iv) $I_{n_k}(x, f) \to f(x)$ if f has modulus of continuity $o\left\{1/\log\log\frac{1}{\delta}\right\}$.

[The proofs resemble that of (7·32). (i) Suppose that $|f| \leqslant 1$. By (7·24), with $n = \nu = n_k$, we have

(v) $$\int_0^{2\pi} \exp \lambda_0 \,|\, I_{n_k}(x, f) \,|\, dx \leqslant \mu_0.$$

Hence the set E_{n_k} of points where $|I_{n_k}| \geqslant (2/\lambda_0)\log\log n_k$ has measure $|E_{n_k}| \leqslant \mu_0(\log n_k)^{-2} = O(k^{-2})$.
(iii) By (v),

(vi) $$\int_0^{2\pi} \exp \lambda_0 \,|\, I_{n_k}[f] - f \,|\, dx \leqslant e^{\lambda_0}\mu_0$$

provided that $|f| \leqslant 1$. Write $f_n = f - \sigma_n[f]$. Since, by Chapter III, (3·16), $f_n = o(1/\log\log n)$ and since

$$I_n[f] - f = I_n[f_n] - f_n,$$

taking a fixed $\epsilon > 0$ we obtain from (vi) that

$$\int_0^{2\pi} \exp\{\epsilon^{-1}\log\log n_k . \,|\, I_{n_k}[f] - f \,|\}\, dx \leqslant e^{\lambda_0}\mu_0$$

for k large enough. This implies that $\limsup |I_{n_k}[f] - f| \leqslant 2\epsilon$ almost everywhere, and so also that $I_{n_k}[f] \to f$ almost everywhere.]

13. Given $\epsilon_1 > \epsilon_2 > \ldots \to 0$, there is a continuous f and a lacunary $\{n_k\}$ such that for almost all x

$$I_{n_k} \neq O(\epsilon_{n_k} \log\log n_k).$$

[The proof is analogous to that of (8·28). Denoting successive primes by ρ_ν, we choose the p_k so that $2p_k + 1$ runs through the primes of the form $\rho_j, \rho_{2j}, \rho_{4j}, \rho_{8j}, \ldots$. It is enough to use the estimate (8·32) for the primes.]

14. There is a function f with modulus of continuity $O(1/\log\log\delta^{-1})$ and a lacunary $\{n_k\}$ such that $\{I_{n_k}[f]\}$ diverges almost everywhere.
[The proof is similar to that of (8·30).]

CHAPTER XI

DIFFERENTIATION OF SERIES. GENERALIZED DERIVATIVES

1. Cesàro summability of differentiated series

In Chapter III, § 7, we investigated the Abel summability of $S'[f]$. We shall now consider the Cesàro summability of repeatedly differentiated series.

Suppose that a function $f(x)$ is defined in the neighbourhood of a point x_0 and that there exist constants $\alpha_0, \alpha_1, \ldots, \alpha_r$ such that for small $|t|$

$$f(x_0+t)=\alpha_0+\alpha_1 t+\ldots+\alpha_{r-1}\frac{t^{r-1}}{(r-1)!}+(\alpha_r+\epsilon_t)\frac{t^r}{r!}, \tag{1·1}$$

where ϵ_t tends to 0 with t. We then say† that f has a *generalized r-th derivative* $f_{(r)}(x_0)$ at x_0 and define $f_{(r)}(x_0)=\alpha_r$. Clearly $\alpha_0=f(x_0)$.

If $f_{(r)}(x_0)$ exists, so does $f_{(r-1)}(x_0)$. The definition of $f_{(1)}(x_0)$ coincides with that of the ordinary derivative $f'(x_0)$. If $f^{(r)}(x_0)$ exists, so does $f_{(r)}(x_0)$ and they have the same value; but the converse is not true for $r>1$, since then the equation (1·1) need not even imply the continuity of f in the neighbourhood of x_0.

The above definition is due to Peano. For applications to trigonometric series a certain modification of it, due to de la Vallée-Poussin, is of importance. We define it separately for r even and odd. Write

$$\chi_{x_0}(t)=\tfrac{1}{2}\{f(x_0+t)+f(x_0-t)\},$$

$$\psi_{x_0}(t)=\tfrac{1}{2}\{f(x_0+t)-f(x_0-t)\}.$$

Suppose first r even. If there are constants $\beta_0, \beta_2, \beta_4, \ldots, \beta_r$ such that

$$\chi_{x_0}(t)=\beta_0+\beta_2\frac{t^2}{2!}+\ldots+\beta_{r-2}\frac{t^{r-2}}{(r-2)!}+(\beta_r+\epsilon_t)\frac{t^r}{r!}, \tag{1·2}$$

where ϵ_t tends to 0 with t, we call β_r the *r-th generalized symmetric derivative*—or *r-th symmetric derivative* for short—of f at x_0. We denote it by the same symbol $f_{(r)}(x_0)$ as the *unsymmetric* (Peano) derivative and shall take care to avoid possible confusion. The definition of the rth symmetric derivative for r odd is similar, except that instead of (1·2) we consider the formula

$$\psi_{x_0}(t)=\beta_1 t+\beta_3\frac{t^3}{3!}+\ldots+\beta_{r-2}\frac{t^{r-2}}{(r-2)!}+(\beta_r+\epsilon_t)\frac{t^r}{r!}. \tag{1·3}$$

† This definition may be slightly modified so as to permit infinite derivatives. Suppose that there are constants $\alpha_0, \alpha_1, \ldots, \alpha_{r-1}$ such that the function $\alpha_r(t)$ defined by

$$f(x_0+t)=\alpha_0+\alpha_1 t+\ldots+\alpha_{r-1}\frac{t^{r-1}}{(r-1)!}+\alpha_r(t)\frac{t^r}{r!}$$

tends to a limit α_r (finite or infinite) as $t\to 0$. We can then call α_r the rth generalized derivative. We shall, however, consider only finite $f_{(r)}(x_0)$.

Taking the semi-sum and semi-difference of (1·1) for $\pm t$, we see that the existence of the unsymmetric derivative implies the existence of the symmetric one of the same order, and equality of both. For symmetric derivatives the existence of $f_{(r)}(x)$ implies that of $f_{(r-2)}(x)$ but not necessarily that of $f_{(r-1)}(x)$. The symmetric derivatives of orders 1 and 2 have already been considered in Chapter I, § 10, and are given by the formulae

$$f_{(1)}(x_0) = \lim_{t\to 0} \frac{f(x_0+\frac{1}{2}t)-f(x_0-\frac{1}{2}t)}{t}, \tag{1·4}$$

$$f_{(2)}(x_0) = \lim_{t\to 0} \frac{f(x_0+t)-2f(x_0)+f(x_0-t)}{t^2}. \tag{1·5}$$

In Chapter I, p. 22, we denoted these limits by $D_1 f(x_0)$ and $D_2 f(x_0)$ and called $D_2 f(x_0)$ the second Riemann (or Schwarz) derivative.†

The Riemann derivative of order r is defined by the formula

$$D_r f(x_0) = \lim_{t\to +0} \frac{\Delta^r f(x_0, t)}{t^r}, \tag{1·6}$$

where $\Delta^r f(x_0,t) = f(x_0+\frac{1}{2}tr) - \binom{r}{1} f(x_0+\frac{1}{2}t(r-2)) + \ldots + (-1)^r f(x_0-\frac{1}{2}tr)$

is the rth symmetric difference of f. It is immediately evident that (1·1) implies $D_r f(x_0) = \alpha_r$, and that also both (1·2) and (1·3) imply $D_r f(x_0) = \beta_r$. Thus, generally, if the symmetric derivative $f_{(r)}(x_0)$ exists, then $D_r f(x_0)$ exists and equals $f_{(r)}(x_0)$. The converse is false except for $r=1$ and $r=2$, in which cases the definitions of $f_{(r)}$ and $D_r f$ coincide.

(1·7) Theorem. *If the symmetric derivative $f_{(r)}(x_0)$ exists, then $S^{(r)}[f]$ is summable (C,α) at x_0 to sum $f_{(r)}(x_0)$, provided $\alpha > r$.*

We may suppose that $r < \alpha \leqslant r+1$. Denote the (C,α) kernel, that is, the nth (C,α) mean of the series $\frac{1}{2}+\cos x+\cos 2x+\ldots$, by $K_n^\alpha(x)$. We have

$$K_n^\alpha(t) = \frac{1}{A_n^\alpha}\sum_{\nu=0}^{n} A_{n-\nu}^{\alpha-1} \frac{\sin(\nu+\frac{1}{2})t}{2\sin\frac{1}{2}t} = \mathscr{I}\left\{\frac{e^{i(n+\frac{1}{2})t}}{2A_n^\alpha \sin\frac{1}{2}t}\sum_{\nu=0}^{n} A_\nu^{\alpha-1} e^{-i\nu t}\right\}. \tag{1·8}$$

We first show that, denoting by C constants independent of n and t,

$$\left|\frac{d^r}{dt^r} K_n^\alpha(t)\right| \leqslant C n^{r+1} \quad (0\leqslant t\leqslant \pi), \tag{1·9}$$

$$\left|\frac{d^r}{dt^r} K_n^\alpha(t)\right| \leqslant \frac{C}{n^{\alpha-r} t^{\alpha+1}} \quad \left(\frac{1}{n}\leqslant t\leqslant \pi\right) \tag{1·10}$$

for $0\leqslant\alpha\leqslant r+1$, $n=1,2,\ldots$.‡ (The inequalities hold for $-1<\alpha\leqslant r+1$, but we do not need this.)

Write

$$u(\beta,n,t) = \sum_{\nu=0}^{n} A_\nu^\beta e^{-i\nu t}.$$

† In Chapter IX we used the notation $D^2 f$ for $D_2 f$.

‡ (1·9) and (1·10) can be combined in one inequality:

$$\left|\frac{d^r}{dt^r} K_n^\alpha(t)\right| \leqslant \frac{C n^{r+1}}{(1+nt)^{\alpha+1}} \quad (0\leqslant t\leqslant\pi).$$

Summation by parts gives

$$u(\beta, n, t) = \{-A_n^\beta e^{-i(n+1)t} + u(\beta-1, n, t)\}(1-e^{-it})^{-1},$$

and so
$$u(\beta, n, t) = -e^{-i(n+1)t}\sum_{j=1}^{s} A_n^{\beta-j+1}(1-e^{-it})^{-j} + u(\beta-s, n, t)(1-e^{-it})^{-s}, \qquad (1\cdot11)$$

for $s = 1, 2, \ldots$. Hence, as is easily seen,

$$K_n^\alpha(t) = \frac{1}{A_n^\alpha}\mathscr{I}\left\{-\frac{e^{-\frac{1}{2}it}}{2\sin\frac{1}{2}t}\sum_{j=1}^{s}\frac{A_n^{\alpha-j}}{(1-e^{-it})^j} + \frac{e^{i(n+\frac{1}{2})t}}{(1-e^{-it})^\alpha 2\sin\frac{1}{2}t} - \frac{\sum_{\nu=n+1}^{\infty} A_\nu^{\alpha-s-1}e^{-i(\nu-n-\frac{1}{2})t}}{(1-e^{-it})^s 2\sin\frac{1}{2}t}\right\}, \qquad (1\cdot12)$$

provided the last series converges.

Take s so large that the last series termwise differentiated r times is absolutely convergent; it is enough to suppose that $s > \alpha + r$. Since $A_n^\delta = O(n^\delta)$, it is easy to see that the rth derivative of the expression in curly brackets is less in absolute value than the sum of the three expressions

$$C\sum_{j=1}^{s}\frac{n^{\alpha-j}}{t^{j+1+r}}, \quad C\sum_{\mu=0}^{r}\frac{n^\mu}{t^{\alpha+1+r-\mu}}, \quad C\sum_{\mu=0}^{r}\frac{n^{\alpha-s+\mu}}{t^{s+r-\mu+1}}, \qquad (1\cdot13)$$

using for the last estimate the fact that

$$\sum_{\nu=n+1}^{\infty} A_\nu^{\alpha-s-1}(\nu-n-\tfrac{1}{2})^\mu = \sum_{\nu=n+1}^{\infty} O(\nu^{\alpha-s+\mu-1}) = O(n^{\alpha-s+\mu}).$$

We now make use of the inequality $\alpha \leqslant r+1$. If $nt \geqslant 1$, the terms of the three sums (1·13) are, respectively, less than

$$n^\alpha(nt)^{-j}t^{-1-r} \leqslant n^\alpha(nt)^{-1}t^{-1-r} \leqslant n^\alpha(nt)^{r-\alpha}t^{-1-r},$$
$$n^\alpha(nt)^{\mu-\alpha}t^{-1-r} \leqslant n^\alpha(nt)^{r-\alpha}t^{-1-r},$$
$$n^\alpha(nt)^{\mu-s}t^{-1-r} \leqslant n^\alpha(nt)^{r-s}t^{-1-r} \leqslant n^\alpha(nt)^{r-\alpha}t^{-1-r},$$

and collecting results we see that $|\{K_n^\alpha(t)\}^{(r)}| \leqslant C(nt)^{r-\alpha}t^{-1-r}$, which is (1·10).

To prove (1·9) we note that if $\alpha \geqslant 0$ then $|\{K_n^\alpha(t)\}^{(r)}|$ is

$$\left|\left\{\tfrac{1}{2}A_n^\alpha + \sum_{\nu=1}^{r} A_{n-\nu}^\alpha \cos\nu t\right\}^{(r)}\right| \Big/ A_n^\alpha \leqslant 1^r + 2^r + \ldots + n^r \leqslant n^{r+1}.$$

To complete the proof of (1·7) suppose that r is even; the argument for r odd is similar. If the (C, α) means of $\mathsf{S}[f]$ are $\sigma_n^\alpha(x)$, the (C, α) means of $\mathsf{S}^{(r)}[f]$ are $\{\sigma_n^\alpha(x)\}^{(r)}$, and

$$\{\sigma_n^\alpha(x_0)\}^{(r)} = \frac{1}{\pi}\int_{-\pi}^{\pi} f(x_0+t)\{K_n^\alpha(t)\}^{(r)}\,dt = \frac{2}{\pi}\int_0^\pi \chi_{x_0}(t)\{K_n^\alpha(t)\}^{(r)}\,dt. \qquad (1\cdot14)$$

Since (1·7) is obvious if f is a trigonometric polynomial $T(x)$, and since, given any $2s+1$ numbers $\xi_0, \xi_1, \ldots, \xi_{2s}$, we can always find a T such that $T^{(j)}(x_0) = \xi_j$ for $j = 0, 1, \ldots, 2s$ (by writing T in the complex form and solving the equations for the coefficients), we may subtract a suitable T from f and suppose that $\beta_r = \beta_{r-2} = \ldots = 0$ in (1·2). Then by (1·9) and (1·10), the last member in (1·14) is

$$\int_0^{1/n} o(t^r)\,O(n^{r+1})dt + \int_{1/n}^{\pi} o(t^r)\,O\{n^{r-\alpha}t^{-\alpha-1}\}\,dt = o(1) + o(1) = o(1), \qquad (1\cdot15)$$

and (1·7) is proved.

Remark. It is not difficult to construct examples showing that (1·7) is false for $\alpha=r$ (see vol. I, p. 314, Example 2). It is, however, useful to observe that, *if* $f=0$ *near* $x=x_0$, *then* $S^{(r)}[f]$ *is summable* (C, r) *at* x_0 *to sum* 0.

This is an immediate consequence of the localization theorem (9·20) of Chapter IX, but it can also be obtained directly from (1·12) and the resulting estimates for $\{K_n^r(t)\}^{(r)}$. First observe that if $f=0$ for $|x-x_0|<\delta$, then in the last integral (1·14) we actually integrate over $\delta\leqslant t\leqslant\pi$. If we drop the curly brackets in (1·12), the contributions of the first and last terms on the right to $\{K_n^r(t)\}^{(r)}$ are $o(1)$ in (δ,π). Although the contribution of the middle term on the right is apparently only $O(1)$, an application of the Riemann-Lebesgue theorem (Chapter II, (4·4)) shows that this term is $o(1)$.

Suppose that f, defined in the neighbourhood of x_0, has $r-1$ unsymmetric derivatives $\alpha_0, \alpha_1, \ldots, \alpha_{r-1}$, and define $\omega_r(x_0,t)$ by

$$f(x_0+t)=\alpha_0+\alpha_1 t+\ldots+\alpha_{r-1}\frac{t^{r-1}}{(r-1)!}+\omega_r(x_0,t)\frac{t^r}{r!}. \tag{1·16}$$

If $\omega_r(x_0,t)$ has a limit as $t\to 0$, f has also an rth derivative $f_{(r)}(x_0)$. It can happen that $\omega_r(x_0,t)$ has different limits for $t\to+0$ and $t\to-0$. In this case we may consider

$$\delta_r(x_0)=\lim_{t\to+0}\delta_r(x_0,t), \quad \text{where} \quad \delta_r(x_0,t)=\omega_r(x_0,t)-\omega_r(x_0,-t)$$

as the *jump* of $f_{(r)}$ at x_0, even if $f_{(r)}(x)$ is not defined near x_0. Clearly if f has at x_0 ordinary right-hand and left-hand rth derivatives $f_+^{(r)}(x_0)$ and $f_-^{(r)}(x_0)$, then $\delta_r(x_0)$ exists and equals $f_+^{(r)}(x_0)-f_-^{(r)}(x_0)$.

If r is odd, (1·16) gives

$$\chi_{x_0}(t)=\alpha_0+\alpha_2\frac{t^2}{2!}+\ldots+\alpha_{r-1}\frac{t^{r-1}}{(r-1)!}+\tfrac{1}{2}\delta_r(x_0,t)\frac{t^r}{r!}; \tag{1·17}$$

and if r is even,

$$\psi_{x_0}(t)=\alpha_1 t+\alpha_3\frac{t^3}{3!}+\ldots+\alpha_{r-1}\frac{t^{r-1}}{(r-1)!}+\tfrac{1}{2}\delta_r(x_0,t)\frac{t^r}{r!}. \tag{1·18}$$

Suppose now, without assuming anything about $\omega_r(x_0,t)$, that there exist constants α_j such that we have either (1·17) or (1·18), according as r is odd or even, and that

$$\delta_r(x_0)=\lim_{t\to+0}\delta_r(x_0,t) \tag{1·19}$$

exists. Then $\delta_r(x_0)$ may be thought of as a jump of the rth derivative, even if this derivative does not exist near x_0.

If $r=0$, (1·18) may be interpreted as $\psi_{x_0}(t)=\tfrac{1}{2}\delta_0(x_0,t)$, that is,

$$\delta_0(x_0)=\lim_{t\to+0}\{f(x_0+t)-f(x_0-t)\},$$

and $\delta_0(x_0)=0$ means that f is symmetrically continuous at x_0. If $r=1$, (1·17) gives

$$\delta_1(x_0,t)=\frac{f(x_0+t)+f(x_0-t)-2f(x_0)}{t},$$

and $\delta_1(x_0)=0$ means that f is smooth at x_0 (Chapter II, § 3), a property which serves as a substitute† for the continuity of the first derivative at x_0.

† The following observation may be helpful. When investigating the existence or behaviour of a symmetric $f_{(r)}(x_0)$ we usually presuppose the existence of symmetric $f_{(r-2)}(x_0)$, $f_{(r-4)}(x_0)$, ...; the symmetric $f_{(r-1)}(x_0)$, $f_{(r-3)}(x_0)$, ... need not exist. (For example, the existence of a symmetric $f_{(1)}(x_0)$ does not presuppose that $f(x_0)$ is defined.) The existence of $\delta_r(x_0)$, however, presupposes that of these last derivatives. Hence the existence of a symmetric $f_{(r)}(x_0)$ and of $\delta_r(x_0)$ implies the existence of an unsymmetric $f_{(r-1)}(x_0)$.

It is natural to call the condition $\delta_r(x_0)=0$ the *smoothness of order r* of f at x_0.

In Chapter III, § 9, we showed that the existence of the jump $\delta_0(x_0)$ is reflected in the behaviour of the terms of $S'[f]$ at x_0. We generalize this result.

(1·20) THEOREM. *Let $f\sim\Sigma A_n(x)$ and suppose that $\delta_r(x_0)$ exists and is finite. Then, for any $\alpha>r+1$,*

$$\delta_r(x_0)=\pi(\mathrm{C},\alpha)\lim A_n^{(r+1)}(x_0). \tag{1·21}$$

Suppose, say, that r is even and consider the function $g(x)=\Sigma n^{-1}\sin n(x-x_0)$, which has jump π at x_0, and its rth integral

$$G(x)=(-1)^{\frac{1}{2}r}\,\Sigma n^{-r-1}\sin n(x-x_0).$$

G has at x_0 ordinary right-hand and left-hand rth derivatives, whose difference is π, and we easily verify the theorem for $f=G$. By subtracting $\pi^{-1}\delta_r(x_0)\,G(x)$ from $f(x)$ we reduce the general case to that when $\delta_r(x_0)=0$. Further, by subtracting from f a trigonometric polynomial (for which the theorem is obvious), we may suppose that

$$f_{(r-1)}(x_0)=f_{(r-3)}(x_0)=\ldots=0,$$

so that, finally,

$$\psi_{x_0}(t)=o(t^r). \tag{1·22}$$

Now the (C,α) means of $\{A_n^{(r+1)}(x_0)\}$ are

$$\frac{1}{A_n^\alpha}\frac{1}{\pi}\int_{-\pi}^{\pi}f(x_0+t)\left\{\tfrac{1}{2}A_n^{\alpha-1}+\sum_{\nu=1}^{n}A_{n-\nu}^{\alpha-1}\cos\nu t\right\}^{(r+1)}dt=\frac{2}{\pi}\int_0^{\pi}\psi_{x_0}(t)\frac{A_n^{\alpha-1}}{A_n^\alpha}\{K_n^{\alpha-1}(t)\}^{(r+1)}dt, \tag{1·23}$$

$K_n^\alpha(t)$ denoting the (C,α) kernel considered above. We may suppose that $r+1<\alpha\leqslant r+2$ and apply (1·9) and (1·10) with $\alpha-1$ for α and $r+1$ for r. Since $A_n^{\alpha-1}/A_n^\alpha=O(1/n)$, we immediately see, using estimates analogous to (1·15), that the last member in (1·23) is $o(1)$. This completes the proof of (1·20).

We now consider the problem of the summability of $\tilde{S}^{(r)}[f]$. We recall that

$$\tilde{f}(x)=-\frac{2}{\pi}\int_0^\infty\frac{\psi_x(t)}{t}\,dt \tag{1·24}$$

for almost all x (Chapter III, (3·23), Chapter II, (7·6)).

(1·25) THEOREM. *If f satisfies $\delta_r(x_0)=0$ (in particular if an unsymmetric $f_{(r)}(x_0)$ exists), and if $r<\alpha\leqslant r+1$, then*

$$\{\tilde{\sigma}_n^\alpha(x_0)\}^{(r)}-\frac{(-1)}{\pi}\int_{1/n}^{\infty}\frac{\delta_r(x_0,t)}{t}\,dt\to 0 \tag{1·26}$$

as $n\to\infty$. In particular, $\tilde{S}^{(r)}[f]$ is summable (C,α) *at x_0 if and only if the integral*

$$-\frac{1}{\pi}\int_0^\infty\frac{\delta_r(x_0,t)}{t}\,dt=-\frac{1}{\pi}\lim_{\epsilon\to+0}\int_\epsilon^\infty \tag{1·27}$$

exists; and if the integral does exist it represents the (C,α) *sum of $\tilde{S}^{(r)}[f]$ at x_0.*

The restriction $\alpha\leqslant r+1$ can be dropped, but it saves calculation.

For $r=0$ this theorem is included in Theorem (5·8) of Chapter III. The integral (1·27) may be called the *conjugate of f of order r*. It converges to $(d^r\tilde{f}/dx^r)_{x=x_0}$ if, for example, f has r continuous derivatives and $f^{(r)}$ is in Λ_β for some $\beta>0$ (since in this case $\tilde{f}^{(r)}$ exists and is continuous).

If $r \geqslant 1$ the integral on the right-hand side of (1·27) converges absolutely for each $\epsilon > 0$. For (1·17) and (1·18) imply that, for $t > \epsilon$,

$$t^{-1}\delta_r(x_0, t) = O(t^{-2}) + O\left\{\frac{|f(x_0+t)| + |f(x_0-t)|}{t^{r+1}}\right\}.$$

Suppose, for example, that r is even. Since after subtracting from f a suitable trigonometric polynomial we have $\psi_{x_0}(t) = o(t^r)$, it is enough to prove (1·25) in two special cases: (i) f is a mononomial $e^{i\nu x}$, (ii) f satisfies $\psi_{x_0}(t) = o(t^r)$.

(i) Clearly $\{\tilde{\sigma}_n^\alpha\}^{(r)} \to \tilde{f}^{(r)}$, and, by (1·24),

$$\tilde{f}^{(r)}(x_0) = -\frac{2}{\pi}\int_0^\infty t^{-1}\frac{d^r}{dt^r}\psi_{x_0}(t)\,dt = -\frac{1}{\pi}\int_0^\infty t^{-1}\frac{d^r}{dt^r}\left\{\delta_r(x_0, t)\frac{t^r}{r!}\right\}dt$$

$$= -\frac{1}{\pi}\int_0^\infty \frac{\delta_r(x_0, t)}{t}\,dt,$$

since $\psi_{x_0}(t)$ and $\frac{1}{2}\delta_r(x_0, t)\,t^r/r!$ differ by a polynomial of degree $r-1$ in t; that the integrated terms in the integration by parts are all 0 follows, if at $t = \infty$ we use the fact that $\delta_r t^r$ is the sum of a trigonometric polynomial and a power polynomial of degree $r-1$, and at $t = 0$ the fact that $\delta_r(x_0, t)$ is odd in t and so, in our case, tends to 0 with t.

(ii) Let $\tilde{K}_n^\alpha(t)$ be the conjugate (C, α) kernel, and

$$H_n(t) = \tfrac{1}{2}\cot\tfrac{1}{2}t - \tilde{K}_n^\alpha(t).$$

If $0 \leqslant \alpha \leqslant r+1$, we have $$|\{\tilde{K}_n^\alpha(t)\}^{(r)}| \leqslant Cn^{r+1} \quad (0 \leqslant t \leqslant \pi), \tag{1·28}$$

$$|H_n^{(r)}(t)| \leqslant Cn^{r-\alpha}t^{-\alpha-1} \quad (1/n \leqslant t \leqslant \pi), \tag{1·29}$$

inequalities analogous to (1·9) and (1·10) and proved in the same way. (We obtain $\tilde{K}_n^\alpha(t)$ by replacing $\sin(\nu+\frac{1}{2})t$ by $1 - \cos(\nu+\frac{1}{2})t$ in the second member of (1·8), so that $H_n(t)$ is obtained by replacing $\mathscr{I}$ by $\mathscr{R}$ in the member that follows.) Since r is even,

$$\{\tilde{\sigma}_n^\alpha(x_0)\}^{(r)} = -\frac{2}{\pi}\int_0^\pi \psi_{x_0}(t)\{\tilde{K}_n^\alpha(t)\}^{(r)}\,dt,$$

$$\{\tilde{\sigma}_n^\alpha(x_0)\}^{(r)} - \left\{-\frac{2}{\pi}\int_{1/n}^\pi \psi_{x_0}(t)\,(\tfrac{1}{2}\cot\tfrac{1}{2}t)^{(r)}\,dt\right\}$$

$$= -\frac{2}{\pi}\int_0^{1/n}\psi_{x_0}(t)\{\tilde{K}_n^\alpha(t)\}^{(r)}\,dt + \frac{2}{\pi}\int_{1/n}^\pi \psi_{x_0}(t)\,H_n^{(r)}(t)\,dt, \tag{1·30}$$

and the hypothesis $\psi_{x_0}(t) = o(t^r)$, together with (1·28) and (1·29), shows that the terms on the right in (1·30) are $o(1)$.

Hence it is enough to show that the difference between

$$-\frac{2}{\pi}\int_{1/n}^\pi \psi_{x_0}(t)\,(\tfrac{1}{2}\cot\tfrac{1}{2}t)^{(r)}\,dt = -\frac{1}{\pi}\int_{1/n \leqslant |t| \leqslant \pi} f(x_0+t)\,(\tfrac{1}{2}\cot\tfrac{1}{2}t)^{(r)}\,dt \tag{1·31}$$

and the second term in (1·26) tends to 0 as $n \to \infty$. We may suppose that $r \geqslant 1$, since if $r = 0$ we know the theorem to be true. Using the formula

$$(\tfrac{1}{2}\cot\tfrac{1}{2}t)^{(r)} = \left\{\sum_{-\infty}^{+\infty}\frac{1}{t+2\pi\nu}\right\}^{(r)} = (-1)^r r!\,\Sigma\frac{1}{(t+2\pi\nu)^{r+1}}$$

and the periodicity of f, we write the right-hand side of (1·31) in the form

$$-\frac{1}{\pi}\int f(x_0+t)\frac{dt}{t^{r+1}},$$

where the integral is taken over the complement of the sum of the intervals

$$I_\nu = (2\pi\nu - 1/n,\, 2\pi\nu + 1/n).$$

In view of the absolute integrability of $f(x_0+t)\,t^{-r-1}$ over $|t| \geqslant \pi$, the integral of the latter function over $\sum_{\nu\neq 0} I_\nu$ tends to zero as $n\to\infty$. Hence the difference between (1·31) and

$$-r!\frac{1}{\pi}\int_{|t|\geqslant 1/n}\frac{f(x_0+t)}{t^{r+1}}\,dt = -\frac{1}{\pi}\int_{1/n}^{\infty}\frac{\delta_r(x_0,t)}{t}\,dt$$

tends to zero and the theorem is established.

Remarks. (*a*) If $f=0$ near x_0, then $\tilde{S}^{(r)}[f]$ is summable (C, r) at x_0 to sum (1·27). This follows either by appealing to Theorem (9·20) of Chapter IX or by using an argument parallel to the one indicated in the remark on p. 62.

(*b*) Suppose that f is periodic and absolutely continuous, and has an rth conjugate function, $\tilde{f}_r$, say, at x_0. Then, by hypothesis, the derivatives $f_{(j)}(x_0)=\alpha_j$ exist for $j<r$, and if, for example, r is even, then

$$\tilde{f}_r(x_0) = -r!\frac{2}{\pi}\int_0^\infty t^{-r-1}\left\{\tfrac{1}{2}[f(x_0+t)-f(x_0-t)] - \sum_j \frac{\alpha_j}{j!}t^j\right\}dt, \tag{1·32}$$

where $j=1, 3, \ldots, r-1$.

Since the expression in curly brackets is $o(t^r)$ for $t\to 0$, and $O(t^{r-1})$ for $t\to\infty$, integration by parts shows that

$$\tilde{f}_r(x_0) = (r-1)!\frac{2}{\pi}\int_0^\infty t^{-r}\left\{\tfrac{1}{2}[f'(x_0+t)+f'(x_0-t)] - \Sigma\frac{\alpha_j}{(j-1)!}t^{j-1}\right\}dt. \tag{1·33}$$

(Somewhat more generally, if f is smooth of order r at x_0, the integrals (1·32) and (1·33) are equiconvergent at $t=0$.)

We might write (1·33) in the form $\tilde{f}_r = (\widetilde{f'})_{r-1}$; but this is not strictly true, since f' need not have generalized derivatives in the present sense. (None the less, the relation $f_{(j)} = (f')_{(j-1)}$ does hold, in some sense, at least almost everywhere in x; see Theorem (4·26) below.) The interpretation is correct if we suppose that f has an ordinary $(r-1)$th derivative and the latter is absolutely continuous. We may then repeat the argument r times and come to the following conclusion:

If f has an ordinary $(r-1)$-th derivative which is absolutely continuous, then $\tilde{f}_r$ exists almost everywhere and is equal to the conjugate function of $f^{(r)}$.

2. Summability C of Fourier series

In Chapter II we obtained a number of tests for the convergence of the Fourier series of a function f at a given point. All those tests represent sufficient conditions only, and the problem of finding a non-trivial *necessary and sufficient* condition is still open. It is conceivable that no such condition exists, and that the convergence of the integral

$$\frac{1}{\pi}\int_{-\epsilon}^{\epsilon} f(x_0+t)\frac{\sin nt}{t}\,dt$$

to $f(x_0)$ as $n\to\infty$ is a property *sui generis* not expressible in terms of simpler properties.

The situation is similar if instead of ordinary convergence we consider summability by an assigned Cesàro mean, for example, summability (C, 1): Fejér's fundamental theorem (Chapter III, (3·4)) gives a sufficient condition only. We therefore change the problem and ask, not when $S[f]$ is summable by some particular Cesàro mean, but when it is summable by *some mean or another*, that is, when it is summable C. We first prove the following result:

(2·1) THEOREM. *Let* $\alpha > -1$, *and suppose that*

$$\tfrac{1}{2}a_0 + \sum_{n=1}^{\infty}(a_n\cos nx + b_n\sin nx) = \sum_{n=0}^{\infty} A_n(x) \tag{2·2}$$

is summable (C, α) *at* x_0 *to a finite sum s. Let* $r > \alpha + 1$, *and suppose that* (2·2) *integrated termwise r times converges in the neighbourhood of* x_0 *to sum* $F(x)$. (*This is certainly the case if* $|a_n|+|b_n| = o(n^{\alpha})$.) *Then the symmetric derivative* $F_{(r)}(x_0)$ *exists and equals s.*

We consider separately the cases (i) α integral, (ii) α fractional. Only (i) is important for our immediate purposes, but (ii) is of sufficient independent interest.

(i) We may suppose that $r = \alpha + 2$. Suppose also that r is even; the proof for r odd is similar. Then

$$F(x) = \tfrac{1}{2}a_0\frac{x^r}{r!} + (-1)^{\frac{1}{2}r}\sum_{1}^{\infty} n^{-r}(a_n\cos nx + b_n\sin nx). \tag{2·3}$$

Without loss of generality we may suppose that $x_0 = 0$, $s = 0$, $a_0 = 0$, and also that (2·2) is a purely cosine series, since the contribution of the sine part to the rth symmetric derivative at 0 is 0. We denote the kth Cesàro sums of $0 + a_1 + a_2 + \dots$ by s_n^k, and write

$$\gamma(t) = \frac{\cos t}{t^r},$$

$$\Delta u_n = \Delta^1 u_n = u_n - u_{n+1}, \quad \Delta^j u_n = \Delta(\Delta^{j-1}u_n)$$

for any sequence $\{u_n\}$.

Summing by parts $r - 1 = \alpha + 1$ times we have

$$F(t) = (-1)^{\frac{1}{2}r}t^r\sum_{1}^{\infty} a_n\gamma(nt) = (-1)^{\frac{1}{2}r}t^r\sum s_n^{\alpha}\Delta^{\alpha+1}\gamma(nt). \tag{2·4}$$

It is well known that if $u(x)$ is differentiable j times, then on writing $u_m = u(x_0 + mh)$, where $h > 0$, we have

$$\Delta^j u_n = (-1)^j h^j u^{(j)}(x_0 + nh + \theta jh) \quad (0<\theta<1). \tag{2·5}$$

Let
$$P(x) = \sum_{0}^{\frac{1}{2}r-1}(-1)^{\nu}\frac{x^{2\nu}}{(2\nu)!}, \quad \lambda(x) = \frac{\cos x - P(x)}{x^r}.$$

Then $\gamma(nt) = \lambda(nt) + P(nt)(nt)^{-r}$ and

$$F(t) = \sum_{\nu=0}^{\frac{1}{2}r-1}\frac{D_\nu}{(2\nu)!}t^{2\nu} + t^r R(t), \tag{2·6}$$

where
$$D_\nu = (-1)^{\frac{1}{2}r+\nu}\sum s_n^{\alpha}\Delta^{\alpha+1}n^{2\nu-r}, \quad R(t) = (-1)^{\frac{1}{2}r}\sum s_n^{\alpha}\Delta^{\alpha+1}\lambda(nt).$$

The series defining the D_ν converge absolutely since $s_n^\alpha = o(n^\alpha)$ and, by (2·5),

$$\Delta^{\alpha+1} n^{2\nu-r} = O(n^{2\nu-r-\alpha-1}) = O(n^{-\alpha-3}).$$

The theorem will be established if we show that $R(t) \to 0$ as $t \to 0$. We write $N = N(t) = [1/t]$ for $0 < t \leqslant 1$. Then

$$|R(t)| \leqslant \sum_1^\infty |s_n^\alpha \Delta^{\alpha+1}\lambda(nt)| = \sum_1^N + \sum_{N+1}^\infty = U + V. \tag{2·7}$$

The function $\lambda(u)$ being indefinitely differentiable for all u, an application of (2·5) shows that

$$|\Delta^{\alpha+1}\lambda(nt)| \leqslant Ct^{\alpha+1} \quad (0 \leqslant n \leqslant N),$$

$$U \leqslant Ct^{\alpha+1} \sum_1^N |s_n^\alpha| = Ct^{\alpha+1} o(N^{\alpha+1}) = o(1), \tag{2·8}$$

C denoting constants independent of n and t. It is also easily seen that $|\gamma^{(\alpha+1)}(u)| \leqslant Cu^{-r}$, and hence $|\lambda^{(\alpha+1)}(u)| \leqslant Cu^{-r}$, for $u \geqslant 1$. Using (2·5) again we therefore have

$$V \leqslant Ct^{\alpha+1} \sum_{N+1}^\infty o(n^\alpha) \frac{1}{n^r t^r} \leqslant t^{-1} \sum_{N+1}^\infty o(n^{-2}) = o(1).$$

Hence $V = o(1)$, $U + V = o(1)$, and the theorem follows.

(ii) We may suppose that $\alpha < r-1 < \alpha+1$, so that

$$r - 1 = \alpha + \delta \quad (0 < \delta < 1), \tag{2·9}$$

and also (as before) that r is even, and that $x_0 = 0$, $s = 0$, $a_0 = 0$. We keep the previous definitions of $\gamma(t)$, $P(t)$, $\lambda(t)$.

Summing by parts $r-1$ times we find that $F(t)$ is equal to

$$(-1)^{\frac{1}{2}r} \sum_{n=1}^\infty s_n^{r-2} \Delta^{r-1} \frac{\cos nt}{n^r} = (-1)^{\frac{1}{2}r} t^r \sum_1^\infty s_n^{r-2} \Delta^{r-1}\lambda(nt) + (-1)^{\frac{1}{2}r} \sum_1^\infty s_n^{r-2} \Delta^{r-1} \frac{P(nt)}{n^r}.$$

The last term is an even polynomial of degree $r-2$ (whose coefficients are convergent infinite series), and for the proof of the theorem it is enough to show that

$$\sum_{n=1}^\infty s_n^{r-2} \Delta^{r-1}\lambda(nt) = o(1) \quad \text{as} \quad t \to 0. \tag{2·10}$$

We express s_n^{r-2} in terms of the s_ν^α. After (2·9),

$$s_n^{r-2} = \sum_{\nu=0}^n s_\nu^\alpha A_{n-\nu}^{-2+\delta}, \tag{2·11}$$

and the left-hand side of (2·10) is, after changing the order of summation,

$$\sum_{\nu=1}^\infty s_\nu^\alpha \sum_{n=\nu}^\infty A_{n-\nu}^{-2+\delta} \Delta^{r-1}\lambda(nt) = \sum_{\nu=1}^\infty s_\nu^\alpha \sum_{n=0}^\infty A_n^{-2+\delta} \Delta^{r-1}\lambda((n+\nu)t) = \sum_{\nu=1}^\infty s_\nu^\alpha \xi_\nu(t), \tag{2·12}$$

say. The interchange of the order of summation is legitimate since the double series resulting from the substitution of (2·11) into the series (2·10) converges absolutely: to see this, observe that, since $\alpha > 0$, the sum (2·11) is majorized by†

$$\sum_{\nu=0}^{n} |s_\nu^\alpha| \, |A_{n-\nu}^{-2+\delta}| \leqslant Cn^\alpha \sum_{\nu=0}^{n} |A_{n-\nu}^{-2+\delta}| \leqslant Cn^\alpha, \tag{2·13}$$

and that $\Delta^r \lambda(nt) = O(n^{-r})$ and $r - \alpha > 1$.

Since $\cos u - P(u)$ vanishes at 0 together with its first $r-1$ derivatives, we have successively

$$\cos u - P(u) = \frac{(-1)^{\frac{1}{2}r}}{(r-1)!} \int_0^u (u-v)^{r-1} \cos v \, dv,$$

$$\frac{\cos nt - P(nt)}{n^r} = \frac{(-1)^{\frac{1}{2}r}}{(r-1)!} \int_0^t (t-w)^{r-1} \cos nw \, dw = \frac{(-1)^{\frac{1}{2}r}}{(r-1)!} \mathscr{R} \int_0^t (t-w)^{r-1} e^{inw} \, dw,$$

$$\Delta^{r-1} \lambda(nt) = \frac{(-1)^{\frac{1}{2}r}}{(r-1)! \, t^r} \mathscr{R} \int_0^t (t-w)^{r-1} (1 - e^{iw})^{r-1} e^{inw} \, dw,$$

$$\begin{aligned} \xi_\nu(t) &= \frac{(-1)^{\frac{1}{2}r}}{(r-1)! \, t^r} \mathscr{R} \int_0^t (t-w)^{r-1} (1 - e^{iw})^{r-1} \sum_{n=\nu}^{\infty} A_{n-\nu}^{-2+\delta} e^{inw} \, dw \\ &= \frac{(-1)^{\frac{1}{2}r}}{(r-1)! \, t^r} \mathscr{R} \int_0^t (t-w)^{r-1} (1 - e^{iw})^{\alpha+1} e^{i\nu w} \, dw. \end{aligned} \tag{2·14}$$

The proof of (ii) will be complete if we prove the two estimates

$$|\xi_\nu(t)| \leqslant C \frac{t^{1-\delta}}{\nu^{r-1}}, \quad |\xi_\nu(t)| \leqslant C \frac{t^{-\delta}}{\nu^r}, \tag{2·15}$$

since then, with $N = [1/t]$,

$$\begin{aligned} \sum_1^\infty s_\nu^\alpha \xi_\nu(t) &= \sum_1^N o(\nu^\alpha) \, O\!\left(\frac{t^{1-\delta}}{\nu^{r-1}}\right) + \sum_{N+1}^\infty o(\nu^\alpha) \, O\!\left(\frac{t^{-\delta}}{\nu^r}\right) \\ &= O(t^{1-\delta}) \sum_1^N o(\nu^{-\delta}) + O(t^{-\delta}) \sum_{N+1}^\infty o(\nu^{-\delta-1}) = o(1). \end{aligned}$$

To prove the first estimate (2·15), integrate the last integral (2·14) by parts $r-1$ times. Since $r - 1 < \alpha + 1$ the integrated terms all vanish, and using the formula for the $(r-1)$th derivative of a product we find

$$\begin{aligned} \pm (i\nu)^{-(r-1)} \int_0^t e^{i\nu w} \{(t-w)^{r-1} (1 - e^{iw})^{\alpha+1}\}^{(r-1)} \, dt &= \nu^{-(r-1)} \int_0^t \sum_{j=0}^{r-1} O(w^{\alpha+1-j}) \, O(t^j) \, dt \\ &= O(t^{\alpha+2} \nu^{-(r-1)}), \end{aligned} \tag{2·16}$$

which gives the first formula (2·15). To prove the second formula, we integrate the first integral in (2·16) by parts once more and obtain

$$(i\nu)^{-r} O(t^{\alpha+1}) + (i\nu)^{-r} \int_0^t \sum_{j=0}^{r-1} O(w^{\alpha-j}) \, O(t^j) \, dt = O(\nu^{-r} t^{\alpha+1}),$$

the needed estimate. This completes the proof of Theorem (2·1).

† If $-1 < \alpha < 0$ (i.e. $r = 1$) we have for the first member in (2·13) the estimate

$$\sum_0^n \leqslant Cn^{-2+\delta} \sum_1^{[\frac{1}{2}n]} o(\nu^\alpha) + o(n^\alpha) \sum_{[\frac{1}{2}n]+1}^{n} |A_{n-\nu}^{-2+\delta}| = o(n^{-2+\delta+\alpha+1}) + o(n^\alpha) = o(n^\alpha).$$

If $\Sigma A_n(x)$ is of power series type, (2·1) can be strengthened by considering unsymmetric derivatives.

(2·17) THEOREM. *Let* $\alpha > -1$ *and suppose that*

$$\sum_0^\infty c_n e^{inx} \tag{2·18}$$

is summable (C, α) *at* x_0 *to a finite sum* s. *Let* $r > \alpha + 1$ *and suppose that*

$$c_0 \frac{x^r}{r!} + \sum_1^\infty \frac{c_n}{(in)^r} e^{inx} \tag{2·19}$$

converges near x_0 *to sum* $F(x)$. *Then the unsymmetric* $F_{(r)}(x)$ *exists and equals* s.

The proof runs parallel to that of Theorem (2·1).

The following result is a corollary of (2·1):

(2·20) THEOREM. *If a series* $\Sigma A_n(x)$ *has coefficients* $o(n^\alpha)$ *and is summable* (C, α) *at* x_0 *to sum* s, *then it is also summable at* x_0 *by the Riemann method* R_r *to sum* s, *provided* $r - \alpha > 1$. *By this we mean that*

$$\lim_{h\to 0} \left\{ \tfrac{1}{2}a_0 + \sum_{n=1}^\infty (a_n \cos nx_0 + b_n \sin nx_0) \left(\frac{\sin nh}{nh} \right)^r \right\} = s. \tag{2·21}$$

In particular, a series summable by some Cesàro mean of negative order is also summable by the method of Lebesgue (which is the method R_1; see Chapter IX, § 2), and a series summable (C, α), $\alpha < 1$, is summable R_2. The latter result generalizes the classical Theorem (2·4) of Chapter IX.

The special case $r = 1$ of (2·17) deserves attention.

(2·22) THEOREM. *Suppose that* $\Sigma A_n(x)$ *has coefficients* $O(n^k)$ *for some* k. *Then a necessary and sufficient condition that the series should be summable* C *at* x_0 *to sum* s *is that there should exist an integer* $r > 0$ *such that the function* $F(x)$ *obtained by integrating* $\Sigma A_n(x)$ *termwise* r *times should have a symmetric* $F_{(r)}(x_0)$. *If the series is of power series type, then in the above statement we may replace the symmetric* $F_{(r)}(x_0)$ *by the unsymmetric one.*

This is an immediate consequence of (1·7) and of (2·1) and (2·17).

When the series concerned is a Fourier series, Theorem (2·22) can be stated in a slightly different form.

Given a function $g(t)$ defined to the right of $t = 0$, we say that the number s is the (C, r) *limit* of $g(t)$ as $t \to +0$ if

$$rt^{-r} \int_0^t g(u)(t-u)^{r-1} du \to s \quad \text{as} \quad t \to 0 \quad (r > 0). \tag{2·23}$$

This is an extension of the notion of Cesàro summability from sequences to functions. We write (2·23) in the form $(C, r)\, g(t) \to s$. If $(C, \alpha)\, g(t) \to s$ for some α, we write $(C)\, g(t) \to s$.

(2·24) THEOREM. *A necessary and sufficient condition for* $S[f]$ *to be summable* C *at* x_0 *to sum* s *is that* $(C)\, \chi_{x_0}(t) \to s$.

Since $S[f]$ at $x = x_0$ is the same thing as $S[\chi_{x_0}(t)]$ at $t = 0$, we may suppose that $x_0 = 0$, that f is even, and that $s=0$. Fourier series may be integrated termwise and so, if $F(x)$ is the result of integrating $S[f]$ r times, we have

$$\frac{1}{(r-1)!}\int_0^t f(u)(t-u)^{r-1}\,du = F(t)+P(t), \tag{2·25}$$

where $P(t)$ is a polynomial of degree less than r. From this we see that if $(C, r)\,f(t)\to 0$ as $t\to 0$, then $F_{(r)}(0)$ exists and equals 0.

Conversely, if $F_{(r)}(0)$ exists and is equal to 0, then $F(t)=o(t^r)$ plus a polynomial of degree $r-2$; since the left-hand side of (2·25) is in any case $O(t^{r-1})$, it must be $o(t^r)$, so that $(C, r)\,f(t)\to 0$. To finish the proof of (2·24) we apply (2·22).

Theorem (2·24) can be completed as follows:

(2·26) THEOREM. (i) *If f is non-negative and if $S[f]$ is summable C at a point x, then $S[f]$ is summable (C, ϵ) at that point, for each $\epsilon>0$.*

(ii) *If f is non-negative, a necessary and sufficient condition for $S[f]$ to be summable C at a point x to sum s is that $(C, 1)\ \chi_x(t)\to s$.*

We base the proof of (2·26) on the following lemma:

(2·27) LEMMA. *If Σu_n is bounded (C, α) and summable (C, β), $\beta>\alpha>-1$, then it is summable $(C, \alpha+\delta)$ for each $\delta>0$.*

We may suppose that $\beta=\alpha+1$, $0<\delta<1$, since the general result follows by repeated application of this special case. We may also suppose that the sum of Σu_n is 0, and we have to prove that $s_n^{\alpha+\delta}/A_n^{\alpha+\delta}\to 0$, where $s_n^{\alpha+\delta}$ denote the Cesàro sums for Σu_n. Now

$$s_n^{\alpha+\delta}=\sum_{k=0}^{n} A_{n-k}^{\delta-1}s_k^{\alpha}=\sum_{0}^{[n\theta]}+\sum_{[n\theta]+1}^{n}=P_n+Q_n,$$

where $\frac{1}{2}<\theta<1$. Denoting by C constants independent of n and θ, we have $|s_k^\alpha|\leqslant Ck^\alpha$ for $k\geqslant 1$, and

$$|Q_n|\leqslant Cn^\alpha \sum_{k=[n\theta]+1}^{n} A_{n-k}^{\delta-1}=Cn^\alpha A_{n-[n\theta]-1}^{\delta}\simeq Cn^{\alpha+\delta}(1-\theta)^\delta.$$

(Since $\theta>\frac{1}{2}$, the first inequality holds also for $\alpha<0$.) Hence, if θ is sufficiently close to 1, we have $|Q_n|/A_n^{\alpha+\delta}<\frac{1}{2}\epsilon$, where ϵ is arbitrarily given and $n>n_0$. Having fixed θ we have $P_n=o(n^{\alpha+\delta})$; for, summing by parts,

$$\begin{aligned}|P_n| &\leqslant \left|\sum_{k=0}^{[n\theta]} A_{n-k}^{\delta-2}s_k^{\alpha+1}\right|+o(n^{\alpha+\delta})\\ &\leqslant C[n(1-\theta)]^{\delta-2}\sum_{k=0}^{[n\theta]} o(k^{\alpha+1})+o(n^{\alpha+\delta})\\ &= C[n(1-\theta)]^{\delta-2}\,o(n^{\alpha+2})+o(n^{\alpha+\delta})<\tfrac{1}{2}\epsilon A_n^{\alpha+\delta},\end{aligned}$$

for $n>n_1$. Hence $|s_n^{\alpha+\delta}|/A_n^{\alpha+\delta}<\epsilon$ for $n>\max(n_0, n_1)$, and the lemma follows.

Return to (2·26)(i). We have (2·23), with $\chi_x(u)$ for $g(u)$, for some $r>1$. Since $\chi_x(u)\geqslant 0$, the left-hand side of (2·23) is not less than

$$rt^{-r}\int_0^{\frac{1}{2}t}\chi_x(u)(t-u)^{r-1}\,du\geqslant r2^{1-r}t^{-1}\int_0^{\frac{1}{2}t}\chi_x(u)\,du,$$

so that

$$\int_0^t \chi_x(u)\,du=O(t). \tag{2·28}$$

Write $$\eta_x(t)=t^{-1}\int_0^t |f(x+u)+f(x-u)-2s|\,du.$$

$S[f]$ is summable (C,α), $\alpha>0$, to s at each point x where $\eta_x(t)=o(1)$ (Chapter III, (5·1)), and the same argument shows that $S[f]$ is bounded (C,α) at x if $\eta_x(t)=O(1)$. Since (2·28) and $\chi_x(t)\geqslant 0$ imply $\eta_x(t)=O(1)$, $S[f]$ is, in our case, bounded (C,α) and so, by (2·27), summable $(C,\alpha+\delta)$ for each $\alpha>0$, $\delta>0$. Writing $\alpha+\delta=\epsilon$ we obtain part (i) of (2·26).

Passing to part (ii), we write $\chi_x(t)=\chi(t)=X_0(t)$ and denote by $X_k(t)$ the integral of $X_{k-1}(u)$ over $0\leqslant u\leqslant t$. The relation (2·23) with $\chi(u)$ for $g(u)$ may be written† $X_r(t)\simeq st^r/r!$, and to prove (ii) we have to show that $X_1(t)\simeq st$. Since $X_k(t)$ is a non-decreasing function of t for $k>0$, (ii) follows by repeated application of the following lemma:

(2·29) Lemma. *If $s(t)$, $t\geqslant 0$, is differentiable, $s'(t)$ non-decreasing, and $s(t)\simeq st^\alpha$ as $t\to 0$, then $s'(t)\simeq s\alpha t^{\alpha-1}$.*

Let $0<\theta<1$ be fixed; by the mean-value theorem,

$$(1-\theta)\,ts'(\theta t)\leqslant s(t)-s(\theta t)\leqslant (1-\theta)\,ts'(t). \tag{2·30}$$

Since $s(t)-s(\theta t)\simeq s(1-\theta^\alpha)\,t^\alpha$, this implies

$$\liminf_{t\to 0}\frac{s'(t)}{t^{\alpha-1}}\geqslant s\,\frac{1-\theta^\alpha}{1-\theta}, \tag{2·31}$$

and $$\limsup_{t\to 0}\frac{s'(\theta t)}{(\theta t)^{\alpha-1}}\leqslant s\,\frac{1-\theta^\alpha}{(1-\theta)\,\theta^\alpha},$$

or $$\limsup_{t\to 0}\frac{s'(t)}{t^{\alpha-1}}\leqslant s\,\frac{1-\theta^\alpha}{(1-\theta)\,\theta^\alpha}. \tag{2·32}$$

Taking θ arbitrarily close to 1 we deduce from (2·31) and (2·32) that $s'(t)/t^{\alpha-1}\simeq s\alpha$. This completes the proof of (2·29), and so also of (2·26) (ii).

It is obvious that (2·26) holds if f is merely bounded below, and in particular if f is bounded.

3. A theorem on differentiated series

It is natural to compare the summability of $S^{(r)}[F]$ with the summability of the Fourier series of the difference ratio whose limit is $F^{(r)}$. We consider only the case $r=1$.

(3·1) Theorem. *Let $\alpha\geqslant 0$,*

$$F(x)\sim \tfrac{1}{2}a_0+\sum_{n=1}^{\infty}(a_n\cos nx+b_n\sin nx), \tag{3·2}$$

and suppose that, for a given x_0, the periodic function

$$g(t)=\frac{F(x_0+t)-F(x_0-t)}{4\tan\frac{1}{2}t}=g_{x_0}(t) \tag{3·3}$$

is integrable L. Then a necessary and sufficient condition for $S'[F]$ to be summable $(C,\alpha+1)$ at x_0 to sum s ($\neq\pm\infty$), is that $S[g]$ be summable (C,α) at $t=0$ to s.

† If $s=0$, this is to be understood as $X_r(t)=o(t^r)$. The same remark applies to Lemma (2·29).

The case $\alpha=0$ is of special interest. By Theorem (5·4) below, if F is differentiable in a set E of positive measure, then $S'[F]$ is summable (C, 1) almost everywhere in E. Hence, by (3·1), $S[g]$ converges at $t=0$ for almost all $x_0 \in E$. In particular, if F is absolutely continuous, $S[g]$ converges at $t=0$ for almost all x_0 (though, by Theorem (3·1) of Chapter VIII, $S[F']$ may diverge at almost every point x_0).

(3·4) LEMMA. *Suppose two sequences $u_0, u_1, \ldots$ and $v_0, v_1, \ldots$ satisfy*

$$v_n=(n+1)(u_n-u_{n+1}) \quad (n=0,1,\ldots).$$

Let $\beta>-1$. Then, if $\sum_0^\infty u_n$ is summable (C, β) to sum s, $\sum_0^\infty v_n$ is summable (C, $\beta+1$) to s. Conversely, if Σv_n is summable (C, $\beta+1$), and the (C, $\beta+1$) means of Σu_n are $o(n)$, then Σu_n is summable (C, β).

Denote the Cesàro sums of order β for Σu_n and Σv_n by U_n^β and V_n^β, and the Cesàro means by σ_n^β and τ_n^β, respectively. Hence (see Chapter III, § 1)

$$\sigma_n^\beta=U_n^\beta/A_n^\beta, \quad \tau_n^\beta=V_n^\beta/A_n^\beta,$$

where $$A_n^\beta=\frac{(\beta+1)(\beta+2)\ldots(\beta+n)}{n!}\simeq\frac{n^\beta}{\Gamma(\beta+1)}.$$

Then $$V_n^{\beta+1}=\sum_{\nu=0}^{n} A_{n-\nu}^{\beta+1}v_\nu=\sum_{\nu=0}^{n} A_{n-\nu}^{\beta+1}(\nu+1)u_\nu-\sum_{\nu=0}^{n+1} A_{n-\nu+1}^{\beta+1}\nu u_\nu$$

$$=(\beta+2)\sum_{\nu=0}^{n} A_{n-\nu}^{\beta+1}u_\nu-\sum_{\nu=0}^{n+1} u_\nu[(\beta+1-\nu)A_{n-\nu}^{\beta+1}+\nu A_{n-\nu+1}^{\beta+1}],$$

where $A_{-1}^{\beta+1}=0$. We easily verify that the expression in square brackets equals $(n+1)A_{n-\nu+1}^\beta$, and we get

$$V_n^{\beta+1}=(\beta+2)U_n^{\beta+1}-(n+1)U_{n+1}^\beta, \tag{3·5}$$

$$\tau_n^{\beta+1}=(\beta+2)\sigma_n^{\beta+1}-(\beta+1)\sigma_{n+1}^\beta. \tag{3·6}$$

The first part of the theorem is a corollary of (3·6).

Using the equation $U_{n+1}^\beta=U_{n+1}^{\beta+1}-U_n^{\beta+1}$, we write (3·5) in the form

$$(n+\beta+3)U_n^{\beta+1}-(n+1)U_{n+1}^{\beta+1}=V_n^{\beta+1},$$

or $$\frac{\sigma_n^{\beta+1}}{n+\beta+2}-\frac{\sigma_{n+1}^{\beta+1}}{n+\beta+3}=\frac{\tau_n^{\beta+1}}{(n+\beta+2)(n+\beta+3)}. \tag{3·7}$$

For the proof of the second part of the lemma we may suppose, by changing u_0, that $\tau_n^{\beta+1}\to 0$. The right-hand side of (3·7) is then $o(n^{-2})$, and the remainders of a series with such terms are $o(1/n)$. Hence, using the fact that $\sigma_n^{\beta+1}=o(n)$, we get from (3·7), by addition,

$$\frac{\sigma_n^{\beta+1}}{n+\beta+2}=o\left(\frac{1}{n}\right),$$

that is, $\sigma_n^{\beta+1}\to 0$.

Substituting this in (3·6), we find that $\sigma_n^\beta\to 0$. This completes the proof of the lemma.

To prove the theorem, write

$$v_{n-1}=n(b_n \cos nx_0 - a_n \sin nx_0) \quad (n=1,2,\ldots),$$

$$\sum_{\nu=n}^{N}\frac{v_\nu}{\nu+1}=\sum_{\nu=n+1}^{N+1}(b_\nu \cos \nu x_0 - a_\nu \sin \nu x_0)$$

$$=\frac{2}{\pi}\int_0^\pi \frac{F(x_0+t)-F(x_0-t)}{4\sin\frac{1}{2}t}\{\cos(n+\tfrac{1}{2})t-\cos(N+\tfrac{3}{2})t\}\,dt,$$

and denote the limit of the last expression, as $N\to+\infty$, by u_n. Since the fraction preceding the curly brackets is integrable, we see from the Riemann-Lebesgue theorem that

$$u_n=\sum_{\nu=n}^{\infty}\frac{v_\nu}{\nu+1}=\frac{2}{\pi}\int_0^\pi \frac{F(x_0+t)-F(x_0-t)}{4\sin\frac{1}{2}t}\cos(n+\tfrac{1}{2})t\,dt$$

$$=\frac{2}{\pi}\int_0^\pi g(t)\cos nt\,dt-\frac{2}{\pi}\int_0^\pi [F(x_0+t)-F(x_0-t)]\tfrac{1}{4}\sin nt\,dt$$

$$=u_n'-u_n'',$$

say. Since g is integrable, $\Sigma u_n''$ converges and its sum is $\frac{1}{2}u_0'$. It follows that $u_0+u_1+\ldots$ and $\frac{1}{2}u_0'+u_1'+\ldots$ are simultaneously summable (C,α) or not, and if they are summable their sums are equal.

Observe now that $v_n=(n+1)(u_n-u_{n+1})$, that $\frac{1}{2}u_0'+u_1'+\ldots$ is $S[g]$ at $t=0$, and that $v_0+v_1+\ldots$ is $S'[F]$ (without its constant term) at x_0. Therefore, by the lemma, if $S[g]$ is summable (C,α) at $t=0$ to sum s, then $S'[F]$ is summable $(C,\alpha+1)$ at x_0 to s. The converse is also true, since the partial sums of the series Σu_n, and so also its $(C,\alpha+1)$ means, are $o(n)$†.

4. Theorems on generalized derivatives

In this section $B^{(k)}$ and $C^{(k)}$ denote the classes of functions having respectively a bounded or a continuous kth derivative. More precisely, $f\in B^{(k)}$ means that $f^{(k-1)}$ exists and is in Λ_1, $k=1,2,\ldots$. We consider functions in the interval $0\leqslant x\leqslant 2\pi$.

Let E be a measurable set of positive measure contained in $(0,2\pi)$. A measurable function $f(x)$ will be said to possess in E the *property* β_k, if for each x in E the (unsymmetric) derivatives $f_{(1)}, f_{(2)}(x), \ldots, f_{(k-1)}(x)$ exist, and the expression $\omega(x,t)$ defined by

$$f(x+t)=f(x)+f_{(1)}(x)\,t+\ldots+f_{(k-1)}(x)\frac{t^{k-1}}{(k-1)!}+\omega(x,t)\frac{t^k}{k!} \tag{4·1}$$

remains bounded (not necessarily uniformly in x) as $t\to 0$. Clearly if $f\in B^{(k)}$ then we have (4·1) with $\omega(x,t)$ uniformly bounded in x, t.

The following result is basic for applications of generalized derivatives:

(4·2) THEOREM. *If $f(x)$ has the property β_k in a set E, $|E|>0$, then there is a perfect set $\Pi\subset E$, of measure arbitrarily close to that of E, and a decomposition*

$$f(x)=g(x)+h(x) \tag{4·3}$$

satisfying the following conditions:

(i) $g(x)\in B^{(k)}$;

† In the preceding argument we dropped the (zero) constant term from $S'[F]$. It is, however, easy to see that if $c_0+c_1+c_2+\ldots$ is summable (C,γ) to s, so is $0+c_0+c_1+\ldots$, and conversely.

(ii) *if* $\Delta_1, \Delta_2, \ldots$ *are the intervals contiguous to* Π, *and* $\chi(x)$ *is the distance of* x *from* Π, *we have*

$$|h(x)| \leqslant C\chi^k(x) \tag{4·4}$$

for all x, *except perhaps for those situated in a finite number of the intervals* Δ_i, C *being independent of* x (*in particular*, $h=0$ *in* Π *and* $|h| \leqslant C|\Delta_i|^k$ *for* $x \in \Delta_i$ *and* i *large enough*).

For any integer $n \geqslant 1$ consider the polynomial

$$w_n(x) = \int_0^x t^{n-1}(1-t)^{n-1}\,dt \Big/ \int_0^1 t^{n-1}(1-t)^{n-1}\,dt$$

of degree $2n-1$; in particular $w_1(x)=x$. We have

$$w_n(0)=0, \quad w_n(1)=1; \quad w_n'(0)=w_n'(1)=w_n''(0)=w_n''(1)=\ldots=w_n^{(n-1)}(1)=0. \tag{4·5}$$

Also $$w_n^{(j)}(x)=O(x^{n-j}), \quad w_n^{(j)}(x)=O\{(1-x)^{n-j}\} \quad (j=1, 2, \ldots, n). \tag{4·6}$$

Consider a function $p(x)$ defined on a perfect set P with end-points a, b. By a *parabolic extension* of order n of $p(x)$ we shall mean the function $\pi(x)$, $a \leqslant x \leqslant b$, such that $\pi(x)=p(x)$ on P and

$$\pi(x)=p(\alpha)+w_n\left(\frac{x-\alpha}{\delta}\right)[p(\beta)-p(\alpha)] \quad (\delta=\beta-\alpha), \tag{4·7}$$

in each interval $\Delta=(\alpha,\beta)$ contiguous to P. If $n=1$, π is linear in each Δ.

(4·8) LEMMA. *If* $p(x)$ *is defined on a perfect set* P *and satisfies*

$$|p(x'')-p(x')| \leqslant M|x''-x'|^n \quad (x' \in P, x'' \in P), \tag{4·9}$$

with M *independent of* x', x'', *then the parabolic extension* π *of order* n *of* p *has* $n-1$ *continuous derivatives all of which vanish on* P, *and* $\pi^{(n-1)} \in \Lambda_1$.

Consider first the case $n=1$. By (4·9),

$$|\pi(x'')-\pi(x')| \leqslant M|x''-x'| \tag{4·10}$$

for x' and x'' in P. Since the slope of π outside P numerically does not exceed M, (4·10) holds if x' and x'' belong to the same interval contiguous to P. If (x', x'') contains points of P in its interior, denote by x_1 and x_2 the first and last of those points, so that $x' \leqslant x_1 < x_2 \leqslant x''$, say. Since the analogue of (4·10) is valid for each of the pairs x', x_1; x_1, x_2; x_2, x'', it holds by addition for x', x''. (This type of argument will be repeatedly used below.) Hence the lemma is proved for $n=1$.

Suppose now that $n \geqslant 2$. Denote by $p_P'(x)$ the derivative of p relative to P, that is, the limit of $\{p(x+h)-p(x)\}/h$ for x and $x+h$ in P. By (4·9) we have $\pi_P'(x)=p_P'(x)=0$ in P. By (4·7), (4·9) and (4·6),

$$\pi'(x)=\frac{1}{\delta}w_n'\left(\frac{x-\alpha}{\delta}\right)[p(\beta)-p(\alpha)]=O\{(x-\alpha)^{n-1}\} \tag{4·11}$$

uniformly in all the intervals $\Delta=(\alpha,\beta)$ contiguous to P. We show that $\pi'(x)$ exists and is 0 in P. It is enough to consider, for instance, the right-hand derivative of π.

Let $x_0 \in P$, $h>0$, and denote by x_0+h' the last point of P in (x_0, x_0+h). (If no such point exists, then x_0 is the left-hand end of an interval contiguous to P and the result is immediate from (4·11).) Then

$$\frac{\pi(x_0+h)-\pi(x_0)}{h}=\frac{\pi(x_0+h')-\pi(x_0)}{h'}\frac{h'}{h}+\frac{\pi(x_0+h)-\pi(x_0+h')}{h}. \tag{4·12}$$

The first term on the right tends to 0 with h since $\pi'_P(x_0)=0$ and $h'/h<1$. The second term on the right is $O\{(h-h')^n h^{-1}\}=o(1)$, by the mean-value theorem and (4·11). Hence $\pi'(x_0)$ exists and is 0; clearly $\pi'(x)$ exists outside P.

If, for example, $n=2$, to complete the proof of the theorem we still have to show that $\pi' \in \Lambda_1$, that is, that

$$|\pi'(x'')-\pi'(x')| \leqslant M_1 |x''-x'|. \tag{4·13}$$

This is obvious if x' and x'' are in P, since $\pi'=0$ there; and it is also immediate if x' and x'' are in the same interval $\Delta=(\alpha,\beta)$ contiguous to P, by an application to π' of the mean-value theorem and the estimate

$$\pi''(x)=\delta^{-2}w_2''\left(\frac{x-\alpha}{\delta}\right)[p(\beta)-p(\alpha)]=O(1) \quad (x\in\Delta). \tag{4·14}$$

If (x',x'') contains points of P in its interior, it is enough to consider the extreme points x_1 and x_2 of P in (x',x'') and argue as in the case $n=1$. This proves the lemma for $n=2$.

If $n\geqslant 3$ the argument is similar and we give it in full. Suppose that $n\geqslant 3$, that $\pi', \pi'', \dots, \pi^{(k)}$ exist, and that they are 0 in P. It is enough to show that (i) if $k<n-1$ then $\pi^{(k+1)}$ exists, and $\pi^{(k+1)}=0$ in P; (ii) if $k=n-1$, then $\pi^{(k)}\in\Lambda_1$.

(i) By hypothesis, $\pi^{(k)}=0$ in P. Hence $\pi_P^{(k+1)}=(\pi^{(k)})'_P$ is 0 in P. If $x_0\in P$, we consider a formula analogous to (4·12) with $\pi^{(k)}$ for π. From this, using the estimate

$$\pi^{(k+1)}(x)=\delta^{-(k+1)}w_n^{(k+1)}\left(\frac{x-\alpha}{\delta}\right)[p(\beta)-p(\alpha)]=O\{(x-\alpha)^{n-(k+1)}\} \tag{4·15}$$

in the intervals $\Delta=(\alpha,\beta)$ contiguous to P, we find that $\pi^{(k+1)}(x_0)$ exists and is 0; the existence of $\pi^{(k+1)}$ outside P is obvious.

(ii) We have to show that

$$|\pi^{(n-1)}(x'')-\pi^{(n-1)}(x')| \leqslant M_1 |x''-x'|. \tag{4·16}$$

This is obvious if x' and x'' are in P, and also if x' and x'' are in the same interval $\Delta=(\alpha,\beta)$ contiguous to P, by the mean-value theorem and the estimate $\pi^{(n)}(x)=O(1)$ obtained by taking $k+1=n$ in (4·15). Hence (4·16) holds for general (x',x'') and the lemma is established.

Return to the theorem. Since $\omega(x,t)=O(1)$ for each $x\in E$ and $t\to 0$, there is a perfect set $\Pi\subset E$ with $|E-\Pi|$ arbitrarily small, and two positive numbers M and d such that

$$|\omega(x,t)|<M \quad \text{for} \quad x\in\Pi,\ |t|<d.$$

This Π is the set Π of the theorem.

If we make the special hypothesis that

$$f(x)=f_{(1)}(x)=f_{(2)}(x)=\dots=f_{(k-1)}(x)=0 \quad \text{in} \quad \Pi, \tag{4·17}$$

(4·1) becomes

$$f(x+t)=\omega(x,t)\frac{t^k}{k!}, \tag{4·18}$$

where $|\omega|<M$ for $x\in\Pi$, $|t|<d$. Hence, writing $C=M/k!$, we have $|f|<C\chi^k$ in each interval Δ_i contiguous to Π and of length $<d$. Since there can be only a finite number of Δ's of length not less than d, we obtain (4·2) with $g=0$, $h=f$.

For the proof of the theorem it is therefore enough to show that if we subtract from f a suitable function in $B^{(k)}$ the difference (again denoted by f) satisfies (4·17). We shall do the subtraction in k stages.

In the proof we use higher forward differences of f,

$$\Delta_t^m f(x) = \sum_{j=0}^{m} (-1)^{m-j} \binom{m}{j} f(x+jt),$$

and the formula $$\Delta_t^{m+1} f(x) = \Delta_t^m f(x+t) - \Delta_t^m f(x).$$

Suppose that instead of (4·17) we have only

$$f_{(k-i)}(x) = f_{(k-i+1)}(x) = \ldots = f_{(k-1)}(x) = 0 \quad \text{on} \quad \Pi, \tag{4·19}$$

for some i, $0 \leqslant i \leqslant k-1$ ($i=0$ means that we do not assume any of the conditions (4·17)), so that

$$f(x+t) = f(x) + f_{(1)}(x)\,t + \ldots + \frac{f_{(k-i-1)}(x)}{(k-i-1)!}\,t^{k-i-1} + \omega(x,t)\frac{t^k}{k!} \quad (x \in \Pi). \tag{4·20}$$

Then for $x \in \Pi$ and $|t| < d/k$ we have

$$|\Delta_t^{k-i} f(x)| \leqslant \lambda_{k,i} M |t|^k, \tag{4·21}$$

$$\Delta_t^{k-i-1} f(x) = f_{k-i-1}(x)\,t^{k-i-1} + \theta M \lambda'_{k,i} |t|^k, \tag{4·22}$$

where $-1 < \theta < 1$, and $\lambda_{k,i}$ and $\lambda'_{k,i}$ are constants. If also $x+t \in \Pi$, we may substitute $x+t$ for x in (4·22). Subtracting the result from (4·22) and using (4·21) we get

$$|f_{(k-i-1)}(x+t) - f_{(k-i-1)}(x)| \leqslant M(\lambda_{k,i} + 2\lambda'_{k,i})\,|t|^{i+1}, \tag{4·23}$$

provided $x \in \Pi$, $x+t \in \Pi$, $|t| < d/k$.

Consider any portion P of Π of diameter less than d/k and apply Lemma (4·8) to $p(x) = f_{(k-i-1)}(x)$. Piecing together a finite number of parabolic extensions of p of order $i+1$, we obtain a function $\pi(x)$ differentiable i times, with

$$\pi' = \pi'' = \ldots = \pi^{(i)} = 0 \quad \text{on} \quad \Pi,$$

and with $\pi^{(i)} \in \Lambda_1$.

Denote by $I(x)$ the $(k-i-1)$th indefinite integral of $\pi(x)$. Then $I'(x), I''(x), \ldots, I^{(k-1)}(x)$ exist, $I^{(k-1)}$ is in Λ_1, and

$$I^{(k-i-1)}(x) = f_{(k-i-1)}(x), \quad I^{(k-i)}(x) = \ldots = I^{(k-1)}(x) = 0 \quad \text{on} \quad \Pi.$$

Hence, subtracting I from f and denoting $f - I$ by f again, we obtain (4·20), where now $f_{(k-i-1)}(x) = 0$ on Π. Starting from (4·1) and repeating the argument k times we arrive at (4·18). This completes the proof of the theorem.

Theorem (4·2) has a number of generalizations and consequences.

(*a*) Property β_k is almost everywhere equivalent to the existence of the kth unsymmetric derivative. To see this it is enough to prove the following theorem.

(4·24) Theorem. *If f has the property β_k in E, then the unsymmetric derivative $f_{(k)}$ exists almost everywhere in E.*

Consider the decomposition $f = g + h$ of (4·2). Since functions in Λ_1 are differentiable almost everywhere, $g^{(k)}$, and so also $g_{(k)}$, exists almost everywhere. We show that $h_{(k)}(x)$

exists almost everywhere in Π. This is immediate, since (4·4) implies that at every point x of density of Π we have

$$h(x+t)=o(t^k)$$

as $t \to 0$, that is, $h_{(k)}$ exists and is 0. (Incidentally this shows that $f_{(j)}=g^{(j)}$, $j=1, 2, \ldots, k$, almost everywhere in Π.) Hence $f_{(k)}=g_{(k)}+h_{(k)}$ exists almost everywhere in Π, and so also almost everywhere in E.

(*b*) The following generalization of (4·2) is of interest, though no more effective in applications than the original:

(4·25) THEOREM. *The function g in (4·2) can be made to belong not only to* $\mathrm{B}^{(k)}$ *but also to* $\mathrm{C}^{(k)}$.

The proof runs parallel to that of (4·2), and it is enough to indicate the idea. First, by (*a*), $f_{(k)}(x)$ exists almost everywhere in E, so that in (4·1) we may replace $\omega(x,t)$ by $f_{(k)}(x,t)+\epsilon(x,t)$, where ϵ tends to 0 with t for each x in E. Since the measurability of f implies that of $f_{(1)}, f_{(2)}, \ldots, f_{(k)}$ on the set where they exist, and the latter set is measurable (remembering that the $f_{(j)}$ are limits of certain ratios involving only the function f), we may suppose that $f, f_{(1)}, \ldots, f_{(k)}$ are all continuous relative to Π. From now on we proceed as in the proof of (4·2). We consider the parabolic extension of order 1 of $f_{(k)}$. The resulting function is continuous, its kth integral $I(x)$ is in $\mathrm{C}^{(k)}$, and subtracting $I(x)$ from $f(x)$ we can make f satisfy the condition $f_{(k)}=0$ on Π. Next we consider the parabolic extension of order 2 of $f_{(k-1)}$ and subtract from f the $(k-1)$th integral of that extension, and so on.

(*c*) In Chapter IX, § 2, we introduced the notion of approximate derivative f'_{ap}. We define $f^{(k)}_{\mathrm{ap}}=\{f^{(k-1)}_{\mathrm{ap}}\}'_{\mathrm{ap}}$. Using these notions we can establish relations between successive generalized derivatives $f_{(j)}$.

(4·26) THEOREM. *If $f_{(1)}, f_{(2)}, \ldots, f_{(k)}$ exist in a set E, $|E|>0$, then*

$$f_{(j)}=\{f_{(j-1)}\}'_{\mathrm{ap}} \quad (j=2, 3, \ldots, k) \tag{4·27}$$

almost everywhere in E.

Consider the decomposition $f=g+h$ of (4·2). Since $h=0$ in Π, h has at every point of density of Π approximate derivatives of all orders, all equal to 0. Since $g', g'', \ldots, g^{(k-1)}$ exist everywhere, and $g^{(k)}$ almost everywhere, it follows that $f'_{\mathrm{ap}}, f''_{\mathrm{ap}}, \ldots, f^{(k)}_{\mathrm{ap}}$ exist almost everywhere in Π. Moreover, almost everywhere in Π we have

$$f^{(j)}_{\mathrm{ap}}=g^{(j)}_{\mathrm{ap}}=g^{(j)}=\{g^{(j-1)}\}'=\{g^{(j-1)}\}'_{\mathrm{ap}}=\{f_{(j-1)}\}'_{\mathrm{ap}}$$

for $j=2, 3, \ldots, k$. This proves (4·27) almost everywhere in Π, and so also almost everywhere in E.

(*d*) Though we only consider measurable functions, it is of interest to observe that (4·24) holds without assuming that either f or E are measurable, and that this more general result can be deduced from the special case proved above.

Denote by $f^*(x)$ and $f_*(x)$ respectively the upper and lower bounds of f at the point x. (Thus $f^*(x)$, for instance, is the limit, for $h \to 0$, of the upper bound of f in the interval $(x-h, x+h)$.) The function f^* is upper semi-continuous and f_* is lower semi-continuous; hence both are B-measurable. We have $f_* \leqslant f \leqslant f^*$ everywhere, and $f_*=f=f^*$ at every point of continuity of f, in particular in E.

Denote by E^* the set of all points where f^* has an unsymmetric $(k-1)$th derivative and

$$f^*(x+t)=f^*(x)+f^*_{(1)}(x)\,t+\ldots+f^*_{(k-1)}(x)\frac{t^{k-1}}{(k-1)!}+O(t^k), \tag{4·28}$$

and by E_* the corresponding set for f_*. Both E^* and E_* are B-measurable, and, by the theorem already proved, $f^*_{(k)}$ exists almost everywhere in E^*, and $f_{*(k)}$ almost everywhere in E_*.

Since $f(x)=f^*(x)$ in E, it is geometrically clear that (4·28) holds in E†. Hence $E\subset E^*$, and similarly $E\subset E_*$, so that $E\subset E^*E_*$.

It follows that both $f^*_{(k)}$ and $f_{*(k)}$ exist in a set $E-Z$, where $|Z|=0$. In other words, for x in $E-Z$ we may replace the last term in (4·28) by $\{f^*_{(k)}(x)+o(1)\}\,t^k/k!$, and similarly in the formula for $f_*(x+t)$.

Suppose now that x is in $E-Z$, and that $x+t$ is in $E-Z$ for infinitely many t's tending to 0; only a denumerable subset D of $E-Z$ can fail to have the latter property. Since $f_*(x+t)=f(x+t)=f^*(x+t)$ for such t's, we find that $f_{*(j)}=f_{(j)}=f^*_{(j)}$ in $E-Z-D$ for $j=1, 2, \ldots, k$, and in particular $f_{(k)}$ exists almost everywhere in E.

(*e*) For applications of Theorem (4·2) to Fourier series we must show that it holds for periodic functions. Suppose that f is periodic of period 2π. The set Π defined on p. 75 is then periodic. Take any interval $(\alpha, \alpha+2\pi)$ with α (and so also $\alpha+2\pi$) not in Π, and consider the decomposition $f=g+h$ with properties described in (4·2). By suitably modifying g near the points α and $\alpha+2\pi$, we obtain a new g which when continued periodically is still in $B^{(k)}$, and the new $h=f-g$ will still satisfy the required conditions.

(*f*) Theorem (4·24) can be considerably generalized, at least for measurable functions. Let $\Delta^k f(x,h)$ denote symmetric kth differences; if

$$h^{-k}\Delta^k f(x,h)=O(1) \tag{4·29}$$

for each $x\in E$ as $h\to 0$ (in particular, if $D_k f$ exists in E, and *a fortiori* if a symmetric $f_{(k)}$ exists in E), then the unsymmetric $f_{(k)}(x)$ exists almost everywhere in E. For applications to Fourier series the case $k=2$ is the most interesting, and we confine our attention to it.

(4·30) THEOREM. *Suppose that f is measurable and that*

$$\frac{f(x+h)+f(x-h)-2f(x)}{h^2}=O(1) \tag{4·31}$$

for each x in E as $h\to 0$. Then the unsymmetric derivative $f_{(2)}(x)$ exists almost everywhere in E.

We split the proof into a series of lemmas:

(4·32) LEMMA. *Suppose that 0 is a point of density of a set $\mathscr{E}$. Then for each sufficiently small u (positive or negative) we can find in the interval $(u, 2u)$ a number v such that*

$$\text{(i) } v\in\mathscr{E};\quad \text{(ii) } \tfrac{1}{2}(u+v)\in\mathscr{E};\quad \text{(iii) } u+v\in\mathscr{E}.$$

† If $f(x_0+t)=a_0+a_1t+\ldots+a_{k-1}t^{k-1}+O(t^k)$ for $t\to 0$, the graph of $f(x)$ near $x=x_0$ is contained in domains limited by parabolic curves of order k, and so the same holds for the graph of $f^*(x)$. In particular, $a_j=f^*_{(j)}(x_0)/j!$ for $j<k$.

Suppose, for example, that $u>0$. Let $\gamma(x)$ be the characteristic function of $\mathscr{E}$, and write $\Gamma(x)=\int_0^x \gamma(t)\,dt$. The sets of points v in $(u, 2u)$ satisfying conditions (i), (ii) and (iii) have measures respectively

$$\left.\begin{aligned} \int_u^{2u} \gamma(v)\,dv &= \Gamma(2u)-\Gamma(u), \\ \int_u^{2u} \gamma\{\tfrac{1}{2}(u+v)\}\,dv &= 2\{\Gamma(\tfrac{3}{2}u)-\Gamma(u)\}, \\ \int_u^{2u} \gamma(u+v)\,dv &= \Gamma(3u)-\Gamma(2u). \end{aligned}\right\} \tag{4·33}$$

Since $\Gamma(u)\simeq u$ for small u, each of the expressions on the right-hand side in (4·33) is asymptotically equal to u, that is, the length of $(u, 2u)$, and it is clear that if u is small enough there is a v satisfying conditions (i), (ii) and (iii).

(4·34) LEMMA. *Under the hypothesis of* (4·30), *f is bounded in the neighbourhood of almost any point of E.*

Denote by E_m^* the set of x such that $|f(x)|<m$ and $|\Delta^2 f(x,h)|<m$ for $0<h<1/m$. Since $E\subset E_1^*+E_2^*+\dots$, it is enough to show that f is bounded near each point of density of a given E_m^*.

Suppose, for example, that $x=0$ is a point of density of E_m^* and that u is small; in particular that $0<u<1/m$. Apply (4·32) with $\mathscr{E}=E_m^*$. Then $\frac{1}{2}(u+v)\in E_m^*$, $v-u<1/m$, so that

$$|f(u)-2f(\tfrac{1}{2}u+\tfrac{1}{2}v)+f(v)|=|\Delta^2 f(\tfrac{1}{2}u+\tfrac{1}{2}v, \tfrac{1}{2}v-\tfrac{1}{2}u)|<m. \tag{4·35}$$

Since v and $\frac{1}{2}(u+v)$ are in E_m^*, the values of f at those points are numerically less than m, and (4·35) implies that $|f(u)|<4m$. This completes the proof of the lemma. (In the argument we did not need property (iii) of (4·32).)

(4·36) LEMMA. *Under the hypothesis of* (4·30) *we have*

$$\frac{\Delta^2 f(x+h,h)}{h^2}=\frac{f(x+2h)-2f(x+h)+f(x)}{h^2}=O(1) \tag{4·37}$$

at almost all points of E.

Denote by E_m the set of points x such that

$$|h^{-2}\Delta^2 f(x,h)|<m \quad \text{for} \quad 0<h<1/m.$$

We have $E\subset E_1+E_2+\dots$, and it is enough to prove that (4·37) holds if x is a point of density of any E_m.

Suppose, for instance, that $x=0$ is a point of density of E_m and that $h>0$. Consider the expressions

$$\delta_1=f(2u)-2f(u)+f(0),$$

$$\delta_2=f(u+v)-2f(\tfrac{1}{2}u+\tfrac{1}{2}v)+f(0),$$

$$\delta_3=f(2v)-2f(v)+f(0),$$

where u is positive and small, and v satisfies the conditions of Lemma (4·32) with $\mathscr{E}=E_m$.

Since $\frac{1}{2}(u+v)$ and v are in E_m, we have $\delta_2 = O\{(u+v)^2\} = O(u^2)$, and $\delta_3 = O(v^2) = O(u^2)$. On the other hand,

$$\delta_1 - 2\delta_2 + \delta_3 = \Delta^2 f(u+v, v-u) - 2\Delta^2 f(\tfrac{1}{2}u + \tfrac{1}{2}v, \tfrac{1}{2}v - \tfrac{1}{2}u),$$

and since $u+v$ and $\frac{1}{2}(u+v)$ are in E_m, the right-hand side is $O\{(u-v)^2\} = O(u^2)$. But δ_2 and δ_3 are also $O(u^2)$. Hence $\delta_1 = O(u^2)$ and the lemma follows.

(4·38) LEMMA. *Under the hypothesis of* (4·30), $f'(x)$ *exists almost everywhere in* E.

Consider a point $x \in E$ such that f is bounded near x and (4·37) holds. Almost all $x \in E$ have these properties; suppose that $x=0$ is one of them and suppose also that $f(0)=0$. Then

$$f(h) - 2f(\tfrac{1}{2}h) = O(h^2), \quad f(\tfrac{1}{2}h) - 2f(\tfrac{1}{4}h) = O\{(\tfrac{1}{2}h)^2\}, \quad \dots,$$

$$f\left(\frac{1}{2^{n-1}}h\right) - 2f\left(\frac{1}{2^n}h\right) = O\left\{\left(\frac{1}{2^{n-1}}h\right)^2\right\}, \quad \dots.$$

If we multiply the first equation by 1, the second by 2, ..., the nth by 2^{n-1}, and add, we get

$$f(h) - 2^n f\left(\frac{h}{2^n}\right) = O(h^2)\,(1 + 2^{-1} + \dots + 2^{-(n-1)}) = O(h^2). \tag{4·39}$$

Suppose that f is bounded for $|h| \leqslant \epsilon$. We confine $|h|$ to the interval $(\frac{1}{2}\epsilon, \epsilon)$ and write $h/2^n = h'$. As $n \to \infty$, h' takes all sufficiently small (non-zero) values. By (4·39),

$$h\frac{f(h')}{h'} = f(h) + O(h^2) = O(1),$$

and hence (since h stays away from 0), $f(h')/h' = O(1)$ as $h' \to 0$.

It follows that f satisfies condition β_1 at almost all points of E, and therefore, by (4·24), f' exists almost everywhere in E.

It is now easy to complete the proof of (4·30). Consider a point $x \in E$ where we have (4·37) and f' exists. Without loss of generality we may suppose that $x=0$ and that $f(0) = f'(0) = 0$. Keeping h fixed in (4·39) and making $n \to \infty$ we see that $f(h) = O(h^2)$. Hence f satisfies condition β_2 at almost all points of E, and, by (4·24), the unsymmetric derivative $f_{(2)}$ exists almost everywhere in E.

5. Applications of Theorem (4·2) to Fourier series

These applications also require Theorem (2·1) of Chapter IV, which asserts that if Π is a periodic perfect set and $\chi(x) = \chi_\Pi(x)$ is the distance of x from Π, then

$$\int_{-\pi}^{\pi} \frac{\chi^\lambda(x+t)}{|t|^{\lambda+1}}\,dt \tag{5·1}$$

is finite almost everywhere in Π for all $\lambda > 0$.

(5·2) THEOREM. *If f is periodic and integrable, and has an unsymmetric derivative $f_{(r)}(x)$, $r \geqslant 1$, in a set E, then the r-th conjugate function*

$$\tilde{f}_r(x) = -\frac{1}{\pi}\int_0^\infty \frac{\delta_r(x,t)}{t}\,dt \tag{5·3}$$

(cf. (1·27)) *exists almost everywhere in* E.

Consider the decomposition $f=g+h$ of (4·2), where g is periodic and in $B^{(r)}$. Since $\tilde{g}_r$ exists almost everywhere (see Remark (*b*) on p. 65), it is enough to show that $\tilde{h}_r$ exists almost everywhere in Π. This is, however, immediate if we observe that

$$|h(x+t)| \leqslant C\chi^r(x+t)$$

for each x in Π and $|t|$ sufficiently small, so that at each point of Π where

$$h=h_{(1)}=\ldots=h_{(r-1)}=h_{(r)}=0$$

(that is, almost everywhere in Π) the integral

$$\tilde{h}_r(x)=-\frac{1}{\pi}\int_0^\infty \frac{h(x+t)\pm h(x-t)}{t^{r+1}}\,dt$$

is majorized near $t=0$ by

$$C\int_0^\infty \frac{\chi^r(x+t)+\chi^r(x-t)}{t^{r+1}}\,dt,$$

an expression finite almost everywhere in Π (see (5·1)).

(5·4) THEOREM. *If f has an unsymmetric derivative $f_{(r)}$, $r\geqslant 1$, in a set E, then both $S^{(r)}[f]$ and $\tilde{S}^{(r)}[f]$ are summable* (C, r) *almost everywhere in E.*

Consider again the decomposition $f=g+h$ of Theorem (4·2), where g is periodic and in $B^{(r)}$. By Theorem (3·23) of Chapter III, $S^{(r)}[g]=S[g^{(r)}]$ is summable (C, 1), and so also summable (C, r), for almost all x. The first part of the theorem will therefore be established if we show that the (C, r) means of $S^{(r)}[h]$ converge almost everywhere in Π. These means are

$$(-1)^r\frac{1}{\pi}\int_{-\pi}^{\pi} h(x+t)\frac{d^r}{dt^r}K_n^r(t)\,dt. \tag{5·5}$$

By (1·9) and (1·10),

$$\left|\frac{d^r}{dt^r}K_n^r(t)\right| \leqslant \frac{C}{|t|^{r+1}} \quad (|t|\leqslant\pi), \tag{5·6}$$

so that (5·5) is majorized by

$$C\int_{-\pi}^{\pi} |t|^{-r-1}\,|h(x+t)|\,dt. \tag{5·7}$$

This integral is finite at almost all points of Π, since $|h(x+t)|\leqslant C\chi^r(x+t)$ for x in Π and $|t|$ small enough, and (5·1) with $\lambda=r$ is finite almost everywhere in Π.

At each point of Π where (5·7) is finite, the means (5·5) are not only bounded but even convergent. For replacing h by 0 outside a sufficiently small neighbourhood of x does not affect the (C, r) summability of $S^{(r)}[h]$ (see the Remark on p. 62), and at the same time makes (5·7) arbitrarily small. This completes the proof of the summability of $S^{(r)}[f]$. By (1·7), the (C, r) sum of $S^{(r)}[f]$ must be $f_{(r)}$ almost everywhere in E.

The proof of the summability of $\tilde{S}^{(r)}[f]$ is similar. Here again $\tilde{S}^{(r)}[g]=\tilde{S}[g^{(r)}]$ is summable (C, 1), and so also (C, r), almost everywhere, and the problem reduces to showing that $\tilde{S}^{(r)}[h]$ is summable (C, r) at almost all points of Π. If we use the fact that the rth conjugate function $\tilde{h}_r$ exists almost everywhere in Π, the proof of the (C, r) summability of $\tilde{S}^{(r)}[h]$ is parallel to the proof of the summability of $S^{(r)}[h]$, and we may omit the details. By (1·25), the (C, r) sum of $\tilde{S}^{(r)}[f]$ is $\tilde{f}_r$ at almost all points of E.

The function conjugate to an absolutely continuous function may be unbounded in every interval. Hence if f is differentiable in a set of positive measure, $\tilde{f}$ need not have the same property. However, we have the following

(5·8) THEOREM. *Let f be integrable, and F the indefinite integral of f, both periodic. If f has an r-th unsymmetric derivative in a set E, the conjugate function $\tilde{F}$ of F has an $(r+1)$-th unsymmetric derivative almost everywhere in E.*

Let $f=g+h$ be the decomposition of (4·2), and let G and H be indefinite integrals of g and h. We may suppose that $F=G+H$. By modifying h suitably in an interval contiguous to Π we may suppose that the constant term of $S[h]$ is 0. Hence the constant term of $S[g]=S[f]-S[h]$ is also 0, and both G and H are periodic. It follows that $S[G]$ is obtained by termwise integration $r+1$ times of $S[G^{(r+1)}]=S[g^{(r)}]$, and consequently $S[\tilde{G}]=\tilde{S}[G]$ is obtained by integrating $r+1$ times $\tilde{S}[g^{(r)}]$, which is a Fourier series. This implies that $\tilde{G}$ has almost everywhere an ordinary derivative of order $r+1$, and it will be enough to show that an unsymmetric $\tilde{H}_{(r+1)}$ exists almost everywhere in Π.

It simplifies the argument slightly if instead of $\tilde{H}(x)$ we consider

$$\bar{H}(x)=\frac{1}{\pi}\int_{-\pi}^{\pi}\frac{H(t)}{x-t}dt, \tag{5·9}$$

which (since $1/u-\frac{1}{2}\cot\frac{1}{2}u$ is regular for $|u|<2\pi$) differs from $\tilde{H}(x)$ by a function regular in the interior of $(-\pi,\pi)$.

We know that

$$h(x+t)=o(t^r), \tag{5·10}$$

$$\int_{-\pi}^{\pi}\frac{|h(x+t)|}{|t|^{r+1}}dt<\infty \tag{5·11}$$

almost everywhere in Π. We show that $\bar{H}_{(r+1)}$ exists at each $x\in\Pi$ where we have (5·10), (5·11), and $H'(x)=0$ $(=h(x))$. Without loss of generality we may suppose that $x=0$ is such a point. By (5·10), h is bounded near the origin, H satisfies condition Λ_1 there, and the integral (5·9) converges near $x=0$. Finally, we may suppose that $H(x)=\int_0^x h\,dt$, so that, by (5·10),

$$H(t)=o(t^{r+1}). \tag{5·12}$$

Write

$$\bar{H}(x)=-\frac{1}{\pi}\int_{-\pi}^{\pi}H(t)\left[\frac{1}{t}+\frac{x}{t^2}+\dots+\frac{x^{r+1}}{t^{r+2}}+\frac{x^{r+2}}{t^{r+2}}\frac{1}{t-x}\right]dt$$

$$=-\frac{1}{\pi}\sum_{k=0}^{r+1}x^k\int_{-\pi}^{\pi}H(t)\,t^{-(k+1)}dt+\frac{x^{r+2}}{\pi}\int_{-\pi}^{\pi}\frac{H(t)}{t^{r+2}(x-t)}dt. \tag{5·13}$$

By (5·12), the coefficients of x^k are absolutely convergent integrals provided $k\leqslant r$. If $0<\epsilon<\pi$, integration by parts gives

$$-(r+1)\int_{\pm\epsilon}^{\pm\pi}H(t)\,t^{-(r+2)}dt=[H(t)\,t^{-(r+1)}]_{\pm\epsilon}^{\pm\pi}-\int_{\pm\epsilon}^{\pm\pi}h(t)\,t^{-(r+1)}dt,$$

and, by (5·11) and (5·12), the terms on the right have limits as $\epsilon\to+0$. Hence all the integrals in (5·13) have meaning, and the theorem will be established if we show that the last integral in (5·13)—call it $\rho(x)$—is $o(1/x)$ as $x\to 0$.

Suppose, for example, that $x \to +0$, fix a small $\epsilon > 0$, and write

$$\rho(x) = \int_{-\pi}^{\pi} \frac{H(t)}{t^{r+2}(x-t)}\,dt = \int_{\epsilon \leqslant |t| \leqslant \pi} + \int_{-\epsilon}^{0} + \int_{0}^{\frac{1}{2}x} + \int_{\frac{1}{2}x}^{\frac{3}{2}x} + \int_{\frac{3}{2}x}^{\epsilon}$$
$$= \sum_{j=1}^{5} I_j, \tag{5·14}$$

say. Clearly $I_1 = O(1) = o(1/x)$. An application of the second mean-value theorem to the (Riemann) integrals I_2, I_3, I_5 shows that they are respectively

$$\frac{1}{x}\int_{-\theta_1\epsilon}^{0} \frac{H(t)}{t^{r+2}}\,dt, \quad \frac{2}{x}\int_{\frac{1}{2}\theta_2 x}^{\frac{1}{2}x} \frac{H(t)}{t^{r+2}}\,dt, \quad -\frac{2}{x}\int_{\frac{3}{2}x}^{\theta_3\epsilon} \frac{H(t)}{t^{r+2}}\,dt,$$

where $0 < \theta_j < 1$, and so, in view of the convergence of $\int_{-\pi}^{\pi} Ht^{-r-2}\,dt$, are arbitrarily small in comparison with $1/x$ provided ϵ is small enough. Finally write

$$I_4 = -\int_{\frac{1}{2}x}^{\frac{3}{2}x} \frac{1}{t^{r+2}} \frac{H(t)-H(x)}{t-x}\,dt - H(x)\int_{\frac{1}{2}x}^{\frac{3}{2}x} \frac{1}{t^{r+2}} \frac{dt}{t-x} = I_{4,1} + I_{4,2},$$

the last integral being taken in the 'principal value' sense. Since $[H(t)-H(x)]/(t-x)$ is contained between the upper and lower bounds of h in (x,t), (5·10) implies that

$$I_{4,1} = o(x^r)\int_{\frac{1}{2}x}^{\frac{3}{2}x} t^{-r-2}\,dt = o(x^{-1}).$$

By (5·12) and the mean-value theorem,

$$I_{4,2} = o(x^{r+1})\int_{0}^{\frac{1}{2}x} \left\{\frac{1}{(x-t)^{r+2}} - \frac{1}{(x+t)^{r+2}}\right\}\frac{dt}{t}$$
$$= o(x^{r+1})\int_{0}^{\frac{1}{2}x} \frac{(r+2)\,2t}{(x-\theta_4 t)^{r+3}}\frac{dt}{t} = o(x^{r+1})\,O(x^{-r-3})\,\tfrac{1}{2}x = o(1/x).$$

Collecting results we see that $\rho(x) = o(1/x)$, and (5·8) is established.

(5·15) THEOREM. *Let f be integrable, F the integral of f, both periodic. Then $\tilde{F}$ has almost everywhere an approximate derivative equal to $\tilde{f}$.*

This is a corollary of (5·8). For let Φ be the integral of F, and suppose that Φ is periodic. Since F is differentiable almost everywhere, the unsymmetric $\tilde{\Phi}_{(2)}$ exists almost everywhere. By (1·7), $S''[\tilde{\Phi}] = \tilde{S}''[\Phi] = \tilde{S}[f]$ is summable (C, 3) to sum $\tilde{\Phi}_{(2)}$. Hence $\tilde{\Phi}_{(2)} = \tilde{f}$ almost everywhere. Since, by (4·26), we also have $\tilde{\Phi}_{(2)} = \{\tilde{\Phi}'\}'_{\mathrm{ap}} = \tilde{F}'_{\mathrm{ap}}$, almost everywhere, the theorem follows.

6. The Integral M and Fourier series

The notion of integral which we systematically use is that of Lebesgue; only in very few instances have we considered other definitions (see Chapter V, (1·8); Chapter VII, § 4).

The Lebesgue integral has a number of simple properties. For example, if f is integrable, so is $|f|$; an integrable f is almost everywhere the derivative of its indefinite integral; integration by parts holds; finally, we can, under very general conditions, pass to the limit under the sign of integral. These properties, when applied to Fourier series, give a theory of satisfactory simplicity and generality. No other integral (we disregard Stieltjes type generalizations) has all the properties just listed. Adopting a definition of integral more general than Lebesgue's, we may strengthen individual results, at the expense, however, of the generality and coherence of the theory.

Generalizations of the Lebesgue integral are, nevertheless, of interest for Fourier series, and we consider two of them.

The first generalization is usually called *Denjoy's special integral* (to distinguish it from *Denjoy's general integral*, which does not concern us here), or *Perron's integral*. We shall base it on the notions of major and minor functions, and call it the M-*integral*. There are other, equivalent, definitions, but this one will have the advantage for setting a perspective for another extension. We recall the basic definitions and properties.†

Let $f(x)$ be a real-valued, not necessarily finite-valued, function defined in a finite interval $a \leqslant x \leqslant b$. By a *major function* for f in (a, b), we mean any $\Psi(x)$, $a \leqslant x \leqslant b$, which is continuous, is 0 at $x = a$, and whose Dini numbers at each x are not less than $f(x)$; wherever $f = -\infty$, we require that the Dini numbers of Ψ shall be greater than $-\infty$. A *minor function* $\Phi(x)$ is defined correspondingly; in particular the Dini numbers of Φ shall not be greater than f, and shall be less than $+\infty$ wherever $f = +\infty$. Since the Dini numbers of $\Psi - \Phi$ are everywhere non-negative, $\Psi(x) - \Phi(x)$ is non-decreasing; in particular $\Psi(b) \geqslant \Phi(b)$. If

$$\inf_{\Psi} \Psi(b) = \sup_{\Phi} \Phi(b), \tag{6·1}$$

we say that f is M-*integrable* over (a, b), call the common value of both sides of (6·1) the M-*integral* of f over (a, b), and denote it by

$$(\mathrm{M}) \int_a^b f(x)\, dx.$$

We list a few properties of this integral.

(i) Every f integrable M over (a, b) is necessarily measurable, and is integrable M over every subinterval of (a, b). The function

$$F(x) = (\mathrm{M}) \int_a^x f(t)\, dt$$

is continuous, and at almost all points of (a, b) the derivative $F'(x)$ exists and equals $f(x)$.

(ii) The M-integrability, and the value of the integral, of an f are not affected if we change the values of f in a set of measure 0.

(iii) The M-integral is additive for a pair of adjacent intervals. The integral of a sum of two functions is equal to the sum of the integrals.

(iv) Every function integrable L over (a, b) is integrable M, and the values of both integrals are the same (this is Lemma (3·8) of Chapter IX); for $f \geqslant 0$ integrabilities L and M are equivalent.

(v) If f is integrable M over (a, b), then there is a dense family of subintervals of (a, b) over each of which f is integrable L.

(vi) Every function integrable R (we include here functions having improper Riemann integrals) is integrable M, and both integrals have the same value.

† For the proofs and more details we refer the reader to Saks's *Theory of the integral*.

(vii) If f is M-integrable over (a,b), and g is of bounded variation over (a,b), then fg is integrable over (a,b), and we have the formula for integration by parts

$$\int_a^b fg\,dx = [Fg]_a^b - \int_a^b F\,dg.$$

We apply these properties to Fourier series confining our attention to a few problems only.

(*a*) If f is periodic and integrable M over a period, it has, by (vii), Fourier coefficients

$$a_n = \frac{1}{\pi}(\mathrm{M})\int_0^{2\pi} f(x)\cos nx\,dx, \quad b_n = \frac{1}{\pi}(\mathrm{M})\int_0^{2\pi} f(x)\sin nx\,dx,$$

and a Fourier series $\quad \frac{1}{2}a_0 + \sum_1^\infty (a_n \cos na + b_n \sin nx) = \sum_0^\infty A_n(x).$

Suppose that $a_0 = 0$. Then the integral $F(x)$ of f is periodic and continuous, and integration by parts shows that $S[f] = S'[F]$. Hence, by Theorem (5·4) in the case $r=1$, *the Fourier series of a periodic and* M*-integrable f is almost everywhere summable* (C, 1) *to sum $f(x)$*.

(*b*) The Riemann-Lebesgue theorem, an important tool for Fourier series, fails for the integral M: the coefficients a_n, b_n need not tend to 0. But the relation $S[f] = S'[F]$, valid if $a_0 = 0$, shows that, in any case,

$$a_n = o(n), \quad b_n = o(n). \tag{6·2}$$

These estimates cannot be improved (see Theorem (6·4) below); in particular we cannot replace summability (C, 1) in (*a*) by summability (C, k), $k<1$.

(*c*) Suppose that $a_0 = 0$. The relation $S[f] = S'[F]$ can also be written $\tilde{S}[f] = \tilde{S}'[F]$.

(*d*) Suppose that f is periodic and integrable M, and that f is integrable L over a subinterval (a,b) of a period. Let f^* be the periodic L-integrable function which coincides with f in (a,b), and is 0 elsewhere, mod 2π. In view of (6·2) and Theorem (9·23) of Chapter IX, the differences $S[f] - S[f^*]$ and $\tilde{S}[f] - \tilde{S}[f^*]$ are uniformly summable (C, 1) in each $(a+\epsilon, b-\epsilon)$, the former to 0. Hence in the intervals where f is integrable L, the behaviour of $S[f]$ and $\tilde{S}[f]$ can be read off from that of $S[f^*]$ and $\tilde{S}[f^*]$.

(*e*) In view of (i), (vii), and Theorem (3·3) of Chapter IV, if f is periodic and integrable M, the conjugate function

$$\tilde{f}(x) = -\frac{1}{\pi}\lim_{\epsilon\to+0}\frac{1}{\pi}\int_\epsilon^\pi \frac{f(x+t)-f(x-t)}{2\tan\frac{1}{2}t}\,dt = -\frac{1}{\pi}\lim_{\epsilon\to+0}\int \frac{F(x+t)+F(x-t)-2F(x)}{(2\sin\frac{1}{2}t)^2}\,dt \tag{6·3}$$

exists almost everywhere. By (5·4) and (1·25), if *f is periodic and integrable* M, $\tilde{S}[f]$ *is summable* (C, 1) *almost everywhere to sum* (6·3).

(*f*) *If a $\Sigma A_n(x)$, with coefficients $o(n)$ is summable* A *to an $f(x)$ finite and integrable* M, *then the series is* $S[f]$. For f integrable L, this is Theorem (7·4) of Chapter IX, and for f integrable M the proof remains the same. Corresponding results hold if we consider the limits of indetermination of partial sums and Abel means.

While Riemann's proper integral is less general than Lebesgue's, no comparison is possible for Riemann's improper integral. The latter is, however, less general than the integral M, and the results (*a*) to (*f*) above hold for Riemann's improper integral. For the same reason, the assertion that the estimates (6·2) are best possible for M-integrals is a corollary of the following theorem

(6·4) THEOREM. *Given any sequence of positive numbers $\lambda_n = o(n)$ there is an R-integrable f whose sine coefficients are greater than λ_n for infinitely many n.*

Let $\lambda_k = \epsilon_k k$, $\epsilon_k \to 0$. We shall define a sequence of non-overlapping intervals $I_k = (\frac{1}{2}\alpha_k, \alpha_k)$ approaching the point 0 from the right. Let

(A) $f(x) = c_k \sin n_k x$ for $x \in I_k$ $(k = 1, 2, \ldots)$, $f = 0$ elsewhere. The c_k and the integers $n_1 < n_2 < \ldots$ are to satisfy a number of conditions, in particular:

(B) $n_k \alpha_k$ are integral multiples of 4π (hence f is continuous for $x > 0$, and the integral of f over I_k is 0);

(C) $c_k/n_k = 1/k \to 0$ (this implies that f is integrable R over $(0, \pi)$).

Let $n_1 = 4$, $c_1 = 4$, $I_1 = (\frac{1}{2}\pi, \pi)$, and suppose the n_i, c_i, α_i have been defined for $i < k$, and consequently $f(x)$ has been defined for $\frac{1}{2}\alpha_{k-1} \leqslant x \leqslant \pi$. Let $\alpha_k = 4\pi/p$, p being the smallest integer such that $\alpha_k \leqslant 1/n_{k-1}$. Hence

(D) $$1/n_{k-1} \geqslant \alpha_k \geqslant 1/2n_{k-1}.$$

Having defined I_k, we take n_k so large that

(E) $$\int_{I_k} \sin^2 n_k x \, dx \geqslant \tfrac{1}{4} |I_k| = \tfrac{1}{8}\alpha_k \quad \text{(the integral tends to } \tfrac{1}{2}|I_k|\text{)},$$

(F) $$\left| \int_{\alpha_k}^{\pi} f(x) \sin n_k x \, dx \right| < 1,$$

(G) $$\epsilon_{n_k} < 1/64 k n_{k-1}.$$

Write $$\int_0^{\pi} f(x) \sin n_k x \, dx = \int_0^{\alpha_{k+1}} + \int_{\frac{1}{2}\alpha_k}^{\alpha_k} + \int_{\alpha_k}^{\pi} = P_k + Q_k + R_k.$$

By (F), $|R_k| < 1$. Since, by (D), $\sin n_k x$ increases in $(0, \alpha_{k+1})$, the second mean-value theorem gives $P_k \to 0$. Finally, by (A), (E), (C), (D), (G),

$$Q_k > \tfrac{1}{8} c_k \alpha_k = n_k \alpha_k / 8k > n_k / 16 k n_{k-1} > 4\epsilon_{n_k} n_k = 4\lambda_{n_k}.$$

It follows that $$\pi b_{n_k} > 4\lambda_{n_k} - 1 - o(1),$$
that is $b_{n_k} > \lambda_{n_k}$ for k large.

7. The integral M²

Integral M was introduced by Denjoy (in a way different from that of § 6) to integrate an exact derivative, and to show that an everywhere finitely differentiable function is the indefinite integral of the derivative. We now turn to a somewhat similar problem, and show that, with a suitable definition of an integral, an everywhere convergent trigonometric series is the Fourier series of its sum.

There are certain difficulties here. On the one hand, if an everywhere convergent trigonometric series $\Sigma A_n(x)$ is the Fourier series of its sum $f(x)$, it is natural to expect that the series $\frac{1}{2}a_0 x - \Sigma B_n(x)/n$ obtained by termwise integration represents the indefinite integral of f. On the other hand, there are everywhere convergent series $\Sigma A_n(x)$ which after termwise integration cease to converge everywhere;

$$\Sigma (\log n)^{-1} \sin nx$$

is an instance in point, and starting from it we can, by the process of condensation of singularities, construct an everywhere convergent $\Sigma A_n(x)$ such that $\Sigma B_n(x)/n$ diverges in an infinite set, even a set of the power of the continuum (though, of course, it must be of measure 0, since $\Sigma B_n(x)/n$ has coefficients $o(1/n)$). Hence we must expect that the indefinite integral of f need not be defined everywhere, and that the integrability of f over a period need not imply integrability over every interval.

However, if $\Sigma A_n(x)$ converges everywhere to sum $f(x)$, the series $\frac{1}{4}a_0x^2 - \Sigma A_n(x)/n^2$ obtained by two successive integrations converges uniformly, and we may expect that its sum $F(x)$ represents a second integral of f. Hence we may try to deal directly with second integrals, without defining first integrals, and in what follows we shall follow this course.

Consider an $f(x)$, $a \leqslant x \leqslant b$, real-valued, possibly $\pm\infty$ at some points. A function $\Psi(x)$ will be called a *major function* (*of order* 2, to distinguish it from the major functions considered in § 6, which we shall occasionally call major functions of order 1) for f if Ψ is continuous, is 0 at $x=a$ and $x=b$, and if

$$\underline{D}^2\Psi(x) = \liminf_{h\to 0} \frac{\Psi(x+h)+\Psi(x-h)-2\Psi(x)}{h^2} \geqslant f(x) \tag{7·1}$$

at each x interior to (a,b); wherever $f(x) = -\infty$ we require strict inequality in (7·1). A *minor function* $\Phi(x)$ is defined correspondingly, by replacing in (7·1) $\underline{D}^2$ by $\overline{D}^2$ and lim inf by lim sup, reversing the sign of inequality, and requiring strict inequality wherever $f(x) = +\infty$. In what follows, $\Psi(x)$ (with or without indices) will always denote a major function, $\Phi(x)$ a minor function.

If $\Omega(x) = \Psi(x) - \Phi(x)$, the hypotheses imply that $\underline{D}^2\Omega \geqslant \underline{D}^2\Psi - \overline{D}^2\Phi \geqslant 0$, so that Ω is convex (Chapter I, (10·7)) and, since $\Omega(a) = \Omega(b) = 0$, non-positive in (a,b). Hence, if $a \leqslant c \leqslant b$ we always have $\Psi(c) \leqslant \Phi(c)$, and in particular

$$\sup_{\Psi} \Psi(c) \leqslant \inf_{\Phi} \Phi(c).$$

If we have equality here for some c, then it holds for any other c, since if a sequence of convex functions $\Omega_n(x)$ equal to 0 at $x=a$ and $x=b$ tends to 0 at some point interior to (a,b), then it tends to 0, even uniformly, in the whole interval (a,b); this fact will be repeatedly used below. In this case we say that f is M^2-*integrable* over (a,b), and denoting the common value of $\sup_{\Psi} \Psi(x)$ and $\inf_{\Phi} \Phi(x)$ by $F(x)$ $(a \leqslant x \leqslant b)$, we call $F(x)$ the *normalized* (by the condition $F(a) = F(b) = 0$) *second indefinite integral* of f in (a,b); in symbols

$$F(x) = (\mathrm{M}^2)\int_{(a,\,b,\,x)} f(y)\,dy. \tag{7·2}$$

If we add to $F(x)$ any linear function, we call the sum *a second indefinite integral* of f in (a,b).

We establish a number of properties of $F(x)$.

(i) *For any* $\Phi(x)$, $\Psi(x)$, *we have*

$$\Psi(x) \leqslant F(x) \leqslant \Phi(x),$$

and both differences $\Psi(x) - F(x)$ *and* $F(x) - \Phi(x)$ *are convex in* (a,b).

Let $\{\Phi_n(x)\}$ and $\{\Psi_n(x)\}$ be such that $\Psi_n(x) - \Phi_n(x) \to 0$ in (a,b). Hence $\Phi_n(x) \to F(x)$, $\Psi_n(x) \to F(x)$. Since $\Psi_n(x) - \Phi(x)$ and $\Psi(x) - \Phi_n(x)$ are convex and non-positive in (a,b), the same follows for the limit functions $F(x) - \Phi(x)$ and $\Psi(x) - F(x)$.

(ii) *If f is integrable* M^2 *in* (a,b) *it is integrable* M^2 *in every subinterval* (a',b'). Let Ψ_n and Φ_n be major and minor functions for f such that $\Psi_n - \Phi_n \to 0$, and let $L_n^{(1)}$ and $L_n^{(2)}$ be linear functions taking at a' and b' the same values as Ψ_n and Φ_n respectively.

Then $L_n^{(1)}$ and $L_n^{(2)}$ tend to a common limit $L(x)$, $\Psi_n^* = \Psi_n - L_n^{(1)}$ and $\Phi_n^* = \Phi_n - L_n^{(2)}$ are respectively major and minor functions for f in (a', b'), and $\Psi_n^* - \Phi_n^* \to 0$ in (a', b').

Incidentally, $\int_{(a,b,x)} f(y)\,dy - \int_{(a',b',x)} f(y)\,dy$ is linear in (a', b').

(iii) *If f is integrable* M (*in particular integrable* L) *in* (a, b), *it is also integrable* M^2.

Let $\psi_n(x)$ and $\phi_n(x)$ be major and minor functions of order 1 for f in (a, b), such that $\psi_n(x) - \phi_n(x)$ tends (uniformly) to 0, and let

$$\Psi_n(x) = \int_a^x \psi_n(y)\,dy + L_n^{(1)}(x), \quad \Phi_n(x) = \int_a^x \phi_n(y)\,dy + L_n^{(2)}(x),$$

where $L_n^{(1)}$ and $L_n^{(2)}$ are linear functions such that Ψ_n and Φ_n vanish at $x = a$ and $x = b$. Clearly Ψ_n and Φ_n are major and minor functions of order 2 for f in (a, b) and since $L_n^{(1)}$ and $L_n^{(2)}$ tend to a common limit $L(x)$, we also have $\Psi_n(x) - \Phi_n(x) \to 0$ in (a, b), and (iii) follows. We also have

$$(\mathrm{M}^2) \int_{(a,b,x)} f\,dy = \int_a^x \left\{(\mathrm{M}) \int_a^y f\,dt\right\} dy + L(x). \tag{7·3}$$

(iv) *If f_1 and f_2 are integrable* M^2 *in* (a, b), *and c_1, c_2 are constants, then $f = c_1 f_1 + c_2 f_2$ is integrable* M^2 *in* (a, b) *and*

$$\int_{(a,b,x)} f(y)\,dy = c_1 \int_{(a,b,x)} f_1(y)\,dy + c_2 \int_{(a,b,x)} f_2(y)\,dy.$$

This is obvious when $f = c_1 f_1$ (consider separately the cases $c_1 > 0$ and $c_1 < 0$) and when $f = f_1 + f_2$, and the general case follows from these two.

(v) *If f is integrable* M^2 *in* (a, b), *then* $F(x) = \int_{(a,b,x)} f(y)\,dy$ *is continuous, and, almost everywhere, a finite $D^2F(x)$ exists and equals $f(x)$.*

Let $G(x)$ be convex and continuous in $\alpha \leqslant x \leqslant \beta$; we shall find estimates for the measure of the set of points where $D^2G(x) \geqslant k > 0$.

Suppose first that G is non-decreasing, and denote by $g(x)$ the right-hand side derivative of $G(x)$; in every (α', β') interior to (α, β), $g(x)$ is non-decreasing and bounded and $g'(x)$ exists almost everywhere and is L-integrable. We have

$$G(\beta') - G(\alpha') = \int_{\alpha'}^{\beta'} g(x)\,dx \geqslant \int_{\alpha'}^{\beta'} \{g(x) - g(\alpha')\}\,dx$$

$$\geqslant \int_{\alpha'}^{\beta'} \left\{\int_{\alpha'}^{x} g'(y)\,dy\right\} dx = \int_{\alpha'}^{\beta'} (\beta' - y)\,g'(y)\,dy,$$

and, making $\alpha' \to \alpha$, $\beta' \to \beta$, we obtain

$$G(\beta) - G(\alpha) \geqslant \int_{\alpha}^{\beta} (\beta - y)\,g'(y)\,dy \geqslant \int_{\alpha+\epsilon}^{\beta} (\beta - y)\,g'(y)\,dy$$

for any $\epsilon < \beta - \alpha$. It follows that the set of points in $(\alpha + \epsilon, \beta)$ where $g' \geqslant k$ is of measure not exceeding $\{G(\beta) - G(\alpha)\}/k\epsilon$, and hence the subset of (α, β) where $g' \geqslant k$ has measure not exceeding $\epsilon + \{G(\beta) - G(\alpha)\}/k\epsilon$, a result valid even if $\epsilon \geqslant \beta - \alpha$.

A similar estimate, with $G(\alpha) - G(\beta)$ for $G(\beta) - G(\alpha)$, holds for $G(x)$ convex and non-increasing. Since for a general convex G we can split (α, β) into two subintervals in each of which G is monotone, we see that for any $\epsilon > 0$, $k > 0$ the subset of (α, β) where $g' \geq k$ has measure not exceeding

$$2\epsilon + 2\omega/k\epsilon,$$

where ω is the oscillation of G in (a, b). Since $D^2G(x) = g'(x)$ wherever $g'(x)$ exist, as we easily see from the formula

$$\frac{G(x+h)+G(x-h)-2G(x)}{h^2} = \frac{1}{h^2}\int_0^h [g(x+t)-g(x-t)]\,dt$$

(in particlar, D^2G exists almost everywhere), we come to the following conclusion: *if the oscillation of a convex $G(x)$ in (α, β) is small, $D^2G(x)$ is small except in a set of small measure.*

Return to (iv), and consider a major function Ψ for f. In the decomposition $F = (F - \Psi) + \Psi$ the function $F - \Psi$ is convex and, for a suitable Ψ, uniformly small. Hence $D^2(F - \Psi)$ is small except in a small set. Since $\underline{D}^2\Psi(x) \geq f(x)$ and $\underline{D}^2\Psi(x) > -\infty$ for all x, it easily follows that $\underline{D}^2F(x) \geq f(x)$, $\underline{D}^2F(x) > -\infty$ at almost all points in (α, β). Using minor functions we similarly obtain that $\bar{D}^2F(x) \leq f(x)$, $\bar{D}^2F(x) < +\infty$ almost everywhere in (α, β), and (v) follows.

(vi) *If $f(x)$ is* M^2*-integrable in (α, β), and if $f_1 \equiv f$, then f_1 is also* M^2*-integrable and*

$$\int_{(a,\,b,\,x)} f(y)\,dy = \int_{(a,\,b,\,x)} f_1(y)\,dy.$$

We first consider the case when f_1 can differ from f only in the set E of points where $f = +\infty$. Hence $f_1 \leq f$, and any major function Ψ for f is a major function for f_1.

Since, by (v), $|E| = 0$, the function $h(x)$ equal to $-\infty$ in E and to 0 elsewhere is, by (ii), M^2 integrable over (a, b) and, by (7·3), its normalized second indefinite integral is 0 identically. Let Φ^* be a minor function for h; hence $\bar{D}^2\Phi^* \leq 0$ everywhere, $\bar{D}^2\Phi^* = -\infty$ in E. If Φ is a minor function for f, the sum $\Phi + \Phi^*$ is a minor function for f_1; for

$$\bar{D}^2(\Phi + \Phi^*) \leq \bar{D}^2\Phi + \bar{D}^2\Phi^*,$$

where the two terms on the right can never be $+\infty$; hence

$$\bar{D}^2(\Phi + \Phi^*) = -\infty \leq f_1 \quad \text{in } E,$$

$$\bar{D}^2(\Phi + \Phi^*) \leq \bar{D}^2\Phi \leq f = f_1 < +\infty \quad \text{outside } E.$$

Since $\Psi - \Phi$ and Φ^* can be made arbitrarily small, the same holds for $\Psi - (\Phi + \Phi^*)$ and (vi) is established in the case considered.

Similarly, we can prove (vi) in the case when f_1 can differ from f only in the set where $f = -\infty$. Passing to the general case, let $g(x) = f(x)$ wherever f is finite, and $g(x) = 0$ elsewhere. The function g is finite-valued, the decomposition $f_1 = (f_1 - g) + g$ has meaning, the normalized second indefinite integral for g is the same as for f, the normalized integral for $f_1 - g$ ($\equiv 0$) is identically 0, and it is enough to apply (iv).

We can now speak of M^2-integrals of functions defined only almost everywhere, since we may complete the definition of the function in the exceptional set of measure 0 in an arbitrary way.

(vii) *Suppose that the partial sums of a trigonometric series $\Sigma A_n(x)$ are bounded at each x, and let*

$$F(x) = \tfrac{1}{4}a_0x^2 - \sum_1^\infty A_n(x)/n^2.$$

Then $D^2F(x)$ exists and is finite almost everywhere, and is M² *integrable in any finite interval (a, b), and $F(x)$ is a second indefinite integral of D^2F.*

The hypothesis implies that the coefficients of $\Sigma A_n(x)$ are bounded; hence $F(x)$ is continuous. We also have (Chapter IX, (2·7))

$$\frac{F(x+h)+F(x-h)-2F(x)}{h^2}=O(1) \quad (h\to 0)$$

at each x, which, by (4·30), implies that $D^2F(x)$ exists and is finite almost everywhere.

The rest of the assertion follows from the fact that if $L(x)$ is the linear function coinciding with $F(x)$ at $x=a$ and $x=b$, then $F(x)-L(x)$ is both a major and minor function for $f(x)=\underline{D}^2F(x)$.

So far we have only considered real-valued functions. If f is complex-valued, $f=f_1+if_2$, we may say that f is M²-integrable over (a, b), if f_1 and f_2 are, and define the (normalized) second indefinite integral of f as F_1+iF_2, where F_k is the (normalized) second integral of f_k $(k=1, 2)$.

It is clear that (vii) holds for a complex-valued trigonometric series $\Sigma c_k e^{ikx}$ if its (symmetric) partial sums are bounded at each x, and $F(x)$ is defined as

$$\tfrac{1}{2}c_0x^2-\Sigma' c_k e^{ikx}/k^2.$$

Before we prove our final result, we must express the Fourier coefficients of a function in terms of M²-integrals. Consider first the Fourier series

$$\sum_{-\infty}^{+\infty} c_k e^{ikx}$$

of an $f\in$ L. We have

$$c_0=\frac{1}{2\pi}\int_{\alpha}^{\alpha+2\pi} f(x)\,dx$$

for any α. Integrating this over $-2\pi\leqslant\alpha\leqslant 0$, and denoting by $F(x)$ a second integral of f, we obtain

$$c_0=\frac{1}{4\pi^2}\{F(2\pi)+F(-2\pi)-2F(0)\},$$

a result independent of the arbitrary linear component of $F(x)$.

Given any f, if F is its second indefinite integral, we shall write

$$F(x+h)+F(x-h)-2F(x)=\mathrm{V}(f;\, x, h).$$

With this notation, the formula for c_0 becomes

$$c_0=\frac{1}{4\pi^2}\mathrm{V}(f;\, 0, 2\pi). \tag{7·4}$$

(7·5) THEOREM. *Suppose that $\Sigma c_k e^{ikx}$ has bounded partial sums at each point x, and write $F(x)=\frac{1}{2}c_0x^2-\Sigma' c_k k^{-2} e^{ikx}$. Then*

$$f(x)=D^2F(x)$$

exists almost everywhere, the products

$$f(x)\,e^{-imx} \quad (m=0, \pm 1, \pm 2, \dots)$$

are M^2*-integrable over every finite interval, and*

$$c_m = \frac{1}{4\pi^2} V(f e^{-imx}; 0, 2\pi) \tag{7·6}$$

for each m.

By (vii) (in the complex case), $f = D^2 F$ exists almost everywhere and is M^2-integrable over any finite interval, and F is a second indefinite integral of f. From $F(x) = \frac{1}{2}c_0 x^2 - \Sigma' c_k k^{-2} e^{ikx}$ we immediately deduce that

$$\frac{1}{4\pi^2}[F(2\pi) + F(-2\pi) - 2F(0)] = c_0,$$

and (7·6) follows from $m = 0$.

Consider the formal product $\Sigma c_{k+m} e^{ikx}$ of $\Sigma c_k e^{ikx}$ and e^{-imx}. The partial sums of $\Sigma c_{k+m} e^{ikx}$ are bounded at each x. It follows that $\Sigma c_{k+m} e^{ikx}$ is summable R almost everywhere to a sum $g(x)$, and that

$$c_m = \frac{1}{4\pi^2} V(g; 0, 2\pi).$$

It is enough to show that $g = f e^{-imx}$ almost everywhere.

To this end observe that, by (1·7), we may take for f and g the (C, 3) sums of $\Sigma c_k e^{ikx}$ and $\Sigma c_{k+m} e^{ikx}$ respectively. Also observe that the two sequences

$$c_k e^{ikx}, \quad c_{-k} e^{-ikx} \quad (k = 0, 1, 2, \ldots) \tag{7·7}$$

are summable (C, 1) almost everywhere to 0. This follows from the fact that the two series $\sum_0^\infty (k+1)^{-1} c_k e^{ikx}$ and $\sum_0^\infty (k+1)^{-1} c_{-k} e^{-ikx}$ have coefficients $O(1/k)$, and so are convergent wherever they are summable (C, 1) (Chapter III, (1·26)); it is now enough to apply the fact that, if Σu_k converges then $u_0 + 2u_1 + \ldots + (k+1) u_k = o(k)$.

By the final remarks of Chapter IX, at the points where the two sequences (7·7) are summable (C, 1) to 0 the series $\sum_{k=-\infty}^{+\infty} c_{k+m} e^{ikx}$ is equisummable (C, 1), and so also (C, 3), with $e^{-imx} \sum_{k=-\infty}^{\infty} c_k e^{ikx}$. Hence $g = f e^{-imx}$ almost everywhere and (7·5) is proved completely.

Remark. The example of the series $\frac{1}{2} + \cos x + \cos 2x + \ldots$ shows that, if the partial sums of $\Sigma c_k e^{ikx}$ are allowed to be unbounded at a single point, (7·6) need not be true. If, however, $c_k \to 0$, the conclusions of (7·5) hold even if the partial sums of $\Sigma c_k e^{ikx}$ cease to be bounded at a denumerable set of points. The function F is then smooth, and the exceptional denumerable set does not affect the validity of the result; we omit the details, which are not difficult.

MISCELLANEOUS THEOREMS AND EXAMPLES

1. Let $\chi_x(t) = \frac{1}{2}\{f(x+t) + f(x-t)\}$.

(i) If $\beta > \alpha \geq 0$, and if (C, α) $\chi_{x_0}(t) \to s$ as $t \to +0$, then $S[f]$ is summable (C, β) at $x = x_0$ to sum s.

(ii) If $\beta > -1$, $\alpha > \beta + 1$, and if $S[f]$ is summable (C, β) at $x = x_0$ to sum s, then (C, α) $\chi_{x_0}(t) \to s$ as $t \to 0$. (Bosanquet [2].)

2. If f is bounded and $f \sim \Sigma A_n(x)$, then $\Sigma B_n(x_0)$ is summable C if and only if it is summable (C, ϵ) for every $\epsilon > 0$; $\Sigma B_n(x_0)$ is summable C if and only if the integral

$$-\frac{1}{\pi}\int_0^\pi [f(x_0+t)-f(x_0-t)]\,\tfrac{1}{2}\cot\tfrac{1}{2}t\,dt$$

exists; and if the integral exists, it represents the C-sum of $\Sigma B_n(x_0)$. (Prasad [2], Hardy and Littlewood [26].)

3. Suppose that $F(x)$ is non-decreasing, and $F(x+2\pi)-F(x)$ is constant.

(i) If $S[dF]$ is summable C at x, then it is summable (C, ϵ) at that point, for each $\epsilon > 0$.

(ii) A necessary and sufficient condition for $S[dF]$ to be summable C at x to sum s is that

$$\lim_{t\to 0}\frac{F(x+t)-F(x-t)}{2t}=s$$

[The proof is similar to that of (2·26).]

4. If $\sum_0^\infty c_n e^{inx}$ is everywhere summable $(C, -\epsilon)$, $0<\epsilon<1$, to a finite sum $f(x)$, then $f(x)$ is M-integrable, and $\Sigma c_n e^{inx}$ is the Fourier series of f.

CHAPTER XII

INTERPOLATION OF LINEAR OPERATIONS. MORE ABOUT FOURIER COEFFICIENTS

1. The Riesz-Thorin theorem

In this section we prove a general theorem on linear operations which, when applied to trigonometric series, leads to a number of interesting results. The proof of the theorem is based on the following maximum principle of Phragmén and Lindelöf:

(1·1) THEOREM. *Suppose that $f(z)$, $z=x+iy$, is continuous and bounded in the closed strip*

$$B: \quad \alpha \leqslant x \leqslant \beta,$$

and regular in the interior of B. If $|f(z)| \leqslant M$ on the lines $x=\alpha$ and $x=\beta$, then $|f(z)| \leqslant M$ also in the interior of B. If, in addition, $|f(z_0)|=M$ at a point z_0 interior to B, then $f(z)$ is constant.

Suppose first that

$$f(x+iy) \to 0 \tag{1·2}$$

as $y \to \pm\infty$, uniformly in $\alpha \leqslant x \leqslant \beta$. If $z_0=x_0+iy_0$ is in the interior of B, the inequality $|f(z_0)| \leqslant M$ follows from the classical maximum principle applied to f in the rectangle $\alpha \leqslant x \leqslant \beta$, $|y| \leqslant \eta$, where η is so large that $\eta > |y_0|$ and $|f(x \pm i\eta)| \leqslant M$ for $\alpha \leqslant x \leqslant \beta$.

If (1·2) does not hold, we consider the function

$$f_n(z) = f(z)\, e^{z^2/n} = f(z)\, e^{(x^2-y^2)/n}\, e^{2ixy/n} \quad (n=1,2,\ldots),$$

which satisfies (1·2), and on the lines $x=\alpha$ and $x=\beta$ does not exceed $M e^{\gamma^2/n}$, where $\gamma = \max(|\alpha|, |\beta|)$. Hence

$$|f_n(z_0)| \leqslant M e^{\gamma^2/n},$$

and on making $n \to \infty$ we get $|f(z_0)| \leqslant M$.

If $|f(z_0)|=M$, then f is constant, for otherwise in the neighbourhood of z_0 we should have points z such that $|f(z)| > |f(z_0)| = M$, which is impossible.

It is convenient to state the Phragmén-Lindelöf principle in a slightly stronger form, which, however, is a corollary of (1·1).

(1·3) THEOREM. *Suppose that $f(z)$ is continuous and bounded in the strip B, and regular in the interior of B, and that*

$$|f(\alpha+iy)| \leqslant M_1, \quad |f(\beta+iy)| \leqslant M_2 \tag{1·4}$$

for all y. Then for every $z_0=x_0+iy_0$ in the interior of B we have

$$|f(x_0+iy_0)| \leqslant M_1^{L(x_0)}\, M_2^{1-L(x_0)}, \tag{1·5}$$

where $L(t)$ is the linear function taking the values 1 and 0 for $t=\alpha$ and $t=\beta$ respectively. If we have equality in (1·5), then

$$F(z) = f(z)/M_1^{L(z)}\, M_2^{1-L(z)}$$

is a constant of absolute value 1.

It is enough to observe that $F(z)$ satisfies the hypotheses of (1·1) with $M=1$.

Theorem (1·3) connects the upper bounds of $|f|$ on the three lines $x=\alpha$, $x=x_0$, $x=\beta$, and is often called the *three-line theorem.*

We now introduce some notions from the theory of linear operations.

Let R be any *measure space*, i.e. a space on which a non-negative and totally additive measure (mass distribution) $\mu(E)$ is defined, at least for some ('measurable') subsets E of R. For our purposes it is enough to assume that R is an n-dimensional Euclidean space, or a subset of it. The notion of measure leads to that of the Lebesgue-Stieltjes integral

$$\int_R f d\mu \tag{1·6}$$

of a complex-valued function f. The most elementary properties of this integral we take for granted. The important special cases—and the only ones interesting in our applications—are when

(i) $\mu(E)$ is the Lebesgue measure of E; or

(ii) the mass μ is concentrated at a denumerable sequence of points $a_1, a_2, \ldots$.

In the first case, (1·6) is the ordinary Lebesgue integral. In the second, (1·6) may be defined as

$$\Sigma f(a_i)\,\mu_i,$$

where μ_i is the mass concentrated at a_i and the series converges absolutely. The reader unfamiliar with the Lebesgue-Stieltjes integral may still proceed provided he interprets (1·6) in one of these two ways.

We consider only functions measurable with respect to μ, write

$$\| f \|_{r,\mu} = \left(\int_R |f|^r d\mu\right)^{1/r} \tag{1·7}$$

for $0<r<\infty$, and denote by

$$\| f \|_{\infty,\mu}$$

the essential upper bound of $|f|$ on R, that is, the least number M such that $|f| \leqslant M$ except on a set E for which $\mu(E)=0$. We denote by $\mathrm{L}^{r,\mu}$ the class of functions such that $\| f \|_{r,\mu}$ is finite. If no confusion arises, we write $\| f \|_r$ and L^r instead of $\| f \|_{r,\mu}$ and $\mathrm{L}^{r,\mu}$. In the rest of this section we always have $1 \leqslant r \leqslant \infty$.

We call *simple* any function taking only a finite number of values and (if $\mu(R)$ is infinite) vanishing outside a subset of R of finite measure. The set of all simple functions will be denoted by S. It follows from the definition of the integral that the set S is dense in every L^r, $1 \leqslant r < \infty$, though not necessarily in L^∞ unless $\mu(R)$ is finite.

We need the following two results, in which $r' = r/(r-1)$:

$$\left|\int_R fg\,d\mu\right| \leqslant \| f \|_r \| g \|_{r'} \quad (1 \leqslant r \leqslant \infty), \tag{1·8}$$

$$\| f \|_r = \sup_g \left|\int_R fg\,d\mu\right| \quad (g \in S,\ \| g \|_{r'} = 1,\ 1 \leqslant r \leqslant \infty). \tag{1·9}$$

The first is Hölder's inequality, whose proof in the general case is the same as for one-dimensional Lebesgue integrals (see Chapter I, § 9). The proof of the second, again in the case of one-dimensional integrals and for g not necessarily simple, was given in Chapter I, p. 19. The extension to Lebesgue-Stieltjes integrals does not require new ideas. That for g we can take simple functions is easily verified directly when $r'=\infty$, and follows when $r'<\infty$ from the fact that S is dense in $L^{r'}$.

Denote by R_1 and R_2 two measure spaces with measures μ and ν respectively. Let

$$h=Tf$$

be a linear operation from R_1 to R_2. By this we mean that

$$T(\alpha_1 f_1+\alpha_2 f_2)=\alpha_1 Tf_1+\alpha_2 Tf_2$$

for all complex numbers α_1, α_2 and all complex-valued functions f on R_1 with $\|f\|_r$ finite, and that the function h is defined on R_2. We say that the operation T is of *type* (r, s), where $1\leqslant r\leqslant\infty$, $1\leqslant s\leqslant\infty$, if

$$\|h\|_{s,\nu}\leqslant M\|f\|_{r,\mu}. \tag{1·10}$$

The least value of M here is the *norm* of the operation (see Chapter IV, § 9).

If Tf is initially defined for simple functions only, and if $1\leqslant r<\infty$, then there is a unique extension of it, as a linear operation and with the same value of M in (1·10), to the whole of $L^{r,\mu}$ (see Chapter IV, (9·3)).

Our main result can now be stated as follows:

(1·11) THEOREM OF M. RIESZ-THORIN. *Let R_1 and R_2 be two measure spaces with measures μ and ν respectively. Let T be a linear operation defined for all simple functions f on R_1. Suppose that T is simultaneously of type* $(1/\alpha_1, 1/\beta_1)$ *and* $(1/\alpha_2, 1/\beta_2)$, *i.e. that*

$$\|Tf\|_{1/\beta_1}\leqslant M_1\|f\|_{1/\alpha_1},\quad \|Tf\|_{1/\beta_2}\leqslant M_2\|f\|_{1/\alpha_2}, \tag{1·12}$$

the points (α_1,β_1) *and* (α_2,β_2) *belonging to the square*

$$0\leqslant\alpha\leqslant 1,\quad 0\leqslant\beta\leqslant 1.$$

Then T is also of the type $(1/\alpha, 1/\beta)$ *for all*

$$\alpha=(1-t)\alpha_1+t\alpha_2,\quad \beta=(1-t)\beta_1+t\beta_2\quad (0<t<1), \tag{1·13}$$

and

$$\|Tf\|_{1/\beta}\leqslant M_1^{1-t}M_2^t\|f\|_{1/\alpha}. \tag{1·14}$$

In particular, if $\alpha>0$ the operation T can be uniquely extended to the whole space $L^{1/\alpha,\mu}$, preserving (1·14).

We fix t in (1·13), and so also the numbers α, β, and consider the functions

$$\alpha(z)=(1-z)\alpha_1+z\alpha_2,\quad \beta(z)=(1-z)\beta_1+z\beta_2,$$

which for $z=0$, $z=1$, $z=t$ reduce to α_1,β_1; α_2,β_2; α,β respectively. We consider z in the strip $0\leqslant x\leqslant 1$, which we call B. For any simple f,

$$\|Tf\|_{1/\beta}=\sup_g\left|\int_{R_2} Tf.g\,d\nu\right|\quad (g\in S,\ \|g\|_{1/(1-\beta)}=1). \tag{1·15}$$

We may suppose that $\|f\|_{1/\alpha}=1$. Fixing f and g, consider the integral

$$I=\int_{R_2} Tf.g\,d\nu.$$

We write

$$f=|f|e^{iu},\quad g=|g|e^{iv},$$

and introduce the functions

$$F_z = |f|^{\alpha(z)/\alpha} e^{iu}, \tag{1·16}$$

$$G_z = |g|^{(1-\beta(z))/(1-\beta)} e^{iv}, \tag{1·17}$$

where we temporarily suppose that $\alpha > 0$ and $\beta < 1$. The integral

$$\Phi(z) = \int_{R_2} TF_z . G_z d\nu \tag{1·18}$$

reduces to I for $z = t$.

If $f = 0$, F_z is to be taken as 0 whatever the value of the exponent; similarly for G_z. Thus, if $c_1, c_2, \ldots$ are the distinct non-zero values of f, and $\chi_1, \chi_2, \ldots$ the characteristic functions of the sets where these values are taken, and if $c_j = |c_j| e^{iu_j}$ we have

$$F_z = \Sigma e^{iu_j} |c_j|^{\alpha(z)/\alpha} \chi_j.$$

A similar formula, with exponents $(1-\beta(z))/(1-\beta)$, holds for G_z. If we replace χ_j by $T\chi_j$ in the sum defining F_z, we get TF_z. Substituting this in (1·18), we see that $\Phi(z)$ is a finite linear combination, with constant coefficients, of exponentials a^z with $a > 0$. In particular, $\Phi(z)$ is bounded in the strip B.

Consider now any z whose real part is 0. The real part of $\alpha(z)$ is then α_1. Hölder's inequality applied to (1·18) gives

$$|\Phi(z)| \leq \| TF_z \|_{1/\beta_1} \| G_z \|_{1/(1-\beta_1)} \leq M_1 \| F_z \|_{1/\alpha_1} \| G_z \|_{1/(1-\beta_1)}. \tag{1·19}$$

But it follows from (1·16) that

$$\| F_z \|_{1/\alpha_1} = \| \, |f|^{\alpha_1/\alpha} \|_{1/\alpha_1} = \| f \|_{1/\alpha}^{\alpha_1/\alpha} = 1^{\alpha_1/\alpha} = 1,$$

the second equality being valid both for $\alpha_1 > 0$ and $\alpha_1 = 0$. Similarly

$$\| G_z \|_{1/(1-\beta_1)} = \| \, |g|^{(1-\beta_1)/(1-\beta)} \|_{1/(1-\beta_1)} = \| g \|_{1/(1-\beta)}^{(1-\beta_1)/(1-\beta)} = 1.$$

Hence $|\Phi(z)| \leq M_1$ for $x = 0$. Similarly $|\Phi(z)| \leq M_2$ for $x = 1$. It follows from (1·3) that

$$|I| = |\Phi(t)| \leq M_1^{1-t} M_2^t.$$

By (1·15), $\| Tf \|_{1/\beta} = \sup_g |I|$ satisfies (1·14).

The two exceptional cases $\alpha = 0$ and $\beta = 1$ cannot occur simultaneously. If $\beta = 1$, then also $\beta_1 = \beta_2 = 1$, and $\alpha > 0$. We define F_z as before and set $G_z = g$; thus G_z is independent of z. The rest of the proof is simplified, since in (1·19) we may replace $\| G_z \|_{1/(1-\beta_1)}$ by $\| g \|_{1/(1-\beta_1)} = 1$. If $\alpha = 0$, and so $\alpha_1 = \alpha_2 = 0$, $\beta < 1$, we set $F_z = f$ and keep the old definition of G_z. (It may be also observed that if $\alpha_1 = \alpha_2$, then (1·14) is a consequence of Hölder's inequality.)

The extensibility of T to the whole of $L^{1/\alpha}$ has already been discussed.

It is sometimes convenient to be able to take $M_1 = M_2 = 1$. This we can always do; if $\beta_1 \neq \beta_2$ then we need only multiply T and $d\nu$ by suitable constants, while if $\alpha_1 \neq \alpha_2$ we may multiply f and $d\mu$ by constants.

It is natural to ask if we can have equality in (1·14) for some $f \not\equiv 0$ in $L^{1/\alpha}$. It turns out that if

$$\| Tf \|_{1/\beta} = M_1^{1-t} M_2^t \| f \|_{1/\alpha} \tag{1·20}$$

and $\beta > 0$, then f must satisfy a certain functional relation from which we can often deduce characteristic properties of f. We may suppose that $\| f \|_{1/\alpha} = 1$. We write

$$g = \left(\frac{|Tf|}{\| Tf \|_{1/\beta}} \right)^{(1-\beta)/\beta} \overline{\operatorname{sign}(Tf)}. \tag{1·21}$$

Hence $\| g \|_{1/(1-\beta)} = 1$ and

$$\| Tf \|_{1/\beta} = \int_{R_2} Tf . g d\nu.$$

Supposing that $\alpha > 0$, $\beta < 1$, we define F_z and G_z by (1·16) and (1·17) and consider the function

$$\Phi(z) = \int_{R_2} TF_z . G_z d\nu. \tag{1·22}$$

This function is defined for each z in the strip B. For since F_z is in $L^{1/\alpha(x)}$, TF_z is in $L^{1/\beta(x)}$; and since G_z is in $L^{1/(1-\beta(x))}$, the integral in (1·22) converges. Moreover $\Phi(z)$ is bounded in B, since

$$| \Phi(z) | \leqslant \| TF_z \|_{1/\beta(x)} \| G_z \|_{1/(1-\beta(x))} \leqslant M_1^{1-x} M_2^x \| F_z \|_{1/\alpha(x)} . 1$$
$$= M_1^{1-x} M_2^x, \tag{1·23}$$

by Theorem (1·11).

We show that Φ is regular in the interior of B.

For each $m = 1, 2, \ldots$ we consider a simple function f_m defined on R_1 and having the following properties:

(i) $| f_m | \leqslant | f |$,

(ii) $| f_m - f | < 1/m$ whenever $| f | \leqslant m$,

(iii) $f_m = 0$ whenever $| f | > m$.

Clearly $f_m \to f$ as $m \to \infty$. Similarly, we define on R_2 simple functions $g_1, g_2, \ldots$, corresponding to g. Denote by $F_{m,z}$, $G_{m,z}$, $\Phi_m(z)$ the functions F_z, G_z, $\Phi(z)$ formed with f_m, g_m. Hence

$$\Phi_m(z) = \int_{R_2} TF_{m,z} . G_{m,z} d\nu$$

is an entire function. An argument similar to (1·23), coupled with (i) and its analogue for g, shows that $| \Phi_m(z) | \leqslant M_1^{1-x} M_2^x$ for $0 \leqslant x \leqslant 1$; in particular, the Φ_m are uniformly bounded in B. If we show that $\Phi_m(z) \to \Phi(z)$ for each z with $0 < x < 1$, the regularity of Φ in the interior of B will follow.

We fix $z = x + iy$, where $0 < x < 1$; hence $\alpha(x) > 0$, $1 - \beta(x) > 0$. It is enough to show that each of the integrals

$$\Delta_1 = \int_{R_2} T\{F_{m,z} - F_z\} . G_{m,z} d\nu,$$

$$\Delta_2 = \int_{R_2} TF_z . \{G_{m,z} - G_z\} d\nu$$

tends to 0 as $m \to \infty$. Now

$$| \Delta_1 | \leqslant \| T\{F_{m,z} - F_z\} \|_{1/\beta(x)} \| G_{m,z} \|_{1/(1-\beta(x))}$$
$$\leqslant M_1^{1-x} M_2^x \| F_{m,z} - F_z \|_{1/\alpha(x)} \| G_z \|_{1/(1-\beta(x))} = M_1^{1-x} M_2^x \| F_{m,z} - F_z \|_{1/\alpha(x)}.$$

Observing that $F_{m,z} \to F_z$ at each point of R_1, and that, by (i), $| F_{m,z} - F_z |^{1/\alpha(x)}$ is majorized by the integrable function $2^{1/\alpha(x)} | f |^{1/\alpha}$, we immediately see that $\| F_{m,z} - F_z \|_{1/\alpha(x)} \to 0$. Hence $\Delta_1 \to 0$. Similarly $\Delta_2 \to 0$. It follows that Φ is regular in the interior† of B.

The function $\Phi(z)/M_1^{1-z} M_2^z$ is regular in the interior of B and, by (1·23), is not greater than 1 in absolute value there. If its modulus is 1 at some point interior to B, the function is constant in B. In particular, *if we have* (1·20) *for some* f *with* $\| f \|_{1/\alpha} = 1$, *and if* $\beta > 0$, *then we have the relation*

$$\int_{R_2} T(| f |^{\alpha(z)/\alpha} \operatorname{sign} f) . (| g |^{(1-\beta(z))/(1-\beta)} \operatorname{sign} g)\, d\nu = M_1^{1-z} M_2^z \quad (0 < x < 1), \qquad \textbf{(1·24)}$$

where g *is given by* (1·21) *and satisfies* $\| g \|_{1/(1-\beta)} = 1$. This is the functional relation alluded to above.

In the argument above we supposed that $\alpha > 0$, $\beta < 1$. If $\alpha = 0$ or $\beta = 1$, we have to modify the F_z and G_z, as we did in the proof of Theorem (1·11). These cases have no interesting applications.

We pass to a generalization of (1·11). Denote by B the strip

$$B: 0 \leqslant x \leqslant 1$$

in the plane of the complex variable $z = x + iy$. A function $\Phi(z)$, continuous in B and regular in the interior of B, will be said to satisfy *condition* E if

$$\log | \Phi(x+iy) | \leqslant A\, e^{a|y|} \quad (0 \leqslant x \leqslant 1), \qquad \textbf{(1·25)}$$

where A and a are positive constants and

$$a < \pi.$$

We consider a whole family of linear operations T_z depending on a complex parameter $z = x + iy$; for our purposes it is enough to suppose that z is confined to the strip B. Keeping the previous notation, we call such a family $\{T_z\}$ *analytic* if, for any simple f and g,

$$\Phi(z) = \int_{R_2} T_z f . g\, d\nu \qquad \textbf{(1·26)}$$

is continuous in B and regular in the interior of B. (The constants a, A in (1·25) may then depend on f and g.) We say that $\{T_z\}$ *satisfies condition* E, if each Φ in (1·26) does.

(1·27) LEMMA. *Suppose that* $\Phi(z)$ *is any function continuous in* B, *regular in the interior of* B, *satisfying condition* E; *and that*

$$| \Phi(iy) | \leqslant M_1(y), \quad | \Phi(1+iy) | \leqslant M_2(y) \qquad \textbf{(1·28)}$$

where $\log M_1(y)$ *and* $\log M_2(y)$ *are* $O(e^{a|y|})$, $a < \pi$. *Then for each* $0 < x < 1$ *we have*

$$| \Phi(x) | \leqslant A_x = A_x(M_1, M_2), \qquad \textbf{(1·29)}$$

where A_x *depends only on* x, *and the functions* M_1, M_2, *and is bounded in* $0 < x < 1$ *if* M_1 *and* M_2 *are fixed.*

† If $\alpha(x)$ stays away from 0, and $\beta(x)$ away from 1, for $0 \leqslant x \leqslant 1$, then the above argument easily shows that $\Phi_m(z)$ tends uniformly to $\Phi(z)$ for z remaining within any fixed bounded subset of B. It follows that if neither of the numbers α_1, α_2 is 0, and neither of the numbers β_1, β_2 is 1, then $\Phi(z)$ is continuous in the closed strip B.

This lemma is easily deducible by conformal mapping from results obtained in Chapter VII, § 7. We recall these results.

Suppose that $\Psi(\zeta)$ is regular in $|\zeta|<1$. Then if $|\zeta|=\rho<R<1$, we have

$$\log|\Psi(\rho e^{i\theta})| \leqslant \frac{1}{\pi}\int_{-\pi}^{\pi} \log|\Psi(R e^{i\phi})|\, P\left(\frac{\rho}{R}, \phi-\theta\right) d\phi. \tag{1·30}$$

If $\log|\Psi(R e^{i\phi})|$ is majorized (algebraically) by an integrable function of ϕ, independent of R, and if the radial limits $\Psi(e^{i\phi})=\lim \Psi(R e^{i\phi})$ exist almost everywhere, we may make $R\to 1$ in (1·30) and obtain

$$\log|\Psi(\rho e^{i\theta})| \leqslant \frac{1}{\pi}\int_{-\pi}^{\pi} \log|\Psi(e^{i\phi})|\, P(\rho, \phi-\theta)\, d\phi. \tag{1·31}$$

In particular, (1·31) holds if $\Psi(\zeta)$ is continuous in $|\zeta|\leqslant 1$ except for a finite number of points ζ_0 on $|\zeta|=1$ in the neighbourhood of each of which

$$\log|\Psi(\zeta)| \leqslant O\{|\zeta-\zeta_0|^{-k}\} \quad (k<1). \tag{1·32}$$

Suppose now that $\Phi(z)$ satisfies the hypotheses of the lemma, and consider a conformal mapping $z=h(\zeta)$ of the circle

$$\Gamma: \quad |\zeta|\leqslant 1$$

onto the strip B. Considering first the mapping $w=i(1+\zeta)/(1-\zeta)$ of Γ onto the half-plane $\mathscr{I}w\geqslant 0$, and then the mapping $z=(\pi i)^{-1}\log w$ of that half-plane onto B, we arrive at the relations

$$z=\frac{1}{\pi i}\log\left\{i\frac{1+\zeta}{1-\zeta}\right\}=h(\zeta), \tag{1·33}$$

$$\zeta=\frac{e^{\pi i z}-i}{e^{\pi i z}+i}. \tag{1·34}$$

In this mapping the points $y=+\infty$ and $y=-\infty$ correspond to $\zeta=-1$ and $\zeta=+1$ respectively, and the segment $0\leqslant x\leqslant 1$, $y=0$ of B to the diameter $(i, -i)$ of Γ.

Write

$$\Phi(z)=\Phi(h(\zeta))=\Psi(\zeta).$$

The function $\Psi(\zeta)$ is regular in $|\zeta|<1$ and continuous in Γ, except possibly for the points $\zeta_0=\pm 1$; and we easily verify that (1·25) leads to (1·32) for $\zeta_0=\pm 1$, with $k=a/\pi<1$. Hence $\Psi(\zeta)$ satisfies (1·31). Going back to the variable z on the left-hand side of (1·31), and observing that to the segment $0<x<1$ corresponds the argument $\theta=\pm\frac{1}{2}\pi$ in Γ, for which the right-hand side of (1·31) is bounded above in $0\leqslant\rho<1$, we obtain (1·29).

Remark. It is not difficult to obtain a precise value of A_x (though this is not important for us). First, by (1·34), if $z=x$ then

$$\zeta=\frac{e^{\pi i x}-i}{e^{\pi i x}+i}=-i\,\frac{\cos\pi x}{1+\sin\pi x}.$$

Suppose, say, that $0\leqslant x\leqslant\frac{1}{2}$. Then

$$\rho=|\zeta|=\frac{\cos\pi x}{1+\sin\pi x}, \tag{1·35}$$

and $\theta=-\frac{1}{2}\pi$ in (1·31), so that

$$P(\rho,\phi-\theta)=P(\rho,\phi+\tfrac{1}{2}\pi)=\tfrac{1}{2}\,\frac{1-\rho^2}{1+2\rho\sin\phi+\rho^2}. \tag{1·36}$$

If $-\pi<\phi<0$, then z is on the line $x=0$, and from

$$e^{i\phi}=\frac{e^{-\pi y}-i}{e^{-\pi y}+i}$$

(see (1·34)) we deduce the two relations

$$\sin\phi=-\frac{1}{\cosh\pi y},\quad d\phi=-\frac{\pi}{\cosh\pi y}dy. \tag{1·37}$$

These equations hold, without the minus signs, on the arc $0<\phi<\pi$ of $|\zeta|=1$, corresponding to the line $x=1$ of B. Combining this with (1·35) and (1·36) we deduce from (1·31) that

$$\log|\Phi(x)|\leqslant\tfrac{1}{2}\sin\pi x\left\{\int_{-\infty}^{+\infty}\frac{\log|\Phi(iy)|}{\cosh\pi y-\cos\pi x}dy+\int_{-\infty}^{+\infty}\frac{\log|\Phi(1+iy)|}{\cosh\pi y+\cos\pi x}dy\right\}.$$

The same argument proves the formula for $\frac{1}{2}\leqslant x<1$. If we replace here $|\Phi(iy)|$ and $|\Phi(1+iy)|$ by $M_1(y)$ and $M_2(y)$ respectively, we obtain (1·29) with

$$\log A_x=\tfrac{1}{2}\sin\pi x\left\{\int_{-\infty}^{+\infty}\frac{\log M_1(y)}{\cosh\pi y-\cos\pi x}dy+\int_{-\infty}^{+\infty}\frac{\log M_2(y)}{\cosh\pi y+\cos\pi x}dy\right\}. \tag{1·38}$$

(1·39) THEOREM. *Let $\{T_z\}$ be an analytic family of linear operators defined for all simple functions on R_1 and satisfying condition* E. *Let (α_1,β_1) and (α_2,β_2) be two points of the square*

$$0\leqslant\alpha\leqslant1,\quad 0\leqslant\beta\leqslant1,$$

and let (α,β) be given by (1·13). *Suppose, finally, that*

$$\|T_{iy}f\|_{1/\beta_1}\leqslant M_1(y)\|f\|_{1/\alpha_1}, \tag{1·40}$$

$$\|T_{1+iy}f\|_{1/\beta_2}\leqslant M_2(y)\|f\|_{1/\alpha_2} \tag{1·41}$$

for each simple f, where

$$\log M_k(y)\leqslant A\,e^{a|y|}\quad(a<\pi;\ k=1,2). \tag{1·42}$$

Then

$$\|T_t f\|_{1/\beta}\leqslant A_t\|f\|_{1/\alpha}, \tag{1·43}$$

where A_t is the same as in Lemma (1·27).

The proof is so like that of (1·11) that it is enough to indicate its main points. Let f and g be any two simple functions on R_1 and R_2 respectively, satisfying

$$\|f\|_{1/\alpha}=\|g\|_{1/(1-\beta)}=1.$$

It is enough to show that

$$\left|\int_{R_2}T_t f.g\,dv\right|\leqslant A_t. \tag{1·44}$$

Supposing first that $\alpha>0$, $\beta<1$, take the functions F_z and G_z defined by (1·16) and (1·17), and consider the function

$$\Phi(z)=\int_{R_2}T_zF_z.G_z\,dv$$

analogous to (1·18). Then using the analytic character of T_z we easily verify that $\Phi(z)$ is continuous in B and regular in the interior of B, and satisfies condition E. Moreover (see (1·19))

$$|\Phi(iy)|\leqslant\|T_{iy}F_{iy}\|_{1/\beta_1}\|G_{iy}\|_{1/(1-\beta_1)}\leqslant M_1(y)\|F_{iy}\|_{1/\alpha_1}\leqslant M_1(y),$$

and similarly

$$|\Phi(1+iy)|\leqslant M_2(y).$$

Therefore, by Lemma (1·27), $|\Phi(t)|\leqslant A_t$ and (1·44) follows. The cases $\alpha=0$ and $\beta=1$ are treated as in the proof of (1·11). Hence (1·39) is established. We easily see that it reduces to (1·11) if T_z is independent of z.

2. The theorems of Hausdorff-Young and F. Riesz

In what follows we denote by p and q numbers satisfying the inequalities

$$1 < p \leqslant 2, \quad 2 \leqslant q < \infty. \tag{2·1}$$

Thus, with the standard notation $r' = r/(r-1)$, every p' is a q, and every q' is a p.

Let $f(t)$ be a function defined in a fixed interval (a, b). For every $r > 0$ we write

$$\| f \|_r = \left\{ \int_a^b | f(t) |^r \, dt \right\}^{1/r} = \mathfrak{M}_r[f].$$

Similarly, for any sequence $c = \{c_n\}$ of complex numbers we write

$$\| c \|_r = (\Sigma \, | c_n |^r)^{1/r}.$$

For functions f defined in $(0, 2\pi)$ we shall also use the notation

$$\mathfrak{A}_r[f] = \left\{ \frac{1}{2\pi} \int_0^{2\pi} | f(t) |^r \, dt \right\}^{1/r}.$$

It has been proved (see Chapter II, (1·12) and Chapter IV, § 1) that for every f with Fourier coefficients c_n we have

$$\frac{1}{2\pi} \int_0^{2\pi} | f |^2 \, dt = \Sigma \, | c_n |^2. \tag{2·2}$$

This formula contains two propositions. First, if $f \in L^2$, then the series on the right converges and its sum is equal to the integral on the left (*Parseval*). Secondly, if $\{c_n\}$ is any two-way infinite sequence such that $\Sigma \, | c_n |^2 < \infty$, then there is an $f \in L^2$ having c_n for its Fourier coefficients and satisfying (2·2) (*Riesz-Fischer*). It is natural to inquire if these results can be extended to exponents other than 2. It turns out that a partial extension at least is possible.

(2·3) Theorem of Hausdorff-Young. *Let* $1 < p \leqslant 2$. (i) *Suppose that* $f(t) \in L^p(0, 2\pi)$ *and*

$$c_n = \frac{1}{2\pi} \int_0^{2\pi} f(t) \, e^{-int} \, dt \quad (n = 0, \pm 1, \pm 2, \ldots). \tag{2·4}$$

Then

$$\| c \|_{p'} \leqslant \mathfrak{A}_p[f]. \tag{2·5}$$

(ii) *Given any two-way infinite sequence* $\{c_n\}$ *of complex numbers with* $\| c \|_p < \infty$, *there is an* $f \in L^{p'}(0, 2\pi)$ *satisfying* (2·4) *and*

$$\mathfrak{A}_{p'}[f] \leqslant \| c \|_p. \tag{2·6}$$

Part (i) is an extension of Parseval's theorem, with '=' replaced by '$\leqslant$'. Part (ii) extends the Riesz-Fischer theorem. In both (i) and (ii) the argument goes from p to p', that is, from the smaller to the larger index. The results become false if we replace p by q. For

(*a*) there is a continuous f (so that $f \in L^r$ for all $r > 0$) such that $\| c \|_p = \infty$ for all $p < 2$; the series

$$\Sigma \pm \frac{\cos nx}{n^{\frac{1}{2}} \log^2 n}$$

is, for a suitable choice of signs, a case in point (see Chapter V, (8·34); also Chapter V, (4·11));

(*b*) there is a series $\Sigma c_n e^{inx}$ which is not a Fourier series although $\| c \|_q < \infty$ for every $q > 2$. As an example we may take the series

$$\Sigma n^{-\frac{1}{2}} \cos 2^n x,$$

or the series

$$\Sigma \pm n^{-\frac{1}{2}} \cos nx,$$

again with suitable signs (Chapter V, (6·4), (8·14)).

It is apparent that between the two parts of the Hausdorff-Young theorem there is a certain dualism. Part (ii) is obtained from (i) if the function f depending on the variable t is replaced by the function c depending on the variable n, integration is replaced by summation and vice versa. This dualism can be detected in various parts of the theory of Fourier series and is an important guide in the search for new results.

Theorem (2·3) is a special case of the following result about any system of functions $\phi_n(t)$, $n = 1, 2, \ldots$, orthonormal and uniformly bounded over an interval (a, b):

$$| \phi_n(t) | \leqslant M \tag{2·7}$$

for t in (a, b) and all n.

(2·8) Theorem of F. Riesz. *Let* $1 < p \leqslant 2$. (i) *If* $f \in L^p(a, b)$, *then the Fourier coefficients*

$$c_n = \int_a^b f \bar{\phi}_n \, dt \tag{2·9}$$

satisfy the inequality

$$\| c \|_{p'} \leqslant M^{(2/p)-1} \| f \|_p. \tag{2·10}$$

(ii) *Given any sequence* $c_1, c_2, \ldots$ *with* $\| c \|_p$ *finite, there is an* $f \in L^{p'}(a, b)$ *satisfying* (2·9) *for all n and*

$$\| f \|_{p'} \leqslant M^{(2/p)-1} \| c \|_p. \tag{2·11}$$

(2·3) is clearly a corollary of (2·8). If (a, b) is finite, then $f \in L^p$ implies $f \in L$, and this together with (2·7) shows the existence of the integrals (2·9). The latter exist, however, even if (a, b) is infinite. For

$$\int_a^b | \phi_n |^{p'} dt \leqslant M^{p'-2} \int_a^b | \phi_n |^2 dt = M^{p'-2}. \tag{2·12}$$

Hence $\phi_n \in L^{p'}$ and $f \bar{\phi}_n$ is integrable.

We prove (2·10). Consider the numbers c_n in (2·9). We have

$$\Sigma | c_n |^2 \leqslant \int_a^b | f |^2 dt, \tag{2·13}$$

$$\sup | c_n | \leqslant M \int_a^b | f | \, dt. \tag{2·14}$$

The first is Bessel's inequality, and the second follows from (2·7) and (2·9). Let μ denote the ordinary Lebesgue measure, and let ν be the additive measure assigning value 1 to the sets consisting of a single point $x = n$, $n = 1, 2, \ldots$, and vanishing for sets not containing any such point. If $c(x) = c_n$ for $x = n$, $n = 1, 2, \ldots$, and is arbitrary elsewhere, (2·13) and (2·14) can be written

$$\| c \|_{2,\nu} \leqslant \| f \|_{2,\mu}, \tag{2·15}$$

$$\| c \|_{\infty,\nu} \leqslant M \| f \|_{1,\mu}. \tag{2·16}$$

Thus, using the terminology of § 1, the linear operation $c(x) = Tf$ is simultaneously of types (2, 2) and (1, ∞), with norms $M_1 \leqslant 1$ and $M_2 \leqslant M$ respectively. It is defined for functions which are not necessarily simple, but if we confine our attention to the latter and use (1·11), we see that T is also of type (p, p'), where $p = 1/\alpha$, $p' = 1/(1-\alpha)$ and $\frac{1}{2} \leqslant \alpha \leqslant 1$. The norm corresponding to type (p, p') is, by (1·14), not greater than

$$M_1^{(1-\alpha)/(1-\frac{1}{2})} M_2^{(\alpha-\frac{1}{2})/(1-\frac{1}{2})} \leqslant 1^{2-2\alpha} M^{2\alpha-1} = M^{(2/p)-1}.$$

Hence
$$\| c \|_{p',\nu} \leqslant M^{(2/p)-1} \| f \|_{p,\mu},$$
which is exactly (2·10).

Part (ii) can be established by a similar argument (which is left to the reader), but it is simpler and more instructive to deduce it from part (i) by an argument which shows the mutual relation of the parts.

Let $c_1, c_2, \ldots$ be given, with $\| c \|_p < \infty$, and let

$$f_n = c_1 \phi_1 + c_2 \phi_2 + \ldots + c_n \phi_n,$$

where $n = 1, 2, \ldots$. We know that $f_n \in L^{p'}$, by (2·12). For any $g \in L^p$ with Fourier coefficients $d_1, d_2, \ldots$ we have

$$\left| \int_a^b \bar{f}_n g \, dx \right| = \left| \sum_1^n \bar{c}_k d_k \right| \leqslant (\Sigma \mid c_k \mid^p)^{1/p} (\Sigma \mid d_k \mid^{p'})^{1/p'}$$
$$\leqslant \| c \|_p M^{(2/p)-1} \| g \|_p,$$

by (i) of (2·8). The upper bound of the left-hand side, for all g with $\| g \|_p = 1$, is $\| \bar{f}_n \|_{p'} = \| f_n \|_{p'}$, so that
$$\| f_n \|_{p'} \leqslant M^{(2/p)-1} \| c \|_p. \qquad \textbf{(2·17)}$$

Since $\| c \|_p < \infty$ implies $\| c \|_2 < \infty$, the series $c_1 \phi_1 + c_2 \phi_2 + \ldots$ is the Fourier series of an f such that $\| f_n - f \|_2 \to 0$ (Chapter IV, (1·1)). Hence $f_n \to f$ almost everywhere, if n tends to $+\infty$ through a sequence of values (Chapter I, (11·6)). Applying Fatou's lemma to (2·17) we obtain (2·11).

The function f of (2·8) (ii) is the function of the Riesz-Fischer theorem. The preceding argument shows that it is not only in L^2 but also in $L^{p'}$. If $m < n$, (2·17) gives

$$\| f_m - f_n \|_{p'} \leqslant M^{(2/p)-1} \left(\sum_{m+1}^{n} \mid c_k \mid^p \right)^{1/p} \to 0$$

as $m, n \to \infty$. Hence, if $\| c \|_p < \infty$, f_n tends to f in $L^{p'}$.

In a similar way we could deduce (i) from (ii), so that both parts of (2·8) are, in a way, equivalent.

In the foregoing proof of (i) we could avoid the use of the Lebesgue-Stieltjes integral. For if we set $c(x) = c_n$ for $n-1 \leqslant x < n$ the inequalities (2·13) and (2·14) can be written

$$\int_0^\infty \mid c(x) \mid^2 dx \leqslant \int_a^b \mid f \mid^2 dx, \quad \sup \mid c(x) \mid \leqslant M \int_a^b \mid f \mid dx,$$

and we apply (1·11) in the case when μ and ν are ordinary Lebesgue measures.

We now investigate the conditions under which we have equality in (2·10) or (2·11). For the sake of simplicity we assume that $\{\phi_n\}$ is complete. Since we always have equality if $p = 2$, we may suppose that $1 \leqslant p < 2$.

(2·18) THEOREM. (i) *A necessary condition for equality in* (2·10) *is that the Fourier series of* f *be finite:*

$$f(t)=\sum_{k=1}^{N} c_{n_k}\phi_{n_k} \quad (n_1<n_2<\ldots<n_N). \tag{2·19}$$

For such functions we have equality in (2·10) *if and only if*

(*a*) $|c_{n_1}|=|c_{n_2}|=\ldots=|c_{n_N}|$;

(*b*) $|f(t)|$ *is constant in a set* E *of measure* $1/NM^2$, *and* $f=0$ *outside* E.

(ii) *A necessary condition for equality in* (2·11) *is that only a finite number of the* c's, *say* $c_{n_1}, c_{n_2}, \ldots, c_{n_N}$, *are distinct from* 0 *and that they satisfy* (*a*). *The function* f *is then of the form* (2·19), *and a necessary and sufficient condition for equality in* (2·11) *is that* f *satisfy* (*b*).

(i) The proof is based on the formula (1·24). We may suppose that $f \not\equiv 0$, for otherwise the result is obvious. Consider the transformation $c(x)=Tf$, and the measures μ and ν appearing in (2·15) and (2·16); and suppose that for an f with $\|f\|_p=1$ we have $\|c\|_{p'}=M^{(2/p)-1}$. If

$$(\alpha_1,\beta_1)=(\tfrac{1}{2},\tfrac{1}{2}), \quad (\alpha_2,\beta_2)=(1,0), \quad (\alpha,\beta)=(1/p,1/p'),$$

then $\alpha(z)=1-\beta(z)=\frac{1}{2}(1+z)$. In our case $M_1=1$, $M_2\leqslant M$. Since the mass ν is concentrated at the points $n=1,2,\ldots$, integration in (1·24) is actually summation, and the relation may be written

$$\sum_{n=1}^{\infty}|d_n|^{\frac{1}{2}p(1+z)}\operatorname{sign} d_n\int_a^b |f(t)|^{\frac{1}{2}p(1+z)}(\operatorname{sign} f)\,\bar{\phi}_n\,dt=M^z, \tag{2·20}$$

where, by (1·21),
$$d_n=\left(\frac{|c_n|}{\|c\|_{p'}}\right)^{p'-1}\operatorname{sign}\bar{c}_n.$$

Write
$$|f|^p=F, \quad |d_n|^p=D_n, \quad \operatorname{sign} f(t)=\eta(t), \quad \operatorname{sign} d_n=\epsilon_n.$$

Hence
$$F\geqslant 0, \quad D_n\geqslant 0, \quad \int_a^b F\,dt=1, \quad \Sigma D_n=1,$$

and
$$\Sigma D_n^{\frac{1}{2}(1+z)}\epsilon_n\int_a^b F^{\frac{1}{2}(1+z)}\eta\bar{\phi}_n\,dt=M^z. \tag{2·21}$$

The left-hand side here is a regular function of z for $0<x<1$. It is also continuous for $0\leqslant x\leqslant 1$, since

(A) each term of the series is a continuous function of z in $0\leqslant x\leqslant 1$, the integrand being majorized by the integrable function MF wherever $F\geqslant 1$, and by the integrable function $F^{\frac{1}{2}}|\phi_n|$ elsewhere;

(B) each integral is numerically bounded by a constant independent of z and n, in consequence of the estimates above;

(C) the series (2·21) converges absolutely and uniformly in the strip $0\leqslant\mathscr{R}z\leqslant 1$ (apply Hölder's first inequality with exponents $2/(1+x)$ and $2/(1-x)$ to the series, and then (2·8) (i)).

Hence (2·21) holds for $z=1$:

$$\Sigma D_n\epsilon_n\int_a^b F\eta\bar{\phi}_n\,dt=M.$$

This and the relations
$$\Sigma D_n=1, \quad \left|\int_a^b F\eta\bar{\phi}_n\,dt\right|\leqslant M\int_a^b F\,dt=M$$

imply that *for each* n *with* $D_n\neq 0$ *we have*

$$\epsilon_n\int_a^b F\eta\bar{\phi}_n\,dt=M,$$

or
$$\int_a^b F\operatorname{sign} f.\bar{\phi}_n\operatorname{sign}\bar{c}_n\,dt=M. \tag{2·22}$$

By the Riemann-Lebesgue theorem (Chapter II, (4·4)) for $\{\phi_n\}$, the integral here tends to 0. Hence there are at most a finite number of D_n distinct from 0, and so only a finite number of c_n distinct from 0. The system $\{\phi_n\}$ being complete, we have (2·19).

The equation (2·22) is valid for $n=n_1,n_2,\ldots,n_N$. Hence the set E where F (or f) is distinct from 0 has the following properties:

$$\operatorname{sign} f=\operatorname{sign}(c_n\phi_n), \quad |\phi_n|=M,$$

for $n=n_1,\ldots,n_N$ and almost all points of E. Together with (2·19) this gives

$$|f(t)|=M\Sigma|c_{n_k}|. \tag{2·23}$$

Thus $|f|$ is constant almost everywhere in E (see (b)).

The equation (2·22) leads to

$$\left|\int_a^b |f|^p \operatorname{sign} f . \bar{\phi}_n dt\right| = M,$$

and since $|f|$ is given by (2·23) on E and is 0 elsewhere in (a, b) we have

$$\left|\int_a^b f\bar{\phi}_n dt\right| = \left|\int_E f\bar{\phi}_n dt\right| = M^{2-p}(\Sigma|c_{n_k}|)^{1-p}$$

for $n = n_1, \ldots, n_N$. This proves the necessity of condition (a).

It remains to find $|E|$. From (2·23) and Parseval's formula we have

$$\int_a^b |f|^2 dt = \int_E |f|^2 dt = M^2(\Sigma|c_{n_k}|)^2 |E| = M^2N^2|c_{n_1}|^2|E|,$$

$$\int_a^b |f|^2 dt = \Sigma|c_{n_k}|^2 = N|c_{n_1}|^2,$$

so that $|E| = 1/NM^2$.

This proves the necessity of the conditions. To prove the sufficiency, suppose that f is of the form (2·19) and satisfies (a) and (b). If c^* is the common value of the $|c_{n_k}|$, and f^* the value of $|f|$ in E, (2·10) can be written $MNc^* \leqslant f^*$. Since the opposite inequality is an immediate consequence of (2·19) we have $MNc^* = f^*$, the sign of equality in (2·10), and part (i) of (2·18) is established.

Passing to part (ii), suppose that we have equality in (2·11), and that $\|c\|_p \neq 0$ (otherwise the result is obvious). Since $\|f\|_{p'} < \infty$, there is a g with $\|g\|_p = 1$, and with Fourier coefficients d_n, such that $\|f\|_{p'} = \int_a^b \bar{f} g\, dt$. Also $\|f - f_m\|_{p'} \to 0$, where $f_m = c_1\phi_1 + \ldots + c_m\phi_m$ (p. 103). Hence

$$\|f\|_{p'} = \int_a^b \bar{f} g\, dt = \lim \int_a^b \bar{f}_m g\, dt = \lim \sum_1^m \bar{c}_n d_n = \Sigma \bar{c}_n d_n$$

$$\leqslant \|c\|_p \|d\|_{p'} \leqslant M^{(2/p)-1}\|c\|_p. \qquad (2·24)$$

The extreme terms here are equal. Hence $\|d\|_{p'} = M^{(2/p)-1}$. It follows from (i) that

$$g = d_{n_1}\phi_{n_1} + \ldots + d_{n_N}\phi_{n_N}, \quad \text{where} \quad |d_{n_1}| = \ldots = |d_{n_N}|,$$

and that $|g|$ is constant in a set E of measure $1/NM^2$, and is 0 outside E. Moreover, Hölder's inequality in (2·24) degenerates into equality, which is only possible if the $|c_n|^p$ and $|d_n|^{p'}$ are proportional. Thus $|c_{n_1}| = \ldots = |c_{n_N}|$, and the remaining c_n are 0. The first equation in (2·24) shows that $|f|^{p'}$ and $|g|^p$ are proportional. Hence $|f|$ is constant in E, and is 0 elsewhere. This proves the necessity of the conditions in (ii), and the sufficiency is easily verified.

(2·25) THEOREM. (i) *The sign of equality occurs in* (2·5) *if and only if* $f(t) = A\, e^{ikt}$, *where* A *is a constant and* k *an integer.*

(ii) *The sign of equality occurs in* (2·6) *if and only if all the* c_n, *except possibly one, are equal to* 0.

(i) It is enough to apply (2·18) (i) to the system $\{(2\pi)^{-\frac{1}{2}} e^{inx}\}$, orthonormal on $(0, 2\pi)$. If we have equality in (2·5) for an $f \not\equiv 0$, then $f = \Sigma c_n e^{inx}$ is a polynomial of, say, N terms, equal to 0 outside a set $E \subset (0, 2\pi)$ of measure $2\pi/N$. It follows that $N = 1$, since otherwise we would have $f \equiv 0$ (Chapter X, (1·7)). This proves (i), and (ii) is proved similarly.

3. Interpolation of operations in the classes H^r

The inequalities (2·13) and (2·14) from which we deduced Theorem (2·8) are valid for general functions f. There are, however, inequalities which are only satisfied by functions of special types. For example, the inequality

$$\sum_{\nu=0}^{\infty} \frac{|c_\nu|}{\nu+1} \leqslant \tfrac{1}{2}\int_0^{2\pi} |f|\, dt \qquad (3·1)$$

is valid for functions of the class H (see Chapter VII, § 8) but not necessarily for general integrable functions. Thus if we wanted to interpolate between (3·1) and

$$\sum_{\nu=0}^{\infty} |c_\nu|^2 = \frac{1}{2\pi}\int_0^{2\pi} |f|^2\,dt, \tag{3·2}$$

we could not apply Theorem (1·11). This brings us to the problem of interpolation of operations in the classes H^r.

We begin by proving an auxiliary result about the interpolation of *multilinear* operations, that is, of operations

$$h = T[f_1, f_2, \ldots, f_n]$$

linear in each f_j.

(3·3) THEOREM. *Let E and $E_1, E_2, \ldots, E_n$ be measure spaces with measures ν and $\mu_1, \mu_2, \ldots, \mu_n$ respectively. Let $h = T[f_1, f_2, \ldots, f_n]$ be a multilinear operation defined for simple functions f_j on E_j, $j = 1, \ldots, n$, where the function h is defined on E. Suppose that T is simultaneously of types $(1/\alpha_1^{(1)}, \ldots, 1/\alpha_n^{(1)}, 1/\beta^{(1)})$ and $(1/\alpha_1^{(2)}, \ldots, 1/\alpha_n^{(2)}, 1/\beta^{(2)})$, that is,*

$$\| T[f_1, f_2, \ldots, f_n] \|_{1/\beta^{(k)}} \leqslant M_k \| f_1 \|_{1/\alpha_1^{(k)}} \cdots \| f_n \|_{1/\alpha_n^{(k)}} \quad (k = 1, 2), \tag{3·4}$$

where $\quad 0 \leqslant \beta^{(k)} \leqslant 1, \quad 0 \leqslant \alpha_j^{(k)} \leqslant 1 \quad (k = 1, 2;\ j = 1, 2, \ldots, n).$

Then T is also of the type $(1/\alpha_1, \ldots, 1/\alpha_n, 1/\beta)$ for

$$\left.\begin{aligned} \alpha_j &= (1-t)\,\alpha_j^{(1)} + t\alpha_j^{(2)}, \\ \beta &= (1-t)\,\beta^{(1)} + t\beta^{(2)}, \end{aligned}\right\} \quad (j = 1, 2, \ldots, n;\ 0 < t < 1), \tag{3·5}$$

and the inequality

$$\| T[f_1, f_2, \ldots, f_n] \|_{1/\beta} \leqslant M_1^{1-t} M_2^t \| f_1 \|_{1/\alpha_1} \cdots \| f_n \|_{1/\alpha_n} \tag{3·6}$$

holds.

Moreover, if all the α_j are positive, T can be extended by continuity to

$$L^{1/\alpha_1, \mu_1} \times \ldots \times L^{1/\alpha_n, \mu_n},$$

preserving (3·6).

The proof is similar to that of Theorem (1·11). We first suppose that $\alpha_1, \alpha_2, \ldots, \alpha_n$ are positive and that $\beta < 1$. Fix simple functions $f_1, f_2, \ldots, f_n$ with

$$\| f_j \|_{1/\alpha_j} = 1 \quad (j = 1, 2, \ldots, n),$$

and a simple function g with $\| g \|_{1/(1-\beta)} = 1$. We fix t in (3·5) and consider the functions

$$\beta(z) = \beta^{(1)}(1-z) + \beta^{(2)}z,$$

$$\alpha_j(z) = \alpha_j^{(1)}(1-z) + \alpha_j^{(2)}z,$$

reducing to β, α_j for $z = t$. Writing

$$f_j = |f_j|\, e^{iu_j}, \quad g = |g|\, e^{iv},$$

we consider the integral

$$\Phi(z) = \int_E T[|f_1|^{\alpha_1(z)/\alpha_1} e^{iu_1}, \ldots, |f_n|^{\alpha_n(z)/\alpha_n} e^{iu_n}]\, |g|^{(1-\beta(z))/(1-\beta)} e^{iv}\, d\nu, \tag{3·7}$$

which for $z = t$ reduces to $\quad I = \int_E T[f_1, \ldots, f_n]\, g\, d\nu.$

Since g and the f_j are simple, $\Phi(z)$ is a linear combination of exponentials λ^z, $\lambda > 0$. For $x=0$, Hölder's inequality gives (as in (1·19))

$$|\Phi(z)| \leqslant \| \, |g|^{(1-\beta(1))/(1-\beta)} \|_{1/(1-\beta(1))} \| T[\ldots, |f_j|^{\alpha_j(z)/\alpha_j} e^{iu_j}, \ldots] \|_{1/\beta(1)}$$
$$\leqslant 1 . M_1 \prod_j \| \, |f_j|^{\alpha_j^{(1)}/\alpha_j} \|_{1/\alpha_j^{(1)}} = M_1.$$

Similarly $|\Phi(z)| \leqslant M_2$ for $x=1$. Hence

$$|I| = |\Phi(t)| \leqslant M_1^{1-t} M_2^t.$$

Since $\| T[f_1, \ldots, f_n] \|_{1/\beta}$ is the upper bound of $|I|$ for all simple g's with $\|g\|_{1/(1-\beta)} = 1$, (3·6) follows when all the norms on the right are 1, and so also in the general case.

The exceptional case $\beta = 1$ is treated (as in the proof of (1·11)) by replacing $|g|^{(1-\beta(z))/(1-\beta)} e^{iv}$ in (3·7) by g. Similarly, if some of the α_j are zero, we replace the corresponding $|f_j|^{\alpha_j(z)/\alpha_j} e^{iu_j}$ in (3·7) by f_j.

It remains to show that if all the α_j are positive and if (3·6) is valid for simple f_j, then T can be extended by continuity to $\mathrm{L}^{1/\alpha_1} \times \ldots \times \mathrm{L}^{1/\alpha_n}$. This follows from the inequality

$$\| T[f_1^{(1)}, \ldots, f_n^{(1)}] - T[f_1^{(2)}, \ldots, f_n^{(2)}] \|_{1/\beta}$$
$$\leqslant M_1^{1-t} M_2^t (\sum_j \| f_j^{(1)} - f_j^{(2)} \|_{1/\alpha_j}) (\sup_{j,k} \| f_j^{(k)} \|_{1/\alpha_j})^{n-1}$$

which, in turn, is a consequence of the inequality

$$\| T[f_1^{(1)}, f_2^{(1)}, \ldots, f_n^{(1)}] - T[f_1^{(2)}, f_2^{(2)}, \ldots, f_n^{(2)}] \|_{1/\beta}$$
$$\leqslant \| T[f_1^{(1)}, f_2^{(1)}, \ldots, f_n^{(1)}] - T[f_1^{(2)}, f_2^{(1)}, f_3^{(1)}, \ldots, f_n^{(1)}] \|_{1/\beta}$$
$$+ \| T[f_1^{(2)}, f_2^{(1)}, \ldots, f_n^{(1)}] - T[f_1^{(2)}, f_2^{(2)}, \ldots, f_n^{(1)}] \|_{1/\beta}$$
$$\cdots\cdots\cdots\cdots\cdots\cdots\cdots\cdots\cdots\cdots\cdots\cdots$$
$$+ \| T[f_1^{(2)}, f_2^{(2)}, \ldots, f_{n-1}^{(2)}, f_n^{(1)}] - T[f_1^{(2)}, f_2^{(2)}, \ldots, f_n^{(2)}] \|_{1/\beta}. \tag{3·8}$$

Thus Theorem (3·3) is established.

We now fix $r > 0$ and consider the class H^r of all functions

$$F(z) = \sum_0^\infty c_\nu z^\nu,$$

regular for $|z| < 1$, such that the expression

$$\mathfrak{A}_r(\rho) = \mathfrak{A}_r(\rho, F) = \left\{ \frac{1}{2\pi} \int_0^{2\pi} |F(\rho e^{i\theta})|^r \, d\theta \right\}^{1/r}$$

is bounded as $\rho \to 1$ (Chapter VII, § 7). We define the norm $\|F\|_r$ by the formula

$$\|F\|_r = \lim_{\rho \to 1} \mathfrak{A}_r(\rho) = \left\{ \frac{1}{2\pi} \int_0^{2\pi} |F(e^{i\theta})|^r \, d\theta \right\}^{1/r},$$

where

$$F(e^{i\theta}) = \lim_{\rho \to 1} F(\rho e^{i\theta}).$$

We know that H^r is a metric space if the distance between two points F and G is defined as $\|F - G\|_r$ for $r \geqslant 1$, and as $\|F - G\|_r^r$ for $0 < r < 1$. We also know that H^r is a complete and separable space (Chapter VII, p. 284).

We shall consider a linear operation

$$h = TF,$$

defined for all F in H^r, yielding $h \in \mathrm{L}^s$ in some measure space E and satisfying an inequality

$$\| h \|_s \leqslant M \| F \|_r,$$

with M independent of F. If T is initially defined on a linear subset of functions F dense in H^r and satisfies the last inequality, we can, by continuity, extend T to the whole of H^r preserving the constant M. The polynomials

$$P(z) = c_0 + c_1 z + \dots + c_n z^n$$

are an example of a linear subset dense in every H^r.

(3·9) THEOREM. *Let (α_1, β_1) and (α_2, β_2) be two points of the strip*

$$0 < \alpha < +\infty, \quad 0 \leqslant \beta \leqslant 1. \tag{3·10}$$

Let T be a linear operation defined for all polynomials $P(z)$, and let TP be defined on some measure space E. Suppose that

$$\| TP \|_{1/\beta_1} \leqslant M_1 \| P \|_{1/\alpha_1}, \quad \| TP \|_{1/\beta_2} \leqslant M_2 \| P \|_{1/\alpha_2}. \tag{3·11}$$

Then for every point (α, β) of the segment

$$\alpha = \alpha_1(1-t) + \alpha_2 t, \quad \beta = \beta_1(1-t) + \beta_2 t \quad (0 < t < 1),$$

we have the inequality

$$\| TP \|_{1/\beta} \leqslant K M_1^{1-t} M_2^t \| P \|_{1/\alpha}, \tag{3·12}$$

K denoting a constant depending on α_1, α_2 only.

In particular, T can be extended to the whole of $\mathrm{H}^{1/\alpha}$ preserving (3·12).

Proof. (i) The inequalities (3·11) show that we can extend T in both H^{1/α_1} and H^{1/α_2}, while preserving M_1 and M_2. *The extension TF is the same both in* H^{1/α_1} *and* H^{1/α_2}. For suppose that

$$\alpha_1 \leqslant \alpha_2.$$

Then $\mathrm{H}^{1/\alpha_1} \subset \mathrm{H}^{1/\alpha_2}$. If $F \in \mathrm{H}^{1/\alpha_1}$ and $\| P_n - F \|_{1/\alpha_1} \to 0$, then also $\| P_n - F \|_{1/\alpha_2} \to 0$. The inequalities (3·11) extended to H^{1/α_1} and H^{1/α_2} show that TP_n tends to limits both in L^{1/β_1} and L^{1/β_2}. These limits must be the same since they are, almost everywhere, ordinary limits of a suitable subsequence of $\{TP_n\}$.

(ii) Suppose again that $\alpha_1 \leqslant \alpha_2$. Let $n > 0$ be an integer such that $\alpha_2 < n$. Hence also $\alpha_1 < n$. For any system of n complex-valued simple functions $g_1, g_2, \dots, g_n$ defined on $(0, 2\pi)$ we define an operation T^* by the formula

$$T^*[g_1, g_2, \dots, g_n] = T[F_1 F_2 \dots F_n], \tag{3·13}$$

where

$$F_j(z) = \frac{1}{2\pi} \int_0^{2\pi} \frac{e^{it} + z}{e^{it} - z} g_j(t)\, dt \quad (j = 1, 2, \dots, n). \tag{3·14}$$

We know (Chapter VII, (7·2), and the inequality following it) that

$$\| F_j \|_r \leqslant A_r \| g_j \|_r \quad (1 < r < \infty). \tag{3·15}$$

Hence $F_j \in \mathrm{H}^{n/\alpha_1}$. By Hölder's inequality, $F_1 F_2 \dots F_n \in \mathrm{H}^{1/\alpha_1}$, and so the left-hand side of (3·13) is defined. It is additive in each g_j. By (3·11), extended to the whole of H^{1/α_k},

$$\| T^*[g_1, \dots, g_n] \|_{1/\beta_k} \leqslant M_k \| F_1 \dots F_n \|_{1/\alpha_k} \leqslant M_k \| F_1 \|_{n/\alpha_k} \cdots \| F_n \|_{n/\alpha_k}.$$

Hence, using (3·15),

$$\| T^*[g_1, \ldots, g_n] \|_{1/\beta_k} \leqslant M_k A^n_{n/\alpha_k} \| g_1 \|_{n/\alpha_k} \cdots \| g_n \|_{n/\alpha_k} \tag{3·16}$$

for $k=1, 2$. It follows that T^* *is a multilinear operation defined for all simple functions* $g_1, g_2, \ldots, g_n$. By (3·3),

$$\| T^*[g_1, g_2, \ldots, g_n] \|_{1/\beta} \leqslant (A^{1-t}_{n/\alpha_1} A^t_{n/\alpha_2})^n M_1^{1-t} M_2^t \prod_j \| g_j \|_{n/\alpha}. \tag{3·17}$$

(iii) The formula (3·13) defines T^* when $g_1, \ldots, g_n$ are simple. The formula (3·16) shows that T^* can be extended to $\mathrm{L}^{n/\alpha_2} \times \mathrm{L}^{n/\alpha_2} \times \ldots \times \mathrm{L}^{n/\alpha_2}$, preserving (3·17). But if g_j is in L^{n/α_2}, then F_j in (3·14) belongs to H^{n/α_2}, and so $F_1 \ldots F_n$ is in H^{1/α_2}. Thus the right-hand side of (3·13) is defined. We shall show that *the equation* (3·13) *still holds in this case.*

If the g_j belong to H^{n/α_2}, and if the g_j^m are simple and satisfy $\| g_j^m - g_j \|_{n/\alpha_2} \to 0$ as $m \to \infty$, then

$$\| T^*[g_1^m, \ldots, g_n^m] - T^*[g_1, \ldots, g_n] \|_{1/\beta_2} \to 0, \tag{3·18}$$

by an argument similar to (3·8). On the other hand, if F_j^m is derived from g_j^m by means of (3·14) we have

$$\| F_j^m - F_j \|_{n/\alpha_2} \to 0, \quad \| F_j^m \|_{n/\alpha_2} \leqslant A_{n/\alpha_2} \| g_j^m \|_{n/\alpha_2} = O(1),$$

so that, by an argument similar to (3·8) but using (3·15), we have

$$\| T[F_1^m \ldots F_n^m] - T[F_1 \ldots F_n] \|_{1/\beta_2} \to 0.$$

This and (3·18) give (3·13) in the case considered.

It follows that (3·17) *holds if all the* g_j *belong to* $\mathrm{L}^{n/\alpha}$. For then g_j is in L^{n/α_2}, since $\alpha_1 \leqslant \alpha \leqslant \alpha_2$.

(iv) Given any polynomial P, we write

$$P(z) = B(z)\, G(z),$$

where $B(z)$ is the Blaschke product of P, and $G(z)$ (also a polynomial) has no zeros for $|z| < 1$. Hence

$$P = F_1 F_2 \ldots F_n, \quad \text{where} \quad F_1 = BG^{1/n}, \quad F_2 = \ldots = F_n = G^{1/n}.$$

Multiplying P by a number of absolute value 1, we may suppose that $P(0)$ is real. Since $G(0) > 0$ (Chapter VII, p. 275), $B(0)$ is real. Taking the main branch of $G^{1/n}$, we see that $F_j(0)$ is real for all j. This and the boundedness of F_j in $|z| < 1$ imply that F_j is of the form (3·14), with $g_j \in \mathrm{L}^{n/\alpha}$ and real-valued. Hence (3·13) holds and, by (3·3),

$$\| TP \|_{1/\beta} = \| T[F_1 \ldots F_n] \|_{1/\beta} = \| T^*[g_1, \ldots, g_n] \|_{1/\beta}$$
$$\leqslant (A^{1-t}_{n/\alpha_1} A^t_{n/\alpha_2})^n M_1^{1-t} M_2^t \prod_j \| g_j \|_{n/\alpha}.$$

The last product does not exceed

$$\prod_j \left\{ \int_0^{2\pi} | F_j(e^{it}) |^{n/\alpha} dt \right\}^{\alpha/n} = \prod_j \left\{ \int_0^{2\pi} | G(e^{it}) |^{1/\alpha} dt \right\}^{\alpha/n} = (2\pi)^\alpha \| P \|_{1/\alpha},$$

which gives (3·12) with $\quad K = (2\pi)^\alpha \max \{A^n_{n/\alpha_1}, A^n_{n/\alpha_2}\}.$

(3·19) THEOREM OF HARDY AND LITTLEWOOD. (i) *Suppose that*

$$f(x) \sim \sum_{-\infty}^{+\infty} c_n e^{inx}$$

is in L^p, $1 < p \leqslant 2$. *Then*

$$\{\sum_k |c_k|^p (|k|+1)^{p-2}\}^{1/p} \leqslant A_p \left\{\int_0^{2\pi} |f|^p dt\right\}^{1/p} \qquad (3\cdot20)$$

(ii) *Let* $\ldots c_{-1}, c_0, c_1, \ldots$ *be complex numbers such that* $\Sigma |c_k|^q (|k|+1)^{q-2}$ *is finite,* $q \geqslant 2$. *Then the* c_k *are the Fourier coefficients of an* f *in* L^q, *and*

$$\left\{\int_0^{2\pi} |f|^q dt\right\}^{1/q} \leqslant A_q \{\Sigma |c_k|^q (|k|+1)^{q-2}\}^{1/q}. \qquad (3\cdot21)$$

We first prove the following theorem which is partly more and partly less general than (i), and from which (i) follows without difficulty:

(3·22) THEOREM. *If* F *is in* L^p, $1 \leqslant p \leqslant 2$, *and* $S[F]$ *is of power series type*, $F \sim \sum_0^\infty c_k e^{ikx}$, *then*

$$\left\{\sum_0^\infty |c_k|^p (k+1)^{p-2}\right\}^{1/p} \leqslant A \left\{\frac{1}{2\pi}\int_0^{2\pi} |F|^p dx\right\}^{1/p}, \qquad (3\cdot23)$$

where A *is an absolute constant.*

For any polynomial $P(z) = \sum_0^n c_k z^k$ we have

$$\left.\begin{aligned} \Sigma |c_k|^2 &= \frac{1}{2\pi}\int_0^{2\pi} |P(e^{ix})|^2 dx, \\ \Sigma \frac{|c_k|}{k+1} &\leqslant \tfrac{1}{2}\int_0^{2\pi} |P(e^{ix})| dx, \end{aligned}\right\} \qquad (3\cdot24)$$

these being Parseval's formula and the inequality (3·1) respectively. We define an operation $h = TP$ by setting h equal to $c_k(k+1)$ at the points $k = 0, 1, 2, \ldots$, with h arbitrary elsewhere. If ψ is the additive measure assigning the value $(k+1)^{-2}$ to the set consisting of the single point k and the value 0 to the sets not containing any of the points k, (3·24) can be written

$$\|h\|_{2,\psi} = \|P\|_2, \quad \|h\|_{1,\psi} \leqslant \pi \|P\|_1.$$

An application of Theorem (3·9) then gives $\|h\|_{p,\psi} \leqslant A \|P\|_p$, that is,

$$\sum_0^n |c_k|^p (k+1)^{p-2} \leqslant \frac{A^p}{2\pi}\int_0^{2\pi} |P(e^{ix})|^p dx, \qquad (3\cdot25)$$

where A is an absolute constant.

If $F(R,x) = \Sigma c_k R^k e^{ikx}$ is the Poisson integral of the F in Theorem (3·22), we apply (3·25) to the nth partial sum of $F(R,x)$. Making first n tend to $+\infty$ and then R tend to 1, we obtain (3·23).

Return to (3·19). In the proof of (i) we may suppose that f is real-valued. Let $F(z)$ be the analytic function whose real part is the Poisson integral of f, and whose imaginary part vanishes at $z = 0$. Since $c_{-k} = \bar{c}_k$, the left-hand side of (3·20) does not exceed

$$\left\{2\sum_0^\infty |c_k|^p (k+1)^{p-2}\right\}^{1/p} \leqslant 2^{1/p} A \|F(e^{ix})\|_p \leqslant A_p \left\{\int_0^{2\pi} |f|^p dx\right\}^{1/p},$$

by (3·23) and Chapter VII, (2·4). This proves (i).

Part (ii) of (3·19) is dual to part (i), and can be obtained from it by an argument similar to that used in the proof of (2·8) (ii). We shall do this in a more general case in § 5.

The constant A_p in (3·20) tends to $+\infty$ as $p \to 1$; for otherwise, making $p \to 1$ through a sequence of p's for which A_p is bounded, we would deduce (3·20) for $p=1$, which is not true. Similarly the constant A_q in (3·21) tends to $+\infty$ as $q \to \infty$ (even if $c_{-1}=c_{-2}=\ldots=0$). For then the left-hand side tends to the essential upper bound of $|f|$, and $\{\Sigma |c_k|^q (|k|+1)^{q-2}\}^{1/q}$ tends to the upper bound of the numbers $|c_k|(|k|+1)$. If A_q remained bounded for some sequence of values of q tending to ∞, a function with coefficients $O(1/n)$ would necessarily be bounded, which of course is not true.

4. Marcinkiewicz's theorem on the interpolation of operations

An operation $h=Tf$ will be called *quasi-linear* if $T(f_1+f_2)$ is uniquely defined whenever Tf_1 and Tf_2 are defined, and if

$$|T(f_1+f_2)| \leqslant \kappa(|Tf_1|+|Tf_2|), \tag{4·1}$$

where κ is a constant independent of f_1 and f_2. If $\kappa=1$, we call T *sublinear*.

We fix two measure spaces R_1 and R_2, with measures μ and ν respectively, and consider a quasi-linear operation $h=Tf$, where f is defined over R_1 and h over R_2. Suppose that

$$1 \leqslant r \leqslant \infty, \quad 1 \leqslant s \leqslant \infty. \tag{4·2}$$

As in the linear case (§ 1), T is said to be *of type* (r,s), if Tf is defined in $L^{r,\mu}$ and if

$$\| h \|_{s,\nu} \leqslant M \| f \|_{r,\mu}, \tag{4·3}$$

with M independent of f; the least admissible value of M is the (r,s) *norm* of T.

Denote by $E_y[h]$ the set of points of R_2 where

$$|h| > y > 0.$$

If $s<\infty$, (4·3) implies that

$$\nu(E_y[h]) \leqslant \left(\frac{M}{y}\| f \|_r\right)^s. \tag{4·4}$$

A quasi-linear operation T which satisfies (4·4), with an M independent of f and y, will be said to be of *weak type* (r,s); the least value of M in (4·4) may be called the *weak* (r,s) *norm* of T. For the sake of emphasis, operations of type (r,s) will be occasionally called of *strong type* (r,s).

There exist operations which are of weak type (r,s) without being of strong type (r,s). By Theorem (3·16) of Chapter IV, the operation $Tf=\tilde{f}$, the conjugate function of f, is of weak type (1, 1), though it is not of type (1, 1). Another example is the operation $Tf=\theta_{|f|}(x)$ defined in Chapter I, § 13. It is sublinear and is of weak, but not of strong, type (1, 1). Other examples will be found in § 5 below.

We have defined weak type (r,s) for $s<\infty$. We define weak type (r,∞) as identical with strong type (r,∞). Hence T is of (weak, strong) type (r,∞) if

$$\operatorname{ess\,sup} |h| \leqslant M \| f \|_r \quad (1 \leqslant r \leqslant \infty). \tag{4·5}$$

(4·6) Theorem of Marcinkiewicz. *Let (α_1, β_1) and (α_2, β_2) be any two points of the triangle*

$$\Delta: \quad 0 \leqslant \beta \leqslant \alpha \leqslant 1,$$

such that $\beta_1 \neq \beta_2$. Suppose that a quasi-linear operation $h = Tf$ is simultaneously of weak types $(1/\alpha_1, 1/\beta_1)$ and $(1/\alpha_2, 1/\beta_2)$, with norms M_1 and M_2 respectively. Then for any point (α, β) with

$$\alpha = (1-t)\,\alpha_1 + t\alpha_2, \quad \beta = (1-t)\,\beta_1 + t\beta_2 \quad (0 < t < 1)$$

the operation T is of strong type $(1/\alpha, 1/\beta)$, and we have

$$\| h \|_{1/\beta} \leqslant K M_1^{1-t} M_2^t \| f \|_{1/\alpha}, \tag{4·7}$$

where $K = K_{t, \kappa, \alpha_1, \beta_1, \alpha_2, \beta_2}$ is independent of f, and is bounded if $\alpha_1, \beta_1, \alpha_2, \beta_2$ are fixed and t stays away from 0 and 1.

We postpone comments on Theorem (4·6), and in particular a comparison with the Riesz-Thorin theorem (1·11), until after the proof.

We begin with the following general remark. Consider any non-negative and μ-measurable function f defined over R_1. Then for any $p \geqslant 1$ we have

$$\int_{R_1} f^p \, d\mu = -\int_0^\infty y^p \, dm(y) = p \int_0^\infty y^{p-1} m(y) \, dy, \tag{4·8}$$

where $m(y)$ is the distribution function of f, that is, the μ-measure of the set of points where $f > y$.

The first equation (4·8) follows immediately from the definition of the Lebesgue-Stieltjes integral. (Compare a similar argument in Chapter I, § 13, for $p = 1$.) The second integral is meant as $\lim_{\omega \to \infty} \lim_{\epsilon \to 0} \int_\epsilon^\omega$, and the equation is valid whether $\int_{R_1} f^p \, d\mu$ is finite or not. The equality of the second and third integrals may be seen by observing that if either of them is finite then $y^p m(y) \to 0$ as y tends to 0 or ∞, and that an integration by parts transforms one integral into the other. If both integrals are infinite there is nothing to prove. Thus (4·8) is established.

In what follows we systematically write $1/\alpha = a$, $1/\beta = b$, $1/\alpha_1 = a_1$, ...

Return to (4·6). We may suppose that

$$\alpha_1 \leqslant \alpha_2.$$

Let $f \in L^{1/\alpha}$. Write $f = f' + f''$, where $f' = f$ whenever $|f| \leqslant 1$, and $f' = 0$ otherwise; thus $|f''| > 1$ or else $f'' = 0$. The condition $f \in L^{1/\alpha}$ implies that $f' \in L^{1/\alpha_1}$, $f'' \in L^{1/\alpha_2}$. Hence Tf' and Tf'' exist, by hypothesis, and so also does $Tf = T(f' + f'')$.

We have to show that $h = Tf$ satisfies (4·7). We first consider the case when both β_1 and β_2 are different from 0. This implies that $\alpha_1 \neq 0$, $\alpha_2 \neq 0$.

Denote by $m(y)$ and $n(y)$ the distribution functions of $|f|$ and $|h|$. Then

$$\| h \|_b^b = b \int_0^\infty y^{b-1} n(y) \, dy = (2\kappa)^b \, b \int_0^\infty y^{b-1} n(2\kappa y) \, dy, \tag{4·9}$$

κ being the same as in (4·1). For a fixed $z > 0$ we consider the decomposition

$$f = f_1 + f_2, \tag{4·10}$$

in which $f_1=f$ when $|f|\leqslant z$ and $f_1=e^{i\arg f}z$ elsewhere. It follows that

$$|f_1|=\min(|f|,z),\quad |f|=|f_1|+|f_2|. \tag{4·11}$$

Set $h_1=Tf_1$, $h_2=Tf_2$. The inequality (4·1) indicates that $|h|>2\kappa y$ at those points, at most, at which either $|h_1|>y$ or $|h_2|>y$. Denote by $m_1(y)$, $m_2(y)$, $n_1(y)$, $n_2(y)$ the distribution functions of $|f_1|$, $|f_2|$, $|h_1|$, $|h_2|$ respectively. Then

$$\begin{aligned} n(2\kappa y) &\leqslant n_1(y)+n_2(y) \\ &\leqslant M_1^{b_1}y^{-b_1}\|f_1\|_{a_1}^{b_1}+M_2^{b_2}y^{-b_2}\|f_2\|_{a_2}^{b_2}, \end{aligned} \tag{4·12}$$

by an application of (4·4) to f_1 and f_2. The right-hand side here depends on z, and the main idea of the proof consists in defining z as a suitable monotone function of y, $z=z(y)$, to be determined later.

By (4·11),

$$\begin{aligned} m_1(y)&=m(y) && \text{for}\quad 0<y\leqslant z, \\ m_1(y)&=0 && \text{for}\quad y>z, \\ m_2(y)&=m(y+z) && \text{for}\quad y>0; \end{aligned}$$

and an application of (4·12) shows that the last integral in (4·9) does not exceed

$$\begin{aligned} M_1^{b_1}\int_0^\infty & y^{b-b_1-1}\left\{\int_{R_1}|f_1|^{a_1}d\mu\right\}^{b_1/a_1}dy+M_2^{b_2}\int_0^\infty y^{b-b_2-1}\left\{\int_{R_1}|f_2|^{a_2}d\mu\right\}^{b_2/a_2}dy \\ &=M_1^{b_1}a_1^{k_1}\int_0^\infty y^{b-b_1-1}\left\{\int_0^z t^{a_1-1}m(t)\,dt\right\}^{k_1}dy \\ &\quad+M_2^{b_2}a_2^{k_2}\int_0^\infty y^{b-b_2-1}\left\{\int_z^\infty (t-z)^{a_2-1}m(t)\,dt\right\}^{k_2}dy, \end{aligned} \tag{4·13}$$

where $$k_1=b_1/a_1,\quad k_2=b_2/a_2$$

are not less than 1, by hypothesis.

Initially, instead of $\alpha_1\leqslant\alpha_2$ we make the stronger assumption $\alpha_1<\alpha_2$ (that is, $a_2<a_1$) and consider separately the two cases

$$\text{(i)}\quad \beta_1<\beta_2,\qquad \text{(ii)}\quad \beta_2<\beta_1.$$

Case (i). We have $b_2<b<b_1$, and we set

$$z=(y/A)^\xi,$$

where A and ξ are positive numbers to be determined later.

Denote by P and Q the two integrals on the right-hand side of (4·13). Then†

$$\left.\begin{aligned} P^{1/k_1}&=\sup_\chi\int_0^\infty y^{b-b_1-1}\left\{\int_0^z t^{a_1-1}m(t)\,dt\right\}\chi(y)\,dy && \text{for}\quad \int_0^\infty y^{b-b_1-1}\chi^{k_1'}(y)\,dy\leqslant 1; \\ Q^{1/k_2}&=\sup_\omega\int_0^\infty y^{b-b_2-1}\left\{\int_z^\infty (t-z)^{a_2-1}m(t)\,dt\right\}\omega(y)\,dy && \text{for}\quad \int_0^\infty y^{b-b_2-1}\omega^{k_2'}(y)\,dy\leqslant 1. \end{aligned}\right\} \tag{4·14}$$

† The computation which follows is designed to lead to the estimate (4·17) below. In the case (important in applications) $\alpha_1=\beta_1$, $\alpha_2=\beta_2$, when $k_1=k_2=1$, the computation simplifies, and instead of using (4·14) and the degenerate form of Hölder's inequality it is enough to interchange the order of integration in (4·13).

The integral under the first 'sup' is

$$\int_0^\infty t^{a_1-1}m(t)\left\{\int_{At^{1/\xi}}^\infty y^{b-b_1-1}\chi(y)\,dy\right\}dt$$
$$\leqslant \int_0^\infty t^{a_1-1}m(t)\left\{\int_{At^{1/\xi}}^\infty y^{b-b_1-1}\,dy\right\}^{1/k_1}\left\{\int_{At^{1/\xi}}^\infty y^{b-b_1-1}\chi^{k_1'}(y)\,dy\right\}^{1/k_1'}dt$$
$$\leqslant A^{(b-b_1)/k_1}(b_1-b)^{-1/k_1}\int_0^\infty t^{a_1-1+(b-b_1)/k_1\xi}m(t)\,dt, \tag{4·15}$$

by Hölder's inequality and the condition for χ, remembering that $b_1>b$. Similarly, substituting t^{a_2-1} for $(t-z)^{a_2-1}$ and using Hölder's inequality and the condition for ω, we find that the integral under the second 'sup' in (4·14) does not exceed

$$\int_0^\infty t^{a_2-1}m(t)\left\{\int_0^{At^{1/\xi}} y^{b-b_2-1}\omega(y)\,dy\right\}dt$$
$$\leqslant \int_0^\infty t^{a_2-1}m(t)\left\{\int_0^{At^{1/\xi}} y^{b-b_2-1}\,dy\right\}^{1/k_2}\left\{\int_0^{At^{1/\xi}} y^{b-b_2-1}\omega^{k_2'}(y)\,dy\right\}^{1/k_2'}dt$$
$$\leqslant A^{(b-b_2)/k_2}(b-b_2)^{-1/k_2}\int_0^\infty t^{a_2-1+(b-b_2)/k_2\xi}m(t)\,dt. \tag{4·16}$$

Collecting results we find that

$$\|h\|_b^b \leqslant (2\kappa)^b b\left\{M_1^{b_1}a_1^{k_1}\frac{A^{b-b_1}}{b_1-b}\left(\int_0^\infty t^{a_1-1+(b-b_1)/k_1\xi}m(t)\,dt\right)^{k_1}\right.$$
$$\left.+M_2^{b_2}a_2^{k_2}\frac{A^{b-b_2}}{b-b_2}\left(\int_0^\infty t^{a_2-1+(b-b_2)/k_2\xi}m(t)\,dt\right)^{k_2}\right\}. \tag{4·17}$$

We now select ξ so that the exponent of t in both integrals is $a-1$. This is possible, and we find

$$\xi=\frac{\alpha(\beta-\beta_1)}{\beta(\alpha-\alpha_1)}=\frac{\alpha(\beta-\beta_2)}{\beta(\alpha-\alpha_2)}.$$

Next† we set
$$A=M_1^\rho M_2^\sigma\|f\|_a^\tau,$$

and select ρ, σ, τ so that both terms on the right of (4·17) contain the same powers of M_1, M_2 and $\|f\|_a$. A simple computation shows that

$$A=M_1^{b_1/(b_1-b_2)}M_2^{b_2/(b_2-b_1)}\|f\|_a^{a(k_2-k_1)/(b_2-b_1)},$$

and that both terms in (4·17) contain $(M_1^{1-t}M_2^t\|f\|_a)^b$; thus we obtain (4·7) with

$$K^b=(2\kappa)^b b\left\{\frac{(a_1/a)^{k_1}}{b_1-b}+\frac{(a_2/a)^{k_2}}{b-b_2}\right\}. \tag{4·18}$$

Case (ii). We now have $b_1<b<b_2$. We set, as before, $z=(y/A)^\xi$, where A is positive and ξ *is negative*. In the inner integrals in (4·15) and (4·16) the intervals of integration $(At^{1/\xi},\infty)$ and $(0, At^{1/\xi})$ are then interchanged, but otherwise the proof remains the same, and we again arrive at (4·7) with K^b given by (4·18), but with $b-b_1$ and b_2-b for b_1-b and $b-b_2$ respectively in the denominators.

† Since $\beta_1 \neq \beta_2$ we can make $M_1=M_2=1$ by multiplying T and $d\nu$ by suitable constants (cf. p. 96). If also T is positive homogeneous—that is, if $|T(kf)|=|k|\,|Tf|$ for all constants k—then we can choose $\|f\|_a=1$, $A=1$, and the proof in case (i) is already complete. The same remark applies to the remaining cases.

We now consider the case $\alpha_1=\alpha_2$. (This has no interesting applications, nor does the case $\beta_1=\beta_2$ not covered by the theorem.) Suppose that $\alpha_1=\alpha=\alpha_2$ and that, for example, $\beta_1<\beta<\beta_2$. By hypothesis,

$$n(y)\leqslant(M_1y^{-1}\|f\|_a)^{b_1},\quad n(y)<(M_2y^{-1}\|f\|_a)^{b_2}, \tag{4·19}$$

where $n(y)=\nu(E_y[h])$. We split the first integral (4·9) into two, extended over $(0,A)$ and (A,∞), where $0<A<\infty$. If we apply to these two integrals the two inequalities (4·19) respectively, and observe that $b_2<b<b_1$, we see that $\|h\|_b^b$ is finite. Setting $A=M_1^\rho M_2^\sigma\|f\|_a^\tau$ and selecting ρ, σ, τ so that the exponents of $M_1, M_2, \|f\|_a$ in both integrals are the same, we arrive at (4·7).

It remains to consider the case when one of the numbers β_1, β_2 is 0. Suppose that $\beta_1=0$. The proof of the theorem requires that after the decomposition (4·10) we estimate $n_1(y)$ in terms of $\|h_1\|_{b_1}=\operatorname{ess\,sup}|h_1|$. In general this cannot be done unless we know that $|h_1|\leqslant y$, in which case $n_1(y)=0$. Thus we must choose z so that $|h_1|\leqslant y$. To be more specific, let us confine our attention to the case $0\leqslant\alpha_1<\alpha_2$ $(\beta_2>\beta_1=0)$, and consider separately the two subcases (*a*) $\alpha_1=0$, (*b*) $\alpha_1>0$.

(*a*) Return to (4·10), (4·11) and (4·12), and take $z=y/M_1$. Then

$$\operatorname{ess\,sup}|h_1|\leqslant M_1\operatorname{ess\,sup}|f_1|\leqslant M_1(y/M_1)=y.$$

It follows that $n_1(y)=0$ in (4·12), and so of the two terms in (4·17) only the second remains. Setting there $A=M_1$, we arrive at (4·7), where again K is given by (4·18) but with the first term in curly brackets omitted.

We can easily verify that the choice of z and A here conforms to the same pattern as in the general argument above.

(*b*) We again select z so that $n_1(y)=0$, after which the proof proceeds as before. Suppose that

$$z=(y/A)^\xi,\quad\text{where}\quad \xi=a_1/(a_1-a),\quad A=\lambda M_1\|f\|_a^{a/a_1}$$

and λ is a numerical factor to be determined presently. Except for the presence of this factor, z, ξ and A are given by the formulae above simplified by the hypothesis that $\beta_1=0$. It is therefore enough to show that for a suitable λ we have $\operatorname{ess\,sup}|h_1|\leqslant y$.

By hypothesis,

$$\operatorname{ess\,sup}|h_1|\leqslant M_1\|f_1\|_{a_1}=M_1\left\{a_1\int_0^z t^{a_1-1}m(t)\,dt\right\}^{1/a_1}.$$

It follows that we certainly have $\operatorname{ess\,sup}|h_1|\leqslant y$ if

$$M_1^{a_1}a_1\int_0^z t^{a_1-1}m(t)\,dt\leqslant(Az^{1/\xi})^{a_1}=A^{a_1}z^{a_1-a},$$

and *a fortiori* (since $a_1>a$) if

$$M_1^{a_1}a_1z^{a_1-a}\int_0^\infty t^{a-1}m(t)\,dt\leqslant\lambda^{a_1}M_1^{a_1}\|f\|_a^a z^{a_1-a}.$$

Since the integral on the left is $a^{-1}\|f\|_a^a$, the inequality is satisfied if $\lambda^{a_1}\geqslant a_1/a$. This completes the proof in the subcase (*b*). The value of K is easily found.

Theorem (4·6) is therefore established.

Remarks. (*a*) The constant K in (4·7) tends, in general, to ∞ as t tends to 0 (or 1), for otherwise the operation would be of strong type $(1/\alpha_1, 1/\beta_1)$, which need not be the case. The drawback of the proof given above is that, even if T is linear and of strong types $(1/\alpha_1, 1/\beta_1)$ and $(1/\alpha_2, 1/\beta_2)$, we cannot show that K is bounded as t tends to 0 or 1. (There is one exception: if $\beta_1 = 0$, the first term in curly brackets in (4·18) is absent, and K remains bounded as $\beta \to \beta_1$.) Hence Theorem (4·6) is not a complete substitute for the Riesz-Thorin theorem. In compensation, (4·6) applies to a number of cases when (1·11) does not; and then additional devices (if a direct appeal to (1·11) fails) may show that K is bounded.

We illustrate this by the example $Tf = \tilde{f}$. By Theorem (3·16) of Chapter IV, T is of weak type $(1, 1)$. T is also of strong type $(2, 2)$. Hence, by (4·6), T is of strong type (p_0, p_0) for each $1 < p_0 < 2$. By the very elementary Theorem (2·21) of Chapter VII, T is also of type (p_0', p_0') and so, again by (4·6), of type (p, p), if $p_0 < p < p_0'$. The norm A_p in the inequality

$$\| \tilde{f} \|_p \leqslant A_p \| f \|_p \tag{4·20}$$

is bounded in every interval $p_0 + \epsilon \leqslant p \leqslant p_0' - \epsilon$ $(\epsilon > 0)$, and so also in every interval $1 + \epsilon \leqslant p \leqslant 1/\epsilon$. The inequality (4·20) which was proved in Chapter VII, § 2, by complex methods can therefore be obtained by means of real variable ones.

(*b*) From (4·18) we see that in any case

$$K^b = O\left(\frac{1}{t}\right), \quad K^b = O\left(\frac{1}{1-t}\right), \tag{4·21}$$

for t tending to 0 and 1 respectively.

(*c*) In a number of problems we are led to consider integrals of the type $\int \phi(|f|)\, d\mu$, where ϕ is not necessarily a power. Theorem (4·6) makes it possible to 'interpolate' the function ϕ. Without striving for too much generality we may consider here some special cases which are both illustrative and useful.

(4·22) THEOREM. *Suppose that $\mu(R_1) < \infty$, $\nu(R_2) < \infty$, and that a quasi-linear operation $h = Tf$ is of weak types (a, a) and (b, b), where $1 \leqslant a < b < \infty$. Suppose also that $\phi(u)$, $u \geqslant 0$, is a continuous increasing function satisfying the conditions $\phi(0) = 0$ and*

$$\phi(2u) = O\{\phi(u)\}, \tag{4·23}$$

$$\int_u^\infty \frac{\phi(t)}{t^{b+1}}\, dt = O\left\{\frac{\phi(u)}{u^b}\right\}, \tag{4·24}$$

$$\int_1^u \frac{\phi(t)}{t^{a+1}}\, dt = O\left\{\frac{\phi(u)}{u^a}\right\}, \tag{4·25}$$

for $u \to \infty$. Then $h = Tf$ is defined for every f with $\phi(|f|)$ integrable, and

$$\int_{R_1} \phi(|h|)\, d\nu \leqslant K \int_{R_1} \phi(|f|)\, d\mu + K, \tag{4·26}$$

where K is independent of f.

The function
$$\phi(u) = u^c \psi(u), \tag{4·27}$$

where $a < c < b$ and $\psi(u)$ is a slowly varying function (Chapter V, § 2), gives an example satisfying the above conditions. The validity of (4·23) is obvious. To prove (4·24), take $0<\eta<b-c$; then, for u large enough,

$$\int_u^\infty \frac{\phi(t)}{t^{b+1}}dt=\int_u^\infty \frac{\psi(t)}{t^\eta}\frac{dt}{t^{b-c-\eta+1}}\leqslant\frac{\psi(u)}{u^\eta}\int_u^\infty\frac{dt}{t^{b-c-\eta+1}}$$

$$=\frac{\psi(u)}{u^\eta}O\left\{\frac{1}{u^{b-c-\eta}}\right\}=O\left\{\frac{\phi(u)}{u^b}\right\}.$$

The inequality (4·25) is proved similarly.

Return to (4·22). The proof that Tf exists is the same as in Theorem (4·6), and we only have to establish (4·26). The proof is similar to that of (4·7), but is in some respects simpler, since we are dealing with points of the hypotenuse of the triangle Δ. However, to avoid having to justify inverting the order of repeated Stieltjes integrals, we confine ourselves to sums rather than integrals.

Let $n(y)$ be the distribution function of $|h|$. We have

$$\int_{R_2}\phi(|h|)\,d\nu=-\int_0^\infty\phi(y)\,dn(y)=\int_0^\infty n(y)\,d\phi(y), \tag{4·28}$$

the passage from the second to the third integral being justified as in (4·8).

We write $\lambda=2\kappa$ and denote by K any positive constant independent of f. If η_j is the measure of the set in which $|h|\geqslant\lambda 2^j$, $j=0,1,2,\ldots$, then from (4·28) and $\nu(R_2)<\infty$ we deduce

$$\int_{R_2}\phi(|h|)\,d\nu\leqslant K+\sum_{j=0}^\infty\eta_j\{\phi(\lambda 2^{j+1})-\phi(\lambda 2^j)\}=K+\sum_0^\infty\eta_j\delta_j, \tag{4·29}$$

say. For any fixed j we write $f=f_1+f_2$, where f_1 equals f or 0 according as $|f|\leqslant 2^j$ or $|f|>2^j$. At the points where $|h|>\lambda 2^j$ we have either $|h_1|>2^j$ or $|h_2|>2^j$, where $h_i=Tf_i$. Since f_1 and f_2 are in L^b and L^a respectively, we have

$$\eta_j\leqslant K\left\{2^{-jb}\int_{R_1}|f_1|^b\,d\mu+2^{-ja}\int_{R_1}|f_2|^a\,d\mu\right\}$$

$$\leqslant K\left\{2^{-jb}\sum_0^j 2^{ib}\epsilon_i+2^{-ja}\sum_{j+1}^\infty 2^{ia}\epsilon_i\right\},$$

where ϵ_i, $i=1,2,\ldots$, denotes the μ-measure of the set in which $2^{i-1}<|f|\leqslant 2^i$, and ϵ_0 that of the set where $|f|\leqslant 1$. If we substitute this estimate of η_j in (4·29), and interchange the order of summation, we are led to (4·26), provided we can prove that each of the sums

$$\sum_{i=0}^\infty 2^{ib}\epsilon_i\sum_{j=i}^\infty\delta_j 2^{-jb},\quad \sum_{i=1}^\infty\epsilon_i 2^{ia}\sum_{j=0}^{i-1}\delta_j 2^{-ja} \tag{4·30}$$

is majorized by $K+K\int_{R_1}\phi(|f|)\,d\mu$.

We may suppose that $\lambda\geqslant 1$. Since $\delta_j\leqslant\phi(\lambda 2^{j+1})$, we have

$$\sum_{j=i}^\infty\delta_j 2^{-jb}\leqslant K\sum_{j=i}^\infty 2^{-j(b+1)}\int_{\lambda 2^{j+1}}^{\lambda 2^{j+2}}\phi(u)\,du\leqslant K\sum_{j=i}^\infty\int_{\lambda 2^{j+1}}^{\lambda 2^{j+2}}u^{-b-1}\phi(u)\,du$$

$$\leqslant K\int_{2^i}^\infty u^{-b-1}\phi(u)\,du\leqslant K\phi(2^i)\,2^{-ib}, \tag{4·31}$$

by (4·24). Hence the first sum (4·30) does not exceed

$$K\sum_{i=0}^{\infty}\epsilon_i\phi(2^i)\leqslant K\epsilon_0\phi(1)+K\sum_{i=1}^{\infty}\epsilon_i\phi(2^{i-1})\leqslant K+K\int_{R_1}\phi(|f|)\,d\mu, \tag{4·32}$$

by an application of (4·23). Using (4·23) and (4·25) we obtain a similar estimate for the second sum (4·30), and the theorem is proved.

(4·33) THEOREM. *If $\phi(u)=u^r\psi(u)$, where $1<r<\infty$ and $\psi(u)$ is a positive slowly varying function, then*

$$\int_0^{2\pi}\phi(|\tilde f|)\,dx\leqslant K\int_0^{2\pi}\phi(|f|)\,dx+K.$$

This is an obvious corollary of (4·20) and Theorem (4·22).

The theorem which follows is a modification of (4·22) in the case when $a=1$ and (4·25) does not necessarily hold, that is, when the growth of $\phi(u)$ may be 'close' to that of u.

(4·34) THEOREM. *Suppose that $\mu(R_1)$ and $\nu(R_2)$ are finite, that $1<b<\infty$, and that $h=Tf$ is a quasilinear operation which is simultaneously of weak types $(1,1)$ and (b,b). Let $\chi(u)$, $u\geqslant 0$, be equal to 0 in a right-hand neighbourhood of $u=0$, say for $u\leqslant 1$, positive and increasing elsewhere; and suppose that $\chi(2u)=O\{\chi(u)\}$ for large u. Write*

$$\phi(u)=u\int_0^u t^{-2}\chi(t)\,dt \tag{4·35}$$

and suppose that $\phi(u)$ satisfies (4·24). Then $h=Tf$ is defined for all f such that $\phi(|f|)$ is integrable, and we have

$$\int_{R_2}\chi(|h|)\,d\nu\leqslant K\int_{R_1}\phi(|f|)\,d\mu+K, \tag{4·36}$$

where K is independent of f.

The proof is similar to that of (4·22) and we shall be brief. First, we verify that $\chi(2u)=O\{\chi(u)\}$ implies $\phi(2u)=O\{\phi(u)\}$. Secondly,

$$\phi(u)\geqslant u\int_{\frac12 u}^u t^{-2}\chi(t)\,dt\geqslant u\chi(\tfrac12 u)\int_{\frac12 u}^u t^{-2}dt=\chi(\tfrac12 u),$$

which in conjunction with $\chi(u)=O\{\chi(\frac12 u)\}$ implies that

$$\chi(u)=O\{\phi(u)\} \tag{4·37}$$

for large u. Next, (4·35) shows that $\phi(u)/u$ is bounded away from 0 as $u\to\infty$; hence if $\phi(|f|)$ is integrable, so is $|f|$, and so Tf is defined. We have now only to prove (4·36).

Substitute χ for ϕ in (4·29), so that now $\delta_j=\chi(\lambda 2^{j+1})-\chi(\lambda 2^j)$. Arguing as before, we shall have proved (4·36) if we can show that each of the two sums (4·30), where now $a=1$, is majorized by $K+K\int_{R_1}\phi(|f|)\,d\mu$. We again have (4·31), with χ for ϕ in the first three sums; and, in view of (4·37) and (4·32), the first sum (4·30) does not exceed $K\int_{R_1}\phi(|f|)\,d\mu+K$. An analogous argument shows that the second sum (4·30) is majorized by a similar expression. (It is here that we have to define ϕ in terms of χ as we did.)

The most interesting special case is when $\chi = 0$ for $u \leqslant 1$ and $\chi = u$ otherwise. Then $\phi(u) = u \log^+ u$, and (4·36) may be written

$$\int_{R_2} |h| \, d\nu \leqslant K \int_{R_1} |f| \log^+ |f| \, d\mu + K. \tag{4·38}$$

More generally, the hypotheses concerning the behaviour of χ and ϕ for large u are satisfied if, for such u, $\chi(u) = u\psi(u)$, where ψ is a positive slowly varying function. Hence we have the following result:

(4·39) THEOREM. *Suppose that $\chi(u) = u\psi(u)$, where $\psi(u)$ is a slowly varying function, positive except in the neighbourhood of $u = 0$ where it is 0. Then, with ϕ defined by* (4·35), *we have*

$$\int_0^{2\pi} \chi(|\tilde{f}|) \, dx \leqslant K \int_0^{2\pi} \phi(|f|) \, dx + K. \tag{4·40}$$

Suppose that the operation $h = Tf$ is of type $(1/\alpha, 1/\beta)$ for (α, β) interior to a segment l with end-points (α_1, β_1) and (α_2, β_2), so that

$$\| h \|_{1/\beta} \leqslant M_{\alpha\beta} \| f \|_{1/\alpha}$$

for such (α, β). The norm $M_{\alpha\beta}$ usually tends to ∞ as (α, β) approaches an end-point of l. If $M_{\alpha\beta}$ does not increase too rapidly, we can easily obtain additional information about the degree of integrability of h for f in L^{1/α_1} or L^{1/α_2}. For the sake of brevity we confine our attention to special, but fairly typical, cases which have some bearing on results obtained previously. We suppose that T is linear, that both $\mu(R_1)$ and $\nu(R_2)$ are finite, and that l is on the line $\alpha = \beta$.

(4·41) THEOREM. *Suppose that $h = Tf$ is of type (r, r) for each $1 < r < \infty$, so that*

$$\| h \|_r \leqslant A_r \| f \|_r \quad (1 < r < \infty).$$

Let $\rho > 0$. Then,

(i) *if*
$$A_r = O(r^\rho) \quad (r \to \infty),$$

there exist positive constants λ, K such that

$$\int_{R_2} \exp(\lambda |h|^{1/\rho}) \, d\nu \leqslant K \tag{4·42}$$

for each f with $|f| \leqslant 1$;

(ii) *if*
$$A_r = O\{(r-1)^{-\rho}\} \quad (r \to 1),$$

then Tf is uniquely defined for each f such that $|f| (\log^+ |f|)^\rho$ is integrable, and we have

$$\int_{R_2} |h| \, d\nu \leqslant K \int_{R_1} |f| (\log^+ |f|)^\rho \, d\mu + K, \tag{4·43}$$

with K independent of f.

Proof. (i) By hypothesis, $A_r \leqslant Ar^\rho$ for, say, $r \geqslant 2$. Hence

$$\int_{R_2} |h|^{k/\rho} \, d\nu \leqslant A^{k/\rho} (k/\rho)^k \int_{R_1} |f|^{k/\rho} \, d\mu \leqslant A^{k/\rho} (k/\rho)^k \mu(R_1), \tag{4·44}$$

for k integral and not less than 2ρ. We multiply (4·44) by $\lambda^k/k!$ and sum over all such values of k. Since $k^k/k! \leqslant e^k$, the series on the right converges if we set $\lambda A^{1/\rho} e/\rho = \frac{1}{2}$, and so, denoting by $P(u)$ a suitable polynomial of degree less than 2ρ, we have

$$\int_{R_2} \{\exp(\lambda |h|^{1/\rho}) - P(|h|)\} \, d\nu \leqslant K. \tag{4·45}$$

The inequality (4·42) follows from (4·45). For at the points where $|h|$ is sufficiently large the integrand in (4·42) is majorized by, say, twice the integrand in (4·45), and the contribution of the remaining values of h in (4·42) is $O(1)$.

(ii) For many operations T, the inequality (4·43) is 'dual' to (4·42) and is obtainable from the latter by means of Young's inequality; but we give a direct proof.

Suppose first that f is simple. Hence Tf is defined. For each $k=1, 2, \ldots$ we define f_k as the function equal to f wherever $2^{k-1} \leqslant |f| < 2^k$, and equal to 0 elsewhere; by f_0 we mean the function equal to f wherever $|f|<1$, and equal to 0 elsewhere. Then $f=\Sigma f_k$, the number of terms being finite. Correspondingly, $h=\Sigma h_k$, where $h_k = Tf_k$. Denote by ϵ_k the μ-measure of the set where $f_k \neq 0$. We have

$$\int_{R_2} |h_k|\, d\nu \leqslant \nu^{1/r'}(R_2)\, \| h_k \|_r \leqslant KA_r \| f_k \|_r \leqslant KA_r 2^k \epsilon_k^{1/r} \tag{4·46}$$

for $1<r<\infty$. By hypothesis, $A_r \leqslant K(r-1)^{-\rho}$ for, say, $1<r<2$; and if we substitute $r=1+1/(k+1)$ in (4·46) we obtain

$$\int_{R_2} |h_k|\, d\nu \leqslant K(k+1)^\rho\, 2^k \epsilon_k^{(k+1)/(k+2)},$$

$$\int_{R_2} |h|\, d\nu \leqslant \Sigma \int_{R_2} |h_k|\, d\nu \leqslant K \sum_0^\infty (k+1)^\rho\, 2^k \epsilon_k^{(k+1)/(k+2)}.$$

Observe now that those terms of the last series in which $\epsilon_k \leqslant 3^{-k}$ have a finite sum, and that for the remaining k we have $\epsilon_k^{(k+1)/(k+2)} \leqslant K\epsilon_k$. Hence

$$\int_{R_2} |h|\, d\nu \leqslant K + K \sum_0^\infty (k+1)^\rho\, 2^k \epsilon_k \leqslant K + K \int_{R_1} |f|\, (\log^+ |f|)^\rho\, d\mu,$$

and (4·43) is proved if f is simple.

Now for any $\omega>0$ we can apply (4·43) to ωf and obtain

$$\int_{R_2} |h|\, d\nu \leqslant K \int_{R_1} |f|\, \{\log^+(\omega |f|)\}^\rho\, d\nu + K/\omega,$$

so that K/ω is arbitrarily small if ω is large enough. It follows that, for any f such that $|f|\,(\log^+|f|)^\rho$ is integrable, if f_n is a sequence of simple functions such that $\int_{R_1} |f-f_n|\,(\log^+ \omega\, |f-f_n|)^\rho\, d\mu \to 0$ for each $\omega>0$, then $\| h_m - h_n \|_1 \to 0$ as $m, n\to\infty$. Hence Tf_m tends in L to a limit h which may be taken as Tf and which satisfies (4·43).

This completes the proof of (4·41). It is clear that the requirement that T should be of type (r, r) is needed in (i) only for $r>r_0$, and in (ii) only for r near 1.

5. Paley's theorems on Fourier coefficients

In this section we shall extend Theorem (3·19) to general systems $\{\phi_n\}$, $n=1, 2, \ldots$, of functions orthonormal and uniformly bounded,

$$|\phi_n(x)| \leqslant M,$$

over an interval (a, b). Given a sequence of numbers $c_1, c_2, \ldots$ we write

$$\mathfrak{B}_r[c] = \{\sum_k |c_k|^r k^{r-2}\}^{1/r}$$

(5·1) THEOREM OF PALEY. (i) *If* $f \in L^p$, $1 < p \leqslant 2$, *and if* $c_1, c_2, \ldots$ *are the Fourier coefficients of* f *with respect to* $\phi_1, \phi_2, \ldots$, *then* $\mathfrak{B}_p[c]$ *is finite and*

$$\mathfrak{B}_p[c] \leqslant A_p M^{(2-p)/p} \| f \|_p. \tag{5·2}$$

(ii) *If given numbers* $c_1, c_2, \ldots$ *satisfy the condition* $\mathfrak{B}_q[c] < \infty$ *for some* $q \geqslant 2$, *then there is an* $f \in L^q$ *having the* c_n *as its Fourier coefficients with respect to* $\{\phi_n\}$, *and such that*

$$\| f \|_q \leqslant A'_q M^{(q-2)/q} \mathfrak{B}_q[c]. \tag{5·3}$$

The function f *is the limit, in* L^q, *of* $s_n = c_1\phi_1 + \ldots + c_n\phi_n$ *as* $n \to \infty$.

(iii) *Moreover, we may take*

$$A'_q = A_{q'}. \tag{5·4}$$

Let μ be the ordinary Lebesgue measure in (a, b); by $\| f \|_r$ we mean a norm with respect to μ. Let ν give measure $1/n^2$ to the set consisting of the single point n, $n = 1, 2, \ldots$, and measure 0 to a set which does not contain any of these points. The linear operation

$$h = Tf = \{nc_n\} = \left\{ n \int_a^b f \bar{\phi}_n \, dx \right\}$$

is defined everywhere in $L^{r,\mu}$, $1 \leqslant r \leqslant 2$, and

$$\| h \|_{r,\nu} = \{\sum_n | nc_n |^r n^{-2}\}^{1/r} = \mathfrak{B}_r[c].$$

Bessel's inequality $\mathfrak{B}_2[c] \leqslant \| f \|_2$ implies that T is of type $(2, 2)$. It is generally not of type $(1, 1)$ (except in special cases; see §3). We show that T *is of weak type* $(1, 1)$; more precisely we show that

$$\nu\{E_y[h]\} \leqslant \frac{2M}{y} \| f \|_1. \tag{5·5}$$

The left-hand side here is Σn^{-2} extended over those n for which $| nc_n | > y$. For such n,

$$y < | nc_n | \leqslant n \int_a^b | f \bar{\phi}_n | \, dx \leqslant nM \| f \|_1,$$

that is,

$$n > y/M \| f \|_1.$$

If we set $\omega = y/M \| f \|_1$ and suppose that $\omega \geqslant 1$, then

$$\sum_{n > \omega} n^{-2} \leqslant 2\omega^{-1} = 2My^{-1} \| f \|_1,$$

which is (5·5). The latter is obvious $\left(\text{since } \sum_1^\infty n^{-2} = \tfrac{1}{6}\pi^2\right)$ if $\omega < 1$.

By Theorem (4·6), with $(\alpha_1, \beta_1) = (1, 1)$, $(\alpha_2, \beta_2) = (\tfrac{1}{2}, \tfrac{1}{2})$, $(\alpha, \beta) = (1/p, 1/p)$, $M_1 = 2M$, $M_2 = 1$, we have

$$\mathfrak{B}_p[c] = \| Tf \|_{p,\nu} \leqslant A_p M^{(2-p)/p} \| f \|_p,$$

which is (i).

We deduce (ii) from (i). Write $q = p'$ and consider any g with $\| g \|_p \leqslant 1$. Let $d_1, d_2, \ldots$ be the Fourier coefficients of g with respect to $\phi_1, \phi_2, \ldots$, which exist by (2·12). For the s_n in (ii) we have

$$\left| \int_a^b \bar{s}_n g \, dx \right| = \left| \sum_1^n \bar{c}_k d_k \right| = \left| \sum_1^n \bar{c}_k k^{(q-2)/q} \,.\, d_k k^{(p-2)/p} \right|$$

$$\leqslant \left(\sum_1^n | c_k |^q k^{q-2} \right)^{1/q} \left(\sum_1^n | d_k |^p k^{p-2} \right)^{1/p} \leqslant A_p M^{(2-p)/p} \left(\sum_1^n | c_k |^q k^{q-2} \right)^{1/q}, \tag{5·6}$$

by Hölder's inequality and (5·2) applied to g. The upper bound of the left-hand side here for all permissible g is $\| s_n \|_q$. Thus

$$\| s_n \|_q \leqslant A_p M^{(q-2)/q} \left(\sum_1^n | c_k |^q k^{q-2} \right)^{1/q}. \tag{5·7}$$

This inequality applied to $s_n - s_m$ gives

$$\| s_n - s_m \|_q \leqslant A_p M^{(q-2)/q} \left(\sum_{m+1}^n | c_k |^q k^{q-2} \right)^{1/q} \to 0$$

as $m, n \to \infty$. Hence there is an $f \in L^q$ such that $\| s_n - f \|_q \to 0$. This and (5·7) lead to (5·3), and also to (5·4).†

Condition $\mathfrak{B}_q [c] < \infty$ implies $\|c\|_2 < \infty$ since, by Hölder's inequality,

$$\Sigma | c_k |^2 = \Sigma | c_k |^2 k^{2(q-2)/q} . k^{-2(q-2)/q}$$
$$\leqslant (\Sigma | c_k |^q k^{q-2})^{2/q} (\Sigma k^{-2})^{(q-2)/q}.$$

It follows (Chapter IV, § 1) that $\Sigma c_k \phi_k$ is the Fourier series of an $f_1 \in L^2$ such that $\| s_n - f_1 \|_2 \to 0$. Hence $f \equiv f_1$ and the proof of (ii) is complete.

We add that (ii) cannot be obtained by interpolating between $q = 2$ and $q = \infty$. For the operation $f = Th$ transforming a sequence $c_1, c_2, \ldots$ into a function $f \sim \Sigma c_k \phi_k$ is not of type (∞, ∞): the finiteness of $\sup | kc_k |$ does not imply the boundedness of f.

By Remark (*b*) on p. 116 we have $A_p = O\{(p-1)^{-1}\}$ as $p \to 1$. By the Riesz-Thorin theorem (1·11), $A_p = O(1)$ as $p \to 2$ (compare Remark (*a*) on p. 115). Hence, using also (5·4), we may take

$$A_p \leqslant A/(p-1), \quad A'_q \leqslant Aq, \tag{5·8}$$

where A is an absolute constant.

It is obvious that by applying (5·1) to the normalized system 1, e^{ix}, e^{-ix}, e^{2ix}, e^{-2ix}, ... we obtain Theorem (3·19).

Consider two finite sequences $a_1, a_2, \ldots, a_n$ and $b_1, b_2, \ldots, b_n$ of non-negative numbers and set

$$S = a_1 b_1 + a_2 b_2 + \ldots + a_n b_n.$$

Suppose that $\{a_k\}$ is monotone, either non-increasing or non-decreasing. Then, *rearranging b_k in all possible manners we get for S the largest possible value if b_k varies in the same sense as a_k*, that is, if both are non-increasing or both non-decreasing; *S is a minimum if they vary in opposite senses.* Suppose, for example, that $a_1 \geqslant a_2 \geqslant \ldots$ and that $b_1 \leqslant b_2$. Then replacing $a_1 b_1 + a_2 b_2$ by $a_1 b_2 + a_2 b_1$ we increase S by $(a_1 - a_2)(b_2 - b_1)$. Similarly for the other case.

Now consider the case when $\{a_k\}$ and $\{b_k\}$ are infinite, and suppose that $\{a_k\}$ is monotone and $b_k \to 0$. The terms b_k distinct from 0 can be rearranged into a non-increasing sequence. If it is finite, we complete it by 0's. The resulting sequence will be denoted by $\{b_k^*\}$ and called the *non-increasing rearrangement of* $\{b_k\}$. It is not difficult to see that if $\{a_k\}$ increases, then $\delta = \Sigma a_k b_k$ is a minimum if the b_k are arranged in decreasing order; if $\{a_k\}$ decreases, δ is a maximum for this rearrangement. For if all b_k are distinct from 0, the argument above remains valid. If some of the b_k are 0, we drop in S the corresponding terms $a_k b_k$ and maximize or minimize the remainder by rearranging the $b_k \neq 0$ into a descending order. The sum we get will be $\Sigma a_k b_k^*$.

† We are not interested in the least values of A_p and A'_q. If, however, A_p and A'_q are to have the least values, then the foregoing argument only gives $A'_q \leqslant A_{q'}$, and to obtain equality we must also use an argument dual to (5·6).

Given a sequence of complex numbers $c_k \to 0$, we denote by $\{c_k^*\}$ the sequence $|c_1|, |c_2|, \ldots$ rearranged in decreasing order. Writing $\mathfrak{B}_r^*[c] = \mathfrak{B}_r[c^*]$, we have

$$\mathfrak{B}_q^*[c] \leqslant \mathfrak{B}_q[c], \quad \mathfrak{B}_p^*[c] \geqslant \mathfrak{B}_p[c]. \tag{5·9}$$

(5·10) Theorem. (i) *Under the hypotheses of* (5·1) (i) *we have*

$$\mathfrak{B}_p^*[c] \leqslant A_p M^{(2-p)/p} \| f \|_p.$$

(ii) *If $c_1, c_2, \ldots$ are complex numbers tending to 0 such that $\mathfrak{B}_q^*[c]$ is finite, then the c_n are the Fourier coefficients, with respect to the $\{\phi_n\}$, of an $f \in L^q$ satisfying*

$$\| f \|_q \leqslant A_q' M^{(q-2)/q} \mathfrak{B}_q^*[c].$$

Part (i) follows from (5·1) (i) if we rearrange the ϕ_k corresponding to the c_k distinct from 0. (Observe that (2·8) (i) implies that $c_k \to 0$.)

Conversely, suppose that for a given $\{c_n\} \to 0$ we have $\mathfrak{B}_q^*[c] < \infty$. Let $|c_{n_1}|, |c_{n_2}|, \ldots$ be the numbers which after rearrangement go into $c_1^*, c_2^*, \ldots$. By (5·1) (ii), $\Sigma c_{n_k} \phi_{n_k}$ is the Fourier series of an $f \in L^q$ having c_{n_k} as its Fourier coefficient with respect to ϕ_{n_k} and satisfying $\| f \|_q \leqslant A_q' M^{(q-2)/q} \mathfrak{B}_q^*[c]$; moreover, f is the limit in L^2 of the partial sums of $\Sigma c_{n_k} \phi_{n_k}$. The latter result shows that the Fourier coefficients of f with respect to the ϕ_k different from $\phi_{n_1}, \phi_{n_2}, \ldots$ are 0. This completes the proof of (ii).

It is interesting to observe that Paley's theorem in the form (5·10) contains the F. Riesz theorem (2·8) as a special case, though in a slightly less sharp form, the right-hand sides of the inequalities in question,

$$\| c \|_{p'} \leqslant M^{(2-p)/p} \| f \|_p, \quad \| f \|_{p'} \leqslant M^{(2-p)/p} \| c \|_p,$$

having an extra factor β_p depending on p only.

It is sufficient to show that

$$\mathfrak{B}_p^*[c] \geqslant \gamma_p' \| c \|_{p'}, \tag{5·11}$$

$$\mathfrak{B}_q^*[c] \leqslant \gamma_q \| c \|_{q'}, \tag{5·12}$$

where γ_p' depends only on p, and γ_q only on q. We prove the second of these inequalities only, the proof of the first being similar. We use the fact that $(x+y+\ldots)^r \geqslant x^r + y^r + \ldots$ for $x, y, \ldots$ non-negative and $r \geqslant 1$. Then

$$\begin{aligned}
\sum_{n=1}^{\infty} c_n^{*q} n^{q-2} = \sum_{\nu=0}^{\infty} \sum_{n=2^\nu}^{2^{\nu+1}-1} c_n^{*q} n^{q-2} &\leqslant 2^{q-2} \sum_{\nu=0}^{\infty} c_{2^\nu}^{*q} 2^{\nu(q-1)} \\
&= 2^{q-2} \sum_{\nu=0}^{\infty} (c_{2^\nu}^{*q'} 2^\nu)^{q-1} \\
&\leqslant 2^{2q-3} \left(c_1^{*q'} + \sum_{\nu=1}^{\infty} c_{2^\nu}^{*q'} 2^{\nu-1} \right)^{q-1} \\
&\leqslant 2^{2q-3} \left(c_1^{*q'} + \sum_{\nu=1}^{\infty} \sum_{n=2^{\nu-1}}^{2^\nu-1} c_n^{*q'} \right)^{q-1} \\
&\leqslant 2^{2q-3} \left(2 \sum_{n=1}^{\infty} c_n^{*q'} \right)^{q-1} = 2^{3q-4} \| c_n \|_{q'}^q,
\end{aligned}$$

and (5·12) is established.

This result might suggest that perhaps the criteria of Paley and Riesz are, roughly speaking, equally powerful. That this is not so can be seen by the example

$$\phi_n(x) = \cos nx, \quad (a,b) = (0,\pi), \quad c_n = n^{-\frac{3}{4}} \log^{-\frac{3}{4}}(n+1)$$

for $n = 1, 2, \ldots$. Then $\mathfrak{B}_4^*[c] = \mathfrak{B}_4[c]$ is finite and, by (5·10) (ii), $\Sigma c_n \cos nx$ is in L^4. Since $\| c \|_{\frac{4}{3}} = \infty$, this result is not a consequence of (2·8) (ii).

We now prove results dual to (5·1) and (5·10), in which the roles of f and c_n are reversed. We need the following lemma, in which f^* denotes the non-increasing rearrangement of $|f|$ (see Chapter I, § 13).

(5·13) Lemma. *If $f(x)$ and $g(x)$ are non-negative, and the latter also non-increasing, in a finite interval (a,b), then*

$$\int_a^b gf\,dx \leqslant \int_a^b gf^*\,dx. \tag{5·14}$$

We first observe that if $f_n(x) \to f(x)$ almost everywhere, then $f_n^*(x) \to f^*(x)$, except possibly at a denumerable set of points. For $| E(f_n > y) | \to | E(f > y) |$ for each y which is not taken by $f(x)$ in a set of positive measure, that is, $m_n(y) \to m(y)$ for such y, if $m_n(y)$ and $m(y)$ denote the distribution functions of f_n and f. Considering the inverse functions $f_n^*(x)$ and $f^*(x)$ of $m_n(y)$ and $m(y)$, it is intuitive geometrically that $f_n^*(x) \to f^*(x)$ for each x which does not correspond to a stretch of constancy of $m(y)$.

Next, if $\{f_n\}$ is monotone and tends to the limit f, and if (5·14) holds for each f_n, then it also holds for f. This follows from the preceding remark and from Lebesgue's theorem on the integration of monotone sequences.

Finally, (5·14) is true if (a,b) can be decomposed into a finite number of intervals of equal length in each of which f, and so also f^*, is a constant. For then the integrals (5·14) reduce to sums, a case discussed previously. Since starting with such functions we can, by monotone passages to limits, obtain any measurable function f (or, rather, a function equivalent to f), (5·14) is established.

We now pass to the duals of (5·1) and (5·10). It will simplify the proof slightly if we assume that (a,b) is finite, for example of the form $(0,h)$, but the proof for the general case is much the same (see also the Remark on p. 126). By f^* we now denote the function non-increasing and equimeasurable with $|f|$. We also write

$$\mathfrak{U}_r[f] = \left(\int_0^h |f|^r x^{r-2}\,dx\right)^{1/r},$$

$$\mathfrak{U}_r^*[f] = \mathfrak{U}_r[f^*] = \left(\int_0^h f^{*r} x^{r-2}\,dx\right)^{1/r}.$$

The following theorem corresponds to (5·10):

(5·15) Theorem. (i) *If for a sequence $c_1, c_2, \ldots$ we have $\| c \|_p < \infty$, then the c_n are the Fourier coefficients with respect to the ϕ_n of an f with*

$$\mathfrak{U}_p^*[f] \leqslant A_p \mathfrak{M}^{(2-p)/p} \| c \|_p. \tag{5·16}$$

(ii) *If $\mathfrak{U}_q^*[f]$ is finite, and if c_n are the Fourier coefficients of f, then $\| c \|_q$ is finite and*

$$\| c \|_q \leqslant A_q' \mathfrak{M}^{(q-2)/q} \mathfrak{U}_q^*[f] \tag{5·17}$$

with $A_q' = A_{q'}$.

We begin by proving (5·16) in a weaker form with $\mathfrak{U}$ for $\mathfrak{U}^*$.

Let μ be the additive measure giving value 1 to the sets consisting of the single points $x=n$, $n=1, 2, \ldots$, and 0 to sets not containing any of these points. By ν we shall mean the measure defined on $(0, h)$ by the condition $d\nu = x^{-2}dx$. Then

$$\| c \|_{r,\mu} = \| c \|_r, \quad \| g \|_{r,\nu} = \mathfrak{U}_r[f]$$

for $g = xf(x)$. The inequality (5·16) holds for $p=2$, with $A_2=1$, and it will be proved generally if we show that the operation

$$g = Tc = xf(x) = x\Sigma c_n \phi_n$$

is of weak type $(1, 1)$. We shall show that

$$\nu(E(| g | > y)) \leqslant My^{-1} \| c \|_1. \tag{5·18}$$

The left-hand side here is $\int x^{-2}dx$ extended over the set where $x | f(x) | > y$. For such x,

$$y < x | f(x) | = x | \Sigma c_n \phi_n | \leqslant xM \| c \|_1,$$

whence $x > y/M \| c \|_1 = \omega$, say. If $\omega \geqslant h$, there is nothing to prove. If $\omega < h$, the left-hand side of (5·18) is less than $\int_\omega^\infty x^{-2}dx = My^{-1} \| c \|_1$.

We now prove the weaker form of (5·17). Let $p=q'$. We fix $N>0$ and consider all sequences $d_1, d_2, \ldots, d_N$ with $\| d \|_q = 1$. Write $g = d_1\phi_1 + \ldots + d_N\phi_N$. Then

$$\left(\sum_1^N | c_n |^q\right)^{1/q} = \sup_d \left| \sum_1^N c_n \bar{d}_n \right| = \sup_d \left| \int_0^h f\bar{g}\,dx \right|$$

$$= \sup_d \left| \int_0^h fx^{(q-2)/q} . \bar{g}x^{(p-2)/p}\,dx \right| \leqslant \mathfrak{U}_q[f] \sup_d \mathfrak{U}_p[g]$$

$$\leqslant A_p M^{(2-p)/p} \mathfrak{U}_q[f],$$

by (5·16). On making $N \to \infty$ we get (5·17) with $A'_q \leqslant A_{q'}$.

We now prove the actual inequality (5·17), with $\mathfrak{U}^*$. We first suppose that f is a step function. Rearranging the order of the intervals of constancy of f, we transform $| f |$ into f^*. At the same time $f(x)$ is transformed into a function $F(x)$, and the ϕ_n are transformed into functions ψ_n again forming an ortho-normal system. Since the coefficient of f with respect to ϕ_n is equal to the coefficient of F with respect to ψ_n, (5·17) follows from the weaker inequality previously established.

To prove (5·17) in the general case, let $\{f_k\}$ be a sequence of functions for each of which (5·17) is true. Since any bounded f is almost everywhere the limit of a uniformly bounded sequence $\{f_k\}$ of step functions, and since $c_n^k \to c_n$, $f_k^* \to f^*$ as $k \to \infty$, (5·17) holds with f_k^*, c_n^k replaced by f^*, c_n. If f is arbitrary, we set $f_k(x) = f(x)$ wherever $| f(x) | \leqslant k$, and $f(x) = 0$ elsewhere. Here again $c_n^k \to c_n$, and f_k^* tends increasingly to f^*; and since the f_k are bounded, (5·17) holds for f.

To prove (5·16) we fix $N>0$, set $f_N = c_1\phi_1 + \ldots + c_N\phi_N$, and observe that

$$\mathfrak{U}_p[f_N^*] = \sup_g \int_0^h f_N^* g\,dx$$

for all $g \geq 0$ with $\mathfrak{U}_q[g] \leq 1$; g may further be restricted to the class of step functions. Then $\int_0^h f_N^* g\,dx = \int_0^h \bar{f}_N \gamma\,dx$, where the absolute value of $\gamma(x) = \gamma(x; g, N)$ is equidistributed with $|g|$. Denoting by d_n the Fourier coefficients of γ we have

$$\begin{aligned}\mathfrak{U}_p[f_N^*] = \sup_g \int_0^h f_N^* g\,dx = \sup_g \left|\sum_1^N \bar{c}_n d_n\right| &\leq \sup_g \| c \|_p \| d \|_q \\ &\leq \| c \|_p \sup_g \{A'_q M^{(q-2)/q} \mathfrak{U}_q[\gamma^*]\} \\ &\leq A'_q M^{(2-p)/p} \| c \|_p, \end{aligned} \tag{5·19}$$

since
$$\mathfrak{U}_q[\gamma^*] = \mathfrak{U}_q[g^*] \leq \mathfrak{U}_q[g] \leq 1.$$

Since the finiteness of $\| c \|_p$ implies that of $\| c \|_2$, there is a sequence $\{f_{N_k}\}$ converging to f almost everywhere. It follows that $f_{N_k}^* \to f^*$ outside a denumerable set (cf. p. 124). Comparing the extreme terms in (5·19) and making $N = N_k \to \infty$, we get (5·16), with $A_p = A'_{p'}$.

Remark. In the case $(a, b) = (-\infty, +\infty)$ it is convenient to define f^* as the function equidistributed with $|f|$, even, and non-increasing in $(0, \infty)$. We may then set

$$\mathfrak{U}_r^*[f] = \left(\int_{-\infty}^{+\infty} f^{*r} |x|^{r-2}\,dx\right)^{1/r}.$$

Compare Theorems (2·8) and (5·10). Their first parts assert that both $\| c \|_{p'}$ and $\mathfrak{B}_p^*[c]$ are less than fixed multiples of $\| f \|_p$. We now prove a result containing these two as special cases. A similar unification is given for the second parts of (2·8) and (5·10).

(5·20) THEOREM. (i) *Suppose that*

$$p \leq r \leq p', \quad \lambda = \frac{1}{p} + \frac{1}{r} - 1.$$

Then for the coefficients c_n of $f \in L^p$ with respect to ϕ_n we have

$$\{\Sigma(c_n^* n^{-\lambda})^r\}^{1/r} \leq A_p M^{(2-p)/p} \| f \|_p. \tag{5·21}$$

(ii) *Suppose that*
$$q' \leq s \leq q, \quad \mu = \frac{1}{q} + \frac{1}{s} - 1,$$

and that a sequence $c_n \to 0$ satisfies $\Sigma(c_n^* n^{-\mu})^s < \infty$. Then there is an $f \in L^q$ having the c_n as its coefficients with respect to ϕ_n and such that

$$\| f \|_q \leq A_q M^{(q-2)/q} \{\Sigma(c_n^* n^{-\mu})^s\}^{1/s}. \tag{5·22}$$

Clearly $\lambda \geq 0$, $\mu \leq 0$.

The left-hand side of (5·21) is

$$\{\Sigma(c_n^* n^{1/p'})^r n^{-1}\}^{1/r} \tag{5·23}$$

and may be written $\left\{\int \xi(x)^r\,d\eta(x)\right\}^{1/r}$, where $\eta(x)$ is a non-decreasing step function. Thus the logarithm of (5·23) is a convex function of $1/r$ (Chapter I, (10·12)). But for $r = p$ and $r = p'$ the expression (5·23) reduces to $\mathfrak{B}_p^*[c]$ and to $\| c^* \|_{p'}$ respectively. Since a function convex in an interval attains its maximum at one of the ends of the interval, (5·21) follows from (5·10) (i) and (2·8) (i).

(ii) may be obtained from (i) in the same way as in (5·1). It is enough to prove (5·22) with $|c_n|$ for c_n^*. Then, setting $q=p'$, $s=r'$ and using the same notation as in (5·6), we have

$$\left|\int_a^b \bar{s}_n g\,dx\right| = \left|\sum_1^n \bar{c}_k k^{-\mu}.d_k k^{-\lambda}\right| \leqslant \{\Sigma\,|\,c_k k^{-\mu}\,|^s\}^{1/s}\{\Sigma\,|\,d_k k^{-\lambda}\,|^r\}^{1/r}$$
$$\leqslant A_p M^{(2-p)/p}\left\{\sum_1^n |\,c_k k^{-\mu}|^s\right\}^{1/s},$$

and continuing the argument as before we get $\|f-s_n\|_q \to 0$.

In a similar way we could prove a generalization of (5·15).

The following corollary of (5·20) will be needed in § 9 below.

(5·24) THEOREM. *Suppose that*

$$1<p\leqslant 2\leqslant q<\infty,\quad f\sim\Sigma c_n\phi_n\in L^p,$$

and write

$$\alpha=\frac{1}{p}-\frac{1}{q},\quad g\sim\Sigma c_n n^{-\alpha}\epsilon_n\phi_n,$$

where ϵ_n is any sequence of numbers such that $|\epsilon_n|\leqslant 1$. Then

$$\|g\|_q\leqslant A_{pq}M^{2\alpha}\|f\|_p. \tag{5·25}$$

Take $r=s=2$ in (5·20) and apply (5·21) and (5·22) to f and g respectively, replacing there, as we may, c_n^* by $|c_n|$. Thus

$$(\Sigma\,|\,c_n n^{-\lambda}\,|^2)^{\frac{1}{2}}\leqslant A_p M^{(2-p)/p}\|f\|_p \quad \left(\lambda=\frac{1}{p}-\frac{1}{2}\right),$$
$$\|g\|_q\leqslant A_q M^{(q-2)/q}(\Sigma\,|\,c_n n^{-(\alpha+\mu)}\,|^2)^{\frac{1}{2}} \quad \left(\mu=\frac{1}{q}-\frac{1}{2}\right),$$

remembering that $|\epsilon_n|\leqslant 1$. These inequalities imply (5·25) if $\mu+\alpha=\lambda$, that is, if $\alpha=1/p-1/q$.

Remark. It is easy to see that (5·24) fails in the cases $p<q<2$ and $2<p<q$ (which are equivalent). Suppose that $2<p<q$, that

$$f(x)\sim\Sigma\eta_n\frac{\cos nx}{n^{\frac{1}{2}}\log n}$$

in $(0,\pi)$, and that $\eta_n=\pm 1$. Choosing for $\{\eta_n\}$ a suitable sequence of ± 1, we have $f\in L^p$ for all $p>2$ (Chapter V, (8·16)). Taking $\epsilon_n=\eta_n$, and for α any positive number less than $\frac{1}{2}$, we have

$$g(x)\sim\Sigma\frac{\cos nx}{n^{\frac{1}{2}+\alpha}\log n}.$$

By Chapter V, (2·6), $g(x)$ is exactly of order $x^{-\frac{1}{2}+\alpha}/\log(1/x)$ near $x=0$, and so is not in L^q if $\frac{1}{2}-\alpha>1/q$, and in particular if $\alpha=1/p-1/q$.

6. Theorems of Hardy and Littlewood about rearrangements of Fourier coefficients

The theorems of the previous section, when applied to the orthogonal system $\{e^{inx}\}$, can be stated in a different form and give the solution of an interesting problem. It will be convenient to change the notation slightly.

Given a sequence $c_0, c_1, c_{-1}, c_2, c_{-2}, \ldots$ tending to 0, let $c_0^*\geqslant c_1^*\geqslant c_{-1}^*\geqslant\ldots$ be the sequence $|c_0|, |c_1|, |c_{-1}|, \ldots$ rearranged in descending order of magnitude. Similarly, given a function $f(x)$, $-\pi \leqslant x \leqslant +\pi$, we denote by $f^*(x)$, $-\pi \leqslant x \leqslant \pi$, the function which is even, equimeasurable with $|f(x)|$, and non-increasing in $(0, \pi)$. For $0 \leqslant x \leqslant \pi$, $f^*(x)$ may be defined as the function inverse to $\frac{1}{2}|E(|f| > y)|$. We set

$$\mathfrak{B}_r[c] = \left\{ \sum_{-\infty}^{+\infty} |c_n|^r (|n|+1)^{r-2} \right\}^{1/r}, \tag{6·1}$$

$$\mathfrak{U}_r[f] = \left\{ \int_{-\pi}^{\pi} |f|^r |x|^{r-2} dx \right\}^{1/r}. \tag{6·2}$$

If for the moment we denote the sequence $c_0^*, c_1^*, c_{-1}^*, \ldots$ by $d_1, d_2, d_3, \ldots$, then the ratio

$$\sum_{-\infty}^{\infty} c_n^{*r} (|n|+1)^{r-2} \Big/ \sum_{1}^{\infty} d_n^r n^{r-2}$$

is contained between two positive numbers depending only on r. Hence we see that Theorem (5·10) remains true for the system $1, e^{ix}, e^{-ix}, \ldots$ if $\mathfrak{B}_r$ is given by (6·1). Similarly, Theorem (5·15) holds for this system if the interval $(0, h)$ is replaced by $(-\pi, \pi)$ and $\mathfrak{U}_r$ is defined by (6·2). In the rest of this section we adopt the definitions (6·1) and (6·2).

We know that a necessary and sufficient condition for the numbers $c_0, c_1, c_{-1}, \ldots$ to be the Fourier coefficients of an $f \in L^2$ is that $\Sigma |c_n|^2 < \infty$. This condition bears on the moduli of the c_n only. Hence a necessary and sufficient condition that the complex numbers $c_0, c_1, c_{-1}, \ldots$ should be, *for every variation of their arguments*, the Fourier coefficients of a quadratically integrable function, is again $\Sigma |c_n|^2 < \infty$. One may ask if anything similar holds for other classes L^r. The answer is always negative; for consider the two series

$$\sum_{n=1}^{\infty} n^{-\alpha} e^{inx}, \quad \sum_{n=1}^{\infty} \pm n^{-\alpha} e^{inx} \quad (0 < \alpha < 1).$$

If e.g. $\alpha = \frac{3}{4}$, the first series belongs to L^q for $q < 4$ only (Chapter V, (2·1)), while the second belongs, for a suitable sequence of signs, to every L^q (Chapter V, (8·16)); thus two functions, one of which belongs to L^q and the other does not, can have the same $|c_n|$. If $\alpha = \frac{1}{4}$, the first series belongs to L^p, $p < \frac{4}{3}$, while the second need not be a Fourier series.

These facts suggest a change in the problem. We shall now vary not only the arguments of the c_n but also their order, and we ask when the new sequences will be those of Fourier coefficients, with respect to the system $1, e^{ix}, e^{-ix}, \ldots$, of functions belonging to L^r. The results which follow are due to Hardy and Littlewood.

(6·3) THEOREM. (i) *A necessary and sufficient condition that numbers $c_n \to 0$ should be, for every variation of their arguments and arrangement, the Fourier coefficients of a function $f \in L^q$, is that $\mathfrak{B}_q^*[c] < \infty$. If the condition is satisfied, then*

$$\mathfrak{M}_q[f] \leqslant A_q' \mathfrak{B}_q^*[c] \tag{6·4}$$

for every such f.

(ii) *A necessary and sufficient condition that the c_n should be, for some variation of their arguments and arrangement, the Fourier coefficients of an $f \in L^p$, is that $\mathfrak{B}_p^*[c] < \infty$. Moreover, we have*

$$\mathfrak{B}_p^*[c] \leqslant A_p \mathfrak{M}_p[f] \tag{6·5}$$

for every such f.

The proof is based on the following lemma:

(6·6) LEMMA. (i) *If* $a_1 \geqslant a_2 \geqslant \ldots \to 0$, *a necessary and sufficient condition that the function* $g(x) = \Sigma a_n \cos nx$ *should belong to* L^r, $r > 1$, *is that the sum* $S_r = \Sigma a_n^r n^{r-2}$ *should be finite.*

(ii) *The result holds also for sine series.*

Let $G(x)$ and $H(x)$ denote respectively the integrals of g and $|g|$ over the interval $(0, x)$. Let $A_n = a_1 + a_2 + \ldots + a_n$. By B we shall mean a constant depending at most on r, but not necessarily always the same. If $g \in L$, in particular if $g \in L^r$, the series defining g is $S[g]$ (see Chapter V, § 1) and so

$$G(x) = \int_0^x g\,dt = \sum_{n=1}^{\infty} \frac{a_n}{n} \sin nx,$$

$$G\left(\frac{\pi}{n}\right) = \sum_{m=1}^{n-1} \left(\frac{a_m}{m} - \frac{a_{m+n}}{m+n} + \frac{a_{m+2n}}{m+2n} - \ldots\right) \sin \frac{m\pi}{n} \geqslant \sum_{m=1}^{n-1} \left(\frac{a_m}{m} - \frac{a_{m+n}}{m+n}\right) \sin \frac{m\pi}{n}$$

$$\geqslant B \sum_{[n/3]+1}^{[2n/3]} \left(\frac{a_m}{m} - \frac{a_{m+n}}{m+n}\right) \geqslant B \sum_{[n/3]+1}^{[2n/3]} \frac{a_m}{m}$$

$$\geqslant B a_n,$$

$$\sum_2^{\infty} a_n^r n^{r-2} \leqslant B \sum_2^{\infty} n^{r-2} G^r\left(\frac{\pi}{n}\right) \leqslant B \sum_2^{\infty} n^{r-2} H^r\left(\frac{\pi}{n}\right)$$

$$\leqslant B \sum_2^{\infty} \int_{\pi/n}^{\pi/(n-1)} \left\{\frac{H(x)}{x}\right\}^r dx = B \int_0^{\pi} \left\{\frac{H(x)}{x}\right\}^r dx$$

$$\leqslant B \int_0^{\pi} |g|^r dx,$$

by Chapter I, (9·16). This establishes the necessity of the condition in (6·6) (i).

To show that the condition is sufficient we observe that

$$|g(x)| \leqslant \sum_1^n a_\nu + \left| \sum_{n+1}^{\infty} a_\nu \cos \nu x \right| \leqslant A_n + \pi a_n / x$$

(Chapter I, (2·3)). It follows that $|g(x)| \leqslant BA_n$ for $\pi/(n+1) \leqslant x \leqslant \pi/n$. Hence

$$\int_0^{\pi} |g|^r dx = \sum_1^{\infty} \int_{\pi/(n+1)}^{\pi/n} |g|^r dx \leqslant B \sum A_n^r n^{-2}, \tag{6·7}$$

and it remains to show that the last series converges if $S_r < \infty$.

Let $a(x)$ denote the function equal to a_n for $n-1 \leqslant x < n$ $(n = 1, 2, \ldots)$, and $A(x)$ the integral of $a(t)$ over $(0, x)$. The inequality $S_r < \infty$ implies that $a^r(x)\, x^{r-2}$ is integrable over $(0, \infty)$. So (by Chapter I, (9·16), with $s = r-2$) is the function

$$\{A(x)/x\}^r x^{r-2} = A^r(x)\, x^{-2}.$$

Since the integrability of the latter function is equivalent to the convergence of $\Sigma A_n^r n^{-2}$, Lemma (6·6) (i) follows.

Lemma (6·6) (ii) can be obtained by a similar argument, or, even more simply, deduced from part (i) by using Theorem (2·4) of Chapter VII.

We are now in a position to prove (6·3). That the condition of (i) is sufficient follows from Theorem (5·10) (ii), whence also we can deduce the inequality (6·4). To prove the necessity of the condition, consider the two series $\Sigma c_n^* e^{inx}$ and $\Sigma c_{-n}^* e^{inx}$. Since both of them belong to L^q, so does their sum

$$\sum_{-\infty}^{+\infty} (c_n^* + c_{-n}^*) e^{inx} = 2\left[c_0^* + \sum_1^{\infty} (c_n^* + c_{-n}^*) \cos nx \right],$$

and from (6·6) (i) we see that $\mathfrak{B}_q^*[c] < \infty$.

From (5·10) (i) we see that the condition of Theorem (6·3) (ii) is necessary. That it is also sufficient follows from the fact that $\Sigma c_n^* e^{inx}$ belongs to L^p if $\mathfrak{B}_p^*[c] < \infty$.

The following theorem, in which we consider the rearrangements not of the coefficients but of the function, is a dual of (6·3).

(6·8) THEOREM. (i) *A necessary and sufficient condition that $\| c \|_q$ should be finite for all $f(x)$ having the same $f^*(x)$ is that $\mathfrak{U}_q^*[f]$ should be finite. If $\mathfrak{U}_q^*[f]$ is finite, then*

$$\| c \|_q \leqslant A_q' \mathfrak{U}_q^*[f]. \tag{6·9}$$

(ii) *A necessary and sufficient condition that $\| c \|_p$ should be finite for some $f(x)$ with a given $f^*(x)$ is that $\mathfrak{U}_p^*[f] < \infty$. If this condition is satisfied, then*

$$\mathfrak{U}_p^*[f] \leqslant A_p \| c \|_p \tag{6·10}$$

The proof of (6·8) is similar to that of (6·3), and indeed is slightly easier since $f^*(x)$, unlike c_n^*, is an even function of the argument. The only thing we need is the following lemma.

(6·11) LEMMA. *If a function $g(x)$, $|x| \leqslant \pi$, is non-negative, even, and decreasing in $(0, \pi)$, and if a_n are the cosine coefficients of g, then a necessary and sufficient condition that $\| a \|_r < \infty$, $r > 1$, is that the function $g^r(x) x^{r-2}$ should be integrable.*

We shall only sketch the proof, which follows the same lines as that of (6·6) (i). Denoting by $G(x)$ the integral of g over $(0, x)$, we show that

$$|a_n| \leqslant 2G(\pi/n), \quad A_n \geqslant Bg(\pi/n), \tag{6·12}$$

where $A_n = |a_1| + |a_2| + \ldots + |a_n|$. The first inequality follows from the formula

$$\tfrac{1}{2}\pi a_n = \int_0^{\pi/n} g(x) \cos nx \, dx + \int_{\pi/n}^{\pi} g(x) \cos nx \, dx.$$

For the modulus of the first term on the right is at most $G(\pi/n)$, and, by the second mean-value theorem, the second term on the right is numerically at most

$$g(\pi/n) \, . \, 2/n \leqslant G(\pi/n).$$

To prove the second inequality (6·12), we observe that $\frac{1}{2}a_0 + a_1 + \ldots + \frac{1}{2}a_n$ is equal to

$$\frac{2}{\pi}\int_0^{\pi} g(t) \frac{\sin nt}{2\tan\frac{1}{2}t} dt \geqslant \frac{2}{\pi}\int_0^{\pi/n} \left[\frac{g(t)}{2\tan\frac{1}{2}t} - \frac{g(t+\pi/n)}{2\tan\frac{1}{2}(t+\pi/n)}\right] \sin nt \, dt$$

$$\geqslant B\int_0^{\pi/2n} \frac{g(t)}{2\tan\frac{1}{2}t} \sin nt \, dt \geqslant B\int_0^{\pi/2n} \frac{g(t)}{t} \sin nt \, dt \geqslant Bg(\pi/2n) \geqslant Bg(\pi/n).$$

We now observe that if $g^r x^{r-2}$ is integrable, so is $G^r x^{-2}$ (Chapter I, (9·16)); thus $\Sigma G^r(\pi/n) < \infty$ and, in view of the first inequality (6·12), $\| a \|_r < \infty$. Conversely, if

$\| a \|_r < \infty$, then $\Sigma\{A_n/n\}^r < \infty$ (an easy consequence of Chapter I, (9·16), with $s=0$) and the second inequality in (6·12) give $\Sigma n^{-r} g^r(\pi/n) < \infty$. The latter inequality is equivalent to the integrability of $g^r x^{r-2}$.

7. Lacunary coefficients

We know that a necessary condition for a sequence $\{a_n, b_n\}$ to be that of the Fourier coefficients of an integrable function f is that $|a_n|+|b_n| \to 0$. If a_n, b_n are to be the coefficients of a continuous f, the series $\Sigma(a_n^2+b_n^2)$ must converge. Neither condition is sufficient, but we shall prove that, for some indices n at least, the Fourier coefficients of integrable, or continuous, functions may be prescribed, roughly speaking, arbitrarily.

(7·1) THEOREM. *Let $\{n_j\}$ be a sequence of positive integers such that*

$$n_{j+1}/n_j > \lambda > 1$$

for $j=1, 2, \ldots$, and let $\{x_j, y_j\}$ be a sequence of pairs of real numbers.

(i) *If $\Sigma(x_j^2+y_j^2) < \infty$, then there is a continuous f with coefficients a_n, b_n satisfying*

$$a_{n_j} = x_j, \quad b_{n_j} = y_j \quad (j=1, 2, \ldots). \tag{7·2}$$

(ii) *If $|x_j|+|y_j| \to 0$, then there is an integrable f satisfying* (7·2).

(iii) *If x_j, y_j are bounded, then there is a continuous non-decreasing $F(x)$, $0 \leqslant x \leqslant 2\pi$, with Fourier-Stieltjes coefficients a_n, b_n satisfying* (7·2).

We begin with (iii) and suppose, as we may, that $\rho_j^2 = x_j^2 + y_j^2 \leqslant 1$ for all j. First suppose that $\lambda \geqslant 3$. We set $A_{n_j}(x) = x_j \cos n_j x + y_j \sin n_j x$ and consider the Riesz product

$$\prod_{j=1}^{\infty} \{1 + A_{n_j}(x)\} \tag{7·3}$$

(Chapter V, § 7). Since $\rho_j \leqslant 1$, $\lambda \geqslant 3$, this product, when multiplied out, is the Fourier-Stieltjes series of a continuous, non-decreasing (in general, singular) function with coefficients x_j, y_j at the places n_j.

If $\lambda > 1$, we take k so large that $\lambda^k \geqslant 3$, and split $\{n_j\}$ into k sequences: $n_1, n_{k+1}, n_{2k+1}, \ldots$; $n_2, n_{k+2}, n_{2k+2}, \ldots; \ldots; n_k, n_{2k}, n_{3k}, \ldots$. Consider the sum

$$\prod_{s=0}^{\infty} (1 + A_{n_{sk+1}}) + \prod_{s=0}^{\infty} (1 + A_{n_{sk+2}}) + \ldots + \prod_{s=0}^{\infty} (1 + A_{n_{sk+k}}) \tag{7·4}$$

which is the Fourier-Stieltjes series of a continuous non-decreasing function. We know (Chapter V, Remark (*b*) on p. 211) that if k is large enough the k series originating from the product in (7·4) have no terms in common, and so their sum satisfies the assertion in (iii).

To prove (ii), let $\{\epsilon_k\}$ be a positive *convex* sequence tending to 0 such that the sequences $\{x_j/\epsilon_{n_j}\}$ and $\{y_j/\epsilon_{n_j}\}$ are bounded. (This is possible since $|x_j|+|y_j| \to 0$.) By (iii), there is a Fourier-Stieltjes series

$$\tfrac{1}{2}a_0 + \sum_{1}^{\infty} (a_k \cos kx + b_k \sin kx)$$

such that $a_{n_j} = x_j/\epsilon_{n_j}$, $b_{n_j} = y_j/\epsilon_{n_j}$ for all j. Since multiplying the terms of a Fourier-

Stieltjes series by a convex sequence tending to 0 transforms it into a Fourier series (Chapter IV, (11·10) and Chapter V, (1·5)), the series

$$\tfrac{1}{2}a_0\epsilon_0+\sum_1^\infty \epsilon_k(a_k\cos kx+b_k\sin kx)$$

satisfies the assertion of (ii). Incidentally, the f obtained in this way from (7·3) or (7·4) is non-negative.

The proof of (i) follows the same lines. If in (7·3), or (7·4), we substitute iA_{n_j} for A_{n_j}, we obtain Fourier series of bounded functions (Chapter V, (7·12)), and the imaginary parts of these series have coefficients x_j, y_j at the places n_j. The passage to continuous functions follows the same pattern as in the proof of (ii).

(7·5) THEOREM. *Given an arbitrary function $\phi(u)$ tending to $+\infty$ with u, there exists a continuous function f having coefficients a_n, b_n such that if we set $r_n^2=a_n^2+b_n^2$ the series $\Sigma r_n^2\phi(1/r_n)$ diverges.*

Let $\{\alpha_k,\beta_k\}$ be an arbitrary sequence such that $\Sigma\rho_k^2<\infty$, $\Sigma\rho_k^2\phi(1/\rho_k)=\infty$, where $\rho_k^2=\alpha_k^2+\beta_k^2$. By (i), there is a continuous f such that $a_{2^k}=\alpha_k$, $b_{2^k}=\beta_k$. Since $\Sigma\rho_k^2\phi(1/\rho_k)$ diverges, so does $\Sigma r_k^2\phi(1/r_k)$.

Theorem (4·11) of Chapter V is a corollary of (7·5) (take, for example, $\phi(u)=\log u$).

We know that the Fourier coefficients of an integrable f can tend to 0 arbitrarily slowly (Chapter V, p. 184). This is no longer true for $f\in L^r$, $r>1$, if only because then $\|c\|_{r'}<\infty$ if $r\leqslant 2$. The latter result can be strengthened if we restrict ourselves to lacunary coefficients.

(7·6) THEOREM. *Suppose that $n_{j+1}/n_j>\lambda>1$ for all j. If a_n, b_n are the coefficients of an $f\in L^r$, $r>1$, the series $\Sigma(a_{n_j}^2+b_{n_j}^2)$ converges. The result holds if merely $|f|(\log^+|f|)^{\frac{1}{2}}$ is integrable.*

We fix $N>0$. For a suitable sequence $\alpha_1,\beta_1,\ldots,\alpha_N,\beta_N$ with $\Sigma(\alpha_j^2+\beta_j^2)=1$, we have

$$\left\{\sum_{j=1}^N(a_{n_j}^2+b_{n_j}^2)\right\}^{\frac{1}{2}}=\sum_{j=1}^N(a_{n_j}\alpha_j+b_{n_j}\beta_j)=\frac{1}{\pi}\int_0^{2\pi}fg\,dt, \tag{7·7}$$

where $g=\Sigma(\alpha_j\cos n_jt+\beta_j\sin n_jt)$. By Theorem (8·20) of Chapter V, there are positive constants γ, δ depending on λ only such that $\int_0^{2\pi}e^{\gamma g^2}dx\leqslant\delta$. Let $\Phi(u)=e^{\gamma u^2}-\gamma u^2-1$. The functions Φ and Φ' vanish for $u=0$, and Φ' is strictly increasing for $u\geqslant 0$. Hence $\Phi(u)$ is a Young function (Chapter I, § 9). Its complementary function $\Psi(v)$, as is easily seen, is $O(v\log^{\frac{1}{2}}v)$ for $v\to\infty$. In other words,

$$\Psi(v)\leqslant A_\lambda v(\log^+v)^{\frac{1}{2}}+B_\lambda$$

for $v\geqslant 0$. Young's inequality now shows that the last term in (7·7) is not greater than

$$\frac{1}{\pi}\int_0^{2\pi}\{\Phi(|g|)+\Psi(|f|)\}\,dt\leqslant\int_0^{2\pi}e^{\gamma g^2}dt+A_\lambda\int_0^{2\pi}|f|(\log^+|f|)^{\frac{1}{2}}dt+2B_\lambda,$$

which gives

$$\left\{\sum_{j=1}^N(a_{n_j}^2+b_{n_j}^2)\right\}^{\frac{1}{2}}\leqslant A_\lambda\int_0^{2\pi}|f|(\log^+|f|)^{\frac{1}{2}}dt+A'_\lambda,$$

with $A'_\lambda=2B_\lambda+\delta$. The inequality holds if we replace N by ∞, and so (7·6) is established.

For power series of type H we have a similar result which is neither more nor less general than (7·6).

(7·8) Theorem. *If* $n_{j+1}/n_j > \lambda > 1$ *and* $F(z) = \sum_0^\infty \alpha_n z^n \in \mathrm{H}$, *then* $\sum |\alpha_{n_j}|^2$ *converges.*

By Chapter VII, (7·22), we have $F = F_1 F_2$, where both

$$F_1(z) = \sum_0^\infty \beta_n z^n \quad \text{and} \quad F_2(z) = \sum_0^\infty \gamma_n z^n$$

are in H^2. Write $\sum |\beta_n|^2 = B^2$, $\sum |\gamma_n|^2 = C^2$. Then

$$|\alpha_{n_j}| = \left| \sum_{k=0}^{n_j} \beta_k \gamma_{n_j-k} \right| \leqslant \sum_{k=0}^{n_{j-1}} + \sum_{k=n_{j-1}+1}^{n_j}$$

$$\leqslant B\left(\sum_{n_j-n_{j-1}}^{n_j} |\gamma_k|^2 \right)^{\frac{1}{2}} + C\left(\sum_{n_{j-1}+1}^{n_j} |\beta_k|^2 \right)^{\frac{1}{2}},$$

by Schwarz's inequality. Hence

$$|\alpha_{n_j}|^2 \leqslant 2B^2 \sum_{n_j-n_{j-1}}^{n_j} |\gamma_k|^2 + 2C^2 \sum_{n_{j-1}+1}^{n_j} |\beta_k|^2.$$

Sum these inequalities for $j = 1, 2, \ldots$. Since $n_j - n_{j-1} > (1-\lambda^{-1})\, n_j$, there is a constant $K = K_\lambda$ such that each k belongs to at most K intervals $(n_j - n_{j-1}, n_j)$. It follows that

$$\sum |\alpha_{n_j}|^2 \leqslant 2B^2 K \sum |\gamma_k|^2 + 2C^2 \sum |\beta_k|^2 = 2(K+1)\, B^2 C^2.$$

8. Fractional integration

Suppose that $f(x)$ is integrable in an interval (a, b). Denote by $F_1(x)$ the integral of f over (a, x), and by $F_\alpha(x)$ the integral of $F_{\alpha-1}$ over (a, x), $\alpha = 2, 3, \ldots$. A classical formula, easily verifiable by induction, gives

$$F_\alpha(x) = \frac{1}{\Gamma(\alpha)} \int_a^x (x-t)^{\alpha-1} f(t)\, dt \quad (a \leqslant x \leqslant b), \tag{8·1}$$

where $\Gamma(\alpha) = (\alpha-1)!$. If $\Gamma(\alpha)$ is Euler's gamma function, the formula (8·1) may be taken as a definition of $F_\alpha(x)$ for every $\alpha > 0$.

Let $g(u) = u^{\alpha-1}/\Gamma(\alpha)$ for $u > 0$ and $g(u) = 0$ elsewhere. Then

$$F_\alpha(x) = \int_a^b f(t)\, g(x-t)\, dt$$

is the convolution of the integrable functions f and g, and so exists for almost all x and is itself integrable (Chapter II, (1·5)). If $\alpha > 1$, $F_\alpha(x)$ is even continuous (a result also valid for $\alpha = 1$), since g is then continuous.

This definition of fractional integral is due to Riemann and Liouville. In the theory of trigonometric series it is not entirely satisfactory, since in general $F_\alpha(x)$ is not periodic even if f is. Moreover, it makes $F_\alpha(x)$ depend on a particular choice of a. For this reason we shall consider another definition, introduced by Weyl and more convenient for trigonometric series.

Let $f(x)$ be an integrable function of period 2π. *We suppose once for all that the integral of f over* $(0, 2\pi)$ *is* 0, so that the constant term of $S[f]$ is 0. It follows that the integral

f_1 of f is also periodic, whatever the constant of integration. If we choose this constant of integration in such a way that the integral of f_1 over $(0, 2\pi)$ is 0—in other words, if the constant term of $S[f_1]$ is 0—then the integral f_2 of f_1 will also be periodic, and so on. Generally, having defined $f_1, f_2, \dots, f_{\alpha-1}$, we choose for f_α that primitive of $f_{\alpha-1}$ whose integral over $(0, 2\pi)$ is 0. In other words, if

$$f(x) \sim \Sigma c_n e^{inx}, \quad c_0 = 0, \tag{8·2}$$

then
$$f_\alpha(x) \sim \Sigma c_n \frac{e^{inx}}{(in)^\alpha} = \frac{1}{2\pi}\int_0^{2\pi} f(t)\,\Psi_\alpha(x-t)\,dt, \tag{8·3}$$

where
$$\Psi_\alpha(t) = \Sigma' \frac{e^{int}}{(in)^\alpha} = \Sigma \gamma_n^{(\alpha)} e^{int}, \tag{8·4}$$

say. The function $\Psi_\alpha(t)$ was already considered in Chapter II, §2, where it was denoted by $B_\alpha(t)$. For $0 < t < 2\pi$, it is a polynomial of degree α.

The formula (8·3) may be considered as a definition of f_α for every $\alpha > 0$, provided that in (8·4) we set

$$\gamma_n^{(\alpha)} = (in)^{-\alpha} = |\,n\,|^{-\alpha} \exp(-\tfrac{1}{2}\pi i\alpha \operatorname{sign} n) \quad \text{for} \quad n \neq 0,\ \gamma_0^{(\alpha)} = 0. \tag{8·5}$$

The series (8·4) can then also be written

$$2\cos\tfrac{1}{2}\pi\alpha \sum_1^\infty \frac{\cos nt}{n^\alpha} + 2\sin\tfrac{1}{2}\pi\alpha \sum_1^\infty \frac{\sin nt}{n^\alpha}. \tag{8·6}$$

It follows from Chapter I, (2·4), that this converges for $t \neq 0$ to a sum $\Psi_\alpha(t)$, and from Chapter V, (1·5) and (1·14), that it is $S[\Psi_\alpha]$. Hence the integral (8·3) exists almost everywhere and its value $f_\alpha(x)$ is integrable. By Chapter III, §4, the series in (8·3) converges almost everywhere and is $S[f_\alpha]$.

The series conjugate to (8·6) converges to an integrable sum $\tilde{\Psi}_\alpha(t)$, and is $S[\tilde{\Psi}_\alpha]$. It follows that the conjugate of the series in (8·3) is a convolution of $S[f]$ and $S[\tilde{\Psi}_\alpha]$, and so is also a Fourier series. It converges almost everywhere.

Denote $f_\alpha(x)$ by $I_\alpha[f]$. The first equation (8·3) gives

$$I_\beta[I_\alpha[f]] = I_{\alpha+\beta}[f] \quad (\alpha, \beta > 0).$$

Since f_α coincides for $\alpha = 1, 2, \dots$ with an ordinary integral, the case $0 < \alpha < 1$ is the most interesting one.

We now define f^α, the derivative of f of fractional order α. Supposing first that $0 < \alpha < 1$, we set

$$f^\alpha(x) = \frac{d}{dx} f_{1-\alpha}(x) \quad (0 < \alpha < 1).$$

Consider a special case, namely when $f_{1-\alpha}(x)$ is absolutely continuous. Then

$$S[f^\alpha] = S[f'_{1-\alpha}] = S'[f_{1-\alpha}] = \Sigma (in)^\alpha c_n e^{inx} = \Sigma \gamma_n^{(-\alpha)} c_n e^{inx}. \tag{8·7}$$

If $\alpha > 0$ is arbitrary and n is the least integer greater than α, it is customary to define f^α by the formula

$$f^\alpha = \frac{d^n}{dx^n} f_{n-\alpha}(x).$$

We shall not, however, use this general definition, since we are concerned only with the case $0 < \alpha < 1$.

The function with Fourier coefficients (8·5) was discussed in Chapter II, § 13. From Chapter II, (13·8) and (13·9) we see that

$$\Gamma(\alpha)\,\Psi_\alpha(x)=2\pi\lim_{n\to\infty}\left\{x^{\alpha-1}+(x+2\pi)^{\alpha-1}+\ldots+(x+2\pi n)^{\alpha-1}-(2\pi)^{\alpha-1}\frac{n^\alpha}{\alpha}\right\} \tag{8·8}$$

for $0<\alpha<1$, $0<x<2\pi$.

If we drop the term $x^{\alpha-1}$ on the right of (8·8), the resulting expression will converge uniformly for $x\geq 0$. Hence, since f is periodic and its integral vanishes over a period, we see that

$$\frac{1}{2\pi}\int_0^{2\pi} f(x-t)\,\Psi_\alpha(t)\,dt=\lim_{n\to\infty}\frac{1}{\Gamma(\alpha)}\int_0^{2\pi} f(x-t)\sum_0^n(t+2\pi\nu)^{\alpha-1}dt$$

$$=\frac{1}{\Gamma(\alpha)}\int_0^\infty f(x-t)\,t^{\alpha-1}\,dt,$$

or

$$f_\alpha(x)=\frac{1}{\Gamma(\alpha)}\int_{-\infty}^x f(t)\,(x-t)^{\alpha-1}\,dt, \tag{8·9}$$

It thus turns out that the new definition of fractional integral differs from (8·1) only in that now $a=-\infty$. It must be stressed, however, that the convergence of the integral (8·9) is bound up with the vanishing of the integral of f over a period.

Denote by $\Gamma(\alpha)\,r_\alpha(x)$ the expression resulting from the omission of $x^{\alpha-1}$ in (8·8). Clearly $r_\alpha(x)$ has derivatives of all orders for $x>-2\pi$ (and is actually regular there). Since Ψ_α is periodic we deduce from (8·8) that, for $-2\pi<x<2\pi$,

$$\Psi_\alpha(x)=\psi_\alpha(x)+r_\alpha(x), \tag{8·10}$$

where $\psi_\alpha(x)$ is the function equal to $2\pi x^{\alpha-1}/\Gamma(\alpha)$ for $x>0$ and to 0 for $x<0$. Hence

$$f_\alpha(x)=\frac{1}{\Gamma(\alpha)}\int_0^x f(t)\,(x-t)^{\alpha-1}\,dt+\frac{1}{2\pi}\int_0^{2\pi} f(t)\,r_\alpha(x-t)\,dt$$

for $0<x<2\pi$. In this range the last function has derivatives of all orders, and we see that, from the point of view of differentiation properties, the definition (8·9) is not essentially different from (8·1) with $a=0$, provided x is in the interior of $(0,2\pi)$.

For some problems we may replace the series (8·3) by a somewhat simpler expression. Suppose for simplicity that

$$f(x)\sim\sum_1^\infty(a_n\cos nx+b_n\sin nx)=\sum_1^\infty A_n(x)$$

is real-valued. Then

$$f_\alpha(x)=\cos\tfrac{1}{2}\pi\alpha\,\Sigma n^{-\alpha}A_n(x)+\sin\tfrac{1}{2}\pi\alpha\,\Sigma n^{-\alpha}B_n(x), \tag{8·11}$$

by (8·6). Suppose also, to take an example, that we want to prove that $f_\alpha\in\Lambda_\beta$, $0<\beta<1$. In this case it is the same thing to show that the function

$$\sum_1^\infty n^{-\alpha}(a_n\cos nx+b_n\sin nx) \tag{8·12}$$

belongs to Λ_β. For if it does, so does the conjugate function (see Chapter III, (13·29)), and so also, by (8·11), does f_α. Conversely, if $f_\alpha\in\Lambda_\beta$, then also $\tilde f_\alpha\in\Lambda_\beta$; and writing for $\tilde f_\alpha$ an equation analogous to (8·11), we deduce from these two that (8·12) is in Λ_β.

In this argument we might consider instead of Λ_β any class K of functions such that $\tilde{f}$ belongs to K whenever f does. We might take for K the class L^r, $1<r<\infty$, the class Λ_*, etc. Thus it is sometimes enough to study the series (8·12) instead of (8·11).

(8·13) THEOREM. *Let* $0 \leqslant \alpha < 1$, $\beta > 0$, *and suppose that* $f \in \Lambda_\alpha$. *Then* (i) $f_\beta \in \Lambda_{\alpha+\beta}$ *if* $\alpha+\beta<1$; (ii) $f_\beta \in \Lambda_*$ *if* $\alpha+\beta=1$.

(8·14) THEOREM. *Let* $0<\gamma<\alpha<1$. *Then* (i) $f^\gamma \in \Lambda_{\alpha-\gamma}$ *if* $f \in \Lambda_\alpha$; (ii) $f^\gamma \in \Lambda_{1-\gamma}$ *if* $f \in \Lambda_*$.

These theorems clearly show the effect of fractional integration and differentiation upon the Lipschitz character of the function. They also show that for these operations the class Λ_* is more natural than Λ_1. The corresponding results with λ for Λ throughout are also true and can be proved in the same way (they are needed for (9·1) below).

In the proof we need the inequalities

$$|\Psi_\alpha(t)| \leqslant C_\alpha |t|^{\alpha-1}, \quad |\Psi'_\alpha(t)| \leqslant C_\alpha |t|^{\alpha-2}, \quad |\Psi''_\alpha(t)| \leqslant C_\alpha |t|^{\alpha-3}, \tag{8·15}$$

valid for $0<|t|\leqslant\pi$. Differentiation here is with respect to t, and C_α depends on α only. These inequalities are consequences of (8·10) and the properties of r_α.

We begin with (8·13). Suppose that $f \in \Lambda_\alpha$, $0<\alpha<1$, and that $0<h\leqslant\frac{1}{2}\pi$. Then

$$2\pi f_\beta(x) = \int_{-\pi}^{\pi} f(x-t)\,\Psi_\beta(t)\,dt = \int_{-\pi}^{\pi} [f(x-t)-f(x)]\,\Psi_\beta(t)\,dt,$$

$$2\pi f_\beta(x+h) = \int_{-\pi}^{\pi} [f(x+h-t)-f(x)]\,\Psi_\beta(t)\,dt = \int_{-\pi}^{\pi} [f(x-t)-f(x)]\,\Psi_\beta(t+h)\,dt,$$

$$2\pi[f_\beta(x+h)-f_\beta(x)] = \int_{-\pi}^{\pi} [f(x-t)-f(x)]\,[\Psi_\beta(t+h)-\Psi_\beta(t)]\,dt$$

$$= \int_{|t|\leqslant 2h} + \int_{2h\leqslant|t|\leqslant\pi} = A+B.$$

We have

$$|A| = \int_{-2h}^{2h} O(|t|^\alpha)\{|\Psi_\beta(t+h)|+|\Psi_\beta(t)|\}\,dt$$

$$\leqslant O(h^\alpha)\int_{-3h}^{3h} 2\,|\Psi_\beta(t)|\,dt = O(h^\alpha)\int_0^{3h} O(t^{\beta-1})\,dt = O(h^{\alpha+\beta}), \tag{8·16}$$

and by the mean-value theorem and the second inequality (8·15)

$$|B| \leqslant \int_{2h\leqslant|t|\leqslant\pi} O(|t|^\alpha)\,h\,|\Psi'_\beta(t+\theta h)|\,dt \quad (0<\theta<1),$$

$$\leqslant h\int_{2h\leqslant|t|\leqslant\pi} O(|t|^\alpha)\,O\{(|t|-h)^{\beta-2}\}\,dt = O(h)\int_{2h}^{\infty} t^{\alpha+\beta-2}\,dt = O(h^{\alpha+\beta}),$$

since $\alpha+\beta<1$. Hence $A+B=O(h^{\alpha+\beta})$, and (i) follows.

Passing to (ii), suppose that $0<\alpha<1$, $\alpha+\beta=1$ (the case $\alpha=0$, $\beta=1$ is obvious). Then

$$2\pi\{f_\beta(x+h)+f_\beta(x-h)-2f_\beta(x)\}$$

$$= \int_{-\pi}^{\pi} [f(x-t)-f(x)]\,\{\Psi_\beta(t+h)+\Psi_\beta(t-h)-2\Psi_\beta(t)\}\,dt = \int_{|t|\leqslant 2h} + \int_{2h\leqslant|t|\leqslant\pi} = A+B.$$

Arguing as in (8·16) we have $\quad A = O(h^{\alpha+\beta}) = O(h).$

Clearly $\quad |B| \leqslant \int_{2h\leqslant|t|\leqslant\pi} O(|t|^{\alpha})\, h^2\, |\Psi_{\beta}''(t+\theta h)|\, dt = O(h^2)\int_{2h}^{\infty} t^{\alpha+\beta-3}\, dt = O(h).$

Hence $A+B=O(h)$ and $f_{\beta}\in\Lambda_{*}$.

For the proof of (8·14) (i) we have to show that $g(x)=df_{1-\gamma}(x)/dx$ exists and is in $\Lambda_{\alpha-\gamma}$. Let F be the integral of f. We have

$$2\pi f_{\beta}(x) = -\int_0^{2\pi} \frac{d}{dt}\{F(x-t)-F(x)\}\,\Psi_{\beta}(t)\,dt = \int_0^{2\pi}\{F(x-t)-F(x)\}\,\Psi_{\beta}'(t)\,dt,$$

since the integrated term is 0. We replace here β by $1-\gamma$ and differentiate both sides. Then

$$g(x) = \frac{1}{2\pi}\int_{-\pi}^{\pi}\{f(x-t)-f(x)\}\,\Psi_{1-\gamma}'(t)\,dt$$

exists, since the integral on the right converges absolutely and uniformly. We have

$$2\pi\{g(x+h)-g(x)\} = \int_{-\pi}^{\pi}\Delta(x,h,t)\,\Psi_{1-\gamma}'(t)\,dt,$$

where $$\Delta = f(x+h-t)-f(x+h)-f(x-t)+f(x).$$

Clearly $\Delta=O(|t|^{\alpha})$, and regrouping terms we also find that $\Delta=O(h^{\alpha})$. Applying these estimates we find

$$2\pi\{g(x+h)-g(x)\} = \int_{|t|\leqslant h} O(|t|^{\alpha})\,O(|t|^{-\gamma-1})\,dt + \int_{h\leqslant|t|\leqslant\pi} O(h^{\alpha})\,O(|t|^{-\gamma-1})\,dt = O(h^{\alpha-\gamma}),$$

and (i) follows.

It remains to prove (ii) of (8·14). Let $P_{\alpha}(t)$ and $Q_{\alpha}(t)$ be respectively the coefficients of $\cos\frac{1}{2}\pi\alpha$ and $\sin\frac{1}{2}\pi\alpha$ in (8·6). Since $P_{\alpha}(t)$ and $Q_{\alpha}(t)$ are linear combinations with constant coefficients of $\Psi_{\alpha}(t)$ and $\Psi_{\alpha}(-t)$, it follows that P_{α} and Q_{α} satisfy inequalities analogous to (8·15).

Suppose that $f\in\Lambda_{*}$, and let F be the integral of f. In the equation

$$2\pi f_{1-\gamma}(x) = -\sin\tfrac{1}{2}\pi\gamma\int_{-\pi}^{\pi}\frac{d}{dt}F(x-t)\,P_{1-\gamma}(t)\,dt - \cos\tfrac{1}{2}\pi\gamma\int_{-\pi}^{\pi}\frac{d}{dt}F(x-t)\,Q_{1-\gamma}(t)\,dt,$$

the two integrals on the right are conjugate functions. If then we show that the last has a derivative in $\Lambda_{1-\gamma}$, the same will follow for $f_{1-\gamma}$, and (ii) will be established.

As in (i) this derivative, which we will call $g(x)$, exists and is equal to

$$\frac{1}{2\pi}\int_{-\pi}^{\pi}[f(x-t)-f(x)]\,Q_{1-\gamma}'(t)\,dt = \frac{1}{\pi}\int_0^{\pi}\phi_x(t)\,Q_{1-\gamma}'(t)\,dt,$$

where $\phi_x(t)=\frac{1}{2}\{f(x+t)+f(x-t)-2f(x)\}$. Hence

$$\pi\{g(x+h)+g(x-h)-2g(x)\} = \int_0^{\pi}\Delta(x,h,t)\,Q_{1-\gamma}'(t)\,dt,$$

where $\Delta=\phi_{x+h}(t)+\phi_{x-h}(t)-2\phi_x(t)$. Since, by hypothesis, $\phi_x(t)=O(t)$, it follows that $\Delta=O(t)$. On the other hand, regrouping the terms we find that

$$\Delta = \phi_{x+t}(h)+\phi_{x-t}(h)-2\phi_x(h) = O(h).$$

Hence

$$\pi\{g(x+h)+g(x-h)-2g(x)\} = \int_0^{h} O(t)\,O(t^{-\gamma-1})\,dt + \int_h^{\pi} O(h)\,O(t^{-\gamma-1})\,dt = O(h^{1-\gamma}).$$

By Remark (*d*) on p. 120 of Chapter III (see also footnote in Chapter II, p. 44), this implies that $g \in \Lambda_{1-\gamma}$ and the proof of (8·14) is completed.

The Weierstrass functions

$$w(x;\alpha) = \Sigma\, 2^{-n\alpha}\cos 2^n x, \quad \tilde{w}(x;\alpha) = \Sigma\, 2^{-n\alpha}\sin 2^n x$$

show that in (8·13) (ii) we cannot substitute the class Λ_1 for Λ_*. For $w(x;\alpha)$ and $\tilde{w}(x;\alpha)$ are both in Λ_α if $0<\alpha<1$, and in Λ_* if $\alpha=1$, and these results are best possible (Chapter II, (4·9)). Hence $w_{1-\alpha}(x;\alpha)$, which is a linear combination with constant coefficients of $w(x;1)$ and $\tilde{w}(x;1)$, is in Λ_* (a result which conforms to (ii) of Theorem (8·13)). That $w_{1-\alpha}(x;\alpha)$ is not in Λ_1 is clear since the coefficients of $w_{1-\alpha}(x;\alpha)$ are not $o(1/n)$.

9. Fractional integration (*cont.*)

We now investigate the effect of fractional integration on the classes L^r.

(9·1) THEOREM. *Suppose that $f \in L^r$, $1 \leqslant r < \infty$. If $1/r < \alpha < 1 + 1/r$, then $f_\alpha \in \lambda_{\alpha-1/r}$. If $\alpha = 1 + 1/r$, then $f_\alpha \in \lambda_*$.*

If $r=1$, then $1<\alpha<2$; and since the integral f_1 of f is continuous, the assertion follows from (8·13) (i), (ii) with λ for Λ throughout. Similarly, if $r>1$, it is enough to consider the case $1/r<\alpha<1$, since the remaining case is obtained by combining this particular one with (8·13) (i), (ii).

Suppose then that $r>1$, $1/r<\alpha<1$. By Hölder's inequality,

$$2\pi\,|f_\alpha(x+h)-f_\alpha(x)| = \left|\int_{-\pi}^{\pi} f(x-t)\{\Psi_\alpha(t+h)-\Psi_\alpha(t)\}\,dt\right|$$
$$\leqslant \left(\int_{-\pi}^{\pi}|f(t)|^r\,dt\right)^{1/r}\left\{\int_{-\pi}^{\pi}|\Psi_\alpha(t+h)-\Psi_\alpha(t)|^{r'}\,dt\right\}^{1/r'}. \qquad (9·2)$$

The first factor on the right can be made arbitrarily small by subtracting from f a trigonometric polynomial (for which the result is obvious). It is therefore enough to prove that the last factor is $O(h^{\alpha-1/r})$. Using (8·15) we write

$$\int_{-\pi}^{\pi}|\Psi_\alpha(t+h)-\Psi_\alpha(t)|^{r'}\,dt = \int_{|t|\leqslant 2h} + \int_{2h\leqslant|t|\leqslant\pi} = A+B,$$

where
$$A \leqslant 2^{r'-1}\int_{-2h}^{2h}(|\Psi_\alpha(t+h)|^{r'}+|\Psi_\alpha(t)|^{r'})\,dt$$
$$\leqslant 2^{r'}\int_{-3h}^{3h}|\Psi_\alpha(t)|^{r'}\,dt = \int_{-3h}^{3h}O(|t|^{(\alpha-1)r'})\,dt = O(h^{(\alpha-1)r'+1}),$$
$$B \leqslant \int_{2h\leqslant|t|\leqslant\pi} h^{r'}|\Psi'_\alpha(t+\theta h)|^{r'}\,dt$$
$$= O(h^{r'})\int_{2h}^{\infty} t^{(\alpha-2)r'}\,dt = O(h^{(\alpha-1)r'+1}),$$

so that the last factor in (9·2) is $O(h^{\alpha-1/r})$. The inequalities $(\alpha-1)r'>-1$ and $(\alpha-2)r'<-1$, which we used in estimating A and B, are equivalent to the hypothesis $1/r<\alpha<1+1/r$.

In what follows we shall use complex methods, and for this purpose a modification of the definition (8·9) will be needed.

Suppose that

$$\phi(z) = \sum_0^\infty c_n z^n, \tag{9·3}$$

where $z = \rho e^{i\theta}$, is regular for $|z| < 1$. We define the fractional integral $\Phi_\alpha(z)$ of $\phi(z)$ by the Riemann-Liouville formula (8·1), where now the integral is taken along the segment $[0, z]$. Thus for $\alpha > 0$ and $|z| < 1$ we set

$$\Phi_\alpha(z) = \frac{1}{\Gamma(\alpha)} \int_{[0,z]} (z-\zeta)^{\alpha-1} \phi(\zeta)\, d\zeta = z^\alpha \frac{z^{-1}}{\Gamma(\alpha)} \int_{[0,z]} \left(1 - \frac{\zeta}{z}\right)^{\alpha-1} \phi(\zeta)\, d\zeta$$
$$= z^\alpha \Phi_\alpha^*(z), \tag{9·4}$$

say, where $(z-\zeta)^{\alpha-1}$ and z^α denote the principal values of the powers: $z^\alpha = \exp(\alpha \log z)$ with $-\pi < \mathscr{I} \log z \leqslant \pi$. On setting $\zeta = zt$, $0 \leqslant t \leqslant 1$, and

$$\delta_n^{(\alpha)} = \frac{1}{\Gamma(\alpha)} \int_0^1 (1-t)^{\alpha-1} t^n\, dt = \frac{\Gamma(n+1)}{\Gamma(n+\alpha+1)}, \tag{9·5}$$

we see that

$$\Phi_\alpha^*(z) = \frac{1}{\Gamma(\alpha)} \int_0^1 (1-t)^{\alpha-1} \phi(zt)\, dt = \sum_0^\infty c_n \delta_n^{(\alpha)} z^n \tag{9·6}$$

is a function regular for $|z| < 1$ (since $\delta_n^{(\alpha)} = O(1)$). Thus in any case $|\Phi_\alpha(z)|$ is single-valued for $|z| < 1$. Moreover, the boundary values of $|\Phi_\alpha(z)|$ and $|\Phi_\alpha^*(z)|$ on $|z| = 1$, wherever they exist, are the same. From (9·6) we see that $\Phi_\alpha^*(z)$ has a limit on any radius of the unit circle on which $\phi(z)$ has a limit.

It is useful to observe that the numbers $\delta_n^{(\alpha)}$ are closely related to the Cesàro numbers A_n^α introduced in Chapter III, § 1; in fact

$$\delta_n^{(\alpha)} = \frac{1}{A_n^\alpha \Gamma(\alpha+1)}. \tag{9·7}$$

Thus the numbers $\delta_n^{(\alpha)}$ decrease monotonically to 0 as n increases, and are asymptotically equal to $n^{-\alpha}$. Even more precisely (see Chapter III, (1·18))

$$|\delta_n^{(\alpha)} - n^{-\alpha}| \leqslant A n^{-\alpha-1}. \tag{9·8}$$

Since fractional integration leads in any case to multivalued functions, we may consider from the start, instead of (9·3), the (in general) multivalued function

$$\phi(z) = z^\gamma \sum_0^\infty c_n z^n, \tag{9·9}$$

where, say, $\gamma > -1$, $c_0 \neq 0$. In this case (9·4) leads to

$$\Phi_\alpha(z) = z^{\alpha+\gamma} \sum_0^\infty c_n \frac{\Gamma(n+\gamma+1)}{\Gamma(n+\alpha+\gamma+1)} z^n = z^{\alpha+\gamma} \Phi_\alpha^*(z), \tag{9·10}$$

say,† a formula which immediately shows that

$$\{\Phi_\beta(z)\}_\alpha = \Phi_{\alpha+\beta}(z).$$

Since $\Phi_\alpha(z)$ is an ordinary repeated integral if $\alpha = 1, 2, \ldots$, we see that the case $0 < \alpha < 1$ is again the most interesting one.

† The fact that $\Phi_\alpha^*(z)$ has two meanings should not lead to confusion.

Suppose for the moment that $\phi(z)$ in (9·3) is of class H. Then $\Sigma c_n e^{in\theta}$ is the Fourier series of

$$\phi(e^{i\theta}) = \lim_{\rho\to 1} \phi(\rho e^{i\theta}).$$

Using the formula (1·15) of Chapter III, we verify that

$$\delta_n^{(\alpha)} - \delta_{n+1}^{(\alpha)} = \frac{\alpha}{(n+\alpha+1) A_n^\alpha \Gamma(\alpha+1)}$$

is a decreasing function of n. Hence $\{\delta_n^{(\alpha)}\}$ is a convex sequence and $\Sigma c_n \delta_n^{(\alpha)} e^{in\theta}$ is also a Fourier series (Chapter IV, (11·10), Chapter V, (1·5)). As can be seen from (9·6), it must be the Fourier series of $\Phi_\alpha^*(e^{i\theta}) = \lim_{\rho\to 1} \Phi_\alpha^*(\rho e^{i\theta})$. Thus the latter function is integrable and

$$\Phi_\alpha^*(e^{i\theta}) \sim \sum_0^\infty c_n \delta_n^{(\alpha)} e^{in\theta}. \tag{9·11}$$

If $c_0 = 0$ we may also consider the Weyl fractional integral $\phi_\alpha(\theta)$ of $\phi(e^{i\theta})$. By (8·3),

$$\phi_\alpha(\theta) \sim i^{-\alpha} \sum_1^\infty c_n n^{-\alpha} e^{in\theta} \quad (i^{-\alpha} = \exp(-\tfrac{1}{2}\pi i\alpha)); \tag{9·12}$$

and, by (9·8),

$$|i^{-\alpha}\Phi_\alpha^*(e^{i\theta}) - \phi_\alpha(\theta)| \leqslant A \sum_1^\infty n^{-1-\alpha} . \max |c_n| \leqslant A_\alpha \| \phi(e^{i\theta}) \|_1$$

$$\leqslant A_\alpha \| \phi(e^{i\theta}) \|_s \tag{9·13}$$

for any $s \geqslant 1$. Hence

$$\| i^{-\alpha}\Phi_\alpha^*(e^{i\theta}) - \phi_\alpha(\theta) \|_s \leqslant A_\alpha \| \phi(e^{i\theta}) \|_s \quad (s \geqslant 1). \tag{9·14}$$

Thus in considering the norm there is no essential difference between $\Phi_\alpha^*(e^{i\theta})$ and $\phi_\alpha(\theta)$.

We now consider the fractional integral of the function $\phi(z) = z^\gamma \phi^*(z)$ for ϕ^* in H^r, $r > 0$. The case $\gamma = 0$ is the only one of interest, but the proof requires us to consider the general case. As in Chapter VII, § 7, we write

$$\| \phi \|_r = \lim_{\rho\to 1} \left(\frac{1}{2\pi}\int_0^{2\pi} |\phi(\rho e^{i\theta})|^r d\theta\right)^{1/r} = \left(\frac{1}{2\pi}\int_0^{2\pi} |\phi^*(e^{i\theta})|^r d\theta\right)^{1/r} = \| \phi^* \|_r = \mathfrak{A}_r[\phi^*(e^{i\theta})],$$

and in order to unify notation we temporarily set

$$\| f \|_r = \mathfrak{A}_r[f]$$

for any $f(\theta) \in L^r$.

(9·15) THEOREM. *Set $\phi(z) = z^\gamma \phi^*(z)$, where $\gamma > -1$, $\phi^* \in H^r$, $r > 0$; and let*

$$s > r, \quad \alpha = 1/r - 1/s > 0.$$

Then the Φ_α^* in the equation $\Phi_\alpha(z) = z^{\alpha+\gamma}\Phi_\alpha^*(z)$ is in H^s, and

$$\| \Phi_\alpha^* \|_s \leqslant A_{\gamma,r,s} \| \phi^* \|_r. \tag{9·16}$$

Suppose first that $r = 2$. The co-factor of $c_n z^n$ in (9·10) is $\{1/\Gamma(\alpha)\} \int_0^1 \lambda^{\gamma+n}(1-\lambda)^{\alpha-1} d\lambda$. Suppose we replace γ here by 0 if γ is non-negative, and by -1 otherwise. We obtain

$\delta_n^{(\alpha)}$ and $\delta_{n-1}^{(\alpha)}$ respectively, and in either case we increase the integral. Thus the co-factor is $O(n^{-\alpha})$. Hence, applying Theorem (5·24), we get

$$\| \Phi_\alpha^* \|_{1/(\frac{1}{2}-\alpha)} \leqslant A_{\alpha,\gamma} \| \phi^* \|_2. \tag{9·17}$$

We now pass to the case $r<2$. Since every ϕ^* in H^r is a difference of two functions in H^r without zeros and with norms not exceeding $2\|\phi^*\|_r$, we may suppose that ϕ^* has no zeros. Then $\phi^* = \psi^{*2/r}$, where $\psi^* \in \mathrm{H}^2$ and $\|\psi^*\|_2^2 = \|\phi^*\|_r^r$. Denote by $N(\theta)$ the upper bound of $|\psi^*(z)|$ on the radius $\arg z = \theta$. By Chapter VII, (7·36),

$$\| N \|_2 \leqslant A \| \psi^* \|_2. \tag{9·18}$$

Since $r<2$, we have

$$\begin{aligned} |\Phi_\alpha^*(z)| &\leqslant \frac{1}{\Gamma(\alpha)} \int_0^1 (1-\lambda)^{\alpha-1} \lambda^\gamma |\psi^*(\lambda z)|^{2/r} d\lambda \\ &\leqslant \frac{1}{\Gamma(\alpha)} N^{(2-r)/r}(\theta) \int_0^1 (1-\lambda)^{\alpha-1}\lambda^\gamma |\psi^*(z\lambda)| \, d\lambda. \end{aligned} \tag{9·19}$$

The function $\psi^*(z)$ is the Poisson integral of $\psi^*(e^{i\theta})$, and so $|\psi^*(z)|$ is majorized by the Poisson integral $u(z)$ of the non-negative function $|\psi^*(e^{i\theta})|$. Let

$$\chi(z) = u(z) + iv(z), \quad \text{where} \quad v(0) = 0,$$

be the regular function with real part $u(z)$. By Parseval's formula,

$$\| \chi \|_2^2 \leqslant 2 \| \psi^* \|_2^2. \tag{9·20}$$

The right-hand side of (9·19) is increased if we replace $|\psi^*(z\lambda)|$ by $u(z\lambda)$. Hence

$$|\Phi_\alpha^*(\rho e^{i\theta})| \leqslant N^{(2-r)/r}(\theta) \, |\chi_\alpha^*(\rho e^{i\theta})|.$$

Apply here Hölder's inequality in the form

$$\| f_1 f_2 \|_{1/\beta} \leqslant \| f_1 \|_{1/\beta_1} \| f_2 \|_{1/\beta_2} \quad (\beta_1, \beta_2 > 0;\ \beta = \beta_1 + \beta_2),$$

where

$$f_1 = N^{(2-r)/r}, \quad f_2 = \chi_\alpha^*, \quad \beta_1 = (2-r)/2r, \quad \beta_2 = \tfrac{1}{2} - \alpha.$$

Using, (9·18), (9·17) (with χ for ϕ^*) and (9·20) we have

$$\begin{aligned} \| \Phi_\alpha^*(\rho e^{i\theta}) \|_{1/(\frac{1}{r}-\alpha)} &\leqslant \| N \|_2^{(2-r)/r} \| \chi_\alpha^*(\rho e^{i\theta}) \|_{1/(\frac{1}{2}-\alpha)} \leqslant (A \| \psi^* \|_2)^{(2-r)/r} A_{\alpha,\gamma} \| \chi \|_2, \\ &\leqslant (A \| \psi^* \|_2)^{(2-r)/r} A_{\alpha,\gamma} 2^{\frac{1}{2}} \| \psi^* \|_2 = A_{\alpha,\gamma,r} \| \psi^* \|_2^{2/r} = A_{\alpha,\gamma,r} \| \phi^* \|_r. \end{aligned}$$

If we compare the extreme terms and make $\rho \to 1$, we get (9·16). It must, however, be observed that in applying (9·17) we tacitly assumed that $\alpha < \frac{1}{2}$.

Suppose now that $r>2$; then $\alpha < \frac{1}{2}$. We fix $\rho < 1$. Then

$$\| \Phi_\alpha^*(\rho e^{i\theta}) \|_s = \sup_g \left| \frac{1}{2\pi} \int_0^{2\pi} \Phi_\alpha^*(\rho e^{i\theta}) \bar{g}(\theta) \, d\theta \right|, \tag{9·21}$$

for all g with $\|g\|_{s'} = 1$. Fix $g \sim \sum_{-\infty}^{+\infty} d_n e^{in\theta}$. Denoting by $\delta_n^{\alpha,\gamma}$ the co-factor of $c_n z^n$ in (9·10), we write the last integral in the form

$$\sum_0^\infty c_n \delta_n^{\alpha,\gamma} \bar{d}_n z^n = \frac{1}{2\pi} \int_0^{2\pi} \phi(e^{i\theta}) \bar{\Omega}_\alpha^*(\rho e^{i\theta}) \, d\theta,$$

where $\Omega_\alpha^*(z)$ is associated with the function

$$\omega(z)=z^\gamma\omega^*(z),\quad \omega^*(z)=\sum_0^\infty d_n z^n.$$

We now observe that $\alpha=1/r-1/s=1/s'-1/r'$ and that $1<s'<r'<2$. Applying Hölder's inequality, using (9·16) in the case already established, and applying Chapter VII, (2·28), we thus see that the last integral is numerically majorized by

$$\|\phi\|_r\,\|\Omega_\alpha^*(\rho e^{i\theta})\|_{r'}\leqslant\|\phi\|_r A_{\alpha,\gamma,r}\|\omega^*(\rho e^{i\theta})\|_{s'}$$
$$\leqslant\|\phi\|_r A_{\alpha,\gamma,r}A_{s'}\|g\|_{s'}=A_{\gamma,r,s}\|\phi\|_r.$$

Going back to (9·21) and making $\rho\to1$ we obtain (9·16), where A depends on α,γ,r,s, that is, on γ,r,s.

This completes the proof of (9·16) for $\alpha<\frac12$. The general result follows from it by inserting between $r=r_0$ and $s=r_k$ auxiliary numbers $r_1<r_2<\ldots<r_{k-1}$ such that $1/r_{j-1}-1/r_j<\frac12$, and successively applying the formula $(\Phi_\lambda)_\mu=\Phi_{\lambda+\mu}$. (Observe that the asterisks in (9·16) may be dropped.)

The following result, in which the fractional integrals are meant in the Weyl sense, is a simple corollary of (9·15).

(9·22) *Suppose that*

$$1<r<s<\infty,\quad \alpha=1/r-1/s.$$

Then if $f\in L^r$, we have $f_\alpha\in L^s$ and

$$\|f_\alpha\|_s\leqslant A_{r,s}\|f\|_r. \tag{9·23}$$

The result is false for $r=1$, but it does hold in this case if $S[f]$ is of power series type.

Suppose first that $S[f]$ is of power series type and $r\geqslant1$. Then $f(\theta)=\phi(e^{i\theta})$, with $\phi(z)\in H$. The inequality (9·23) then follows from (9·16) with $\gamma=0$ and from (9·14).

If $r>1$ and $S[f]$ is not of power series type, we may suppose that f is real-valued. Let $\phi(e^{i\theta})=f(\theta)+i\tilde f(\theta)$. Then $S[\phi(e^{i\theta})]$ is of power series type and $\|\phi(e^{i\theta})\|_r\leqslant A_r\|f\|_r$, by Chapter VII, (2·4). Thus

$$\|f_\alpha\|_s\leqslant\|\phi_\alpha\|_s\leqslant A_{r,s}\|\phi\|_r\leqslant A_{r,s}A_r\|f\|_r=A_{r,s}\|f\|_r.$$

To show that (9·22) is false for $r=1$, let

$$f(\theta)\sim\sum_2^\infty\frac{\cos n\theta}{(\log n)^{1-\alpha}}.$$

By Chapter V, (1·5), the series on the right is a Fourier series, and, by (8·11), $S[f_\alpha]$ is

$$\cos\tfrac12\pi\alpha\sum_2^\infty\frac{\cos n\theta}{n^\alpha(\log n)^{1-\alpha}}+\sin\tfrac12\pi\alpha\sum_2^\infty\frac{\sin n\theta}{n^\alpha(\log n)^{1-\alpha}}.$$

By the formulae (2·8) of Chapter V, this sum is exactly of the order of $\left(\theta\log\dfrac1\theta\right)^{\alpha-1}$ as $\theta\to+0$, and so is not in $L^s=L^{1/(1-\alpha)}$.

10. Fourier-Stieltjes coefficients

Denote by

$$c_n=c_n[dF]=\frac1{2\pi}\int_0^{2\pi}e^{-inx}\,dF(x) \tag{10·1}$$

the Fourier-Stieltjes coefficients of a function $F(x)$ of bounded variation over $0 \leqslant x \leqslant 2\pi$. The class of functions F such that $c_n \to 0$ will be denoted by R. We know that every F in R is continuous (Chapter III, (9·6)), but that the converse is false (Chapter V, (7·5); see also §11 below). We shall investigate properties of $F \in$ R, confining our attention, as we may, to real-valued F.

(10·2) THEOREM. *If $F(x)$ belongs to* R, *so does its positive, negative and absolute variation.*

Denote by $P(x)$, $N(x)$, $V(x)$ respectively the positive, negative and absolute variations of F over $(0, x)$. If $c_n \to 0$, then $F(x)$ is continuous, and so also are $P(x)$, $N(x)$, $V(x)$. Owing to the relations

$$P + N = V, \quad P - N = F(x) - F(0),$$

it is enough to prove the part of (10·2) that concerns V. Incidentally, (10·2) implies that *every F from* R *is a difference of two non-decreasing functions from* R.

Denote by b_n, b'_n the Fourier-Stieltjes coefficients of $G(x)$, $H(x)$. The inequality

$$2\pi \,|\, b_n - b'_n \,| = \left| \int_0^{2\pi} e^{-inx}\, d\{G(x) - H(x)\} \right| \leqslant \int_0^{2\pi} |\, d(G - H) \,|$$

shows that if for a given G there exist functions $H \in$ R such that the total variation of $G - H$ is arbitrarily small, then $G \in$ R.

Next we prove that if $F \in$ R, then

$$\int_0^{2\pi} \tau(x)\, e^{-inx}\, dF(x) \to 0 \tag{10·3}$$

for any step-function $\tau(x)$. The relation (10·3) is true if $\tau(x) = e^{imx}$, $m = 0, \pm 1, \ldots$, and so also when τ is any polynomial. Hence it holds for any continuous and periodic τ. If $\tau(x)$ is the characteristic function of an interval (a, b) (closed, open or half-open) and $h(x)$ is the continuous function vanishing outside $(a - \eta, b + \eta)$, equal to 1 in $(a + \eta, b - \eta)$, and linear in the remaining two intervals, then

$$\left| \int_0^{2\pi} \{\tau(x) - h(x)\}\, e^{-inx}\, dF(x) \right| \leqslant \left(\int_{a-\eta}^{a+\eta} + \int_{b-\eta}^{b+\eta} \right) |\, dF \,|$$

is small with η, and (10·3) holds again. Hence it is valid for any step function $\tau(x)$.

Take now any subdivision $0 = x_0 < x_1 < \ldots < x_k = 2\pi$ of $(0, 2\pi)$ such that

$$V(2\pi) - \sum_1^k |\, F(x_j) - F(x_{j-1}) \,| \leqslant \epsilon, \tag{10·4}$$

and let δ_j be defined as $+1$ or -1 according as $F(x_j) - F(x_{j-1})$ is $\geqslant 0$ or < 0. Set $\tau(x)$ equal to δ_j for $x_{j-1} \leqslant x < x_j$, $j = 1, \ldots, k$, and $\tau(2\pi) = \tau(2\pi - 0)$. Set

$$G(x) = \int_0^x \tau(t)\, dF(t), \quad \Delta(x) = V(x) - G(x).$$

Since
$$\int_0^{2\pi} e^{-inx}\, dG(x) = \int_0^{2\pi} e^{-inx}\, \tau(x)\, dF(x) \to 0,$$

(10·2) will follow if we show that the total variation of Δ over $(0, 2\pi)$ does not exceed ϵ.

Take now any subdivision $0 = x'_0 < x'_1 < \ldots < x'_q = 2\pi$ of $(0, 2\pi)$ containing all the points $x_0, x_1, \ldots, x_k$, and let (x'_{p-1}, x'_p) be a subinterval of (x_{j-1}, x_j). The difference

$$\begin{aligned}\Delta(x'_p) - \Delta(x'_{p-1}) &= V(x'_p) - V(x'_{p-1}) - [G(x'_p) - G(x'_{p-1})]\\ &= V(x'_p) - V(x'_{p-1}) - \delta_j[F(x'_p) - F(x'_{p-1})]\end{aligned}$$

is non-negative. Hence

$$\begin{aligned}\Sigma\,|\,\Delta(x'_p) - \Delta(x'_{p-1})\,| &= V(2\pi) - V(0) - \sum_{p,j} \delta_j\{F(x'_p) - F(x'_{p-1})\}\\ &= V(2\pi) - \sum_j |\,F(x_j) - F(x_{j-1})\,|.\end{aligned}$$

By (10·4), the total variation over $(0, 2\pi)$ of the continuous function Δ does not exceed ϵ, and (10·2) follows.

(10·5) THEOREM. *A necessary and sufficient condition for a function $F(x)$ of bounded variation over $0 \leqslant x \leqslant 2\pi$ to belong to* R *is that for the characteristic function $\chi(x)$ of each interval (a, b) (repeated* mod 2π) *we have*

$$\int_0^{2\pi} \chi(nx)\, dF(x) \to \int_0^{2\pi} dF(x) \frac{1}{2\pi}\int_0^{2\pi} \chi(x)\, dx = \frac{b-a}{2\pi}\{F(2\pi) - F(0)\}. \qquad \textbf{(10·6)}$$

For F absolutely continuous, the relation has already been proved (Chapter II, (4·15)). Theorem (10·5) asserts that $F \in \mathrm{R}$ if and only if the mass dF is distributed over $(0, 2\pi)$ with a certain homogeneity. A good illustration is provided by the Cantor-Lebesgue function constant over every interval contiguous to the Cantor ternary set (Chapter V, § 3). Denote by $\chi(x)$ the characteristic function of the middle third of $(0, 2\pi)$. Then $F(x)$ is constant on each interval of the set whose characteristic function is $\chi(3^k x)$. Hence in this case, and for $n = 3^k$, the left-hand side of (10·6) is zero, and so does not tend to the right-hand side. Hence it follows from Theorem (10·5) that the Fourier-Stieltjes coefficients of the Cantor-Lebesgue function cannot tend to zero, a fact already established (Chapter V, (3·6)).

In proving the necessity part of (10·5) we may suppose that F is non-decreasing. If $\Sigma c_m e^{imx}$ is $S[dF]$, then

$$F(x) = c_0 x + d + \sum_{m=-\infty}^{+\infty}{}' (c_m/im)\, e^{imx}.$$

The intervals where $\chi(nx) = 1$ are of the form

$$\left(\frac{u}{n} + \frac{2k\pi}{n} - \frac{\theta}{n},\ \frac{u}{n} + \frac{2k\pi}{n} + \frac{\theta}{n}\right) \quad (k = 1, 2, \ldots, n), \qquad \textbf{(10·7)}$$

2θ being the length of the interval where $\chi(x) = 1$. Since F is continuous, the left-hand side of (10·6) is

$$\begin{aligned}F_n(u, \theta) &= \sum_{k=1}^{n} \left\{F\left(\frac{u}{n} + \frac{2k\pi}{n} + \frac{\theta}{n}\right) - F\left(\frac{u}{n} + \frac{2k\pi}{n} - \frac{\theta}{n}\right)\right\}\\ &= 2c_0\theta + 2\sum_{\lambda=-\infty}^{+\infty}{}' (c_{\lambda n}/\lambda)\, e^{i\lambda u} \sin \lambda\theta. \qquad \textbf{(10·8)}\end{aligned}$$

The right-hand side here is, for fixed u, a non-decreasing function of θ, and is, except for the linear term, a Fourier series in θ. The termwise differentiated series

$$2c_0 + 2\sum_{\lambda=-\infty}^{+\infty}{}' c_{\lambda n}\, e^{i\lambda u} \cos \lambda\theta = 2c_0 + \sum_{\lambda=-\infty}^{+\infty}{}' (c_{\lambda n} e^{i\lambda u} + c_{-\lambda n}\, e^{-i\lambda u})\, e^{i\lambda\theta}$$

is $S[dF_n]$. By hypothesis, $c_m \to 0$. Hence, as $n \to \infty$, the last series converges termwise to $2c_0 + \Sigma' 0 . e^{i\lambda\theta}$. By Chapter IV, (4·22), $F_n(u, \theta)$ converges, for each u, to $2c_0\theta$, which is the right-hand side of (10·6).

Conversely, suppose that (10·6) holds for the intervals (10·7), for every u *and a single* θ, $0 < \theta < \pi$. Disregarding a denumerable set of values of u we may suppose that $\chi(nx)$ and $F(x)$ have no discontinuity in common. Then again the series on the right of (10·8) represents the left-hand side of (10·6). Consider this series *qua* Fourier series in u. By hypothesis, its sum converges (boundedly) to the constant term $2c_0\theta$ for almost all u. Hence the remaining coefficients must tend to 0, and taking $\lambda = \pm 1$ we get $c_n \to 0$.

(10·9) THEOREM. *Suppose that* $F \in \mathrm{R}$, *and denote by* $B(x)$ *any function such that the (Lebesgue-Stieltjes) integral* $J = \int_0^{2\pi} B(x)\, dF$ *exists. Then*

$$J_n = \int_0^{2\pi} e^{-inx} B(x)\, dF \to 0. \tag{10·10}$$

Given an $\epsilon > 0$ we can find a polynomial $T(x)$ such that

$$\int_0^{2\pi} | B(x) - T(x) | \, | dF | < \epsilon.$$

(This inequality is certainly true, and is a consequence of the definition of an integral, if T is a suitable *step function*. Since F is continuous, it is also true for some continuous function T, and so also for some polynomial.) Hence, setting $J'_n = \int_0^{2\pi} e^{-inx} T(x)\, dF$, we have $| J_n - J'_n | < \epsilon$, and since $J'_n \to 0$ we get (10·10).

Remark. For immediate applications we shall only need (10·9) in the case when $B(x)$ is continuous, except possibly at one point, and bounded. Then the integrals involved are all Riemann-Stieltjes integrals, and the proof is elementary.

(10·11) THEOREM. *If* $F \in \mathrm{R}$, *then*

$$I(u) = \int_0^{2\pi} e^{-iux}\, dF(x) \to 0$$

as u *tends continuously to* $\pm\infty$.

Suppose that $I(u) \neq o(1)$; then there is a sequence $\{u_k\}$ tending to infinity such that $| I(u_k) | \geq \delta > 0$. Let $u_k = n_k + \alpha_k$, where $n_k = [u_k]$, $0 \leq \alpha_k < 1$. By considering a subsequence of $\{u_k\}$, we may suppose that α_k tends to a limit α. Then

$$\lim_{k\to\infty} \left\{ I(u_k) - \int_0^{2\pi} e^{-i(n_k+\alpha)x}\, dF \right\} = 0.$$

But, by (10·9),

$$\int_0^{2\pi} e^{-inx} e^{-i\alpha x}\, dF \to 0$$

as $n \to \infty$. Hence $I(u_k) \to 0$, a contradiction.

Remark. The expression $I(u)/2\pi$ is the Fourier-Stieltjes transform of a function equal to $F(x)$ for $0 \leq x \leq 2\pi$, and equal to $F(0)$ and $F(2\pi)$ for $x < 0$ and $x > 2\pi$ respectively.

If F is absolutely continuous, then $c_n[dF]$ tends to 0. The c_n cannot tend to 0 if F is of bounded variation and has non-removable discontinuities. How rapidly can the c_n tend to 0 if F is continuous and singular? It is clear that in any case we have then $\Sigma | c_n |^2 = \infty$.

(10·12) THEOREM. *There is a monotone and singular F such that $c_n[dF] = O(n^{-\frac{1}{2}+\epsilon})$ for every $\epsilon > 0$.*

The idea of the proof will be to consider an $S[dG]$ with coefficients small 'on the average', and then by a simple mapping of the interval $(-\pi, \pi)$ onto itself to obtain an F with coefficients actually 'small'.

Write $n_k = 2^{2^k}$ and consider the Riesz product

$$\prod_{k=1}^{\infty} (1 + \cos n_k x) = 1 + \sum_1^{\infty} \alpha_\nu \cos \nu x = \sum_{-\infty}^{+\infty} \gamma_\nu e^{i\nu x}$$

(see Chapter V, § 7). The series here is an $S[dG]$, where G is increasing, continuous and singular (Chapter V, (7·5), (7·6)). We have $0 \leqslant \gamma_\nu \leqslant 1$ for each ν. For any integer $N > 3$ there is a k such that $n_{k-1} \leqslant N < n_k$. Hence, if $\Pi_k(x)$ is the kth partial product and if $\mu_k = n_k + n_{k-1} + n_{k-2} + \dots$, we have

$$\sum_{-N}^{N} \gamma_\nu \leqslant \sum_{-\mu_k}^{\mu_k} \gamma_\nu = \Pi_k(0) = 2^k \leqslant C \log N,$$

and comparing the extreme terms we see that the γ_ν are small 'on the average'.

Consider now the one-one mapping

$$x = x(t) = \frac{1}{2}\left(t + \frac{t^2}{\pi} \operatorname{sign} t\right) \quad (-\pi \leqslant t \leqslant \pi),$$

of the interval $(-\pi, \pi)$ onto itself, and set $F(x) = G(t)$. Since $x'(t)$ is contained between two positive bounds, $F(x)$ is increasing, continuous and singular. Moreover, by Chapter IV, (8·7),

$$c_n[dF] = \frac{1}{2\pi}\int_{-\pi}^{\pi} e^{-inx}\, dF(x) = \frac{1}{2\pi}\int_{-\pi}^{\pi} e^{-inx(t)}\, dG(t)$$
$$= \sum_{\nu=-\infty}^{+\infty} \lambda_{n,\nu} \gamma_\nu.$$

The series here converges absolutely since the function $e^{-inx(t)}$ has a derivative of bounded variation, and so its Fourier coefficients

$$\lambda_{n,\nu} = \frac{1}{2\pi}\int_{-\pi}^{\pi} e^{-i(nx(t)+\nu t)}\, dt$$

are $O(1/\nu^2)$.

Suppose that $n > 0$. We shall prove in a moment the inequalities

$$|\lambda_{n,\nu}| \leqslant A n^{-\frac{1}{2}} \quad \text{for all } \nu, \tag{10·13}$$

$$|\lambda_{n,\nu}| \leqslant A \nu^{-2} \quad \text{for } |\nu| \geqslant 3n, \tag{10·14}$$

where A is independent of ν and n. Taking these for granted we have

$$|c_n[dF]| \leqslant \sum_{\nu=-\infty}^{+\infty} |\lambda_{n,\nu}| \gamma_\nu = \sum_{|\nu| \leqslant 3n} + \sum_{|\nu| > 3n} = P + Q,$$

$$P \leqslant A n^{-\frac{1}{2}} \sum_{-3n}^{3n} \gamma_\nu \leqslant A n^{-\frac{1}{2}} . C \log 3n = O(n^{-\frac{1}{2}} \log n),$$

$$Q \leqslant A \sum_{|\nu| > 3n} \nu^{-2} . 1 = O(1/n).$$

Hence $$P+Q=O(n^{-\frac{1}{2}}\log n)+O(1/n)=O(n^{-\frac{1}{2}+\epsilon})$$

for each $\epsilon>0$, which proves the theorem.

It remains, however, to prove (10·13) and (10·14). The first inequality follows from Lemma (4·3) of Chapter V, since for the function $f(t)=nx(t)+\nu t$ we have $f''(t)=\pm n\pi^{-1}$. (Since f'' is discontinuous at $t=0$, we must consider the intervals $(-\pi,0)$ and $(0,\pi)$ separately.)

Since $f(t)$ is odd, $\pi\lambda_{n,\nu}$ is

$$\int_0^\pi \cos f\,dt=\int_0^\pi \frac{d\sin f}{f'}=\int_0^\pi \frac{\sin f}{f'^2}f''\,dt=n\pi^{-1}\int_0^\pi \frac{\sin f}{f'^3}f'\,dt.$$

The function $f'=nx'+\nu$ is monotone in $(0,\pi)$, and for $|\nu|\geqslant\frac{3}{2}n$ is of constant sign, since $\frac{1}{2}\leqslant x'\leqslant\frac{3}{2}$. For $|\nu|\geqslant 3n$ we have $|f'|\geqslant\frac{1}{2}|\nu|$, and the second mean-value theorem applied to the factor $(1/f')^3$ shows the integral to be numerically not greater than

$$n\pi^{-1}.(2/|\nu|)^3.2\leqslant A/\nu^2.$$

11. Fourier-Stieltjes coefficients and sets of constant ratio of dissection

We shall now investigate the behaviour of the Fourier-Stieltjes coefficients of certain non-decreasing and continuous functions.

Take $0<\xi<\frac{1}{2}$, and consider a perfect non-dense set $E=E(\xi)$ constructed on $(0,2\pi)$ in the familiar Cantor manner, except that at every stage of construction we remove, not the middle third, but a concentric interval of relative length $1-2\xi$. We considered such sets in Chapter V, § 3, where we called them *sets of constant ratio of dissection.* The Fourier-Stieltjes coefficients of the Lebesgue function $F(x)$ associated with E are given by the formula (Chapter V, (3·5))

$$c_n=(-1)^n(2\pi)^{-1}\prod_{k=1}^{\infty}\cos\{\pi n\xi^{k-1}(1-\xi)\}. \tag{11·1}$$

One of our problems is to characterize the values of ξ for which $c_n\to 0$.

If $1/\xi$ is a positive integer, and in particular if E is Cantor's ternary set, the coefficients c_n do not tend to 0 (Chapter V, (3·6)). For other values of ξ, and especially for ξ irrational, the problem is much more delicate; it reveals, quite unexpectedly, connexions with algebraic number theory.

The proof of (11·1) shows that the formula holds for n non-integral, provided $(-1)^n$ is replaced by $e^{-\pi in}$. Hence, in the light of (10·11), we may state our problem as follows: *for what ξ's does the function*

$$\gamma(u)=\prod_{k=0}^{\infty}\cos(\pi u\xi^k) \tag{11·2}$$

tend to 0 as $u\to\infty$?

In discussing this problem we suppose always that $0<\xi<1$. Although in constructing the set E we had to take $0<\xi<\frac{1}{2}$, the formulae for c_n and $\gamma(u)$ have meaning (and are of interest in the calculus of probability) for $0<\xi<1$. Incidentally, restricting ξ to the interval $(0,\frac{1}{2})$ would not simplify the argument.

Suppose that $\gamma(u)\neq o(1)$ as $u\to\infty$. We shall deduce from this hypothesis certain consequences about ξ.

Let $$\theta=1/\xi\quad(\theta>1).$$

By hypothesis, there is a sequence $u_1 < u_2 < \ldots < u_s < \ldots \to \infty$ such that $|\gamma(u_s)| \geq \delta > 0$. We can write

$$u_s = \lambda_s \theta^{m_s},$$

where the m_s are integers tending monotonically to $+\infty$, and $1 \leq \lambda_s < \theta$. By selecting a subsequence of $\{u_s\}$, we may suppose that $\lambda_s \to \lambda$, $1 \leq \lambda \leq \theta$.

Obviously, $$|\gamma(u_s)| \leq |\cos(\pi\lambda_s) . \cos(\pi\lambda_s\theta) . \cos(\pi\lambda_s\theta^2) \ldots \cos(\pi\lambda_s\theta^{m_s})|.$$

Hence $$\prod_{m=0}^{m_s}\{1 - \sin^2(\pi\lambda_s\theta^m)\} \geq \delta^2,$$

and, using the inequality $e^x \geq 1 + x$,

$$\sum_{m=0}^{m_s} \sin^2(\pi\lambda_s\theta^m) \leq \log(1/\delta^2).$$

Therefore, for $t > s$, $$\sum_{m=0}^{m_s} \sin^2(\pi\lambda_t\theta^m) \leq \log(1/\delta^2).$$

Keeping s fixed and making $t \to \infty$, and then making $s \to \infty$, we obtain from the last inequality the relation

$$\sum_{m=0}^{\infty} \sin^2(\pi\lambda\theta^m) \leq \log(1/\delta^2). \tag{11·3}$$

Thus we have the following theorem:

(11·4) THEOREM. *If the coefficients c_n in* (11·1) *do not tend to* 0, *there is a real $\lambda \neq 0$ such that the series*

$$\Sigma \sin^2(\pi\lambda\theta^m) \tag{11·5}$$

converges.

Denote by $\{a\}$ the distance between a and the nearest integer; thus $0 \leq \{a\} \leq \frac{1}{2}$. The convergence of (11·5) is equivalent to the convergence of

$$\Sigma\{\lambda\theta^m\}^2. \tag{11·6}$$

Suppose that y is an algebraic integer of degree n, so that y satisfies an equation

$$y^n + b_1 y^{n-1} + \ldots + b_n = 0, \tag{11·7}$$

where $b_1, b_2, \ldots, b_n$ are rational integers, and satisfies no equation of this type and of lower degree. We say that y is an S *number*, if $y > 1$ and if all the conjugates of y (other than y) have moduli less than 1.

(11·8) THEOREM. *Suppose that $\theta > 1$. A necessary and sufficient condition that there exists a real $\lambda \neq 0$ such that $\Sigma\{\lambda\theta^m\}^2 < \infty$, is that θ be an* S *number.*

The sufficiency of the condition is immediate. Suppose that $\theta > 1$ is an algebraic integer of degree n with conjugates $\alpha_1, \alpha_2, \ldots, \alpha_{n-1}$ of absolute values less than 1. Then $\theta^m + \alpha_1^m + \alpha_2^m + \ldots + \alpha_{n-1}^m$ is a rational integer c_m such that

$$|c_m - \theta^m| \leq (n-1)\sigma^m \quad (\sigma = \max|\alpha_j|),$$

and the convergence of $\Sigma\{\lambda\theta^m\}$, with $\lambda = 1$, follows.

The proof of the necessity is deeper and is based on two lemmas. The first is the following:

(11·9) THEOREM OF KRONECKER. *A power series*

$$\sum_{m=0}^{\infty} c_m z^m \tag{11·10}$$

represents a rational function regular at the origin if and only if the determinants

$$\Delta_m = \begin{vmatrix} c_0 & c_1 & \dots & c_m \\ c_1 & c_2 & \dots & c_{m+1} \\ \dots & \dots & \dots & \dots \\ c_m & c_{m+1} & \dots & c_{2m} \end{vmatrix}$$

are 0 *for all large enough* m.

Proof. First we note that a power series (11·10) represents a rational function regular at the origin if and only if the c_m satisfy, for all large enough m, a recurrence relation

$$c_m\gamma_0 + c_{m+1}\gamma_1 + \dots + c_{m+k}\gamma_k = 0, \tag{11·11}$$

where $\gamma_0, \gamma_1, \dots, \gamma_k$ are independent of m and not all 0. For, supposing as we may that $\gamma_k \neq 0$, the validity of (11·11) for large enough m means that the product of (11·10) by the polynomial

$$\gamma_0 z^k + \gamma_1 z^{k-1} + \dots + \gamma_k,$$

which does not vanish at the origin, is itself a polynomial, that is, (11·10) represents a rational function regular at the origin.

Suppose now that we have (11·11) for $m > m_0$, and that $\gamma_k = 1$. Then, for $N > m_0 + k$, the last column of Δ_N is a linear combination of the preceding k columns, and so $\Delta_N = 0$. It remains therefore to show that if $\Delta_m = 0$ for all large enough m, then, with suitable $\gamma_0, \gamma_1, \dots, \gamma_k$ not all 0, we have (11·11) for all large enough m.

The special case when $\Delta_0 = \Delta_1 = \dots = 0$ is immediate. For then $c_0 = 0$, and if $c_0 = c_1 = \dots = c_{m-1} = 0$, so that $\Delta_m = (-1)^m c_m^m$, then also $c_m = 0$, that is, (11·10) vanishes identically.

Suppose therefore that

$$\Delta_{m_0} = \Delta_{m_0+1} = \dots = 0, \quad \Delta_{m_0-1} \neq 0, \tag{11·12}$$

for some $m_0 \geq 1$. From $\Delta_{m_0} = 0$ we deduce the existence of constants $\gamma_0, \gamma_1, \dots, \gamma_{m_0}$, not all 0, such that

$$\gamma_0 c_h + \gamma_1 c_{h+1} + \dots + \gamma_{m_0} c_{h+m_0} = 0 \quad (0 \leq h \leq m_0).$$

Since $\Delta_{m_0-1} \neq 0$, we may suppose that $\gamma_{m_0} = 1$. Write

$$C_m = \gamma_0 c_m + \gamma_1 c_{m+1} + \dots + \gamma_{m_0} c_{m+m_0}.$$

Then $C_m = 0$ for all $m \leq m_0$. We wish to show that $C_m = 0$ also for $m > m_0$.

Suppose that for some positive m_1 we have $C_{m_0+m_1} \neq 0$, while $C_m = 0$ for $m < m_0 + m_1$. Consider $\Delta_{m_0+m_1}$. Adding to each of its columns of rank not less than m_0 the m_0 preceding columns multiplied respectively (in order of increasing rank) by $\gamma_0, \gamma_1, \dots, \gamma_{m_0-1}$, we obtain

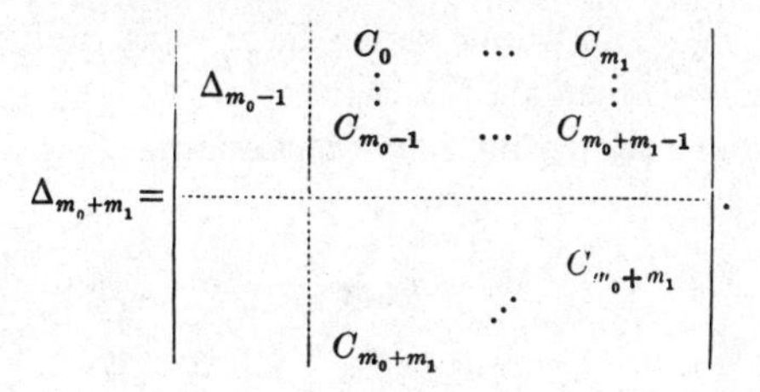

Since the elements in the upper right square are all 0, those above the diagonal of the lower right square are all 0 and those on the diagonal are all $C_{m_0+m_1}$, we have

$$\Delta_{m_0+m_1} = (-1)^{m_1}\Delta_{m_0-1} C_{m_0+m_1}^{m_1+1}.$$

Hence, by (11·12), $C_{m_0+m_1} = 0$, contrary to hypothesis. It follows that $C_m = 0$ for all $m \geq 0$, and the proof of (11·9) is completed.

The second lemma we need is as follows:

(11·13) LEMMA. *Suppose that a power series $\Sigma c_m z^m$ with integral coefficients represents a rational function; then we can write*

$$\Sigma c_m z^m = \frac{a_0 + a_1 z + \ldots + a_h z^h}{1 + b_1 z + \ldots + b_k z^k},$$

where the fraction on the right is irreducible and all the a's and b's are integers.

(i) We have $\Sigma c_m z^m = P/Q$, where

$$P(z) = a_0 + a_1 z + \ldots + a_h z^h, \quad Q(z) = b_0 + b_1 z + \ldots + b_k z^k, \; b_0 \neq 0.$$

Since $b_k c_n + b_{k-1} c_{n+1} + \ldots + b_0 c_{n+k} = 0$ for $n + k > h$, the b's are commensurable; if we suppose, as we may, that the b's are rational, the a's are also rational. We may suppose that the a's and b's are integers and that P and Q are without common roots. We may also suppose that the c's are co-prime (i.e. have no common divisor greater than 1) and that the b's are co-prime. For otherwise, expressing the a's in terms of the b's and c's, we see that the common divisor of the b's would also divide all a's, and could be cancelled out in the equation $\Sigma c_m z^m = P/Q$. It is enough to show that on these assumptions $b_0 = 1$.

(ii) If $\Sigma \gamma_m z^m$ is a product of power series $\Sigma \alpha_m z^m$ and $\Sigma \beta_m z^m$ with integral coefficients, and if all the γ's are divisible by a prime p, then either all the α's or all the β's are necessarily divisible by p. For suppose neither is true and denote by α_s and β_t the first α and β not divisible by p. Then in

$$\gamma_{s+t} = \alpha_0 \beta_{s+t} + \ldots + \alpha_{s-1}\beta_{t+1} + \alpha_s \beta_t + \alpha_{s+1}\beta_{t-1} + \ldots + \alpha_{s+t}\beta_0$$

all the terms on the right except $\alpha_s \beta_t$ are divisible by p, and $\alpha_s \beta_t$ is not. It follows that γ_{s+t} is not divisible by p, contrary to hypothesis.

(iii) If the P and Q of (i) have integral coefficients and no root in common, there are polynomials R and S with integral coefficients such that

$$P(x) R(x) + Q(x) S(x) = N,$$

where N is an integer. Since the power series for P/Q has integral coefficients, the same holds for the power series

$$\Sigma c'_m z^m = \frac{N}{Q} = \frac{P}{Q} R + S.$$

We assumed that all the b's are co-prime. Then, by (ii), each prime factor of N divides all c'_m. Hence we may suppose that $N = 1$. This gives $c'_0 b_0 = 1$, whence $b_0 = 1$, which proves the lemma.

We pass to the proof of necessity in Theorem (11·8). Write $\lambda\theta^m = c_m + \delta_m$, where c_m is an integer and $|\delta_m| \leqslant \frac{1}{2}$. The power series $\Sigma c_m z^m$ has radius of convergence $1/\theta$. We show that it represents a rational function.

By (11·9), it is enough to show that the determinants Δ_m vanish for $m \geqslant m_0$. Now, Δ_m can be written

$$\begin{vmatrix} c_0 & c_1 - \theta c_0 & \cdots & c_m - \theta c_{m-1} \\ c_1 & c_2 - \theta c_1 & \cdots & c_{m+1} - \theta c_m \\ \vdots & & & \\ c_m & c_{m+1} - \theta c_m & \cdots & c_{2m} - \theta c_{2m-1} \end{vmatrix} = \begin{vmatrix} c_0 & \eta_1 & \cdots & \eta_m \\ c_1 & \eta_2 & \cdots & \eta_{m+1} \\ \vdots & & & \\ c_m & \eta_{m+1} & \cdots & \eta_{2m} \end{vmatrix},$$

where
$$\eta_k = c_k - \theta c_{k-1} = \theta\delta_{k-1} - \delta_k.$$

Since
$$\eta_k^2 \leqslant (\theta^2 + 1)(\delta_{k-1}^2 + \delta_k^2),$$

the series $\Sigma\eta_k^2$ converges. Let R_k be the kth remainder of the series. Hadamard's well-known estimate of the absolute value of a determinant gives

$$\Delta_m^2 \leqslant \left(\sum_0^m c_k^2\right)\left(\sum_1^{m+1} \eta_k^2\right)\left(\sum_2^{m+2} \eta_k^2\right)\dots\left(\sum_m^{2m} \eta_k^2\right) \leqslant \left(\sum_0^m c_k^2\right) R_1 R_2 \dots R_m.$$

Now, since $|\lambda\theta^k - c_k| \leqslant \frac{1}{2}$, we have

$$\sum_0^m c_k^2 \leqslant C\theta^{2m},$$

where $C = C(\lambda, \theta)$ is independent of m. Since $R_m \to 0$, it follows that $\Delta_m \to 0$. But Δ_m is a rational integer. Hence $\Delta_m = 0$ for $m \geqslant m_0$.

By Lemma (11·13),

$$\Sigma c_m z^m = \frac{P(z)}{1 + b_1 z + \dots + b_n z^n},$$

where the b's, as well as the coefficients of the polynomial P, are integers. Write

$$\begin{aligned}\sum_0^\infty \delta_m z^m &= \sum_0^\infty \lambda\theta^m z^m - \sum_0^\infty c_m z^m \\ &= \frac{\lambda}{1-\theta z} - \frac{P(z)}{1 + b_1 z + \dots + b_n z^n}.\end{aligned} \tag{11·14}$$

Since $|\delta_m| \leqslant \frac{1}{2}$, the radius of convergence of $\Sigma\delta_m z^m$ is at least 1. Hence $1/\theta$ is a root of $1 + b_1 z + \dots + b_n z^n$, and all other roots have modulus at least 1. Since $\Sigma\delta_m^2 < \infty$, the rational function $\Sigma\delta_m z^m$ cannot have a pole on $|z| = 1$. It follows that

$$1 + b_1 z + \dots + b_n z^n = 0$$

has one root, $1/\theta$, inside the unit circle, and the remaining $n-1$ roots outside the unit circle. Therefore the roots of the reciprocal equation

$$z^n + b_1 z^{n-1} + \dots + b_n = 0 \tag{11·15}$$

are all algebraic integers, and all, except θ, are situated inside the unit circle. Hence θ is an S number (of degree n, since (11·15) is clearly irreducible).

Incidentally, λ belongs to the field generated over the rationals by θ. For if

$$Q = 1 + b_1 z + \dots + b_n z^n,$$

then, by (11·14), $-\lambda/\theta$ is the residue of P/Q at $z = 1/\theta$, so that

$$-\frac{\lambda}{\theta} = \frac{P(1/\theta)}{Q'(1/\theta)}.$$

(11·16) THEOREM. *Suppose that* $0 < \xi < 1$. *A necessary and sufficient condition that the coefficients* c_n *in* (11·1) *do not tend to* 0 *is that* $\theta = 1/\xi$ *should be an* S *number different from* 2.

From (11·4) and (11·8) we see that if $c_n \neq o(1)$, then θ is an S number. Moreover, $\theta \neq 2$, since otherwise $c_n = 0$ for $n \neq 0$ (and $F(x) = x + C$ in $(0, 2\pi)$).

Suppose, conversely, that $\theta \neq 2$ is an S number. It is enough to show that $\gamma(u)$ in (11·2) does not tend to 0 as $u \to \infty$. Set $u = \theta^k$, $k = 0, 1, 2, \dots$. Then

$$|\gamma(\theta^k)| = |\cos(\pi\theta)\cos(\pi\theta^2)\dots\cos(\pi\theta^k)| . |\cos(\pi/\theta)\cos(\pi/\theta^2)\dots|.$$

The proof of the sufficiency part of (11·8) shows that $\Sigma \sin^2 \pi\theta^m < \infty$. Hence the infinite product $\Pi \cos^2 \pi\theta^m$ converges to a number A, which is positive unless for some m we have $\theta^m = j + \frac{1}{2}$ with j integral. The latter is impossible if θ is an S number. Hence

$$|\gamma(\theta^k)| \geqslant A^{\frac{1}{2}} |\cos(\pi/\theta)\cos(\pi/\theta^2)\dots|.$$

The last infinite product converges to a number $B \neq 0$, since $\theta \neq 2$. (If θ is an S number $\theta^m = 2$ is impossible for $m > 1$.) Hence $|\gamma(\theta^k)| \geqslant A^{\frac{1}{2}}|B|$ and (11·16) is established.

The theorem shows that *except for a denumerable set of ξ's we always have $c_n \to 0$.*

If $\xi \neq \frac{1}{2}$ is rational, say $\xi = p/q$, and the fraction is irreducible, then $\theta = q/p$, and $c_n \neq o(1)$ if and only if $p = 1$ cf. (Chapter V, (3·6)).

(11·17) THEOREM. *If $0 < \xi < \frac{1}{2}$ and $\theta = 1/\xi$ is not an S number then the perfect set $E(\xi)$ is a set of multiplicity.*

(11·18) THEOREM. *If $0 < \xi < \frac{1}{2}$ and $\theta = 1/\xi$ is an S number then $E(\xi)$ is a set of uniqueness.*

Theorem (11·17) is a corollary of (11·16) since under the hypothesis of (11·17), if F is the Cantor-Lebesgue function associated with $E(\xi)$, $c_n \to 0$ and, by Theorem (6·8) of Chapter IX, $S[dF]$ converges to 0 outside E without being identically 0. Thus E is a set of multiplicity even in the restricted sense (Chapter IX, p. 348).

Theorem (11·18), on the other hand, does not follow from (11·16). For though, if θ is an S number, $S[dF]$ does not converge to 0 outside E (indeed, it diverges almost everywhere), it is not inconceivable that some other trigonometric series, not necessarily even a Fourier-Stieltjes series, converges to 0 outside E without being identically 0. As a matter of fact, the proof of (11·18) requires some new ideas.

In Chapter IX, § 6, we introduced the notion of sets $H^{(n)}$. We now need a slight generalization of this notion.

Consider vectors $(x_1, x_2, \dots, x_n)$ of n-dimensional Euclidean space R^n. An infinite sequence of vectors $V^{(m)} = (v_1^{(m)}, v_2^{(m)}, \dots, v_n^{(m)})$ is called *normal* if for every non-zero vector $(a_1, a_2, \dots, a_n)$ with integral components we have

$$|v_1^{(m)} a_1 + v_2^{(m)} a_2 + \dots + v_n^{(m)} a_n| \to \infty$$

as $m \to \infty$. A set $E \subset (0, 2\pi)$ will be called an $H_*^{(n)}$ *set* if there is a subdomain Δ of the fundamental cube

$$K: 0 \leqslant x_j < 2\pi \quad (j = 1, 2, \dots, n),$$

and a normal sequence $\{V^{(m)}\}$ such that for each $x \in E$ the point $(xv_1^{(m)}, xv_2^{(m)}, \dots, xv_n^{(m)})$ never enters Δ, mod 2π $(m = 1, 2, \dots)$. If in this definition we consider only vectors $V^{(m)} = (v_1^{(m)}, \dots, v_n^{(m)})$ with integral components, we obtain the sets $H^{(n)}$ of Chapter IX, § 6.

(11·19) LEMMA. *Every set of type $H_*^{(n)}$ is a sum of a finite number of sets $H^{(n)}$.*

It is easy to see that in the definition of a set $H_*^{(n)}$ we may take all the numbers $v_j^{(m)}$ rational, with the same denominator, at the cost possibly of decreasing Δ. Furthermore, since now the fractional parts of the $v_j^{(m)}$ can take only a finite number of values, we may suppose, by considering a subsequence of $\{V^{(m)}\}$, that the fractional parts of the $v_j^{(m)}$ are the same for each j:

$$v_j^{(m)} = n_j^{(m)} + r_j \quad (m = 1, 2, \dots; j = 1, 2, \dots, n), \tag{11·20}$$

where the $n_j^{(m)}$ are integers, and $0 \leqslant r_j < 1$. Without loss of generality we may suppose that Δ is a cube $|x_j - x_j^0| < \delta$, $j = 1, 2, \ldots, n$.

We split $(0, 2\pi)$ into disjoint intervals $I_1, I_2, \ldots, I_p$, each of length less than $\frac{1}{2}\delta$, and write $E_i = EI_i$. Hence $E = \Sigma E_i$, and it is enough to show that each E_i is of type $\mathrm{H}^{(n)}$.

If $x \in E_i$ then, by (11·20),

$$v_j^{(m)} x = n_j^{(m)} x + r_j x,$$

and, since x is in $I_i = (a_i, b_i)$, $r_j x$ is in the interval $(r_j a_i, r_j b_i)$ of length less than $\frac{1}{2}\delta$. It follows that, since the points $(v_1^{(m)} x, \ldots, v_n^{(m)} x)$ do not enter Δ, mod 2π, the points $(n_1^{(m)} x, \ldots, n_n^{(m)} x)$ do not enter the cube $|x_j - x_j^0| < \frac{1}{2}\delta$, $j = 1, 2, \ldots, n$. Hence each E_i is of type $\mathrm{H}^{(n)}$.

Since the closure of a set of type $\mathrm{H}^{(n)}$ is $\mathrm{H}^{(n)}$, we deduce from (11·19) and Theorem (6·15) of Chapter IX that *sets* $\mathrm{H}_*^{(n)}$ *are sets of uniqueness.*

Theorem (11·18) will be established, if we show that, *if θ is an* S *number of degree n, then $E(\xi)$ is of type* $\mathrm{H}_*^{(n)}$.

The proof of this makes essential use of the following lemma:

(11·21) THEOREM OF MINKOWSKI. *Consider n linear homogeneous forms*

$$\xi_r = \alpha_{r1} x_1 + \alpha_{r2} x_2 + \ldots + \alpha_{rn} x_n \quad (r = 1, 2, \ldots, n) \tag{11·22}$$

in $x_1, x_2, \ldots, x_n$, with real coefficients and determinant $\Delta \neq 0$. If $\lambda_1, \lambda_2, \ldots, \lambda_n$ are positive numbers satisfying

$$\lambda_1 \lambda_2 \ldots \lambda_n \geqslant |\Delta|,$$

then there are integers $x_1, x_2, \ldots, x_n$ not all 0 such that

$$|\xi_1| \leqslant \lambda_1, \quad |\xi_2| \leqslant \lambda_2, \quad \ldots, \quad |\xi_n| \leqslant \lambda_n.$$

The conclusion holds if some of the forms have complex coefficients, provided that along with each such ξ its conjugate also appears among the forms (11·22), *and provided that the λ's corresponding to conjugate ξ's are equal.*

We prove this at the end of the section and proceed in the meanwhile with the deduction of (11·18).

Let θ be an S number and $\alpha_1, \alpha_2, \ldots, \alpha_{n-1}$, $|\alpha_j| < 1$, its remaining conjugates. Let $P(z) = z^n + b_1 z^{n-1} + \ldots + b_n$ be the irreducible polynomial with integral coefficients and roots $\theta, \alpha_1, \alpha_2, \ldots, \alpha_{n-1}$. Denote by $Q(z)$ the polynomial reciprocal to $P(z)$:

$$Q(z) = z^n P(1/z) = 1 + b_1 z + \ldots + b_n z^n.$$

Finally, let $R(z) = a_0 z^{n-1} + a_1 z^{n-2} + \ldots + a_{n-1}$ be any polynomial of degree $n-1$ with integral coefficients, and $S(z)$ the reciprocal of $R(z)$. Consider the formulae

$$\frac{R(z)}{Q(z)} = \sum_0^\infty c_m z^m = \frac{\lambda}{1 - \theta z} + \sum_{j=1}^{n-1} \frac{\mu_j}{1 - \alpha_j z}. \tag{11·23}$$

Since the constant term of $Q(z)$ is 1, the c_m are integers.

By changing, if necessary, the sign of R we may suppose that $\lambda > 0$. In view of (11·23),

$$\lambda \theta^m = c_m + \delta_m, \tag{11·24}$$

where

$$\delta_m = -\sum_{j=1}^{n-1} \mu_j \alpha_j^m \to 0, \tag{11·25}$$

$$\sum_0^\infty |\delta_m| \leqslant \sum_{j=1}^{n-1} \frac{|\mu_j|}{1 - |\alpha_j|}. \tag{11·26}$$

Also $$\lambda=\lim_{z\to 1/\theta}(1-\theta z)\frac{R(z)}{Q(z)}=-\theta\frac{R(1/\theta)}{Q'(1/\theta)}=\frac{-\theta R(1/\theta)}{\{z^n P(1/z)\}'_{z=1/\theta}}=\frac{S(\theta)}{P'(\theta)},$$

and since a similar argument gives the μ_j, we have

$$\lambda=\frac{S(\theta)}{P'(\theta)},\quad \mu_j=\frac{S(\alpha_j)}{P'(\alpha_j)}\quad (j=1,2,\dots,n-1). \tag{11·27}$$

The coefficients of $S(z)=a_{n-1}z^{n-1}+\dots+a_1z+a_0$ have so far been arbitrary integers. We now choose them so that

$$\left.\begin{aligned}\lambda=|\lambda|=\left|\frac{S(\theta)}{P'(\theta)}\right|\leqslant \eta\theta^N,\\ |\mu_j|=\left|\frac{S(\alpha_j)}{P'(\alpha_j)}\right|\leqslant\eta\quad (j=1,2,\dots,n-1),\end{aligned}\right\} \tag{11·28}$$

where η and N will be fixed presently. Since $S(\theta)$, $S(\alpha_1)$, ..., $S(\alpha_{n-1})$ are linear forms in the a_j whose determinant (the Vandermonde determinant of θ, α_1, ..., α_{n-1}) is non-zero and depends only on θ, we can apply (11·21); and we easily see that we can satisfy (11·28) provided

$$\eta^n\theta^N\geqslant A, \tag{11·29}$$

where $A=A(\theta)$ depends on θ only.

Write $d_m=\lambda\theta^m/(\theta-1)$. The sequence of vectors

$$V^{(m)}=(d_{m+1}, d_{m+2}, \dots, d_{m+n})\quad (m=0,1,\dots)$$

is normal, since for any non-zero vector $(e_1, e_2, \dots, e_n)$ with integral e's we have

$$\sum_{k=1}^n d_{m+k}e_k=\frac{\lambda\theta^{m+1}}{\theta-1}(e_1+e_2\theta+\dots+e_n\theta^{n-1})\to\pm\infty,$$

since $e_1+e_2\theta+\dots+e_n\theta^{n-1}\neq 0$. We show that there is a subdomain Δ of K such that for no $x\in E$ is any $(d_{m+1}x, \dots, d_{m+n}x)$ in Δ, mod 2π.

We know (Chapter V, §3) that the points of E are given by

$$x=2\pi(\theta-1)(\epsilon_1\theta^{-1}+\epsilon_2\theta^{-2}+\dots),$$

where the ϵ_i are arbitrarily 0 or 1. Hence for any fixed N we have, mod 2π,

$$\begin{aligned}d_m x=\frac{\lambda\theta^m}{\theta-1}x&=2\pi\lambda(\epsilon_{m+1}\theta^{-1}+\epsilon_{m+2}\theta^{-2}+\dots+\epsilon_{m+N}\theta^{-N})\\&\quad+2\pi\lambda(\epsilon_{m+N+1}\theta^{-N-1}+\epsilon_{m+N+2}\theta^{-N-2}+\dots)\\&\quad+2\pi(\epsilon_m\delta_0+\epsilon_{m-1}\delta_1+\dots+\epsilon_1\delta_{m-1}),\\&=U+V+W,\end{aligned} \tag{11·30}$$

say (cf. (11·24)). By (11·28) and (11·26),

$$0\leqslant V\leqslant 2\pi\lambda\theta^{-N}\leqslant 2\pi\eta,$$

$$|W|\leqslant 2\pi\sum_0^\infty|\delta_k|\leqslant 2\pi B\eta,$$

where $B=B(\theta)$. Hence, by (11·30),

$$-C\eta\leqslant d_m x-2\pi\lambda(\epsilon_{m+1}\theta^{-1}+\dots+\epsilon_{m+N}\theta^{-N})\leqslant C\eta \tag{11·31}$$

where $C=B+1$.

Denote by g_m the fractional part of $\lambda(\epsilon_{m+1}\theta^{-1}+\ldots+\epsilon_{m+N}\theta^{-N})$, and by O_m the point $(2\pi g_m, 2\pi g_{m+1}, \ldots, 2\pi g_{m+n-1})$ of K. By (11·31), the point $(d_m x, \ldots, d_{m+n-1}x)$ is situated, mod 2π, in a cube with centre O_m whose sides are parallel to the co-ordinate axes and have length $2C\eta$. But the number of distinct O_m is at most 2^{N+n-1}, since each O_m is determined by the numbers $\epsilon_{m+1}, \epsilon_{m+2}, \ldots, \epsilon_{m+N+n-1}$ and each ϵ_i is either 0 or 1. Hence if, for example,

$$2^{N+n-1}(2C\eta)^n \leqslant \tfrac{1}{2}(2\pi)^n, \tag{11·32}$$

then there is a subdomain of K free of the points $(d_m x, \ldots, d_{m+n-1}x)$ and E is an $H_*^{(n)}$ set.

Write (11·32) in the form

$$2^N \eta^n \leqslant D, \tag{11·33}$$

where D depends on θ and n only. Theorem (11·18) will be established if we show that there exist N and η satisfying both (11·29) and (11·33). Suppose that we have equality in (11·33). Then it is enough to take N so large that $(\frac{1}{2}\theta)^N \geqslant A/D$, and then to determine η from $2^N\eta^n = D$.

Return to the proof of (11·21) and suppose first that all the ξ are real. It is enough to prove the theorem under the hypothesis that $\lambda_1 \ldots \lambda_n$ is strictly greater than $|\Delta|$, for once this is done we may replace λ_1 by $\lambda_1' > \lambda_1$ and then make $\lambda_1' \to \lambda_1$, so obtaining the general result.

Suppose that the inequalities $|\xi_j| \leqslant \lambda_j$, $j = 1, 2, \ldots, n$, have no non-zero integral solution; we shall then come to a contradiction with the hypotheses. Write $\xi_j(x)$ for ξ_j, where $x = (x_1, x_2, \ldots, x_n)$. Let D be the set of x such that

$$|\xi_1(x)| \leqslant \tfrac{1}{2}\lambda_1, \quad |\xi_2(x)| \leqslant \tfrac{1}{2}\lambda_2, \quad \ldots, \quad |\xi_n(x)| \leqslant \tfrac{1}{2}\lambda_n.$$

Denote by D_g the set obtained from D by a parallel translation which moves the origin to the integral point $g = (g_1, g_2, \ldots, g_n)$. Clearly D_g is given by the inequalities $|\xi_j(x-g)| \leqslant \frac{1}{2}\lambda_j$ for all j. If $g' \neq g''$, $D_{g'}$ and $D_{g''}$ have no points in common; for if such a point x^0 existed, the inequalities $|\xi_j(x^0 - g')| \leqslant \frac{1}{2}\lambda_j$ and $|\xi_j(x^0 - g'')| \leqslant \frac{1}{2}\lambda_j$ would lead to

$$|\xi_j(g' - g'')| \leqslant \lambda_j$$

for all j, and the non-zero integral $g = g' - g''$ would satisfy all the inequalities $|\xi_j| \leqslant \lambda_j$, contrary to the assumption that no such solution exists.

We now observe that the measure $|D|$ of D is

$$\int\!\cdots\!\int_D dx_1 \ldots dx_n = \int_{|\xi_1|\leqslant\frac{1}{2}\lambda_1}\!\!\cdots\int_{|\xi_n|\leqslant\frac{1}{2}\lambda_n} \left|\frac{\partial(x_1, \ldots, x_n)}{\partial(\xi_1, \ldots, \xi_n)}\right| d\xi_1 \ldots d\xi_n = \frac{\lambda_1 \ldots \lambda_n}{|\Delta|} > 1,$$

since, by hypothesis, $\lambda_1 \ldots \lambda_n$ exceeds $|\Delta|$. Consider all D_g with g belonging to the cube $|x_j| \leqslant N$, $j = 1, 2, \ldots, n$, where N is an integer. Since the various D_g have no points in common, the measure of the union of those D_g on the one hand is $(2N+1)^n |D|$, and on the other hand does not exceed $(2N+2d)^n$, where d denotes the largest distance of the points of D from the origin. Hence

$$(2N+1)^n |D| \leqslant (2N+2d)^n.$$

Dividing this by N^n and making $N \to \infty$, we obtain $|D| \leqslant 1$, a contradiction with the previous inequality $|D| > 1$. This proves (11·21) when all the ξ are real.

Suppose now that, for example, ξ_1 and ξ_2 are complex conjugates, and replace ξ_1 and ξ_2 by the real forms

$$\xi_1' = \frac{\xi_1 + \xi_2}{2}, \quad \xi_2' = \frac{\xi_1 - \xi_2}{2i},$$

and λ_1, λ_2 by

$$\lambda_1' = \lambda_2' = 2^{-\frac{1}{2}}\lambda_1.$$

We proceed similarly with all conjugate pairs of ξ's, and if ξ_j is real we write $\xi_j' = \xi_j$, $\lambda_j' = \lambda_j$. The determinant Δ' of the new forms satisfies

$$|\Delta'| = 2^{-r}|\Delta|,$$

where r is the number of complex-conjugate pairs of ξ's. Since $|\Delta'| \leqslant \lambda_1' \dots \lambda_n'$, there is an integral non-zero solution x^0 of

$$|\xi_1'| \leqslant \lambda_1', \quad |\xi_2'| \leqslant \lambda_2', \quad \dots, \quad |\xi_n'| \leqslant \lambda_n';$$

and since, for example, $\quad |\xi_1(x^0)|^2 = \xi_1'^2(x^0) + \xi_2'^2(x^0) \leqslant \tfrac{1}{2}\lambda_1^2 + \tfrac{1}{2}\lambda_1^2 = \lambda_1^2,$

we easily see that $|\xi_j(x^0)| \leqslant \lambda_j$ for all j.

MISCELLANEOUS THEOREMS AND EXAMPLES

For Examples 1–4 below see Marcinkiewicz and Zygmund [4].

We suppose that $\{\phi_n\}$ is orthonormal in (a, b) and that

(i) $$\|\phi_n\|_{q_0} = \left(\int_a^b |\phi_n|^{q_0} dx\right)^{1/q_0} \leqslant M_n < \infty$$

for $n = 1, 2, \dots$ and some $2 < q_0 \leqslant \infty$. We write $p_0 = q_0' = q_0/(q_0 - 1)$.

1. Suppose that $p_0 < p \leqslant 2$, and define q by

$$\frac{p_0}{p} + \frac{2 - p_0}{q} = 1.$$

Then the Fourier coefficients c_n of f with respect to ϕ_n satisfy

$$(\Sigma M_n^{q-2} |c_n|^q)^{1/q} \leqslant \|f\|_p,$$

a relation which for $q_0 = \infty$, $M_1 = M_2 = \dots = M$ reduces to (2·10).

[The proof is similar to that of (2·10): we have $|c_n| M_n^{-1} \leqslant \|f\|_{p_0}$ and Bessel's inequality $\|c\|_2 \leqslant \|f\|_2$.]

2. Suppose that $1 \leqslant p \leqslant 2$ and define q by

$$\frac{2 - p_0}{p} + \frac{p_0}{q} = 1.$$

Then if both $\Sigma M_n^{2-p} |c_n|^p$ and $\Sigma |c_n|^2$ are finite, there is an $f \in L^q(a, b)$ whose Fourier coefficients with respect to ϕ_n are c_n, such that

$$\|f\|_q \leqslant (\Sigma M_n^{2-p} |c_n|^p)^{1/p},$$

a relation which reduces to (2·11) when $q_0 = \infty$, $M_1 = M_2 = \dots = M$.

If the M_n are bounded below (which is certainly the case if (a, b) is finite, since then

$$M_n \geqslant (b - a)^{(2 - q_0)/2q_0})$$

the convergence of $\Sigma M_n^{2-p} |c_n|^p$ implies that of $\Sigma |c_n|^2$.

[Consider finite sums $\sum_1^N c_n \phi_n = f$; interpolate between $\|f\|_2 = \|c\|_2$ and

$$\|f\|_{q_0} \leqslant \sum_1^N M_n |c_n|,$$

and in making $N \to \infty$ apply the Riesz-Fischer theorem (Chapter IV, § 1).]

3. Suppose that $p_0 < p < 2$, that $f \in L^p(a, b)$, and that

$$M_1 \leqslant M_2 \leqslant \dots.$$

Then the coefficients c_n of f with respect to ϕ_n satisfy

$$\{\Sigma |c_n|^p M_n^{(p-2)p_0/(2-p_0)} n^{(p-2)/(2-p)}\} \leqslant A_{p, p_0} \|f\|_p,$$

a generalization of (5·2).

[Write Bessel's inequality in the form

$$\Sigma |c_n n^\alpha M_n^\beta|^2 n^{-2\alpha} M_n^{-2\beta} \leqslant \|f\|_2^2,$$

where α, β are positive numbers to be fixed presently. Consider the additive measure ν equal to $n^{-2\alpha}M_n^{-2\beta}$ for the set consisting of the single point $n=1, 2, \ldots$, and equal to 0 for the sets not containing any point n. Consider the linear operation

$$Tf=\{c_n n^\alpha M_n^\beta\}=\left\{n^\alpha M_n^\beta \int_a^b f\bar{\phi}_n\,dx\right\},$$

and the norms $\| Tf \|_{r,\nu}$. From (i) we deduce that $|c_n| \leqslant M_n \| f \|_{p_0}$, and we shall try to choose α, β so that Tf is of weak type (p_0, p_0). The measure ν of the set in which $|Tf|>y>0$ is equal to $\Sigma n^{-2\alpha}M_n^{-2\beta}$ extended over those n for which $|c_n n^\alpha M_n^\beta|>y$. For such n we have, *a fortiori*,

$$\text{(ii)} \qquad n^\alpha M_n^{\beta+1} \| f \|_{p_0} > y.$$

Let n_0 be the least n satisfying (ii). Since $M_1 \leqslant M_2 \leqslant \ldots$, (ii) holds for each $n \geqslant n_0$. We shall see presently that $\alpha \geqslant 1$. Anticipating this, and denoting by A a positive absolute constant, we have

$$\sum_{n\geqslant n_0} n^{-2\alpha}M_n^{-2\beta} \leqslant M_{n_0}^{-2\beta} \sum_{n\geqslant n_0} n^{-2\alpha} \leqslant A n_0^{-2\alpha+1} M_{n_0}^{-2\beta}.$$

Comparing this with (ii), we choose α, β so that $2\alpha-1$ and 2β are proportional to $\alpha, \beta+1$, and that the coefficient of proportionality is p_0. This gives $\alpha=1/(2-p_0)$, $\beta=p_0/(2-p_0)$, and an application of (4·6) easily completes the proof.]

4. Suppose that $2 \leqslant q < q_0$, that $M_1 \leqslant M_2 \leqslant \ldots$, and that $S=\Sigma |c_n|^q M_n^{(q-2)p_0/(2-p_0)} n^{(q-2)/(2-p_0)}$ is finite. Then there is an $f \in L^q(a, b)$ with coefficients c_n, satisfying

$$\| f \|_q \leqslant B_{q,q_0} S^{1/q}.$$

Moreover,

$$B_{q,q_0}=A_{p,p_0}, \quad B_{q,q_0} \leqslant A\frac{q_0-2}{q_0-q}q.$$

In Examples 5–8 which follow, $\{\phi_n\}$ is a uniformly bounded system ($|\phi_n| \leqslant M$) orthonormal on (a, b).

5. If $|c_n| \leqslant 1/n$ for $n=1, 2, \ldots$, the c_n are the coefficients, with respect to $\{\phi_n\}$, of an f such that

$$\int_a^b \{e^{\gamma M^{-1}|f|}-1\}\,dx \leqslant M^{-2}\delta,$$

where γ and δ are positive absolute constants.

[The proof is similar to that of (4·41) (i). Supposing for simplicity that $c_1=0$, we deduce from (2·11) that

$$\frac{\gamma^q}{q!}\int_a^b |f|^q\,dx \leqslant \frac{\gamma^q}{q!}M^{q-2}\left(\sum_{n=2}^{\infty} n^{-q/(q-1)}\right)^{q-1} \leqslant \frac{\gamma^q}{q!}M^{q-2}(q-1)^{q-1}$$

for $q=2, 3, \ldots$, and observe that

$$\exp|u| \leqslant 2\cosh u = 2\sum_0^\infty \frac{u^{2k}}{(2k)!}.$$

We may take for γ any fixed number less than $1/e$.

A somewhat different argument is based on (5·3) and (5·8):

$$\int_a^b |f|^q\,dx \leqslant M^{q-2}A^q q^q \Sigma |nc_n|^q . n^{-2} \leqslant M^{q-2}A^q q^q,$$

from which point we proceed as before.]

6. If $\sum_1^\infty n|c_n|^2 \leqslant 1$, there is an f with coefficients c_n such that

$$\int_a^b \{e^{\gamma M^{-2}|f|^2}-1\}\,dx \leqslant M^{-2}\delta.$$

More generally, if $\Sigma n^k |c_n|^{k+1} \leqslant 1$ for some $k>0$, then

$$\int_a^b \{e^{\gamma M^{-1-1/k}|f|^{1+1/k}}-1\}\,dx \leqslant M^{-2}\delta_k,$$

where γ_k and δ_k are positive and depend on k only.

(The first result has applications to power series. For if $F(z)=\sum_0^\infty c_n z^n$ is regular for $|z|<1$, we have

$$\sum_1^\infty n|c_n|^2=\frac{1}{2\pi}\iint_{|z|<1}|F'(\rho e^{i\theta})|^2\rho\,d\rho\,d\theta.$$

A proof based on complex variable theory gives in this case somewhat better values for γ and δ; see Beurling [2]).

[(i) As in Example 5, $\|f\|_q M^{(2-q)/q}$ does not exceed

$$\|c\|_p=(\Sigma n^{\frac{1}{2}p}|c_n|^p.n^{-\frac{1}{2}p})^{1/p}\leqslant(\Sigma n|c_n|^2)^{\frac{1}{2}}(\Sigma n^{-p/(2-p)})^{(2-p)/2p},$$

and the last factor on the right does not exceed $A(p-1)^{-\frac{1}{2}}\leqslant Aq^{\frac{1}{2}}$.]

7. Suppose that $b-a<\infty$. If $1<p<2$, $\phi(u)=u^p\phi_1(u)$, where ϕ_1 is positive and slowly varying, then the coefficients c_n of f satisfy

$$\Sigma\phi(n|c_n|)n^{-2}\leqslant K\int_a^b\phi(|f|)\,dx+K.$$

If $2<q<\infty$, $\psi(u)=u^q\psi_1(u)$, where ψ_1 is positive and slowly varying, and if $\Sigma\psi(n|c_n|)n^{-2}<\infty$, then the c_n are the coefficients of an f such that

$$\int_a^b\psi(|f|)\,dx\leqslant K\Sigma\psi(n|c_n|)n^{-2}+K.$$

8. Suppose that $|f|(\log^+|f|)^\alpha\in L(a,b)$, $b-a<\infty$, $\alpha>0$. Then

$$\text{(i)}\qquad \sum_2^\infty\frac{|c_n|}{n}(\log n)^{\alpha-1}\leqslant K\int_a^b|f|(\log^+|f|)^\alpha\,dx+K,$$

$$\text{(ii)}\qquad \Sigma\frac{|c_n|}{n}\left(\log\frac{1}{|c_n|}\right)^{\alpha-1}\leqslant K\int_a^b|f|(\log^+|f|)^\alpha\,dx+K,$$

$$\text{(iii)}\qquad \Sigma\exp(-k|c_n|^{-1/\alpha})<\infty,$$

for each $k>0$, with K independent of f.

[By (4·34) applied to $Tf=\{nc_n\}$, we have (if $\alpha\geqslant1$; for $0<\alpha<1$ we replace $\log(n|c_n|)$ on the left by $\log(2+n|c_n|)$)

$$\text{(iv)}\qquad \Sigma\frac{|c_n|}{n}\{\log(n|c_n|)\}^{\alpha-1}\leqslant K\int_a^b|f|(\log^+|f|)^\alpha\,dx+K.$$

For the n such that $n^{-\frac{1}{2}}\leqslant|c_n|\leqslant n^{\frac{1}{2}}$, $\log(n|c_n|)$ is exactly of order $\log n$, and since the contribution of the remaining terms in (i) is $\leqslant K$, (i) follows from (iv). (For $\alpha=1$, (i) also follows from Example 5 and Young's inequality.) Since $|c_n|\geqslant1/n$ implies $\log 1/|c_n|\leqslant\log n$, (i) implies (ii) for $\alpha\geqslant1$. If $0<\alpha<1$, we must modify (ii) by either omitting the terms on the left with, say, $|c_n|\geqslant\frac{1}{2}$ or replacing $\log(1/|c_n|)$ by $\log(2+1/|c_n|)$ for all n; in either case the contribution of those terms on the left of (i) for which $|c_n|\geqslant1/n$ is less than

$$K\Sigma|c_n|^2=K\int_a^b|f|^2\,dx\leqslant K\int_a^b|f|\log^+|f|\,dx+K.$$

To prove (iii), observe that (i) holds with c_n^* for c_n, and so $c_n^*=o\{(\log n)^{-\alpha}\}$.]

9. If $1<r<s<\infty$, $\alpha=1/r-1/s$, and if $f\in L^r$, then the least value of the constant $A_{r,s}$ in the inequality $\|f_\alpha\|_s\leqslant A_{r,s}\|f\|_r$ (cf. (9·23)) satisfies

$$A_{r,s}\leqslant A_r s^{s'/r'}.$$

10. It is easy to deduce from (9·22) that if $f\in L^r$, $r>1$, then $f_{1/r}$ is integrable in every power. But more than this is true: there are positive constants λ, Λ such that if $\|f\|_r=1$, then

$$\int_0^{2\pi}\exp\{\lambda|f_{1/r}|^{r'}\}\,dx\leqslant\Lambda.$$

[Apply an argument similar to that used in the proof of (4·41) (i) to the result of the preceding Example.]

11. The first part of Theorem (9·22) is false for $r=1$; the following result is a substitute. If $0<\alpha<1$, $\beta=1/(1-\alpha)$, $f\in L(\log^+ L)^{1-\alpha}$, then $f_\alpha \in L^\beta$, and

$$\mathfrak{M}_\beta[f_\alpha] \leqslant A_\alpha \int_0^{2\pi} |f|(\log^+|f|)^{1-\alpha}\,dx + A_\alpha.$$

12. (i) If $0<\alpha<1$, $0<\beta<1$, $\alpha+\beta>1$, and if $f\in\Lambda_\alpha$, then f_β has a derivative $f'_\beta \in \Lambda_{\alpha+\beta-1}$.

(ii) If $0<\alpha<\gamma<1$, and if f has a derivative $f'\in\Lambda_\alpha$, then $f^\gamma \in \Lambda_{1+\gamma-\alpha}$.

[Corollaries of (8·13) and (8·14).]

13. Let $f(x)\sim\Sigma c_n e^{inx}$, $g(x)\sim\Sigma d_n e^{inx}$, $h(x)\sim\Sigma c_n d_n e^{inx}$.

(i) If $f\in\Lambda_\alpha$, $0<\alpha<1$, $g\in\Lambda_\beta$, $0<\beta<1$, then $h\in\Lambda_{\alpha+\beta}$, $h\in\Lambda_*$, $h'\in\Lambda_{\alpha+\beta-1}$, according as $\alpha+\beta<1$, $\alpha+\beta=1$, $\alpha+\beta>1$.

(ii) If $f\in\Lambda_\alpha$, $0<\alpha<1$, $g\in\Lambda_*$, then $h'\in\Lambda_\alpha$.

(iii) If $f\in\Lambda_*$, $g\in\Lambda_*$, then $h'\in\Lambda_*$.

[(i) Let $f(r,x)$, $g(r,x)$, $h(r,x)$ be the Poisson integrals of f, g, h. Then

$$h''(r^2,x)=\frac{1}{2\pi}\int_0^{2\pi} f'(r,t)\,g'(r,x-t)\,dt,$$

where differentiation is with respect to the angle.

If $f\in\Lambda_\alpha$, $g\in\Lambda_\beta$, then $f'(r,t)=O(\delta^{\alpha-1})$, $g'(r,t)=O(\delta^{\beta-1})$ ($\delta=1-r$; see Chapter VII, (5·1)(i)); hence $h''(r^2,x)=O(\delta^{\alpha+\beta-2})$, and so also $h''(r,x)=O(\delta^{\alpha+\beta-2})$. If $\alpha+\beta=1$, it is enough to apply Chapter VII, (5·1)(ii). If $\alpha+\beta<1$, Laplace's equation for $h(r,x)$ gives

$$r^{-1}\frac{\partial}{\partial r}\left(r\frac{\partial}{\partial r}h(r,x)\right)=O(\delta^{\alpha+\beta-2}),$$

and integrating with respect to r we obtain successively $\partial h(r,x)/\partial r=O(\delta^{\alpha+\beta-1})$, $h'(r,x)=O(\delta^{\alpha+\beta-1})$, $h(x)\in\Lambda_{\alpha+\beta}$. Similarly for $\alpha+\beta>1$. The conclusion holds if $f\in\Lambda^s_\alpha$, $g\in\Lambda^{s'}_\beta$, $1\leqslant s\leqslant\infty$.]

14. Let $k\geqslant1$, $1\leqslant r\leqslant2$, $1/r+1/r'=1$, $\lambda>0$, $\Sigma|c_\nu|^r<\infty$. If $\phi_1,\phi_2,\dots$ are Rademacher's functions and $f(t)=\Sigma c_\nu\phi_\nu(t)$, then

(i) $$\left(\int_0^1 |f(t)|^k\,dt\right)^{1/k}\leqslant Ak^{1/r'}(\Sigma|c_\nu|^r)^{1/r},$$

(ii) $$\int_0^1 \exp(\lambda|f|^{r'})\,dt<\infty,$$

where A is an absolute constant.

[(i) holds for $r=2$ (Chapter V, § 8) and for $r=1$; apply (1·11). (ii) Apply (4·41)(i).]

15. Results analogous to those of the preceding example hold for lacunary series $\Sigma A_{n_k}(x)$, $n_{k+1}/n_k>q>1$, except that $A=A_q$ in (i).

16. If $g\sim\Sigma\rho_\nu\cos(\nu x+x_\nu)$, $g\in L(\log^+L)^{1/r}$, $r\geqslant2$, and if $n_{k+1}/n_k>q>1$, then

$$(\Sigma\rho^r_{n_k})^{1/r}\leqslant A_{r,q}\int_0^{2\pi} g.(\log^+ g)^{1/r}\,dx+A_{r,q}$$

(cf. (7·6)).

17. If $r\geqslant2$, $\Sigma\rho^r_\nu<\infty$, $\epsilon>0$, then for almost all changes of sign the nth partial sum of

$$\Sigma\pm\rho_\nu\cos(\nu x+x_\nu)\quad\text{is}\quad o\{(\log n)^{1/r'}\}$$

uniformly in x, and the series $\Sigma\pm(\log\nu)^{-\epsilon-1/r'}\rho_\nu\cos(\nu x+x_\nu)$ converges uniformly [cf. Chapter V, (8·34)].

18. If $f\sim\Sigma c_n e^{inx}$, $g\sim\Sigma c'_n e^{inx}$, $f\in L^p$, $g\in L^{p'}$, $1<p<\infty$, then

$$\frac{1}{2\pi}\int_0^{2\pi} fg\,dx=\sum_{-\infty}^{+\infty} c_n c'_{-n},$$

where the series on the right converges (Chapter VII, (6·12)). The convergence is absolute if $p=2$, but for no other value of p. (M. Riesz [1].)

[Suppose, for example, that $1<p<2$. Let $0<\alpha<\frac{1}{2}, p<1/(1-\alpha)$. There is an $h \in \Lambda_{\frac{1}{2}}$, $h \sim \Sigma d_n e^{inx}$, such that $\Sigma |d_n| = \infty$ (Chapter VI, (3·1)). Write $d_n = |n|^{-\alpha} . d_n |n|^{\alpha} = c_n c_n'$, say, $(n \neq 0)$. Then $\Sigma c_n e^{inx} \in L^p$ (Chapter V, (2·1)), and $\Sigma c_n' e^{inx} = S[g]$, where g is continuous (cf. (8·14)) and so also in $L^{p'}$, and $\Sigma |c_n c_{-n}'| = \Sigma |d_n| = \infty$.]

A set E is said to be *a set of multiplicity in the restricted sense*, or set M_0, if there is a Fourier-Stieltjes series converging to 0 outside E, but not everywhere. A set which is not M_0 will be called *a set of uniqueness in the wide sense*, or set U_0; a series $\Sigma A_n(x)$ converging to 0 outside such a set is either identically 0, or else is not a Fourier–Stieltjes series. Every B-measurable set which is a U_0 is necessarily of measure 0.

19. (i) If E is M_0, then there is a perfect subset of E which is also M_0.

(ii) If E is an M_0, then there is a non-decreasing F, $F \not\equiv$ const., such that $S[dF]$ converges to 0 outside E.

(iii) If Borel sets $E_1, E_2, \dots, E_n, \dots$ are U_0, their sum $E = \Sigma E_i$ is also U_0.

(iv) If E is a U_0, and if an $S[dF]$ converges, or is only summable A, outside E to a finite integrable function f, then $S[dF] = S[f]$; in particular, if $S[dF]$ is summable A to 0 outside E, then $F =$ const.

[(i) We may suppose that $|E| = 0$, since otherwise the assertion is obvious. Suppose that an $S[dF] = \Sigma A_n(x)$ ($F \not\equiv$ const.) converges to 0 outside E. Hence F is singular (and continuous, since the coefficients of $\Sigma A_n(x)$ tend to 0). Let $\mathscr{E}$ be the set of points where F' exists and is $\pm\infty$. Since $S[dF]$ is summable A to $\pm\infty$ in $\mathscr{E}$ (Chapter III, § 7), $\mathscr{E}$ is a subset of E. It is well known (see, for example, Saks, *Theory of the integral*, p. 125) that the total variation of F over $\mathscr{E}$ is not 0. Let P be a perfect subset of $\mathscr{E}$ such that $\int_P |dF| \neq 0$, χ the characteristic function of P, and $G = \int_0^x \chi dF$. G is not constant, the coefficients of dG tend to 0 (cf. (10·9)), $S[dG]$ converges to 0 outside P.

(ii) Let G_1 and G_2 be the positive and negative variations of the G in (i); consider $S[dG_1]$ and $S[dG_2]$.

(iii) If E is M_0, then there is a perfect subset P of E, and an $S[dG]$, $G \not\equiv$ const., converging to 0 outside P. Since $P = PE = \Sigma PE_i$, there is an i_0 such that $\int_{PE_{i_0}} |dG| \neq 0$. Let P' be a perfect subset of PE_{i_0} such that $\int_{P'} |dG| \neq 0$. If χ' is the characteristic function of P', then $H(x) = \int_0^x \chi' dG$ is not constant, $S[dH]$ converges to 0 outside P', and so also outside $E_{i_0} \supset PE_{i_0} \supset P'$, which contradicts the hypothesis that E_{i_0} is a U_0.

(iv) If F is not absolutely continuous, the set $\mathscr{E}$ of points where $F' = \pm\infty$ is not empty, and $\int_{\mathscr{E}} |dF| \neq 0$. If P is a perfect subset of $\mathscr{E}$ such that $\int_P |dF| \neq 0$, χ the characteristic function of P, $G = \int_0^x \chi dF \not\equiv$ const., then $S[dG]$ converges to 0 outside P, and so also outside E, which contradicts the hypothesis that E is U_0.]

20. If $c_n \to 0$, then the function $\Phi(x) = \frac{1}{2}c_0 x^2 - \Sigma' c_n n^{-2} e^{inx}$, obtained by integrating $\Sigma c_n e^{inx}$ termwise twice, has the following property: for each u and any $\theta \neq 0$ the expression

$$(*)\ \left(\frac{4\theta}{N}\right)^{-1} \sum_{k=1}^{N} \left\{\Phi\left(u + \frac{2k\pi}{N} + \frac{2\theta}{N}\right) + \Phi\left(u + \frac{2k\pi}{N} - \frac{2\theta}{N}\right) - 2\Phi\left(u + \frac{2k\pi}{N}\right)\right\}$$

tends to $c_0 \theta$ as $N \to \infty$. Conversely, if $\Sigma c_n e^{inx}$ is such that the periodic part of the series defining $\Phi(x)$ is the Fourier series of a continuous function, and if $(*)$ tends to 0 for some $\theta \not\equiv 0 \bmod \pi$ and all u (not necessarily uniformly in u), then $c_n \to 0$. This is an analogue of (10·5) for general trigonometric series.

CHAPTER XIII

CONVERGENCE AND SUMMABILITY ALMOST EVERYWHERE

1. Partial sums of $S[f]$ for $f \in L^2$

In Chapter II, § 11, we showed that if $f \in L$, then the partial sums $S_n(x) = S_n(x; f)$ of the Fourier series

$$\tfrac{1}{2}a_0 + \sum_{1}^{\infty} (a_k \cos kx + b_k \sin kx) = \sum_{0}^{\infty} A_k(x) \qquad (1\cdot1)$$

of f are $o(\log n)$ almost everywhere. For $f \in L^2$, this estimate can be strengthened:

(1·2) THEOREM. *If $f \in L^2$, then $S_n(x) = o\{(\log n)^{\frac{1}{2}}\}$ almost everywhere. Furthermore, the function*

$$S^*(x) = \sup_{n \geqslant 2} \{|S_n(x)|/(\log n)^{\frac{1}{2}}\} \qquad (1\cdot3)$$

is in L^2 and

$$\mathfrak{M}_2[S^*] \leqslant A\mathfrak{M}_2[f]. \qquad (1\cdot4)$$

Here and in the rest of this section A denotes a positive absolute constant not necessarily always the same. The condition $n \geqslant 2$ in (1·3) can be omitted if we replace $\log n$ by $\log(n+2)$.

The method used to prove the theorem is of considerable intrinsic interest, and can be applied in other instances.

We begin with (1·4). Let $S_N^*(x)$ be defined like $S^*(x)$ except that n is bounded by the number N. Since $S_N^*(x)$ increases and tends to $S^*(x)$ as $N \to \infty$, it is enough to prove (1·4) with S^* replaced by S_N^*.

Let $n(x)$ be any step function taking integral values and such that $2 \leqslant n(x) \leqslant N$. Let $\lambda(x) = 1/\log n(x)$. It is enough to prove that

$$\left\{\int_{-\pi}^{\pi} \lambda(x)\, S_{n(x)}^2(x)\, dx\right\}^{\frac{1}{2}} \leqslant A\mathfrak{M}_2[f], \qquad (1\cdot5)$$

since

$$S_N^*(x) = \lambda^{\frac{1}{2}}(x)\,|S_{n(x)}(x)|$$

for a suitable $n(x)$.

We show first that for the Dirichlet kernel $D_n(x)$ we have

$$\int_{-\pi}^{\pi} |D_{n(x)}(x)|\, dx \leqslant A \log N, \qquad (1\cdot6)$$

a generalization of the fact that the Nth Lebesgue constant is $O(\log N)$ (Chapter II, (12·1)). Since $|D_{n(x)}(x)|$ is majorized by both AN and $A/|x|$, the integral in (1·6) does not exceed

$$\int_{|x| \leqslant 1/N} AN\, dx + \int_{1/N \leqslant |x| \leqslant \pi} A\,|x|^{-1}\, dx \leqslant A + A \log N \leqslant A \log N.$$

Returning to (1·5), we use the fact that the left-hand side is

$$\int_{-\pi}^{\pi} \lambda^{\frac{1}{2}}(x)\, S_{n(x)}(x)\, \phi(x)\, dx$$

for some ϕ with $\mathfrak{M}_2[\phi]=1$. Let $\lambda^{\frac{1}{2}}(x)\,\phi(x)=\psi(x)$. Then the left-hand side of (1·5) is

$$\begin{aligned}\int_{-\pi}^{\pi} S_{n(x)}(x)\,\psi(x)\,dx &= \int_{-\pi}^{\pi}\psi(x)\,dx\left\{\frac{1}{\pi}\int_{-\pi}^{\pi} f(t)\,D_{n(x)}(x-t)\,dt\right\}\\ &= \int_{-\pi}^{\pi} f(t)\,dt\left\{\frac{1}{\pi}\int_{-\pi}^{\pi}\psi(x)\,D_{n(x)}(x-t)\,dx\right\}\\ &\leqslant \mathfrak{M}_2[f]\,.\,\mathfrak{M}_2\left\{\frac{1}{\pi}\int_{-\pi}^{\pi}\psi(x)\,D_{n(x)}(x-t)\,dx\right\},\end{aligned}$$

and it remains to show that the last factor does not exceed A.

Its square can be written

$$\begin{aligned}&\int_{-\pi}^{\pi}\left\{\frac{1}{\pi}\int_{-\pi}^{\pi}\psi(x)\,D_{n(x)}(x-t)\,dx\right\}\left\{\frac{1}{\pi}\int_{-\pi}^{\pi}\psi(y)\,D_{n(y)}(y-t)\,dy\right\}dt\\ &\quad=\frac{1}{\pi}\int_{-\pi}^{\pi}\int_{-\pi}^{\pi}\psi(x)\,\psi(y)\left\{\frac{1}{\pi}\int_{-\pi}^{\pi}D_{n(x)}(x-t)\,D_{n(y)}(y-t)\,dt\right\}dx\,dy\\ &\quad=\frac{1}{\pi}\int_{-\pi}^{\pi}\int_{-\pi}^{\pi}\psi(x)\,\psi(y)\,D_{n(x,y)}(x-y)\,dx\,dy,\end{aligned}\tag{1·7}$$

where $$n(x,y)=\min\{n(x),n(y)\}.$$

The last integral does not exceed

$$\frac{1}{2\pi}\int_{-\pi}^{\pi}\int_{-\pi}^{\pi}\psi^2(x)\,|\,D_{n(x,y)}(x-y)\,|\,dx\,dy+\frac{1}{2\pi}\int_{-\pi}^{\pi}\int_{-\pi}^{\pi}\psi^2(y)\,|\,D_{n(x,y)}(x-y)\,|\,dx\,dy=J_1+J_2.$$

Integrating in J_1 with respect to y and using (1·6), we get

$$J_1\leqslant\frac{1}{2\pi}\int_{-\pi}^{\pi}\psi^2(x)\,A\log n(x)\,dx=A\int_{-\pi}^{\pi}\phi^2(x)\,dx=A.$$

Similarly, $J_2\leqslant A$. Thus the right-hand side of (1·7) does not exceed A and (1·5) is established.

The inequality (1·4) shows that almost everywhere we have $S^*(x)<\infty$, that is, $S_n(x)=O\{(\log n)^{\frac{1}{2}}\}$. To refine the '$O$' to '$o$' we have to show that the function

$$S_*(x)=\limsup_{n\to\infty}\{|\,S_n(x)\,|/(\log n)^{\frac{1}{2}}\}$$

is zero almost everywhere. Since $S_*\leqslant S^*$, the inequality (1·4) gives $\mathfrak{M}_2[S_*]\leqslant A\mathfrak{M}_2[f]$. But $\mathfrak{M}_2[f]$ may be made arbitrarily small, without changing S_*, by subtracting from f a suitable polynomial. Thus $\mathfrak{M}_2[S_*]=0$, whence $S_*=0$ almost everywhere.

(1·8) THEOREM. *If $\Sigma A_n(x)$ is the Fourier series of an $f\in L^2$, then the partial sums $s_n(x)$ of*

$$\sum_{2}^{\infty}(a_n\cos nx+b_n\sin nx)/(\log n)^{\frac{1}{2}}\tag{1·9}$$

converge almost everywhere. Moreover, for $s^*(x) = \sup |s_n(x)|$ *we have*

$$\mathfrak{M}_2[s^*] \leqslant A\mathfrak{M}_2[f]. \tag{1·10}$$

Let $l_0 = l_1 = 0$, $l_n = (\log n)^{-\frac{1}{2}}$ for $n \geqslant 2$. If the $\sigma_n(x)$ are the (C, 1) means of $S[f]$ and

$$\sigma^*(x) = \sup_n |\sigma_n(x)|,$$

repeated summation by parts gives

$$s_n = \sum_0^n (S_k - S_{k-1}) l_k = \sum_0^{n-1} S_k \Delta l_k + S_n l_n = \sum_0^{n-2} (k+1)\sigma_k \Delta^2 l_k + n\sigma_{n-1}\Delta l_{n-1} + S_n l_n, \tag{1·11}$$

$$|s_n(x) - S_n(x) l_n| \leqslant \sigma^*(x)\left\{\sum_0^{n-2}(k+1)|\Delta^2 l_k| + n\Delta l_{n-1}\right\} \leqslant A\sigma^*(x),$$

since $\Delta^2 l_k \geqslant 0$ for $k \geqslant 2$, $\Sigma(k+1)\Delta^2 l_k$ converges and $n\Delta l_{n-1} \to 0$ (Chapter III, (4·1)). Thus

$$s^* \leqslant S^* + A\sigma^*,$$

$$\mathfrak{M}_2[s^*] \leqslant \mathfrak{M}_2[S^*] + A\mathfrak{M}_2[\sigma^*] \leqslant A\mathfrak{M}_2[f] + A\mathfrak{M}_2[f] = A\mathfrak{M}_2[f],$$

since (Chapter IV, § 7)

$$\mathfrak{M}_2[\sigma^*] \leqslant A\mathfrak{M}_2[f]. \tag{1·12}$$

This proves (1·10). The convergence of (1·9) follows from the estimate $S_n = o\{(\log n)^{\frac{1}{2}}\}$ and from Chapter III, (4·3).

(1·13) Theorem. *If* $\Sigma(\alpha_k^2 + \beta_k^2)\log k$ *is finite, the partial sums* t_n *of*

$$\sum_2^\infty (\alpha_k \cos kx + \beta_k \sin kx) = S[f]$$

converge almost everywhere and $t^*(x) = \sup_n |t_n(x)|$ *satisfies the inequality* $\mathfrak{M}_2[t^*] \leqslant A\mathfrak{M}_2[f]$.

This is an obvious variant of (1·8). It generalizes Theorem (1·8) of Chapter IV. It can also be stated in the following equivalent form.

(1·14) Theorem. *If* $f \in L$ *and the function*

$$g(x) = \left\{\int_0^\pi \frac{[f(x+t) - f(x-t)]^2}{t} dt\right\}^{\frac{1}{2}} \tag{1·15}$$

belongs to L^2, *then* $S[f]$ *converges almost everywhere.*

Let α_n, β_n be the coefficients of f. By Parseval's formula,

$$\mathfrak{M}_2^2[g] = \int_0^\pi \frac{dt}{t}\int_{-\pi}^\pi [f(x+t) - f(x-t)]^2 dx = 4\pi\int_0^\pi \frac{dt}{t}\sum_1^\infty (\alpha_n^2 + \beta_n^2)\sin^2 nt$$

$$= 4\pi\sum_1^\infty (\alpha_n^2 + \beta_n^2)\int_0^\pi \frac{\sin^2 nt}{t} dt,$$

and the convergence of $S[f]$ will follow from (1·13) if we show that the last integral is exactly of order $\log n$; actually, it is

$$\frac{1}{2}\int_0^\pi \frac{1-\cos 2nt}{t} dt = \frac{1}{2}\int_0^\pi \tilde{D}_{2n}^*(t)\,dt + O(1) \simeq \frac{1}{\pi}\log n,$$

by Chapter II, (12·3).

Incidentally, we have shown that if $g \in L^2$, then also $f \in L^2$.

The following corollary of (1·14) will be needed later:

(1·16) Theorem. *If $\alpha > \frac{1}{2}$ and if*

$$|f(x+h)-f(x)| \leqslant A(\log 1/|h|)^{-\alpha} \quad (0 \leqslant x \leqslant 2\pi),$$

for, say, $|h| \leqslant \frac{1}{2}$, then $S[f]$ converges almost everywhere.

For in this case g is bounded and accordingly in L^2.

The problem of what happens in the case when $\alpha = \frac{1}{2}$ is open. It is well to remember that, even for $\alpha = 1$, $S[f]$ may diverge at some points, though its partial sums are uniformly bounded (Chapter VIII, (2·1)). For $\alpha > 1$, $S[f]$ is, of course, uniformly convergent (Chapter II, (6·8)).

Remark. While the estimate $S_n(x) = o(\log n)$ holds at every point of the Lebesgue set (Chapter II, (11·9)), and in particular at every point of continuity of f, the estimate $S_n(x) = o\{(\log n)^{\frac{1}{2}}\}$ of Theorem (1·2) has only been shown to hold almost everywhere, and it is not known what property of the function (non-trivially) guarantees this estimate at a given point. That mere continuity is not enough, follows from Chapter VIII, (1·2).

It is conceivable that the estimate $S_n = o\{(\log n)^{\frac{1}{2}}\}$ in (1·2) is best possible. This is a rather strong conjecture, since it implies in particular the existence of an $f \in L^2$ with $S[f]$ divergent almost everywhere. The problem is open. It is easy, however, to prove the following result:

(1·17) Theorem. *Let $f \in L^2$.*

(i) *If $0 < n_1 < n_2 < \ldots$ is a sequence of indices satisfying a condition $n_{k+1}/n_k > q > 1$ for all k, then the partial sums S_{n_k} of $S[f]$ converge almost everywhere.*

(ii) *The function $S^*(x) = \sup_k |S_{n_k}(x)|$ satisfies an inequality*

$$\mathfrak{M}_2[S^*] \leqslant A_q \mathfrak{M}_2[f], \tag{1·18}$$

with A_q depending on q only.

The proof of (ii) is based on the following corollary of (1·12) and Chapter III, (1·29):

(1·19) Lemma. *If the Fourier series of an integrable function $f(x)$ possesses infinitely many gaps $m_k < j \leqslant m_k'$ with $m_k'/m_k > q > 1$, then the partial sums S_{m_k} (and $S_{m_k'}$) converge almost everywhere to $f(x)$. The functions $s^*(x) = \sup |S_{m_k}(x)|$ and $\sigma^*(x) = \sup |\sigma_n(x)|$ satisfy an inequality*

$$s^*(x) \leqslant A_q \sigma^*(x). \tag{1·20}$$

Returning to (i), let $n_0 = 0$, and let

$$\Delta_0 = A_0(x), \quad \Delta_{k+1} = \sum_{n_k+1}^{n_{k+1}} A_j(x) \quad (k = 0, 1, 2, \ldots),$$

where $A_j(x)$ are the terms of $S[f]$. We split $S[f]$ into the series

$$T_1 = \Delta_0 + \Delta_2 + \Delta_4 + \ldots, \quad T_2 = \Delta_1 + \Delta_3 + \ldots.$$

By the Riesz-Fischer theorem, T_1 and T_2 are Fourier series of functions f' and f'', say. For T_1, n_k is then either of type m_k or of type m_k' above, and similarly for T_2. By (1·19), the partial sums S'_{n_k} and S''_{n_k} of T_1 and T_2 converge almost everywhere. The same holds for $S_{n_k} = S'_{n_k} + S''_{n_k}$, and (i) is established.

Passing to (ii), let $f', f'', S'_{n_k}, S''_{n_k}$ have the same meaning as before. Observing that if the function f in (1·19) is in L^2 the inequalities (1·12) and (1·20) also imply

$$\mathfrak{M}_2[s^*] \leqslant A_q \mathfrak{M}_2[\sigma^*] \leqslant A_q \mathfrak{M}_2[f], \tag{1·21}$$

we have

$$S^* \leqslant \sup |S'_{n_k}| + \sup |S''_{n_k}|,$$

$$\mathfrak{M}_2[S^*] \leqslant A_q \mathfrak{M}_2[f'] + A_q \mathfrak{M}_2[f''] \leqslant A_q \mathfrak{M}_2[f],$$

since, by Parseval's formula, $\mathfrak{M}_2[f]$ majorizes both $\mathfrak{M}_2[f']$ and $\mathfrak{M}_2[f'']$. This completes the proof of (ii).

It may be added that, since the hypotheses of (1·17) imply that $\mathfrak{M}_2[f - S_n] \to 0$, the existence of some sequence $\{S_{n_k}\}$ converging almost everywhere follows already from Chapter I, (11·6). The point of (1·17) is that in our case we have an $\{n_k\}$ independent of f.

The theorem which follows throws some light upon the still unsolved problem of the existence of an $f \in L^2$ with $S[f]$ divergent almost everywhere.

If $S[f]$ converges almost everywhere, then the function

$$s^*(x) = \sup_n |S_n(x)|,$$

where S_n is the nth partial sum of $S[f]$, is finite almost everywhere. However, more than that is true:

(**1·22**) Theorem. *If every* $S[f]$, $f \in L^2$, *converges almost everywhere, then the operation* $Tf = s^*$ *is of weak type* (2, 2) (Chapter XII, § 4), *that is, if* $E(y) = E(y, f)$ *is the set of points where* $s^*(x) > y > 0$, *then*

$$|E(y)| \leqslant Ay^{-2} \|f\|_2^2, \tag{1·23}$$

where A *is an absolute constant.*

Suppose that T is not of weak type (2, 2). Then for each $N = 1, 2, \ldots$ there is a function $p_N(x)$ and numbers $\omega(N)$ and y_N such that

$$\|p_N\|_2 = 1, \quad \omega(N) \to +\infty, \quad |E(y_N, p_N)| > \omega(N)\, y_N^{-2}.$$

It is not difficult to see that for the p_N we may take polynomials.

From $\omega(N) \to \infty$ we deduce that $y_N \to \infty$. Hence we can find a non-decreasing sequence $\{N_k\}$ such that

$$\Sigma y_{N_k}^{-2} < \infty, \quad \Sigma \omega(N_k)\, y_{N_k}^{-2} = \infty.$$

The last equation implies that $\Sigma\, |E(y_{N_k}, p_{N_k})|$ diverges.

Given any (periodic) set $\mathscr{E}$ we denote by $\mathscr{E}^x$ the translate of $\mathscr{E}$ by x. We need the following lemma.

(**1·24**) Lemma. *Given any sequence* $\{\mathscr{E}_k\}$ *of sets such that* $\Sigma\, |\mathscr{E}_k| = \infty$, *there exist numbers* x_k *such that almost every point belongs to infinitely many* $\mathscr{E}_k^{x_k}$.

We take this lemma temporarily for granted, and apply it to $\{\mathscr{E}_k\} = \{E(y_{N_k}, p_{N_k})\}$. Choose integers m_k which increase so fast that the polynomials $e^{im_k x} P_{N_k}(x)$ do not overlap. Then

$$\Sigma\, e^{im_k x} p_{N_k}(x - x_k)\, y_{N_k}^{-1} \tag{1·25}$$

converges almost everywhere to an $f \in L^2$:

$$\|\Sigma\, e^{im_k x} p_{N_k}(x - x_k)\, y_{N_k}^{-1}\|_2^2 = \Sigma\, \|p_{N_k}(x - x_k)\|_2^2\, y_{N_k}^{-2} = \Sigma y_{N_k}^{-2} < \infty.$$

Written out in full, (1·25) is $S[f]$. If $x - x_k \in E(y_{N_k}, p_{N_k})$, then a connected block of terms of $S[f]$ exceeds, numerically, $y_{N_k}\, y_{N_k}^{-1} = 1$. Hence $S[f]$ diverges at each point belonging to infinitely many $E^{x_k}(y_{N_k}, p_{N_k})$, that is diverges almost everywhere.

Corollaries. (i) If for some $0 < \epsilon < 2$ the integral

$$\int_0^{2\pi} \{s^*(x)\}^{2-\epsilon}\, dx \tag{1·26}$$

is not uniformly bounded for all f with $\|f\|_2 = 1$, then there is a Fourier series of the class L^2 diverging almost everywhere.

To see this, observe that the integral (1·26) is $(2-\epsilon)\int_0^\infty |E(y)|\,y^{1-\epsilon}\,dy$ (cf. Chapter XII, (4·8)). Since $y^{1-\epsilon}$ is integrable over $0<y\leq 1$, the unboundedness of (1·26) implies that of

$$\int_1^\infty |E(y)|\,y^2\frac{dy}{y^{1+\epsilon}},$$

and since $y^{-1-\epsilon}$ is integrable over $(1,\infty)$, $|E(y)|\,y^2$ is unbounded in y and f, $\|f\|_2=1$.

(ii) The case $\epsilon=1$ is of interest. The unboundedness of (1·26) in f, $\|f\|_2=1$, is then equivalent to that of

$$\int_0^{2\pi} S_{n(x)}(x;f)\,dx,$$

where $n(x)$ is any non-negative integral-valued function of x, taking, say, a finite number of values only. The last integral is equal to

$$\int_0^{2\pi} f(t)\left\{\frac{1}{\pi}\int_0^{2\pi} D_{n(x)}(x-t)\,dx\right\}dt=\int_0^{2\pi} f(t)\,I(t)\,dt,$$

say, and its unboundedness is, by Theorem (9·14) of Chapter I, equivalent to that of

$$\|I\|_2^2=\int_0^{2\pi}\left\{\frac{1}{\pi}\int_0^{2\pi} D_{n(x)}(x-t)\,dx\right\}^2 dt=\frac{1}{\pi}\int_0^{2\pi}\int_0^{2\pi} D_{n(x,y)}(x-y)\,dx\,dy, \tag{1·27}$$

where $n(x,y)=\min\{n(x),n(y)\}$. Hence, *if the last integral in* (1·27) *is not uniformly bounded for all* $n(x)$, *there is an* $\mathsf{S}[f]$, $f\in L^2$, *diverging almost everywhere.*

We now pass to the proof of Lemma (1·24). If $C\mathscr{E}_k^{x_k}$ is the complement of $\mathscr{E}_k^{x_k}$, then the set of points which belong to a finite number of $\mathscr{E}_k^{x_k}$ only is

$$\sum_{p=1}^{\infty} C\mathscr{E}_p^{x_p}\,.\,C\mathscr{E}_{p+1}^{x_{p+1}}\,.\,C\mathscr{E}_{p+2}^{x_{p+2}}\ldots, \tag{1·28}$$

and so is contained in

$$\prod_1^{p_1} C\mathscr{E}_k^{x_k}+\prod_{p_1+1}^{p_2} C\mathscr{E}_k^{x_k}+\prod_{p_2+1}^{p_3} C\mathscr{E}_{pk}^{x_k}+\ldots, \tag{1·29}$$

where $p_1<p_2<p_3<\ldots$ will be chosen in a moment. Let $\chi_k(t)$ be the characteristic function of $C\mathscr{E}_k$. Then the characteristic function of the first product in (1·29) is $\chi_1(t+x_1)\,\chi_2(t+x_2)\ldots\chi_{p_1}(t+x_k)$. Since

$$\frac{1}{(2\pi)^{p_1}}\int_0^{2\pi}\ldots\int_0^{2\pi}\left\{\int_0^{2\pi}\chi_1(t+x_1)\ldots\chi_{p_1}(t+x_{p_1})\,dt\right\}dx_1\ldots dx_{p_1}=\prod_1^{p_1}\frac{1}{2\pi}\int_0^{2\pi}\chi_k(t)\,dt=\prod_1^{p_1}\left(1-\frac{|\mathscr{E}_k|}{2\pi}\right)$$

and since, by hypothesis, $\Sigma|\mathscr{E}_k|=\infty$, the last product can be made less than $\frac{1}{2}$ by taking p_1 large enough. It follows that there are $x_1, x_2, \ldots, x_{p_1}$ such that the measure of the first term in (1·29) is less than $\frac{1}{2}$. Similarly, we can select p_2 and $x_{p_1+1}, \ldots, x_{p_2}$ so that the second term in (1·29) has measure less than $(\frac{1}{2})^2$, and so on. Since now the measures of the terms of the series (1·29) have a finite sum, we easily see that (1·28) is of measure 0, and (1·24) follows.

2. Order of magnitude of S_n for $f\in L^p$

Theorems (1·2) and (1·17) have analogues for functions in L^p. In this section we consider only generalizations of (1·2); for generalizations of (1·17) see Chapter XV, (4·4). The main results can be stated as follows:

(2·1) THEOREM. *Let* $1<p\leq 2$, $f\in L^p$. *Then*

$$S_n(x)=o\{(\log n)^{1/p}\} \tag{2·2}$$

almost everywhere. Moreover, the function

$$S^*(x)=\sup_{n\geq 2}\{|S_n(x)|/(\log n)^{1/p}\} \tag{2·3}$$

satisfies an inequality $$\mathfrak{M}_p[S^*] \leqslant A_p \mathfrak{M}_p[f]. \tag{2·4}$$

The inequality (2·4) *fails for* $p=1$, *but is valid if* $S[f]$ *is of power series type, and then we may also replace the* A_p, $1 \leqslant p \leqslant 2$, *of* (2·4) *by an absolute constant* A.

(2·5) THEOREM. *If* $f \in L^q$, $2 \leqslant q \leqslant \infty$, *the function*

$$s^*(x) = \sup_{n \geqslant 2} |S_n(x)|/(\log n)^{1/p} \quad (p=q'),$$

satisfies the inequality $$\mathfrak{M}_q[s^*] \leqslant A \mathfrak{M}_q[f]. \tag{2·6}$$

Whether or not (2·1) holds for $p > 2$ is an open problem. It is not inconceivable that it does not, and that the estimate $S_n = o\{(\log n)^{\frac{1}{2}}\}$, valid almost everywhere for f in L^2, cannot be improved for f in L^q, $q > 2$, or even for f continuous.

The proof of (2·5) is immediate. The theorem is valid for $q=2$ and also for $q=\infty$, since

$$|S_n(x)| \leqslant A \log n$$

if $|f| \leqslant 1$ and $n \geqslant 2$ (Chapter II, § 12). Fix any step-function $n(x)$, $2 \leqslant n(x) \leqslant N$, taking integral values only and consider the linear operation

$$g = Tf = S_{n(x)}(x)/\log n(x). \tag{2·7}$$

The inequality $$\left\{\int_0^{2\pi} |g(x)|^q \log n(x)\, dx\right\}^{1/q} \leqslant A_q \mathfrak{M}_q[f] \tag{2·8}$$

is valid for $q=2$ and $q=\infty$. On the left we have a norm $\| g \|_q$ taken with respect to the measure $d\mu = \log n(x)\, dx$. By Theorem (1·11) of Chapter XII, (2·8) holds for $2 < q < \infty$, with $A_q \leqslant \max(A_2, A_\infty)$. By choosing $n(x)$ suitably, we deduce from (2·8) the inequality (2·6) with s^* replaced by s_N^*, the latter function being defined like s^* except that n is to be bounded by N. The result now follows on making N tend to infinity.

Passing to (2·1), we first observe that we could prove (2·4) in exactly the same way as we proved (2·6), by an interpolation of the linear operation (2·7) between $p=1$ and $p=2$, if the inequality (2·4) were true for $p=1$. But, as we shall show below, this is not the case. If we temporarily take for granted, however, that (2·4) *holds for* $p=1$ *and* $S[f]$ *of power series type*, an application of the interpolation theorem (3·9) of Chapter XII gives (2·4) for such series, even with A_p replaced by A.

Now let f be any real-valued function in L^p, $1 < p \leqslant 2$. We set $F = f + i\tilde{f}$, where $\tilde{f}$ is the conjugate of f. Then (2·4) holds for F, since $S[F]$ is of power series type. Since the function S^* for f is majorized by the function S^* for F, and since $\mathfrak{M}_p[F] \leqslant A_p \mathfrak{M}_p[f]$ (Chapter VII, (2·4)), the inequality (2·4) for f follows.

It remains therefore to prove (2·4) for $p=1$ and

$$f \sim c_0 + c_1 e^{ix} + c_2 e^{i2x} + \ldots.$$

The function $$F(z) = c_0 + c_1 z + c_2 z^2 + \ldots$$

is regular for $|z| < 1$ and of class H, and its boundary values $F(e^{ix})$ coincide almost everywhere with $f(x)$. Without loss of generality we may suppose that F has no zeros for $|z| < 1$ (by Chapter VII, (7·23), F is a sum of functions F_1 and F_2 without zeros, such that $\mathfrak{M}[F_1(e^{ix})]$ and $\mathfrak{M}[F_2(e^{ix})]$ do not exceed $2\mathfrak{M}[F(e^{ix})]$). Then $F(z) = G^2(z)$, where

$$G(z) = d_0 + d_1 z + d_2 z^2 + \ldots \tag{2·9}$$

is of class H^2. We shall have to consider the Cesàro sums and means of the series (2·9),

$$S_n^\alpha(z) = S_n^\alpha(z;\, G) = \sum_{\nu=0}^{n} A_{n-\nu}^\alpha d_\nu z^\nu,$$

$$\tau_n^\alpha(z) = S_n^\alpha(z)/A_n^\alpha,$$

where
$$A_n^\alpha = \frac{(\alpha+1)(\alpha+2)\dots(\alpha+n)}{n!} \simeq \frac{n^\alpha}{\Gamma(\alpha+1)} \tag{2·10}$$

(see Chapter III, § 1). Observe that if Σw_n is the Cauchy product of Σu_n and Σv_n, and if W_n^γ, U_n^γ, V_n^γ denote the (C, γ) sums of the series, then the equations

$$\sum_0^\infty W_n^{\alpha+\beta+1} z^n = \frac{1}{(1-z)^{\alpha+\beta+2}} \sum_0^\infty w_n z^n = \frac{1}{(1-z)^{\alpha+1}} \sum_0^\infty u_n z^n \, \frac{1}{(1-z)^{\beta+1}} \sum_0^\infty v_n z^n$$

imply that

$$W_n^{\alpha+\beta+1} = \sum_{\nu=0}^{n} U_\nu^\alpha V_{n-\nu}^\beta.$$

Hence the relation $F = G^2$ leads to the equation

$$S_n(e^{ix};\, F) = \sum_{\nu=0}^{n} S_\nu^{-\frac12}(e^{ix};\, G)\, S_{n-\nu}^{-\frac12}(e^{ix};\, G). \tag{2·11}$$

(2·12) LEMMA. *For any $G(z) \in H^2$ we have*

$$\int_0^{2\pi} \left\{ \sum_{n=0}^{\infty} \frac{|\tau_n^{-\frac12}(e^{ix}) - \tau_n^{\frac12}(e^{ix})|^2}{(n+1)\log(n+2)} \right\} dx \leqslant A \int_0^{2\pi} |G(e^{ix})|^2\, dx. \tag{2·13}$$

We first verify, using (2·10), that

$$\tau_n^{\alpha-1}(z) - \tau_n^\alpha(z) = \frac{1}{\alpha A_n^\alpha} \sum_{\nu=0}^{n} \nu A_{n-\nu}^{\alpha-1} d_\nu z^\nu.$$

Then
$$\frac{1}{2\pi} \int_0^{2\pi} |\tau_n^{-\frac12}(e^{ix}) - \tau_n^{\frac12}(e^{ix})|^2\, dx = \frac{4}{(A_n^{\frac12})^2} \sum_{\nu=0}^{n} \nu^2 (A_{n-\nu}^{-\frac12})^2 |d_\nu|^2. \tag{2·14}$$

Let $l_n = 1/\log(n+2)$. Since A_n^α is exactly of order n^α, the left-hand side of (2·13) does not exceed

$$A \sum_{n=0}^{\infty} \frac{l_n}{(n+1)^2} \sum_{\nu=0}^{n} |d_\nu|^2 \nu^2 (n-\nu+1)^{-1} = A \sum_{\nu=1}^{\infty} \nu^2 |d_\nu|^2 \sum_{n=\nu}^{\infty} \frac{l_n}{(n+1)^2 (n-\nu+1)}.$$

We split the inner sum on the right into two, extended over the ranges $\nu \leqslant n \leqslant 2\nu$ and $n > 2\nu$. Correspondingly, the whole expression is split into two parts, P and Q, where

$$P \leqslant A \sum_{\nu=1}^{\infty} |d_\nu|^2 \nu^2 l_\nu (\nu+1)^{-2} \sum_{n=\nu}^{2\nu} \frac{1}{n-\nu+1} \leqslant A \sum_{\nu=1}^{\infty} |d_\nu|^2,$$

$$Q \leqslant A \sum_{\nu=1}^{\infty} |d_\nu|^2 \nu^2 \sum_{n=2\nu+1}^{\infty} \frac{l_n}{n^3} \leqslant A \sum_{\nu=1}^{\infty} |d_\nu|^2,$$

since $l_n < 1$ for $n > 1$. Thus the left-hand side of (2·13), which is majorized by $P + Q$, does not exceed $A\Sigma |d_\nu|^2$, and the lemma is established.

An application of Schwarz's inequality to (2·11) gives

$$| S_n(e^{ix}; F) | \leqslant \sum_{\nu=0}^{n} | S_\nu^{-\frac{1}{2}}(e^{ix}; G) |^2 = \sum_{\nu=0}^{n} | \tau_\nu^{-\frac{1}{2}}(e^{ix}; G) A_\nu^{-\frac{1}{2}} |^2$$

$$\leqslant A \sum_{\nu=0}^{n} \frac{| \tau_\nu^{-\frac{1}{2}} - \tau_\nu^{\frac{1}{2}} |^2}{\nu+1} + A \sum_{\nu=0}^{n} \frac{| \tau_\nu^{\frac{1}{2}} |^2}{\nu+1}$$

$$= U_n(x) + V_n(x). \tag{2·15}$$

Now let
$$\psi(x) = \sup_n | \tau_n^{\frac{1}{2}}(e^{ix}; G) |,$$

and let $\phi^2(x)$ be the integrand on the left of (2·13). From Chapter IV, (7·8), we have

$$\mathfrak{M}_2[\psi] \leqslant A \mathfrak{M}_2[G(e^{ix})], \tag{2·16}$$

and (2·13) means that
$$\mathfrak{M}_2[\phi] \leqslant A \mathfrak{M}_2[G(e^{ix})]. \tag{2·17}$$

On the other hand, clearly,

$$U_n(x) \leqslant A\phi^2(x) \log(n+2) \tag{2·18}$$

and
$$V_n(x) \leqslant A\psi^2(x) \sum_0^n \frac{1}{\nu+1} \leqslant A\psi^2(x) \log(n+2). \tag{2·19}$$

From (2·15), (2·18), and (2·19) we therefore get

$$S^*(x) = \sup_{n \geqslant 2} \frac{| S_n(e^{ix}; F) |}{\log n} \leqslant A \sup_{n \geqslant 0} \frac{|S_n(e^{ix}; F)|}{\log(n+2)}$$

$$\leqslant A[\phi^2(x) + \psi^2(x)],$$

and so, by (2·16) and (2·17),

$$\mathfrak{M}[S^*] \leqslant A \mathfrak{M}_2^2[G(e^{ix})] = A \mathfrak{M}[F(e^{ix})].$$

This completes the proof of the inequality (2·4) for $1 < p \leqslant 2$. It implies that at almost every point we have (2·2), with 'O' instead of 'o'. The passage from 'O' to 'o' is the same as in the case $p = 2$.

That we have (2·2) almost everywhere, for $f \in L$ and $p = 1$, is a classical result (Chapter II, (11·9)). That in this case, however, the function

$$S^*(x) = \sup_n | S_n(x) | / \log n \tag{2·20}$$

need not be integrable, can be shown by the example of the function

$$f(x) \sim \tfrac{1}{2}\lambda_0 + \sum_1^\infty \lambda_n \cos nx,$$

where $\lambda_n = 1/\log\log n$ for n large and is positive, decreasing and convex (see Chapter V, (1·5)). Summing by parts twice we get

$$S_n(x) = \sum_0^{n-2} (m+1) \Delta^2\lambda_m K_m(x) + n\Delta\lambda_{n-1} K_{n-1}(x) + \lambda_n D_n(x).$$

The first two terms on the right are non-negative and the third, divided by $\log n$, is

$$(\log n \log\log n)^{-1} \frac{\sin nx}{x} + o(1),$$

where the $o(1)$ is uniform in x. Since for $n=[\pi/2x]$ this expression is exactly of order

$$\frac{1}{x\log(1/x)\log\log(1/x)}$$

as $x\to+0$, the function (2·20) is not integrable.

(2·21) Theorem. *If f is in L^p, $1<p\leqslant 2$, and has coefficients a_n, b_n, then the partial sums $s_n(x)$ of*

$$\sum_{k=2}^{\infty}(a_k\cos kx+b_k\sin kx)/(\log k)^{1/p} \tag{2·22}$$

converge almost everywhere, and the function $s^*(x)=\sup_n|s_n(x)|$ satisfies an inequality

$$\mathfrak{M}_p[s^*]\leqslant A_p\mathfrak{M}_p[f]. \tag{2·23}$$

This follows from (2·1) exactly as (1·8) followed from (1·2).

If $f\in L$, the series (2·22), with $p=1$, converges almost everywhere (Chapter III, (4·4)), but (2·23) no longer holds (see also Example 4 on p. 197).

3. A test for the convergence of S[f] almost everywhere

The Dini-Lipschitz theorem asserts that S[f] converges, even uniformly, if

$$f(x+h)-f(x)=o\left\{\frac{1}{\log 1/|h|}\right\}\quad(h\to 0), \tag{3·1}$$

uniformly in x. We have already pointed out (Chapter VIII, p. 303) that (3·1) may hold at an individual point without S[f] converging there. We now show that if (3·1) holds at every point of a set E, not necessarily uniformly in x, then S[f] converges almost everywhere in E. More generally:

(3·2) Theorem. *Suppose that an $f\in L$ satisfies for every $x\in E$ the condition*

$$\frac{1}{h}\int_0^h|f(x+t)-f(x)|\,dt=O\left\{\frac{1}{\log 1/|h|}\right\}\quad(h\to 0). \tag{3·3}$$

Then S[f] converges almost everywhere in E.

It is easily verified that at an individual point x the relation

$$f(x+h)-f(x)=O\left\{\frac{1}{\log 1/|h|}\right\}, \tag{3·4}$$

—and *a fortiori* (3·1)—implies (3·3).

It may be observed—a result which we shall need below—that if (3·4) holds uniformly in x, the convergence almost everywhere of S[f] follows from Theorem (1·16). Theorem (3·2) will therefore be established if we show that under its hypotheses

$$f(x)=\phi(x)+\psi(x), \tag{3·5}$$

where ϕ satisfies a condition analogous to (3·4) uniformly in x, and ψ satisfies the Dini condition (see Chapter II, (6·1), or (3·8) below) in a subset of E whose measure (for a suitable decomposition (3·5)) is arbitrarily close to $|E|$.

From the hypotheses of (3·2), there follows the existence of a perfect set $P \subset E$, with $|E-P|$ arbitrarily small, and of two positive numbers M and δ such that $\delta < 1$ and

$$\left|\frac{1}{h}\int_0^h |f(x+t)-f(x)|\,dt\right| \leqslant \frac{M}{\log 1/|h|} \quad \text{for} \quad x \in P,\ |h| < \delta. \tag{3·6}$$

For if E_n denotes the subset of E in which the inequality (3·6) holds for $M=n$, $\delta = 1/n$, then $E_n \subset E_{n+1}$, $E = \Sigma E_n$, so that $|E_n| \to |E|$. Since the E_n are measurable, it is enough to take n_0 large and select for P a 'large' subset of E_{n_0}. We shall then have (3·6) with $M = n_0$, $\delta = 1/n_0$. We may suppose that $n_0 \geqslant 3$.

Our next step is to deduce from (3·6) the existence of a number M_1 such that for any two points x, y in P we have

$$|f(x)-f(y)| \leqslant \frac{M_1}{\log 1/h}, \quad \text{where} \quad h = |y-x| < \delta. \tag{3·7}$$

We may suppose that $x < y$. Fix x and y and let

$$\xi(u) = |f(u)-f(x)| \log \frac{1}{u-x},$$
$$I = (x + \tfrac{1}{3}h, x + \tfrac{2}{3}h).$$

The set H of points $u \in I$ at which $\xi(u)$ exceeds a number N is of measure

$$\begin{aligned} |H| &\leqslant N^{-1}\int_I \xi(u)\,du \leqslant N^{-1}|\log \tfrac{1}{3}h| \left|\int_I |f(u)-f(x)|\,du\right| \\ &\leqslant N^{-1}|\log \tfrac{1}{3}h| \left|\int_0^h |f(x+t)-f(x)|\,dt\right| \leqslant \frac{h}{N} M \frac{\log \tfrac{1}{3}h}{\log h} \\ &< \frac{h}{N} 2M = \frac{6M}{N}|I|, \end{aligned}$$

since, by hypothesis, $h < \frac{1}{3}$.

Similarly, we prove that the subset H_1 of I in which $|f(u)-f(y)| \log 1/(y-u)$ exceeds N is of measure less than $6M|I|/N$. Thus for $N = 12M$ there is in I a point u_0 which belongs neither to H nor to H_1, such that

$$|f(x)-f(u_0)| \leqslant N/\log\frac{1}{u_0-x}, \quad |f(y)-f(u_0)| \leqslant N/\log\frac{1}{y-u_0};$$

and by addition we have (3·7) with $M_1 = 2N = 24M$.

Now let $\phi(x)$ be the function equal to $f(x)$ in P and linear in the closure of each interval contiguous to P. Let $\psi(x)$ be defined by (3·5). We shall show that ϕ and ψ have the required properties.

First, if x and y, $x < y$, are in P, (3·7) holds with f replaced by ϕ. If x and y are in the same interval contiguous to P, the linearity of ϕ makes (3·7) still valid for ϕ. In the remaining case, inserting between x and y terminal points of the contiguous intervals which contain x and y respectively, and using the previous cases, we again get (3·7) for ϕ, with M_1 replaced by $3M_1$. Thus ϕ satisfies a condition analogous to (3·4), uniformly in x.

We show finally that ψ satisfies the Dini condition

$$\int_{-\pi}^{\pi} \frac{|\psi(t)|}{|x-t|}\,dt < +\infty \tag{3·8}$$

for almost all $x \in P$. Clearly $\psi = 0$ in P. Let $d_1, d_2, \ldots$ denote the intervals contiguous to P, and also their lengths. Let $\chi(t)$ denote the distance of the point t from P. This function was considered in Chapter IV, § 2, where we proved that

$$\int_{-\pi}^{\pi} \frac{\{\log 1/\chi(t)\}^{-1}}{|x-t|} dt < +\infty \tag{3·9}$$

almost everywhere in P. It is enough to show that (3·8) holds at every point of density of P at which we have (3·9). Let x be such a point.

Now for any $d_j = (a_j, b_j)$ of length less than δ,

$$\int_{d_j} |\psi(t)|\, dt = \int_{d_j} |\psi(t) - \psi(a_j)|\, dt \leqslant \int_{d_j} |f(t) - f(a_j)|\, dt + \int_{d_j} |\phi(t) - \phi(a_j)|\, dt$$
$$\leqslant M d_j (\log 1/d_j)^{-1} + 3M_1 d_j (\log 1/d_j)^{-1} < AM d_j (\log 1/d_j)^{-1},$$

by (3·6) and the analogue of (3·7) for ϕ. Let d_j be so close to x that the length of d_j is less than the distance ρ_j of x from d_j. Then

$$\int_{d_j} \frac{|\psi(t)|}{|x-t|} dt \leqslant \frac{1}{\rho_j} \int_{d_j} |\psi(t)|\, dt < \frac{AM}{\rho_j} d_j (\log 1/d_j)^{-1}$$
$$\leqslant AM \int_{d_j} \frac{\{\log 1/\chi(t)\}^{-1}}{|x-t|} dt.$$

Comparing the extreme terms and summing over all the intervals d_j sufficiently close to x, we see that the part of the integral in (3·8) extended over a sufficiently small neighbourhood of x does not exceed a fixed multiple of the corresponding part of the integral (3·9). Thus (3·8) holds at the point x, and the proof of (3·2) is completed.

Condition (3·3) expresses a sort of 'average continuity' of f at x. The following theorem, in which we use the integral modulus of continuity of f,

$$\omega_1(\delta) = \omega_1(\delta; f) = \sup_{0<t\leqslant\delta} \int_{-\pi}^{\pi} |f(x+t) - f(x)|\, dx$$

(see Chapter II, § 3), is much more on the surface.

(3·10) Theorem. *If*

$$\int_0^{\pi} t^{-1}\omega_1(t)\, dt < \infty,$$

then $\mathsf{S}[f]$ *converges almost everywhere.*

Since

$$\int_{-\pi}^{\pi} dx \int_0^{\pi} \frac{|f(x \pm t) - f(x)|}{t} dt = \int_0^{\pi} \frac{dt}{t} \int_{-\pi}^{\pi} |f(x \pm t) - f(x)|\, dx \leqslant \int_0^{\pi} \frac{\omega_1(t)}{t} dt < \infty,$$

by hypothesis, the inner integral on the left is finite for almost all x; thus f satisfies Dini's condition almost everywhere, and (3·10) follows.

In particular, $\mathsf{S}[f]$ converges almost everywhere, provided

$$\int_{-\pi}^{\pi} |f(x+t) - f(x)|\, dt = O\left(\log \frac{1}{t}\right)^{-1-\epsilon} \quad (\epsilon > 0), \tag{3·11}$$

as $t \to +0$. Whether or not the result holds for $\epsilon = 0$ is an open problem, though we do know that in this case $\mathfrak{M}_1[f - S_n] \to 0$ (Chapter IV, p. 180, Example 7).

4. Majorants for the partial sums of $S[f]$ and $\tilde{S}[f]$

Consider the series

$$\tfrac{1}{2}a_0 + \sum_{k=1}^{\infty} (a_k \cos kx + b_k \sin kx) \equiv \sum_{k=0}^{\infty} A_k(x) \tag{4·1}$$

and its conjugate

$$\sum_{k=1}^{\infty} (a_k \sin kx - b_k \cos kx) \equiv \sum_{k=1}^{\infty} B_k(x). \tag{4·2}$$

Using a notation slightly different from that of preceding sections, we denote the partial sums and (C, 1) means of (4·1) by s_n, σ_n, and those of (4·2) by $\tilde{s}_n$, $\tilde{\sigma}_n$.

An interesting phenomenon in the theory of trigonometric series is that one-sided estimates for the s_n can lead to two-sided ones, and also to estimates for the $\tilde{s}_n$. This topic is discussed in the present section and the two following. The proofs are based on the formulae

$$\tfrac{1}{2}\{s_n(x+\alpha)+s_n(x-\alpha)\}$$
$$= \sum_{k=0}^{n-2} (k+1)\,\sigma_k(x)\,\Delta^2 \cos k\alpha + n\sigma_{n-1}(x)\,\Delta \cos (n-1)\,\alpha + s_n(x) \cos n\alpha, \tag{4·3}$$

$$-\tfrac{1}{2}\{s_n(x+\alpha)-s_n(x-\alpha)\}$$
$$= \sum_{k=1}^{n-2} (k+1)\,\tilde{\sigma}_k(x)\,\Delta^2 \sin k\alpha + n\tilde{\sigma}_{n-1}(x)\,\Delta \sin (n-1)\,\alpha + \tilde{s}_n(x) \sin n\alpha, \tag{4·4}$$

which follow from the obvious relations

$$\tfrac{1}{2}\{s_n(x+\alpha)+s_n(x-\alpha)\} = \sum_{k=0}^{n} A_k(x) \cos k\alpha,$$

$$-\tfrac{1}{2}\{s_n(x+\alpha)-s_n(x-\alpha)\} = \sum_{k=1}^{n} B_k(x) \sin k\alpha,$$

on summing by parts twice. (We have used (4·3) and (4·4) already in Chapter III, § 12.)

(4·5) THEOREM. *Suppose that* $\Sigma A_k(x)$ *is an* $S[f]$, *and that there is a* $\phi(x)$ *such that*

$$s_n(x) \geqslant \phi(x) \quad \textit{for} \quad n=0, 1, 2, \ldots. \tag{4·6}$$

(i) *If* f *and* ϕ *are both in* L^r, $1<r<\infty$, *there exist functions* Φ, ψ, Ψ, *also in* L^r, *such that*

$$s_n(x) \leqslant \Phi(x), \tag{4·7}$$

$$\psi(x) \leqslant \tilde{s}_n(x) \leqslant \Psi(x), \tag{4·8}$$

for $n=0, 1, \ldots$.

(ii) *If* $f \log^+ |f|$ *and* $\phi \log^+ |\phi|$ *are integrable, we still have* (4·7) *and* (4·8), *but with* Φ, ψ, Ψ *in* L.

(iii) *If* f *and* ϕ *are in* L, *we have* (4·7) *and* (4·8) *with* Φ, ψ, Ψ *in* L^α *for every* $0<\alpha<1$.

We denote by A a positive absolute constant. Write

$$\sup_n |\sigma_n(x)| = \sigma^*(x), \quad \sup_n |\tilde{\sigma}_n(x)| = \tilde{\sigma}^*(x),$$

$$f^*(x) = \sup_{0<|h|\leqslant\pi} \left\{\frac{1}{h}\int_0^h |f(x+t)|\,dt\right\}.$$

We know (Chapter IV, (7·8)) that
$$\sigma^*(x) \leqslant Af^*(x). \tag{4·9}$$

Since†
$$|\Delta \cos k\alpha| \leqslant |\alpha|, \quad |\Delta^2 \cos k\alpha| \leqslant \alpha^2,$$
we obtain from (4·3) and (4·6)
$$-s_n(x)\cos n\alpha \leqslant -\tfrac{1}{2}\{\phi(x+\alpha)+\phi(x-\alpha)\}+\sigma^*(x)\left[\alpha^2\sum_0^{n-2}(k+1)+n|\alpha|\right],$$
or, using (4·9) and taking $|\alpha| \leqslant A/n$,
$$-s_n(x)\cos n\alpha \leqslant \tfrac{1}{2}\{|\phi(x+\alpha)|+|\phi(x-\alpha)|\}+Af^*(x).$$

Integrate this over the interval $\pi/2n \leqslant \alpha \leqslant \pi/n$. The resulting inequality will not be affected if the integral on the right is replaced by $\int_0^{\pi/n}$; and defining ϕ^* similarly to f^* we get
$$s_n(x) \leqslant \frac{n}{2}\int_0^{\pi/n}\{|\phi(x+\alpha)|+|\phi(x-\alpha)|\}\,d\alpha + Af^*(x),$$
$$s_n(x) \leqslant A\{\phi^*(x)+f^*(x)\}. \tag{4·10}$$

We know (Chapter I, (13·14)) that according as $|f|^r$, $f\log^+|f|$ or f is integrable, the function f^* is in L^r, L or L^α respectively. The same holds for ϕ^*, and denoting the right-hand side of (4·10) by $\Phi(x)$, we get (4·7) in all cases (i), (ii) and (iii).

Since we now have two-sided estimates for $s_n(x)$, it would seem natural to apply an argument similar to the preceding one to (4·4) in order to obtain (4·8). But except in case (i), the integrability of Φ is not as good as that of ϕ, and the argument would lead to functions ψ and Ψ not as good as Φ. For this reason a slight modification of the proof is called for.

Let R_1 and R_2 denote the right-hand sides in (4·3) and (4·4), without the last terms $s_n \cos n\alpha$ and $\tilde{s}_n \sin n\alpha$. If $|\alpha| \leqslant A/n$, the functions $|R_1|$ and $|R_2|$ are majorized by $A\sigma^*(x)$ and $A\tilde{\sigma}^*(x)$ respectively. Thus, subtracting (4·3) from (4·4), we get successively
$$\tilde{s}_n(x)\sin n\alpha - s_n(x)\cos n\alpha = R_1 - R_2 - s_n(x+\alpha),$$
$$\tilde{s}_n(x)\sin n\alpha \leqslant |s_n(x)| + A\{\sigma^*(x)+\tilde{\sigma}^*(x)\} - \phi(x+\alpha),$$
$$\tilde{s}_n(x)\sin n\alpha \leqslant |\phi(x)|+|\Phi(x)|+A\{f^*(x)+\tilde{\sigma}^*(x)\}+|\phi(x+\alpha)|.$$

Integrating the last inequality over the interval $0<\alpha\leqslant\pi/2n$ if $\tilde{s}_n(x)\geqslant 0$, and over $-\pi/2n\leqslant\alpha<0$ if $\tilde{s}_n(x)<0$, we obtain
$$|\tilde{s}_n(x)| \leqslant A\{|\phi(x)|+|\Phi(x)|+f^*(x)+\phi^*(x)+\tilde{\sigma}^*(x)\}. \tag{4·11}$$

Let $\Psi(x)$ be the right-hand side here. We have then (4·8) with $\psi=-\Psi$, and it remains to verify that Ψ satisfies the required conditions. In view of what has already been said of Φ and f^*, it is enough to observe that, for f satisfying the hypotheses of (i), (ii) or (iii) respectively, the function $\tilde{\sigma}^*$ belongs to L^r, L or L^α (Chapter VII, (7·42)).

We shall say that a sequence of functions $f_n(x)$ is *uniformly semi-convergent to $f(x)$ from below* if, for any $\epsilon>0$,
$$f_n(x)-f(x) > -\epsilon$$
for all x, provided $n \geqslant n_0(\epsilon)$.

† Similar estimates hold for $\sin k\alpha$ and will be used below.

(4·12) THEOREM. (i) *If f is bounded, $|f| \leqslant M$, and if there is a constant a such that*

$$s_n(x) \geqslant -a \quad (n=0, 1, \ldots),$$

then there is another constant $b = b(a, M)$ such that

$$s_n(x) \leqslant b \quad (n=0, 1, \ldots). \tag{4·13}$$

(ii) *If f is continuous, and $S[f]$ is uniformly semi-convergent to f from below, then $S[f]$ converges uniformly.*

(i) Let R_1 be defined as before and let $\alpha = \pi/n$ in (4·3). We get

$$-s_n(x) + R_1 \geqslant -a,$$

$$s_n(x) \leqslant |R_1| + a \leqslant A\sigma^*(x) + a$$

$$\leqslant AM + a = b,$$

and (4·13) is established.

(ii) We note that $b = AM + a$ in (i), and that if we assume $s_n(x) \geqslant -a$ for $n = k, k+1, \ldots$ only, we have (4·13) for $n \geqslant k$.

Suppose that

$$s_n - f \geqslant -\epsilon \tag{4·14}$$

for $n \geqslant n_0$, and let T be a polynomial of order m such that $|f - T| < \epsilon$. We write (4·14) in the form

$$(s_n - T) - (f - T) \geqslant -\epsilon,$$

and note that $s_n - T$ is the nth partial sum of $S[f-T]$ provided $n \geqslant m$. Applying (i) to $f - T$ with $M = \epsilon$, $a = 2\epsilon$ we therefore have, for $n \geqslant \max(n_0, m)$,

$$-2\epsilon \leqslant s_n - T \leqslant AM + a = (A+2)\epsilon,$$

$$|s_n - f| \leqslant |s_n - T| + \epsilon \leqslant (A+3)\epsilon,$$

and (ii) follows.

5. Behaviour of the partial sums of $S[f]$ and $\tilde{S}[f]$

From the convergence of $S[f]$ we can sometimes deduce that of $\tilde{S}[f]$. For example, if $S[f]$ converges uniformly then $\tilde{S}[f]$ converges almost everywhere (Chapter VII, (6·14)). The theorem that follows is of a similar nature though considerably stronger. It applies to general trigonometric series, but the special case of Fourier series is the most interesting one.

(5·1) THEOREM. *If the series*

$$\tfrac{1}{2}a_0 + \sum_{k=1}^{\infty} (a_k \cos kx + b_k \sin kx) = \sum_{k=0}^{\infty} A_k(x) \tag{5·2}$$

converges in a set E, and if the conjugate series

$$\sum_{k=1}^{\infty} (a_k \sin kx - b_k \cos kx) = \sum_{k=1}^{\infty} B_k(x) \tag{5·3}$$

is summable (C, 1) *almost everywhere in E, then it converges almost everywhere in E. In particular, for any $f \in L$ the sets of the points of convergence of $S[f]$ and $\tilde{S}[f]$ are the same, except for a set of measure zero.*

We may suppose that $|E|>0$, and that $\Sigma B_k(x)$ is summable (C, 1) everywhere in E.

Let s_n, $\tilde{s}_n$, σ_n, $\tilde{\sigma}_n$ denote the partial sums and the (C, 1) means of (5·2) and (5·3) respectively, and let

$$\sigma(x)=\lim \sigma_n(x), \quad \tilde{\sigma}(x)=\lim \tilde{\sigma}_n(x),$$

wherever the limits exist. By Theorem (12·15) of Chapter III,

$$\tfrac{1}{2}[s_n(x+\alpha_n)-s_n(x-\alpha_n)]+[\tilde{s}_n(x)-\tilde{\sigma}(x)]\sin n\alpha_n \to 0 \tag{5·4}$$

for $x \in E$ and any sequence of numbers $\alpha_n=O(1/n)$. (This is also an easy consequence of (4·4).) The whole subsequent argument will be based on this relation.

Let $\mathscr{E} \subset E$ be a set of positive measure in which $s_n(x)$ converges uniformly. (By the theorem of Egorov, $|E-\mathscr{E}|$ can be made arbitrarily small.) In particular, $\sigma(x)$ is continuous on $\mathscr{E}$. Let ξ be a point in $\mathscr{E}$ which is at the same time a point of density of $\mathscr{E}$, and let $\mu(h)$ be the measure of $\mathscr{E}(\xi, \xi+h)$, the part of $\mathscr{E}$ situated in $(\xi, \xi+h)$. Then

$$\mu(h)=h+o(h), \quad \mu(2h)=2h+o(h). \tag{5·5}$$

The resulting relation $\quad \mu(2h)-\mu(h)=h+o(h)$

shows that the average density of $\mathscr{E}$ in the interval $(\xi+h, \xi+2h)$ tends to 1 as $h \to 0$. Since the same holds for the interval $(\xi-2h, \xi-h)$, we obtain, by taking $h=1/n$, the following conclusion: *for all large enough n there is a number μ_n, $1 \leqslant \mu_n \leqslant 2$, such that $\xi \pm \mu_n/n \in \mathscr{E}$.*

We now apply (5·4) with $x=\xi$, $\alpha_n=\mu_n/n$. By the uniform convergence of $\{s_n(x)\}$ on $\mathscr{E}$, the first term in square brackets tends to 0 as $n \to \infty$. Observing that

$$\sin n\alpha_n=\sin \mu_n$$

stays away from zero, we get $\quad \tilde{s}_n(\xi)-\tilde{\sigma}(\xi) \to 0.$

Since almost all points of $\mathscr{E}$ are points of density of $\mathscr{E}$, and since $|E-\mathscr{E}|$ may be arbitrarily small, (5·1) is established.

Remarks. (i) If $\Sigma B_k(x)$ is summable (C, 1) in E, and if a sequence $\{s_{n_k}(x)\}$ of the partial sums of $\Sigma A_k(x)$ converges in E, then $\{\tilde{s}_{n_k}(x)\}$ converges almost everywhere in E.

(ii) Let $\rho_1<\rho_2<\ldots<\rho_n<\ldots\to\infty$. If $\Sigma B_k(x)$ is summable (C, 1) in E, and if $s_n(x)=O(\rho_n)$ at each point of E, then $\tilde{s}_n(x)=O(\rho_n)$ at almost any point of E. The result holds if 'O' is replaced by 'o' throughout.

The proofs are similar to that of (5·1). Remark (i) was already used in Chapter V, §7.

A similar argument gives information about the limits of indetermination of the partial sums of (5·2) and (5·3). With the notation

$$s^*(x)=\limsup s_n(x), \quad s_*(x)=\liminf s_n(x), \tag{5·6}$$

we have the following result:

(5·7) THEOREM. (i) *If $\Sigma A_k(x)$ is summable* (C, 1) *for $x \in E$ to sum $\sigma(x)$, and if $s^*(x)<+\infty$ in E, then*

$$s_*(x)>-\infty, \quad \sigma(x)=\tfrac{1}{2}[s^*(x)+s_*(x)] \tag{5·8}$$

almost everywhere in E.

(ii) *If $\Sigma A_k(x)$ is summable* (C, 1) *in E, and if $s^*(x)=+\infty$ in E, then $s_*(x)=-\infty$ almost everywhere in E.*

This theorem shows that at almost all points at which $\Sigma A_k(x)$ is summable (C, 1), the s_n oscillate symmetrically about $\sigma(x) = \lim \sigma_n(x)$. In particular, if $\Sigma A_k(x)$ is an S[f], or $\tilde{\mathrm{S}}[f]$, the oscillation is symmetric with respect to $f(x)$ or $\tilde{f}(x)$ as the case may be. That a Fourier series may oscillate finitely almost everywhere is shown by Theorem (3·14) of Chapter VIII.

Since in (i) the inequalities $s^*(x) < +\infty$ and $s_*(x) > -\infty$ are obviously interchangeable, (ii) is an immediate corollary of (i). We may suppose that $|E| > 0$.

To prove (i), we note first that, by Chapter III, (12·9),

$$\tfrac{1}{2}\{s_n(x+\alpha_n)+s_n(x-\alpha_n)\}-\{s_n(x)-\sigma(x)\}\cos n\alpha_n \to \sigma(x) \tag{5·9}$$

for $x \in E$, $\alpha_n = O(1/n)$ (a simple consequence also of (4·3)). Next, by hypothesis, $s^*(x)$ is finite in E. The sequence of measurable functions

$$t_n(x) = \max\{s_n(x), s^*(x)\}$$

converges to $s^*(x)$ in E. We can therefore find a set $\mathscr{E} \subset E$, with $|E - \mathscr{E}|$ arbitrarily small, such that the convergence is uniform on $\mathscr{E}$. It follows that there exist positive numbers $\epsilon_n \to 0$ such that

$$s_n(x) \leqslant s^*(x) + \epsilon_n \quad \text{for} \quad x \in \mathscr{E}.$$

Since $s^*(x)$ is measurable, we may also suppose (by further reduction of $\mathscr{E}$, if necessary) that s^* is continuous on $\mathscr{E}$.

We shall now need a result which is a refinement of one contained in the proof of (5·1). It will be used repeatedly and we state it as a separate lemma:

(5·10) LEMMA. *Let ξ be a point of density of a measurable set $\mathscr{E}$ and γ a fixed positive number. Then there is a sequence of numbers $\mu_n \to \gamma$ such that $\xi \pm \mu_n/n \in \mathscr{E}$ for all large enough n.*

An argument based on equations analogous to (5·5) shows that if $\alpha < \beta$ are two positive numbers, then the average density of $\mathscr{E}$ in the interval $(\xi + \alpha h, \xi + \beta h)$ tends to 1 as $h \to 0$. Hence taking $\alpha = \gamma$, $\beta = \gamma + \eta$, where η is a fixed positive number, we deduce that there is a sequence of numbers λ_n contained between γ and $\gamma + \eta$ such that $\xi \pm \lambda_n/n \in \mathscr{E}$ for n large enough. By making η tend to 0, and piecing together the corresponding sequences $\{\lambda_n\}$, we can obtain a $\{\mu_n\}$ satisfying (5·10).

Returning to (5·7), let $x = \xi$ be a point of density of $\mathscr{E}$ belonging to $\mathscr{E}$, and let $\alpha_n = \mu_n/n$, where $\{\mu_n\}$ satisfies (5·10) with $\gamma = \pi$. We may suppose that $\sigma(\xi) = 0$, for otherwise we subtract $\sigma(\xi)$ from the constant term of $\Sigma A_k(x)$. Using (5·9) and the continuity of s^* over $\mathscr{E}$ we get

$$s_n(\xi)\cos n\alpha_n \leqslant \tfrac{1}{2}\{s^*(\xi+\alpha_n)+s^*(\xi-\alpha_n)\}+o(1) \leqslant s^*(\xi)+o(1),$$

and so

$$-s_*(\xi) \leqslant s^*(\xi), \tag{5·11}$$

since $\cos n\alpha_n \to -1$. This proves the first inequality (5·8) for $x = \xi$ and so also almost everywhere in E.

In the general case when $\sigma(\xi) \neq 0$, the inequality (5·11) is

$$\tfrac{1}{2}\{s^*(\xi) + s_*(\xi)\} \geqslant \sigma(\xi). \tag{5·12}$$

If we started with the function s_*, which we now know to be finite almost everywhere in E, we should obtain (5·11), or (5·12), with reversed inequality signs. This completes the proof of the second relation (5·8).

(5·13) Theorem. *If $\Sigma A_k(x)$ is summable* (C, 1) *in a set E to a finite sum $\sigma(x)$, and if a sequence $\{s_{n_k}(x)\}$ of partial sums of the series converges in E to a limit $s(x)$, then $s(x)=\sigma(x)$ almost everywhere in E.*

We may suppose that $|E|>0$ and (by reducing $|E|$ arbitrarily little) that $\{\sigma_n\}$ and $\{s_{n_k}\}$ converge uniformly on E; in particular $s(x)$ is continuous on E. Let $\xi\in E$ be a point of density of E, and let $\{\alpha_n\}$ be such that $\xi\pm\alpha_n\in E$, $n\alpha_n\to\pi$. From (5·9) with $x=\xi$, $n=n_k$ we deduce

$$s(\xi)-\{s(\xi)-\sigma(\xi)\}(-1)=\sigma(\xi),$$

that is, $s(\xi)=\sigma(\xi)$.

The theorem which follows is stated without proof. It generalizes Chapter VIII, (3·14).

(5·14) Theorem. *Let $\phi(x)$ be periodic, measurable, non-negative, not necessarily finite almost everywhere. There is an $\mathsf{S}[f]$ such that, for almost all x,*

$$s_*(x)=f(x)-\phi(x), \quad s^*(x)=f(x)+\phi(x).$$

6. Theorems on the partial sums of power series

Consider the power series

$$\sum_{k=0}^{\infty} c_k e^{ikx} \tag{6·1}$$

in $z=e^{ix}$. The partial sums and the (C, 1) means of the series will be denoted by $t_n(x)$ and $\tau_n(x)$ respectively. For each x we denote by $L(x)$ the set of all limit points (including the point at infinity) of the sequence $\{t_n(x)\}$. $L(x)$ is a closed set which reduces to a single point if (6·1) either converges or else diverges to ∞.

Wherever the limit

$$\tau(x)=\lim \tau_n(x)$$

exists, we write $m(x)=\liminf|\tau(x)-t_n(x)|$, $M(x)=\limsup|\tau(x)-t_n(x)|$.

We introduce some geometric terminology. By $A(\zeta_0;\alpha,\beta)$, where $0\leqslant\alpha\leqslant\beta\leqslant\infty$, we mean the annulus $\alpha\leqslant|\zeta-\zeta_0|\leqslant\beta$ of the complex plane. The disk $A(\zeta_0;0,\beta)$ and the circumference $A(\zeta_0;\beta,\beta)$ will be denoted by $D(\zeta_0,\beta)$ and $C(\zeta_0,\beta)$ respectively. The interiors of A and D will be denoted by A^0 and D^0.

We say that a set Z in the complex plane is *of circular structure* if it is a union of a finite or infinite (denumerable or not) family of circumferences with common centre ζ_0, the *centre* of Z. If Z is closed, the radii of the smallest and largest circumferences (with centre ζ_0) that it contains will be called the *extreme radii* of Z.

(6·2) Theorem. *If $\Sigma c_k e^{ikx}$ is summable* (C, 1) *for $x\in E$, then for almost all $x\in E$ the set $L(x)$ is of circular structure with centre $\tau(x)$ and extreme radii $m(x)$ and $M(x)$.*

Thus if, for fixed x, we consider the terms of the sequence $\{t_n(x)\}$ as successive positions of a moving point, then for almost all $x\in E$ the terms $t_n(x)$ move, roughly speaking, on the circumferences constituting $L(x)$. The terms may jump from one circumference to another, but accumulate to every point of each circumference.

If $c_n\to 0$, then $t_n(x)-t_{n-1}(x)\to 0$, and $L(x)$ must, for almost all $x\in E$, be the annulus $A(\tau(x);m(x),M(x))$. It need not be so in the general case; we give examples below.

If the real part of (6·1) is an $S[f]$, then $\int_0^{2\pi} |\tau - t_n|^\mu dx \to 0$ for $0<\mu<1$ (Chapter VII, (6·8)), and so there is a sequence $\{t_{n_k}(x)\}$ converging to $\tau(x)$ almost everywhere. In this case $m(x)=0$ and, for almost all $x \in E$, $L(x)$ is a circular disk with centre $\tau(x)$.

The proof of (6·2) is based on the following lemma, in which $L_1(x)$ denotes the set of limit points of the sequence $\{t_n(x)-\tau(x)\}$, that is, the translate of $L(x)$ by $-\tau(x)$:

(6·3) LEMMA. *Let E be the set of x such that* (6·1) *is summable* (C, 1), *and that $L_1(x)$ does not contain any point of a fixed disk $D^0(\zeta, r)$ with $|\zeta| \geqslant r$. Then for almost all $x \in E$ the set $L(x)$ does not contain any point of the annulus $A^0(\tau(x); |\zeta|-r, |\zeta|+r)$.*

First, by Chapter III, (12·16),

$$t_n(x+\alpha_n) - [t_n(x)-\tau(x)]\, e^{in\alpha_n} \to \tau(x) \tag{6·4}$$

for $x \in E$, $\alpha_n = O(1/n)$.

Let $r_k = r(1-1/k)$ for $k=2, 3, \ldots$, and let $E_{k,N}$ be the set of x at which (6·1) is summable (C, 1) and such that all $t_n(x)-\tau(x)$ are outside $D^0(\zeta, r_k)$ for $n \geqslant N$. Then

$$E = \prod_k \sum_N E_{k,N},$$

and it is enough to show that for almost all x in each $E_{k,N}$ the set $L(x)$ is disjoint with

$$A^0(\tau(x); |\zeta|-r_k, |\zeta|+r_k).$$

The functions $t_n(x)$ being continuous and $\tau(x)$ measurable (on the set where it exists), each $E_{k,N}$ is measurable.

We may suppose that $|E_{k,N}|>0$. Let $\xi \in E_{k,N}$ be a point of density of $E_{k,N}$. The lemma will be established if we show that $L(\xi)$ is disjoint with $A^0(\tau(\xi); |\zeta|-r_k, |\zeta|+r_k)$. Supposing, as we may, that $\tau(\xi)=0$, we get from (6·4)

$$t_n(\xi) = t_n(\xi+\alpha_n)\, e^{-in\alpha_n} + o(1). \tag{6·5}$$

Given any real number γ, let $\{\alpha_n\}$ be such that $n\alpha_n \to \gamma$ and $\xi+\alpha_n \in E_{k,N}$ for all large n. It follows from (6·5) that $L(\xi)=L_1(\xi)$ does not contain any point of $D^0(\zeta e^{-i\gamma}, r_k)$ and so, since γ is arbitrary, any point of $A^0(0; |\zeta|-r_k, |\zeta|+r_k)$.

We proceed to deduce (6·2) from (6·3). We denote by X the set of $x \in E$ such that $L_1(x)$ is not of circular structure with centre 0. The sets L_1 being closed, for every $x \in X$ there is a disk $D^0(\zeta, r) = D^0(\xi+i\eta, r)$ with ξ, η and r rational, $|\zeta| \geqslant r$, such that

(i) $L_1(x)$ contains no point of $D^0(\zeta, r)$;

(ii) $L_1(x)$ does contain points of $A^0(0; |\zeta|-r, |\zeta|+r)$.

Let $X_{\zeta,r}$ denote the set of $x \in X$ satisfying (i) and (ii). By (6·3), $|X_{\zeta,r}|=0$ for each pair ζ, r. Since $X=\Sigma X_{\zeta,r}$, where the summation is extended over all rational ξ, η, r, it follows that $|X|=0$, and (6·2) is established.

Now consider together with s^*, s_* (see (5·6)) the functions

$$\tilde{s}^*(x) = \limsup \tilde{s}_n(x), \quad \tilde{s}_*(x) = \liminf \tilde{s}_n(x).$$

(6·6) THEOREM. (i) *Suppose that both $\Sigma A_k(x)$ and $\Sigma B_k(x)$ are summable* (C, 1) *for $x \in E$ to sums $\sigma(x)$ and $\tilde{\sigma}(x)$ respectively, and that $s_*(x) > -\infty$ in E. Then at almost all points $x \in E$ the four functions s^*, s_*, $\tilde{s}^*$, $\tilde{s}_*$, are finite and satisfy the relations*

$$\tfrac{1}{2}\{s^*(x)+s_*(x)\} = \sigma(x), \tag{6·7}$$

$$\tfrac{1}{2}\{\tilde{s}^*(x)+\tilde{s}_*(x)\} = \tilde{\sigma}(x), \tag{6·8}$$

$$s^*(x)-s_*(x) = \tilde{s}^*(x)-\tilde{s}_*(x). \tag{6·9}$$

(ii) *If the condition on s_* is replaced by $s_*(x) = -\infty$ in E, then almost everywhere in E we have*

$$s^*(x) = \tilde{s}^*(x) = +\infty, \quad \tilde{s}_*(x) = -\infty. \tag{6·10}$$

(i) The relation (6·7) is contained in (5·7). Let $\Sigma A_k(x)$ and $\Sigma B_k(x)$ be the real and imaginary parts of $\sum_0^\infty c_k e^{ikx}$, which is then summable (C, 1) in E. Thus, at almost all $x \in E$, $L(x)$ is of circular structure with centre $\tau(x) = \sigma(x) + i\tilde{\sigma}(x)$, and so is bounded since $s_*(x) > -\infty$ in E; and the relations (6·8) and (6·9) follow immediately.

(ii) By (5·7), $s^*(x) = +\infty$ almost everywhere in E. By (i) both $\tilde{s}^*$ and $\tilde{s}_*$ are infinite almost everywhere in E, for otherwise s_* would be finite almost everywhere in E. Since $\Sigma B_k(x)$ is summable (C, 1) in E, $\tilde{s}^*$ and $\tilde{s}_*$ must be infinites of opposite sign, so that $\tilde{s}_* = -\infty$, $\tilde{s}^* = +\infty$ almost everywhere in E.

Examples for (6·2). Let

$$G(z) = 1 + z + z^2 + \dots.$$

(i) The partial sums $t_n(x)$ of $G(e^{ix})$ are bounded, and the (C, 1) means $\tau_n(x)$ converge to $\tau(x) = (1 - e^{ix})^{-1}$, for each $x \neq 0 \pmod{2\pi}$. Here

$$t_n(x) - \tau(x) = -e^{i(n+1)x}(1 - e^{ix})^{-1},$$

so that all $t_n(x)$ are situated on the circumference with centre $\tau(x)$ and radius $|1 - e^{ix}|^{-1}$. If x is incommensurable with π, $L(x)$ coincides with this circumference.

(ii) For the series

$$\alpha G(e^{ix}) + \beta G(e^{2ix}) = (\alpha + \beta) + \alpha e^{ix} + (\alpha + \beta) e^{2ix} + \dots,$$

the $t_n(x)$ are situated on two circumferences with common centre

$$\alpha(1 - e^{ix})^{-1} + \beta(1 - e^{2ix})^{-1}$$

and radii

$$|1 - e^{ix}|^{-1} |\alpha + \beta(1 + e^{\pm ix})^{-1}|.$$

If β/α is not real, these two circumferences are different. If x is incommensurable with π, $L(x)$ is the union of these circumferences.

(iii) Finally, consider

$$G(z) + \alpha G(e^{i\lambda} z) \quad (z = e^{ix}),$$

where $0 < \alpha < 1$, and λ is incommensurable with π. If x and λ are linearly independent, the fractional parts of $(n+1)x/2\pi$ and $(n+1)\lambda/2\pi$ can be arbitrarily close to any preassigned numbers of $(0, 1)$. Using this, we easily show that, except for a denumerable set of x's, $L(x)$ is an annulus not reducing to a disk or circumference.

By combining examples (ii) and (iii), we get series for which $L(x)$ consists of several concentric annuli.

(iv) For the power series $\Sigma c_k e^{ikx}$ whose real part is the $S[f]$ of (5·14), the set $L(x)$ is for almost all x the disk with centre $f(x) + i\tilde{f}(x)$ and radius $\phi(x)$.

7. Strong summability of Fourier series. The case $f \in L^r$, $r > 1$

Let $q > 0$. A sequence $s_0, s_1, \dots$, or a series with partial sums $s_0, s_1, \dots$, will be called summable H_q to limit (sum) s if

$$\frac{|s_0 - s|^q + |s_1 - s|^q + \dots + |s_n - s|^q}{n+1} \to 0 \quad \text{as} \quad n \to \infty. \tag{7·1}$$

This kind of summability was first considered by Hardy and Littlewood, and is, for $q \geqslant 1$, a generalization of the method of the first arithmetic mean. For clearly summability H_1 implies (C, 1), and Hölder's inequality shows that if (7·1) is true for some q it holds for any smaller q. Summability H_1 indicates that the mean value of $s_k - s$ tends to zero, not because of the cancellation of positive and negative terms, but because the indices k for which $|s_k - s|$ is not small are sparse (see (7·2) below).

We shall also speak of (7·1) as the *strong summability* (C, 1)—or merely *strong summability*—of $\{s_k\}$.

Let
$$(S) \quad n_1 < n_2 < \ldots < n_k < \ldots$$

be an increasing sequence of positive integers, and let $\nu(N)$ denote the number of the $n_k \leqslant N$. The number $d_N = \nu(N)/(N+1)$ is the density of S in the interval $(0, N)$, and the limit $d = \lim d_N$, if it exists, will be called the *density* of S. The sequence of positive integers complementary to S has then the density $1-d$. Only the cases $d=0$ and $d=1$ will be of interest to us. It is easily verified that $d=1$ if and only if $k/n_k \to 1$, and $d=0$ if and only if $k/n_k \to 0$.

We shall say that a sequence $\{s_k\}$, or a series with partial sums s_k, is *almost convergent* to limit (sum) s if there is a sequence $\{n_k\}$ of density 1 such that $s_{n_k} \to s$. Since the intersection of two sequences $\{n_k\}$ of density 1 has infinitely many terms (indeed, is of density 1), the number s is determined uniquely.

(7·2) THEOREM. (i) *If $\{s_n\}$ is summable H_q to limit s, then $\{s_n\}$ is almost convergent to s.*

(ii) *Conversely, if $\{s_n\}$ is almost convergent to s, and is bounded, then, for any $q > 0$, $\{s_n\}$ is summable H_q to s.*

We may suppose that $s=0$. We first prove the following lemma: *A necessary and sufficient condition for an $\{s_n\}$ to be almost convergent to 0 is that for any $\epsilon > 0$ the n's such that $|s_n| \leqslant \epsilon$ have density* 1.

Only the sufficiency needs a proof. Let S_m be the set of indices n such that

$$|s_n| \leqslant 1/m \quad (m = 1, 2, \ldots).$$

Then $S_1 \supset S_2 \supset \ldots \supset S_m \supset \ldots$, and each S_m has density 1. Define a sequence

$$N_1 < N_2 < \ldots < N_m < \ldots$$

such that S_m has density $> 1 - 1/m$ in $(0, N)$ provided $N > N_m$. Let S be the sequence defined as follows. It consists of the elements of S_1 which do not exceed N_2, the elements of S_2 which are between N_2 and N_3, the elements of S_3 which are between N_3 and N_4, and so on. Then if $N_m < N \leqslant N_{m+1}$, the density of S in $(0, N)$ is not less than the density of S_m in $(0, N)$, and so exceeds $1 - 1/m$. Hence S is of density 1, and clearly $s_n \to 0$ as $n \to \infty$ in S. This proves the lemma.

Return to (7·2) (i). Fixing an $\epsilon > 0$, let $\nu(N)$ be the number of $k \leqslant N$ with $|s_k| > \epsilon$. The left-hand side of (7·1), with $s=0$, is then not less than $\epsilon^q \nu(n)/(n+1)$. Hence $\nu(n) = o(n)$, and an application of the lemma yields the result.

The proof of (7·2) (ii) is left to the reader.

The rest of this section is devoted to the strong summability of Fourier series.

If $f \in L^r$, $r \geqslant 1$, we shall write

$$\Phi_{x,r}(h) = \int_0^h |\phi_x(t)|^r dt = \int_0^h |\tfrac{1}{2}\{f(x+t)+f(x-t)\}-f(x)|^r dt.$$

It was shown in Chapter II, § 11, that $\Phi_{x,r}(h) = o(h)$ almost everywhere.

(7·3) THEOREM. (i) *If $f \in L^r$, $r > 1$, then $S[f]$ is summable H_q to sum $f(x)$ at every point x at which $\Phi_{x,r}(h) = o(h)$.*

(ii) *If $f \in L$ and f is continuous at every x, $a \leqslant x \leqslant b$, then $S[f]$ is uniformly summable H_q to $f(x)$ over (a, b).*

That $S[f]$ is almost everywhere summable H_q for functions f merely integrable, is a deeper result requiring different methods; it will be proved in the next section.

Let $s_\nu(x) = S_\nu(x; f)$. It is sufficient to prove the relation

$$\frac{1}{n+1}\sum_{\nu=0}^{n} |s_\nu(x) - f(x)|^q \to 0 \tag{7·4}$$

for $q = r/(r-1) = r'$. For $\{h^{-1}\Phi_{x,r}(h)\}^{1/r}$ is a non-decreasing function of r, and so if $\Phi_{x,r}(h) = o(h)$ for some r, the relation holds for any smaller r. Taking r sufficiently close to 1 we obtain q as large as we please. It is also sufficient to prove (7·4) for the modified partial sums s_ν^* (Chapter II, § 5), since $|s_\nu - f|^q$ differs from $|s_\nu^* - f|^q$ by an amount tending uniformly to zero.

If $0 < \nu \leqslant n$, we have

$$s_\nu^*(x) - f(x) = \frac{2}{\pi}\int_0^\pi \phi_x(t)\frac{\sin \nu t}{2\tan\frac{1}{2}t}dt = \frac{2}{\pi}\left(\int_0^{1/n} + \int_{1/n}^\pi\right)$$
$$= \alpha_\nu^{(n)} + \beta_\nu^{(n)},$$

$$\left\{\frac{1}{n+1}\sum_{1}^{n}|s_\nu^* - f|^q\right\}^{1/q} \leqslant \left\{\frac{1}{n+1}\sum_{\nu=1}^{n}|\alpha_\nu^{(n)}|^q\right\}^{1/q} + \left\{\frac{1}{n+1}\sum_{\nu=1}^{n}|\beta_\nu^{(n)}|^q\right\}^{1/q},$$

and (i) will be established if we show that each term on the right in the last inequality is $o(1)$.

Clearly
$$|\alpha_\nu^{(n)}| \leqslant 2\pi^{-1}\nu\Phi_{x,1}(1/n) \leqslant \nu\Phi_{x,1}(1/\nu) = o(1),$$

since $\Phi_{x,r}(h) = o(h)$ implies $\Phi_{x,1}(h) = o(h)$, and

$$\left\{\frac{1}{n+1}\sum_{\nu=1}^{n}|\alpha_\nu^{(n)}|^q\right\}^{1/q} \to 0. \tag{7·5}$$

The β's are Fourier coefficients of the function equal to $\phi_x(t)\cot\frac{1}{2}t$ for $1/n \leqslant t \leqslant \pi$ and to 0 in $(-\pi, 1/n)$. Hence applying the Hausdorff-Young inequality (Chapter XII, (2·3)), and supposing, as we may, that $r \leqslant 2$, we get

$$\left\{\frac{1}{n+1}\sum_{\nu=1}^{n}|\beta_\nu^{(n)}|^q\right\}^{1/q} \leqslant \frac{1}{(n+1)^{1/q}}\left(\frac{1}{\pi}\int_{1/n}^{\pi}\left|\frac{\phi_x(t)}{\tan\frac{1}{2}t}\right|^r dt\right)^{1/r}, \tag{7·6}$$

where $q = r'$. Replacing $\tan\frac{1}{2}t$ by $\frac{1}{2}t$ and integrating by parts, we see that the right-hand side of (7·6) does not exceed a fixed multiple of

$$\frac{1}{(n+1)^{1/q}}\left\{\left[\frac{\Phi_{x,r}(t)}{t^r}\right]_{1/n}^{\pi} + r\int_{1/n}^{\pi}\frac{\Phi_{x,r}(t)}{t^{r+1}}dt\right\}^{1/r}$$
$$\leqslant \frac{1}{(n+1)^{1/q}}\left\{O(1) + \int_{1/n}^{\pi} o(t^{-r})\,dt\right\}^{1/r}$$
$$= (n+1)^{-1/q}\{o(n^{r-1}) + o(n^{r-1})\}^{1/r} = o(1).$$

Hence the left-hand side of (7·6) tends to 0 and this, together with (7·5), proves (7·3) (i).

A curious feature of the above argument is that the less we assume about the function—the smaller the number r is—the larger the value for q we obtain. The argument, however, breaks down for $r=1$.

If $f\in L^r$, $r>1$, the proof of (ii) is essentially the same as that of (i). We need only observe that if $a\leqslant x\leqslant b$, then $\Phi_{x,r}(h)=o(h)$, $\Phi_{x,1}(h)=o(h)$ uniformly in x, and that the estimates we obtain are also uniform in x. In particular, *if the function $f(x)$ is everywhere continuous*, (7·4) *holds uniformly for all x.*

Suppose now that $f\in L$ in (ii). We can then find an interval (a_1, b_1), $a_1<a\leqslant b<b_1$, such that f is bounded in (a_1, b_1). Let $f=f'+f''$, where $f'=f$ in (a_1, b_1) and $f'=0$ elsewhere. For the partial sums s'_ν and s''_ν of $S[f']$ and $S[f'']$ we have $s_\nu=s'_\nu+s''_\nu$, and

$$\left\{\frac{1}{n+1}\sum_{\nu=0}^{n}|s_\nu-f|^q\right\}^{1/q}\leqslant\left\{\frac{1}{n+1}\sum_{\nu=0}^{n}|s'_\nu-f'|^q\right\}^{1/q}+\left\{\frac{1}{n+1}\sum_{\nu=0}^{n}|s''_\nu-f''|^q\right\}^{1/q}$$

The first term on the right tends uniformly to zero for $a\leqslant x\leqslant b$, since f' is bounded and so is in every L^r. Since $f''=0$ in (a_1, b_1), $|s''_\nu-f''|^q$ tends uniformly to zero in (a, b). Hence also the second term on the right of the last inequality tends uniformly to 0 in (a, b), and the proof of (ii) is completed.

Clearly, (7·4) is true if f is in L and is continuous at the point x. (The argument applies when (a, b) reduces to a point.) The result actually holds if f has a simple discontinuity at x and $2f(x)=f(x+0)+f(x-0)$.

For $r=q=2$, Theorem (7·3) (i) has an analogue for general orthogonal systems.

(7·7) THEOREM. *Let $\phi_0(x)$, $\phi_1(x)$, ... be an orthonormal system in the interval (a, b). Then the series*

$$c_0\phi_0(x)+c_1\phi_1(x)+\dots, \tag{7·8}$$

with $\Sigma|c_k|^2<\infty$, is summable H_2 at almost all points at which it is summable (C, 1).

We need the following lemma:

(7·9) LEMMA. *If $s_n(x)$ and $\sigma_n(x)$ are the partial sums and the* (C, 1) *means of* (7·8), *with $\Sigma|c_k|^2<\infty$, we have*

$$\sum_{n=1}^{\infty}\frac{|s_n-\sigma_n|^2}{n}<+\infty \tag{7·10}$$

almost everywhere in (a, b).

The lemma will follow if we show that the termwise integral of the series (7·10) over (a, b) is finite. But

$$\sum_{n=1}^{\infty}\frac{1}{n}\int_a^b|s_n-\sigma_n|^2dx=\sum_{n=1}^{\infty}\frac{1}{n(n+1)^2}\sum_{k=1}^{n}k^2|c_k|^2$$

$$=\sum_{k=1}^{\infty}k^2|c_k|^2\sum_{n=k}^{\infty}\frac{1}{n(n+1)^2}\leqslant\sum_{k=1}^{\infty}k^2|c_k|^2\cdot\frac{1}{k^2}=\sum_{k=1}^{\infty}|c_k|^2,$$

and (7·9) follows.

Let now E be the set of points in (a, b) where $s(x)=\lim\sigma_n(x)$ exists. Recalling that for every convergent Σu_n we have $u_1+2u_2+\dots+nu_n=o(n)$ (Chapter III, (1·25)), we deduce from (7·10) that

$$\frac{1}{n+1}\sum_{k=1}^{n}|s_k-\sigma_k|^2\to 0$$

for almost all x. Theorem (7·7) follows from this and the fact that in the inequality

$$\left\{\frac{1}{n+1}\sum_{k=0}^{n}|s_k-s|^2\right\}^{\frac{1}{2}} \leqslant \left\{\frac{1}{n+1}\sum_{k=0}^{n}|s_k-\sigma_k|^2\right\}^{\frac{1}{2}} + \left\{\frac{1}{n+1}\sum_{k=0}^{n}|\sigma_k-s|^2\right\}^{\frac{1}{2}}$$

the second term on the right is $o(1)$ for $x \in E$.

8. Strong summability of $S[f]$ and $\tilde{S}[f]$ in the general case

(8·1) THEOREM. *If $f \in L$, then for every $q>0$ both $S[f]$ and $\tilde{S}[f]$ are summable H_q almost everywhere.*

We may suppose that $q \geqslant 1$. Since the (C, 1) sums of $S[f]$ and $\tilde{S}[f]$ are f and $\tilde{f}$ almost everywhere, the sums in (8·1) must also be f and $\tilde{f}$ almost everywhere.

It is no longer true that $S[f]$ is summable H_q at every point where $\Phi_{x,1}(h)=o(h)$. Hence the proof of (8·1) must be different from that of (7·3). It is also much more difficult. We shall still use the Hausdorff-Young theorem; but since f need not belong to any L^p, $p>1$, it will be necessary to deal not with f itself but with its Poisson integral $U(\rho,\theta)$, or even with the analytic function whose real part is U, and then to make ρ tend to 1.

We may suppose that $f \geqslant 0$. Let f_1 be a bounded function coinciding with f on a perfect set $E \subset (0, 2\pi)$ and equal to 0 elsewhere. $S[f_1]$ and $\tilde{S}[f_1]$ are, by (7·3), summable H_q almost everywhere in E, and if we prove this also for $f-f_1$ it will be true for f; then (8·1) will follow since $|E|$ may be arbitrarily close to 2π. Thus the problem is reduced to proving that, if an $f \in L$ is non-negative, and vanishes in a perfect set E, then $S[f]$ and $\tilde{S}[f]$ are both summable H_q at almost all points of E.

Next a further reduction. Let $F(x)$ be the indefinite integral of f. At almost all points x in E the ratio

$$R(x,h)=\frac{F(x+h)-F(x)}{h}$$

tends to 0 with h. Let E_m be the set of $x \in E$ such that $|R(x,h)| \leqslant m$ for $|h| \leqslant \pi$. The sets E_m are closed and E differs from ΣE_m by a set of measure 0. It is therefore enough to prove that $S[f]$ and $\tilde{S}[f]$ are summable H_q almost everywhere in each E_m. Fixing m and writing E for E_m we thus have

$$\int_x^{x+h} f(t)\,dt \leqslant m|h| \quad \text{for} \quad x \in E,\ |h| \leqslant \pi, \tag{8·2}$$

$$\int_x^{x+h} f(t)\,dt = o(h) \quad \text{for almost all } x \in E. \tag{8·3}$$

Let $U(\rho,\theta)$ be the Poisson integral of f, and let $\delta = 1-\rho$. We know (Chapter III, (7·10)) that for $x \in E$, $|h| \leqslant \pi$, (8·2) implies

$$0 \leqslant U(\rho, x+h) \leqslant Am\left(1+\frac{|h|}{\delta}\right), \tag{8·4}$$

$$\left|\int_0^h U(\rho, x+t)\,dt\right| \leqslant Am|h|. \tag{8·5}$$

Here and in the rest of the section A denotes an absolute constant, not always the same.

We take for the q of (8·1) $q=p'=p/(p-1)$, where p satisfies $1<p<2$ but is otherwise at our disposal; by taking $p-1$ sufficiently small we get q arbitrarily large.

If $\chi(x)$ is the distance of x from the closed set E, then, by Chapter IV, (2·1),

$$\int_{-\pi}^{\pi} f(x+t)\frac{\chi^{p-1}(x+t)}{|t|^p}\,dt<\infty \tag{8·6}$$

for almost all $x\in E$. Also, $S[f]$ and $\tilde{S}[f]$ are summable (C, 1) almost everywhere in E.

Almost all $x\in E$ have *all* the properties enumerated above. If we show that at every such x both $S[f]$ and $\tilde{S}[f]$ are summable H_q, (8·1) will be established. Without loss of generality we may suppose that $x=0$ is such a point.

We now consider the power series

$$\tfrac{1}{2}a_0+\sum_{\nu=1}^{\infty}(a_\nu-ib_\nu)z^\nu \quad (z=\rho e^{ix}), \tag{8·7}$$

which for $z=e^{ix}$ reduces to $S[f]+i\tilde{S}[f]$. Denote the sum of the series for $|z|<1$ by $F(z)$ and the partial sums and the (C, 1) means for $z=e^{ix}$ by $t_n(x)$ and $\tau_n(x)$ respectively. Since by hypothesis $\lim\tau_n(0)$ exists, it is enough to show that

$$\sum_{\nu=1}^{n}|t_\nu-\tau_\nu|^q=o(n), \tag{8·8}$$

where $t_\nu=t_\nu(0)$, $\tau_\nu=\tau_\nu(0)$.

This in turn will follow if we show that

$$\sum_{\nu=1}^{\infty}(\nu+1)^q|t_\nu-\tau_\nu|^q\rho^{\nu q}=o(\delta^{-q-1}), \tag{8·9}$$

where $\delta=1-\rho\to 0$. For if we set $\delta=1/n$, retain on the left only the terms given by $\nu\leqslant n$, and observe that $(1-1/n)^\nu\geqslant(1-1/n)^n>e^{-1}$ for such ν, we get

$$\sum_{\nu=1}^{n}(\nu+1)^q|t_\nu-\tau_\nu|^q=o(n^{q+1}),$$

a relation from which (8·8) easily follows by summation by parts.

Returning to (8·9) we note that

$$(n+1)(t_n-\tau_n)=\sum_{0}^{n}\nu(a_\nu-ib_\nu)$$

is the nth partial sum of $\Sigma(a_\nu-ib_\nu)\nu z^\nu$ for $z=1$. Hence, by Chapter III, (1·7),

$$\sum_{\nu=0}^{\infty}(\nu+1)(t_\nu-\tau_\nu)z^\nu=\frac{zF'(z)}{1-z}.$$

By the Hausdorff-Young theorem,

$$\left\{\sum_{\nu=0}^{\infty}(\nu+1)^q|t_\nu-\tau_\nu|^q\rho^{\nu q}\right\}^{1/q}\leqslant\left\{\frac{\rho^p}{2\pi}\int_{-\pi}^{\pi}\left|\frac{F'(\rho e^{i\psi})}{1-\rho e^{i\psi}}\right|^p d\psi\right\}^{1/p}, \tag{8·10}$$

and (8·9) will be established if we show that

$$\int_{-\pi}^{\pi}\left|\frac{F'(\rho e^{i\psi})}{1-\rho e^{i\psi}}\right|^p d\psi=o(\delta^{1-2p}). \tag{8·11}$$

Finally, recalling that, by Chapter VII, (2·31),

$$|F'(z)| \leqslant \frac{A}{\delta} U(\rho, x),$$

where U is the Poisson integral of f, we see that (8·11) will follow if we show that

$$I(\rho) = \int_{-\pi}^{\pi} \frac{U^p(\rho, \psi)}{\Delta^{\frac{1}{2}p}(\rho, \psi)} d\psi = o(\delta^{1-p}), \tag{8·12}$$

where $$\Delta(\rho, \psi) = 1 - 2\rho \cos \psi + \rho^2.$$

The rest of this section is devoted to the proof of (8·12).

First of all we recall the obvious estimates

$$(a) \quad \Delta(\rho, \psi) \geqslant \delta^2, \qquad (b) \quad \Delta(\rho, \psi) \geqslant A\psi^2. \tag{8·13}$$

We also have $$\int_{-\pi}^{\pi} \frac{d\psi}{\Delta^\alpha(\rho, \psi)} \leqslant A_\alpha \delta^{1-2\alpha}, \qquad (\alpha > \tfrac{1}{2}) \tag{8·14}$$

which follows immediately if for $|\psi| \leqslant \delta$ we apply to the integrand the first, and for the remaining intervals of ψ the second, of the inequalities (8·13).

Next we note that the inequalities (8·4) and (8·5) with $x = 0$ become

$$0 \leqslant U(\rho, \psi) \leqslant Am\left(1 + \frac{|\psi|}{\delta}\right), \tag{8·15}$$

$$\left| \int_0^h U(\rho, \psi)\, d\psi \right| \leqslant Am\,|h|. \tag{8·16}$$

Finally, writing $U^p = U^{p-1}U$, we have (see (8·12))

$$\begin{aligned} I(\rho) &= \int_{-\pi}^{\pi} \frac{U^{p-1}(\rho, \psi)}{\Delta^{\frac{1}{2}p}(\rho, \psi)} d\psi \frac{1}{\pi} \int_{-\pi}^{\pi} f(u)\, P(\rho, \psi - u)\, du \\ &= \frac{1}{\pi} \int_{-\pi}^{\pi} f(u) du \int_{-\pi}^{\pi} \frac{U^{p-1}(\rho, \psi)\, P(\rho, \psi - u)}{\Delta^{\frac{1}{2}p}(\rho, \psi)} d\psi. \end{aligned} \tag{8·17}$$

We shall estimate $I(\rho)$ by splitting the last integral into a few parts, appealing in each case to appropriate hypotheses about the behaviour of f at $x = 0$.

We first split the range $|u| \leqslant \pi$ of the integral last written into two parts, $|u| \leqslant \delta$ and the remainder. Correspondingly,

$$I(\rho) = I_1(\rho) + I_2(\rho). \tag{8·18}$$

Using (8·13) (a),

$$\begin{aligned} I_1(\rho) &\leqslant A\delta^{-p} \int_{|u| \leqslant \delta} f(u)\, du \left\{ \frac{1}{\pi} \int_{-\pi}^{\pi} U^{p-1}(\rho, \psi)\, P(\rho, \psi - u)\, d\psi \right\} \\ &\leqslant A\delta^{-p} \int_{|u| \leqslant \delta} f(u)\, du \left\{ \frac{1}{\pi} \int_{-\pi}^{\pi} U(\rho, \psi)\, P(\rho, \psi - u)\, d\psi \right\}^{p-1} \\ &= A\delta^{-p} \int_{|u| \leqslant \delta} f(u)\, U^{p-1}(\rho^2, u)\, du, \end{aligned} \tag{8·19}$$

by Jensen's inequality (since $0 < p - 1 < 1$).

Using (8·15) and (8·3) we have

$$I_1(\rho) \leqslant Am^{p-1}\delta^{-p}\int_{|u|\leqslant\delta} f(u)\,du \leqslant Am^{p-1}\delta^{-p}\,.\,o(\delta)$$
$$= o(\delta^{1-p}). \tag{8·20}$$

Next

$$I_2(\rho) = I_{2,1}(\rho) + I_{2,2}(\rho), \tag{8·21}$$

where

$$I_{2,1}(\rho) = \int_{\delta\leqslant|u|\leqslant\pi} f(u)\,du \int_{|\psi|\leqslant\frac{1}{2}|u|} \frac{U^{p-1}(\rho,\psi)\,P(\rho,\psi-u)}{\Delta^{\frac{1}{2}p}(\rho,\psi)}\,d\psi,$$

$$I_{2,2}(\rho) = \int_{\delta\leqslant|u|\leqslant\pi} f(u)\,du \int_{\frac{1}{2}|u|\leqslant|\psi|\leqslant\pi} \frac{U^{p-1}(\rho,\psi)\,P(\rho,\psi-u)}{\Delta^{\frac{1}{2}p}(\rho,\psi)}\,d\psi.$$

The inner integral in $I_{2,1}(\rho)$ does not exceed

$$P(\rho,\tfrac{1}{2}u)\int_{|\psi|\leqslant\frac{1}{2}|u|} \frac{U^{p-1}(\rho,\psi)}{\Delta^{\frac{1}{2}p}(\rho,\psi)}\,d\psi$$
$$\leqslant P(\rho,\tfrac{1}{2}u)\left(\int_{|\psi|\leqslant\frac{1}{2}|u|} U(\rho,\psi)\,d\psi\right)^{p-1}\left(\int_{|\psi|\leqslant\frac{1}{2}|u|} \Delta^{-p/(2(2-p))}\,d\psi\right)^{2-p},$$

by Hölder's inequality, and so, by (8·16) and (8·14), does not exceed

$$A\delta u^{-2}(Am\,|\,u\,|)^{p-1}\,(A_p\,\delta^{1-p/(2-p)})^{2-p} \leqslant A_p\,m^{p-1}\,|\,u\,|^{p-3}\delta^{3-2p}.$$

Hence

$$I_{2,1}(\rho) \leqslant A_p\,m^{p-1}\delta^{3-2p}\int_{\delta\leqslant|u|\leqslant\pi} \frac{f(u)}{u^{3-p}}\,du \leqslant A_p\,m^{p-1}\delta^{3-2p}\,.\,o(\delta^{p-2})$$
$$= o(\delta^{1-p}), \tag{8·22}$$

by integration by parts and application of (8·3) with $x=0$.

Finally,

$$I_{2,2}(\rho) \leqslant \int_{\delta\leqslant|u|\leqslant\pi} \frac{f(u)\,du}{\Delta^{\frac{1}{2}p}(\rho,\frac{1}{2}u)} \int_{-\pi}^{\pi} U^{p-1}(\rho,\psi)\,P(\rho,\psi-u)\,d\psi$$
$$\leqslant A\int_{\delta\leqslant|u|\leqslant\pi} \frac{f(u)}{|\,u\,|^p}\,U^{p-1}(\rho^2,u)\,du,$$

by an argument similar to that in (8·19).

The inequality (8·4) shows that

$$U(\rho^2,u) \leqslant Am\left\{1+\frac{\chi(u)}{1-\rho^2}\right\} \leqslant Am\left\{1+\frac{\chi(u)}{\delta}\right\},$$

$\chi(u)$ being the distance of u from E. Hence

$$I_{2,2}(\rho) \leqslant Am^{p-1}\int_{\delta\leqslant u\leqslant\pi} |\,u\,|^{-p} f(u)\,du + Am^{p-1}\delta^{1-p}\int_{\delta\leqslant|u|\leqslant\pi} |\,u\,|^{-p} f(u)\,\chi^{p-1}(u)\,du$$
$$\leqslant o(\delta^{1-p}) + Am^{p-1}\,\delta^{1-p}\int_{-\pi}^{\pi} |\,u\,|^{-p} f(u)\,\chi^{p-1}(u)\,du. \tag{8·23}$$

From (8·18), (8·20), (8·21), (8·22) and (8·23), we get

$$I(\rho) \leqslant o(\delta^{1-p}) + Am^{p-1}\delta^{1-p}\int_{-\pi}^{\pi} f(u)\,\frac{\chi^{p-1}(u)}{|\,u\,|^p}\,du, \tag{8·24}$$

and so, by the hypothesis about the integral (8·6) for $x=0$, we certainly have

$$I(\rho) = O(\delta^{1-p}).$$

For the refinement

$$I(\rho) = o(\delta^{1-p}), \tag{8·25}$$

we set $f=f_1+f_2$, where $f_1=f$ in $(-2\eta, 2\eta)$ and $f_1=0$ elsewhere. The value of m for f_1 is not increased and, by (8·24), $I(\rho, f_1)\leqslant \epsilon\delta^{1-p}$ for any fixed ϵ and δ sufficiently small, provided η is small enough. It is therefore enough to show that $I(\rho, f_2)=o(\delta^{1-p})$. Let U_2 be the Poisson integral of f_2. Then $I(\rho, f_2)$ is given by an integral analogous to (8·12) with U_2 for U. The part of this integral extended over $(-\eta, \eta)$ tends to 0 with δ since $U_2(\rho, \psi)$ tends to 0 on $-\eta\leqslant\psi\leqslant\eta$ (cf. (8·14)). The remaining part does not exceed

$$O(1)\int_{\eta\leqslant|\psi|\leqslant\pi} U_2^p(\rho,\psi)\,d\psi = O\{\operatorname*{Max}_{\psi} U_2^{p-1}(\rho,\psi)\}\int_{-\pi}^{\pi} U_2(\rho,\psi)\,d\psi = o(\delta^{p-1})\int_{-\pi}^{\pi} f_2\,d\psi = o(\delta^{p-1}).$$

9. Almost convergence of $S[f]$ and $\tilde{S}[f]$

The following result is an immediate consequence of (7·2), (7·3) and (8·1).

(9·1) THEOREM. *For any $f\in L$ and almost all x both $S[f]$ and $\tilde{S}[f]$ are almost convergent to sums $f(x)$ and $\tilde{f}(x)$ respectively. If $f\in L^r$, $r>1$, $S[f]$ is almost convergent to $f(x)$ at every point at which $\Phi_{x,r}(h)=o(h)$.*

Applying (7·9) to the trigonometric system, we prove the following theorem:

(9·2) THEOREM. *If $f\in L^2$, then for almost all x the sequence $1, 2, 3, \ldots$ can be broken up into two complementary sequences $\{m_k\}$ and $\{n_k\}$* (generally depending on x) *such that*

$$s_{m_k}(x)\to f(x), \quad \Sigma 1/n_k<\infty. \tag{9·3}$$

Let x be a point at which simultaneously

$$\Sigma \frac{|s_n-\sigma_n|^2}{n}<\infty \tag{9·4}$$

and $\sigma_n\to f$. For a fixed $\epsilon>0$, the sequence $\{\nu_k\}$ of indices n such that $|s_n(x)-\sigma_n(x)|\geqslant\epsilon$ satisfies, therefore, the condition $\Sigma 1/\nu_k<\infty$. By omitting the first few terms we may make $\Sigma 1/\nu_k$ arbitrarily small.

Take now $\epsilon=2^{-1}$. For N_1 large enough, we can split the sequence $N_1+1, N_1+2, \ldots$ into two complementary subsequences $\nu_1^{(1)}, \nu_2^{(1)}, \ldots$ and $\mu_1^{(1)}, \mu_2^{(1)}, \ldots$ such that $\Sigma 1/\nu_k^{(1)}<2^{-1}$ and $|s_n(x)-\sigma_n(x)|<2^{-1}$ for $n\in\{\mu_k^{(1)}\}$. Similarly, taking $\epsilon=2^{-2}$, we can find $N_2>N_1$ so large that the sequence $N_2+1, N_2+2, \ldots$ can be split into two, $\{\nu_k^{(2)}\}$ and $\{\mu_k^{(2)}\}$, such that $\Sigma 1/\nu_k^{(2)}<2^{-2}$ and $|s_n(x)-\sigma_n(x)|<2^{-2}$ for $n\in\{\mu_k^{(2)}\}$, and so on.

Now let $\{n_k\}$ be the sequence of all $\nu_j^{(s)}$ arranged, without repetition, in increasing order, and let $\{m_k\}$ be the remaining positive integers. Then

$$\Sigma 1/n_k\leqslant\Sigma 1/\nu_k^{(1)}+\Sigma 1/\nu_k^{(2)}+\ldots<2^{-1}+2^{-2}+\ldots=1.$$

Also, for any fixed s, all the $m_k>N_s$ must be in the sequence $\mu_1^{(s)}, \mu_2^{(s)}, \ldots$, and hence

$$|s_{m_k}(x)-\sigma_{m_k}(x)|<2^{-s}.$$

Since $\sigma_{m_k}(x)\to f(x)$, it follows that $s_{m_k}(x)\to f(x)$, and (9·2) is established.

A sequence $\{n_k\}$ satisfying $\Sigma 1/n_k<\infty$ is necessarily of density 0, and so the complementary sequence $\{m_k\}$ is of density 1.

Theorem (9·2) holds for $f \in L^p$, $p > 1$, and even for general $S[f]$ of power series type, but the proof is then considerably more difficult (see Chapter XV, § 1). The result, however, does not hold for f merely integrable.

10. Theorems on the convergence of orthogonal series

The theorem which follows will be needed only in the case of the trigonometric system, but even in this case we are obliged to consider a general, uniformly bounded, orthonormal system.

We denote such a system by $\phi_1(x)$, $\phi_2(x)$, ..., suppose that it is orthonormal on (a, b), and that

$$|\phi_\nu| \leqslant M \quad (\nu = 1, 2, \ldots).$$

By q we denote a number strictly greater than 2. Finally, for every sequence $c_1, c_2, \ldots$ tending to 0, where $c_\nu \neq 0$ for all ν, we denote by $c_1^*, c_2^*, \ldots$ the sequence $|c_1|, |c_2|, \ldots$ rearranged in descending order of magnitude, where if several $|c_n|$ are equal we rearrange them in order of increasing index n.

(10·1) THEOREM OF MENŠOV-PALEY. *If for a sequence* $c_1, c_2, \ldots$ *the expression*

$$\mathfrak{B}_q^*[c] = \left(\sum_{\nu=1}^{\infty} c_\nu^{*q} \nu^{q-2} \right)^{1/q} \quad (q > 2) \tag{10·2}$$

is finite, then the series

$$\sum_1^\infty c_\nu \phi_\nu(x) \tag{10·3}$$

converges almost everywhere. Moreover the function

$$S^*(x) = \sup_n \left| \sum_{\nu=1}^{n} c_\nu \phi_\nu(x) \right| \tag{10·4}$$

is in L^q *and satisfies an inequality*

$$\mathfrak{M}_q[S^*] \leqslant A_q M^{(q-2)/q} \mathfrak{B}_q^*[c]. \tag{10·5}$$

For the present A_q denotes a positive constant depending on q only.

We begin with a few remarks.

(*a*) We proved in Chapter XII, § 5, that if $\mathfrak{B}_q^*[c]$ is finite, then (10·3) is the Fourier series of an $f \in L^q$ and

$$\mathfrak{M}_q[f] \leqslant A_q M^{(q-2)/q} \mathfrak{B}_q^*[c]. \tag{10·6}$$

The proof of (10·6) gave no information about the convergence of (10·3).

(*b*) An inequality weaker† than (10·5), namely,

$$\mathfrak{M}_q[S^*] \leqslant A_q M^{(q-2)/q} \mathfrak{B}_q[c], \tag{10·7}$$

is much easier to prove. The generalization from (10·7) to (10·5) is the main difficulty in the proof of the theorem. This generalization is quite important since we have the inequality

$$(\Sigma |c_\nu|^p)^{1/p} \geqslant A_q \mathfrak{B}_q^*[c] \quad \left(p = q' = \frac{q}{q-1}\right)$$

(see p. 123), so that Theorem (10·1) holds if we replace $\mathfrak{B}_q^*[c]$ by $(\Sigma |c_\nu|^p)^{1/p}$ throughout, a result which cannot be deduced from (10·7).

† We know that $\mathfrak{B}_q^*[c] \leqslant \mathfrak{B}_q[c]$; see p. 123.

(c) The convergence of (10·3) is a consequence of (10·5). For suppose that we replace the first k coefficients of (10·3) by 0's and denote the resulting function $S^*(x)$ by $S_k^*(x)$. An application of (10·5) then shows that

$$\int_a^b S_k^{*q}(x)\,dx \to 0$$

as $k \to \infty$. Hence for each $\epsilon > 0$ the measure of the set of points where $S_k^*(x) > \epsilon$ tends to 0 as $k \to \infty$. This shows that $\Sigma c_\nu \phi_\nu$ converges almost everywhere. It is therefore enough to prove (10·5).

(d) A corollary of (10·1) is that if $\mathfrak{B}_q^*[c]$ is finite, then any rearrangement of $\Sigma c_\nu \phi_\nu$ converges almost everywhere.

We now pass to the proof of (10·5).

(10·8) LEMMA. *Let*
$$\Phi(x) = \sum_1^{2^\mu} d_\nu \phi_\nu(x).$$
The function
$$\Phi^*(x) = \sup_{n \leqslant 2^\mu} \left| \sum_1^n d_\nu \phi_\nu(x) \right|$$
satisfies an inequality
$$\mathfrak{M}_q[\Phi^*] \leqslant A_q M^{(q-2)/q} 2^{\mu(q-2)/q} \left(\sum_1^{2^\mu} |d_\nu|^q \right)^{1/q}. \tag{10·9}$$

With Φ for Φ^* on the left the inequality is an immediate consequence of (10·6).

Corresponding to each λ of $0 \leqslant \lambda \leqslant \mu$ we consider a splitting of Φ into 2^λ successive blocks 'of rank λ', each containing $2^{\mu-\lambda}$ terms, and denote the kth block and the 'greatest' block by

$$\Phi_{k,\lambda} = \sum_{(k-1)2^{\mu-\lambda}+1}^{k2^{\mu-\lambda}} d_\nu \phi_\nu, \tag{10·10}$$

$$\Phi_\lambda^*(x) = \sup_k |\Phi_{k,\lambda}(x)|, \tag{10·11}$$

where $k = 1, 2, \ldots, 2^\lambda$.

Considering the dyadic development of an integer n, $1 \leqslant n \leqslant 2^\mu$, we see that the nth partial sum of Φ is the sum of a number of blocks of different ranks. Hence

$$\Phi^* \leqslant \sum_{\lambda=0}^{\mu} \Phi_\lambda^*, \tag{10·12}$$

and therefore
$$\mathfrak{M}_q[\Phi^*] \leqslant \sum_{\lambda=0}^{\mu} \mathfrak{M}_q[\Phi_\lambda^*]. \tag{10·13}$$

But, by (10·10) and (10·11),

$$\int_a^b \Phi_\lambda^{*q}\,dx \leqslant \int_a^b \sum_{k=1}^{2^\lambda} |\Phi_{k,\lambda}|^q\,dx = \sum_{k=1}^{2^\lambda} \int_a^b |\Phi_{k,\lambda}|^q\,dx. \tag{10·14}$$

Now an application of (10·6) to (10·10) shows that the first integral in (10·14) does not exceed

$$\sum_{k=1}^{2^\lambda} A_q^q M^{q-2} 2^{(\mu-\lambda)(q-2)} \sum_{(k-1)2^{\mu-\lambda}+1}^{k2^{\mu-\lambda}} |d_\nu|^q = A_q^q M^{q-2} 2^{(\mu-\lambda)(q-2)} \sum_1^{2^\mu} |d_\nu|^q.$$

Substituting this in (10·13) we have

$$\mathfrak{M}_q[\Phi^*] \leqslant \sum_{\lambda=0}^{\mu} A_q M^{(q-2)/q} 2^{(\mu-\lambda)(q-2)/q} \left(\sum_1^{2^\mu} |d_\nu|^q \right)^{1/q}$$
$$\leqslant A_q M^{(q-2)/q} 2^{\mu(q-2)/q} \left(\sum_1^{2^\mu} |d_\nu|^q \right)^{1/q},$$

which proves the lemma.

We now set
$$\epsilon_\mu = \sum_{2^{\mu-1}}^{2^\mu - 1} c_\nu^{*q} \nu^{q-2},$$

where $\mu = 1, 2, \ldots$, and consider the terms $c_l \phi_l$ such that $|c_l|$ is one of the c_ν^* in ϵ_μ. We denote the sum of these terms by $\Phi_\mu(x)$, and suppose that the order of terms within each Φ_μ is the same as in (10·3). Let $\Phi_\mu^*(x)$ be the greatest absolute value of the partial sums of $\Phi_\mu(x)$. (Φ_μ has $2^{\mu-1}$ terms; thus Φ_μ and Φ_μ^* may be considered as the functions Φ and Φ^* of the lemma.) Since each partial sum of (10·3) is obtained by the addition of a (finite) number of the partial sums of various Φ_μ, we have
$$S^*(x) \leqslant \Phi_1^*(x) + \Phi_2^*(x) + \ldots. \tag{10·15}$$

An estimate of $\mathfrak{M}_q[\Phi_\mu^*]$ is given by Lemma (10·8), and it is natural to combine this with Minkowski's inequality applied to (10·15). Unfortunately, the resulting estimate for $\mathfrak{M}_q[S^*]$ is not good enough to give (10·5) and we must proceed differently.

We first show that
$$\int_a^b (\Phi_\lambda^* \Phi_\mu^*)^{\frac{1}{2}q}\, dx \leqslant A_q^q M^{q-2} \epsilon_\lambda^{\frac{1}{2}} \epsilon_\mu^{\frac{1}{2}} 2^{-\alpha_q |\lambda-\mu|}, \tag{10·16}$$

where α_q is positive and depends on q only.

We may suppose that $\lambda \leqslant \mu$. Writing
$$(\Phi_\lambda^* \Phi_\mu^*)^{\frac{1}{2}q} = (\Phi_\lambda^{*\frac{1}{2}q \cdot 4/(q+2)})(\Phi_\lambda^{*\frac{1}{2}q \cdot (q-2)/(q+2)} \Phi_\mu^{*\frac{1}{2}q}),$$

and applying Hölder's inequality with exponents $\frac{1}{2}(q+2)$ and $(q+2)/q$ we have
$$\int_a^b (\Phi_\lambda^* \Phi_\mu^*)^{\frac{1}{2}q}\, dx \leqslant \left(\int_a^b \Phi_\lambda^{*q}\, dx\right)^{2/(q+2)} \left(\int_a^b \Phi_\lambda^{*\frac{1}{2}q-1} \Phi_\mu^{*\frac{1}{2}q+1}\, dx\right)^{q/(q+2)}$$
$$= P^{2/(q+2)} Q^{q/(q+2)}, \tag{10·17}$$

say. Lemma (10·8) applies to P, and
$$P \leqslant A_q^q M^{q-2} 2^{\lambda(q-2)} \sum_{2^{\lambda-1}}^{2^\lambda - 1} c_\nu^{*q} \leqslant A_q^q M^{q-2} \sum_{2^{\lambda-1}}^{2^\lambda - 1} c_\nu^{*q} \nu^{q-2}$$
$$= A_q^q M^{q-2} \epsilon_\lambda, \tag{10·18}$$
$$Q \leqslant \sup_x \{\Phi_\lambda^*(x)\}^{\frac{1}{2}q-1} \int_a^b \Phi_\mu^{*\frac{1}{2}q+1}\, dx.$$

Lemma (10·8) is applicable to the integral since $\frac{1}{2}q + 1 > 2$, and we have
$$Q \leqslant M^{\frac{1}{2}q-1} \left(\sum_{2^{\lambda-1}}^{2^\lambda - 1} c_\nu^*\right)^{\frac{1}{2}q-1} M^{\frac{1}{2}q-1} A_{\frac{1}{2}q+1}^{\frac{1}{2}q+1} 2^{(\mu-1)(\frac{1}{2}q-1)} \sum_{2^{\mu-1}}^{2^\mu - 1} c_\nu^{*\frac{1}{2}q+1}. \tag{10·19}$$

Observe now that
$$\sum_{2^{\lambda-1}}^{2^\lambda - 1} c_\nu^* = \sum_{2^{\lambda-1}}^{2^\lambda - 1} c_\nu^* \nu^{(q-2)/q} \cdot \nu^{-(q-2)/q} \leqslant \epsilon_\lambda^{1/q} \left(\sum_{2^{\lambda-1}}^{2^\lambda - 1} \nu^{-(q-2)/(q-1)}\right)^{(q-1)/q}$$
$$\leqslant \epsilon_\lambda^{1/q} (2^{\lambda-1} \cdot 2^{-(\lambda-1)(q-2)/(q-1)})^{(q-1)/q}$$
$$= \epsilon_\lambda^{1/q} 2^{(\lambda-1)/q},$$

and similarly
$$\sum_{2^{\mu-1}}^{2^\mu - 1} c_\nu^{*\frac{1}{2}q+1} = \sum_{2^{\mu-1}}^{2^\mu - 1} c_\nu^{*\frac{1}{2}q+1} \nu^{(q-2)(q+2)/2q} \cdot \nu^{-(q-2)(q+2)/2q}$$
$$\leqslant \left(\sum_{2^{\mu-1}}^{2^\mu - 1} c_\nu^{*q} \nu^{q-2}\right)^{(q+2)/2q} \left(\sum_{2^{\mu-1}}^{2^\mu - 1} \nu^{-(q+2)}\right)^{(q-2)/2q}$$
$$\leqslant \epsilon_\mu^{(q+2)/2q} \cdot (2^{-(\mu-1)(q+1)})^{(q-2)/2q}.$$

Substituting these estimates in (10·19) and replacing there $A^{\frac{1}{2}q+1}_{\frac{1}{2}q+1}$ by A^q_q (merely a matter of notation), we easily deduce from (10·17), (10·18) and (10·19) the inequality (10·16) with $\alpha_q = \frac{1}{2}(q-2)/(q+2)$.

Return to (10·15) and denote by r the least integer not less than q. Then

$$\mathfrak{M}^q_q[S^*] \leqslant \int_a^b (\Sigma\, \Phi^*_\lambda)^q\, dx = \int_a^b \{(\Sigma \Phi^*_\lambda)^r\}^{q/r}\, dx$$

$$= \int_a^b \{\sum_{\lambda_1} \dots \sum_{\lambda_r} \Phi^*_{\lambda_1} \dots \Phi^*_{\lambda_r}\}^{q/r}\, dx$$

$$\leqslant \sum_{\lambda_1} \dots \sum_{\lambda_r} \int_a^b (\Phi^*_{\lambda_1} \dots \Phi^*_{\lambda_r})^{q/r}\, dx. \tag{10·20}$$

Writing

$$\Phi^*_{\lambda_1}\Phi^*_{\lambda_2} \dots \Phi^*_{\lambda_r} = \{(\Phi^*_{\lambda_1}\Phi^*_{\lambda_2})(\Phi^*_{\lambda_1}\Phi^*_{\lambda_3}) \dots (\Phi^*_{\lambda_1}\Phi^*_{\lambda_r}) \dots (\Phi^*_{\lambda_{r-1}}\Phi^*_{\lambda_r})\}^{1/(r-1)},$$

where the number of bracketed factors is $R = \frac{1}{2}r(r-1)$, and applying Hölder's inequality with exponents R we have, by (10·16),

$$\int_a^b (\Phi^*_{\lambda_1} \dots \Phi^*_{\lambda_r})^{q/r}\, dx \leqslant \prod_{1 \leqslant i < j \leqslant r} \left\{\int_a^b (\Phi^*_{\lambda_i}\Phi^*_{\lambda_j})^{\frac{1}{2}q}\, dx\right\}^{1/R}$$

$$\leqslant A^q_q M^{(q-2)} \prod_{1 \leqslant i < j \leqslant r} (\epsilon_{\lambda_i}\epsilon_{\lambda_j})^{1/2R}\, 2^{-\alpha_q |\lambda_i - \lambda_j|/R}$$

$$= A^q_q M^{(q-2)} \prod_{i=1}^{r} \epsilon^{1/r}_{\lambda_i} \prod_{j=1}^{r}{}^{(i)} \{2^{-\frac{1}{2}\alpha_q |\lambda_i - \lambda_j|/R}\},$$

where the upper suffix i indicates that the factor $j = i$ (which factor, by the way, is equal to 1) is omitted. Substituting this in the last sum in (10·20), and using Hölder's inequality again, we have

$$\mathfrak{M}^q_q[S^*] \leqslant A^q_q M^{q-2} \prod_{i=1}^{r} \left\{\sum_{\lambda_1, \dots, \lambda_r} \epsilon_{\lambda_i} \prod_{j=1}^{r}{}^{(i)}\, 2^{-\alpha_q |\lambda_i - \lambda_j|/(r-1)}\right\}^{1/r}.$$

Consider the multiple sum in curly brackets. Summing first with respect to $\lambda_1, \dots, \lambda_{i-1}, \lambda_{i+1}, \dots, \lambda_r$ and then with respect to λ_i, we see that the sum under consideration, and so also the product $\prod_{i=1}^{r}\{\dots\}^{1/r}$, does not exceed

$$\left(\sum_{\lambda=1}^{\infty} \epsilon_\lambda\right)\left\{\sum_{\nu=-\infty}^{+\infty} 2^{-\alpha_q|\nu|/(r-1)}\right\}^{r-1}.$$

Since $r - 1 < q \leqslant r$, we have

$$\mathfrak{M}^q_q[S^*] \leqslant A^q_q M^{q-2} \left(\sum_{\lambda=1}^{\infty} \epsilon_\lambda\right)(\Sigma 2^{-\alpha_q|\nu|/q})^q$$

$$= A^q_q M^{q-2} \mathfrak{B}^{*q}_q[c],$$

and (10·5) is established. This also completes the proof of Theorem (10·1).

Theorem (10·1) fails for $q = 2$. There are a number of substitute results in this case; we only prove the following, in which $\phi_1, \phi_2, \dots$ is an arbitrary (not necessarily uniformly bounded) system orthonormal on (a, b).

(10·21) THEOREM OF MENŠOV-RADEMACHER. *If*

$$\Sigma |c_\nu|^2 \log^2 \nu < \infty, \tag{10·22}$$

the series $\Sigma c_\nu \phi_\nu$ converges almost everywhere in (a, b). Moreover, the function $S^*(x)$ of (10·4) *satisfies the inequality*

$$\mathfrak{M}_2[S^*] \leqslant A \left\{ \sum_{\nu=1}^{\infty} |c_\nu|^2 \log^2 (\nu+1) \right\}^{\frac{1}{2}}. \tag{10·23}$$

The first part of the theorem extends Theorem (1·8) of Chapter IV to the most general orthogonal series. On the other hand, Theorem (1·13) asserts that for the trigonometric system the factor $\log^2 \nu$ in (10·22) can be replaced by $\log \nu$.

As in Theorem (10·1) it is enough to prove (10·23).

(10·24) LEMMA. *For the most general orthonormal system the function Φ^* of Lemma* (10·8) *satisfies*

$$\mathfrak{M}_2[\Phi^*] \leqslant A(\mu+1) \left(\sum_{1}^{2^\mu} |d_\nu|^2 \right)^{\frac{1}{2}}. \tag{10·25}$$

We keep the notations (10·10) and (10·11). The inequalities (10·13) and (10·14) hold for $q = 2$ (the factor M^{q-2} dropping out and A_q becoming an absolute constant A) so that

$$\int_a^b \Phi_\lambda^{*2} dx \leqslant A^2 \sum_{1}^{2^\mu} |d_\nu|^2,$$

$$\mathfrak{M}_2[\Phi^*] \leqslant \sum_{\lambda=0}^{\mu} \mathfrak{M}_2[\Phi_\lambda^*] \leqslant A(\mu+1) \left(\sum_{1}^{2^\mu} |d_\nu|^2 \right)^{\frac{1}{2}}.$$

(10·26) LEMMA. *If $\Sigma |c_\nu|^2 \log \nu$ converges, then*

(i) *the partial sums $S_{2^\mu} = \sum_{1}^{2^\mu} c_\nu \phi_\nu$ of $\Sigma c_\nu \phi_\nu$ converge almost everywhere;*

(ii) *the function* $$S_*(x) = \sup_\mu |S_{2^\mu}(x)|$$

satisfies

$$\mathfrak{M}_2[S_*] \leqslant A \left\{ \sum_{1}^{\infty} |c_\nu|^2 \log (\nu+1) \right\}^{\frac{1}{2}}. \tag{10·27}$$

Only (ii) is needed for the proof of (10·23).

By the theorem of Riesz-Fischer (Chapter IV, § 1), $\Sigma c_\nu \phi_\nu$ is the Fourier series of an $f \in L^2$ and

$$\int_a^b |S_{2^\mu} - f|^2 dx = \sum_{2^\mu+1}^{\infty} |c_\nu|^2,$$

$$\int_a^b \sum_{\mu=0}^{\infty} |S_{2^\mu} - f|^2 dx = \sum_{\mu=0}^{\infty} \sum_{\nu=2^\mu+1}^{\infty} |c_\nu|^2 = \sum_{\nu=2}^{\infty} |c_\nu|^2 \sum_{\mu=0}^{[\log(\nu-1)/\log 2]} 1$$

$$\leqslant A \sum_{\nu=2}^{\infty} |c_\nu|^2 \log (\nu+1). \tag{10·28}$$

Thus, under the hypotheses of (10·26), $\Sigma |S_{2^\mu} - f|^2$ converges almost everywhere, and, in particular, $S_{2^\mu} \to f$ almost everywhere. This proves (i).

Since $|S_{2^\mu}|^2 \leqslant 2(|f|^2 + |S_{2^\mu} - f|^2)$, we also have

$$S_*^2 \leqslant 2|f|^2 + 2\Sigma |S_{2^\mu} - f|^2.$$

By (10·28), the integral over (a, b) of the last sum does not exceed $A\Sigma |c_\nu|^2 \log(\nu+1)$, and since

$$\int_a^b |f|^2 dx = \Sigma |c_\nu|^2 \leqslant A\Sigma |c_\nu|^2 \log(\nu+1),$$

(10·27) follows.

We can now prove (10·23). Setting

$$S_\mu^*(x) = \sup_{2^{\mu-1}<n<2^\mu} \left| \sum_{2^{\mu-1}+1}^{n} c_\nu \phi_\nu(x) \right|,$$

we have
$$S^* \leqslant S_* + \sup_\mu S_\mu^*,$$
$$S^{*2} \leqslant 2S_*^2 + 2\Sigma S_\mu^{*2},$$

and so, by (10·27) and (10·25),

$$\int_a^b S^{*2} dx \leqslant A^2 \sum_1^\infty |c_\nu|^2 \log(\nu+1) + A^2 \sum_{\mu=1}^\infty \mu^2 \sum_{2^{\mu-1}}^{2^\mu-1} |c_\nu|^2$$
$$\leqslant A^2 \sum_1^\infty |c_\nu|^2 \log^2(\nu+1) + A^2 \sum_{\mu=1}^\infty \sum_{2^{\mu-1}}^{2^\mu-1} |c_\nu|^2 \log^2(\nu+1)$$
$$= A^2 \sum_{\nu=1}^\infty |c_\nu|^2 \log^2(\nu+1).$$

11. Capacity of sets and convergence of Fourier series

Theorems about the convergence, or summability, almost everywhere of a trigonometric series $\Sigma A_n(x)$ can, in some cases, be refined by giving additional information about the exceptional sets of measure 0. There are a number of classifications of sets of measure 0; here we consider only one, based on the notion of 'capacity'. We confine our attention to Borel sets.

We fix a function $\lambda(x)$, periodic, integrable, real-valued, even, continuous for $x \neq 0$, and tending to $+\infty$ as $x \to 0$; in particular, $\lambda(x)$ is bounded below. Consider a set $E \subset (0, 2\pi)$, and let $d\mu$ be any (non-negative) mass distribution concentrated on E, of total mass 1; hence $\int_0^{2\pi} d\mu = \int_E d\mu = 1$. The convolution

$$l(x) = l(x, \mu) = \frac{1}{\pi} \int_0^{2\pi} \lambda(x-t)\, d\mu(t) \tag{11·1}$$

has a definite value, finite or $+\infty$; it is uniformly bounded below for all μ. There are two possibilities: (i) $l(x)$ is unbounded above for all μ, (ii) $l(x)$ is bounded above for some μ. Case (i) occurs, for example, if E is finite or denumerably infinite; case (ii) if E is of positive Lebesgue measure. (If $\chi(x)$ is the characteristic function of E, and $d\mu(x) = |E|^{-1} \chi(x)\, dx$, (11·1) is bounded.) In case (i) we may consider the set as 'small', in case (ii) as 'large'. In case (i) we say that E is of $\{\lambda\}$-capacity 0, in case (ii) that E has a positive $\{\lambda\}$-capacity.

Suppose, in addition to the previous hypotheses, that the mean value of λ over a period is non-negative. The same then holds for $l(x)$, and $L(\mu) = \sup_x l(x, \mu)$ is also non-negative. The non-negative number

$$\mathscr{C}(E) = \{\inf_\mu L(\mu)\}^{-1} \tag{11·2}$$

will be called the *$\{\lambda\}$-capacity of E*.

If $E \subset E_1$, then $\mathscr{C}(E) \leqslant \mathscr{C}(E_1)$. Hence we may think of $\mathscr{C}(E)$ as a sort of measure of E; but it has the character of an 'inner measure' since it is defined by means of masses situated in E. We might apply the definition (11·2) to open and closed sets, and inquire about the conditions for E to have $\inf \mathscr{C}(O) = \sup \mathscr{C}(F)$, where O are open sets containing E, and F closed subsets of E. We are, however, interested in sets of capacity 0 only, and we define sets of *outer* $\{\lambda\}$-*capacity* 0 as those which can be covered by open sets O with $\mathscr{C}(O)$ arbitrarily small.

If E_1 and E_2 are sets of outer $\{\lambda\}$-capacity 0, *so is* $E_1 + E_2$. For suppose that $O_1 \supset E_1$ and $O_2 \supset E_2$ are open sets such that $\mathscr{C}(O_1) < \epsilon$ and $\mathscr{C}(O_2) < \epsilon$. Let $d\mu$ be any mass distribution of total mass 1, concentrated in $O_1 + O_2$ and suppose that $\int_{O_1} d\mu \geqslant \frac{1}{2}$. The first term on the right in

$$\int_{O_1+O_2} \lambda(x-t)\, d\mu(t) = \int_{O_1} \lambda(x-t)\, d\mu(t) + \int_{O_2 - O_1 O_2} \lambda(x-t)\, d\mu(t)$$

is then at least $1/2\epsilon$ for some x, and the second is uniformly bounded below, and since ϵ can be arbitrarily small, the assertion follows.

Of main interest in applications are the cases

$$\lambda(x) \sim \sum_1^\infty \frac{\cos nx}{n} = \log \frac{1}{|\,2 \sin \frac{1}{2}x\,|},$$

$$\lambda(x) \sim \sum_1^\infty \frac{\cos nx}{n^{1-\alpha}} \simeq C_\alpha \,|\, x\,|^{-\alpha} \quad (x \to 0;\ 0 < \alpha < 1)$$

(see Chapter I, (2·8) and Chapter V, (2·1)). In the first case we also speak of the *logarithmic capacity*, in the second of α-*capacity*. The logarithmic capacity is a limiting case ($\alpha = 0$) of α-capacity. It is not difficult to see that if E is of outer α-capacity 0, then it is of outer α' capacity 0 for $0 \leqslant \alpha < \alpha' < 1$.

(11·3) THEOREM. (i) *If* $\Sigma n(a_n^2 + b_n^2) < \infty$, *then the set of points of divergence of the trigonometric series* $\Sigma A_n(x)$ *is of outer logarithmic capacity* 0.

(ii) *If* $0 < \alpha < 1$, $\Sigma n^\alpha (a_n^2 + b_n^2) < \infty$, *then the set of the points of divergence of* $\Sigma A_n(x)$ *is of outer* $(1-\alpha)$-*capacity* 0.

(i) If $\omega(n)$ tends monotonically to $+\infty$, then at the points where $\Sigma A_n(x)$ diverges the partial sums of $\Sigma \omega(n)\, A_n(x)$ are unbounded (Chapter I (2·4)). Taking $\omega(n)$ such that

$$\Sigma \omega^2(n)\, n(a_n^2 + b_n^2) < \infty,$$

we see that it is enough to show that the set of the points where the partial sums $s_n(x)$ of $\Sigma A_n(x)$ are unbounded is of outer logarithmic capacity 0. It is even enough to show this for the set E of points where the $s_n(x)$ are unbounded above.

The proof of the following lemma resembles that of Theorem (1·2).

(11·4) LEMMA. *Let O be an open set, and $d\mu$ a mass distribution concentrated in O. If*

$$\frac{1}{2\pi} \int_0^{2\pi} \log \frac{1}{|\,2 \sin \frac{1}{2}(x-y)\,|}\, d\mu(y) \leqslant M \tag{11·5}$$

for all x, if $\Sigma n(a_n^2+b_n^2)\leqslant 1$, and if $n(x)$ is any Borel measurable function taking only non-negative integral values, then the partial sums $s_n(x)$ of $\sum_1^\infty A_n(x)$ satisfy

$$\frac{1}{\pi}\int_0^{2\pi} s_{n(x)}(x)\,d\mu(x)\leqslant M+A, \tag{11·6}$$

where A is an absolute constant.

Let $G_n(x)$ and $H_n(x)$ be the partial sums of $\Sigma n^{-\frac{1}{2}}\cos nx$ and $\Sigma n^{-1}\cos nx$ respectively. By hypothesis, $\Sigma n^{\frac{1}{2}}A_n(x)$ is an $S[F]$, $F\in L^2$, and we have

$$\frac{1}{\pi}\int_0^{2\pi} s_{n(x)}(x)\,d\mu(x)=\frac{1}{\pi}\int_0^{2\pi}\left\{\frac{1}{\pi}\int_0^{2\pi} F(t)\,G_{n(x)}(x-t)\,dt\right\}d\mu(x)$$

$$=\frac{1}{\pi}\int_0^{2\pi} F(t)\left\{\frac{1}{\pi}\int_0^{2\pi} G_{n(x)}(x-t)\,d\mu(x)\right\}dt=\frac{1}{\pi}\int_0^{2\pi} F(t)\,I(t)\,dt,$$

say. By Schwarz's inequality, and the hypothesis $\Sigma n(a_n^2+b_n^2)\leqslant 1$,

$$\left\{\frac{1}{\pi}\int_0^{2\pi} F(t)\,I(t)\,dt\right\}^2\leqslant\frac{1}{\pi}\int_0^{2\pi} F^2(t)\,dt\,.\,\frac{1}{\pi}\int_0^{2\pi} I^2(t)\,dt$$

$$\leqslant 1\,.\,\frac{1}{\pi}\int_0^{2\pi}\left\{\int_0^{2\pi} G_{n(x)}(x-t)\,d\mu(x)\right\}\left\{\int_0^{2\pi} G_{n(y)}(y-t)\,d\mu(y)\right\}dt$$

$$=\int_0^{2\pi}\int_0^{2\pi}\left\{\frac{1}{\pi}\int_0^{2\pi} G_{n(x)}(x-t)\,G_{n(y)}(y-t)\,dt\right\}d\mu(x)\,d\mu(y)=\int_0^{2\pi}\int_0^{2\pi} H_{n(x,y)}(x-y)\,d\mu(x)\,d\mu(y)$$

where $n(x,y)=\min\{n(x),n(y)\}$.

Since, by Chapter V, (2·28),

$$H_n(x)\leqslant\log|\tfrac{1}{2}\operatorname{cosec}\tfrac{1}{2}x|+A, \tag{11·7}$$

using this estimate for $H_{n(x,y)}(x-y)$ in the last integral, integrating first with respect to y, and applying (11·5), we obtain (11·6).

Fix a positive integer N. Since $\{s_n\}$ is unbounded above at each point of E, using the continuity of the s_n we can associate with every $x\in E$ an integer $n=n_x$ and an open neighbourhood such that $s_n>N$ in this neighbourhood. We can cover E by a denumerable family of the neighbourhoods. Their union O_N is an open set in which we have a Borel-measurable function $n(x)$ such that $s_{n(x)}(x)\geqslant N$.

Suppose now that E is not of outer logarithmic capacity 0. Then there is an M such that for any open $O\supset E$ and some $d\mu$, of total mass 1, concentrated in O, we have (11·5), and so also (11·6). But for $O=O_N$ and the $n(x)$ just mentioned

$$\int_{O_N} s_{n(x)}(x)\,d\mu\geqslant N\int_{O_N} d\mu=N,$$

which contradicts (11·6) for N large enough. This proves (i).

(ii) We modify Lemma (11·4) as follows. If $\Sigma n^\alpha(a_n^2+b_n^2)\leqslant 1$, and if on the left of (11·5) we replace the cofactor of $d\mu(y)$ by $H(x)=\Sigma n^{-\alpha}\cos nx$, we still have (11·6), with $A_\alpha M+A_\alpha$ on the right. The proof is the same as before, if we define $G_n(x)$ and $H_n(x)$

as partial sums of $\Sigma n^{-\frac{1}{2}\alpha}\cos nx$ and $\Sigma n^{-\alpha}\cos nx$ respectively, and $S[F]$ as $\Sigma n^{\frac{1}{2}\alpha}A_n(x)$; instead of (11·7) we now have

$$H_n(x) \leqslant A_\alpha H(x) + A_\alpha,$$

an immediate consequence of Chapter V, (2·26) and the fact that $H(x)$ is exactly of order $|x|^{-(1-\alpha)}$ for $x\to 0$. The rest of the proof is unchanged.

MISCELLANEOUS THEOREMS AND EXAMPLES

1. Let $f\in L^2$, and let $E_1, E_2, \ldots, E_n, \ldots$ be any increasing or decreasing sequence of sets in $(0, 2\pi)$. Then

$$\sum_{k=2}^{\infty} (\log k)^{-1}\left\{\left(\int_{E_k} f\cos kx\,dx\right)^2+\left(\int_{E_k} f\sin kx\,dx\right)^2\right\}\leqslant A\,\|f\|_2^2,$$

where A is an absolute constant. (Salem [2].)

[The theorem is equivalent to (1·10); cf. Chapter I, (9·14)]

2. Let $0<n_1<n_2<\ldots<n_s<\ldots$, $n_{s+1}/n_s>q>1$, and let $E_1, E_2, \ldots$ be any increasing or decreasing sequence of sets in $(0, 2\pi)$ such that $E_k=E_{n_s}$ for $n_{s-1}<k\leqslant n_s$ $(s=1, 2, \ldots)$. If $f\in L^2$, then

$$\sum_{k=1}^{\infty}\left\{\left(\int_{E_k} f\cos kx\,dx\right)^2+\left(\int_{E_k} f\sin kx\,dx\right)^2\right\}\leqslant A_q\,\|f\|_2^2.$$

[See (1·10).]

3. Let $\{a_n\}$ be positive decreasing, such that $\{na_n\}$ is monotone and $\Sigma a_n/n<\infty$. If $s_n(x)$ and $t_n(x)$ are respectively the partial sums of $\Sigma a_n\cos nx$ and $\Sigma a_n\sin nx$, then the functions $\sup_n|s_n(x)|$ and $\sup_n|t_n(x)|$ are both integrable.

4. If $f\sim\Sigma A_n(x)$, $\delta>0$, then the partial sums of both $\Sigma(\log n)^{-1-\delta}A_n(x)$ and $\Sigma(\log n)^{-1-\delta}B_n(x)$ can be majorized by integrable functions. For $\delta=0$ this is no longer true.

5. If $a_k\geqslant 0$ for $k=1, 2, \ldots$, and if $\Sigma a_k\sin kx$ is the Fourier series of a bounded function, then partial sums of the series are uniformly bounded; if f is continuous, the series converges uniformly. (Paley [7].)

[Let s_n and σ_n be the partial sums and (C, 1) means of the series. If, for example, $|f|\leqslant M$, then $|\sigma_{2n}(x)|\leqslant M$, $|\sigma'_{2n}(x)|\leqslant 2nM$ and, for $x=0$,

$$\sum_{k=1}^{2n}\left(1-\frac{k}{2n+1}\right)ka_k\leqslant 2nM,\quad \sum_{k=1}^{n} ka_k\leqslant 4nM,\quad |s_n(x)-\sigma_n(x)|\leqslant 4M.]$$

6. If $\Sigma A_k(x)$ is summable H_q, $q\geqslant 1$, in a set E, and if $\Sigma B_k(x)$ is summable (C, 1) in E, then $\Sigma B_k(x)$ is summable H_q almost everywhere in E. (Marcinkiewicz and Zygmund [7].)

[The proof resembles that of (5·1); the result holds without the hypothesis that $\Sigma B_k(x)$ is summable (C, 1) in E, but the proof is then more difficult.]

7. Let $f\sim\Sigma A_n(x)$. If

$$\frac{1}{h}\int_0^h [f(x+t)-f(x-t)]\,dt = o\left\{\frac{1}{\log 1/h}\right\}\quad (h\to+0)$$

uniformly in $a\leqslant x\leqslant b$, then $s_n(x)-\sigma_n(x)$ tends uniformly to 0 in every subinterval $(a+\epsilon, b-\epsilon)$ of (a, b) and, in particular, $s_n(x)$

(i) converges uniformly in $(a+\epsilon, b-\epsilon)$ if f is continuous in (a, b);

(ii) converges almost everywhere in (a, b) if f is integrable. (Salem [14].)

8. Let $I_n(x)=I_n(x, f)$ be the Lagrange interpolating polynomials with $2n+1$ equidistant fundamental points (Chapter X, § 1). If f is integrable R, and if $\liminf I_n(x)>-\infty$ for $x\in E$, then $\limsup I_n(x)<+\infty$ almost everywhere in E, and

$$f(x)=\tfrac{1}{2}\{\liminf I_n(x)+\limsup I_n(x)\}$$

almost everywhere in E. (Marcinkiewicz [10].)

[The proof is analogous to that of (5·7).]

9. Let $I_n(x)$ be the interpolating polynomials (Chapter X, § 1) of an $f \in \mathrm{R}$. If $\lim I_n(x)$ and $\lim \hat{f}_n(x)$ (see Chapter X, § 10) exist and are finite for $x \in E$, $|E| > 0$, then $\lim \tilde{I}_n(x)$ exists, and equals $\hat{f}(x)$, almost everywhere in E.

Both here and in the previous example we have analogous results for $I_{n,\nu}$, $\tilde{I}_{n,\nu}$.

10. Let $\Sigma(a_k \cos n_k x + b_k \sin n_k x)$ be a finite lacunary polynomial, $n_{k+1}/n_k > q > 1$; let $\lambda > 1$ and $E \subset (0, 2\pi)$, $|E| > 0$. Then there is a number $\nu_0 = \nu_0(q, \lambda, E)$ such that

$$\int_E S^{*2} dx \leqslant \tfrac{1}{2}\lambda \, |E| \, \Sigma(a_k^2 + b_k^2) \quad (S^*(x) = \max_k |S_{n_k}(x)|),$$

provided $n_1 > \nu_0$. The result holds for infinite $\Sigma A_{n_k}(x)$, provided $\Sigma(a_k^2 + b_k^2) < \infty$.

CHAPTER XIV

MORE ABOUT COMPLEX METHODS

1. Boundary behaviour of harmonic and analytic functions

The results we are going to prove in this section are not only of intrinsic interest but also of importance for the study of trigonometric series. Though the behaviour of functions $f(z)$ regular for $|z|<1$ is our primary concern, it will be convenient to assume merely that the analytic functions we consider are meromorphic there and do not reduce to constants.

Let D denote the unit circle $|z|<1$ and C the circumference $|z|=1$. By a *triangular neighbourhood* $T(\theta_0)$ of a point $e^{i\theta_0}\in C$ we mean any open triangle contained in D and having $e^{i\theta_0}$ as a vertex. If $T(\theta_0)$ is isosceles and bisected by the radius to $e^{i\theta_0}$, we call the neighbourhood *symmetric*.

A function $f(z)$ defined in D will be said to satisfy condition B *at* θ_0, if there is some $T(\theta_0)$ such that $f(T(\theta_0))$ is bounded. (For any set Z, $f(Z)$ will mean the set of numbers $f(z)$, $z\in Z$.) We say that f satisfies condition B *in a set* $E\subset(0,2\pi)$ if it satisfies condition B at every point $\theta_0\in E$. The triangles $T(\theta_0)$ need not remain congruent, nor the bounds for $f(T(\theta_0))$ the same, as θ_0 runs through E.

(1·1) THEOREM. *If a function $u(z)$ harmonic in $|z|<1$ satisfies condition* B *in a set E, then $u(z)$ has a non-tangential limit at almost all points of E.*

This result is a generalization of the fact that a function harmonic and bounded in D has a non-tangential limit almost everywhere on C (the function u being then the Poisson integral of a bounded function); and this special result will be used in the proof of it. Owing to the importance of the result we give two different proofs.

First proof. This will be based on conformal mapping.

For any $0<\delta<1$, let C_δ denote the circumference $|z|=\delta$. By Ω_δ we mean the open region bounded by the two tangents from $z=1$ to C_δ and by the more distant arc of C_δ between the points of contact. The set Ω_δ increases monotonically with δ and tends to D as $\delta\to 1$.

By $\Omega_\delta(\theta)$ we mean the domain Ω_δ rotated through an angle θ around $z=0$. If there is no confusion, we shall write $\Omega(\theta)$ instead of $\Omega_\delta(\theta)$.

We first consider the somewhat simpler case when all the $T(\theta_0)$ are symmetric. Then for every $\theta_0\in E$ we can find an integer $n=n(\theta_0)$ such that

$$|u(z)|\leqslant n \quad \text{for} \quad z\in\Omega_{1/n}(\theta_0). \tag{1·2}$$

The set of all points θ_0 in $(0,2\pi)$ for which (1·2) holds we call E_n. Clearly,

$$E\subset E_2+E_3+\dots,$$

and it is enough to show that u has a non-tangential limit almost everywhere in each E_n.

We therefore fix n and write $E_n=P$, $\Omega_{1/n}(\theta)=\Omega(\theta)$. As can be seen from (1·2), the set P is closed. The intervals contiguous to it we denote by (α_k,β_k). Let

$$U=\sum_{\theta\in P}\Omega(\theta) \tag{1·3}$$

be the union of all $\Omega(\theta)$ with $\theta \in P$. The set U is open. Since we may restrict θ in (1·3) to a dense subset of P, containing, say, the points α_k, β_k, we immediately see that the boundary B of U is a simple closed curve consisting (see the figure) of

(i) the set Π of points $e^{i\theta}$, $\theta \in P$;

(ii) the rectilinear sides a_k, b_k of similar curvilinear triangles constructed on the circular arcs $(e^{i\alpha_k}, e^{i\beta_k})$ contiguous to Π; and possibly

(iii) arcs of $C_{1/n}$. For convenience we shall always ignore the third possibility.

We inscribe in B a simple closed polygonal line consisting of a finite number of the pairs a_k, b_k and of chords of C joining points of Π. Since the length of $a_k + b_k$ does not exceed a fixed multiple of $\beta_k - \alpha_k$, it follows from the definition of the length of a curve that B is rectifiable. The same argument shows that Π has the same length whether as part of B or of C.

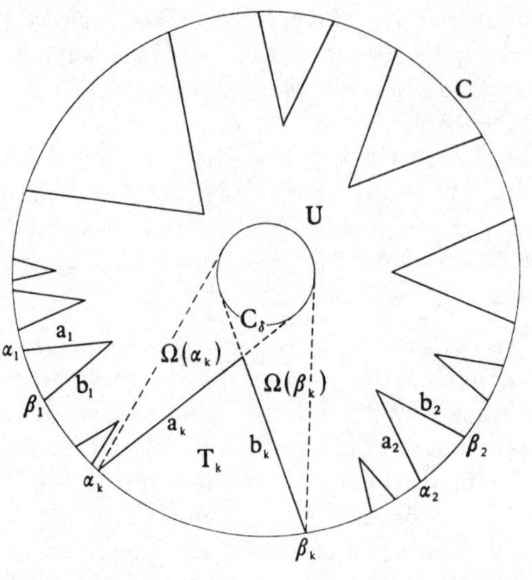

Clearly u is bounded ($|u| \leq n$) in U. Therefore, if $z = \phi(\zeta)$ maps conformally the circle D ($|\zeta| < 1$) onto U, the function

$$u(\phi(\zeta)) = u^*(\zeta) \tag{1·4}$$

is harmonic and bounded in D. Hence $u^*(\zeta)$ has a non-tangential limit almost everywhere on C ($|\zeta| = 1$). The exceptional set on C is mapped onto a set of measure zero on B (Chapter VII, (10·17)). If, in addition to the latter, we disregard on B the set, of measure zero, of points at which B has no tangent, then at the remaining points the mapping is conformal (Chapter VII, (10·13)) and so the function $u(z)$ has a non-tangential limit.

In particular $u(z)$ has at almost all points of Π a non-tangential limit as z approaches the point from inside U. But a tangent to B at a point of Π is also a tangent to C, and the non-tangential approaches through U and D mean the same thing.

Thus $u(z)$ has a non-tangential limit in Π, except for a set Z of length zero on B. The set Z is also of length zero on C, and (1·1) is established, under the assumption that all the neighbourhoods T considered are symmetric.

We shall now remove the latter restriction. The triangles $T(\theta)$ are characterized by the two sides emanating from $e^{i\theta}$, by the angle between these sides and by the angle that the bisector of $T(\theta)$ at $e^{i\theta}$ makes with the radius to $e^{i\theta}$. By reducing $T(\theta)$ if necessary, we may suppose that the two sides in question are rational and the two angles commensurable with π. We may also select bounds for $u(T(\theta))$ from among positive integers. Thus, by splitting E into a denumerable family of subsets, we may reduce the general case to a special one in which all the triangles $T(\theta)$, $\theta \in E$, are congruent and identically situated (with respect to the radii to $e^{i\theta}$), and the function u is uniformly bounded in them. As before, we may suppose that E is closed.

Now let θ_0 be any point of density of E. We shall show that u is bounded in a symmetric neighbourhood $T^*(\theta_0)$. For let Δ be any angle with vertex at $e^{i\theta_0}$ bisected by the radius to $e^{i\theta_0}$. Then θ_0 being a point of density, if an arc (α_k, β_k) contiguous to E approaches θ_0, its length becomes infinitesimally small in comparison with its distance from θ_0. It follows easily geometrically that if θ approaches θ_0 in E the triangles $T(\theta)$ cover all the points of Δ which are sufficiently close to $e^{i\theta_0}$. Thus u is bounded in a certain symmetric neighbourhood $T^*(\theta_0)$.

Since almost all points of E are points of density, u has a non-tangential limit at almost all points of E by the case previously considered, and (1·1) is proved in full generality.

Second proof. This is independent of conformal mapping and can be applied in more general situations when conformal mapping is not available.

It will be convenient to modify our definitions and denote by $T(\theta)$ a curvilinear triangle limited by an arc of some circle $|z| = \rho < 1$ and by the two rectilinear segments joining $e^{i\theta}$ to the end-points of the arc. As before, we may restrict ourselves to *symmetric* $T(\theta)$, bisected by the radius to $e^{i\theta}$. Furthermore, as in the preceding proof, we may reduce the problem to the case when E—henceforth denoted by P—is closed, the $T(\theta)$ are all identical (i.e. congruent), and u is uniformly bounded, say $|u| \leqslant 1$, in all $T(\theta)$, $\theta \in P$. From now on, by $T(\theta)$ we denote therefore a definite triangle whose curvilinear part is on a fixed circle $|z| = 1 - \delta_0$. By $T_1(\theta)$ we denote the part of $T(\theta)$ situated in the ring $1 - \frac{1}{2}\delta_0 < |z| < 1$, and we write

$$V = \sum_{\theta \in P} T_1(\theta).$$

Hence V is an open set, not necessarily connected. We denote the boundary of V by B. Clearly $|u| \leqslant 1$ in V.

Let $\rho_n = 1 - 1/n$ for $n = 2, 3, \ldots$, and let $P^{(n)}$ be the set of θ's such that $\rho_n e^{i\theta}$ is in V. $P^{(n)}$ consists of a finite number of open arcs and contains P if n is large enough; we consider only such values of n. Let $\phi_n(z)$ be the Poisson integral of the function equal to $u(\rho_n e^{i\theta})$ in $P^{(n)}$ and to 0 in the complementary set $Q^{(n)}$, and let $\psi_n(z)$ be the Poisson integral of the function equal to 0 in $P^{(n)}$ and to $u(\rho_n e^{i\theta})$ in $Q^{(n)}$. Thus

$$u(\rho_n z) = \phi_n(z) + \psi_n(z) \quad (|z| < 1). \tag{1·5}$$

Since $|\phi_n| \leqslant 1$ for all n, the ϕ_n are equi-continuous in each circle $|z| \leqslant 1 - \epsilon$.† It

† Our hypotheses imply that in the formula

$$\phi_n(\rho e^{i\theta}) = \sum_{-\infty}^{+\infty} c_k^{(n)} e^{ik\theta} \rho^{|k|}$$

the c's are numerically not greater than 1, so that $\partial\phi_n/\partial\rho$ and $\partial\phi_n/\partial\theta$ are uniformly bounded—and *a fortiori* the ϕ_n are equicontinuous—for $\rho \leqslant 1 - \epsilon$.

follows that there is a subsequence $\{\phi_{n_k}(z)\}$ converging uniformly in each circle $|z| \leqslant 1-\epsilon$ to a harmonic limit $\phi(z)$ ($|z|<1$). By (1·5), $\{\psi_{n_k}(z)\}$ also converges to a harmonic limit $\psi(z)$, and we have a decomposition

$$u(z) = \phi(z) + \psi(z). \tag{1·6}$$

Since $|\phi| \leqslant 1$, ϕ has a non-tangential limit at almost all points of C, and it remains to show that ψ has a non-tangential limit at almost all points of Π, where Π denotes the set of points $e^{i\theta}$, $\theta \in P$. More precisely, we show that this limit is 0 almost everywhere in Π. (Observe that all ψ_n are 0 on Π.)

This will be established if we prove the existence of a positive function $\chi(z)$ which majorizes $|\psi|$ in V, and which tends non-tangentially to 0 at almost all points of Π. For let θ_0 be a point of density of P such that χ has a non-tangential limit 0 at $e^{i\theta_0}$. Consider any fixed triangular neighbourhood N of $e^{i\theta_0}$. Since θ_0 is a point of density for P, the union of all $T_1(\theta)$ with θ in P and sufficiently close to θ_0 contains all points of N which are sufficiently close to $e^{i\theta_0}$. Hence $|\psi(z)| \leqslant \chi(z)$ near $e^{i\theta_0}$ in N, and since $\chi(z)$ tends to 0 as z in N approaches $e^{i\theta_0}$, the same holds for $\psi(z)$. It follows that ψ has a non-tangential limit 0 at $e^{i\theta_0}$.

To construct the required χ, we denote by $\chi_1(z)$ the Poisson integral of the characteristic function of the set Q complementary to P, and consider the positive harmonic function

$$\chi_2(z) = P(r, x) + \chi_1(z) \quad (z = r\,e^{ix}), \tag{1·7}$$

where $P(r, x)$ is the Poisson kernel. Clearly χ_2 has a non-tangential limit 0 at almost all points of Π. We assert that *at each point z_0 of the boundary B of V which is not on $|z|=1$ the function χ_2 stays above a fixed positive number depending only on our (standard) neighbourhood T.*

This is obvious if $|z_0| = 1-\frac{1}{2}\delta_0$, since $\chi_2(z) > P(r, x)$†. Suppose now that z_0 is in the ring $1-\frac{1}{2}\delta_0 < |z| < 1$. For any z in the ring, we call the largest (open) arc (α, β) such that z is in $T(\theta)$ for all θ in (α, β) *the arc associated with z*. The geometric interpretation of the Poisson integral of the characteristic function of (α, β) (see Chapter III, (6·18)) shows that the value of this integral at z exceeds a fixed positive number which depends only on T. But if $z = z_0$ is on B, the associated arc (α, β) is contained in Q, and $\chi_2(z_0)$ is not less than the Poisson integral of the characteristic function of (α, β). This completes the proof of the assertion.

Next we show that $|\psi_n(z)| \leqslant 2$ on B. (Observe that ψ_n is continuous and vanishes on Π.) It is enough to prove that $|\psi_n(z)| \leqslant 2$ in V. If z_0 is in V, z_0 is in some $T_1(\theta_0)$, $\theta_0 \in P$, and, since

$$\rho_n > 1 - \tfrac{1}{2}\delta_0 > \frac{1-\delta_0}{1-\frac{1}{2}\delta_0},$$

$\rho_n z_0$ is in $T(\theta_0)$, which, in view of (1·5) and $|\phi_n| \leqslant 1$, gives $|\psi_n(z_0)| \leqslant 2$.

Hence χ_2 stays away from 0 on $B - \Pi$, while the ψ_n are uniformly bounded there. There exists therefore an $M > 0$ such that $M\chi_2 \pm \psi_n \geqslant 0$ on $B - \Pi$. If z in V tends to a point of Π, we have

$$\liminf \{M\chi_2 \pm \psi_n\} = \liminf M\chi_2 \geqslant 0,$$

† Since χ_1 is strictly positive in $|z|<1$, the conclusion holds if we omit the term $P(r, x)$ in (1·7), and take $\chi_2 = \chi_1$. The definition (1·7) will, however, be convenient to us later (Chapter XVII, §4).

and the maximum principle for harmonic functions shows that $M\chi_2 \pm \psi_n \geqslant 0$, that is, $|\psi_n| \leqslant M\chi_2$, in V. Finally, making $n \to \infty$, we get

$$|\psi(z)| \leqslant M\chi_2(z) \quad \text{in } V. \tag{1·8}$$

Hence $\chi = M\chi_2$ has the required properties and the second proof of (1·1) is completed.

Theorem (1·1) holds if $u(z)$ is a function regular in $|z|<1$. For we may either apply (1·1) to the real and imaginary parts of u, which are harmonic functions, or repeat the first proof with u regular. In the latter case, the argument works even if $u(z)$ is meromorphic in $|z|<1$ (that is, with poles as its only singularities there). We briefly recapitulate the proof. Arguing as before we reach the stage when θ is confined to a closed set P, and there are positive numbers δ and η independent of θ such that u is uniformly bounded at all the points of $\Omega_\delta(\theta)$, $\theta \in P$, which are distant from $e^{i\theta}$ by not more than η. We can no longer assert the boundedness of u in the whole of $\Omega_\delta(\theta)$. It follows that the singularities of u in the domain U defined by (1·3) have a distance from Π exceeding a fixed positive number, and so are situated in a circle $|z| \leqslant 1-\epsilon$. Since u is meromorphic, we can have at most a finite number of such singularities, and multiplying u by a suitable polynomial we obtain a function $u_1(z)$ regular and bounded in U. Considering the function $u_1^*(\zeta) = u_1(\phi(\zeta))$ of (1·4), we see that $u_1(z)$ has a non-tangential limit almost everywhere in Π, and the same holds for $u(z)$.

We have implicitly proved the following theorem:

(1·9) THEOREM. *If a function $u(z)$ meromorphic (in particular, regular) for $|z|<1$ has a non-tangential limit zero in a set of positive measure on $|z|=1$, then $u(z) \equiv 0$.*

For we may suppose that our set P is of positive measure. The function $u_1^*(\zeta)$ is then regular and bounded in $|\zeta|<1$ and has non-tangential limit zero in a set of positive measure on $|\zeta|=1$. Hence $u_1^*(\zeta) \equiv 0$ (Chapter VII, (7·25)), $u(z) \equiv 0$.

COROLLARY. *If two functions u_1 and u_2, meromorphic in $|z|<1$, have the same non-tangential limits in a set of positive measure on $|z|=1$, then $u_1 \equiv u_2$.*

Theorem (1·9) easily leads to a more general result.

We shall say that a function $u(z)$, meromorphic in $|z|<1$, behaves *restrictedly* at θ_0 if there exists a triangular neighbourhood $T(\theta_0)$ such that $u(T(\theta_0))$ is not dense in the whole plane, i.e. is wholly situated outside a certain circle. Otherwise we shall say that u behaves *unrestrictedly* at θ_0.

(1·10) THEOREM. *Let $E \subset (0, 2\pi)$ be a set of points at which a meromorphic function $u(z)$, $|z|<1$, behaves restrictedly. Then at almost all points $\theta_0 \in E$ the function u has a finite non-tangential limit.*

Thus at almost all $\theta \in (0, 2\pi)$ there are only two possibilities: either u has a finite non-tangential limit there or else $u(T(\theta))$ is dense in the complex plane for every $T(\theta)$.

The proof is immediate. With every $\theta \in E$ we can associate a $T(\theta)$ and a circle $|\zeta - \alpha| < \rho$ such that $u(T(\theta))$ has no point in common with the circle. Taking α, ρ rational we may, by splitting E into a denumerable family of subsets, suppose that α and ρ remain unchanged throughout E. The meromorphic function

$$1/\{u(z) - \alpha\}$$

is therefore bounded in some $T(\theta)$, for each $\theta \in E$, and so has a non-tangential limit almost everywhere in E. By (1·9), this limit is distinct from zero, and so the non-tangential limit of $u(z)$ exists and is finite, at almost all points of E.

The following special cases of (1·10) deserve separate mention:

(i) No meromorphic (in particular, regular) function in $|z|<1$ can have an infinite non-tangential limit on $|z|=1$ in a set of positive measure. (This also follows from (1·9): consider $1/u(z)$.) The same holds for harmonic functions.

(ii) If a harmonic function has a non-tangential limit in a set E of positive measure, the conjugate harmonic function has a non-tangential limit almost everywhere in E.

(1·11) THEOREM. *Let $f_1(z)$ and $f_2(z)$ be regular in the rectangles $-a<x<a$, $0<y<b$ and $-a<x<a$, $0>y>-b$ respectively. If $f_1(z)-f_2(\bar{z})$ has a non-tangential limit as z approaches non-tangentially, say from the upper half-plane, any point of a set E situated on the interval $(-a,a)$ of the real axis, then both $f_1(z)$ and $f_2(z)$ have non-tangential limits at almost all points of E.*

Let $f_1=u_1+iv_1$, $f_2=u_2+iv_2$, $y>0$; the hypothesis is that the harmonic functions $u_1(x,y)-u_2(x,-y)$, $v_1(x,y)-v_2(x,-y)$ have non-tangential limits at E. It follows that the conjugate harmonic functions $v_1(x,y)+v_2(x,-y)$, $-u_1(x,y)-u_2(x,-y)$ have non-tangential limits at almost all points of E, and hence the same holds for $u_1(x,y)$, $v_1(x,y)$, $u_2(x,-y)$, $v_2(x,-y)$.

The result which follows shows that for functions harmonic or analytic in $|z|<1$ the radial behaviour may be totally different from the non-tangential behaviour.

(1·12) THEOREM. *Let $g(z)$ be any function continuous in $|z|<1$, and E any set of the first category on $|z|=1$. Then there is a function $f(z)$ regular in $|z|<1$ and such that along each radius terminating in E*

$$\lim_{r\to 1}\{f(re^{ix})-g(re^{ix})\}=0.$$

The proof is based on the approximation of continuous functions by polynomials and will not be given here. Since E can be of measure 2π, the theorem shows that, from the point of view of radial behaviour almost everywhere in $|z|<1$, there is no difference between functions which are regular and functions which are merely continuous. The following special cases may be mentioned: (*a*) A function f (regular in $|z|<1$) may tend to 0 along almost every radius without vanishing identically. (*b*) f may be bounded along almost every radius (of course not uniformly) and have a limit only on a set of radii of measure 0. (*c*) The real part of f may be bounded and the imaginary part unbounded along almost every radius.

(1·13) THEOREM. *If*

$$u(z)=u(r,x)=\sum_0^\infty A_k(x)\,r^k, \tag{1·14}$$

harmonic in $|z|<1$, satisfies for a given (α,β)

$$\int_\alpha^\beta |u(r,x)|\,dx\leqslant M<\infty \quad (0\leqslant r<1), \tag{1·15}$$

then

$$u(z)=\phi(z)+\psi(z), \tag{1·16}$$

where $\phi(z)$ is a Poisson-Stieltjes integral and $\psi(z)$ tends uniformly to 0 in every interval $(\alpha+\epsilon,\beta-\epsilon)$ $(\epsilon>0)$ as $r\to 1$. In particular, $u(z)$ has a non-tangential limit almost everywhere on (α,β).

The proof resembles the second proof of (1·1) but is easier.

Let $r_n = 1 - 1/n$, $n = 2, 3, \ldots$, and let $\phi_n(e^{ix})$ be the function equal to $u(r_n e^{ix})$ in $\alpha \leqslant x \leqslant \beta$ and to 0 elsewhere; and $\psi_n(e^{ix})$ the function equal to 0 in $\alpha \leqslant x \leqslant \beta$ and to $u(r_n e^{ix})$ elsewhere. If $\phi_n(z)$ and $\psi_n(z)$ are the Poisson integrals of $\phi_n(e^{ix})$ and $\psi_n(e^{ix})$, then

$$u(r_n z) = \phi_n(z) + \psi_n(z). \tag{1·17}$$

Since, by (1·15), $\mathfrak{M}[\phi_n(e^{ix})] \leqslant M$, the Fourier coefficients of the $\phi_n(e^{ix})$ are uniformly bounded, the functions $\phi_n(z)$ are equicontinuous in every circle $|z| \leqslant 1 - \epsilon$, and we can find a sequence $\{\phi_{n_k}(z)\}$ converging in $|z| < 1$ to a harmonic $\phi(z)$. By (1·17), $\{\psi_{n_k}(z)\}$ converges in $|z| < 1$ to a harmonic $\psi(z)$, and we have $u(z) = \phi(z) + \psi(z)$. We will show that ϕ and ψ have the required properties.

Since $\mathfrak{M}[\phi_n(e^{ix})] \leqslant M$, we have $\mathfrak{M}[\phi_n(r e^{ix})] \leqslant M$ for $r < 1$. Making $n = n_k \to \infty$, we obtain $\mathfrak{M}[\phi(r e^{ix})] \leqslant M$ for $r < 1$. By Chapter IV, (6·5), $\phi(z)$ is a Poisson-Stieltjes integral (of a function constant in the complement of (α, β)).

Let $\alpha < \alpha' < \alpha'' < \beta'' < \beta' < \beta$, and let S, S', S'' be central sectors of $|z| < 1$ supported respectively by the arcs (α, β), (α', β'), (α'', β''); we include in the sectors the terminal radii but exclude the arcs on $|z| = 1$. Let $P(r, x)$ be the Poisson kernel. If we show that, for $z \in S'$ and A large enough,

$$|\psi(z)| \leqslant A[P(r, x - \alpha') + P(r, x - \beta')],$$

it will follow that $\psi(z) \to 0$ for $z \in S''$ and $r \to 1$, and the theorem will be proved. It is enough to show that

$$|\psi_n(z)| \leqslant A[P(r, x - \alpha') + P(r, x - \beta')] \quad \text{for} \quad z \in S', \tag{1·18}$$

with A independent of n, and then make $n = n_k \to \infty$.

First we show that if $z \in S'$ then

$$\psi_n(z) = O\{1/(1-r)\}, \quad \text{uniformly in } n. \tag{1·19}$$

This will follow from (1·17) if we prove the same for $\phi_n(z)$ and $u(r_n z)$. For $\phi_n(z)$, the result is obvious (for all x) since the Fourier coefficients of $\phi_n(e^{ix})$ are uniformly bounded. As to u, observe that if we multiply (1·15) by r and integrate over $1 - 2\delta \leqslant r \leqslant 1$ we obtain

$$\int_\alpha^\beta \int_{1-2\delta}^1 |u(r, x)|\, r\, dr\, dx \leqslant 2M\delta. \tag{1·20}$$

Suppose now that $z_0 \in S'$ and write $|z_0| = 1 - \delta$. If δ is so small that the disk $|z - z_0| < \delta$ is in S, then,† denoting by $d\sigma$ the element of area and using (1·20),

$$|u(z_0)| \leqslant \frac{1}{\pi\delta^2} \iint_{|z-z_0|<\delta} |u|\, d\sigma \leqslant \frac{1}{\pi\delta^2} 2M\delta \leqslant \frac{M}{1 - |z_0|}.$$

Hence, with a suitable M', $|u(z_0)| \leqslant M'/(1 - |z_0|)$ for *all* $z_0 \in S'$. It follows that $|u(r_n z_0)| \leqslant M'/(1 - |r_n z_0|) \leqslant M'/(1 - |z_0|)$ for $z_0 \in S'$, and (1·19) is established.

We can now prove (1·13). If $0 \leqslant r < 1$, and $z = r e^{ix}$ is on either of the radii bounding S', then $\psi_n(z)$ satisfies (1·19), and the sum in square brackets in (1·18) is at least $\frac{1}{2} + r + r^2 + \ldots > 1/2\,(1 - r)$. It follows that (1·18) holds for such z if A is large enough.

† The first inequality follows if we multiply the inequality $|u(z_0)| \leqslant \mathfrak{M}[u(z_0 + \rho e^{it})]$ by ρ and integrate over $0 < \rho < \delta$.

If z approaches, from S', any point of the circular arc bounding S', then $\phi_n(z) \to 0$ (observe that $\psi_n(e^{it}) = 0$ on (α, β)) and $\liminf\{P(r, x-\alpha') + P(r, x-\beta')\} \geqslant 0$. By the maximum principle for harmonic functions (compare a similar argument in the second proof of (1·1)) we have (1·18), and (1·13) is established.

The preceding argument shows that if we replace (1·15) by

$$\mathfrak{M}[u(r', x) - u(r'', x); \alpha, \beta] \to 0 \quad (r', r'' \to 1),$$

then the ϕ in (1·16) is the Poisson integral of a function f vanishing outside (α, β). The conclusion holds if instead of (1·15) we have $\mathfrak{M}_p[u; \alpha, \beta] = O(1)$ for some $p > 1$; f is then in L^p. If $u(r\,e^{ix})$ is bounded for $\alpha \leqslant x \leqslant \beta$, $r < 1$, $\phi(z)$ is bounded in $|z| < 1$.

In the remainder of the section we consider the problem of the convergence of a sequence of functions regular in $|z| < 1$.

(1·21) THEOREM. *Let $F_1(z), F_2(z), \ldots$ be regular in $|z| < 1$ and satisfy*

$$\int_0^{2\pi} \log^+ |F_n(\rho\,e^{ix})|\,dx \leqslant M < \infty \quad (0 \leqslant \rho < 1;\ n = 1, 2, \ldots). \tag{1·22}$$

Then, if the boundary values $F_n(e^{ix})$ converge on a set of positive measure, the sequence $\{F_n(z)\}$ converges uniformly in every circle $|z| < 1 - \epsilon$ $(\epsilon > 0)$.

In view of (1·22), $F_n(e^{ix}) = \lim F_n(r\,e^{ix})$ exists almost everywhere (Chapter VII, (7·25)). Hence, for each n, we can find a set $Z_n \subset (0, 2\pi)$ of measure arbitrarily small, such that $F_n(r\,e^{ix})$ is bounded on the set of radii terminating outside Z_n. By the theorem of Egorov, $\{F_n(e^{ix})\}$ converges uniformly over subsets of E of measure arbitrarily close to $|E|$. Hence, choosing the Z_n so that $\Sigma |Z_n| < \frac{1}{2}|E|$, we deduce the existence of a set $\mathscr{E} \subset E$, $|\mathscr{E}| > 0$, with the following properties

(i) $\{F_n(e^{ix})\}$ converges uniformly on $\mathscr{E}$;

(ii) For each n, $F_n(r\,e^{ix})$ is bounded for $0 \leqslant r < 1$, $x \in \mathscr{E}$.

Fix m, n, and write $F_{m,n} = F_m - F_n$, $\mu_{m,n} = \sup |F_{m,n}(e^{it})|$ for $t \in \mathscr{E}$. Since for $r < R < 1$ we have

$$\log |F_{m,n}(r\,e^{ix})| \leqslant \frac{1}{\pi} \int_0^{2\pi} \log |F_{m,n}(R\,e^{it})|\, P\left(\frac{r}{R}, t - x\right) dt \tag{1·23}$$

(Chapter VII, (7·11)), and since (1·22) implies

$$\int_0^{2\pi} \log^+ |F_{m,n}(\rho\,e^{it})|\,dt \leqslant M' < \infty \quad (0 \leqslant \rho < 1),$$

where M' is independent of m, n, splitting the integral (1·23) into two, one extended over $\mathscr{E}$ and the other over $(0, 2\pi) - \mathscr{E}$, we have

$$\log |F_{m,n}(r\,e^{ix})| \leqslant \frac{1}{\pi} \int_{\mathscr{E}} \log |F_{m,n}(R\,e^{it})|\, P\left(\frac{r}{R}, t - x\right) dt + \frac{M'}{\pi} \cdot \frac{1}{1 - r/R}.$$

By (ii), $\log |F_{m,n}(R\,e^{it})|$ is bounded above on $\mathscr{E}$. Hence making $R \to 1$ we obtain

$$\log |F_{m,n}(r\,e^{ix})| \leqslant \log \mu_{m,n} \cdot \frac{1}{\pi} \int_{\mathscr{E}} P(r, t - x)\,dt + \frac{M'}{\pi} \cdot \frac{1}{1 - r}. \tag{1·24}$$

If $r \leqslant 1 - \epsilon$, the last term is bounded. If m and n tend to ∞, the preceding term tends to $-\infty$, since the co-factor of $\log \mu_{m,n}$, the Poisson integral of the characteristic function of $\mathscr{E}$, is bounded away from 0 for $r \leqslant 1 - \epsilon$. Hence $F_{m,n}(r\,e^{ix}) \to 0$ and (1·21) is established.

Condition (1·22) is satisfied if the $F_n(z)$ are uniformly bounded in $|z|<1$. In this case the theorem may be strengthened as follows.

(1·25) THEOREM. *If $F_n(z)$ $(n=1,2,\dots)$ are regular and uniformly bounded in $|z|<1$, and if $\{F_n(e^{ix})\}$ converges at every point of a set E, $|E|>0$, then almost all points x_0 of E have the property that $\{F_n(z)\}$ converges uniformly in $\Omega_\sigma(x_0)$ for all $\sigma<1$.*

We may suppose that $|F_n(z)|\leqslant\frac{1}{2}$ for all n, so that $|F_{m,n}|=|F_m-F_n|\leqslant 1$. Let $\mathscr{E}$, $|\mathscr{E}|>0$, be a subset of E on which $\{F_n(e^{ix})\}$ converges uniformly, and let m, n be so large that the upper bound of $|F_{m,n}(e^{it})|$ on $\mathscr{E}$ is less than $\eta<1$. Then, making $R\to 1$ in (1·23), we have

$$\log|F_{m,n}(re^{ix})|\leqslant\frac{1}{\pi}\int_0^{2\pi}\log|F_{m,n}(e^{it})|\,P(r,t-x)\,dt\leqslant-\log\frac{1}{\eta}.\frac{1}{\pi}\int_{\mathscr{E}}P(r,t-x)\,dt.$$

The last factor, the Poisson integral of the characteristic function of $\mathscr{E}$, tends to 1 at every point of density x_0 of $\mathscr{E}$ as z tends to e^{ix_0} through $\Omega_\sigma(x_0)$ (Chapter III, (7·6)). Hence, if $z\in\Omega_\sigma(x_0)$ the integral stays above a positive quantity and $F_{m,n}(z)$ tends uniformly to 0 as $m,n\to\infty$. This completes the proof of (1·25).

Remark. Let $\mathscr{E}'$ be a subset of $\mathscr{E}$ in which the density of $\mathscr{E}$ is uniformly 1 (that is, the density of $\mathscr{E}$ in (x_0-h,x_0+h) tends to 1 as $h\to 0$, uniformly in $x_0\in\mathscr{E}'$). The preceding argument shows that $\{F_n(z)\}$ converges uniformly in the union of all $\Omega_\sigma(x_0)$, $x_0\in\mathscr{E}'$, for a fixed $\sigma<1$. The measure of $\mathscr{E}-\mathscr{E}'$, and so also that of $E-\mathscr{E}'$, can be arbitrarily small.

2. The function $s(\theta)$

Let C, C_δ, D, $\Omega_\delta(\theta)$ have the same meanings as in the preceding section.

For any function

$$F(z)=F(x+iy)=F(re^{it})$$

regular in D, consider the integral

$$s(\theta)=s_\delta(\theta,F)=\left\{\iint_{\Omega_\delta(\theta)}|F'(z)|^2\,d\sigma\right\}^{\frac{1}{2}}\quad(d\sigma=dx\,dy). \tag{2·1}$$

It is a non-negative, possibly infinite, function defined for $0\leqslant\theta\leqslant 2\pi$, and $s^2(\theta)$ is the area of the image of $\Omega(\theta)$ by F. For the finiteness of $s(\theta)$ only the part of $\Omega(\theta)$ situated in an arbitrarily small neighbourhood of $e^{i\theta}$ is relevant. The main result of this section will be that at almost all points θ the finiteness of $s(\theta)$ is equivalent to the existence of the non-tangential limit of F. More precisely,

(2·2) THEOREM. (i) *If $F(z)$ has a non-tangential limit in a set $E\subset(0,2\pi)$, then $s(\theta)$ is finite at almost all points of E for every $\delta<1$.*

(ii) *Conversely, if for every $\theta\in E$ there is a $\delta=\delta(\theta)$ such that $s_\delta(\theta)<\infty$, then $F(z)$ has a non-tangential limit at almost all points of E.*

For the proof of (i) we need the following result:

(2·3) LEMMA. *Suppose that $F(z)=\Sigma c_m z^m$ is of class H^2. Then for any fixed $\delta<1$,*

$$\int_0^{2\pi}s_\delta^2(\theta)\,d\theta\leqslant A_\delta\Sigma|c_m|^2. \tag{2·4}$$

In particular, for almost all θ, $s_\delta(\theta)$ is finite for every $\delta<1$.

Here and hereafter A_δ denotes a constant depending only on δ.

Let $\chi_\theta(z)$ be the characteristic function of the set $\Omega(\theta)$. The left-hand side of (2·4) is

$$\int_0^{2\pi} d\theta \int_{|z|<1} |F'(z)|^2 \chi_\theta(z)\, d\sigma = \int_{|z|<1} |F'(z)|^2 \left\{\int_0^{2\pi} \chi_\theta(z)\, d\theta\right\} d\sigma. \tag{2·5}$$

We fix z, and consider $\chi_\theta(z)$ as a function of θ. We denote the integral in curly brackets by $I(z)$, and distinguish the two cases

$$(a)\quad |z|<\delta, \qquad (b)\quad |z| \geqslant \delta.$$

In case (a), $\chi_\theta(z)=1$ for all θ, and $I(z)=2\pi$.

In case (b), $\chi_\theta(z)=0$, except for the θ's on an arc γ of C cut out by the angle opposite to the one formed by the tangents from z to C_δ. As $|z|\to 1$, the length of γ is asymptotically proportional to $1-|z|$. It follows that the length of γ is always contained between two fixed multiples (depending on δ) of $1-|z|$. Hence

$$I(z) \leqslant A_\delta(1-|z|), \tag{2·6}$$

an inequality which holds also in case (a).

Thus the right-hand side of (2·5) does not exceed A_δ times

$$\begin{aligned}\int_{|z|<1} (1-r)\,|F'(z)|^2\, d\sigma &= \int_0^1\int_0^{2\pi} (1-r)\,|F'(re^{it})|^2\, r\,dr\,dt\\ &= 2\pi\int_0^1 (1-r)\,\Sigma m^2 |c_m|^2 r^{2m-1}\,dr\\ &= 2\pi\Sigma m^2|c_m|^2 \int_0^1 (1-r)\, r^{2m-1}\,dr\\ &= 2\pi\sum_1^\infty \frac{m^2}{2m(2m+1)}|c_m|^2\\ &\leqslant 2\pi\Sigma |c_m|^2,\end{aligned} \tag{2·7}$$

and (2·3) is established.

Remarks. (a) Since $2m(2m+1) \leqslant 2m\,.\,3m = 6m^2$, the left-hand side of (2·7) is not less than a fixed multiple of the right, and so is finite if and only if $F \in \mathrm{H}^2$.

(b) Let

$$S(r) = S(r, F) = \int_{|z|\leqslant r} |F'(z)|^2\, d\sigma$$

be the area of the image of the circle $|z|\leqslant r$ by F. Changing the order of integration we get

$$\int_0^1 S(r)\,dr = \int_{|z|<1} (1-r)\,|F'(z)|^2\, d\sigma, \tag{2·8}$$

and so the integral on the left is finite if and only if $F\in \mathrm{H}^2$.

(c) In view of our observation about the length of γ, the inequality opposite to (2·6) is also true. Since the special nature of F is irrelevant for the validity of (2·5) we have implicitly arrived at the following result, which will be used later.

(2·9) Lemma. *For any function $G(z)$ non-negative in D $(G \not\equiv 0)$, the ratio*

$$\int_0^{2\pi} d\theta \int_{\Omega_\delta(\theta)} G(z)\, d\sigma \Big/ \int_{|z|<1} (1-r)\, G(z)\, d\sigma$$

is contained between two positive constants depending on δ only.

Returning to (2·2) (i), it is enough to prove it for a fixed $\delta<1$. The result will follow if we show that we have $s(t)<\infty$ for almost all $t\in H$, where H is the set of all t such that F is bounded in $\Omega(t)$. Clearly $H=H_1+H_2+\dots$, where H_n consists of all t's such that $|F|\leqslant n$ in $\Omega(t)$. The sets H_n are closed and it is enough to show that $s(t)<\infty$ almost everywhere in each H_n.

We fix n and write $H_n=P$. Let U be the union of all $\Omega(t)$ with t in P; this set was already considered in the preceding section. Let $z=\phi(\zeta)$, $\zeta=\xi+i\eta$, map the circle D ($|\zeta|<1$) conformally onto U. Then

$$\Phi(\zeta)=F(\phi(\zeta)) \tag{2·10}$$

is regular and bounded for $|\zeta|<1$. Hence, by (2·3), $s_\delta(\theta,\Phi)$ is finite for almost all θ and all $\delta<1$.

Let $z^*=e^{it^*}$ be the general point of Π, and let $z^*=\phi(\zeta^*)$, $\zeta^*=e^{i\theta^*}$. Almost all points z^* have the following properties:

(*a*) The boundary B of U has a tangent at z^*, and so the mapping $z=\phi(\zeta)$ is conformal at ζ^* (Chapter VII, (10·13));

(*b*) $s_\eta(\theta^*,\Phi)<\infty$ for all $\eta<1$.

It is enough to show that, for every $z^*=e^{it^*}$ satisfying (*a*) and (*b*), we have $s(t^*,F)<\infty$. Now, if $\overline{\Omega}$ is the set of the ζ mapped by the function $z=\phi(\zeta)$ onto $\Omega(t^*)$, we have

$$\int_{\overline{\Omega}}|\Phi'(\zeta)|^2\,d\xi\,d\eta=\int_{\Omega(t^*)}\left|\frac{d}{dz}\Phi(\zeta).\frac{dz}{d\zeta}\right|^2\left|\frac{d\zeta}{dz}\right|^2dx\,dy=\int_{\Omega(t^*)}|F'(z)|^2\,dx\,dy.$$

The left-hand side here is finite, since the part of $\overline{\Omega}$ which is in a small neighbourhood of $\zeta=\zeta^*=e^{i\theta^*}$ is contained in $\Omega_\eta(\theta^*)$, provided $\eta>\delta$ (condition (*a*)), and since $s_\eta(\theta^*,\Phi)<\infty$ (condition (*b*)). Hence also the right-hand side is finite, and the proof of (2·2)(i) is completed.

Passing to part (ii), let E be the set of all t such that $s_\delta(t,F)$ is finite for some $\delta=\delta(t)>0$. Then $E=E_2+E_3+\dots$, where E_n consists of the t such that $s_{1/n}(t,F)\leqslant n$. We fix n and write $E_n=P$. It is enough to show that F has a non-tangential limit at almost all points of P.

By Fatou's lemma (Chapter I, (11·2)), the set P is closed. We consider once more the union U of all $\Omega_{1/n}(t)$, $t\in P$, and a function $z=\phi(\zeta)$ mapping conformally D ($|\zeta|<1$) onto U. This time, however, we take a special ϕ, with $\phi(0)=0$. Since $|\phi(\zeta)|<1$, Schwarz's lemma† gives

$$|\phi(\zeta)|\leqslant|\zeta|. \tag{2·11}$$

If we show that the function $\Phi(\zeta)=F(\phi(\zeta))$ has a non-tangential limit at almost all points θ, an argument already used (in the proof of (1·1)) will show that F has a non-tangential limit at almost all points of P.

It is enough to show that $\Phi\in H^2$ or, what is the same thing, by Remark (*b*) on p. 208, that

$$\int_0^1 S(r,\Phi)\,dr<\infty. \tag{2·12}$$

Let $U(r)$ be the part of U situated in the circle $|z|\leqslant r$, and let $U_*(r)$ be the image under $z=\phi(\zeta)$ of the circle $|\zeta|\leqslant r$. The inequality (2·11) implies that

$$U_*(r)\subset U(r).$$

† Schwarz's lemma asserts (2·11) for any function $\phi(\zeta)$ regular in $|\zeta|<1$ and satisfying the conditions $\phi(0)=0$, $|\phi(\zeta)|\leqslant1$. For the ratio $\phi(\zeta)/\zeta$ is regular for $|\zeta|<1$, and numerically does not exceed $1/(1-\epsilon)$ for $|\zeta|=1-\epsilon$, and so also in $|\zeta|\leqslant1-\epsilon$; and it is enough to make $\epsilon\to0$.

Let $\chi(z)$ denote the characteristic function of U. Then

$$S(r,\Phi)=\int_{|\zeta|\leqslant r}|\Phi'(\zeta)|^2\,d\sigma=\int_{U_*(r)}|F'(z)|^2\,d\sigma$$

$$\leqslant\int_{U(r)}|F'|^2\,d\sigma=\int_{|z|\leqslant r}|F'|^2\chi\,d\sigma$$

and, by (2·9),

$$\int_0^1 S(r,\Phi)\,dr\leqslant\int_{|z|<1}(1-r)\,|F'(re^{it})|^2\,\chi(re^{it})\,d\sigma$$

$$\leqslant A_\delta\int_0^{2\pi}dt\int_{\Omega(t)}|F'|^2\chi\,d\sigma.$$

It is therefore enough to show that the inner integral on the right is bounded as a function of t.

This is clear if $t\in P$, since the integral is then $s^2(t,F)\leqslant n^2$. Suppose now that t is on an interval (α_k,β_k) contiguous to P. It is geometrically immediate that $\Omega(t)$ is then contained in the union of $\Omega(\alpha_k)$, $\Omega(\beta_k)$ and the curvilinear triangle T_k whose sides are a_k, b_k and the arc $(e^{i\alpha_k}, e^{i\beta_k})$ (see the figure on p. 200). Since $\chi=0$ in T_k,

$$\int_{\Omega(t)}|F'|^2\chi\,d\sigma\leqslant\int_{\Omega(\alpha_k)}+\int_{\Omega(\beta_k)}\leqslant 2n^2.$$

This completes the proof of (2·2) (ii).

Remark. The conclusion of (2·2) (ii) holds if we suppose that for every $\theta\in E$ there is a triangular neighbourhood $T(\theta)$, not necessarily symmetric, such that

$$\int_{T(\theta)}|F'|^2\,d\sigma<\infty.$$

The proof follows the same lines as before (though the geometric details are a little awkward) if instead of U we consider another domain, analogous to U, obtained from the union of various $T(\theta)$.

3. The Littlewood-Paley function $g(\theta)$

For any function $$F(z)=\sum_0^\infty c_m z^m\quad(z=re^{i\theta})$$
regular in $|z|<1$ we set

$$g(\theta)=g(\theta,F)=\left\{\int_0^1(1-r)\,|F'(re^{i\theta})|^2\,dr\right\}^{\frac12}.$$

If the real part of F is the Poisson integral of a function $f(\theta)$, we shall also write $g(\theta,f)$ for $g(\theta,F)$.

The function g, introduced by Littlewood and Paley, will be used in the next chapter to obtain a number of important results. It has no obvious geometric significance, but it is majorized by a function which has one, as we now show.

(3·1) Theorem. $$g(\theta)\leqslant A_\delta s_\delta(\theta),$$

where $s_\delta(\theta)$ is defined in (2·1) and A_δ depends on δ only.

We may suppose that $\theta=0$. Let $\rho_n=1-2^{-n}$, $n=0,1,\ldots$, and let r_n be a point on (ρ_n,ρ_{n+1}) at which $|F'(r)|$ attains its maximum. Then

$$g^2(0)\leqslant\sum_{n=0}^{\infty}|F'(r_n)|^2\int_{\rho_n}^{\rho_{n+1}}(1-\rho)\,d\rho\leqslant\sum_{n=0}^{\infty}\left|\frac{F'(r_n)}{2^n}\right|^2. \tag{3·2}$$

By Cauchy's theorem, if $\rho<1-r$,

$$|F'(r)|=\left|\frac{1}{2\pi i}\int_{|\zeta-r|=\rho}\frac{F'(\zeta)}{\zeta-r}\,d\zeta\right|\leqslant\frac{1}{2\pi}\int_0^{2\pi}|F'(r+\rho e^{it})|\,dt,$$

$$|F'(r)|^2\leqslant\frac{1}{2\pi}\int_0^{2\pi}|F'(r+\rho e^{it})|^2\,dt. \tag{3·3}$$

Let $D(r,R)$ denote the circle $|z-r|\leqslant R$, and suppose that it is contained in $|z|<1$. Multiplying (3·3) by ρ and integrating over $0\leqslant\rho\leqslant R$, we get

$$\tfrac{1}{2}R^2|F'(r)|^2\leqslant\frac{1}{2\pi}\int_{D(r,R)}|F'(z)|^2\,d\sigma\quad(d\sigma=dx\,dy).$$

We set here $r=r_n$, $R=\eta 2^{-n}$, η being a constant. If $\eta=\eta(\delta)$ is small enough, all the circles $D(r_{2k},\eta 2^{-2k})$ are in $\Omega=\Omega_\delta(0)$ and no two have points in common. Hence

$$\pi\eta^2\sum_{k=0}^{\infty}\left|\frac{F'(r_{2k})}{2^{2k}}\right|^2\leqslant\int_\Omega|F'(z)|^2\,d\sigma. \tag{3·4}$$

This inequality is still true if $2k$ is replaced by $2k+1$. Adding the new inequality to (3·4), and using (3·2), we get (3·1) for $\theta=0$.

Combining (2·2) (i) and (3·1) we see that $g(\theta)$ is finite at almost all θ for which F has a non-tangential limit.

(3·5) THEOREM. *Suppose that $F\in \mathrm{H}^\lambda$, $\lambda>0$, and let $F(e^{i\theta})=\lim_{r\to1}F(re^{i\theta})$. Then*

$$\left\{\int_0^{2\pi}g^\lambda(\theta)\,d\theta\right\}^{1/\lambda}\leqslant A_\lambda\left\{\int_0^{2\pi}|F(e^{i\theta})|^\lambda\,d\theta\right\}^{1/\lambda}, \tag{3·6}$$

where A_λ depends on λ only.

It is enough to prove (3·6) for an F regular in $|z|\leqslant1$. For if $0<R<1$, $g_R=g(\theta,F(Rz))$, then

$$g_R^2(\theta)=R^2\int_0^1(1-\rho)\,|F'(R\rho e^{i\theta})|^2\,d\rho\leqslant R^2\int_0^1(1-\rho R)\,|F'(R\rho e^{i\theta})|^2\,d\rho$$

$$=R\int_0^R(1-\rho)\,|F'(\rho e^{i\theta})|^2\,d\rho\leqslant g^2(\theta).$$

Thus
$$g_R(\theta)\leqslant g(\theta),\quad g_R(\theta)\to g(\theta)\quad\text{as}\quad R\to1,$$

and if (3·6) is valid with g and $F(z)$ replaced by g_R, $F(Rz)$, we obtain (3·6) in full generality by making $R\to1$.

The proof of (3·5) consists of three parts. (*a*) First we establish it for $\lambda=2$. (*b*) Next we show that if (3·6) is true for any particular λ it holds for all smaller λ. (*c*) Having thus established (3·6) for $\lambda\leqslant2$, we pass from this to $\lambda>2$ by a conjugacy argument.

(*a*) Arguing as in (2·7) we have

$$\int_0^{2\pi} g^2(\theta)\,d\theta = \int_0^{2\pi}\int_0^1 (1-\rho)\,|\,F'(\rho e^{i\theta})\,|^2\,d\rho\,d\theta$$

$$= 2\pi \sum_{m=1}^{\infty} |\,c_m\,|^2 \frac{m^2}{(2m-1)\,2m} \leqslant 2\pi \sum |\,c_m\,|^2$$

$$= \int_0^{2\pi} |\,F(e^{i\theta})\,|^2\,d\theta.$$

The argument does not require F to be regular on $|\,z\,|=1$.

(*b*) Suppose that (3·6) holds for some λ, and let $0<\kappa<\lambda$. Suppose that $F\in \mathrm{H}^\kappa$ and assume temporarily that F has no zero for $|\,z\,|<1$. Consider a branch of $F_1 = F^{\kappa/\lambda}$. Then

$$F_1^\lambda = F^\kappa, \quad F_1 \in \mathrm{H}^\lambda.$$

Set
$$F_1^*(\theta) = \sup_{\rho<1} |\,F_1(\rho e^{i\theta})\,|,$$

$$g(\theta) = g(\theta, F), \quad g_1(\theta) = g(\theta, F_1).$$

By Chapter VII, (7·36),
$$\mathfrak{M}_\lambda[F_1^*] \leqslant C_\lambda\, \mathfrak{M}_\lambda[F_1(e^{i\theta})], \tag{3·7}$$

and
$$g^2(\theta) = (\lambda/\kappa)^2 \int_0^1 (1-\rho)\,|\,F_1\,|^{2(\lambda-\kappa)/\kappa}\,|\,F_1'\,|^2\,d\rho$$

$$\leqslant (\lambda/\kappa)^2 \{F_1^*(\theta)\}^{2(\lambda-\kappa)/\kappa} g_1^2(\theta),$$

$$\int_0^{2\pi} g^\kappa\,d\theta \leqslant (\lambda/\kappa)^\kappa \int_0^{2\pi} F_1^{*\lambda-\kappa} g_1^\kappa\,d\theta$$

$$\leqslant (\lambda/\kappa)^\kappa\, \mathfrak{M}_\lambda^{\lambda-\kappa}[F_1^*]\, \mathfrak{M}_\lambda^\kappa[g_1].$$

Since (3·6) is supposed true for our particular λ, we have, using (3·7),

$$\mathfrak{M}_\kappa[g] \leqslant (\lambda/\kappa)\,\{C_\lambda\,\mathfrak{M}_\lambda[F_1(e^{i\theta})]\}^{(\lambda-\kappa)/\kappa} A_\lambda\, \mathfrak{M}_\lambda[F_1(e^{i\theta})] = A_\kappa\, \mathfrak{M}_\kappa[F(e^{i\theta})],$$

where
$$A_\kappa = (\lambda/\kappa)\, A_\lambda\, C_\lambda^{(\lambda-\kappa)/\kappa}. \tag{3·8}$$

Suppose now that $F(z)\,(\not\equiv 0)$, does have zeros for $|\,z\,|\leqslant 1$. By Chapter VII, (7·23), we have $F = F_1 + F_2$, where F_1 and F_2 have no zeros and $\mathfrak{M}_\kappa[F_j(e^{ix})] \leqslant 2\mathfrak{M}_\kappa[F(e^{ix})]$ for $j=1,2.$† If we set $g_j = g(\theta, F_j)$, then $g \leqslant g_1 + g_2$, by Minkowski's inequality, and

$$\mathfrak{M}_\kappa[g] \leqslant \mathfrak{M}_\kappa[g_1] + \mathfrak{M}_\kappa[g_2]$$

$$\leqslant A_\kappa\{\mathfrak{M}_\kappa[F_1] + \mathfrak{M}_\kappa[F_2]\} \leqslant 4A_\kappa\, \mathfrak{M}_\kappa[F].$$

Here again, the regularity of F on $|\,z\,|=1$ is not called on.

(*c*) The proof here is rather intricate and it is worth while to observe that for a number of important results the simple case $0<\lambda\leqslant 2$ already proved is sufficient.

In view of (*b*) it is enough to prove (3·6) for $\lambda \geqslant 4$. Let μ be the exponent conjugate to $\frac{1}{2}\lambda$, so that $1<\mu\leqslant 2$. Let $\xi(\theta)$ be any non-negative function such that $\mathfrak{M}_\mu[\xi]\leqslant 1$. Then, with $g = g(\theta, F)$,

$$\mathfrak{M}_\lambda^2[g] = \mathfrak{M}_{\frac{1}{2}\lambda}[g^2] = \sup_\xi \int_0^{2\pi} g^2 \xi\,d\theta. \tag{3·9}$$

We may even restrict ξ to trigonometric polynomials. Fix such a polynomial and

† Of course this F_1 has nothing to do with the F_1 considered above.

write $\gamma(\theta)=g(\theta,\xi)$. Let $\zeta(z)$ be a power polynomial with $\mathscr{I}\zeta(0)=0$, having $\xi(\theta)$ as its real part for $z=e^{i\theta}$. By Chapter VII, (2·27),

$$\mathfrak{M}_\mu[\zeta(e^{i\theta})]\leqslant R_\mu \mathfrak{M}_\mu[\xi]\leqslant R_\mu, \tag{3·10}$$

with a constant R_μ depending on μ only. Since (3·6) has already been established for $1\leqslant\lambda\leqslant 2$,

$$\mathfrak{M}_\mu[\gamma]\leqslant A\mathfrak{M}_\mu[\zeta(e^{i\theta})]\leqslant AR_\mu, \tag{3·11}$$

where $A=\sup\limits_{1\leqslant\kappa\leqslant 2} A_\kappa$ is finite, by (3·8) with $\lambda=2$.

Return to the integral on the right of (3·9); we write

$$\int_0^{2\pi} g^2(\theta)\,\xi(\theta)\,d\theta=\int_0^1(1-\rho)\left\{\int_0^{2\pi}|F'(\rho e^{i\theta})|^2\,\xi(\theta)\,d\theta\right\}d\rho, \tag{3·12}$$

and our next step is to show that

$$\int_0^{2\pi} g^2(\theta)\,\xi(\theta)\,d\theta\leqslant 4\int_0^1\rho(1-\rho)\left\{\int_0^{2\pi}|F'(\rho e^{i\theta})|^2\,\xi(\rho,\theta)\,d\theta\right\}d\rho, \tag{3·13}$$

where $\xi(\rho,\theta)$ is the Poisson integral of ξ. (Thus $\xi(\rho,\theta)\geqslant 0$.)

Now the right-hand side of (3·12) is (replacing ρ by ρ^2)

$$2\int_0^1(1-\rho^2)\left\{\int_0^{2\pi}|F'(\rho^2e^{i\theta})|^2\,\xi(\theta)\,d\theta\right\}\rho\,d\rho\leqslant 4\int_0^1\rho(1-\rho)\left\{\int_0^{2\pi}|F'(\rho^2e^{i\theta})|^2\,\xi(\theta)\,d\theta\right\}d\rho.$$

Let

$$w(z)=|F'(z)|^2.$$

Since the modulus of a harmonic function ($F'^2(\rho z)$ in our case) is majorized by the Poisson integral of the values that this modulus takes on $|z|=1$, we have

$$w(\rho^2e^{i\theta})\leqslant\frac{1}{\pi}\int_0^{2\pi}w(\rho e^{iu})\,P(\rho,\theta-u)\,du,$$

$$\int_0^{2\pi}w(\rho^2e^{i\theta})\,\xi(\theta)\,d\theta\leqslant\int_0^{2\pi}\xi(\theta)\left\{\frac{1}{\pi}\int_0^{2\pi}w(\rho e^{iu})\,P(\rho,\theta-u)\,du\right\}d\theta$$

$$=\int_0^{2\pi}w(\rho e^{iu})\,\xi(\rho,u)\,du$$

(on inverting the order of integration), and (3·13) is established.

In (3·13) we should like to integrate by parts in the inner integral on the right, so as to move the operation of differentiation from F to ξ and bring in the function $\gamma(\theta)$ about which we know something (see (3·11)). The only way to do this seems to be by using the formula

$$4|F'|^2=\Delta(|F|^2)$$

established in Chapter VII, (3·6). The symbol ΔU here stands for the Laplacian of U. In polar co-ordinates,

$$\Delta U=\rho^{-1}(\rho U_\rho)_\rho+\rho^{-2}U_{\theta\theta}=\rho^{-1}U_\rho+U_{\rho\rho}+\rho^{-2}U_{\theta\theta}. \tag{3·14}$$

This formula implies that for any $a(\rho,\theta)$, $b(\rho,\theta)$,

$$\Delta(ab)=a\Delta b+b\Delta a+2(a_\rho b_\rho+\rho^{-2}a_\theta b_\theta).$$

Thus, taking $a=|F|^2$, $b=\xi(\rho,\theta)$ and observing that $\Delta\xi=0$, we have

$$4|F'|^2\xi=\Delta(|F|^2\xi)-2\{|F^2|_\rho\,\xi_\rho+\rho^{-2}|F^2|_\theta\,\xi_\theta\},$$

where $\xi = \xi(\rho, \theta)$. It follows that the right-hand side of (3·13) does not exceed

$$\int_0^{2\pi}\int_0^1 \rho(1-\rho)\,\Delta(|F|^2\xi)\,d\rho\,d\theta + 4\int_0^{2\pi}\int_0^1 (1-\rho)\,|F|\,(|F_\rho|\,|\xi_\rho| + \rho^{-2}|F_\theta|\,|\xi_\theta|)\,d\rho\,d\theta$$
$$= I_1 + I_2, \quad (3\cdot15)$$

say. Since $\quad |F(\rho e^{i\theta})| \leqslant F^*(\theta) = \sup\limits_{\rho<1} |F(\rho e^{i\theta})|, \quad |F_\rho| = |F'|,$

$$|\rho^{-1}F_\theta| = |F'|, \quad |\xi_\rho| \leqslant |\zeta'|, \quad |\rho^{-1}\xi_\theta| \leqslant |\zeta'|,$$

we find $\quad I_2 \leqslant 8\int_0^{2\pi} F^*(\theta)\,d\theta\left\{\int_0^1 (1-\rho)\,|F'|\,|\zeta'|\,d\rho\right\} \leqslant 8\int_0^{2\pi} F^*(\theta)\,g(\theta)\,\gamma(\theta)\,d\theta$

$$\leqslant 8\mathfrak{M}_\lambda[F^*]\,\mathfrak{M}_\lambda[g]\,\mathfrak{M}_\mu[\gamma] \leqslant 8AC_\lambda R_\mu \mathfrak{M}_\lambda[g]\,\mathfrak{M}_\lambda[F(e^{i\theta})], \quad (3\cdot16)$$

using successively Schwarz's inequality, Hölder's inequality with three indices λ, λ, μ, the inequality (3·7) applied to F, and (3·11).

To I_1 we apply the first equation in (3·14), noting that by periodicity the integral of $(|F|^2\xi)_{\theta\theta}$ over $(0, 2\pi)$ is zero. Thus, integrating by parts,

$$I_1 = \int_0^{2\pi} d\theta \int_0^1 (1-\rho)\frac{\partial}{\partial\rho}\left(\rho\frac{\partial}{\partial\rho}|F|^2\xi\right)d\rho = \int_0^{2\pi} d\theta\int_0^1 \rho\frac{\partial}{\partial\rho}(|F|^2\xi)\,d\rho$$
$$= \int_0^{2\pi}\left\{|F(e^{i\theta})|^2\xi(\theta) - \int_0^1 |F(\rho e^{i\theta})|^2\xi(\rho,\theta)\,d\rho\right\}d\theta$$
$$\leqslant \int_0^{2\pi} |F(e^{i\theta})|^2\xi(\theta)\,d\theta \leqslant \mathfrak{M}_\lambda^2[F(e^{i\theta})]\,\mathfrak{M}_\mu[\xi]$$
$$\leqslant \mathfrak{M}_\lambda^2[F(e^{i\theta})]. \quad (3\cdot17)$$

From (3·9), (3·13), (3·15), (3·16) and (3·17) we get

$$\mathfrak{M}_\lambda^2[g] \leqslant 8AC_\lambda R_\mu \mathfrak{M}_\lambda[F(e^{i\theta})]\,\mathfrak{M}_\lambda[g] + \mathfrak{M}_\lambda^2[F(e^{i\theta})].$$

Let $\quad X = \mathfrak{M}_\lambda[g]/\mathfrak{M}_\lambda[F(e^{i\theta})].$

Since we are supposing F regular on $|z| = 1$, X is well defined and does not exceed the largest root of the equation

$$X^2 = 8AC_\lambda R_\mu X + 1. \quad (3\cdot18)$$

This proves (3·6) for $\lambda \geqslant 4$ and so also for all $\lambda > 0$.

There is a converse of (3·5), at least for $\lambda > 1$.

(3·19) THEOREM. *Suppose that F is regular for $|z| < 1$, that $F(0) = 0$ and that $g(\theta, F) \in L^\lambda$, $\lambda > 1$. Then $F \in H^\lambda$ and*

$$\left\{\int_0^{2\pi} |F(e^{i\theta})|^\lambda\,d\theta\right\}^{1/\lambda} \leqslant B_\lambda\left\{\int_0^{2\pi} g^\lambda(\theta)\,d\theta\right\}^{1/\lambda}. \quad (3\cdot20)$$

(Since g is independent of $F(0)$, (3·20) cannot be true without some assumption about $F(0)$.) As in the proof of (3·5) we may suppose that F is regular for $|z| \leqslant 1$.

(3·21) LEMMA. *Let $f_1(\theta) \sim \sum_{-\infty}^{+\infty} c_k' e^{ik\theta}$, $f_2(\theta) \sim \sum_{-\infty}^{+\infty} c_k'' e^{ik\theta}$ be continuous functions, and suppose that at least one of the numbers c_0', c_0'' is zero. Then, with $g_k = g(\theta, f_k)$,*

$$\left|\int_0^{2\pi} f_1 f_2\,d\theta\right| \leqslant 4\int_0^{2\pi} g_1 g_2\,d\theta. \quad (3\cdot22)$$

Let $f_k(\rho, \theta)$ be the Poisson integral of f_k, and let $F_k(z)$ be the function regular for $|z| < 1$, with real part $f_k(\rho, \theta)$ and with $\mathscr{I} F_k(0) = 0$. We have

$$\frac{1}{2\pi}\int_0^{2\pi} f_1 f_2 \, d\theta = \Sigma c_k' c_{-k}'' = \lim_{r \to 1} \Sigma c_k' c_{-k}'' r^{|2k|},$$

the middle series converging absolutely. Since $c_0' c_0'' = 0$, termwise integration shows the last series to be

$$4\int_0^r \frac{d\sigma}{\sigma}\int_0^\sigma \left(\sum_{-\infty}^{+\infty} |k| c_k' \rho^{|k|-1} |k| c_{-k}'' \rho^{|k|-1}\right) \rho \, d\rho$$

$$= 4\int_0^r \frac{d\sigma}{\sigma}\int_0^\sigma \left(\frac{1}{2\pi}\int_0^{2\pi} \frac{\partial f_1(\rho,\theta)}{\partial \rho}\frac{\partial f_2(\rho,\theta)}{\partial \rho} d\theta\right) \rho \, d\rho.$$

Therefore, since $\sigma > \rho$,

$$\left|\frac{1}{2\pi}\int_0^{2\pi} f_1 f_2 \, d\theta\right| \leqslant \frac{4}{2\pi}\int_0^1 d\sigma \int_0^\sigma \int_0^{2\pi} |f_{1\rho} f_{2\rho}| \, d\rho \, d\theta$$

$$= \frac{4}{2\pi}\int_0^{2\pi} \left\{\int_0^1 |f_{1\rho} f_{2\rho}| \, d\rho \int_\rho^1 d\sigma\right\} d\theta$$

$$= \frac{4}{2\pi}\int_0^{2\pi} d\theta \left\{\int_0^1 (1-\rho) |f_{1\rho} f_{2\rho}| \, d\rho\right\}$$

$$\leqslant \frac{4}{2\pi}\int_0^{2\pi} d\theta \left\{\int_0^1 (1-\rho) |F_1'(\rho e^{i\theta})| \, |F_2'(\rho e^{i\theta})| \, d\rho\right\}$$

$$\leqslant \frac{4}{2\pi}\int_0^{2\pi} g_1(\theta) g_2(\theta) \, d\theta,$$

by Schwarz's inequality; and (3·22) is established.

We now set

$$h(\theta) = |F(e^{i\theta})|^{\lambda-1} \overline{\operatorname{sign} F(e^{i\theta})}, \quad \gamma(\theta) = g(\theta, h),$$

and denote by $H(z)$ the function regular for $|z| < 1$, with $\mathscr{I} H(0) = 0$, the real part of which is the Poisson integral of $h(\theta)$. Then

$$\begin{aligned} \mathfrak{M}_{\lambda'}[\gamma] &\leqslant A_{\lambda'} \mathfrak{M}_{\lambda'}[H(e^{i\theta})] \leqslant A_{\lambda'} R_{\lambda'} \mathfrak{M}_{\lambda'}[h] \\ &= A_{\lambda'} R_{\lambda'} \mathfrak{M}_\lambda^{\lambda-1}[F(e^{i\theta})], \end{aligned} \tag{3·23}$$

using (3·5) and an inequality analogous to (3·10). Hence

$$\mathfrak{M}_\lambda^\lambda[F(e^{i\theta})] = \int_0^{2\pi} |F(e^{i\theta})|^\lambda \, d\theta$$

$$= \int_0^{2\pi} F(e^{i\theta}) h(\theta) \, d\theta \leqslant 4\int_0^{2\pi} g(\theta) \gamma(\theta) \, d\theta$$

$$\leqslant 4\mathfrak{M}_\lambda[g] \mathfrak{M}_{\lambda'}[\gamma] \leqslant 4A_{\lambda'} R_{\lambda'} \mathfrak{M}_\lambda[g] \mathfrak{M}_\lambda^{\lambda-1}[F(e^{i\theta})]$$

by (3·22) and (3·23). A comparison of the extreme members gives (3·20) with $B_\lambda = 4A_{\lambda'} R_{\lambda'}$.

The following generalization of the inequality $\mathfrak{M}_2[g] \leqslant A\mathfrak{M}_2[F]$ will find an application in Chapter XV, § 6.

(3·24) Theorem. *If $f \in L^q$, $2 \leqslant q < \infty$, and $f(\rho, \theta)$ is the Poisson integral of f, then*

$$\int_0^1 \int_0^{2\pi} (1-\rho)^{q-1} \left| \rho^{-1} \frac{\partial}{\partial \theta} f(\rho, \theta) \right|^q d\rho\, d\theta \leqslant A^q \int_0^{2\pi} |f|^q d\theta, \tag{3·25}$$

where A is an absolute constant.

The inequality can be written $\| h \|_q \leqslant A\mathfrak{M}_q[f]$, where $h = (1-\rho)\ \rho^{-1} f_\theta(\rho, \theta)$ and the norm of h is taken with respect to the measure $d\mu = (1-\rho)^{-1} d\rho\, d\theta$ in the unit circle.

After Theorem (1·11) of Chapter XIII, it is enough to show that the operation $h = Tf$ is both of type (2, 2) and of type (∞, ∞). The former is a consequence of (2·7) (observe that $|\rho^{-1} f_\theta| \leqslant |F'|$, where $F(z)$ is the function whose real part is f, and that $\mathfrak{M}_2[F(e^{ix})] \leqslant 2\mathfrak{M}_2[f]$), and to prove the latter we verify that

$$(1-\rho)\rho^{-1} \left| \frac{\partial}{\partial \theta} P(\rho, \theta) \right| \leqslant AP(\rho, \theta),$$

so that $\sup_{\rho<1} |h(\rho, \theta)| \leqslant A \sup |f(\theta)|$.

As a special case of (3·25) we have

$$\int_0^1 \int_0^{2\pi} (1-\rho)^{q-1} |F'(\rho e^{i\theta})|^q d\rho\, d\theta \leqslant A \int_0^{2\pi} |F(e^{i\theta})|^q d\theta \tag{3·26}$$

for $F \in H^q$.

4. Convergence of conjugate series

In §§ 4, 5, and 6 of the preceding chapter we investigated the behaviour of the partial sums of a trigonometric series under the hypothesis that the series was summable (C, 1). The results could therefore be applied to Fourier series and their conjugates. Some of our theorems are, however, valid without the (C, 1) hypothesis, though the proofs become less elementary. We shall confine our attention to a generalization of Theorem (5·1) of Chapter XIII.

(4·1) Theorem. *If a series*

$$\tfrac{1}{2}a_0 + \sum_{n=1}^{\infty} (a_n \cos nx + b_n \sin nx) \equiv \sum_{n=0}^{\infty} A_n(x) \tag{4·2}$$

converges in a set E, then the conjugate series

$$\sum_{n=1}^{\infty} (a_n \sin nx - b_n \cos nx) \equiv \sum_{n=1}^{\infty} B_n(x) \tag{4·3}$$

converges almost everywhere in E.

The proof requires a series of lemmas.

A sequence of functions $s_n(x)$, or a series with partial sums $s_n(x)$, defined in the neighbourhood of $x = x_0$, will be said to converge *stably* at x_0 to limit, or sum, s if

$$s_n(x_0 + h_n) \to s \quad \text{for every sequence} \quad h_n = O(1/n).$$

This is a modification (considered already in Chapter III, § 12) of the notion of uniform convergence *at* x_0; in the latter, h_n was merely supposed to tend to zero. An equivalent definition of stable convergence would be that for any positive A, ϵ there exists an n_0 such that

$$|s_n(x_0+h)-s|<\epsilon \quad \text{for} \quad n \geqslant n_0, \ |h| \leqslant A/n.$$

(4·4) Lemma. *A necessary and sufficient condition that*

$$\tfrac{1}{2}a_0 + \sum_{n=1}^{\infty} a_n \cos nx \tag{4·5}$$

converge stably at $x=0$ to sum s is that $\frac{1}{2}a_0+a_1+\ldots$ converge to s.

The necessity is obvious. The sufficiency follows from Chapter III, (12·16).

(4·6) Lemma. *A necessary and sufficient condition for*

$$\sum_{n=1}^{\infty} b_n \sin nx \tag{4·7}$$

to converge stably at $x=0$ (to sum 0) is that

$$t_n = b_1 + 2b_2 + \ldots + nb_n = o(n). \tag{4·8}$$

Sufficiency. Suppose that $|\alpha_n| \leqslant A/n$, $t_n = \epsilon_n n$, $\epsilon_n \to 0$. Then with $\xi(x) = (\sin x)/x$ we have, summing by parts,

$$s_n(\alpha_n) = \alpha_n \sum_{\nu=1}^{n} \nu b_\nu \xi(\nu\alpha_n) = \alpha_n \left\{ \sum_{\nu=1}^{n-1} t_\nu \Delta\xi(\nu\alpha_n) + t_n \xi(n\alpha_n) \right\},$$

$$|s_n(\alpha_n)| \leqslant A \left\{ \sum_{\nu=1}^{n-1} |\epsilon_\nu| \, |\Delta\xi(\nu\alpha_n)| + |\epsilon_n| \, |\xi(n\alpha_n)| \right\}. \tag{4·9}$$

Since ξ is of bounded variation in every finite interval, the right-hand side in (4·9) is a linear transformation of $\{|\epsilon_\nu|\}$ with matrix satisfying conditions (i) and (ii) of regularity (Chapter III, § 1). Hence $s_n(\alpha_n) \to 0$.

Necessity. Let $\eta(x) = x/\sin x$, $\alpha_n = 1/n$. Then

$$\frac{t_n}{n} = \sum_{\nu=1}^{n} b_\nu \sin \nu\alpha_n \, \eta(\nu\alpha_n) = \sum_{\nu=1}^{n-1} s_\nu(\alpha_n) \, \Delta\eta(\nu\alpha_n) + s_n(\alpha_n)\, \eta(n\alpha_n),$$

$$\left|\frac{t_n}{n}\right| \leqslant \sum_{\nu=1}^{n-1} \epsilon'_\nu \, |\Delta\eta(\nu\alpha_n)| + \epsilon'_n \eta(n\alpha_n),$$

where

$$\epsilon'_\nu = \max_{m \geqslant \nu} |s_\nu(\alpha_m)| \to 0, \tag{4·10}$$

by hypothesis. Hence $t_n/n \to 0$ for the same reason as above.

Remark. Selecting $\alpha_n = 1/n$ in this argument was partly arbitrary. Any $\{\alpha_n\}$ with

$$0 < \delta \leqslant n\alpha_n \leqslant \pi - \delta \quad (\delta > 0) \tag{4·11}$$

would do. We shall use this remark in a moment.

(4·12) Lemma. *The series $\Sigma A_\nu(x)$ converges stably at x_0 to sum s if and only if*

(i) $A_0(x_0)+A_1(x_0)+A_2(x_0)+\ldots$ *converges to s; and*

(ii) $B_1(x_0)+2B_2(x_0)+\ldots+nB_n(x_0)=o(n)$.

We may suppose that $x_0=0$. Taking the semi-sum and semi-difference of $\Sigma A_\nu(x)$ at the points $\pm x$, we see that the series is stably convergent to s at $x=0$ if and only if

$\frac{1}{2}a_0+\Sigma a_\nu \cos \nu x$ and $\Sigma b_\nu \sin \nu x$ converge stably at $x=0$ to sums s and 0 respectively, and it is enough to apply Lemmas (4·4) and (4·6).

(4·13) LEMMA. *A necessary and sufficient condition for $\Sigma A_n(x)$ to converge stably at x_0 to sum s is that*

(i) *the series converges at x_0 to sum s; and*

(ii) *there exists a sequence $\{\alpha_n\}$, satisfying $0<\delta\leqslant n\alpha_n\leqslant\pi-\delta$, such that*

$$s_\nu(x_0+\alpha_n)\to s \quad \textit{for} \quad n\geqslant\nu\to\infty.$$

The necessity being obvious, we suppose (i) and (ii) satisfied and take $x_0=0$. From (i) and Lemma (4·4) we see that the cosine part of $\Sigma A_n(x)$ converges stably at $x=0$ to sum s. Hence, by (ii),

$$\sum_{k=1}^{\nu} b_k \sin k\alpha_n \to 0,$$

which, by virtue of the Remark to Lemma (4·6), leads to $t_n=o(n)$, and so to the stable convergence of $\Sigma b_k \sin kx$ at $x=0$. Thus $\Sigma A_k(x)$ is stably convergent at $x=0$ to sum s.

Theorem (1·34) of Chapter III asserts that if a power series $\Sigma a_n z^n$ converges at a point z_0 with $|z_0|=1$ to sum s, then the function

$$F(z)=\sum_0^\infty a_n z^n \quad (|z|<1)$$

tends to s as z tends to z_0 non-tangentially. The corresponding result is false for trigonometric series $\Sigma A_n(x)$; such a series may well converge at a point x_0 while the harmonic function

$$\tfrac{1}{2}a_0+\sum_{n=1}^\infty (a_n \cos nx + b_n \sin nx)\, r^n \tag{4·14}$$

does not tend to any limit as $re^{ix}\to e^{ix_0}$ non-tangentially. (Consider the series $\Sigma n^{-1} \sin nx$ with $x_0=0$.) However, we have the following lemma:

(4·15) LEMMA. *If $\Sigma A_n(x)$ converges stably at x_0 to sum s, the harmonic function $\Sigma A_n(x)\, r^n$ tends to s as $r\, e^{ix_0}$ tends to e^{ix_0} non-tangentially.*

Let $x_0=0$. Then $\frac{1}{2}a_0+a_1+\dots$ converges to s, and (4·8) holds. It is enough to show that

$$u(r,x)=\tfrac{1}{2}a_0+\sum_{\nu=1}^\infty a_\nu r^\nu \cos \nu x\to s, \tag{4·16}$$

$$v(r,x)=\sum_{\nu=1}^\infty b_\nu r^\nu \sin \nu x \to 0, \tag{4·17}$$

as z tends non-tangentially to 1. Since $u(r,x)=\mathscr{R}\left\{\frac{1}{2}a_0+\sum_{\nu=1}^\infty a_\nu z^\nu\right\}$, (4·16) follows from Theorem (1·34) of Chapter III. But

$$v(r,x)=\int_0^x \left(\sum_1^\infty \nu b_\nu r^\nu \cos \nu t\right) dt = \int_0^x \mathscr{R}\left\{\sum_1^\infty \nu b_\nu \zeta^\nu\right\} dt$$

$$=\int_0^x \mathscr{R}\left\{(1-\zeta)\sum_0^\infty t_\nu \zeta^\nu\right\} dt,$$

where $\zeta=r\,e^{it}$ (see Chapter III, (1·7)), and since $t_\nu=o(\nu)$, $|1-\zeta|\leqslant C(1-r)$ we have

$$|v(r,x)|\leqslant|x|.C(1-r).o(1-r)^{-2}=|x|.o(1-r)^{-1}=o(1),$$

which is (4·17).

(4·18) Lemma. *If $\Sigma A_n(x)$ converges for $x \in E$, it converges stably at almost all points of E.*

It is enough to show that if $\Sigma A_n(x)$ converges uniformly on $\mathscr{E} \subset E$, then it converges stably at every point $\xi \in \mathscr{E}$ which is a point of density of $\mathscr{E}$.

Let $\{\alpha_n\}$ be such that $0 < n\alpha_n \to \frac{1}{2}\pi$ and $\xi + \alpha_n \in \mathscr{E}$ for all n large enough (see (5·10) of Chapter XIII). Since

$$s_\nu(\xi + \alpha_n) - s(\xi + \alpha_n) \to 0, \quad s(\xi + \alpha_n) - s(\xi) \to 0$$

for $n, \nu \to \infty$, it follows that $s_\nu(\xi + \alpha_n) \to s(\xi)$, and it is enough to apply Lemma (4·13).

The proof of (4·1) can now be completed as follows. If $\Sigma A_n(x)$ converges on E, it converges stably on a set $E_1 \subset E$ of the same measure as E. By (4·15), the harmonic function $\Sigma A_n(x) r^n$ has a non-tangential limit at every point of E_1. By Theorem (1·10), the conjugate harmonic function

$$v(r, x) = \sum_{n=1}^{\infty} (a_n \sin nx - b_n \cos nx) r^n$$

has a non-tangential limit in a set $E_2 \subset E_1 \subset E$ of the same measure as E. In particular, $\Sigma B_n(x)$ is Abel summable in E_2. But, by (4·12),

$$B_1(x) + 2B_2(x) + \ldots + nB_n(x) = o(n)$$

in $E_1 \supset E_2$. It follows that $\Sigma B_\nu(x)$ converges in E_2, since by the theorem of Tauber (Chapter III, (1·36)), if Σu_ν is Abel summable and $u_1 + 2u_2 + \ldots + nu_n = o(n)$, then Σu_ν converges. Thus (4·1) is established.

As a corollary of (4·1) and of Theorem (2·27) of Chapter IX we obtain the following theorem which was initially stated without proof (Chapter IX, (2·28)).

(4·19) Theorem. *If the series $\Sigma A_n(x)$ converges in a set E of positive measure to sum $s(x)$, then at almost all points of E the sum*

$$F(x) = \tfrac{1}{2}a_0 x + \sum_{n=1}^{\infty} (a_n \sin nx - b_n \cos nx)/n$$

of the termwise integrated series has an approximate derivative equal to $s(x)$.

5. The Marcinkiewicz function $\mu(\theta)$

We look for an analogue of the Littlewood-Paley function $g(\theta)$ defined without entering the interior of the unit circle, or in terms of real variables.

We might, for example, consider the function

$$\nu(\theta) = \nu(\theta, f) = \left\{\int_0^\pi t \left|\frac{f(\theta+t) - f(\theta-t)}{2t}\right|^2 dt\right\}^{\frac{1}{2}}$$

$$= \tfrac{1}{2}\left\{\int_0^\pi \frac{|f(\theta+t) - f(\theta-t)|^2}{t} dt\right\}^{\frac{1}{2}},$$

whose definition has a certain analogy with that of g. (We did consider this function on p. 163 in a different context.) On closer inspection, however, it turns out that $\nu(\theta)$

does not have the required properties. For example, $\nu(\theta)$ may be infinite for all θ even if f is everywhere continuous.†

Following Marcinkiewicz we modify the above definition and replace

$$\frac{f(\theta+t)-f(\theta-t)}{2t} \quad \text{by} \quad \frac{F(\theta+t)+F(\theta-t)-2F(\theta)}{t^2},$$

where F is the indefinite integral of f.

(5·1) Theorem. *Suppose that f is periodic and integrable, and F the indefinite integral of f. Then*

$$\mu(\theta)=\mu(\theta,F)=\left\{\int_0^\pi \frac{|F(\theta+t)+F(\theta-t)-2F(\theta)|^2}{t^3}dt\right\}^{\frac{1}{2}} \tag{5·2}$$

is finite almost everywhere.

(5·3) Theorem. *If F is any function in L^2, periodic, and differentiable in a set E, then $\mu(\theta,F)$ is finite almost everywhere in E.*

Clearly (5·3) contains (5·1); but for the proof of (5·3) we must first establish (5·1) in the special case when $f \in L^2$. We show that then $\mu(\theta) \in L^2$.

Write

$$f \sim \Sigma c_n e^{in\theta}, \quad \Delta(\theta,t)=F(\theta+t)+F(\theta-t)-2F(\theta).$$

Then

$$\Delta(\theta,t)=i\Sigma' \frac{c_n}{n} e^{in\theta}(2\sin\tfrac{1}{2}nt)^2,$$

$$\frac{1}{2\pi}\int_{-\pi}^{\pi}|\Delta(\theta,t)|^2 d\theta=\Sigma'\frac{|c_n|^2}{n^2}(2\sin\tfrac{1}{2}nt)^4,$$

$$\int_0^\pi t^{-3}dt\int_{-\pi}^{\pi}|\Delta(\theta,t)|^2 d\theta=32\pi\Sigma'|c_n|^2\int_0^\pi \frac{(\sin\frac{1}{2}nt)^4}{n^2t^3}dt.$$

The integrals on the right are all less than $\int_0^\infty u^{-3}(\sin\frac{1}{2}u)^4\,du$, a finite quantity. Hence the double integral on the left, which is $\mathfrak{M}_2^2[\mu]$, is also finite. In particular, $\mu(\theta)<\infty$ almost everywhere.

We now prove (5·3), for which we may suppose that F is real-valued. The proof resembles the proof of the existence of the conjugate function in Chapter IV, § 3.

Denote by E_n the set of points θ in E such that

$$\left|\frac{F(\theta+h)-F(\theta)}{h}\right| \leqslant n \quad \text{for} \quad |h| \leqslant 1/n. \tag{5·4}$$

Then $E=E_1+E_2+\dots$. Fix n and denote by P a perfect subset of E_n. Theorem (5·3) will follow if we show that $\mu(\theta,F)<\infty$ almost everywhere in P.

Let $G(\theta)$ be the function which coincides with F on P and is linear in the closure of each interval contiguous to P. It is easy to see that G satisfies a Lipschitz condition of order 1, and in particular is the indefinite integral of a function in L^2. It follows that $\mu(\theta,G)<\infty$ almost everywhere.

† In Chapter IV, p. 133, we constructed a continuous f such that $\int_0^\pi t^{-1}|f(\theta+t)-f(\theta-t)|\,dt=+\infty$ everywhere. The same construction gives a continuous f for which $\nu(\theta)=+\infty$ everywhere.

Write $F=G+H$; hence $H=0$ in P. It is enough to show that $\mu(\theta,H)<\infty$ almost everywhere in P. Denote by $\chi(\theta)$ the distance of θ from P. For $\theta\in P$, we have by (5·4),

$$|H(\theta+t)|\leqslant B\chi(\theta+t) \quad \text{for} \quad |t|\leqslant 1/n,$$

where B is independent of θ. Hence

$$\mu^2(\theta,H)\leqslant\int_0^\pi t^{-3}\{|H(\theta+t)|+|H(\theta-t)|\}^2\,dt$$
$$\leqslant 2B^2\left\{\int_0^{1/n}\frac{\chi^2(\theta+t)+\chi^2(\theta-t)}{t^3}\,dt\right\}+2n^3\int_0^{2\pi}H^2(\theta+t)\,dt.$$

Since the first integral on the right is finite almost everywhere in P (Chapter IV, (2·1)), the same holds for $\mu(\theta,H)$ and the proof of (5·3) is completed.

The following theorem, which is an analogue of (3·5) and (3·19), is stated without proof:

(5·5) THEOREM. *If f is periodic and in* L^r, $1<r<\infty$, *and F is the indefinite integral of f, then $\mu(\theta)=\mu(\theta,F)$ satisfies*

$$\mathfrak{M}_r[\mu]\leqslant A_r\mathfrak{M}_r[f], \tag{5·6}$$

$$\mathfrak{M}_r[f]\leqslant B_r\mathfrak{M}_r[\mu], \tag{5·7}$$

provided, in the case of (5·7), *that the constant term of* $\mathsf{S}[f]$ *is* 0.

MISCELLANEOUS THEOREMS AND EXAMPLES

1. Suppose that f is periodic, of the class L^2 and that the generalized derivative $f_{(k)}(x)$ (Chapter XI, § 1) exists at each point of a set E:

$$f(x+t)=\sum_{\nu=0}^{k-1}f_{(\nu)}(x)\frac{t^\nu}{\nu!}+\omega_k(x,t)\frac{t^k}{k!},$$

where $\omega_k(x,t)\to f_{(k)}(x)$ for each $x\in E$. Then the integral

$$\int_0^\pi\frac{[\omega_k(x,t)-\omega_k(x,-t)]^2}{t}\,dt$$

is finite for almost all points of E.

[If $k=1$, this is (5·3); the proof for general k follows a similar line.]

2. (i) If $f(z)$ is regular for $|z|<1$ and has a non-tangential limit in a set E on $|z|=1$, then, for almost all $x\in E$,

$$(*)\quad \int_0^r|f'(\rho e^{ix})|\,d\rho=o\left\{\log^{\frac{1}{2}}\frac{1}{1-r}\right\}\quad (r\to 1).$$

(ii) The result is best possible, in the sense that given any function $\epsilon(r)$, $0<r<1$, positive and tending to 0 as $r\to 1$, there is an $f(z)$ having a non-tangential limit almost everywhere and such that for almost all x the left-hand side of (*) is not $O\{\epsilon(r)\log^{\frac{1}{2}}1/(1-r)\}$.

[(i) The function g of § 3 exists almost everywhere in E. (ii) Consider $f(z)=\Sigma k^{-1}z^{n_k}$ where the n_k increase fast enough.]

3. Theorem (3·5) can be completed as follows. There are positive constants α, β such that for any $F(z)$ regular in $|z|<1$ and such that $|\mathscr{R}F(z)|<1$, we have

$$\int_0^{2\pi}\exp\{\alpha g(\theta,F)\}\,d\theta\leqslant\beta.$$

CHAPTER XV

APPLICATIONS OF THE LITTLEWOOD-PALEY FUNCTION TO FOURIER SERIES

1. General remarks

In § 1 of Chapter XIII we proved that if

$$n_{k+1}/n_k > q > 1$$

for all k, the partial sums of order n_k of $S[f]$ converge almost everywhere, provided $f \in L^2$. One of the main results of this chapter is that the theorem holds for $f \in L^p$, $p > 1$. (It was observed in Chapter VIII, p. 308, that it fails for $p = 1$.) The proof is now much less elementary than for $p = 2$. It employs complex methods and in particular the function $g(\theta)$ introduced in § 3 of Chapter XIV.

It is easier to apply this function if we work not with general trigonometric series but with power series

$$\sum_{\nu=0}^{\infty} c_\nu e^{i\nu\theta} \tag{1·1}$$

in $e^{i\theta}$, or, what amounts to the same thing, with functions

$$F(z) = \sum_{\nu=0}^{\infty} c_\nu z^\nu \tag{1·2}$$

regular for $|z| < 1$.

We shall also have to introduce various auxiliary functions related to $g(\theta)$ which give information about the behaviour of $\Sigma c_\nu e^{i\nu\theta}$.

Let

$$n_0 = 0, \quad n_1 = 1 < n_2 < n_3 < \dots$$

be a fixed sequence of indices satisfying one or both of the two conditions

$$(a) \quad n_{k+1}/n_k > \alpha > 1, \qquad (b) \quad n_{k+1}/n_k < \beta < \infty \quad (k = 1, 2, \dots) \tag{1·3}$$

and let

$$\Delta_0 = c_0, \quad \Delta_k(\theta) = \sum_{\nu=n_{k-1}+1}^{n_k} c_\nu e^{i\nu\theta} \quad (k = 1, 2, \dots) \tag{1·4}$$

be blocks of successive terms of $\Sigma c_\nu e^{i\nu\theta}$, so that formally the series is

$$\sum_{k=0}^{\infty} \Delta_k(\theta). \tag{1·5}$$

The first auxiliary function we introduce is

$$\gamma(\theta) = \gamma(\theta, F) = \left(\sum_0^\infty |\Delta_k(\theta)|^2 \right)^{\frac{1}{2}}. \tag{1·6}$$

Denote by $t_n(\theta)$ and $\tau_n(\theta)$ the partial sums and the (C, 1) means of $\Sigma c_\nu e^{i\nu\theta}$. We shall also consider the function

$$\gamma_1(\theta) = \gamma_1(\theta, F) = \left(\sum_{k=1}^\infty |t_{n_k}(\theta) - \tau_{n_k}(\theta)|^2 \right)^{\frac{1}{2}} = \left(\sum_{k=1}^\infty \frac{|t'_{n_k}(\theta)|^2}{(n_k+1)^2} \right)^{\frac{1}{2}}. \tag{1·7}$$

Both γ and γ_1 depend on $\{n_k\}$. This does not apply to our third auxiliary function

$$\gamma_2(\theta)=\gamma_2(\theta, F)=\left(\sum_{\nu=1}^{\infty}\frac{|t_\nu(\theta)-\tau_\nu(\theta)|^2}{\nu}\right)^{\frac{1}{2}}=\left(\sum_{\nu=1}^{\infty}\frac{|t'_\nu(\theta)|^2}{\nu(\nu+1)^2}\right)^{\frac{1}{2}}. \tag{1·8}$$

The significance of γ_1 is easy to grasp. For suppose that $\Sigma c_\nu e^{i\nu\theta}$ is an $S[f]$, where $f\in L^r$, $r\geqslant 1$. Then $\tau_n\to f$ almost everywhere. If $\gamma_1(\theta)$ is finite almost everywhere, we have $t_{n_k}(\theta)-\tau_{n_k}(\theta)\to 0$, and so also

$$t_{n_k}(\theta)\to f(\theta), \tag{1·9}$$

almost everywhere.

To obtain another application, suppose that $\Sigma c_\nu e^{i\nu\theta}$ is in L^r, where $r>1$, and that also $\gamma_1(\theta)$ is in L^r. Since the $\tau_n(\theta)$ are all majorized by a function in L^r (see Chapter IV, (7·8)), the same holds for the $t_{n_k}(\theta)$, as we see from the inequalities

$$|t_{n_k}|\leqslant|\tau_{n_k}|+|t_{n_k}-\tau_{n_k}|\leqslant|\tau_{n_k}|+\gamma_1. \tag{1·10}$$

The function γ_2 was already considered in Chapter XIII, Lemma (7·9).

It is not immediately obvious why the indices $\{n_k\}$ in (1·6) and (1·7) should be subjected to either of the conditions (1·3). This will appear from later considerations. Here we call attention to a purely heuristic argument which makes at least plausible a connexion of, say, the function $\gamma(\theta)$ with

$$g(\theta)=\left\{\int_0^1(1-\rho)\,|F'(\rho e^{i\theta})|^2\,d\rho\right\}^{\frac{1}{2}}, \tag{1·11}$$

and so also with the function $s(\theta)$ defined in Chapter XIV, § 2.

The starting-point is the observation that in certain cases the behaviour of the nth partial sum s_n of an infinite series Σu_ν is not unlike the behaviour of the power series

$$F(r)=\Sigma u_\nu r^\nu$$

for $r=r_n=1-1/n$. (Compare, for example, the proof of Tauber's theorem in Chapter III. Though other instances of a similar nature can be quoted, it does not seem possible to cover them by any precise theorem; we merely state a principle which may be a helpful guide.) This, in turn, leads us to compare a block s_m-s_n of successive terms of Σu_ν with $F(r_m)-F(r_n)=\int_{r_n}^{r_m}F'(\rho)\,d\rho$. If we apply the idea to the series $\Sigma c_\nu e^{i\nu\theta}$ and suppose for simplicity that $n_k=2^k$, we are led to expect that the function $\gamma^2(\theta)$ is, in some sense, majorized by

$$\begin{aligned}\sum_k\left|\int_{r_{2^k}}^{r_{2^{k+1}}}\frac{d}{d\rho}F(\rho e^{i\theta})\,d\rho\right|^2&\leqslant\sum_k(r_{2^{k+1}}-r_{2^k})\int_{r_{2^k}}^{r_{2^{k+1}}}|F'(\rho e^{i\theta})|^2\,d\rho\\&=\sum_k(1-r_{2^{k+1}})\int_{r_{2^k}}^{r_{2^{k+1}}}|F'(\rho e^{i\theta})|^2\,d\rho\\&\leqslant\sum_k\int_{r_{2^k}}^{r_{2^{k+1}}}(1-\rho)\,|F'(\rho e^{i\theta})|^2\,d\rho\leqslant g^2(\theta),\end{aligned}$$

so that γ is majorized by g.

We shall see that this is actually true, though the proof is far from simple and bears little resemblance to the heuristic argument.

The next three sections will be devoted to the case when $\Sigma c_\nu e^{i\nu\theta}$ is an $\mathsf{S}[f]$ with $f \in \mathrm{L}^r$, $1 < r < \infty$. This case yields the most complete and clear-cut results; the extreme cases $r = 1$ and $r = \infty$ will be discussed later.

By A_α, $A_{\alpha,\beta}$, etc. (occasionally by A'_α, $B_{\alpha,\beta}$, etc.) we mean constants, not necessarily always the same, depending exclusively (except when otherwise stated) on the parameters shown as subscripts. By A (without subscript) we denote an absolute constant.

2. Functions in L^r, $1 < r < \infty$

The main result of this and the next two sections is:

(2·1) THEOREM. *Let $1 < r < \infty$. Suppose that*

$$F(z) = \sum_0^\infty c_\nu z^\nu \tag{2·2}$$

is regular for $|z| < 1$, and that

$$\sum_0^\infty c_\nu e^{i\nu\theta} \tag{2·3}$$

is the Fourier series of $\quad f(\theta) = F(e^{i\theta}) = \lim_{\rho \to 1} F(\rho e^{i\theta}).$

Then (cf. (1·3))

$$\mathfrak{M}_r[\gamma] \leqslant A_{\alpha,r} \mathfrak{M}_r[f], \tag{2·4}$$

$$\mathfrak{M}_r[\gamma_1] \leqslant A_{\alpha,r} \mathfrak{M}_r[f], \tag{2·5}$$

$$\mathfrak{M}_r[\gamma_2] \leqslant A_r \mathfrak{M}_r[f]. \tag{2·6}$$

We also have opposite inequalities

$$\mathfrak{M}_r[\gamma] \geqslant A_{\alpha,r} \mathfrak{M}_r[f], \tag{2·7}$$

$$\mathfrak{M}_r[\gamma_1] \geqslant A_{\beta,r} \mathfrak{M}_r[f], \tag{2·8}$$

$$\mathfrak{M}_r[\gamma_2] \geqslant A_r \mathfrak{M}_r[f], \tag{2·9}$$

provided we suppose† *in the case of* (2·8) *and* (2·9) *that $c_0 = F(0) = 0$.*

Remark. The last three inequalities are understood to mean that if any one of the functions γ, γ_1, γ_2 is in L^r, then $\Sigma c_\nu e^{i\nu\theta}$ is an $\mathsf{S}[f]$, where $f(\theta) = \lim F(\rho e^{i\theta})$ satisfies the corresponding inequality.

In this section we prove (2·5), (2·6), and also (2·4) in a somewhat weaker form (with $A_{\alpha,\beta,r}$ for $A_{\alpha,r}$). We need a number of lemmas.

(2·10) LEMMA. *Suppose that $\mathfrak{S}$ is a linear subspace of functions $\phi(x)$ in $\mathrm{L}^r(a, b)$, $1 \leqslant r < \infty$, and*

$$\psi = T\phi \quad (\psi = \psi(y),\ a' \leqslant y \leqslant b')$$

an additive operation defined for $\phi \in \mathfrak{S}$ and satisfying the conditions

(i) *if ϕ is real-valued so is ψ;*

(ii) *$\mathfrak{M}_r[\psi] \leqslant M \mathfrak{M}_r[\phi]$, with M independent of ϕ.*

Let $\phi_1, \phi_2, \ldots, \phi_n$ be a set of functions in $\mathfrak{S}$, and write $\psi_j = T\phi_j$,

$$\Phi = (\Sigma |\phi_j|^2)^{\frac{1}{2}}, \quad \Psi = (\Sigma |\psi_j|^2)^{\frac{1}{2}}.$$

† Since γ_1 and γ_2 are independent of c_0, it is clear that (2·8) and (2·9) cannot be true without some assumption about c_0.

Then $$\mathfrak{M}_r[\Psi] \leqslant M \mathfrak{M}_r[\Phi], \tag{2·11}$$
with the same M *as in* (ii).

Suppose first that the ϕ_j are real-valued. Let $\alpha_1, \alpha_2, \ldots, \alpha_n$ be a fixed set of direction angles in n-dimensional Euclidean space. Then $(\cos\alpha_1, \ldots, \cos\alpha_n)$ is a point of the n-dimensional unit sphere Σ. If $\phi = \Sigma\phi_j \cos\alpha_j$, $\psi = \Sigma\psi_j \cos\alpha_j$, then $\psi = T\phi$ and, by (ii),

$$\int_{a'}^{b'} |\Sigma\psi_j \cos\alpha_j|^r \, dy \leqslant M^r \int_a^b |\Sigma\phi_j \cos\alpha_j|^r \, dx. \tag{2·12}$$

The integrand on the right is $|\Phi(x)\cos\delta|^r$, δ denoting the angle between the vectors $(\phi_1, \ldots, \phi_n)$ and $(\cos\alpha_1, \ldots, \cos\alpha_n)$. Similarly, the integrand on the left is $|\Psi(y)\cos\delta'|^r$, δ' being the angle between $(\psi_1, \ldots, \psi_n)$ and $(\cos\alpha_1, \ldots, \cos\alpha_n)$. We now integrate (2·12) over Σ, interchange the order of integration on each side and observe that $\int_\Sigma |\cos\delta|^r d\sigma$ is independent of x and equal to $\int_\Sigma |\cos\delta'|^r d\sigma$, which similarly is independent of y; cancelling this factor on both sides of the inequality we arrive at (2·11).

If the $\phi_j = \phi_j' + i\phi_j''$ are complex-valued, then $\psi_j = \psi_j' + i\psi_j''$, where $\psi_j' = T\phi_j'$, $\psi_j'' = T\phi_j''$, and the equations
$$|\phi_j|^2 = \phi_j'^2 + \phi_j''^2, \quad |\psi_j|^2 = \psi_j'^2 + \psi_j''^2$$
reduce this case to the previous one, with n replaced by $2n$.

Making $n \to \infty$ we see that (2·11) holds in the case when $\{\phi_j\}$ is an infinite set.

(2·13) LEMMA. *Let* $f_1, f_2, \ldots, f_N$ *be periodic, complex-valued and of class* L^r, $1 < r < \infty$. *Let* $\tilde f_j$ *be the conjugate of* f_j *and write*
$$\Phi = (\Sigma |f_j|^2)^{\frac12}, \quad \Psi = (\Sigma |\tilde f_j|^2)^{\frac12}.$$
Then $$\mathfrak{M}_r[\Psi] \leqslant A_r \mathfrak{M}_r[\Phi]. \tag{2·14}$$

This follows from (2·10) and the inequality $\mathfrak{M}_r[\tilde f] \leqslant A_r \mathfrak{M}_r[f]$ (see Chapter VII, (2·4)).

(2·15) LEMMA. *Let* $f_1, f_2, \ldots, f_N$ *be periodic and in* L^r, $1 < r < \infty$. *Let* $s_{n,k}$ *be the k-th partial sum of* $S[f_n]$, *and* $k = k_n$ *a function of n. Then*
$$\int_0^{2\pi} \left(\sum_{n=1}^N |s_{n,k_n}|^2\right)^{\frac12 r} d\theta \leqslant A_r^r \int_0^{2\pi} \left(\sum_{n=1}^N |f_n|^2\right)^{\frac12 r} d\theta. \tag{2·16}$$

Let $g_{n,k}$ and $h_{n,k}$ denote respectively $f_n \cos k\theta$ and $f_n \sin k\theta$. Then (compare Chapter VII, (6·2))
$$s_{n,k}(\theta) = \tilde g_{n,k}(\theta)\sin k\theta - \tilde h_{n,k}(\theta)\cos k\theta + \alpha_{n,k}(\theta),$$
where
$$\alpha_{n,k}(\theta) = \frac{1}{2\pi}\int_0^{2\pi} f_n(t)\cos k(t-\theta)\, dt.$$

Denote the functions g_{n,k_n}, h_{n,k_n}, α_{n,k_n} by g_n, h_n, α_n, and let I be the integral on the left of (2·16). We have, successively,
$$|s_{n,k_n}| \leqslant |\tilde g_n| + |\tilde h_n| + |\alpha_n|,$$
$$(\Sigma |s_{n,k_n}|^2)^{\frac12} \leqslant (\Sigma |\tilde g_n|^2)^{\frac12} + (\Sigma |\tilde h_n|^2)^{\frac12} + (\Sigma |\alpha_n|^2)^{\frac12},$$
$$I \leqslant 3^{r-1}\left\{\int_0^{2\pi} (\Sigma |\tilde g_n|^2)^{\frac12 r} d\theta + \int_0^{2\pi} (\Sigma |\tilde h_n|^2)^{\frac12 r} d\theta + \int_0^{2\pi} (\Sigma |\alpha_n|^2)^{\frac12 r} d\theta\right\}. \tag{2·17}$$

From (2·14) we have

$$\int_0^{2\pi} (\Sigma \,|\, \tilde{g}_n \,|^2)^{\frac{1}{2}r}\, d\theta \leqslant A_r^r \int_0^{2\pi} (\Sigma \,|\, f_n \,|^2)^{\frac{1}{2}r}\, d\theta, \tag{2·18}$$

and the same inequality holds for the $\tilde{h}_n$. Since

$$|\, \alpha_n(\theta) \,|^2 \leqslant \left(\frac{1}{2\pi}\int_0^{2\pi} |\, f_n(t) \,|\, dt\right)^2,$$

we have $$(\Sigma \,|\, \alpha_n(\theta) \,|^2)^{\frac{1}{2}} \leqslant \frac{1}{2\pi}\left\{\Sigma\left(\int_0^{2\pi} |\, f_n \,|\, dt\right)^2\right\}^{\frac{1}{2}} \leqslant \frac{1}{2\pi}\int_0^{2\pi} (\Sigma \,|\, f_n \,|^2)^{\frac{1}{2}}\, dt,$$

by Minkowski's inequality, so that

$$\int_0^{2\pi} (\Sigma \,|\, \alpha_n(\theta) \,|^2)^{\frac{1}{2}r}\, d\theta \leqslant \int_0^{2\pi} (\Sigma \,|\, f_n(\theta) \,|^2)^{\frac{1}{2}r}\, d\theta. \tag{2·19}$$

Collecting estimates, we obtain from (2·17)

$$I \leqslant 3^{r-1}(2A_r^r + 1)\int_0^{2\pi} (\Sigma \,|\, f_n(\theta) \,|^2)^{\frac{1}{2}r}\, d\theta \leqslant A_r^r \int_0^{2\pi} (\Sigma \,|\, f_n(\theta) \,|^2)^{\frac{1}{2}r}\, d\theta,$$

and (2·16) is proved.

For any periodic f we denote by $f(\rho, \theta)$ the Poisson integral of f.

(2·20) Lemma. *Let $f_1, f_2, \ldots, f_N$ be periodic and in L^r, $1 < r < \infty$. Let $k = k_n$ and $\rho = \rho_n$, $0 \leqslant \rho_n < 1$, be functions of n. Then*

$$\int_0^{2\pi} \left(\sum_{n=1}^{N} |\, s_{n,k_n}(\rho_n, \theta) \,|^2\right)^{\frac{1}{2}r} d\theta \leqslant A_r^r \int_0^{2\pi} (\sum_{n=1}^{N} |\, f_n(\theta) \,|^2)^{\frac{1}{2}r}\, d\theta. \tag{2·21}$$

Summing by parts and applying Jensen's inequality we have

$$\begin{aligned}|\, s_{n,k}(\rho_n, \theta) \,|^2 &= \left|(1-\rho_n)\sum_{\nu=0}^{k-1} s_{n,\nu}(\theta)\,\rho_n^\nu + s_{n,k}(\theta)\,\rho_n^k\right|^2 \\ &\leqslant (1-\rho_n)\sum_{\nu=0}^{k-1} |\, s_{n,\nu}(\theta) \,|^2 \rho_n^\nu + |\, s_{n,k}(\theta) \,|^2 \rho_n^k.\end{aligned}$$

Let I be the left-hand side of (2·21). Then

$$\begin{aligned} I &\leqslant \int_0^{2\pi} \left\{\sum_{n=1}^{N}\sum_{\nu=0}^{k_n-1} (1-\rho_n)\,|\, s_{n,\nu}(\theta) \,|^2 \rho_n^\nu + \sum_{n=1}^{N} |\, s_{n,k_n}(\theta) \,|^2 \rho_n^{k_n}\right\}^{\frac{1}{2}r} d\theta \\ &\leqslant A_r^r \int_0^{2\pi} \left\{\sum_{n=1}^{N}\sum_{\nu=0}^{k_n-1} (1-\rho_n)\,|\, f_n(\theta) \,|^2 \rho_n^\nu + \sum_{n=1}^{N} |\, f_n(\theta) \,|^2 \rho_n^{k_n}\right\}^{\frac{1}{2}r} d\theta \\ &= A_r^r \int_0^{2\pi} \left(\sum_{n=1}^{N} |\, f_n(\theta) \,|^2\right)^{\frac{1}{2}r} d\theta, \end{aligned}$$

by Lemma (2·15), and (2·21) is established.

(2·22) Lemma. *Let $0 \leqslant \rho_n < 1$ for $n = 1, 2, \ldots, N$, and let δ_n denote any subinterval of $(\rho_n, 1)$. Then, under the hypotheses of Lemma* (2·20),

$$\int_0^{2\pi} \left(\sum_{n=1}^{N} |\, s_{n,k_n}(\rho_n, \theta) \,|^2\right)^{\frac{1}{2}r} d\theta \leqslant A_r^r \int_0^{2\pi} \left(\sum_{n=1}^{N} \frac{1}{|\,\delta_n\,|}\int_{\delta_n} |\, f_n(\rho, \theta) \,|^2\, d\rho\right)^{\frac{1}{2}r} d\theta, \tag{2·23}$$

where the constant A_r is the same as in (2·21).

We may suppose without loss of generality that no δ_n contains the point $\rho=1$. If ρ_n' is any number between ρ_n and 1, (2·21) gives

$$\int_0^{2\pi}\left(\sum_{n=1}^{N}|s_{n,k_n}(\rho_n,\theta)|^2\right)^{\frac{1}{2}r}d\theta\leqslant A_r^r\int_0^{2\pi}\left(\sum_{n=1}^{N}|f(\rho_n',\theta)|^2\right)^{\frac{1}{2}r}d\theta.$$

We split each interval δ_n into m equal parts and denote by $\rho_n^{(i)}$, where $i=1,2,\dots,m$, the left-hand ends of the subintervals of δ_n thus obtained. The last inequality holds if we replace each term $|s_{n,k_n}(\rho_n,\theta)|^2$ on the left by m terms equal to $m^{-1}|s_{n,k_n}(\rho_n,\theta)|^2$, and simultaneously replace the term $|f(\rho_n',\theta)|^2$ on the right by $m^{-1}\sum_{i=1}^{m}|f_n(\rho_n^{(i)},\theta)|^2$. Making $m\to\infty$, we obtain (2·23).

We now pass to the proof of the inequalities (2·4), (2·5) and (2·6). It is enough to prove them with

$$g(\theta)=\left\{\int_0^1(1-\rho)\,|F'(\rho e^{i\theta})|^2\,d\rho\right\}^{\frac{1}{2}}$$

instead of $f(\theta)=\lim F(\rho e^{i\theta})$ on the right. The conclusion then follows from the basic inequality

$$\mathfrak{M}_r[g]\leqslant A_r\mathfrak{M}_r[f] \tag{2·24}$$

established in Chapter XIV, § 3. Of the chain of lemmas proved in this section we need only the last, namely (2·22).

We begin with (2·6). Our starting point is the formula

$$t_n(\theta)-\tau_n(\theta)=-i\frac{t_n'(\theta)}{n+1}=-\frac{i}{n+1}\sum_{\nu=0}^{n}i\nu\,c_\nu e^{i\nu\theta}\rho^\nu.\rho^{-\nu}. \tag{2·25}$$

Summing by parts we have

$$t_n'(\theta)=\rho^{-n}t_n'(\rho,\theta)-(1-\rho)\sum_{\nu=0}^{n-1}\rho^{-\nu-1}t_\nu'(\rho,\theta)$$

for $0<\rho<1$, the dash denoting differentiation with respect to θ. Hence

$$\begin{aligned}|t_n'(\theta)|^2&\leqslant 2\left\{\rho^{-2n}|t_n'(\rho,\theta)|^2+(1-\rho)^2\left(\sum_{\nu=0}^{n-1}\rho^{-\nu-1}|t_\nu'(\rho,\theta|\right)^2\right\}\\&\leqslant 2\left\{\rho^{-2n}|t_n'(\rho,\theta)|^2+\frac{1-\rho}{\rho^n}\cdot\sum_{\nu=0}^{n-1}\rho^{-\nu-1}|t_\nu'(\rho,\theta)|^2\right\}.\end{aligned} \tag{2·26}$$

Let
$$\rho=\rho_n=1-\frac{1}{n+1},\quad \delta_n=(\rho_n,\rho_{n+1}). \tag{2·27}$$

Then $\rho_n^\nu>e^{-1}$ for $0\leqslant\nu\leqslant n$. Using (1·8) and applying Lemma (2·22) we have

$$\begin{aligned}\mathfrak{M}_r^r[\gamma_2]&\leqslant 2^{\frac{1}{2}r}\int_0^{2\pi}\left\{\sum_{n=1}^{\infty}\frac{|t_n'(\rho_n,\theta)|^2}{n^3\rho_n^{2n}}+\sum_{n=1}^{\infty}\frac{1-\rho_n}{n^3\rho_n^n}\sum_{\nu=0}^{n-1}\rho_n^{-\nu-1}|t_\nu'(\rho_n,\theta)|^2\right\}^{\frac{1}{2}r}d\theta\\
\leqslant 2^{\frac{1}{2}r}e^rA_r^r&\int_0^{2\pi}\left\{\sum_{n=1}^{\infty}\frac{1}{n^3|\delta_n|}\int_{\delta_n}\left|\frac{d}{d\theta}F(\rho e^{i\theta})\right|^2d\rho+\sum_{n=1}^{\infty}\frac{1}{n^4}\sum_{\nu=0}^{n-1}\frac{1}{|\delta_n|}\int_{\delta_n}\left|\frac{d}{d\theta}F(\rho e^{i\theta})\right|^2d\rho\right\}^{\frac{1}{2}r}d\theta\\
&\leqslant(2^{\frac{1}{2}}eA_r)^r\int_0^{2\pi}\left\{\sum_{n=1}^{\infty}\frac{2(n+1)(n+2)}{n^3}\int_{\delta_n}|F'(\rho e^{i\theta})|^2d\rho\right\}^{\frac{1}{2}r}d\theta\\
&\leqslant(2^{\frac{1}{2}}eA_r)^r\int_0^{2\pi}\left\{\sum_{n=1}^{\infty}\frac{2(n+1)(n+2)^2}{n^3}\int_{\delta_n}(1-\rho)|F'(\rho e^{i\theta})|^2d\rho\right\}^{\frac{1}{2}r}d\theta\\
&\leqslant(10eA_r)^r\int_0^{2\pi}g^r(\theta)\,d\theta.\end{aligned} \tag{2·28}$$

It follows that $$\mathfrak{M}_r[\gamma_2] \leqslant A_r \mathfrak{M}_r[g],$$
and (2·6) is established (see (2·24)).

It is worth noting that in the definition (2·27) of ρ_n we applied the principle used in the heuristic argument of § 1.

We pass to the proof of (2·5), which is largely similar to that of (2·6). Suppose that $\{n_k\}$ satisfies the condition (1·3) (a). Observing that

$$\frac{n_{k+1}+1}{n_k+1} \geqslant \frac{\alpha n_k+1}{n_k+1} \geqslant \tfrac{1}{2}(\alpha+1) = \alpha' > 1$$

for $k=1, 2, \ldots$, we set

$$\rho_k = 1 - \frac{1}{n_k+1}, \quad \delta_k = \left(\rho_k,\ 1 - \frac{1}{\alpha'(n_k+1)}\right). \tag{2·29}$$

No two δ_k overlap. If we replace n by n_k in (2·26), we have

$$\begin{aligned}
\mathfrak{M}_r^r[\gamma_1] &\leqslant 2^{\frac{1}{2}r} e^r \int_0^{2\pi} \left\{\sum_{k=1}^{\infty} [n_k^{-2} \,|\, t'_{n_k}(\rho_k, \theta) \,|^2 + n_k^{-3} \sum_{\nu=0}^{n_k-1} |\, t'_\nu(\rho_k, \theta) \,|^2]\right\}^{\frac{1}{2}r} d\theta \\
&\leqslant 2^{\frac{1}{2}r} e^r A_r^r \int_0^{2\pi} \left\{2 \sum_{k=1}^{\infty} n_k^{-2} \frac{1}{|\,\delta_k\,|} \int_{\delta_k} \left|\frac{d}{d\theta} F(\rho e^{i\theta})\right|^2 d\rho\right\}^{\frac{1}{2}r} d\theta \\
&\leqslant (2eA_r)^r \int_0^{2\pi} \left\{\frac{\alpha'^2}{\alpha'-1} \sum_{k=1}^{\infty} \left(\frac{n_k+1}{n_k}\right)^2 \int_{\delta_k} (1-\rho) \,|\, F'(\rho e^{i\theta}) \,|^2 d\rho\right\}^{\frac{1}{2}r} d\theta.
\end{aligned}$$

whence $$\mathfrak{M}_r[\gamma_1] \leqslant \frac{4\alpha' e A_r}{(\alpha'-1)^{\frac{1}{2}}} \left\{\int_0^{2\pi} g^r(\theta)\, d\theta\right\}^{1/r},$$

where $\alpha' = \frac{1}{2}(1+\alpha)$, and (2·5) follows by an application of (2·24).

We now pass to the proof of (2·4) in which we temporarily require that $\{n_k\}$ satisfies both conditions (1·3). For $0 \leqslant m < n$, summing by parts, we have

$$\begin{aligned}
i \sum_{m+1}^{n} c_\nu e^{i\nu\theta} &= \sum_{m+1}^{n} i\nu c_\nu e^{i\nu\theta} \frac{1}{\nu} \\
&= -\frac{t'_m}{m+1} + \frac{t'_n}{n+1} + \sum_{m+1}^{n} \frac{t'_\nu}{\nu(\nu+1)}.
\end{aligned}$$

Hence, with $m = n_{k-1}$, $n = n_k$ (see (1·4))

$$|\,\Delta_k\,|^2 \leqslant 3\left\{\left|\frac{t'_{n_{k-1}}}{n_{k-1}+1}\right|^2 + \left|\frac{t'_{n_k}}{n_k+1}\right|^2 + \left|\sum_{n_{k-1}+1}^{n_k} \frac{t'_\nu}{\nu(\nu+1)}\right|^2\right\}. \tag{2·30}$$

Since $g(\theta)$ is independent of c_0, and

$$\mathfrak{M}_r[\Delta_0] = (2\pi)^{1/r} \,|\, c_0 \,| \leqslant (2\pi)^{1/r} \frac{1}{2\pi} \int_0^{2\pi} |\, f \,|\, d\theta \leqslant \mathfrak{M}_r[f], \tag{2·31}$$

it is enough to prove (2·4) under the hypothesis $F(0)=0$.

From (2·30) and the equation $|\, t'_\nu \,|/(\nu+1) = |\, t_\nu - \tau_\nu \,|$ we deduce

$$\begin{aligned}
\gamma^2 = \sum_1^\infty |\,\Delta_k\,|^2 &\leqslant 6\gamma_1^2 + 3 \sum_{k=1}^{\infty} \left|\sum_{n_{k-1}+1}^{n_k} \frac{t'_\nu}{\nu(\nu+1)}\right|^2 \\
&= 6\gamma_1^2 + 3\gamma_3^2,
\end{aligned}$$

say, where $\gamma_3 \geqslant 0$; and it remains to show that

$$\mathfrak{M}_r[\gamma_3] \leqslant A_{\alpha,\beta,r}\,\mathfrak{M}_r[f]. \tag{2·32}$$

Now
$$\left|\sum_{n_{k-1}+1}^{n_k} \frac{t'_\nu}{\nu(\nu+1)}\right|^2 \leqslant \frac{1}{(n_{k-1}+1)^4}\left(\sum_{n_{k-1}+1}^{n_k} |t'_\nu|\right)^2 \leqslant \frac{n_k}{(n_{k-1}+1)^4}\sum_{n_{k-1}+1}^{n_k} |t'_\nu|^2$$
$$\leqslant \frac{\beta^4}{n_k^3}\sum_{n_{k-1}+1}^{n_k} |t'_\nu|^2.$$

By (2·26) with ν for n, μ for ν, and ρ_κ for ρ, where $\rho_\kappa = 1-(n_k+1)^{-1}$, we obtain

$$\mathfrak{M}_r^r[\gamma_3] \leqslant \beta^{2r}\int_0^{2\pi}\left\{\sum_{k=1}^{\infty}\frac{1}{n_k^3}\sum_{n_{k-1}+1}^{n_k} |t'_\nu(\theta)|^2\right\}^{\frac{1}{2}r} d\theta$$
$$\leqslant 2^{\frac{1}{2}r}\beta^{2r}e^r\int_0^{2\pi}\left\{\sum_{k=1}^{\infty}\frac{1}{n_k^3}\left[\sum_{\nu=n_{k-1}+1}^{n_k}\left(|t'_\nu(\rho_k,\theta)|^2 + \frac{1}{n_k}\sum_{\mu=0}^{\nu-1}|t'_\mu(\rho_k,\theta)|^2\right)\right]\right\}^{\frac{1}{2}r} d\theta.$$

By Lemma (2·22) the last inequality holds if we replace each $|t'_\mu|^2$ on the right by $|\delta_k|^{-1}\int_{\delta_k} |F'(\rho e^{i\theta})|^2\,d\rho$, provided we simultaneously multiply the right-hand side by A_r^r. A computation analogous to the one used in estimating $\mathfrak{M}_r^r[\gamma_1]$ gives

$$\mathfrak{M}_r^r[\gamma_3] \leqslant (2\beta^2 e A_r)^r\int_0^{2\pi}\left\{\sum_{k=1}^{\infty}\frac{1}{n_k^2}\frac{1}{|\delta_k|}\int_{\delta_k} |F'(\rho e^{i\theta})|^2\,d\rho\right\}^{\frac{1}{2}r} d\theta,$$
$$\mathfrak{M}_r[\gamma_3] \leqslant \frac{4\beta^2 e\alpha' A_r}{(\alpha'-1)^{\frac{1}{2}}}\,\mathfrak{M}_r[g],$$

where $\alpha' = \frac{1}{2}(1+\alpha)$, and (2·32) follows by an application of (2·24). This completes the proof of (2·4) with $A_{\alpha,\beta,r}$ for $A_{\alpha,r}$ (that the constant is actually independent of β will be shown in §4 below).

3. Functions in L^r, $1<r<\infty$ (*cont.*)

We now prove the inequalities (2·8) and (2·9), beginning with the latter.

(3·1) LEMMA. *For each* θ,

$$g(\theta) \leqslant A\gamma_2(\theta). \tag{3·2}$$

We have
$$|F'(\rho e^{i\theta})| = \left|\sum_1^\infty \nu c_\nu \rho^{\nu-1} e^{i\nu\theta}\right| = (1-\rho)\left|\sum_1^\infty t'_\nu \rho^{\nu-1}\right|,$$

where $t'_\nu = t'_\nu(\theta)$. If $\rho_n = 1-1/n$ we can write

$$g^2(\theta) = \sum_{n=1}^{\infty}\int_{\rho_n}^{\rho_{n+1}} (1-\rho)\,|F'(\rho e^{i\theta})|^2\,d\rho$$
$$\leqslant \sum_{n=1}^{\infty} n^{-3}\left\{(1-\rho_n)\sum_1^\infty |t'_\nu|\,\rho_{n+1}^{\nu-1}\right\}^2$$
$$\leqslant 2\sum_{n=1}^{\infty} n^{-5}\left(\sum_{\nu=1}^{n} |t'_\nu|\,\rho_{n+1}^{\nu-1}\right)^2 + 2\sum_{n=1}^{\infty} n^{-5}\left(\sum_{\nu=n+1}^{\infty} |t'_\nu|\,\rho_{n+1}^{\nu-1}\right)^2$$
$$= P+Q,$$

say. From the estimates

$$P \leqslant 2 \sum_{n=1}^{\infty} n^{-5} \left(\sum_{\nu=1}^{n} |t'_\nu|^2 \right) \left(\sum_{\nu=1}^{n} 1^2 \right) = 2 \sum_{\nu=1}^{\infty} |t'_\nu|^2 \sum_{n=\nu}^{\infty} n^{-4}$$

$$\leqslant A \sum_{\nu=1}^{\infty} |t'_\nu|^2 \nu^{-3},$$

$$Q \leqslant 2 \sum_{n=1}^{\infty} n^{-5} \left(\sum_{\nu=n+1}^{\infty} |t'_\nu|^2 \nu^{-4} \right) \left(\sum_{\nu=n+1}^{\infty} \nu^4 \rho_{n+1}^{2(\nu-1)} \right)$$

$$\leqslant 2 \sum_{n=1}^{\infty} n^{-5} \left(\sum_{\nu=n+1}^{\infty} |t'_\nu|^2 \nu^{-4} \right) \left(\sum_{\nu=1}^{\infty} \nu^4 \rho_{n+1}^{2(\nu-1)} \right)$$

$$\leqslant A \sum_{n=1}^{\infty} n^{-5} (1-\rho_{n+1})^{-5} \sum_{\nu=n}^{\infty} |t'_\nu|^2 \nu^{-4} = A \sum_{\nu=1}^{\infty} |t'_\nu|^2 \nu^{-3},$$

we deduce
$$g^2(\theta) \leqslant P + Q \leqslant A \sum_{\nu=1}^{\infty} |t'_\nu|^2 \nu^{-3} \leqslant A \gamma_2^2(\theta),$$

and the lemma is established.

The inequality (2·9) follows from (3·2) and from the basic inequality

$$\mathfrak{M}_r[F(\rho e^{i\theta})] \leqslant A_r \mathfrak{M}_r[g] \quad (F(0)=0) \tag{3·3}$$

(Chapter XIV, (3·19)).

The inequality (2·8) is a simple consequence of (2·9) and Lemma (2·15) (with $N=\infty$):

$$\mathfrak{M}_r^r[\gamma_2] = \int_0^{2\pi} \left\{ \sum_{k=1}^{\infty} \sum_{\nu=n_{k-1}+1}^{n_k} \frac{|t'_\nu|^2}{\nu(\nu+1)^2} \right\}^{\frac{1}{2}r} d\theta \leqslant A_r^r \int_0^{2\pi} \left\{ \sum_{k=1}^{\infty} |t'_{n_k}|^2 \sum_{n_{k-1}+1}^{n_k} \frac{1}{\nu(\nu+1)^2} \right\}^{\frac{1}{2}r} d\theta$$

$$\leqslant A_r^r \int_0^{2\pi} \left\{ \sum_{k=1}^{\infty} |t'_{n_k}|^2 \frac{n_k}{(n_{k-1}+1)^3} \right\}^{\frac{1}{2}r} d\theta \leqslant A_r^r \int_0^{2\pi} \left\{ \sum_{k=1}^{\infty} |t'_{n_k}|^2 \frac{\beta^3}{(n_k+1)^2} \right\}^{\frac{1}{2}r} d\theta$$

$$= A_r^r \beta^{\frac{3}{2}r} \mathfrak{M}_r^r[\gamma_1],$$

so that
$$\mathfrak{M}_r[\gamma_2] \leqslant A_{\beta,r} \mathfrak{M}_r[\gamma_1], \tag{3·4}$$

and (2·8) follows from (2·9).

The inequality (2·7) will be proved in the next section.

4. Theorems on the partial sums of $S[f]$, $f \in L^r$, $1<r<\infty$

We first consider the case of an $S[f]$ of power series type:

$$f(\theta) \sim \sum_0^{\infty} c_\nu e^{i\nu\theta} \in L^r \quad (1<r<\infty), \tag{4·1}$$

so that, with the previous notation,

$$f(\theta) = F(e^{i\theta}) = \lim_{\rho \to 1} F(\rho e^{i\theta}),$$

where
$$F(z) = \sum_0^{\infty} c_\nu z^\nu \tag{4·2}$$

is in H^r. The partial sums and (C, 1) means of the series (4·1) will again be denoted by t_ν and τ_ν.

(4·3) Theorem. *Given an* $f(\theta) \in L^r$, *we can, for almost all* θ, *decompose the sequence* $1, 2, \ldots$ *into two complementary sequences* $\{p_k\}$ *and* $\{q_k\}$ *(depending in general on* θ) *such that*

(i) $t_{p_k}(\theta) \to f(\theta)$;

(ii) $\Sigma 1/q_k < \infty$.

The case $r=2$ was proved in Chapter XIII, § 9. The proof was based exclusively on the finiteness of $\gamma_2(\theta)$ almost everywhere, and so is valid for general $r>1$ in view of (2·6).

We show in § 5 (see (5·10)) that (4·3) holds for $r=1$.

(4·4) Theorem. *Suppose that the indices* n_k *satisfy the conditions*

$$n_0=0, \quad n_1=1, \quad n_{k+1}/n_k > \alpha > 1 \quad (k=1, 2, \ldots), \tag{4·5}$$

and set

$$t^*(\theta) = \sup_k |t_{n_k}(\theta)|.$$

Then for $f \in L^r$, *we have*

(i) $t_{n_k}(\theta) \to f(\theta)$ *almost everywhere;*

(ii) $\mathfrak{M}_r[t^*] \leqslant A_{\alpha,r} \mathfrak{M}_r[f]$.

By (2·5), $\gamma_1(\theta) < \infty$ almost everywhere, and this implies (i) (see (1·9)).

The function $\tau^*(\theta) = \sup_\nu |\tau_\nu(\theta)|$ satisfies an inequality $\mathfrak{M}_r[\tau^*] \leqslant A_r \mathfrak{M}_r[f]$ (Chapter IV, (7·8)), and since, by (1·10),

$$t^* \leqslant \tau^* + \gamma_1, \quad \mathfrak{M}_r[t^*] \leqslant \mathfrak{M}_r[\tau^*] + \mathfrak{M}_r[\gamma_1], \tag{4·6}$$

(ii) follows from (2·5).

(4·7) Theorem. *Suppose that* $\{n_k\}$ *satisfies the conditions*

$$n_0=0, \; n_1=1, \quad n_{k+1}/n_k > \alpha > 1 \quad (k=1, 2, \ldots), \tag{4·8}$$

and that $\{\epsilon_k\}$ *is any sequence of numbers* ± 1. *Then, for* $f \in L^r$, *the series (cf.* (1·4))

$$\sum_0^\infty \epsilon_k \Delta_k \tag{4·9}$$

is (when written at length as a trigonometric series) the Fourier series of an $f_1 \in L^r$, *and*

$$A_{\alpha,r} \mathfrak{M}_r[f] \leqslant \mathfrak{M}_r[f_1] \leqslant A_{\alpha,r} \mathfrak{M}_r[f]. \tag{4·10}$$

(4·11) Theorem. *Suppose that* f *and* $\{n_k\}$ *are the same as in* (4·7) *and that* $\{\eta_k\}$ *is any sequence consisting entirely of the numbers* 0, 1. *Then*

$$\sum_0^\infty \eta_k \Delta_k \tag{4·12}$$

is the Fourier series of an $f_2 \in L^r$ *and*

$$\mathfrak{M}_r[f_2] \leqslant A_{\alpha,r} \mathfrak{M}_r[f]. \tag{4·13}$$

We first deduce (4·11) from (4·7). We have $\eta_k = \frac{1}{2}(1+\epsilon_k)$, where $\epsilon_k = \pm 1$. Thus, with the notation of the previous theorem,

$$f_2 = \tfrac{1}{2}(f+f_1),$$

and (4·12) is in L^r. The inequality $\mathfrak{M}_r[f_2] \leqslant \frac{1}{2}\{\mathfrak{M}_r[f] + \mathfrak{M}_r[f_1]\}$, together with the second inequality (4·10), gives (4·13).

It is instructive to compare the last two theorems with Theorems (8·12) and (8·41) of Chapter V.

Before we pass to (4·7) we prove the following theorem.

(4·14) THEOREM. *Suppose that $\lambda_0, \lambda_1, \ldots$ is a sequence of numbers such that*

$$|\lambda_\nu| \leqslant M, \quad \sum_{2^\nu}^{2^{\nu+1}-1} |\lambda_j - \lambda_{j+1}| \leqslant M \quad (\nu = 0, 1, \ldots). \tag{4·15}$$

Then, under the hypothesis (4·1),

$$\Sigma c_\nu \lambda_\nu e^{i\nu\theta} \tag{4·16}$$

is the Fourier series of an $h(\theta) \in L^r$ and

$$\mathfrak{M}_r[h] \leqslant A_r M \mathfrak{M}_r[f]. \tag{4·17}$$

Since $\mathfrak{M}_r[c_0] \leqslant \mathfrak{M}_r[f]$ (see (2·31)), we may suppose without loss of generality that $c_0 = 0$.

We take

$$n_k = 2^k - 1 \quad (k = 0, 1, \ldots), \tag{4·18}$$

define the Δ_k for f accordingly, and set

$$\left.\begin{aligned} \Delta_{k,s} &= \sum_{n_{k-1}+1}^{s} c_\nu e^{i\nu\theta} \\ \Delta'_k &= \sum_{n_{k-1}+1}^{n_k} c_\nu \lambda_\nu e^{i\nu\theta} \end{aligned}\right\} \quad (s \geqslant n_{k-1}+1;\ k = 1, 2, \ldots). \tag{4·19}$$

Summation by parts and Schwarz's inequality give

$$\Delta'_k = \sum_{s=n_{k-1}+1}^{n_k} \Delta_{k,s}(\lambda_s - \lambda_{s+1}) + \Delta_k \lambda_{n_k+1},$$

$$|\Delta'_k|^2 \leqslant \left(\sum_{s=n_{k-1}+1}^{n_k} |\lambda_s - \lambda_{s+1}| + |\lambda_{n_k+1}|\right)\left(\sum_{s=n_{k-1}+1}^{n_k} |\Delta_{k,s}|^2 |\lambda_s - \lambda_{s+1}| + |\Delta_k|^2 |\lambda_{n_k+1}|\right)$$

$$\leqslant 2M\left(\sum_{s=n_{k-1}+1}^{n_k} |\Delta_{k,s}|^2 |\lambda_s - \lambda_{s+1}| + |\Delta_k|^2 |\lambda_{n_k+1}|\right).$$

Hence, by Lemma (2·15),

$$\int_0^{2\pi} \left(\sum_{k=1}^{\infty} |\Delta'_k|^2\right)^{\frac{1}{2}r} d\theta \leqslant (2M)^{\frac{1}{2}r} A_r^r \int_0^{2\pi} \left\{\sum_{k=1}^{\infty} |\Delta_k|^2 \left(\sum_{n_{k-1}+1}^{n_k} |\lambda_s - \lambda_{s+1}| + |\lambda_{n_k+1}|\right)\right\}^{\frac{1}{2}r} d\theta,$$

$$\int_0^{2\pi} \left(\sum_{k=1}^{\infty} |\Delta'_k|^2\right)^{\frac{1}{2}r} d\theta \leqslant (2MA_r)^r \int_0^{2\pi} \left(\sum_{k=1}^{\infty} |\Delta_k|^2\right)^{\frac{1}{2}r} d\theta. \tag{4·20}$$

The sequence (4·18) satisfies the condition $2 \leqslant n_{k+1}/n_k \leqslant 3$, and Theorem (4·14) follows from (4·20) and the inequality $\mathfrak{M}_2[\gamma] \leqslant A_{\alpha,\beta,r} \mathfrak{M}_r[f]$ established in § 2, p. 228.

Example. If δ is real and non-zero, the numbers $\lambda_\nu = \nu^{i\delta}$ satisfy (4·15).

Theorem (4·7) is a corollary of (4·14). If $\{n_k\}$ satisfies (4·8), the sequence $\{\lambda_j\}$ defined by the conditions

$$\lambda_0 = \epsilon_0, \quad \lambda_j = \epsilon_k \quad \text{for} \quad n_{k-1} < j \leqslant n_k \quad (k = 1, 2, \ldots)$$

satisfies (4·15) with $M = M_\alpha$, and the second inequality (4·10) is a consequence of (4·17). Since f and f_1 in (4·7) play symmetric roles, the first inequality (4·10) is a consequence of the second. This completes the proof of (4·7).

Let now the $\{\epsilon_k\}$ in (4·7) be the sequence of Rademacher functions $r_k(t)$ (Chapter I, § 3). If t is not a diadic rational, we have

$$A^r_{\alpha,r}\mathfrak{M}^r_r[f] \leqslant \int_0^{2\pi} |\Sigma r_k(t)\Delta_k(x)|^r dx \leqslant A^r_{\alpha,r}\mathfrak{M}^r_r[f].$$

Integrate this over $0 \leqslant t \leqslant 1$ and interchange the order of integration. Observing that for any $s = \Sigma a_k r_k(t)$, with $S = (\Sigma |a_k|^2)^{\frac{1}{2}}$ finite we have

$$A_r S \leqslant \mathfrak{M}_r[s] \leqslant B_r S \quad (r > 0), \tag{4·21}$$

(Chapter V, (8·4)), we immediately obtain

$$A^r_{\alpha,r}\mathfrak{M}^r_r[f] \leqslant \int_0^{2\pi} (\Sigma |\Delta_k(x)|^2)^{\frac{1}{2}r} dx \leqslant A^r_{\alpha,r}\mathfrak{M}^r_r[f].†$$

The second inequality here is the same as (2·4) (we had already proved it in § 2, with $A_{\alpha,\beta,r}$ for $A_{\alpha,r}$), and the first is the same as (2·7). Thus Theorem (2·1) is proved completely.

The following result is a simple consequence of (2·4) and (2·7):

(4·22) THEOREM. *Suppose that* $f \in L^r$ *and that* $n_0 = 0$, $n_1 = 1$, $n_{k+1}/n_k > \alpha > 1$. *Then*

(i) *if* $2 \leqslant r < \infty$, *we have*

$$\left\{\int_0^{2\pi} (\Sigma |\Delta_k|^r)\, d\theta\right\}^{1/r} \leqslant A_{\alpha,r}\mathfrak{M}_r[f];$$

(ii) *if* $1 < r \leqslant 2$ *we have*

$$\left\{\int_0^{2\pi} (\Sigma |\Delta_k|^r)\, d\theta\right\}^{1/r} \geqslant A_{\alpha,r}\mathfrak{M}_r[f].$$

Case (i) follows from (2·4) and the inequality

$$(\Sigma |\Delta_k|^2)^{\frac{1}{2}r} \geqslant \Sigma |\Delta_k|^r \tag{4·23}$$

valid for $r \geqslant 2$. Similarly (ii) follows from (2·7) and the inequality opposite to (4·23), valid for $r \leqslant 2$.

Similar conclusions could be derived from the remaining inequalities of Theorem (2·1).

The results obtained for $S[f]$ of power series type have analogues for general $S[f] \in L^r$, $1 < r < \infty$. As an example we shall consider the case of the inequalities (2·4) and (2·7).

Suppose that $\phi(\theta) \in L^r$ is real-valued and

$$\phi(\theta) \sim \tfrac{1}{2}a_0 + \Sigma(a_\nu \cos \nu\theta + b_\nu \sin \nu\theta).$$

Then $\tilde{\phi} \in L^r$, and

$$f = \phi + i\tilde{\phi} \sim \tfrac{1}{2}a_0 + \sum_1^\infty (a_\nu - ib_\nu)\, e^{i\nu\theta} = \sum_0^\infty c_\nu e^{i\nu\theta}.$$

Retaining the previous definition of Δ_k (see (1·4)), where $\{n_k\}$ satisfies the conditions (4·8), we set $\Delta_k = \delta_k + i\tilde{\delta}_k$. Thus

$$\delta_0 = \tfrac{1}{2}a_0, \quad \delta_k = \sum_{n_{k-1}+1}^{n_k} (a_\nu \cos \nu\theta + b_\nu \sin \nu\theta) \quad (k = 1, 2, \ldots).$$

(4·24) THEOREM. *If* $\phi \in L^r$, $1 < r < \infty$, *then*

$$A_{\alpha,r}\mathfrak{M}_r[\phi] \leqslant \left\{\int_0^{2\pi} (\Sigma\delta_k^2)^{\frac{1}{2}r}\, d\theta\right\}^{1/r} \leqslant A_{\alpha,r}\mathfrak{M}_r[\phi]. \tag{4·25}$$

† This argument presupposes the finiteness of $\Sigma |\Delta_k(x)|^2$ almost everywhere. It is enough to apply it to the partial sum of order n_k of $S[f]$ and then make $k \to \infty$.

Let $\gamma' = (\Sigma\delta_k^2)^{\frac{1}{2}}$, $\gamma'' = (\Sigma\tilde{\delta}_k^2)^{\frac{1}{2}}$. Then

$$\gamma^2 = \gamma'^2 + \gamma''^2, \quad \mathfrak{M}_r[\gamma] \leqslant \mathfrak{M}_r[\gamma'] + \mathfrak{M}_r[\gamma''].$$

The second inequality (4·25) is immediate since the middle term in (4·25) does not exceed

$$\mathfrak{M}_r[\gamma] \leqslant A_{\alpha, r}\,\mathfrak{M}_r[f] \leqslant A_{\alpha, r}\,\mathfrak{M}_r[\phi].$$

On the other hand,

$$\mathfrak{M}_r[\phi] \leqslant \mathfrak{M}_r[f] \leqslant A_{\alpha, r}\,\mathfrak{M}_r[\gamma] \leqslant A_{\alpha, r}(\mathfrak{M}_r[\gamma'] + \mathfrak{M}_r[\gamma'']),$$

and the first inequality (4·25) will follow if we show that

$$\mathfrak{M}_r[\gamma''] \leqslant A_r \mathfrak{M}_r[\gamma']. \tag{4·26}$$

It is enough to prove (4·26) when ϕ is a polynomial of order n_k (and then make k tend to ∞). But in this case (4·26) is a consequence of Lemma (2·10), since $\tilde{\phi} = T\phi$ is a linear operation satisfying the hypotheses of the lemma (with $M = A_r$).

5. The limiting case $r = 1$

Most (though not all) of the results obtained in the preceding sections fail in the limiting cases $r = 1$ and $r = \infty$, the first of which we now discuss. Our discussion will be less complete than in the case $1 < r < \infty$, and we shall almost exclusively confine our attention to analogues of the first three inequalities of Theorem (2·1). As before, we write

$$f(\theta) = F(e^{i\theta}) \sim \sum_0^\infty c_\nu e^{i\nu\theta}, \tag{5·1}$$

where $F(z) = \Sigma c_\nu z^\nu$ is regular for $|z| < 1$.

(5·2) Theorem. *If* $f \sim \sum_0^\infty c_\nu e^{i\nu\theta} \in L$, *then*

$$\mathfrak{M}_\mu[\gamma] \leqslant A_{\alpha, \beta, \mu}\mathfrak{M}[f], \tag{5·3}$$

$$\mathfrak{M}_\mu[\gamma_1] \leqslant A_{\alpha, \mu}\mathfrak{M}[f], \tag{5·4}$$

$$\mathfrak{M}_\mu[\gamma_2] \leqslant A_\mu \mathfrak{M}[f], \tag{5·5}$$

for every $0 < \mu < 1$. *In particular the functions* $\gamma, \gamma_1, \gamma_2$ *are finite almost everywhere.*

(5·6) Theorem. *If* $f \sim \sum_0^\infty c_\nu e^{i\nu\theta}$ *satisfies* $f \log^+ |f| \in L$, *then*

$$\mathfrak{M}[\gamma] \leqslant A_{\alpha, \beta}\int_0^{2\pi} |f| \log^+ |f|\, d\theta + A_{\alpha, \beta}, \tag{5·7}$$

$$\mathfrak{M}[\gamma_1] \leqslant A_\alpha\int_0^{2\pi} |f| \log^+ |f|\, d\theta + A_\alpha, \tag{5·8}$$

$$\mathfrak{M}[\gamma_2] \leqslant A\int_0^{2\pi} |f| \log^+ |f|\, d\theta + A. \tag{5·9}$$

These inequalities are the main results of this section. Before we pass to their proofs we give a few corollaries.

(5·10) Theorem. *Theorem* (4·3) *holds for* $r = 1$.

This follows from the finiteness of γ_2 almost everywhere.

Since the convergence of Σu_ν implies that $u_1+2u_2+\ldots+nu_n=o(n)$, we deduce from (5·5) that $S[f]$ *is summable* H_2 *almost everywhere* (see Chapter XIII, § 8).

This result is weaker than Theorem (8·1) of Chapter XIII, if only because the latter theorem was proved for $S[f]$ not necessarily of power series type.

(5·11) THEOREM. *Let $f\sim\sum_0^\infty c_\nu e^{i\nu\theta}$ and suppose that $n_{k+1}/n_k>\alpha>1$ for all k. Then $t_{n_k}(\theta)\to f(\theta)$ almost everywhere. Moreover, for $t^*(\theta)=\sup_k |t_{n_k}(\theta)|$ we have*

$$\mathfrak{M}_\mu[t^*]\leqslant A_{\alpha,\mu}\mathfrak{M}[f] \quad (0<\mu<1), \tag{5·12}$$

$$\mathfrak{M}[t^*]\leqslant A_\alpha\int_0^{2\pi}|f|\log^+|f|\,d\theta+A_\alpha. \tag{5·13}$$

The proof of the relation $t_{n_k}\to f$ is the same as in Theorem (4·4). The inequalities (5·12) and (5·13) follow from (5·4), (5·8), the first inequality (4·6), and the estimates

$$\mathfrak{M}_\mu[\tau^*]\leqslant A_\mu\mathfrak{M}[f],\quad \mathfrak{M}[\tau^*]\leqslant A\int_0^{2\pi}|f|\log^+|f|\,d\theta+A$$

(see Chapter IV, § 7).

Return to Theorems (5·2) and (5·6). For their proofs we need lemmas analogous to those used in §§ 2 and 3. But this time we shall not be able to use the direction angles for the purpose of averaging (see Lemma (2·10)); less precise results obtained by means of Rademacher functions will, however, be adequate for our purposes.

(5·14) LEMMA. *Let $r_\nu(t)$ be Rademacher functions,*

$$s(t)=\Sigma a_\nu r_\nu(t),\quad S=(\Sigma|a_\nu|^2)^{\frac12}<\infty.$$

Then

$$A_r S\leqslant\mathfrak{M}_r[s]\leqslant B_r S \quad (r>0), \tag{5·15}$$

$$AS\log^+S-A'\leqslant\int_0^1|s|\log^+|s|\,dt\leqslant AS\log^+S+A'. \tag{5·16}$$

Here (5·15) is the same as (4·21).

In order to prove (5·16) we set $\chi(u)=\log(u+e)$, observe that $\chi^2(u)$ is concave and apply the inequalities of Schwarz and Jensen:

$$\int_0^1|s|\log^+|s|\,dt\leqslant\int_0^1|s|\,\chi(|s|)\,dt\leqslant\left(\int_0^1|s|^2dt\right)^{\frac12}\left(\int_0^1\chi^2(|s|)\,dt\right)^{\frac12}$$

$$\leqslant\left(\int_0^1|s|^2dt\right)^{\frac12}\left(\chi^2\left(\int_0^1|s|\,dt\right)\right)^{\frac12}\leqslant S\chi(S)\leqslant AS\log^+S+A'.$$

The first inequality (5·16) is a consequence of Theorem (8·25) of Chapter V, according to which there are two positive absolute constants ϵ,η such that $|s(t)|\geqslant\eta S$ in a set of measure at least ϵ. Thus

$$\int_0^1|s|\log^+|s|\,dt\geqslant\epsilon\eta\,S\log^+(\eta S). \tag{5·17}$$

We may suppose that $\eta\leqslant1$. The right-hand side in (5·17) is $\epsilon\eta S(\log^+S-\log 1/\eta)$ if $S>\eta^{-1}$, and is 0 if $S\leqslant\eta^{-1}$. Hence the first inequality (5·16) holds in every case.

The two lemmas which follow are analogues of Lemma (2·10), from which we retain some notation.

(5·18) LEMMA. *Let* $0 < s \leqslant r$ *and let* $\mathfrak{S}$ *be a linear subspace of* $L^r(a, b)$. *Suppose that* $\psi = T\phi$ *is an additive operation defined for* $\phi \in \mathfrak{S}$ ($\psi = \psi(y)$, $a' \leqslant y \leqslant b'$), *such that*

$$\mathfrak{M}_s[\psi] \leqslant M\mathfrak{M}_r[\phi] \quad (\phi \in \mathfrak{S}).$$

Then there is a constant $A_{r,s}$ *such that*

$$\mathfrak{M}_s[\Psi] \leqslant MA_{r,s}\mathfrak{M}_r[\Phi]. \tag{5·19}$$

If
$$\phi = \phi_t(x) = \Sigma r_\nu(t)\,\phi_\nu(x), \quad \psi = \psi_t(y) = \Sigma r_\nu(t)\,\psi_\nu(x),$$

then $\psi = T\phi$. Integrating the inequality $\mathfrak{M}_s^s[\psi] \leqslant M^s\mathfrak{M}_r^s[\phi]$ with respect to t, and applying Hölder's inequality on the right, we have

$$\int_0^1 dt \int_{a'}^{b'} |\, \Sigma r_\nu(t)\,\psi_\nu(y)\,|^s\, dy \leqslant M^s \left\{ \int_0^1 dt \int_a^b |\, \Sigma r_\nu(t)\,\phi_\nu(x)\,|^r\, dx \right\}^{s/r}.$$

If we now change the order of integration on both sides and use (5·15) we get

$$A_s^s \int_{a'}^{b'} \Psi^s(y)\, dy \leqslant M^s \left\{ B_r^r \int_a^b \Phi^r(x)\, dx \right\}^{s/r},$$

which is (5·19) with $A_{r,s} = B_r/A_s$.

(5·20) LEMMA. *Suppose that* $\mathfrak{S}$ *is a linear subspace of* $L(a, b)$, *with* $b - a < \infty$, *and that* $\psi = T\phi$ *is an additive operation defined for* $\phi \in \mathfrak{S}$ ($\psi = \psi(y)$, $a' \leqslant y \leqslant b'$), *such that*

$$\int_{a'}^{b'} |\,\psi\,|\, dy \leqslant M \int_a^b |\,\phi\,| \log^+ |\,\phi\,|\, dx + M.$$

Then
$$\int_{a'}^{b'} \Psi\, dy \leqslant M' \int_a^b \Phi \log^+ \Phi\, dx + M',$$

where M' *depends on* M *and* $b - a$.

It is enough to integrate both sides of the inequality

$$\int_{a'}^{b'} |\, \Sigma r_\nu(t)\,\psi_\nu(y)\,|\, dy \leqslant M \int_a^b |\, \Sigma r_\nu(t)\,\phi_\nu(x)\,| \log^+ |\, \Sigma r_\nu(t)\,\phi_\nu(x)\,|\, dx + M$$

with respect to t, change the order of integration, and use (5·16).

We shall use Lemmas (5·18) and (5·20) with $T\phi = \tilde{\phi}$.

(5·21) LEMMA. *With the notation of Lemma* (2·15) *we have*

$$\int_0^{2\pi} (\Sigma\,|\, s_{n,k_n}\,|^2)^{\frac{1}{2}\mu}\, d\theta \leqslant \left\{ A_\mu \int_0^{2\pi} (\Sigma\,|\, f_n\,|^2)^{\frac{1}{2}}\, d\theta \right\}^\mu \quad (0 < \mu < 1), \tag{5·22}$$

$$\int_0^{2\pi} (\Sigma\,|\, s_{n,k_n}\,|^2)^{\frac{1}{2}}\, d\theta \leqslant A \int_0^{2\pi} (\Sigma\,|\, f_n\,|^2)^{\frac{1}{2}} \log^+ (\Sigma\,|\, f_n\,|^2)^{\frac{1}{2}}\, d\theta + A. \tag{5·23}$$

The proofs resemble so closely that of Lemma (2·15) that we need indicate only the main points. We begin with (5·22), and denote the left-hand side by I. Starting with the inequality immediately preceding (2·17), we get instead of (2·17) the inequality

$$I \leqslant \int_0^{2\pi} (\Sigma\,|\, \tilde{g}_n\,|^2)^{\frac{1}{2}\mu}\, d\theta + \int_0^{2\pi} (\Sigma\,|\, \tilde{h}_n\,|^2)^{\frac{1}{2}\mu}\, d\theta + \int_0^{2\pi} (\Sigma\,|\, \alpha_n\,|^2)^{\frac{1}{2}\mu}\, d\theta,$$

instead of (2·18) the inequality

$$\int_0^{2\pi}(\Sigma|\tilde{g}_n|^2)^{\frac{1}{2}\mu}\,d\theta\leqslant\left\{A_\mu\int_0^{2\pi}(\Sigma|f_n|^2)^{\frac{1}{2}}\,d\theta\right\}^\mu$$

(a corollary of Lemma (5·18) when $T\phi=\tilde{\phi}$, $r=1$, $s=\mu$), and instead of (2·19) the inequality

$$\int_0^{2\pi}(\Sigma|\alpha_n|^2)^{\frac{1}{2}\mu}\,d\theta\leqslant(2\pi)^{1-\mu}\left\{\int_0^{2\pi}(\Sigma|f_n|^2)^{\frac{1}{2}}\,d\theta\right\}^\mu.$$

Collecting estimates we arrive at (5·22). The proof of (5·23) is similar.

(5·24) Lemma. *With the notation of Lemma* (2·20) *we have*

$$\int_0^{2\pi}(\Sigma|s_{n,k_n}(\rho_n,\theta)|^2)^{\frac{1}{2}\mu}\,d\theta\leqslant\left\{A_\mu\int_0^{2\pi}(\Sigma|f_n|^2)^{\frac{1}{2}}\,d\theta\right\}^\mu\quad(0<\mu<1),\tag{5·25}$$

$$\int_0^{2\pi}(\Sigma|s_{n,k_n}(\rho_n,\theta)|^2)^{\frac{1}{2}}\,d\theta\leqslant A\int_0^{2\pi}(\Sigma|f_n|^2)^{\frac{1}{2}}\log^+(\Sigma|f_n|^2)^{\frac{1}{2}}\,d\theta+A.\tag{5·26}$$

The proof is similar to that of Lemma (2·20).

(5·27) Lemma. *With the notation of Lemma* (2·22) *we have*

$$\int_0^{2\pi}(\Sigma|s_{n,k_n}(\rho_n,\theta)|^2)^{\frac{1}{2}\mu}\,d\theta\leqslant\left\{A_\mu\int_0^{2\pi}\left(\Sigma|\delta_n|^{-1}\int_{\delta_n}|f_n(\rho,\theta)|^2\,d\rho\right)^{\frac{1}{2}}d\theta\right\}^\mu,\tag{5·28}$$

$$\int_0^{2\pi}(\Sigma|s_{n,k_n}(\rho_n,\theta)|^2)^{\frac{1}{2}}\,d\theta\leqslant A\int_0^{2\pi}\left(\Sigma|\delta_n|^{-1}\int_{\delta_n}|f_n(\rho,\theta)|^2\,d\rho\right)^{\frac{1}{2}}\log^+(\Sigma\ldots)^{\frac{1}{2}}\,d\theta+A.\tag{5·29}$$

The proof is similar to that of Lemma (2·22).

Passing to the proof of Theorem (5·2), we observe that it is enough to prove the inequalities with $\mathfrak{M}[g]$ instead of $\mathfrak{M}[f]$ on the right, since by Theorem (3·5) of Chapter XIV we have

$$\mathfrak{M}[g]\leqslant A\mathfrak{M}[f].$$

We begin with the proof of (5·5), and again we may be brief, since the proof runs parallel to that of (2·6). Starting with (2·25) we have, instead of the first inequality (2·28),

$$\mathfrak{M}_\mu^\mu[\gamma_2]\leqslant 2^{\frac{1}{2}\mu}\int_0^{2\pi}\{\ldots\}^{\frac{1}{2}\mu}\,d\theta;$$

and a series of inequalities, parallel to those in (2·28) but based on (5·28), gives $\mathfrak{M}_\mu[\gamma_2]\leqslant A_\mu\mathfrak{M}[g]$. This completes the proof of (5·5).

The proof of (5·4) resembles that of (2·5) and does not require additional explanations. Similarly the proof of (5·3) resembles that of (2·4) with $A_{\alpha,\beta,r}$ for $A_{\alpha,r}$ (p. 228).

It remains to prove Theorem (5·6). We confine our attention to the inequality (5·9), the proof of the remaining inequalities being analogous.

If we start out from the first inequality (2·28), where now $r=1$, and use (5·29), we arrive at the inequality

$$\int_0^{2\pi}\gamma_2\,d\theta\leqslant A\int_0^{2\pi}g\log^+(A'g)\,d\theta\leqslant A\int_0^{2\pi}g\log^+g\,d\theta+A.$$

Hence (5·9) will be established if we prove the following lemma:

(5·30) LEMMA. *If* $f \sim \sum_0^\infty c_\nu e^{i\nu\theta} \in L\log^+ L$, *we have*

$$\int_0^{2\pi} g\log^+ g\,d\theta \leqslant A\int_0^{2\pi} |f|\log^+|f|\,d\theta + A. \tag{5·31}$$

We may suppose that c_0 is real. Consider the inequality

$$\mathfrak{M}_r[g] \leqslant A_r \mathfrak{M}_r[f],$$

valid for $F \in H^r$, $r > 0$. Suppose now that $r > 1$, and set

$$f(\theta) = u(\theta) + iv(\theta).$$

Then $$\mathfrak{M}_r[g] \leqslant A_r \mathfrak{M}_r[u], \tag{5·32}$$

since $\mathfrak{M}_r[f] \leqslant A_r \mathfrak{M}_r[u]$ (Chapter VII, (2·4)). The function $u(\theta)$ is an arbitrary (periodic) function in L^r and $F(z)$ is uniquely determined by $u(\theta)$. Hence

$$g(\theta) = \left\{\int_0^1 (1-\rho)\,|F'(\rho e^{i\theta})|^2\,d\rho\right\}^{\frac{1}{2}} = Tu \tag{5·33}$$

is an operation defined for $u \in L^r$. By Minkowski's inequality, T is a sublinear operation and, by (5·32), is of type (r, r) for each $r > 1$ (Chapter XII, pp. 111, 95). By† Theorem (4·22) of Chapter XII we have

$$\int_0^{2\pi} g^2\log^+ g\,d\theta \leqslant A\int_0^{2\pi} u^2\log^+ u\,d\theta + A,$$

and, *a fortiori*, $$\int_0^{2\pi} g^2\log^+ g\,d\theta \leqslant A\int_0^{2\pi} |f|^2\log^+|f|\,d\theta + A. \tag{5·34}$$

In order to deduce (5·31) from (5·34) we suppose first that F has no zeros in $|z| < 1$, and set $F = F_1^2$. Then denoting by g_1 the g corresponding to F_1, we have

$$g(\theta) = 2\left\{\int_0^1 (1-\rho)\,|F_1|^2\,|F_1'|^2\,d\rho\right\}^{\frac{1}{2}} \leqslant 2f_1^*(\theta)\,g_1(\theta),$$

where $f_1^*(\theta) = \sup_\rho |F_1(\rho e^{i\theta})|$. Hence

$$\begin{aligned}\int_0^{2\pi} g\log^+ g\,d\theta &\leqslant \int_0^{2\pi} (2f_1^* g_1)\log^+(2f_1^* g_1)\,d\theta \\ &\leqslant 2\int_0^{2\pi} f_1^{*2}\log^+(2f_1^{*2})\,d\theta + 2\int_0^{2\pi} g_1^2\log^+(2g_1^2)\,d\theta + A \\ &\leqslant A\int_0^{2\pi} f_1^{*2}\log^+ f_1^*\,d\theta + A\int_0^{2\pi} g_1^2\log^+ g_1\,d\theta + A.\end{aligned} \tag{5·35}$$

By (5·34) applied to F_1,

$$\int_0^{2\pi} g_1^2\log^+ g_1\,d\theta \leqslant A\int_0^{2\pi} |f_1|^2\log^+|f_1|\,d\theta + A \leqslant A\int_0^{2\pi} |f|\log^+|f|\,d\theta + A.$$

From this and (5·35) we obtain (5·31), provided we show that

$$\int_0^{2\pi} f_1^{*2}\log^+ f_1^*\,d\theta \leqslant A\int_0^{2\pi} |f_1|^2\log^+|f_1|\,d\theta + A \leqslant A\int_0^{2\pi} |f|\log^+|f|\,d\theta + A. \tag{5·36}$$

† We set there $\phi(u) = u^2\log^+ u$; the fact that ϕ is strictly increasing for $u \geqslant 1$ only, is of no importance.

Only the first inequality requires proof. Considering $F(z)$ as the Poisson integral of f, we immediately find that $f^*(\theta)=\sup_\rho |F(\rho e^{i\theta})|=Tf$ is a sublinear operation, of type (r,r) for $r>1$ (Chapter IV, (7·8)). Hence (5·36) follows from Theorem (4·22) of Chapter XII.

If F has zeros in $|z|<1$ we have a decomposition $F=F_1+F_2$, where F_1 and F_2 are without zeros and $|f_1|\leqslant 2|f|$, $|f_2|\leqslant 2|f|$ (Chapter VII, (7·23)). If g_1 and g_2 are the g for F_1 and F_2, then $g\leqslant g_1+g_2$ and

$$\begin{aligned}\int_0^{2\pi} g\log^+ g\,d\theta &\leqslant \int_0^{2\pi}(2g_1)\log^+(2g_1)\,d\theta+\int_0^{2\pi}(2g_2)\log^+(2g_2)\,d\theta\\ &\leqslant A\int_0^{2\pi} g_1\log^+ g_1 d\theta + A\int_0^{2\pi} g_2\log^+ g_2 d\theta + A\\ &\leqslant A\int_0^{2\pi}|f_1|\log^+|f_1|\,d\theta + A\int_0^{2\pi}|f_2|\log^+|f_2|\,d\theta + A\\ &\leqslant A\int_0^{2\pi}|f|\log^+|f|\,d\theta + A,\end{aligned}$$

which completes the proof of (5·9).

6. The limiting case $r=\infty$

This time, and contrary to what precedes, we get the best results by considering general Fourier series

$$\tfrac{1}{2}a_0+\sum_{\nu=1}^{\infty}(a_\nu\cos\nu\theta+b_\nu\sin\nu\theta)=\sum_{\nu=0}^{\infty}A_\nu(\theta) \tag{6·1}$$

and their conjugates

$$\sum_{\nu=1}^{\infty}(a_\nu\sin\nu\theta-b_\nu\cos\nu\theta)=\sum_{\nu=1}^{\infty}B_\nu(\theta). \tag{6·2}$$

The partial sums and the (C, 1) means of (6·1) are denoted by $s_n(\theta)$ and $\sigma_n(\theta)$, and of (6·2) by $\tilde{s}_n(\theta)$ and $\tilde{\sigma}_n(\theta)$.

Suppose that

$$n_0=0,\quad n_1=1<n_2<n_3<\dots,\quad n_{k+1}/n_k>\alpha>1. \tag{6·3}$$

If $\Sigma A_\nu(\theta)$ is the development of a bounded function, then both $\Sigma A_\nu(\theta)$ and $\Sigma B_\nu(\theta)$ are in L^r, $r>1$, so that, by (4·4), the sequences $\{s_{n_k}\}$ and $\{\tilde{s}_{n_k}\}$ converge almost everywhere, and the functions

$$s^*(\theta)=\sup_k|s_{n_k}(\theta)|,\quad \tilde{s}^*(\theta)=\sup_k|\tilde{s}_{n_k}(\theta)| \tag{6·4}$$

are in L^r. It is the latter fact which we are now going to generalize.

(6·5) Theorem. *Suppose that $\Sigma A_\nu(\theta)$ is an $S[f]$, where $|f|\leqslant 1$, and that $\{n_k\}$ satisfies (6·3). There are positive constants λ and μ, depending on α only, such that we have*

$$\int_0^{2\pi}\exp(\lambda s^*)\,d\theta\leqslant\mu,\quad \int_0^{2\pi}\exp(\lambda\tilde{s}^*)\,d\theta\leqslant\mu. \tag{6·6}$$

If f is continuous, the integrals (6·6) are finite no matter how large λ is.

The part concerning continuous f is a corollary of the result about bounded functions, as is seen from the decomposition $f=f_1+f_2$, where f_1 is a polynomial and $\max|f_2|$

arbitrarily small. It is therefore enough to prove (6·6), and we begin with the first inequality. The proof has some points in common with earlier ones, but is shorter and more elementary, no use being made of the deeper inequalities for the function g.

(6·7) LEMMA. *If $\{n_k\}$ satisfies (6·3) and if $\Sigma A_\nu(\theta)$ is an $S[f]$, where $f \in L^q$, $q \geqslant 2$, then*

$$\int_0^{2\pi} \left(\sum_{k=1}^{\infty} | s_{n_k} - \sigma_{n_k} |^q \right) d\theta \leqslant A_{q,\alpha}^q \int_0^{2\pi} | f |^q \, d\theta, \tag{6·8}$$

where

$$A_{q,\alpha} \leqslant q A_\alpha. \tag{6·9}$$

Except for the estimate (6·9), which is important, (6·8) is contained in (2·5) in view of the obvious inequality $\Sigma | a_n |^q \leqslant (\Sigma | a_n |^2)^{\frac{1}{2}q}$; but the following proof is on different lines.

From the equations $s_n - \sigma_n = \tilde{s}_n'/(n+1)$ we deduce (compare a similar argument in (2·25) and (2·26))

$$| s_n(\theta) - \sigma_n(\theta) | \leqslant \frac{1}{n+1} \left| (1-\rho) \sum_{\nu=1}^{n-1} \tilde{s}_\nu'(\rho, \theta) \rho^{-\nu-1} \right| + \frac{1}{n+1} | \tilde{s}_n'(\rho, \theta) | \rho^{-n}.$$

Hence, if $n > 0$,

$$\begin{aligned} | s_n(\theta) - \sigma_n(\theta) |^q &\leqslant \frac{2^{q-1}}{n^q} (1-\rho)^q \left| \sum_{\nu=1}^{n-1} \tilde{s}_\nu'(\rho, \theta) \rho^{-\nu-1} \right|^q + \frac{2^{q-1}}{n^q} | \tilde{s}_n'(\rho, \theta) |^q \rho^{-nq} \\ &\leqslant \frac{2^{q-1}}{n^q} (1-\rho)^q \left(\sum_{\nu=1}^{n-1} | s_\nu'(\rho, \theta) |^q \right) \left(\sum_{\nu=0}^{n-1} \rho^{-(\nu+1)q'} \right)^{q-1} + \frac{2^{q-1}}{n^q} | \tilde{s}_n'(\rho, \theta) |^q \rho^{-nq} \\ &\leqslant \frac{2^{q-1}}{n^q} \rho^{-nq} (1-\rho) \sum_{\nu=1}^{n-1} | s_\nu'(\rho, \theta) |^q + \frac{2^{q-1}}{n^q} \rho^{-nq} | \tilde{s}_n'(\rho, \theta) |^q. \end{aligned} \tag{6·10}$$

By the second inequality (6·5) of Chapter VII, we have

$$\int_0^{2\pi} | \tilde{s}_\nu'(\rho, \theta) |^q \, d\theta \leqslant A_q^q \int_0^{2\pi} | f'(\rho, \theta) |^q \, d\theta, \tag{6·11}$$

where, as is easily verified (see Chapter VII, (6·7), (6·3), and (3·8)),

$$A_q \leqslant A q. \tag{6·12}$$

Setting $\rho = \rho_n = 1 - 1/n$ in (6·10) and applying (6·11) and (6·12), we obtain

$$\int_0^{2\pi} | s_n - \sigma_n |^q \, d\theta \leqslant \frac{A^q q^q}{n^q} \int_0^{2\pi} | f'(\rho_n, \theta) |^q \, d\theta.$$

The integral $\int_0^{2\pi} | f'(\rho, \theta) |^q \, d\theta$ is an increasing function of ρ. Hence, setting

$$\rho_{n_k} = 1 - 1/n_k, \quad \delta_k = (1 - 1/n_k, 1 - 1/\alpha n_k),$$

we have

$$\begin{aligned} &\int_0^{2\pi} (\Sigma | s_{n_k} - \sigma_{n_k} |^q) \, d\theta \\ &\leqslant A^q q^q \sum_{k=1}^{\infty} \frac{1}{n_k^q} \frac{1}{| \delta_k |} \int_0^{2\pi} d\theta \int_{\delta_k} | f'(\rho, \theta) |^q \, d\rho \\ &\leqslant \frac{(Aq\alpha)^q}{\alpha - 1} \sum_{k=1}^{\infty} \int_0^{2\pi} d\theta \int_{\delta_k} | f'(\rho, \theta) |^q (1-\rho)^{q-1} \, d\rho \\ &\leqslant \frac{(Aq\alpha)^q}{\alpha - 1} \int_0^{2\pi} \int_0^1 | f'(\rho, \theta) |^q (1-\rho)^{q-1} \, d\rho \, d\theta, \end{aligned}$$

and (6·8) follows if we observe that the last integral does not exceed $A^q\int_0^{2\pi}|f(\theta)|^q\,d\theta$ (Chapter XIV, (3·25)).

We can now prove the first inequality (6·6). Set

$$\chi(u)=e^u-u-1=\frac{u^2}{2!}+\frac{u^3}{3!}+\ldots.$$

Multiplying (6·8) by $(2\lambda)^q/q!$, summing the results over $q=2,3,\ldots$, using (6·9), and observing that $q^q/q!\leqslant e^q$, we find

$$\int_0^{2\pi}\sum_{k=1}^{\infty}\chi(2\lambda\,|\,s_{n_k}-\sigma_{n_k}|)\,d\theta\leqslant 2\pi\sum_{q=2}^{\infty}\frac{(2\lambda A_\alpha q)^q}{q!}<A_{\lambda,\alpha}, \tag{6·13}$$

provided $2A_\alpha\lambda e<1$.

Now, we have successively

$$|\,s_{n_k}|\leqslant|\,\sigma_{n_k}|+|\,s_{n_k}-\sigma_{n_k}|\leqslant 1+|\,s_{n_k}-\sigma_{n_k}|, \tag{6·14}$$

$$s^*\leqslant 1+\sup_k|\,s_{n_k}-\sigma_{n_k}|, \tag{6·15}$$

$$\chi(\lambda s^*)\leqslant\chi(2\lambda)+\Sigma\chi(2\lambda\,|\,s_{n_k}-\sigma_{n_k}|). \tag{6·16}$$

Hence, fixing λ and observing that

$$\int_0^{2\pi}\exp(\lambda s^*)\,d\theta\leqslant A\int_0^{2\pi}\chi(\lambda s^*)\,d\theta+A,$$

we obtain the first inequality (6·6) from (6·16) and (6·13).

The proof of the second inequality (6·6) is similar. Clearly (6·8) and (6·9) hold if we substitute $\tilde{s}_{n_k}$, $\tilde{\sigma}_{n_k}$ for s_{n_k}, σ_{n_k}, and the same applies to (6·13). Instead of (6·14) and (6·15), we now have

$$\bar{s}^*\leqslant\bar{\sigma}^*+\sup_k|\,\tilde{s}_{n_k}-\tilde{\sigma}_{n_k}|,$$

$$\chi(\lambda\bar{s}^*)\leqslant\chi(2\lambda\bar{\sigma}^*)+\Sigma\chi(2\lambda\,|\,\tilde{s}_{n_k}-\tilde{\sigma}_{n_k}|),$$

where $\bar{\sigma}^*=\sup|\,\tilde{\sigma}_n|$. Since $\chi(2\lambda\bar{\sigma}^*)\leqslant\exp(2\lambda\bar{\sigma}^*)$, it is enough to show, writing 2λ for λ, that

$$\int_0^{2\pi}\exp(\lambda\bar{\sigma}^*)\,d\theta\leqslant A_\lambda \tag{6·17}$$

for λ sufficiently small. This in turn will follow if we show that $\mathfrak{M}_q[\bar{\sigma}^*]\leqslant Aq$ for $q\geqslant 2$. But

$$\mathfrak{M}_q[\bar{\sigma}^*]\leqslant A\mathfrak{M}_q[\tilde{f}]\leqslant Aq\mathfrak{M}_q[f]\leqslant Aq,$$

by the inequalities (7·8) of Chapter IV, and (2·4) and (3·8) of Chapter VII.

CHAPTER XVI

FOURIER INTEGRALS

1. General remarks

Given a function $f(x)$ in $(-\infty, \infty)$ and a number $\omega > 0$, consider the integrals

$$S_\omega(x) = S_\omega(x, f) = \frac{1}{\pi}\int_{-\infty}^{+\infty} f(x+t)\frac{\sin \omega t}{t}\,dt, \tag{1·1}$$

$$\tilde{S}_\omega(x) = \tilde{S}_\omega(x, f) = -\frac{1}{\pi}\int_{-\infty}^{+\infty} f(x+t)\frac{1-\cos \omega t}{t}\,dt. \tag{1·2}$$

If f is periodic and integrable over a period, the integral (1·1) exists and represents the sum of the terms of $S[f]$ of rank not exceeding ω, with the understanding that, if ω is an integer, only half of the term of rank ω is taken; (1·2) bears the same relation to $\tilde{S}[f]$ (see Chapter II, § 7).

In this section we consider the integrals (1·1) and (1·2) for general (non-periodic) functions in $(-\infty, +\infty)$. We suppose that f is integrable over every finite interval, and that the behaviour of f near $\pm\infty$ is such that (1·1) and (1·2) converge, say absolutely. This is certainly the case if $|f(x)|/(1+|x|)$ is integrable over $(-\infty, +\infty)$, and so, in particular, if $f \in L(-\infty, +\infty)$ or (using Hölder's inequality) if $f \in L^r(-\infty, +\infty)$ for some $r > 1$.

It is an important fact that if f satisfies locally some convergence test for Fourier series, then $S_\omega(x, f) \to f(x)$ as $\omega \to \infty$. This result is known as *the representation of f by Fourier's single integral*, and is a corollary of the following theorem, which also contains a parallel result for the integral (1·2).

(1·3) THEOREM. *Suppose that $|f(t)|/(1+|t|)$ is integrable over $(-\infty, +\infty)$, and let $f_a(t)$ be the periodic function coinciding with f in the interior of a given interval $J_a = (a, a+2\pi)$. Let $s_n(x)$ and $\tilde{s}_n(x)$ be the partial sums of $S[f_a]$ and $\tilde{S}[f_a]$. Then in each interval J'_a totally interior to J_a, the differences*

$$S_\omega(x) - s_{[\omega]}(x), \quad \tilde{S}_\omega(x) - \tilde{s}_{[\omega]}(x)$$

tend uniformly to limits as $\omega \to \infty$, the first limit being 0.

We shall only discuss $S_\omega - s_{[\omega]}$, the argument applying without change to $\tilde{S}_\omega - \tilde{s}_{[\omega]}$. For the sake of clarity we first consider only pointwise convergence. Suppose that $x_0 \in J'_a$, and that $\delta > 0$ is so small that $x_0 \pm \delta \in J_a$. From (1·1), using the Riemann-Lebesgue theorem in the form (4·6) of Chapter II, we deduce

$$S_\omega(x_0) = \frac{1}{\pi}\int_{-\delta}^{\delta} f(x_0+t)\frac{\sin \omega t}{t}\,dt + o(1), \tag{1·4}$$

since $|f(x_0+t)t^{-1}|$ is integrable over $|t| \geq \delta$. Write $\omega' = [\omega] + \frac{1}{2}$, so that $-\frac{1}{2} \leq \omega - \omega' \leq \frac{1}{2}$. The difference between the last integral and the integral in the formula

$$s_{[\omega]}(x_0) = \frac{1}{\pi}\int_{-\delta}^{\delta} f(x_0+t)\frac{\sin \omega' t}{t}\,dt + o(1) \tag{1·5}$$

(Chapter II, (7·1)) is

$$\frac{\omega-\omega'}{\pi}\int_{-\delta}^{\delta} f(x_0+t)\frac{2\sin\frac{1}{2}(\omega-\omega')t}{(\omega-\omega')t}\cos\tfrac{1}{2}(\omega+\omega')t\,dt=\frac{\omega-\omega'}{\pi}\int_{-\delta'}^{\delta''} f(x_0+t)\cos\tfrac{1}{2}(\omega+\omega')t\,dt \tag{1·6}$$

(where $0<\delta'<\delta$, $0<\delta''<\delta$), by an application of the second mean-value theorem to the factor $2\sin\frac{1}{2}(\omega-\omega')t/(\omega-\omega')t$, which is even, positive and decreasing in $(0,\delta)$. By the Riemann-Lebesgue theorem, the last integral is $o(1)$, and the relation $S_\omega(x_0)-s_{[\omega]}(x_0)\to 0$ follows.

It remains to show that the $o(1)$ in the preceding argument is uniform in $x_0\in J'_a$.

Split the integral in (1·1) into three, extended respectively over $|t|\leqslant\delta$, $\delta\leqslant|t|\leqslant\Delta$ and $|t|\geqslant\Delta$. If Δ is large enough, the integral over $|t|\geqslant\Delta$ is arbitrarily small uniformly in x_0 (and ω). Fix Δ and apply the second mean-value theorem to the factor $1/t$ in the integrals over $(-\Delta,-\delta)$ and (δ,Δ); we immediately see, by the Riemann-Lebesgue theorem, that the integrals are $o(1)$ uniformly in x_0.† Hence the $o(1)$ in (1·4) is uniform in x_0. That the $o(1)$ in (1·5) is uniform was shown in Chapter II, § 7, and that the last integral in (1·6) tends uniformly to 0, is obvious. Hence $S_\omega(x)-s_{[\omega]}(x)=o(1)$ uniformly in $x_0\in J'_\alpha$, and the proof of (1·3) is completed.

Theorem (1·3) enables us to deduce tests for the pointwise convergence of $S_\omega(x)$ from the corresponding tests for Fourier series. The same applies to uniform convergence over *finite* intervals since such intervals can be split into a finite number of intervals of length less than 2π. We omit specific applications.

Return for a moment to the difference $\tilde{S}_\omega-\tilde{s}_{[\omega]}$. Suppose that the integral defining the conjugate function $\tilde{f}_a$ converges at an $x_0\in J'_a$. Then the integral

$$-\frac{1}{\pi}\int_0^\infty\frac{f(x_0+t)-f(x_0-t)}{t}dt=-\frac{1}{\pi}\int_{-\infty}^{+\infty}\frac{f(t)}{t-x_0}dt=-\frac{1}{\pi}\lim_{\epsilon\to+0}\int_{|t-x_0|\geqslant\epsilon}\frac{f(t)}{t-x_0}dt$$

also converges; for near $t=\pm\infty$ it converges absolutely. It is called the *function conjugate to f*, or the *Hilbert transform* of f, and is denoted‡ by $\tilde{f}(x)$. *It exists for almost all x_0 in $(-\infty,+\infty)$.* The proof of (1·3) is easily modified to show that

$$\tilde{S}_\omega(x_0)-\tilde{f}(x_0)-\{\tilde{s}_{[\omega]}(x_0)-\tilde{f}_a(x_0)\}\to 0, \tag{1·7}$$

so that $\tilde{S}_\omega(x_0)\to\tilde{f}(x_0)$, if $\tilde{s}_n(x_0)\to\tilde{f}_a(x_0)$.

(1·8) THEOREM. *The conclusions of Theorem* (1·3) *hold if f is integrable over every finite interval and if, in addition, $f(t)/t$ tends to 0 as $t\to\pm\infty$ and is of bounded variation in the neighbourhood of $t=\pm\infty$.*

We confine our attention to $S_\omega-s_{[\omega]}$. The last hypothesis of (1·8) means that there is a $B>0$ such that $f(t)/t$ is of bounded variation in $(-\infty,-B)$ and $(B,+\infty)$.

We may suppose that $(-B,B)$ contains J_a in its interior. Let $g(x)=f(x)$ in $(-B,B)$ and $g(x)=0$ elsewhere. Then $g_a=f_a$ and, by (1·3), $S_\omega(x_0,g)-s_{[\omega]}(x_0;f_a)\to 0$ uniformly in $x_0\in J'_a$. If we write $f=g+h$ and show that $S_\omega(x_0,h)\to 0$ uniformly in $x_0\in J'_a$, then by combining this relation with the preceding one we obtain the required result.

† We apply Theorem (4·6) of Chapter II to the family of functions $f(x_0+t)$ depending on the parameter x_0, but the change of variable $x_0+t=t'$ moves the parameter from the integrand to the limits of integration.

‡ We use the same notation for the conjugate function in the periodic and non-periodic case.

Now $h=0$ in $(-B, B)$ and $h=f$ elsewhere, so that

$$S_\omega(x_0, h) = \frac{1}{\pi}\left(\int_{-\infty}^{-B} + \int_{B}^{+\infty}\right)\frac{f(t)}{t}\frac{t}{t-x_0}\sin\omega(t-x_0)\,dt. \tag{1·9}$$

We may suppose that $f(t)/t$ is monotonically decreasing to 0 in $(-\infty, -B)$ and $(B, +\infty)$ as $t \to \pm\infty$; for otherwise we can replace $f(t)/t$ by a difference of two functions with this property. The integrals are limits, for $B' \to +\infty$, of integrals extended over $(-B', -B)$ and (B, B') respectively. Applying the second mean-value theorem to the monotone factors $f(t)/t$ and $t/(t-x_0)$, and then making $B' \to \infty$, we find that (1·9) tends uniformly to 0 for $x_0 \in J'_a$, and the proof of (1·8) is completed.

Suppose that $f(t)$ is integrable over $(-\infty, +\infty)$. Then

$$S_\omega(x) = \frac{1}{\pi}\int_{-\infty}^{+\infty} f(t)\,dt \int_0^\omega \cos s(x-t)\,ds = \frac{1}{\pi}\int_0^\omega ds \int_{-\infty}^{+\infty} f(t)\cos s(x-t)\,dt, \tag{1·10}$$

the inversion of the order of integration being justified by the absolute convergence. Hence $S_\omega(x)$ is a partial integral of the infinite integral

$$\frac{1}{\pi}\int_0^\infty ds \int_{-\infty}^{+\infty} f(t)\cos s(x-t)\,dt \tag{1·11}$$

$$= \int_0^\infty \{a(s)\cos sx + b(s)\sin sx\}\,ds = \int_{-\infty}^{+\infty} c(s)\,e^{isx}\,ds, \tag{1·12}$$

where
$$a(s) = \frac{1}{\pi}\int_{-\infty}^{+\infty} f(t)\cos st\,dt, \quad b(s) = \frac{1}{\pi}\int_{-\infty}^{+\infty} f(t)\sin st\,dt, \tag{1·13}$$

$$c(s) = \frac{1}{2\pi}\int_{-\infty}^{+\infty} f(t)\,e^{-ist}\,dt = \tfrac{1}{2}\{a(s) - ib(s)\}. \tag{1·14}$$

The integrals $a(s)$, $b(s)$, $c(s)$ are analogues of Fourier coefficients; they are called the *Fourier transforms* (cosine, sine, complex) of f. We considered them earlier in Chapter I, § 4. The integrals (1·12) are analogues of Fourier series; they are called *Fourier integrals*; sometimes, in view of the representation (1·11), *Fourier's repeated integrals.* We may also consider general integrals of the form (1·12), where the coefficient functions $a(s)$, $b(s)$, $c(s)$ are not necessarily Fourier transforms; such integrals are called *trigonometric integrals.* If $a(s)$ and $b(s)$ are real-valued and $\gamma(s) = a(s) - ib(s)$, (1·12) is the real part of the *Laplace integral*

$$\int_0^\infty \gamma(s)\,e^{isz}\,ds \quad (z = x+iy), \tag{1·15}$$

for $z = x$. The imaginary part of (1·15) for $z = x$ is the integral

$$\int_0^\infty \{a(s)\sin sx - b(s)\cos sx\}\,ds = -i\int_{-\infty}^{+\infty} c(s)\,(\operatorname{sign} s)\,e^{isx}\,ds, \tag{1·16}$$

conjugate to (1·12). The integral (1·15) converges, under very general conditions, in the upper half-plane $y > 0$ to a function regular there.

Given an $f(t)$, $-\infty < t < +\infty$, such that the integrals (1·13) have a meaning, we may ask in what sense (1·12) represents $f(x)$. If the inversion of the order of integration in (1·10) is justified for all ω, then the partial integrals of (1·12), $S_\omega(x)$ say, are given by (1·1). The problem then reduces to that of the representation of f by Fourier's single integral, and so (in view of Theorem (1·3)) to that of the representation of a function by its Fourier series. The formula (1·10) is, however, true only under certain conditions on the behaviour of $f(t)$ near $t = \pm\infty$. Consequently the range of application of Fourier's repeated integral is more restricted than that of Fourier's single integral.

The formula (1·10) holds if $f \in L(-\infty, +\infty)$, and the problem of the convergence of (1·12) and (1·16) may therefore be considered as settled in this case by Theorem (1·3).

The case when $f(t)$ is not necessarily integrable in the neighbourhood of $t = \pm\infty$, but is of bounded variation there and tends to 0 with $1/t$, has applications. We may suppose that $f(t)$ tends monotonically to 0 as $t \to \pm\infty$, $|t| \geq B > 0$. The second mean-value theorem shows then that the transforms $a(s)$ and $b(s)$ converge for $s \neq 0$, and that the convergence is uniform for $|s| \geq \delta > 0$. Hence the following version of (1·10) holds:

$$\frac{1}{\pi}\int_\delta^\omega ds \int_{-\infty}^{+\infty} f(t)\cos s(x-t)\,dt = \frac{1}{\pi}\int_{-\infty}^{+\infty} f(t)\,\frac{\sin\omega(x-t)}{x-t}\,dt - \frac{1}{\pi}\int_{-\infty}^{+\infty} f(t)\,\frac{\sin\delta(x-t)}{x-t}\,dt. \quad (1·17)$$

We show that the last integral tends to 0 with δ. Split the interval of integration $(-\infty, +\infty)$ into $(-\infty, -A)$, $(-A, A)$, $(A, +\infty)$. If $A > B$ is large enough, the second mean-value theorem applied to f shows that the first and third integrals are arbitrarily small. If A is fixed, the second integral tends to 0 with δ. This proves the assertion, and we have the following theorem:

(1·18) THEOREM. *If f is of bounded variation near $t = \pm\infty$ and tends to 0 with $1/t$, we still have* (1·10), *provided the outer integral on the right is taken as* $\lim_{\delta\to+0}\int_\delta^\omega$.

That the latter restriction is essential, and that $g(s) = \int_{-\infty}^{+\infty} f(t)\cos s(x-t)\,dt$, *qua* function of s, need not be Lebesgue-integrable near $s = 0$, can be seen from the following example. Take a sequence $a_1 > a_2 > \ldots \to 0$ such that the sum of $\Sigma a_n \cos ns$ is not integrable near $s = 0$ (see Chapter V, (1·9)). Let $x = 0$, and let $f(t)$ be even and satisfy

$$f(t) = 0 \quad \text{for} \quad 0 \leq t < \tfrac{1}{2},$$

$$f(t) = a_n \quad \text{for} \quad n - \tfrac{1}{2} \leq t < n + \tfrac{1}{2} \quad (n = 1, 2, \ldots).$$

Then

$$\frac{s}{4\sin\frac{1}{2}s}\,g(s) = \Sigma a_n \cos ns,$$

and $g(s)$ is not integrable near $s = 0$.

This example shows that *under the hypothesis of* (1·18) *the outer integral in Fourier's repeated integral must be taken as* $\lim_{\omega\to+\infty}\lim_{\delta\to+0}\int_\delta^\omega$.

Return to (1·3), and for each $\omega \geq 0$ denote the sum of terms of rank not exceeding ω in $S[f_a]$ and $\tilde{S}[f_a]$ by s_ω and $\tilde{s}_\omega$ respectively. Thus $s_\omega = s_{[\omega]}$, $\tilde{s}_\omega = \tilde{s}_{[\omega]}$. By (1·3), $S_\omega(x) - s_\omega(x)$ tends uniformly to 0 in J'_a. It follows that

$$\frac{1}{w}\int_0^w S_\omega(x)\,d\omega - \frac{1}{w}\int_0^w s_\omega(x)\,d\omega \quad (1·19)$$

tends uniformly to 0 in J'_a as $w \to \infty$. Denote the second term by $\sigma_w(x)$; it is a continuous analogue of the discontinuous (C, 1) mean $(s_0 + \ldots + s_n)/(n+1)$, and the two are equivalent (see Chapter III, p. 83). Integrating (1·1) over $0 \leq \omega \leq w$ and inverting the order of integration in the repeated integral we find for the (C, 1) means of $S_\omega(x)$ the expression

$$\frac{1}{\pi}\int_{-\infty}^{+\infty} f(x+t)\,\frac{2\sin^2\frac{1}{2}wt}{wt^2}\,dt, \quad (1·20)$$

which is an analogue of Fejér's integral for Fourier series. Using basic results about the summability (C, 1) of Fourier series (see Chapter III, (3·7), (3·9)) we arrive at the following theorem:

(1·21) THEOREM. *If* $|f(t)|/(1+|t|)$ *is integrable over* $(-\infty, +\infty)$, *the* (C, 1) *means* (1·20) *of* $S_\omega(x, f)$ *converge to* f *almost everywhere. In particular they converge to*

$$\tfrac{1}{2}\{f(x+0)+f(x-0)\}$$

at each point where $f(x\pm 0)$ *exist. The convergence is uniform over each finite and closed interval of continuity of* f.†

If we integrate (1·7) over $0 \leqslant \omega \leqslant w$ and observe that the continuous (C, 1) means $\tilde{\sigma}_w(x)$ of $\tilde{s}_\omega(x)$ tend to $\tilde{f}_a(x)$ almost everywhere (Chapter III, (3·23)) we obtain similarly:

(1·22) THEOREM. *If* $|f(t)|/(1+|t|)$ *is integrable over* $(-\infty, +\infty)$, *the* (C, 1) *means*

$$-\frac{1}{\pi}\int_{-\infty}^{+\infty} f(x+t)\left\{\frac{1}{t}-\frac{\sin wt}{wt^2}\right\}dt$$

of $\tilde{S}_\omega(x)$ *converge to the Hilbert transform* $\tilde{f}(x)$ *almost everywhere.*

2. Fourier transforms

In § 1 we investigated the representability of functions by their Fourier integrals, single and repeated, and proved that, locally at least, the problem is reducible to that of the representability of periodic functions by their Fourier series. We shall now study in greater detail properties of Fourier transforms as such. The latter are analogues of Fourier coefficients, but there is now more symmetry in the situation, since f and its Fourier transform are both functions of a continuous variable.

Given an $f(x)$ in $(-\infty, +\infty)$, consider its Fourier transform

$$c(x) = \frac{1}{2\pi}\int_{-\infty}^{+\infty} f(y)\, e^{-ixy}\, dy. \qquad (2·1)$$

If f is in $L(-\infty, +\infty)$, $c(x)$ is defined everywhere, is a continuous function of x, and tends to 0 as $x \to \pm\infty$; the latter is the Riemann-Lebesgue theorem (Chapter II, (4·6)). The representability of f by its Fourier transform means, as we proved in the preceding section, that

$$f(x) = \int_{-\infty}^{+\infty} c(y)\, e^{ixy}\, dy, \qquad (2·2)$$

in the sense that the partial integrals $\int_{-\omega}^{+\omega}$ of the integral on the right are summable (C, 1) to $f(x)$ almost everywhere.

We may also define $c(x)$ in certain cases when f is not absolutely integrable over $(-\infty, +\infty)$, but then the validity of (2·2) must be investigated afresh. If, for example, f is of bounded variation over $(-\infty, +\infty)$, then $c(x)$ exists for $x \neq 0$, and we have (2·2)

† It may be instructive for the reader to prove this theorem (and Theorem (1·22) below) directly, without using Fourier series. It is not difficult to see that the integral (1·20) has meaning, and converges almost everywhere to $f(x)$ as $w \to \infty$, if $f(t)/(1+t^2) \in L(-\infty, +\infty)$.

for all x, provided $f(x)=\frac{1}{2}\{f(x+0)+f(x-0)\}$, and provided the integral (2·2) is understood as

$$\lim_{\omega\to+\infty}\lim_{\delta\to+0}\int_{\delta\leqslant|y|\leqslant\omega};$$

this is a consequence of (1·8) and (1·18).

Remark. Suppose that $\gamma(\lambda)$ is integrable over every finite interval of λ (but not necessarily over $(-\infty, +\infty)$) and that

$$g(x)=\int_{-\infty}^{+\infty}\gamma(\lambda)\,e^{ix\lambda}\,d\lambda,$$

where the integral converges uniformly over every finite interval of x. We have then

$$\frac{1}{2\pi\omega}\int_0^{\omega}dw\int_{-w}^{w}g(x)\,e^{-ix\lambda}\,dx=\frac{1}{\pi}\int_{-\infty}^{+\infty}\gamma(\mu)\,\frac{2\sin^2\frac{1}{2}(\mu-\lambda)}{\omega(\mu-\lambda)^2}\,d\mu.$$

The formula is easy to verify by substituting $g(x)=\int_{-\infty}^{+\infty}\gamma(\mu)\,e^{ix\mu}\,d\mu$ into the left-hand side, and integrating first with respect to x and w, which is justified by uniform convergence. The right-hand side of the formula tends to $\gamma(\mu)$ for almost all λ under rather general conditions for γ (it is sufficient, for example, to assume that

$$\gamma(\mu)/(1+\mu^2)\in L(-\infty,+\infty)),$$

and we have then

$$\gamma(\lambda)=\frac{1}{2\pi}\int_{-\infty}^{+\infty}g(x)\,e^{-ix\lambda}\,dx,$$

where the integral is meant in the (C, 1) sense. This observation will be used in § 10 below.

It is desirable to make the definition of the Fourier transform more symmetric by changing the factor $(2\pi)^{-1}$ in (2·1) to $(2\pi)^{-\frac{1}{2}}$. The Fourier transform of f, thus modified, will be denoted systematically by $\hat{f}$, and the relations (2·1) and (2·2) can be written

$$\hat{f}(x)=\frac{1}{\sqrt{(2\pi)}}\int_{-\infty}^{+\infty}f(y)\,e^{-ixy}\,dy, \tag{2·3}$$

$$f(x)=\frac{1}{\sqrt{(2\pi)}}\int_{-\infty}^{+\infty}\hat{f}(y)\,e^{ixy}\,dy. \tag{2·4}$$

These two relations are sometimes called *Fourier inversion formulae*.

In the special case when f is even, $f(-x)=f(x)$, (2·3) and (2·4) can be written

$$\hat{f}(x)=\sqrt{\frac{2}{\pi}}\int_0^{\infty}f(y)\cos xy\,dy, \tag{2·5}$$

$$f(x)=\sqrt{\frac{2}{\pi}}\int_0^{\infty}\hat{f}(y)\cos xy\,dy, \tag{2·6}$$

so that $\hat{f}(x)$ is also even. If f is odd, $f(-x)=-f(x)$, then

$$i\hat{f}(x)=\sqrt{\frac{2}{\pi}}\int_0^{\infty}f(y)\sin xy\,dy, \tag{2·7}$$

$$-i f(x)=\sqrt{\frac{2}{\pi}}\int_0^{\infty}\hat{f}(y)\sin xy\,dy, \tag{2·8}$$

and $\hat{f}(x)$ is odd.

If f is defined in $(0, \infty)$ only, the right-hand sides of (2·5) and (2·7) will be called the *cosine* and *sine* transforms of f and denoted by $\hat{f}_c$ and $\hat{f}_s$ respectively.

If $f \in L(-\infty, +\infty)$, $\hat{f}(x)$ is continuous and tends to 0 as $x \to \pm\infty$. If, in addition, $\hat{f} \in L(-\infty, +\infty)$, the right-hand side of (2·4) is a continuous function, which we may identify with the initial function f by modifying the latter, if necessary, in a set of measure 0. Hence if both f and $\hat{f}$ are in $L(-\infty, +\infty)$, then both are continuous and tend to 0 at infinity.

But the integrability of f need not imply that of $\hat{f}$. Hence, in spite of the formal resemblance of the relations (2·3) and (2·4), there is a basic asymmetry in the metric properties of f and $\hat{f}$, if f is merely integrable over $(-\infty, +\infty)$. It turns out that this asymmetry disappears if we confine our attention to functions f in $L^2 = L^2(-\infty, \infty)$, and that in this case we also have the basic *Parseval-Plancherel formula*

$$\int_{-\infty}^{+\infty} |\hat{f}(x)|^2\, dx = \int_{-\infty}^{+\infty} |f(x)|^2\, dx. \tag{2·9}$$

But of course we must first define $\hat{f}$ for $f \in L^2$.

Let S be the class of all step functions in $(-\infty, +\infty)$ which are 0 near $\pm\infty$. We define the Fourier transforms of such f directly by (2·3). For all other $f \in L^2$, we shall define $\hat{f}$ indirectly. Let $f \in S$. Then, if $0 < \omega < \infty$,

$$\left.\begin{aligned} \int_{-\omega}^{+\omega} |\hat{f}(x)|^2\, dx &= \int_{-\omega}^{\omega} dx \frac{1}{2\pi} \int_{-\infty}^{+\infty} f(y)\, e^{-ixy}\, dy \int_{-\infty}^{+\infty} \bar{f}(y')\, e^{ixy'}\, dy' \\ &= \int_{-\infty}^{+\infty}\int_{-\infty}^{+\infty} f(y)\bar{f}(y')\, dy\, dy' \frac{1}{2\pi}\int_{-\omega}^{\omega} e^{ix(y'-y)}\, dx \\ = \frac{1}{\pi}\int_{-\infty}^{+\infty}\int_{-\infty}^{+\infty} f(y)\bar{f}(y') &\frac{\sin \omega(y-y')}{y-y'}\, dy\, dy' = \int_{-\infty}^{+\infty} f(y)\, S_\omega(y, \bar{f})\, dy, \end{aligned}\right\} \tag{2·10}$$

where S_ω is given by (1·1). The transformations are legitimate since all integrations are actually over finite intervals.

Observe now that, in the case under consideration, $S_\omega(y, \bar{f})$ is uniformly bounded in y and ω, and tends to $\bar{f}(y)$ as $\omega \to +\infty$, except possibly at a finite number of points. It is enough to consider the case when f is the characteristic function of an interval; the result then follows from the formula

$$\int_0^\infty \frac{\sin \lambda t}{t}\, dt = \tfrac{1}{2}\pi \operatorname{sign} \lambda,$$

and the fact that the partial integrals of this integral are uniformly bounded. Comparing the extreme terms of (2·10) and making $\omega \to \infty$, we obtain (2·9) by Lebesgue's theorem on the termwise integration of bounded sequences.

The formula (2·3) defines a linear operation $\hat{f} = Tf$ for all f in the set S, which is dense in L^2. Since (2·9) holds for $f \in S$, the operation Tf can be extended by continuity to all $f \in L^2$; this extension is unique (see Chapter IV, (9·3), and satisfies $\| Tf \| \leqslant \| f \|$, where

$$\| f \| = \| f \|_2 = \left(\int_{-\infty}^{+\infty} |f|^2\, dx \right)^{\frac{1}{2}}.$$

We show now that we actually have $\| Tf \| = \| f \|$. For let $f \in L^2$, and let $f_1, f_2, \ldots$ in S be such that $\| f - f_n \| \to 0$. In particular, by Minkowski's inequality, $\| f_n \| \to \| f \|$. Since

$$\| Tf_n - Tf \| = \| T(f_n - f) \| \leqslant \| f_n - f \| \to 0,$$

we also have $\| Tf_n \| \to \| Tf \|$, and making $n \to \infty$ in the equation $\| Tf_n \| = \| f_n \|$, we obtain $\| Tf \| = \| f \|$, which is (2·9). Collecting results we see that there is a linear operation $\hat{f} = Tf$ defined for all $f \in L^2$, satisfying (2·9) and given by (2·3) for $f \in S$.

Next we show that if $f \in L^2$ is 0 outside some interval $(-A, A)$, then $\hat{f} = Tf$ is still given by the formula (2·3). For denote by $f^*(x)$ the value of the integral (2·3); we have to show that $\hat{f} = f^*$ almost everywhere. Take a sequence of step functions $f_1, f_2, \ldots$ vanishing outside $(-A, A)$ and such that $\| f - f_n \| \to 0$. Hence $\| \hat{f} - \hat{f}_n \| \to 0$, and in particular

$$\int_{-\omega}^{+\omega} | \hat{f} - \hat{f}_n |^2 \, dx \to 0 \tag{2·11}$$

for each $\omega > 0$. Schwarz's inequality shows that $\hat{f}_n$ tends uniformly to f^* over every finite interval, and in particular $\int_{-\omega}^{+\omega} | \hat{f}_n - f^* |^2 \, dx \to 0$ as $n \to \infty$. This and (2·11) show that $f^* = \hat{f}$ almost everywhere in $(-\omega, \omega)$, and so also almost everywhere in $(-\infty, \infty)$.

Finally, for an arbitrary $f \in L^2$, write

$$\hat{f}_\omega(x) = \frac{1}{\sqrt{(2\pi)}} \int_{-\omega}^{+\omega} f(y) \, e^{-ixy} \, dy. \tag{2·12}$$

Then $\hat{f}_\omega = Tf_\omega$, where f_ω coincides with f in $(-\omega, \omega)$ and is 0 elsewhere. Since

$$\| \hat{f}_\omega - \hat{f} \| = \| T(f_\omega - f) \| = \| f_\omega - f \| \to 0$$

as $\omega \to \infty$, we see that for each $f \in L^2$, *the integral* (2·3) *converges in* L^2 *to* $\hat{f}$, that is, $\| \hat{f}_\omega - \hat{f} \| \to 0$.

The last relation implies that there is a sequence ω_k such that $\hat{f}_{\omega_k} \to \hat{f}$ almost everywhere. In particular, *if the right-hand side of* (2·3) *converges almost everywhere, it converges to* $\hat{f}$.

We now prove (2·4) for $f \in L^2$. It can be written

$$f = T^* Tf, \tag{2·13}$$

where T^* denotes the operation obtained from T by changing e^{-ixy} to e^{ixy} in (2·3) $(T^* f = \overline{T\bar{f}})$. Since T and T^* are continuous in L^2, so is T^*T, and it is enough to prove (2·13) when $f \in S$, or even when f is the characteristic function of an interval (a, b). In this case $\hat{f}(x) = i(e^{-ibx} - e^{-iax})(2\pi)^{-\frac{1}{2}} x^{-1}$, and

$$\frac{1}{\sqrt{(2\pi)}} \int_{-\infty}^{+\infty} \hat{f}(y) \, e^{ixy} \, dy = \frac{1}{\pi} \int_0^\infty \frac{\sin (x-a) y}{y} \, dy + \frac{1}{\pi} \int_0^\infty \frac{\sin (b-x) y}{y} \, dy$$

$$= \tfrac{1}{2} \{ \operatorname{sign} (x-a) + \operatorname{sign} (b-x) \}. \tag{2·14}$$

Hence the left-hand side is 1 for $a < x < b$, and is 0 for $x < a$ and $x > b$. Hence $T^*Tf = f$, and the proof of (2·4) is completed.

We can represent $\hat{f}$ in a different form. Let $\Phi(x)$ and $\Phi_\omega(x)$ be the integrals of $\hat{f}$ and $\hat{f}_\omega$ respectively, from 0 to x. By Schwarz's inequality,

$$| \Phi(x) - \Phi_\omega(x) | = \left| \int_0^x (\hat{f} - \hat{f}_\omega) \, dy \right| \leqslant | x |^{\frac{1}{2}} \| \hat{f} - \hat{f}_\omega \| = o(1)$$

as $\omega \to \infty$, so that $\Phi(x) = \lim_{\omega} \Phi_\omega(x)$ for each x. Since, by (2·12),

$$\Phi_\omega(x) = \frac{1}{\sqrt{(2\pi)}} \int_{-\omega}^{\omega} f(y) \frac{e^{-ixy} - 1}{-iy} \, dy,$$

and $\hat{f}(x) = \Phi'(x) = (d/dx) \lim \Phi_\omega(x)$ almost everywhere, it follows that almost everywhere we have

$$\hat{f}(x) = \frac{d}{dx} \left\{ \frac{1}{\sqrt{(2\pi)}} \int_{-\infty}^{+\infty} f(y) \frac{e^{-ixy} - 1}{-iy} dy \right\}, \tag{2·15}$$

and correspondingly for (2·4),

$$f(x) = \frac{d}{dx} \left\{ \frac{1}{\sqrt{(2\pi)}} \int_{-\infty}^{+\infty} \hat{f}(y) \frac{e^{ixy} - 1}{iy} \, dy \right\}. \tag{2·16}$$

For f in L^2, the results about $\hat{f}$ obtained so far can be summarized in the following theorem:

(2·17) THEOREM OF PLANCHEREL. *For each f in $L^2(-\infty, \infty)$ the integral (2·3) converges in L^2 to a function $\hat{f} \in L^2$; that is to say, $\| \hat{f} - \hat{f}_\omega \| \to 0$, where $\hat{f}_\omega$ is given by (2·12). Similarly the integral (2·4) converges in L^2 to f. The functions f and $\hat{f}$ satisfy the Parseval formula $\| \hat{f} \| = \| f \|$, and for almost all x the relations (2·15) and (2·16) also.*

If f_1 and f_2 are in L^2, then

$$\int_{-\infty}^{+\infty} f_1 \bar{f}_2 dx = \int_{-\infty}^{+\infty} \hat{f}_1 \bar{\hat{f}}_2 dx. \tag{2·18}$$

For applying (2·9) to $f = f_1 + f_2$ and using the equations $\| \hat{f}_j \| = \| f_j \|$, $j = 1, 2$, we find that the real parts of the two sides of (2·18) are equal. Hence, replacing f_1 by if_1, the imaginary parts must also be equal, and (2·18) follows.

The formula (2·18), which generalizes (2·9), is the source of many important identities. It shows in particular that *if $\{f_n\}$ is an orthonormal system on $(-\infty, +\infty)$, so is $\{\hat{f}_n\}$; if $\{f_n\}$ is complete, $\{\hat{f}_n\}$ is also complete.*

The formula (2·4) tells us that each f in L^2 is the Fourier transform of a $g \in L^2$, namely of $g(x) = \hat{f}(-x)$.

If $\hat{f}_\omega$ is given by (2·12), and $0 < w < \infty$, we have

$$\frac{1}{\sqrt{(2\pi)}} \int_{-w}^{w} \hat{f}_\omega(y) \, e^{ixy} \, dy = \frac{1}{\pi} \int_{-\omega}^{\omega} f(y) \frac{\sin w(x-y)}{x-y} dy.$$

Since $\hat{f}_\omega$ tends to $\hat{f}$ in L^2, we can make $\omega \to \infty$ and obtain

$$\frac{1}{\sqrt{(2\pi)}} \int_{-w}^{w} \hat{f}(y) \, e^{ixy} \, dy = \frac{1}{\pi} \int_{-\infty}^{+\infty} f(y) \frac{\sin w(x-y)}{x-y} dy. \tag{2·19}$$

Substituting here $\hat{f}(y)$ for $f(y)$ on the right and $f(-y)$ for $\hat{f}(y)$ on the left, we find the formula

$$\frac{1}{\sqrt{(2\pi)}} \int_{-w}^{w} f(y) \, e^{-ixy} \, dy = \frac{1}{\pi} \int_{-\infty}^{+\infty} \hat{f}(y) \frac{\sin w(x-y)}{x-y} dy \tag{2·20}$$

or

$$\hat{f}_w(x) = S_w(x, \hat{f}).$$

From this we can deduce the following consequences (for $f \in L^2$):

(*a*) *The integral (2·3) is summable* (C, 1) *almost everywhere to $\hat{f}$.* For $|\hat{f}(t)|/(1 + |t|)$ is in L, and it is enough to apply (1·21).

(*b*) *We have*

$$\hat{f}_\omega(x) = o\{(\log \omega)^{\frac{1}{2}}\}$$

for almost all x. For (1·3) is applicable, and it is enough to use the fact that the nth partial sum of a Fourier series in L^2 is $o\{(\log n)^{\frac{1}{2}}\}$ almost everywhere (Chapter XIII, § 1).

(*c*) *If* $\omega_1 < \omega_2 < \dots$, $\omega_{k+1}/\omega_k > q > 1$, *then* $\hat{f}_{\omega_k}(x) \to \hat{f}(x)$ *almost everywhere*. The argument is the same as in (*b*), except that now we use the fact that lacunary partial sums of an $S[f]$ in L^2 converge almost everywhere (see Chapter XIII, (1·17)).

Whether or not we have $\hat{f}_\omega(x) \to \hat{f}(x)$ almost everywhere for f in L^2 is an open question. In view of (1·3) and the fact that $\hat{f}$ in (2·20) is the most general function in L^2, the problem is equivalent to the problem whether the Fourier series of a general function in L^2 converges almost everywhere.

We consider a few examples of Fourier transforms.

(*a*) If $f(x)$ is the characteristic function of an interval $(a-\lambda, a+\lambda)$, then

$$\hat{f}(x) = \left(\frac{2}{\pi}\right)^{\frac{1}{2}} e^{-ixa} \frac{\sin \lambda x}{x}. \tag{2·21}$$

An application of (2·9) to this f gives the basic formula

$$\int_{-\infty}^{+\infty} \frac{\sin^2 \lambda x}{\lambda x^2} = \pi \tag{2·22}$$

(see also Chapter III, (11·4)).

Consider a sequence of numbers $n_1, n_2, \dots$, not necessarily integers, such that the intervals $(n_k - \lambda, n_k + \lambda)$ do not overlap. Then the characteristic functions of these intervals are mutually orthogonal. By (2·18), the same holds for their transforms, a result which, in view of (2·21) and (2·22), can also be stated in the following form:

If the intervals $(n_k - \lambda, n_k + \lambda)$, $k = 1, 2, \dots$, *do not overlap, then the functions* $e^{in_1x}, e^{in_2x}, \dots$ *form an orthonormal system over* $(-\infty, +\infty)$ *with respect to the weight*

$$d\omega(x) = \frac{\sin^2 \lambda x}{\pi \lambda x^2}$$

(see Chapter I, p. 6).

(*b*) Let $\chi(x) = \chi_h(x)$ be the 'triangular' function defined by the conditions that $\chi(x) = 0$ for $|x| \geqslant h$, $\chi(0) = 1$, $\chi(x)$ is linear in the intervals $(-h, 0)$ and $(0, h)$. Then

$$\hat{\chi}(x) = \left(\frac{2}{\pi}\right)^{\frac{1}{2}} \frac{2 \sin^2 \frac{1}{2}hx}{hx^2}, \tag{2·23}$$

a relation which can also be written

$$\frac{1}{\pi} \int_{-\infty}^{+\infty} e^{-iyx} \frac{2 \sin^2 \frac{1}{2}hy}{hy^2} dy = \left(1 - \frac{|x|}{h}\right)^+. \tag{2·24}$$

(*c*) Let $f(x) = e^{-ax}$, $x > 0$, where $a > 0$. Then for the cosine and sine transforms of f we have

$$f_c(x) = \left(\frac{2}{\pi}\right)^{\frac{1}{2}} \frac{a}{x^2 + a^2}, \tag{2·25}$$

$$f_s(x) = \left(\frac{2}{\pi}\right)^{\frac{1}{2}} \frac{x}{x^2 + a^2}. \tag{2·26}$$

The inversion formula for (2·25) leads to the familiar equation

$$\int_0^{\infty} \frac{\cos xy}{y^2 + a^2} dy = \tfrac{1}{2} \frac{\pi}{a} e^{-a|x|}.$$

(*d*) Let $0 < s < 1$, $f(x) = (x^+)^{s-1}$. Then

$$\hat{f}(x) = (2\pi)^{-\frac{1}{2}} |x|^{-s} \Gamma(s) \exp\{-\tfrac{1}{2}\pi s \operatorname{sign} x\}$$

(*e*) Let
$$f(x)=\frac{1}{\sqrt{(2\pi)}\,i}\frac{1}{x-\zeta},$$
where $\zeta=\xi+i\eta$ is complex $(\eta\neq 0)$. Then, if $\eta>0$, we have
$$\hat{f}(x)=0 \quad\text{for}\quad x>0, \qquad \hat{f}(x)=e^{-ix\zeta} \quad\text{for}\quad x<0;$$
if $\eta<0$, we have
$$\hat{f}(x)=-e^{-ix\zeta} \quad\text{for}\quad x>0, \qquad \hat{f}(x)=0 \quad\text{for}\quad x<0.$$

(*f*) The function
$$g(x)=\frac{1}{\sqrt{(2\pi)}}\,e^{-\frac{1}{2}x^2}$$
is its own Fourier transform, that is to say, $\hat{g}(x)=g(x)$.

Clearly
$$\hat{g}(y)=\frac{1}{2\pi}\int_{-\infty}^{+\infty}e^{-\frac{1}{2}x^2-ixy}\,dx=e^{-\frac{1}{2}y^2}\frac{1}{2\pi}\int_{-\infty}^{+\infty}e^{-\frac{1}{2}(x+iy)^2}\,dx.$$
If we show that the last integral is independent of y, that is equal to $\int_{-\infty}^{+\infty}e^{-\frac{1}{2}x^2}dx=(2\pi)^{\frac{1}{2}}$, the assertion will follow. To show this independence it is enough to apply Cauchy's theorem to the function
$$e^{-\frac{1}{2}z^2}=e^{-\frac{1}{2}(x+iy)^2}$$
and the rectangle with vertices $\pm R$, $\pm R+iy$, and then make $R\to\infty$.

We conclude this section with a few results about Fourier transforms. We consider only functions which belong either to L or to L^2, though the formulae hold under more general conditions. We occasionally write Tf for $\hat{f}$.

The first formula is
$$Tf(x+a)=e^{ixa}\,Tf(x). \tag{2·27}$$

If $f\in$ L, the formula is an immediate consequence of the equation
$$\int_{-\omega}^{\omega}f(y+a)\,e^{-ixy}\,dy=e^{ixa}\int_{-\omega+a}^{\omega+a}f(y)\,e^{-ixy}\,dy.$$
If $f\in$ L^2, it is enough to observe that, in view of (2·9), the integral $\int_{\omega}^{\omega+a}f(y)\,e^{-ixy}\,dy$ tends to 0 in L^2 as $\omega\to\infty$, and the same holds for $\int_{-\omega-a}^{-\omega}$.

Given two functions g and h in $(-\infty,+\infty)$, we define their *convolution* $g{*}h$ (compare Chapter II, § 1) by the formula
$$g{*}h=\frac{1}{\sqrt{(2\pi)}}\int_{-\infty}^{+\infty}g(y)\,h(x-y)\,dy. \tag{2·28}$$
We want to show that
$$T(g{*}h)=Tg\,.\,Th. \tag{2·29}$$

As in the proof of Theorem (1·15) of Chapter II, if $g\in$ L^s, $h\in$ L^t and
$$\frac{1}{r}=\frac{1}{s}+\frac{1}{t}-1, \tag{2·30}$$
then $f=g{*}h$ exists almost everywhere as an absolutely convergent integral, is in L^r, and satisfies
$$\left(\frac{1}{\sqrt{(2\pi)}}\int_{-\infty}^{+\infty}|f|^r\,dx\right)^{1/r}\leqslant\left(\frac{1}{\sqrt{(2\pi)}}\int_{-\infty}^{+\infty}|g|^s\,dx\right)^{1/s}\left(\frac{1}{\sqrt{(2\pi)}}\int_{-\infty}^{+\infty}|h|^t\,dx\right)^{1/t}. \tag{2·31}$$
Observe now that Tf is defined if r is either 1 or 2, and so if either both s and t are 1,

or one of them is 1 and the other 2. In the first case, (2·29) is justified by the following argument; analogous to the one we used for Fourier coefficients (Chapter II, § 1):

$$\hat{f}(x)=\frac{1}{\sqrt{(2\pi)}}\int_{-\infty}^{+\infty} e^{-ixy}\,dy\,\frac{1}{\sqrt{(2\pi)}}\int_{-\infty}^{+\infty} g(u)\,h(y-u)\,du$$

$$=\frac{1}{\sqrt{(2\pi)}}\int_{-\infty}^{+\infty} g(u)\,e^{-ixu}\,du\,\frac{1}{\sqrt{(2\pi)}}\int_{-\infty}^{+\infty} h(y-u)\,e^{-ix(y-u)}\,dy$$

$$=\hat{g}(x)\,\hat{h}(x).$$

In the second case, suppose, for example, that $s=1$, $t=2$. Thus f is in L^2. Consider a sequence $h_1, h_2, \ldots$ of step functions vanishing near $\pm\infty$ and such that $\|h-h_n\|_2\to 0$. Write $f_n=g*h_n$. By (2·31), with $r=2$, $s=1$, $t=2$, we find that

$$\|f_n-f\|_2=\|g*(h-h_n)\|_2\to 0.$$

Now

$$\hat{f}_n=\hat{g}\hat{h}_n \tag{2·32}$$

by the case already dealt with. Since

$$\|\hat{f}-\hat{f}_n\|_2=\|f-f_n\|_2\to 0,$$

and

$$\|\hat{g}\hat{h}_n-\hat{g}\hat{h}\|_2\leqslant(\max|\hat{g}|)^{\frac{1}{2}}\,\|\hat{h}-\hat{h}_n\|_2\to 0,$$

(2·32) leads to (2·29).

If g and h are in L^2, their convolution is bounded and continuous. Since bounded functions in general have no Fourier transforms, we cannot expect (2·29) to be valid in the same sense as before. If we observe, however, that

$$\frac{1}{\sqrt{(2\pi)}}\int_{-\infty}^{+\infty} g(y)\,h(x-y)\,dy=\frac{1}{\sqrt{(2\pi)}}\int_{-\infty}^{+\infty}\hat{g}(y)\,\hat{h}(y)\,e^{ixy}\,dy, \tag{2·33}$$

a formula easily deducible from (2·18) and (2·27), we may interpret this as (2·29).

The inversion formulae (2·3) and (2·4) can be represented in various forms. One of these, particularly useful for functions of a complex variable, will be given here.

Suppose that $f(x)$ is defined for $0<x<\infty$. Consider the integral

$$F(s)=\int_0^{\infty} x^{s-1}f(x)\,dx. \tag{2·34}$$

The function F is usually called the *Mellin transform* of f. The familiar formula

$$\Gamma(s)=\int_0^{\infty} e^{-x}\,x^{s-1}\,dx$$

shows that $\Gamma(s)$ is the Mellin transform of e^{-x}. Though f is a function of a real variable, it is important to consider s complex, $s=\sigma+it$.

Set $x=e^{-y}$. We can then write (2·34) in the form

$$F(\sigma+it)=\int_{-\infty}^{+\infty} f(e^{-y})\,e^{-y\sigma}\,e^{-iyt}\,dy, \tag{2·35}$$

and so, formally (compare (2·3) and (2·4)),

$$f(e^{-y})\,e^{-y\sigma}=\frac{1}{2\pi}\int_{-\infty}^{+\infty} F(\sigma+it)\,e^{iyt}\,dt=\frac{1}{2\pi}\lim_{\omega\to\infty}\int_{-\omega}^{+\omega}, \tag{2·36}$$

or
$$f(x)=\frac{1}{2\pi i}\int_{\sigma-i\infty}^{\sigma+i\infty}F(s)\,x^{-s}ds=\frac{1}{2\pi i}\lim_{\omega\to+\infty}\int_{\sigma-i\omega}^{\sigma+i\omega}. \tag{2·37}$$

The pair (2·34) and (2·37) are called *Mellin inversion formulae*. The preceding argument was formal, but since we know conditions for the validity of Fourier's inversion formulae, these conditions can easily be adjusted to the Mellin case. For example, *if $x^{\sigma-1}f(x)\in L(0,\infty)$, and f is of bounded variation near x, then*

$$\tfrac{1}{2}[f(x+0)+f(x-0)]=\frac{1}{2\pi i}\lim_{\omega\to+\infty}\int_{\sigma-i\omega}^{\sigma+i\omega}F(s)\,x^{-s}ds. \tag{2·38}$$

For the conditions on f guarantee the absolute convergence of the integral in (2·35) and the validity of (2·36), with the left-hand side replaced by half the sum of its limits from the left and right.

Since Mellin's transform is merely another version of Fourier's, derived by a change of variable, all results for Fourier transforms can be stated in the language of Mellin transforms.

We add that in general the domain of convergence of the integral (2·34) $\left(\text{considered as }\lim_{\omega\to+\infty}\lim_{\epsilon\to+0}\int_{\epsilon}^{\omega}\right)$ is a strip $\alpha<\sigma<\beta$, and $F(s)$ is regular in it. For it is enough to write the integral (2·35) in the form $\int_{-\infty}^{0}+\int_{0}^{\infty}$ and observe that $\int_{0}^{\infty}$ converges in a half-plane $\sigma>\alpha$, and $\int_{-\infty}^{0}$ in a half-plane $\sigma<\beta$.

3. Fourier transforms (*cont.*)

In the previous section we investigated the Fourier transforms

$$\hat{f}(x)=\frac{1}{\sqrt{(2\pi)}}\int_{-\infty}^{+\infty}f(y)\,e^{-ixy}\,dy \tag{3·1}$$

in the cases when either $f\in L$ or $f\in L^2$. In the former, the integral converges absolutely; in the latter, it converges in L^2. It is natural to ask if the integral (3·1) has meaning for f in other L^r.

In this section we use the notation

$$\|f\|_r=\left\{\int_{-\infty}^{+\infty}|f(x)|^r\,dx\right\}^{1/r}.$$

(3·2) THEOREM. *If $f\in L^p$, $1<p\leqslant 2$, the right-hand side of* (3·1) *converges in $L^{p'}=L^{p/(p-1)}$ to an $\hat{f}\in L^{p'}$. The function $\hat{f}$ satisfies the inequality*

$$\left(\frac{1}{\sqrt{(2\pi)}}\int_{-\infty}^{+\infty}|\hat{f}(x)|^{p'}dx\right)^{1/p'}\leqslant\left(\frac{1}{\sqrt{(2\pi)}}\int_{-\infty}^{+\infty}|f(x)|^{p}dx\right)^{1/p}, \tag{3·3}$$

the equations (2·15) *and* (2·16) *for almost all x, and we have*

$$f(x)=\frac{1}{\sqrt{(2\pi)}}\int_{-\infty}^{+\infty}\hat{f}(y)\,e^{ixy}\,dy, \tag{3·4}$$

where the right-hand side converges in L^p.

The inequality (3·3) is an extension to Fourier transforms of the theorem of Hausdorff-Young (see Chapter XII, (2·3)). It asserts that if f is in L^p, then $\hat{f}$ exists and is in $L^{p'}$, and conversely each f in L^p is the Fourier transform of some function from $L^{p'}$.

For the proof of (3·2) we observe that (3·1) defines a linear operation $\hat{f}=Tf$ if for example, f is a *simple* function (one taking only a finite number of values and vanishing near $\pm\infty$). Using the terminology of Chapter XII, § 1, we may say that T is of types $(1,\infty)$ and $(2,2)$, and that the norms of T satisfy

$$M_{1,\infty}=(2\pi)^{-\frac{1}{2}}, \quad M_{2,2}=1.$$

Hence, by the fundamental Theorem (1·11) of Chapter XII, the operation T is (uniquely) extensible to all f in L^p and is of type (p,p'), with $M_{p,p'}\leqslant(2\pi)^{\frac{1}{2}-1/p}$. The last inequality is (3·3).

Given an f, let f_ω denote (as in the preceding section) the function equal to f in $(-\omega,\omega)$ and to 0 elsewhere. If $f\in L^p$, then $f_\omega\in L$, so that $\hat{f}_\omega=Tf_\omega$ satisfies (3·1). Since, by (3·3),

$$\|\hat{f}-\hat{f}_\omega\|_{p'}\leqslant\|f-f_\omega\|_p\to 0 \quad (\omega\to\infty),$$

the right-hand side of (3·1) converges in $L^{p'}$ to an $\hat{f}\in L^{p'}$.

The proof of (2·15) for $f\in L^p$, $1<p<2$, is the same as in the case $p=2$, except that instead of Schwarz's inequality we now use Hölder's.

The proof of (2·16) is based on the following remark. Suppose that $g\in L$, and consider the Fourier repeated integral of g, that is,

$$\frac{1}{\sqrt{(2\pi)}}\int_{-\infty}^{+\infty}\hat{g}(y)\,e^{ixy}\,dy. \tag{3·5}$$

Integrating this formally over a finite interval (α,β), we obtain the integral of g over (α,β), that is,

$$\int_\alpha^\beta g(x)\,dx=\frac{1}{\sqrt{(2\pi)}}\int_{-\infty}^{+\infty}\hat{g}(y)\frac{e^{i\beta y}-e^{i\alpha y}}{iy}\,dy, \tag{3·6}$$

where the integral on the right converges. This could be proved in many ways. The shortest is to assume, as we may, that $\beta-\alpha<2\pi$, and observe that the partial integrals

$$\frac{1}{\sqrt{(2\pi)}}\int_{-\omega}^{\omega}\hat{g}(y)\,e^{ixy}\,dy \tag{3·7}$$

of (3·5) are, by (1·3), uniformly equiconvergent over (α,β) with the partial sums of the Fourier series of a function coinciding with g in $(\alpha-\epsilon,\beta+\epsilon)$; since Fourier series when integrated termwise give the value of the integral of the function, (3·6) then follows.

Hence, observing that $f_\omega\in L$, we have for $|x|<\omega$:

$$\int_0^x f(t)\,dt=\int_0^x f_\omega(t)\,dt=\frac{1}{\sqrt{(2\pi)}}\int_{-\infty}^{+\infty}\hat{f}_\omega(y)\frac{e^{ixy}-1}{iy}\,dy.$$

Since $\|\hat{f}-\hat{f}_\omega\|_{p'}\to 0$, Hölder's inequality shows that

$$\int_0^x f(t)\,dt=\frac{1}{\sqrt{(2\pi)}}\int_{-\infty}^{+\infty}\hat{f}(y)\frac{e^{ixy}-1}{iy}\,dy,$$

and (2·16) follows.

It remains to show that the right-hand side of (3·4) converges in L^p to f. This is an analogue of Theorem (6·4) of Chapter VII, which asserts that the partial sums of an $S[f]$ in L^p converge to f in L^p. The proof is also analogous, and is based on the following:

(3·8) THEOREM. *If $f \in L^r$, $r > 1$, the Hilbert transform*

$$\tilde{f}(x) = \frac{1}{\pi}\int_{-\infty}^{+\infty} \frac{f(t)}{x-t}\,dt = -\lim_{\epsilon\to 0}\frac{1}{\pi}\int_{\epsilon}^{\infty}\frac{f(x+t)-f(x-t)}{t}\,dt \tag{3·9}$$

exists almost everywhere, and

$$\|\tilde{f}\|_r \leqslant A_r \|f\|_r, \tag{3·10}$$

where A_r depends on r only.

That $\tilde{f}$ exists was pointed out in §1. To prove (3·10), write

$$g_n(x) = \frac{1}{2\pi n}\int_{-\pi n}^{\pi n} f(t)\cot\frac{x-t}{2n}\,dt \quad (|x| < \pi),$$

and consider the difference $\delta_n(x) = \tilde{f}(x) - g_n(x)$. Clearly $\delta_n(x)$ is

$$\frac{1}{2\pi n}\int_{-\pi n}^{\pi n} f(t)\left\{\frac{2n}{x-t} - \cot\frac{x-t}{2n}\right\}dt + \frac{1}{\pi}\int_{-\infty}^{-\pi n}\frac{f(t)}{x-t}\,dt + \frac{1}{\pi}\int_{\pi n}^{+\infty}\frac{f(t)}{x-t}\,dt = \alpha_n + \beta_n + \gamma_n.$$

Hölder's inequality shows that $\beta_n \to 0$, $\gamma_n \to 0$ for each x. If x is fixed and n large enough we have

$$|\alpha_n| \leqslant An^{-1}\int_{-\pi n}^{\pi n} |f|\,dt \leqslant An^{-1}\left(\int_{-\pi n}^{\pi n} |f|^r\,dt\right)^{1/r}(2\pi n)^{1/r'} = o(1),$$

where A is an absolute constant. Thus $\delta_n \to 0$.

The functions g_n and $\tilde{f}$ are defined at exactly the same points inside $(-\pi n, \pi n)$, and $g_n \to \tilde{f}$ wherever $\tilde{f}$ exists. Observe that $g_n(nx)$ is the conjugate function of $f(nx)$ in $|x| < \pi$. Using the inequality (2·5) of Chapter VII we have

$$\left(\int_{-\pi n}^{\pi n} |g_n|^r\,dx\right)^{1/r} \leqslant A_r\left(\int_{-\pi n}^{\pi n} |f|^r\,dx\right)^{1/r} \leqslant A_r\left(\int_{-\infty}^{+\infty} |f|^r\,dx\right)^{1/r},$$

whence, by Fatou's lemma (Chapter I, (11·2)),

$$\left(\int_{-w}^{w} |\tilde{f}|^r\,dx\right)^{1/r} \leqslant A_r\left(\int_{-\infty}^{+\infty} |f|^r\,dx\right)^{1/r}$$

for each $w > 0$, and making $w \to \infty$ we obtain (3·10).

Consider now the formula (see (2·19))

$$\frac{1}{\sqrt{(2\pi)}}\int_{-w}^{w} \hat{f}(y)\,e^{ixy}\,dy = \frac{1}{\pi}\int_{-\infty}^{+\infty} f(y)\frac{\sin w(x-y)}{x-y}\,dy. \tag{3·11}$$

This was established for f in L^2, but holds also for $f \in L^p$, $1 < p < 2$. (It is enough to use, in the proof, Hölder's inequality instead of Schwarz's.) If

$$f_1(y) = f(y)\cos wy, \quad f_2(y) = f(y)\sin wy,$$

then the right-hand side of (3·11) is $\tilde{f}_1(x)\sin wx - \tilde{f}_2(x)\cos wx$, and, denoting the left-hand side by $f^w(x)$, we have, by (3·10),

$$\|f^w\|_p \leqslant 2A_p\|f\|_p. \tag{3·12}$$

It remains to show that $\| f - f^w \|_p \to 0$ as $w \to \infty$. It is now enough to prove this for a dense set of functions, for example for step functions. This will follow if we verify the result for the characteristic function of an interval (a, b). Then

$$\hat{f}(y) = (2\pi)^{-\frac{1}{2}} y^{-1} i (e^{-iyb} - e^{-iya}),$$

$$f^w(x) = \frac{1}{\pi}\left[\int_0^{(x-a)w} \frac{\sin y}{y}\,dy + \int_0^{(b-x)w} \frac{\sin y}{y}\,dy\right]. \tag{3·13}$$

Since $f^w(x)$ is uniformly bounded in x and w, and tends to $f(x)$ for $x \neq a, b$ as $w \to \infty$, the relation $\| f^w - f \|_p \to 0$ will be established if we show that the integral of $|f^w|^p$ over the set $|x| > A$ is small for A large. This is immediate if we observe that, by the second mean-value theorem, the right-hand side of (3·13) is $w^{-1}O(1/x)$ for $|x|$ large. This completes the proof of (3·2).

The following result partly generalizes (3·2):

(3·14) Theorem. *Suppose that $f \in L^p$, $1 \leqslant p < 2$. Then the Fourier transform*

$$\hat{f}(x) = \lim_{\omega \to \infty} \hat{f}_\omega(x) = \lim_{\omega \to \infty} \frac{1}{\sqrt{(2\pi)}} \int_{-\omega}^{\omega} f(y)\, e^{-ixy}\, dy \tag{3·15}$$

converges almost everywhere. Moreover, the function

$$\xi(x) = \sup_{\alpha, \beta \geqslant 0} \left| \int_{-\alpha}^{\beta} f(y)\, e^{-ixy}\, dy \right| \tag{3·16}$$

is in $L^{p'}$, and

$$\| \xi \|_{p'} \leqslant A_p \| f \|_p. \tag{3·17}$$

We may suppose that $1 < p < 2$. It is enough to prove (3·17), from which the existence almost everywhere of the limit (3·15) follows easily. To see this, write $f = g + h$, where g is a step function vanishing near $\pm\infty$ and $\| h \|_p$ is small. Then $\hat{f}_\omega = \hat{g}_\omega + \hat{h}_\omega$, $\hat{g}_\omega$ tends to a limit at each point, and, by (3·17) applied to h, $\limsup \hat{h}_\omega - \liminf \hat{h}_\omega$ is small except in a set of small measure. Since the latter difference is equal to $\limsup \hat{f}_\omega - \liminf \hat{f}_\omega$, the existence of $\lim \hat{f}_\omega$ almost everywhere follows.

We may suppose that $f(x) = 0$ for $x < 0$, and that $\alpha = 0$ in (3·16). We may also suppose that $f(x) = 0$ for x large enough, $x > \gamma$ say. For then (with A_p independent of γ) the general result follows by a passage to the limit.

We fix a $\lambda > 0$ and denote by $\xi^\lambda(x)$ the function analogous to $\xi(x)$, except that β now runs through integral multiples of λ. It is clear that $\xi^\lambda(x) \to \xi(x)$ as $\lambda \to 0$.

Consider also the series

$$\sum_{k=0}^{\infty} a_k^\lambda e^{-ik\lambda x}, \quad \text{where} \quad a_k^\lambda = \int_{k\lambda}^{(k+1)\lambda} f(y)\, dy. \tag{3·18}$$

Denote the partial sums of the series by $\eta_n^\lambda(x)$, and let $\eta^\lambda(x) = \sup_n |\eta_n^\lambda(x)|$. Since the functions $\phi_n(x) = (\frac{1}{2}\lambda/\pi)^{\frac{1}{2}} e^{-in\lambda x}$ form an orthonormal system on $(-\pi/\lambda, \pi/\lambda)$, and $|\phi_n| \leqslant (\frac{1}{2}\lambda/\pi)^{\frac{1}{2}}$, Theorem (10·1) of Chapter XIII and the inequality (5·12) of Chapter XII show that

$$\left\{\int_{-\pi/\lambda}^{\pi/\lambda} \left\{\eta^\lambda(x) \left(\frac{\lambda}{2\pi}\right)^{\frac{1}{2}}\right\}^{p'} dx\right\}^{1/p'} \leqslant A_p \left(\frac{\lambda}{2\pi}\right)^{\frac{1}{2}(2-p)/p} \left(\sum_k |a_k^\lambda|^p\right)^{1/p}. \tag{3·19}$$

Now $$\left(\sum_k |a_k^\lambda|^p\right)^{1/p} \leqslant \left\{\sum_k \left(\int_{k\lambda}^{(k+1)\lambda} |f|\, dy\right)^p\right\}^{1/p} \leqslant \left\{\sum_k \int_{k\lambda}^{(k+1)\lambda} |f|^p\, dy\right\}^{1/p} \lambda^{1/p'},$$

by Hölder's inequality, and (3·19) reduces to

$$\left\{\int_{-\pi/\lambda}^{\pi/\lambda} \{\eta^\lambda(x)\}^{p'}\,dx\right\}^{1/p'} \leqslant A_p \| f \|_p, \tag{3·20}$$

with a different A_p.

Fix a $w>0$ and suppose that $\pi/\lambda > w$. By (3·20),

$$\left\{\int_{-w}^{w} \{\eta^\lambda(x)\}^{p'}\,dx\right\}^{1/p'} \leqslant A_p \| f \|_p. \tag{3·21}$$

Supposing that $n\lambda \leqslant \gamma$ and $|x| \leqslant w$, we have

$$\left|\eta^\lambda_{n-1}(x) - \int_0^{n\lambda} f(y)\, e^{-ixy}\,dy\right| \leqslant \sum_{k=0}^{n-1}\int_{k\lambda}^{(k+1)\lambda} |f(y)|\,|e^{-ik\lambda x} - e^{-ixy}|\,dy$$

$$\leqslant \lambda\,|x| \int_0^\gamma |f|\,dy \leqslant \lambda w \| f \|_p \gamma^{1/p'},$$

so that
$$\xi^\lambda(x) \leqslant \eta^\lambda(x) + \lambda w \| f \|_p \gamma^{1/p'},$$

and, by (3·21), $\left\{\int_{-w}^{w} (\xi^\lambda)^{p'}\,dx\right\}^{1/p'} \leqslant A_p \| f \|_p + (2w)^{1/p'} \lambda w \| f \|_p \gamma^{1/p'}.$

Making here first $\lambda \to 0$ and then $w \to \infty$ we obtain (3·17). This completes the proof of Theorem (3·14).

Let $2 < q < \infty$. It is not true in general that functions f in L^q have Fourier transforms. Consider the following example. Let $\alpha_1, \alpha_2, \ldots$ be a sequence of numbers such that $\Sigma |\alpha_n|^q < \infty$, $\Sigma |\alpha_n|^2 = \infty$ (take, for instance, $\alpha_n = n^{-\frac{1}{2}}$), and let $f(x)$ be a function such that $f(x) = \alpha_n$ for $2^n - \frac{1}{2} \leqslant |x| \leqslant 2^n + \frac{1}{2}$ $(n = 1, 2, \ldots)$, $f(x) = 0$ elsewhere. Then

$$\hat{f}_\omega(x) = \frac{2}{\sqrt{(2\pi)}}\frac{2\sin\frac{1}{2}x}{x}\sum_{2^n \leqslant \omega} \alpha_n \cos 2^n x + o(1),$$

where the $o(1)$ is uniform in x. Since $\Sigma\alpha_n \cos 2^n x$ is not summable by any linear method in any set of positive measure (Chapter V, (6·4)), $\hat{f}(x)$ cannot be defined as a (generalized) limit of $\hat{f}_\omega(x)$.

Nor does the formula (2·15) help here, though the integral on the right converges absolutely and represents a continuous function if $f \in L^q$. We show that, for the f just defined, this function is differentiable only in a set of measure zero. It is equal, except for a numerical factor, to

$$\sum_1^\infty \alpha_n \int_{2^n-\frac{1}{2}}^{2^n+\frac{1}{2}} \frac{\sin xy}{y}\,dy = \Sigma \alpha_n \int_{2^n-\frac{1}{2}}^{2^n+\frac{1}{2}} \left(\frac{1}{y} - \frac{1}{2^n}\right) \sin xy\,dy + \Sigma \frac{\alpha_n}{2^n} \int_{2^n-\frac{1}{2}}^{2^n+\frac{1}{2}} \sin xy\,dy = G_1(x) + G_2(x),$$

say. We easily verify that the termwise differentiated series for G_1 has terms $O(2^{-n})$, so that G_1' exists everywhere. But

$$G_2(x) = \frac{2\sin\frac{1}{2}x}{x}\Sigma\alpha_n \frac{\sin 2^n x}{2^n},$$

and since the function $\Sigma\alpha_n 2^{-n} \sin 2^n x$ has a derivative in a set of measure zero only (otherwise the series $\Sigma\alpha_n \cos 2^n x$ would be summable by the method of Lebesgue in a set of positive measure), $G_1 + G_2$ is differentiable only in a set of measure zero.

4. Fourier-Stieltjes transforms

Let $F(x)$ be of bounded variation in $(-\infty, +\infty)$. The function

$$\phi(x) = \frac{1}{\sqrt{(2\pi)}}\int_{-\infty}^{+\infty} e^{-ixy}\,dF(y) \tag{4·1}$$

is called the *Fourier-Stieltjes transform* of F, or the *Fourier transform* of dF. We shall also denote it by $\hat{dF}$. If F is the integral of an $f \in L(-\infty, +\infty)$, then $\hat{dF} = \hat{f}$.

The integral (4·1) converges absolutely and uniformly, and $\phi(x)$ is a bounded and continuous function. The example $F(x) = \operatorname{sign} x$ shows that $\phi(x)$ need not tend to 0 as $x \to \pm\infty$. This F is discontinuous, but there are continuous F with ϕ not tending to 0. For example, if $F(x)$ coincides on $(0, 2\pi)$ with the Cantor-Lebesgue function (Chapter V, § 3), and is equal to $F(0)$ for $x \leqslant 0$, and to $F(2\pi)$ for $x \geqslant 2\pi$, then $(2\pi)^{-\frac{1}{2}}\phi(n)$ is the nth Fourier-Stieltjes coefficient of $F(x)$, $0 \leqslant x \leqslant 2\pi$, and so does not tend to 0 (Chapter V, (3·6)); and *a fortiori* $\phi(x)$ does not tend to 0.

Let $\lambda_1, \lambda_2, \dots$ be all the discontinuities of F, and $c_1, c_2, \dots$ the corresponding jumps:

$$c_n = F(\lambda_n + 0) - F(\lambda_n - 0).$$

We call F a *function of jumps*, if

$$F(y) - F(-\infty) = \sum_{\lambda_n \leqslant y}{}' c_n, \tag{4·2}$$

where the dash indicates that if y coincides with a λ_m, then c_m in the sum is to be replaced by $F(\lambda_m) - F(\lambda_m - 0)$. The series in (4·2) converges absolutely, and (4·1) can be written

$$(2\pi)^{\frac{1}{2}}\phi(x) = \Sigma c_n e^{-i\lambda_n x}. \tag{4·3}$$

The case $\lambda_n = n$, $n = 0, \pm 1, \pm 2, \dots$, is of special interest, and the series (4·3) then becomes a general, absolutely convergent, trigonometric series

$$\sum_{n=-\infty}^{+\infty} c_n e^{-inx}. \tag{4·4}$$

We now show that ϕ determines F, apart from an arbitrary additive constant.

(4·5) Theorem. *If F is of bounded variation in $(-\infty, +\infty)$ and satisfies*

$$F(x) = \tfrac{1}{2}\{F(x+0) + F(x-0)\} \tag{4·6}$$

for all x, then

$$F(x) - F(0) = \frac{1}{\sqrt{(2\pi)}}\int_{-\infty}^{+\infty} \phi(\xi)\frac{e^{i\xi x} - 1}{i\xi}\,d\xi = \frac{1}{\sqrt{(2\pi)}}\lim_{\omega\to+\infty}\int_{-\omega}^{\omega}. \tag{4·7}$$

Fix ρ, write

$$\phi(\xi)\,e^{i\rho\xi} = \frac{1}{\sqrt{(2\pi)}}\int_{-\infty}^{+\infty} e^{-i\xi(x-\rho)}\,dF(x) = \frac{1}{\sqrt{(2\pi)}}\int_{-\infty}^{+\infty} e^{-i\xi x}\,dF(x+\rho),$$

and subtract (4·1) from this. Then, writing

$$G(x) = F(x+\rho) - F(x),$$

we have

$$\phi(\xi)\,(e^{i\rho\xi} - 1) = \frac{1}{\sqrt{(2\pi)}}\int_{-\infty}^{+\infty} e^{-i\xi x}\,dG(x). \tag{4·8}$$

The function G is of bounded variation and integrable over $(-\infty, +\infty)$. To verify the latter fact we may suppose that F is non-decreasing. If, for example, $\rho \geqslant 0$, so that $G \geqslant 0$, we have

$$\int_{-\omega}^{\omega} G(x)\,dx = \int_{-\omega}^{\omega}\{F(x+\rho) - F(x)\}\,dx = \left(\int_{\omega}^{\omega+\rho} - \int_{-\omega}^{-\omega+\rho}\right)F(x)\,dx \to \rho[F(+\infty) - F(-\infty)] \tag{4·9}$$

as $\omega \to \infty$, which proves the integrability of G.

Since $G(x)\to 0$ as $x\to\pm\infty$, we obtain, on integrating by parts,

$$\phi(\xi)\,(e^{i\rho\xi}-1)=\frac{1}{\sqrt{(2\pi)}}\,i\xi\int_{-\infty}^{+\infty}G(x)\,e^{-i\xi x}\,dx,$$

so that, in view of (4·6),

$$G(x)=\frac{1}{\sqrt{(2\pi)}}\int_{-\infty}^{+\infty}\phi(\xi)\frac{e^{i\rho\xi}-1}{i\xi}\,e^{i\xi x}\,d\xi=\frac{1}{\sqrt{(2\pi)}}\lim_{\omega\to+\infty}\int_{-\omega}^{\omega}\tag{4·10}$$

for each x. In particular, taking $x=0$, we deduce (4·7) with ρ for x.

The integral (4·7) converges in general only conditionally (and symmetrically). But it converges boundedly over each finite interval of x; this is a consequence of (1·3) and of the bounded convergence of Fourier series of functions of bounded variation (Chapter II, (8·1)).

From (4·7) we deduce that

$$F(x)-F(-x)=\frac{2}{\sqrt{(2\pi)}}\int_{-\infty}^{+\infty}\phi(\xi)\frac{\sin\xi x}{\xi}\,d\xi.$$

If we integrate this with respect to x and change the order of integration on the right, we get the formula

$$\int_0^x[F(y)-F(-y)]\,dy=\frac{2}{\sqrt{(2\pi)}}\int_{-\infty}^{+\infty}\phi(\xi)\frac{1-\cos\xi x}{\xi^2}\,d\xi,\tag{4·11}$$

in which the integral converges absolutely, and so is easier to deal with than that of (4·7).

The behaviour of $\phi(x)$ as $x\to\pm\infty$ is to a certain degree influenced by the discontinuities of F. The result which we are going to prove is analogous to the formula (9·7) of Chapter III. Without loss of generality we may suppose that F satisfies (4·6).

Let $h>0$. By (4·7),

$$F(x+h)-F(x-h)=\frac{2}{\sqrt{(2\pi)}}\int_{-\infty}^{+\infty}\phi(\xi)\,e^{i\xi x}\frac{\sin\xi h}{\xi}\,d\xi.\tag{4·12}$$

Hence, by the Parseval-Plancherel formula (2·9),

$$\frac{1}{h}\int_{-\infty}^{+\infty}|F(x+h)-F(x-h)|^2\,dx=4\int_{-\infty}^{+\infty}|\phi(\xi)|^2\frac{\sin^2\xi h}{\xi^2 h}\,d\xi,\tag{4·13}$$

provided either side is finite (and so also in the general case.)

The left-hand side is finite. For, using (4·9) for non-decreasing F, we find that for general F we have

$$\int_{-\infty}^{+\infty}|F(x+\rho)-F(x)|\,dx\leqslant V\,|\rho|,\tag{4·14}$$

$$\int_{-\infty}^{+\infty}|F(x+h)-F(x-h)|\,dx\leqslant 2Vh,\tag{4·15}$$

where V is the total variation of F. The last inequality and the boundedness of F imply that the left-hand side of (4·13) is finite.

Let $\lambda_1,\lambda_2,\ldots$ be all the discontinuities of F, and $d_k=F(\lambda_k+0)-F(\lambda_k-0)$ the corresponding jumps. We prove the following formula:

$$\lim_{h\to+0}\int_{-\infty}^{+\infty}|\phi(\xi)|^2\frac{\sin^2 h\xi}{h\xi^2}\,d\xi=\tfrac{1}{2}\Sigma\,|d_k|^2.\tag{4·16}$$

In view of (4·13) we have to show that

$$\frac{1}{h}\int_{-\infty}^{+\infty} |F(x+h)-F(x-h)|^2\,dx \to 2\Sigma\,|d_k|^2. \tag{4·17}$$

(*a*) This is certainly true if F is everywhere continuous. It is then uniformly continuous in $(-\infty, \infty)$, and denoting its modulus of continuity by $\omega(\delta)$ we see that the left-hand side of (4·17) does not exceed

$$\omega(2h)\,h^{-1}\int_{-\infty}^{+\infty} |F(x+h)-F(x-h)|\,dx \leqslant 2V\omega(2h) \tag{4·18}$$

(see (4·15)), and so tends to 0 with h.

(*b*) The formula (4·17) is true if F is a step function having jumps $d_1, d_2, \ldots, d_k$ at the points $\lambda_1, \lambda_2, \ldots, \lambda_k$ and continuous elsewhere. For if h is small enough,

$$F(x+h)-F(x-h)$$

in (4·17) is equal to d_j for x inside $(\lambda_j-h, \lambda_j+h)$, $j=1, 2, \ldots, k$, and is 0 outside these intervals, so that the left-hand side is $2(d_1^2+\ldots+d_k^2)$, and the formula follows.

(*c*) In the general case write $F=F_1+F_2$, where F_1 is a step function of the type just considered, formed with the first k discontinuities of F, and k is so large that all the discontinuities of F_2 are numerically less than a given ϵ. If $\omega(\delta)$ is the modulus of continuity of F_2, then $\omega(2h)<\epsilon$ if h is small enough. Hence the left-hand side of (4·17) with F_2 for F can be made arbitrarily small if k is large enough.

Denote the square root of the left-hand side of (4·17) by $\alpha_h(F)$. Since $\alpha_h(F)$ is contained between $\alpha_h(F_1)\pm\alpha_h(F_2)$, and since for k large enough and h sufficiently small $\alpha_h(F_2)$ can be made arbitrarily small and $\alpha_h(F_1)$ arbitrarily close to $(2\Sigma d_j^2)^{\frac{1}{2}}$, (4·17) and (4·16) follow.

(4·19) THEOREM. *Each of the following two conditions is both necessary and sufficient for a function $F(x)$ of bounded variation to be continuous:*

$$\int_{-\omega}^{\omega} |\phi(\xi)|^2\,d\xi = o(\omega) \tag{4·20}$$

$$\int_{-\omega}^{\omega} |\phi(\xi)|\,d\xi = o(\omega) \tag{4·21}$$

$$(\omega\to\infty).$$

This is an analogue of Theorem (9·6) of Chapter III and the observation following it, and the proof follows the same pattern.

In Chapter IV, § 4, we discussed a kind of convergence of a sequence of *non-decreasing* functions $F_1(x), F_2(x), \ldots, 0\leqslant x\leqslant 2\pi$. We say that such a sequence converges if there is a (non-decreasing) $F(x)$ such that $F_n(x)\to F(x)$ at each point of continuity of F.

Denote the Fourier coefficients of dF_n by c_ν^n. We showed that if the F_n are also uniformly bounded, then they converge in the sense just described if and only if $\lim_n c_\nu^n$ exists for each ν. Moreover, if $c_\nu=\lim_n c_\nu^n$ does exist for each ν, then the c_ν are the Fourier-Stieltjes coefficients of $F=\lim F_n$.

Consider now functions $F_1(x), F_2(x), \ldots$, non-decreasing and bounded in $(-\infty, +\infty)$, and denote the Fourier transform of dF_n by $\phi_n(x)$. The following two examples show that results valid for the finite closed interval $(0, 2\pi)$ do not extend to $(-\infty, +\infty)$.

(*a*) Let $F_n(x)=0$ for $x<n$, $F_n(x)=1$ for $x \geqslant n$. Then $F_n(x)\to 0$ for each x, and yet $\phi_n(x)=(2\pi)^{-\frac{1}{2}}e^{-inx}$ does not tend to a limit as $n\to+\infty$, except at $x\equiv 0 \pmod{2\pi}$.

(*b*) Let $F_n(x)$ be continuous, equal to 0 for $x\leqslant -n$, equal to 1 for $x\geqslant n$, and linear in $(-n, +n)$. Then $F_n(x)\to\frac{1}{2}$ for each x; but though $\phi_n(x)=(2\pi)^{-\frac{1}{2}}(\sin nx)/nx$ has a limit $(2\pi)^{-\frac{1}{2}}$ at 0, 0 elsewhere, this limit is not the Fourier-Stieltjes transform of $\lim F_n$.

The reason of this failure is that the interval $(-\infty, +\infty)$ is not compact; and the situation can be restored if we confine our attention to non-decreasing F which have fixed limits at $\pm\infty$, for example, such that

$$F(-\infty)=0, \quad F(+\infty)=1. \tag{4·22}$$

Such non-decreasing functions are called *distribution functions*. It is sometimes required that F be continuous on a definite side, but we do not impose this restriction.

The Fourier-Stieltjes transform of a distribution function F is called the *characteristic function* of F, but for historical reasons the factor $(2\pi)^{-\frac{1}{2}}$ in (4·1) is then omitted and the characteristic function is defined by

$$\phi(x)=\int_{-\infty}^{+\infty} e^{-ixy}\,dF(y), \tag{4·23}$$

a rule to which we shall adhere through the rest of the section.

Thus a characteristic function is continuous, does not exceed 1 in absolute value, and takes the value 1 at the origin. It determines F uniquely at the points of continuity of F.

(4·24) THEOREM. *Let $F_1, F_2, \ldots$ be distribution functions and $\phi_1, \phi_2, \ldots$ their characteristic functions. Then*

(i) *if F_n converges to a distribution function F (at the points of continuity of the latter), and if ϕ is the characteristic function of F, then*

$$\phi_n(x)\to\phi(x), \tag{4·25}$$

and the convergence is uniform over each finite interval.

(ii) *Conversely, suppose that the characteristic functions $\phi_n(x)$ tend to a limit at each x, and that the limit function $\phi(x)$ is continuous at the particular point $x=0$. Then the F_n converge to a distribution function F whose characteristic function is ϕ.*

Example (*b*) above shows that the continuity of ϕ at 0 is important for the validity of (ii). This condition is certainly satisfied if the convergence of $\{\phi_n\}$ is uniform near 0.

(i) Given an $\epsilon>0$, select an $\omega>0$ such that F is continuous at $\pm\omega$ and

$$F(-\omega)<\epsilon, \quad 1-F(\omega)<\epsilon.$$

Then F_n satisfies the same inequalities for n large enough. Now $\phi(x)-\phi_n(x)$ can be written

$$\int_{|y|\leqslant\omega} e^{-ixy}\,d\{F(y)-F_n(y)\}+\int_{|y|\geqslant\omega} e^{-ixy}\,dF(y)-\int_{|y|\geqslant\omega} e^{-ixy}\,dF_n(y)=P+Q+R,$$

say. Clearly $|Q|\leqslant\{1-F(\omega)\}+F(-\omega)<2\epsilon$ for all x, and similarly $|R|<2\epsilon$ for n large enough.

Integrating by parts we find that

$$P = \{[F(y) - F_n(y)]\, e^{-ixy}\}_{-\omega}^{+\omega} + ix \int_{|y| \leqslant \omega} [F(y) - F_n(y)]\, e^{-ixy}\, dy.$$

Since $F_n(\pm\omega) \to F(\pm\omega)$, the first term on the right tends to 0 uniformly in x. The second term does not exceed

$$|x| \int_{|y| \leqslant \omega} |F(y) - F_n(y)|\, dy$$

in absolute value, and so tends to 0 uniformly over each finite interval.

Collecting results, we find that $|\phi_n(x) - \phi(x)| \leqslant 5\epsilon$ for x in a fixed interval and n large enough. This proves (i).

(ii) The F_n being monotone and uniformly bounded, we can find a sequence $F_{n_k}(x)$ converging to a non-decreasing $F(x)$ (see Chapter IV, (4·6)). It is enough to show that F is a distribution function, and that the F_n tend to F at each point of continuity of F.

Apply the formula (4·11) to the F_{n_k} and recall that we are using the definition (4·23). We find

$$\int_0^h \{F_{n_k}(y) - F_{n_k}(-y)\}\, dy = \frac{1}{\pi} \int_{-\infty}^{+\infty} \phi_{n_k}(\xi) \frac{1 - \cos \xi h}{\xi^2}\, d\xi, \tag{4·26}$$

where $h > 0$. Since the ϕ_n are uniformly bounded and tend to ϕ, (4·26) gives

$$h^{-1} \int_0^h \{F(y) - F(-y)\}\, dy = \frac{1}{\pi} \int_{-\infty}^{+\infty} \phi(\xi) \frac{1 - \cos \xi h}{\xi^2 h}\, d\xi = \frac{1}{\pi} \int_{-\infty}^{+\infty} \phi\left(\frac{\xi}{h}\right) \frac{1 - \cos \xi}{\xi^2}\, d\xi. \tag{4·27}$$

Make $h \to \infty$. Since $\phi(t)$ is continuous at 0, $\phi(t/h)$ converges to 1 uniformly in every finite interval, and from the absolute integrability of the function $(1 - \cos \xi)/\xi^2$ and the formula (2·22) we deduce that the last term of (4·27) tends to 1. But the first term of (4·27) clearly tends to $F(+\infty) - F(-\infty)$. Since $0 \leqslant F(x) \leqslant 1$, this implies that $F(-\infty) = 0$, $F(+\infty) = 1$, that is, that F is a distribution function.

Suppose now that ξ is a point of continuity of F and $\{F_n(\xi)\}$ does not converge to $F(\xi)$. We can then find a sequence $\{F_{n_k}\}$ which converges everywhere to a distribution function $F^*(x)$ such that $F^*(\xi) \neq F(\xi)$, say $F^*(\xi) > F(\xi)$. In view of the continuity of F at ξ, we have $F^* > F$ in an interval $(\xi, \xi + \eta)$, which is impossible since, by (i), the characteristic function of F^* is also ϕ, and the characteristic function defines the distribution function essentially uniquely.

5. Applications to trigonometric series

Theorem (4·24) plays a fundamental role in the calculus of probability, especially in the study of the behaviour of sums of large numbers of independent random variables. In this section we make an analogous application of (4·24) to lacunary trigonometric series (Chapter V, § 6) whose behaviour in many respects resembles that of series of independent random variables.

Consider the function

$$G(x) = \frac{1}{\sqrt{(2\pi)}} \int_{-\infty}^{x} e^{-\frac{1}{2}y^2}\, dy. \tag{5·1}$$

It is continuous and increasing, and satisfies the conditions $G(-\infty) = 0$, $G(+\infty) = 1$;

thus it is a distribution function. It is called the *(standard) Gauss distribution function*. In view of example (*f*) on p. 252, the characteristic function of $G(x)$ is

$$\gamma(x) = e^{-\frac{1}{2}x^2}. \tag{5·2}$$

Let $f(x)$ be a (measurable) function defined on a set E of finite positive measure. For each y denote by $\mathscr{E}(f \leqslant y; E)$, or $\mathscr{E}(f \leqslant y)$, the subset of E where $f \leqslant y$. The ratio

$$F(y) = \frac{|\mathscr{E}(f \leqslant y)|}{|E|}$$

is obviously a distribution function; it is called the *distribution function of f on E*.

If we have a sequence of functions f_n on E, and if their distribution functions on E, F_n say, converge to a distribution function F, we say that the f_n have *asymptotically* the distribution function F. In particular, if $F = G$ is given by (5·1), we say that the f_n are asymptotically *Gauss distributed*.

Consider a lacunary trigonometric series

$$\sum_{k=1}^{\infty} (a_k \cos n_k x + b_k \sin n_k x) = \sum r_k \cos(n_k x + x_k) = \sum A_{n_k}(x), \tag{5·3}$$

where $n_{k+1}/n_k \geqslant q > 1$, and $r_k \geqslant 0$, for all k. Write

$$S_k(x) = \sum_{j=1}^{k} (a_j \cos n_j x + b_j \sin n_j x),$$

$$A_k = \left\{\tfrac{1}{2} \sum_{j=1}^{k} (a_j^2 + b_j^2)\right\}^{\frac{1}{2}}.$$

(Note that A_m and $A_m(x)$ mean different things.) We confine our attention primarily to series such that

$$\text{(i) } A_k \to \infty, \quad \text{(ii) } r_k/A_k \to 0. \tag{5·4}$$

We know that if $A_k \to \infty$, then the $S_k(x)$ are unbounded at almost all points (Chapter V, (6·10)); the theorem which follows makes this fact in some respects more precise. The second condition (5·4) is satisfied if, for example, the r_k are bounded but $\sum A_{n_k}(x)$ is not in L^2. It is not satisfied if the r_k increase too rapidly, for example if $r_k = 2^{k\alpha}$, $\alpha > 0$.

(5·5) THEOREM. *Under the hypothesis* (5·4), *the functions* $S_k(x)/A_k$ *are asymptotically Gauss distributed on* $(0, 2\pi)$. *More generally, they are asymptotically Gauss distributed on each set* $E \subset (0, 2\pi)$ *of positive measure, that is*

$$|E|^{-1} |\mathscr{E}(S_k/A_k \leqslant y; E)| \to \frac{1}{\sqrt{(2\pi)}} \int_{-\infty}^{y} e^{-\frac{1}{2}x^2} dx. \tag{5·6}$$

In view of (4·24) and (5·2), it is enough to show that the characteristic function of the left-hand side of (5·6) tends to $e^{-\frac{1}{2}x^2}$. For simplicity of notation we suppose that (5·3) is a pure cosine series, but the argument is unchanged if we replace $a_j \cos n_j x$ by $r_j \cos(n_j x + x_j)$. We first suppose that $q \geqslant 3$, in which case the proof is somewhat simpler.

Denote the left-hand side of (5·6) by $F_k(y)$ and its characteristic function by $\phi_k(x)$,

$$\phi_k(x) = \int_{-\infty}^{+\infty} e^{-ixy}\, dF_k(y).$$

From the definition of the Lebesgue integral, applied to the real and imaginary parts of the right-hand side, we find that

$$\phi_k(x) = |E|^{-1} \int_E \exp\{-ixS_k(t)/A_k\}\, dt$$

$$= |E|^{-1} \int_E \exp\left\{-ixA_k^{-1} \sum_{j=1}^{k} a_j \cos n_j t\right\} dt. \tag{5·7}$$

It is not difficult to see that (5·4) is equivalent to

$$(\max_{1 \leqslant j \leqslant k} |a_j|)/A_k \to 0. \tag{5·8}$$

Hence, using the fact that

$$\exp z = (1+z)\exp\{\tfrac{1}{2}z^2 + O(|z|^3)\} = (1+z)\exp\{\tfrac{1}{2}z^2 + o(|z|^2)\}$$

for $z \to 0$, we can write (5·7) in the form

$$|E|^{-1} \int_E e^{o(1)} \prod_{j=1}^{k} \left\{(1 - ixa_j A_k^{-1} \cos n_j t) \exp\left(-\frac{x^2 a_j^2}{2A_k^2} \cos^2 n_j t\right)\right\} dt, \tag{5·9}$$

where the term $o(1)$ in $e^{o(1)}$ tends to 0 uniformly in t as $k \to \infty$, provided $x = O(1)$, which we assume from now on.

Observe now (since $1 + u \leqslant e^u$) that

$$\left|\prod_{j=1}^{k} (1 - ixa_j A_k^{-1} \cos n_j t)\right| \leqslant \left\{\prod_{j=1}^{k} (1 + x^2 a_j^2 A_k^{-2})\right\}^{\frac{1}{2}} \leqslant e^{x^2}, \tag{5·10}$$

and that if we write

$$\sum_{j=1}^{k} \frac{a_j^2}{A_k^2} \cos^2 n_j t = 1 + \sum_{j=1}^{k} \frac{a_j^2}{2A_k^2} \cos 2n_j t = 1 + \xi_k(t),$$

then the measure of the set of points in E where $|\xi_k(t)| \geqslant \eta > 0$ is less than

$$\eta^{-2} \int_E \xi_k^2(t)\, dt \leqslant \eta^{-2} \int_0^{2\pi} \xi_k^2(t)\, dt = \tfrac{1}{4}\pi\eta^{-2} \left(\sum_{j=1}^{k} a_j^4\right) A_k^{-4}$$

$$\leqslant \tfrac{1}{2}\pi\eta^{-2} (\max_{1 \leqslant j \leqslant k} a_j^2)/A_k^2,$$

and so, by (5·8), tends to 0 as $k \to \infty$. Since $|\xi_k(t)| \leqslant 1$, it follows that (5·9) is

$$|E|^{-1} e^{-\frac{1}{2}x^2} \int_E \prod_{j=1}^{k} (1 - ixa_j A_k^{-1} \cos n_j t)\, dt + o(1), \tag{5·11}$$

where the $o(1)$ is uniform in x. Denote the last integral by I_k. It is enough to show that $I_k \to |E|$.

Write

$$\prod_{j=1}^{k} (1 - ixa_j A_k^{-1} \cos n_j t) = \alpha_0^{(k)} + \sum_{\nu \geqslant 1} \alpha_\nu^{(k)} \cos \nu t, \tag{5·12}$$

$$\epsilon_j^{(k)} = |x|\,|a_j|/A_k, \tag{5·13}$$

so that the $\alpha_\nu^{(k)}$ depend also on x. The numbers α which actually occur in the sum (that is, those such that $\alpha_\nu^{(k)} \neq 0$) correspond to the indices $\nu \geqslant 0$ of the form

$$n_{i_1} \pm n_{i_2} \pm \dots \pm n_{i_r} \quad (i_1 > i_2 > \dots > i_r). \tag{5·14}$$

If $q \geqslant 3$, such a representation, if it exists, is unique (see Chapter V, p. 209). In particular, $\alpha_0^{(k)} = 1$, so that

$$I_k = |E| + \pi \sum_{\nu \geqslant 1} h_\nu \alpha_\nu^{(k)},$$

where the h_ν are the cosine coefficients of the characteristic function of E. (Thus if $E = (0, 2\pi)$, then $I_k = |E|$ and the proof is completed.) If a positive ν is of the form (5·14), then

$$|\alpha_\nu^{(k)}| \leqslant 2^{1-r} \epsilon_{j_1}^{(k)} \epsilon_{j_2}^{(k)} \dots \epsilon_{j_r}^{(k)} \tag{5·15}$$

and, in view of (5·8), each $\alpha_\nu^{(k)}$ tends to 0 as $k \to \infty$, and $\nu = 1, 2, \dots$. Hence, if ν_0 is fixed,

$$\sum_{\nu \geqslant 1} |h_\nu \alpha_\nu^{(k)}| = \sum_{\nu=1}^{\nu_0} + \sum_{\nu_0+1}^{\infty} \leqslant o(1) + \left(\sum_{\nu_0+1}^{\infty} h_\nu^2\right)^{\frac{1}{2}} \left(\sum_{\nu_0+1}^{\infty} |\alpha_\nu^{(k)}|^2\right)^{\frac{1}{2}}$$

The first factor in the last product is arbitrarily small if ν_0 is large enough (since $\Sigma h_\nu^2 < \infty$) and the last factor is bounded, by (5·12), (5·10), and Parseval's formula. Hence $I_k \to |E|$, and the theorem is proved for $q \geqslant 3$.

The proof in the case $1 < q < 3$ is similar, except that we operate not with individual terms of $\Sigma A_{n_k}(x)$ but with blocks of them. We may suppose that $n_{k+1}/n_k > q$ for all k.

Let r be an integer such that $q^r \geqslant Q \geqslant 3$, where Q will be chosen later. Write $\Sigma A_{n_k}(x)$ in the form $\Delta_1 + \Delta_2 + \dots$, where

$$\Delta_j(t) = \sum_{i=(j-1)r+1}^{jr} a_i \cos n_i t \quad (j = 1, 2, \dots).$$

In view of (5·4), it is enough to prove (5·6) when k is an even multiple of r, $k = 2Nr$. Decompose $S_k = S_{2Nr}$ into two sums, one consisting of the Δ_j with j odd, the other of those with j even. Applying the estimate $e^z = (1+z)\exp(\frac{1}{2}z^2 + o(|z|^2))$ to the blocks Δ_j and using (5·4), we obtain for ϕ_k the expression (see (5·9) and (5·11))

$$|E|^{-1} \int_E e^{o(1)}$$

$$\times \prod_{j=1}^{N} \{(1 - ix\Delta_{2j-1} A_k^{-1}) \exp(-\tfrac{1}{2}x^2 A_k^{-2} \Delta_{2j-1}^2)(1 - ix\Delta_{2j} A_k^{-1}) \exp(-\tfrac{1}{2}x^2 A_k^{-2} \Delta_{2j}^2)\}\, dt$$

$$= o(1) + |E|^{-1} \int_E \prod_{j=1}^{N} \{(1 - ix\Delta_{2j-1} A_k^{-1})(1 - ix\Delta_{2j} A_k^{-1})\} \exp(-\tfrac{1}{2}x^2 A_k^{-2} \sum_1^{2N} \Delta_j^2)\, dt. \tag{5·16}$$

In this argument we used the fact that the product $\Pi\{\ \}$ is bounded (compare (5·10)).

Write

$$A_k^{-2} \sum_{j=1}^{2N} \Delta_j^2 = 1 + \xi_N(t).$$

If we show that

$$\int_0^{2\pi} \xi_N^2\, dt \to 0, \tag{5·17}$$

then, as before, (5·16) will reduce to

$$o(1) + |E|^{-1} e^{-\frac{1}{2}x^2} \int_E \prod_{j=1}^{N} \{(1 - ixA_k^{-1} \Delta_{2j-1})(1 - ixA_k^{-1} \Delta_{2j})\}\, dt,$$

and it will be enough to show that the last integral, I_N say, tends to $|E|$.

Take (5·17) temporarily for granted, and write

$$\prod_{j=1}^{N} (1 - ixA_k^{-1}\Delta_{2j-1}) = \alpha_0^{(N)} + \sum_1^{\infty} \alpha_\mu^{(N)} \cos \mu t, \tag{5·18}$$

$$\prod_{j=1}^{N} (1 - ixA_k^{-1}\Delta_{2j}) = \beta_0^{(N)} + \sum_1^{\infty} \beta_\nu^{(N)} \cos \nu t. \tag{5·19}$$

The μ and ν which actually occur here are of the form (5·14), where now $i_1 - i_2 > r$, $i_2 - i_3 > r$, ... Such a μ is contained between

$$n_{i_1}(1 - Q^{-1} - Q^{-2} - \ldots) \quad \text{and} \quad n_{i_1}(1 + Q^{-1} + Q^{-2} + \ldots),$$

that is, between $n_{i_1}(Q-2)/(Q-1)$ and $n_{i_1}Q/(Q-1)$. Since $Q \geq 3$, there is no cancellation of terms in the products (5·18) and (5·19), and in particular $\alpha_0^{(N)} = \beta_0^{(N)} = 1$.

The condition $n_{i+1}/n_i > q > 1$ implies that the intervals $(n_i q^{-\frac{1}{2}}, n_i q^{\frac{1}{2}})$ have no points in common. Take r so large that

$$q^{-\frac{1}{2}} < (Q-2)/(Q-1) < Q/(Q-1) < q^{\frac{1}{2}}. \tag{5·20}$$

Hence the μ and ν which actually occur in (5·18) and (5·19) are confined to the intervals $(n_i q^{-\frac{1}{2}}, n_i q^{\frac{1}{2}})$, where the n_i come respectively from blocks Δ with odd or even indices. This implies, in particular, that the series (5·18) and (5·19) do not overlap, except for the constant terms. (Hence, if $|E| = (0, 2\pi)$, then $I_N = |E|$, and the theorem is established once we prove (5·17).) Indeed, more than mere non-overlapping is true; if μ and ν actually occur in (5·18) and (5·19) respectively, then either

$$\mu/\nu \leq q^{-\frac{1}{2}} \quad \text{or} \quad \mu/\nu \geq q^{\frac{1}{2}}. \tag{5·21}$$

If h_λ are the cosine coefficients of the characteristic function of E, then

$$I_N = |E| + \tfrac{1}{2}\pi \sum_{\substack{\mu,\nu \geq 0 \\ \mu+\nu > 0}} \alpha_\mu^{(N)} \beta_\nu^{(N)} (h_{\mu+\nu} + h_{\mu-\nu}). \tag{5·22}$$

We prove, as before, that $\alpha_\mu^{(N)}$ and $\beta_\nu^{(N)}$ tend to 0 as $N \to \infty$, for each $\mu, \nu = 1, 2, \ldots$. Hence to prove $I_N \to |E|$ we need only show that

$$\sigma = \sum_{\mu+\nu \geq \rho} |\alpha_\mu^{(N)} \beta_\nu^{(N)} h_{\mu \pm \nu}| \tag{5·23}$$

can be made arbitrarily small if ρ is large enough.

Now using the observation (5·21) we easily find that each integer $n \neq 0$ can be represented in at most a limited number of ways, ω say, in the form $\mu \pm \nu$ (compare the argument in Chapter V, p. 203). By Schwarz's inequality,

$$\sigma^2 \leq \sum_{\mu+\nu \geq \rho} |\alpha_\mu^{(N)} \beta_\nu^{(N)}|^2 \sum_{\mu+\nu \geq \rho} h_{\mu \pm \nu}^2 \leq \left(\sum_0^{\infty} |\alpha_\mu^{(N)}|^2\right) \left(\sum_0^{\infty} |\beta_\nu^{(N)}|^2\right) \omega \sum_{n \geq n_0} h_n^2,$$

where n_0 is the least integer representable in the form $\mu \pm \nu$ with $\mu + \nu \geq \rho$. It is not difficult to see that n_0 is large with ρ. Hence the last factor is arbitrarily small if ρ is large enough. But the preceding two factors are bounded, in view of the boundedness of (5·18) and (5·19). Hence $I_N \to |E|$.

It remains to prove (5·17). Write $\delta_j = A_k^{-2}\Delta_j^2$. The non-constant terms of δ_j are

$$A_k^{-2}\sum_{i'<i}^{(j)} a_i a_{i'} \cos(n_i - n_{i'})t + A_k^{-2} \sum_{i'<i}^{(j)} a_i a_{i'} \cos(n_i + n_{i'})t$$

$$+ \tfrac{1}{2}A_k^{-2}\sum^{(j)} a_i^2 \cos 2n_i t, \tag{5·24}$$

where the superscript j means that the n_i come from Δ_j. The integral over $(0, 2\pi)$ of the square of the last sum is $\pi\sum^{(j)} a_i^4$. It is clear that if i and i' differ by not less than a certain $d = d(q)$, then the numbers $n_i + n_{i'}$ of the second sum are all distinct, from which we easily deduce that if we square the sum and integrate we obtain $C_q \sum^{(j)} a_i^4$ at most. The same holds for the first sum; and collecting results we deduce that

$$\int_0^{2\pi} \delta_j^2\, dt \leqslant C_q A_k^{-4} \sum^{(j)} a_i^4. \tag{5·25}$$

Similarly, if $|j - j'| \geqslant d$ then δ_j and $\delta_{j'}$ are mutually orthogonal. Hence

$$\int_0^{2\pi} \left(\sum_{j=1}^{2N} \delta_j\right)^2 dt = \sum_{|j-j'|<d} \int_0^{2\pi} \delta_j \delta_{j'}\, dt \leqslant \sum_{|j-j'|<d} \int_0^{2\pi} \tfrac{1}{2}(\delta_j^2 + \delta_{j'}^2)\, dt \leqslant d \sum_{j=1}^{2N} \int_0^{2\pi} \delta_j^2\, dt,$$

and combining this with (5·25) we get

$$\int_0^{2\pi} \xi_N^2\, dt \leqslant C_q A_k^{-4} \sum_1^k a_i^4, \tag{5·26}$$

which, as before, leads to (5·17).

This completes the proof of (5·5).

The theorem which follows indicates that if $A_k \to \infty$ then the second condition (5·4) is indispensable for the validity of (5·5).

(5·27) THEOREM. *Suppose that $A_k \to \infty$ and that the distribution functions $F_k(y)$ of the ratios S_k/A_k formed for $\Sigma A_{n_k}(x)$ tend to a distribution function F which is not constant outside a finite interval (that is, either $F(y) > 0$ for all y or $F(y) < 1$ for all y). Then* (5·4) (ii) *must hold.*

Suppose e.g. that $F(y) < 1$ for all y, and that (5·4) (ii) does not hold. There is then an $\epsilon > 0$ such that $r_k/A_k > 2^{\frac{1}{2}}\epsilon$ for infinitely many k; consider only such k. Let $E_k(y)$ denote the subset of E where $S_k/A_k \leqslant y$. From $A_{k-1}^2 + \frac{1}{2}r_k^2 = A_k^2$ we deduce that

$$A_{k-1}/A_k < (1-\epsilon^2)^{\frac{1}{2}}.$$

Since the last term in

$$\frac{S_k}{A_k} = \frac{S_{k-1}}{A_{k-1}}\frac{A_{k-1}}{A_k} + \frac{a_k \cos n_k x + b_k \sin n_k x}{A_k}$$

does not exceed $\sqrt{2}$, it follows that if $y > 0$ then at each point of $E_{k-1}(y)$ we have $S_k/A_k \leqslant y(1-\epsilon^2)^{\frac{1}{2}} + \sqrt{2}$. It follows that $E_{k-1}(y) \subset E_k(y(1-\epsilon^2)^{\frac{1}{2}} + \sqrt{2})$, so that

$$F_{k-1}(y) \leqslant F_k(y(1-\epsilon^2)^{\frac{1}{2}} + \sqrt{2}).$$

Let y be a point of continuity of F. Making $k \to \infty$ we obtain

$$F(y) \leqslant F(y(1-\epsilon^2)^{\frac{1}{2}} + \sqrt{2}). \tag{5·28}$$

But from the hypothesis that $F(y) < 1$ for all y and the fact that $F(y) \to 1$ as $y \to \infty$ it follows that there are points of continuity $y > 0$ such that

$$F(y) > F(y(1-\epsilon^2)^{\frac{1}{2}} + \sqrt{2}).$$

This contradicts (5·28) and completes the proof of (5·27).

If the series (5·3) is an $\mathrm{S}[f], f \in \mathrm{L}^2$, then it converges almost everywhere (see Chapter V, (6·3)). Let $R_k(x) = f(x) - S_{k-1}(x)$. The proof of the following result follows the same pattern as that of (5·5):

(5·29) THEOREM. *Suppose that* $\Sigma A_{n_k}(x)$ *is an* $\mathrm{S}[f], f \in \mathrm{L}^2$, *and that*

$$r_k/B_k \to 0, \quad \textit{where} \quad B_k = \left(\tfrac{1}{2} \sum_{j=k}^{\infty} r_j^2\right)^{\frac{1}{2}}. \tag{5·30}$$

Then the distribution function of $R_k(x)/B_k$ *over each set* E *of positive measure is asymptotically Gaussian.*

Theorem (5·5) and (5·29) have analogues for general linear methods of summability, but the case of main interest is that of summability A. The following result corresponds to (5·5):

(5·31) THEOREM. *Suppose that* (5·4) *is satisfied. Then on each set of positive measure the distribution function of*

$$\Sigma \,(a_k \cos n_k x + b_k \sin n_k x)\, r^{n_k} \Big/ \left\{\sum_1^{\infty} \tfrac{1}{2}(a_k^2 + b_k^2)\, r^{2n_k}\right\}^{\frac{1}{2}}$$

is asymptotically Gaussian as $r \to 1$.

Theorem (5·29) has a similar analogue.

6. Applications to trigonometric series (*cont.*)

Denote by S a series

$$\sum_1^{\infty} (a_n \cos nx + b_n \sin nx), \tag{6·1}$$

and by S_t the series

$$\sum_1^{\infty} (a_n \cos nx + b_n \sin nx)\, \psi_n(t), \tag{6·2}$$

where $\psi_1(t), \psi_2(t), \ldots$ are the Rademacher functions (Chapter I, § 3). In Chapter V, § 8, we saw that for almost all t the series S_t converges or diverges almost everywhere in $(0, 2\pi)$ according as $\Sigma(a_n^2 + b_n^2)$ is finite or not. We shall now study the distribution functions of the partial sums $S_{k,t}(x)$ of (6·2) *qua* functions of x.

We shall consider functions $\Omega(u)$, $u > 0$, which are positive, monotonically increasing to $+\infty$, and such that $u/\Omega(u)$ also monotonically increases to $+\infty$, and

$$\int_1^{\infty} u^{-2}\Omega(u)\, du < \infty. \tag{6·3}$$

For example, we may take $\Omega(u) = u^{\alpha}$, where $0 < \alpha < 1$.

We write

$$r_k = (a_k^2 + b_k^2)^{\frac{1}{2}}, \quad A_k = \left(\tfrac{1}{2} \sum_1^{k} r_j^2\right)^{\frac{1}{2}}$$

(6·4) Theorem. *Suppose that*

$$A_k \to \infty, \quad r_k^2 = O\{\Omega(A_k^2)\}, \tag{6·5}$$

where $\Omega(u)$ satisfies the conditions just stated. Then for almost all t the functions $S_{k,t}(x)/A_k$ are asymptotically Gauss distributed on $(0, 2\pi)$.

The conditions (6·5) are equivalent to

$$\max_{1 \leqslant j \leqslant k} r_j^2 = O\{\Omega(A_k^2)\}. \tag{6·6}$$

They are satisfied if, for example, $r_k = O(1)$, $A_k \to \infty$. Since (6·5) implies that $r_k = o(A_k)$, the condition (6·5) is stronger than (5·4).

Denote the distribution function of $S_{k,t}/A_k$ by $F_{k,t}(y)$, and the characteristic function of $F_{k,t}$ by $\phi_{k,t}$; we shall also use the abbreviated notation F_k, ϕ_k, and shall denote the partial sums of (6·2) by S_k; no confusion should arise from this. We suppose for simplicity that S is a cosine series, so that $r_k = |a_k|$.

We have (compare (5·7) and (5·9))

$$\phi_{k,t}(\lambda) = \frac{1}{2\pi}\int_0^{2\pi} \exp\{-i\lambda S_{k,t}(x)/A_k\}\,dx$$
$$= \frac{1}{2\pi}\int_0^{2\pi} e^{o(1)} \prod_{j=1}^{k} \left\{(1 - i\lambda a_j A_k^{-1}\psi_j(t)\cos jx)\exp\left(-\frac{\lambda^2 a_j^2}{2A_k^2}\cos^2 jx\right)\right\} dx,$$

where the $o(1)$ is uniform in $\lambda = O(1)$.

Observe now that

$$\left|\prod_{j=1}^{k}(1 - i\lambda a_j A_k^{-1}\psi_j(t)\cos jx)\right| \leqslant \prod_{j=1}^{k}(1 + \lambda^2 a_j^2 A_k^{-2})^{\frac{1}{2}} \leqslant e^{\lambda^2},$$

and that, if $$\sum_1^k a_j^2 A_k^{-2}\cos^2 jx = 1 + \sum_1^k \tfrac{1}{2} a_j^2 A_k^{-2}\cos 2jx = 1 + \xi_k(x),$$

then the measure of the set of points where $|\xi_k(x)| > \delta > 0$ does not exceed

$$\delta^{-2}\int_0^{2\pi} \xi_k^2\,dx = \tfrac{1}{4}\pi\delta^{-2}(a_1^4 + \ldots + a_k^4)\,A_k^{-4},$$

and so tends to 0 as $N \to \infty$.

Hence, with an error tending to 0 as $k \to \infty$ (uniformly in $\lambda = O(1)$), $\phi_k(\lambda)$ is equal to

$$e^{-\frac{1}{2}\lambda^2}\frac{1}{2\pi}\int_0^{2\pi}\prod_{j=1}^{k}(1 - i\lambda a_j A_k^{-1}\psi_j(t)\cos jx)\,dx,$$

and it remains to show that

$$\frac{1}{2\pi}\int_0^{2\pi}\prod_{j=1}^{k}(1 - i\epsilon_j\psi_j(t)\cos jx)\,dx \to 1 \tag{6·7}$$

for almost all t, where $\epsilon_j = \epsilon_j(k) = \lambda a_j A_k^{-1}$.

Denote the last integrand by $\Pi(x)$ and write

$$J_k(t) = \frac{1}{2\pi}\int_0^{2\pi}\Pi(x)\,dx - 1 = \frac{1}{2\pi}\int_0^{2\pi}\{\Pi(x) - 1\}\,dx.$$

Then $$|J_k(t)|^2 = \frac{1}{4\pi^2}\int_0^{2\pi}\int_0^{2\pi}\{\Pi(x)-1\}\{\overline{\Pi}(y)-1\}\,dx\,dy,$$

$$\int_0^1 |J_k(t)|^2\,dt = \frac{1}{4\pi^2}\int_0^{2\pi}\int_0^{2\pi} dx\,dy\int_0^1 \{\Pi(x)-1\}\{\overline{\Pi}(y)-1\}\,dt.$$

Since $$\int_0^1 \Pi(x)\,dt = \int_0^1 \overline{\Pi}(y)\,dt = 1$$

and $$\int_0^1 \Pi(x)\,\overline{\Pi}(y)\,dt = \int_0^1 \prod_1^k \{1+\epsilon_j^2\cos jx\cos jy + i\epsilon_j(\cos jx - \cos jy)\,\psi_j(t)\}\,dt$$

$$= \prod_1^k (1+\epsilon^2\cos jx\cos jy),$$

we deduce that

$$\int_0^1 |J_k(t)|^2\,dt = \frac{1}{4\pi^2}\left\{\int_0^{2\pi}\int_0^{2\pi}\prod_{j=1}^k (1+\epsilon_j^2\cos jx\cos jy)\,dx\,dy\right\} - 1$$

$$\leqslant \frac{1}{4\pi^2}\int_0^{2\pi}\int_0^{2\pi}\exp\left(\sum_1^k \epsilon_i^2\cos jx\cos jy\right)dx\,dy - 1.$$

If we now apply the equation $e^u = 1+u+\frac{1}{2}u^2e^{\eta u}$, $0<\eta<1$, to

$$u = \sum_1^k \epsilon_j^2 \cos jx\cos jy,$$

and observe that $$|u| \leqslant \sum_1^k \epsilon_j^2 = 2\lambda^2, \qquad \int_0^{2\pi}\int_0^{2\pi} u\,dx\,dy = 0,$$

we get $$\int_0^{2\pi} |J_k(t)|^2\,dt \leqslant \frac{e^{2\lambda^2}}{8\pi^2}\int_0^{2\pi}\int_0^{2\pi}\left(\sum_1^k \epsilon_j^2\cos jx\cos jy\right)^2 dx\,dy = \tfrac{1}{8}e^{2\lambda^2}\sum_1^k \epsilon_j^4$$

$$= \tfrac{1}{8}\lambda^4 e^{2\lambda^2} A_k^{-4}\sum_1^k a_j^4,$$

which, in view of (6·6), gives $$\int_0^1 |J_k(t)|^2\,dt = O\{\omega(A_k^2)\}, \tag{6·8}$$

where $\omega(u) = \Omega(u)/u$ decreases monotonically to 0.

We now fix a $\theta>1$ and denote by n_j the first integer satisfying $\theta^j \leqslant A_{n_j}^2 < \theta^{j+1}$. Such an integer exists for all large enough j, for otherwise for infinitely many j and suitable n we should have $\theta^j \leqslant A_n^2 < \theta^{j+1}$, $A_{n-1}^2 < \theta^{j-1}$, that is,

$$\tfrac{1}{2}a_n^2 > \theta^j - \theta^{j-1}, \quad \tfrac{1}{2}a_n^2/A_n^2 > (\theta^j - \theta^{j-1})/\theta^{j+1} = (\theta-1)/\theta^2,$$

contradicting the relation $a_n^2 = o(A_n^2)$. Thus, by (6·8),

$$\int_0^1 |J_{n_j}(t)|^2\,dt = O\{\omega(\theta^j)\}. \tag{6·9}$$

Since the condition (6·3) is equivalent to the convergence of $\Sigma\omega(n)/n$, and so also to that of $\Sigma\omega(\theta^n)$, (6·9) implies that $J_{n_j}\to 0$, that is,

$$\frac{1}{2\pi}\int_0^{2\pi} e^{-i\lambda S_{n_j}/A_{n_j}}\,dx \to e^{-\frac{1}{2}\lambda^2} \tag{6·10}$$

for almost all t, uniformly over each finite interval of λ.

Consider now any integer m such that $n_j \leqslant m < n_{j+1}$, and let

$$\Delta = \frac{1}{2\pi}\int_0^{2\pi} e^{-i\lambda S_m/A_m}\,dx - \frac{1}{2\pi}\int_0^{2\pi} e^{-i\lambda S_{n_j}/A_{n_j}}\,dx. \tag{6·11}$$

Using the estimate $|e^{iv} - e^{iv'} - i(v-v')| \leqslant \frac{1}{2}(v-v')^2$ and the fact that the integrals of S_m and S_{n_j} over $(0, 2\pi)$ are 0, we find

$$\begin{aligned} |\Delta| &\leqslant \frac{\lambda^2}{4\pi}\int_0^{2\pi}\left(\frac{S_m}{A_m} - \frac{S_{n_j}}{A_{n_j}}\right)^2 dx \\ &\leqslant \frac{\lambda^2}{2\pi}\int_0^{2\pi}\frac{(S_m - S_{n_j})^2}{A_m^2}\,dx + \frac{\lambda^2}{2\pi}\int_0^{2\pi} S_{n_j}^2\left(\frac{1}{A_m} - \frac{1}{A_{n_j}}\right)^2 dx \\ &\leqslant \lambda^2\frac{A_{n_{j+1}}^2 - A_{n_j}^2}{A_{n_j}^2} + \lambda^2\frac{(A_{n_{j+1}} - A_{n_j})^2}{A_{n_j}^2} \\ &\leqslant 2\lambda^2\frac{A_{n_{j+1}}^2 - A_{n_j}^2}{A_{n_j}^2} \leqslant 2\lambda^2\frac{\theta^{j+2} - \theta^j}{\theta^j}, \end{aligned}$$

that is,

$$|\Delta| \leqslant 2\lambda^2(\theta^2 - 1).$$

Since θ can be arbitrarily close to 1, we deduce from this, (6·11) and (6·10) that

$$\frac{1}{2\pi}\int_0^{2\pi} e^{-i\lambda S_m/A_m}\,dx \to e^{-\frac{1}{2}\lambda^2}$$

for almost all t, and (6·4) is established.

A modification of the proof shows that the functions $S_{k,t}(x)/A_k$ are, for almost all t, asymptotically Gauss distributed not only on $(0, 2\pi)$ but on each subset of $(0, 2\pi)$ of positive measure, so that (6·4) is a complete analogue of (5·5). There are corresponding analogues of (5·29) and (5·31), and their proofs do not require new ideas.

7. The Paley-Wiener theorem

We shall now characterize the Fourier transforms of functions which are in L^2 and vanish outside a finite interval; for the sake of simplicity we assume the interval to be symmetric with respect to the origin. It turns out that the problem has connexions with the theory of *integral functions of exponential type*, that is to say, functions $F(z)$ regular in the complex plane and such that

$$F(z) = O(e^{a|z|}) \quad (z \to \infty) \tag{7·1}$$

for some positive a. The lower bound σ of such a is called the *type* of F; it is necessarily non-negative. The class of integral functions of type at most σ will be denoted by E^σ. The functions F in E^σ satisfy

$$F(z) = O\{e^{(\sigma+\epsilon)|z|}\} \quad (z \to \infty)$$

for each $\epsilon > 0$.

(7·2) THEOREM OF PALEY-WIENER. *Let $\sigma > 0$. We have*

$$F(x) = \int_{-\sigma}^{\sigma} f(\xi)\, e^{i\xi x}\, d\xi \tag{7·3}$$

for some f in $\mathrm{L}^2(-\sigma, \sigma)$ if and only if $F(x)$ is in $\mathrm{L}^2(-\infty, +\infty)$ and can be extended to the complex plane as a function in E^σ.

It is immediate that if we have (7·3) where f is in L^2 (and so also in L), then

$$F(z)=\int_{-\sigma}^{\sigma} f(\xi)\, e^{i\xi z}\, d\xi \tag{7·4}$$

is an integral function of a complex variable $z=x+iy$ and

$$|F(z)| \leqslant e^{\sigma|z|}\int_{-\sigma}^{\sigma} |f(\xi)|\, d\xi,$$

so that $F(z)\in \mathrm{E}^\sigma$; clearly $F(x)\in \mathrm{L}^2(-\infty, +\infty)$.

The converse lies deeper. We have to show that if an $F(z)$ in E^σ is in L^2 on the real axis, then the function

$$f(x)=\int_{-\infty}^{+\infty} F(\xi)\, e^{-i\xi x}\, d\xi \tag{7·5}$$

is 0 for almost all x outside $(-\sigma, \sigma)$; the integral (7·5) is meant as the limit in L^2 of the partial integrals $\int_{-\omega}^{\omega}$ for $\omega\to\infty$.

Suppose that $F(z)\in \mathrm{E}^\sigma$. Consider the function

$$g(z;\, \theta)=\int_0^{\infty} F(\zeta)\, e^{-\zeta z}\, d\zeta \quad (z=x+iy), \tag{7·6}$$

where the integral is taken along the ray $\arg\zeta=-\theta$. We show that

(i) *the integral* (7·6) *converges absolutely and uniformly in each half-plane contained together with its boundary in the half-plane*

$$x\cos\theta+y\sin\theta>\sigma.$$

This latter is that half-plane bounded by the tangent to $|\zeta|=\sigma$ at $e^{i\theta}$ which does not contain the origin; call it H_θ.

It is clear that at $\zeta=\rho e^{-i\theta}$ the integrand of (7·6) is majorized by

$$O\{\exp((\sigma+\epsilon)\rho-\mathscr{R}((x+iy)\rho e^{-i\theta}))\}=O\{\exp((\sigma+\epsilon)\rho-\rho(x\cos\theta+y\sin\theta))\}$$

for each $\epsilon>0$, and so tends to 0 exponentially if $z\in H_\theta$; from this (i) follows.

(ii) *If $0<|\theta''-\theta'|<\pi$, the functions $g(z;\, \theta')$ and $g(z;\, \theta'')$ coincide in the intersection of $H_{\theta'}$ and $H_{\theta''}$; hence each function is an analytic continuation of the other.*

Suppose that $0<\theta''-\theta'<\pi$, and that $z\in H_{\theta'}H_{\theta''}$. Fix z and consider the integrand $G(\zeta)$ of (7·6) as a function of ζ alone. It is geometrically obvious that z belongs to all H_θ, $\theta'\leqslant\theta\leqslant\theta''$, and that $G(\zeta)$ tends exponentially to 0 as $\zeta\to\infty$ in the angle $(-\theta'', -\theta')$. By Cauchy's theorem, we can rotate the ray of integration within this angle without changing the value of the integral. This proves (ii).

Write $g_0(z)=g(z;\, 0)$, $g_1(z)=g(z;\, \pi)$.

(iii) *The functions g_0 and g_1 are regular in the half-planes $x>0$ and $x<0$ respectively, and are analytic continuations of each other across the segments $y>\sigma$ and $y<-\sigma$ of the imaginary axis.* (In particular, g_0 and g_1 define jointly a function regular in the complex plane cut along the segment $-\sigma\leqslant y\leqslant\sigma$ of the imaginary axis.)

If $x \geqslant \epsilon > 0$, then

$$|g_0(z)| = \left|\int_0^\infty F(\xi)\, e^{-\xi(x+iy)}\, d\xi\right| \leqslant \left(\int_0^\infty |F(\xi)|^2\, d\xi\right)^{\frac{1}{2}} \left(\int_0^\infty e^{-2\epsilon\xi}\, d\xi\right)^{\frac{1}{2}},$$

which shows that g_0 is regular for $x > 0$. Similarly, we prove the regularity of g_1 for $x < 0$.

Consider now $g(z; \frac{1}{2}\pi)$. It is regular in $H_{\frac{1}{2}\pi}$ and, by (ii), coincides with g_0 in the subset $\mathscr{R}z > \sigma$ of $H_{\frac{1}{2}\pi}$. Since g_0 is regular in the whole half-plane $x > 0$, it follows that $g(z; \frac{1}{2}\pi)$ is the analytic continuation of g_0 across the segment $y > \sigma$ of the imaginary axis. Similarly, $g(z; \frac{1}{2}\pi)$ is the analytic continuation of g_1 across that same segment, so that g_0 and g_1 are analytic continuations of each other across the segment. A similiar argument holds for the segment $y < -\sigma$, and (iii) is established.

Consider the integral

$$g_0(x+iy) = \int_0^\infty F(\xi)\, e^{-\xi x}\, e^{-i\xi y}\, d\xi \tag{7·7}$$

for $x > 0$. It is, except for the factor $(2\pi)^{-\frac{1}{2}}$, the Fourier transform of the function equal to $F(\xi)\, e^{-\xi x}$ for $\xi > 0$, and to 0 elsewhere. Since

$$\int_0^\infty |F(\xi)\,(1 - e^{-\xi x})|^2\, d\xi$$

tends to 0 as $x \to +0$, (7·7) tends in L^2 to

$$\int_0^\infty F(\xi)\, e^{-i\xi y}\, d\xi. \tag{7·8}$$

Similarly, $$g_1(x+iy) = \int_0^{-\infty} F(\xi)\, e^{-\xi(x+iy)}\, d\xi = -\int_{-\infty}^0 F(\xi)\, e^{-\xi(x+iy)}\, d\xi \tag{7·9}$$

tends in L^2, as $x \to -0$, to

$$-\int_{-\infty}^0 F(\xi)\, e^{-i\xi y}\, d\xi. \tag{7·10}$$

Hence as $x \to +0$, the difference $g_0(x+iy) - g_1(-x+iy)$ tends in L^2 to

$$\int_{-\infty}^{+\infty} F(\xi)\, e^{-i\xi y}\, d\xi. \tag{7·11}$$

But, by (iii), this difference tends pointwise to 0 if $|y| > \sigma$. Hence the function (7·11) is 0 for almost all such y. This completes the proof of (7·2).

The argument just concluded holds for $\sigma = 0$, and from (7·4) we deduce that $F \equiv 0$ is the only function from E^0 which is in L^2 on the real axis.

From (7·2) and the Riemann-Lebesgue theorem we deduce that if an $F(z) \in E^\sigma$ is in L^2 on the real axis, then $F(x) \to 0$ as $x \to \pm\infty$.

Theorem (7·2) leads to interesting representations of functions from E^σ which are in L^2, or are merely bounded, on the real axis.

We have already observed (see § 2) that if $\{\phi_n\}$ is an orthonormal system on $(-\infty, +\infty)$, then the Fourier transforms $\{\hat{\phi}_n\}$ are also an orthonormal system on $(-\infty, +\infty)$; if $\{\phi_n\}$ is complete, so is $\{\hat{\phi}_n\}$.

Consider the functions ϕ_n defined by **(7·12)**

$$\phi_n(x) = (2\pi)^{-\frac{1}{2}}\, e^{inx} \quad (|x| \leqslant \pi), \qquad \phi_n(x) = 0 \quad (|x| > \pi),$$

where $n=0, \pm 1, \pm 2, \ldots$. This system is orthonormal over $(-\infty, +\infty)$. It is not complete: the Fourier coefficients of any function vanishing in $(-\pi, \pi)$ are all 0. It is easy to see that

$$\hat{\phi}_n(x) = (-1)^n \frac{\sin \pi x}{\pi(x-n)}. \tag{7·13}$$

Given any $F \in L^2(-\infty, +\infty)$, consider its Fourier series with respect to the system (7·13):

$$F(x) \sim \Sigma a_n \hat{\phi}_n = \Sigma a_n (-1)^n \frac{\sin \pi x}{\pi(x-n)}, \tag{7·14}$$

where a_n are the Fourier coefficients of F. The system $\{\hat{\phi}_n\}$ is not complete, so that in general the series (7·14) does not represent F. We may ask which functions F are representable by their Fourier series, that is, which F satisfy

$$\left\| F - \sum_{-N}^{N} a_n \hat{\phi}_n \right\| \to 0 \tag{7·15}$$

as $N \to +\infty$. $\left(\text{We write } \|f\| \text{ for } \left(\int_{-\infty}^{+\infty} |f|^2 dx\right)^{\frac{1}{2}}.\right)$

We know (Chapter IV, § 1) that a necessary and sufficient condition for the validity of (7·15) is that for each $\epsilon > 0$ we can find a finite linear combination $\Sigma \alpha_n \hat{\phi}_n$ with constant coefficients such that

$$\| F - \Sigma \alpha_n \hat{\phi}_n \| < \epsilon. \tag{7·16}$$

Let f be the function whose Fourier transform is F. Then (7·16) is equivalent to

$$\| f - \Sigma \alpha_n \phi_n \| < \epsilon. \tag{7·17}$$

Since the ϕ_n are all 0 outside $(-\pi, \pi)$, we can have (7·17) if and only if $f \equiv 0$ outside $(-\pi, \pi)$. Hence, by (7·2), we have (7·15) if and only if F is in E^π and in $L^2(-\infty, +\infty)$.

Suppose these conditions satisfied. Observing that Σa_n^2 and $\Sigma (\sin \pi x)^2/(x-n)^2$ converge, and applying Schwarz's inequality, we find that the series (7·14) converges uniformly over $(-\infty, +\infty)$. Since F is continuous, we can replace the sign '$\sim$' in (7·14) by '$=$':

$$F(x) = \frac{\sin \pi x}{\pi} \Sigma a_n \frac{(-1)^n}{x-n}. \tag{7·18}$$

Setting here $x = n$ we find the interesting fact that

$$a_n = F(n) \quad (n = 0, \pm 1, \pm 2, \ldots).$$

The last series converges also in the complex plane, uniformly in each band $-a \leqslant \mathcal{J} z \leqslant a$. Hence it represents an integral function and we may replace x by z in (7·18). Thus we have the following theorem:

(7·19) THEOREM. *If F is in E^π and its restriction to the real axis is in L^2, we have the interpolation formula*

$$F(z) = \frac{\sin \pi z}{\pi} \sum_{-\infty}^{+\infty} (-1)^n \frac{F(n)}{z-n}. \tag{7·20}$$

From this we can easily obtain interpolation formulae for functions F in E^π which are merely bounded on the real axis. For then $G(z) = \{F(z) - F(0)\}/z$ is both in E^π and in $L^2(-\infty, +\infty)$, and applying (7·20) to G we find

$$F(z) = F(0) + \frac{\sin \pi z}{\pi} \left[F'(0) + \sum_{n=-\infty}^{+\infty}{}' (-1)^n z \frac{F(n) - F(0)}{n(z-n)} \right]. \tag{7·21}$$

If we use the development

$$\frac{\pi}{\sin \pi z} = \frac{1}{z} + \sum_{-\infty}^{+\infty}{}' \left(\frac{1}{z-n} + \frac{1}{n}\right)(-1)^n = \frac{1}{z} + z \sum_{-\infty}^{+\infty}{}' \frac{(-1)^n}{n(z-n)},$$

we may rewrite (7·21) as follows:

$$F(z) = \frac{\sin \pi z}{\pi}\left\{F'(0) + \frac{F(0)}{z} + \sum_{-\infty}^{+\infty}{}' (-1)^n F(n)\left(\frac{1}{z-n} + \frac{1}{n}\right)\right\}. \tag{7·22}$$

Both (7·20) and (7·21) presuppose that $F \in \mathrm{E}^{\pi}$. If F is in E^{σ}, $\sigma > 0$, then $F(z\pi/\sigma)$ is in E^{π}, and we may apply previous results and obtain modifications of preceding formulae. In concrete problems, however, it is easier to reverse the procedure and reduce the case E^{σ} to E^{π}.

In Chapter X, § 3, we proved a number of inequalities for the derivatives of trigonometric polynomials $T(x) = \sum_{-n}^{+n} c_k e^{ikx}$. Typical, and the most important, of these results is *Bernstein's inequality*

$$\max_x |T'(x)| \leqslant n \max_x |T(x)|, \tag{7·23}$$

where we actually have strict inequality unless T is a monomial $A \cos(nx + \alpha)$. Since $T(x)$ is the restriction to the real axis of the integral function $\sum_{-n}^{n} c_k e^{ikz}$ which belongs to E^n and is bounded on the real axis, (7·23) is a special case of the following result likewise due to Bernstein:

(7·24) THEOREM. *If F is in E^{σ}, and bounded on the real axis, and if $M = \sup |F(x)|$ then*

$$|F'(x)| \leqslant \sigma M \quad (-\infty < x < +\infty), \tag{7·25}$$

the sign of equality being possible if and only if

$$F(z) = a e^{i\sigma z} + b e^{-i\sigma z}, \tag{7·26}$$

where a and b are arbitrary constants.

We may suppose that $\sigma > 0$, for when this case is proved, (7·25) with $\sigma = 0$ follows by taking limits. (Hence for $\sigma = 0$ constants are the only admissible functions.) Furthermore, we may suppose that $\sigma = \pi$, for otherwise we take $F(z\pi/\sigma)$ instead of $F(z)$.

The termwise differentiation of (7·22) leads to a series converging uniformly for all real z. Hence, denoting by $F_1(z)$ the contents of the curly brackets in (7·22), we have

$$F'(x) = \cos \pi x F_1(x) + \frac{\sin \pi x}{\pi} \sum_{n=-\infty}^{+\infty} \frac{(-1)^{n-1} F(n)}{(x-n)^2},$$

and taking $x = \frac{1}{2}$ we find

$$F'(\tfrac{1}{2}) = \frac{4}{\pi} \sum_{-\infty}^{+\infty} \frac{(-1)^{n-1} F(n)}{(2n-1)^2}, \tag{7·27}$$

$$|F'(\tfrac{1}{2})| \leqslant \frac{4}{\pi} M \sum_{-\infty}^{+\infty} \frac{1}{(2n-1)^2} = \frac{4}{\pi} M \cdot \tfrac{1}{4}\pi^2 = M\pi. \tag{7·28}$$

Take an x_0 and consider the function $G(z)=F(x_0+z-\frac{1}{2})$, which is in E^π and satisfies $|G|\leqslant M$ on the real axis. By the result just obtained,

$$|F'(x_0)|=|G'(\tfrac{1}{2})|\leqslant M\pi,$$

and (7·25) follows.

Applying (7·27) to G and replacing x_0 by x we obtain

$$F'(x)=\frac{4}{\pi}\sum_{-\infty}^{+\infty}\frac{(-1)^n F(x+n+\frac{1}{2})}{(2n+1)^2}. \tag{7·29}$$

This is a generalization of a formula for the derivative of a trigonometric polynomial (see Chapter X, (3·11)).

Suppose now that we have equality in (7·25) for $x=x_1$. It follows from (7·29) that then

$$F(x_1+n+\tfrac{1}{2})(-1)^n=M\,e^{i\alpha}\quad (n=0,\pm 1,\pm 2,\ldots).$$

Set $F(x_1+z+\frac{1}{2})=H(z)$. Then $H(n)(-1)^n=M\,e^{i\alpha}=H(0)$. If we apply (7·22) to $H(z)$ and use the formula

$$\pi\cot\pi z=\frac{1}{z}+\sum_{-\infty}^{+\infty}{}'\left(\frac{1}{z-n}+\frac{1}{n}\right),$$

we find that

$$H(z)=\frac{\sin\pi z}{\pi}\{H'(0)+\pi H(0)\cot\pi z\}=A\cos\pi z+B\sin\pi z=A_1\,e^{i\pi z}+B_1\,e^{-i\pi z},$$

which, since $F(z)=H(z-x_1-\frac{1}{2})$, leads to (7·26) with $\sigma=\pi$.

It is immediate that for the function (7·26) we actually have equality in (7·25) for some x. For as x increases the arguments of the numbers $a\,e^{i\sigma x}$ and $b\,e^{-i\sigma x}$ vary in opposite directions, so that for some $x=x_1$ the arguments differ by 2π, and

$$|F'(x_1)|=\sigma(|a|+|b|)=\sigma\max_x F(x)$$

Concerning the cases of equality in (7·25) for F in E^σ, it is interesting to note that the equality

$$\sup_x|F'(x)|=\sigma\sup|F(x)| \tag{7·30}$$

can occur for functions other than (7·26). For example,

$$F(z)=\cos\sqrt{(1+z^2)} \tag{7·31}$$

is a function in E^1 for which $\max|F(x)|=\sup|F'(x)|=1$, so that (7·30) holds with $\sigma=1$. But for this F we have $|F'(x)|<1$ for all x.

The formula (7·29) is a source of a number of inequalities analogous to those we proved for the trigonometric polynomials. We state only one, an analogue of the formula (3·17) of Chapter X.

(7·32) THEOREM. *Suppose that F is in E^σ and is bounded on the real axis. Then for any $\omega(u)$ which is convex, non-negative, and non-decreasing we have*

$$\int_{-\infty}^{+\infty}\omega(\sigma^{-1}|F'(x)|)\,dx\leqslant\int_{-\infty}^{+\infty}\omega(|F(x)|)\,dx.$$

The case $\omega(u)=u^p$, $p\geqslant 1$, is the most interesting.

8. Riemann theory of trigonometric integrals

In this and the next section we consider general trigonometric integrals

$$\int_{-\infty}^{+\infty} e^{i\lambda x}\, d\chi(\lambda), \tag{8·1}$$

where χ is of bounded variation in each finite interval and the integral is meant as the limit, ordinary or generalized, of the symmetric partial integrals $\int_{-\omega}^{\omega}$ as $\omega \to +\infty$; the latter are meant in the Riemann-Stieltjes sense. Without loss of generality we may assume that χ has no removable discontinuities.

Since $e^{i\lambda x}$ is of bounded variation over each finite interval of λ, the partial integrals $\int_{-\omega}^{\omega}$ exist if χ is merely continuous, or even merely bounded and with no more than a denumerable set of discontinuities. We do not investigate this generalization systematically (though occasionally we have to consider it), but concentrate on the case when χ is locally of bounded variation.

Integrals

$$\int_{-\infty}^{+\infty} e^{i\lambda x}\, h(\lambda)\, d\lambda, \tag{8·2}$$

where h is integrable over each finite interval, and series

$$\sum_{-\infty}^{+\infty} c_n\, e^{inx} \tag{8·3}$$

are both special cases of (8·1), the latter when χ is a step function discontinuous at most at the points $\lambda = n$.

In Chapter IX we discussed Riemann's theory of trigonometric series, especially for those series with coefficients tending to 0. In this section we prove a number of analogous results for the integrals (8·1). In spite of certain dissimilarities (see below), the two theories run parallel and no basically new ideas are involved. Our approach will, however, be different and, as in § 1 of this chapter, our main purpose will be to show that under certain conditions the general integral (8·1) is, in each interval of length less than 2π, uniformly equiconvergent with a certain series $\Sigma\, c_n e^{inx}$ having coefficients tending to 0. This will enable us to translate results about trigonometric series into results about integrals (8·1).

While the terms of a convergent series $u_0 + u_1 + \ldots + u_n + \ldots$ must necessarily tend to 0, an integral $\int_0^{\infty} u(x)\, dx = \lim \int_0^{\omega}$ may converge without $u(x)$ tending to 0. This is the main difference in the behaviour of trigonometric series and integrals, and the only one which requires serious attention.

The theorem of Cantor-Lebesgue, which asserts that if $\Sigma c_n\, e^{inx}$ converges in a set of positive measure then $c_n \to 0$ as $n \to \pm\infty$ (Chapter IX, (1·2)), has the following analogue for trigonometric integrals:

(8·4) THEOREM. *If* (8·1) *converges in a set E of positive measure, then*

$$\lim_{\lambda \to \pm\infty} \{ \sup_{0 \leqslant h \leqslant 1} | \chi(\lambda + h) - \chi(\lambda) | \} = 0. \tag{8·5}$$

The integral (8·1) can be written

$$\int_0^\infty \{\cos\lambda x\, d\chi_1(\lambda) + \sin\lambda x\, d\chi_2(\lambda)\}, \tag{8·6}$$

where $\chi_1(\lambda) = \chi(\lambda) - \chi(-\lambda), \quad \chi_2(\lambda) = i[\chi(\lambda) + \chi(-\lambda)],$

and it is enough to show that χ_1 and χ_2 satisfy conditions analogous to (8·5) for $\lambda \to +\infty$. Without loss of generality we may suppose that χ_1 and χ_2 are real-valued, otherwise we consider the real and imaginary parts of (8·6) separately.

Let $\mathscr{E}$ be a closed subset of E, of positive measure, such that (8·6) converges uniformly on $\mathscr{E}$. Suppose first that $x=0$ is a point of density of $\mathscr{E}$, and so also belongs to $\mathscr{E}$. Substituting $-x$ for x in (8·6) and taking half the sum and half the difference of the integrals, we see that both integrals

$$\int_0^\infty \cos\lambda x\, d\chi_1(\lambda), \quad \int_0^\infty \sin\lambda x\, d\chi_2(\lambda) \tag{8·7}$$

converge uniformly on a closed set $\mathscr{E}^*$ having 0 as a point of density. ($\mathscr{E}^*$ is the intersection of $\mathscr{E}$ with its reflexion in $x=0$.) If we set $x=0$ in the first integral, we see that $\int_0^\infty d\chi_1$ converges, and the condition for χ_1 follows.

The uniform convergence of the second integral (8·7) on $\mathscr{E}^*$ implies that

$$\int_u^{u+h} \sin\lambda x\, d\chi_2(\lambda) \tag{8·8}$$

tends to 0 as $u \to \pm\infty$, uniformly in $x \in \mathscr{E}^*$ and $0 \leqslant h \leqslant 1$. Since 0 is a point of density for $\mathscr{E}^*$, each interval $(\frac{1}{6}\pi u^{-1}, \frac{1}{3}\pi u^{-1})$ contains a point $x = x_u$ of $\mathscr{E}^*$, provided u is large enough. It follows that

$$\int_u^{u+h} \sin\lambda x_u\, d\chi_2(\lambda) \tag{8·9}$$

tends to 0, uniformly in $0 \leqslant h \leqslant 1$. Observing that $\lambda x_u = u x_u + (\lambda - u) x_u$ is in the interval $(\frac{1}{6}\pi, \frac{1}{2}\pi)$ for u large enough and $u \leqslant \lambda \leqslant u+1$, we see that the factor $\sin\lambda x_u$ in (8·9) is monotonically increasing. By the second mean-value theorem,

$$\chi_2(u+h) - \chi_2(u) = \int_u^{u+h} \frac{1}{\sin\lambda x_u} \sin\lambda x_u\, d\chi_2(\lambda) = \frac{1}{\sin u x_u}\int_u^{u+h'} \sin\lambda x_u\, d\chi_2(\lambda),$$

where $0 < h' < h$; and since the last integral, being of type (8·9), tends to 0, we find that χ_2 satisfies a condition analogous to (8·5).

Consider now the general case when 0 is not necessarily a point of density of $\mathscr{E}$. Let x_0 be a point of density of $\mathscr{E}$. Making the substitution $x = x' + x_0$, we write (8·6) in the form

$$\int_0^\infty \{\cos\lambda x'\, d\chi_1^*(\lambda) + \sin\lambda x'\, d\chi_2^*(\lambda)\}, \tag{8·10}$$

where

$$\chi_1^*(\lambda) = \int_0^\lambda \{\cos\mu x_0\, d\chi_1(\mu) + \sin\mu x_0\, d\chi_2(\mu)\},$$

$$\chi_2^*(\lambda) = \int_0^\lambda \{-\sin\mu x_0\, d\chi_1(\mu) + \cos\mu x_0\, d\chi_2(\mu)\}.$$

Since (8·10) converges uniformly in a closed set having $x'=0$ as a point of density, the expressions

$$\chi_1^*(u+h)-\chi_1^*(u)=\int_u^{u+h}\{\cos\mu x_0\,d\chi_1(\mu)+\sin\mu x_0\,d\chi_2(\mu)\}, \tag{8·11}$$

$$\chi_2^*(u+h)-\chi_2^*(u)=\int_u^{u+h}\{-\sin\mu x_0\,d\chi_1(\mu)+\cos\mu x_0\,d\chi_2(\mu)\} \tag{8·12}$$

tend to 0 as $u\to+\infty$, uniformly in $0\leqslant h\leqslant 1$. The same holds if we multiply the integrand of (8·11) by $\cos\mu x_0$ and that of (8·12) by $-\sin\mu x_0$, since both multipliers have only a finite number of maxima and minima in $u\leqslant\mu\leqslant u+1$, and it is enough to apply the second mean-value theorem to each interval in which the multiplier is monotone. Adding the two resulting integrals we obtain

$$\int_u^{u+h}d\chi_1(\mu)=\chi_1(u+h)-\chi_1(u),$$

which proves the condition for χ_1. It we multiply the integrands in (8·11) and (8·12) by $\sin\mu x_0$ and $\cos\mu x_0$ respectively and add, we obtain the result for χ_2. This completes the proof of (8·4).

For the integrals (8·2) the condition (8·5) is equivalent to

$$\lim_{u\to\pm\infty}\left\{\max_{0\leqslant h\leqslant 1}\left|\int_u^{u+h}h(\lambda)\,d\lambda\right|\right\}=0,$$

and for the series (8·3) to the condition $c_n\to 0$ (the conclusion of the Cantor-Lebesgue theorem). We call the condition (8·5) *condition* N_0, and consider almost exclusively integrals (8·1) satisfying this condition.

From (8·11), (8·12) and the second mean-value theorem we deduce that condition N_0 remains invariant under translation $x\to x+x_0$ of the variable; and the condition is satisfied uniformly in x_0 if $x_0=O(1)$.

It is also easily seen that if the integral

$$\int_0^\infty e^{i\lambda x}\,d\chi(\lambda)$$

(an analogue of a power series) converges at a single point x_0, then χ satisfies condition N_0.

We immediately see that condition N_0 implies

$$\chi(u)=o(|\,u\,|)$$

as $u\to\pm\infty$; more generally

$$\int_0^u e^{i\lambda x_0}\,d\chi\,(\lambda)=o(|\,u\,|)$$

uniformly in $x_0=O(1)$.

In Chapter IX we associated with a trigonometric series $\Sigma c_n e^{inx}$ having coefficients tending to 0 a continuous function

$$F(x)=\tfrac{1}{2}c_0x^2-\sum_{-\infty}^{+\infty}{}' c_n n^{-2}e^{inx},$$

obtained by integrating the series twice. Integrating (8·1) formally twice we introduce in the integrand the factor $-\lambda^{-2}$, which is unbounded near $\lambda=0$ and may result in the divergence of the new integral. To avoid this difficulty we define the function F for (8·1) somewhat differently and write

$$F(x) = -\int_{|\lambda|\leqslant 1} \frac{e^{i\lambda x}-1-i\lambda x}{\lambda^2}\,d\chi(\lambda) - \int_{|\lambda|\geqslant 1} \frac{e^{i\lambda x}}{\lambda^2}\,d\chi(\lambda). \tag{8·13}$$

Formal differentiation of the right-hand side twice still gives (8·1), and the integral converges uniformly over each finite interval of x, provided χ satisfies condition N_0. To see this we write

$$\chi^*(\lambda) = \int_0^\lambda e^{i\mu x}\,d\chi(\mu) = \chi_x(\lambda),$$

integrate by parts and use the fact that $\chi^*(\lambda) = o(|\lambda|)$ uniformly in x over each finite interval.

The two theorems of Riemann about the function F (see Chapter IX, (2·4), (2·8)) have the following analogues:

(8·14) THEOREM. *If χ satisfies condition N_0 and (8·1) converges at a point x_0 to value s, then*

$$D^2F(x_0) = \lim_{h\to 0} \frac{F(x_0+2h)+F(x_0-2h)-2F(x_0)}{4h^2} \tag{8·15}$$

exists and is equal to s.

(8·16) THEOREM. *If χ satisfies condition N_0, then F is smooth, that is,*

$$\frac{F(x+2h)+F(x-2h)-2F(x)}{4h} \to 0, \tag{8·17}$$

as $h\to 0$, for each x.

To prove (8·14) write

$$\chi^*(\lambda) = \int_{-\lambda}^{\lambda} e^{ix_0\mu}\,d\chi(\mu). \tag{8·18}$$

We may suppose that χ is continuous at 0. (Otherwise we subtract from χ a fixed multiple of the function sign λ, for which (8·1) is identically 2, $F(x)$ is x^2, and (8·15) is obvious.) The ratio in (8·15) is then equal to

$$\int_{-\infty}^{+\infty} e^{ix_0\lambda}\left(\frac{\sin\lambda h}{\lambda h}\right)^2 d\chi(\lambda) = \int_0^\infty \left(\frac{\sin\lambda h}{\lambda h}\right)^2 d\chi^*(\lambda) = -\int_0^\infty \chi^*(\lambda)\frac{d}{d\lambda}\left(\frac{\sin\lambda h}{\lambda h}\right)^2 d\lambda,$$

since the integrated terms in the integration by parts vanish. Substituting

$$\chi^*(\lambda) = s+\epsilon(\lambda), \quad \text{where} \quad \epsilon(\lambda)\to 0,$$

we split the last integral into two, the first of which is s; then we have only to show that the second tends to 0 with h. This is immediate if we observe that the total variation of $(\sin\lambda h)^2/\lambda^2h^2$ over each finite interval tends to 0 with h, that it is bounded (constant) over $(0, \infty)$ and that $\epsilon(\lambda)$ is arbitrarily small outside a sufficiently large interval.

To prove (8·16) we use the notation (8·18) and write the ratio (8·17) in the form

$$\int_0^\infty \frac{\sin^2\lambda h}{\lambda^2 h}\,d\chi^*(\lambda) = \sum_{n=0}^\infty \int_{n\pi/h}^{(n+\frac{1}{2})\pi/h} + \sum_{n=0}^\infty \int_{(n+\frac{1}{2})\pi/h}^{(n+1)\pi/h} = P+Q,$$

say. The function $\sin^2\lambda h$ is increasing in each integral in P and decreasing in each integral in Q. Since χ^* satisfies condition N_0, the oscillation of χ^* over each interval occurring either in P or Q is $o(1/h)$ as $h\to 0$. Hence, applying the second mean-value theorem to the factors $\sin^2 \lambda h$ and λ^{-2} in each integral of P except the first, in which we apply it instead to the (decreasing) ratio $(\sin \lambda h)^2/\lambda^2$, we find that

$$P = h^2 . h^{-1} . o(h^{-1}) + \sum_{n=1}^{\infty} \frac{1^2}{(n\pi/h)^2} h^{-1} o(h^{-1}) = o(1).$$

The relation $Q = o(1)$ is proved similarly, and (8·17) is established. It holds uniformly over any finite interval of x.

The integral

$$-i \int_{-\infty}^{+\infty} (\operatorname{sign} \lambda)\, e^{i\lambda x}\, d\chi(\lambda) \tag{8·19}$$

may be called *conjugate* to (8·1). Since sign λ is discontinuous at $\lambda = 0$, and the integrals are taken in the Riemann-Stieltjes sense, we may suppose for the sake of simplicity that χ is continuous at 0.

In the rest of this section we consider the 'formal multiplication' of trigonometric integrals. It is analogous to the formal multiplication of trigonometric series discussed in Chapter IX, though the details are somewhat less simple. We restrict ourselves to results which will be useful in the proofs of the main theorems of the next section.

Given two integrals

$$\int_{-\infty}^{+\infty} e^{i\lambda x}\, d\phi(\lambda), \tag{8·20}$$

$$\int_{-\infty}^{+\infty} e^{i\lambda x}\, d\psi(\lambda), \tag{8·21}$$

we define their *formal product* as

$$\int_{-\infty}^{+\infty} e^{i\lambda x}\, d\chi(\lambda), \tag{8·22}$$

where

$$\chi(\lambda) = \int_{-\infty}^{+\infty} \{\phi(\lambda - \mu) - \phi(\lambda_0 - \mu)\}\, d\psi(\mu), \tag{8·23}$$

λ_0 being any fixed number. If the integral (8·23) exists for some λ_0 (and all λ), then it exists for any other λ_0, and the two χ's differ by an additive constant. We assume once for all that ϕ and ψ (but not necessarily χ) are of bounded variation over each finite interval.

The integral (8·23) converges, even absolutely, if ϕ satisfies condition N_0 and ψ is of bounded variation over $(-\infty, +\infty)$; moreover, χ satisfies condition N_0. For if $0 \leqslant h \leqslant 1$, then

$$|\chi(\lambda + h) - \chi(\lambda)| \leqslant \int_{-\infty}^{+\infty} |\phi(\lambda + h - \mu) - \phi(\lambda - \mu)|\, |d\psi(\mu)| = \int_{-M}^{M} + \int_R,$$

where R is the complement of $(-M, M)$. Since the last integral is arbitrarily small if M is large enough, and the preceding integral tends to 0 as $\lambda \to \pm\infty$ and M remains fixed, the assertion easily follows. The hypotheses do not guarantee, however, that χ is of bounded variation in any interval.

Under the same hypotheses we may interchange the roles of ϕ and ψ in (8·23). For, integrating by parts and making the substitution $\lambda - \mu = \nu$, we find for the partial integral of (8·23) over $A \leqslant \mu \leqslant B$ the value

$$\{\phi(\lambda - B) - \phi(\lambda_0 - B)\}\, \psi(B) - \{\phi(\lambda - A) - \phi(\lambda_0 - A)\}\, \psi(A)$$
$$+ \int_{\lambda - B}^{\lambda - A} \psi(\lambda - \nu)\, d\phi(\nu) - \int_{\lambda_0 - B}^{\lambda_0 - A} \psi(\lambda_0 - \nu)\, d\phi(\nu). \tag{8·24}$$

The integrated terms tend to 0 as $A \to -\infty$, $B \to +\infty$, since the expressions in curly brackets tend to 0 and ψ is bounded. Supposing ϕ and ψ real-valued, and decomposing ψ into a difference of two non-decreasing positive and bounded functions, we find that if we omit λ, λ_0 in the limits of the last two integrals we commit errors $o(1)$. Hence the difference between (8·24) and

$$\int_{-B}^{-A} \{\psi(\lambda-\nu)-\psi(\lambda_0-\nu)\}\, d\phi(\nu)$$

tends to 0, and the symmetry of χ with respect to ϕ and ψ is proved.

If ϕ satisfies condition N_0 and $\int_{-\infty}^{+\infty} |\mu|\,|d\psi(\mu)| < \infty$, we may take for χ the integral

$$\int_{-\infty}^{+\infty} \phi(\lambda-\mu)\, d\psi(\mu), \tag{8·25}$$

which converges absolutely since $\phi(u)=o(|u|)$ for large $|u|$, and which differs from (8·23) by an additive constant.

If the two functions in (8·23) have a common discontinuity, the integral does not exist in the Riemann-Stieltjes sense. We may then either use the Lebesgue-Stieltjes definition or assume that one of the functions ϕ, ψ is continuous. The latter course is sufficient for most purposes since in our main applications (8·21) will be the Fourier integral of a function $L(x)$ vanishing outside a finite interval and having any prescribed number of continuous derivatives.† In this case (8·21) is

$$\int_{-\infty}^{+\infty} e^{i\lambda x} g(\lambda)\, d\lambda, \tag{8·26}$$

where
$$g(\lambda)=\frac{1}{2\pi}\int_{-\infty}^{+\infty} e^{-i\lambda x}\, L(x)\, dx. \tag{8·27}$$

Such a g is even analytic, and integration by parts shows that if L has k continuous derivatives then each derivative of g is $O(|\lambda|^{-k})$ at infinity.

It is useful to have a simple sufficient condition which guarantees that χ is of bounded variation over each finite interval. Suppose that ϕ satisfies condition N_0, and denote the indefinite integral of ϕ by Φ. If $d\psi(\lambda)=g(\lambda)\,d\lambda$, and if $\lambda g'(\lambda) \in L(-\infty, +\infty)$, then, as we easily see, $g(\lambda)=o(1/\lambda)$ and, integrating by parts, we can write (8·23) in the form

$$\int_{-\infty}^{+\infty} [\Phi(\lambda-\mu)-\Phi(\lambda_0-\mu)]\, g'(\mu)\, d\mu.$$

This function is in Λ_1 (and in particular is of bounded variation) over each finite interval, since the formally differentiated integral is a locally bounded function.

It is also easy to verify that, if ϕ satisfies a condition stronger than N_0, namely

$$\int_{\lambda}^{\lambda+1} |d\phi(\mu)| = o(1) \quad (\lambda \to +\infty),$$

and if ψ is of bounded variation over $(-\infty, +\infty)$, then the χ in (8·23) is of bounded variation over each finite interval (and the result holds if $o(1)$ is replaced by $O(1)$).

† Using the Lebesgue-Stieltjes definition, it is not difficult to see that if ϕ satisfies condition N_0 and ψ is of bounded variation over $(-\infty, +\infty)$, then the χ in (8·23) is bounded over each finite interval and is continuous except possibly at the points $\xi+\eta$, where ξ and η are discontinuities of ϕ and ψ respectively.

The following result is an analogue of Theorem (4·20) of Chapter IX:

(8·28) THEOREM. *Suppose that ϕ satisfies condition N_0, that* $\int_{-\infty}^{+\infty}|\lambda||d\psi(\lambda)|<\infty$, *and that* (8·22) *is the formal product of* (8·20) *and* (8·21). *Then, if $L(x)$ is the value of* (8·21), *the two differences*

$$\Delta_\omega(x)=\int_{-\omega}^{\omega}e^{ix\lambda}\,d\chi(\lambda)-L(x)\int_{-\omega}^{\omega}e^{ix\lambda}\,d\phi(\lambda), \tag{8·29}$$

$$\tilde{\Delta}_\omega(x)=\int_{-\omega}^{\omega}(-i\operatorname{sign}\lambda)\,e^{ix\lambda}\,d\chi(\lambda)-L(x)\int_{-\omega}^{\omega}(-i\operatorname{sign}\lambda)\,e^{ix\lambda}\,d\phi(\lambda) \tag{8·30}$$

converge uniformly over each finite interval as $\omega\to+\infty$, the former to limit 0.

We consider only (8·29), the proof for (8·30) being similar, and take first, for simplicity, $x=0$. We may define χ by (8·25). Then

$$\begin{aligned}\Delta_\omega(0)&=\chi(\omega)-\chi(-\omega)-L(0)\{\phi(\omega)-\phi(-\omega)\}\\&=\int_{-\infty}^{+\infty}\{\phi(\omega-\mu)-\phi(\omega)\}\,d\psi(\mu)-\int_{-\infty}^{+\infty}\{\phi(-\omega-\mu)-\phi(-\omega)\}\,d\psi(\mu),\end{aligned} \tag{8·31}$$

and it is enough to show that the last two integrals tend to 0. This, in turn, will follow if we show that given an $\epsilon>0$ we have

$$|\phi(\omega-\mu)-\phi(\omega)|<\epsilon|\mu|,\quad |\phi(-\omega-\mu)-\phi(-\omega)|<\epsilon|\mu| \tag{8·32}$$

for all μ, provided ω is large enough. It is enough to consider the first inequality.

We observe that if μ' and μ'' tend both to $+\infty$ or to $-\infty$, then $\phi(\mu'')-\phi(\mu')$ is $o(|\mu''-\mu'|+1)$. It follows that

$$\phi(\omega-\mu)-\phi(\omega)=o(|\mu|) \tag{8·33}$$

if $\omega\to\infty$, $\mu\leqslant\frac{1}{2}\omega$, for then both ω and $\omega-\mu$ tend to $+\infty$. We also have (8·33) if $\omega\to\infty$, $\mu\geqslant 2\omega$, for then the left-hand side is $o(\mu-\omega)+o(\omega)=o(\mu)$. Finally, if $\frac{1}{2}\omega\leqslant\mu\leqslant 2\omega$, the left-hand side is $o(\omega)=o(\mu)$. Hence we have (8·33) for $\omega\to\infty$, uniformly in μ, and the relation $\Delta_\omega(0)\to 0$ follows.

Passing to general x, write

$$\int_0^u e^{ixt}\,d\phi(t)=\phi_x(u),$$

and define $\psi_x(u)$, $\chi_x(u)$ correspondingly. The first term on the right of (8·29) is $\chi_x(\omega)-\chi_x(-\omega)$, and is equal to

$$\begin{aligned}&\chi(\omega)\,e^{ix\omega}-\chi(-\omega)\,e^{-ix\omega}-ix\int_{-\omega}^{\omega}e^{ix\lambda}\,\chi(\lambda)\,d\lambda\\&=\int_{-\infty}^{+\infty}e^{ix(\omega-\mu)}\,\phi(\omega-\mu)\,e^{ix\mu}\,d\psi(\mu)-\int_{-\infty}^{+\infty}e^{ix(-\omega-\mu)}\,\phi(-\omega-\mu)\,e^{ix\mu}\,d\psi(\mu)\\&\qquad-ix\int_{-\omega}^{\omega}d\lambda\int_{-\infty}^{+\infty}e^{ix(\lambda-\mu)}\,\phi(\lambda-\mu)\,e^{ix\mu}\,d\psi(\mu).\end{aligned}$$

The inner integral in the repeated integral converges absolutely. If we interchange the order of integration and combine the three integrals in one we obtain

$$\int_{-\infty}^{\infty}\{\phi_x(\omega-\mu)-\phi_x(-\omega-\mu)\}\,d\psi_x(\mu).$$

Since the last term in (8·29) is $\{\phi_x(\omega)-\phi_x(-\omega)\}\int_{-\infty}^{+\infty} d\psi_x(\mu)$, we easily obtain the formula

$$\Delta_\omega(x)=\int_{-\infty}^{+\infty}\{\phi_x(\omega-\mu)-\phi_x(\omega)\}\,d\psi_x(\mu)-\int_{-\infty}^{+\infty}\{\phi_x(-\omega-\mu)-\phi_x(-\omega)\}\,d\psi_x(\mu), \quad \textbf{(8·34)}$$

analogous to (8·31). Since $\phi_x(\lambda)$ satisfies condition N_0 uniformly in each finite interval of x, and $|d\psi_x|=|d\psi|$, the previous argument shows that $\Delta_\omega(x)$ tends to 0 uniformly over each finite interval. This completes the proof of (8·28).

The lemma which follows will be used in the next section. It is obviously a special case of a more general result which, however, is not needed.

(8·35) LEMMA. *If ϕ is bounded and*

$$\sup_{0\leqslant h\leqslant 1}|\phi(u+h)-\phi(u)|=o(u^{-2}) \quad (u\to\pm\infty), \quad \textbf{(8·36)}$$

$$\int_{-\infty}^{+\infty} u^2\,|d\psi(u)|<\infty, \quad \psi(\pm\infty)=0, \quad \textbf{(8·37)}$$

then

$$\chi(u)=\int_{-\infty}^{+\infty}\phi(u-v)\,d\psi(v)=o(u^{-2}) \quad \textbf{(8·38)}$$

as $u\to\pm\infty$.

We first show that, if $\epsilon_p=o(p^{-2})$, and $\eta(u)$ is bounded and $o(u^{-2})$, then

$$\sum_{p=-\infty}^{+\infty}\epsilon_p\,\eta(u-p)=o(u^{-2}). \quad \textbf{(8·39)}$$

We suppose that $u\to+\infty$, and split the sum into two parts, A and B, extended over $|p|\leqslant\frac{1}{2}u$ and $|p|>\frac{1}{2}u$ respectively. Then

$$|A|\leqslant\sup_{v\geqslant\frac{1}{2}u}|\eta(v)|\sum_{|p|\leqslant\frac{1}{2}u}|\epsilon_p|=o(u^{-2}).O(1)=o(u^{-2}),$$

$$|B|\leqslant\sup_{|p|>\frac{1}{2}u}|\epsilon_p|\sum_{|p|>\frac{1}{2}u}|\eta(u-p)|=o(u^{-2}).O(1)=o(u^{-2}),$$

and (8·39) follows.

Return to (8·38), and again suppose that $u\to+\infty$. We have

$$\chi(u)=-\int_{-\infty}^{+\infty}\psi(v)\,d_v\phi(u-v), \quad \textbf{(8·40)}$$

the integrated term being 0 since $\phi=O(1)$, $\psi(\pm\infty)=0$. Split the integral into two, extended over $v\geqslant 0$ and $v\leqslant 0$ respectively; it is enough to show that the first of them is $o(u^{-2})$.

We may suppose that ϕ and ψ are real-valued, and also that $\psi(u)$ is monotonically decreasing to 0 in $(0,\infty)$; for otherwise we decompose ψ into a difference of two such functions, the positive and negative variations of ψ, which also satisfy (8·37). Then if, for instance, $p>0$, (8·37) implies

$$\psi(p)=-\int_p^\infty d\psi(u)\leqslant\frac{1}{p^2}\int_p^\infty u^2|\,d\psi(u)|=o(p^{-2}).$$

Denote the upper bound of the left-hand side of (8·36) for $w\leqslant u\leqslant w+1$ by $\alpha(w)$;

clearly $\alpha(w)=o(w^{-2})$. By the second mean-value theorem, denoting by θ_p numbers between 0 and 1, we have

$$\left|\int_0^\infty \psi(v)\,d_v\phi(u-v)\right| = \left|\sum_{p=1}^\infty \psi(p-1)\int_{p-1}^{p-\theta_p} d_v\phi(u-v)\right| \leqslant \sum_{p=1}^\infty \psi(p-1)\,\alpha(u-p).$$

The last sum is $o(u^{-2})$ by the previous remark about $\{\epsilon_p\}$ and $\eta(u)$.

9. Equiconvergence theorems

In Chapter IX, § 4, we proved that if $\Sigma c_n e^{inx}$ is any series with coefficients tending to 0, $F(x)$ the function obtained by integrating the series termwise twice, and $L(x)$ a periodic sufficiently differentiable function equal to 1 in an interval J, $a \leqslant x \leqslant b$, and to 0 outside an interval J', $a' \leqslant x \leqslant b'$, where $a' < a < b < b'$, then the differences

$$\sum_{|n|\leqslant N} c_n e^{inx} - \frac{1}{\pi}\int_{J'} F(t)\,L(t)\,D_N''(x-t)\,dt, \tag{9·1}$$

$$\sum_{|n|\leqslant N} (-i\,\mathrm{sign}\,n)\,c_n e^{inx} - \frac{1}{\pi}\int_{J'} F(t)\,L(t)\,\tilde{D}_N''(x-t)\,dt \tag{9·2}$$

converge uniformly in J', the limit of the first being 0.

We need a corresponding result for integrals

$$\int_{-\infty}^{+\infty} e^{i\lambda x}\,d\chi(\lambda). \tag{9·3}$$

We denote by F the function (8·13).

(9·4) THEOREM. *If χ satisfies condition N_0, and $L(x)$ is differentiable five times, is equal to 1 in the interval J, $a \leqslant x \leqslant b$, and to 0 outside the interval J', $a' \leqslant x \leqslant b'$, then the differences*

$$\int_{-\omega}^{\omega} e^{i\lambda x}\,d\chi(\lambda) - \frac{1}{\pi}\int_{J'} F(t)\,L(t)\frac{d^2}{dt^2}\frac{\sin\omega(x-t)}{x-t}\,dt, \tag{9·5}$$

$$\int_{-\omega}^{\omega} (-i\,\mathrm{sign}\,\lambda)\,e^{i\lambda x}\,d\chi(\lambda) - \frac{1}{\pi}\int_{J'} F(t)\,L(t)\frac{d^2}{dt^2}\frac{1-\cos\omega(x-t)}{x-t}\,dt \tag{9·6}$$

converge uniformly in J, the limit of (9·5) *being* 0.

We consider only (9·5). We may suppose that χ is constant in $(-1, +1)$. For if (9·3) is of the form $\int_{-1}^{+1} e^{i\lambda x}\,d\chi(\lambda)$, then the corresponding F is an entire function, and the second integral in (9·5) is, after integration by parts,

$$\pi^{-1}\int_{J'} (FL)''\frac{\sin\omega(x-t)}{x-t}\,dt,$$

and so converges uniformly to $\int_{-1}^{+1} e^{i\lambda x}\,d\chi(\lambda)$ in J.

Under our hypothesis, therefore,

$$F(x) = \int_{-\infty}^{+\infty} (-e^{i\lambda x})\,\lambda^{-2}\,d\chi(\lambda). \tag{9·7}$$

It is natural to expect that if I_3 is the formal product of the trigonometric integrals I_1 and I_2, then

$$I_3'' = I_1''I_2 + 2I_1'I_2' + I_1I_2'', \tag{9·8}$$

where products are formal products, and dashes denote formal differentiation. Take this formula temporarily for granted.

Denote the integral (9·7) by $\mathfrak{J}$ and write it as

$$\int_{-\infty}^{+\infty} e^{i\lambda x}\,d\mathrm{X}(\lambda), \tag{9·9}$$

where

$$\mathrm{X}(\lambda)=-\int_{-\infty}^{\lambda} \mu^{-2}\,d\chi(\mu). \tag{9·10}$$

Denote the Fourier integral of L by $\mathfrak{F}[L]$. Then

$$\mathfrak{F}[L]\equiv\int_{-\infty}^{+\infty} e^{i\lambda x} g(\lambda)\,d\lambda, \tag{9·11}$$

where

$$g(\lambda)=\frac{1}{2\pi}\int_{-\infty}^{+\infty} L(x)\,e^{-i\lambda x}\,dx=\frac{1}{2\pi}\int_{J'} L(x)\,e^{-i\lambda x}\,dx. \tag{9·12}$$

If $\mathfrak{P}$ is the formal product of $\mathfrak{J}$ and $\mathfrak{F}[L]$, we have, by (9·8),

$$\mathfrak{P}''=\mathfrak{J}''\mathfrak{F}[L]+2\mathfrak{J}'\mathfrak{F}[L']+\mathfrak{J}\mathfrak{F}[L''], \tag{9·13}$$

since $\mathfrak{F}^{(k)}[L]=\mathfrak{F}[L^{(k)}]$ for $k=1,2$. Observing that $L=1$, $L'=L''=0$ in J, and that the Fourier transforms of L, L', L'' are $O(\lambda^{-5})$, $O(\lambda^{-4})$, $O(\lambda^{-3})$ respectively, we obtain, by an application of (8·29),

$$\mathfrak{P}''_\omega-\mathfrak{J}''_\omega\to 0$$

uniformly in J, where $\mathfrak{J}_\omega$ and $\mathfrak{P}_\omega$ are symmetric partial integrals of $\mathfrak{J}$ and $\mathfrak{P}$. Since $\mathfrak{J}''_\omega$ is the first integral (9·5), the theorem will be proved if we show that

$$\mathfrak{P}_\omega=\frac{1}{\pi}\int_{J'} F(t)\,L(t)\,\frac{\sin\omega(x-t)}{x-t}\,dt. \tag{9·14}$$

It is enough to show that, under our hypotheses,

$$\mathfrak{P}\equiv\int_{-\infty}^{+\infty} e^{i\lambda x} p(\lambda)\,d\lambda, \tag{9·15}$$

where p is continuous and

$$p(\lambda)=o(\lambda^{-2}). \tag{9·16}$$

For the last relation implies that $p(\lambda)\in\mathrm{L}(-\infty,+\infty)$, and since, by (8·29), $\mathfrak{P}_\omega$ converges everywhere to $F(x)\,L(x)$, the equations (9·15) and

$$p(\lambda)=\frac{1}{2\pi}\int_{-\infty}^{+\infty} F(x)L(x)\,e^{-i\lambda x}\,dx \tag{9·17}$$

lead to (9·14).

Now, in any case,

$$\mathfrak{P}\equiv\int_{-\infty}^{+\infty} e^{i\lambda x}\,dP(\lambda),$$

where

$$P(\lambda)=\int_{-\infty}^{+\infty} g(\lambda-\mu)\,\mathrm{X}(\mu)\,d\mu.$$

Since $\mathrm{X}(\mu)=O(1)$, $g(\mu)=O(\mu^{-5})$, the last integral converges absolutely and uniformly provided λ remains in a finite interval. The same holds if we substitute $g'(\lambda-\mu)$ for

$g(\lambda-\mu)$ since, by (9·12), $g'(\lambda)$ is the Fourier transform of $-ixL(x)$. Hence P has a continuous derivative

$$p(\lambda)=P'(\lambda)=\int_{-\infty}^{+\infty} g'(\lambda-\mu)\,\mathrm{X}(\mu)\,d\mu=\int_{-\infty}^{+\infty} \mathrm{X}(\lambda-\mu)\,g'(\mu)\,d\mu,$$

and, by Lemma (8·35), $p(\lambda)=o(\lambda^{-2})$.

Hence (9·4) is established. The estimate (9·16) was much better than we actually needed, but the equation

$$\frac{1}{2\pi}\int_{-\infty}^{+\infty} F(t)\,L(t)\,e^{-i\lambda t}\,dt=o(\lambda^{-2}), \qquad \textbf{(9·18)}$$

which follows from (9·16) and (9·17), will be useful later.

Return to (9·8). This is an analogue of a formula for the formal product of trigonometric series (see Chapter IX, (4·32)), but the proof for integrals is somewhat less simple. While $I_1 I_2$, and so also $(I_1 I_2)''$, has meaning if, for example, the ϕ in $I_1=\int_{-\infty}^{+\infty} e^{i\lambda x}\,d\phi$ satisfies condition N_0 and the ψ in $I_2=\int_{-\infty}^{+\infty} e^{i\lambda x}\,d\psi$ is of bounded variation over $(-\infty, +\infty)$, the right-hand side of (9·8) requires stronger conditions on ϕ and ψ to be meaningful; we prove (9·8) under the hypotheses (8·36) and (8·37), so that all terms in (9·8) have meaning.

We define χ by (8·23), write

$$\chi_k(\lambda)=\int_{\lambda_0}^{\lambda} \mu^k\,d\chi(\mu)$$

for $k=1,2$, and define $\phi_k(\lambda)$, $\psi_k(\lambda)$ similarly. The right-hand side of (9·8) is $i^2\int_{-\infty}^{+\infty} e^{i\lambda x}\,d\chi_*(\lambda)$, where $\chi_*(\lambda)$ is

$$\begin{aligned}\int_{-\infty}^{+\infty} \{[\phi_2(\lambda-\mu)-\phi_2(\lambda_0-\mu)]&+2[\phi_1(\lambda-\mu)-\phi_1(\lambda_0-\mu)]\,\mu\\ &+[\phi(\lambda-\mu)-\phi(\lambda_0-\mu)]\,\mu^2\}\,d\psi(\mu)\\ &=\int_{-\infty}^{+\infty}\left\{\int_{\lambda_0-\mu}^{\lambda-\mu} (t^2+2t\mu+\mu^2)\,d\phi(t)\right\}d\psi(\mu)\\ &=\int_{-\infty}^{+\infty}\left\{\int_{\lambda_0}^{\lambda} t^2\,d_t\phi(t-\mu)\right\}d\psi(\mu)\\ &=\int_{-\infty}^{+\infty}\left\{\lambda^2[\phi(\lambda-\mu)-\phi(\lambda_0-\mu)]\right.\\ &\qquad\left.-2\int_{\lambda_0}^{\lambda} t[\phi(t-\mu)-\phi(\lambda_0-\mu)]\,dt\right\}d\psi(\mu).\end{aligned}$$

If we split the last integral into two and invert the order of integration (this being justified by the hypotheses on ϕ and ψ), we obtain

$$\lambda^2\chi(\lambda)-2\int_{\lambda_0}^{\lambda} t\chi(t)\,dt=\int_{\lambda_0}^{\lambda} t^2\,d\chi(t),$$

so that $\chi_*(\lambda)=\chi_2(\lambda)$, which implies (9·8).

The following two theorems are the main results of this section:

(9·19) FIRST EQUICONVERGENCE THEOREM. *If χ satisfies condition N_0, then for any interval J of length less than 2π there is a trigonometric series $\Sigma c_n e^{inx}$ with coefficients tending to 0, such that the two differences*

$$\int_{-\omega}^{\omega} e^{i\lambda x}\, d\chi(\lambda) - \sum_{|n|\leqslant\omega} c_n e^{inx}, \tag{9·20}$$

$$\int_{-\omega}^{\omega} (-i \operatorname{sign} \lambda)\, e^{i\lambda x}\, d\chi(\lambda) - \sum_{|n|\leqslant\omega} (-i \operatorname{sign} n)\, c_n e^{inx} \tag{9·21}$$

tend uniformly to limits in J as $\omega \to +\infty$, the first limit being 0.

(9·22) SECOND EQUICONVERGENCE THEOREM. *If χ satisfies condition N_0, then for any interval J of finite length there is an integral $\int_{-\infty}^{+\infty} c(\lambda)\, e^{i\lambda x}\, d\lambda$, where $c(\lambda)$ is continuous and tends to 0 with $1/\lambda$, such that*

$$\int_{-\omega}^{+\omega} e^{i\lambda x}\, d\chi(\lambda) - \int_{-\omega}^{\omega} e^{i\lambda x} c(\lambda)\, d\lambda \tag{9·23}$$

and

$$\int_{-\omega}^{+\omega} (-i \operatorname{sign} \lambda)\, e^{i\lambda x}\, d\chi(\lambda) - \int_{-\omega}^{\omega} (-i \operatorname{sign} \lambda)\, e^{i\lambda x} c(\lambda)\, d\lambda \tag{9·24}$$

tend uniformly to limits in J as $\omega \to +\infty$, the first limit being 0.

We consider only the differences (9·20) and (9·23) and begin with (9·20). Since χ satisfies condition N_0, it is enough to take $\omega = N + \frac{1}{2}$, where N is an integer.

Let J' be an interval of length 2π containing J in its interior. Let $L(x)$ be a function having five continuous derivatives, and equal to 1 in J and to 0 outside J'. We have (9·18) and, in particular,

$$\frac{1}{2\pi} \int_{J'} F(x) L(x)\, e^{-inx}\, dx = o(n^{-2}), \tag{9·25}$$

so that the Fourier series of the periodic function coinciding with FL in J' has coefficients $o(n^{-2})$.

Let $\Sigma c_n e^{inx}$ be obtained by differentiating that Fourier series twice; then $c_n \to 0$ and, incidentally, $c_0 = 0$. Clearly

$$\sum_{|n|\leqslant N} c_n e^{inx} = \frac{1}{\pi} \int_{J'} F(t)\, L(t) \frac{d^2}{dt^2} \frac{\sin (N+\frac{1}{2})(x-t)}{2 \sin \frac{1}{2}(x-t)}\, dt. \tag{9·26}$$

By Theorem (9·4) with $\omega = N + \frac{1}{2}$,

$$\int_{-(N+\frac{1}{2})}^{(N+\frac{1}{2})} e^{i\lambda x}\, d\chi(\lambda) - \frac{1}{\pi} \int_{J'} F(t)\, L(t) \frac{d^2}{dt^2} \frac{\sin (N+\frac{1}{2})(x-t)}{x-t}\, dt \tag{9·27}$$

tends uniformly to 0 in J as $N \to +\infty$. If we show that

$$\int_{J'} F(t)\, L(t) \frac{d^2}{dt^2} \left[\left\{ \frac{1}{2 \sin \frac{1}{2}(x-t)} - \frac{1}{x-t} \right\} \sin (N+\tfrac{1}{2})(x-t) \right] dt \tag{9·28}$$

tends uniformly to 0 in J, the assertion of Theorem (9·19) concerning (9·20) will follow.

Fix an x in J and denote the difference in curly brackets in (9·28) by $\Delta(x-t)$. Performing the differentiation, we express (9·28) as a sum of three terms, one of which is

$$-(N+\tfrac{1}{2})^2\int_{J'} F(t)\,L(t)\,\Delta(x-t)\sin(N+\tfrac{1}{2})(x-t)\,dt \tag{9·29}$$

and the other two contain lower powers of $(N+\frac{1}{2})$ (and higher derivatives of Δ). We show that the terms tend to 0; it is enough to consider (9·29).

We need only observe that $\Delta(u)$ is analytic in $-2\pi<u<2\pi$, so that $L_x(t)=L(t)\,\Delta(x-t)$ has, like $L(t)$, five continuous derivatives; the fact that (9·29) tends to 0 then becomes a consequence of the general relation (9·18). Moreover, if $x\in J$ and $t\in J'$, then $x-t$ stays in an interval $(-2\pi+\epsilon, 2\pi-\epsilon)$, and all estimates for $L_x(t)$ are uniform in x, with the result that (9·29) tends uniformly to 0 in J. We may consider the proof of (9·19) as completed.

The proof of (9·22) is even easier. By (9·18) the Fourier transform of FL is $o(\lambda^{-2})$. Hence, if $\int_{-\infty}^{+\infty} c(\lambda)\,e^{i\lambda x}\,d\lambda$ is obtained by differentiating the Fourier integral of FL twice, we have

$$\int_{-\omega}^{\omega} c(\lambda)\,e^{i\lambda x}\,d\lambda=\frac{1}{\pi}\int_{J'} F(t)\,L(t)\frac{d^2}{dt^2}\frac{\sin\omega(x-t)}{x-t}\,dt,$$

and since, by (9·4),

$$\int_{-\omega}^{\omega} e^{i\lambda x}\,d\chi(\lambda)-\frac{1}{\pi}\int_{J'} F(t)\,L(t)\frac{d^2}{dt^2}\frac{\sin\omega(x-t)}{x-t}\,dt$$

tends uniformly to 0 in J, the difference (9·23) also tends uniformly to 0 in J.

Incidentally, not only is the $c(\lambda)$ in (9·22) continuous but it may have as many derivatives as we please—even infinitely many if L is differentiable infinitely many times.

(9·30) Theorem. *Suppose that χ satisfies condition* N_0 *and that the F in* (8·13) *satisfies in an interval J of length less than 2π an equation*

$$F(x)=\int_{x_0}^{x} du\int_{x_0}^{u} f(t)\,dt+Ax+B, \tag{9·31}$$

where A, B are constants, x_0 is a point of J, and f is integrable over J. Then, if J' is any interval of length 2π containing J in its interior, f^* a periodic function coinciding with f in J', and $\Sigma c_n^*\,e^{inx}$ the Fourier series of f^*, the differences

$$\int_{-\omega}^{\omega} e^{i\lambda x}\,d\chi(\lambda)-\sum_{|n|\leqslant\omega} c_n^*\,e^{inx}, \tag{9·32}$$

$$\int_{-\omega}^{\omega}(-i\,\mathrm{sign}\,\lambda)\,e^{i\lambda x}\,d\chi(\lambda)-\sum_{|n|\leqslant\omega}(-i\,\mathrm{sign}\,n)\,c_n^*\,e^{inx} \tag{9·33}$$

converge uniformly in every interval J'' interior to J, the limit of the first being 0.

This is a consequence of (9·4). Consider, for example, (9·5). Integrating the second integral by parts, we find that $\int_{-\infty}^{+\infty} e^{i\lambda x}\,d\chi(\lambda)$ is uniformly equiconvergent on J with the Fourier integral of the function $(FL)''$, which on J is equal to f and so also to f^*. Since, by (1·3), the Fourier integral of $(FL)''$ is uniformly equiconvergent on J'' with $S[f^*]$, the assertion follows.

Suppose that χ satisfies condition N_0, and consider the function

$$\Phi(z) = \int_0^\infty e^{-\lambda z}\, d\chi(\lambda) \tag{9·34}$$

of the complex variable $z = x + iy$. Integrating by parts we easily find that the integral (9·34) converges uniformly in each bounded closed subdomain of the half-plane $x > 0$; in particular, $\Phi(z)$ is regular in this half-plane.

The following theorem generalizes the theorem of Fatou-Riesz on the convergence of power series on the circle of convergence (see Chapter IX, (5·7)):

(9·35) Theorem. *If χ satisfies condition N_0 and the $\Phi(z)$ in* (9·34) *is analytically continuable across a finite closed segment J, $a \leqslant y \leqslant b$, of the imaginary axis, then the integral* (9·34) *converges uniformly on J.*

This is a simple consequence of (9·30). We may suppose that $\chi = 0$ for $0 \leqslant \lambda \leqslant 1$, and that $b - a < 2\pi$. Denote J by J'' and reserve the notation J for a slightly larger closed segment across which Φ is still continuable. Consider the function

$$\Psi(z) = \int_1^\infty \lambda^{-2} e^{-\lambda z}\, d\chi(\lambda). \tag{9·36}$$

It is regular in the half-plane $x > 0$ and it is easy to see that the integral (9·36) converges uniformly in each bounded subset of the closed half-plane $x \geqslant 0$. For $x > 0$ we have $\Psi'' = \Phi$, so that Ψ is continuable across J, and $-\Psi(iy)$ is a second integral of $\Phi(iy)$ on $a \leqslant y \leqslant b$. Since $-\Psi(iy)$ is the function F formed for the integral $\int_1^\infty e^{-i\lambda y}\, d\chi(\lambda)$, and since the Fourier series of a periodic function which coincides with $\Phi(iy)$ for $a \leqslant y \leqslant b$ is obviously uniformly convergent on J'', the theorem follows.

We have already observed (p. 280) that if the integral (9·34) has at least one point of convergence on the imaginary axis, then χ must satisfy condition N_0.

10. Problems of uniqueness

We shall now investigate the uniqueness of representation of functions by integrals

$$\int_{-\infty}^{+\infty} e^{i\lambda x}\, d\phi(\lambda) \tag{10·1}$$

and, in particular, by integrals

$$\int_{-\infty}^{+\infty} e^{i\lambda x} c(\lambda)\, d\lambda, \tag{10·2}$$

where $c(\lambda)$ is integrable over each finite interval. We may assume that ϕ has only regular discontinuities.

(10·3) Theorem. *Suppose that* (10·2) *converges everywhere to a function $f(x)$ which is finite and integrable over each finite interval. Then*

$$c(\lambda) = (\mathrm{C}, 1)\frac{1}{2\pi}\int_{-\infty}^{+\infty} f(x)\, e^{-i\lambda x}\, dx = \lim_{\omega\to+\infty} \frac{1}{2\pi\omega}\int_0^\omega dw \int_{-w}^{w} f(x)\, e^{-i\lambda x}\, dx \tag{10·4}$$

for almost all λ.

In view of Theorem (8·4), the function $\phi(\lambda)=\int_0^\lambda c(\mu)\,d\mu$ satisfies condition N_0:

$$\max_{0\leqslant h\leqslant 1}\left|\int_\lambda^{\lambda+h} c(\mu)\,d\mu\right|=o(1)\quad(\lambda\to\pm\infty). \tag{10·5}$$

We begin by making a stronger assumption, from which we free ourselves later, namely that $c(\lambda)\to 0$ as $\lambda\to\pm\infty$.

The theorem is obvious if $c(\lambda)$ vanishes outside a finite interval. We may therefore suppose without loss of generality that $c(\lambda)=0$ in $(-1,1)$. We write

$$F(x)=-\int_{-\infty}^{+\infty}\mu^{-2}c(\mu)\,e^{i\mu x}\,d\mu. \tag{10·6}$$

By Theorem (9·19), in each interval J of length less than 2π the integral (10·2) is uniformly equiconvergent with a series $\Sigma c_n e^{inx}$ having coefficients tending to 0 (and depending on J):

$$\int_{-\omega}^{\omega} c(\lambda)\,e^{i\lambda x}\,d\lambda-\sum_{|n|\leqslant\omega} c_n e^{inx}\to 0\quad(x\in J). \tag{10·7}$$

We may suppose that $c_0=0$ (see p. 289). If we integrate (10·7) twice over a subinterval (x_0,x) of J we get

$$\int_{-\omega}^{\omega}(-\lambda^{-2})\,c(\lambda)\,e^{i\lambda x}\,d\lambda-\sum_{|n|\leqslant\omega}{}'\,(-n^{-2})\,c_n e^{inx}-A_\omega x-B_\omega\to 0\quad(x\in J), \tag{10·8}$$

where A_ω and B_ω are constants depending on ω and x_0 only. It is clear that A_ω and B_ω must tend to limits as $\omega\to+\infty$. By Theorem (2·4), and Lemma (3·13) of Chapter IX, the series $-\Sigma' n^{-2}c_n e^{inx}$ converges in J to a second integral of f. Hence

$$F(x)=\int_0^x dy\int_0^y f(t)\,dt+Ax+B, \tag{10·9}$$

for x in J. It might appear that the constants A and B may vary with J, but using the continuity and smoothness of F (see (8·17)) we find that A and B are independent of J and, since J is arbitrary, (10·9) holds for $-\infty<x<+\infty$.†

The hypothesis $c(\lambda)=o(1)$ implies that $c(\lambda)\,\lambda^{-2}$ is in $L(-\infty,+\infty)$, and the Riemann Lebesgue theorem applied to (10·6) shows that $F(x)\to 0$ as $x\to\pm\infty$. Inverting (10·6) we find that

$$-\frac{c(\lambda)}{\lambda^2}=(C,1)\frac{1}{2\pi}\int_{-\infty}^{+\infty}F(x)\,e^{-i\lambda x}\,dx=\lim_{\omega\to\infty}\frac{1}{2\pi\omega}\int_0^\omega dw\int_{-w}^{w}F(x)\,e^{-i\lambda x}\,dx \tag{10·10}$$

for almost all λ.

Now

$$\int_{-w}^{w}F(x)\,e^{-i\lambda x}\,dx=\left[iF(x)\frac{e^{-i\lambda x}}{\lambda}+F'(x)\frac{e^{-i\lambda x}}{\lambda^2}\right]_{x=-w}^{w}-\frac{1}{\lambda^2}\int_{-w}^{w}f(x)\,e^{-i\lambda x}\,dx \tag{10·11}$$

if $\lambda\neq 0$. We have just observed that $F(\pm\infty)=0$, and integration by parts gives

$$(C,1)\lim_{w\to+\infty}F'(\pm w)\,e^{\mp i\lambda w}=0. \tag{10·12}$$

From this and the preceding two formulae we get (10·4).

† We could have obtained (10·9) without using equiconvergence with trigonometric series, by appealing directly to Theorem (8·14) of this chapter and Theorem (2·4) of Chapter IX. We wanted, however, to indicate a method which might be advantageous in other cases.

Remark. If $c(\lambda)$ is continuous, or has only regular discontinuities of the first kind, then (10·4) holds for all λ. For $\lambda \neq 0$ this follows from the preceding argument. To prove the result for $\lambda=0$ we have to show, since $c(\lambda)=0$ near $\lambda=0$, that

$$\omega^{-1}\int_0^\omega dw \int_{-w}^{w} f(t)\,dt \to 0 \quad (\omega \to +\infty).$$

But, in view of (10·9), the left-hand side is

$$\omega^{-1}\int_0^\omega [F'(w)-F'(-w)]\,dw = \omega^{-1}[F(\omega)+F(-\omega)-2F(0)] = o(1),$$

which proves the desired result.

Before we proceed with the proof we apply the results above to the integral (10·1). Suppose that (10·1) converges everywhere to an $f(x)$ integrable over each finite interval. Fix a $\rho \neq 0$. We have

$$\int_{-\infty}^{+\infty} e^{ix\lambda}\,d\phi(\lambda) = f(x), \qquad \int_{-\infty}^{+\infty} e^{i\lambda x}\,d\phi(\lambda+\rho) = f(x)\,e^{-ix\rho}.$$

Subtracting these equations and integrating by parts we have

$$\int_{-\infty}^{+\infty} [\phi(\lambda+\rho)-\phi(\lambda)]\,e^{ix\lambda}\,d\lambda = f(x)\frac{e^{-i\rho x}-1}{-ix}, \tag{10·13}$$

since the integrated term is 0 because of the condition N_0 for ϕ. Hence, by the result already established,

$$\phi(\lambda+\rho)-\phi(\lambda) = (\mathrm{C},1)\frac{1}{2\pi}\int_{-\infty}^{+\infty} f(x)\,e^{-i\lambda x}\frac{e^{-ix\rho}-1}{-ix}\,dx, \tag{10·14}$$

since ϕ has only regular discontinuities. We come therefore to the following conclusion:

(10·15) Theorem. *If* (10·1) *converges everywhere to an* $f(x)$ *integrable over each finite interval, we have* (10·14) *for all* λ *and* ρ.

Return to (10·3). It remains to remove the hypothesis $c(\lambda)=o(1)$. This was used to show that we have (10·10) and that the integrated terms in (10·11) tend to 0 as $w \to +\infty$. The latter need not be true under the hypothesis (10·5), but for our purposes it is enough to show that these terms tend to 0 by the method of the first arithmetic mean. We have to show that

$$\int_0^\omega F(w)\,e^{-i\lambda w}\,dw = o(\omega), \tag{10·16}$$

$$\int_0^\omega F'(w)\,e^{-i\lambda w}\,dw = o(\omega), \tag{10·17}$$

for each $\lambda \neq 0$, and that this still holds if w is replaced by $-w$ in the integrand. It is enough to prove (10·16) and (10·17). Integration by parts shows that (10·17) is a consequence of (10·16) and the relation

$$F(w) = o(w) \quad (w \to \pm\infty), \tag{10·18}$$

and we confine our attention to the latter two relations.

To prove (10·18), we may suppose that $w \to +\infty$ and that $c(\lambda)=0$ in $(-k,k)$, where

k is arbitrarily large but fixed (by the Riemann-Lebesgue theorem the contribution of the interval $(-k, k)$ to $F(w)$ is $o(1)$ as $w \to \infty$). Write (10·6) in the form

$$F(w) = -\sum_{n=0}^{\infty} \int_{\pi n w^{-1} \leqslant |\mu| \leqslant \pi(n+1) w^{-1}} \mu^{-2} c(\mu)\, e^{i\mu w}\, d\mu. \tag{10·19}$$

Since $c(\lambda) = 0$ in $(-k, k)$, it is enough to sum over those n such that $\pi(n+1) w^{-1} > k$. Let n_0 be the smallest n satisfying this inequality; clearly $n_0 \simeq kw/\pi$. In each interval of integration in (10·19) the real and imaginary parts of $e^{i\mu w}$ have at most one extremum. Applying the second mean-value theorem and observing that the real and imaginary parts of ϕ satisfy condition N_0, we find that the series (10·19) is majorized in absolute value by a fixed multiple of

$$\sum_{n=n_0}^{\infty} (\pi n w^{-1})^{-2} . O(1) = O(w^2) \sum_{n \geqslant n_0} n^{-2} = O(n_0^{-1} w^2) = O(w/k), \tag{10·20}$$

and this proves (10·18).

Passing to (10·16), we observe that it is obviously true if $c(\mu)$ vanishes outside a finite interval, since in this case $F(w) = o(1)$. Fixing our λ, we may therefore suppose that $c(\mu) = 0$ for $|\mu| \leqslant |\lambda| + 1$. By (10·6) we may write the integral (10·16) in the form

$$i \int_{-\infty}^{+\infty} \mu^{-2} c(\mu) \frac{e^{i(\mu-\lambda)\omega} - 1}{\mu - \lambda}\, d\mu = i\, e^{-i\lambda\omega} \int_{-\infty}^{+\infty} \mu^{-2} c^*(\mu)\, e^{i\mu\omega}\, d\mu + O(1),$$

where $c^*(\mu) = c(\mu)/(\mu - \lambda)$. By the second mean-value theorem, $c^*(\mu)$ satisfies a condition analogous to (10·5), so that, applying to the last integral the decomposition of (10·19), with ω for w, and using estimates analogous to those of (10·20), we obtain (10·16).

It remains to prove (10·10). Since the integral representing $F(x)$ converges uniformly in every finite interval, we have

$$\frac{1}{2\pi\omega} \int_0^{\omega} dw \int_{-w}^{w} F(x)\, e^{-ix\lambda}\, dx = \frac{1}{\pi} \int_{-\infty}^{+\infty} \left(-\frac{c(\mu)}{\mu^2} \right) \frac{2 \sin^2 \frac{1}{2}\omega(\mu - \lambda)}{\omega(\mu - \lambda)^2}\, d\mu$$

(see the remark on p. 247). If we show that the right-hand side tends to $-c(\lambda)/\lambda^2$ for almost every λ, the proof of (10·3) will be completed. Since this is true for functions $c(\lambda)$ vanishing outside a finite interval, it is enough to show that if λ is fixed and if $c(\mu) = 0$ for $-k \leqslant \mu \leqslant k$, where, say, $k \geqslant |\lambda| + 1$, then the integral is small for k large, but fixed, and $\omega \to \infty$. Writing $c_*(\mu) = -c(\mu)/(\mu - \lambda)^2$, we present the integral in the form

$$\frac{2}{\omega} \cdot \sum_{n \geqslant n_0} \int_{\pi n \omega^{-1} \leqslant |\mu| \leqslant \pi(n+1)\omega^{-1}} c_*(\mu)\, \mu^{-2} \sin^2 \tfrac{1}{2}\omega(\mu - \lambda)\, d\mu,$$

and the argument follows the same line as in (10·20).

Remarks. (*a*) If the f of Theorem (10·3) is identically 0, then $c(\lambda) \equiv 0$. In this special case the proof given above simplifies somewhat. A similar remark applies to Theorem (10·15).

(*b*) If instead of assuming the convergence of (10·2) we merely suppose that ϕ satisfies condition N_0 and the two functions

$$f_*(x) = \liminf_{\omega \to +\infty} \int_{-\omega}^{\omega} e^{i\lambda x} c(\lambda)\, d\lambda, \quad f^*(x) = \limsup_{\omega \to +\infty} \int_{-\omega}^{+\omega} e^{i\lambda x} c(\lambda)\, d\lambda$$

are finite except for a denumerable set of x, and one of them is integrable over each finite interval, then $f_* = f^*$ almost everywhere and we have (10·4) for almost all λ, where $f(x)$ is the common value of $f_*(x)$ and $f^*(x)$. This follows from the preceding argument and the corresponding result for Fourier series (Chapter IX, (3·25)).

We recall certain definitions and results from Chapter IX. A point set E is said to be a set of *uniqueness* for trigonometric series, or a U-set, if each series

$$\sum_{-\infty}^{+\infty} c_n e^{inx} \tag{10·21}$$

which converges to 0 in the complement of E, vanishes identically. A set which is not U is called a set of *multiplicity*, or an M-set.

Similar definitions may be introduced for trigonometric integrals

$$\int_{-\infty}^{+\infty} c(\lambda)\, e^{i\lambda x}\, d\lambda, \tag{10·22}$$

or integrals

$$\int_{-\infty}^{+\infty} e^{i\lambda x}\, d\chi(\lambda). \tag{10·23}$$

A set E is a U-set for integrals (10·22), if the convergence of an integral (10·22) in the complement of E to 0 implies that $c(\lambda) \equiv 0$; for integrals (10·23) we replace the latter condition by $\chi = \text{const}$. Sets which are not U are called M-sets. To avoid misunderstanding, U-sets for expressions (10·21), (10·22), and (10·23) will be denoted respectively by U_s, U_i, and $U_{i'}$, and the corresponding M-sets by M_s, M_i, $M_{i'}$. Sets U_s and M_s, if considered on $(-\infty, +\infty)$, are periodic, of period 2π; the other sets are, in general, non-periodic.

We confine our attention to measurable sets. Sets U_s are then of measure 0 (Chapter IX, § 6). The converse is not true: among sets of measure 0 there are M_s-sets (see Chapter IX, § 6, and Chapter XII, (11·17), (11·18)). Sets U_i and $U_{i'}$ are also of measure 0; the proof is the same as for sets U_s.

Formal multiplication of trigonometric series shows that

(*a*) *If a U_s-set E is contained in an open interval J, and if a series $\Sigma\, c_n e^{inx}$ converges to 0 in $J - E$, then the series converges to 0 in J.*

Let $\lambda(x)$ have three continuous derivatives, be distinct from 0 in J, and equal to 0 elsewhere. The formal product of $\Sigma c_n e^{inx}$ by $S[\lambda]$ converges to 0 in the complement of E, and so everywhere. It follows that $\Sigma c_n e^{inx}$ converges to 0 in J.

From this it follows at once that

(*b*) *If a set E is locally U_s (that is, if for every x there exists a neighbourhood N_x of x such that EN_x is U_s), then E is a U_s-set.*

Results (*a*) and (*b*) hold for sets U_i and $U_{i'}$, and the proofs are essentially unchanged.

(10·24) THEOREM. *Sets U_s, U_i, $U_{i'}$ are locally the same.*

It is clear that each set $U_{i'}$ is both U_s and U_i, and the theorem will be established if we show that

(i) every set U_s of diameter less than 2π is $U_{i'}$;

(ii) every set U_i of finite diameter is $U_{i'}$.

(i) Suppose that E is a U_s-set of diameter less than 2π, and that an integral (10·23) converges to 0 in the complement of E. Let J be an open interval containing E and of length less than 2π. By (9·19), there is a series $\Sigma c_n e^{inx}$, with coefficients tending to 0, equiconvergent with the integral on J. By proposition (*a*) above, the series converges to 0 in J. Hence the same holds for the integral, and since the latter, by hypothesis, converges to 0 outside J, it converges to 0 everywhere, and so is identically 0. This gives (i), and the proof of (ii) is similar, except that instead of (9·19) we use (9·22).

Proposition (i) shows that every set E which is an $\mathrm{M}_{i'}$ and has diameter less than 2π is an M_s. But the argument just used gives also a series of multiplicity for E. For if an integral (10·23) converges to 0 in the complement of E but is not identically 0, and if J and $\Sigma c_n e^{inx}$ have the same meaning as above, then the latter series converges to 0 in $J-E$ but not at each point of J. Hence the formal product of $\Sigma c_n e^{inx}$ by $\mathsf{S}[L]$, where L, sufficiently differentiable, is 0 outside J and different from 0 in J, is a trigonometric series which converges to 0 in the complement of E but not everywhere. A similar remark applies to (ii).

Given a set E and a number l, we denote by E_l the set of points lx, where $x \in E$.

We now prove the following result which was stated in Chapter IX, (6·18):

(10·25) THEOREM. *If E is contained in J_0: $0 \leqslant x < 2\pi$, and is a* U_s*-set, then each E_l, $l>0$, which is contained in J_0 is also a* U_s*-set.*

We may suppose that $x=0$ is not in E, since otherwise we first prove the theorem for the set E minus the point 0, and then add $x=0$ to E_l (since adding a point to a U_s-set does not affect the U_s-character of the set). Then both E and E_l are interior to $(0, 2\pi)$.

Consider a series $\Sigma c_n e^{inx}$, S say, converging to 0 at each point of $(0, 2\pi)$ which is not in E_l. Let x_0 be a point of E; it is enough to show that S converges to 0 at lx_0.

Denote by J an open interval containing x_0; then J_l contains lx_0. We take J so small that both J and J_l are contained, together with their endpoints, in the interior of $(0, 2\pi)$. Since S converges to 0 in $J_l - E_l$, the series

$$\Sigma c_n e^{inlx} \tag{10·26}$$

converges to 0 in $J-E$. Treating (10·26) as an integral $\int_{-\infty}^{+\infty} e^{i\lambda x}\, d\chi(\lambda)$ and applying (9·19), we find a trigonometric series $\Sigma c_n' e^{inx}$ with coefficients tending to 0, S' say, equiconvergent with (10·26) on J. By proposition (*a*) above, S' converges to 0 in J. Hence (10·26) converges to 0 in J, and in particular at x_0. It follows that S converges to 0 at lx_0, and the proof of (10·25) is completed.

The fact that if E is a U_i-set, then each E_l is also a U_i-set, is much more obvious and follows by a change of variable in the integral. Similarly it is immediate that if E is a $\mathrm{U}_{i'}$-set, so is E_l.

A set E of period 2π and measure 0 will be called a U_s^*-set, if every trigonometric series converging in the complement of E to a finite and integrable function f is necessarily $\mathsf{S}[f]$.† Every U_s^*-set is obviously a U_s-set; whether the converse is true seems not to be known except in the case when E is closed. For, by the principle of localization, if $\Sigma c_n e^{inx}$ converges outside a closed set E, $|E|=0$, to an $f \in \mathrm{L}$, then the difference between the series and $\mathsf{S}[f]$ converges to 0 outside E, and so is identically 0 if E is U.

(10·27) THEOREM. *If a set E is locally* U_s^**, then it is* U_s^*.

Suppose a trigonometric series S converges in the complement of E to a finite and integrable function f. Consider any point x_0 and an open interval J containing x_0

† Cf. Chapter IX, p. 350.

such that JE is U_s^*. Let J' be an interval containing x_0 and totally interior to J, $L(x)$ a sufficiently differentiable function equal to 1 on J' and to 0 outside J, and $T = SS[L]$. Clearly $T = S[fL]$, and since S and T are uniformly equiconvergent on J', the function F obtained by integrating S twice is, on J', a second integral of f:

$$F(x) = \int_0^x dy \int_0^y f(t)\, dt + Ax + B \quad (x \in J'). \tag{10·28}$$

By the Heine-Borel theorem, and the continuity and smoothness of F, (10·28) is valid for all x, with A and B independent of x. Hence $S = S[f]$.

A set $E \subset (-\infty, +\infty)$ of measure 0 will be called a U_i^*-set, if every trigonometric integral (10·2) converging outside E to a finite function $f(x)$ integrable over each finite interval is the Fourier integral of f in the sense of (10·4).

(10·29) THEOREM. *Every set E which is locally* U_s^*, *is* U_i^*. *Every set E which is* U_i^* *and of diameter less than* 2π, *is* U_s^*.

Suppose that a set $E \subset (-\infty, +\infty)$ is locally U_s^*, and that an integral (10·22), I say, converges outside E to a finite $f(x)$ integrable over each finite interval. Let J be an interval of length less than 2π, and S a trigonometric series with coefficients tending to 0 uniformly equiconvergent with I on J. As in the proof of (10·27), we find that S when integrated twice represents in J a second integral of f. Hence by integrating I formally twice we also obtain a function which on J is a second integral of f. This leads to the formula (10·9), valid for all x, with A and B independent of x, and the Fourier character of I follows as in the proof of Theorem (10·3).

The proof of the second part of (10·29) is similar.

The idea of extending results from trigonometric series to trigonometric integrals by using the equiconvergence theorems of the preceding section can be applied in many other cases. For example, we may obtain an analogue of Theorem (10·25) for U_s^*-sets; we may obtain uniqueness theorems for integrals (10·22) or (10·23) summable by Abel's method; we may consider limits of indetermination for integrals, etc. No new difficulties appear if it is only a matter of translating results from series to integrals (or conversely), and there is no point in considering such results in detail.

MISCELLANEOUS THEOREMS AND EXAMPLES

1. If $\phi \in L^2(-\infty, +\infty)$, the Hilbert transform $\tilde{\phi}$ satisfies

$$\| \tilde{\phi} \|_2 = \| \phi \|_2.$$

2. If $\{\phi_\nu\}$ is orthonormal over $(-\infty, +\infty)$, so is $\{\tilde{\phi}_\nu\}$; if $\{\phi_\nu\}$ is, in addition, complete, so is $\{\tilde{\phi}_\nu\}$.

3. Let $\psi_0, \psi_1, \psi_2, \ldots$ be orthonormal and complete over a finite interval $(-\sigma, \sigma)$, and let $\tilde{\psi}_\nu$ be the Hilbert transform of the function equal to ψ_ν in $(-\sigma, \sigma)$ and to 0 elsewhere. Then the system $\{\tilde{\psi}_\nu\}$ is orthonormal over $(-\infty, +\infty)$, and is complete with respect to functions which are in $L^2(-\infty, +\infty)$ and restrictions to the real axis of functions from E^σ (p. 272).

In particular, if $P_0(x)$, $P_1(x)$, ... are Legendre polynomials, the functions

$$Q_n(x) = \frac{1}{\pi} \int_{-1}^{+1} \frac{P_n(t)}{x-t}\, dt$$

are orthogonal over $(-\infty, +\infty)$, and form a system complete with respect to functions which are both in E^1 and $L^2(-\infty, +\infty)$.

4. Let $a_\nu^2+b_\nu^2\leqslant 1$, $0<2h\leqslant\lambda_1<\lambda_2<\dots$, $\lambda_{\nu+1}/\lambda_\nu\geqslant 3$ for $\nu=1,2,\dots$, and let

$$F_N(x)=\frac{1}{\pi}\int_{-\infty}^{x}\frac{\sin^2 yh}{y^2h}\prod_{\nu=1}^{N}(1+a_\nu\cos\lambda_\nu y+b_\nu\sin\lambda_\nu y)\,dy.$$

Then $F_N(x)$ is a distribution function (p. 262) and tends, uniformly over any finite interval, as $N\to\infty$, to a continuous distribution function $F(x)$ satisfying

$$\tfrac{1}{2}a_\nu=\int_{-\infty}^{+\infty}\cos\lambda_\nu x\,dF(x),\quad \tfrac{1}{2}b_\nu=\int_{-\infty}^{+\infty}\sin\lambda_\nu x\,dF(x)$$

($F(x)$ is an analogue of the Riesz product; Chapter V, § 7). If, in addition, $\Sigma(a_\nu^2+b_\nu^2)=\infty$, F is singular. Similarly, if $\Sigma(a_\nu^2+b_\nu^2)<\infty$ and if the λ_ν satisfy the same hypotheses as before, the function

$$f(x)=\frac{\sin^2 hx}{x^2h}\prod_{\nu=1}^{\infty}(1+ia_\nu\cos\lambda_\nu x+ib_\nu\sin\lambda_\nu x)=g+ih$$

is bounded, is in $L(-\infty,+\infty)$ and satisfies

$$\tfrac{1}{2}a_\nu=\frac{1}{\pi}\int_{-\infty}^{+\infty}h(x)\cos\lambda_\nu x\,dx,$$

with a similar formula for b_ν. The functions F and f can be used to obtain for Fourier integrals theorems analogous to those of Chapter XII, § 7.

5. If $\Sigma|a_\nu|^2<\infty$, $0<\lambda_1<\lambda_2<\dots$, $\lambda_{\nu+1}/\lambda_\nu>q>1$ then $\Sigma a_\nu e^{i\lambda_\nu x}$ converges almost everywhere. (Kac [1].)

[Let δ be so small that the intervals $(\lambda_\nu-\delta,\lambda_\nu+\delta)$ do not overlap, and let

$$f(x)=a_\nu\quad\text{for}\quad x\in(\lambda_\nu-\delta,\lambda_\nu+\delta),\quad f(x)=0\ \text{elsewhere}.$$

Then $f\in L^2$, $\int_0^{\lambda_n}f(t)\,e^{itx}(1-t/\lambda_n)\,dt$ tends to a limit almost everywhere, and since the hypotheses imply that $\lambda_n^{-1}\int_0^{\lambda_n}t\,|f(t)|\,dt\to 0$,

$$\lim_{n\to\infty}\int_0^{\lambda_n}f(t)\,e^{itx}\,dt=\frac{2\sin\delta x}{x}\lim\sum_{\nu=1}^{n}a_\nu e^{i\lambda_\nu x}$$

exists almost everywhere.]

6. If $0<\lambda_1<\lambda_2<\dots$, $\lambda_{\nu+1}/\lambda_\nu>q>1$, $\lambda_{-\nu}=-\lambda_\nu$, and if $\Sigma c_\nu e^{i\lambda_\nu x}$ converges in a set of positive measure, then $\Sigma|c_\nu|^2<\infty$. (Kac [1], Hartman [1].)

[The proof is analogous to that of Theorem (6·4) of Chapter V; we use the fact that the system $\{e^{i\lambda_\nu x}\}$ is orthogonal over $(-\infty,+\infty)$ with respect to a weight function $x^{-2}\sin^2\delta x$.]

7. If the λ_ν satisfy the hypotheses of Example 6, if $c_{-\nu}=\bar{c}_\nu$, and if the symmetric partial sums of $\Sigma c_\nu e^{i\lambda_\nu x}$ are bounded below in a set of positive measure, then $\Sigma|c_\nu|^2<\infty$.

8. Suppose that $f(x)$ is integrable over every finite interval and that $\Sigma\left(\int_n^{n+1}|f(y)|\,dy\right)^2$ converges. If $f_\omega(x)$ is equal to $f(x)$ in $(-\omega,\omega)$ and to 0 elsewhere, and if $\hat{f}_\omega$ is the Fourier transform of f_ω, then there is a function $\hat{f}$ such that $\hat{f}(x)/(1+|x|)\in L^2(-\infty,+\infty)$ and

$$(\dagger)\qquad \int_{-\infty}^{+\infty}\frac{|\hat{f}_\omega(x)-\hat{f}(x)|^2}{1+x^2}\,dx\to 0\quad(\omega\to\infty).$$

In particular, $\hat{f}$ converges to $\hat{f}_\omega$ in L^2 over every finite interval. (Wiener [5].)

[If $g\in L(-\infty,+\infty)$ and G is the indefinite integral of g, then

$$(*)\qquad \frac{\sin hx}{hx}\int_{-\infty}^{+\infty}g(t)\,e^{-ixt}\,dt=\int_{-\infty}^{+\infty}\frac{G(x+h)-G(x-h)}{2h}\,e^{-ixt}\,dt.$$

Denoting by F and F_ω the indefinite integrals of f and f_ω we therefore have (*) with f_ω, F_ω for g, G. The hypotheses about f imply that $\phi_h(x)=\{F(x+h)-F(x-h)\}/2h$ is in $L^2(-\infty,+\infty)$ and that $\|\phi_h-\phi_{h,\omega}\|_2\to 0$ as $\omega\to\infty$, where $\phi_{h,\omega}$ is the ϕ_h for F_ω. Since

$$(**)\qquad \left\|\frac{\sin hx}{hx}(\hat{f}_\omega(x)-\hat{f}_{\omega'}(x))\right\|_2=\|\hat{\phi}_{h,\omega}-\hat{\phi}_{h,\omega'}\|_2=\|\phi_{h,\omega}-\phi_{h,\omega'}\|_2$$

and since the right-hand side tends to 0 as $\omega, \omega' \to \infty$, it follows that $\int_{-\pi/2h}^{\pi/2h} |\hat{f}_\omega - \hat{f}_{\omega'}|^2 dx \to 0$ for each $h > 0$, and so there is an $\hat{f}$ such that $\hat{f}_\omega$ tends to $\hat{f}$ in L^2 over every interval.

To obtain the stronger form of the assertion we observe that $\hat{f}_\omega(x + \frac{1}{2}\pi)$ is the Fourier transform of $e^{-\frac{1}{2}\pi it} f_\omega(t)$. Hence, from (**) with $h = 1$ we deduce that

$$\int_{-\infty}^{+\infty} \left(\frac{\sin x}{x}\right)^2 \left|\hat{f}_\omega\left(x + \frac{\pi}{2}\right) - \hat{f}_{\omega'}\left(x + \frac{\pi}{2}\right)\right|^2 dx = \int_{-\infty}^{+\infty} \frac{\cos^2 x}{(x - \frac{1}{2}\pi)^2} |\hat{f}_\omega(x) - \hat{f}_{\omega'}(x)|^2 dx \to 0,$$

as $\omega, \omega' \to \infty$. Considering the intervals $-\pi \leqslant x \leqslant \pi$ and its complement separately, using the weaker result obtained above and the fact that $\cos^2 x + \sin^2 x = 1$, we arrive at the desired conclusion.

If f satisfies the hypothesis that $\Sigma\left(\int_n^{n+1} |f(y)|\, dy\right)^p < \infty$, where $1 < p < 2$, then

$$\hat{f}(x)/(1 + |x|) \in L^{p'}(-\infty, +\infty),$$

and we have (†) with exponents 2 replaced by p' throughout.]

9. Let $\zeta = \rho e^{i\theta}$, $z = x + iy$. A function $f(\zeta)$, regular in $|\zeta| < 1$, is said to belong to the class H^r, $r > 0$, if there is a finite μ such that

$$\frac{1}{2\pi} \int_0^{2\pi} |f(\rho e^{i\theta})|^r d\theta \leqslant \mu^r \quad (0 < \rho < 1)$$

(cf. Chapter VII, § 7). Correspondingly, a function $F(z)$, regular in the half-plane $x > 0$ will be said to belong to the class $\mathfrak{H}^r$, if there is a finite M such that

$$\frac{1}{2\pi} \int_{-\infty}^{+\infty} |F(x + iy)|^r dy \leqslant M^r \quad (0 < x < \infty).$$

Consider the correspondence $\qquad \zeta = \dfrac{z-1}{z+1}, \quad z = \dfrac{1+\zeta}{1-\zeta}$

between the circle and the half-plane.

(i) A necessary and sufficient condition that a function $F(z)$ regular for $x > 0$ should belong to $\mathfrak{H}^r$ is that

$$f(\zeta) = 2^{1/r}(1 - \zeta)^{-2/r} F(z)$$

should belong to H^r.

(ii) A necessary and sufficient condition that a function $f(\zeta)$ regular for $|\zeta| < 1$ should belong to H^r is that

$$F(z) = 2^{1/r}(1 + z)^{-2/r} f(\zeta)$$

should belong to $\mathfrak{H}^r$.

(iii) If for μ and M in the definitions above we take the least admissible values, then $\mu = M$ in (i) and (ii).

10. If $f \in L^p(-\infty, +\infty)$, $1 < p < \infty$ and

$$\phi(x) = \sup_{\epsilon > 0} \left| -\frac{1}{\pi} \int_\epsilon^\infty \frac{f(x+t) - f(x-t)}{t} dt \right|,$$

then $\|\phi\|_p \leqslant A_p \|f\|_p$. (Cf. Chapter VII, (7·38).)

CHAPTER XVII

A TOPIC IN MULTIPLE FOURIER SERIES

1. General remarks

The notions of orthogonality of functions, of Fourier coefficients, and of Fourier series, initially defined for functions of a single variable, extend without difficulty to functions of several variables.

If D is a set in the m-dimensional Euclidean space E^m, functions $\phi_1(p), \phi_2(p), \ldots$ of a point p in D are said to form an orthogonal system in D if

$$\int_D \phi_j(p)\,\bar{\phi}_k(p)\,dp \begin{cases} =0 & (j \neq k), \\ =\lambda_j > 0 & (j=k), \end{cases}$$

for all j, k; here dp denotes the element of m-dimensional volume. If

$$c_n = \frac{1}{\lambda_n}\int_D f(p)\,\bar{\phi}_n(p)\,dp$$

are the Fourier coefficients of f with respect to $\{\phi_n\}$, $\Sigma c_n \phi_n$ is the Fourier series of f, and we write

$$f(p) \sim \Sigma c_n \phi_n(p). \qquad (1\cdot1)$$

Results obtained for general orthogonal series of a single variable extend to the case of several variables, and the proofs remain in general the same. (Replacing single by m-dimensional integration is purely formal.) In particular, we still have Bessel's inequality, the Riesz-Fischer theorem, the equivalence of the notions of a complete and closed orthogonal system (Chapter IV, § 1), etc.

Suppose that for each $j=1, 2, \ldots, m$ the system

$$\phi_1^j(x), \quad \phi_2^j(x), \quad \ldots, \quad \phi_n^j(x), \quad \ldots \qquad (1\cdot2)$$

is orthogonal in an interval (a_j, b_j). It is clear that the functions

$$\phi_{n_1, n_2, \ldots, n_m}(x_1, x_2, \ldots, x_m) = \phi_{n_1}^1(x_1)\,\phi_{n_2}^2(x_2)\ldots\phi_{n_m}^m(x_m) \qquad (1\cdot3)$$

are then orthogonal in the m-dimensional interval (parallelopiped)

$$a_j \leqslant x_j \leqslant b_j \quad (j=1, 2, \ldots, m),$$

and the λ's associated with (1·3) are the products of the corresponding λ's for (1·2). Finally, if all the systems (1·2) are complete, so is (1·3).†

The system $\{e^{inx}\}_{n=0, \pm1, \pm2, \ldots}$ is orthogonal and complete over $(-\pi, \pi)$, and all the λ's are equal to 2π. Hence the system

$$e^{i(n_1x_1+n_2x_2+\cdots+n_mx_m)} \quad (-\infty < n_j < +\infty; j=1, 2, \ldots, m) \qquad (1\cdot4)$$

is orthogonal and complete over the m-dimensional cube

$$Q: -\pi \leqslant x_j \leqslant \pi \quad (j=1, 2, \ldots, m);$$

† For $m=2$ the proof is indicated in Chapter I, p. 34, Example 8.

the Fourier coefficients of an f integrable over Q are

$$c_{n_1,\ldots,n_m} = (2\pi)^{-m}\int_Q f(y_1, \ldots, y_m)\, e^{-i(n_1y_1+\cdots+n_my_m)}\, dy_1 \ldots dy_m; \tag{1·5}$$

and we write
$$f(x_1, \ldots, x_m) \sim \Sigma c_{n_1,\ldots,n_m} e^{i(n_1x_1+\cdots+n_mx_m)}. \tag{1·6}$$

We denote the series by $S[f]$. We do not arrange the terms of $S[f]$ linearly (as in (1·1)), but treat it as a multiple series. We may consider f as defined over E^m and periodic (i.e. periodic in each x_j).

The completeness of (1·4) implies (see Chapter IV, § 1) the validity of the Parseval formula

$$\Sigma\, |c_{n_1,\ldots,n_m}|^2 = (2\pi)^{-m}\int_Q |f(x_1, \ldots, x_m)|^2\, dx_1 \ldots dx_m. \tag{1·7}$$

It often simplifies formulae if we use vector notation. We denote points or vectors $(x_1, \ldots, x_m)$, $(y_1, \ldots, y_m)$, ... by the corresponding bold letters $\mathbf{x}$, $\mathbf{y}$, ... Points $(n_1, \ldots, n_m)$ with integral co-ordinates are denoted by $\mathbf{n}$, and the point $(0, 0, \ldots, 0)$ by $\mathbf{0}$ and $(1, 1, \ldots, 1)$ by $\mathbf{1}$. The symbols $k\mathbf{x}$ (where k is real) and $\mathbf{x}+\mathbf{y}$ mean the vectors $(kx_1, \ldots, kx_m)$ and $(x_1+y_1, \ldots, x_m+y_m)$ respectively, and $(\mathbf{xy})$ stands for the scalar product $x_1y_1+\ldots+x_my_m$. Finally, we write $|\mathbf{x}|$ for $(x_1^2+\ldots+x_m^2)^{\frac{1}{2}}$, and $d\mathbf{x}$ for the element of volume $dx_1\,dx_2 \ldots dx_m$.

With this notation, (1·4) is $e^{i(\mathbf{nx})}$, and (1·5), (1·6) and (1·7) take respectively the forms

$$c_{\mathbf{n}} = (2\pi)^{-m}\int_Q f(\mathbf{y})\, e^{-i(\mathbf{ny})}\, d\mathbf{y},$$

$$f(\mathbf{x}) \sim \Sigma c_{\mathbf{n}}\, e^{i(\mathbf{nx})},$$

$$\Sigma\, |c_{\mathbf{n}}|^2 = (2\pi)^{-m}\int_Q |f(\mathbf{x})|^2\, d\mathbf{x}.$$

The theory of multiple Fourier series (1·6) is vast, but much of it is a straightforward extension of results for a single variable. Only those results are of interest, of course, which have no counterpart for a single variable, or whose proofs require essentially new ideas. In this chapter we confine our attention to a rather special topic, namely the rectangular summability of Fourier series (see below) and its applications, and in this section we collect a few simple facts and definitions.

The proof of the Riemann-Lebesgue theorem (Chapter II, (4·4)) is based on Bessel's inequality and holds for the system (1·4): *the coefficients $c_{\mathbf{n}}$ of any integrable function f tend to* 0 *as* $|\mathbf{n}| \to \infty$; this means that only a finite number of the $c_{\mathbf{n}}$ exceed numerically a given $\epsilon > 0$. A slightly more general version, analogous to Theorem (4·6) of Chapter II and proved in the same way, asserts that if f is integrable over E^m, then

$$\int_{E^m} f(\mathbf{y})\, e^{i(\mathbf{xy})}\, d\mathbf{y}$$

tends to 0 as $|\mathbf{x}| \to \infty$.

If D is any bounded domain in E^m and $f \sim \Sigma c_{\mathbf{n}}\, e^{i(\mathbf{nx})}$, a sum

$$\sum_{\mathbf{n}\in D} c_{\mathbf{n}}\, e^{i(\mathbf{nx})}$$

is a *partial sum* of $S[f]$; we denote it by $S_D(\mathbf{x})$, or by $S_D(\mathbf{x}; f)$. If we have a sequence of domains $D_1, D_2, \ldots$ such that each $\mathbf{n}$ belongs to all D_j with j large enough, we may ask if $S_{D_j}(\mathbf{x})$ converges to f (pointwise or in some norm) as $j \to \infty$. This problem has many aspects since the D_j may have various shapes. In general, however, partial sums are not adequate to represent the function and, as in the case of a single variable, we may have to consider various means of partial sums.

The most important special cases are when the D_j are either m-dimensional spheres or m-dimensional intervals, with centre at $\mathbf{0}$. *Spherical* partial sums are

$$\sum_{|\mathbf{n}|\leqslant R} c_{\mathbf{n}}\, e^{i(\mathbf{nx})}, \tag{1·8}$$

and the *rectangular* ones can be written

$$\sum_{|n_j|\leqslant N_j} c_{\mathbf{n}}\, e^{i(\mathbf{nx})}. \tag{1·9}$$

We shall also use for (1·9) the notations $S_{N_1,\ldots,N_m}(\mathbf{x})$ or $S_{\mathbf{N}}(\mathbf{x})$ or $S_{\mathbf{N}}[f]$.

The spherical and rectangular partial sums of an $S[f]$ behave in many respects quite differently. Each type requires a different technique and makes appeal to different properties of f. If we consider $S[f]$ primarily as an orthogonal series, it is natural to arrange its terms according to the magnitude of $|\mathbf{n}|$, and dealing with spherical partial sums has advantages. On the other hand, there are problems where rectangular partial sums, and their means, are indispensable. This is, in particular, true of the problems of the behaviour of multiple power series

$$\Sigma c_{n_1,\ldots,n_m} z_1^{n_1} \ldots z_m^{n_m}$$

near the boundary of the domain of convergence.

In this chapter we consider only rectangular partial sums of $S[f]$, and their means. We shall be concerned exclusively with the (C, 1) (that is, the first arithmetic) means

$$\sigma_{\mathbf{n}}(\mathbf{x}) = \sigma_{n_1,\ldots,n_m}(\mathbf{x}) = \frac{1}{(n_1+1)\ldots(n_m+1)} \sum_{0}^{n_1,\ldots,n_m} S_{k_1,\ldots,k_m}(\mathbf{x}), \tag{1·10}$$

and the Abel means

$$f(\mathbf{r},\mathbf{x}) = \Sigma c_{n_1,\ldots,n_m} r_1^{|n_1|} \ldots r_m^{|n_m|} e^{i(n_1x_1+\cdots+n_mx_m)}. \tag{1·11}$$

While for one-dimensional series the method (C, 1) is stronger than ordinary convergence, and the method A stronger than (C, 1), and while under certain conditions the same holds in the multi-dimensional case, these conditions are quite restrictive and complicated, and we shall not be concerned with them here. The methods (C, 1) and A have interesting features: the $\sigma_{\mathbf{n}}$ are trigonometric polynomials in m variables, and the A-means are harmonic functions of each pair of variables (r_j, x_j), and so also of their totality. As with one-dimensional series, the methods run largely parallel, and in general it is enough to prove results for one of them only.

It is easy to see that the partial sums of $S[f]$ are given by

$$\begin{aligned} S_{\mathbf{n}}(\mathbf{x}) &= \pi^{-m}\int_Q f(x_1+t_1,\ldots,x_m+t_m)\prod_{j=1}^{m} D_{n_j}(t_j)\,dt_1\ldots dt_m \\ &= \pi^{-m}\int_Q f(\mathbf{x}+\mathbf{t})\,D_{\mathbf{n}}(\mathbf{t})\,d\mathbf{t} = \pi^{-m}\int_Q f(\mathbf{t})\,D_{\mathbf{n}}(\mathbf{x}-\mathbf{t})\,d\mathbf{t}, \end{aligned} \tag{1·12}$$

using the obvious abbreviation

$$D_{\mathbf{n}}(\mathbf{t}) = D_{n_1}(t_1)\ldots D_{n_m}(t_m). \tag{1·13}$$

Similarly, if $K_n(t)$ and $P(r,t)$ are the Féjer and Poisson kernels, and we set

$$K_{\mathbf{n}}(\mathbf{t}) = K_{n_1}(t_1) \dots K_{n_m}(t_m), \tag{1·14}$$

$$P(\mathbf{r},\mathbf{t}) = P(r_1 . t_1) \dots P(r_m, t_m), \tag{1·15}$$

we have

$$\sigma_{\mathbf{n}}(\mathbf{x}) = \pi^{-m} \int_Q f(\mathbf{x}+\mathbf{t})\, K_{\mathbf{n}}(\mathbf{t})\, d\mathbf{t}, \tag{1·16}$$

$$f(\mathbf{r},\mathbf{x}) = \pi^{-m} \int_Q f(\mathbf{x}+\mathbf{t})\, P(\mathbf{r},\mathbf{t})\, d\mathbf{t}. \tag{1·17}$$

The right-hand side of (1·17) is the *Poisson integral* of f.

It is useful to observe that if $f(x_1, \dots, x_m) = f_1(x_1) \dots f_m(x_m)$ we have

$$S_{\mathbf{n}}(\mathbf{x}; f) = S_{n_1}(x_1; f_1) \dots S_{n_m}(x_m; f_m), \tag{1·18}$$

and corresponding formulae hold for $\sigma_{\mathbf{n}}(\mathbf{x})$ and $f(\mathbf{r},\mathbf{x})$.

We have to discriminate between various kinds of convergence of multiple sequences. We say that $\mathbf{n} = (n_1, \dots, n_m)$ tends to $+\infty$ if each n_j tends to $+\infty$, and we say that $\{s_{\mathbf{n}}\}$ converges to limit s if for each $\epsilon > 0$ we have $|s_{\mathbf{n}} - s| < \epsilon$ provided that all n_j are large enough. This definition, as opposed to one with $|\mathbf{n}| \to \infty$, imposes no condition on infinitely many of the $s_{\mathbf{n}}$, and as a result there are occasionally difficulties in operating with it. Sometimes we have $s_{\mathbf{n}} \to s$ as $|\mathbf{n}| \to \infty$; this happens, for example, in the Riemann-Lebesgue theorem.

In defining summability (C, 1) we suppose that the (C, 1) means $\sigma_{\mathbf{n}}$ tend to a limit as $\mathbf{n} \to \infty$.

In Abel summability we have a variable $\mathbf{r} = (r_1, \dots, r_m)$ tending continuously to $\mathbf{1} = (1, \dots, 1)$, and the preceding remarks apply also to this case. If $\mathbf{z} = (z_1, \dots, z_m)$, $\mathbf{z}^0 = (z_1^0, \dots, z_m^0)$ and if $|z_j| < 1$, $|z_j^0| = 1$ for each j, we say that $\mathbf{z}$ tends *non-tangentially* to $\mathbf{z}^0$, if each z_j tends non-tangentially to z_j^0.

If

$$\Sigma \gamma_{n_1, \dots, n_m} r_1^{|n_1|} \dots r_m^{|n_m|} e^{i(n_1 x_1 + \dots + n_m x_m)}$$

tends to a limit s as $(r_1 e^{ix_1}, \dots, r_m e^{ix_m})$ tends non-tangentially to $(e^{ix_1^0}, \dots, e^{ix_m^0})$, we say that $\Sigma \gamma_{n_1, \dots, n_m} e^{i(n_1 x_1^0 + \dots + n_m x_m^0)}$ is *summable* A* to sum s.

Return to $\sigma_{\mathbf{n}}(\mathbf{x})$ and $f(\mathbf{r},\mathbf{x})$. By (1·14) and (1·15), we have

$$\pi^{-m} \int_Q K_{\mathbf{n}}(\mathbf{t})\, d\mathbf{t} = 1, \quad \pi^{-m} \int_Q P(\mathbf{r},\mathbf{t})\, d\mathbf{t} = 1. \tag{1·19}$$

From this, the positiveness of $K_{\mathbf{n}}(\mathbf{t})$ and $P(\mathbf{r},\mathbf{t})$, and (1·16) and (1·17), we see that if $m \leqslant f \leqslant M$, both $\sigma_{\mathbf{N}}$ and $f(\mathbf{r},\mathbf{x})$ satisfy the same inequality.

Let $r \geqslant 1$. Applying Jensen's inequality to (1·16), and using the first equation (1·19) we obtain, successively,

$$|\sigma_{\mathbf{n}}(\mathbf{x})|^r \leqslant \pi^{-m} \int_Q |f(\mathbf{x}+\mathbf{t})|^r K_{\mathbf{n}}(\mathbf{t})\, d\mathbf{t},$$

$$\int_Q |\sigma_{\mathbf{n}}(\mathbf{x})|^r d\mathbf{x} \leqslant \int_Q |f(\mathbf{x})|^r d\mathbf{x}.$$

The inequalities hold for $f(\mathbf{r},\mathbf{x})$.

The following theorem is an extension of the classical theorems of Fejér and Poisson (see Chapter III, (3·4), (6·11)) to multiple Fourier series.

(1·20) THEOREM. *If f is bounded, then $S[f]$ is summable both* (C, 1) *and* A* *at each point of continuity, and the summability is uniform over any closed set of points of continuity.*

We shall see presently that the condition that f should be bounded is essential. Consider summability (C, 1). If $0<\delta<\pi$, then, by (1·14),

$$\int_{Q, |\mathbf{t}|\geqslant\delta} K_{\mathbf{n}}(\mathbf{t})\,d\mathbf{t}\to 0 \tag{1·21}$$

as $\mathbf{n}\to\infty$, since the domain of integration can be split into a finite number of subdomains in each of which at least one of the t_j satisfies $\delta m^{-\frac{1}{2}}\leqslant |t_j|\leqslant\pi$, so that the corresponding $K_{n_j}(t_j)$ tends uniformly to 0, while the integral of the product of the remaining K, even if taken over the whole of Q, remains bounded. Suppose now that $\mathbf{x}_0$ is a point of continuity of f. Write

$$\sigma_{\mathbf{n}}(\mathbf{x}_0)-f(\mathbf{x}_0)=\pi^{-m}\int_Q [f(\mathbf{x}_0+\mathbf{t})-f(\mathbf{x}_0)]\,K_{\mathbf{n}}(\mathbf{t})\,d\mathbf{t}, \tag{1·22}$$

and split the integral on the right into two, one extended over a small sphere $|\mathbf{t}|\leqslant\delta$, and the other over the remainder of Q. The first integral is small with δ, for all $\mathbf{n}$, and if δ is fixed the second is small as $\mathbf{n}\to\infty$. (Observe that $|f(\mathbf{x}_0+\mathbf{t})-f(\mathbf{x}_0)|\leqslant 2M$, where $M=\sup|f|$, and use (1·21).) Hence $\sigma_{\mathbf{n}}(\mathbf{x}_0)-f(\mathbf{x}_0)$ tends to 0, and uniformly in $\mathbf{x}_0$ if $\mathbf{x}_0$ is in a closed set of points of continuity of f. The argument is the same for summability A and, with minor modifications, for summability A*.

(1·23) THEOREM. *If $f\in L^p$, $1\leqslant p<\infty$, then $\|f-\sigma_{\mathbf{n}}\|_p\to 0$ as $\mathbf{n}\to\infty$; the result holds also for the Abel means.*

Here

$$\|f\|_p=\left\{\int_Q |f(\mathbf{x})|^p\,d\mathbf{x}\right\}^{1/p}.$$

Suppose that ϕ is continuous, periodic and such that $\|f-\phi\|_p<\epsilon$. Then

$$\|f-\sigma_{\mathbf{n}}[f]\|_p\leqslant\|f-\phi\|_p+\|\phi-\sigma_{\mathbf{n}}[\phi]\|_p+\|\sigma_{\mathbf{n}}[\phi-f]\|_p, \tag{1·24}$$

and the first and last terms on the right are less than ϵ, while the middle term tends to 0 as $\mathbf{n}\to\infty$. Hence $\|f-\sigma_{\mathbf{n}}[f]\|_p\to 0$.

We now consider problems of localization. Suppose that $f=0$ in a neighbourhood $|\mathbf{x}|\leqslant\delta$ of $\mathbf{0}$. We show by an example that this does not necessarily imply that $S_{\mathbf{n}}(\mathbf{0};f)$ converges to 0.

Suppose, for simplicity, that $m=2$ and set $f(\mathbf{x})=f_1(x_1)\,f_2(x_2)$, where f_1 and f_2 are continuous, $f_1(x)=0$ for $|x|\leqslant\delta$, and $S_n(0;f_2)$ is unbounded (Chapter VIII, § 1). Then $f(\mathbf{x})=0$ for $|\mathbf{x}|\leqslant\delta$, and

$$S_{\mathbf{n}}(\mathbf{0};f)=S_{n_1}(0;f_1)\,S_{n_2}(0;f_2). \tag{1·25}$$

Considering suitable f_1 we may suppose that $S_{n_1}(0;f_1)\neq 0$ for infinitely many n_1, so that, if n_2 increases sufficiently rapidly, the hypothesis $S_{n_2}(0;f_2)\neq O(1)$ implies that (1·25) is unbounded for $\mathbf{n}\to\infty$. Hence, even in the case of continuous functions, the principle of localization does not hold for spherical neighbourhoods.

Clearly, the principle of localization does hold for continuous, or even only bounded, functions and spherical neighbourhoods, provided we consider not the partial sums but the (C, 1) or Abel means; this is a consequence of Theorem (1·20). In the above assertion the condition of boundedness is essential and cannot be omitted. Take the example $f(\mathbf{x}) = f_1(x_1) f_2(x_2)$, but this time with f_2 integrable and with $\sigma_n(0; f_2) \neq O(1)$ (take say $f_2(x) = \Sigma n^{-1} \cos nx$). Then

$$\sigma_{\mathbf{n}}(\mathbf{0}, f) = \sigma_{n_1}(0, f_1)\, \sigma_{n_2}(0, f_2) \neq O(1),$$

though $f = 0$ near $\mathbf{x} = \mathbf{0}$.

This example shows that the condition of boundedness in (1·20) cannot be dropped.

Given an $\mathbf{x}^0 = (x_1^0, \ldots, x_n^0)$ and a $\delta > 0$, we call the set of $\mathbf{x}$ satisfying at least one of the inequalities

$$|x_1 - x_1^0| < \delta, \quad \ldots, \quad |x_n - x_n^0| < \delta \tag{1·26}$$

a *cross-neighbourhood* of $\mathbf{x}^0$. (For $m = 2$ it actually has a cross-like shape.) Its diameter does not tend to 0 with δ.

(1·27) THEOREM. *If $f = 0$ in a cross-neighbourhood of $\mathbf{x}_0$, then $S_{\mathbf{n}}(\mathbf{x}_0; f)$, $\sigma_{\mathbf{n}}(\mathbf{x}_0, f)$ and $f(\mathbf{r}, \mathbf{x}_0)$ all tend to 0 as $\mathbf{n} \to \infty$ or $\mathbf{r} \to \mathbf{1}$.*

For $S_{\mathbf{n}}$ the result is a consequence of the Riemann-Lebesgue theorem. For the $\sigma_{\mathbf{n}}$ and $f(\mathbf{r}, \mathbf{x})$ it is a consequence of the fact that $K_{\mathbf{n}}(\mathbf{t})$ and $P(\mathbf{r}, \mathbf{t})$ tend to 0 uniformly outside every cross-neighbourhood of $\mathbf{0}$. It is easy to see that the convergence of $\sigma_{\mathbf{N}}$ and $f(\mathbf{r}, \mathbf{x})$ is uniform in every m-dimensional interval $|x_j - x_j^0| \leqslant \delta'$, $j = 1, \ldots, m$, $\delta' < \delta$; this result holds also for the $S_{\mathbf{n}}$, but requires a somewhat more delicate proof, similar to that of Theorem (6·3) of Chapter II.

Theorem (1·27) expresses the principle of localization for cross-neighbourhoods.

2. Strong differentiability of multiple integrals and its applications

The problems of summability of Fourier series have close connexions with the problems of the differentiability of integrals. The differentiability of multiple integrals has features which do not appear in the one-dimensional case.

Let $f(\mathbf{x}) = f(x_1, \ldots, x_m)$ be integrable over the cube

$$Q: 0 \leqslant x_j \leqslant 1 \quad (j = 1, \ldots, m).$$

The integral of f over any subset E of Q is denoted by $F(E)$. We shall denote subintervals $a_j \leqslant x_j \leqslant b_j$, $j = 1, \ldots, m$, of Q by I.

We say that I *tends to* $\mathbf{x}$ if I contains $\mathbf{x}$ and if all dimensions $b_j - a_j$ tend to 0. If, in addition, the ratio of each two dimensions remains bounded, we say that I tends *restrictedly* to $\mathbf{x}$.

The classical theorem of Lebesgue asserts that for almost all x the ratio

$$\frac{F(I)}{|I|} = \frac{\int_I f(\mathbf{y})\, d\mathbf{y}}{|I|} \tag{2·1}$$

converges to $f(\mathbf{x})$ if I tends restrictedly to $\mathbf{x}$. Examples show that the condition of restrictedness is essential and cannot be omitted. (For a stronger statement see part (ii) of Theorem (2·2) below.)

If (2·1) has a finite limit as I tends unrestrictedly to $\mathbf{x}$, we say that F is *strongly differentiable* at $\mathbf{x}$. In view of the theorem of Lebesgue, if F is strongly differentiable in a set E, the limit of (2·1) must be $f(\mathbf{x})$ almost everywhere in E.

We write systematically

$$f^*(\mathbf{x}) = \sup_{I \ni \mathbf{x}} \frac{\int_I |f(\mathbf{y})|\, d\mathbf{y}}{|I|}.$$

For positive f the modulus sign in the numerator can be omitted.

(2·2) THEOREM. (i) *If* $|f|(\log^+|f|)^{m-1}$ *is integrable (in particular, if* f *is in any* L^p, $p > 1$), *the integral* F *of* f *is strongly differentiable almost everywhere.*

(ii) *Given any* $\phi(u)$, $u \geq 0$, *positive, increasing, and such that* $\phi(u) = o(u \log^{m-1} u)$ *for* $u \to \infty$, *there is an integrable* $f \geq 0$ *such that* $\phi(f)$ *is also integrable and* (2·1) *is unbounded at each* $\mathbf{x}$ *as* I *tends to* $\mathbf{x}$.

(iii) *With the norm* $\|f\|_p = \left(\int_Q |f|^p\, d\mathbf{y}\right)^{1/p}$, *we have*

$$\|f^*\|_p \leq A_{p,m} \|f\|_p \quad (p > 1), \tag{2·3}$$

$$\|f^*\|_1 \leq A_m \int_Q |f|(\log^+|f|)^m\, d\mathbf{y} + A_m, \tag{2·4}$$

$$\|f^*\|_\mu \leq A_{\mu,m} \int_Q |f|(\log^+|f|)^{m-1}\, d\mathbf{y} + A_{\mu,m} \quad (0 < \mu < 1). \tag{2·5}$$

We do not prove (ii); it is included here only to show that (i) cannot be improved. It is enough in (i) and (iii) to consider non-negative f.

We begin with (iii) and recall the following results (Chapter I, Theorem (13·15)). If $g(x)$, $0 \leq x \leq 1$, is integrable and non-negative, and $g^*(x) = \sup_{h \neq 0} \left\{h^{-1}\int_x^{x+h} g\, dt\right\}$†, then

$$\|g^*\|_p \leq A_p \|g\|_p \quad (p > 1), \tag{2·6}$$

$$\|g^*\|_1 \leq A\int_0^1 g \log^+ g\, dt + A, \tag{2·7}$$

$$\|g^*\|_\mu \leq A_\mu \|g\|_1 \quad (0 < \mu < 1). \tag{2·8}$$

From (2·7) we shall deduce that for $r \geq 2$ we have

$$\int_0^1 g^* . (\log^+ g^*)^{r-1}\, dt \leq A_r \int_0^1 g . (\log^+ g)^r\, dt + A_r. \tag{2·9}$$

It is enough to prove this for g non-increasing (see Chapter I, (13·4)), when

$$g^*(x) = x^{-1}\int_0^x g\, dt.$$

Since $\psi(u) = u(\log^+ u)^{r-1}$ is non-decreasing and convex, Jensen's inequality, together with (2·7), gives

$$\int_0^1 \psi(g^*)\, dx = \int_0^1 \psi\left(\frac{1}{x}\int_0^x g(t)\, dt\right) dx \leq \int_0^1 \left\{x^{-1}\int_0^x \psi(g)\, dt\right\} dx \leq A\int_0^1 \psi(g)\log^+\psi(g)\, dx + A$$

$$= A\int_0^1 g(\log^+ g)^{r-1}\log^+\{g . (\log^+ g)^{r-1}\}\, dx + A \leq A_r\int_0^1 g . (\log^+ g)^r\, dx + A_r,$$

and (2·9) follows.

† We use a different notation now; the present g^* is the Θ_g of Chapter I, § 13.

Given any integrable $f(\mathbf{x}) = f(x_1, \ldots, x_m)$ we consider, for each $j = 1, 2, \ldots, m$, functions

$$M_j\{f\} = M_j(\mathbf{x}, f) = \sup_{h \neq 0} \left\{ h^{-1} \int_{x_j}^{x_j+h} f(x_1, \ldots, x_{j-1}, t, x_{j+1}, \ldots, x_m)\, dt \right\}. \qquad (2\cdot10)$$

The M_j are measurable and, if $I \ni \mathbf{x}$, the ratio (2·1) does not exceed $M_m M_{m-1} \ldots M_1\{f\}$ taken at $\mathbf{x}$. In particular

$$f^* \leqslant M_m M_{m-1} \ldots M_1\{f\}. \qquad (2\cdot11)$$

Each M_j, being of the type g^*, satisfies with respect to x_j inequalities analogous to (2·6), (2·7), (2·8) and (2·9). From this (iii) follows easily. Take $m = 2$, a case entirely typical. By (2·11) and (2·6),

$$\begin{aligned} \| f^* \|_p^p &\leqslant \int_0^1 dx_1 \int_0^1 (M_2 M_1\{f\})^p\, dx_2 \leqslant A_p^p \int_0^1 dx_1 \int_0^1 M_1^p\{f\}\, dx_2 \\ &= A_p^p \int_0^1 dx_2 \int_0^1 M_1^p\{f\}\, dx_1 \leqslant A_p^{2p} \int_0^1 dx_2 \int_0^1 f^p\, dx_1 \\ &= A_p^{2p} \| f \|_p^p, \end{aligned}$$

which gives (2·3). We prove (2·4) similarly. As regards (2·5) for, say, $m = 2$ we have, using Hölder's inequality at the appropriate place,

$$\begin{aligned} \| f^* \|_\mu^\mu &\leqslant \int_0^1 dx_1 \int_0^1 (M_2 M_1\{f\})^\mu\, dx_2 \leqslant A_\mu^\mu \int_0^1 dx_1 \left(\int_0^1 M_1\{f\}\, dx_2 \right)^\mu \\ &\leqslant A_\mu^\mu \left(\int_0^1 dx_1 \int_0^1 M_1\{f\}\, dx_2 \right)^\mu = A_\mu^\mu \left(\int_0^1 dx_2 \int_0^1 M_1\{f\}\, dx_1 \right)^\mu \\ &\leqslant A_\mu^\mu \left\{ \int_0^1 dx_2 \left(A \int_0^1 f \log^+ f\, dx_1 + A \right) \right\}^\mu. \end{aligned}$$

This gives (2·5) and completes the proof of (iii).

Part (i) is a simple corollary of (2·5). If we apply it to kf, where k is a positive constant, we obtain

$$\| f^* \|_\mu \leqslant A_{\mu, m} \int_Q |f| (\log^+ |kf|)^{m-1} d\mathbf{y} + k^{-1} A_{\mu, m}.$$

The last term is arbitrarily small for k large enough. Fix k and write $f = f_1 + f_2$, where f_1 is continuous and the integral of $|f_2| (\log^+ |kf_2|)^{m-1}$ small. By the last inequality with, say, $\mu = \frac{1}{2}$, f_2^* is small except in a set E_2 of small measure. In particular, if $\mathbf{x}$ is not in E_2 and I contains $\mathbf{x}$, $F_2(I)/|I|$ is small. On the other hand, $F_1(I)/|I|$ converges uniformly to $f_1(\mathbf{x})$ as I converges to $\mathbf{x}$; and f_1 differs little from f, except in a set E_1 of small measure. Hence, if $\mathbf{x}$ is not in the small set $E_1 + E_2$,

$$\frac{F(I)}{|I|} = \frac{F_1(I)}{|I|} + \frac{F_2(I)}{|I|}$$

differs little from $f(\mathbf{x})$, provided I contains $\mathbf{x}$ and is small enough. This completes the proof of (i).

Clearly (2·2) holds for functions defined in any finite interval, though the constants A in (2·4) and (2·5) will then depend also on the interval.

Consider a set E in E^m. We say that $\mathbf{x}$ is a point of *strong density* for E, if $|EI|/|I|$ tends to 1 when I tends unrestrictedly to $\mathbf{x}$. Theorem (2·2) (i) applied to the characteristic function of E implies that *almost all points of E are points of strong density for E.*

We now apply (2·2) to Fourier series. Given a periodic $f(\mathbf{x})=f(x_1, \ldots, x_m)$, denote by $\sigma_{\mathbf{N}}$ the (C, 1) means of $S[f]$ and write

$$\sigma_*(\mathbf{x})=\sup_{\mathbf{N}} |\sigma_{\mathbf{N}}(\mathbf{x})|. \tag{2·12}$$

Also, fixing a δ with $0<\delta<1$, denote by $\Omega_\delta(x)$, or simply by $\Omega(x)$, where $0 \leqslant x \leqslant 2\pi$, the domain limited by the two tangents from e^{ix} to the circle $|z|=\delta$ and the more distant arc of the circle, and by $\Omega(\mathbf{x})$ the Cartesian product of the domains $\Omega(x_1), \ldots, \Omega(x_m)$. We write

$$f_*(\mathbf{x})=\sup_{(\mathbf{r},\mathbf{y})\in\Omega(\mathbf{x})} |f(\mathbf{r},\mathbf{y})|, \tag{2·13}$$

where $f(\mathbf{r},\mathbf{y})$ is the Abel mean of $S[f]$.

(Note that f^* and f_* have totally different meanings.)

(2·14) THEOREM. (i) *If* $|f|(\log^+|f|)^{m-1}$ *is integrable (in particular if* $f\in L^p, p>1$), *then* $S[f]$ *is summable both* (C, 1) *and* A* *at almost all* $\mathbf{x}$.

(ii) *Given any* $\phi(u)$, $u\geqslant 0$, *positive, increasing and* $o(u\log^{m-1}u)$ *for* $u\to\infty$, *there is a periodic and integrable* $f\geqslant 0$ *such that* $\phi(f)$ *is integrable and* $S[f]$ *is nowhere summable either* (C, 1) *or* A.

(iii) *We have*

$$\|\sigma_*\|_p \leqslant A_{p,m}\|f\|_p \quad (p>1), \tag{2·15}$$

$$\|\sigma_*\|_1 \leqslant A_m\int_Q |f|(\log^+|f|)^m d\mathbf{x}+A_m, \tag{2·16}$$

$$\|\sigma_*\|_\mu \leqslant A_{\mu,m}\int_Q |f|(\log^+|f|)^{m-1} d\mathbf{x}+A_{\mu,m} \quad (0<\mu<1); \tag{2·17}$$

and analogous inequalities hold for f_* *(with A's depending also on* δ).

We begin with (ii). Suppose that the f of (2·2) (ii) is initially defined for $-\pi\leqslant x_j<\pi$, $j=1, \ldots, m$, and then continued periodically. Let $\mathbf{x}^0$ be interior to Q. There is a sequence of intervals I tending to $\mathbf{x}$ such that $F(I)/|I|$ tends to $+\infty$. We may suppose that the I's have $\mathbf{x}^0$ for centre and, moreover, are of the form

$$I_{n_1,\ldots,n_m}: \quad -1/n_j\leqslant t_j-x_j^0\leqslant 1/n_j \quad (j=1, \ldots, m),$$

with n_j integral. Since $K_{\mathbf{n}}\geqslant 0$, and $K_n(t)\geqslant An$ for $|t|\leqslant 1/n$, we deduce from (1·16) that

$$\sigma_{n_1,\ldots,n_m}(\mathbf{x}^0)\geqslant A_m n_1\ldots n_m\int_{I_{n_1\ldots n_m}} f(\mathbf{t})\,d\mathbf{t}=A_m\frac{F(I_{n_1,\ldots,n_m})}{|I_{n_1,\ldots,n_m}|}. \tag{2·18}$$

Hence $\sigma_{\mathbf{n}}(\mathbf{x}^0)$ is unbounded as $\mathbf{n}\to\infty$, and $S[f]$ is not summable (C, 1) at $\mathbf{x}_0$. Clearly if $g(\mathbf{x})=f(\mathbf{x})+f(\mathbf{x}+\pi\mathbf{1})$, $S[g]$ is not summable (C, 1) anywhere, not even on the boundary of Q.

The same argument works for summability A.

In the proof of (iii) we may suppose that $f\geqslant 0$. We recall (Chapter IV, (7·8)) that if the σ_n are the (C, 1) means for a $g(x)\geqslant 0$, then

$$\sigma_*(x)\leqslant Ag^*(x), \tag{2·19}$$

where now $g^*(x)=\sup_{|h|\leqslant\pi} h^{-1}\int_x^{x+h} g\,dt$; this g^* satisfies inequalities analogous to (2·6)–(2·8), with norms taken over $(0, 2\pi)$.

Suppose now that we modify the definition (2·10) of M_j by replacing the condition $h \neq 0$ by $|h| \leqslant \pi$. Then starting from

$$\sigma_{\mathbf{n}}(\mathbf{x}) = \pi^{-m}\int_{-\pi}^{\pi} \dots \int_{-\pi}^{\pi} f(x_1+t_1, \dots, x_m+t_m) K_{n_1}(t_1) \dots K_{n_m}(t_m)\, dt_1 \dots dt_m$$

and using (2·19) we deduce the inequality

$$\sigma_* \leqslant A^m M_m M_{m-1} \dots M_1 f,$$

analogous to (2·11), and the proof of (iii) runs parallel to that of (2·2) (iii) (the new M's behave like the old ones, with new constants). Similarly for the method A* (see Chapter IV, (7·8)).

For the proof of (i) we write $f = f' + f''$, where f' is a trigonometric polynomial and $\| f'' \|_1$ is small (see (1·23)). Then

$$|\sigma_{\mathbf{n}}(\mathbf{x}; f) - \sigma_{\mathbf{n}'}(\mathbf{x}; f)| \leqslant |\sigma_{\mathbf{n}}(\mathbf{x}; f'') - \sigma_{\mathbf{n}'}(\mathbf{x}; f'')| + o(1) \leqslant 2\sigma_*(\mathbf{x}; f'') + o(1),$$

where the $o(1)$ is uniform in $\mathbf{x}$. Since, by (2·17), $\sigma_*(\mathbf{x}; f'')$ is small except in a set of small measure, the convergence of $\{\sigma_{\mathbf{n}}(\mathbf{x}; f)\}$ almost everywhere follows without difficulty. Similarly for the method A*.

It must be stressed that, in contrast to the case of a single variable (see Chapter III, (3·9), (7·9)), Theorem (2·14) (i) does not specify the points at which summability (C, 1), or A, happens. It is possible to state conditions, satisfied almost everywhere, which guarantee summability C, or A, at individual points, but they are inevitably rather complicated in the absence of a theorem of localization.

3. Restricted summability of Fourier series

We say that $\mathbf{n} = (n_1, \dots, n_m)$ tends *restrictedly* to ∞, if the n_j tend to $+\infty$ in such a way that all the ratios n_j/n_k $(j, k = 1, \dots, m)$ remain bounded. Similarly $\mathbf{r} = (r_1, \dots, r_m)$ tends restrictedly to $\mathbf{1}$, if each r_j tends to 1 and all the ratios $(1-r_j)/(1-r_k)$ remain bounded.

Let $\sigma_{\mathbf{n}} = \sigma_{n_1, \dots, n_m}$ be the (C, 1) means of a series S. We say that S is *restrictedly summable* (C, 1) to sum s, if $\sigma_{\mathbf{n}}$ tends to s as $\mathbf{n}$ tends restrictedly to ∞. Similarly, we define *restricted summability* A. If, in the definition of summability A* in § 1, we introduce the condition that all the ratios $(1-r_j)/(1-r_k)$ are to remain bounded, we obtain the definition of *restricted summability* A*.

(3·1) Theorem. *Each* $S[f]$ *is almost everywhere restrictedly summable, both* (C, 1) *and* A*, *to sum* f.

Restricted summability of series is analogous to restricted differentiability of integrals, and one might expect that (3·1) would follow fairly easily from the restricted differentiability of indefinite integrals. Actually the proof of (3·1) is not so easy. We give it in the sufficiently typical case $m=2$. We need first some lemmas. By rectangles, we shall mean rectangles with sides parallel to the axes.

(3·2) Lemma. *Let* $h(t)$ *and* $k(t)$ *be two continuous functions of* $t \geqslant 0$, *strictly increasing to* $+\infty$ *and* 0 *at* $t=0$. *Let* E *be a plane set of finite positive outer measure. Suppose that with each* (x, y) *in* E *we associate a rectangle* $R = R_{x,y}$ *with centre* (x, y) *and sides* $2h(t)$, $2k(t)$, *where t depends on (x, y). Then we can find a finite number of disjoint rectangles $R_j = R_{x_j, y_j}$, $j = 1, 2, \ldots, n$, such that*

$$\sum_1^n |R_j| > \tfrac{1}{26}|E|. \tag{3·3}$$

The proof which follows effectively shows that the $\frac{1}{26}$ in (3·3) can be replaced by any constant less than $\frac{1}{9}$, but this is of no significance; the point is that the numerical factor is independent of the functions h, k. For applications to (3·1) we need only the case $h(t) = t$, $k(t) = \alpha t$, where α is a constant, but the proof is no simpler in this case.

Denote by K_1 the family of all R's associated with E, and let t_1^* be the upper bound of the corresponding t's. We may suppose that $t_1^* < \infty$, for otherwise it is enough to select a single R with $|R| > \frac{1}{26}|E|$. Take an R_1 such that if t_1 is the corresponding value of t, then

$$h(t_1) > \tfrac{1}{2}h(t_1^*), \quad k(t_1) > \tfrac{1}{2}k(t_1^*).$$

Denote by K_2' the family of those R in K_1 which have points in common with R_1, and by K_2 the set of the remaining R; thus $K_1 = K_2' + K_2$. It is easily seen that the rectangle $\bar{R}_1$ obtained by expanding R_1 about its centre by a factor 5 contains all the R's from K_2'.

We repeat the argument, starting this time from K_2. Let t_2^* be the upper bound of all relevant t's, and select from K_2 an R_2 such that if t_2 is the corresponding t we have

$$h(t_2) > \tfrac{1}{2}h(t_2^*), \quad k(t_2) > \tfrac{1}{2}k(t_2^*).$$

Denote by K_3' the set of those $R \in K_2$ which have points in common with R_2, and by K_3 the set of the remaining $R \in K_2$. Hence $K_2 = K_3' + K_3$, and all the R in K_3' are contained in the rectangle $\bar{R}_2$ obtained by expanding R_2 likewise by a factor 5.

The construction is now clear. We obtain a sequence of numbers $t_1^*, t_2^*, \ldots$, finite or not (in the former case K_j is empty for some j), and a sequence of disjoint rectangles $R_1, R_2, \ldots$. Suppose first that $\{t_j^*\}$ is infinite. Since $t_1^* \geq t_2^* \geq t_3^* \geq \ldots$, there are two possibilities: (i) the t_m^* stay above a positive number, (ii) $t_m^* \to 0$.

In case (i), (3·3) is obvious for n large enough. In case (ii) it is easy to see that each R in K_1 is contained in some $\bar{R}_j$. For otherwise a certain rectangle R' would belong to K_j for each j, which is impossible since the dimensions of the rectangles from K_j do not exceed $2h(t_j^*)$, $2k(t_j^*)$, and so tend to 0 as $j \to \infty$.

Since E is contained in the union of the $R \in K_1$, E is contained in $\bar{R}_1 + \bar{R}_2 + \ldots$. Hence

$$|E| \leq \sum_1^\infty |\bar{R}_j| = 25 \sum_1^\infty |R_j|, \tag{3·4}$$

which gives (3·3) for n large enough.

If $\{R_j\}$ is finite and ends with R_n, we have (3·4) with n for ∞, and (3·3) holds *a fortiori*. This completes the proof of (3·2).

(3·5) Lemma. *Let Q be the square $|x| \leq \pi$, $|y| \leq \pi$, and Q' the square $|x| \leq 2\pi$, $|y| \leq 2\pi$. Suppose that f is integrable over Q', and for each (x, y) in Q write*

$$f_*(x, y) = \sup_t \frac{1}{4hk} \int_{-h}^{h} \int_{-k}^{k} |f(x+u, y+v)|\, du\, dv, \tag{3·6}$$

where $h(t)$ and $k(t)$ are the functions of Lemma (3·2) and t is so small that the domain of *integration is contained in Q'. Then the set $\mathscr{E}_*(\xi)$ of points of Q for which $f_*(x,y)>\xi$ satisfies*

$$|\mathscr{E}_*(\xi)| \leqslant 26\xi^{-1}\int_{Q'} |f(x,y)|\,dx\,dy. \tag{3·7}$$

If (x,y) is in $\mathscr{E}_*(\xi)$, there is a rectangle R with centre (x,y) and sides $2h(t)$, $2k(t)$ such that the ratio on the right of (3·6) exceeds ξ. By (3·2), there are a finite number of such rectangles disjoint and satisfying (3·3) with $E=\mathscr{E}_*(\xi)$; and the lemma follows from

$$\int_{Q'} |f|\,dx\,dy \geqslant \sum_1^n \int_{R_j} |f|\,dx\,dy > \tfrac{1}{26}\xi\,|\mathscr{E}_*(\xi)|.$$

(3·8) Lemma. *Let $h(t)$, $k(t)$, f be the functions of Lemma (3·5), let α and β be fixed positive numbers, and let $f_*^{\alpha,\beta}(x,y)$ be the function f_* of Lemma (3·5) with $\alpha h(t)$, $\beta k(t)$ for $h(t)$, $k(t)$. For (x,y) in Q let*

$$f^*(x,y) = \sup_{i,j}\{f_*^{2^i,2^j}(x,y)\,2^{-\frac{1}{2}(i+j)}\} \quad (i,j=0,1,2,\ldots). \tag{3·9}$$

Then for the set $\mathscr{E}^*(\xi)$ of points (x,y) in Q at which $f^*(x,y)>\xi$ we have

$$|\mathscr{E}^*(\xi)| \leqslant A\xi^{-1}\int_{Q'} |f|\,dx\,dy. \tag{3·10}$$

Let $\mathscr{E}_*^{\alpha,\beta}(\xi)$ be the set $\mathscr{E}_*(\xi)$ of Lemma (3·5) with $\alpha h(t)$, $\beta k(t)$ for $h(t)$, $k(t)$. Since $f^*(x,y)>\xi$ if and only if $f_*^{2^i,2^j}(x,y)>\xi\,2^{\frac{1}{2}(i+j)}$ for some non-negative integers i, j, the set $\mathscr{E}^*(\xi)$ is included in the union of all $\mathscr{E}_*^{2^i,2^j}(\xi\,2^{\frac{1}{2}(i+j)})$, and

$$|\mathscr{E}^*(\xi)| \leqslant \sum_{i,j} |\mathscr{E}_*^{2^i,2^j}(\xi\,2^{\frac{1}{2}(i+j)})| \leqslant 26\xi^{-1}\Big(\sum_{i,j} 2^{-\frac{1}{2}(i+j)}\Big)\int_{Q'} |f|\,dx\,dy,$$

which gives (3·10).

(3·11) Lemma. *Let $\sigma_{\mu\nu}$ be the* (C, 1) *mean of* S[*f*], *and let*

$$\sigma_*^\alpha(x,y) = \sup_{\mu,\nu} |\sigma_{\mu\nu}(x,y)| \quad \textit{for} \quad \alpha^{-1}\leqslant \nu/\mu \leqslant \alpha. \tag{3·12}$$

Then the set of points (x,y) of Q for which $\sigma_*^\alpha(x,y)\geqslant\xi$ has measure not greater than

$$A_\alpha\xi^{-1}\int_Q |f|\,dx\,dy.$$

In the proof, f_* and f^* denote the functions of Lemmas (3·5) and (3·8) for $h(t)=k(t)=t$. In view of Lemma (3·8), it is enough to show that

$$\sigma_*^\alpha(x,y) \leqslant A_\alpha f^*(x,y). \tag{3·13}$$

Clearly $|\sigma_{\mu\nu}(x,y)|$ is majorized by a sum of four integrals of which a typical one is

$$\tau_{\mu\nu}(x,y) = \frac{1}{\pi^2}\int_0^\pi\int_0^\pi |f(x+u,y+v)|\,K_\mu(u)\,K_\nu(v)\,du\,dv, \tag{3·14}$$

and for the proof of (3·13) it is enough to show that

$$\tau_{\mu\nu}(x,y) \leqslant A_\alpha f^*(x,y). \tag{3·15}$$

Let $\lambda=\min(\mu,\nu)\geqslant 1$. Since $K_n(t)$ is majorized in $(0,\pi)$ by n and by $An^{-1}t^{-2}$, both $K_\mu(t)$ and $K_\nu(t)$ are majorized, with a suitable A_α, by either of the expressions

$$A_\alpha\lambda, \quad A_\alpha\lambda^{-1}t^{-2}. \tag{3·16}$$

Split the domain of integration in (3·14) into four rectangles by the lines $u=1/\lambda$, $v=1/\lambda$. Using both estimates (3·16), we easily see that $\tau_{\mu\nu}(x,y)$ is majorized by

$$A_\alpha\int_0^{1/\lambda}du\int_{1/\lambda}^{\pi}v^{-2}|f|\,dv+A_\alpha\int_0^{1/\lambda}dv\int_{1/\lambda}^{\pi}u^{-2}|f|\,du$$

$$+A_\alpha\lambda^{-2}\int_{1/\lambda}^{\pi}\int_{1/\lambda}^{\pi}u^{-2}v^{-2}|f|\,du\,dv+A_\alpha\lambda^2\int_0^{1/\lambda}\int_0^{1/\lambda}|f|\,du\,dv$$

$$=A_\alpha P_\lambda+A_\alpha Q_\lambda+A_\alpha R_\lambda+A_\alpha S_\lambda,$$

say, where f stands for $f(x+u,y+v)$. The inequality (3·15) will be proved if we show that P_λ, Q_λ, R_λ, S_λ are all majorized by $Af^*(x,y)$.

Let L be the integer satisfying $\pi\leqslant 2^L/\lambda<2\pi$. Then

$$P_\lambda(x,y)\leqslant\sum_{j=1}^{L}\int_0^{1/\lambda}du\int_{2^{j-1}/\lambda}^{2^j/\lambda}v^{-2}|f|\,dv\leqslant 4\lambda^2\sum_{j=1}^{L}2^{-2j}\int_{-1/\lambda}^{1/\lambda}du\int_{-2^j/\lambda}^{2^j/\lambda}|f|\,du$$

$$\leqslant 16\sum_{j=1}^{L}2^{-j}f_*^{1,2^j}(x,y)\leqslant 16f^*(x,y)\sum_{j=1}^{\infty}2^{-\frac{1}{2}j},$$

so that $P_\lambda\leqslant Af^*(x,y)$, and it is clear that the same inequality holds for Q_λ. Similarly

$$R_\lambda(x,y)\leqslant\lambda^{-2}\sum_{i,j=1}^{L}\int_{2^{i-1}/\lambda}^{2^i/\lambda}\int_{2^{j-1}/\lambda}^{2^j/\lambda}u^{-2}v^{-2}|f|\,du\,dv$$

$$\leqslant 16\lambda^2\sum_{i,j=1}^{L}2^{-2(i+j)}\int_{-2^i/\lambda}^{2^i/\lambda}\int_{-2^j/\lambda}^{2^j/\lambda}|f|\,du\,dv$$

$$\leqslant 64\sum_{i,j=1}^{L}2^{-(i+j)}f_*^{2^i,2^j}(x,y)$$

$$\leqslant 64f^*(x,y)\sum_{i,j=1}^{\infty}2^{-\frac{1}{2}(i+j)}=Af^*(x,y).$$

Since $S_\lambda(x,y)\leqslant 4f_*(x,y)\leqslant 4f^*(x,y)$, the lemma follows.

By the standard decomposition $f=f'+f''$, where f' is a trigonometric polynomial and $\|f''\|_1$ is small (cf. (1·23)), we deduce from (3·11) that $\sigma_{\mu\nu}(x,y)$ converges almost everywhere as $\mu,\nu\to\infty$, provided $\alpha^{-1}\leqslant\nu/\mu\leqslant\alpha$. Taking $\alpha=1,2,3,\ldots$ we see that $\sigma_{\mu\nu}(x,y)$ converges restrictedly almost everywhere. By (1·23) it must converge almost everywhere to limit f. This completes the proof of the part of (3·1) concerning summability (C, 1).

The argument goes without change for summability A since the Poisson kernel satisfies the same inequalities as Fejér's, and the extension to A* does not introduce new difficulties.

Let $F(E)$ be a countably additive function of sets defined for subsets of the semi-open cube (torus)

$$Q_0:\quad -\pi\leqslant x_j<\pi\quad (j=1,2,\ldots,m).$$

If desirable, we can extend $F(E)$ to the whole E^m by assigning identical distributions to all cubes obtained from Q_0 by translations $2\pi\mathbf{n}$. It is well known that $F(E)$ has at almost all points $\mathbf{x}$ a restricted derivative $F'(\mathbf{x})$, by which we mean the limit of $F(I)/|I|$ as I tends restrictedly to $\mathbf{x}$. The derivative of $F(E)$ is almost everywhere the same as the derivative of the absolutely continuous component of $F(E)$.

We may consider the Fourier series of $F(d\mathbf{x})$ (a *Fourier-Stieltjes* series)†

$$F(d\mathbf{x}) \sim \Sigma c_{\mathbf{n}} e^{i(\mathbf{nx})}, \tag{3·17}$$

where

$$c_{\mathbf{n}} = (2\pi)^{-m} \int_{Q_0} e^{-i(\mathbf{ny})} F(d\mathbf{y})$$

are the Fourier coefficients of $F(d\mathbf{x})$. The following theorem generalizes (3·1).

(3·18) THEOREM. *The series* (3·17) *is almost everywhere restrictedly summable both* (C, 1) *and* A* *to sum* $F'(\mathbf{x})$.

In view of (3·1) it is enough to prove (3·18) in the case when $F(E)$ is purely singular, that is, when $F'(\mathbf{x}) = 0$ almost everywhere. We may also confine our attention to summability (C, 1). Going over the proofs of Lemmas (3·5), (3·8) and (3·11), we verify that the measure of the set of points where $\sigma_*^\alpha(x, y) > \xi$ does not exceed $A_\alpha \xi^{-1} \int_{Q_0} F(dx\,dy)$, so that $\sigma_{\mu\nu}(x, y) = O(1)$ at almost all points if (μ, ν) tends restrictedly to ∞. The proof, however, that $\sigma_{\mu\nu}(x, y)$ tends restrictedly to 0 almost everywhere is somewhat elaborate, and the argument in the case $m = 2$ would no longer be typical. We take therefore general m, and prove the following lemma:

(3·19) LEMMA. *If* $F(E) = 0$ *for all* E *contained in an interval*

$$I: \quad |x_j - x_j^0| \leqslant a_j \quad (j = 1, \dots, m),$$

then the (C, 1) *means* $\sigma_{\mathbf{n}}$ *of* (3·17) *tend restrictedly to* 0 *at almost all points of* I.

Denote by I_δ the interval I with $a_j - \delta$ for a_j, where $\delta > 0$ and $j = 1, \dots, m$. It is enough to prove that the $\sigma_{\mathbf{n}}$ tend restrictedly to 0 almost everywhere in I_δ. We may suppose that $F \geqslant 0$. If $\mathbf{x} \in I_\delta$, then in

$$\sigma_{\mathbf{n}}(\mathbf{x}) = \pi^{-m} \int_{Q_0} K_{\mathbf{n}}(\mathbf{x} - \mathbf{t})\, F(d\mathbf{t}) \tag{3·20}$$

we need only integrate over the union of m (overlapping) sets each characterized by a single inequality $|x_j - t_j| \geqslant \delta$, $j = 1, \dots, m$. Since $F \geqslant 0$, it is enough to show that the integral extended over one of these sets, say over $|x_1 - t_1| \geqslant \delta$, tends restrictedly to 0 almost everywhere in I_δ. Denote by Q_0' the $(m-1)$-dimensional interval $-\pi \leqslant x_j < \pi$, $j = 2, \dots, m$, and for $E' \subset Q_0'$ write $\Phi(E') = F(E)$, where E is the set of points of Q_0 that project on to points of E'. Then the integral in question is majorized by

$$\pi^{-m} \max_{\delta \leqslant t \leqslant \pi} K_{n_1}(t) \int_{Q_0'} K_{\mathbf{n}'}(\mathbf{x}' - \mathbf{y}')\, \Phi(d\mathbf{y}'),$$

where $\mathbf{n}' = (n_2, \dots, n_m)$, $\mathbf{x}' = (x_2, \dots, x_m)$, $\mathbf{y}' = (y_2, \dots, y_m)$. We already know that the last integral is bounded almost everywhere in Q_0' as $\mathbf{n}'$ tends restrictedly to ∞, and since the preceding factor tends to 0 with $1/n_1$, the product tends to 0 and the lemma follows.

We now complete the proof of (3·18) for summability (C, 1); we may suppose that $m = 2$ and that F is singular. Then given any $\epsilon > 0$ we can find an open set $O \subset Q_0$, with $|Q_0 - O|$ arbitrarily small, such that $\int_O |F(d\mathbf{x})| < \epsilon$. Let $F_1(E)$ and $F_2(E)$ denote respectively $F(OE)$ and $F(E - O)$, and let $\sigma'_{\mu\nu}$ and $\sigma''_{\mu\nu}$ be the (C, 1) means of the Fourier series of

† Since $F(E)$ is a set function, the notation $F(d\mathbf{x})$ seems preferable to $dF(\mathbf{x})$.

F_1 and F_2. By (3·19), the $\sigma''_{\mu\nu}$ tend restrictedly to 0 almost everywhere in O, and, as we know, the measure of the subset of Q_0 where the upper bound of the $|\sigma'_{\mu\nu}|$, subject to $\alpha^{-1} \leqslant \nu/\mu \leqslant \alpha$, exceeds $\epsilon^{\frac{1}{2}}$, is less than $A_\alpha \epsilon^{-\frac{1}{2}} \int_{Q_0} |F_1(d\mathbf{x})| < A_\alpha \epsilon^{\frac{1}{2}}$. From this we easily deduce that $\sigma_{\mu\nu} = \sigma'_{\mu\nu} + \sigma''_{\mu\nu}$ tends restrictedly to 0 almost everywhere.

Let $\sigma_\mathbf{n}$ and $f(\mathbf{r}, \mathbf{x})$ be the (C, 1) and Abel means of a series $\Sigma c_\mathbf{n} e^{i(\mathbf{nx})}$. The theorem which follows will be needed in § 5 below.

(3·21) THEOREM. (i) *Either of the conditions*

$$\int_Q |\sigma_\mathbf{n}(\mathbf{x})| \, d\mathbf{x} \leqslant M < \infty, \quad \int_Q |f(\mathbf{r}, \mathbf{x})| \, d\mathbf{x} \leqslant M < \infty \tag{3·22}$$

is both necessary and sufficient for $\Sigma c_\mathbf{n} e^{i(\mathbf{nx})}$ to be the Fourier series of a mass distribution.

(ii) *Either of the conditions*

$$\sigma_\mathbf{n}(\mathbf{x}) \geqslant 0, \quad f(\mathbf{r}, \mathbf{x}) \geqslant 0 \tag{3·23}$$

is both necessary and sufficient for $\Sigma c_\mathbf{n} e^{i(\mathbf{nx})}$ to be an $S[dF]$ with $dF \geqslant 0$.

Consider, for example, the $\sigma_\mathbf{n}$. If we have (3·20), then

$$|\sigma_\mathbf{n}(\mathbf{x})| \leqslant \pi^{-m} \int_{Q_0} K_\mathbf{n}(\mathbf{x} - \mathbf{y}) \, |F(d\mathbf{y})|.$$

Integrating this with respect to $\mathbf{x}$ over Q, and interchanging the order of integration on the right, we get

$$\int_Q |\sigma_\mathbf{n}(\mathbf{x})| \, d\mathbf{x} \leqslant \int_{Q_0} |F(d\mathbf{y})| \, \pi^{-m} \int_Q K_\mathbf{n}(\mathbf{x} - \mathbf{y}) \, d\mathbf{x} = \int_{Q_0} |F(d\mathbf{y})|.$$

This gives the necessity part of (i). The necessity in (ii) is proved in the same way as for ordinary Fourier series, from (3·20).

Conversely, consider the first inequality (3·22) and denote by $F_\mathbf{N}(E)$ the mass distribution with density $\sigma_\mathbf{N}(\mathbf{x})$. Then the total mass of $F_\mathbf{N}(E)$ over Q does not exceed M. Therefore† we can find m sequences $N_{1,k}, N_{2,k}, \ldots, N_{m,k}$, $k = 1, 2, \ldots$, each tending to ∞, and a mass distribution $F(E)$ such that for every m-dimensional interval $I \subset Q$ we have

$$F(I) = \lim_{k \to \infty} F_{\mathbf{N}_k}(I)$$

provided the total mass of F over the boundary of I is 0. Now, if $|n_i| \leqslant N_i$,

$$c_{n_1, \ldots, n_m} \left(1 - \frac{|n_1|}{N_1 + 1}\right) \cdots \left(1 - \frac{|n_m|}{N_m + 1}\right) = \pi^{-m} \int_Q e^{-i(\mathbf{ny})} \sigma_\mathbf{N}(\mathbf{y}) \, d\mathbf{y} = \pi^{-m} \int_Q e^{-i(\mathbf{ny})} F_\mathbf{N}(d\mathbf{y})$$

and the exponential function is continuous. Hence, decomposing Q into a large number of non-overlapping intervals I with the property just stated,‡ substituting $N_{1,k}, \ldots, N_{m,k}$ for $N_1, \ldots, N_m$, and making $k \to \infty$, we find that the $c_\mathbf{n}$ are the Fourier coefficients of F. This proves the sufficiency part of (i), and the sufficiency part of (ii) follows if we note that $\sigma_\mathbf{N} \geqslant 0$ implies $F_\mathbf{N} \geqslant 0$.

† We use here the m-dimensional analogue of Theorem (4·6) of Chapter IV. Cf. the references in the notes to Chapter XVII at the end of the volume.

‡ If the boundary of Q carries a non-zero mass we have to replace Q by a suitable translation of Q.

4. Power series of several variables

In this and the next section we consider functions of m complex variables $z_1, z_2, \ldots, z_m$. We write $z_k = x_k + iy_k$, $z_k = r_k e^{i\theta_k}$ (and, if no confusion arises, also $z_k = r_k e^{ix_k}$) and $\mathbf{z} = (z_1, \ldots, z_m)$. The set of all points $\mathbf{z}$ is denoted by Z_m.

The (open) subset of Z_m defined by the inequalities

$$|z_1| < 1, \quad |z_2| < 1, \quad \ldots, \quad |z_m| < 1$$

will be called the *unit m-cylinder* and denoted by Δ_m; it is the Cartesian product of m open unit disks and has dimension $2m$. The *boundary* B_m of Δ_m has dimension $2m-1$ and is the union, for $j = 1, \ldots, m$, of the Cartesian products of the circumferences $|z_j| = 1$ and the disks $|z_k| \leqslant 1$, $k \neq j$. The set

$$|z_1| = 1, \quad |z_2| = 1, \quad \ldots, \quad |z_m| = 1$$

is a subset of B_m; it has dimension m and will be called the *extremal boundary* of Δ_m; we denote† it by Γ_m. In certain problems it plays a more significant role than the full boundary B_m.

A function $u(z_1, \ldots, z_m)$ defined and continuous in a domain $D \subset Z_m$ will be called *m-harmonic* if it is harmonic in each pair of variables x_k, y_k; it is then a *a fortiori* harmonic in the totality of variables $x_1, \ldots, y_m$.

It is clear that each $u(z_1, \ldots, z_m)$ which is representable in Δ_m by an absolutely convergent series

$$\sum_{-\infty}^{+\infty} c_{n_1, \ldots, n_m} r_1^{|n_1|} \ldots r_m^{|n_m|} e^{i(n_1\theta_1 + \ldots + n_m\theta_m)} \tag{4·1}$$

is m-harmonic in Δ_m. Conversely, if $u(z_1, \ldots, z_m)$ is m-harmonic in Δ_m, it is representable by a series (4·1) absolutely convergent in Δ_m. For suppose, for simplicity, that $m = 2$ and $u = u(z_1, z_2)$. Then for each fixed z_2, $|z_2| < 1$, u is of the form

$$\sum_{j=-\infty}^{+\infty} c_j(z_2) r_1^{|j|} e^{ij\theta_1}, \quad \text{where} \quad c_j(z_2) = \frac{1}{2\pi r_1^{|j|}} \int_0^{2\pi} u(r_1 e^{is}, z_2) e^{-ijs} ds. \tag{4·2}$$

It is obvious that each $c_j(z_2)$ is harmonic in x_2, y_2, so that

$$c_j(z_2) = \sum_{k=-\infty}^{+\infty} c_{jk} r_2^{|k|} e^{ik\theta_2}, \tag{4·3}$$

where

$$c_{jk} = \frac{1}{2\pi r_2^{|k|}} \int_0^{2\pi} c_j(r_2 e^{it}) e^{-ikt} dt$$

$$= \frac{1}{4\pi^2 r_1^{|j|} r_2^{|k|}} \int_0^{2\pi} \int_0^{2\pi} u(r_1 e^{is}, r_2 e^{it}) e^{-i(js+kt)} ds\, dt. \tag{4·4}$$

Substituting (4·3) in the series (4·2) we obtain a series (4·1) (with $m = 2$) which converges absolutely in Δ_2; for if we fix any $\rho_1 < 1$, $\rho_2 < 1$ and observe that, by (4·4),

$$|c_{jk}| \leqslant M\rho_1^{-|j|}\rho_2^{-|k|}, \quad \text{where} \quad M = M(\rho_1, \rho_2),$$

we see that (4·1) converges absolutely for $r_1 < \rho_1$, $r_2 < \rho_2$, and so also in Δ_2.

† Γ_m is sometimes called the *edge* of B_m, or the *distinguished boundary* of Δ_m.

A function $\phi(\mathbf{z})=\phi(z_1, \ldots, z_m)$ will be called *regular* in Δ_m if it is representable in Δ_m by an absolutely convergent power series

$$\sum_{n_1, \ldots, n_m \geqslant 0} c_{n_1, \ldots, n_m} z_1^{n_1} \ldots z_m^{n_m} = \sum_{n_1, \ldots, n_m \geqslant 0}^{\infty} c_{n_1, \ldots, n_m} r_1^{n_1} \ldots r_m^{n_m} e^{i(n_1\theta_1 + \ldots + n_m\theta_m)}. \tag{4·5}$$

We say that ϕ belongs to the class H^p, $p>0$, if the integral

$$\int_Q |\phi(r_1 e^{i\theta_1}, \ldots, r_m e^{i\theta_m})|^p \, d\theta_1 \ldots d\theta_m \tag{4·6}$$

is bounded for $r_1<1, \ldots, r_m<1$. We say that ϕ belongs to the class N if

$$\int_Q \log^+ |\phi(r_1 e^{i\theta_1}, \ldots, r_m e^{i\theta_m})| \, d\theta_1 \ldots d\theta_m$$

is bounded, and to the class N_α (where $\alpha>0$) if

$$\int_Q \log^+ |\phi(r_1 e^{i\theta_1}, \ldots)| \, \{\log^+ \log^+ |\phi(r_1 e^{i\theta_1}, \ldots)|\}^\alpha \, d\theta_1 \ldots d\theta_m$$

is bounded, for $r_1<1, \ldots, r_m<1$. It is clear that if $q>p$, then

$$\mathrm{H}^q \subset \mathrm{H}^p \subset \mathrm{N}_\alpha \subset \mathrm{N}.$$

We have studied the classes H^p and N in the case $m=1$ (see Chapter VII, § 7). This case is rather exceptional and many results facilitating its study are false when $m>1$. For example, the zeros of regular functions of a single variable are isolated; and if we divide any f from H^p or N by the corresponding Blaschke product we do not alter the class of the function, so that we can thereby reduce the general case to that of a function without zeros. The zeros of regular functions of several variables, on the other hand, form continua, and no analogue of the Blaschke product exists, with the result that the theory of classes H^p and N is much less complete.

In view of the Parseval formula (1·7), the function (4·5) is in the class H^2 if and only if $\Sigma |c_\mathbf{n}|^2$ is finite, that is to say, if and only if $\Sigma c_\mathbf{n} e^{i(\mathbf{nx})}$ is a Fourier series of class L^2. Hence, by (2·14) (i), a function $\phi(\mathbf{z})$ from H^2 has a non-tangential limit at almost all points of the extremal boundary Γ_m. This result will be generalized considerably (see Theorem (4·8) below), but the following very special case of it is already of interest.

(4·7) THEOREM. *If $\phi(z_1, \ldots, z_m)$ is regular and bounded in Δ_m, then ϕ has a non-tangential limit at almost all points of Γ_m.*

This result can be generalized as follows:

(4·8) THEOREM. *If $\phi(z_1, \ldots, z_m)$ is in N_{m-1} (in particular if $\phi \in \mathrm{H}^p$), then ϕ has a non-tangential limit at almost all points of Γ_m.*

We fix a number $0<\lambda<1$, and consider the function

$$\phi^\lambda(z_1, \ldots, z_m) = \phi(\lambda z_1, \ldots, \lambda z_m),$$

which is continuous in the closure of Δ_m. We recall that if $\psi(z)$ is regular for $|z| \leqslant 1$, then $\log^+ |\psi(z)|$ is majorized in $|z| \leqslant 1$ by the Poisson integral of the function $\log^+ |\psi(e^{it})|$ (Chapter VII, p. 273). Applying this to $\phi^\lambda(z_1, \ldots, z_m)$ *qua* function of z_1, we find that

$$\log^+ |\phi^\lambda(r_1 e^{ix_1}, z_2, \ldots, z_m)| \leqslant \frac{1}{\pi} \int_0^{2\pi} \log^+ |\phi^\lambda(e^{iy_1}, z_2, \ldots, z_m)| \, P(r_1, y_1 - x_1) \, dy_1 \tag{4·9}$$

for $r_1 < 1$, $|z_2| \leqslant 1, \ldots, |z_m| \leqslant 1$. Similarly

$$\log^+ |\phi^\lambda(r_1 e^{ix_1}, r_2 e^{ix_2}, z_3, \ldots, z_m)| \leqslant \frac{1}{\pi}\int_0^{2\pi} \log^+ |\phi^\lambda(r_1 e^{ix_1}, e^{iy_2}, z_3, \ldots, z_m)|\, P(r_2, y_2 - x_2)\, |\, dy_2$$

for $r_1 < 1$, $r_2 < 1$, $|z_3| \leqslant 1, \ldots, |z_m| \leqslant 1$. This combined with (4·9), gives

$$\log^+ |\phi^\lambda(r_1 e^{ix_1}, r_2 e^{ix_2}, z_3, \ldots, z_m)| \leqslant \frac{1}{\pi^2}\int_0^{2\pi}\int_0^{2\pi} \log^+ |\phi^\lambda(e^{iy_1}, e^{iy_2}, z_3, \ldots, z_m)|\, P(r_1, y_1 - x_1)\, P(r_2, y_2 - x_2)\, dy_1 dy_2,$$

an inequality immediately extensible to

$$\log^+ |\phi^\lambda(r_1 e^{ix_1}, \ldots, r_m e^{ix_m})| \leqslant \frac{1}{\pi^m}\int_Q \log^+ |\phi^\lambda(e^{iy_1}, \ldots, e^{iy_m})| \prod_{j=1}^m P(r_j, y_j - x_j)\, dy_j. \quad \textbf{(4·10)}$$

Hence $\log^+ |\phi^\lambda(\mathbf{z})|$ is majorized in Δ_m by the Poisson integral of the values taken on Γ_m by $\log^+ |\phi^\lambda|$.

We fix a δ with $0 < \delta < 1$, and consider the domain $\Omega(x) = \Omega_\delta(x)$ described on p. 199, and the Cartesian product $\Omega(\mathbf{x})$ of domains $\Omega(x_1), \ldots, \Omega(x_m)$. Let $u^\lambda(\mathbf{z}) = u^\lambda(z_1, \ldots, z_m)$ be the Poisson integral of the (non-negative) function

$$\log^+ |\phi^\lambda(e^{iy_1}, \ldots, e^{iy_m})|.$$

Denote by $u_*^\lambda(\mathbf{x})$ the upper bound of $u^\lambda(\mathbf{z})$, by $v_*^\lambda(\mathbf{x})$ the upper bound of $\log^+ |\phi^\lambda(\mathbf{z})|$, and by $v_*(\mathbf{x})$ the upper bound of $\log^+ |\phi(\mathbf{z})|$, all for $\mathbf{z} \in \Omega(\mathbf{x})$ (see (2·13)). By Theorem (2·14), we have for $0 < \mu < 1$

$$\left\{\int_Q [u_*^\lambda(\mathbf{x})]^\mu\, d\mathbf{x}\right\}^{1/\mu} \leqslant A_{\delta,\mu,m}\int_Q \log^+ |\phi^\lambda(e^{ix_1}, \ldots)|\, \{\log^+ \log^+ |\phi^\lambda(e^{ix_1}, \ldots)|\}^{m-1}\, d\mathbf{x} + A_{\delta,\mu,m}$$

$$= A_{\delta,\mu,m}\int_Q \log^+ |\phi(\lambda e^{ix_1}, \ldots)|\, \{\log^+ \log^+ |\phi(\lambda e^{ix_1}, \ldots)|\}^{m-1}\, d\mathbf{x} + A_{\delta,\mu,m}. \quad \textbf{(4·11)}$$

By hypothesis, the last integral stays below a certain bound M independent of λ. By (4·10), we have $v_*^\lambda(\mathbf{x}) \leqslant u_*^\lambda(\mathbf{x})$, and in (4·11) we may replace u_*^λ by v_*^λ. It is also easily seen that $v_*^\lambda(\mathbf{x}) \to v_*(\mathbf{x})$ for each $\mathbf{x}$ as $\lambda \to 1$. Hence (4·11) leads to

$$\left\{\int_Q [v_*(\mathbf{x})]^\mu\, d\mathbf{x}\right\}^{1/\mu} \leqslant A_{\delta,\mu,m} M + A_{\delta,\mu,m}, \quad \textbf{(4·12)}$$

which implies in particular that $v_*(\mathbf{x})$ is finite at almost all points of Γ_m.

The function $v_*(\mathbf{x})$ depends on the parameter δ. Making δ tend to 1 through a sequence of values, we see that $\log^+ |\phi(\mathbf{z})|$ is non-tangentially bounded at almost all points of Γ_m. Hence $\phi(\mathbf{z})$ is non-tangentially bounded at almost all points of Γ_m.

The rest of the proof of Theorem (4·8) consists in showing that we can refine the non-tangential boundedness of ϕ to non-tangential convergence.

Given a point z_0, $|z_0| = 1$, consider a curvilinear triangle limited by an arc of a circle $|z| = \rho < 1$ and by the two segments joining the end-points of the arc to z_0. The interior of such a triangle will be called a *triangular neighbourhood* of z_0 and will be denoted by $T(z_0)$ (see p. 201). If $\mathbf{z}=(z_1, \ldots, z_m)$ is on Γ_m, the Cartesian product of triangular neighbourhoods $T(z_j)$, $j=1, \ldots, m$, will be called a *triangular neighbourhood* of $\mathbf{z}$ and denoted by $T(\mathbf{z})$; the $T(z_j)$ may of course vary with j.

A function $u(\mathbf{z})$ defined in Δ_m will be said to satisfy *condition* B at a point $\mathbf{z}_0 \in \Gamma_m$, if u is bounded in some triangular neighbourhood of $\mathbf{z}_0$.

The proof of Theorem (4·8) will be completed if we establish the following m-dimensional analogue of Theorem (1·1) of Chapter XIV.

(4·13) THEOREM. *If $u(\mathbf{z})=u(z_1, \ldots, z_m)$ is m-harmonic in Δ_m, and satisfies condition* B *at each point of a set $E \subset Q$, then u has a non-tangential limit at almost all points of E.*

This theorem gives slightly more than we actually need, since the ϕ of Theorem (4·8) is bounded not only in triangular neighbourhoods $T(\mathbf{z})$ but in larger domains $\Omega(\mathbf{z})$. The proof of (4·13) runs parallel to the second proof of Theorem (1·1) of Chapter XIV, but certain details require more elaboration. We may suppose that u is real-valued.

Suppose first that all $T(\mathbf{z})$ in (4·13) are *symmetric*, that is that each $T(z_j)$ is bisected by the radius to z_j. We may further suppose that $T(z_1), \ldots, T(z_m)$ are congruent. Two neighbourhoods $T(\mathbf{z}')$ and $T(\mathbf{z}'')$ will be called *similar* if $T(z_j')$ and $T(z_j'')$ are congruent for all j. The same argument as in the case of a single variable shows that under the hypotheses of (4·13) we can find a perfect subset P of E with $|E-P|$ arbitrarily small, and similar neighbourhoods $T(\mathbf{z}^0)=T(z_1^0)\times \ldots \times T(z_m^0)$, such that if $\mathbf{z}^0$ is in Π, where Π denotes the set of points $(e^{ix_1}, \ldots, e^{ix_m})$ with $\mathbf{x}$ in P, then u is uniformly bounded, say $|u| \leqslant 1$, in $T(\mathbf{z}^0)$. From now on, $T(\mathbf{z}^0)$ denotes this uniquely defined neighbourhood.

Suppose that the curvilinear parts of the $T(\mathbf{z}^0)$ are on the circle $|z|=1-\delta_0$ and denote by $T_1(z^0)$ the intersection of $T(z^0)$ with the ring $1-\frac{1}{2}\delta_0<|z|<1$. Write

$$T_1(\mathbf{z}^0)=T_1(z_1^0)\times \ldots \times T_1(z_m^0),$$

and

$$V=\sum_{\mathbf{z}^0 \in \Pi} T_1(\mathbf{z}^0). \tag{4·14}$$

The set V is open, but not necessarily connected, and $|u| \leqslant 1$ in V. Denote the boundary of V by B.

Write $\rho_n=1-1/n$, $n=2, 3, \ldots$, and denote by $P^{(n)}$ the set of $\mathbf{x}=(x_1, \ldots, x_m)$ such that $(\rho_n e^{ix_1}, \ldots, \rho_n e^{ix_m})$ is in V. $P^{(n)}$ is open, and contains P for n large enough; only such n are considered. Let $\phi_n(\mathbf{z})$ be the Poisson integral of the function equal to $u(\rho_n e^{ix_1}, \ldots, \rho_n e^{ix_m})$ in $P^{(n)}$ and to 0 in the complementary set $Q^{(n)}$, and let $\psi_n(\mathbf{z})$ be the Poisson integral of the function equal to 0 in $P^{(n)}$ and to $u(\rho_n e^{ix_1}, \ldots, \rho_n e^{ix_m})$ in $Q^{(n)}$; then ψ_n is continuous in the closure of V and is 0 on Π. Since $u(\rho_n \mathbf{z})$ is the Poisson integral of the values it takes on Γ_m, we have

$$u(\rho_n \mathbf{z})=\phi_n(\mathbf{z})+\psi_n(\mathbf{z}). \tag{4·15}$$

The ϕ_n are Poisson integrals of a function numerically not exceeding 1, and so are equicontinuous in the set $|z_1| \leqslant 1-\epsilon, \ldots, |z_m| \leqslant 1-\epsilon$. We can therefore find a sequence $\{\phi_{n_k}\}$ converging in Δ_m, and uniformly in each set just written. The function $\phi(\mathbf{z})=\lim \phi_{n_k}(\mathbf{z})$ is m-harmonic, with $|\phi(\mathbf{z})| \leqslant 1$, in Δ_m. By (4·15), $\{\psi_{n_k}(\mathbf{z})\}$ converges in Δ_m to an m-harmonic function $\psi(\mathbf{z})$, and

$$u(\mathbf{z})=\phi(\mathbf{z})+\psi(\mathbf{z}). \tag{4·16}$$

Since ϕ is bounded in Δ_m, and thus is the Poisson integral of a bounded function, ϕ has a non-tangential limit at almost all points of Γ_m. For the proof of (4·13) it is therefore enough to show that ψ has a non-tangential limit almost everywhere in Π; we show that this limit is 0 almost everywhere in Π. For simplicity of notation we suppose that $m=2$, but the proof is general.

It is enough to prove that there exists a positive $\chi(\mathbf{z})$ in Δ_m which majorizes $|\psi(\mathbf{z})|$ in V and tends non-tangentially to 0 at almost all points of Π. For consider an $\mathbf{x}^0=(x_1^0, x_2^0)$ in P which is a point of *strong* density for P (see p. 307) and at which χ tends non-tangentially to 0; almost all $\mathbf{x}^0 \in P$ have this property. The hypothesis that $\mathbf{x}^0$ is a point of strong density implies that if $0<\lambda<1$ is *fixed*, and (x_1, x_2) is any point sufficiently close to (x_1^0, x_2^0), then the rectangle (of the points (ξ_1, ξ_2))

$$|\xi_1-x_1| \leqslant \lambda |x_1-x_0^1|, \quad |\xi_2-x_2| \leqslant \lambda |x_2-x_2^0| \tag{4·17}$$

must necessarily contain points of P. This in turn implies that if N is any fixed triangular neighbourhood of $\mathbf{z}^0=(e^{ix_1^0}, e^{ix_2^0})$, then the part of it which is sufficiently close to $\mathbf{z}^0$ is totally covered by the $T(\mathbf{z})$ with $\mathbf{z}$ in Π†. Hence $\chi(\mathbf{z})$ majorizes $|\psi(\mathbf{z})|$ for those $\mathbf{z}$ in N which are close to $\mathbf{z}^0$, and ψ has a non-tangential limit 0 at $\mathbf{z}^0$.

It remains to construct the required χ.

Let $h(x_1, x_2)$ be the characteristic function of the open set Q complementary to P, let

$$\chi_1(\mathbf{z})=\frac{1}{\pi}\int_{-\pi}^{\pi} P(r_1, t_1-x_1)\,dt_1\left\{\frac{1}{\pi}\int_{-\pi}^{\pi} h(t_1, t_2)\,P(r_2, t_2-x_2)\,dt_2\right\} \tag{4·18}$$

be the Poisson integral of h, and let

$$\chi_2(\mathbf{z})=P(r_1, x_1)+P(r_2, x_2)+\chi_1(\mathbf{z}), \tag{4·19}$$

where $\mathbf{z}=(r_1 e^{ix_1}, r_2 e^{ix_2})$. Clearly χ_2 has a non-tangential limit 0 at almost all points of Π.

For each n, $\psi_n(\mathbf{z})$ is a continuous function in the closure $\bar{V}$ of V. If we show that there is an $M>0$ independent of n such that, for each n, as z in V approaches a point $\mathbf{z}^0$ on the boundary of V, we have

$$-M \limsup_{\mathbf{z}\to\mathbf{z}^0} \chi_2(\mathbf{z}) \leqslant \psi_n(\mathbf{z}^0) \leqslant M \limsup_{\mathbf{z}\to\mathbf{z}^0} \chi_2(\mathbf{z}), \tag{4·20}$$

then the maximum principle for harmonic functions (each m-harmonic function is also harmonic) will imply that

$$|\psi_n(\mathbf{z})| \leqslant M\chi_2(\mathbf{z}) \quad (\mathbf{z}\in V),$$

† Since the situation is geometrically somewhat less simple than in the one-dimensional case, we add a few words of explanation. Instead of a bi-cylinder we may consider the Cartesian product of the half-planes $-\infty<x_1<+\infty$, $y_1>0$ and $-\infty<x_2<+\infty$, $y_2>0$. Let P be a closed set of positive measure in the plane $y_1=0$, $y_2=0$, and $T(\xi_1, \xi_2)$ a fixed triangular neighbourhood

$$h>y_1>\alpha|x_1-\xi_1|, \quad h>y_2>\alpha|x_2-\xi_2|$$

of a variable point $(\xi_1, \xi_2)\in P$. We have to show that if (x_1^0, x_2^0) is a point of strong density for P, $\beta>0$ is fixed, and y_1, y_2 are arbitrary positive and sufficiently small, then any point $(x_1, y_1)\times(x_2, y_2)$ satisfying

$$(*) \quad y_1>\beta|x_1-x_1^0|, \quad y_2>\beta|x_2-x_2^0|$$

belongs to some $T(\xi_1, \xi_2)$, with $(\xi_1, \xi_2)\in P$, that is

$$(**) \quad y_1>\alpha|x_1-\xi_1|, \quad y_2>\alpha_2|x_2-\xi_2| \quad ((\xi_1, \xi_2)\in P).$$

We may suppose that $\beta<\alpha$, for otherwise the result is obvious. But if we have (*), then the $(\xi_1, \xi_2)\in P$ satisfying (4·17) with $\lambda=\beta/\alpha$ satisfy also (**).

and, making $n \to \infty$, $$|\psi(\mathbf{z})| \leqslant M\chi_2(\mathbf{z}) \quad (\mathbf{z} \in V),$$

so that $\chi = M\chi_2$ will have the desired properties.

We recall that $|\psi_n| \leqslant 2$ in V, and $\psi_n = 0$ in Π. Let $z_0 \in B$ and consider various special cases:

(i) $|z_1^0| = 1$, $|z_2^0| = 1$. (4·20) is then obvious;

(ii) $|z_1^0| = 1 - \frac{1}{2}\delta_0$ or $|z_2^0| = 1 - \frac{1}{2}\delta_0$. (4·20) is again obvious;

(iii) both z_1^0 and z_2^0 are in the ring

$$R: \quad 1 - \tfrac{1}{2}\delta_0 < |z| < 1.$$

Given a $z \in R$, consider the largest open arc (α, β) such that z is in $T_1(e^{ix})$ for x in (α, β); we call this the arc *associated* with z. If $z_1 \in R$, $z_2 \in R$, the point $\mathbf{z} = (z_1, z_2)$ is in V if and only if the Cartesian product of the arc (α_1, β_1) associated with z_1 and the arc (α_2, β_2) associated with z_2 contains points of P. Hence, under the hypothesis of (iii), if (α_j, β_j) is associated with z_j^0, $j = 1, 2$, the Poisson integral of h is at z^0 not less than the Poisson integral of the characteristic function of $(\alpha_1, \beta_1) \times (\alpha_2, \beta_2)$, which is the product of the one-dimensional Poisson integrals of the characteristic functions of (α_j, β_j), $j = 1, 2$. But the Poisson integral of the characteristic function of (α_j, β_j) exceeds at z_j^0 a positive number depending on T only (Chapter III, (6·18)). Hence $\chi_2(\mathbf{z}^0)$ is bounded away from 0, and since $|\psi_n(\mathbf{z}^0)| \leqslant 2$ we again have (4·20).

(iv) $z_1^0 \in R$, $|z_2^0| = 1$ (or, vice versa, $z_2^0 \in R$, $|z_1^0| = 1$). Write $z_j^0 = r_j^0 e^{ix_j^0}$ and consider the formula (4·18). Fix t_1 in the inner integral. Then $h(t_1, t_2)$ becomes a function of t_2, continuous at each point where h takes the value 1. It follows that as $z_2 \to z_2^0$ the inner integral approaches $h(t_1, x_2^0)$ if $h(t_1, x_2^0) = 1$, and has non-negative inferior limit if $h(t_1, x_2^0) = 0$. Hence in all cases the inferior limit of the inner integral as $z_2 \to z_2^0$ is not less than $h(t_1, x_2^0)$. This gives

$$\liminf_{\mathbf{z} \to \mathbf{z}^0} \chi_2(\mathbf{z}) \geqslant P(r_1^0, x_1^0) + \frac{1}{\pi} \int_{-\pi}^{\pi} h(t_1, x_2^0)\, P(r_1, t_1 - x_1^0)\, dt_1. \tag{4·21}$$

Denote the right-hand side by $\chi_2(z_1^0; x_2^0)$, and replace z_1^0 by general z, $|z| < 1$. The resulting function $\chi_2(z; x_2^0)$ is a one-dimensional analogue of $\chi_2(\mathbf{z})$ and has been discussed in detail in Chapter XIV, p. 202. Clearly $h(t_1, x_2^0)$ is the characteristic function of a one-dimensional open (possibly empty) set $Q(x_2^0)$. The complementary (one-dimensional) set $P(x_2^0)$ is closed, and non-empty, and consists of points t_1 such that $(t_1, x_2^0) \in P$. Let $V(x_2^0)$ be the union of $T_1(e^{it_1})$ for $t_1 \in P(x_2^0)$, and let $\Pi(x_2^0)$ be the set of points e^{it_1}, $t_1 \in P(x_2^0)$.

Consider also $\psi_n(z, e^{ix_2^0})$ *qua* function of z. It is harmonic and numerically not greater than 2 in $V(x_2^0)$, continuous in the closure of $V(x_2^0)$, and 0 in $\Pi(x_2^0)$. Therefore (see Chapter XIV, (1·8)) there exists a constant M, depending on T only, such that $|\psi_n(z, e^{ix_2^0})| \leqslant M\chi_2(z; x_2^0)$ for $z \in V(x_2^0)$, from which we deduce, by (4·21), that

$$|\psi_n(\mathbf{z}^0)| = |\psi_n(z_1^0, z_2^0)| \leqslant M\chi_2(z_1^0; x_2^0) \leqslant M \limsup_{\mathbf{z} \to \mathbf{z}^0} \chi_2(\mathbf{z}).$$

This completes the proof of (4·20), and so also of Theorem (4·13) when the neighbourhoods T are symmetrical. The general case is reducible to this, since if we fix T and a closed set P as in the symmetric case, then at every point of strong density of P we have a symmetric neighbourhood in which u is bounded.

The preceding proof also shows that the conclusion of (4·13) holds if u is m-harmonic not in the whole of Δ_m but only in triangular neighbourhoods of each point of E. We shall use this remark presently.

If u is regular in Δ_m, (4·13) can be strengthened. We say that $\Phi(\mathbf{z})$, regular in Δ_m, behaves *restrictedly* at a point $\mathbf{z}^0 = (e^{ix_1^0}, \ldots, e^{ix_m^0})$, if there is a triangular neighbourhood $T(\mathbf{z}^0)$ such that $\Phi\{T(\mathbf{z}^0)\}$ is not dense in the complex plane.

(4·22) THEOREM. *If $w = \Phi(\mathbf{z}) = \Phi(z_1, \ldots, z_m)$ is regular in Δ_m and behaves restrictedly at each point of a set $E \subset \Gamma_m$, then Φ has a finite non-tangential limit almost everywhere in E.*

We may suppose, without loss of generality, that each $\Phi\{T(\mathbf{z}^0)\}$, $\mathbf{z}^0 \in E$, is disjoint with a fixed circle $|w - w_0| < r$, so that $u = 1/\{\Phi(\mathbf{z}) - w_0\}$ is regular and bounded in each $T(\mathbf{z}^0)$, $\mathbf{z}^0 \in E$. Hence u has a non-tangential limit almost everywhere in E. It follows that Φ has a non-tangential limit almost everywhere in E, and we have to show that this limit can be infinite only in a set of measure 0.

For a fixed choice of real parameters $\alpha_1, \alpha_2, \ldots, \alpha_m$ consider the function

$$\Psi(z) = \Phi(z\,e^{i\alpha_1}, z\,e^{i\alpha_2}, \ldots, z\,e^{i\alpha_m}) \tag{4·23}$$

regular in $|z| < 1$. Let E^∞ be the set of points where $\Phi(\mathbf{z})$ has an infinite non-tangential limit. The set of t's such that $\Psi(z)$ has an infinite non-tangential limit at e^{it} is of measure 0 (Chapter XIV, (1·11)) and contains the t's such that $(t+\alpha_1, \ldots, t+\alpha_m) \in E^\infty$. Hence the intersection of E^∞ with the straight line $x_j = t + \alpha_j$, $j = 1, \ldots, m$, has linear measure 0. The α's being arbitrary and E^∞ measurable, we have $|E^\infty| = 0$.

(4·24) THEOREM. *If $\Phi(\mathbf{z})$ is regular in Δ_m and not identically 0, then the set E^0 of Γ_m, where Φ has a non-tangential limit 0, is of measure 0.*

Consider the function (4·23). It cannot be identically 0 for any choice of the α's. Hence, arguing as before, the intersection of E^0 with any straight line $x_j = t + \alpha_j$, $j = 1, 2, \ldots, m$, is of measure 0 (see Chapter XIV, (1·9)) which implies that $|E^0| = 0$.

5. Power series of several variables (*cont.*)

We shall now study the boundary behaviour of functions ϕ regular in Δ_m and of class N, that is, such that

$$\int_Q \log^+ |\phi(r_1 e^{ix_1}, \ldots, r_m e^{ix_m})|\, dx_1 \ldots dx_m \leqslant A \tag{5·1}$$

for some A and all $r_j < 1$.

(5·2) THEOREM. *If $\phi \in \mathrm{N}$, then ϕ has a restricted non-tangential limit at almost all points of Γ_m.*

We first show that if $\phi \in \mathrm{N}$ then $\log^+ |\phi(\mathbf{z})|$ is majorized in Δ_m by a non-negative m-harmonic function $u(\mathbf{z})$.

Consider the Poisson integral

$$u_{R_1 \ldots R_m}(z_1, \ldots, z_m) = \pi^{-m} \int_Q \log^+ |\phi(R_1 e^{it_1}, \ldots, R_m e^{it_m})| \prod_{j=1}^m P(r_j/R_j, t_j - x_j)\, dt_1 \ldots dt_m, \tag{5·3}$$

where $R_j < 1$ for all j. The right-hand side is m-harmonic in the m-cylinder $|z_1| < R_1, \ldots, |z_m| < R_m$, and when extended by continuity to its closure coincides with $\log^+|\phi|$ for $|z_1| = R_1, \ldots, |z_m| = R_m$.

Since ϕ is a regular function of z_1 for $z_2, \ldots, z_m$ fixed, the integral

$$\pi^{-1}\int_0^{2\pi} \log^+|\phi(R_1 e^{it_1}, \ldots)|\, P(r_1/R_1, t_1 - x_1)\, dt_1 \tag{5·4}$$

is a non-decreasing function of R_1 for $r_1 < R_1$ (Chapter VII, p. 273)†. If we multiply (5·4) by $P(r_2/R_2, x_2 - t_2) \ldots P(r_m/R_m, x_m - t_m)$ and integrate over $(0, 2\pi)$ with respect to $t_2, \ldots, t_m$, we still have a non-decreasing function of R_1. It follows that, for any $\mathbf{z}$ in Δ_m, $u_{R_1 \ldots R_m}(\mathbf{z})$ is a non-decreasing function of $R_1, \ldots, R_m$ provided that $|z_1| < R_1, \ldots, |z_m| < R_m$. Hence at each $\mathbf{z} \in \Delta_m$ the function $u_{R_1 \ldots R_m}(\mathbf{z})$, which is defined if $R_1, \ldots, R_m$ are sufficiently close to 1, tends, increasing, to a limit $u(\mathbf{z}) = u(z_1, \ldots, z_m)$.

This limit is finite, since if $|z_j| \leqslant R < R' \leqslant R_j < 1$ for all j, then, by (5·3),

$$|u_{R_1 \ldots R_m}(\mathbf{z})| \leqslant \left(\frac{1}{2\pi}\frac{R'+R}{R'-R}\right)^m \int_Q \log^+|\phi(R_1 e^{it_1}, \ldots R_m e^{it_m})|\, dt_1 \ldots dt_m,$$

so that the left-hand side is uniformly bounded.

A similar argument, under the same hypotheses concerning the z_j and R_j, shows that the partial derivatives of $u_{R_1 \ldots R_m}$ with respect to r_j and x_j are also uniformly bounded. Hence the convergence of $u_{R_1 \ldots R_m}(\mathbf{z})$ to $u(\mathbf{z})$ is uniform in every m-cylinder $|z_1| \leqslant R, \ldots, |z_m| \leqslant R$, $R < 1$, which shows that $u(\mathbf{z})$ is m-harmonic in Δ_m.

Since ϕ is regular in z_1, we have

$$\log^+|\phi(r_1 e^{ix_1}, z_2, \ldots, z_m)| \leqslant \pi^{-1}\int_0^{2\pi} \log^+|\phi(R_1 e^{it_1}, z_2, \ldots, z_m)|\, P(r_1/R_1, x_1 - t_1)\, dt_1$$

for $r_1 < R_1 < 1$ (Chapter VII, p. 273). Since $\phi(r_1 e^{ix_1}, z_2, \ldots, z_m)$ is regular in z_2, we have

$$\log^+|\phi(r_1 e^{ix_1}, r_2 e^{ix_2}, z_3, \ldots, z_m)| \leqslant \pi^{-1}\int_0^{2\pi} \log^+|\phi(r_1 e^{ix_1}, R_2 e^{it_2}, z_3, \ldots, z_m)|\, P(r_2/R_2, x_2 - t_2)\, dt_2$$

for $r_2 < R_2 < 1$. This and the previous inequality give

$$\log^+|\phi(r_1 e^{ix_1}, r_2 e^{ix_2}, z_3, \ldots, z_m)| \leqslant \pi^{-2}\int_0^{2\pi}\int_0^{2\pi} \log^+|\phi(R_1 e^{it_1}, R_2 e^{it_2}, z_3, \ldots, z_m)|\, P(r_1/R_1, x_1 - t_1)\, P(r_2/R_2, x_2 - t_2)\, dt_1 dt_2$$

for $r_1 < R_1$, $r_2 < R_2$. Proceeding in this way we see that $\log^+|\phi(r_1 e^{ix_1}, \ldots, r_m e^{ix_m})|$ does not exceed (5·3), and making $R_1, \ldots, R_m$ tend to 1 we find that $\log^+|\phi(\mathbf{z})|$ is majorized by a non-negative m-harmonic function $u(\mathbf{z})$, as asserted.

† Let $\phi(z)$ be regular for $|z| < 1$, and let $0 < R_1 < R_2 < 1$. In the argument above we are using the fact that, if $r < R_1$, then

$$\frac{1}{\pi}\int_0^{2\pi} \log^+|\phi(R_1 e^{it})|\, P\left(\frac{r}{R_1}, x - t\right) dt \leqslant \frac{1}{\pi}\int_0^{2\pi} \log^+|\phi(R_2 e^{it})|\, P\left(\frac{r}{R_2}, x - t\right) dt.$$

To prove this, denote the right-hand side by $u(z)$, where $z = re^{ix}$. Since u is harmonic in $|z| < R_2$, the right-hand side does not change for $|z| < R_1$, if $\log^+|\phi(R_2 e^{it})|$ is replaced by $u(R_1 e^{it})$ and $P(r/R_2, x - t)$ by $P(r/R_1, x - t)$. Since, by Chapter VII, (7·10), we have $\log^+|\phi(R_1 e^{it})| \leqslant u(R_1 e^{it})$, the required inequality follows.

By (3·21) and (3·18), u has a restricted non-tangential limit at almost all points of Γ_m. Hence ϕ is restrictedly non-tangentially bounded at almost all points of Γ_m. This gives (5·2) as a corollary if we prove the following general result:

(5·5) THEOREM. *Suppose that $\phi(\mathbf{z})$ is regular in Δ_m, and that at each point of a set E situated on the extremal boundary Γ_m, ϕ is restrictedly non-tangentially bounded. Then ϕ has a restricted non-tangential limit almost everywhere in E.*

The main idea of the proof consists in reducing the case of restricted non-tangential boundedness to that of ordinary (unrestricted) non-tangential boundedness by a suitable change of variables. The geometric aspect of the change of variables is slightly easier to grasp if we consider functions not in Δ_m but in the Cartesian product of half-planes; these two cases are clearly equivalent through linear transformations of the individual variables.

Write $z_j = x_j + iy_j$, and suppose that $\phi(\mathbf{z}) = \phi(z_1, \ldots, z_m)$ is regular in the domain

$$y_1 > 0, \quad \ldots, \quad y_m > 0, \tag{5·6}$$

and that, as $\mathbf{z}$ restrictedly and non-tangentially approaches any point $\mathbf{x}^0 = (x_1^0, \ldots, x_m^0)$ of a set E situated on the extremal boundary

$$y_1 = 0, \quad \ldots, \quad y_m = 0 \tag{5·7}$$

of (5·6), the function ϕ remains bounded (not necessarily uniformly in $\mathbf{x}^0$).

The hypothesis that ϕ is bounded as $\mathbf{z}$ restrictedly and non-tangentially approaches an $\mathbf{x}^0 \in E$ means that ϕ remains bounded as $\mathbf{z}$ tends to $\mathbf{x}^0$ in such a way that

$$\max_{j,k} (y_k/y_j) \leqslant \alpha \quad (j, k = 1, \ldots, m), \tag{5·8}$$

$$|x_j - x_j^0| \leqslant \beta y_j \quad (j = 1, \ldots, m), \tag{5·9}$$

where α and β are any fixed but arbitrary constants. Of course the upper bound of ϕ may depend also on α and β.

Condition (5·8) expresses the restrictedness, and (5·9) the non-tangential character, of the approach. If we omit (5·8) and merely require that the y's tend to 0 through positive values we obtain the unrestricted non-tangential approach.

We fix an ϵ, $0 < \epsilon < 1$, and consider the transformation

$$\left.\begin{aligned} z_1 &= w_1 + \epsilon w_2 + \ldots + \epsilon w_m, \\ z_2 &= \epsilon w_1 + w_2 + \ldots + \epsilon w_m, \\ &\ldots\ldots\ldots\ldots\ldots\ldots\ldots\ldots \\ z_m &= \epsilon w_1 + \epsilon w_2 + \ldots + w_m. \end{aligned}\right\} \tag{5·10}$$

On setting $w_j = u_j + iv_j$ we reduce this to two transformations

$$\left.\begin{aligned} x_1 &= u_1 + \epsilon u_2 + \ldots + \epsilon u_m, \\ &\ldots\ldots\ldots\ldots\ldots\ldots\ldots\ldots \\ x_m &= \epsilon u_1 + \epsilon u_2 + \ldots + u_m, \end{aligned}\right\} \tag{5·11}$$

and
$$\left.\begin{aligned} y_1 &= v_1+\epsilon v_2+\ldots+\epsilon v_m, \\ &\ldots\ldots\ldots\ldots\ldots\ldots\ldots\ldots \\ y_m &= \epsilon v_1+\epsilon v_2+\ldots+ v_m, \end{aligned}\right\} \qquad (5\cdot12)$$
either of which we denote by T.

Let X, Y, U, V be the spaces of points $\mathbf{x}=(x_1, \ldots, x_m)$, $\mathbf{y}=(y_1, \ldots, y_m)$, etc. Let Y^+ be the 'positive octant' of Y, that is, the (open) set of all $\mathbf{y}$'s with strictly positive co-ordinates; we define V^+ similarly.

Since T is non-singular, it establishes a one-one correspondence between the points of X and U. The map of V^+ by T is an 'angle' A_ϵ, a conical domain whose closure, except for the origin, is interior to Y^+. (If the v_j are non-negative but not all 0, the y_j are positive.) The map of A_ϵ by T is another angle, which we denote by B_ϵ, whose closure, except for the origin, is interior to A_ϵ. For obviously $B_\epsilon \subset A_\epsilon$, and if some boundary point of B_ϵ were on the boundary of A_ϵ, it would mean that there were two points p and q on the boundary of the positive octant such that $T^2p=Tq$, that is, $q=Tp$, which is impossible unless $p=q=0$.

We now prove the following two facts:

(i) If the point $(w_1, \ldots, w_m)$, where all the w_j have positive imaginary parts, approaches a point $(u_1^0, \ldots, u_m^0)$ non-tangentially, then $(z_1, \ldots, z_m)$ approaches the corresponding point $(x_1^0, \ldots, x_m^0)$ restrictedly and non-tangentially.

(ii) If $(z_1, \ldots, z_m)$ approaches $(x_1^0, \ldots, x_m^0)$ non-tangentially and in such a way that $(y_1, \ldots, y_m) \in B_\epsilon$, then $(w_1, \ldots, w_m)$ approaches $(u_1^0, \ldots, u_m^0)$ non-tangentially.

Observe that (ii) is, in a sense, a converse of (i). For in the first place, if $(y_1, \ldots, y_m)$ is in B_ϵ, and so also in A_ϵ, then we have (5·8) for some $\alpha=\alpha_\epsilon$. Conversely, since T tends to the identity transformation as $\epsilon\to 0$, and so A_ϵ and B_ϵ exhaust the positive octant as $\epsilon \to 0$, it follows that if we have (5·8) for some α, then $(y_1, \ldots, y_m)$ is in B_ϵ for ϵ small enough.

In view of the homogeneity of T we may suppose that
$$(x_1^0, \ldots, x_m^0)=(u_1^0, \ldots, u_m^0)=(0, \ldots, 0).$$

To prove (i), suppose that $|u_j| \leqslant Av_j$ for all j. Then, by (5·11) and (5·12), $|x_j| \leqslant Ay_j$. Moreover, by (5·12), for all $(y_1, \ldots, y_m) \in A_\epsilon$ we have
$$\epsilon \max v_j \leqslant y_k \leqslant (1+n\epsilon)\max v_j \quad (k=1, \ldots, m), \quad n=m-1, \qquad (5\cdot13)$$
which shows that the ratio of any two y's does not exceed $(1+n\epsilon)/\epsilon$; hence (i) is established.

To prove (ii), suppose that $|x_j| \leqslant Ay_j$ for all j. Solving the equations (5·11) with respect to $u_1, \ldots, u_m$, we obtain the u_k as linear functions of $x_1, \ldots, x_m$. Hence $|u_k| \leqslant B \max |x_j|$ for all k, where B is a constant. This, combined with the preceding inequality, shows that $|u_k| \leqslant AB \max y_j$, for all k, and so, by (5·13),
$$|u_k| \leqslant AB(1+n\epsilon)\max v_j. \qquad (5\cdot14)$$
Finally, since $(y_1, \ldots, y_m)$ is in B_ϵ, it follows that $(v_1, \ldots, v_m)$ is in A_ϵ, and since the ratio of any two v's in A_ϵ does not exceed $(1+n\epsilon)/\epsilon$, (5·14) implies that
$$|u_k| \leqslant AB(1+n\epsilon)^2 \epsilon^{-1} v_k \quad (k=1, \ldots, m),$$
which proves (ii).

Return to the function $\phi(z_1, \ldots, z_m)$, and, substituting for $z_1, \ldots, z_m$ the values (5·10), write

$$\phi(z_1, \ldots, z_m) = \psi(w_1, \ldots, w_m). \tag{5·15}$$

The function ψ is defined and regular if all the v_j are positive. Let E^* be the set in the U-space obtained from E by the transformation (5·11); the measures of E^* and E differ by a constant factor different from 0. If $(w_1, \ldots, w_m)$ approaches non-tangentially a point $(u_1^0, \ldots, u_m^0) \in E^*$, then, by (i), $(z_1, \ldots, z_m)$ approaches restrictedly and non-tangentially the corresponding point $(x_1^0, \ldots, x_m^0) \in E$ and, by (5·15) and the hypothesis of (5·5), ψ remains bounded. Hence, after (4·13), ψ has a non-tangential limit almost everywhere in E^*. This in turn implies, by (ii), that if z approaches non-tangentially a point $(x_1^0, \ldots, x_m^0) \in E$ in such a way that $(y_1, \ldots, y_m)$ stays in B_ϵ, then $\phi(z_1, \ldots, z_m)$ tends to a limit; the exceptional set depends on ϵ, and we call it N_ϵ. Take now a sequence $\epsilon_1, \epsilon_2, \ldots \to 0$, and write $N = \Sigma N_{\epsilon_i}$. Clearly $|N| = 0$, and since B_ϵ exhausts the positive octant as $\epsilon \to 0$, we immediately see that ϕ has a restricted non-tangential limit at all points of $E - N$.

This concludes the proof of (5·5), and so also of (5·2).

The account just given may be completed by the following remarks.

Theorem (5·5) is an analogue of (4·13), but while in (4·13) it was enough to assume that the function was m-harmonic, in (5·5) we imposed on ϕ a much stronger condition, that of regularity. It is, however, easy to see that (5·5) holds if ϕ is the real part of a function regular in Δ_m.

The restricted non-tangential approach in Theorem (5·5) may be replaced by a slightly weaker hypothesis, namely, that $\mathbf{z}$ approaches $\mathbf{x}^0$ restrictedly through a triangular neighbourhood $T(\mathbf{x}^0)$. The condition of boundedness may be replaced by the requirement that ϕ does not take values inside some circle; the circle may vary with $\mathbf{x}^0$, and even with the upper bound of the ratios $(1-r_j)/(1-r_k)$ when all the r's tend to 1. It can then be shown (we omit the proof since it does not require new ideas) that under these weaker conditions ϕ is restrictedly non-tangentially bounded at almost all points of E, and so the conclusion of (5·5) still holds.

Finally, if ϕ is regular in Δ_m and tends restrictedly and non-tangentially to 0 in a set E of positive measure situated on Γ_m, then $\phi \equiv 0$. For 'transplanting' ϕ from Δ_m to the Cartesian product of half-planes and using (5·15) we find, after (4·24), that $\psi \equiv 0$, and so $\phi \equiv 0$.

The rest of this section is devoted to the study of the behaviour of $\phi(\mathbf{z}) = \phi(z_1, \ldots, z_m)$ when only one of the variables z_j tends to the boundary. By $\chi(t)$ we denote a function which is non-decreasing and convex for $t \geq 0$, satisfies $\chi(0) = 0$, but is not identically 0. Then $\chi(t)/t > c > 0$ for t large enough.

(5·16) THEOREM. *Suppose that $\phi(\mathbf{z}) = \phi(z_1, \ldots, z_m)$ is regular in Δ_m, and that*

$$\int_0^{2\pi} \cdots \int_0^{2\pi} \log^+ |\phi(r_1 e^{ix_1}, \ldots, r_m e^{ix_m})| \, dx_1 \ldots dx_m < M \tag{5·17}$$

or some M and all $r_j < 1$. Then, for almost all x_1 in $(0, 2\pi)$ and all $z_2, \ldots, z_m$ of modulus less than 1, the function $\phi(z_1, z_2, \ldots, z_m)$ converges to a limit as z_1 tends non-tangentially to e^{ix_1}. The convergence is uniform over every set

$$|z_2| \leq 1 - \eta, \quad \ldots, \quad |z_m| \leq 1 - \eta \quad (\eta > 0),$$

so that

$$\phi^{x_1}(z_2, \ldots, z_m) = \lim_{z_1 \to e^{ix_1}} \phi(z_1, z_2, \ldots, z_m)$$

is a regular function of $z_2, \ldots, z_m$ in Δ_{m-1}.

If instead of (5·17) *we have*

$$\int_0^{2\pi}\cdots\int_0^{2\pi}\chi\{\log^+|\phi(r_1e^{ix_1},\ldots,r_me^{ix_m})|\}\,dx_1\ldots dx_m<M, \tag{5·18}$$

then
$$\int_0^{2\pi}\cdots\int_0^{2\pi}\chi\{\log^+|\phi^{x_1}(r_2e^{ix_2},\ldots,r_me^{ix_m})|\}\,dx_2\ldots dx_m<M(x_1)<\infty \tag{5·19}$$

for almost all x_1.

In particular, if $\phi(z_1,\ldots,z_m)$ *is in* H^α, *then almost all functions* $\phi^{x_1}(z_2,\ldots,z_m)$ *are in* H^α, $\alpha>0$.

Since $\chi(t)/t>c>0$ for large t, (5·18) clearly implies (5·17) (with a new M).

Consider
$$I_\rho(z)=\int_0^{2\pi}\cdots\int_0^{2\pi}\log^+|\phi(z,\rho e^{ix_2},\ldots,\rho e^{ix_m})|\,dx_2\ldots dx_m, \tag{5·20}$$

where $z=re^{ix}$, $r<1$, $\rho<1$. Fix an $R<1$ and suppose that $r<R$. We have

$$\log^+|\phi(re^{ix},\rho e^{ix_2},\ldots,\rho e^{ix_m})| \leqslant \frac{1}{\pi}\int_0^{2\pi}\log^+|\phi(Re^{it},\rho e^{ix_2},\ldots,\rho e^{ix_m})|\,P(r/R,x-t)\,dt, \tag{5·21}$$

and, furthermore, the right-hand side is a non-decreasing function of R in $(r,1)$. Integrating this over $0\leqslant x_2\leqslant 2\pi,\ \ldots,\ 0\leqslant x_m\leqslant 2\pi$, and interchanging the order of integration on the right, we obtain the inequality

$$I_\rho(re^{ix})\leqslant\frac{1}{\pi}\int_0^{2\pi}I_\rho(Re^{it})\,P(r/R,x-t)\,dt\quad(r<R). \tag{5·22}$$

Denote the right-hand side here by $u_R(re^{ix})$. It also is a non-decreasing function of R for $r<R<1$. Moreover, it is non-negative and harmonic in re^{ix}, and the limit

$$u(re^{ix})=\lim_{R\to1}u_R(re^{ix}),$$

finite or infinite, exists in the interior of the unit circle.

If we integrate $u_R(re^{ix})$ with respect to x over $(0,2\pi)$ and apply (5·20) and (5·17), we get

$$\int_0^{2\pi}u_R(re^{ix})\,dx=\int_0^{2\pi}I_\rho(Re^{it})\,dt$$
$$=\int_0^{2\pi}\cdots\int_0^{2\pi}\log^+|\phi(Re^{it},\rho e^{ix_2},\ldots,\rho e^{ix_m})|\,dt\,dx_2\ldots dx_m<M. \tag{5·23}$$

Now the limit of a non-decreasing sequence of harmonic functions is either harmonic or $+\infty$ (uniformly in every smaller domain). The latter is impossible since in that case the first integral (5·23) would tend to $+\infty$. Hence u is harmonic, and using (5·22) we come to the following conclusion: *under the hypothesis* (5·17) *the function* $I_\rho(z)$ *is majorized in* $|z|<1$ *by a non-negative harmonic function* $u(z)$.†

Denote by $M_\delta(x_1)$ the upper bound of $u(z)$ in the domain $\Omega_\delta(e^{ix_1})$. (For the definition of Ω_δ see p. 199.) Since a non-negative harmonic function is the Poisson-Stieltjes integral of a non-negative mass distribution, $M_\delta(x_1)$ is finite for almost all x_1 and all $\delta<1$. Suppose that $M_\delta(x_1^0)$ is finite for all $\delta<1$, and that $|z_2|\leqslant r_0,\ldots,|z_m|\leqslant r_0$, where

† Our $u(z)$ is an increasing function of the parameter ρ, but making $\rho\to1$ we can obtain a $u(z)$ independent of ρ.

$r_0 < \rho$. Then, since $\log^+ |\phi(z_1, \ldots, z_m)|$ is majorized by its Poisson integral with respect to any of the variables $z_1, \ldots, z_m$, we have

$$\log^+ |\phi(z_1, \ldots, z_m)| \leqslant \frac{1}{2\pi}\frac{\rho+r_0}{\rho-r_0}\int_0^{2\pi} \log^+ |\phi(z_1, \rho e^{ix_2}, z_3, \ldots, z_m)|\, dx_2$$

$$\leqslant \left(\frac{1}{2\pi}\right)^2 \left(\frac{\rho+r_0}{\rho-r_0}\right)^2 \int_0^{2\pi}\int_0^{2\pi} \log^+ |\phi(z_1, \rho e^{ix_2}, \rho e^{ix_3}, \ldots, z_m)|\, dx_2 dx_3$$

....

$$\leqslant \left(\frac{1}{2\pi}\right)^{m-1} \left(\frac{\rho+r_0}{\rho-r_0}\right)^{m-1} \int_0^{2\pi}\ldots\int_0^{2\pi} \log^+ |\phi(z_1, \rho e^{ix_2}, \ldots, \rho e^{ix_m})|\, dx_2 \ldots dx_m$$

$$= \left(\frac{1}{2\pi}\right)^{m-1} \left(\frac{\rho+r_0}{\rho-r_0}\right)^{m-1} I_\rho(z_1),$$

and we get $\qquad \log^+ |\phi(z_1, z_2, \ldots, z_m)| \leqslant \pi^{1-m}(\rho - r_0)^{1-m} M_\delta(x_1^0)$

for $\qquad z_1 \in \Omega_\delta(x_1^0), \quad |z_2| \leqslant r_0, \ldots, |z_m| \leqslant r_0.$

Hence $\phi(z_1, z_2, \ldots, z_m)$ is bounded for $z_1 \in \Omega_\delta(x_1^0)$ and $|z_2|, \ldots, |z_m|$ not exceeding r_0, for each $r_0 < 1$ and $\delta < 1$. It follows that *$\phi(z_1, z_2, \ldots, z_m)$ is equicontinuous in the variables $z_2, z_3, \ldots, z_m$, provided these variables numerically do not exceed an $r_0 < 1$ and the parameter z_1 stays within an $\Omega_\delta(x_1^0)$ such that $M_\delta(x_1^0)$ is finite.*

We can now prove the existence and regularity of $\phi^{x_1}(z_2, \ldots, z_m)$ for almost all x_1. It is enough to show that there is a sequence of points $(z_2^k, \ldots, z_m^k)$, $k = 1, 2, \ldots$, dense in Δ_{m-1} and such that $\phi(z_1, z_2^k, \ldots, z_m^k)$ has, for each k, a non-tangential limit at almost all points e^{ix_1}. For there will then exist a set E of measure 2π, such that for $x_1^0 \in E$ all functions $\phi(z_1, z_2^k, \ldots, z_m^k)$ have non-tangential limits at $e^{ix_1^0}$. Using the equicontinuity just established, we come to the following conclusion: if $x_1^0 \in E$ and $|z_2| \leqslant 1 - \eta, \ldots, |z_m| \leqslant 1 - \eta$, the function $\phi(z_1, z_2, \ldots, z_m)$ converges to a limit as z_1 tends non-tangentially to $e^{ix_1^0}$, and the convergence is uniform in $z_2, \ldots, z_m$. Hence $\phi^{x_1^0}(z_2, \ldots, z_m)$ exists, and is regular in Δ_{m-1}.

The existence of the required sequence $\{(z_2^k, \ldots, z_m^k)\}$ will be established if we show that for each system of radii $r_2^0, \ldots, r_m^0$ less than 1 almost all systems of amplitudes $(x_2^0, \ldots, x_m^0)$ have the property that $\phi(z_1, r_2^0 e^{ix_2^0}, \ldots, r_m^0 e^{ix_m^0})$ has a non-tangential limit almost everywhere. By (5·17),

$$\int_0^{2\pi}\ldots\int_0^{2\pi}\left\{\int_0^{2\pi} \log^+ |\phi(r_1 e^{ix_1}, r_2^0 e^{ix_2}, \ldots, r_m^0 e^{ix_m})|\, dx_1\right\} dx_2 \ldots dx_m < M.$$

The inner integral being a non-decreasing function of r_1, it follows that at almost all points $(x_2^0, \ldots, x_m^0)$ we have

$$\int_0^{2\pi} \log^+ |\phi(r_1 e^{ix_1}, r_2^0 e^{ix_2^0}, \ldots, r_m^0 e^{ix_m^0})|\, dx_1 = O(1)$$

as $r_1 \to 1$, so that $\phi(z_1, r_2^0 e^{ix_2^0}, \ldots, r_m^0 e^{ix_m^0})$ has a non-tangential limit almost everywhere.

Suppose now we have (5·18). We write it as

$$\int_0^{2\pi}\left\{\int_0^{2\pi}\ldots\int_0^{2\pi} \chi\{\log^+ |\phi(r_1 e^{ix_1}, \ldots, r_m e^{ix_m})|\}\, dx_2 \ldots dx_m\right\} dx_1 < M.$$

Making $r_1 \to 1$ and keeping $r_2, \ldots, r_m$ fixed we obtain

$$\int_0^{2\pi}\left\{\int_0^{2\pi}\cdots\int_0^{2\pi}\chi\{\log^+|\phi^{x_1}(r_2e^{ix_2}, \ldots, r_me^{ix_m})|\}\,dx_2\ldots dx_m\right\}dx_1 \leqslant M.$$

Since the inner (repeated) integral is a non-decreasing function of each of the radii $r_2, \ldots, r_m$, we deduce (5·19) with $M(x_1)$ finite almost everywhere. This completes the proof of (5·16).

If $\phi(z_1, \ldots, z_m)$ is in N, then, by (5·19) with $\chi(t)=t$, the function $\phi^{x_1}(z_2, \ldots, z_m)$ is in N for almost all x_1. Hence the iterated limit

$$\phi^{x_1x_2}(z_2, \ldots, z_m) = \lim_{z_2\to e^{ix_2}} \phi^{x_1}(z_2, \ldots, z_m)$$

exists and is in N for almost all points (x_1, x_2). Repeating the argument we see that if $\phi(z_1, \ldots, z_m)$ is in N then the iterated limit

$$\phi^{x_1x_2\ldots x_m} = \lim_{z_m\to e^{ix_m}} \{\ldots\{\lim_{z_1\to e^{ix_1}} \phi(z_1, \ldots, z_m)\}\}$$

exists at almost all points $(x_1, \ldots, x_m)$.

If ϕ is in N_{m-1} then, by (4·8), an ordinary non-tangential limit, which we may denote by $\phi(e^{ix_1}, \ldots, e^{ix_m})$, also exists almost everywhere.

(5·24) THEOREM. *If $\phi(z_1, \ldots, z_m)$ is in N_{m-1}, then*

$$\phi(e^{ix_1}, \ldots, e^{ix_m}) = \phi^{x_1x_2\ldots x_m} \tag{5·25}$$

at almost all points $(x_1, \ldots, x_m)$. In particular, at almost all points $(x_1, \ldots, x_m)$ the value of the iterated limit of ϕ is independent of the order of the passage to the limit.

Suppose, for example, that $m=2$. The existence of $\phi(e^{ix_1}, e^{ix_2})$ almost everywhere implies that, given an $\epsilon>0$, we can find a set E in the square $Q(0\leqslant x_1\leqslant 2\pi, 0\leqslant x_2\leqslant 2\pi)$ such that $|Q-E|$ is arbitrarily small and

$$|\phi(r_1e^{ix_1}, r_2e^{ix_2}) - \phi(e^{ix_1}, e^{ix_2})| \leqslant \epsilon \tag{5·26}$$

for $\qquad (x_1, x_2)\in E, \quad 1-r_1\leqslant\delta, \quad 1-r_2\leqslant\delta, \quad \delta=\delta(\epsilon).$

(We use here only the existence of the radial limit.) Hence, making $r_1\to 1$ in (5·26), we see that

$$|\phi^{x_1}(r_2e^{ix_2}) - \phi(e^{ix_1}, e^{ix_2})| \leqslant \epsilon \quad (1-r_2\leqslant\delta)$$

for almost all x_1 such that (x_1, x_2) is in E, that is, for almost all (x_1, x_2) in E. Now, making $r_2\to 1$, we see that $|\phi^{x_1x_2} - \phi(e^{ix_1}, e^{ix_2})| \leqslant \epsilon$ almost everywhere in E. Since ϵ and $|Q-E|$ are arbitrarily small, the theorem follows.

MISCELLANEOUS THEOREMS AND EXAMPLES

1. If the integral of $|f|$ is strongly differentiable in E, the integral of f is strongly differentiable almost everywhere in E.

[For each $n=1, 2, \ldots$ write $f=g_n+h_n$, where $g_n(\mathbf{x})$ is either $f(\mathbf{x})$ or 0, according as $|f(\mathbf{x})|\leqslant n$ or $|f(\mathbf{x})|>n$. Then $|f|=|g_n|+|h_n|$, $F=G_n+H_n$, $\Phi=\Gamma_n+\mathrm{X}_n$, where F, G_n, H_n, Φ, Γ_n, X_n denote respectively the integrals of f, g_n, h_n, $|f|$, $|g_n|$, $|h_n|$. Since g_n is bounded, it follows that G_n and Γ_n are strongly differentiable almost everywhere, $\mathrm{X}_n=\Phi-\Gamma_n$ is strongly differentiable almost everywhere in E, and $\mathrm{X}_n'=|h_n|$ almost everywhere in E. Let E_n be the subset of E where (i) $h_n=0$,

(ii) the strong derivative of X_n exists and equals $|h_n|=0$, (iii) the strong derivative of G_n exists and equals g_n. Then, if $\mathbf{x}_0 \in E_n$ and I is an interval converging to $\mathbf{x}_0$, the right-hand side of

$$F(I)\,|\,I\,|^{-1} = G_n(I)\,|\,I\,|^{-1} + H_n(I)\,|\,I\,|^{-1}$$

tends to $g_n(\mathbf{x}_0)+0=f(\mathbf{x}_0)$. Observe that $|E-E_n|\to 0$ as $n\to\infty$.]

2. There is an $f\in \mathrm{L}$ such that the integral of f is strongly differentiable almost everywhere, and the integral of $|f|$ almost nowhere. (Papoulis [1].)

3. (i) Suppose that $f(\mathbf{x})=f(x_1,\ldots,x_m)$ is 0 near $\mathbf{x}=\mathbf{x}^0$ and that each of the m functions

$$g_j(x_1,\ldots,x_{j-1},x_{j+1},\ldots,x_m)=\int_{-\pi}^{\pi}|\,f(x_1,\ldots,x_{j-1},t,x_{j+1},\ldots,x_m)\,|\,dt$$

is bounded in $\mathbf{x}$. Then $\mathsf{S}[f]$ is summable to 0, both $(C,1)$ and A^*, near $\mathbf{x}=\mathbf{x}^0$. (ii) If $m=2$, the result holds if g_1 and g_2 are bounded *near* $\mathbf{x}^0$.

4. (i) If the integral of $f(x_1,x_2)$ is strongly differentiable in E, then $\mathsf{S}[f]$ is summable A^* almost everywhere in E. (ii) If the integral of $|f(x_1,x_2)|$ is strongly differentiable in E, then $\mathsf{S}[f]$ is summable $(C,1)$ almost everywhere in E. (iii) If $f(x_1,x_2)$ is 0 in a two-dimensional interval I, then $\mathsf{S}[f]$ is summable to 0, both $(C,1)$ and A^*, almost everywhere in I.

5. The analogue of Example 4 (iii) is false for $m\geq 3$: there is an $f(\mathbf{x})=f(x_1,x_2,x_3)$ which is 0 in a neighbourhood N of $(0,0,0)$ and such that $\mathsf{S}[f]$ is summable neither $(C,1)$ nor A at any point of N.

[Let $f_1(x_1,x_2)$ be periodic, non-negative and such that, say, the Abel means of $\mathsf{S}[f_1]$ are unbounded at each point (x_1,x_2) (cf. (2·14) (ii)). Let $f(x_1,x_2,x_3)$ be equal to $f(x_1,x_2)$ for $\frac{1}{2}\pi\leq|x_3|\leq\pi$, and to 0 elsewhere. Then, if $|x_3|\leq\frac{1}{4}\pi$ and (x_1,x_2) is fixed, the Abel mean of $\mathsf{S}[f]$ at (x_1,x_2,x_3) exceeds

$$P(r_3,\tfrac{1}{4}\pi)\,\pi^{-3}\int_{-\pi}^{\pi}\int_{-\pi}^{\pi}f_1(x_1+t_1,x_2+t_2)\,P(r_1,t_1)\,P(r_2,t_2)\,dt_1\,dt_2,$$

and so is unbounded if $1-r_3$ tends to 0 slowly enough in comparison with $1-r_1$ and $1-r_2$.]

6. Suppose that $f(\mathbf{z})=f(z_1,\ldots,z_n)\in \mathrm{N}$. If the integrals

$$(*)\qquad \int_0^{x_1}\cdots\int_0^{x_m}\log^+|\,f(r_1e^{iy_1},\ldots,r_m e^{iy_m})\,|\,dy_1\ldots dy_m$$

are uniformly absolutely continuous (that is, if the integral of $\log^+|f|$ taken over sets $E\subset Q$ is small with $|E|$, uniformly in $\mathbf{r}$), then for any convex non-decreasing $\chi(u)$, $u\geq 0$, we have

$$\int_Q \chi\{\log^+|\,f(r_1e^{ix_1},\ldots,r_me^{ix_m})\,|\}\,d\mathbf{x}\leq\int_Q\chi\{\log^+|\,f(e^{ix_1},\ldots,e^{ix_m})\,|\}\,d\mathbf{x},$$

where $f(e^{ix_1},\ldots,e^{ix_m})$ denotes the restricted radial limit of f.

In particular, the conclusion holds if there is a non-negative convex $\chi(u)$ such that $\chi(u)/u\to\infty$ with u, and the integral $\int_Q\chi\{\log^+|\,f(r_1e^{ix_1},\ldots,r_me^{ix_m})\,|\}\,d\mathbf{x}$ is bounded.

[If $r_j<\rho<1$ for all j we have (cf. pp. 321–2)

$$\log^+|\,f(r_1e^{ix_1},\ldots,r_me^{ix_m})\,|\leq\pi^{-m}\int_Q\log^+|\,f(\rho e^{it_1},\ldots,\rho e^{it_m})\,|\prod_{j=1}^{m}P\left(\frac{r_j}{\rho},x_j-t_j\right)d\mathbf{t}.$$

Using the hypothesis of uniform absolute continuity, we can make $\rho\to 1$ on the right, and we obtain

$$\log^+|\,f(r_1e^{ix_1},\ldots,r_me^{ix_m})\,|\leq\pi^{-m}\int_Q\log^+|\,f(e^{it_1},\ldots,e^{it_m})\,|\prod_{j=1}^{m}P(r_j,x_j-t_j)\,d\mathbf{t}.$$

By Jensen's inequality, we may replace $\log^+|f|$ by $\chi(\log^+|f|)$ throughout, and integrating both sides of the resulting inequality with respect to $\mathbf{x}$ over Q, and changing the order of integration on the right, we obtain the desired inequality.]

7. If $f(\mathbf{z})\in \mathrm{H}^\alpha$, and if the boundary values of f belong to L^β, $\beta>\alpha$, then $f(\mathbf{z})\in\mathrm{H}^\beta$.

8. If $f(\mathbf{z}) \in \mathrm{N}$, and if the integral (*) in Example 6 is a uniformly absolutely continuous function of $(x_1, \ldots, x_m)$ then, for almost all x_1,

$$\int_0^{x_2} \cdots \int_0^{x_m} \log^+ | f^{x_1}(r_2 e^{iy_2} \ldots, r_m e^{iy_m}) | \, dy_2 \ldots dy_m$$

is an absolutely continuous function of $(x_2, \ldots, x_m)$.

[The hypothesis is equivalent to that of the existence of a $\chi(u)$, $u \geqslant 0$, non-negative, convex, non-decreasing, satisfying $\chi(u)/u \to \infty$ as $u \to \infty$, and such that the integral

$$\int_Q \chi\{\log^+ | f(r_1 e^{ix_1}, \ldots) |\} \, d\mathbf{x}$$

is bounded in $\mathbf{r}$. Apply (5·16).]

9. Let $f(\mathbf{z}) \in \mathrm{H}^\alpha$, and let $g_\delta(\mathbf{x})$ be the upper bound of $|f|$ in the domain $\Omega(\mathbf{x}) = \Omega_\delta(\mathbf{x})$ considered on p. 317. Show that if $\int_Q | f(\mathbf{z}) |^\alpha \, d\mathbf{x} \leqslant M$ for $\mathbf{z} \in \Delta_m$, then

(i) $\int_Q | g_\delta(\mathbf{x}) |^\alpha \, d\mathbf{x} \leqslant A_{\alpha,\delta} M,$

(ii) $\int_Q | f(r_1 e^{ix_1}, \ldots, r_m e^{ix_m}) - f(e^{ix_1}, \ldots, e^{ix_m}) |^\alpha \, d\mathbf{x} \to 0 \quad (\text{as } \mathbf{r} \to \mathbf{1}).$

10. If $f(\mathbf{z}) \in \mathrm{N}_1$ (that is, if the integral of $\log^+ | f(\mathbf{z}) | \log\log^+ | f(\mathbf{z}) |$ over Q is bounded in $\mathbf{r}$), then

$$f^{x_1 x_2}(z_3, \ldots, z_m) = f^{x_2 x_1}(z_3, \ldots, z_m)$$

for almost all (x_1, x_2). If $f(\mathbf{z}) \in \mathrm{N}_2$, then $f^{x_1 x_2 x_3}(z_4, \ldots, z_m)$ is almost always independent of the permutation of (x_1, x_2, x_3), etc.

NOTES

CHAPTER X

§ 1. There are two expository articles about trigonometric interpolation: Burkhardt, Trigonometrische Interpolation, *Enz. der Math. Wiss.* II, A, **9***a*, pp. 642–93, covering the period up to the end of the nineteenth century, and E. Feldheim [1], presenting results prior to 1938.

The case of non-equidistant fundamental points has been studied extensively by Fejér; a list of his papers will be found in Feldheim [1]; see also Erdös and Turàn [1].

In using the Stieltjes notation we follow Marcinkiewicz [7].

§ 2. Theorem (2·10) seems to go back to Bessel; see Burkchardt, *loc. cit.* p. 648 sqq.

§ 3. For (3·6) and (3·11) see M. Riesz [3]. Theorem (3·13) of Bernstein [1] has considerable literature; see, for example, an expository article of Schaeffer [2]. An important generalization will be found in Schaeffer [3]. (3·16) will be found in Zygmund [10]; see also E. Stein [1]. For (3·19) see Szegö [3].

§ 4. For (4·8) see de la Vallée-Poussin [3]. (4·9) goes back to Euler and Gauss; cf. Burkchardt, *loc. cit.* p. 651.

§ 5. Convergence of interpolating polynomials is studied in de la Vallée-Poussin [3] and Faber [2], where we find (5·6) in the case $\nu = n$. It seems that Jackson [5] was the first to consider the convergence of $I_{n,\nu}$ for $\nu \neq n$, but he limited himself to Theorem (5·6), and apparently did not recognize the generality of the problem.

For (5·5) see Natanson [1], for (5·13) and (5·16) Faber [2]. (5·17) (i) is due to Marcinkiewicz [9]; the extension to (ii) was communicated to us in conversation by P. Erdös; see also Erdös [2].

§ 6. Polynomials J_n were introduced by Jackson [3]; the $B_{n,n}$ by Marcinkiewicz [6], [10]. Bernstein [3], was the first to point out that the J_n are Hermite interpolation polynomials. (6·12) is due to Fejér [13]; for (6·14), (6·15), (6·18) see Bernstein [5], [6]; for (6·14) see also Offord [1], Klein [1].

§ 7. In connexion with (7·1) and (7·4) see Erdös and Turàn [1. I]. (7·5) is due to Marcinkiewicz [9] (see also Marcinkiewicz and Zygmund [6]; Lozinski [1]); an analogue of (7·6) for Tchebyshev interpolation was proved by Erdös and Feldheim [1]; interesting extensions to derivatives will be found in Zarantonello [1]. For (7·10), (7·23), (7·26), (7·27), (7·29) see Marcinkiewicz and Zygmund [6]; (7·30) is an earlier result of Bernstein [6]. (7·32) was proved by Erdös [3]. That $(I_0 + I_1 + \ldots + I_n)/(n+1)$ (and *a fortiori* the ξ_n of (7·32)) can be unbounded at some points, even if f is continuous, was shown by Marcinkiewicz [10].

§ 8. (8·6) is due to Faber; it seems to be still an open problem whether for any sequence of systems of fundamental points there is a continuous function for which the interpolating polynomials diverge at some point. For (8·8) see Faber [3], Marcinkiewicz [9]. (8·14) was proved independently by Grünwald [1] and Marcinkiewicz [7]. (8·25) is proved in Marcinkiewicz [7] (the fact that there are continuous f such that $S[f]$ converges uniformly, and $\{I_n[f]\}$ diverges at some points, was observed by Faber [2]), and so are (8·28) and (8·30).

§ 9. (9·1) is essentially a result of Grünwald [2] and Marcinkiewicz [8], though both of them actually proved (9·15); the proof of (9·1) given in the text follows the argument of Marcinkiewicz [8]. For (9·16) see Gosselin [1].

CHAPTER XI

§ 1. The definitions of generalized symmetric derivatives appear in de la Vallée-Poussin [2]. For (1·7) and (1·20) see Gronwall [4], Zygmund [22]. (1·25) is stated in Plessner [5] and his *Trigonometrische Reihen.*

§ 2. For (2·24) and (2·26) see Hardy and Littlewood [12], [8]; they were the first to formulate the problem of necessary and sufficient conditions for the summability C of a Fourier series. For (2·1) see Plessner, *Trigonometrische Reihen*, p. 1381, who considers only integral α (the case $-1<\alpha<0$, $r=1$, was proved by Hardy and Littlewood [26]). For (2·20) and (2·22) see Kogbetliantz [1], Verblunsky [4]. See also Bosanquet [2], Obrechkoff [1].

§ 3. For (3·1) see Zygmund [12; § 3], where also earlier literature is indicated. A generalization will be found in Bosanquet [1].

§ 4. The basic theorem (4·2) is proved in Marcinkiewicz [2]; (4·30) is a special case of a general theorem proved in Marcinkiewicz and Zygmund [2].

§ 5. For (5·2) see Plessner [5], where complex methods are used; the proof of the text is from Marcinkiewicz [2]; (5·4) and (5·8) will be found in Marcinkiewicz [2] and Marcinkiewicz and Zygmund [2] respectively. (5·15) is due to Titchmarsh [1].

§ 6. References to the M-integral are scattered in the literature on Fourier series; the existence of $\tilde{f}$ for $f \in \mathrm{M}$ is proved in Plessner [5]. That Fourier-Riemann coefficients need not tend to 0 was known to Riemann [1]; another example is indicated in the first edition of this book, p. 19. For (6·4) see Titchmarsh [7].

§ 7. Basic work here is due to Denjoy; see his *Calcul des coefficients d'une série trigonométrique*, where we also find references to his earlier papers. Marcinkiewicz and Zygmund [2], have shown that, suitably defined, major and minor functions of order 1 are sufficient to prove the Fourier character of an everywhere convergent $\Sigma A_n(x)$; in using major and minor functions of order 2 we follow Gage and James [1], and James [1]. Further bibliographic references (especially to the work of Burkill and Verblunsky) will be found in James [2] and Jeffery's *Trigonometric series*; see also Taylor [1].

CHAPTER XII

§ 1. A theorem equivalent to (1·11) was first proved by M. Riesz [4] for (α, β) in the triangle $0 \leqslant \beta \leqslant \alpha \leqslant 1$, and extended to the square $0 \leqslant \alpha \leqslant 1$, $0 \leqslant \beta \leqslant 1$ by Thorin [1]. In Riesz's paper we find for the first time the idea of an 'interpolation of linear operations', which has proved very fruitful since. Riesz's proof was real-variable, Thorin used complex methods. Further simplifications of proof will be found in Thorin [2], Tamarkin and Zygmund [1], Calderón and Zygmund [1] (the argument of the text is that of the latter paper). An extension of the argument to 'sublinear' operations (for the definition see § 4 below) will be found in Calderón and Zygmund [3]. (1·39) is due to E. Stein [2]; see also Hirschman [1]. Interpolation of measures has been considered in Stein [2], Stein and Weiss [1]. The observation on p. 96 that we can take $M_1 = M_2 = 1$ in (1·11) is due to G. Weiss.

For other proofs and extensions, see Paley [4] (his argument is reproduced in the first edition of the book), L. C. Young [1], Salem [16], [17], Calderón and Zygmund [4].

§ 2. (2·3) was proved by W. H. Young [3], [4], for $p'=4, 6, 8, \ldots$, and by Hausdorff [1] in the general case; (2·8) is due to F. Riesz [8]; the idea of proving (2·8) by using interpolation of linear operations is due to M. Riesz [4]. (2·25) was proved by Hardy and Littlewood [15]; the extension to (2·18) is due to Verblunsky [3]. See also Hardy and Littlewood [23]; Mulholland [1].

§ 3. A particular instance of interpolation in the classes H^r (a proof of (3·22)) will be found in Thorin [2]; general theorems were obtained by Salem and Zygmund [6], Calderón and Zygmund [1] (whence we take the argument of the text) and G. Weiss [2].

§ 4. The basic theorems (4·6), (4·22) and (4·39) are due to Marcinkiewicz [14]; see also Zygmund [28]. The argument giving (4·41) (i) is not novel (see, for example, Zygmund [4]) but the theorem has not been formulated explicitly; for (4·41) see Yano [1] (also Titchmarsh [3]).

Recently Cotlar and Bruschi [1] have shown that the proof of (4·6) can be modified so as to give the Thorin-Riesz theorem (1·11) for the triangle $0 \leqslant \beta \leqslant \alpha$, with an extra constant on the right of (1·14).

§ 5. The main results here are due to Paley [3], who extended (3·19) to uniformly bounded orthonormal systems (later, certain extensions which do not require new methods were also formulated, independently, by Verblunsky [3], and in the first edition of this book). See also Zygmund [28]. For further generalizations see Pitt [1], Stein and Weiss [1], Littlewood [5].

§ 6. See Hardy and Littlewood [15], [22], Gabriel [1], [2].

§ 7. For (7·1) see Banach [1], Sidon [4], [5]; the first edition of this book reproduced the proof of (7·1) from Verblunsky [3]; the present argument is from Salem and Zygmund [1]. (7·5) is due to Gronwall [3] (see also Paley [6]); for (7·6) see Zygmund [8] (which contains a slip easy to correct: the class conjugate to $\exp L^2$ is $L(\log^+ L)^{\frac{1}{2}}$); for (7·8)—Paley [5], Hardy and Littlewood [24].

§§ 8, 9. For the definition (8·3) see Weyl [1]. The main results here are due to Hardy and Littlewood [$9_{\text{I, II}}$], [19] (for limiting cases see also Zygmund [5], [1]).

§ 10. For (10·2) and (10·9) see Milicer-Grużewska [1] ((10·9) is stated there for $B(x)$ bounded only; that the result is valid in the general case was observed by Pyatetski-Shapiro [1], and Salem (unpublished)). The proof of (10·9) given there is shorter but less elementary. For (10·5) compare Rajchman [7]. (10·11) will be found in Salem [12_{I}].

(10·12) is due to Wiener and Wintner [2], who generalized an earlier result of Littlewood [4]; see also Schaeffer [1]. That the mass carried by F can be confined to a perfect non-dense set of measure 0, was shown by Salem [9], [11]. Ivašev-Musatov [1] showed recently that the c_n in (10·12) can be made $O(n^{-\frac{1}{2}})$, $O\{(n \log n)^{-\frac{1}{2}}\}$, $O\{(n \log n \log\log n)^{-\frac{1}{2}}\}$, ..., etc.

§ 11. Important work in the theory of numbers S has been done by Pisot [1]; see also his expository article [2] (to the literature mentioned there one should add Hardy [12]). In calling the numbers S we follow Pisot's terminology. The importance of numbers S for the problems of uniqueness was first realized by Salem [$12_{\text{I, II}}$] (see also an earlier paper of Erdös [4]), where we find (11·17) and (11·18). Salem's proof of (11·18) contains a slip, as stated by Salem himself in [12_{III}]. However, the theorem is correct, and was proved finally in Salem and Zygmund [8], [9], by using an idea contained in Salem [12_{III}] together with the important notion of sets $H^{(n)}$.

Sets $H^{(n)}$ were introduced by Pyatetski-Shapiro [1], [2]; he also showed that for each n there are sets $H^{(n)}$ which are not denumerable sums of sets $H^{(n-1)}$.

Sets $H^{(n)}_*$, which appear in the proof of (11·18) could be avoided, and we could deal directly with sets $H^{(n)}$, by borrowing a little more from the theory of algebraic numbers; see Salem and Zygmund [9].

In [2], Pyatetski-Shapiro shows also that there are perfect sets M which are not M in the restricted sense (for the definition of sets M in the restricted sense see Chapter IX, p. 348).

Example 12, p. 159. R. O'Neil has suggested the following argument, somewhat simpler than that indicated in the text. Suppose, for example, that $f \in \Lambda_\alpha$, $0<\alpha<1$, $g \in \Lambda_\beta$, $0<\beta<1$, $\alpha+\beta<1$. We can easily verify that

$$h(x+2u)+h(x-2u)-2h(x)=\frac{1}{2\pi}\int_0^{2\pi}[f(t+u)-f(t-u)]\,[g(x-t+u)-g(x-t-u)]\,dt.$$

The right-hand side is clearly $O(|u|^{\alpha+\beta})$, and so $h \in \Lambda_{\alpha+\beta}$ (cf. footnote in Chapter II, p. 44). The argument works in other cases.

CHAPTER XIII

§ 1. (1·2) and (1·8) are final forms of results obtained successively by Kolmogorov and Seliverstov [1], [2], Plessner [3], Hardy and Littlewood [21]. For (1·14) see Plessner [3]; (1·17) is a corollary of results of Kolmogorov [5] and Hardy and Littlewood [1]. (1·22) was communicated to us by A. P. Calderón.

§ 2. (2·1), (2·5) and (2·21) are due to Littlewood and Paley, [1]. The original proof of (2·1) was quite difficult; simplifications are from Salem and Zygmund [6]. See also Marcinkiewicz [13].

§ 3. (3·2) is due to Marcinkiewicz [5]; in paper [2] he shows that the $O\{(\log 1/h)^{-1}\}$ in (3·3) cannot be replaced by anything tending less rapidly to 0. See also Salem [14].

§ 4. See Paley and Zygmund [2].

§ 5. (5·1) is due to Kuttner [2]; the proof of the text is from Marcinkiewicz and Zygmund [2], where also (5·7) will be found. (5·13) was communicated to us by Mrs M. Weiss. (5·14) is due to Menšov [3]; see also his papers [4], [5].

§ 6. See Marcinkiewicz and Zygmund [7].

§ 7. For (7·3) see Hardy and Littlewood [4], Sutton [1], Carleman [3]; for (7·7)—Borgen [1], Zygmund [23].

§8. (8·1) was proved by Marcinkiewicz [10], for $q=2$; for general q see Zygmund [21_{II}]. That $S[f]$ need not be summable H_1 at the points where $\phi_{x,1}(h)=o(h)$, was shown by Hardy and Littlewood [6]; see also Wang [1], Hardy and Littlewood [14].

§10. That $\Sigma c_\nu \phi_\nu$ converges almost everywhere (for *any* orthonormal $\{\phi_\nu\}$) if $\Sigma |c_\nu|^p < \infty$ $(p<2)$, was proved by Menšov [6_{III}]; the rest of (10·1) is in Paley [3]. For (10·21) see Menšov [6_{I}], Rademacher [1]; a different proof will be found in Salem [17]. For more results about the convergence and summability of orthogonal series see Kaczmarz and Steinhaus, *Orthogonalreihen*.

§11. For (11·3) (i) see Beurling [3]; for (ii)—Salem and Zygmund [10], Broman [1].

CHAPTER XIV

§1. (1·1), proved by the method of conformal mapping, and (1·9), are in Privalov [1]; see also Lusin and Privalov [1]. The second proof of (1·1) is due to Calderón [2]. (1·10) was proved by Plessner [5]; for further study see Collingwood and Cartwright [1]. (1·11) seems to be new; a simplification in its proof we owe to P. Cohen. For (1·12) see Bagemihl and Seidel [1], Rudin [1]. (1·21) is due to Ostrowski [1] and (in the case of uniformly bounded functions) Khintchin [1]. (1·25) will be found in Zygmund [25]; the condition of the uniform boundedness of the F_n can be replaced by a weaker one (*loc. cit.*), but not by the condition that the integrals $\int_0^{2\pi} \log^+ |F_n(re^{ix})|\,dx$ are uniformly bounded (a counter example can be easily constructed by means of the function $\exp\{(1+z)/(1-z)\}$), though the latter condition implies the existence of a subsequence $\{F_{n_k}(z)\}$ converging uniformly in each $\Omega_\sigma(x_0)$ for almost every $x_0 \in E$; see Tumarkin [1].

§2. For (2·3) see Lusin [5]; parts (i) and (ii) of (2·2) will be found respectively in Marcinkiewicz and Zygmund [8] and Spencer [1]; see also Calderón [3]. The case of integrals of $|F'|^2$ over domains tangent in the unit circle is discussed in Piranian and Rudin [1].

§3. For (3·5) and (3·15) see Littlewood and Paley [$1_{\text{I, II}}$]; the proof of (3·15) given in the text is new and was communicated to us by J. E. Littlewood. Partial extensions of (3·15) to the case $\lambda \leq 1$ will be found in Flett [2]. Analogues of (3·5) and (3·15) for the function s_δ are discussed in Marcinkiewicz and Zygmund [8]. For (3·24) see Littlewood and Paley [1_{II}].

§4. For (4·1) see Plessner [4]; the proof of the text is from Marcinkiewicz and Zygmund [7]. The result admits of an extension to summability (C, α) (*loc. cit.*), but not to summability A (cf. (1·12)).

§5. For (5·1) and (5·3) see Marcinkiewicz [1]; for (5·5)—Zygmund [20].

CHAPTER XV

Most of the results of this chapter will be found in Littlewood and Paley [1]; they mostly confined themselves to functions in L^p, $p>1$. Extensions of results to other classes, and simplifications of proofs, will be found in Zygmund [21, I]. (4·14) is due to Marcinkiewicz [12]. The function γ_2 is essentially the same as the function g^* introduced by Littlewood and Paley (*loc. cit.*), and important in their work; this fact was noticed rather late; see Sunouchi [2], Zygmund [27].

For (2·10) see also Marcinkiewicz and Zygmund [5], Boas and Bochner [1].

Extension of the theory to classes H^r, $r<1$, is not yet complete; see Zygmund [21_{II}] (Theorem 4 of the paper contains a misprint: the denominator $(\log n)^{1/\alpha}$ should be replaced by $(\log n)^{1/\lambda}$), [20], [27], Sunouchi [1], [2], Stein and Weiss [2]. Here we only mention the following result: *If* $F(z)=\Sigma c_n z^n \in H^r$, $r<1$, *then* $\Sigma c_n e^{in\theta}$ *is summable* $(C, r^{-1}-1)$ *almost everywhere* (Zygmund [21_{II}]). The example of the function $F(z)=(1-z)^{-1/r}=\Sigma A_n^{(1/r)-1} z^n$, which belongs to every $H^{r-\epsilon}$ $(\epsilon>0)$, shows that summability $(C, r^{-1}-1)$ is a best possible result.

CHAPTER XVI

§1. In this chapter, after a general introduction, we consider only a few selected problems on Fourier integrals. For other aspects of the theory, and bibliography, see Titchmarsh, *Fourier integrals*; Bochner, *Vorlesungen über Fouriersche Integrale*; Wiener, *The Fourier integral*; Paley and Wiener, *Fourier transforms in the complex domain*; and relevant chapters of Schwarz's *Théorie des distributions*. For older literature see Burkchardt, *Trigonometrische Reihen und Integrale*.

A brief account of the summability (C, k) of integrals can be found in the first edition of this book.

§ 2. For (2·17) see Plancherel [2], [3].

§ 3. For (3·2) see Titchmarsh [6]; also M. Riesz [4], Hille and Tamarkin [3]. Hewitt and Hirschman [1] have shown that we have strict inequality in (3·3) unless $f \equiv 0$. For (3·8) see M. Riesz [1] (a limiting case is in Kober [1]); for (3·14)—Zygmund [29], Kac [2].

The important problem of the harmonic analysis of bounded functions has been started by Beurling. A brief account of the results obtained, together with bibliographic references, will be found in Pollard [1].

§ 4. For (4·19) see Wiener [4]; the essence of (4·24) is a classical result of the calculus of probability, in a form strengthened by Cramér.

§ 5. See Ferrand and Fortet [1]; Salem and Zygmund [7].

§ 6. Salem and Zygmund [2].

§ 7. More details about the theory functions of exponential type will be found in Boas's *Entire functions*, to which we also refer for bibliography. In connexion with (7·19) see Hardy [11]. The example (7·31) was communicated to us by R. P. Boas.

§ 8. For (8·4) see Wolf [3].

§ 9. The case of integrals $\int e^{i\lambda x} d\chi(\lambda)$ with χ absolutely continuous is discussed in Zygmund [30]; for the general case see Wolf [3], Zygmund [31], where also applications are given. All these papers also treat the case of summable integrals.

§ 10. (10·3) is due to Offord [2], who also considers the more general case of integrals summable (C, 1). The latter case can also be reduced, by means of equisummability theorems, to that of series summable (C, 1) (treated in Chapter IX, §§ 7, 8).

CHAPTER XVII

§ 1. An elementary introduction to double Fourier series can be found in Tonelli's *Serie trigonometriche*. For bibliographic references to older literature see Geiringer [1].

Concerning spherical summability of multiple Fourier series see Bochner [3], E. Stein [2], Chandrasekharan and Minakshisundaram, *Typical means*.

§ 2. For (2·2) (i), (iii) and (2·14) (i), (iii) see Jessen, Marcinkiewicz and Zygmund [1] (also Burkill [1]). A proof of (2·2) (ii) will be found in Saks [2].

For certain aspects of rectangular summability see also Herriot [1].

§ 3. Restricted (C, 1) summability of double Fourier series was first considered by Moore [2]; for the main results of this section see Marcinkiewicz and Zygmund [9], Zygmund [24].

The m-dimensional analogue of Theorem (4·6) of Chapter IV which we use in the proof of (3·21) can be found in de la Vallée-Poussin [4] or Frostman [1].

§ 4. For (4·7) see Zygmund [26] (where, however, only the case of radial approach is explicitly stated), and for (4·8)—Zygmund [25], where a different proof is given; the present proof uses ideas from Calderón [2]. Calderón's paper also contains (4·13), (4·22) and (4·24).

For related problems see also Bochner [2], Bergman and Marcinkiewicz [1], Bers [1].

§ 5. See Calderón and Zygmund [2].

BIBLIOGRAPHY

A. BOOKS

AKHIESER, N., *Lectures on the theory of approximation* (in Russian) (Moscow, 1947), pp. 1–307; English translation, *Theory of Approximation* (New York, 1956), pp. 1–307; German translation, *Vorlesungen über Approximationstheorie* (Berlin, 1953), pp. 1–309.

BANACH, S., *Opérations linéaires*, Monografje Matematyczne (Warsaw, 1932), pp. vii + 254.

BOAS, R. P., *Entire functions* (New York, 1954), pp. 1–276.

BOCHNER, S., *Vorlesungen über Fouriersche Integrale* (Leipzig, 1932), pp. 1–227.

BURKHARDT, H., Trigonometrische Reihen und Integrale (bis etwa 1850), *Enz. d. Math. Wiss.* II, A 12, 819–1354. Trigonometrische Interpolation, *Enz. d. Math. Wiss.* II, A 9*a*, 642–93.

CHANDRASEKHARAN, K. and MINAKSHISUNDARAM, S., *Typical means* (Oxford, 1952), pp. 1–139.

DENJOY, A., *Calcul des coefficients d'unes érie trigonométrique*, four parts (Paris, 1941–9), pp. 1–701.

EVANS, G. C., *The logarithmic potential*, American Math. Soc. Coll. Publ. VI (1927), 1–150.

GRENANDER, U. and SZEGÖ, G., *Toeplitz forms and their applications* (Berkeley, Los Angeles. 1958), vii + 245.

HARDY, G. H., *Divergent Series* (Oxford, 1949), pp. xiv + 396.

HARDY, G. H., LITTLEWOOD, J. E. and PÓLYA, G., *Inequalities* (second edition) (Cambridge, 1952), pp. xii + 324.

HARDY, G. H. and ROGOSINSKI, W. W., *Fourier Series*, Cambridge Tracts, no. 38 (1950), 1–100.

HILB, E. and RIESZ, M., Neuere Untersuchungen über trigonometrische Reihen, *Enz. d. Math. Wiss.* II, C 10, 1189–1228.

HILLE, E., WALSH, J. L. and SHOHAT, J. A., *A bibliography on orthogonal polynomials*, Bull. of the Nat. Research Council, no. 103, Nat. Acad. of Sciences, Washington, 1940.

JACKSON, D., *The theory of approximation*, American Math. Soc. Coll. Publ., XI, 1930, 1–178.

JEFFERY, R. L., *Trigonometric series*, Canadian Math. Congress (Toronto, 1956), pp. 1–39.

KACZMARZ, S. and STEINHAUS, H., *Theorie der Orthogonalreihen*, Monografje Matematyczne (Warsaw, 1935), pp. vi + 294.

LEBESGUE, H., *Leçons sur les séries trigonométriques* (Paris, 1906), pp. 1–128.

LEVINSON, N., *Gap and density theorems*, American Math. Soc. Coll. Publ. XXVI, 1940, viii + 246.

LITTLEWOOD, J. E., *Lectures on the theory of functions* (Oxford, 1944), pp. 1–243.

MANDELBROJT, S., *Séries de Fourier et classes quasi-analytiques de fonctions* (Paris, 1935), pp. viii + 156.

NEVANLINNA, R., *Eindeutige analytische Funktionen*, 2nd edition (Springer, 1953), pp. 1–379.

PALEY, R. E. A. C. and WIENER, N., *Fourier transforms in the complex domain*, American Math. Soc. Coll. Publ. XIX, 1934, 1–184.

PLESSNER, A., *Trigonometrische Reihen*, in Pascal's *Repertorium d. höheren Analysis* (I, 3), 1325–96.

RADO, T., *Subharmonic functions, Ergebnisse d. Mathematik* (Berlin, 1937), pp. 1–56.

RIESZ, F. and SZ. B. NAGY, *Leçons d'analyse fonctionnelle* (Budapest, 1952), pp. viii + 448.

SCHWARTZ, L., *Théorie des distributions*, vol. I (2nd edition), Actualités Scientifiques et Industrielles, no. 1245; vol. II, *ibidem*, no. 1122.

SZEGÖ, G., *Orthogonal polynomials*, American Math. Soc. Coll. Publ. XXIII, 1939, vii + 401.

DE LA VALLÉE-POUSSIN, CH. J., *Leçons sur l'approximation des fonctions d'une variable réelle* (Paris, 1919), pp. vi + 150.

TONELLI, L., *Serie trigonometriche* (Bologna, 1928), pp. viii + 526.

WIENER, N., *The Fourier Integral and certain of its applications* (Cambridge, 1933), pp. xi + 201.

B. ORIGINAL PAPERS

Abbreviations of titles of periodicals

Acc. Lincei = Rendiconti della Accademia Nazionale dei Lincei. *A.E.N.S.* = Annales de l'Ecole Normale Supérieure. *A.J.M.* = American Journal of Mathematics. *A.M.* = Acta Mathematica. *Ann. Math.* = Annals of Mathematics. *A.S.* = Acta Scientiarum Mathematicarum (Szeged). *B.A.M.S.* = Bulletin of the American Mathematical Society. *B.A.P.* = Bulletin de l'Académie

Polonaise. *B.S.M.* = Bulletin des Sciences Mathématiques. *B.S.M.F.* = Bulletin de la Société Mathématique de France. *C.R.* = Comptes Rendus des Séances de l'Académie des Sciences (Paris). *Doklady* = Doklady Akademii Nauk SSSR (Moscow). *Duke J.* = Duke Mathematical Journal. *F.M.* = Fundamenta Mathematicae. *Izvestia* = Izvestia Akademii Nauk SSSR; Seriya Matematičeskaya. *J.f.M.* = Journal für die reine und angewandte Mathematik. *J.L.M.S.* = Journal of the London Mathematical Society. *M.A.* = Mathematische Annalen. *M.f.M.* = Monatshefte für Mathematik. *M.M.* = Messenger for Mathematics. *M.Z.* = Mathematische Zeitschrift. *P.L.M.S.* = Proceedings of the London Mathematical Society. *Proc.A.M.S.* = Proceedings of the American Mathematical Society. *Q.J.* = The Quarterly Journal of Mathematics (Oxford). *R.P.* = Rendiconti del Circolo Matematico di Palermo. *S.M.* = Studia Mathematica. *Sbornik* = Matematičeskiĭ Sbornik (Moscow).

Akhieser, N. and Krein, M., [1] On the best approximation of periodic functions by trigonometric sums (in Russian), *Doklady*, **15** (1937), 107–11.

Agnew, R. P., [1] Rogosinski-Bernstein trigonometric summability methods and modified arithmetic means, *Ann. Math.* **56** (1952), 53–79.

Alexits, G., [1] Sur l'ordre de grandeur de l'approximation d'une fonction par les moyennes de sa série de Fourier, *Mat. Fiz. Lapok*, **48** (1941), 410–22.

[2] Über eine Hadamardsche Fragestellung, *M.Z.* **30** (1929), 43–6.

Aliančić, S., Bojanić, R. and Tomić, M., [1] Sur l'intégrabilité de certaines séries trigonométriques, *Publ. Inst. Math. Acad. Serbe*, **8** (1955), 67–84.

[2] Sur le comportement asymptotique au voisinage de zéro des séries trigonométriques de sinus à coefficients mononotes, *ibid.* **10** (1956), 101–20.

Arbault, J., [1] Sur l'ensemble de convergence absolue d'une série trigonométrique, *B.S.M.F.* **80** (1952), 253–317.

Bagemihl, F. and Seidel, W., [1] Some boundary properties of analytic functions, *M.Z.* **61** (1954), 186–99.

Banach, S., [1] Über eine Eigenschaft der lakunären trigonometrischen Reihen, *S.M.* **2** (1930), 207–20.

Banach, S. and Steinhaus, H., [1] Sur le principe de condensation de singularités, *F.M.* **9** (1927), 50–61.

Bary, N., [1] Generalization of inequalities of S. N. Bernstein and A. A. Markov (in Russian), *Izvestia*, **18** (1954), 159–76.

[2] Sur l'unicité du développement trigonométrique, *F.M.* **9** (1927), 62–118.

[3] The uniqueness problem of the representation of functions by trigonometric series (in Russian), *Uspekhi Matematičeskich Nauk*, **4** (1949), no. 3, 3–68; English translation, *American Math. Soc.* no. 52 (New York, 1951), pp. 1–89.

[4] Sur l'unicité du développement trigonométrique, *C.R.* **177** (1923), 1195–7.

[5] Primitive functions and trigonometric series (in Russian), *Sbornik*, **31** (1952), 687–702.

Bary, N. and Stečkin, S. B., [1] Best approximation and differential properties of two conjugate functions (in Russian), *Trudy Moskovskogo Matematičeskogo Obščestva*, **5** (1956), 483–522.

Bergman, S. and Marcinkiewicz, J., [1] Sur les fonctions analytiques de deux variables complexes, *F.M.* **33** (1939), 75–94.

Bernstein, S., [1] Sur l'ordre de la meilleure approximation des fonctions continues par des polynômes de degré donné, *Mém. Acad. Roy. Belgique*, 2me série, **4** (1912), 1–104.

[2] Sur la convergence absolue des séries trigonométriques, *C.R.* **158** (1914), 1661–4.

[3] Sur la convergence absolue des séries trigonométriques (in Russian), *Comm. Soc. Math. Kharkov*, 2me série, **14** (1914), 139–44.

[4] Sur un procédé de sommation des séries trigonométriques, *C.R.* **191** (1930), 976–9.

[5] Sur une modification de la formule d'interpolation de Lagrange, *Comm. de la Soc. Math. de Kharkov*, **5** (1931), 49–57.

[6] Sur une classe de formules d'interpolation, *Izvestia*, no. 9 (1931), pp. 1151–61.

Bers, L., [1] Bounded analytic functions of two complex variables, *A.J.M.* **64** (1942), 514–29.

Besicovitch, A. S., [1] Sur la nature des fonctions à carré sommable et des ensembles mesurables, *F.M.* **4** (1923), 172–95.

[2] On a general metric property of summable functions, *J.L.M.S.* **1** (1926), 120–8.

Beurling, A., [1] Sur les intégrales de Fourier absolument convergentes, *Neuvième Congrès des mathématiciens scandinaves* (Helsingfors, 1938), pp. 345–66.

BEURLING, A., [2] Etude sur un problème de majoration, *Diss.* (Uppsala, 1933), pp. 1–109.
[3] Sur les ensembles exceptionnels, *A.M.* **72** (1940), 1–13.

BILLIK, M., [1] Orlicz spaces (M.A. thesis submitted at the Mass. Inst. of Technology), 1957.

BIRNBAUM, Z. and ORLICZ, W., [1] Über die Verallgemeinerung des Begriffes der zueinander konjugierten Potenzen, *S.M.* **3** (1931), 1–67.

BOAS, R. P., [1] Integrability of trigonometric series, (I) *Duke J.* **18** (1951), 787–93; (II) *M.Z.* **55** (1952), 183–6; (III) *Q.J.* **3** (1952), 217–21.

BOAS, R. P. and BOCHNER, S., [1] On a theorem of M. Riesz for Fourier series, *J.L.M.S.* **14** (1939), 62–73.

BOCHNER, S., [1] Über Faktorenfolgen für Fouriersche Reihen, *A.S.* **4** (1929), 125–9.
[2] Boundary values of analytic functions in several variables and almost periodic functions, *Ann. Math.* **45** (1944), 708–22.
[3] Summation of multiple Fourier series by spherical means, *T.A.M.S.* **40** (1936), 175–207.

BOHR, H., [1] Über einen Satz von J. Pál, *A.S.* **7** (1935), 129–35.

BOKS, T. J., [1] Sur le rapport entre les méthodes d'intégration de Riemann et de Lebesgue, *R.P.* **45** (1921), 211–64.

BORGEN, S., [1] Über (C, 1) Summierbarkeit von Reihen orthogonaler Funktionen, *M.A.* **98** (1928), 125–50.

BOSANQUET, L. S., [1] Note on differentiated Fourier series, *Q.J.* **10** (1939), 67–74.
[2] On the summability of Fourier series, *P.L.M.S.* **31** (1930), 144–64.

BROMAN, A., [1] On two classes of trigonometrical series, *Diss.* (Uppsala, 1947), pp. 1–51.

BRUSCHI, M. and COTLAR, M., *see* Cotlar, M. and Bruschi, M.

BURKILL, J. C., [1] On the differentiability of multiple integrals, *J.L.M.S.* **26** (1951), 244–9.

CALDERÓN, A. P., [1] On theorems of M. Riesz and Zygmund, *Proc.A.M.S.* **1** (1950), 533–5.
[2] On the behaviour of harmonic functions on the boundary, *T.A.M.S.* **68** (1950), 47–54.
[3] On a theorem of Marcinkiewicz and Zygmund, *ibid.* 55–61.

CALDERÓN, A. P., GONZÁLEZ-DOMÍNGUEZ, A. and ZYGMUND, A., [1] Nota sobre los valores límites de funciones analíticas, *Revista de la Unión Matemática Argentina*, **14** (1949), 16–19.

CALDERÓN, A. P. and ZYGMUND, A., [1] On the theorem of Hausdorff-Young and its extensions, *Ann. Math. Studies*, **25** (1950), 166–88.
[2] Note on the boundary values of functions of several complex variables, *ibid.* 144–65.
[3] A note on the interpolation of sublinear operations, *A.J.M.* **78** (1956), 282–8.
[4] A note on the interpolation of linear operations, *S.M.* **12** (1951), 194–204.
[5] On singular integrals, *A.M.* **88** (1952), 85–139.

CANTOR, G., [1] Über die Ausdehnung eines Satzes aus der Theorie der trigonometrischen Reihen, *M.A.* **5** (1872), 123–32.

CARATHÉODORY, C., [1] Über die Variabilitätsbereich der Koeffizienten von Potenzreihen die gegebene Werte nicht annehmen, *M.A.* **64** (1907), 95–115.
[2] Über die Fouriersche Koeffizienten monotoner Funktionen, *Sitzungsberichte d. Preuss. Akad. der Wiss.* 1920, pp. 559–73.

CARLEMAN, T., [1] Sur les équations intégrales singulières à noyau réel et symétrique, *Uppsala Universitets Årsskrift*, 1923, pp. 1–228.
[2] Über die Fourierkoeffizienten einer stetigen Funktion, *A.M.* **41** (1918), 377–84.
[3] A theorem concerning Fourier Series, *P.L.M.S.* **21** (1923), 483–92.

CARLESON, L., [1] Sets of uniqueness for functions regular in the unit circle, *A.M.* **87** (1952), 325–45.

CARTWRIGHT, M., [1] On analytic functions regular in the unit circle, *Q.J.* **4** (1933), 246–57.

CARTWRIGHT, M. and COLLINGWOOD, E. F., *see* Collingwood, E. F. and Cartwright, M.

CHAUNDY, T. W. and JOLLIFFE, A. E., [1] The uniform convergence of a certain class of trigonometrical series, *P.L.M.S.* **15** (1916), 214–16.

COLLINGWOOD, E. F. and CARTWRIGHT, M., [1] Boundary theorems for a function regular in the unit circle, *A.M.* **87** (1952), 83–146.

VAN DER CORPUT, J. G., [1] Zahlentheoretische Abschätzungen, *M.A.* **84** (1921), 53–79.

COTLAR, M. and BRUSCHI, M., [1] On the convexity theorems of Riesz-Thorin and Marcinkiewicz, *Revista, Universidad Nacional de la Plata, Publicaciones de la Facultad de Ciencias Fisicomatematicas*, **5** (1956), 162–72.

CRAMÉR, H., [1] Etudes sur la sommation des séries de Fourier, *Arkiv för Matematik, Astronomi och Fysik*, **13** (1919), No. 20, 1–21.

DENJOY, A., [1] Sur l'absolue convergence des séries trigonométriques, *C.R.* **155** (1912), 135–6.
[2] Sur les ensembles parfaits présentant le caractère (A), *Acc. Lincei*, $\mathbf{29}_2$ (1920), 316–18.
[3] Sur l'intégration riemannienne, *C.R.* **169** (1919), 219–21.

DIEUDONNÉ, J. M., [1] Sur les fonctions univalentes, *C.R.* **192** (1931), 1148–50.

DOOB, J. L., [1] The boundary values of analytic functions, *T.A.M.S.* **34** (1932), 153–70.

EDMONDS, S., [1] The Parseval formula for monotonic functions, Part I, *Proc. Camb. Phil. Soc.* **43** (1947), 289–306; Parts II and III, *ibid.* **46** (1950), 231–48, 249–67.

ERDÖS, P., [1] On the convergence of trigonometric series, *J. Math. Phys.* **22** (1943), 37–9.
[2] Corrections to two of my papers, *Ann. Math.* **44** (1943), 647–51.
[3] Some theorems and remarks on interpolation, *A.S.* **12** (1950), 11–17.
[4] On a family of symmetric Bernoulli convolutions, *A.J.M.* **61** (1939), 974–6.

ERDÖS, P. and FELDHEIM, E., [1] Sur le mode de convergence dans l'interpolation de Lagrange, *C.R.* **203** (1936), 913–15.

ERDÖS, P. and GÁL, I., [1] Law of the iterated logarithm, *Indagationes Math.* **17** (1955), 65–84.

ERDÖS, P. and TURÀN, P., [1] On interpolation, (I) Quadrature and mean convergence in the Lagrange interpolation, *Ann. Math.* **38** (1937), 142–55; (II) On the distribution of the fundamental points of Lagrange and Hermite interpolation, *ibid.* **39** (1938), 703–24; (III) Interpolatory theory of polynomials, *ibid.* **41** (1940), 510–53.

FABER, G., [1] Über das Verhalten analytischer Funktionen an Verzweigungsstellen, *Münchener Sitzungsberichte*, 1917, pp. 263–84.
[2] Über stetige Funktionen, II, *M.A.* **69** (1910), 372–443.
[3] Über die interpolatorische Darstellung stetiger Funktionen, *Jahresbericht d. Deutschen Math. Vereinigung*, **23** (1914), 192–210.

FATOU, P., [1] Séries trigonométriques et séries de Taylor, *A.M.* **30** (1906), 335–400.
[2] Sur la convergence absolue des séries trigonométriques, *B.S.M.F.* **41** (1913), 47–53.

FAVARD, J., [1] Sur les meilleurs procédés d'approximation, *B.S.M.* **61** (1937), 209–24, 243–56.

FEJÉR, L., [1] Lebesguesche Konstanten und divergente Fourier-reihen, *J.f.M.* **139** (1910), 22–53.
[2] Über die Bestimmung des Sprunges einer Funktion aus ihrer Fourierreihe, *ibid.* **142** (1913), 165–8.
[3] Untersuchungen über Fouriersche Reihen, *M.A.* **58** (1904), 501–69.
[4] Neue Eigenschaften der Mittelwerte bei den Fourierreihen, *J.L.M.S.* **8** (1933), 53–62.
[5] Über gewisse durch die Fouriersche und Laplacesche Reihe definierte Mittelkurven und Mittelflächen, *R.P.* **38** (1914), 79–97.
[6] Über die positivität von Summen die nach trigonometrischen und Legendreschen Funktionen fortschreiten, *A.S.* **2** (1925), 75–86.
[7] Über Potenzreichen deren Summe im abgeschlossenen Konvergenzkreise überall stetig ist, *Münchener Sitzungsberichte*, 1917, pp. 33–50.
[8] Sur les singularités de la série de Fourier des fonctions continues, *A.E.N.S.* **28** (1911), 63–103.
[9] Über konjugierte trigonometrische Reihen, *J.f.M.* **144** (1913), 48–56.
[10] Über gewisse Minimumprobleme der Funktionentheorie, *M.A.* **97** (1926), 104–23.
[11] Trigonometrische Reihen und Potenzreihen mit mehrfach monotoner Koeffizientenfolge, *T.A.M.S.* **39** (1936), 18–59.
[12] La convergence sur son cercle de convergence d'une série de puissances effectuant une représentation conforme du cercle sur le plan simple, *C.R.* **156** (1913), 46–9.
[13] Die Abschätzungen eines Polynomes in einem Intervalle, *M.Z.* **32** (1930), 426–57.

FEJÉR, L. and RIESZ, F., [1] Über einige Funktionentheoretische Ungleichungen, *M.Z.* **11** (1921), 305–14.

FEKETE, M., [1] Über die Faktorenfolgen welche die 'Klasse' einer Fourierschen Reihe unverändert lassen, *A.S.* **1** (1923), 148–66.

FELDHEIM, E., [1] Théorie de la convergence des procédés d'interpolation et de quadrature mécanique, *Mémorial des Sciences Math.* **95** (1939), 1–90.

FERRAND, J. and FORTET, R., [1] Sur des suites arithmétiques équiréparties, *C.R.* **224** (1947), 516–18.

FICHTENHOLZ, G., [1] Sur l'intégrale de Poisson et quelques questions qui s'y rattachent, *F.M.* **13** (1929), 1–33.

FINE, N., [1] On the Walsh functions, *T.A.M.S.* **65** (1949), 372–414.
[2] Cesàro summability of Walsh-Fourier series, *Proc. Nat. Acad. U.S.A.* **41** (1955), 588–91.

FISCHER, E., [1] Sur la convergence en moyenne, *C.R.* **144** (1907), 1022–4.

FLETT, T. M., [1] Some remarks on a maximal theorem of Hardy and Littlewood, *Q.J.* **6** (1955), 275–82.

[2] On some theorems of Littlewood and Paley, *J.L.M.S.* **31** (1956), 336–44.

FROSTMAN, O., [1] Potentiel d'équilibre et capacité des ensembles avec quelques applications à la théorie des fonctions, *Comm. of the Math. Seminar of the U. of Lund*, **3**, 1–118.

GABRIEL, R. M., [1] The rearrangement of positive Fourier coefficients, *P.L.M.S.* **33** (1932), 32–51.

[2] A star inequality for harmonic functions, *ibid.* **34** (1932), 305–13.

GAGE, W. H. and JAMES, R. D., [1] A generalized integral, *Proc. Roy. Soc. Canada*, **40** (1946), III, 25–36.

GATTEGNO, C. and OSTROWSKI, A., [1] Représentation conforme à la frontière, (I) Domaines généraux, *Mémorial d. Sci. Math.* **109** (1949), 1–58; (II) Domaines particuliers, *ibid.* **110** (1949), 1–56.

GEIRINGER, H., [1] Trigonometrische Doppelreihen, *M.f.M.* **29** (1918), 65–144.

GERGEN, J. J., [1] Convergence and summability criteria for Fourier series, *Q.J.* **1** (1930), 252–75.

GOSSELIN, R., [1] On the convergence behavior of trigonometric interpolating polynomials, *Pacific J. Math.* **5** (1955), 915–22.

GRONWALL, T. H., [1] Über die Gibbssche Erscheinung etc., *M.A.* **72** (1912), 228–61.

[2] Zur Gibbsschen Erscheinung, *Ann. Math.* **31** (1930), 232–40.

[3] On the Fourier coefficients of a continuous function, *B.A.M.S.* **27** (1921), 320–1.

[4] Über eine Summationsmethode und ihre Anwendung auf die Fouriersche Reihe, *J.f.M.* **147** (1916), 16–35.

GROSZ, W., [1] Zur Poissonschen Summierung, *Wiener Berichte*, **124** (1915), 1017–37.

GRÜNWALD, G., [1] Über Divergenzerscheinungen der Lagrangeschen Interpolationspolynome, *A.S.* **7** (1935), 207–21.

[2] Über Divergenzerscheinungen der Lagrangeschen Interpolationspolynome der stetigen Funktionen, *Ann. Math.* **37** (1936), 908–18.

HARDY, G. H., [1] Weierstrass's non-differentiable function, *T.A.M.S.* **17** (1916), 301–25.

[2] Notes on some points in the integral calculus (LXIV), *M.M.* **57** (1928), 12–16.

[3] On the summability of Fourier series, *P.L.M.S.* **12** (1913), 365–72.

[4] Notes on some points in the integral calculus (LVI), *M.M.* **52** (1922), 49–53.

[5] Notes on some points in the integral calculus (LV), *M.M.* **51** (1922), 186–92.

[6] A new proof of the functional equation of the Zeta-function, *Matematisk Tidskrift*, B, 1922, pp. 71–3.

[7] A theorem concerning Taylor's series, *Q.J.* **44** (1913), 147–60.

[8] Remarks on three recent notes in the Journal, *J.L.M.S.* **3** (1928), 166–9.

[9] The mean value of the modulus of an analytic function, *P.L.M.S.* **14** (1914), 269–77.

[10] The multiplication of conditionally convergent series, *P.L.M.S.* **6** (1908), 410–23.

[11] Notes on special systems of orthogonal functions (IV): the orthogonal functions of Whittaker's cardinal series, *Proc. Camb. Phil. Soc.* **37** (1941), 331–48.

[12] A problem of Diophantine approximation, *J. Indian Math. Soc.* **11** (1919), 162–6.

HARDY, G. H. and LITTLEWOOD, J. E., [1] A maximal theorem with function-theoretic applications, *A.M.* **54** (1930), 81–116.

[2] Some new convergence criteria for Fourier series, *J.L.M.S.* **7** (1932), 252–6.

[3] Some new convergence criteria for Fourier series, *Annali di Pisa*, **3** (1934), 43–62.

[4] Sur la série de Fourier d'une fonction à carré sommable, *C.R.* **156** (1913), 1307–9.

[5] On the absolute convergence of Fourier series, *J.L.M.S.* **3** (1928), 250–3.

[6] On the strong summability of Fourier series, *F.M.* **25** (1935), 162–89.

[7] The allied series of a Fourier series, *P.L.M.S.* **24** (1925), 211–46.

[8] The Fourier series of a positive function, *J.L.M.S.* **1** (1926), 134–8.

[9] Some properties of fractional integral (I), *M.Z.* **28** (1928), 565–606; (II), *ibid.* **34** (1931–2), 403–39.

[10] A convergence criterion for Fourier series, *M.Z.* **28** (1928), 612–34.

[11] Some problems of Diophantine approximation: A remarkable trigonometrical series, *Proc. Nat. Acad. U.S.A.* **2** (1916), 583–6.

[12] Solution of the Cesàro summability problem for power series and Fourier series, *M.Z.* **19** (1923), 67–96.

[13] Some properties of conjugate functions, *J.f.M.* **167** (1932), 405–23.

[14] Some theorems on Fourier series and Fourier power series, *Duke J.* **2** (1936), 354–81.

[15] Some new properties of Fourier constants, *M.A.* **97** (1926), 159–209.

[16] Two theorems concerning Fourier series, *J.L.M.S.* **1** (1926), 19–25.

[17] A further note on the converse of Abel's theorem, *P.L.M.S.* **25** (1926), 219–36.

[18] Tauberian theorems concerning power series and Dirichlet series whose coefficients are positive, *P.L.M.S.* **13** (1914), 174–91.

[19] Theorems concerning mean values of analytic or harmonic functions, *Q.J.* **12** (1942), 221–56.

[20] Some new cases of Parseval's theorem, *M.Z.* **34** (1932), 620–33.

[21] On the partial sums of Fourier series, *Proc. Camb. Phil. Soc.* **40** (1944), 103–7.

[22] Some new properties of Fourier constants, *J.L.M.S.* **6** (1931), 3–9.

[23] A problem concerning majorants of Fourier series, *Q.J.* **6** (1935), 304–15.

[24] Generalizations of a theory of Paley, *Q.J.* **8** (1937), 161–71.

[25] On the Fourier series conjugate to the Fourier series of a bounded function, *J.L.M.S.* **6** (1931), 278–86.

[26] On Young's convergence criterion for Fourier series, *P.L.M.S.* **28** (1928), 301–11.

Hardy, G. H. and Rogosinski, W. W., [1] On the Gibbs phenomenon, *J.L.M.S.* **18** (1943), 83–7.

[2] Asymptotic expressions for sums of certain trigonometric series, *Q.J.* **16** (1945), 49–58.

[3] On sine series with positive coefficients, *J.L.M.S.* **18** (1943), 50–7.

Hartman, Ph., [1] The divergence of non-harmonic gap series, *Duke J.* **9** (1942), 404–5.

Hartman, Ph. and Wintner, A., [1] A sine series with monotonic coefficients, *J.L.M.S.* **28** (1953), 102–4.

Hausdorff, F., [1] Eine Ausdehnung des Parsevalschen Satzes über Fourier-reihen, *M.Z.* **16** (1923), 163–9.

Helson, H., [1] Proof of a conjecture of Steinhaus, *Proc. Nat. Acad. U.S.A.* **40** (1954), 205–6.

[2] Fourier transforms on perfect sets, *S.M.* **14** (1954), 209–13.

Herglotz, G., [1] Über Potenzreihen mit positivem reellem Teil im Einheitskreise, *Leipziger Berichte*, **63** (1911), 501–11.

Herriot, J. G., [1] Nörlund summability of multiple Fourier series, *Duke J.* **11** (1944), 735–54.

Herzog, F. and Piranian, J., [1] Sets of convergence of Taylor series (I), *Duke J.* **16** (1949), 529–34; (II), *ibid.* **20** (1953), 41–54.

Hewitt, E. and Hirschman, I. I., [1] A maximum problem in harmonic analysis, *A.J.M.* **76** (1954), 839–51.

Heywood, Ph., [1] A note on a theorem of Hardy on trigonometrical series, *J.L.M.S.* **29** (1954), 373–8.

[2] On the integrability of functions, defined by trigonometric series, *Q.J.* **5** (1954), 71–6.

Hille, E., [1] On the analytical theory of semi-groups, *Proc. Nat. Acad. U.S.A.* **28** (1942), 421–4.

[2] Note on a power series considered by Hardy and Littlewood, *J.L.M.S.* **4** (1929), 176–82.

[3] On functions of bounded deviation, *P.L.M.S.* **31** (1930), 165–73.

Hille, E. and Tamarkin, J. D., [1] On the summability of Fourier series: (I), *T.A.M.S.* **34** (1932), 757–83; (II), *Ann. Math.* **34** (1933), 329–44 and 602–5; (III), *M.A.* **108** (1933), 525–77.

[2] Remarks on a known example of a monotone function, *American Math. Monthly*, **36** (1929), 255–64.

[3] On the theory of Fourier transforms, *B.A.M.S.* **39** (1933), 768–74.

Hirschman, I. I., [1] A convexity theorem for certain groups of transformations, *J. d'Analyse Math.* **2** (1953), 209–18.

[2] Fractional integration, *A.J.M.* **75** (1953), 531–46.

Hobson, E. W., [1] On the integration of trigonometrical series, *J.L.M.S.* **2** (1927), 164–6.

[2] On the uniform convergence of Fourier series, *P.L.M.S.* **5** (1907), 275–89.

Hyltén-Cavallius, C., [1] Geometrical methods applied to trigonometrical series, *Comm. of the Math. Seminar of the U. of Lund*, 1950.

Ingham, A. E., [1] On the 'high-indices' theorem of Hardy and Littlewood, *Q.J.* **8** (1937), 1–7.

[2] Note on a certain power series, *Ann. Math.* **31** (1930), 241–5.

[3] Some trigonometric inequalities with applications to the theory of series, *M.Z.* **41** (1936), 367–79.

IVAŠEV-MUSATOV, O. S., [1] The coefficients of trigonometric null series (in Russian), *Izvestia*, **21** (1957), 559–78.

IZUMI, S., [1] A simple proof of Littlewood's tauberian theorem, *Proc. Japanese Acad.* **30** (1954), 927–9.

JACKSON, D., [1] Über eine trigonometrische Summe, *R.P.* **32** (1911), 257–62.

[2] On approximations by trigonometrical sums and polynomials, *T.A.M.S.* **13** (1912), 491–515.

[3] A formula for trigonometric interpolation, *R.P.* **37** (1914), 371–5.

[4] On the order of magnitude of coefficients in trigonometric interpolation, *T.A.M.S.* **21** (1920), 321–32.

[5] Some notes on trigonometric interpolation, *Amer. Math. Monthly*, **34** (1927), 401–5.

JAMES, R. D., [1] A generalized integral (II), *Canad. J. Math.* **2** (1950), 297–306.

[2] Integrals and summable trigonometric series, *B.A.M.S.* **61** (1955), 1-15.

JESSEN, B., [1] On the approximation of Lebesgue integrals by Riemann sums, *Ann. Math.* **35** (1934), 248–51.

JESSEN, B., MARCINKIEWICZ, J. and ZYGMUND, A., [1] Note on the differentiability of multiple integrals, *F.M.* **25** (1935), 217–34.

KAC, M., [1] Convergence and divergence of non-harmonic gap series, *Duke J.* **8** (1941), 541–5.

[2] On a theorem of Zygmund, *Proc. Camb. Phil. Soc.* **47** (1951), 475–6.

KACZMARZ, S., [1] Über ein Orthogonalsystem, *Comptes Rendus du I Congrès des mathématiciens des pays slaves* (Warszawa, 1929), pp. 189–92.

[2] Integrale vom Dinischen typus, *S.M.* **3** (1931), 189–99.

[3] The divergence of certain integrals, *J.L.M.S.* **7** (1932), 218–22.

[4] On some classes of Fourier series, *J.L.M.S.* **8** (1933), 39–46.

KACZMARZ, S. and MARCINKIEWICZ, J., [1] Sur les multiplicateurs des séries orthogonales, *S.M.* **7** (1938), 73–81.

KAHANE, J. P., [1] Généralisation d'un théorème de S. Bernstein, *B.S.M.F.* **85** (1957), 221–8.

KAHANE, J. P. and SALEM, R., [1] Sur les ensembles linéaires non portant pas de pseudomesures, *C.R.* **243** (1956), 1185–7.

[2] Construction de pseudomesures sur les ensembles parfaits symétriques, *ibid.* 1986–8.

KARAMATA, J., [1] Über die Hardy-Littlewoodsche Umkehrungen des Abelschen Stetigkeitssatzes *M.Z.* **32** (1930), 319–20.

[2] Sur un mode de croissance régulière, *B.S.M.F.* **61** (1933), 55–62.

[3] Sur la sommabilité de S. Bernstein et quelques procédés qui s'y rattachent, *Sbornik*, **21** (1947), 13–22.

[4] Über die Beziehung zwischen dem Bernsteinschen und Cesàroschen Limitierungsverfahren, *M.Z.* **52** (1949), 305–6.

[5] Suites des fonctionnelles linéaires et facteurs de convergence des séries de Fourier, *Journal de Math.* **35** (1956), 87–95.

[6] Remarque relative à la sommation des séries de Fourier par le procédé de Nörlund, *Publ. Sci. de l'Université d'Alger, Sciences Mathématiques*, **1** (1954), 7–14.

KARAMATA, J. and TOMIĆ, M., [1] Sur la sommation des séries de Fourier des fonctions continues, *Publ. Inst. Math. Acad. Serbe*, **8** (1955), 123–38.

KHINTCHIN, J., [1] Sur les suites de fonctions analytiques bornées dans leur ensemble, *F.M.* **4** (1923), 72–5.

KHINTCHIN, J. and KOLMOGOROV, A. N., [1] Über Konvergenz von Reihen deren Glieder durch den Zufall bestimmt werden, *Sbornik*, **32** (1925), 668–77.

KLEIN, G., [1] A note on interpolation, *P.A.M.S.* **1** (1950), 695–702.

KOBER, H., [1] A note on Hilbert transforms, *J.L.M.S.* **18** (1943), 66–71.

KOGBETLIANTZ, E., [1] Recherches sur l'unicité des séries ultra-sphériques, *J.M.* **5** (1924), 125–96.

[2] Analogies entre les séries trigonométriques et les séries sphériques, *A.E.N.S.* **40** (1923), 259–323.

KOLMOGOROV, A. N., [1] Zur Grössenordnung des Restgliedes Fourierscher Reihen differenzierbarer Funktionen, *Ann. Math.* **36** (1935), 521–6.

[2] Sur les fonctions harmoniques conjuguées et les séries de Fourier, *F.M.* **7** (1925), 23–8.

[3] Sur l'ordre de grandeur des coefficients de la série de Fourier-Lebesgue, *B.A.P.* (1923), 83–6.

[4] Sur la possibilité de la définition générale de la dérivée, de l'intégrale et de la sommation des séries divergentes, *C.R.* **180** (1925), 362–4.

[5] Une contribution à l'étude de la convergence des séries de Fourier, *F.M.* **5** (1924), 96–7.

[6] Une série de Fourier-Lebesgue divergente presque partout, *F.M.* **4** (1923), 324–8.

[7] Une série de Fourier-Lebesgue divergente partout, *C.R.* **183** (1926), 1327–8.

KOLMOGOROV, A. N. and SELIVERSTOV, G., [1] Sur la convergence des séries de Fourier, *C.R.* **178** (1925), 303–5.

[2] Sur la convergence des séries de Fourier, *Acc. Lincei*, **3** (1926), 307–10.

KORN, A., [1] Über Minimal-flächen deren Randcurven wenig von ebenen Kurven abweichen, *Abh. K. Preussischen Akademie der Wiss., Phys.-Math. Klasse* (1909), pp. 1–37.

KRYLOV, V., [1] Functions regular in a half-plane (in Russian), *Sbornik*, **6** (1939), 95–138.

KUTTNER, B., [1] Some relations between different kinds of Riemann summability, *P.L.M.S.* **40** (1936), 524–40.

[2] A theorem on trigonometric series, *J.L.M.S.* **10** (1935), 131–40.

LANDAU, E., [1] Über eine trigonometrische Ungleichung, *M.Z.* **37** (1933), 36.

[2] Abschätzungen der Koeffizientensumme einer Potenzreihe, *Archiv. Math. Physik*, **21** (1913), 42–50.

LEBESGUE, H., [1] Sur la représentation trigonométrique approchée des fonctions satisfaisant à une condition de Lipschitz, *B.S.M.F.* **38** (1910), 184–210.

[2] Recherches sur la convergence des séries de Fourier, *M.A.* **61** (1905), 251–80.

LÉVY, P., [1] L'espace de répartitions linéaires, *B.S.M.* **62** (1938), 305–20, 324–37.

[2] Sur la convergence absolue des séries de Fourier, *Compositio Math.* **1** (1934), 1–14.

LITTLEWOOD, J. E., [1] On bounded bilinear forms in an infinite number of variables, *Q.J.* **1** (1930), 164–74.

[2] On a theorem of Fatou, *J.L.M.S.* **2** (1927), 172–6.

[3] The converse of Abel's theorem on power series, *P.L.M.S.* **9** (1910), 434–48.

[4] On Fourier coefficients of functions of bounded variation, *Q.J.* **7** (1936), 219–26.

[5] On a theorem of Paley, *J.L.M.S.* **29** (1954), 387–95.

[6] On mean values of power series, *P.L.M.S.* **25** (1924), 328–37.

[7] On mean values of power series, *J.L.M.S.* **5** (1930), 179–82.

LITTLEWOOD, J. E. and PALEY, R. E. A. C., [1] Theorems on Fourier series and power series, (I) *J.L.M.S.* **6** (1931), 230–3; (II) *P.L.M.S.* **42** (1936), 52–89; (III) *ibid.* **43** (1937), 105–26.

LOOMIS, L., [1] A note on Hilbert's transform, *B.A.M.S.* **52** (1946), 1082–6.

LOZINSKI, S., [1] On convergence and summability of Fourier series and interpolation processes, *Sbornik*, **14** (1944), 175–263.

LUKÁCS, F., [1] Über die Bestimmung des Sprunges einer Funktion aus ihrer Fourierreihe, *J.f.M.* **150** (1920), 107–12.

LUSIN, N., [1] *Integral and trigonometric series* (in Russian) (Moscow, 1915); second edition (Moscow, 1951) (with critical and historical annotations by N. Bary and D. E. Menšov); also reprinted in Lusin's *Collected Works*.

[2] Sur l'absolue convergence des séries trigonométriques, *C.R.* **155** (1912), 580–2.

[3] Über eine Potenzreihe, *R.P.* **32** (1911), 386–90.

[4] On the localization of the principle of finite area, *Doklady*, **56** (1947), 447–50.

[5] Sur une propriété des fonctions à carré sommable, *Bull. Calcutta Math. Soc.* **20** (1930), 139–54.

LUSIN, N. and PRIVALOV, I., [1] Sur l'unicité et la multiplicité des fonctions analytiques, *A.E.N.S.* **42** (1925), 143–91.

LUXEMBURG, W. A. J., [1] Banach function spaces, *Diss.* (Delft, 1955), pp. 1–70.

MALLIAVIN, P., [1] Sur la convergence absolue des séries trigonométriques, *C.R.* **228** (1949), 1467–9.

MANDELBROJT, S., [1] Modern researches on the singularities of functions defined by Taylor's series, *Rice Institute Pamphlets*, **14** (1927), no. 4, 225–32.

[2] Some theorems connected with the theory of infinitely differentiable functions, *Duke J.* **11** (1944), 341–9.

MARCINKIEWICZ, J., [1] Sur quelques intégrales du type de Dini, *Ann. de la Soc. Polonaise de Math.* **17** (1938), 42–50.

[2] Sur les séries de Fourier, *F.M.* **27** (1936), 38–69.

[3] Quelques théorèmes sur les séries de fonctions, *Bull. du Séminaire Math. de l'Université de Wilno*, **1** (1937).

Marcinkiewicz, J., [4] On Riemann's two methods of summation, *J.L.M.S.* **10** (1935), 268–72

[5] On the convergence of Fourier series, *ibid.* pp. 264–8.

[6] A new proof of a theorem on Fourier series, *J.L.M.S.* **8** (1933), 179.

[7] On interpolating polynomials (in Polish), *Wiadomości Matematyczne*, **39** (1935), 85–125.

[8] Sur la divergence des polynomes d'interpolation, *A.S.* **8** (1937), 131–5.

[9] Quelques remarques sur l'interpolation, *ibid.* pp. 127–30.

[10] Sur l'interpolation, *S.M.* **6** (1936), 1–17, 67–81.

[11] Sur la sommabilité forte des séries de Fourier, *J.L.M.S.* **14** (1939), 162–8.

[12] Sur les multiplicateurs des séries de Fourier, *S.M.* **8** (1939), 78–91.

[13] Sur une nouvelle condition sur la convergence presque partout des séries de Fourier, *Ann. Pisa*, **8** (1939), 139–40.

[14] Sur l'interpolation d'opérations, *C.R.* **208** (1939), 1272–3.

Marcinkiewicz, J. and Zygmund, A., [1] Quelques théorèmes sur les fonctions indépendantes, *S.M.* **7** (1938), 104–20.

[2] On the differentiability of functions and summability of trigonometrical series, *F.M.* **26** (1936), 1–43.

[3] Two theorems on trigonometrical series, *Sbornik*, **2** (1937), 733–8.

[4] Some theorems on orthogonal systems, *F.M.* **28** (1937), 309–35.

[5] Quelques inégalités pour les opérations linéaires, *F.M.* **32** (1939), 115–21.

[6] Mean values of trigonometric polynomials, *F.M.* **28** (1937), 131–66.

[7] On the behavior of trigonometric series and power series, *T.A.M.S.* **50** (1941), 407–53.

[8] A theorem of Lusin, *Duke J.* **4** (1938), 473–85.

[9] On the summability of double Fourier series, *F.M.* **32** (1939), 112–32.

Mazurkiewicz, S., [1] Sur l'intégrale $\int_0^1 \frac{f(x+t)+f(x-t)-2f(x)}{t}\,dt$, *S.M.* **3** (1931), 114–18.

[2] Sur les séries de puissances et les séries trigonométriques non sommables (in Polish, with a French summary), *Prace Matematyczno-Fizyczne*, **28** (1919), 109–18.

[3] Sur les séries de puissances, *F.M.* **3** (1922), 52–8.

Menšov, D., [1] Sur la convergence uniforme des séries de Fourier (in Russian), *Sbornik*, **11** (1942), 67–96.

[2] Sur l'unicité du développement trigonométrique, *C.R.* **163** (1916), 433–6.

[3] On limits of indeterminacy of Fourier series (in Russian), *Sbornik*, **30** (1950), 601–50.

[4] On limits of indeterminacy in measure of partial sums of trigonometric series (in Russian), *Sbornik*, **34** (1954), 557–74.

[5] On the convergence of trigonometric series (in Russian), *A.S.* **12** (1950), 170–84.

[6] Sur les séries de fonctions orthogonales, (I), *F.M.* **4** (1923), 82–105; (II), *F.M.* **8** (1926), 56–108; (III), *F.M.* **10** (1927), 375–420.

Milicer-Grużewska, H., Sur les fonctions à variation bornée et à l'écart Hadamardien nul, *C.R. Soc. Sci. Varsovie*, **21** (1928), 67–78.

Mirimanoff, D., [1] Remarque sur la notion d'ensemble parfait de 1-re espèce, *F.M.* **4** (1923), 122–3.

Moore, C. N., [1] On the application of Borel's method to the summation of Fourier series, *Proc. Nat. Acad. U.S.A.* **11** (1925), 284–7.

[2] On the summability of double Fourier series of discontinuous functions, *M.A.* **74** (1913), 555–78.

Morgenthaler, G. W., [1] On Walsh-Fourier series, *T.A.M.S.* **84** (1957), 472–507.

Morse, M. and Transue, W., [1] Functionals F bilinear over the product $A \times B$ of two pseudo-normed vector spaces; II. Admissible spaces A, *Ann. Math.* **51** (1950), 576–614.

Mulholland, H. P., [1] Concerning the generalization of the Young-Hausdorff theorem, *P.L.M.S.* **35** (1933), 257–93.

Natanson, I. P. [1] On the convergence of trigonometrical interpolation at equidistant knots, *Ann. Math.* **45** (1944), 457–71.

Neder, A., [1] Zur theorie der trigonometrischen Reihen, *M.A.* **84** (1921), 117–36.

Nevanlinna, R., [1] Über die Anwendung des Poissonschen Integrals zur Untersuchung der Singularitäten analytischer Funktionen, *Verh. des 5. Math. Congress, Helsingfors* (1923), pp. 273–89.

Niemytski, V., [1] Sur quelques classes d'ensembles linéaires avec applications aux séries trigonométriques absolument convergentes (in Russian, with a French summary), *Sbornik*, **33** (1926), 5–32.

NIKOLSKY, S. M., [1] Inequalities for entire functions of finite type and their applications in the theory of differentiable functions of several variables (in Russian), *Trudy Mat. Inst. Steklov*, **38** (1951), 244–78.

[2] Sur l'allure asymptotique du reste dans l'approximation au moyen des sommes de Fejér des fonctions vérifiant la condition de Lipschitz (in Russian, with a French summary), *Doklady*, **4** (1940), 501–8.

OBRECHKOFF, N., [1] Sur la sommation des séries trigonométriques de Fourier par les moyennes arithmétiques, *B.S.M.F.* **62** (1934), 84–109, 167–84.

OFFORD, A. C., [1] Approximation of functions by trigonometric polynomials, *Duke J.* **6** (1940), 505–10.

[2] On the uniqueness of the representation of a function by a trigonometric integral, *P.L.M.S.* **42** (1937), 422–80.

ORLICZ, W., [1] Über eine gewisse Klasse von Räumen von Typus B, *B.A.P.* 1932, pp. 207–20.

[2] Über Räume L^M, *ibid.* (1936), pp. 93–107.

OSTROW, E. H. and STEIN, E. M., [1] A generalization of lemmas of Marcinkiewicz and Fine with applications to singular integrals, *Annali di Pisa*, **11** (1957), 117–35.

OSTROWSKI, A., [1] Über die Bedeutung der Jensenschen Formel für einige Fragen der Komplexen Funktionentheorie, *A.S.* 1 (1922–3), 80–7.

PALEY, R. E. A. C., [1] A remarkable system of orthogonal functions, *P.L.M.S.* **34** (1932), 241–79.

[2] On lacunary power series, *Proc. Nat. Acad. U.S.A.* **19** (1933), 271–2.

[3] Some theorems on orthogonal functions, *S.M.* **3** (1931), 226–45.

[4] A proof of a theorem on bilinear forms, *J.L.M.S.* **6** (1931), 226–30.

[5] On the lacunary coefficients of power series, *Ann. Math.* **34** (1933), 615–16.

[6] A note on power series, *J.L.M.S.* **7** (1932), 122–30.

[7] On Fourier series with positive coefficients, *P.L.M.S.* **7** (1932), 205–8.

PALEY, R. E. A. C. and ZYGMUND, A., [1] On some series of functions, (I) *Proc. Camb. Phil. Soc.* **26** (1930), 337–57; (II), *ibid.* 458–74; (III), *ibid.* **28** (1932), 190–205.

[2] On the partial sums of Fourier series, *S.M.* **2** (1930), 221–7.

[3] A note on analytic functions inside the unit circle, *Proc. Camb. Phil. Soc.* **28** (1932), 266–72.

PAPOULIS, A., [1] On the strong differentiability of the indefinite integral, *T.A.M.S.* **69** (1950), 130–41.

PHRAGMÉN, E., [1] Några reflexioner i anknytning till Dr Riesz's föredrag, *3rd Scandinavian Math. Congress, Kristiania* (1913), pp. 129–42.

PIRANIAN, G. and RUDIN, W., [1] Lusin's theorem on areas of conformal maps, *Michigan Math. J.* **3** (1955–6), 191–9.

PISOT, CH., [1] Sur la répartition modulo 1, *Annali di Pisa*, **7** (1938), 205–48.

[2] Sur une famille remarquable d'entiers algébriques formant un ensemble fermé, *Colloque sur la théorie des nombres, Bruxelles* (1955), pp. 77–83.

PITT, H. R., [1] Theorems on Fourier series and power series, *Duke J.* **3** (1937), 747–55.

PLANCHEREL, M., [1] Le développement de la théorie des séries trigonométriques dans le dernier quart de siècle, *Enseignement Mathématique*, **24** (1925).

[2] Contribution à l'étude de la représentation d'une fonction arbitraire par des intégrales definies, *R.P.* **30** (1910), 289–335.

[3] Sur la convergence et la sommabilité par les moyennes de Cesàro de $\lim_{z=\infty}\int_a^z f(x)\cos xy\,dx$, *M.A.* **76** (1915), 315–26.

PLESSNER, A., [1] Eine Kennzeichnung der totalstetigen Funktionen, *J.f.M.* **160** (1929), 26–32.

[2] Zur Theorie der konjugierten trigonometrischen Reihen, *Mitt. Math. Seminar Universität Giessen*, **10** (1923), 1–36.

[3] Über die Konvergenz von trigonometrischen Reihen, *J.f.M.* **155** (1926), 15–25.

[4] Über konjugierte trigonometrische Reihen, *Doklady*, **4** (1935), 251–3.

[5] Über die Verhalten analytischer Funktionen am Rande ihres Definitionsbereiches, *J.f.M.* **159** (1927), 219–27.

POLLARD, H., [1] The harmonic analysis of bounded functions, *Duke J.* **20** (1953), 499–512.

PRASAD, B. N., [1] On the summability of Fourier series and the bounded variation of power series, *P.L.M.S.* **35** (1933), 407–24.

[2] A theorem on the summability of the allied series of a Fourier series, *J.L.M.S.* **6** (1931), 274–8.

PRIVALOV, I. I., [1] Intégrale de Cauchy (in Russian), *Saratov* (1919), pp. 1–104.
[2] Sur les fonctions conjuguées, *B.S.M.F.* **44** (1916), 100–3.

PYATETSKI-SHAPIRO, I. I., [1] On the problem of uniqueness of expansion of a function in a trigonometric series (in Russian), *Učenye Zapiski Moskovskogo Gosudarstvennogo Universiteta*, 155, Matematika, **5** (1952), 54–72.
[2] Supplement to the work 'On the problem, etc.', *ibid.* 165, **7** (1954), 78–97.

QUADE, E. S., [1] Trigonometric approximation in the mean, *Duke J.* **3** (1937), 529–43.

RADEMACHER, H., [1] Einige Sätze über Reihen von allgemeinen Orthogonalfunktionen, *M.A.* **87** (1922), 112–38.

RAJCHMAN, A., [1] Séries trigonométriques sommables par le procédé de Poisson (in Polish, French summary), *Prace Matematyczno-Fizyczne*, **30** (1919), 19–88.
[2] Sur l'unicité de développement trigonométrique, *F.M.* **3** (1922), 286–302.
[3] Sur le principe de localisation de Riemann (in Polish, French summary), *Comptes Rendus de la Soc. Sci. de Varsovie*, **11** (1918), 115–22.
[4] Sur la multiplication des séries trigonométriques et sur une classe d'ensembles fermés. *M.A.* **95** (1926), 388–408.
[5] Sur la convergence multiple, *C.R.* **181** (1925), 172–4.
[6] Une classe de séries trigonométriques qui convergent presque partout vers zéro, *M.A.* **101** (1929), 686–700.
[7] Sur une classe de fonctions à variation bornée, *C.R.* **187** (1928), 1026–8.

RAJCHMAN, A. and ZYGMUND, A., [1] Sur la possibilité d'appliquer la méthode de Riemann aux séries trigonométriques sommables par le procédé de Poisson, *M.Z.* **25** (1926), 261–73.
[2] Sur la relation du procédé de sommation de Cesàro et celui de Riemann, *B.A.P.* (1925), pp. 69–80.

RANDELS, W. C., [1] On an approximate functional equation of Paley, *T.A.M.S.* **43** (1938), 102–25.

REITER, H., [1] On a certain class of ideals in the L^1-algebra of a locally compact abelian group, *T.A.M.S.* **75** (1953), 505–9.

RIEMANN, B., [1] Über die Darstellbarkeit einer Funktion durch eine trigonometrische Reihe, *Ges. Werke*, 2 Aufl., Leipzig (1892), pp. 227–71.

RIESZ, F., [1] Sur les polynômes trigonométriques, *C.R.* **158** (1914), 1657–61.
[2] Über orthogonale Funktionensysteme, *Göttinger Nachrichten* (1907), pp. 116–22.
[3] Sur certains systemes singuliers d'équations intégrales, *A.E.N.S.* **28** (1911), 33–62.
[4] Eine Ungleichung für harmonische Funktionen, *M.f.M.* **43** (1936), 401–6.
[5] Über ein Problem von Carathéodory, *J.f.M.* **146** (1916), 83–7.
[6] Über die Fourierkoeffizienten einer stetigen Funktion von beschränkter Schwankung, *M.Z.* **2** (1918), 312–15.
[7] Über die Randwerte einer analytischen Funktion, *M.Z.* **18** (1923), 87–95.
[8] Über eine Verallgemeinerung des Parsevalschen Formel, *M.Z.* **18** (1923), 117–24.

RIESZ, F. and M., [1] Über Randwerte einer analytischen Funktion, *Quatrième Congrès des mathématiciens scandinaves, Stockholm* (1916), pp. 27–44.

RIESZ, M., [1] Sur les fonctions conjuguées, *M.Z.* **27** (1927), 218–44.
[2] Sur la sommation des séries de Fourier, *A.S.*, **1** (1923), 104–13.
[3] Eine trigonometrische Interpolationsformel und einige Ungleichungen für Polynome, *Jahresbericht d. Deutschen Math. Ver.* **23** (1914), 354–68.
[4] Sur les maxima des formes bilinéaires et sur les fonctionnelles linéaires, *A.M.* **49** (1926), 465–97.
[5] Neuer Beweis des Fatouschen Satzes, *Göttinger Nachrichten* (1916), pp. 62–5.
[6] Sätze über Potenzreihen, *Arkiv for Matematik, Astronomi och Fysik*, **11** (1916), no. 12.
[7] Summierbare trigonometrische Reihen, *M.A.* **71** (1911), 54–75.

ROGOSINSKI, W., [1] Über positive harmonische Entwickelungen und typisch reelle Potenzreihen, *M.Z.* **35** (1932), 93–121.
[2] Über die Abschnitte trigonometrischer Reihen, *M.A.* **95** (1925), 110–34.
[3] Reihensummierung durch Abschnittskoppelungen, *M.Z.* **25** (1926), 132–49.
[4] Abschnittsverhaltungen bei trigonometrischen und insbesondere Fourierschen Reihen, *M.Z.* **41** (1936), 75–136.

ROGOSINSKI, W. and SZEGÖ, G., [1] Über die Abschnitte von Potenzreihen die in einem Kreise beschränkt bleiben, *M.Z.* **28** (1928), 73–94.

RUDIN, W., [1] Radial cluster sets of analytic functions, *B.A.M.S.* **60** (1954), 545, Abstract no. 718.

SAKS, S., [1] On some functionals, (I) *T.A.M.S.* **35** (1933), 549–56; (II) *ibid.* **41** (1937), 160–70.

[2] On the strong derivatives of functions of intervals, *F.M.* **25** (1935), 245–52.

SALEM, R., [1] On the absolute convergence of trigonometric series, *Duke J.* **8** (1941), 317–34.

[2] Essais sur les séries trigonométriques, *Actualités scientifiques et industrielles*, no. 862 (Paris, 1940), pp. 1–85.

[3] On a theorem of Bohr and Pàl, *B.A.M.S.* **50** (1944), 579–80.

[4] Sur les transformations des séries de Fourier, *F.M.* **33** (1939), 108–14.

[5] A singularity of the Fourier series of a continuous function, *Duke J.* **10** (1943), 711–16.

[6] On some properties of symmetrical perfect sets, *B.A.M.S.* **47** (1941), 820–8.

[7] On a theorem of Zygmund, *Duke J.* **10** (1943), 23–31.

[8] On sets of multiplicity for trigonometric series, *A.J.M.* **64** (1942), 69–82.

[9] On singular monotonic functions of Cantor type, *J. Math. Physics*, **21** (1942), 69–82.

[10] On a problem of Smithies, *Indagationes Math.* **16** (1954), 403–7.

[11] On singular monotonic functions whose spectrum has a given dimension, *Arkiv for Mat.* **1** (1951), 353–65.

[12] (I) Sets of uniqueness and sets of multiplicity, *T.A.M.S.* **54** (1943), 218–28; (II) *ibid.* **56** (1944), 32–49; (III) Rectification to the papers: Sets of uniqueness and sets of multiplicity, I and II, *ibid.* **63** (1948), 595–8.

[13] A remarkable class of algebraic integers, *Duke J.* **11** (1944), 103–8.

[14] New theorems on the convergence of Fourier series, *Koninklijke Nederlandse Akad. Wettenschappen, Math. Sci.* (1954), pp. 550–7.

[15] Convexity theorems, *B.A.M.S.* **55** (1949), 851–60.

[16] Sur une extension du théorème de convexité de M. Marcel Riesz, *Colloquium Math.* **1** (1947), 6–8.

[17] A new proof of a theorem of Menchoff, *Duke J.* **8** (1941), 269–72.

SALEM, R. and ZYGMUND, A., [1] On a theorem of Banach, *Proc. Nat. Acad. U.S.A.* **33** (1947), 293–5.

[2] Trigonometric series whose terms have random sign, *A.M.* **91** (1954), 245–301.

[3] The approximation by partial sums of Fourier series, *T.A.M.S.* **59** (1946), 14–22.

[4] Lacunary power series and Peano curves, *Duke J.* **12** (1945), 569–78.

[5] La loi du logarithme itéré pour les séries trigonométriques lacunaires, *B.S.M.* **74** (1950), 209–24.

[6] A convexity theorem, *Proc. Nat. Acad. U.S.A.* **34** (1948), 443–7.

[7] On lacunary trigonometric series, (I) *Proc. Nat. Acad. U.S.A.* **33** (1947), 333–8; (II), *ibid.* **34** (1948), 54–62.

[8] Sur un théorème de Piatetski-Shapiro, *C.R.* **240** (1955), 2040–2.

[9] Sur les ensembles parfaits dissymétriques à rapport constant, *ibid.* 2281–3.

[10] Capacity of sets and Fourier series, *T.A.M.S.* **59** (1946), 23–41.

SCHAEFFER, A. C., [1] The Fourier-Stieltjes coefficients of a function of bounded variation, *A.J.M.* **61** (1939), 934–40.

[2] Inequalities of A. Markoff and S. Bernstein for polynomials and related functions, *B.A.M.S.* **47** (1941), 565–77.

[3] Entire functions and trigonometric polynomials, *Duke J.* **20** (1953), 77–88.

SCHMETTERER, L., [1] Zur Fourierentwickelung des Produktes zweier Funktionen, *M.f.M.* **53** (1949), 53–62.

[2] Bemerkungen zur Multiplikation unendlicher Reihen, *M.Z.* **54** (1951), 102–14.

SCHOENBERG, I., [1] Über die asymptotische Verteilung reeller Zahlen mod 1, *M.Z.* **28** (1928), 171–99.

SCHUR, I., [1] Über lineare Transformationen in der Theorie der unendlichen Reihen, *J.f.M.* **151** (1921), 79–111.

SCHUR, I. and SZEGÖ, G., [1] Über die Abschnitte einer im Einheitskreise beschränkter Potenzreihen, *Sitzungsberichte d. Berliner Akad.* (1925), pp. 545–60.

SEIDEL, V., [1] On the distribution of values of bounded analytic functions, *T.A.M.S.* **36** (1934), 200–26.

SIDON, S., [1] Über die Fourierkoeffizienten einer stetigen Funktion von beschränkter Schwankung, *A.S.* **2** (1924), 43–6.

[2] Reihentheoretische Sätze und ihre Anwendungen in der Theorie der Fourierschen Reihen, *M.Z.* **10** (1921), 121–7.

SIDON, S., [3] Verallgemeinerung eines Satzes über die absolute Konvergenz von Fourierreihen mit Lücken, *M.A.* **97** (1927), 675–6.

[4] Einige Sätze und Fragestellungen über Fourierkoeffizienten, *M.Z.* **34** (1932), 477–80.

[5] Ein Satz über trigonometrische Polynome und ihre Anwendungen in der Theorie der Fourierreihen, *M.A.* **106** (1932), 536–9.

SIERPIŃSKI, W., [1] Sur l'ensemble des points de convergence d'une suite de fonctions continues. *F.M.* **2** (1921), 41–9.

SMIRNOV, V., [1] Sur les valeurs limites des fonctions analytiques, *C.R.* **188** (1929), 131–3.

[2] Sur les valeurs limites des fonctions régulières à l'intérieur d'un cercle, *Journal de la Soc. Physico-Mathématique de Leningrad*, **2** (1929), 22–37.

ŠNEIDER, A. A., [1] On the uniqueness of expansions in Walsh functions (in Russian), *Sbornik*, **24** (1949), 279–300.

[2] On the convergence of subsequences of the partial sums of Fourier series of Walsh functions, *Doklady*, **70** (1950), 969–71.

SPENCER, D., [1] A function theoretic identity, *A.J.M.* **65** (1943), 147–60.

STEČKIN, S. B., [1] On the absolute convergence of Fourier series (in Russian), *Izvestia*, **17** (1953), 499–512; (II) *ibid.* **19** (1955), 221–46; (III) *ibid.* **20** (1956), 385–412.

STEIN, E., [1] Functions of exponential type, *Ann. of Math.* **65** (1957), 582–92.

[2] Interpolation of linear operators, *T.A.M.S.* **83** (1956), 482–92.

[3] A maximal function with applications to Fourier series, *Ann. Math.* (1958).

STEIN, E. and WEISS, G., [1] Interpolation of operators with change of measures, *T.A.M.S.* **87** (1958), 159–72.

[2] On the interpolation of analytic families of operators acting on H^p-spaces, *Tôhoku Mathematical Journal*, **9** (1957), 318–39.

[3] An extension of a theorem of Marcinkiewicz and some of its applications, *Journal of Math. and Mechanics*, **8** (1959).

STEIN, P., [1] On a theorem of M. Riesz, *J.L.M.S.* **8** (1933), 242–7.

STEINHAUS, H., [1] Sur le développement du produit de deux fonctions en une série de Fourier, *Bull. de l'Ac. de Cracovie* (1913), pp. 113–16.

[2] Sur quelques propretiés de séries trigonométriques et celles de puissances (in Polish, with a French summary), *Rozprawy Akademji Umiejętności* (Cracow, 1915), pp. 175–225.

[3] Über die Wahrscheinlickheit dafür, dass der Konvergenzkreis einer Potenzreihe ihre natürliche Grenze ist, *M.Z.* **21** (1929), 408–16.

[4] A new property of G. Cantor's set (in Polish), *Wektor*, **7** (1917).

[5] Sur les distances des points des ensembles de mesure positive, *F.M.* **1** (1920), 93–104.

[6] Sur la convergence non-uniforme des séries de Fourier, *Bull. de l'Ac. de Cracovie* (1913), pp. 145–60.

[7] A generalization of G. Cantor's theorem on trigonometric series (in Polish), *Wiadomości Matematyczne*, **24** (1920), 197–201.

[8] Une série trigonométrique partout divergente, *Comptes Rendus de la Soc. Sci. Varsovie* (1912), pp. 219–29.

[9] A divergent trigonometrical series, *J.L.M.S.* **4** (1929), 86–8.

[10] Sur un problème de MM. Lusiñ et Sierpiński, *Bull. de l'Acad. de Cracovie*, 1913, 335–350.

SUNOUCHI, G., [1] On the summability of power series and Fourier series, *Tôhoku Math. Journal*, **7** (1955), 96–109.

[2] Theorems on power series of the class H^p, *ibid.* **8** (1956), 125–46.

SUTTON, O. G., [1] On a theorem of Carleman, *P.L.M.S.* **23** (1925), XLVIII–LI.

SZASZ, O., [1] Über den Konvergenzexponent der Fourierschen Reihen, *Münchener Sitzungsberichte* (1922), pp. 135–50.

[2] Über die Fourierschen Reihen gewisser Funktionenklassen, *M.A.* **100** (1928), 530–6.

SZEGÖ, G., [1] Über die Lebesgueschen Konstanten bei den Fourierreihen, *M.Z.* **9** (1921), 163–6.

[2] Über die Randwerte einer analytischen Funktion, *M.A.* **84** (1921), 232–44.

[3] Über einen Satz des Herrn S. Bernstein, *Schriften d. Koenigsberger gelehrten Ges.* **5** (1928), 59–70.

SZEGÖ, G. and ZYGMUND, A., [1] On certain mean values of polynomials, *Journal d'Analyse Math.* **3** (1953/4), 225–44.

SZ. NAGY, B., [1] Sur une classe générale de procédés de sommation pour les séries de Fourier, *Hungarica Acta Math.* $\mathbf{1}_3$ (1948), 14–52.

[2] Approximation von Funktionen durch die arithmetische Mittel ihrer Fourierschen Reihen, *A.S.* **11** (1946), 71–84.

[3] Séries et intégrales de Fourier des fonctions monotones non-bornées, *A.S.* **13** (1948), 118–35.

TAMARKIN, J. D., [1] Remarks on the theory of conjugate functions, *P.L.M.S.* **34** (1932), 379–91.

TAMARKIN, J. D. and ZYGMUND, A., [1] Proof of a theorem of Thorin, *B.A.M.S.* **50** (1944), 279–82.

TAYLOR, S. J., [1] An integral of Perron's type, *Q.J.* **6** (1955), 255–74.

THORIN, G. O., [1] An extension of a convexity theorem due to M. Riesz, *Kungl. Fysiografiska Saellskapet i Lund Forhaendlinger*, **8** (1939), no. 14.

[2] Convexity theorems, *Diss. Lund* (1948), pp. 1–57.

TITCHMARSH, E. C., [1] On conjugate functions, *P.L.M.S.* **29** (1928), 49–80.

[2] The convergence of certain integrals, *ibid.* **24** (1925), 347–58.

[3] Additional note on conjugate functions, *J.L.M.S.* **4** (1929), 204–6.

[4] Reciprocal formulae for series and integrals, *M.Z.* **25** (1926), 321–47.

[5] A theorem on Lebesgue integrals, *J.L.M.S.* **2** (1927), 36–7.

[6] A contribution to the theory of Fourier transforms, *P.L.M.S.* **23** (1924), 279–89.

[7] The order of magnitude of the coefficients in a generalized Fourier series, *P.L.M.S.* **22** (1925), xxiv–xxvi.

TOEPLITZ, O., [1] Über allgemeine lineare Mittelbildungen, *Prace Matematyczno-Fizyczne*, **22** (1911), 113–19.

[2] Über die Fouriersche Entwickelungen positiver Funktionen, *R.P.* **32** (1911), 191–2.

[3] Zur Theorie der quadratischen und bilinearen Formen von unendlich vielen Veränderlichen (I), *M.A.* **70** (1911), 351–76.

TUMARKIN, G. C., [1] On uniform convergence of certain sequences of functions (in Russian), *Doklady*, **105** (1955), 1151–4.

TURÁN, P., [1] Uber die partielle Summen der Fourierreihen, *J.L.M.S.* **13** (1938), 278–82.

ULYANOV, P. L., [1] Application of *A*-integration to a class of trigonometric series, *Sbornik*, **35** (1954), 469–90.

DE LA VALLÉE-POUSSIN, CH. J., [1] Sur l'unicité du développement trigonométrique, *Bull. de l'Acad. Royale de Belgique* (1912), pp. 702–18.

[2] Sur l'approximation des fonctions d'une variable réelle et leurs dérivées par les polynômes et les suites limitées de Fourier, *ibid.* (1908), pp. 193–254.

[3] Sur la convergence des formules d'interpolation entre ordonnées équidistantes, *ibid.* (1908), pp. 319–410.

[4] Extension de la méthode du balayage de Poincaré et problème de Dirichlet, *Ann. de l'Institut H. Poincaré*, **2** (1932), 169–232.

VERBLUNSKY, S., [1] On the theory of trigonometric series, (I) *P.L.M.S.* **34** (1932), 441–56; (II), *ibid.* 457–91; (III), *ibid.* 526–60; (IV), *ibid.* **35** (1933), 445–87; (V), *F.M.* **21** (1933), 168–210.

[2] Note on summable trigonometric series, *J.L.M.S.* **6** (1931), 106–12.

[3] Fourier constants and Lebesgue classes, *P.L.M.S.* **39** (1935), 1–31.

[4] The relation between Riemann's method of summation and Cesàro's, *Proc. Camb. Phil. Soc.* **26** (1930), 34–42.

WALSH, J. L., [1] A closed set of normal orthogonal functions, *A.J.M.* **55** (1923), 5–24.

WANG, FU TRAING, [1] Strong summability of Fourier series, *Duke J.* **12** (1945), 77–87.

WARASZKIEWICZ, Z., [1] Sur un théorème de M. Zygmund, *B.A.P.* (1929), pp. 275–9.

WARSCHAWSKI, S., [1] Über einige Kongvergenzsätze aus der Theorie der konformen Abbildung, *Göttinger Nachrichten* (1930), pp. 344–69.

WEISS, G., [1] A note on Orlicz spaces, *Portugaliae Mat.* **15** (1956), 35–47.

[2] An interpolation theorem for sublinear operations on H^p spaces, *Proc. A.M.S.* **8** (1957), 92–9.

WEISS, MARY, [1] On the law of the iterated logarithm for lacunary trigonometric series, *T.A.M.S.* (1959).

[2] On series of Hardy and Littlewood, *T.A.M.S.* (1959).

[3] On a problem of Littlewood, *J.L.M.S.* **34** (1959).

WEYL, H., [1] Bemerkungen zum Begriff der Differentialquotienten gebrochener Ordnung, *Vierteljahrschrift d. Naturforscher Gesellschaft in Zürich*, **62** (1917), 296–302.

[2] Über die Gleichverteilung von Zahlen mod. Eins, *M.A.* **77** (1916), 313–52.

WIELANDT, H., [1] Zur Umkehrung des Abelschen Stetigkeitssatzes, *M.Z.* **56** (1952), 206–7.

WIENER, N., [1] The quadratic variation of a function and its Fourier coefficients, *Massachusetts J. of Math.* **3** (1924), 72–94.

[2] A class of gap theorems, *Annali di Pisa*, **3** (1934), 367–72.

[3] Tauberian theorems, *Ann. Math.* **33** (1932), 1–100.

[4] Generalized harmonic analysis, *A.M.* **55** (1930), 117–258.

[5] On the representation of functions by trigonometrical integrals, *M.Z.* **24** (1925), 575–616.

WIENER, N. and WINTNER, A., [1] On singular distributions, *J. Math. Physics*, **17** (1938), 233–46.

[2] Fourier-Stieltjes transforms and singular infinite convolutions, *A.J.M.* **40** (1938) 513–22.

WILTON, J. R., [1] An approximate functional equation of a simple type: Applications to a certain trigonometrical series, *J.L.M.S.* **9** (1934), 247–54.

WOLF, F., [1] On summable trigonometrical series, *P.L.M.S.* **45** (1939), 328–56.

[2] The Poisson integral. A study in the uniqueness of harmonic functions, *A.M.* **74** (1941), 65–100.

[3] Contributions to the theory of summable trigonometric integrals, *Univ. of California, Publ. Math.* **1** (1947), 159–227.

YANO, S., [1] An extrapolation theorem, *J. Math. Soc. Japan*, **3** (1951), 296–305.

YOUNG, G. C. and W. H., [1] On the theorem of Riesz and Fischer, *Q.J.* **44** (1913), 49–88.

YOUNG, L. C., [1] On an inequality of Marcel Riesz, *Ann. Math.* **40** (1939), 367–74.

YOUNG, W. H., [1] On the integration of Fourier series, *P.L.M.S.* **9** (1911), 449–62.

[2] Sur la généralisation du théorème de Parseval, *C.R.* **155** (1912), 30–3.

[3] On the multiplication of successions of Fourier constants, *Proc. Roy. Soc.* A, **87** (1912), 331–9.

[4] On the determination of the summability of a function by means of its Fourier constants, *P.L.M.S.* **12** (1913), 71–88.

[5] Konvergenzbedingungen für die verwandte Reihe einer Fourierschen Reihe, *Münchener Sitzungsberichte*, **41** (1911), 361–71.

[6] On the mode of oscillation of a Fourier series and of its allied series, *P.L.M.S.* **12** (1913), 433–52.

[7] On successions with subsequences converging to an integral, *P.L.M.S.* **24** (1926), 1–20.

[8] On the Fourier series of bounded functions, *P.L.M.S.* **12** (1913), 41–70.

[9] On the ordinary convergence of restricted Fourier series, *Proc. Roy. Soc.* A, **93** (1917), 276–92.

[10] On restricted Fourier series and the convergence of power series, *P.L.M.S.* **17** (1918), 353–66.

[11] On the connexion between Legendre series and Fourier series, *P.L.M.S.* **18** (1920), 141–62.

ZAANEN, A. C., [1] On a certain class of Banach spaces, *Ann. Math.* **47** (1946), 657–66.

ZALCWASSER, Z., [1] Sur le phénomène de Gibbs dans la théorie des séries de Fourier des fonctions continues, *F.M.* **12** (1928), 126–51.

ZAMANSKY, M., [1] Classes de saturation de certains procédés d'approximation des séries de Fourier, *A.E.N.S.* **66** (1949), 19–93.

[2] Classes de saturation des procédés de sommation des séries de Fourier, *A.E.N.S.* **67** (1950), 161–98.

[3] Sur l'approximation des fonctions continues, *C.R.* **228** (1949), 460–1.

ZARANTONELLO, E. H., [1] On trigonometric interpolation, *Proc. A.M.S.* **3** (1952), 770–82.

ZELLER, K., [1] Über Konvergenzmengen der Fourierreihen, *Archiv der Math.* **6** (1955), 335–40.

ZYGMUND, A., [1] Smooth functions, *Duke J.* **12** (1945), 47–76.

[2] Sur la sommation des séries conjuguées aux séries de Fourier, *B.A.P.* (1924), pp. 251–8.

[3] On the degree of approximation of functions by their Fejér means, *B.A.M.S.* **51** (1945), 274–8.

[4] Sur les fonctions conjuguées, *F.M.* **13** (1929), 284–303; Corrigenda, *ibid.* **18** (1932), 312.

[5] Some points in the theory of trigonometric and power series, *T.A.M.S.* **36** (1934), 586–617.

[6] Sur un théorème de M. Fekete, *B.A.P.* (1927), pp. 343–7.

[7] On a theorem of Hadamard, *Ann. de la Soc. Polonaise de Math.* **21** (1948), 52–70; Errata, *ibid.* 357–9.

[8] On the convergence of lacunary trigonometric series, *F.M.* **16** (1930), 90–107.
[9] On lacunary trigonometric series, *T.A.M.S.* **34** (1932), 435–46.
[10] A remark on conjugate series, *P.L.M.S.* **34** (1932), 392–400.
[11] Sur la convergence absolue des séries de Fourier, *J.L.M.S.* **3** (1928), 194–6.
[12] Quelques théorèmes sur les séries trigonométriques et celles de puissances, *S.M.* **3** (1931), 77–91.
[13] On a theorem of Littlewood, *Summa Brasiliensis Mathematicae*, **2** (1949), Fasc. 5.
[14] Sur la théorie riemannienne des séries trigonométriques, *M.Z.* **24** (1926), 47–104.
[15] Sur un théorème de M. Fejér, *Bull. du Séminaire Math. de l'Université de Wilno*, **2** (1939), 3–12.
[16] An example in Fourier series, *S.M.* **10** (1948), 113–19.
[17] Note on the formal multiplication of trigonometrical series, *Bull. du Séminaire Math. de l'Univ. de Wilno*, **2** (1939), 62–6.
[18] Sur la théorie riemannienne de certains systèmes orthogonaux, (I) *S.M.* **2** (1930), 97–170; (II) *Prace Matematyczno-Fizyczne*, **39** (1932), 73–117.
[19] Sur les séries trigonométriques sommables par le procédé de Poisson, *M.Z.* **25** (1926), 274–90.
[20] On certain integrals, *T.A.M.S.* **55** (1944), 170–204.
[21] On the convergence and summability of power series on the circle of convergence, (I), *F.M.* **30** (1928), 170–96; (II), *P.L.M.S.* **47** (1941), 326–50.
[22] Sur un théorème de M. Gronwall, *B.A.P.* (1925), pp. 207–17.
[23] Sur l'application de la première moyenne arithmétique, *F.M.* **10** (1926), 356–62.
[24] On the summability of multiple Fourier series, *A.J.M.* **69** (1947), 836–50.
[25] On the boundary values of functions of several complex variables, *F.M.* **36** (1949), 207–35.
[26] On the differentiability of multiple integrals, *F.M.* **23** (1934), 143–9.
[27] On the Littlewood-Paley function $g^*(\theta)$, *Proc. Nat. Acad. U.S.A.* **42** (1956), 208–12.
[28] On a theorem of Marcinkiewicz concerning interpolation of operations, *Journal de Math.* **35** (1956), 223–48.
[29] A remark on Fourier transforms, *Proc. Camb. Phil. Soc.* **32** (1936).
[30] Über die Beziehungen der Eindeutigkeitsfragen in der Theorien der trigonometrischen Reihen und Integrale, *M.A.* **99** (1928), 562–89.
[31] Trigonometric integrals, *Ann. Math.* **48** (1947), 393–440.

INDEX

Numbers in Roman type refer to Vol. I; those in italics to Vol. II.
$a(b)$ refers to item b on page a; n. refers to a footnote.